A TEXTBOOK OF BOTANY

Volume II

A TEXTBOOK
OF
BOTANY

FOR DEGREE STUDENTS

Volume II

Kashinath Bhattacharya MSc, PhD, FIAS, FICAAI, FAScT

Professor, Department of Botany
Visva-Bharati University, Santiniketan, West Bengal

Gopinath Hait MSc, PhD

Principal, Gobardanga Hindu College, North 24 Pgs, West Bengal
Formerly Reader, Dept of Botany, City College, Kolkata, West Bengal

Ashim Kumar Ghosh MSc, PhD

Reader and Head, Department of Botany
City College, Kolkata, West Bengal

New Central Book Agency (P) Ltd

Co-Published by

Bijay Publications

E-mail : bijaypublications@gmail.com

Exclusively Distributed by

Bijay Pustak Bhandar (P) Ltd

90/6A M. G. Road, Bhawani Dutta Lane, Kolkata-700007
✆ : (033) 4605-7856, 8777898576, 8274916104
E-mail : bijaypustakbhandar@gmail.com

NCBA

REGD OFFICE
8/1 Chintamoni Das Lane, Kolkata 700 009, India
email: ncbakolkata@gmail.com

OVERSEAS
NCBA (UK) Ltd, 149 Park Avenue North, Northampton, NN3 2HY, UK
email: ncbauk@yahoo.com

CORPORATE OFFICE
212 Shahpur Jat, New Delhi 110 049
email: ncbaexports@ncbapvtltd.com

BIJAY PUBLICATIONS
90/6A M. G. Road, Bhawani Dutta Lane, Kolkata-7
E-mail : bijaypublications@gmail.com

BIJAY PUSTAK BHANDAR (P) LTD
90/6A M. G. Road, Bhawani Dutta Lane, Kolkata-7
E-mail : bijaypustakbhandar@gmail.com

A TEXTBOOK OF BOTANY [Volume II] • Kashinath Bhattacharya, Gopinath Hait, Ashim Kumar Ghosh

First Published: September 2007
Reprinted: 2009, 2011, 2013, 2014, 2015, 2016, 2017, 2018, 2019, 2020, 2021, 2024

PUBLISHER • TYPESETTER • PRINTER
New Central Book Agency (P) Ltd
8/1 Chintamoni Das Lane, Kolkata 700 009

CO-PUBLISHED BY
Bijay Publications
90/6A M G Road, Bhawani Dutta Lane, Kolkata 700007

TECHNICAL EDITOR : Dr Mita Sen

PROJECT TEAM : Subas Maitra, Joydeep Ghosh and Arun Kabri

PROJECT SUPERVISOR : Ashit Kr. Ghosh

ISBN: 978 81 7381 555 3

Price ₹ 850.00

Dedicated
to our beloved

Abir Bhattacharya
Arunima and Madhurima Hait
&
Ayan Ghosh

Contents

Preface

Like Volume I, the Volume II of 'A Textbook of Botany' is brought out in due time and we hope that this volume will be received with the same degree of enthusiasm by the undergraduate students of various universities of our country. This book, comprising of spermatophytes, is divided into five chapters dealing with Gymnosperms and Palaeobotany, Angiosperm Morphology, Embryology, Taxonomy, and Plant Anatomy. The book is well-illustrated and documented and also possesses comprehensive accounts for undergraduate students of Botany and allied branches.

We express our sincere thanks to all our friends and colleagues who have extended their cooperation and help during the period the book was in preparation. We are obliged to mention about the help rendered by Prof. P. K. Pal of Burdwan University; Prof. S. K. Sen, Dr. Subrata Mandal and Dr. C. H. Rahaman of Visva-Bharati; Dr. Bratati Dey and Dr. Subir Bera of Calcutta University; Dr. S. Chattopadhyay of City College; Dr. Udayan Sarkar, Principal and Dr. Prasenjit Sharma of B. B. College, Asansol; Dr. Prakash Karmakar, Vidyasagar University; Dr. Arati Malakar, Tamluk College; Dr. Swapna Banik, Basirhat College, Dr. Satyabrata Ghosh and Dr. Pampa Chakraborty of Habra Srichaitanya College; Dr. Kashinath Ghosh, Dr. Sampa Mitra and Prof. Swapna Datta of Scottish Church College; Dr. Debashis Jana, Jiyaganj College, Dr. Satadip Sarkar, Dr. Madhubrata Chowdhury, Dr. Jukta Adhikari, Dr. Tarit Baran Sandhu and Prof. Satyaranjan Saha of Presidency College; Dr. Haradhan Saha, DumDum Motijheel College; Dr. Ponty Roy, Principal, Sarojini Naidu College; Dr. Vivekananda Mukherjee, Ranaghat College; Dr. Asok Das, Medinipur College; Dr. Fazlul Rahaman and Dr. Samiran Kundu, Suri Vidyasagar College; Dr. Bholanath Chakraborty and Dr. Subrata Roy of Kalna College, and others.

We are greatly thankful to Mr. Amitabha Sen, Director of New Central Book Agency (P) Ltd, for undertaking the publication of the book. We are also thankful to Dr. (Mrs.) Mita Sen and Mr. Subas Maitra for thorough proof-checking of the final print-outs.

We are thankful to Mr. Sunil Pal, Mr. Narayan Sur and Mr. Jaydeep Ghosh for the illustrations. Thanks are due to Mr. Ashit Kr. Ghosh of New Central Book Agency (P) Ltd for his continuous help in composing, typesetting, drawing and overall supervision throughout the progress of the book. Thanks are also due to Mr. Prabhat Jas and Mr. Soumen Paul of New Central for their help in various ways.

We welcome comments, criticisms and suggestions for future improvement of the book in the subsequent editions.

Kolkata
4th September 2007
Janmastami

Kashinath Bhattacharya
Gopinath Hait
Ashim Kumar Ghosh

1 Gymnosperms and Palaeobotany

The term '**gymnosperm**' was coined by Theophrastus (372-286B.C.), one of Aristotle's (384-322B.C.) pupils for plants whose seeds are unprotected. Gymnosperms (*gymnos* = naked, *sperma* = seed) include the group of flowering plants with naked seeds. Unlike angiosperms, gymnosperm seeds are not enclosed in a ovary. Gymnosperm seeds (ovules) are with a single integument. The unique features of angiosperms i.e., the double fertilisation is completely absent in gymnosperms (except, *Ephedra* often shows double fertilisation).

The gymnosperms originated in the Devonian period of the Palaeozoic Era and formed the supreme vegetation in the Mesozoic Era. At present, gymnosperms are represented by about 72 genera and about 650 species and are widely distributed from North temperate to tropical regions of the world. Gymnosperms form one of the major forest types of our planet having the greatest economic importance as a source of timber, lumber and woodpulp.

MAIN CHARACTERISTIC FEATURES OF GYMNOSPERMS

1. Plants are woody and evergreen, either trees, shrubs or lianes without having any herbaceous member.
2. Plants have a tap root system, generally persists for a long time.
3. Xylem is made up of tracheids, parenchyma and rays. Vessels are only present in the member's Gnetales which have evolved from pitted tracheids, as shown by intermediate stages between pits and perforations.
4. Phloem consists of sieve cells only where sieve areas commonly occur on the radial wall. The companion cells are absent.
5. Secondary growth is present in all members, where mature metaxylem shows bordered pits (In *Stangeria* and *Zamia*, metaxylem shows scalariform thickening).
6. The anther has an exothecium which produces numerous anemophilous pollen grains. Pollen exine is alveolar or granular.
7. Prothallial cell is present in male gametophyte, male gametes rarely motile.
8. Ovule (seed) consists of a single integument (unitegmic) which is exposed (naked).
9. The development of female gametophyte shows a prolonged free nuclear phase and there is a long interval between pollination and fertilisation.
10. Endosperm is haploid (n) and produced before fertilisation.
11. There is a free nuclear stage in the development of proembryo.

CLASSIFICATION OF GYMNOSPERMS

As flowering plants, gymnosperms have been considered as a co-associate group of angiosperms. Gymnosperms have been classified by various workers based mainly on the morphological, anatomical and developmental aspects. It was Robert Brown (in 1827) who first recognised gymnosperms as a separate entity among plant kingdom. Bentham and Hooker (1883) in 'Genera Plantarum' considered the gymnosperms as a coordinate group with the angiosperms and placed them in-between the dicotyledons and monocotyledons. van Tiegham (1898) considered the gymnosperms as one of the two major divisions of Spermetophyta (seed plants) : the Gymnosperms (Astigmatae) and the Angiosperms (Stigmatae).

Coulter and Chamberlain (1917) divided the gymnosperms into different orders — mainly based on the habit, stem anatomy and leaf characteristics. Later, Chamberlain (1935) divided the gymno-sperms into two main groups : Cycadophytes having three orders and Coniferophytes having four orders :

Chamberlain (1935) :

Gymnosperms

A. Cycadophytes (simple stem with massive cortex and thin wood. Sporophylls simple)
   1. Order : Cycadofilicales
   2. Order : Bennettitales
   3. Order : Cycadales

B. Coniferophytes (profused branched stem with thin cortex and massive wood. Sporophylls complex)
   1. Order : Cordaitales
   2. Order : Ginkgoales
   3. Order : Coniferales
   4. Order : Gnetales

On the basis of the origin of ovules (axial or foliar) Sahni (1920) divided gymnosperms into two groups : Phyllospermae (leaves large and compound, ovules borne on leaves or organs) and Stachyospermae (leaves small and simple, ovules borne on stem).

Sahni (1920) :

Gymnosperms

A. Phyllospermae
   1. Pteridospermae
   2. Cycadales
   3. Bennettitales

B. Stachyospermae
   1. Cordaitales
   2. Ginkgoales
   3. Coniferales
   4. Gnetales

Tippo (1942) arbitrarily placed Gymnosperms in Class 2 before Class 3 Angiospermae under the sub-phylum Pteropsida. He divided gymnosperms into two sub-classes : Cycadophytae (including Cycadofilicales, Bennettitales and Cycadales) and Coniferophytae (including Cordaitales, Ginkgoales, Coniferales and Gnetales).

Tippo (1942) :

Sub-kingdom : Embryophyta

Phylum : Bryophyta

Phylum : Tracheophyta

Sub-phylum : Psilopsida

Sub-phylum : Lycopsida

Sub-phylum : Pteropsida

Class 1 : Filicinae

Class 2 : Gymnospermae

Sub-class : Cycadophytae

Order 1 : Cycadofilicales

Order 2 : Bennettitales

Order 3 : Cycadales

Sub-class : Coniferophytae

Order 1 : Cordaitales

Order 2 : Ginkgoales

Order 3 : Coniferales

Order 4 : Gnetales

Class 3 : Angiospermae

Arnold (1948), for the first time, gave a palaeobotanical foundation (based on fossil evidences) to the Chamberlain classification. He interpreted the class Gymnospermae as a polyphyletic group where seeds have originated independently in Cycadophyta and Coniferophyta. He recognised three separate phyla : Cycadophyta, Coniferophyta and Chlamydospermophyta.

Arnold (1948) :

Gymnosperms

Phylum I : Cycadophyta

Order 1 : Pteridospermales

Order 2 : Cycadeoidales

Order 3 : Cycadales

Phylum II : Coniferophyta

Order 1 : Cordaitales

Order 2 : Ginkgoales

Order 3 : Taxales
Order 4 : Coniferales
Phylum III : Chlamydospermophyta
Order 1 : Ephedrales
Order 2 : Gnetales

Takhtajan (1950) treated gymnosperms as class Gymnospermae under the phylum Pteropsida and divided gymnosperms into major sub-classes : Pteridosperminae (one order), Phyllosperminae (2 orders), Stachyosperminae (3 orders) and Chlamydosperminae (3 orders).

Takhtajan (1950) :

Division : Telemophyta
Phylum IV : Pteropsida
Class A : Filicinae
Class B : Gymnospermae
Sub-class 1 : Pteridosperminae
Order 1 : Cycadofilicales
Sub-class 2 : Phyllosperminae
Order 1 : Cycadales
Order 2 : Bennettitales
Sub-class 3 : Stachyosperminae
Order 1 : Cordaitales
Order 2 : Ginkgoales
Order 3 : Coniferales
Sub-class 4 : Chlamydospermi-nae
Order 1 : Ephedrales
Order 2 : Welwitschiales
Order 3 : Gnetales
Class C : Angiospermae

Johansen (1951) classified gymnosperms into five separate phylla with a total of 10 orders.

Johansen (1951) :

Gymnosperms
Phylum 1 : Pteridospermophyta
Order : Cycadofilicales
Phylum 2 : Cycadophyta
Order 1 : Bennettitales
Order 2 : Cycadales
Order 3 : Nilssoniales
Order 4 : Caytoniales
Phylum 3 : Ginkgophyta
Order 1 : Cordaitales
Order 2 : Ginkgoales
Phylum 4 : Coniferophyta
Order 1 : Voltziales
Order 2 : Coniferales
Phylum 5 : Ephedrophyta
Order : Gnetales

In the classification proposed by Pilger and Melchior in the 1954 edition of Engler's *Syllabus der Pflanzanfamilien,* gymnosperms have been divided into four classes where cycads and pteridosperms were merged together to place them under the class Cycadopsida. The order Taxales was segregated out from conifers to create a separate class Taxopsida.

1954 Edition of Engler's Syllabus :

Division XV : Gymnospermae
Class 1 : Cycadopsida
Order 1 : Pteridospermales
Order 2 : Caytoniales
Order 3 : Cycadales
Order 4 : Nilssoniales
Order 5 : Bennettitales
Order 6 : Pentoxylales
Order 7 : Ginkgoales
Class 2 : Coniferopsida
Order 1 : Cordaitales
Order 2 : Coniferales
Class 3 : Taxopsida
Order : Taxales
Class 4 : Chlamydospermae
Order : Gnetales

Pant (1957) classified gymnosperms into three divisions which is simply a modified version of Arnold's (1948) classification.

Pant (1957) :

Gymnosperms
Division I : Cycadophyta
Class 1 : Pteridospermopsida

Order 1 : Lyginopteridales

Order 2 : Medullosales

Order 3 : Glossopteridales

Order 4 : Peltaspermales

Order 5 : Corystospermales

Order 6 : Caytoniales

Class 2 : Cycadopsida

Order 1 : Cycadales

Class 3 : Pentoxylopsida

Order 1 : Pentoxylales

Class 4 : Bennettitopsida

Order 1 : Bennettitales

Division II : Chlamydospermophyta

Class 1 : Gnetopsida

Order 1 : Gnetales

Order 2 : Welwitschiales

Division III : Coniferophyta

Class 1 : Coniferopsida

Order 1 : Cordaitales

Order 2 : Coniferales

Order 3 : Ginkgoales

Class 2 : Ephedropsida

Order 1 : Ephedrales

Class 3 : Czekanowskiopsida

Order 1 : Czekanowskiales

Class 4 : Taxopsida

Order 1 : Taxales

K. R. Sporne (1965) proposed a classification of gymnosperms in his book *"The Morphology of Gymnosperms"* which is substantially the same as that proposed by Pilger and Melchior in the 1954 edition of Engler's *"Syllabus der Pflanzenfamilien"*.

Sporne proposed the diphyletic origin of gymnosperms. He divided gymnosperms into 3 classes, comprised of 9 orders. Except for Gnetales (having practically no fossil record and quite different from others), the other orders fall into two main classes (either Cycadopsida or Coniferopsida) based on wood anatomy, leaf form and seed structure. Sporne emphasised that the correlation between wood, leaf and seed is so strong that some morphologists interpreted gymnosperms as diphyletic and the two major groups, cycads and conifers evolved from different pteridophyte ancestors. This has been evidenced by the fact that both the groups appear almost simultaneously in the Palaeozoic period.

The main disadvantage of this classification is that it did not provide sufficient information regarding affinity and phylogeny of the group. Moreover, the placement and position of the 3rd class, Gnetopsida, has not been highlighted.

K. R. Sporne (1965) :

Gymnosperms

A. Class Cycadopsida (Manoxylic wood, large frond-like leaves which are basically pinnate in form or in venation; seeds radially symmetrical)

Order 1 : Pteridospermales

Family 1 : Lyginopteridaceae

Family 2 : Medullosaceae

Family 3 : Calamopityaceae

Family 4 : Glossopteridaceae

Family 5 : Peltaspermaceae

Family 6 : Corystospermaceae

Family 7 : Caytoniaceae

Order 2 : Bennettitales

Family 1 : Williamsoniaceae

Family 2 : Wielandiellaceae

Family 3 : Cycadeoidaceae

Order 3 : Pentoxylales

Family 1 : Pentoxylaceae

Order 4 : Cycadales

Family 1 : Nilssoniaceae

Family 2 : Cycadaceae

B. Class : Coniferopsida (Pycnoxylic wood; needle-shaped, paddle-shaped or fan-shaped leaves which are basically dichotomous in form or in venation; seeds bilaterally symmetrical)

Order 1 : Cordaitales

Family 1 : Eristophytaceae

Family 2 : Cordaitaceae

Family 3 : Poroxylaceae

Order 2 : Coniferales

Family 1 : Lebachiaceae
Family 2 : Voltziaceae
Family 3 : Palissyaceae
Family 4 : Pinaceae
Family 5 : Taxodiaceae
Family 6 : Cupressaceae
Family 7 : Podocarpaceae
Family 8 : Cephalotaxaceae
Family 9 : Araucariaceae

Order 3 : Taxales

Family 1 : Taxaceae

Order 4 : Ginkgoales

Family 1 : Trichopityaceae
Family 2 : Ginkgoaceae

C. Class Gnetopsida (Secondary wood with vessels; Flowers in inflorescence, male flowers with perianth; microphyle projecting as a long tube)

Order 1 : Gnetales

Family 1 : Gnetaceae
Family 2 : Welwitschiaceae
Family 3 : Ephedraceae

Cronquist, Takhtajan and Zimmermann (1965) classified gymnosperms into three subdivisions — Cycadicae (including Pteridosperms and Cycads), Pinicae (Ginkgos and Conifers) and Gneticae (Chlamydosperms).

Cronquist, Takhtajan and Zimmermann (1965):

Sub-kingdom : Embryobionta

Division 7 : Pinophyta

Sub-division 1 : Cycadicae

Class 1 : Lyginopteridatae
Class 2 : Cycadatae
Class 3 : Bennettitatae

Sub-division 2 : Pinicae

Class 1 : Ginkgoatae
Class 2 : Pinatae

Sub-class 1 : Cordaitidae
Sub-class 2 : Pinidae

Sub-division 3 : Gneticae

Class 1 : Gnetitae

Sub-class 1 : Ephedridae
Sub-class 2 : Welwitschiidae
Sub-class 3 : Gnetidae

Bierhorst (1971) proposed a classification of gymnosperms. He divided gymnosperms into three classes : Cycadopsida (including 4 orders), Coniferopsida (including 4 orders) and Gnetopsida (including 3 orders). Bierhorst also followed more or less the same pattern of segregation of major groups of gymnosperms as proposed by Pilger and Melchior (1954) and Sporne (1965).

Bierhorst (1971) :

Gymnosperms

Class I : Cycadopsida

Order 1 : Pteridospermales
Order 2 : Cycadales
Order 3 : Cycadeoidales
Order 4 : Caytoniales

Class II : Coniferopsida

Order 1 : Cordaitales
Order 2 : Coniferales
Order 3 : Taxales
Order 4 : Ginkgoales

Class III : Gnetopsida

Order 1 : Ephedrales
Order 2 : Gnetales
Order 3 : Welwitschiales

S. V. Meyen (1987) proposed a classification of gymnosperms, past and present, based on the origin and evolution of seeds in his book, *"Fundamentals of Paleobotany"*. Meyen used different morphological terminology for the description of gymnosperms. This classification did not imply the common origin of certain types, but merely aimed to refine and clarify the usage of descriptive terminology. Gymnosperms are linked by transitional forms with Pteridophytes and Angiosperms. Some of these forms gave evidence of phylogenetic links, such as Progymnosperms that phylogenetically linked Pteridophytes with Gymnosperms. In other cases, only certain organs are transitional, whereas similarities evolved independently in various phylogenetic lineages. From successive comparison of all the best-studied gymnosperm genera, it became apparent that the major phylogenetic lineages of gymnosperms and the respective taxa of higher ranks are best reflected in the evolution of seeds and cupules, At the phylogenetic roots of gymnosperm there were two different types of seeds, namely **Platyspermic** [Calamopityales] and **Radiospermic** [Lagenostomales with groups bilaterally symmetrical (BS) and radially symmetrical (RS)]. Both orders are linked to Progymnosperms by forms transitional in the anatomical structure of the stems. Group BS gave rise to Cordaitanthales, from which conifers evolved [**Secondary Platyspermic**], whereas group RS gave rise to Trigonocarpales, which were probably progenitors of Cycads and Bennettites.

The classification of Meyen (1987) differs from those of other earlier systems on the basis of the following characteristics :

1. This system is based on fundamental distinctions established (not postulated) between the major types of gymnosperm seeds—Platyspermic, Radiospermic and Secondary Platyspermic. Taken together with other pertinent characteristics, this allows for the precise discrimination of the classes. Hence this is a significant departure from the preceding classifications.
2. Gymnosperms are monophyletic taxa originating from one of the progymnospermic orders, probably Archaeopteridales.
3. The group of Pteridosperms proved to be heterogeneous, both morphologically and phylogenetically. It can be retained only as an informal unit comprising the genera of uncertain affinities.
4. Ginkgos are again affiliated with Leptostrobans (Czekanowskians), but can be allied with Peltasperms, not Conifers (their similarity is chiefly convergent).
5. The systems revealed the apparent parallelism between the classes in many characteristics, especially of the pollen and leaf structure. This makes it easier to explain the diversity of gymnosperms.

Criticism

Meyen (1987) placed Gnetales and Welwitschiales under the class Cycadopsida (the reason for this treatment has not been stated) and Ephedrales under the class Ginkgoopsida (because of its Platyspermic seed and striated pollen). In most of the classifications those three members were placed in a separate order because of their many common characteristics showing angiospermous affinity. Ignoring the other important characteristics (vegetative and other reproductive structures) this system is primarily based on the seed characteristics. So in spite of having sufficient evidences, this system has many flaws.

S. V. Meyen (1987) :

A. Class. Ginkgoopsida [Primary Platyspermic, non-cupular seeds; male fructifications are leafless branching microsporoclads or epiphyllous or forming synangia borne on underside of unmodified pinnules; stems showing eustele with one ring of protoxylem strand.

Order 1 : Calamopityales (1 family)
Order 2 : Callistophytales (1 family)
Order 3 : Peltaspermales (4 families)
Order 4 : Ginkgoales (1 family)
Order 5 : Leptostrobales (1 family)
Order 6 : Caytoniales (1 family)
Order 7 : Gigantonomiales or Gigantopteridales (1 family)
Order 8 : Arberiales or Glossopteridales ? (1 family)
Order 9 : Pentoxylales ? (1 family)
Order10 : Ephedrales (1 family)

B. Class. Cycadopsida [Radiospermic seeds; cupule in more advanced order undergo transformation into an outer integument; fern-like habit of the leaves which was inherited from Progymnosperm]

Order 1 : Lagenostomales (1 family)
Order 2 : Trigonocarpales (1 family)
Order 3 : Cycadales (1 family)
Order 4 : Bennettitales (2 families)
Order 5 : Gnetales (1 family)
Order 6 : Welwitschiales (1 family)

C. Class. Pinopsida [Secondary Platyspermic seeds; pycnoxylic wood; leaves simple, needle like, scaly, fan-shaped, etc.; male fructifications diversified]

Order 1 : Cordaitanthales (3 familes)

Order 2 : Dicranophyllales (1 family)

Order 3 : Pinales or Conifers (12 families)

Kramer and Green (1990) in their treatise *"The Families and Genera of Vascular Plants"* (edited by Kubitzki, 1990) have divided Gymnosperms into two sub-divisions — Coniferophytina and Cycadophytina. They have included two classes — Ginkgoatae and Pinatae, under Coniferophytina. Ginkgoatae has a single family, Ginkgoaceae, and Pinatae comprises nine families. The authors have divided Cycadophytina division into two classes namely, Cycadatae and Gnetatae. Cycadatae has a single order Cycadales with four families. Gnetatae also contains a single order Gnetales with three families.

This classification is principally based on the living members of gymnosperms where fossil members have totally been ignored. There seems hardly any justification to club Cycadatae and Gnetatae under the division Cycadophytina. The authors themselves have argued that the distinction of orders Pinales and Taxales under the class Pinatae is doubtful due to uncertain position of the Cephalotaxaceae.

Kramer and Green (1990) :

Gymnospermae

Division : Coniferophytina

Class : Ginkgoatae

Order : Ginkgoales

Family : Ginkgoaceae

Class : Pinatae

Order : Pinales

Family : Araucariaceae

Family : Cephalotaxaceae

Family : Cupressaceae

Family : Phyllocladaceae

Family : Pinaceae

Family : Podocarpaceae

Family : Sciadophytaceae

Family : Taxodiaceae

Order : Taxales

Family : Taxaceae

Division : Cycadophytina

Class : Cycadatae

Order : Cycadales

Family : Boweniaceae

Family : Cycadaceae

Family : Stangeriaceae

Family : Zamiaceae

Class : Gnetatae

Order : Gnetales

Family : Ephedraceae

Family : Gnetaceae

Family : Welwitschiaceae

Taylor and Taylor (1993), in the process of establishing major taxa in their book *"The Biology and Evolution of Fossil Plants"*, have raised the important groups to the rank of divisions. They have divided gymnosperms into six major divisions including a separate group 'Other Gymnosperms' covering those taxa which show uncertain affinity, under the Kingdom Plantae. This classification is a synthesis of different natural systems, describing structural features and evolutionary significance of the taxa. The main disadvantage of this classification is that it is based on the fossil members, ignoring the extant taxa.

Taylor and Taylor (1993) :

Kingdom : Plantae

18. Division : Progymnospermopsida
    1. Order : Archaeopteridales
    2. Order : Aneurophytales
    3. Order : Protopityales
    4. "Noeggerathians"
19. Division : Pteridospermophyta
    1. Order : Calamopityales
    2. Order : Buteoxylonales
    3. Order : Lyginopteridales
    4. Order : Medullosales

5. Order : Callistophytales
6. Order : Glossopteridales
7. Order : Caytoniales
8. Order : Corystospermales
9. Order : Peltaspermales

20. Division : Cycadophyta
1. Order : Cycadales
2. Order : Bennettitales (Cycadeoidales)

21. Division : Ginkgophyta
1. Order : Ginkgoales
2. Order : Dicranophyllales

22. Other Gymnosperms of uncertain affinities
1. Order : Czekanowskiales
2. Order : Vojnovskyales
3. Order : Pentoxylales
4. Order : Gigantopteridales

23. Division : Gnetophyta
1. Order : Gnetales

24. Division : Coniferophyta
1. Order : Cordaitales
2. Order : Coniferales

Stewart and Rothwell (1993) proposed a classification of gymnosperms in the book, *"Paleobotany and the Evolution of Plants"* which is a synthesis of several natural systems. They used evidences from the fossil records to establish the natural relationship among the major groups of vascular plants. In this classification, Progymnospermopsida (considered as ancestor of gymnosperms) and Gymnospermopsida (naked seeded plants) were placed as distinct classes under the division Tracheophyta (vascular plants) of the fourth Kingdom Plantae. They divided Progymnospermopsida into three orders and Gymnospermopsida into 13 orders where Gnetales has been placed in the middle as number 8th order in spite of having its advancement. The main advantage of this classification is that it encompasses almost all the fossil records available so far. However, living members mostly have not been enumerated and described. This system of classification is merely a synthesis of different classifications as proposed by earlier workers like Eames (1936), Tippo (1942), Cronquist (1960), etc.

Stewart and Rothwell (1993) :

Kingdom IV. Plantae

Division : Tracheophyta (vascular plants)

Class : Progymnospermopsida (ancestor of Gymnosperms)

Order : Aneurophytales
Order : Archaeopteridales
Order : Protopityales

Class : Gymnospermopsida (plants with naked seeds)

Order : Pteridospermales
Order : Cycadales
Order : Cycadeoidales
Order : Caytoniales
Order : Glossopteridales
Order : Pentoxylales
Order : Czekanowskiales
Order : Gnetales
Order : Ginkgoales
Order : Cordaitales
Order : Voltziales
Order : Coniferales
Order : Taxales

It has been observed that the classification of gymnosperms as proposed by various authors differ widely with reference not only to the criteria adopted for distinction of the major groups of plants, but also to the placement of various families, orders and classes. In this book, we have adopted a classification retaining the Division status for gymnosperms. The major plant types have been divided into six classes where gymnosperms having uncertain affinities are placed under the fourth class "Other Gymnosperms".

The classification followed in this book are :

Division : Gymnospermophyta

Class 1 : Pteridospermopsida

Order 1 : Cycadofilicales (Families : Calamopityaceae, Lyginopteridaceae, Medullosaceae, Callistophytaceae)

Order 2 : Glossopteridales (Family : Glossopteridaceae)

Order 3 : Caytoniales (Families : Caytoniaceae, Peltaspermaceae, Corystospermaceae)

Class 2 : Cycadopsida

Order 1 : Cycadales (Family : Cycadaceae)

Order 2 : Cycadeoidales (= Bennettitales)

(Families : Williamsoniaceae, Cycadeoidaceae)

Class 3 : Ginkgopsida

Order 1 : Ginkgoales (Family : Ginkgoaceae)

Class 4 : Other gymnosperms (Gymnosperms of uncertain affinities)

Order 1 : Czekanowskiales (Family : Czekanowskiaceae)

Order 2 : Pentoxylales (Family : Pentoxylaceae)

Class 5 : Coniferopsida

Order 1 : Cordaitales (Family : Cordaitaceae)

Order 2 : Voltziales (Family: Voltziaceae)

Order 3 : Coniferales (Families : Pinaceae, Taxodiaceae, Cupressaceae, Podocarpaceae, Araucariaceae, Cephalotaxaceae).

Order 4 : Taxales (Family : Taxaceae)

Class 6 : Gnetopsida (= Chlamydospermopsida)

Order 1 : Ephedrales (Family : Ephedraceae)

Order 2 : Welwitschiales (Family : Welwitschiace)

Order 3 : Gnetales (Family : Gnetaceae)

Distinction between Pteridophytes and Gymnosperms

Pteridophytes	**Gymnosperms**
1. Main plant body is sporophyte, gametophyte is independent of sporophyte.	1. Main plant body is sporophyte. There is a reduction in the gametophytic generation which is completely dependent upon the sporophyte.
2. In sporophyte, tap root system gets lost and roots are always adventitious.	2. Sporophyte bears well-developed tap root system.
3. Both homosporous and heterosporous members are present.	3. All the members are heterosporous.
4. Secondary growth is noted in some of the members, cambium is unifacial (only produces secondary xylem inwardly).	4. Secondary growth is present in all the members, cambium is bifacial and produces both secondary xylem and secondary phloem.
5. Male gametophytes develop through endosporic proximal germination of spores.	5. Male gametophytes develop through endosporic distal germination of pollen.
6. Male gametes are motile, bi- or multiflagellated.	6. Male gametes are generally non-motile (multiflagellated, motile sperms are present only in cycads and ginkgos)
7. There is direct access of male gametes to female gametes during fertilisation.	7. During fertilisation, male gametes have to come through pollen tube, hence there is no direct access of male gametes to female gametes.
8. Megasporangium wall is dehiscent in nature, hence pteridophytes are free-sporing.	8. Megasporangium wall is indehiscent in nature, megaspore is retained within megasporangium.
9. Archegonia are comprised of neck canal cells and ventral canal cell.	9. Archegonia lack neck canal cells, and ventral canal cell is frequently eliminated.
10. Absence of seeds.	10. Presence of seeds.

Distinction between Gymnosperms and Angiosperms

Gymnosperms	Angiosperms
1. Plants are woody, either trees, shrubs or lianes without having herbaceous member.	1. Plants are variable, although herbs are predominant.
2. Leaves are evergreen, generally with thick cuticle.	2. Leaves are variable like evergreen, deciduous, semideciduous etc., thickness of cuticle is also variable.
3. They generally lack vessels. Vessels are present only in Gnetales which have evolved from pitted tracheids.	3. All angiosperms bear vessels (except, Tochodendraceae, Winteraceae, etc.) which have evolved from scalariform tracheids.
4. Phloem consists of sieve cells where sieve areas occur on the radial wall. Companion cell is absent.	4. Phloem consists of sieve tube and companion cell. Sieve plates are present on lateral walls.
5. Stem possesses either mesarch or endarch xylem.	5. It is never mesarch.
6. Flowers are naked, i.e. not covered by accessory whorls like sepals or petals (except Gnetales).	6. Flowers are generally covered by accessory whorls.
7. Male gametophyte contains prothallial cells.	7. No prothallial cells are found.
8. Both motile and non-motile male gametes are found.	8. Male gametes are always non-motile.
9. Pollen grains are anemophilous.	9. Pollen are either anemophilous or entomophilous or hydrophilous.
10. Pollen exine (wall) is alveolar or granular.	10. Pollen exine (wall) is tectate.
11. There is a long interval between pollination and fertilisation.	11. The gap between pollination and fertilisation is rather short (except, Amentiferae).
12. Ovules (seeds) consist of a single integument.	12. Ovules (seeds) consist of two integuments (outer and inner).
13. Generally, integument has a vascular supply.	13. The vascular strand terminates at the base of the integuments.
14. The ovules are exposed (naked) on the megasporophylls, hence gymnosperms are called open-seeded plants.	14. Ovules are enclosed in ovary, hence angiosperms are called closed-seeded plants.
15. Endosperm is haploid (n) and produced before fertilisation.	15. Endosperm is triploid (3n) and produced after fertilisation.

CLASS. PTERIDOSPERMOPSIDA

The Carboniferous period is considered as the **'age of ferns'** because of the presence of a large number of primitive ferns (Primofilicales) as well as fern-like fronds of uncertain affinity. It was Stur (in 1883) who first time hypothesised the existence of such type of plants. The systematic position of such fern-like plants were not established. Williamson (1887) recognised the existence of such plants showing combined structural characteristics of both cycads and ferns based on anatomical features. It was H. Potonié in 1899, who, on the basis of anatomical evidence, suggested the existence of a group of plants that was transitional between ferns and seed plants and introduced the concept of Cycadofilicales (Pteridosperms). Finally, two British palaeobotanists, F. W. Oliver and D. H. Scott, in 1904, established the concept of Pteridosperms (seed ferns) based on the detailed work on *Lyginopteris oldhamia* plant.

The Pteridosperms were a very large and diverse assemblage of plants which first appeared in the Upper Devonian, had its climax during the Carboniferous to Triassic and became extinct in the Cretaceous. Due to their wide span of existence, they are conventionally divided into two broad groups : Palaeozoic Pteridosperms (Upper Devonian to Permian) and Mesozoic Pteridosperms (Triassic to Cretaceous).

GENERAL CHARACTERISTICS OF PTERIDOSPERMOPSIDA

1. Some were small trees with upright trunks, others were lianes or vines with scrambling habit.
2. Massive fern-like fronds arranged helically on stem.
3. Primary xylem mesarch (rarely exarch) with variable stelar configurations, ranging from simple protostele to eusteles.
4. Wood is manoxylic and limited in amount. Thin-walled tracheids are with multiseriate pitting on radial walls.
5. Both pollen-bearing organs and ovules (seeds) borne on leaves.
6. Pollen organs aggregated into clusters or arranged into large synangiate organs.
7. Seeds large and solitary, sometimes small and surrounded by cupulate structure.

The Pteridosperms are now divided into three orders :

Class : Pteridospermopsida

Order 1 : Cycadofilicales

Family 1 : Calamopityaceae (U. Devonian to L. Carboniferous)

Family 2 : Lyginopteridaceae (Carboniferous)

Family 3 : Medullosaceae (Carboniferous to Permian)

Family 4 : Callistophytaceae (U. Carboniferous to Permian)

Order 2 : Glossopteridales

Family 1 : Glossopteridaceae (U. Carboniferous to Triassic)

Order 3 : Caytoniales

Family 1 : Caytoniaceae (Triassic to Cretaceous)

Family 2 : Peltaspermaceae (U. Permian to Triassic)

Family 3 : Corystospermaceae (Triassic).

ORDER 1 : CYCADOFILICALES (= PTERIDOSPERMALES)

Family 1 : Calamopityaceae

The family Calamopityaceae is an interesting group of permineralised plants. The family consists entirely of stems and petiole remains where other organs are not known.

The different form-genera so far identified are :

Stems : *Stenomyelon, Calamopitys*

Petiole : *Kalymma*

Stems : The Lower Carboniferous stem (*Stenomyelon*) is considered to be the most primitive member of the family. The beautifully preserved stem (Fig. 1.1A) of *S. tuedianum* shows a primitive type of three lobed protostele dissected by the radiating plates of parenchyma like the Aneurophytales of progymnosperm. Hence this genus is considered to be the ancestor of seed ferns. Metaxylem tracheids exhibit multiseriate bordered pits. In another species, *S. primaevum*, the protoxylem is entirely composed of tracheids with mesarch primary xylem.

Calamopitys is the commonest genus showing a large parenchymatous cortex with sparganum cortical sclerenchyma and leaf traces (Fig. 1.1B). The stele appears roughly triangular which ranges from protosteles with abundant parenchyma to eusteles. *C. americana* is a common species that exceeds 4.0 cm in diameter. The pith is of mixed type, comprised of tracheids and parenchyma.

In both the genera, stems are with compact manoxylic wood, surrounded by primary xylem.

Petiole : The petioles (*Kalymma*) are attached in 2/5 phyllotaxy to the stem of *Stenomyelon*, showing many traces containing two exarch primary xylem. Some petioles show a number of bundles that assume a C-configuration.

Family 2 : Lyginopteridaceae

The family Lyginopteridaceae shows its historical importance in the recognition of the group Pteridosperm (seed fern). There are some characteristic features which can be used to distinguish the family from rest of the members of Pteridosperms :

1. Presence of cauline monostelic vasculature.
2. Presence of axillary branching.
3. Fronds are large with bifurcate rachis.
4. Petioles are small with a single V-, Y- or W-shaped trace that are formed from the fusion of several small traces.
5. Pollen-bearing organs are small, laminar, or terminal in clusteristics on branches of fronds.
6. Pollens are of pre-pollen type with trilete aperture.
7. Ovules are small, radially symmetrical (radiospermic) and borne in cupules having hydrasperman pollen chamber.
8. Nucellus is fused with integument and apex forming a salpinx or lagenostome.

The several important form-genera of Lyginopteridaceae so far identified are :

Stems : *Lyginopteris, Heterangium, Microspermopteris, Schopfiastrum, Tristichia.*

Frond : *Sphenopteris.*

Root : *Kaloxylon.*

Rachis : *Rachiopteris, Lyginorachis.*

Pollen-bearing organs : *Crossotheca, Telangium, Telangiopsis, Feraxotheca.*

Seeds : *Lagenostoma, Physostoma, Stamnostoma, Eurystoma, Lyrasperma, Sphaerostoma, Conostoma.*

Cupule : *Calymmatotheca.*

Lyginopteris oldhamia (= *Calymmatotheca hoeninghausii*) is the most well-known plant proposed by H. Potonié in 1899. But as per rule of priority the correct name for *Lyginopteris* should be *Calymmatotheca hoeninghausii*, as the form-genus *Calymmatotheca* was described earlier by Stur in 1877. The *Lyginopteris* plants were vine-like, growing with the support of neighbouring trees or shrubs. All the parts of the plant (except roots) are provided with **capitate glands**. The different common parts of the plants are :

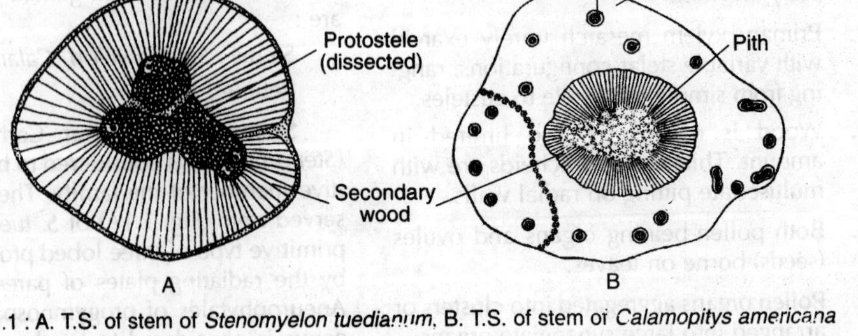

Fig. 1.1 : A. T.S. of stem of *Stenomyelon tuedianum*, B. T.S. of stem of *Calamopitys americana*

Stem : *Lyginopteris oldhamia*
Fond : *Sphenopteris hoeninghausii*
Root : *Kaloxylon hookeri*
Rachis : *Rachiopteris aspera*
Pollen-bearing organ : *Crossotheca hoeninghausii*
Seed : *Lagenostoma lomaxi*
Cupule : *Calymmatotheca hoeninghausii.*

Stems : The frequently branched stem of *Lyginopteris oldhamia* was long, slender (2 mm to 4 cm in diameter). The anatomy of the stem (Fig. 1.2A) shows **eustelic** configuration with broad cortex bounded by a single epidermis. Small multicellular hairs are a common feature of the epidermis. The cortex is divided into two zones : outer cortex with radial patches of sclerenchymatous cells resembling the Roman numerals on a clock face, and inner cortex which is parenchymatous. A ring of secondary wood with radially arranged multiseriate bordered-pitted tracheids is present. The amount of wood is relatively small in proportion to the total size of the stem. Five strands of mesarch primary xylem are present inside the secondary wood. Leaf traces are present outside the stele. The pith is large and parenchymatous and possesses frequent clusters of small sclerotic cells forming **sclerotic nests.**

Heterangium (Fig. 1.3) is another stem genus associated with *Sphenopteris* leaf, reported from Lower and Upper Carboniferous deposits of Scotland. The stem is 2-4 cm in diameter which is rarely branched. The primary vasculature is of **mixed protostele.** A narrow band of secondary xylem is present outside the primary xylem. The secondary wood is similar to *Lyginopteris.* The cortex is also divided into two zones like *Lyginopteris.* The pith is large and reticulate.

Fronds : Bipinnate or tripinnate leaves of *Sphenopteris hoeninghausii* (Fig. 1.2B) are about 0.5m long which are arranged spirally on stem or branches. Leaves are borne at right angles to the rachis which show alternately arranged pinnules with free veins. The leaf anatomy shows a cutinised upper epidermis and mesophyll tissue divisible into palisade and spongy parenchyma. Leaf traces are formed by a tangential division of a primary xylem strand. The trace separates into a pair of strands in the cortex that fuse at the

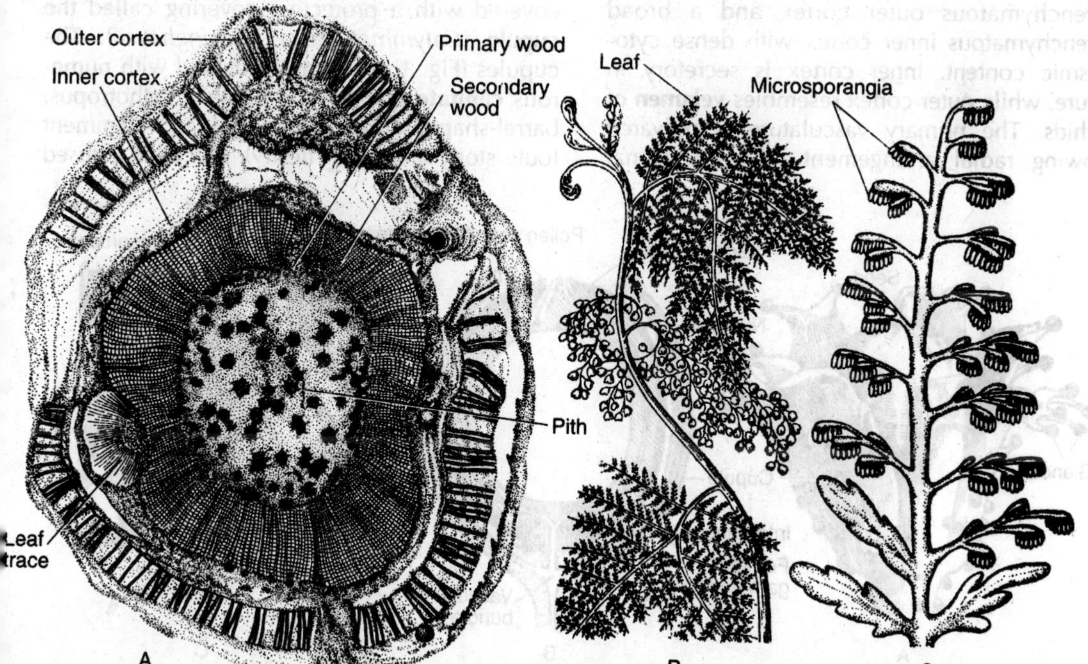

Fig. 1.2 : A. T.S. of *Lyginopteris oldhamia* stem, B. A reconstructed shoot showing arrangement of *Sphenopteris* hoeninghausii leaves, C. *Crossotheca* microsporophyll

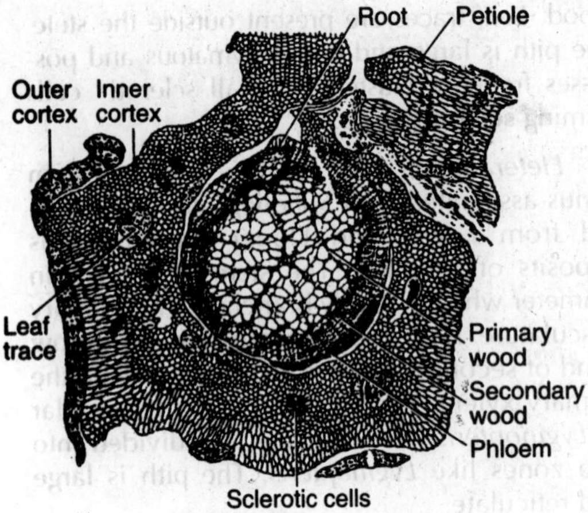

Fig. 1.3 : T.S. of *Heterangium grievii*

base of the petiole (*Rachiopteris aspera*) to form a V-, Y- or W-shaped bundle. Petioles — when preserved as isolated fragments, with a V-, Y- or W-shaped bundle — are referred to as *Lyginorachis.*

Roots : Roots (*Kaloxylon hookeri*) are adventitious, aerial and long. T. S. of root shows an epidermis followed by a well-developed cortex. The cortex is divided into two zones : 2-3 layered parenchymatous outer cortex and a broad parenchymatous inner cortex with dense cytoplasmic content. Inner cortex is secretory in nature, while outer cortex resembles **velamen** of orchids. The primary vasculature is polyarch showing radial arrangement with occasional reports on the occurrence of endodermis and pericycle.

Pollen-bearing organs : The male fructification, *Crossotheca hoeninghausii* (Fig. 1.2C) has been identified in association with the frond, *Sphenopteris hoeninghausii.* The fertile part of *Crossotheca* is spathulate or hastate bearing a number of boat-shaped, pendent, bilocular microsporangia that lack an annulus. The whole structure resembles minute **hair-brush** or **epaulets**. Other pollen-bearing organs of this family are *Telangium, Telangiopsis* and *Feraxotheca.* In these form-genera pollen organs are synangiate with many pollen sacs arranged pinnately.

The **pre-pollen** of Lyginopteridaceae are spherical with trilete aperture, measuring about 70 μm in diameter.

Seeds : A large number of seeds or **pre-ovules** namely, *Lagenostoma* (Fig. 1.4A), *Physostoma, Sphaerostoma, Conostoma, Lyrasperma, Stamnostoma, Eurystoma,* etc. has been identified, of which *Lagenostoma lomaxi* (Fig. 1.4A, B) is better known. *Lagenostoma* seeds are associated with ultimate part of *Lyginopteris* leaves which are about 5.5 mm in length and 4.25 mm in diameter. The seeds are covered with a protective covering called the **cupule** (*Calymmatotheca hoeninghausii*). The cupules (Fig. 1.4A, B) are provided with numerous **capitate glands**. The ovule is orthotropus, barrel-shaped and is covered by an integument (oute stony and inner fleshy). Nucellus is fused

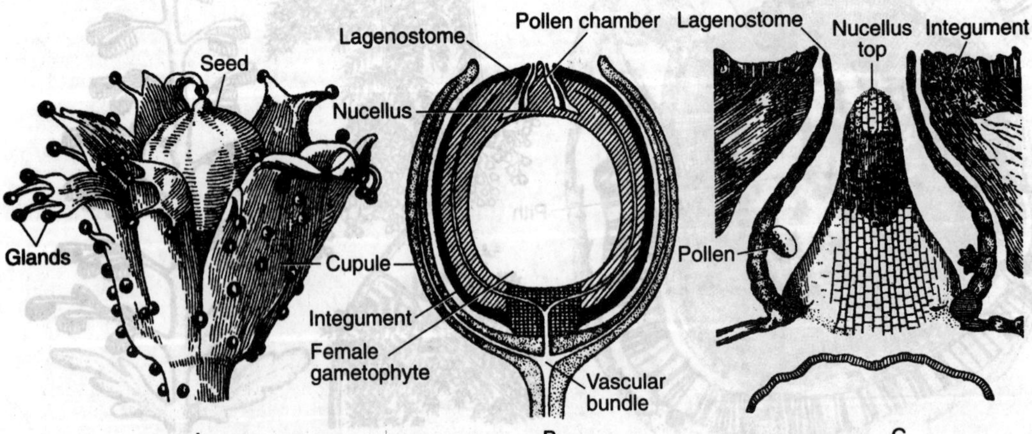

Fig. 1.4 : A. *Lagenostoma lomaxii*, B. Median L.S. of *Lagenostoma*, C. Enlarged view of the top of the *Lagenostoma* showing pollen chamber (lagenostome)

with integument except at the apical chamber. A ring of 8 to 9 vascular bundles are traversing the integument. Tip of the integument is divided in 9 parts, each receiving a vascular strand. The pollen chamber became more complex by the formation of a flask-shaped **lagenostome** due to the extended growth of nucellus (Fig. 1.4C). Trilete pre-pollens are occasionally found in pollen chamber.

Family 3 : Medullosaceae

The members of this family are known to occur in the Mississippian (Lower Carboniferous) extending up to Permian. They were more massive than the Lyginopteridaceae. The genus *Medullosa* was initially established by Cotta in 1832 for Lower Permian stem fossils.

The distinguishing characteristic features of the family :

1. Plants are massive trees.
2. Stems with several segments of vascular tissue, each surrounded by secondary xylem.
3. Fronds with a bifurcate rachis.
4. Massive petioles with large number of scattered traces.
5. Pollen organs synangiate with tubular sporangia.
6. Pollens are of pre-pollen type, large with monolete aperture.
7. Seeds are large, non-cupulate, three-angled.
8. Nucellus free from integument, except at the base.
9. Pollen chambers simple.

The different form-genera of this family so far identified are :

Stems : *Medullosa, Colpoxylon, Sutcliffia, Quaestora.*

Leaves : *Neuropteris, Alethopteris, Reticulopteris, Callipteridium, Mixoneura, Linopteris, Cyclopteris,* etc.

Petiole : *Myeloxylon.*

Pollen organs : *Codonotheca, Whittleseya, Aulacotheca, Dolerotheca, Bernaultia, Potonié,* etc.

Ovules (seeds) : *Pachytesta* (= *Trigonocarpus*), *Stephanospermum, Aetheotesta, Codonospermum,* etc.

Stems : The well-known reconstructed *Medullosa noei* (Stewart and Delevoryas, 1956) shows a general appearance of a tree fern of 3.5 m to 4.5 m height with a crown of leaves. The lower portion of the stem is covered by periderm and adventitious roots. The scars of spirally arranged leaf bases are arranged on the upper portion of the stem. The T.S. of young stem shows two to four segments of vascular tissue. In older stem, a ring of secondary parenchyma is present that superficially resembles another vascular segment (Fig. 1.5A). The number of vascular segments varies not only from level to level, but also within the species. Vascular segment is elliptical or band shaped in cross-section and each segment is **protostelic** in nature, composed of clusters of large metaxylem cells separated by abundant parenchyma. There are one to four mesarch **sympodia** or **axial bundles** near the edge of the primary body. Trachieds of secondary xylem are very long (up to 24 mm) with 12 rows of bordered pits on radial wall. The periderm cells are isodiametric, thick-walled and produced towards inside the stem. The cortical parenchyma and abundant elongated secretory canals are present towards outside. Patches of sclerenchyma fibres form interrupted bands around the periphery of the stem. Most of the geologically younger (Permian) medullosan stems differ markedly from the older (Carboniferous) forms, principally on the basis of the arrangement of vascular segments. *Medullosa anglica* and *M. thompsonii* of Lower-Middle Pennsylvanian are the two other important medullosan stems.

Colpoxylon, recorded from Permian of France, is characterised by the complete absence of central vascular strands in the ground tissue. The leaf traces are small and circular with secondary xylem.

Sutcliffia of Lower Coal Measures of England represents bigger stem (15 cm in diameter) with two large central vascular segments surrounded by many subsidiary segments.

Leaves : Medullosan fronds are very large, dichotomously branched or slightly unequal in lower portion, pinnately compound. The most common leaf-genera are *Neuropteris* and

Alethopteris. Pinnae of *Alethopteris* are attached with *Medullosa noei* stem. Pinnules show a prominent mid-vein from which secondary veins develop at sharp angles. T. S. of pinnule shows its revolute (turned under along the margin) nature and the midrib contains one to four vascular traces. One to two layers of palisade cells are present beneath the upper epidermis, while in the lower part loosely arranged mesophyll cells are found.

In *Neuropteris* (Fig. 1.5B), the pinnules possess a cordate base. The prominent midvein and secondary veins dichotomise several times before entering the pinnule margin. T.S. of *Neuropteris rarinervis* pinnules shows upper epidermis with undulating margins and parenchymatous bundle sheaths.

Other important leaf genera of medullosan members are *Reticulopteris, Mixoneura, Cyclopteris, Odontopteris, Lonchopteris*, etc.

Petiole : The detached petioles of Medullosaceae are called *Myeloxylon* which have numerous secretory canals. T.S. of *Myeloxylon* shows a striking similarity with maize stem in having many scattered vascular bundles and sclerenchymatous fibre at the periphery.

Roots : Roots of *Medullosa* are adventitious. T.S. of roots attached to stems of *Medullosa anglica* are triarch with broad periderm in the older roots. In some Middle Pennsylvanian *Medullosa* roots, xylem exhibit exarch actinostele (**pentarch**) with secondary xylem around the primary body.

Pollen organs : Pollen organs of Medullosaceae are synangiate structure with tubular sporangia showing longitudinal dehiscence. *Bernaultia* is a well-described pollen organ, consists of a hemispherical **campanulum** (bell-shaped structure) with an eccentrically placed pedicel. There are radiating rows of paired sporangia with multicellular hairs. *Whittleseya* (Fig. 1.5B) is a hollow synangiate compression-impression campanulum formed by the fusion of a large number sporangia side by side. *Whittleseya* pollen organs are associated with *Neuropteris* frond. *Codonotheca* is another compression-impression pollen organ, made up of 6 sporangia in a ring that are fused at bases and the apex remains free. A complete fusion of a ring of 6 sporangia is observed in compressed *Aulacotheca*. *Dolerotheca* is a large and complex pollen organ, consisting of four independent radial sporangia that are fused together to form a compound structure. It was observed that *Bernaultia* is associated with *Myeloxylon* frond of *Alethopteris* pinnules.

Pre-Pollen : The dispersed pre-pollen of Medullosaceae are relatively large (100-500 μm), monolete type. These grains are bilaterally symmetrical, round to elliptical in shape. The

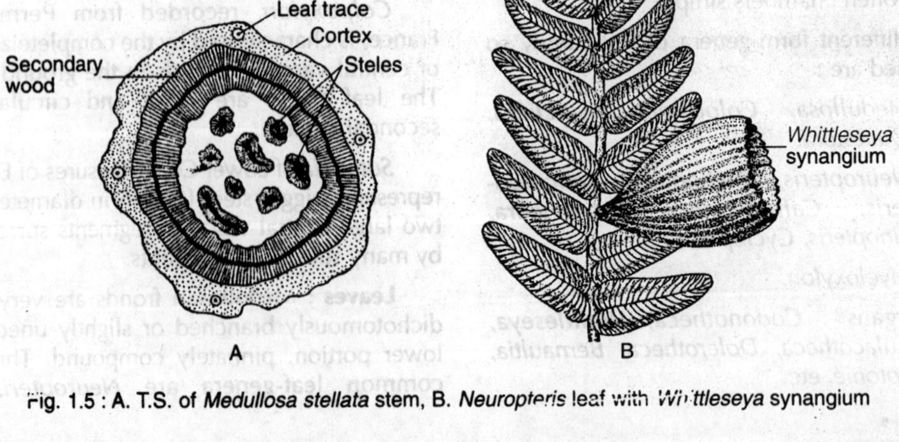

Fig. 1.5 : A. T.S. of *Medullosa stellata* stem, B. *Neuropteris* leaf with *Whittleseya* synangium

proximal aperture is bent with two longitudinal distal grooves.

Ovules (seeds) : The seeds of Medullosaceae are frequently found as casts, although many permineralised seeds have been reported. Brongniart in 1828 proposed *Trigonocarpus* nomenclature for cast specimen of *Medullosa* seeds. It has been observed that permineralised *Medullosan* ovules, *Pachytesta* (Fig. 1.6A–C) are associated with *Alethopteris* fronds. So there has been controversy regarding the validity of the two generic names for the medullosan ovules (seeds). It has now been universally accepted that the name *Trigonocarpus* is to be restricted to those seeds which are preserved as cast or compression-impression fossils and the name *Pachytesta* to be retained for permineralised (petrified) ovules (seeds). Both the genera show **trimerous** organisation. *Panchytesta* ovules are the largest among seed-ferns. *Pachytesta* (Fig. 1.6C) shows its massive nucellus which is free from the integument except at the base. Nucellus forms a simple bell-shaped pollen chamber. The ovule reveals double vascularisation where both the nucellus and integument are supplied with vascular strands. The integument is three-layered (Fig. 1.6B) : an outer fleshy layer (**sarcotesta**), a stony, fibrous, middle layer (**sclerotesta**) and an inner thin layer (**endotesta**). Megagametophytes have been reported to be preserved in *Pachytesta* seeds.

Family 4 : Callistophytaceae

The family Callistophytaceae is probably the best-known group of Palaeozoic Pteridosperms. Callistophytean fossils are principally known from Middle to Upper Pennsylvannia of USA and Upper Carboniferous of France.

General characteristic features of Callistophytaceae :

1. Small shrubby, scrambling plants that branch profusely.
2. Stem eustelic that bears spirally arranged pinnately compound leaves with axillary buds or branches at each node.
3. Young leaves show circinate vernation.

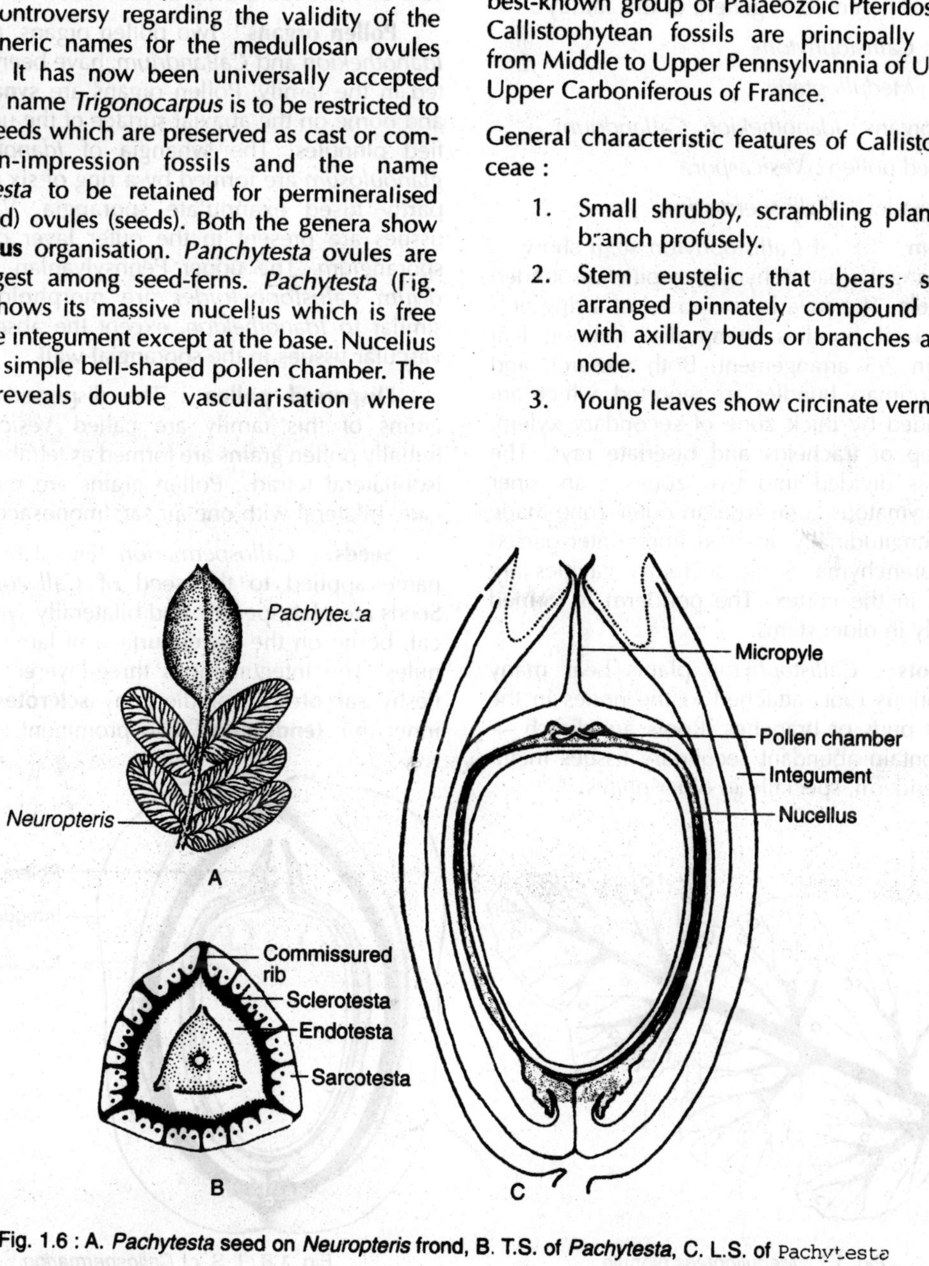

Fig. 1.6 : A. *Pachytesta* seed on *Neuropteris* frond, B. T.S. of *Pachytesta*, C. L.S. of *Pachytesta*

4. Pollen organs synangiate and borne superficially on pinnules of the frond.
5. Pollen grains are small and monosaccate.
6. The seeds are small, platyspermic and non-cupulate.
7. The integument is free from the nucellus except at the base.

The different form genera of this family :

Stems : *Callistophyton*
Frond : *Medullopteris.*

Pollen organs : *Idanothekion, Callandrium*
Dispersed pollen : *Vesicaspora*

Seeds (ovules) : *Callospermarion.*

Stem : T.S. of *Callistophyton* stem shows a central angular parenchymatous pith surrounded by **eustele.** The stele is composed of a ring of 5 to 13 axial bundles (**sympodia**) forming leaf traces in 2/5 arrangement. Both mesarch and exarch primary bundles are reported, which are surrounded by thick zone of secondary xylem, made up of tracheids and biseriate rays. The cortex is divided into two zones : an inner parenchymatous zone, and an outer zone made up of longitudinally-oriented fibres interspersed with parenchyma. Some secretory cavities are present in the cortex. The periderm is formed specially in older stems.

Roots : *Callistophyton* plants bear many adventitious roots attached to the nodes in the axils of buds or branches. Roots are diarch — they contain abundant secondary tissues including periderm, specially in older plants.

Leaves : *Medullopteris* (Fig. 1.7) leaves (up to 30 cm long) are helically arranged on *Callistophyton* stem or branch. Fronds are bi- to quadripinnately compound with prominent secretory cavities. The pinnules are deeply lobed with constricted base. Pinnules adjacent to the tip of the frond are small with blunt tip and wavy margin, while those of more proximal, pinnules are large and lobed. Multicellular hairs are present on the abaxial side of the leaves.

Pollen organs : Two pollen organs, namely *Idanothekion* and *Callandrium,* have been reported in the family. Pollen organs are **synangiate** and borne on the abaxial surface of the unmodified pinnules. The synangia of *Idanothekion glandulosum* are formed by a ring of six to nine partly fused exannulate sporangia. Vascular tissues are present in the outer layer of each sporangium. The upper Pennsylvanian *Callandrium callistophytoides* are morphologically similar to *Idanothekion,* except the absence of vascular tissues in the sporangial wall.

Dispersed pollen : The dispersed pollen grains of this family are called *Vesicaspora.* Initially pollen grains are formed as tetrahedral or isobilateral tetrads. Pollen grains are monosulcate, bilateral with one air sac (monosaccus).

Seeds : *Callospermarion* (Fig. 1.8) is the name applied to the seed of *Callistophyton.* Seeds are platyspermic and bilaterally symmetrical, borne on the lower surface of laminar pinnules. The integument is three-layered : outer fleshy (**sarcotesta**), middle stony (**sclerotesta**) and inner thin (**endotesta**). Two prominent vascular

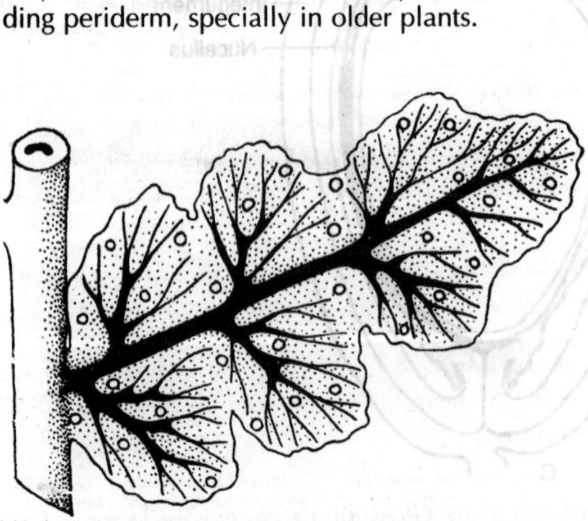

Fig. 1.7 : *Medullopteris* pinnule

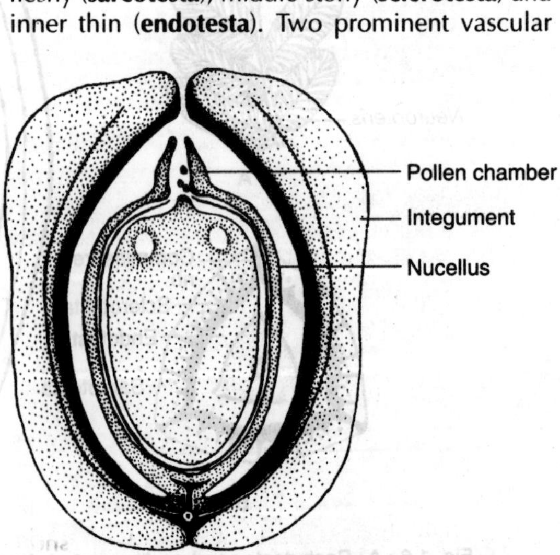

Fig. 1.8 : L.S. of *Callospermarion*

strands are present in the wings of the integument and a small pad of tracheids is present at the base of the nucellus. The integument is free from the nucellus except at the base and the distal end is modified to form a flask-shaped pollen chamber. The ovules were able to trap pollen grains by the production of resinous pollination droplets, extended from the micropylar orifis. This **pollination drop mechanism** of Callistophytaceae suggests its affinity with Cordaitales and modern Conifers.

ORDER 2 : GLOSSOPTERIDALES

Family : Glossopteridaceae

The name *Glossopteris* was first proposed by Brongniart (1828) for certain tongue-shaped leaves with a prominent midrib and reticulate venation. *Glossopteris* occurs in Australia, Africa, South America, Antarctica and India and constitute the Permian Gondwana land floras of Southern Hemisphere that separated from other continents of Northern Hemisphere by the great Tethys sea. *Glossopteris* first appears in the Upper Carboniferous, forms a climax vegetation during the Permian and decline thereafter. Some specimens are also known from the Triassic period of India (Pant and Pant, 1987) and the Jurassic deposits of Mexico (Delevoryas, 1969).

From the evidences obtained from the works of various workers, the Glossopteridales has been characterised by the following features :

1. The *Glossopteris* was a large tree (Fig. 1.9A). The trunk was about 6 m tall with araucarioxylon type of gymnospermous wood. The large number of isolated leaves and, to a much lesser extent, reproductive organs, suggest that the plants were deciduous. This has been established by the occurrence of numerous isolated leaves that are found in autumn/ winter sediments, whereas they are absent in spring/summer sediments.
2. The plant was supported by a root system of *Vertebraria* type.
3. The secondary growth was present in stems, branches and roots showing conspicuous growth ring that indicates climatic variations.
4. Leaves were arranged alternately or in tight helices to form whorls on short shoots.
5. Reproductive organs were either borne on leaves or attached to highly modified leaves. Both male and female fructifications were borne separately. It is not known whether they were monoecious or dioecious ?

Different important form-genera of *Glossopteris* plant so far identified are :

Root : *Vertebraria*

Leaves : *Glossopteris, Gangamopteris, Palaeovittaria, Rubidgea, Euryphyllum, Rhabdotaenia*, etc.

Male fructifications : *Eretmonia, Glossotheca, Kendostrobus.*

Ovule structures : *Lidgettonia, Ottokaria, Scutum, Denkania, Dictyopteridium.*

Root : The detached roots of *Glossopteris* plant are called *Vertebraria* (Fig. 1.9B). The roots are flattened and grooved with wedge-like sectors that radiate from the centre of the axis. The wedges are made up of secondary xylem which covers the central polyarch protoxylem. The pycnoxylic secondary xylem strands alternate with protoxylem strands (Fig. 1.9C). In young stage, the cavities that formed between the radiating arms suggest that the plant grew in semiaquatic environment.

Leaves : *Glossopteris* (Fig. 1.9D) leaves are simple, entire and sessile (rarely petiolate, e.g., *G. petiolata*). Leaves show a great variation in size and shape (linear lanceolate to spathulate ovate). They have a strong midrib from which numerous longitudinally running veins pass out to form a reticulate pattern or remain free. Leaf anatomy shows a typical dorsiventral and **hypostomatic** nature with irregularly dispersed **haplocheilic**, sunken stomata. Some species show a hypodermal layer. Mesophyll is differentiated into palisade and spongy parenchyma. Some scale leaves (rounded or lanceolate) are found to be associated with *Glossopteris* leaves.

Some other less common leaf genera namely, *Gangamopteris, Palaeovittaria, Rubidgea, Euryphyllum, Rhabdotaenia* are also included under Glossopteridaceae family due to their identical form, venation pattern and cuticular structure.

Gangamopteris (Lower Permian) leaf is identical to *Glossopteris* except the absence of well-defined midrib (Fig. 1.9E). *Rhabdotaenia* is

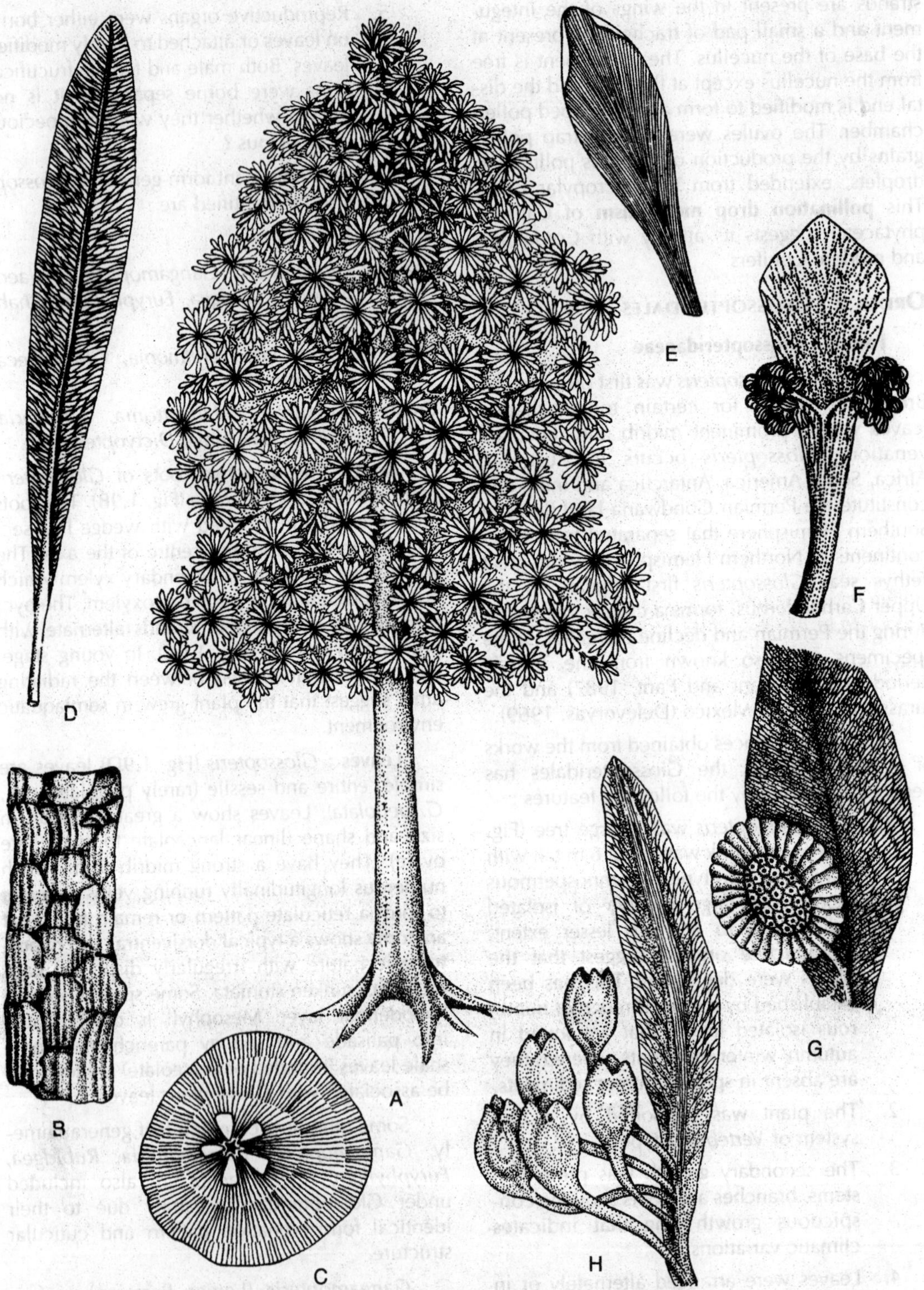

Fig. 1.9 : A. A *Glossopteris* plant, B. *Vertebraria*, C. T.S. of *Vertebraria*, D. *Glossopteris* leaf, E. *Gangamopteris* leaf, F. *Eretmonia*, G. *Scutum leslium* fructification associated with *Glossopteris* leaf, H. *Denkania indica*

distinguished from other leaf genera due to the presence of non-anastomosing lateral veins.

Male fructifications

Eretmonia (Fig. 1.9F) pollen organ consists of a fertile petiolate lamina of triangular to rhomboidal in outline. Two branches bearing whorls of microsporangia are borne on upper part of the lamina. After dehiscence, the sporangia look purse-shaped due to the longitudinal rupture.

The most common pollen organ, *Glossotheca,* is basically similar to *Eretmonia*, except it has 2-3 sporangia-bearing pedicels. Each pedical bifurcates into two daughter branches containing clusters of elongate sporangia at their tips.

Kendostrobus is another pollen organ associated with *Glossopteris* leaf. It is a long, stalked, cylindrical structure having a central axis with spirally arranged naked exannulate sporangia in groups.

The isolated sporangia of *Glossopteris* are known as *Arberiella, Lithangium* and *Polytheca.* Pollen grains are **bisaccate, striated** in *Arberiella,* monolete in *Lithangium* and trilete in *Polytheca.*

Female fructification

The dorsiventral structure bearing seeds of Glossopteridales are variously termed as capitulum, megasporophyll, cupule, fertiliger or cladode.

Scutum (Fig. 1.9G) has been described by Mrs. Plumsted (1952). This lanceolate to ovate fructification is attached by a pedicel to the lower portion of *Glossopteris* leaf midrib. It has two bilaterally concave symmetrical valves. A wing-like expansion is present along the line where the two valves join. Small sac-like structure are present in the concavity of both the halves. The fructification as described by Plumsted is to be bisexual where the lower sacs represent seeds and the upper ones pollen sacs bearing monolete, non-winged pollen. But according to many authors, the *Scutum* is a female fructification where one half of the unit represent a receptacle that contains a large number of ovules on elevated projection. The other half represents a fertile scale that protects the ovules at early age and detaches later to expose micropyle for receiving pollen grains during pollination. Some other authors (Rigby 1978) argued that the ovules of *Scutum* were borne on a single fleshy structure and the appearance of two parted structures is due to the splitting of rock through the middle of the ovule-bearing organ.

Ovulate structure of *Denkania indica* (Fig. 1.9H) was reported from Raniganj stage (Permian) of India (Surange and Chandra 1971). In this organ, six long pedicellate seed-bearing cupules are attached to the adaxial surface of the midrib of *Glossopteris* leaf. Apparently, one seed is produced per cupule comprising of two appressed scale-like valves.

Lidgettonia is another compound ovulate structure consisting of a spathulate leaf with two rows of disc like capitula attached abaxially on the petiole.

Ottokaria is an ovulate fructification attached to *Gangamopteris* leaf by a slender stalk. It consists of a flattened capitulum with a marginal frill bearing a large number of ovules on the under surface.

Some other ovulate structures of glossopterids like *Dictyopteridium, Partha, Rigbya, Plumsteadia, Senotheca, Rusangea,* etc., have been described by various authors.

ORDER 3 : CAYTONIALES

Caytoniales is another group of Pteridosperms which was widespread during the Upper Triassic to Lower Cretaceous. Initially Caytoniales was considered as a close associate of angiosperms because of the apparent closed carpel-like cupulate structure. But now the members of Caytoniales are included under Pteridosperms and are divided into three distinct families namely, Caytoniaceae, Peltaspermaceae and Corystospermaceae.

Family 1 : Caytoniaceae

The family Caytoniaceae had a wide stratigraphic range (U. Triassic to L. Cretaceous). The leaves of this family are placed in form-genus *Sagenopteris,* the male fructification in *Caytonanthus* and the female fructification in *Caytonia.* These form genera were found in close association in the deposits of Greenland, Western Canada, Eastern Russia, England, Siberia, etc.; however, no organic connection could be noticed.

Leaf : *Sagenopteris* leaf (Fig. 1.10A) is palmately compound, consisting of two pairs of four lanceolate leaflets, with a slender petiole. Each leaflet has a prominent midrib showing reticulate venation. **Haplocheilic stomata** are found on the lower surface.

Male fructification : The male fructification, *Caytonanthus* (Fig. 1.10B), is a pinnately branched structure which bears synangia (four fused sporangium) at its ultimate division. A single **synangium** is 1.0 cm long, radially symmetrical, pendent, tubular and quadrilocular (Fig. 1.10C). Pollen grains are small and bisaccate with a distal sulcus aperture, referred to the form-genus *Vitreisporites.* The sporoderm (exine) is bilayered, the inner nexine lamellate and the outer sexine alveolate.

Female fructification : *Caytonia* (Fig. 1.10D) is a flat structure which bears fruit, rachis or megasporophyll. The rachis is considered as megasporophyll, because of its dorsiventral epidermal structure. The seeds are completely enclosed within the fruit-like body or **cupule** with a minute opening or crack below, close to the pedical (Fig. 1.10E). There are a series of transversely arranged thickened bars on the curved inner face of a well-developed cupule (Fig. 1.10F), and the number of these bars corresponds with the number of seeds. Generally 8 to 30 orthotropous seeds are present in single or double rows. Each seed is flattened, oval and unitegmic (with one integument) with a short micropyle pointing towards the opening below. The nucellus is free from the integument. The integument is devoid of any vascular trace.

Caytonia was compared with the angiosperms by the earlier workers (Thomas, 1925) due to the presence of fruit-like structure, the reticulate venation of leaves and the microsporophyll bearing tetralocular sporangia. Hence, Caytoniacene has been considered as the probable ancestor of angiosperms. However, Caytoniaceae exhibit many non-angiospermous characteristics :

(a) Radially symmetrical pollen bearing organ and lack of a filament and connective,

(b) No blind vein ends in leaf,

(c) Bisaccate pollen grains with alveolate sporoderm,

(d) Flate, pinnate megasporophyll,

Fig. 1.10 : A. *Sagenopteris* leaf, B. *Caytonanthus*, C. Cross section of *Caytonanthus* showing the four pollen sacs, D. *Caytonia* fructification showing two rows of cupules, E. V.L.S. of *Caytonia*, F. V.L.S. of fleshy cupule fruit of *Caytonia*

(e) Single integument in ovule,

(f) Pollen reached the micropyle by means of a "pollen drop mechanism" as in gymnosperms.

The Caytoniaceae has now been considered as an unique member of gymnosperms and treated under a separate and independent order Caytoniales of Pteridospermopsida.

Family 2 : Peltaspermaceae

The Peltasperms are a small group of plants known from Upper Triassic deposits of South Africa, Greenland and Sweden. The genus *Lepidopteris* has been designated for the whole plant in spite of having several form genera, because all the form genera have the characteristic blister-like swelling and similar stomatal structures.

The different form-genera so far identified are :

Leaf : *Lepidopteris*

Male fructification : *Antvesia*

Female fructification : *Peltaspermum.*

Leaf : The frond of *Lepidopteris* are bipinnate with broadly based, open-veined pinnules. The rachis and pinnules of the fronds are provided with blister-like swelling.

Male fructification : The pollen organs, *Antvesia,* are pinnately branched which give two to three dichotomously dividing ultimate branches. Each ultimate branch bears two rows of pendulous pollen sacs. Pollen grains are non-saccate, bilaterally symmetrical with a longitudinal furrow.

Female fructification : The ovuliferous structures are very distinctive, called *Peltaspermum.* About 20 ovules (seeds) are borne in a ring on the underside of a stalked, peltate disc. Several discs are helically arranged on a central stalk. Each ovule (seed) is 7 mm long with an elongated and curved micropylar beak.

Family 3 : Corystospermaceae

The leaf genera, *Dicroidium* of Corystospermaceae was first described by H. H. Thomas (1933) from Triassic beds of Natal, South Africa. Most of them were small plants with pinnatified and dichotomously veined leaves and only known from Gondwanaland. The principal and best-known form genera so far reported are :

Leaves : *Dicroidium* (= *Thinnfeldia*), *Pachypteris, Stenopteris.*

Male fructification : *Pteruchus.*

Female fructification : *Umkomasia*

Although no parts of the plants have ever been found attached, they are believed to belong to Corystospermaceae on the evidence of similar epidermal characteristics.

Leaves : All the three leaf genera, (*Dicroidium, Pachypteris, Stenopteris*) show pinnate frond with open venations (Fig. 1.101A, B). Pigg (1988) found organic connection between *Dicroidium* and ovulate cupules from Triassic deposits of Antarctica. Internal structure is similar to that of cycads. A single rachis bears 15 to 20 bundles arranged in a ring of 7 to 8 near abaxial surface and a line of 5 to 8 adaxially. Mesophyll is differentiated into palisade and spongy parenchyma. The epidermis is provided with dicyclic stomata.

Male fructification : The pollen-bearing organ, *Pteruchus* is a small branching system, divided by equal and unequal dichotomies. Each microsporophyll terminates in a flattened, rounded head which bears clusters of elongated pollen sacs on its underside. Pollen grains are bisaccate having a well-defined aperture.

Female fructification : The megasporophyll, *Umkomasia* is situated in the axil of a bract. Each megasporophyll is branched and is subtended by a pair of bracteole. Each branch bears opposite pairs of stalked, recurved, bivalved, **helmet-shaped** (*Corystosperm* means helmet-shaped seed) cupule. Each cupule bears one seed. The micropyle is curved and bifid, projecting beyond the cupule.

CLASS. CYCADOPSIDA

The class Cycadopsida includes two orders : Cycadales and Cycadeoidales. They were the most dominating taxa during mid-Mesozoic Era. However, they were first appeared in the Upper Carboniferous period of Palaeozoic Era and declined in the late Cretaceous.

GENERAL CHARACTERISTICS OF CYCADOPSIDA

1. Plants are generally with stout trunk having manoxylic wood.
2. Leaves are large frond-like. They are basically pinnate in form or in venation.
3. Seeds are large showing radial symmetry.
4. Presence of ciliated sperms.

ORDER. CYCADALES

The members of the Cycadales are woody plants with unbranched stem (occasional with adventitious branching). They bear manoxylic wood with **mucilage canals** in pith and cortex. Leaves are large, pinnate (rarely bipinnate) with **haplocheilic stomala**. Leaf trace is of diploxylic nature (except in Nilssoniaceae). Plants are dioecious and reproductive organs in cones (except female *Cycas*). Megasporophylls bear 2-8 orthotropous ovules with sterile tip. Seeds are large. Microsporophylls are scale-like or peltate with abaxial sporangia. Sperms are multiciliated.

Most authors recognised two families in the order — Nilssoniaceae and Cycadaceae. However, some authors classified the order into four families, such as Nilssoniaceae Cycadaceae, Stangeriaceae and Zamiaceae. Some others preferred to retain a single family Cycadaceae.

The members of Cycadales are commonly called as '**cycads**' and are the only living representatives of the class. The Cycadales probably originated from the Pteridosperm ancestor (*Palaeocycas* megasporophyll with *Taeniopteris* leaf) in the Upper Carboniferous to Permian period (Mamay, 1976). They reached their climax during the Triassic to Jurassic period of Mesozoic Era and mostly declined at the end of Upper Cretaceous period. Thus, the Mesozoic Era is commonly known as "**age of cycads**". At present, they are represented by eleven extant genera and are restricted in the tropical zones of both the Eastern and Western Hemispheres.

Fossil Cycads

The earliest known cycads have been reported mostly as dispersed megasporophylls. *Spermopteris* (Fig. 1.11A) is the first evidence of the fossil cycad from the Upper Pennsylvania of Kansas. It is an ovulate structure of the leaf genus *Taeniopteris* where two rows of ovules attach to the abaxial surface of the leaf.

Plasmatocycas and *Archaeocycas* are the two cycadean megasporophylls reported from the Lowen Permian. *Plasmatocycas* (Fig. 1.11B) megasporophyll has two rows of ovules attached laterally to the stalk of a *Taeniopteris* leaf. *Archaeocycas* (Fig. 1.11C) megasporophyll has an expanded spathulate lamina bearing two rows of ovules on the basal part. The ovules are partly enclosed by the lamina in the basal region.

Leptocycas gracilis is an Upper Triassic cycad, reconstructed by Delevoryas and Hope (1971). The plant has a smooth, slender stem (1.5 m tall) bearing a crown of fronds and a pollen cone.

The reconstruction of *Bjuvia simplex* (Fig. 1.11D), an Upper Triassic cycad, has been made by Florin (1933) based on the *Taeniopteris* leaves and *Palaeocycas* megasprophylls with an imaginary trunk.

Harris (1961) reconstructed a Jurassic cycad bearing *Beania gracilis* ovulate cones (Fig. 1.11E, F) and crown of *Nilssonia* type (Fig. 1.11G) leaves on an imaginary stem. The ovulate cones of *Beania gracilis* resemble seed cone of *Zamia*. The pollen cone, *Androstrobus* (Fig. 1.11H, I) is believed to be associated with the same plant.

Extant cycads

The cycads comprise of eleven (11) living genera of which five (*Microcycas, Dioon, Chigua, Ceratozamia* and *Zamia*) belong to Western Hemisphere and the remaining six (*Cycas, Stangeria, Bowenia, Macrozamia, Lepidozamia* and *Encephalartos*) are confined to Eastern Himesphere. A total of about 100 species of cycads have been reported.

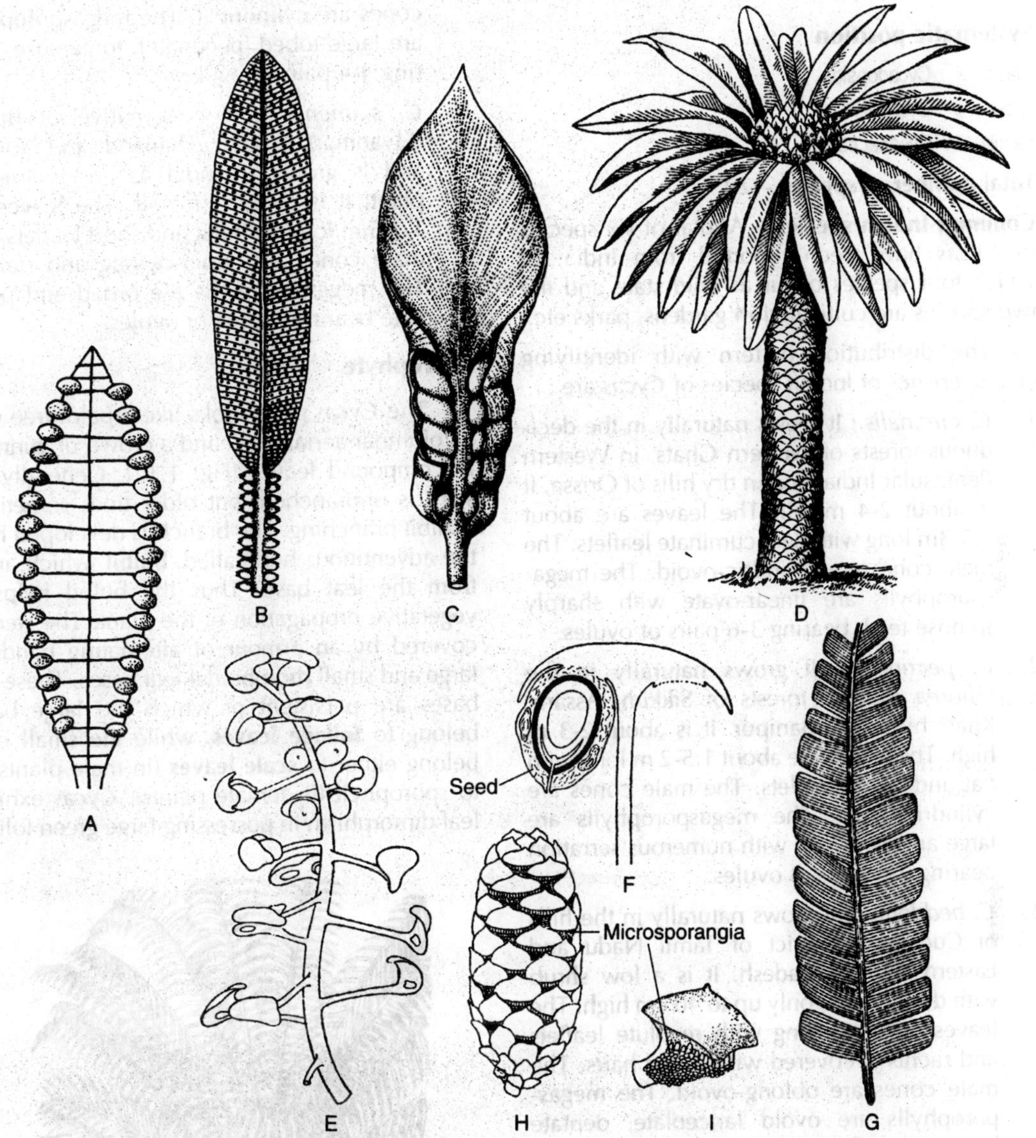

Fig. 1.11 : A. *Spermopteris*, B. *Plasmatocycas*, C. *Archaeocycas*, D. *Bjuvia simplex*, E. *Beania gracilis*, F. L.S. of an ovule of *Beania gracilis*, G. *Nilssonia leaf*, H. Androstrobus manii, I. A microsporophyll of Androstrobus

Worldwide distribution pattern of extant cycads :

Western Hemisphere

Genus	Country of occurrence
Microcycas	: Cuba
Dioon and *Ceratozamia*	: Mexico
Chigua	: Columbia
Zamia	: Mexico, West Indies, North-Western South America, and Florida

Eastern Hemisphere

Genus	Country of occurrence
Cycas	: Indian sub-continent, China, Southern Japan, Madagascar and North Australia
Stangeria and *Encephalartos*	: Tropical and South Africa
Lepidozamia, Macrozamia and *Bowenia*	: Australia

CYCAS

Systematic position

Class. Cycadopsida
Order. Cycadales
Family. Cycadaceae

Total number of species – 20

Common Indian species – A total of six species of *Cycas* have been reported from India, of which four species occur as wild state and the two species are cultivated in gardens, parks etc.

The distribution pattern with identifying characteristics of Indian species of *Cycas* are :

1. *C. circinalis* : It grows naturally in the deciduous forests of Western Ghats, in Western Peninsular India, and in dry hills of Orissa. It is about 2-4 m tall. The leaves are about 1.5-3m long with flat acuminate leaflets. The male cones are cylindric-ovoid. The megasporophylls are linear-ovate with sharply spinose teeth bearing 3-6 pairs of ovules.
2. *C. pectinata* : It grows naturally in the '*Shorea robusta*' forests of Sikkim, Assam, Khasi hills and Manipur. It is about 2-3 m high. The leaves are about 1.5-2 m long with flat and linear leaflets. The male cones are cylindric-ovoid. The megasporophylls are large and pectinate with numerous serration bearing 2-3 pairs of ovules.
3. *C. beddomei* : It grows naturally in the hills of Cuddapah district of Tamil Nadu and Eastern Andhra Pradesh. It is a low shrub with dwarf trunk, only up to 40 cm high. The leaves are 1 m long with revolute leaflets and rachis is covered with tufted hairs. The male cones are oblong-ovoid. The megasporophylls are ovoid lanceolate, dentate bearing 2-3 pairs of ovules.
4. *C. rumphii* : It is found naturally on the beach forests of Andaman and Nicobar Islands. It is a tall tree, up to 6 m height. The leaves are 1-2 m long, with flat and broad (12 mm) leaflets. The male cones are ellipsoidal. The megasporophylls are linear-ovate and dentate, bearing 3-5 pairs of ovules.
5. *C. revoluta* : It is a native of China and Southern Japan and is commonly cultivated in India as an ornamental plant in gardens, parks etc. It is about 3 m tall. The leaves are 1-2 m long with revolute leaflets. The male cones are cylindrical. The megasporophylls are large lobed (pectinate), tomentose, bearing 3-4 pairs of ovules.
6. *C. siamensis* : It is a native of Burma (Myanmar), China, Thailand and Yunnan and is grown in India as an ornamental plant. It is about 3 m tall. The leaves are 1-1.5 m long with flat and short leaflets. The male cones are ovoid-oblong and stalked. The megasporophylls are broad and pectinate bearing 2 pairs of ovules.

Sporophyte

The *Cycas* plant looks like a palm tree with a columnar aerial trunk and a crown of pinnately compound leaves (Fig. 1.12). Generally the stem is unbranched, but older trees sometimes exhibit branching. The branch is developed from the adventitious bud called '**bulbil**' which arises from the leaf base. Thus the bulbil helps in vegetative propagation of the plant. The stem is covered by an armour of alternating bands of large and small rhomboidal leaf bases. These leaf bases are persistent of which the large bases belong to **foliage leaves**, while the small ones belong either to **scale leaves** (in male plants) or to sporophyll (in female plants). *Cycas* exhibits leaf dimorphism in possessing large green foliage

Fig. 1.12 : A female plant of *Cycas circinalis*

leaves and small scale leaves or **cataphylls**. The primary tap root system is replaced by a number of strong, branched adventitious roots arising from the stem.

Root

External morphology : At the beginning, there is a primary root system in *Cycas*. This normal tap root system is short-lived and is replaced by large, fleshy, persistent, branched adventitious roots arising from the stem. Some of the lateral branches grow horizontally in the soil and become **apogeotropic**. They have numerous lenticels on their surface. These roots frequently get inflected with bacteria (*Pseudomonas, Azotobacter*), fungi and blue green algae or **Cyanobacteria** (*Anabaena cycadae, Nostoc punctiforme, Oscillatoria* and *Calothrix*). The infection causes a distortion of their shape. The infected roots repeatedly dichotomise to become a mass of tubercles apparently looks like a coral. These are called **coralloid roots** (Fig. 1.13B) which grow horizontally in soil and get swollen at the tip. The exact role of these endophytic Cyanobacteria is not known. However, it has been reported that these Cyanobacteria can fix nitrogen and promote the growth of *Cycas* plant.

Internal structure

Anatomically, a normal primary root is differentiated into epidermis, cortex and a central vascular system (Fig. 1.13A). **Epidermis** is comprised of single-celled parenchymatous cells with occasional hairs. The **cortex** is multilayered, made up of parenchymatous cells filled with starch. Some cells of cortex contain tannins and sphaeraphides. The single-layered endodermal cells have characteristic carparian bands and the multilayared **pericycle** is made up of parenchymatous cells. The **stele** is diarch with exarch xylem. A mature tap root shows **secondary** growth in thickness both instrastelar and extrastelar.

Anatomically, the **coralloid roots** are more or less identical with the primary tap root except for a well-defined 'algal zone' in the cortex having abundant intercellular spaces occupied by blue-green algae or **Cyanobacteria** (Fig. 1.13C). Besides, some fungi and bacteria are also repor-ted from this zone. The stele is diarch to tetrarch. There is a little or no secondary growth in the coralloid roots.

Stem

External morphology : The *cycas* stem is columnar with a thick armour of **persistent leaf bases**. However, *C. siamensis* has underground short tuberous stem. The stem is usually unbranched, but sometimes it shows branches which are produced from the bulbil. The stem is irregular in outline because of a number of persistent leaf bases.

Internal structure : Internally, the stem shows a similarity with a dicot stem which is differentiated into **epidermis, cortex** and **vascular cylinder** (Fig. 1.14A). The single-layered epidermis in discontinuous due to the presence of leaf-bases. The epidermis is covered with a thick cuticle.

The cortex is massive, made up of parenchymatous cells rich in starch grains. The cortex is traversed by several mucilaginous canals and leaf bases. The endodermal and pericycle layers are not distinct.

In the young stem, the vascular cylinder is very small comprising of several vascular bundles arranged in a ring. Thus the *Cycas* stem shows **ectophloic siphonostelic** configuration. The each **vascular bundle** is conjoint, collateral and open which is separated by parenchymatous **medullary rays**. The xylem is endarch, comprising of tracheids and xylem parenchyma only.

Several leaf-traces are observed in a T.S. of stem. Each leaf receives four traces. Out of the four traces, two traces enter directly into the leaf after arising from the stelar cylinder on the same side are called **direct traces**. The other two traces arise from the stelar cylinder on the opposite side of the leaf and enter the leaf after turning round a semicircle or girdle around the stelar cylinder. These traces are called **girdle traces** (Fig. 1.14C). The girdle of leaf traces is a characteristic of *Cycas* and allied members of Cycadaceae. The two girdle traces join with the direct traces and branches of other girdle traces finally enter the leaf as two traces.

The central part of the stem is occupied by a large parenchymatous pith rich in starch grains. Pith cells also contain tannins and mucilagenous substances.

Secondary growth

The normal intrastelar secondary growth is observed in early stages due to the bifacial acti-

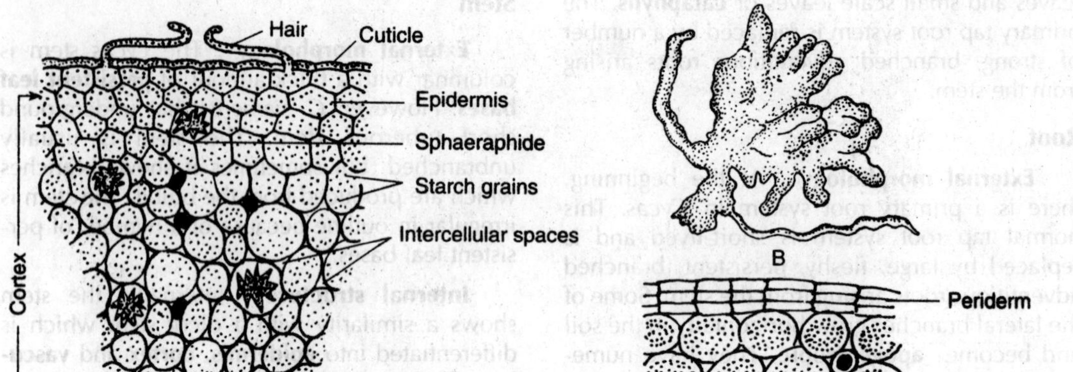

Fig. 1.13 : A. T.S. of primary root of *Cycas revoluta*, B. Coralloid roots, C. T.S. of a part of coralloid root of *Cycas revoluta* showing algal zone

vity of cambium ring. This primary cambium ring is short-lived, thus remains functional for a short time. The secondary xylem thus produced centripetally shows tracheids with 4-5 rows of bordered pits and scalariform thickenings and one row of thin-walled cells in between (Fig. 1.14B). The three types of vascular rays are produced : (i) uniseriate rays, (ii) multiseriate rays, and (iii) foliar multiseriate rays. The phloem is well-developed which may exceed the volume of xylem. The sieve elements contain numerous sieve areas on their radial wall. The companion cells are absent, instead some phloem parenchyma termed **albuminous cells** — are formed, closely associated with sieve cells. The stem does not produce any annual ring.

A peculiarity of *Cycas* stem is that the young stem which is **monoxylic** becomes **polyxylic** at maturity due to the formation of accessory cambium rings in the cortex (Fig. 1.14A). The second cambium ring forms similarly the secondary xylem towards the inner side and secondary phloen towards the outer side. The second cambium ring remains functional for a short time and thereafter its activity ceases. In a similar fashion, successive cambium rings initiate in the cortical region. The first two vascular rings are thicker than the other vascular rings, of which the second vascular ring is wider than the first. The subsequent rings are narrower than these two rings, gradually diminishing towards the periphery. Usually 3-4 rings of wood are produced in *Cycas*, but in *C. pectinata* as many as 20 rings of wood may be formed.

The extrastelar secondary growth has also been observed in the stem of *Cycas* which forms the **periderm.**

Leaf

External morphology : *Cycas* exhibits leaf dimorphism in possessing green foliage leaves and scale leaves.

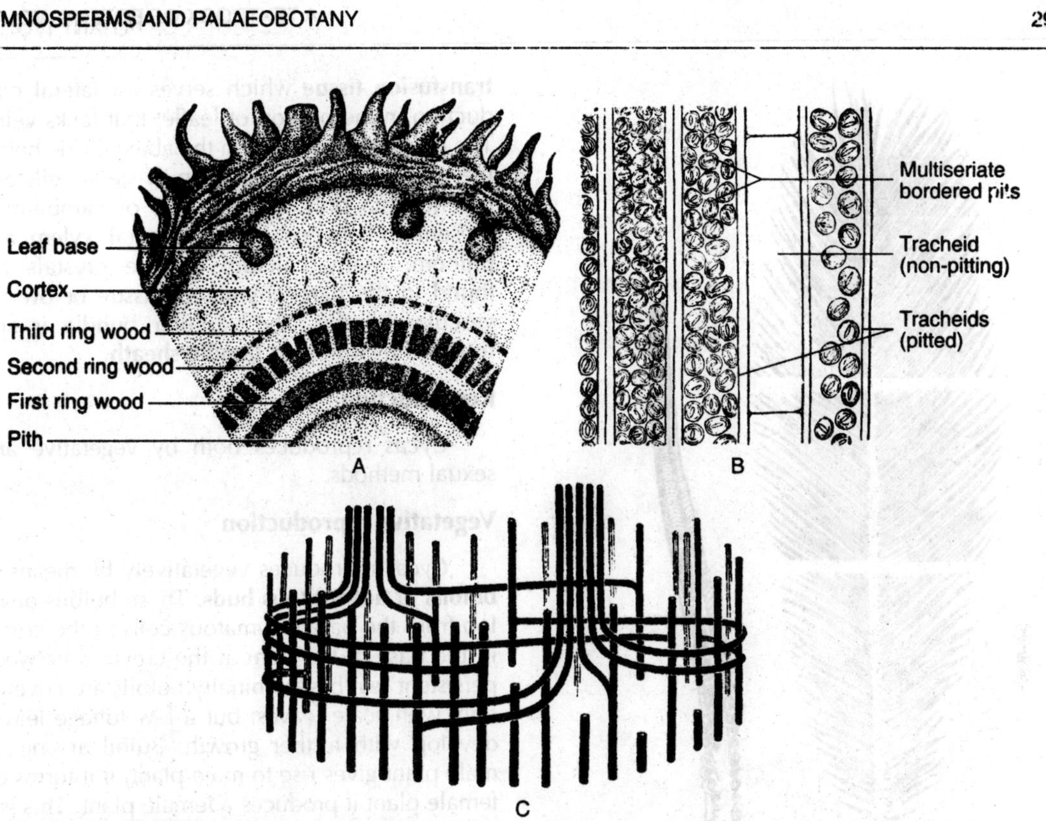

Fig. 1.14 : *Cycas* : A. T.S. of stem, B. L.S. of mature wood, C. Girdling of leaf traces

The **scale leaves** are small, dry, non-green and triangular in shape, covered with ramenta. They have small persistent leaf bases. Their only function is to provide protection to apical meristem and other aerial parts.

The **foliage leaves** are large, unipinnately compound (Fig. 1.15A). A single leaf bears 75-100 pairs of leaflets which are arranged on either side of the rachis in opposite or alternate manner. The leaflets are sessile, elongated having revolute (e.g. *C. revoluta, C beddomei*) or flat (*C. circinalis, C. pectinata, C. rumphii*) margins. Each leaflet is provided with a single unbranched midvein without having any lateral veins. The leaves possess a very long and strong rachis with a short petrole. The base of the petiole is provided with two rows of small stiff spines. The young rachis as well as leaflets are **circinately coiled** like those of ferns (Fig. 1.15B).

Internal structure : The T.S. of rachis shows circular outline. The outermost layer is the **epidermis** which consists of cuticularised thick-walled cells interrupted by **sunken stomata**. The epidermis is followed by **hypodermis,** which is composed of a variable mixture of chlorenchyma and sclerenchyma followed by a parenchymatous ground tissue with many **mucilage canals**. Initially two endarch bundles enter the leaf base and then they spite up into several bundles. These bundles are then dispersed in the ground tissue and are arranged in an inverted Greek letter "**omega**" (Ω) like arc. The vascular bundles are conjoint, collateral and open, bearing both the centrifugal and centripetal xylem. Thus, they are **diploxylic** or **pseudomesarch** in nature. As the bundles move up into pinnae, they become exarch without having any centrifugal xylem. The phloem lies on the side of the protoxylem. The bundles are covered with a bundle sheath.

The *Cycas* leaflet is dorsiventral showing strongly xerophytic characteristics. The T.S. of a leaflet shows a thick-walled, highly cuticularised epidermis. The upper epidermis is continuous, while the lower epidermis is interrupted by stomata (Fig. 1.16). The stomata are **haplocheilic**, situated in pits with overarching rims. The two guard cells are bordered by 8-10 subsidiary cells arranged in a ring. The hypodermis is 1-2 layered thick except for the midrib or at the margins where it becomes 3-4 layered. The hypodermal cells are thick-walled, highly lignified. The mesophylls are differentiated into **palisade**

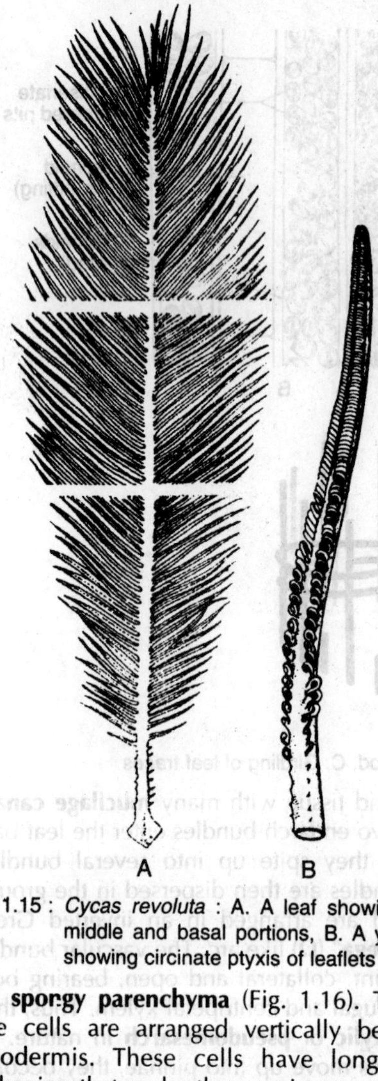

Fig. 1.15 : *Cycas revoluta* : A. A leaf showing apical, middle and basal portions, B. A young leaf showing circinate ptyxis of leaflets

and **spongy parenchyma** (Fig. 1.16). The palisade cells are arranged vertically below the hypodermis. These cells have long fibrous thickening that make them resistant against the external radial and longitudinal pressures. The lower part of the mesophyll consists of spongy parenchymatous cells with intercellular spaces. The vascular bundle is **diploxylic** in nature, made up of a broad triangular centripetal exarch metaxylem and two small patches of centrifugal endarch primary xylem (Fig. 1.16). The **centripetal xylem** is separated from **contrifugal xylem** by a few parenchymatous cells. Some tracheidal cells situated on either side of the centripetal metaxylem are termed **transfusion tissue**. Additionally, there are 3-4 layers of transversly elongated thin-walled colourless tracheidal cells in-between upper palisade layer and lower spongy layer. These cells constitute the **accessory transfusion tissue** which serves for lateral conduction in the portion of leaflet that lacks veins. The phloem is situated on the abaxial side below the xylem which is made up of sieve cells and parenchyma only. A small arc of cambium is present in-between the centripetal xylem and phloem. Many calcium oxalate crystals are found in the parenchymatous tissue below the vascular bundle. The vascular bundle is surrounded by a **fibrous bundle sheath.**

Reproduction

Cycas reproduces both by vegetative and sexual methods.

Vegetative reproduction

Cycas reproduces vegetatively by means of **bulbils** or **adventitous buds**. These bulbils develop from the parenchymatous cells of the cortex in the base of the stem at the crevices between persistent leaf bases. Initially bulbils are covered only with scale leaves, but a few foliage leaves develop with further growth. Bulbil arising on male plant gives rise to male plant, if it forms on female plant it produces a female plant. This is a very common method of vegetative propagation in *C. revoluta.* The vegetative propagation in *C. circinalis* takes place by **suckers** that develop on roots. With further growth they start producing new plants.

Sexual reproduction

Cycas is strickly a **dioecious** plant, but the male and female plants are indistinguishable at the vegetative stage. The compact male cone develops at the apex of the stem in the male plant. However, *Cycas* is the only genus of Cycadaceae which does not produce any female cone. Instead, several megasporophylls arise spirally in acropetal succession around the stem apex of the female plant.

Male cone

The surface of male cone is covered with brown scales at young stage. At maturity, the male cone becomes very large (40-80 cm in length), oval or conical in shape (Fig. 1.17A) which emits odour that can be smelt from quite a moderate distance. The male cone of *Cycas* is the largest among the plant kingdom. The male cone has a central cone axis and numerous microsporophylls are arranged spirally and

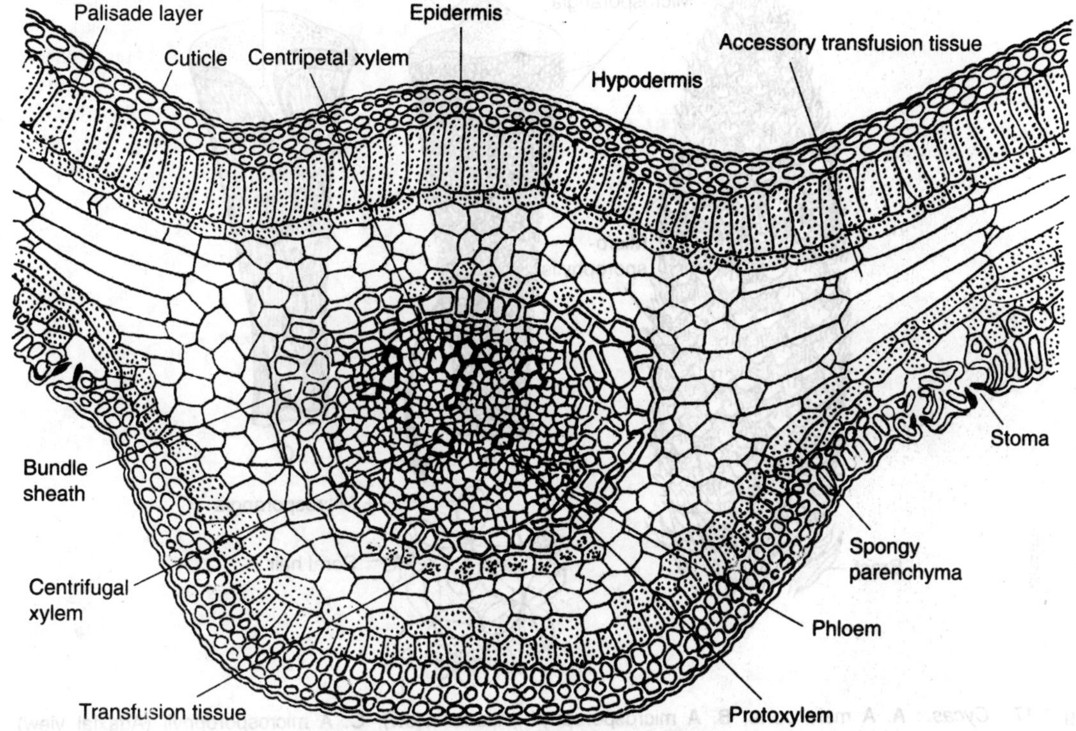

Fig. 1.16 : T.S. of leaflet of *Cycas*

acropetally around the axis. A single **microsporophyll** is a nearly triangular flattened woody structure. It is differentiated into a proximal wedge-shaped fertile part and a distal sterile part, tapering into an upcurved apex called **apophysis** (Fig. 1.17C). Numerous microsporangia (700 in *C. circinalis,* 1100 in *C. media)* are borne on the abaxial (lower) surface of the microsporophyll except at the apex and the base (Fig. 1.17B). Microsporangia are borne in groups of 3-5 forming **sori** that are surrounded by many single-called delicate **hairs** (Fig. 1.17D). Each sporangium is oval or circular in shape having a very short massive stalk. The dehiscence of sporangia takes place by longitudinal slit (Fig. 1.17D). The development of sporangia is of **eusporangiate** type. The sporangial wall is multilayered with a thickened epidermis and a ill-defined tapetum enclosing numerous microspore (pollen) mother cells. Further, microspore mother cells through meiotic division produce numerous microspores or pollen grains. The pollen grain is oval-shaped having a large rounded monosulcate aperture. The pollen is bounded by two concentric wall layers; the outer thick **exine** and the inner thin **intine** (Fig. 1.20A).

Female fructification

The megasporophylls of *Cycas* are not organised into a definite cone. Instead, they arise at the stem apex spirally and acropetally forming a loose crown. The **megasporophylls** are pinnate in nature and are covered with brown hairs called **rementa**. Hence the megasporophylls of *Cycas* are considered to be the modified foliage leaves. **Megasporophylls** are produced every year like the foliage leaves, though they are produced more in numbers than the foliage leaves. A single megasporophyll is a flat dorsiventral structure, measuring up to 30 cm in length. It is differentiated into an upper pinnate lamina and a basal stalk that bears two rows of opposite or sub-opposite, one to six pairs of ovules (Fig. 1.18A–F). There is a great variation in structure of megasporophylls and in number of ovules per megasporophyll in *Cycas* and these criteria can be applied to identify the species of *Cycas*. There is a great reduction in the structure of megasporophylls and in number of ovules among the various species of *Cycas*. *C. revoluta* is the most primitive species where the megasporophyll lamina is much dissected and tapers into a point, bearing 3-4 pairs of ovules (Fig. 1.18A). In *C. pectinata,* the mere

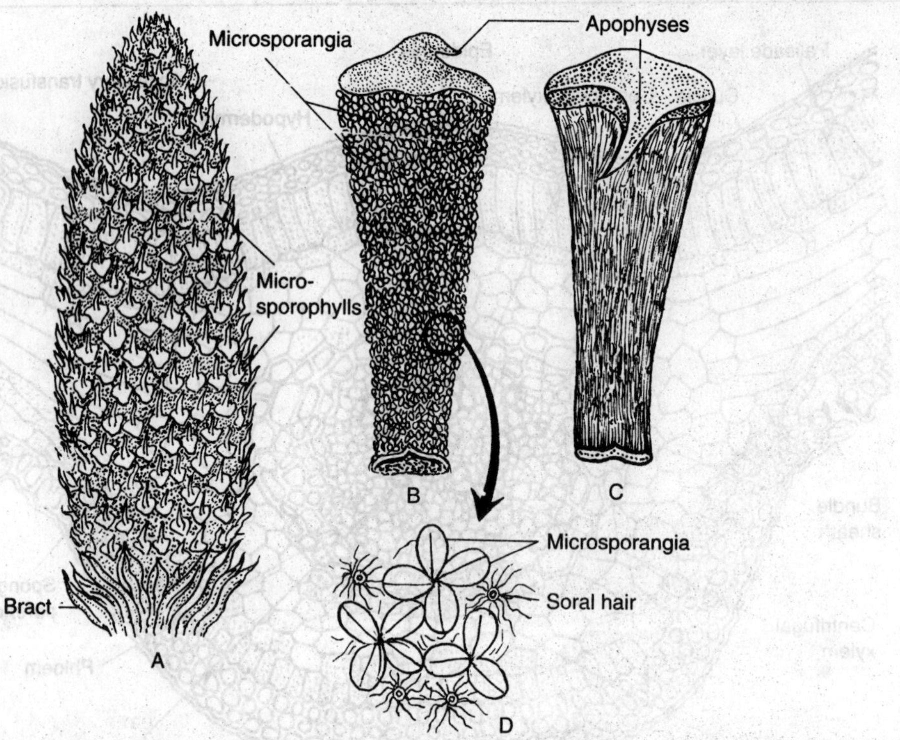

Fig. 1.17 : *Cycas* : A. A male cone, B. A microsporophyll (Abaxial view), C. A microsporophyll (Adaxial view), D. Microsporangia arranged in sori

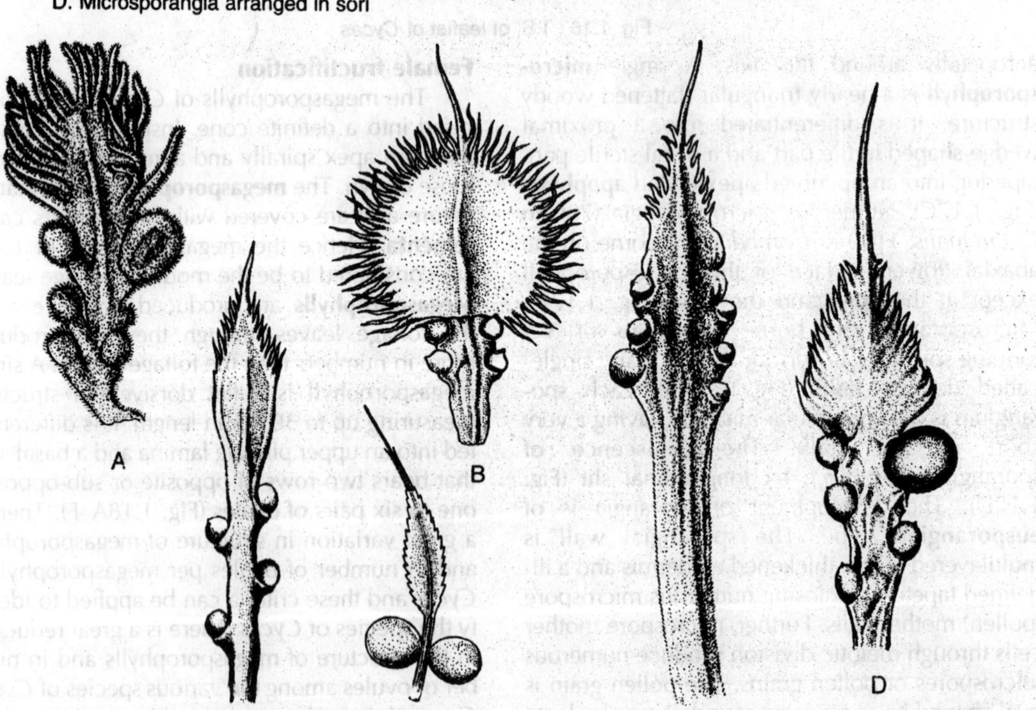

Fig. 1.18 : Megasporophylls : A. *Cycas* revotuta, B. *C. pectinata*, C. *C. circinalis*, D. *C. rumphii*, E. *C. beddomei*, F. *C. normanbyana*

serrated lamina is orbiculate with pectinate margins, bearing 2-3 pairs of ovules (Fig. 1.18B). In *C. circinalis*, the lamina is lanceolate with dentate margins and acuminate apex, bearing 4-6 pairs of ovules (Fig. 1.18C). The further reduction in lamina has been observed in *C. rumphii* (Fig. 1.18D) and *C. beddomei* (Fig. 1.18E) where lamina is ovate-lanceolate with acuminate apex, bearing 2–3 pairs of ovules. The maximum reduction in megasporophyll has been noted in *C. normanbyana*. Its lamina shows a mere serrated margins bearing only one pair of ovule (Fig. 1.18F).

Ovule

The ovules of *Cycas* are **orthotropous** and shortly stalked. The ovules are large, attaining a length of about 6-7 cm with a diameter of 4 cm and is perhaps the largest amongst the plant kingdom. The mature seeds are elliptical, bilobed and slightly flattened — they become fleshy and bright orange or red in colour. There is a single integument which is fused with the nucellus except the top (Fig. 1.19). The integument is differentiated into three layers. The **outer layer** is fleshy and pulpy and becomes variously coloured at maturity. The **middle layer** is very hard and stony, while the **inner layer** is fleshy and becomes papery before maturity. There are frequent mucilage canals and tannin cells in the integument.

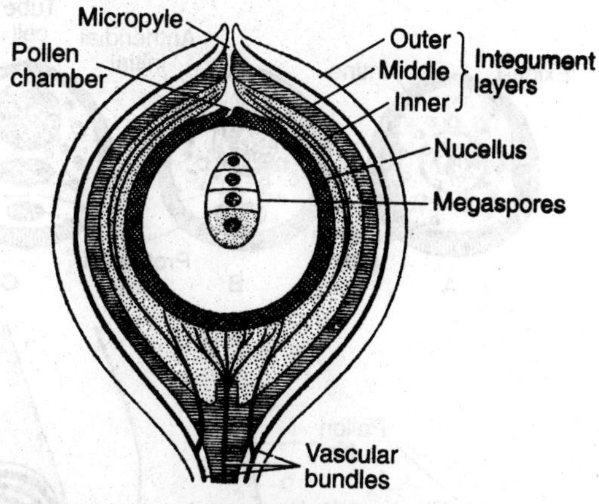

Fig. 1.19 : Vertical median section of *Cycas* ovule

The ovule is supplied with three vascular traces (Fig. 1.19). The median strand supplies the base of the integument, which extends up to the chalazal end of the nucellus and further ramifies abruptly. The two side strands pass to the integument which again divide — one branch supplies to the outer pulpy layer and the other to the inner soft layer.

Megasporogenesis

A deeply situated cell of the nucellus is differentiated into a large megaspore mother cell which undergoes meiotic division to form a **linear tetrad** of four megaspores. Out of the four megaspores the outer three degenerate, while the

Schematic representation in the development of male gametophyte of *Cycas*

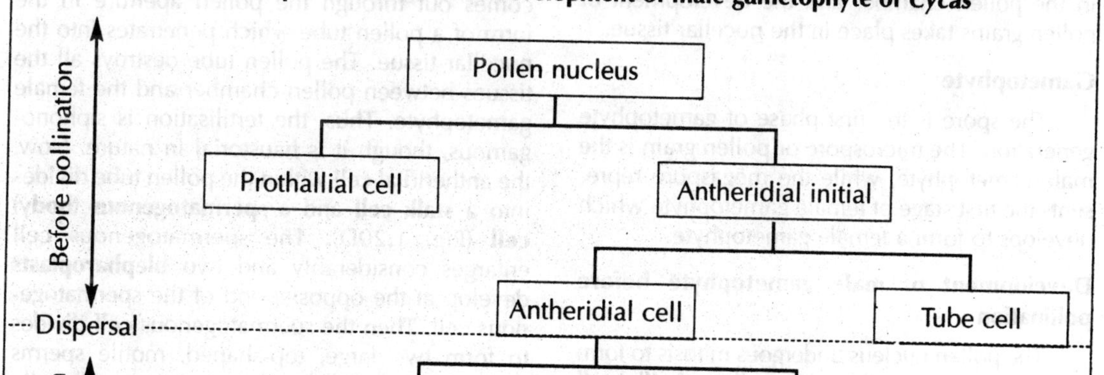

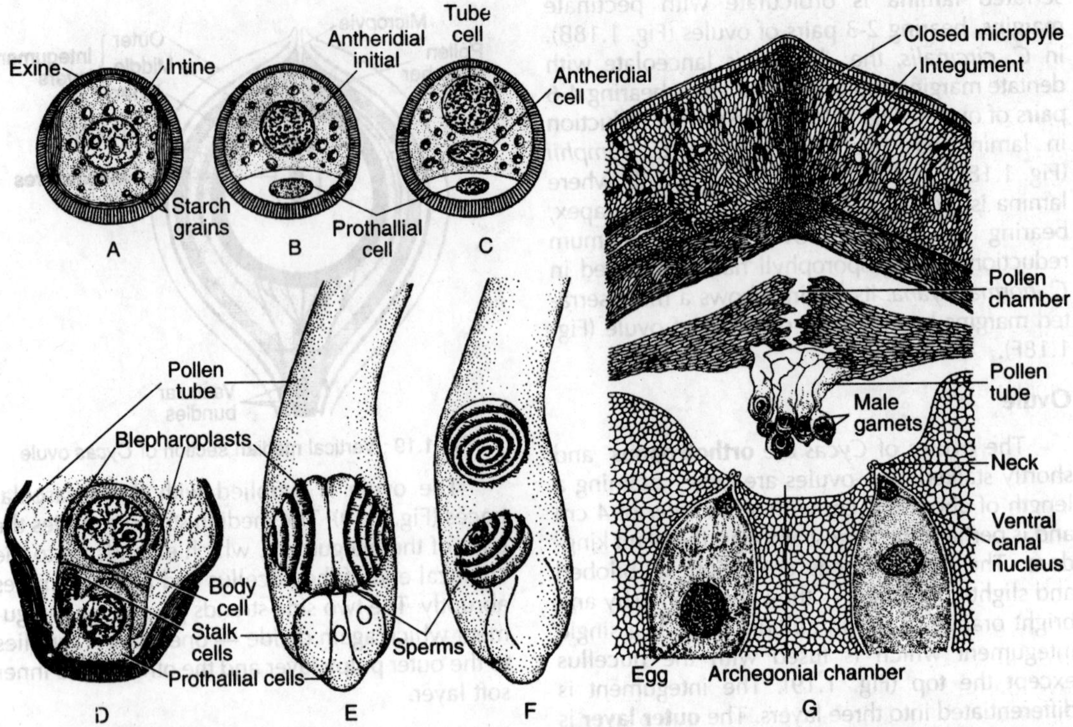

Fig. 1.20 : *Cycas* : A–F. The development of male gametophyte, G. Closed top of ovule after pollination showing course of pollen tubes

lowermost megaspore becomes functional (Fig. 1.21A). The upper free opening of the integument forms the micropyle and a concavity in the top of the nucellar tissue forms the pollen chamber. After pollination, the pollen grains are collected in the pollen chamber and the development of pollen grains takes place in the nucellar tissue.

Gametophyte

The spore is the first phase of gametophyte generation. The microspore or pollen grain is the male gametophyte, while the megaspore represents the first stage of female gametophyte which develops to form a female gametophyte.

Development of male gametophyte before pollination

The pollen nucleus undergoes mitosis to form endosporically a small lens-shaped **prothallial cell** and a large **antheridial initial** (Fig. 1.20B). The prothalial cell does not divide further, while the antheridial initial divides into an **antheridial cell** and a **tube cell** (Fig. 1.20C). The pollen grains are released from the sporangium at the 3-celled stage. After pollination, the further development takes place within the nucellus tissue of the ovule.

Development of male gametophyte after pollination

The further development of male gametophyte takes place within a week after pollination. The tube cell of 3-celled male gametophyte comes out through the pollen aperture in the form of a pollen tube which penetrates into the nucellar tissue. The pollen tube destroys all the tissues between pollen chamber and the female gametophyte. Thus, the fertilisation is siphonogamous, though it is haustorial in nature. Now, the antheridial cell within the pollen tube divides into a **stalk cell** and a **spermatogenous (body) cell** (Fig. 1.20D). The spermatogenous cell enlarges considerably and two **blepharoplasts** develop at the opposite end of the spermatogenous cell. Then the spermatogenous cell divides to form two large, top-shaped, motile sperms with spiral band bearing numerous flagella around the distal end of the sperm (Fig. 1.20E). The sperms of *Cycas* are very large (180-210 μm) and can be visible to the naked eye (Fig. 1.20F).

Development of female gametophyte

The female gametaphyte of *Cycas* develops from the functional megaspore surrounded by a tough membrane that persists in the gametophyte (Fig. 1.21B). The nucleus of the megaspore divides by free nuclear divisions forming a large number of nuclei unaccompanied by wall-formation. In the mean time, the megaspore increases in size and develops a vacuole in the centre which forces the cytoplasm along with nuclei towards the periphery. Thus the nuclei lie in a thin film of cytoplasm around the vacuole (Fig. 1.21B). The cell wall formation starts in a centripetal fashion, from periphery inwards. The wall formation proceeds very rapidly and as a result the central vacuole in obliterated (Fig. 1.21C). The entire gametophyte becomes cellular and the tissue thus formed is called **endosperm**. The peripheral cells of the gametophyte are small, isodiametric devoid of any food reserves, while the inner cells are large, irregular and filled up with starch grains.

Development of archegonia

Two to eight cells of the female gametophyte at the micropylar end enlarge in size and have dense cytoplasm and prominent nuclei. These cells function as **archegonial initials** (Fig. 1.21D). Each archegonial initial divides periclinally to form an outer small **neck initial** and a large **central cell** (Fig. 1.21E). The neck initial divides by a vertical (anticlinal) wall to form two **neck cells** (Fig. 1.21F). A second division takes place prior to fertilisation, thus the neck actually comprises of four cells. The central cell enlarges considerably in size and its nucleus divides into a small ephemeral **ventral canal nucleus** and a large **egg cell** (Fig. 1.20G). The egg of *Cycas* is reported to be the largest among the living plants, thus it can be seen with naked eyes. The nucellar tissue above the archegonia disorganises to form an archegonial chamber (Fig. 1.21G).

Pollination

Cycas is anemophilous i.e., wind pollinated. The airborne pollen grains remain suspended in the air and at the same time some cells of the nucellar beak in the ovule are disorganised to form a viscous fluid. This fluid comes out through the microphyle in the form of a '**pollination drop**'. Thus, some of the airborne pollen grains come in contact with the fluid and are sucked into the pollen chamber through the micropylar canal. The pollen grains are then deposited and concentrated within the pollen chamber as a result of drying off the fluid (Fig. 1.20G). At this stage, the ovule increases in size.

Fertilisation

The sperms and the cytoplasm of pollen tube are released in the archegonial chamber by the rupture of the basal end of pollen tube. The osmotically rich pollen tube cytoplasm causes the rupture of neck cells. The motile sperms enter into the archegonium with a forward and circular motion, ciliary band forming the anterior end. The archegonial chamber is flooded with the fertilisation fluid produced by nucellar cells. The ciliary band of the sperm is left behind on the top of the egg cell. The sperm nucleus fuses with the egg nucleus and thus a **zygote** in formed.

Embryogeny

The zygote enlarges considerably and undergoes numerous free nuclear divisions. As much

Schematic representation in the development of archegonia of *Cycas*

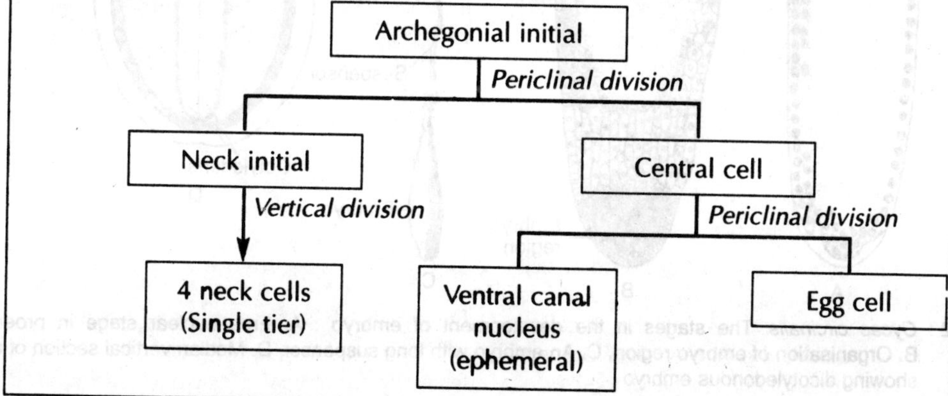

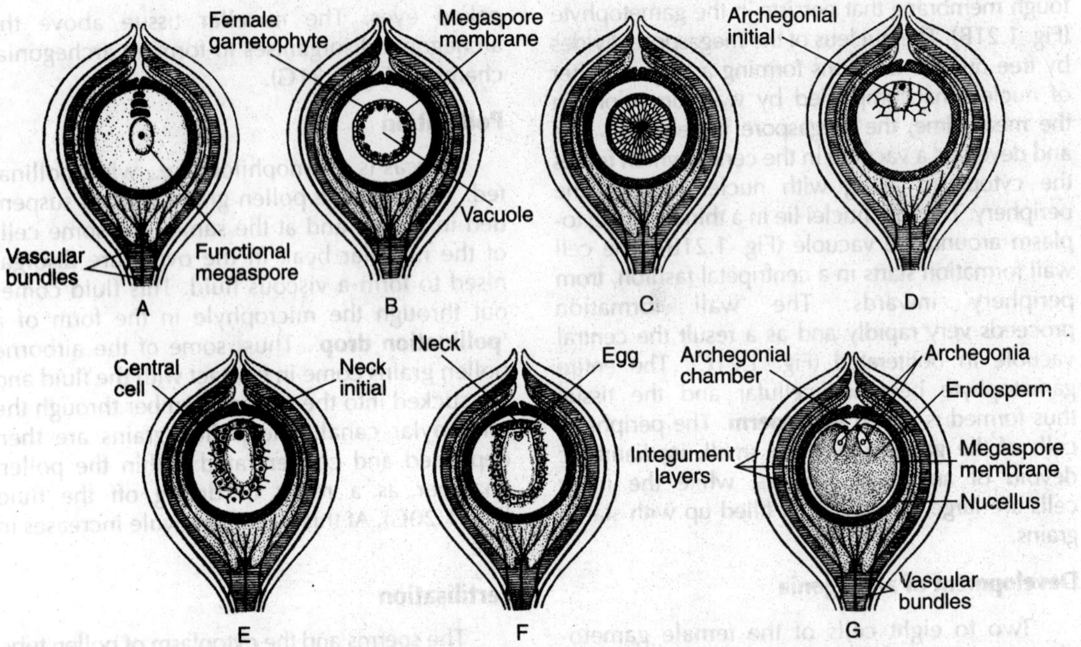

Fig. 1.21 : *Cycas*. Development of female gametophyte within ovule : A. Linear tetrad with one functional megaspore, B. Free nuclear stage, C. Cellular stage, D–F. Stages in development of an archegonium, G. A mature female gametophyte containing two archegonia

as 256 (*C. revoluta*) to 512 (*C. circinalis*) free nuclei have been reported. A large central vacuole is formed and the nuclei are arranged around the central vacuole (Fig. 1.22A). Most of the free nuclei move to the base of the **proembryo**. The cell formation in the proembryo begins from the basal part and extends up the periphery. The entire embryo never becomes cellular. The basal cells are smaller with dense cytoplasm forming the **embryo** (Fig. 1.22B). A dicotylede-

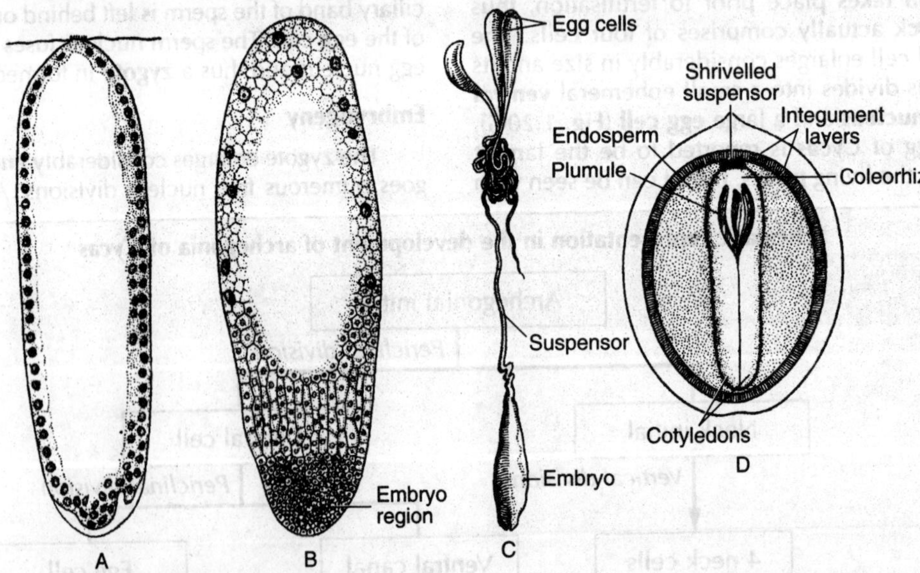

Fig. 1.22 : *Cycas circinalis*. The stages in the development of embryo : A. Free nuclear stage in proembryo, B. Organisation of embryo region, C. An embryo with long suspensor, D. Median vertical section of a seed showing dicotyledonous embryo

nous embryo is developed at the tip. The upper cells elongate greatly to form the **suspensor** which may be coiled and twisted (Fig. 1.22C). Several egg cells may be fertilised to form many zygotes, but ultimately only one embryo in an ovule attains maturity. The layer of cells covering the outer curved face of the embryonal mass is called the **cap**.

Seeds

Cycas seeds are fleshy and remain covered with an orrange red-coloured thick seed coat. The embryo in the seed enlarges much. The nucellus is used up and the seed is covered with the three-layered seed coat (Fig. 1.22D). The coleorrhiza is the first region to differentiate in the embryo proper which is quite hard. The plumule develops later inside the base of the cotyledons. Subsequently the suspensor dries up. The growth of the embryo in seeds is very slow, taking over a year to mature. *Cycas* shows three years reproductive cycle where pollination in first year winter, fertilisation in second year summer and seed shedding in third year summer. The development of embryo continues even after the seeds are shed.

The seeds are dispersed by birds due to their attractive fleshy outer pulp. The germination of the seed is **epigeal**. The first leaves come out at right angle to the cotyledons in a decussated fashion, although the remaining leaves develop spirally (Fig. 1.23).

Figure 1.24 shows the life-cycle of *Cycas*.

Origin and relationships

Cycads probably originated from the Pteridosperm ancestor (*Palaeocycas* megasporophyll with *Taeniopteris* leaf) in the Upper Carboniferous period. Among the order Cycadales, *Cycas* is the most primitive genus because of the absence of female cone. *Cycas* shows relationship with different groups of plants based on their morphological and anatomical features.

Relation to Ferns (Filicopsida)

Cycas resembles ferns in exhibiting the following characteristics :

1. The young rachis as well as leaflets are circinately coiled like those of ferns.
2. Microsporangia are borne in sori on the abaxial surface of the microsporophylls like those of ferns.

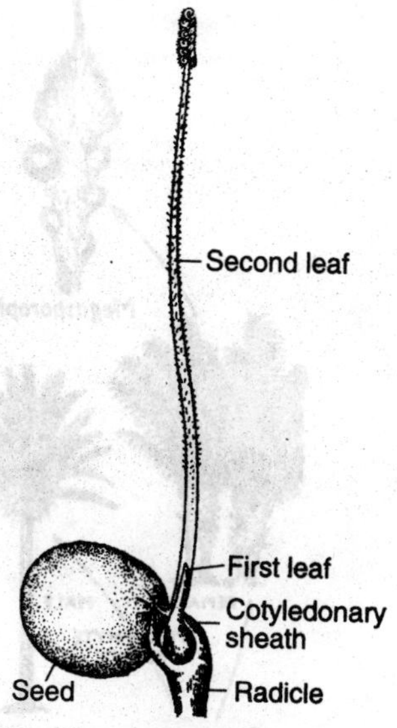

Fig. 1.23 : *Cycas* : The germinating seed

3. The sperms are very large and motile with spiral band, bearing countless cilia like those of ferns.

Relation to Pteridosperms

1. Most of the pteridosperms have stout unbranched columnar stem like those of *Cycas*.
2. Presence of manoxylic wood with large pith and extensive cortex in stem.
3. The leaves are pinnately compound.
4. The ovules have three-layered integument with two systems of vascular bundles.
5. The presence of nucellar beak with a distinct pollen chamber.

Relation to Bennettitales

The points of similarity between *Cycas* and Bennettitales are :

1. The plants are small trees with a barrel-shaped trunk.
2. The stem anatomy shows a very thick cortex, relatiively thin vasculature and large pith. Stele is of siphonostelic with exarch xylem.
3. Presence of manoxylic wood.
4. Leaves are pinnately compound with striking similarities in form, size and venation.
5. The presence of dicotyledonous embryo.
6. The presence of monocolpate (monosulcate) pollen grains.

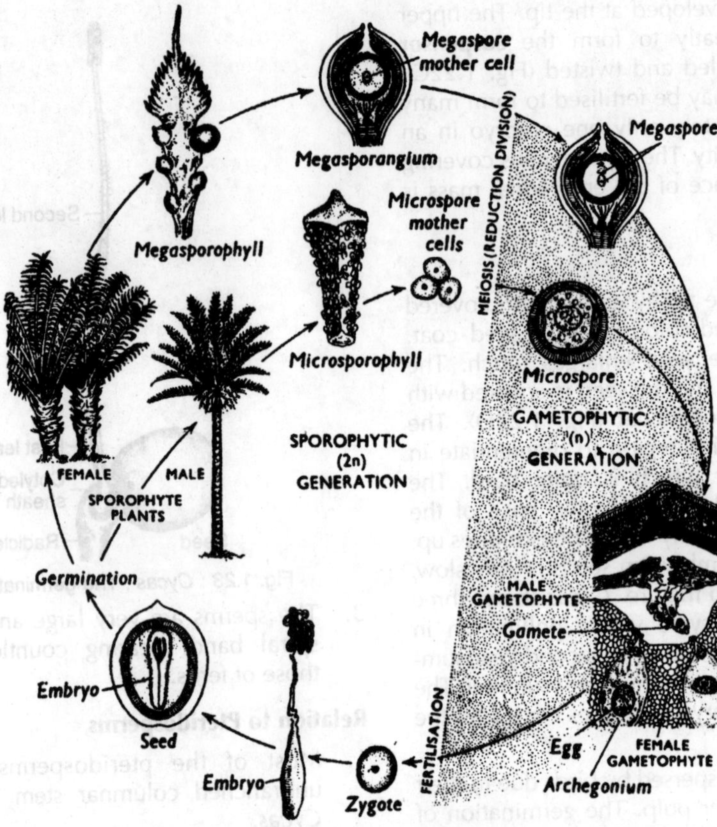

Fig. 1.24 : Life cycle of *Cycas*

Relation to Ginkgoales

There are several common features shared by *Cycas* and *Ginkgo*.

These include :

1. The presence of broad pith, cortex and numerous mucilaginous ducts in stem. Mucilage ducts are also present in leaf, petiole, embryo or even in root.
2. The presence of mesarch xylem in the cotyledon leaves and adult leaves.
3. The presence of nucellar beak with a distinct pollen chamber.
4. Both the micro- and megasporophylls are foliar in nature.
5. The sperms are motile with spiral band bearing numerous cilia.
6. There are extensive free-nuclear divisions during the early stage of embryogeny.
7. The mature seed consists of an endosporic embryo with two cotyledons.

Relation to Cordiatales

The similarities between *Cycas* and Cordiatales include :

1. The presence of large pith and centripetal wood in stem
2. The presence of extensive leaf-sclerenchyma.
3. The seeds are comparatively simple and fairly large with a single integument.
4. The presence of motile sperms.

Relation to Coniferales

Cycas resembles the Coniferales in having the following characteristics :

1. The presence of large pith, broad cortex and centripetal wood in stem.
2. The presence of leaf-sclerenchyma.
3. The presence of haplocheilic sunken stomata in leaves.
4. The seeds are with three-layered integument.

ORDER : CYCADEOIDALES (= BENNETTITALES)

The Cycadeoidales (= Bennettitales) is an extinct group of fossil gymnosperms. The order was named Bennettitales in honour of the English botanist J. J. Bennett. This order caught attention during earlier times by the discovery of silicified trunk known as *Cycadeoidea* from Great Britain. At the same time, *Bennettites* plant was reported from the Lower Cretaceous deposits of Isle of Wight. It becomes apparent that there is no difference between *Cycadeoidea* and *Bennettites*. Thus, according to the Rule of Priority *Cycadeoidea* is a valid name for the genus.

Both the members of Cycadales and Cycadeoidales coexisted during Mesozoic Era. The Mesozoic Era is thus popularly called the "**age of the cycads.**" Cycadeoid leaves are very much similar to cycads in form, structure and venation. However, cycadeoid leaves differ from those of cycads in having **syndetocheilic stomata**, while it is haplocheilic in cycads. Moreover, cycadeoid guard cells are heavily cutinised than that of cycads.

General characteristics of Cycadeoidales

1. In some members, plants are with slender and highly branched stems, or others have trunk-like stems with few branches, or are unbranched.
2. The stem surfaces are covered with spirally arranged persistent leaf bases with or without hairs in-between.
3. The stem anatomy shows a large pith surrounded by a cylinder of wood.
4. The wood is well-developed and compact like that of conifer wood.
5. The primary xylem is endarch. The secondary tracheids either have scalariform pitting on the walls, or have multiseriate bordered pits on their radial wall.
6. Secretory canals are present in pith and cortex.
7. The leaves are pinnately compound (rarely, simple) with paralled venation and syndetocheilic stomata. The guard cells are heavily cutinised and the stomata are oriented at right angles to the veins.
8. The leave traces originate from the primary xylem and pass directly through the secondary xylem to the cortex where they branch forming many traces. Thus, a single leaf base receives many vascular traces.
9. The reproductive structures are bisporangiate (except, Williamsoniaceae where they are monosporangiate) — often called flowers, protected by many bracts.
10. The ovules are borne in large numbers on a dome-shaped, cylindrical or conical receptacle.
11. The ovules are interspersed with vascularized interseminal scales with their enlarged tips fused to form a shield through which micropyles come out.
12. Each ovule has a single integument which is three-layered and the integument protects forward to form a long micropyle.
13. Microsporophylls are in whorls or free or united with numerous elongated microsporangia usually fused into a synangium.
14. The presence of monocolpate (= monosulcate) pollen grains.
15. The embryo is dicotyledonous.

Classification

Sporne (1965) classified the order Cycadeoidales into three families — Williamsoniaceae, Wielandiellaceae and Cycadeoidaceae.

However, in recent times, the Cycadeoidales has been divided into two families — Williamsoniaceae and Cycadeoidaceae (Meyen, 1987; Stewart and Rothwell, 1993).

The order is comprised of four reconstructed members — *Williamsonia, Williamsoniella, Wielandiella* and *Cycadeoidea*. A key has been provided following Darrah (1960) for easy identification of those members (Table 1.1) :

Table 1.1 : Key to the identification of four reconstructed members of Cycadeoidales (After Darrah, 1960)

A. Stem surfaces are covered with persistent leaf bases. Stems are either unbranched or with few branches.

B. Stems unbranched, flowers bisexual, inflorescence partially embedded within leaf base layer .... *Cycadeoidea*

BB. Stems with few branches, flowers unisexual, inflorescence not embedded within leaf base layer ...... *Williamsonia*

AA. Stem smooth, not covered with leaf bases, stems dichotomously branched.

B. Leaves are dissected, and concentrated around the dichasia. Cones are sessile and borne at each dichasium .... *Wielandiella*

BB. Leaves are entire, and are scattered on the axes. Cones are with long, slender peduncle and are borne in the axils of leaves .... *Williamsoniella*

Family : Williamsoniaceae

Several plants under this family were reconstructed. *Williamsonia gigas* was reported by Williamson (1870) for the first time from Jurassic of Yorkshire. The most important plant, *Williamsonia sewardiana* was reconstructed by Prof. Birbal Sahni (1932) from the Jurassic (Upper Gondwana) deposits of the Rajmahal Hills of India. The oldest known member of this family is *Wielandiella* which was recovered from the Upper Triassic (Rhaetian) of Scania in Sweden by Nathorst (1911). Another member, *Williamsoniella* was reported by Thomas (1915) from the Jurassic beds of Yorkshire.

Wielandiella

The plant consists of a very slender, herbaceous stem showing **false dichotomy** (Fig. 1.25). The unipinnately compound leaves are known as *Anomozamites* which are concentrated in the region of bifurcations.

A single sessile bisporangiate cone is borne in the region of each bifurcation. The compression-impression cone shows a central ovulate receptacle bearing ovules, interseminal scales and microsporophylls. The cone is subtended by numerous bracts.

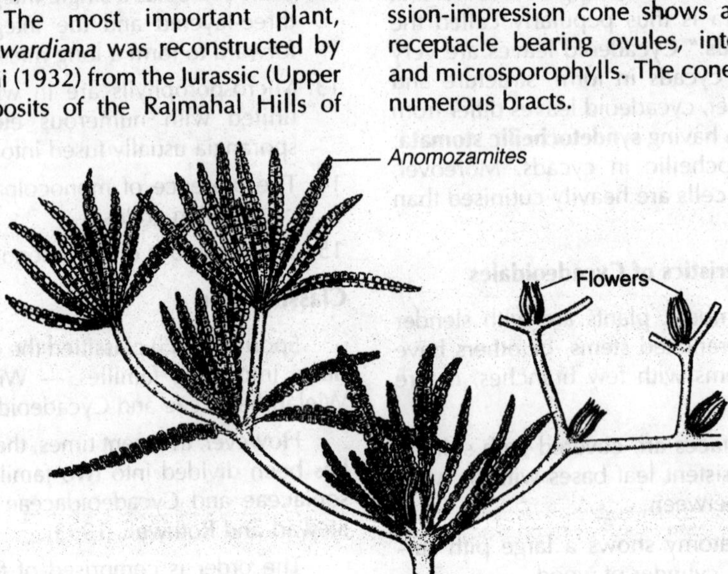

Fig. 1.25 : *Wielandiella angustifolia* with *Anomozamites* leaves

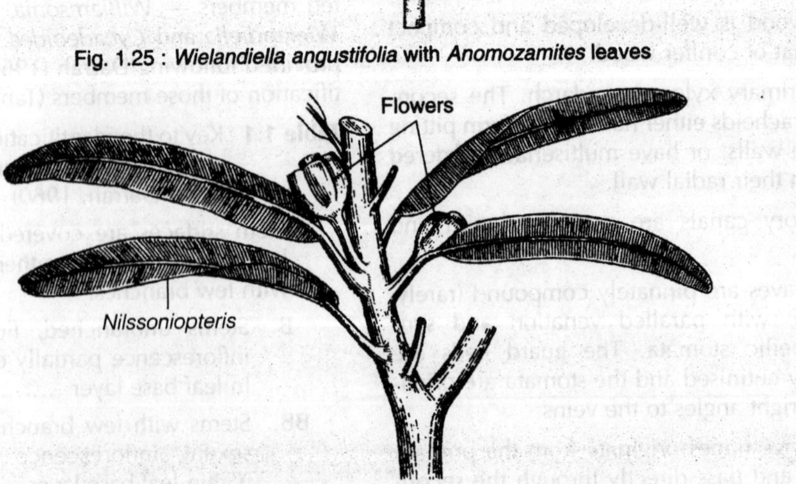

Fig. 1.26 : *Williamsoniella coronata* with *Nilssoniopteris* leaves

Williamsoniella

It is a shrubby plant that bears dichotomously branched axes (Fig. 1.26). The simple, petiolate and lanceolate leaves under the form-genus *Nilssoniopteris* are borne scattered on the axes.

Each bisporangiate cone is borne in the apex of peduncle that develops in the axil of a leaf. There is a centrally elongated ovulate receptacle that bears about 300 ovules and interseminal scales. In the basal region of the receptacle, there are 12 or more microsporophylls and numerous bracts arranged in whorl.

Williamsonia

Williamsonia is the widely studied genus of the family Williamsoniaceae. Several species of *Williamsonia* have been described from different regions of the world. Birbal Sahni (1932) reconstructed *Williamsonia sewardiana* (Fig. 1.30) using *Bucklandia indica* (stem), *Ptilophyllum cutchense* (leaf) and *Williamsonia* (female fructification). T. M. Harris (1969) also made a reconstruction using *Bucklandia pustulosa* (stem), *Ptilophyllum pecten* (leaf) and *Williamsonia leckenbyi* (female fructification). The male fructification of *Williamsonia* is called *Weltrichia* which has never been found in actual connection with the plant. Several species of *Weltrichia* are described from the Jurassic deposits of India. In addition to *Ptilophyllum*, several leaf-genera have also been described showing cycadeoid affinity.

Leaves : Leaves are pinnately compound. Many long slender pinnae are attached to the adaxial surface of the rachis by full width of their broad bases. *Pterophyllum* (Fig. 1.27A, B) leaf is characterised by long, slender pinnae with paralled venation, attached laterally to the rachis. *Ptilophyllum* (Fig. 1.27C, D) leaf has linear to slightly sickle-shaped pinnae, with parallel

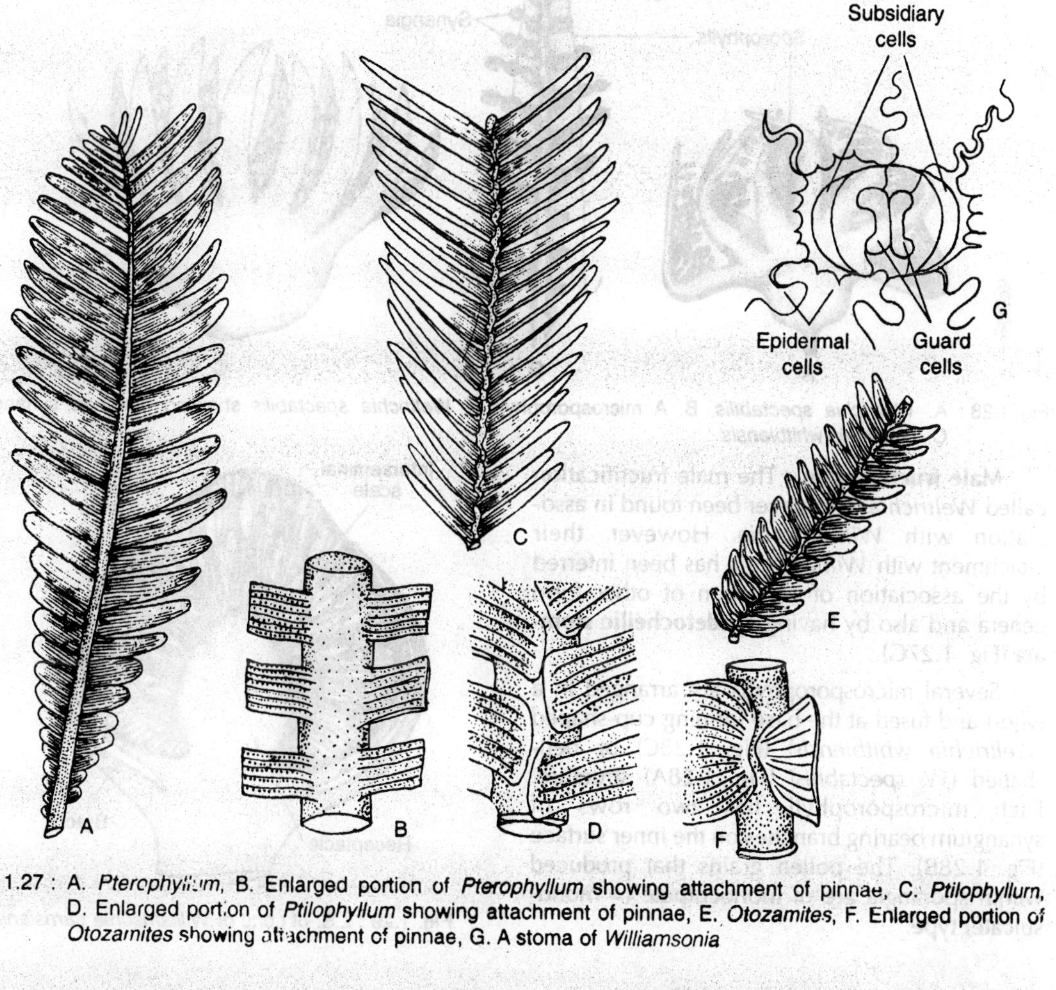

Fig. 1.27 : A. *Pterophyllum*, B. Enlarged portion of *Pterophyllum* showing attachment of pinnae, C. *Ptilophyllum*, D. Enlarged portion of *Ptilophyllum* showing attachment of pinnae, E. *Otozamites*, F. Enlarged portion of *Otozamites* showing attachment of pinnae, G. A stoma of *Williamsonia*

venation attached to the adaxial surface of the rachis. *Otozamites* pinnae (Fig. 1.27E, F) have lobed bases with spreading veins which are attached to the adaxial side of the rachis. *Dictyozamites* leaf is similar to *Otozamites* except that the pinnae have anastomosing network.

Stem : The stem, called *Bucklandia* is columnar in nature, either unbranched or sparsely branched. The surface of the stem is covered with spirally arranged persistent leaf bases. Internal structures are rarely preserved, because most of the specimens were recovered as casts. Anatomically, the stem exhibits a broad **cortex**, **vascular bundles** arranged in a ring and a narrow pith. The wood is formed following **secondary growth** in thickness.

Female fructification : Several species of *Williamsonia* have been described. The female fructification consists of a well-developed central conical **receptacle** which bears numerous stalked ovules and **interseminal scales** arranged spirally on the receptacle (Fig. 1.29). The interseminal scales are vascularised, club-shaped fleshy structures. The distal club-like ends of interseminal scales are fused to form a shield through which ovular micropyles protrude. The female fructification is subtended by spirally arranged perianth-like **bracts** with **scales** at their base and hairs above.

Williamsonia sewardiana

One of the most important contributions made in Palaeobotany is the reconstruction of *Williamsonia sewardiana* by Prof. Birbal Sahni

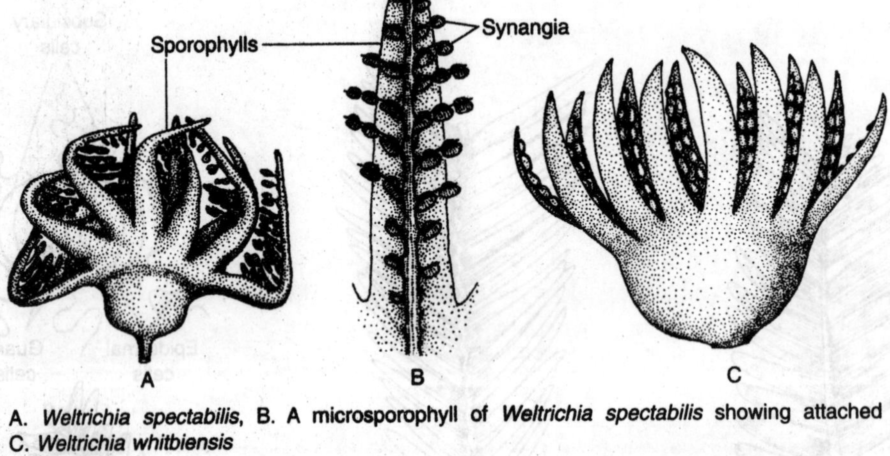

Fig. 1.28 : A. *Weltrichia spectabilis*, B. A microsporophyll of *Weltrichia spectabilis* showing attached synangia, C. *Weltrichia whitbiensis*

Male fructification : The male fructification called *Weltrichia* have never been found in association with *Williamsonia*. However, their attachment with *Williamsonia* has been inferred by the association of a portion of other form genera and also by having **syndetocheilic stomata** (Fig. 1.27G).

Several microsporophylls are arranged in a whorl and fused at the base forming cup-shaped (*Weltrichia whitbiensis*) (Fig. 1.28C) or bell-shaped (*W. spectabilis*) (Fig. 1.28A) structure. Each microsporophyll has two rows of synangium bearing branches on the inner surface (Fig. 1.28B). The pollen grains that produced within sporangia are of monocolpate (= monosulcate) type.

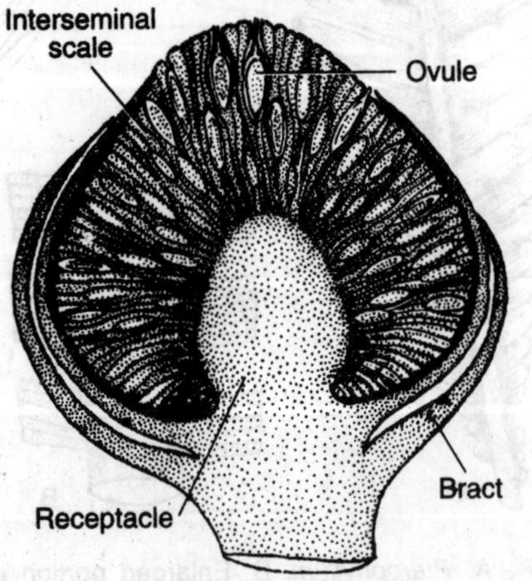

Fig. 1.29 : L.S. of cone of *Williamsonia harrisiana*

(1932) from the Jurassic (Upper Gondwana) deposits of the Rajmahal Hills of India. The plant was named in the honour of two English botanists Prof. Williamson (grand-teacher of Dr. Sahni) and Prof. A. C. Seward (teacher of Dr. Sahni). It was the first reconstructed fossil plant from India. The name of this plant is actually based on the female fructification (*Williamsonia*) rather than the entire plant of which *Bucklandia indica* is the stem and *Ptilophyllum cutchense,* the leaves.

The reconstruction as made by Dr. Sahni shows a small tree-like plant with sparsely branched columnar stem (*Bucklandia indica*) ending in a crown of pinnately compound leaves (*Ptilophyllum cutchense*) and bearing female fructification (*Williamsonia scotica*) on short branched lateral shoot. The plant looks like a miniature *Cycas* (Fig. 1.30).

Bucklandia indica

The stem is armoured with spirally arranged alternating areas of large (foliage) and small (scaly) leaf bases (Fig. 1.30). The branching of the stem is lateral which projects beyond the armour of leaf-curshions. There are two types of branching : the **vegetative** or **sterile shoot** and the **fertile shoot**. There is a prominent constriction at the base of each lateral shoot. It is presumed that the lateral shoots get detached through the constricted base and vegetatively propagate. The main shoot bears both the foliage leaves and scale leaves, while the lateral shoots bear only foliage leaves.

Internally, the stem shows a wide **cortex** and large **pith**. The **vascular bundles** are conjoint, collateral, endarch and open, remain embedded in pith. Many secretory ducts are found in cortex and pith. The secondary xylem, although compact, is manoxylic comprising of scalariform tracheids with multiseriate bordered pits on radial wall. There are uni- to triseriate medullary rays. There are numerous leaf traces. The most striking difference with the *Cycas* stem is the absence of girdling leaf traces in *Bucklandia indica.*

Ptilophyllum cutchense

The fronds have linear to slightly sickle-shaped pinnae which are attached to the adaxial surface of the rachis by full width of their broad and rounded bases (Fig. 1.27C, D). Pinnae exhibit parallel venations.

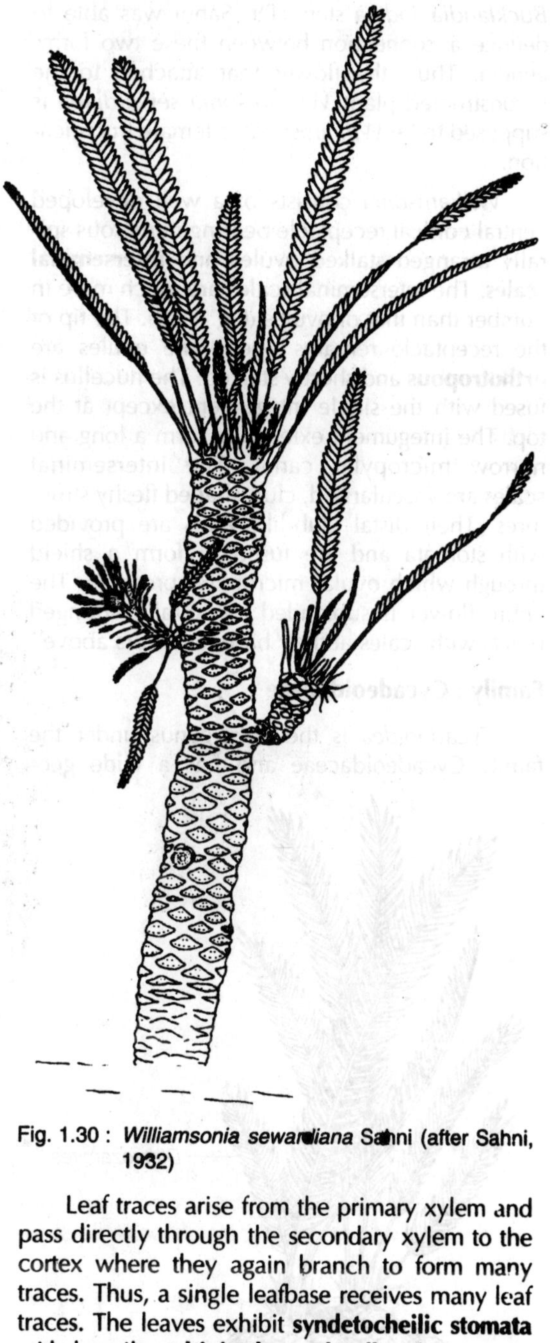

Fig. 1.30 : *Williamsonia sewardiana* Sahni (after Sahni, 1932)

Leaf traces arise from the primary xylem and pass directly through the secondary xylem to the cortex where they again branch to form many traces. Thus, a single leafbase receives many leaf traces. The leaves exhibit **syndetocheilic stomata** with heavily **cultinised guard cells.** The stomate are oriented at right angles to the veins. The epidermal cell walls have wavy outlines (Fig. 1.27G).

Williamsonia

Based on the similar anatomical features in the peduncle of *Willlansonia scotica* cone and

Bucklandia indica stem, Dr. Sahni was able to deduce a connection between these two form-genera. Thus, the flower that attached to the reconstructed plant *Williamsonia sewardiana* is supposed to be *Williamsonia*, a female fructification.

Williamsonia consists of a well developed central conical receptacle bearing numerous spirally arranged stalked ovules and **interseminal scales**. The interseminal scales are much more in number than that of ovules (Fig. 1.29). The tip of the receptacle remains open. The ovules are **orthotropous** and shortly stalked. The nucellus is fused with the single integument except at the top. The integument extends to form a long and narrow micropylar canal. The interseminal scales are vascularised, club-shaped fleshy structures. Their distal club-like ends are provided with stomata and are fused to form a shield through which ovular micropyles protrude. The entire flower is sublanded by spirally arranged bracts with scales at their base and hairs above.

Family : Cycadeoidaceae

Cycadeoidea is the only genus under the family Cycadeoidaceae and has a wide geographical distribution, especially in North America, Europe and India during Upper Jurassic to Upper Cretaceous. The plants have stout, globose trunks with armour of spirally arranged leaf bases embedded in **ramenta** of flat, tongue-shaped scales (Fig. 1.31). The leaves have never been found attached to the stem. It is assumed that *Ptilophyllum* or *Dictyozamites* type of leaves were borne spirally in a crown at the top of the trunk. The flowers were borne in the axil of leaf bases and developed simultaneously. Thus the plants were **monocarpic** i.e., they flowered once during their lifetime.

Anatomically, the trunk exhibits a large pith surrounded by a broad ring of vascular tissue comprising of manoxylic wood and secondary phloem (Fig. 1.32). The vascular ring is dissected by broad rays, either uniseriate or biseriate type. There are scattered secretory ducts in cortex and pith. The leaf traces traversing the cortex are C-shaped and pass directly through the secondary xylem to the cortex where they again divide and arrange themselves in a horse-shoe-shaped fashion without girdling.

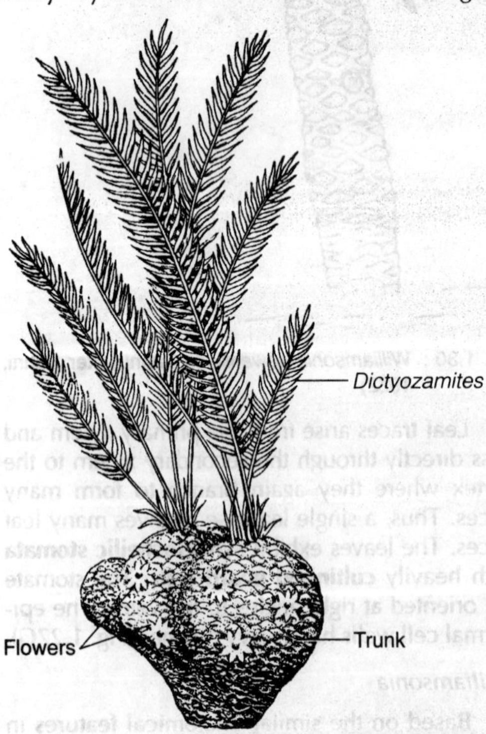

Fig. 1.31 : A reconstructed *Cycadeoidea* plant

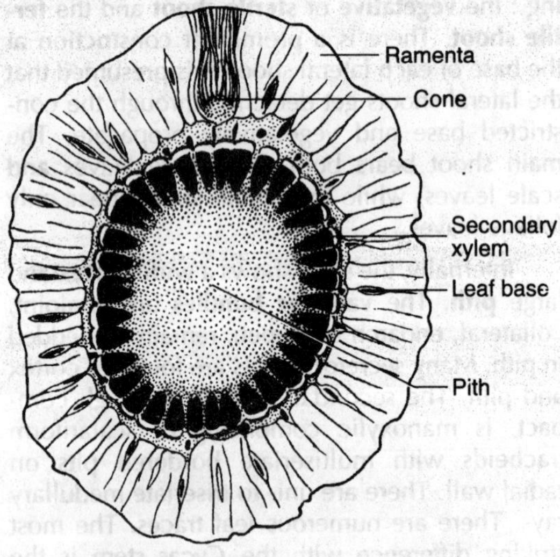

Fig. 1.32 : T.S. of trunk of *Cycadeoidea*

Cycadeoid flowers are bisexual, borne on a short pedical and are covered by about 100 spirally arranged hairy bracts, protected by a **ramentum** or **scales** (Fig. 1.33A). The bracts open up at maturity. There are about 20 pinnate microsporophylls arranged in a whorl at the base of ovuliferous receptacle. Microsporophylls are united at the base and comparable to the

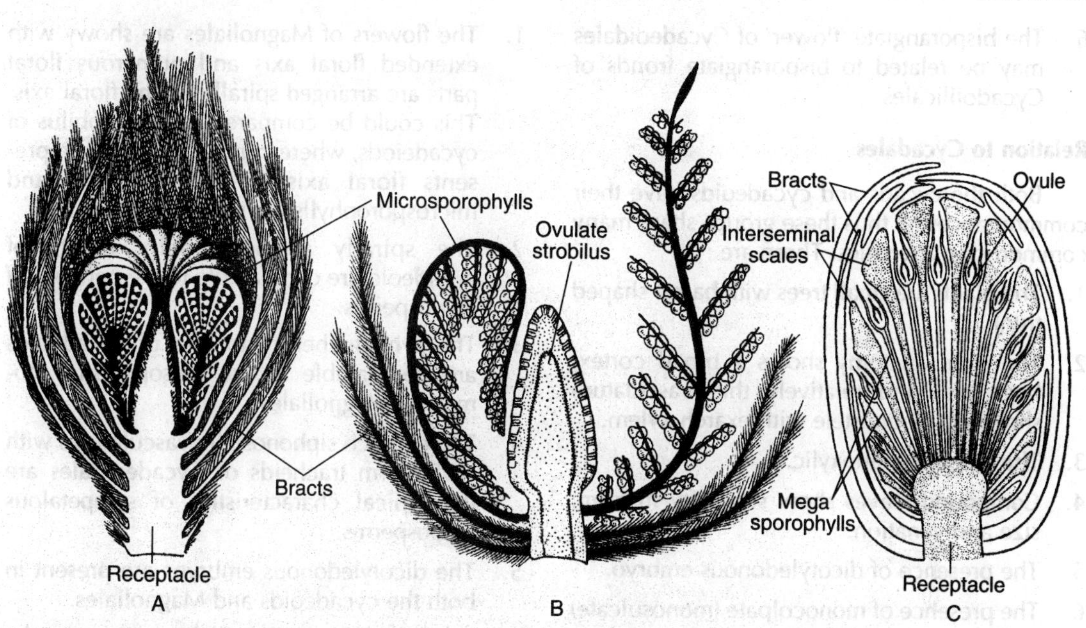

Fig. 1.33 : *Cycadeoidea* : A. An unexpanded flower, B. The same with a microsporophyll expanded, C. A mature flower

segments of an orange. Two rows of kidney-shaped synangia are borne on the inner surface of each pinnule of a microsporophyll (Fig. 1.33B). Each synangium bears 20-30 tubular pollen sacs or sporangia containing monocolpate (monosulcate) pollen grains. As in *Williamsonia*, the ovule-bearing structure is in the form of a conical receptacle on which ovules and interseminal scales are arranged spirally (Fig. 1.33C). The number of ovules is almost equal to the number of **interseminal scales.**

Origin and relationships

Cycadeoidales shows striking similarities with Cycadales. Chamberlain (1920) hypothesised that cycadeoids and cycads are two distinct groups having a common ancestry. Most of the authors suggested that cycadeoids evolved from the pteridosperm complex, probably in the late Carboniferous. Delevoryas (1968) proposed his "foliar theory" in the origin of cycadeoids. This is the most acceptable hypothesis so far proposed. According to this theory, the basal pinna bearing microsporangia and ovules of an axillary shoot of pteridosperm became foreshortened and condensed to form the cycadeoid bisporangiate cone. While, the rest of the sterile frond became unchanged. Thus, the cycadeoid cone is a condensed axillary fertile shoot. The reproductive structure is an unique feature in cycadeoids; however, this group shows similarities with many other groups.

Relation to Ferns

Cycadeoids resemble ferns in the following characteristics :

1. The abundance of multicellular rementa on the plant body.
2. The presence of scalariform tracheids, large pith and direct leaf traces.
3. The pinnatified microsporophyll with lateral rows of synangia in cycadeiods are comparable to the sori of the Marattiales.

Relation to Pteridospermales

Cycadeoids probably originated from pteridosperms, thus they share many common features. These include :

1. The presence of barrel-shaped trunk.
2. The presence of ramental hairs.
3. The presence of direct leaf traces.
4. The stem shows large pith and cortex with manoxylic wood.
5. The presence of leafy microsporophyll bearing synangia.

6. The bisporangiate 'flower' of Cycadeoidales may be related to bisporangiate fronds of Cycadofilicales.

Relation to Cycadales

Both the cycads and cycadeoids have their common ancestry, thus these groups share many common characteristics. These are :

1. The plants are small trees with barrel-shaped trunk.
2. The stem anatomy shows a broad cortex, large pith and relatively thin vasculature showing siphonostele with exarch xylem.
3. The wood is manoxylic.
4. Compound leaves show similarity in form, size and venation.
5. The presence of dicotyledonous embryo.
6. The presence of monocolpate (monosulcate) pollen.

Relation to Angiosperms

Many botanists like Arber and Parkin, Bessey, Hallier, Hutchinson, believed that the Cycadeoidales is the ancestor of angiosperms. This view is based on the similarity in cycadeoid strobili and the flowers of some of the more primitive angiosperms such as Mangoliales and Ranales. There are many other characteristics noted in cycadeoids that closely resemble the primitive angiosperms. These are :

1. The flowers of Magnoliales are showy with extended floral axis and numerous floral parts are arranged spirally on the floral axis. This could be compared to the strobilus of cycadeiods, where conical receptacle represents floral axis on which ovules and microsporophylls are spirally arranged.
2. The spirally arranged hairy bracts of cycadeoid are comparable to the perianth of angiosperms.
3. The syndetocheilic stomata of cycadeoids are comparable to the mesoparasitic stomata of Magnoliales.
4. The endarch siphonostelic vasculatures with scalariform tracheids of Cycadeoidales are the typical characteristics of sympetalous angiosperms.
5. The dicotyledonous embryos are present in both the cycadeoids and Magnoliales.

Pant and Kidwai (1977) disagreed over the above view. Based on the geological occurrence of cycadeoids and angiosperms, they argued that cycadeoid flower represents a parallel line of evolution to angiosperms.

It has been assumed that the homologies between cycadeoids and angiosperms are due to a long course of evolution. The cycadeiods have naked seeds (typical of gymnosperms) with an admixture of sterile interseminal scales. Moreover, wood rays are devoid of marginal cell which is typical of angiosperms.

CLASS. GINKGOPSIDA

The class Ginkgopsida includes a single order Ginkgoales — now represented by a single monotypic extant plant, *Ginkgo biloba*. The members of this group first appeared in the Permian period. They were well-represented all over the world during the Jurassic period and gradually started declining from Cretaceous onwards. The extant plant, *Ginkgo biloba,* has its fossil record as far back as Triassic. The leaf impressions of *Ginkgo biloba* have been identified from the Permian rocks and probably from the Carboniferous rocks. *Ginkgo biloba* has escaped extinction and still exists today being the oldest living seed plant. Thus, *Ginkgo biloba* is referred to as **'living fossil'**.

GENERAL CHARACTERISTICS OF GINKGOPSIDA

1. Mostly tall woody branching trees.
2. Secondary wood pycnoxylic
3. Leaves are leathery, strap-shaped or fan-shaped, often deeply divided with dichotomous venation.
4. Male fructifications are axillary, unbranched, catkin-like bearing microsporangiophores.
5. Each microsporangiophore bears 2-12 pendulous microsporangia.
6. Presence of ciliated sperms.
7. Ovules 2-10, terminal on axillary branching or almost unbranched axes.
8. Seeds are large showing radial symmetry.
9. Seeds are with fleshy outer layer and stony middle layer.

ORDER : GINKGOALES

The order Ginkgoales is comprised of two families — (i) Trichopityaceae, (ii) Ginkgoaceae. The family Trichopityaceae consists of a single genus *Trichopitys* and is probably the oldest known Ginkgophytes, reported as far back as Permian. The family Ginkgoaceae is represented by 14 extinct genera (e.g. *Baiera, Arctobaiera, Sphenobaiera, Ginkgoites* etc.) and by one living genus, *Ginkgo*. The extinct members (Fig. 1.34A, B) showed their worldwide distribution, particularly during Jurassic.

GINKGO

Systematic position

Class. **Ginkgopsida**
Order. **Ginkgoales**
Family. **Ginkgoaceae**

Monotypic species – *Ginkgo biloba.*

Geological distribution – Triassic to present day.

Geographical distribution

Ginkgo biloba is commonly called **Maiden hair tree** because of its leaves that resemble very much with *Adiantum* (called **Maiden hair fern**) leaves with respect to shape and venation. *Ginkgo biloba* is successfully cultivated as an ornamental plant in gardens of North America and Europe. But, in time immemorial Buddhist monks considered *G. biloba* as a sacred plant and took keen interest for its preservation in different places. Thus, unknowingly, this plant

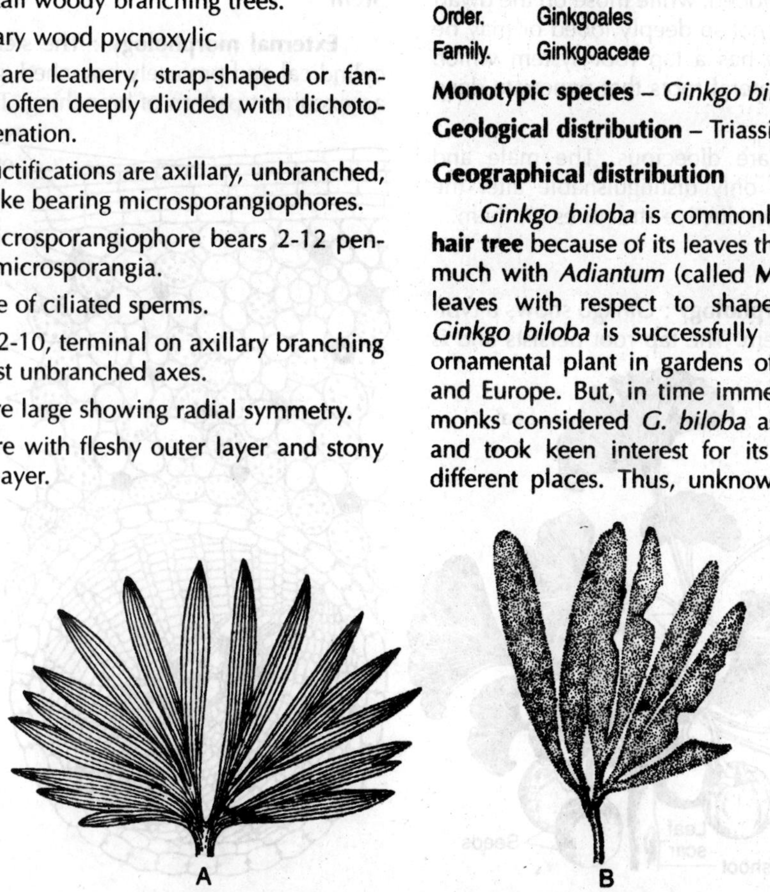

Fig. 1.34 : A. *Baiera gracilis* leaf, B. *Ginkgoites hermelini* leaf

has been preserved and escaped extinction. At present, trees of *G. biloba* are found growing near the Buddhist temples in China, Japan, Thailand, etc. In addition, this plant is cultivated in some places of China and Japan for its edible seeds. It has been reported that *G. biloba* still exits today in South-Eastern China in the wild state.

In India, very few *Ginkgo biloba* trees are found in the Botanic Gardens of Darjeeling, Dehradoon, Mussoorie, Shibpur (Howrah), etc.

Sporophyte

Ginkgo biloba is a tall deciduous tree (up to 30 m height) giving rise to a very irregular pattern of branching. The branching is restricted to the upper part of the stem, thus giving the tree an **excurrent** pattern of growth. The branches are **dimorphic**, bearing two types of shoot : long shoots and dwarf shoots (Fig. 1.35). The leaves also show dimorphism, those present on the long shoot are deeply lobed, while those on the dwarf shoots are either not so deeply lobed or may be entire. *G. biloba* has a tap root system which posseses strong lateral roots that penetrate deep into the soil.

The plants are dioecious. The male and female trees are only distinguishable after the production of reproductive structures on them.

Root

External morphology : *Ginkgo* shows a typical tap root system. The tap root persists and is extensively branched and penetrates deep into the soil. The roots of *Ginkgo biloba* is characterised by vasicular-arbuscular-mycorrhizas (**VAM**) with *Glomus epigaeum* inoculation.

Internal structure : Anatomically, the young root of *Ginkgo* is diarch (Fig. 1.36), but older roots become tetrarch or hexarch. The following structures are discernible in the young root from outside inward : a single-layered **epidermis**; multilayed thin-walled **cortex** containing several tannin-filled cells, calcium oxalate crystals and many mucilage canals; a single-layered **endodermis**; a layer of **pericycle**; diarch radial **vascular bundle**. Xylem is exarch and remains embedded within phloem. The intracellular VAM hyphae are much abundant than the intercellular hyphae.

The normal secondary growth takes place in roots, but the annual rings are not so prominent as in the stem.

Stem

External morphology : The stem is woody, cylindical and profusely branched, giving rise to an excurrent pattern of branching. The branches

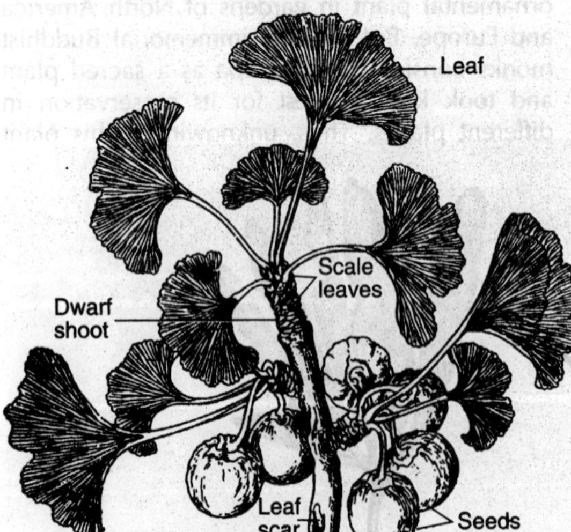

Fig. 1.35 : *Ginkgo biloba.* A shoot bearing dwarf shoots

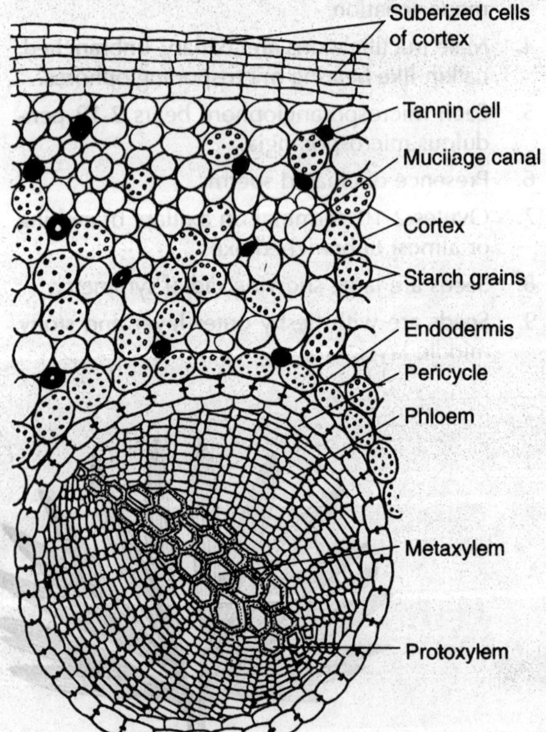

Fig. 1.36 : *Ginkgo biloba.* T.S. of young root

develop on the upper part of the stem. The branches are dimorphic bearing two types of shoots, viz., the **long shoot** and the **dwarf shoot**.

Long shoots are the branches of unlimited growth which gradually become reduced in size, thus giving the tree a pyramid-like or pagoda-like appearance Dwarf shoots are the branches of limited growth which are borne on long shoots. Each dwarf shoot bears **scale leaves**, leaf scars and a crown of foliage leaves (Fig. 1.35). Thus, a dwarf shoot looks like a miniature *Cycas* tree.

Internal structure : Anatomically, the long shoot can easily be distinguished from the dwarf shoot. The long shoot has comparatively narrow cortex and pith with a few mucilage ducts (Fig. 1.37A). The wood is **pycnoxylic**, broad and hard. On the other hand, the dwarf shoot has broad cortex and pith with a large number of mucilage ducts (Fig. 1.37B). The wood is pycnoxylic and less in amount. Moreover, the T.S. of dwarf shoot shows several double leaf traces. On the other-hand, the leaf trace in the T.S. of long shoot can only be visible if the section is cut through the region of the attachment of leaves. The two leaf traces that enter into the petiole of the leaf are developed from separate adjacent axial bundles of the stem.

The stem anatomy in both the long and dwarf shoot shows a heavily cutinised single layered **epidermis**. In order stem, the epidermis is replaced by **periderm.** The epidermis is followed by cortex. The **cortex** is made up parenchymatous cells. The cortex also consists of tanniniferous cells, sphaeraphides, mucilage ducts and few cortical cells contain crystals of calcium oxalate. The cortex of older stem also contains numerous fibres and sclerides. The **endodermis** and **pericycle** are not well-differentiated in long shoot.

In young stem, **vascular bundles** are arranged in a ring. Each vascular bundle is conjoint, collateral and open. Thus, the vascular cylinder becomes **endarch siphonostele** at the onset of secondary growth. The tracheids are arranged in radial rows. The protoxylen tracheids have spiral thickening, while the metaxylem tracheids have reticulate thickening. The sieve cells and parenchyma with occasional albuminous cells constitute the phloem.

Secondary growth in thickness

The cambium is normal in position and normal in function. Thus, the **cambium ring** cuts off a continuous cylinder of secondary xylem inside and secondary phloem outside. The cambium also produces **secondary medullary rays** at places (Fig. 1.38A). The secondary xylem is made up of tracheids whose end almost remain at the same level, without overlapping each other. Thus, wood of *Ginkgo* becomes very brittle and has little commercial value. The secondary phloem is composed of sieve elements, fibres and a few parenchyma.

The R.L.S. of the wood shows (Fig. 1.38C) tracheids with one or two rows of bordered pits. The mature tracheids develop thickened border in the form of crescentic bars called **Bars of Sanio** or **Crassulae** in between the pits. Sometimes, crassulae cross the interior of tracheids. There are uniseriate xylem rays seen in T.L.S. (Fig. 1.38B). The rays vary in height, 1-3 cells high in long shoots and 1-15 cells high in dwarf shoots.

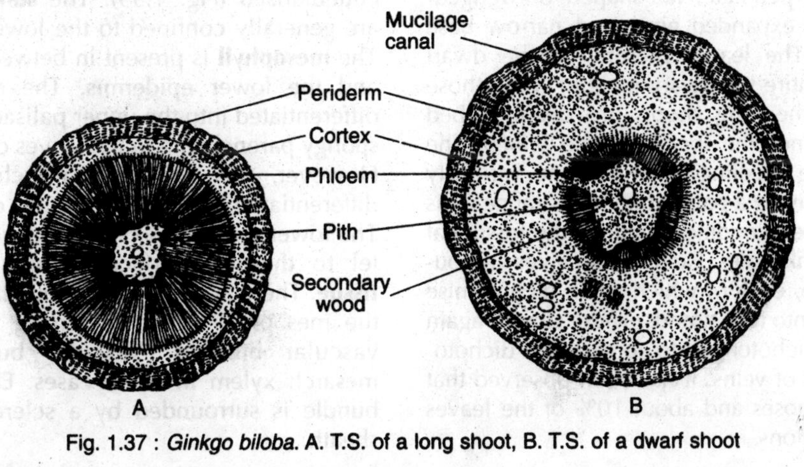

Fig. 1.37 : *Ginkgo biloba*. A. T.S. of a long shoot, B. T.S. of a dwarf shoot

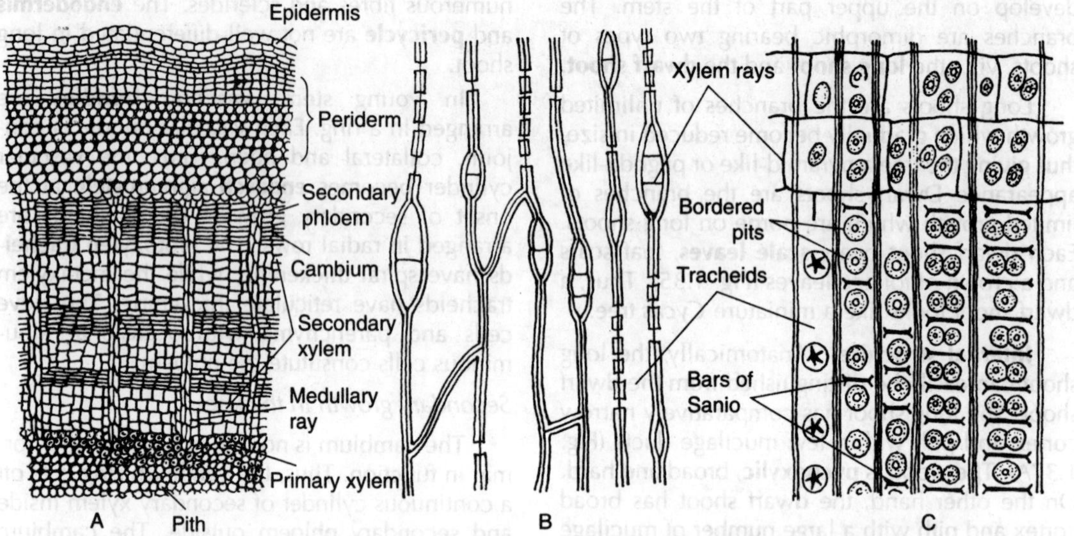

Fig. 1.38 : *Ginkgo biloba*. A. T.S. of long shoot, B. T.L.S. of wood, C. R.L.S. of wood

The unique features in the wood of *G. biloba* that separate them from other gymnosperms are the narrow tracheids and calcium oxalate crystals containing xylem parenchyma. These characters may be used to distinguish *Ginkgo* wood from the *Pinus* wood. The annual ring is not well-defined.

The **extrastelar secondary growth** takes place by the cork cambium which cuts off secondary cortex on the inner side and cork on the outer side. The secondary cortex is well-developed and devoid of mucilage ducts. Later, the epidermis is replaced by the cork.

Leaf

External morphology : The leaves are simple, long petiolate, fan-shaped or wedged-shaped with expanded apex and narrow base (Fig. 1.35). The leaves present on the dwarf shoots are entire with wary margin; while those present on the long shoots are deeply lobed (bilobed), hence the specific name *biloba*. The leaves at the seedling stage become variously lobed, thus giving a palmitified appearance. This variously lobed condition reminds the fossil leaf genera of Ginkgoales. Each leaf receives an double leaf-trace, each of which again dichotomise — resulting into four strands. These strands again repeatedly dichotomise to form open dichotomous system of veins. It has been observed that vein anastomoses and about 10% of the leaves have vein unions.

Internal Structure

Petiole : The T.S. of a petiole shows a thick cuticularised epidermis which is broken by stomata. A few-layered thick hypodermis is present beneath the epidermis. This is followed by a broad cortex where a few mucilage ducts, tannin-filled cells and spaeraphides are irregularly distributed. There are two endarch vascular bundles surrounded by a sclerenchymatous bundle sheath. The cambium is present in-between the phloem and xylem. The bundles are interspersed with some indistinct centripetal xylem called transfusion tracheids.

Lamina : The T.S. of the lamina shows an upper and a lower epidermis which are thickly cuticularised (Fig. 1.39). The **sunken stomata** are generally confined to the lower epidermis. The **mesophyll** is present in between the upper and the lower epidermis. The mesophyll is differentiated into the upper palisade and lower spongy parenchyma in the leaves of long shoot. However, the palisade parenchyma is not differentiated in the leaves of dwarf shoot. The lower mesophyll cells are drawn out parallel to the leaf surface forming **transfusion tissue**. There are many large mucilage canals in the mesophyll region traversing between the vascular bundles. Vascular bundles have mesarch xylem in some cases. Each vascular bundle is surrounded by a sclerenchymatous sheath.

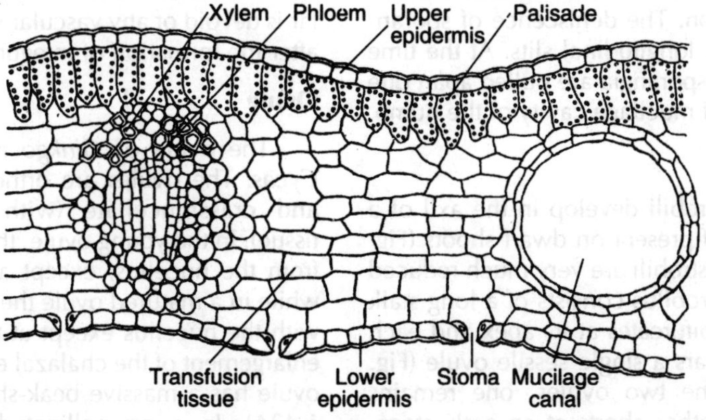

Fig. 1.39 : *Ginkgo biloba*. T.S. of leaf

Reproduction

Ginkgo biloba reproduces sexually. *Ginkgo* is dioecious. Morphologically, the male and the female plants are indistinguishable before the formation of reproductive organs. However, the male and the female trees can be distinguished cytologically at an early stage. In the female plant, out of the 24 chromosomes, four are satellite-bearing chromosomes, while in male plant three chromosomes are with satellite. Like cycads, the reproductive structures of *Ginkgo* are most primitive among living seed plants.

Male strobilus

Male strobili develop in the axil of leaves or on the sides of petiole bases present on dwarf shoots (Fig. 1.40A). The male strobilus is a loose structure and consists of a stalked central axis (2-3 cm in length) on which many microsporophylls are arranged spirally. A male strobilus looks like a catkin inflorescence of angiosperms (Fig. 1.40A).

Each microsporophyll has a long slender stalk terminating into . knob-like portion or **hump**. It bears two pendent microsporangia (Fig. 1.40B). The hump is always provided with a mucilage duct. The morphological nature of the hump is still an open question. According to some botanists, the hump represents an abortive sporangium and the mucilage cavity represents abortive sporogenous tissue. In rare instances, a microsporophyll bears three sporangia.

The microsporangia develops from a hypodermal archesporial cell of the microsporophyll. A mature microsporangium consists of a multi-layered wall, tapetum and microspore mother cells. It is important to note that 1-3 hypodermal layers develop fibrous thickening representing an **endothecium** as in angiosperm. In this point, *Ginkgo* differs from other gymnosperms where the outermost layer of the microsporangium develops fibrous thickening representing an **exothecium**.

Inside the sporangium, each microspore mother cell produces four microspores or pollen grains by meiotic division. The pollen grain is oval-shaped with monosulcate aperture. Each grain is bounded by two concentric wall layers : the outer thick **exine** and the inner thin **intine**.

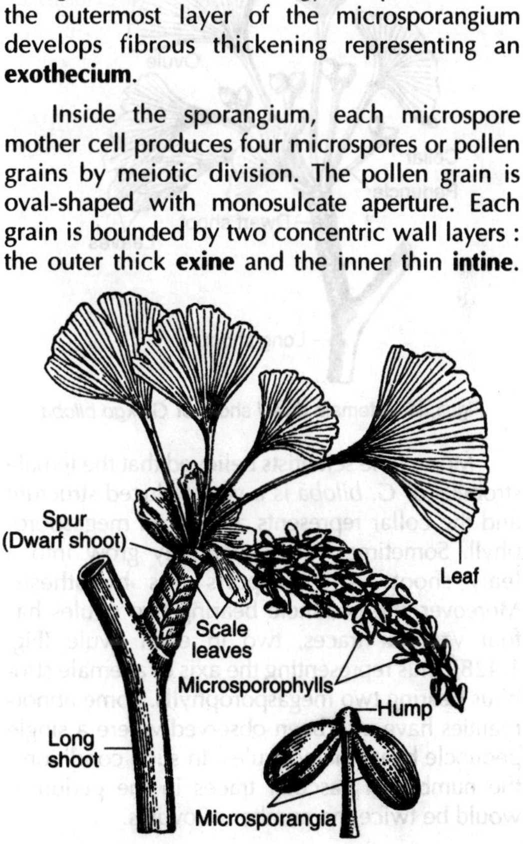

Fig. 1.40 : *Ginkgo biloba* : A. A male dwarf shoot, B. A microsporophyll

The exine is not continuous, being absent at the top apertural region. The dehiscence of sporangia takes place by longitudinal slits. At the time of dehiscence the sporangia are pulled apart due to the shrinkage of mucilage cavity of the hump.

Female strobilus

The female strobili develop in the axil of a leaf or a scale leaf present on dwarf shoots (Fig. 1.41). The female strobili are very much reduced structures. Each strobilus consists of a long stalk or peduncle that bifurcates at its apex and each branch usually bears a single sessile ovule (Fig. 1.42A). Out of the two ovules, one remains viable while the other aborts at an early stage. Sometimes, both the ovules become viable and develop into seeds. The base of each ovule is surrounded by a fleshy cup called **collar**.

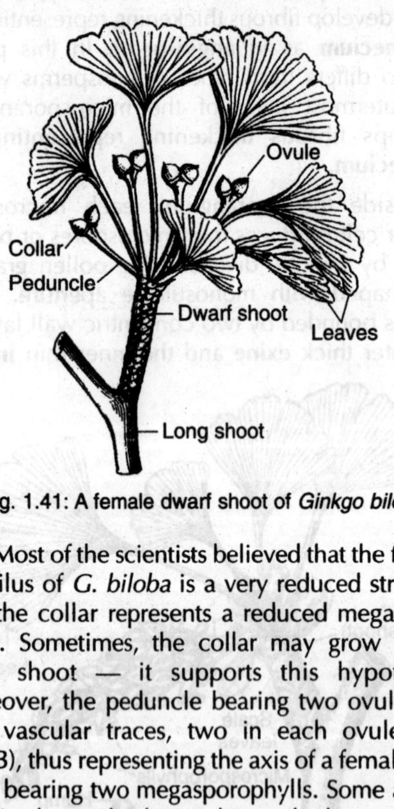

Fig. 1.41: A female dwarf shoot of *Ginkgo biloba*

Most of the scientists believed that the female strobilus of *G. biloba* is a very reduced structure and the collar represents a reduced megasporophyll. Sometimes, the collar may grow into a leafy shoot — it supports this hypothesis. Moreover, the peduncle bearing two ovules has four vascular traces, two in each ovule (Fig. 1.42B), thus representing the axis of a female strobilus bearing two megasporophylls. Some abnormalities have also been observed where a single peduncle bears many ovules. In such conditions, the number of vascular traces in the peduncle would be twice the number of ovules.

Pankow and Sothman (1967) disagreed over the sporophyll nature of the collar. They considered the ovules of *Ginkgo biloba* as cauline and terminal on lateral axis. They argued that the collar is devoid of any vascular supply and develops after the formation of integument.

Ovule

The ovule of *Ginkgo* is similar to that of *Cycas*. The ovules are orthotropous, unitegmic and crassinucellate (with massive nuceller tissue). In the young ovule, the integument is free from the nucellus, except at the chalazal end, while in a matured ovule the integument is fused with the nucellus except at the apex due to the enlargement of the chalazal end (Fig. 1.42B). The ovule has a massive beak-shaped nucellus (Fig. 1.42A). In a pre-pollinated ovule, the apical nucellar cells degenerates to form a deep pollen chamber and the pollination drop.

The single integument is differentiated into three layers : The outer fleshy with numerous mucilage cavities, the middle stony and the inner fleshy, eventually becomes papery. The ovule is supplied with two vascular traces which enter into the inner fleshy layer reaching up to the free part of the nucellus.

Megasporogenesis

A deeply situated cell of the nucellus is differentiated into a large megaspore mother cell.

The megaspore mother cell is surrounded by a well-developed spongy tissue. At a later stage, the spongy tissue cells adjacent to the enlarging gametophyte lose their cell walls. The megaspore mother cell wall becomes thick due to the development of a double layered wall. Now, the megaspore mother cell undergoes meiotic division to form a liner tetrad of four megaspores.

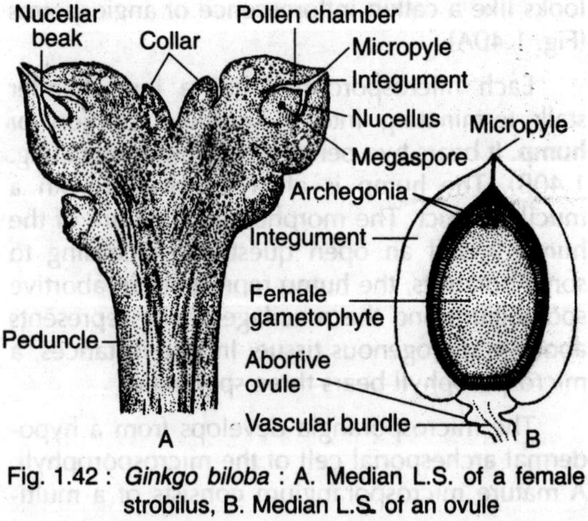

Fig. 1.42 : *Ginkgo biloba* : A. Median L.S. of a female strobilus, B. Median L.S. of an ovule

The upper three megaspores degenerate, while the lowermost one becomes functional.

Gametophyte

The spores (either microspores or megaspores) represent the first phase of gametophyte generation. The microspore or pollen grain is the male gametophyte, while the megaspore is the first stage of female gametophyte which develops into a female gametophyte.

Development of male gametophyte before pollination

The pollen nucleus undergoes mitotic division to produce a small lens-shaped first prothalial cell towards the proximal end (opposite to the aperturalside) and a large **central cell** on the distal end (Fig. 1.43A). The prothallial cell does not divide, whereas the central cell divides to form a second prothallial cell and an **antheridial initial** (Fig. 1.43B). The first prothallial cell is ephemeral, while the second prothallial cell is persistent. The antheridial initial cuts off a small **antheridial cell** and a large **tube cell** (Fig. 1.43C). The pollen grains are released from the microsporangium at this 4-celled stage (2 prothallial cells, an antheridial cell and a tube cell).

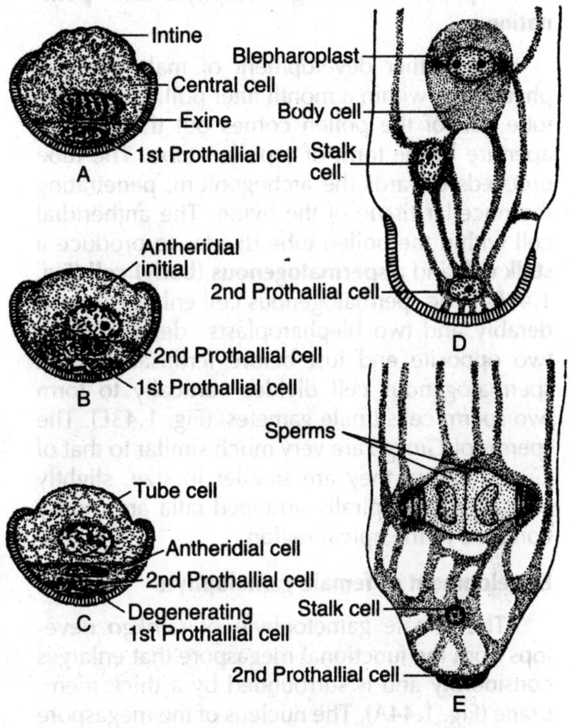

Fig. 1.43 : *Ginkgo biloba* : A–E. The stages in the development of male gametophytes

Schematic representation in the development of male gametophyte of *Ginkgo*

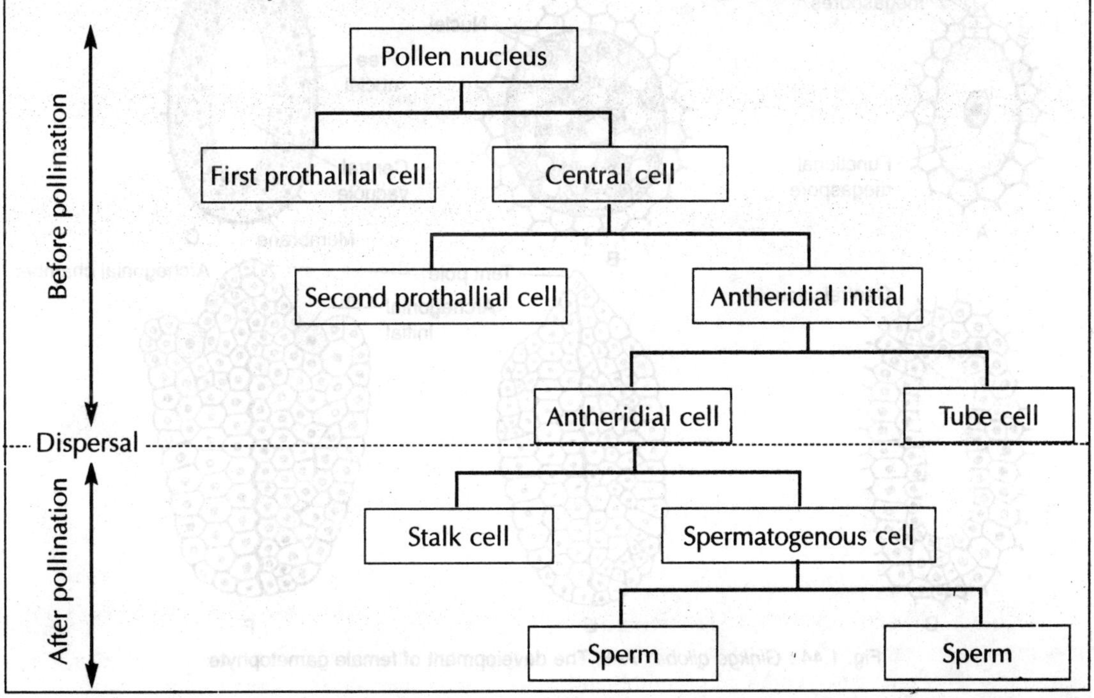

Development of male gametophyte after pollination.

The further development of male gametophyte starts within a month after pollination. The tube cell of the pollen comes out through the aperture in the form of a pollen tube. The tube proceeds towards the archegonium, penetrating the nucellar tissue of the ovule. The antheridial cell within the pollen tube divides to produce a **stalk cell** and a **spermatogenous (body) cell** (Fig. 1.43D). The spermatogenous cell enlarges considerably and two blepharoplasts develop at its two opposite end just before fertilisation. The spermatogenous cell divides vertically to form two sperm cells (male gametes) (Fig. 1.43E). The sperms of *Ginkgo* are very much similar to that of *Cycas,* except they are smaller in size, slightly elongated and spirally arranged cilia are mostly confined to the apical region.

Development of female gamelophyte

The female gametophyte of *Ginkgo* develops from the functional megaspore that enlarges considerbly and is surrounded by a thick membrane (Fig. 1.44A). The nucleus of the megaspore divides mitotically, unaccompanied by wall formation, forming a large number of free nuclei (as much as 8,000 free nuclei) (Fig. 1.44B). The nuclei are restricted to a thin film of cytoplasm at the periphery resulting into the formation of a large central vacuole (Fig. 1.44C). Thereafter, the cell wall formation begins in a centripetal fashion from periphery inwards (Fig. 1.44D), as a result the vacuole is obliterated. The entire gametophyte becomes cellular and the tissue thus formed is called endosperm (Fig. 1.44E). The endosperm cells are haploid in nature, but some polyploid cells are also formed. The cells that contain 2-3 nuclei during wall-formation are transformed to polyploid cells.

The ultrastructural study (Dexheimer, 1973) of the female gametophyte of *Ginkgo* shows four different zonations according to their food reserves. They are : (i) **Lipid zone** (outer 3–4 layers), (ii) **Starch proteiolipid zone** with **large** vacuolated cells, (iii) **Starch zone** consists of large vacuolate cells with peripheral cytoplasm, (iv) **Central zone** comprising of less differentiated cells with very little reserve content.

Development of archegonia

Two to four cells at the micropylar end of the female gemetophyte function as **archegonial initials** (Fig. 1.44F, 1.45A). Each of these cells divides periclinally to form on outer small primary **neck initial** and a large **central cell** (Fig.

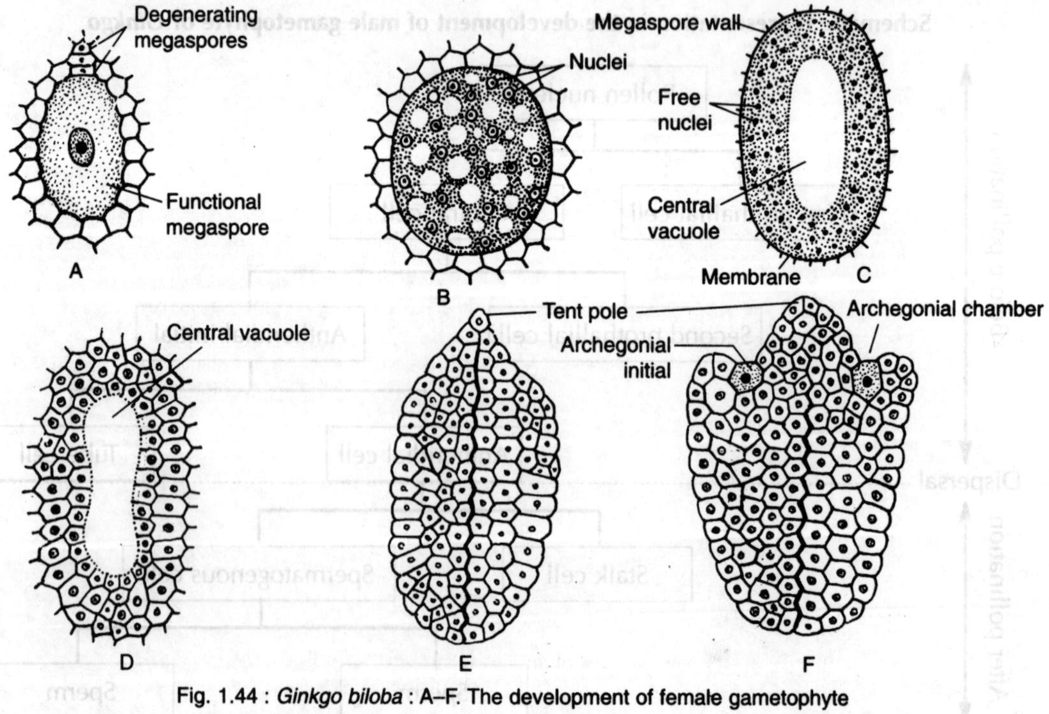

Fig. 1.44 : *Ginkgo biloba* : A–F. The development of female gametophyte

Schematic representation in the development of archegonium in *Ginkgo*

- Archegonial initial
  - Periclinal division →
    - Primary neck initial
      - 2 vertical divisions → 4 neck cells (single tier)
    - Central cell
      - Periclinal division →
        - Ventral canal cell (ephemeral)
        - Egg cell

1.45B). The primary neck initial divides by two vertical walls at right angles to each other resulting in four neck cells arranged in one tier. The nucleus of central cell cuts off an upper ephemeral ventral canal cell and a large egg cells (Fig. 1.45C). With regard to ventral canal cell, *Ginkgo* seems to be more primitive than *Cycas* where a ventral canal nucleus is present (Fig. 1.45D). At this stage, the female gametophyte grows upward forming a beak between the two archegonia (Fig. 1.44F, 1.46). The beak extends upward touching the inner surface of the nucellar beak like a pole in a tent and is known as **'tent pole'**.

Pollination

Like *Cycas*, *Ginkgo* is anemophilous i.e., wind-pollinated. At the time of dispersal of pollen grains, the central strand of elongated nucellar cells in the pre-pollinated ovule disorganises followed by the collapse of epidermis resulting into a narrow and deep pollen-chamber and pollination drop. The viscous fluid oozes out through the microphyle in the form of **pollination drop**. Some of the anemophilous pollen grains are caught in the pollination drop and are brought to the pollen chamber due to the drying-off the fluid. Pollination generally takes place in May. Pollen grains germinate in the pollen chamber, unlike *Cycas* where they germinate in the intermediary chamber. The pollen tube comes out through the aperture. The pollen tube branches freely showing a proximal unbranched tube and a much branched haustorial ramification in between the intercellular cells.

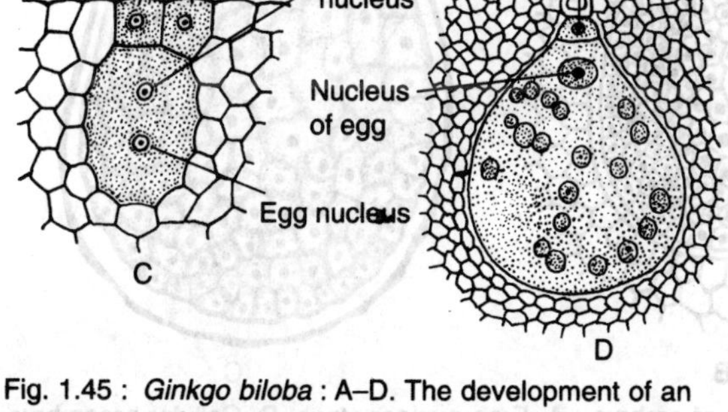

Fig. 1.45 : *Ginkgo biloba* : A–D. The development of an archegonium

Fertilisation

The sperms along with the pollen tube cytoplasm are released in the archegonial chamber by the rupture of the terminal part of pollen tube. The osmotically rich cytoplasm of pollen tube causes the rupture of neck cells, as a result motile sperms enter into the archegonia. The fertilisation takes place in September i.e., four months after pollination. The two sperms proceed towards the archegonium with a forward and circular motion, ciliary band forming the posterior end. However, in *Cycas*, the ciliary band forms the anterior end. The archegonial chamber is filled up with the fertilisation fluid produced by nucellar cells. The ciliary band of the sperm is left behind on the top of the egg cell. The sperm nucleus fuses with the egg nucleus resulting in a zygote.

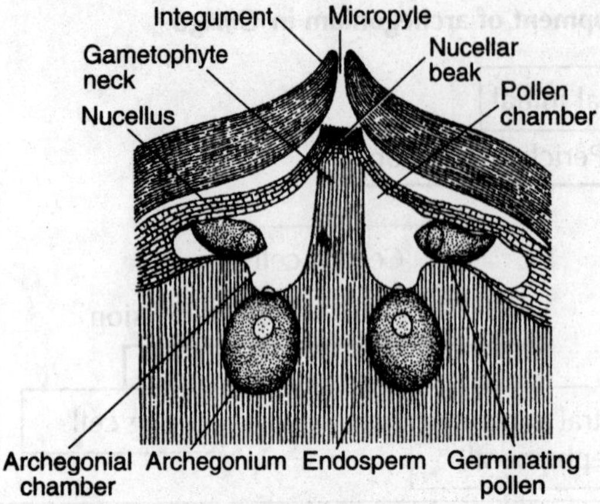

Fig. 1.46 : Ginkgo biloba : Median L.S. of top of ovule during fertilization

Embryogeny

The zygote enlarges considerably and undergoes free nuclear divisions giving rise to 256 nuclei, evenly placed in the cytoplasm of the developing proembryo (Fig. 1.47A). The cell wall formations in the proembryo begin in all the cells simultaneously resulting in a cellular proembryo (Fig. 1.47B). The differentiation of proembryonal cells begins at a later stage. The cells at the micropylar end elongate greatly forming a massive suspensor, while the basal cells (chalazal end) develop into embryonal cells with distinct meristem (Fig. 1.47C). A shoot apex is differentiate at the basal region of the embryo, while a root apex develops at the opposite end. A dicotyledonous embryo (rarely with three cotyledons) with two mesarch bundles is seen in a mature embryo (Fig. 1.48).

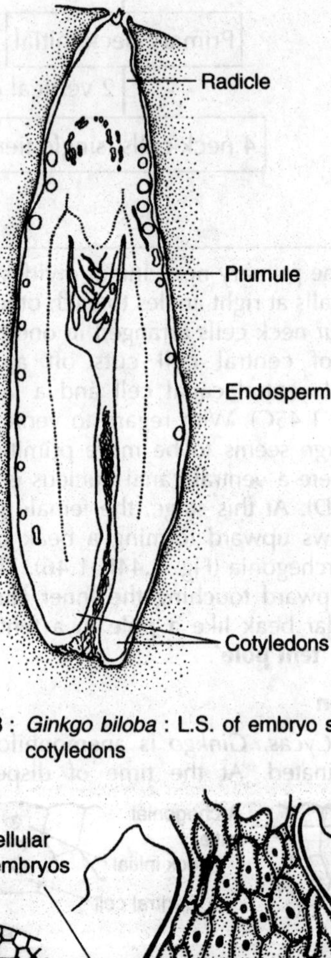

Fig. 1.48 : *Ginkgo biloba* : L.S. of embryo showing two cotyledons

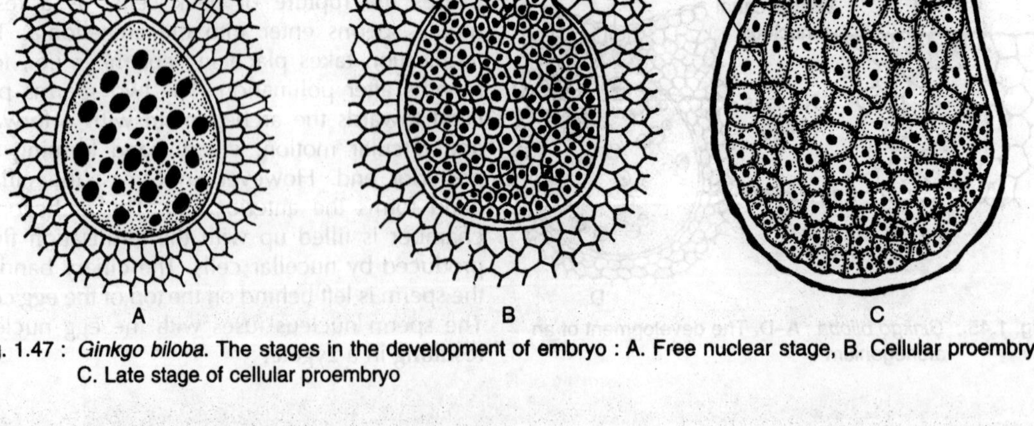

Fig. 1.47 : *Ginkgo biloba*. The stages in the development of embryo : A. Free nuclear stage, B. Cellular proembryo, C. Late stage of cellular proembryo

Seed

The most striking feature in *Ginkgo* is the development of embryo which continues even after the dispersal of seeds. Actually, the major part of the growth of embryo takes place in the detached seeds that are shed before or just after fertilisation. About seven months are required for attaining the maturity of an embryo. Thus, *Ginkgo* shows the two-years reproductive cycle where the pollination takes place in spring, fertilisation in early autumn of the same year and the development of embryo continues even after shedding of seeds. The germination of the seed is **hypogeal.**

Figure 1.49 shows the life cycle of *Ginkgo.*

Origin and relationship

The origin of ginkgos is a matter of controversy. Some botanists hypothesised a cordaitean origin of ginkgos, while some others suggested its pteridospermous origin at the end of the Palaeozoic Era. But in recent palaeobotanical studies it has been postulated that the Ginkgoales was derived from a Palaeozoic ancestor, possibly the progymnosperms which also gave rise to conifers and pteridosperms. Thus, the monotypic genus *Ginkgo biloba* indicates relationship with the ferns, pteridosperms, Cycadales, Cordaitales and Coniferales.

Relation to Ferns

The similarities between *Ginkgo* and ferns are :

1. The leaves of *Ginkgo* shows striking similarity with *Adiantum* (maiden hair fern) in form, shape and venation and hence the name maiden hair tree for *Ginkgo* tree.
2. The sperms are motile and multiciliated.

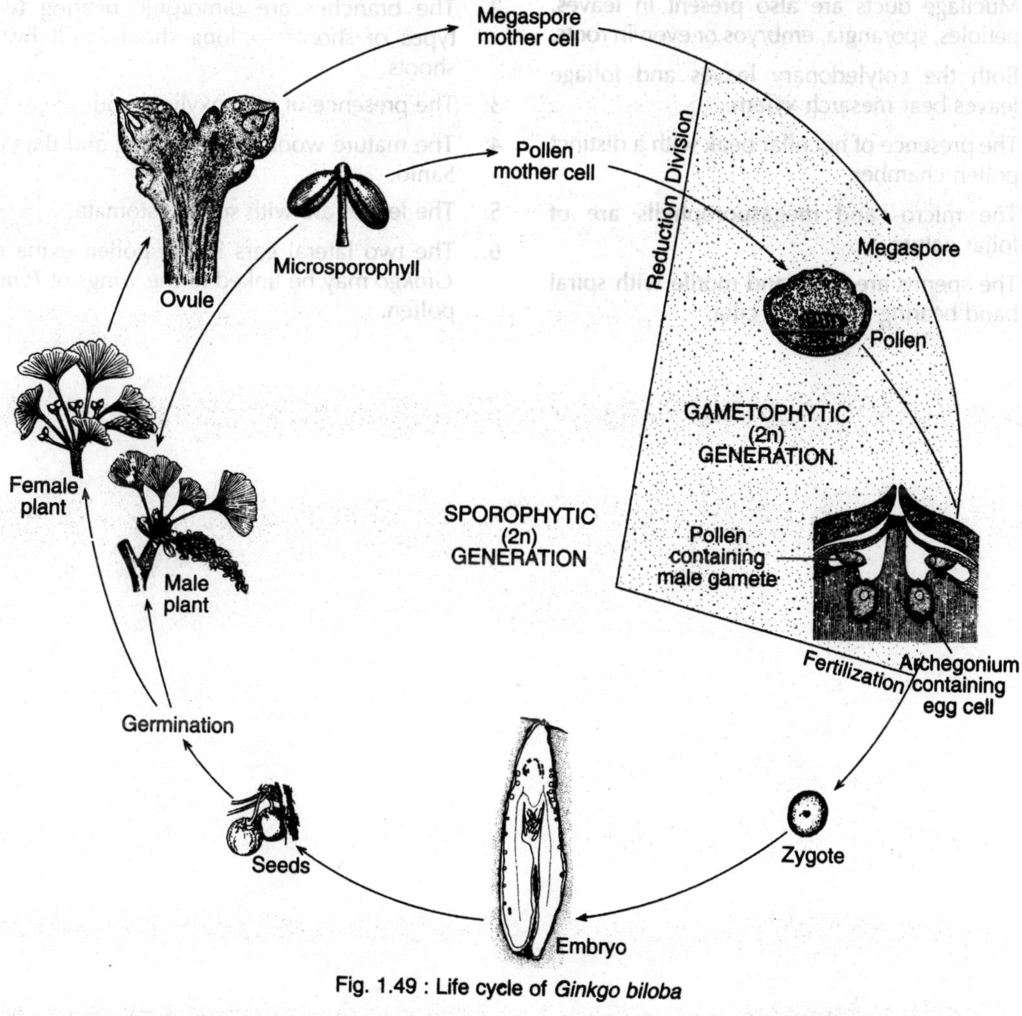

Fig. 1.49 : Life cycle of *Ginkgo biloba*

Relation to Pteridosperms

Ginkgo resembles the pteridosperms in having the following characteristics :

1. The leaf of *Ginkgo* may be linked to the wedge-shaped pinnule of pteridosperm.
2. There is accumulation of abundant reserve food material and lignification in the integument prior to fertilisation.
3. The collar in the ovule of *Ginkgo* may be linked to the cupule of pteridosperm ovule.

Relation to Cycadales

Ginkgo resembles the Cycadales in having the following characteristics :

1. The stem anatomy shows a broad cortex, large pith and numerous mucilage ducts. Mucilage ducts are also present in leaves, petioles, sporangia, embryos or even in roots.
2. Both the cotyledonary leaves and foliage leaves bear mesarch xylem.
3. The presence of nucellar beak with a distinct pollen chamber.
4. The micro- and megasporophylls are of foliar nature.
5. The sperms are large and motile with spiral band bearing numerous cilia.
6. The presence of extensive free-nuclear divisions in the early stage of embryogeny.
7. The embryo with two cotyledons.

Relation to Cordiatales

The similarities between *Ginkgo* and Cordaitales are :

1. The presence of double leaf trace.
2. The presence of motile sperms.
3. The presence of endospermic beak in the mature ovules.

Relation to Coniferales

The resemblance between *Ginkgo* and Coniferales are :

1. The plants are tall trees showing excurrent branching.
2. The branches are dimorphic bearing two types of shoots — long shoots and dwarf shoots.
3. The presence of pycnoxylic wood.
4. The mature wood shows pitting and Bars of Sanio.
5. The leaves are with sunken stomata.
6. The two lateral ears in the pollen exine of *Ginkgo* may be linked to the wings of *Pinus* pollen.

CLASS. CONIFEROPSIDA

The Coniferopsida includes four orders : Cordaitales, Voltziales, Coniferales and Taxales. They probably emerged during the Upper Devonian and were the most dominating taxa during Upper Carboniferous to Triassic. At present, this class is represented by 57 living genera.

GENERAL CHARACTERISTICS OF CONIFEROPSIDA

1. Mostly tall woody plants which are profusely branched.
2. Secondary wood pycnoxylic.
3. There are needle-shaped, paddle-shaped or fan-shaped leaves which are basically dichotomous in form or in venation.
4. Cones compound unisexual.
5. Seeds bilaterally symmetrical.

ORDER 1 : CORDAITALES

The Cordaitales, a group of extinct plants, is reported to be the world's first great forest-maker. This group originated during Upper Devonian and formed the vast forests in the Upper Carboniferous and declined at the beginning of Triassic. The various form genera of Cordaitales have been reported from China, Siberia, North America, Europe, Australia and India. Indian Cordaitales are represented by impression-compression of leaves, seeds and permineralised woods.

Previously, the order Cordaitales was classified into three families — Cordaitaceae, Poroxylaceae, and Pityaceae. Both Poroxylaceae and Pityaceae show their close affinity to pteridosperms. Now, the family Poroxylaceae has been included under the order Pteridospermales and the family Pityaceae shows its close association with pteridosperms. Thus, the order Cordaitales comprises of a single family Cordaitaceae.

GENERAL CHARACTERISTICS OF CORDAITALES

1. The plants are tall trees (Fig. 1.50) with slender branched monopodial stem, attaining a height of about 30 m and bear a dense cluster of branches at the top.
2. The roots are highly branched and swallow, forming stilt roots like those of mangroves.
3. The leaves are simple, sessile, linear or lanceolate or spathulate and are spirally arranged.
4. The leaves are without midribs and with parallel venation. There are double leaf traces.
5. The primary vasculature is of siphonostelic forming a narrow band between the large pith and secondary wood.
6. The secondary wood is pycnoxylic.
7. The presence of discoid pith which shows large air chambers separated by diaphragms.
8. The cones are unisexual and compound consisting of a main (primary) axis bearing several spirally arranged secondary shoots subtended by bracts.
9. Each secondary shoot is fertile, consisting of spirally arranged sterile and fertile scales.

Fig. 1.50 : A *Cordaites* plant

10. There is a homology between male and female cones, the female secondary shoot comprises of 1-4 fertile scales that terminate to 1-4 ovules, and the male secondary shoot with 4-6 terminal pollen sacs.
11. The pollen grains are monosaccate (with one wing).
12. The seeds are bilateral i.e., platyspermic.

Different form genera of this family

Root : *Amyelon*

Stem : *Mesoxylon, Pennsylvanioxylon*

Leaf : *Cordaites*

Fructifications : *Cordaianthus* (Both male and female)

Seed : *Cardiocarpus, Mitrospermum*

Pollen grain : *Florinites* type

The main plant is called *Cordaites*, based on the nomenclature of its leaf genus.

Root

The roots are highly branched and shallow forming stilt roots like those of mangroves. This indicates the halophytic nature of the plants which grew along the sea shores.

Anatomically, the root has a actinostelic structure with diarch to tetrach primary wood which is surrounded by enormous secondary wood (Fig. 1.51). The periderm is well-developed in the cortical zone associated with many lenticels. The cortex may be distinguished into two zones, sometimes infected with endotrophic mycorrhizal fungi. In older roots, a pith is formed at the centre by the medullation of metaxylem, thus the protostelic root transformed into a siphonostelic configuration.

Stem

Though *Cordaites* trees are very large, they bear very slender branches.

Anatomically, the stem shows siphonostelic configuration with a narrow band of primary vasculature in between the large pith and pycnoxylic secondary wood (Fig. 1.52A, B). The primary xylem in *Pennsylvanioxylon* is endarch, while it is mesarch in *Mesoxylon*. The hexagonal bordered pits, restricted to radial walls, are in two or more rows and are very crowded. There are frequent uniseriate rays in the wood (Fig. 1.52C). The secondary phloem is made up of radially arranged sieve cells, parenchyma and

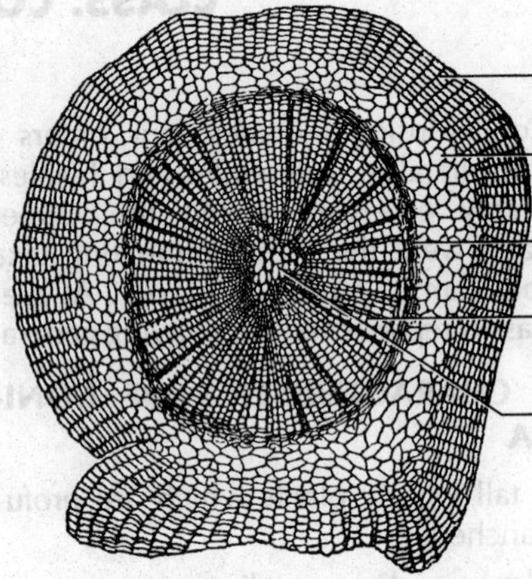

Fig. 1.51: T.S. of *Amyelon* (root).

fibres. The cortex in younger stem exhibits secretary sacs, in older stem the cortex is replaced by periderm (Fig. 1.52B). The pith shows the evidence of large air chambers separated by diaphragms (Fig. 1.52D). During the elongation of stem, the pith cracks transversely with intact periphery, thus appearing like a pile of concave discs. Hence, such a pith is called **discoid pith.**

The stem exhibits many leaf traces which appear to be double, arising from sympodia, and each leaf trace is flanked by two branch traces. In *Pennsylvanioxylon*, the primary xylem in leaf trace is endarch and centripetal, while it is mesarch and both centripetal as well as centrifugal in *Mesoxylon*.

Leaf

The leaves are simple, sessile and long slender, up to 1 m in length (Fig. 1.53A). They are mostly linear, but may be lanceolate, spathulate or obovate. The venation is parallel. The *Cordiates* leaves may be sub-divided into three sub-genera based on the leaf forms, viz., subgenus *Dory-Cordaites* (leaves—strap-like with pointed tip), *Eu-Cordaites* (leaves strap-like with blunt tip) and *Poa-Cordaites* (leaves grass-like).

Anatomically, the leaves exhibit xerophytic characteristics. Both the upper and lower epidermis are heavily cutinised (Fig. 1.53B). The stomata are haplocheilic and mostly confined to the lower epidermis. The stomatal apparatus is comprised of two guard cells (parallel to veins) surrounded by 4–6 subsidiary cells. The palisade and spongy tissue are poorly differentiated with a large number of lacunae in between the veins. The thick-walled sclerenchymatous hypodermal

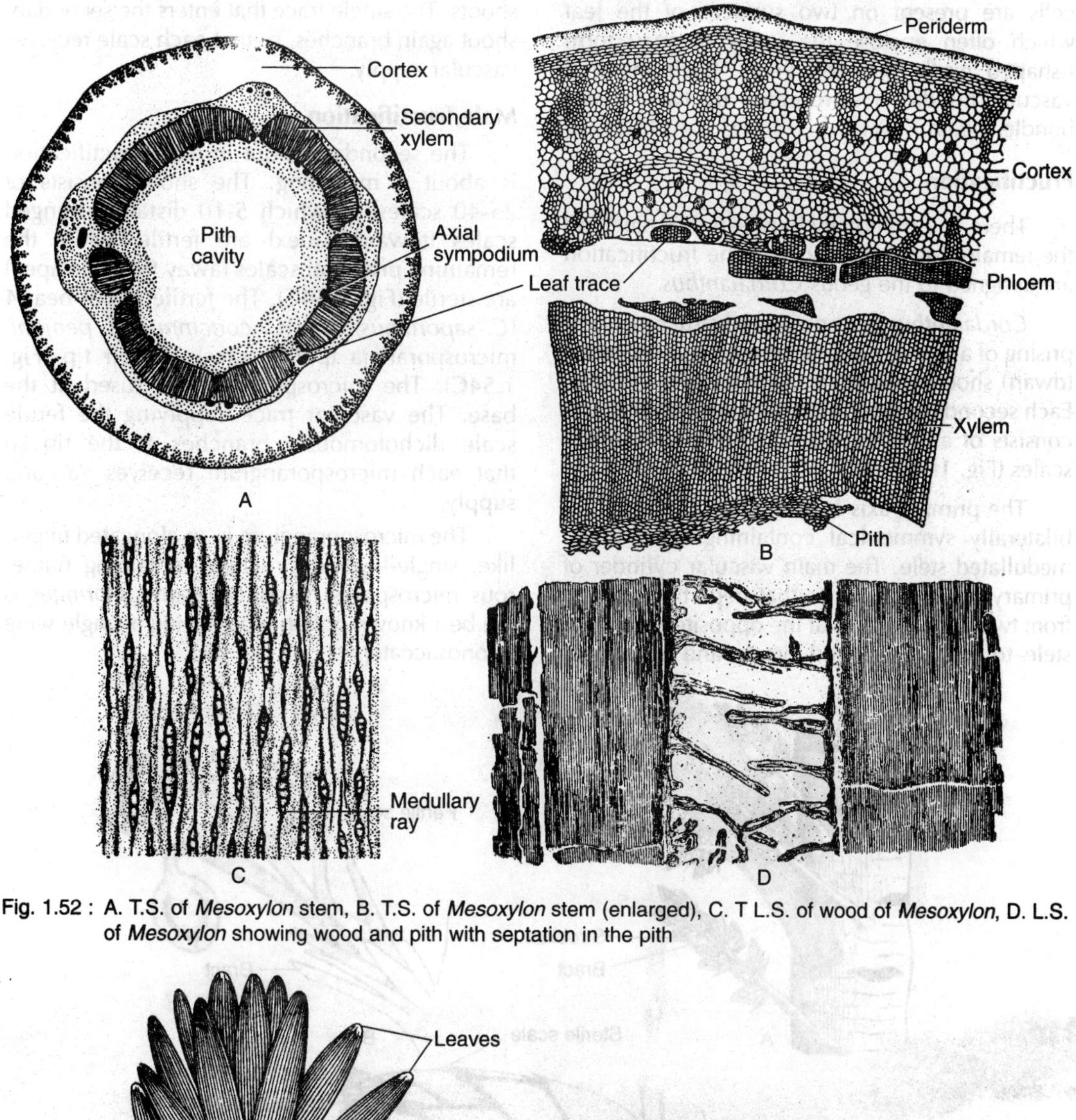

Fig. 1.52 : A. T.S. of *Mesoxylon* stem, B. T.S. of *Mesoxylon* stem (enlarged), C. T L.S. of wood of *Mesoxylon*, D. L.S. of *Mesoxylon* showing wood and pith with septation in the pith

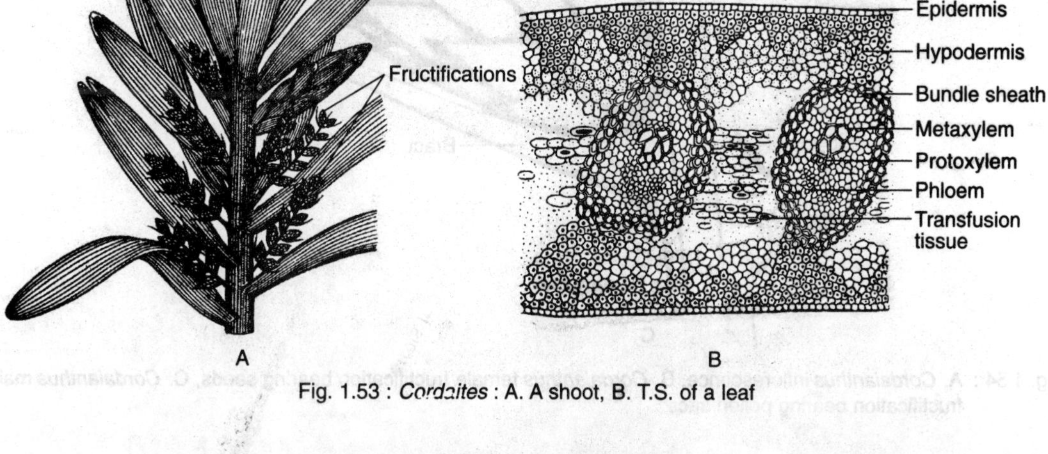

Fig. 1.53 : *Cordaites* : A. A shoot, B. T.S. of a leaf

cells are present on two surfaces of the leaf which often enclose the veins or may form I-shaped girdle between veins. Each mesarch vascular bundle is surrounded by two-layered bundle sheath.

Fructification

There is a homology between the male and the female fructifications; both the fructification are assigned to the genus *Cordaianthus*.

Cordaianthus is a compound structure comprising of a primary axis bearing many secondary (dwarf) shoots subtended by bracts (Fig. 1.54A). Each secondary shoot is radially symmetrical and consists of a central axis with spirally arranged scales (Fig. 1.54B, C).

The primary axis is somewhat flattened and bilaterally symmetrical containing an endarch medullated stele. The main vascular cylinder of primary axis forms traces that depart alternately from two areas situated at the opposite end of the stele to the four ranked bracts and secondary shoots. The single trace that enters the secondary shoot again branches, so that each scale receives vascular supply.

Male fructification

The secondary shoot in male fructification is about 6 mm long. The shoot consists of 25-40 scales, of which 5-10 distally arranged scales (towards apex) are fertile, while the remaining proximal scales (away from the apex) are sterile (Fig. 1.54C). The fertile scales bear 4 (*C. saportanus*) to 6 (*C. concinnus, C. penjoni*) microsporangia (pollen sacs) at their tip (Fig. 1.54C). The microsporangia are fused at the base. The vascular trace supplying the fertile scale dichotomously branches at the tip so that each microsporangium receives vascular supply.

The microsporangium is an elongated finger-like, single-layered structure containing numerous microspores or pollen grams. *Florinites* is the best known pollen type having a single wing (monosaccate).

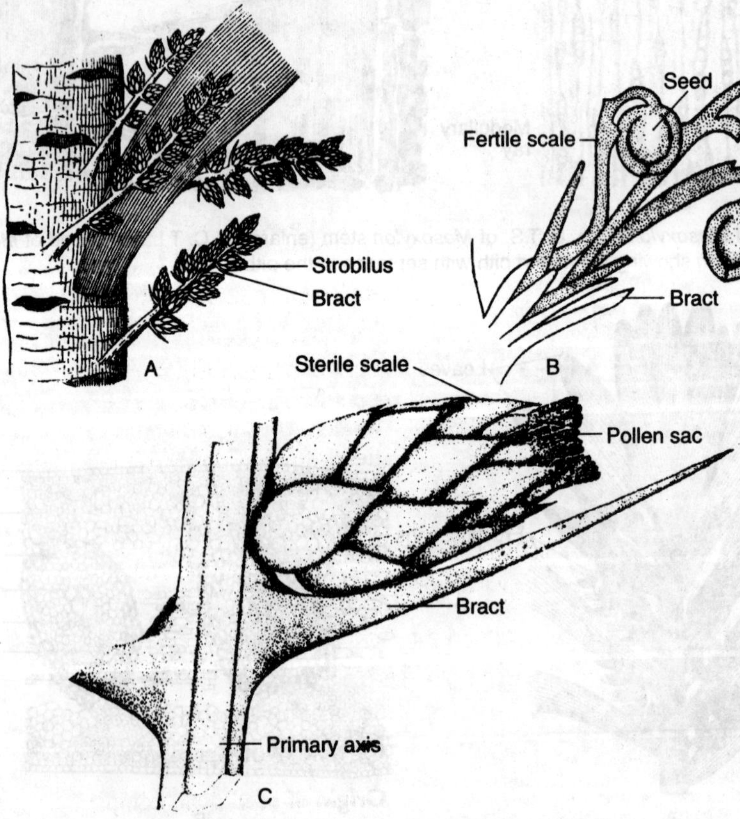

Fig. 1.54 : A. *Cordaianthus* inflorescence, B. *Cordaianthus* female fructification bearing seeds, C. *Cordaianthus* male fructification bearing pollen sacs

Female fructification

The secondary shoot in female fructification consists of 16-20 scales, of which 4-6 distal scales are fertile and the remaining proximal scales are sterile (Fig. 1.54B). Each fertile scale bears a single reflexed (pendulous) ovule at its tip. Sometimes, the fertile scale is forked and contains two to three terminal, pendulous ovules.

Seed

The seeds are cordate, flattened and bilaterally symmetrical (Fig. 1.55A, B). The integument is three-layered : (i) the outer fleshy **sarcotesta** which is laterally expanded into a wing, (ii) the middle **sclerotesta** with spine-like projection into sarcotesta, and (iii) the **endotesta** made up of thin-walled cells. The integument is free from the nucellus except at the base and the nucellar beak of the pollen chamber projects into the microphyle.

There is a single vascular strand that enters the ovule. In *Cardiocarpus*, the single trace enters up to the base of the nucellus, then it divides to form two branches that extend through the inner side of sarcotesta. A second pair of strands passes through the peripheral zone of the nucellus. In *Mitrospermum,* the two lateral strands enter the sarcotesta moving through the sclerotesta. In the sarcotesta, each strand divides further to form several bundles (Fig. 1.55A).

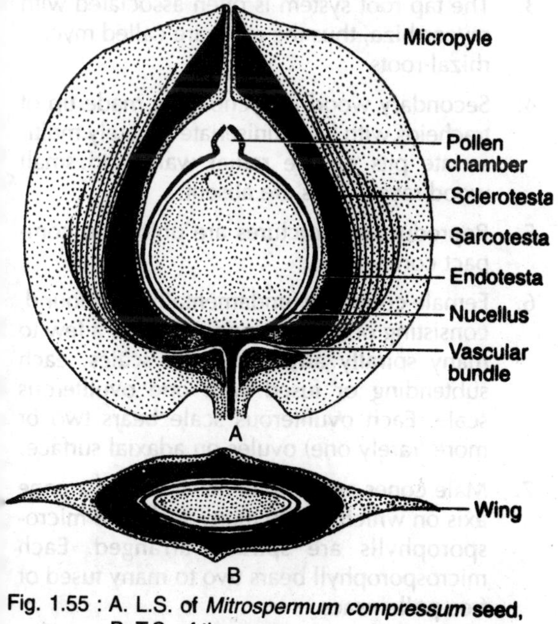

Fig. 1.55 : A. L.S. of *Mitrospermum compressum* seed, B. T.S. of the same

Origin and relationship

The Cordaitales lived side by side with the pteridosperms in Palaeozoic times. It has been assumed that both the groups might have originated from a common ancestral form. Some other botanists postulated that the Cordaitales has evolved from the progymnospermous stock. Anyway, Cordaitales shows its strong affinity with Cycadales, Ginkgoales and Coniferales.

Relation to Cycadales

The members of Cordaitales resemble Cycadales in exhibiting the following characteristics :

1. The presence of large pith and centripetal wood in stem.
2. The presence of a large amount of leaf sclerenchyma.
3. The seeds are comparatively simple, and fairly large with a single three-layered integument.
4. The presence of motile sperms.

Relation to Ginkgoales

The Cordaitales resembles the Ginkgoales in having the following characteristics :

1. The presence of double-leaf traces.
2. The presence of motile sperms.
3. The presence of endospermic beak in the mature ovules.

Relation to Coniferales

The conifers have evolved from the members of Voltziales which are called 'transitional conifers' between the Cordaitales and Coniferales. Thus, Cordaitales shows striking similarities with the Coniferales in exhibiting the following characteristics :

1. The plants are tall branched trees.
2. The leaves are simple with paralled veins.
3. The presence of sclerenchymatous hypodermis in the leaves.
4. The wood is pycnoxylic.
5. The pollen grains are winged.
6. The ovules are bilaterally symmetrical.

Origin of seed-cone complex

It has been established that the cones of *Pinus* are modified compound shoots which

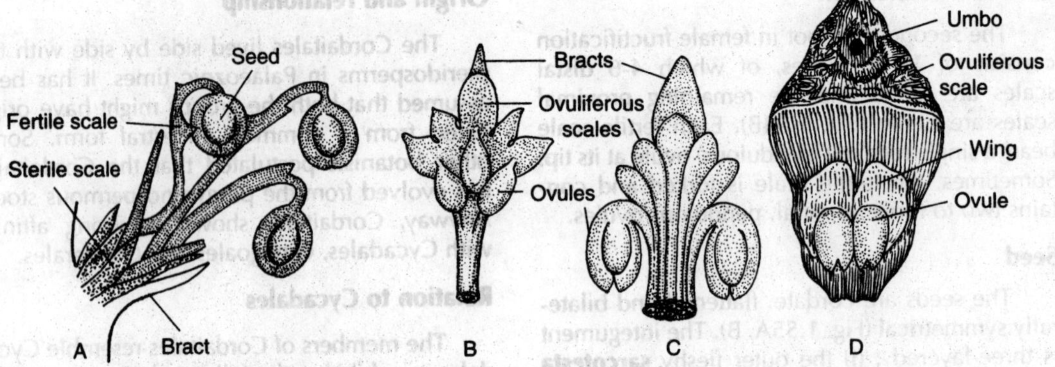

Fig. 1.56. A. *Cordaianthus* female fructification, B. *Pseudovoltzia liebeana*, C. *Glyptolepis longibracteata*, D. An ovuliferous scale of *Pinus* bearing two anatropous seeds

show homology with the *Cordaianthus* fructification. The ovuliferous scales are the modified secondary shoots of *Cordaianthus* and the bract scales of *Pinus* are the modified bracts of *Cordaianthus*.

In the evolutionary stages, the radially symmetical secondary shoots of *Cordaianthus* (Fig. 1.56A) have gradually been planated to become bilateral shoots of Voltziales (e.g., *Ernestiodendron, Wallchiostrobus*). In the next stage of evolution, both the fertile and sterile scales of secondary shoots are laterally fused to form a common platform (ovuliferous scale) bearing ovules (e.g., *Pseudovoltzia, Glyptolepis*) (Fig. 1.56B, C). Thus, an ovuliferous scale is formed which, in turn, is fused with a bract to form a bract-scale complex.

In the present day conifers, the cone axis (= primary axis) bears spirally arranged seed-cone complexes (= secondary shoots). Each seed-cone complex is comprised of ovuliferous scale bea-ring ovulus and bract scale (Fig. 1.56D). During evolution, the female cones of living conifers have undergone further morphological changes.

The evolutionary pathways that followed in the origin and evolution of conifer cones are :

1. Change in the secondary shoots from the radially symmetrical condition to bilateral symmetry.
2. Lateral fusion of fertile or both fertile and sterile scales to form ovuliferous scale.
3. Shifting from erect ovules to inverted ovules.
4. Reduction in the number of ovules.

ORDER 2 : CONIFERALES

The members of the Coniferales are commonly called conifers. They are the largest order among gymnosperms, represented by 52 genera and about 600 species.

General characteristics of conifers

1. They are woody plants showing profused branching often with long and dwarf shoots.
2. The leaves are dimosphic having both the scale leaves and foliar leaves. Foliar leaves are needle-shaped, paddle-shaped or fan-shaped.
3. The tap root system is often associated with mycorrhiza, thus the roots are called mycorrhizal roots.
4. Secondary wood is pycnoxylic, made up of tracheids with large uniseriate or rarely multiseriate pits on the radial walls and small wood rays. Fibres are absent.
5. Reproductive structures are unisexual compact cones.
6. Female cones are fundamentally compound, consisting of a main cone axis with a few to many spirally arranged bract scales, each subtending or fused with one ovuliferous scale. Each ovuliferous scale bears two or more (rarely one) ovules on adaxial surface.
7. Male cones are simple, consisting of a cone axis on which few to many scale like microsporophylls are spirally arranged. Each microsporophyll bears two to many fused or free pollen sacs.

8. Pollen grains are winged or non-winged.
9. Embryos are with two to many cotyledons.

The order coniferales is divided into six families :

Family	Genera
1. Pinaceae : (10 genera)	*Pinus, Picea, Abies, Cedrus, Cathaya, Keeteleria, Larix, Pseudolarix, Tsuga* and *Pseudotsuga.*
2. Taxodiaceae : (10 genera)	*Taxodium, Cryptomeria, Cunninghamia, Sequoia, Sequoiadendron, Taiwania, Glyptostrobus, Metasequoia, Sciadopitys* and *Arthrotaxis.*
3. Cupressaceae : (22 genera)	*Cupressus, Thuja, Biota, Juniperus, Callitris, Libocedrus, Chamaecyparis, Tetraclinis, Actinostrobus, Widdringtonia,* etc.
4. Podocarpaceae : (7 genera)	*Podocarpus, Dacrydium, Microstrobus, Microcachrys, Acmopyle, Phyllocladus* and *Saxegothaea.*
5. Araucariaceae : (2 genera)	*Araucaria* and *Agathis.*
6. Cephalotaxaceae : (1 genus)	*Cephalotaxus.*

Among the 52 genera, about 30 genera are confined to the Northern Hemisphere and 14 genera to the Southern Hemisphere. The rest of the genera are distributed in both the Hemispheres. At present, the conifers form the world's most extensive temperate evergreen forests with exception of a few deciduous taxa like *Larix, Taxodium, Metasequoia* and *Pseudolarix.* The oldest living plants on our planet are found among conifers. A tree of *Pinus longaeva* growing in Eastern Nevada, USA is about 4,900 years old and a tree of *Pinus aristata* in Inyo National Forests of California, USA, has been determined to be more than 4,600 years old. The giant sequoias like *Sequoia sempervirens* and *S. wellingtonia* in the coastal forests of Western North America have been estimated to be over 3,000 years old. A gaint cypress, *Taxodium mucronatum,* near Oaxaca, Mexico has been determined to be more than 2,000 years old.

The giant sequoias (*Sequoia sempervirens* and *S. wellingtonia*) are regarded as the tallest living trees (redwood trees) in the world. Some trees in *Sequoia* National Park, California have been reported to be more than 80 m in height with a base circumference of 285 m.

The conifers have evolved from the Permo-Carboniferous Voltziales members and they reached their climax in mid-Mesozoic forming extensive forests in Northern Europe. Among the living families, Pinaceae and Araucariaceae are the more primitive which have evolved probably in Triassic age; while Cupressaceae and Cephalotaxaceae are comparatively younger, probably appeared during Upper Jurassic to Lower Cretaceous period.

PINUS

Systematic position

Class. Coniferopsida
Order. Coniferales
Family. Pinaceae

Total number of species : 100

Geographical distribution :

Pinus is the most well-known genus of the family Pinaceae and is commonly known as pine. The pines form extensive forests in the Northern Hemisphere including Northern Europe. North and Central America, subtropics of North Africa, the Canary Islands. They are also distributed in the forests of Afghanistan, Pakistan, India, Myanmar (Burma), Philippines and also extend to Indonesia after crossing the equator.

Indian distribution

In India, six species of *Pinus* have been recorded, of which five are restricted to the Himalayas. The distribution pattern of Indian species are :

1. *Pinus roxburghii* (Syn. *P. longifolia*). It is a very common species and popularly called '**Chir pine**'. It grows from 460 to 2,300 m altitude in the outer range of the Himalayas from Afghanistan to Bhutan and Eastern Nepal.

2. *P. wallichiana* (Syn. *P. excelsa, P. griffithii*). It is commonly called '**Blue pine**' which grows from 1,500 to 3,000 m altitude on the Himalayas, frequently in the Kashmir valley, Shimla, Chakrata, Mussoorie and also in

Eastern Nepal. It is almost absent in Sikkim and major parts of Kumaon.

3. *P. gerardiana*. It is commonly known as '**Chilgoza pine**'. It grows at an elevation of 1,830–3,600 m in North-Western Himalayas and in the Kalpa and Pangi districts of Himachal Pradesh.
4. *P. insularis* (Syn. *P. khasya, P. khasiana*). It is called '**Khasi pine**'. It is restricted to the Eastern Himalaya and is found in Garo, Khasi and Jaintia hills and also in Manipur at an elevation of 800 to 2,000 m.
5. *P. armandi*. It is basically a Chinese species which grows wildly in the hills of North-Eastern Himalayas above 1,500 m altitude.
6. *P. merkusii*. It is commonly known as '**Tenasserim Pine**' which is most tropical in nature among the Indian pines. It grows in the hilly slopes of Andaman and Nicobar islands.

Sporophyte

Pinus is a tall evergreen tree giving rise to a series of widespread horizontal branches (Fig. 1.57A). In each year, a whorl of branches is produced in the axil of scale leaves. The branching is restricted to the upper part of the stem, thus giving the tree a pyramid-like appearance. The main stem is cylindrical and covered with scaly bark. The branches are dimorphic, bearing two types of shoots : **long shoots** and **dwarf shoots**, or **spurs** or **brachyblasts** (Fig. 1.57B). *Pinus* exhibits two types of leaves, the scale leaves and the green acicular foliage leaves called **needles** (Fig. 1.57B). The plant has a tap root system which becomes elongated at maturity and posesses strong lateral roots.

The plants are monoecious where the male and female cones are borne on separate branches in the same plant.

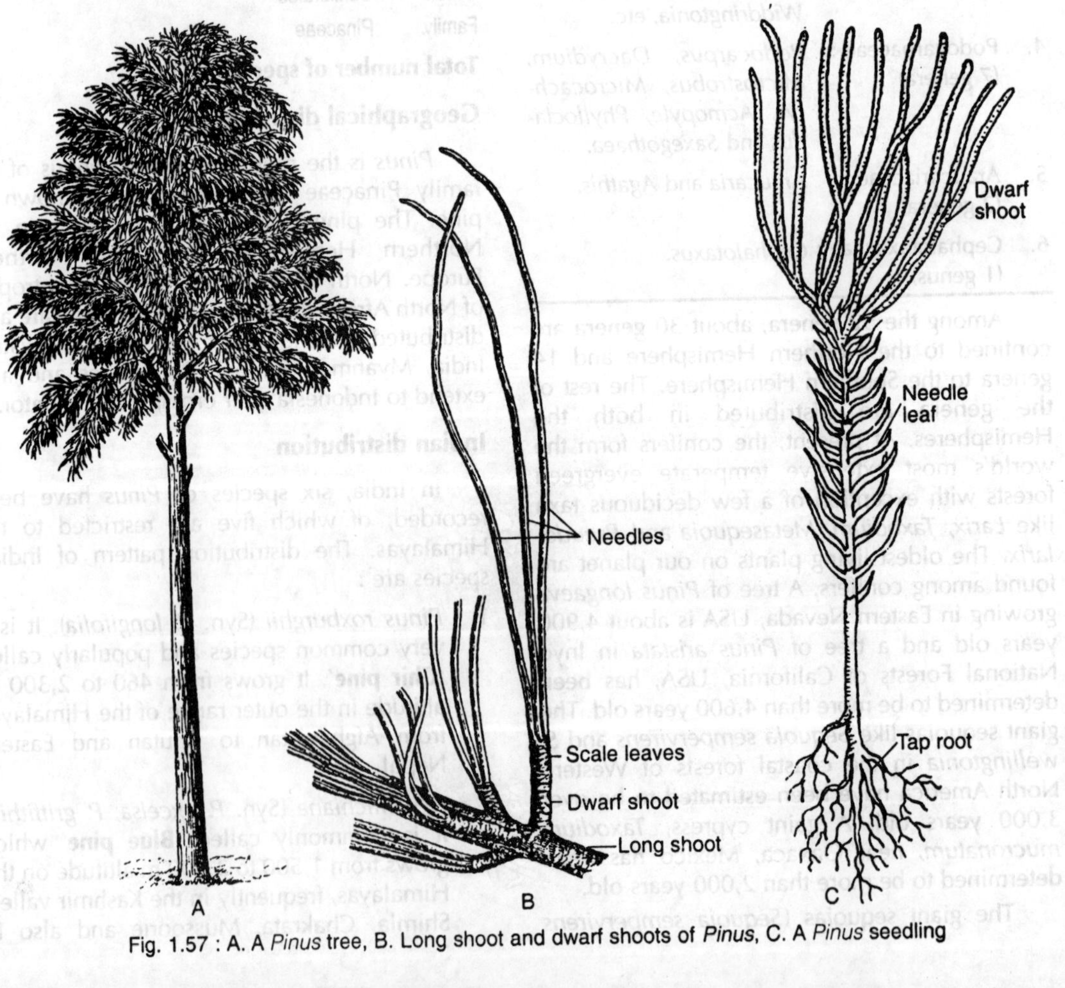

Fig. 1.57 : A. A *Pinus* tree, B. Long shoot and dwarf shoots of *Pinus*, C. A *Pinus* seedling

Root

External morphology : *Pinus* shows a typical tap root system. The tap root (long root) persists but several lateral roots (short or dwarf root) grow extensively with poorly developed root hairs. The roots of *Pinus* is characterised by ectotrophic mycorrhizal association with about 50 different funal species. The predominant species that form **ectotrophic mycorrhiza** belong to *Boletus, Clavaria, Amantia* and *Scleroderma* of the Basidiomycotina. The dwarf roots divide dichotomously and several hyphae penetrate the intercellular spaces in the cortical cells constituting the **Hartig's net** and **mantle.** Thus a symbiotic association is formed between the roots and the fungi where the plant is benefited by increasing the surface of the root for absorption of water and nutrient from soil.

Internal structure : Anatomically, the primary (long) root of *Pinus* is diarch to tetrarch which resembles a typical dicot root. The following structures are discernible from outside inward : **epidermis** with unicellular root hairs, the starch filled parenchymatous **cortex**, a single-layered **endodermis** with typical casparian bands, a multilayered **pericycle**, followed by a diarch or tetrarch radial **vascular bundle** and the starch-filled **pith**. The protoxylem of vascular bundle is bifurcated forming a 'Y'-shaped structure and each arm is associated with a **resin duct**. The metaxylem consists of pitted tracheids. The phloem is present in alternate position to xylem and is comprised of sieve cells and parenchyma.

The secondary growth in root starts very early. The cambium develops below the primary phloem and produces secondary xylem centripetally towards the pith and secondary phloem centrifugally towards the cortex (Fig. 1.58). The cambium in the region of resin ducts cuts off parenchymatous xylem rays (uni- or multiseriate). The root, in a later stage, shows a more or less similar configuration with that of a stem.

The extrastelar secondary growth in root also takes place soon after the commencement of intrastelar secondary growth. The phellogen (cork cambium) which differentiates in the outer region of the pericycle, cuts off phellem (cork cells) towards the outer side and phelloderm (secondary cortex) towards the inner side. Several resin canals are formed in the phelloderm. The cell layers from epidermis to endodermis are sloughed off and eventually the highly suberised and tanniniferous cork becomes the outer protective layer.

Anatomically, the dwarf root is more or less

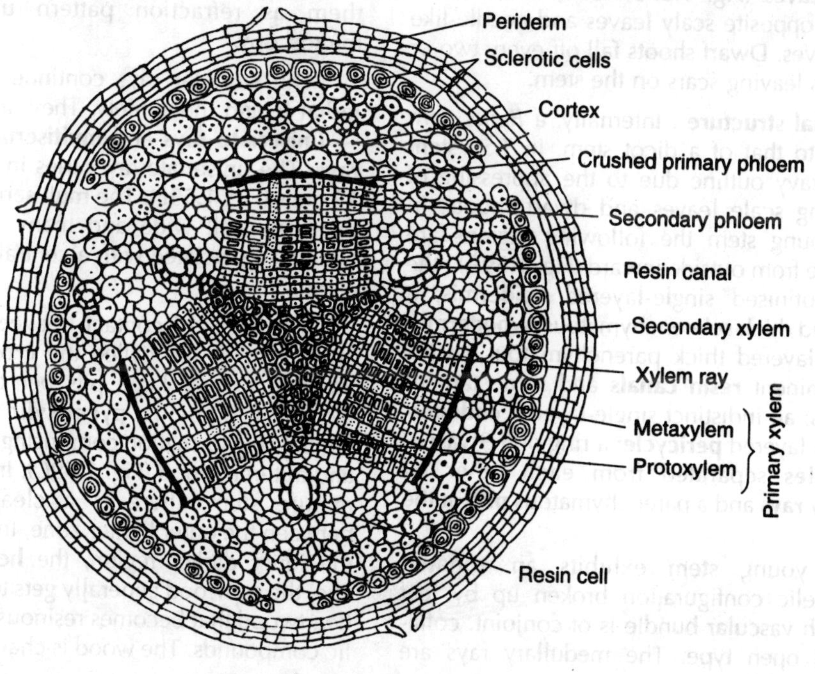

Fig. 1.58 : *Pinus.* T.S. of an old root

similar with the long root. However, the dwarf roots are devoid of root cap, resin ducts, starch in cortical cells and secondary growth.

Stem

External morphology : The stem of *Pinus* is woody, cylindrical and erect with rough uneven scaly bark that peels off. The branches are pseudomonopodial and are restricted to the upper part of the stem. The branches develop spirally in the axil of scale leaves present on the upper part of the stem, giving the tree a pyramid-like or pagoda-like appearance (Fig. 1.57A). There are two types of branches — the long shoot and the dwarf shoot, thus the branches are **dimorphic** (Fig. 1.57B).

Long shoots : These are the branches of unlimited growth bearing an apical bud enclosed in bud scales. Each long shoot develops in the axil of a scale leaf, thus this branch bears only scale leaf. These branches shorten gradually towards the tip, thus giving the tree a pyramid-like or pagoda-like appearance. The long shoot falls off leaving a scar on the stem.

Dwarf shoot : These are the branches of limited growth and also known as **short shoots**, **brachyblasts** or **foliar spurs**. The dwarf shoots are borne on long shoots and develop in the axil of scale leaves (Fig. 1.57B). Each dwarf shoot bears two opposite scaly leaves and needle-like foliage leaves. Dwarf shoots fall off every two or three years leaving scars on the stem.

Internal structure : Internally, a *Pinus* stem is similar to that of a dicot stem. In T.S., stem exhibits wavy outline due to the appressing of surrounding scale leaves and dwarf shoots. In T.S. of young stem the following ragions are discernible from outside inward (Fig. 1.59A) : the heavily cutinised single-layered **epidermis**; a few-layered thick sclerenchymatous **hypodermis**; a several-layered thick parenchymatous **cortex** with prominent **resin canals** and some tannin-filled cells; an indistinct single-layered **endodermis**; a 2-3 layered **pericycle**; a ring of 5-9 **vascular bundles** separated from each other by **medullary rays** and a parenchymatous **pith** at the centre.

The young stem exhibits an endarch siphonostelic configuration broken up by leaf gaps. Each vascular bundle is of conjoint, collateral and open type. The medullary rays are narrow due to the closeness of two adjacent vascular bundles. The tracheids are arranged in radial rows. The protoxylem of tracheids have spiral thickening and the metaxylem have reticulate thickening. The phloem is made up of sieve cells and parenchyma with occasional **albuminous cells**.

Secondary growth in thickness

The fascicular cambium joins with the interfascicular cambium forming a complete cambium ring that cuts off a continuous cylinder of secondary xylem towards the inside and secondary phloem towards the outside (Fig. 1.59B). The cam-bium also produces secondary medullary rays at places. The secondary xylem is comprised of tracheids. The secondary phloem consists of sieve cells arranged in rows. The sieve cell is provided with sieve area on the radial wall throughout its length. The tracheids are more or less quadrilateral in cross-section and are provided with bordered pits on radial wall (Fig. 1.59D). The mature tracheids develop thickened border in the form of crescentic bars in between the pits. These are known as **Crassulae** or **Bars of Sanio** (Fig. 1.59F). According to Sanio (1873) the crassulae are rod-shaped horizontal thickenings in the middle lamella due to the deposition of cellulose or pectin between radial walls of tracheids. However, Mauseth (1988) discribed them as refraction pattern using electron microscopy.

The vascular rays continue to increase in radial length indefinitely. They are of two types — **uniseriate rays** and **multiseriate** or **fusiform rays**. The uniseriate ray varies in height from 1-12 cells (Fig. 1.59E). The multiseriate ray is more than one cell wide and many cells in height which is associated with a centrally-placed resin canal.

The areas of contact between the ray cells and tracheids are called the **cross-fields**. There are one to two large pits per cross-field. The wood of pinus is **pycnoxylic** which shows **annual ring** formation comprising of spring wood and autumn wood (Fig. 1.59C). In the wood of a mature stem, there is a clear demarcation between the outer lighter zone, the **sapwood** and the inner darker region, the **heartwood**. With age, the sapwood generally gets transformed into heartwood that becomes resinous, rich in phenolic compounds. The wood is characterised by the absence of vessels, wood parenchyma and fibres, although fibre tracheids may occur in late wood.

The extrastelar secondary growth takes place by the **cork cambium** or **phellogen** which forms cork on the outer side and secondary cortex on the inner side. Later, the original cork cambium is replaced by successive layers of phellogen deeper and deeper (Fig. 1.59B) into the cortex. So rough uneven scaly bark is formed that peels off.

Anatomically, the dwarf shoot is similar to that of long shoot except for its diameter.

Leaf

Pinus exhibits leaf dimorphism bearing two types of leaves, the **scale leaves** and the green acicular foliage leaves called **needles**.

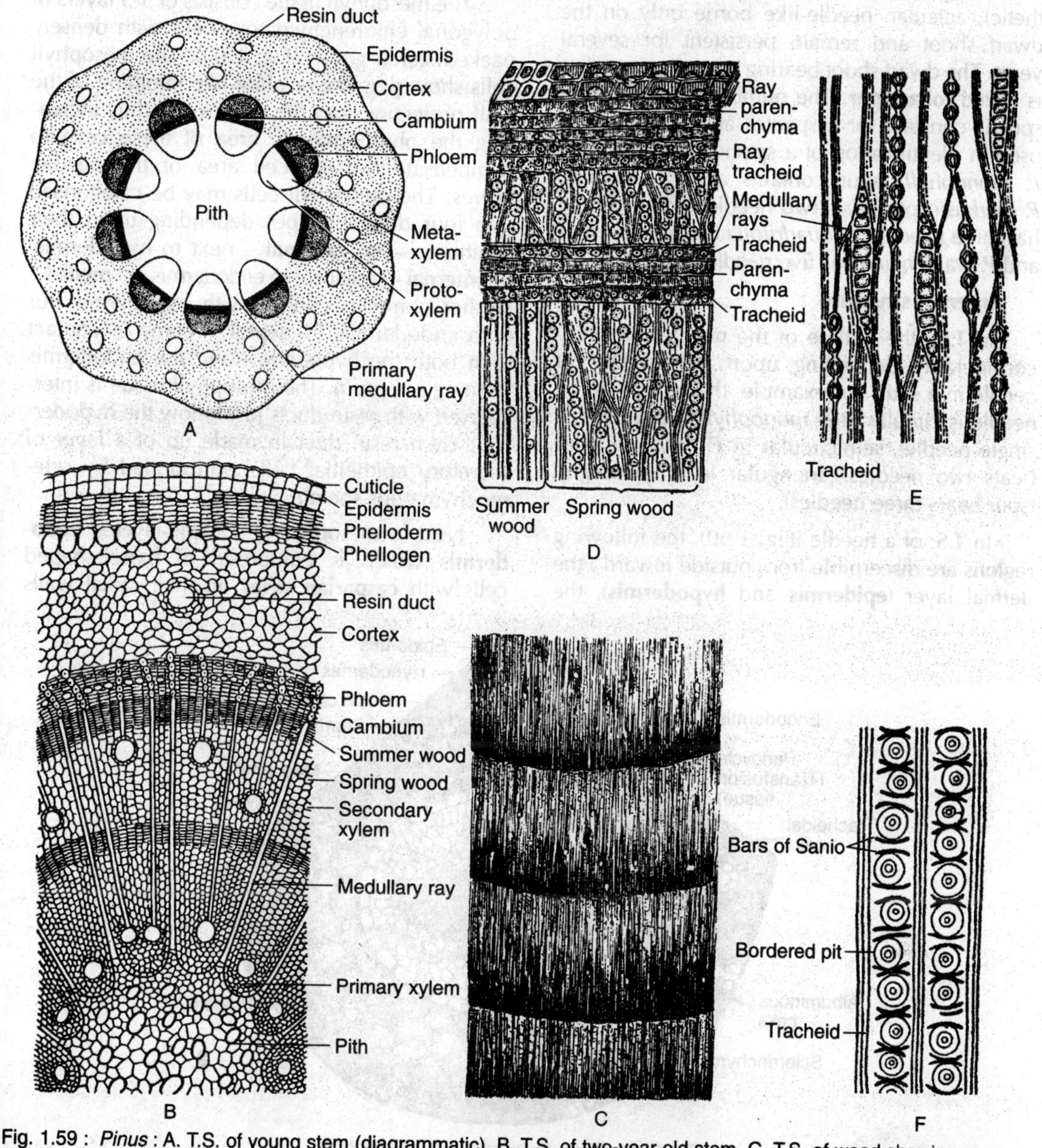

Fig. 1.59 : *Pinus* : A. T.S. of young stem (diagrammatic), B. T.S. of two-year old stem, C. T.S. of wood showing annual rings, D. R.L.S. of secondary wood, E. T.L.S. of secondary wood, F. Tracheids showing Bars of Sanio

External morphology

Scale leaves : These are thin, small and membranous and are dark-brown in colour present on both the long and the dwarf shoots (Fig. 1.57B). They are non-photosynthetic and provide protection to the young buds. The scale leaves that are borne on the dwarf shoot have a distinct midrib and are known as **cataphylls**. At maturity scale leaves fall-off.

Foliage leaves : These are green (photosynthetic), acicular, needle-like borne only on the dwarf shoot and remain persistent for several years. The dwarf shoot bearing a group of needles is called **foliar spur**. The number of needle in a spur is constant for a species and is a criterion used in identification of a species. For example, *P. monophylla* spur contains a single needle, *P. markusii* spur bears two needles, *P. roxburghii* has three needles, *P. quadrifolia* has four needles and *P. wallichiana* has five needles per spur.

Internal structure

In T.S., the outline of the needle may vary considerably depending upon the number of needle in a spur. For example, the outline of the needle is circular in *P. monophylla* (spur bears a single needle), semicircular in *P. markusii* (spur bears two needles), triangular in *P. roxburghii* (spur bears three needles).

In T.S. of a needle (Fig. 1.60), the following regions are discernible from outside inward : the dermal layer (**epidermis** and **hypodermis**), the **mesophyll tissue**, and a centrally-placed monarch or diarch **stele**.

The epidermis (Fig. 1.60) is single-layered, made up of heavily cutinised isodiametric lignified cells. It is interrupted by many **haplocheilic** deeply sunken stomata consisting of two guard cells surrounded by 6-9 subsidiary cells. The hypodermis (Fig. 1.60) is made up of 2-3 layers of thick-walled sclerenchymatous cells which is often interrupted by sub-stomatal cavities.

The mesophyll tissue consists of 3-5 layers of polygonal chlorenchymatous cells with densely packed starch grains (Fig. 1.60). The mesophyll cells show **peg-like** or **plate-like** infolding of the wall projecting into the cell cavity, thus increasing the photosynthetic area of these cells to compensate the reduced area of the needle. leaves. The mesophyll cells may be categorised into four different types depending upon their position, — (a) **external** – next to hypodermis, (b) **internal** – adjacent to endodermis, (c) **medial** – neither come in contact with hypodermis nor with endodermis, (d) **septal** – come in contact with both the hypodermis and the endodermis forming a septum. The mesophyll tissue is interspersed with resin ducts just below the hypodermis. Each resin duct in made up of a layer of secretory epithellial cells, surrounded by sclerenchymatous sheath.

Next to mesophyll is a single-layered **endodermis** which is composed of barrel-shaped cells with **casparian strips** on their radial walls

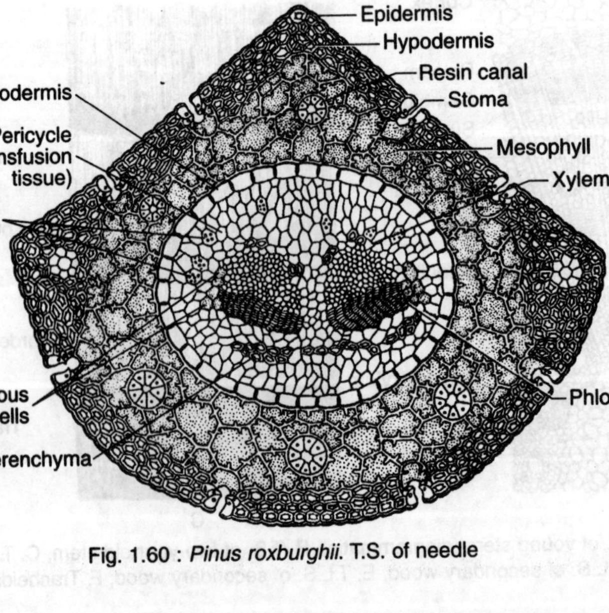

Fig. 1.60 : *Pinus roxburghii*. T.S. of needle

Distinctions between Haploxylon and Diploxylon pines

Haploxylon pines	Diploxylon pines
1. They are the soft pines.	1. They are the hard pines.
2. There is a single vascular bundle in the needle.	2. There are two vascular bundles in the needle.
3. The umbo (tip of the apophysis of ovuliferous scale in the mature cone) is either terminal or dorsal.	3. The umbo is always dorsal.
4. In stem, ray tracheids have smooth walls and tangential pits are developed after one year's growth.	4. In stem, ray tracheids have dentate or reticulate walls and tangential pits are not formed at the end of a year's growth.

(Fig. 1.60). The endodermis is followed by a multi-layered **pericycle** consisting of three types of cells, viz. parenchymatous cells densely filled with starch grains, **albuminous cells** near the phloem filled with proteins and starch grains, and tracheidal cells near the xylem with bordered pits on their tangential and transverse walls (Fig. 1.60). All the cells of pericycle help in conduction of water and minerals, thus they are often termed **trans-fusion tissue**. A sheath of sclerenchyomatous fibres is present is pericycle around the phloem.

The stele of *Pinus* is either monarch or diarch type. On the basis of the number of vascular bundles, *Pinus* may be categorised into two distinct groups — **Haploxylon**, having a single vascular bundle in needle, and **Diploxylon**, having two bundles placed at an angle to each other (Fig. 1.60). The vascular bundles are conjoint,

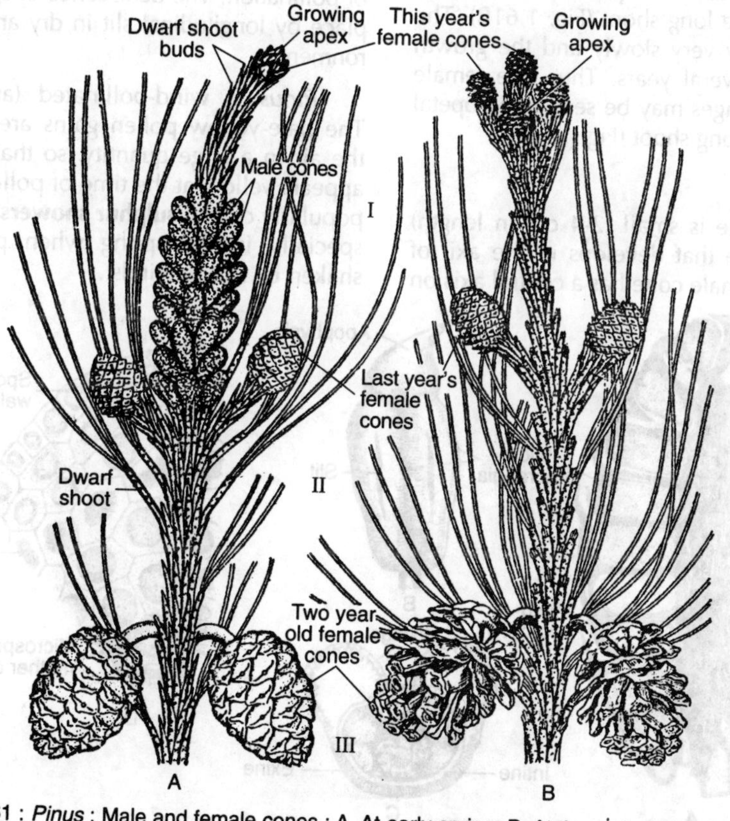

Fig. 1.61 : *Pinus* : Male and female cones : A. At early spring, B. At the end of spring (rains)

collateral and open with adaxial xylem and abaxial phloem. The bundles consist of protoxylem and metaxylem elements. The tracheary elements arrange in radial rows which alternate with rows of parenchymatous cells and albuminous cells. The phloem is composed of sieve cells and parenchyma. The cambium is present in vascular bundle which gives rise to secondary phloem and little or no secondary xylem.

Reproduction

Pinus reproduces sexually. *Pinus* is **monoecious**, but the male and the female cones are produced on separate branches of the same plant. The male cones develop on the lower branches, while the female cones are formed on the upper branches. The male cones, which replace the dwarf shoots, develop in clusters on the base of the current year's long shoot at early spring (Fig. 1.61A). The number of male cones in a cluster varies considerably from 15 (*P. wallichiana*) to 140 (*P. roxburghii*). At the onset of spring, the male cones fall off and simultaneously the young female cones are borne in pairs or in clusters round the tip of the long shoot (Fig. 1.61B). The female cones grow very slowly and the growth may persist for several years. Thus, the female cones of different ages may be seen in acropetal succession in the long shoot (Fig. 1.61B).

Male cone

The male cone is small (2-4 cm in length) and oval in shape that develops in the axil of scale leaves. The male cone has a central axis on which 60-150 **microsporophylls** are spirally arranged around the axis (Fig. 1.62A). A single microsporophyll is a membranous stalked structure with a distal expanded roughly triangular sterile part called **apophysis** (Fig. 1.62B). Each microsporophyll bears two sac-like microsporangia on the abaxial surface.

The development of microsporangia is of eusporangiate type, i.e., it develops from a group of hypodermal cells of the microsporophyll. A mature microsporangium consists of a multilayered wall, tapetum and microspore mother cells (Fig. 1.62D). Each microspore mother cell — by meiotic division — produces four microspores or pollen grains. Thus, at maturity, a single microsporangium contains numerous pale yellow pollen grains. The pollen grains are boat-shaped with monosulcate apertures and are bounded by two concentric wall layers : the outer thick **exine** and the inner thin **intine** (Fig. 1.62C). The exine on the lateral sides of the pollen is expanded to form two **wings** (**sacci**). Thus the pollen grains of *Pinus* are **bisaccate** indicating their anemophilous mode of pollination. The dehiscence of sporangia takes place by longitudinal slit in dry and warm environment.

Pinus is wind-pollinated (anemophilous). The pale-yellow pollen gains are released into the air in a large quantity, so that a pine forest appears yellow at the time of pollination. This is popularly called '**sulphur showers**' which occurs specially in the spring when pine trees are shaken by strong winds.

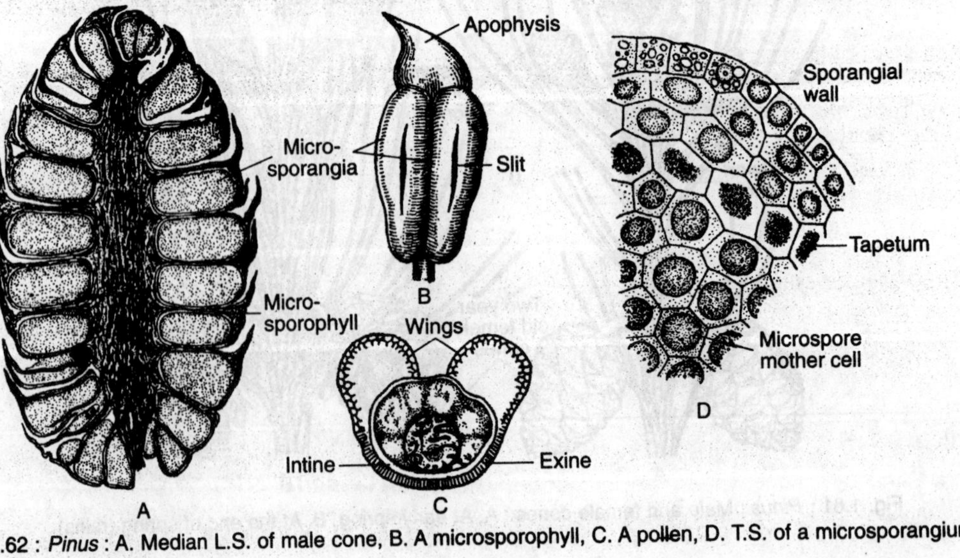

Fig. 1.62 : *Pinus* : A. Median L.S. of male cone, B. A microsporophyll, C. A pollen, D. T.S. of a microsporangium

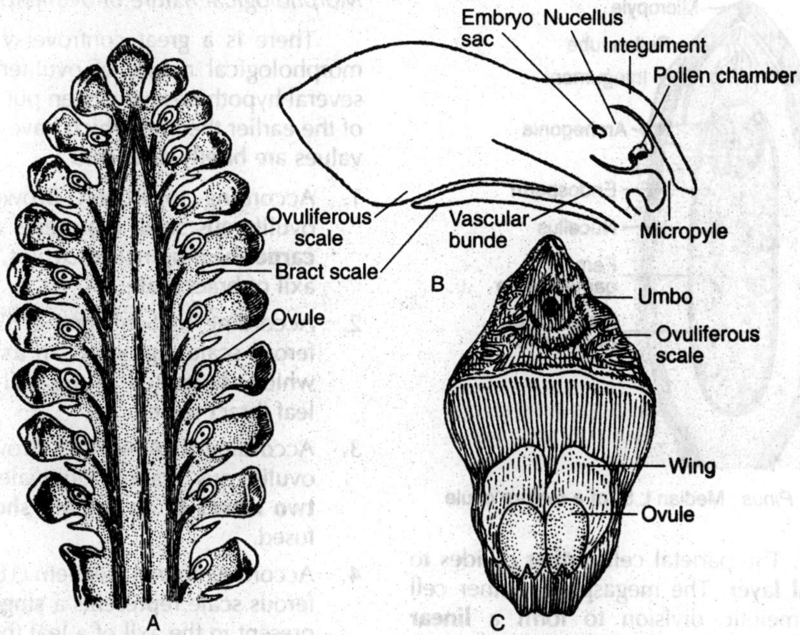

Fig. 1.63 : *Pinus* : A. Median L.S. of female cone, B. V.L.S. of an ovuliferous scale, C. An ovuliferous scale bearing two seeds (fertilized ovules)

Female cone

Female cones are produced in pairs or in clusters in the axil of the scale leaves. The female cones mature very slowly. The first year young cone (Fig. 1.61A, B) is small (1-2 cm in length), soft, compact and red-green in colour. The second year cone (Fig. 1.61A, B) is comparatively large (5-8 cm in length), woody, compact and green in colour. The fully matured third year cone is much larger (15-60 cm in length), woody, loose and brown in colour. Here megasporophylls are separated from each other due to the elongation of the cone axis.

The female cone of *Pinus* represents a compound shoot; it is a complicated structure. The female cone is composed of a central axis on which 80-90 megasporophylls, axillary to bract scale/scale leaves, are arrange spirally (Fig. 1.63A). The bract scale and ovuliferous scale thus form a **seed-scale complex**. A single megasporaphyll consists of two types of scales : (a) a large woody **ovuliferous scale** or **seminiferous scale** bearing two ovules on the adaxial surface, and (b) a **bract scale** or **cone scale** on the abaxial surface (Fig. 1.63B, C). Initially, the ovuliferous scale is much smaller than that of bract scale, but after pollination it becomes larger than the bract scale. The ovuliferous scale is a thick, large, woody, roughly triangular and brownish structure. Its upper thick exposed part is known as **apophysis**. In the mature cone, the tip of the apophysis becomes the '**umbo**' (Fig. 1.63C). There are two separate vascular traces, one supplies to the ovuliferous scale and the other to the bract scale (Fig. 1.63B). There is no separate vascular trace for ovule.

The development of megasporangium (ovule) is of eusporangiate type i.e., an ovule develops from a group of superficial cells of the ovuliferous scale.

Ovule

The ovules of *Pinus* are anatropous, unitegmic and crassinucellate (Fig. 1.64). The single integument is free from the nucellus except at the chalazal end. There is a fairly broad micropylar tube which becomes inwardly curved during pre-pollination stages and becomes outwardly curved at the time of pollination.

The integument is three-layered, the outer fleshy, the middle stony and the inner fleshy.

Megasporogenesis

A hypodermed cell in the nucellar tissue at the micropylar end is differentiated into an **archesporial cell**. It divides periclinally to form an upper **parietal cell** and a lower **megaspore**

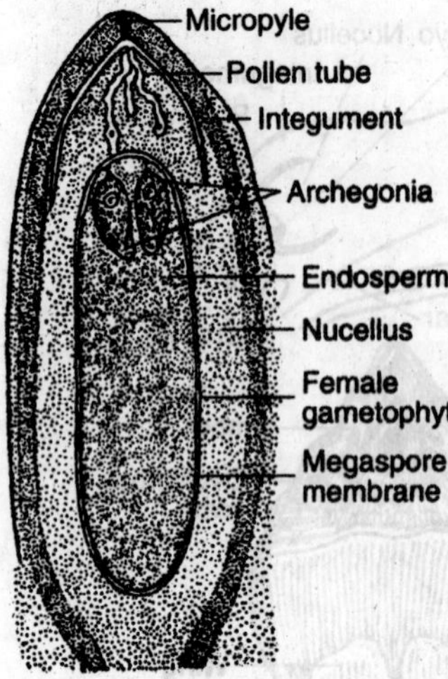

Fig. 1.64 : *Pinus* : Median L.S. of a mature ovule

mother cell. The parietal cell further divides to form **tapetal layer**. The megaspore mother cell undergoes meiotic division to form a **linear tetrad** of four megaspores. The outer three megaspores degenerate, while the lowermost megaspore becomes functional (Fig. 1.66A). The upper free opening of the integument forms the micropyle and a concavity in between the integument and nucellus in the upper part of the ovule forms the pollen chamber. After pollination the pollen grains are stored in the **pollen chamber** and further development of pollen grains takes place in the nucellar tissue.

Morphological nature of ovuliferous scale :

There is a great controversy regarding the morphological nature of ovuliferous scale and several hypotheses have been put forward. Some of the earlier theories which have only a historial values are briefly discussed :

1. According to Robert Brown (1827), the ovuliferous scale represents an **open foliar carpel** bearing naked ovules, present in the axil of bract scale.
2. According to Schleiden (1839), the ovuliferous scale represents an **axillary placenta** which is situated in the axil of an axillary leaf (bract scale).
3. According to Alexander Brown (1842), the ovuliferous scale is equivalent to the **first two leaves of an axillary shoot** which had fused.
4. According to van Tieghem (1869), the ovuliferous scale represents a **single leaf branch**, present in the axil of a leaf (bract).
5. According to the **foliar theory** of Delpino (1889), both the ovuliferous scale and the bract scale are the parts of a tripartite bract, where the two lateral fertile lobes of the bract were fused to form ovuliferous scale and the median sterile part formed the bract scale.
6. According to **ligular** or **excrescence theory** of Sachs (1882) and Eichler (1889), the

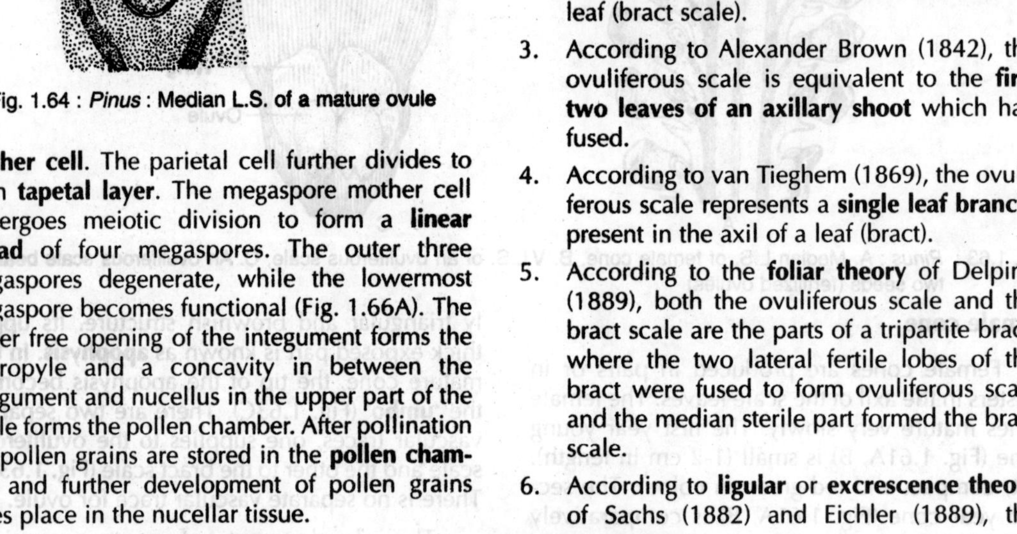

Fig. 1.65 : *Pinus* : A–E. The stages in the development of male gametophyte

female cone represents a simple flower where the cone axis is equivalent to receptacle or thalamus and bract scale to free carpels. The ovuliferous scale represents an outgrowth of the carpel (bract scale) as in ligule of *Selaginella* or the placenta of angiosperms.

7. According to **brachyblast theory** of Braun, the female cone is equivalent to an inflorescence, where ovuliferous scale represents a determinate axillary shoot bearing two fertile leaves with a single ovule on the dorsal surface of each leaf.
8. **Modern hypothesis** : On the basis of the comparative studies of the fossil Cordaitales and Voltziales members, Florin (1951) introduced a terminology **'seed-cone complex'** for ovuliferous scale and bract scale. According to Florin, the female cone of *Pinus* represents an inflorescence where cone axis is the peduncle, bract scale is a true bract and the ovuliferous scale is a rudimentary female flower i.e., a modified reproductive shoot.

This hypothesis has been substantiated by fossil evidences. The occurrence of several Upper Carboniferous to Triassic seed-cone complex in the members of Cordaitales and Voltziales supports the above hypothesis (Fig. 1.56A–D). In *Cordaianthus, Emporia* and *Ernestiodendron* of Upper Carboniferous, the cone axis bears several secondary female reproductive shoots in the axils of bracts. Each secon-dary shoot consists of many spirally-arranged sterile and fertile (with terminal ovule) scales (showing radial symmetry). The further evolution took place in the Triassic Voltziales like *Glyptolepis, Wallchiostrobus*, etc. where the scales were planated (showing bilateral symmetry) and tangentially fused to form a rudimentary ovuliferous scale following Telome hypothesis. Thus, it confirms that the ovuliferous scale is a modified reduced secondary female reproductive shoot and bract scale is actually a reduced bract.

Gametophyte

The spore is the first phase of gametophyte generation. The microspore or pollen grain represents the male gametophyte, while the megaspore represents the first stage of female gametophyte which develops into a female gametophyte.

Schematic representation in the development of male gametophyte of *Pinus*

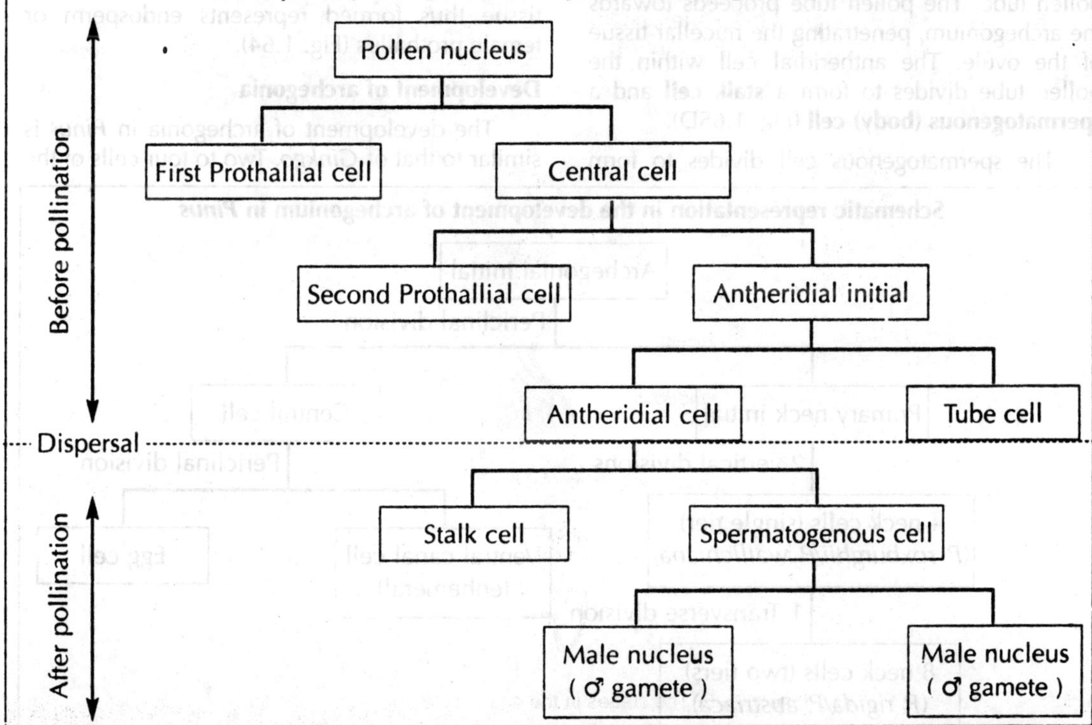

Development of male gametophyte before pollination

The basic pattern of the development of male gametophyte of *Pinus* is similar to that of *Ginkgo*. The pollen grains undergo endosporic development. The pollen nucleus divides mitotically to produce a small lens-shaped **first prothallial cell** towards the proximal end and a large **central cell** on the distal end (Fig. 1.65A). The central cell again cuts off a **second prothallial cell** and an **antheridial initial** (Fig. 1.65B). Both the prothallial cells are ephemeral and the second prothallial cell remains attached to the first prothallial cell. The antheridial initial divides to form a **small antheridial cell** and a large **tube cell** (Fig. 1.65C). The pollen grains are released from the microsporangium at the 4-celled stage (2 prothallial cells, an antheridial cell and a tube cell).

Development of male gametophyte after pollination

After pollination, the 4-celled pollen stores in the pollen chamber and remains ungerminated for about 11 months.The pollen develops the next spring. The tube cell of the pollen comes out through the pollen aperture in the form of a pollen tube. The pollen tube proceeds towards the archegonium, penetrating the nucellar tissue of the ovule. The antheridial cell within the pollen tube divides to form a **stalk cell** and a **spermatogenous (body) cell** (Fig. 1.65D).

The spermatogenous cell divides to form two **male nuclei** just prior to fertilisation (Fig. 1.65E). The male nuclei are actually the male gametes which are non-motile and ephemeral.

Development of female gametophyte

The female gametophyte of *Pinus* develops from the functional megaspore which enlarges considerably (Fig. 1.66A). The nucleus of the megaspore divides mitotically forming a large number of nuclei unaccompanied by wall formation. The number of free nuclei is constant for a particular species, say for example, it is 2,000 for *P. gerardiana* and 2,500 for *P. roxburghii* and *P. wallichiana*. With the increase in size, the megaspore develops a vacuole at the centre which forces by cytoplasm along with nuclei towards the periphery.

Thus, the nuclei lie in a thin film of cytoplasm around the vacuole (Fig. 1.66B). Thereafter, the cell wall formation starts in a centripetal fashion, from periphery inwards. At this stage, numerous radially elongated multinucleate tube-like cells called **alveoli** are formed and the wall formation takes place through alveoli. Each alveolus containing a nucleus at its mouth directs its growth. Then, cross-walls are laid down on each alveolus to form uninucleate cells. In this way, the entire gametophyte becomes cellular and the tissue thus formed represents endosperm or female prothallus (Fig. 1.64).

Development of archegonia

The development of archegonia in *Pinus* is similar to that of *Ginkgo*. Two to four cells of the

Schematic representation in the development of archegonium in *Pinus*

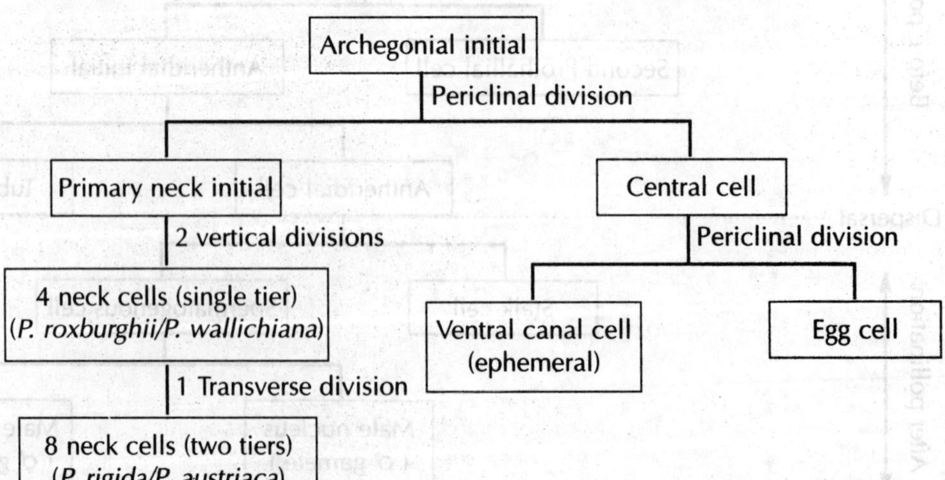

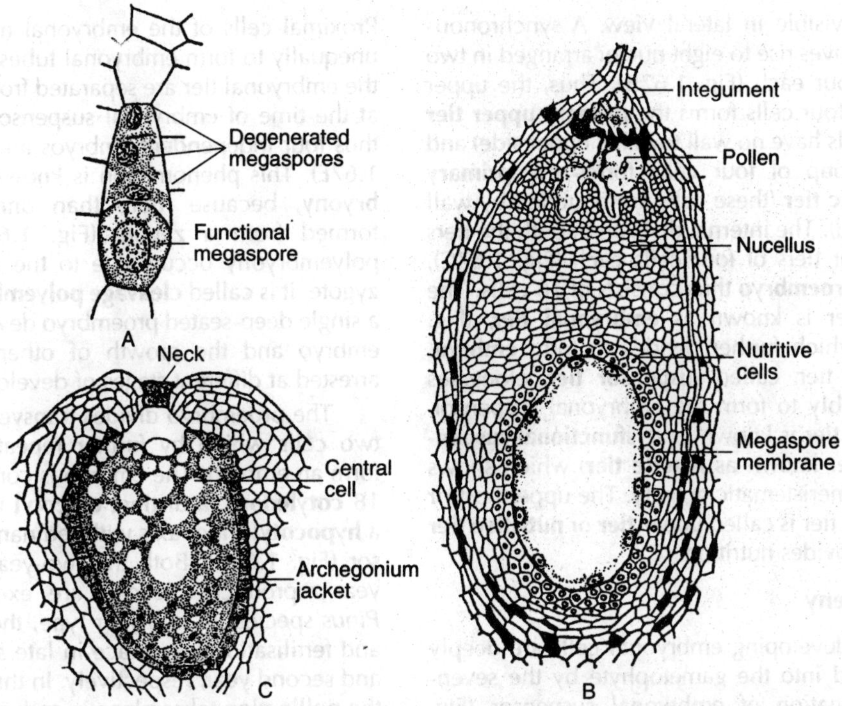

Fig. 1.66 : *Pinus* : A. Linear tetrad showing large functional megaspore, B. Free nuclear stage of female gametophyte, C. An archegonium

female gametophyte at the micropylar end enlarge in size and have dense cytoplasm and prominent nuclei. These cells function as **archegonial initials**. Each archegonial initial divides periclinally to form an outer small **primary neck initial** and a large **central cell**. The primary neck initial divides by two vertical walls at right angles to each other forming a **neck** of four cells. Thus, the four neck cells are arranged in a single tier as in *P. roxburghii* and *P. wallichiana*. However, in *P. rigida, P. austriaca* the four neck cells again divide transversely to form eight cells which are arranged in two tiers. The central cell enlarges very rapidly and its cytoplasm becomes vacuolated (Fig. 1.66C). The nucleus of the central cell divides into an upper ephemeral **ventral canal cell** and a large **egg cell**. A nutritive layer called archegonial jacket is differentiated around the archegonium. The nucellar tissue above the archegonia disorganises to form an **archegonial chamber**.

Pollination

Pinus is anemophilous i.e., wind-pollinated. The pollen grains are dispersed and remain suspended in the air for sometime. At the same time, the nucellar beak in the ovule disorganises forming a viscous sugary liquid containing glucose, fructose and sucrose. This fluid comes out in a cyclic phenomenon (24 hr. cycle) through the micropyle in the form of a **pollination drop** either at night or in the early hours of morning. The pollen grains are caught in the pollination drop and are collected in the pollen chamber as a result of drying off the fluid. The mouth of the micropyle is then sealed from the outer environment.

Fertilisation

The fertilisation takes place after one year of pollination. The pollen tube enters the tip of the archegonium by forcing itself between the cells of the nucellus. The pollen tube wall is disintegrated by the enzymes secreted from the egg and eventually two male nuclei are released. One of the male nuclei fuses with the egg cell and thus a **zygote** is formed.

Development of proembryo

The zygote nucleus divides by two mitotic divisions forming four nuclei which move to the base of zygote (Fig. 1.67A). All the four nuclei are arranged in one plane and only two nuclei are thus visible in lateral view. A synchronous division gives rise to eight nuclei arranged in two tiers of four each (Fig. 1.67B). Thus, the upper group of four cells forms the **primary upper tier** (these cells have no wall towards upper side) and lower group of four cells forms the **primary embryonic tier** (these cells are bounded by wall all around). The internal division in both the tiers forms four tiers of four cells each (Fig. 1.67C), and the **proembryo** thus consists of 16 cells. The lowest tier is known as **embryonal tier,** (Fig. 1.67D) which further divides to form **embryo**. The next tier, called **suspensor tier,** elongates considerably to form the embryonal suspensor. The third tier is known as **dysfunctional suspensor** (earlier known as **rosette tier**) which shows abortive meristematic activity. The uppermost or the fourth tier is called **upper tier** or **nutritive tier** which provides nutrition.

Embryogeny

The developing embryonal cells are deeply embedded into the gametophyte by the sevenfold elongation of embryonal suspensor (Fig. 1.67E). Thus the several **embryonal suspensors** (designated as Es_1, Es_2, Es_3 and so on) are formed. Proximal cells of the embryonal mass elongate unequally to form embryonal tubes. The cells of the embryonal tier are separated from each other at the time of embryonal suspensor elongation, thus four independent embryos are formed (Fig. 1.67E). This phenomenon is known as **polyembryony**, because more than one embryo is formed from a zygote (Fig. 1.67F). As the polyembryony occurs due to the splitting of a zygote, it is called **cleavage polyembryony**. Only a single deep-seated proembryo develops into an embryo and the growth of other embryos is arrested at different stages of development.

The proembryo divides transversely to form two cells which by further repeated divisions form an embryo. The embryo is comprised of 3-18 **cotyledons**, a distinct **epicotyl** root axis and a **hypocotyl** shoot axis with remnants of suspensor (Fig. 1.68A). Both the two-year and three-year reproductive cycles are exemplified by *Pinus* species. In two-year type, the pollination and fertilisation take place in late spring of first and second year, respectively. In three-year type the pollination takes place in spring of first year, and fertilisation in the spring of third year, a lapse of two years. The seeds shed in autumn.

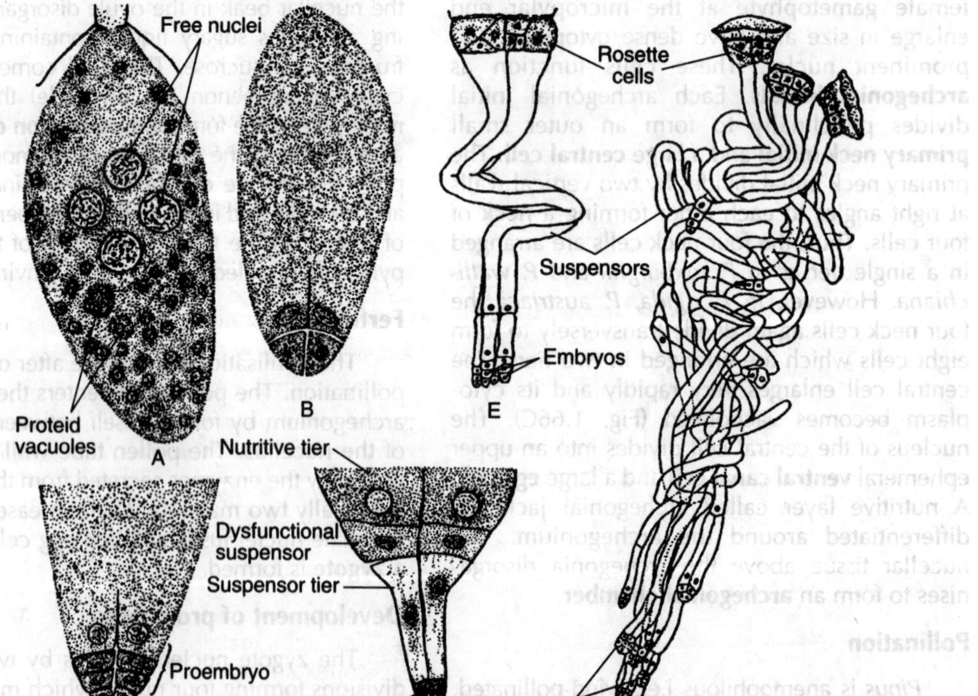

Fig. 1.67 : *Pinus* : A–D. The stages of development of proembryo, E–F. The stages of development of embryo

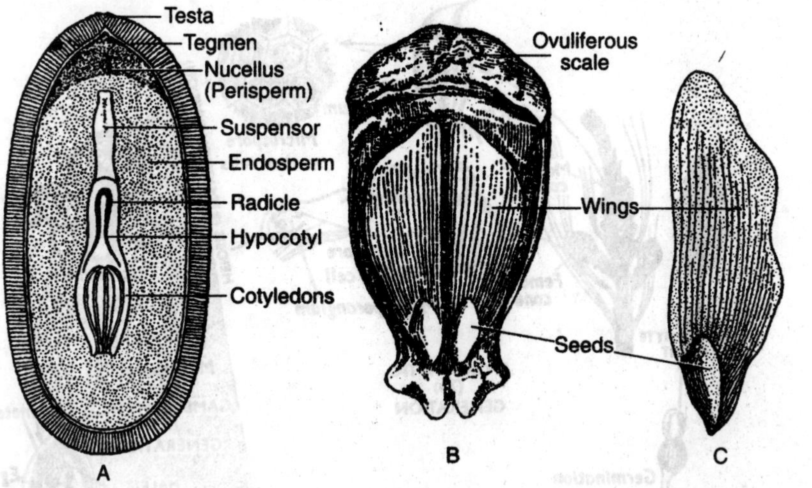

Fig. 1.68 : *Pinus* : A. L.S. of seed, B. An ovuliferous scale bearing seeds, C. A seed with wing

Seeds

The seeds are endowed with a well-developed **wing** which is thin and papery and is easily detachable at maturity (Fig. 1.68B). The outer fleshy layer of integument and part of ovuliferous scale contribute to the wing formation (Fig. 1.68C). The seeds are usually dispersed by wind. The embryo remains embedded within the endosperm.

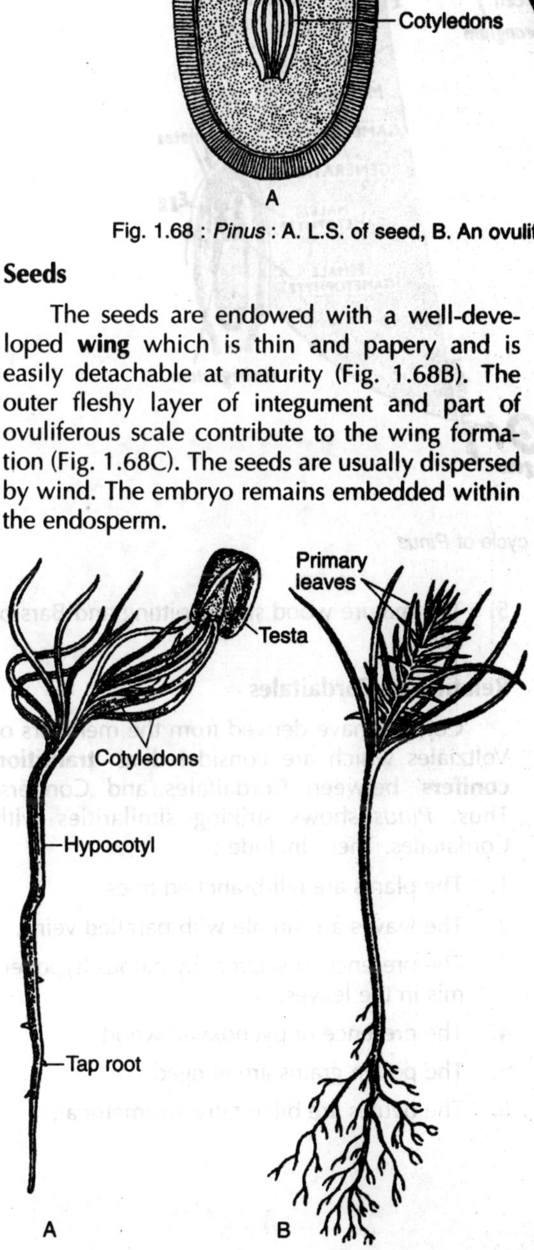

Fig. 1.69 : A–B. The stages in germination of seed

The seeds of *Pinus* remain viable for a long time. The germination of seed is **epigeal** (Fig. 1.69A, B).

Figure 1.70 shows the **life cycle** of *Pinus.*

Origin and relationship

The conifers are large and diversified group of extant gymnosperms. The Pinaceae is the largest family of the modern conifers. The conifers had evolved from the members of Permo-Carboniferous Voltziales, commonly known as **"transition conifers"**. They reached their climax in mid-Mesozoic forming extensive forests in Northern Europe. Among the living families, Pinaceae and Araucariaceae are more primitive, probably evolved during the Triassic age. On the other hand, Cupressaceae and Cephalotaxaceae are comparatively younger which have evolved probably during Upper Jurassic to Lower Cretaceous period.

Pinus is the large and well-representative genus of the family Pinaceae. It indicates relationship with Cycadales, Ginkgoales and Cordaitales.

Relation to Cycadales

The similarities between *Pinus* and Cycadales are :

1. The stem anatomy shows a broad pith, large cortex and centripetal wood.
2. The presence of haplocheilic, sunken stomata in leaves.
3. The presence of leaf sclerenchyma.

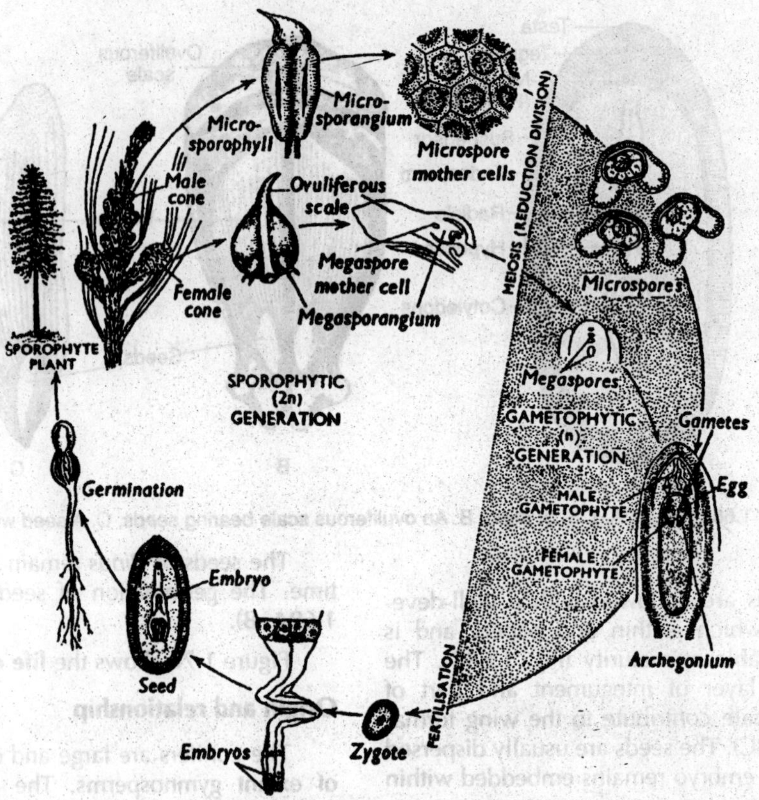

Fig. 1.70 : Life cycle of *Pinus*

4. The seeds are with three-layered integument.
5. The presence of free-nuclear divisions in the development of female gametophyte.

Relation to Ginkgoales

The resemblance between *Pinus* and Ginkgoales are :

1. The plants are profusely branched showing excurrent habit.
2. The shoots are dimorphic, bearing long shoots and dwarf shoots
3. The wood is pycnoxylic.
4. The leaves are with haplocheilic sunken stomata.
5. The mature wood shows pitting and Bars of Sanio.

Relation to Cordaitales

Conifers have derived from the members of Voltziales which are considered as '**transition conifers**' between Cordaitales and Conifers. Thus, *Pinus* shows striking similarities with Cordaitales. These include :

1. The plants are tall-branched trees.
2. The leaves are simple with paralled veins.
3. The presence of sclerenchymatous hypodermis in the leaves.
4. The presence of pycnoxylic wood.
5. The pollen grains are winged.
6. The ovules are bilaterally symmetrical.

CLASS. GNETOPSIDA

The class Gnetopsida is considered to be the highly evolved group of gymnosperms. The class comprises of three distinct orders : Ephedrales, Welwitschiales and Gnetales with monogeneric family in each. However, some scientists preferred to retain them in a single order Gnetales with a monogeneric family in each.

GENERAL CHARACTERISTICS OF THE CLASS GNETOPSIDA

1. Opposite or whorled leaves.
2. Vessels in the secondary xylem.
3. Absence of resin canals.
4. Compound male and female strobili resembling inflorescene.
5. Male flowers with perianth.
6. The microphyle projecting as long tube by prolongation of inner integument.
7. Embryo with two cotyledons.

ORDER : GNETALES

The order Gnetales includes trees or lianes, distributed in the tropical and humid regions of the world. The leaves are large, dorsiventral with unicostate reticulate venation. The shoot apex exhibits a definable tunicea-corpus layers. The female gametophyte is tetrasporic. Archegonia are completely absent. There is no free-nuclear phase in the embryogenesis.

The order Gnetales does not show any fossil counterparts, hence it is considered as a recently evolved group of gymnosperms.

GNETUM

Systematic position

Class. Gnetopsida

Order. Gnetales

Family. Gnetaceae

Total Number of species : About 35

Distribution :

The plants are confined to the moist rain forests of equatorial South America, West Africa, New Guinea, India, South China, Burma (Myanmar), Malaysia, Thailand and Bangladesh. Most of the species are endemic within the areas of their distribution. No species has been reported to be common to both the Hemispheres.

Five species of *Gnetum* are reported from India (Bharadwaj, (1957).

Common Indian Species

1. *Gnetum contractum* : It is a scandent shrub, found in Conoor in Tamil Nadu, Quilon in Kerala and in Nilgiri Hills.
2. *G. ula* : It is a most extensive liana and widely distributed species found in the moist evergreen forests of Western Ghats, Godawari district of Andhra Pradesh and some parts of Orissa.
3. *G. montanum* : It is a robust liana, grows wildly in the tropical Eastern Himalaya from Sikkim eastwards to Darjeeling district, Assam, Khasi and Jaintia Hills, Manipur and also in the Mayur Bhanj in Orissa.
4. *G. latifolium* : It is a scandent shrub with two (out of a total five) varieties growing in India. *G. latifolium* var. *macropodum* is found in Andaman and Nicobar islands, while *G. latifolium* var. *funiculare* is found in South Andaman.
5. *G. gnemon* : It is a tree or erect shrub with two (out of a total of six) varieties growing in India. *G. gnemon* var. *brunonianum* is distributed in Golaghat areas of Assam, South Lusai Hills and Naga Hills. *G. gnemon* var. *griffithii* occurs in Sibsagar and Kungaba areas of Assam and Naga Hills.

Habit :

Most of the species of *Gnetum* are lianes or scandent shrubs. However, *G. gnemon* shows its tree habit and *G. trinerve* is a parasite on *Cinchona* plant.

Sporophyte

Gnetum resembles very much in its characteristics to an angiosperm than a gymnosperm. Based on the studies of 58 characteristics, *Gnetum* shares more than 60% of the characteristics with angiosperm and about 30% characteristics with gymnosperm (Muhammad and Sattler,

1982). *Gnetum* is easily mistaken for a dicot plant unless it is in flowering or fruiting stage. The plant body is differentiated into root, stem and leaves.

Root

External morphology : *Gnetum* shows a typical **tap root** system which is profusely branched. The mature roots show normal secondary growth.

Internal structure : Internally, the root of *Gnetum* resembles the root of angiosperm. The root is differentiated into the outermost layer, **epidermis**, multilayered **cortex** and diarch **vascular cylinder**. The cortex consists of polygonal or oval-shaped parenchymatous cells containing starch grains. The thick-walled **fibre cells** are often present in the cortex. A single-layered **endodermis** encircles a multilayered **pericycle**.

The primary vascular cylinder is diarch, radial and exarch. The **secondary growth** in roots is of normal type. The arcs of cambium form below the phloem groups and above the xylem groups which join together to form a cambium ring. The secondary xylem consists of tracheids possessing uniseriate bordered pits with conspicuous **Bars of Sanio**. The **vessels** are also present. The pits on the vessels are bordered or simple, small and multiseriate with or without inconspicuous Bars of Sanio. The xylem **ray** is composed of thin-walled parenchymatous cells containing starch grains. The phloem consists of sieve cells and parenchyma. The **periderm** is formed due to the extrastelar secondary growth.

Stem

External morphology : Almost all the species of *Gnetum*, except the tree type like *G. gnemon*, exhibits two types of branches viz. **dwarf shoots** or branches of limited growth and **long shoots** or branches of unlimited growth. In climbing and scandent species of *Gnetum*, the stem is articulated with prominent joints. The joint consists of two parts : one just above the node and the other below the node and these two are separated by an **annular groove**. In arboreal species such as *G. gnemon*, the stem exhibits uniform type of branching.

Internal structure : Internally, a *Gnetum* stem is similar to that of a dicot. In T.S. the stem exhibits more or less circular outline (Fig. 1.71A). The following regions are discernible from outside inward : the heavily cutinised single-layered **epidermis** with **sunken stomata**, a several-layered (12-16 layers) thick **cortex** which is differentiated into an outer chlorenchymatous, a middle parenchymatous and an inner sclerenchymatous region. The cells of the inner regions are referred to as **spicular cells** (Fig. 1.71C). Many **fibrous cells** are often present which have branched or unbranched pit canals. The cortex is followed by **endodermis** and **pericycle** layers which are not distinct. Next to pericycle, there is an endarch **siphonostele**.

The vascular cylinder is comprised of 20-24 collateral, open and endarch **vascular bundles**, arranged in a ring (Fig. 1.71A). The vascular bundles are separated from one other by wide **primary medullary rays**. The xylem consists of **tracheids** and few **vessels** (Fig. 1.71B). The protoxylem of tracheids have annular or spiral thickening and the metaxylem have reticulate thickening with uniseriate **bordered pits**. The phloem is composed of **sieve cells** and phloem parechyma. .Laticiferous elements are often found in pith as well as in cortex.

Secondary growth in thickness

In tree species like *G. gnemon*, the secondary growth is of normal type. The fascicular cambium joins with interfascicular cambium to form a complete cambium ring that cuts off a continuous cylinder of secondary xylem towards the inside and secondary phloem towards the outside. The secondary xylem is produced much more than the secondary phloem.

In climbling species (*G. ula, G. africanum*), the **anomalous secondary growth** is noted. Initially, the stem shows normal secondary growth. Later, several successive cambium rings are formed in the cortex. Thus, several extrastellar rings of xylem and phloem are formed which are separated into wedge-shaped bundles because of medullary rays (Fig. 1.71D). Some of these accessory rings remain incomplete and, as a result, the vascular bundles become eccentric with regard to pith (Fig. 1.71E).

The **periderm** is also formed by the activity of phellogen during the third or fourth year of growth of the stem.

The **wood** of *Gnetum* is composed of vessels, tracheids and xylem parenchyma. The vessels are of different sizes and have a single pore on their end walls (Fig. 1.72A). The tracheids are

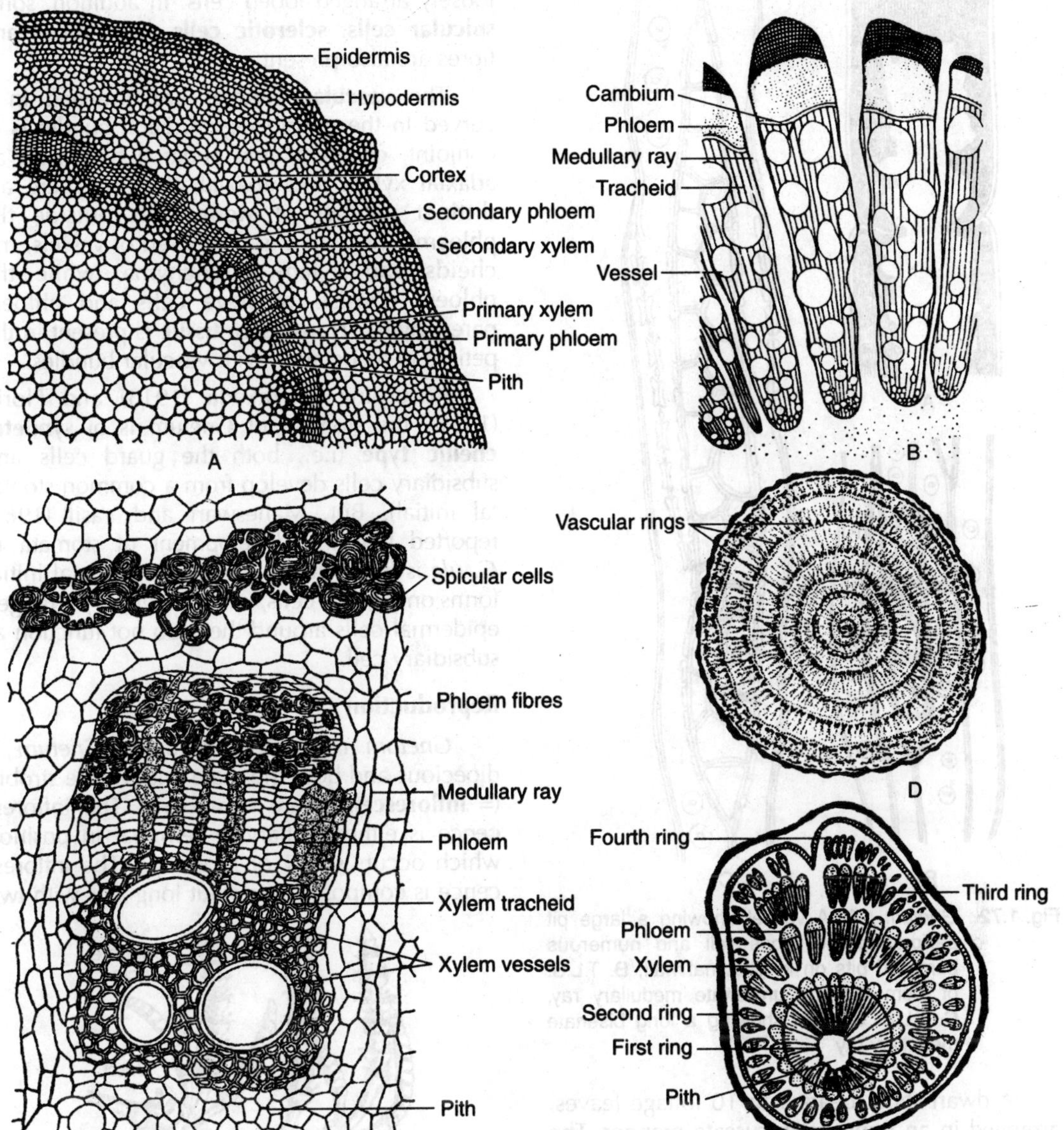

Fig. 1.71 : *Gnetum ula* : A. T.S. of stem, B. Same, a part is enlarged to show vascular bundles, C. Same, one vascular bundle is enlarged, D. T.S. of a mature stem showing polycyclic vascular rings, E. T.S. of a mature stem showing eccentric secondary growth

much longer than the vessels and have bordered pits on both of their radial and tangential walls. The xylem parenchyma cells have simple pits. There are many uni- to multiseriate vascular rays (Fig. 1.72B, C).

The secondary phloem consists of sieve cells and phloem parenchyma. The companion cells are totally absent in *Gnetum,* although some small cells are present near the sieve tube.

The vascular rays are very large and multiseriate which appear boat-shaped in T.L.S. Occasionally, some short biseriate or uniseriate rays are present. The ray cells are thick-walled and pitted.

Leaf

External morphology : *Gnetum* exhibits leaf dimorphism bearing both the **foliage leaves** and **scale leaves.**

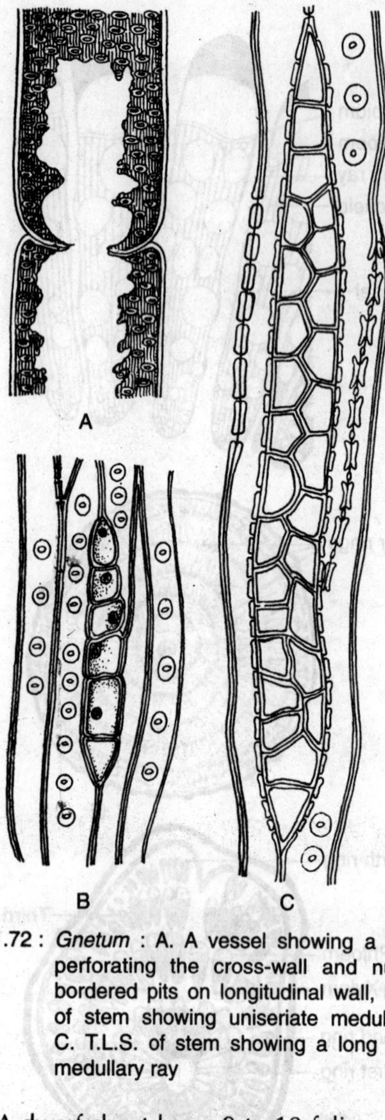

Fig. 1.72 : *Gnetum* : A. A vessel showing a large pit perforating the cross-wall and numerous bordered pits on longitudinal wall, B. T.L.S. of stem showing uniseriate medullary ray, C. T.L.S. of stem showing a long biseriate medullary ray

A dwarf shoot bears 9 to 10 foliage leaves, arranged in an opposite decussate manner. The leaves are large, simple having an oval-shaped broad coriaceous lamina with unicostate reticulate venation. The leaves are exstipulate, shortly petiolate with entire margin. The leaf of *Gnetum* resembles a dicot leaf.

Internal structure : Anatomically, the leaves of *Gnetum* resemble those of dorsiventral dicot leaves. The **epidermis** (both upper and lower) has undulating walls, covered with thick cuticle. The **mesophyll** is differentiated into **palisade** and **spongy** parenchyma. The palisade consists of single-layered compactly arranged elongated cells. The spongy parenchyma is composed of loosely arranged lobed cells. In addition, some **spicular cells**, **sclerotic cells**, latex tubes and fibres are also present.

The **vascular bundles** are arranged in a curved in the midrib region. Each bundle is of conjoint, collateral and endarch type with an adaxial xylem and abaxial phloem. There are distinct patches of **stone cells** outside the phloem. The xylem is composed of vessels, tracheids and xylem parenchyma, while the phloem consists of sieve cells and phloem parenchyma. **Transfusion tissue** is present in the petiole in association with vascular bundles.

According to Takeda (1913) and Florin (1931), the stomata of Gnetum is of **syndetocheilic type** (i.e., both the guard cells and subsidiary cells develop from a common stomatal initial). But, Maheswari and Vasil (1961) reported that the development of stomata in *G. ula* is **haplocheilic** where the stomatal initial forms only guard cells, while irregularly arranged epidermal cells around them do not function as subsidiary cell.

Reproduction

Gnetum reproduces sexually. *Gnetum* is dioecious and both the male and female strobili (= **inflorescence**) are compound. The inflorescence is either axillary or terminal in position which occurs singly or in groups. The inflorescence is composed of a stout long axis with two

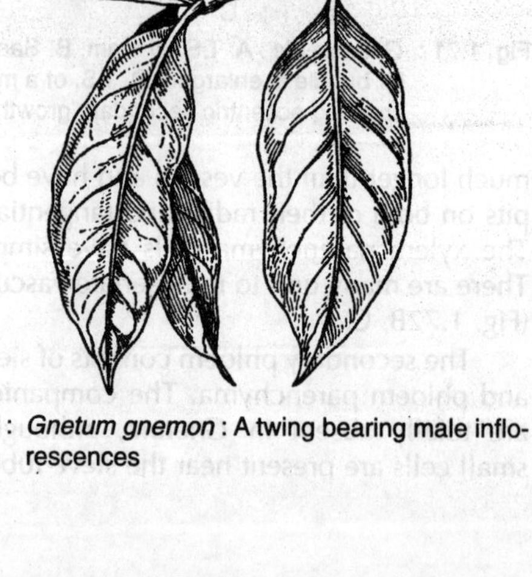

Fig. 1.73 : *Gnetum gnemon* : A twing bearing male inflorescences

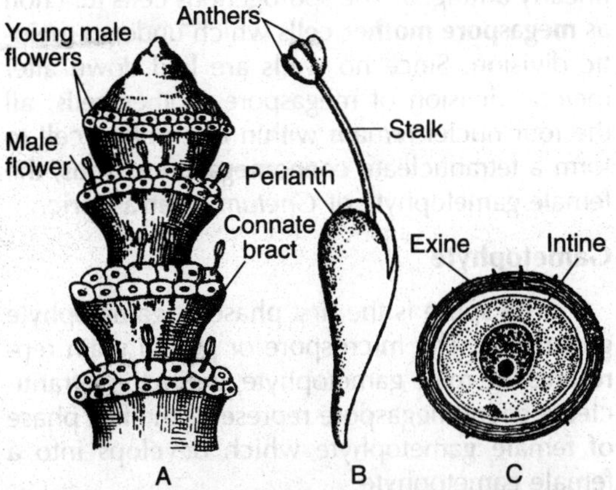

Fig. 1.74 : *Gnetum gnemon* : A. A part of a male inflorescence, B. A male flower, C. A pollen grain

opposite decussate, connate bracts at the base and a series of cup-like bracts called **cupules** or **collars** that are superposed one above the other (Fig. 1.73, 1.74A). The strobilus becomes compact at young stage because of the suppression of internodes. Thus, the collars of a young strobilus appear to be continuous (Fig. 1.73). At maturity, the axis becomes elongated and the collars get separated. There are many rings of flowers in the axil of collars. The collars are developed in acropetal succession and the flowers are initiated as mounds of meristematic cells from the lower surface of a collar.

Male strobilus

The male strobilus is compound and has a long slender axis bearing 10-25 whorl of bracts (collars) (Fig. 1.73). About 12-25 **male flowers** are arranged in three to six rings above each collar (Fig. 1.74A). A single ring of 7-15 imperfect female flowers or abortive ovules is present just above the male flowers.

Male flower : A male flower consists of two unilocular **anthers** on a stalk (**antherophore**) enclosed in a small sheathing **perianth** (Fig. 1.74B). The stalk of the anther elongates rapidly at maturity pushing the anther beyond the collars through a slit in the perianth.

Female strobilus

The female strobilus is very much similar to that of the male strobilus in the young stage. Like male strobilus, the female strobilus consists of an axis bearing several whorl of collars arranged one above the other (Fig. 1.75A). A ring of 4-10 **ovules** (female flowers) is present above each collar (Fig. 1.75B). The male flowers are not found in the female strobilus. The upper few collars are devoid of ovules and are thus sterile (Fig. 1.75B).

Ovule

A single ovule represents a female flower. The ovule is stalked in *G. ula*, but may be subsessile or even sessile. The ovules are orthotropous, crassinucellate (with massive nucellar tissue) and are protected by three envelopes (Fig. 1.75C). The outer envelop which becomes thickened and succulent at maturity is considered to be the perianth corresponding to the perianth of

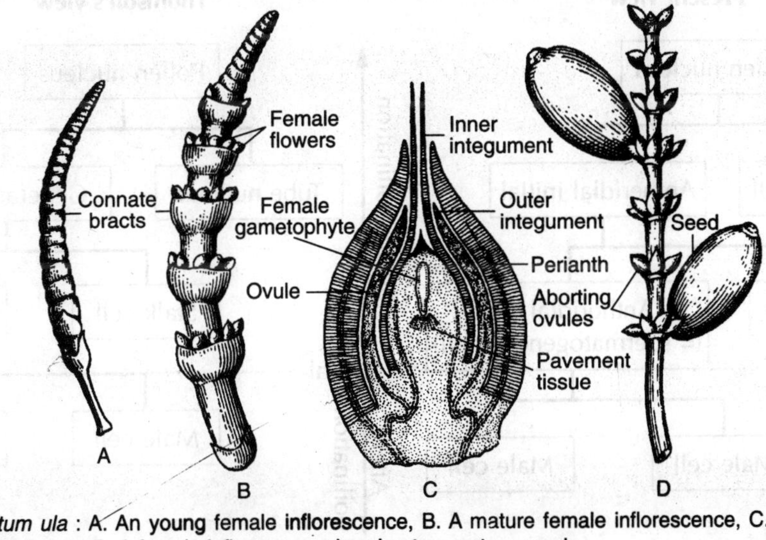

Fig. 1.75 : *Gnetum ula* : A. An young female inflorescence, B. A mature female inflorescence, C. Median L.S. of a female flower, D. A female inflorescence bearing two mature seeds

male flower. The middle and the inner envelopes are actually the **integuments**. Numerous laticiferous ducts and sclerides are present in the perianth with some epidermal stomata. The middle envelop is called the **outer integument** which is anatomically similar to the outer envelop. The inner envelop, i.e., the **inner integument,** elongates far beyond the apical cleft of the perianth and forms a long **micropylar tube** (Fig. 1.75C). Anatomically, the inner integument is different from the other two envelopes, because neither sclerides nor stomata develop in the inner integument. The inner integument is free from the nucellus except at the chalazal end. Two sets of vascular bundles are formed (Fig. 1.75C), of which the outer set passes to the perianth and the inner set again divides and one of its branches passes to the outer integument and the other to the inner integument. All the three envelopes of ovules develop in acropetal manner. A shallow **pollen chamber** is present at the tip of the nucellus.

Megasporogenesis

Two to four hypodermal cells in the nucellar tissue at the micropylar end is differentiated into **primary parietal cells** towards outside and the **primary sporogenous cells** towards inside. The primary parietal cells together with nucellar epidermal cells divide repeatedly to produce a massive nucellus. The primary sporogenous cells divide to form 8-20 **sporogenous cells** which are linearly arranged. The sporogenous cells function as **megaspore mother cells** which undergo meiotic division. Since no walls are laid down after meiotic division of megaspore mother cells, all the four nuclei remain within the mother cell to form a tetranucleate **coenomegaspore**. Thus, the female gametophyte of *Gnetum* is **tetrasporic.**

Gametophyte

The spore is the first phase of gametophyte generation. The microspore or pollen grain represents the male gametophyte, while the tetranucleate coenomegaspore represents the first phase of female gametophyte which develops into a female gametophyte.

Development of male gametophyte before pollination

The pollen grain is **inapeturate** and spherical, bounded by the two concentric wall layers : the outer thick, spiny **exine** and the inner thin **intine** (Fig. 1.74C). The pollen nucleus divides mitotically to produce a small lens-shaped **prothallial cell** and a large **antheridial initial** (Fig. 1.76B). The prothallial cell does not divide further and eventually degenerates. The antheridial initial divides to form an **antheridial cell** and a **tube nucleus** (Fig. 1.76C). The antheridial cell directly functions as **spermatogenous cell.** The pollen grains are released from the microsporangium at this 3-celled stage (one prothallial cell, an antheridial or spermatogenous cell and a tube nucleus).

Schematic representation in the development of male gametophyte of *Gnetum*

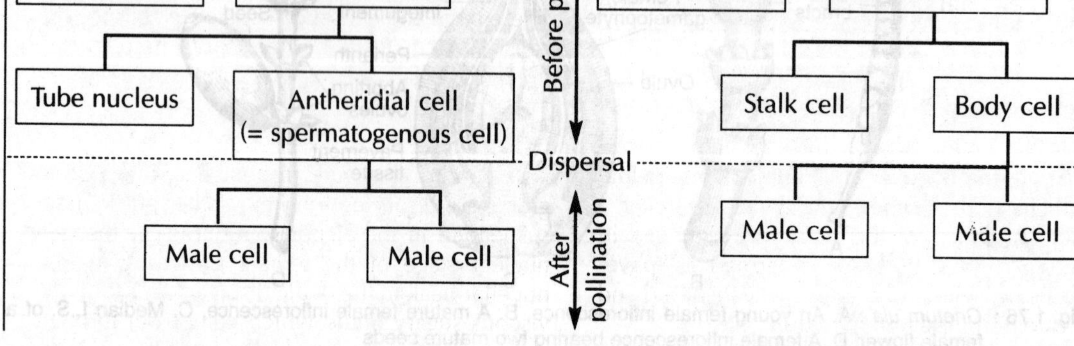

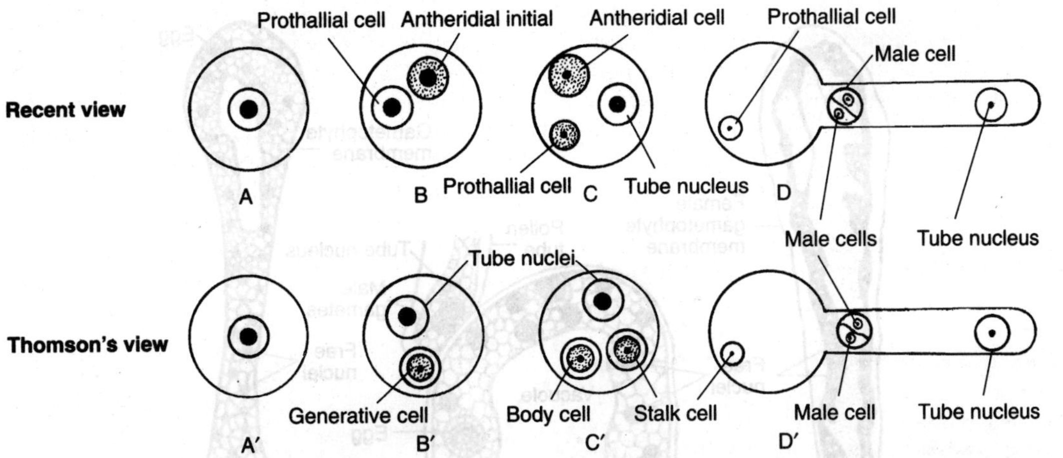

Fig. 1.76 : *Ginkgo biloba*. The development of male gametophyte : A–D. Recent view, A′–D′. Thomson's view

According to Thomson (1961), the prothallial cell is not formed in *Gnetum*. He proposed that the pollen nucleus cuts off a tube nucleus and a **generative cell** (Fig. 1.76B′). The generative cell again divides forming a **stalk cell** and a **body cell** (Fig. 1.76C′). Thus, the pollen grains are released at this 3-celled stage (tube nucleus, stalk cell and body cell).

Development of male gametophyte after pollination

The exine is cast off during pollen germination. The tube cell of the pollen comes out in the form of a **pollen tube** which traverses the nucellar tissue through intercellular spaces. The prothalial cell remains within the pollen grain and eventually disorganises. The spermatogenous cell moves into the pollen tube and subsequently it divides to form two equal (e.g., *G. ula*, *G. gnemon*) or unequal (e.g., *G. africanum*) male cells just prior to fertilisation (Fig. 1.76D). The male cells are actually the **male gametes** which are non-motile. However, according to Thomson (1961) the stalk cell remains within the microspore and eventually degenerates. The body cell moves into the tube where it divides to form two male cells (Fig. 1.76D′).

Development of female gametophyte

At the initial stage, before the gametophyte formation, the nucellar cells immediately below the megaspore mother cell divide to form a tissue. The cells of this tissue are arranged in radiating rows. This tissue is termed as '**pavement tissue**' which eventually gets absorbed and seems to be nutritive in function.

There is a free nuclear division in the coenomegaspore, as a result a large number of free nuclei are formed (Fig. 1.77A). The number of nuclei thus formed varies in different species, viz. 256 in *G. gnemon,* 512 in *G. africanum* and 1500 in *G. ula*. At this stage, a large central vacuole appears and the free nuclei lie in a thin film of cytoplasm around the vacuole towards the periphery (Fig. 1.77B). Later, the nuclei in the peripheral cytoplasm divide repeatedly. At this stage, the upper part of the gametophyte surrounding a vacuole widens, while the lower part of the gametophyte shows accumulation of cytoplasm. Thus, the gametophyte becomes an inverted flask-shaped structure (Fig. 1.77C). The wall formation starts very slowly from the chalazal end towards the micropylar end. Thus, the nuclei remain free at the microphylar end even at the time of fertilisation.

The important characteristic in the female gametophyte of *Gnetum* is the absence of archegonia. One to three nuclei of the gametophyte in the micropylar end enlarge several times and accumulate dense cytoplasm around them. These large and densely cytoplasmic cells are the **eggs** (Fig. 1.77B). It is important to note that all the eggs do not mature simultaneously.

Pollination

Gnetum is wind-pollinated. The pollen grains are dispersed from the anther and remain suspended in the air for some time. At the free nuclear stage of the female gametophyte, the nucellar beak in the ovule disorganises forming a viscous sugary liquid which comes out through

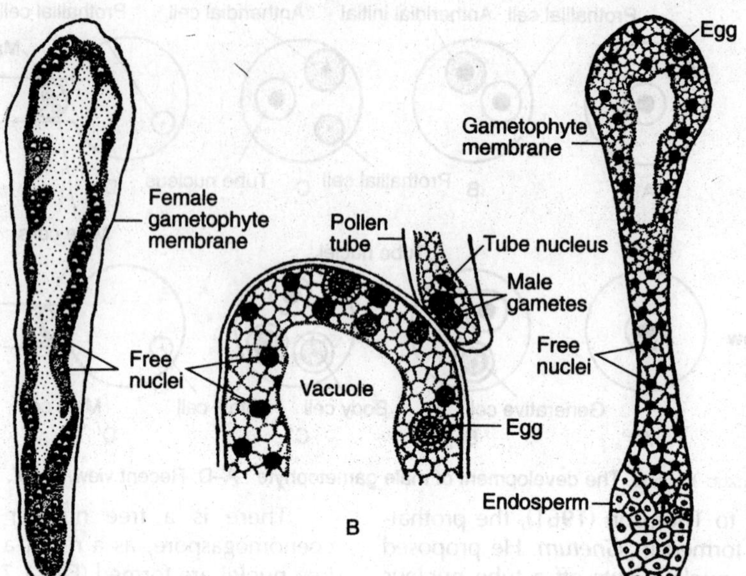

Fig. 1.77 : *Gnetum ula* : A. A female gametophyte in free nuclear stage, B. Top of the female gametophyte showing eggs, C. Female gametophyte showing first part of endosperm at base

the microphyle in the form of a **pollination drop.** The pollen grains are caught in the pollination drop. Due to the drying off of the fluid, the pollen grains are sucked into the micropylar canal and are finally collected in the pollen chamber. The mouth of the micropyle is then sealed from the outer environment due to the development of **flage** (a circular rim or an umbrella-shaped structure develops from the inner integument) and **micropylar closing tissue** (a tissue develops by the proliferation of the inner epidermis of integument at the level of flage).

Fertilisation

The pollen tube enters the female gametophyte and the male gametes move ahead of tube nucleus (Fig. 1.77B). The pollen tube ruptures to discharge the male gametes into the egg cell. The cell sheath of male cell is left outside the egg cell. Usually one of the male nuclei fuses with the egg nucleus and thus a zygote is formed. Sometimes, two male gametes may fuse two different eggs if those eggs are in the vicinity of the pollen tube (Vasil 1959).

Endosperm

In gymnosperms, endosperms are cellular and haploid and are formed before fertilisation. However, in *Gnetum* the development of **endosperms** starts before fertilisation very slowly from lower part of the gametophyte which eventually proceeds upward. After fertilisation, the wall formation starts in such a way that the cytoplasm divides into many multinucleate compartments (Fig. 1.78A). Later, the nuclei in each cell fuse to form a single **polyploid nucleus** (Fig. 1.78B). In this stage, the lower part of the gametophyte becomes cellular, while the upper part remains free nuclear even after fertilisation (Fig. 1.77C). Thus, the development of endosperm takes place even after fertilisation. There is a great variation in the development of endosperm in *Gnetum*. In some cases, the wall formation starts either from the upper part or from the middle part of the gametophyte instead of the lower part and the whole gametophyte may become cellular. Though some portions of the endosperms are formed after fertilisation, the characteristic triploid endosperm through double fertilisation is, however, absent in *Gnetum*.

Embryogeny

In all angiosperms (except, *Paeonia*), the division of zygote is accompanied by wall-formation; while in all gymnosperms (except *Sequoia*, *Welwitschia*), there is a free nuclear phase in the zygote during the development of embryo. However, *Gnetum* occupies an inter-

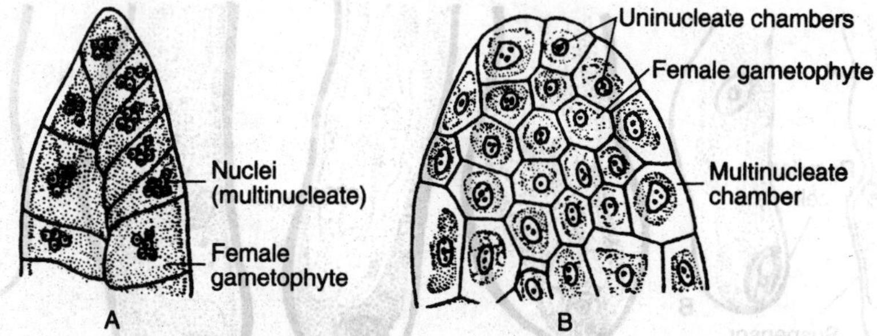

Fig. 1.78 : *Gnetum ula*. Development of endosperm : A. A portion of female gametophyte showing multinucleate endosperm, B. Uninucleate and multinucleate chambers

mediate stage between gymnosperms and angiosperms with regard to embryo development by having both the free-nuclear divisions as well as cell divisions.

There is a great variation in the early development of the embryo in different species of *Gnetum*.

In *G. gnemon*, the zygote develops 1-3 small tubular outgrowths. Only one of the protuberances receives the nucleus and survives, while the remaining protuberances die out (Fig. 1.79A). The surviving tubular outgrowth becomes much elongated and branched and develops in various directions invading the intercellular spaces of the endosperm. These

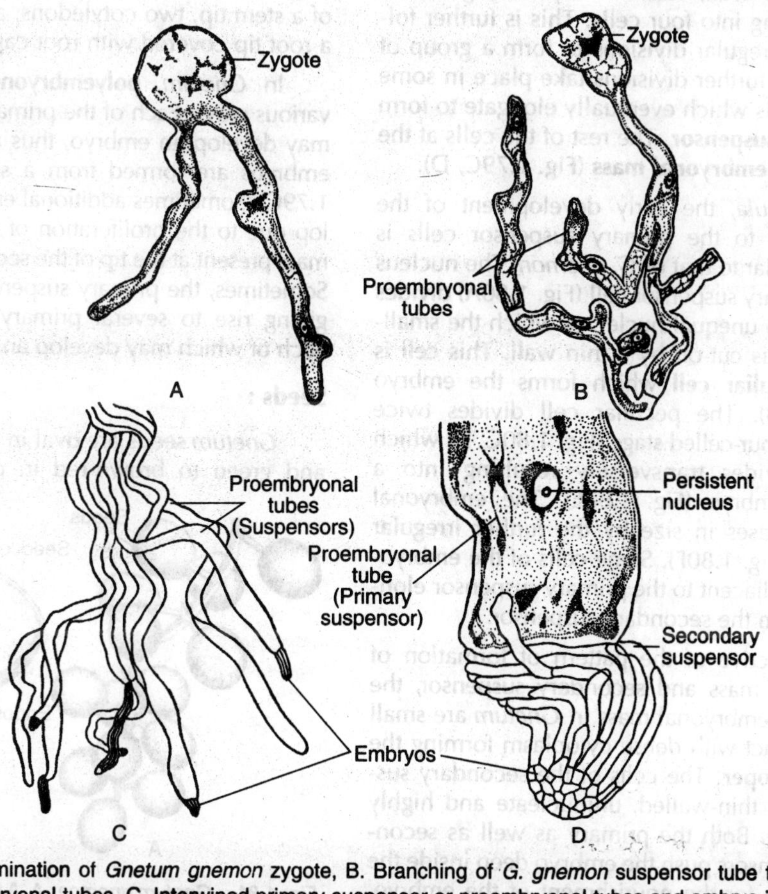

Fig. 1.79 : A. Germination of *Gnetum gnemon* zygote, B. Branching of *G. gnemon* suspensor tube to form several proembryonal tubes, C. Interwined primary suspensor or proembryonal tubes and embryo initials in young seed of *G. ula*, D. Tip of proembryonal tube with growing embryo in *G. ula*

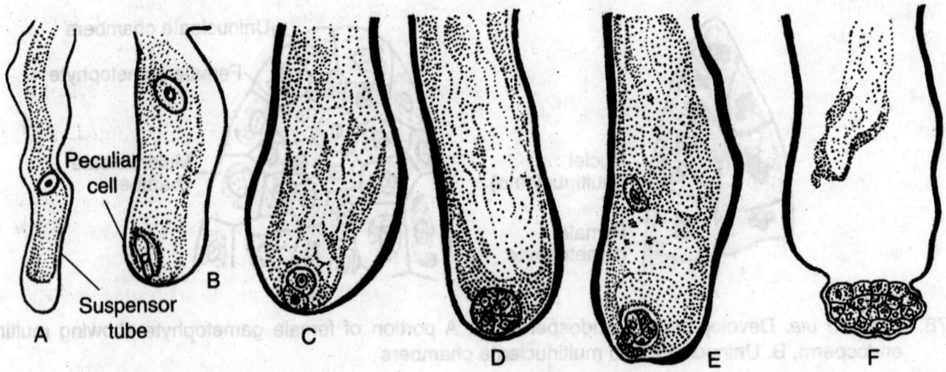

Fig. 1.80 : *Gnetum ula* : A. Tip of primary suspensor tube, B. Formation of peculiar cell, C–E. Peculiar cell dividing to form two (C), four (D) and eight (E) cells, F. Mass of cells formed by the peculiar cell

tubes are called **primary suspensor tubes** or **proembryonal tubes** (Fig. 1.79B). All the primary suspensor tubes remain coiled around each other. At the tip of the primary suspensor tube, a small cell is cut off which eventually divides by a transverse wall, followed by a longitudinal wall resulting into four cells. This is further followed by irregular divisions to form a group of cells. Now, further divisions take place in some of these cells which eventually elongate to form **secondary suspensor**. The rest of the cells at the tip form an **embryonal mass** (Fig. 1.79C, D).

In *G. ula*, the early development of the embryo up to the primary suspensor cells is almost similar to that of *G. gnemon*. The nucleus of the primary suspensor cell (Fig. 1.80A) divides to form two unequal nuclei, of which the smaller nucleus is cut-off by a thin wall. This cell is called **peculiar cell** which forms the embryo (Fig. 1.80B). The peculiar cell divides twice forming a four-celled stage (Fig. 1.80C, D) which further divides transversely resulting into a 8-celled embryo (Fig. 1.80E). The embryonal mass increases in size by the further irregular divisions (Fig. 1.80F). Some cells of the embryonal mass adjacent to the primary suspensor elongate to form the secondary suspensor.

Irrespective of the pattern of formation of embryonal mass and secondary suspensor, the cell of the embryonal mass in *Gnetum* are small and compact with dense cytoplasm forming the **embryo-proper**. The cells of the secondary suspensor are thin-walled, uninucleate and highly vacuolated. Both the primary as well as secondary suspensors push the embryo deep inside the endosperm for the nourishment of the embryo. At the tip of the embryonal cells, a stem tip with two lateral cotyledons is differentiated (Fig. 1.81B). A root tip with a root cap also develops at the opposite end of the stem tip. Simultaneously, a hump-like structure called **feeder** is developed in-between the stem and root tips (Fig. 1.81B). Thus, a mature embryo is composed of a stem tip, two cotyledons, a large feeder and a root tip covered with root cap.

In *Gnetum*, **polyembryony** takes place in various ways. Each of the primary suspensor tube may develop an embryo, thus a large number of embryos are formed from a single zygote (Fig. 1.79C). Sometimes additional embryos may develop due to the proliferation of the proembryonal mass present at the tip of the secondary suspensor. Sometimes, the primary suspensor tube branches giving rise to several primary suspensor tubes, each of which may develop an embryo at its tip.

Seeds :

Gnetum seeds are oval in shape (Fig. 1.81A) and green to brown-red in colour. The seeds

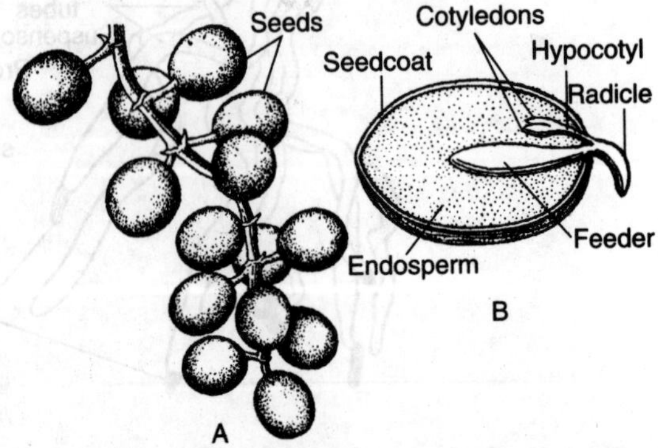

Fig. 1.81 : *Gnetum gnemon* : A. A twig containing seeds, B. Median L.S. of seed about to germinate

remain covered with a three-layered envelop, of which outer is **fleshy**, middle is **stony** and inner is **pepary**. The nucellus is used up and the embryo is embedded within the endosperm. *Gnetum* shows one-year reproductive cycle where pollination, fertilisation and development of embryo take place in one year.

The germination of seed is **epigeal**. The seeds of *G. ula* germinate after one year's of res-ting phase.

Figure 1.82 shows the life **cycle** of *Gnetum.*

Relationship

Gnetum has been included under gymnosperms due to its naked seeds and other gymnospermous characteristics, although it shows several resemblances with angiosperms. Arber and Parkin (1980) argued that both Gnetales (including *Gnetum, Ephedra* and *Welwitschia*) and angiosperms have originated from a common stalk called 'Hemiangiosperm'. But, recent studies have clearly shown that *Gnetum* shares many characteristics with angiosperms due to convergent evolution.

Gymnospermous characteristics of *Gnetum*

1. Wood having metaxylem tracheids with bordered pits.
2. Presence of sieve cells with sieve areas on the radial wall. The companion cells are absent.
3. The anther has an exothecium which produces anemophilous pollen grains.
4. Pollen exine is alveolar or granular.
5. Presence of prothallial cell in male gametophyte.
6. The ovules are naked (exposed).
7. The development of female gametophyte shows a free nuclear phase.
8. Pollination takes place through 'pollination drop mechanism'.
9. Presence of cleavage polyembryony.

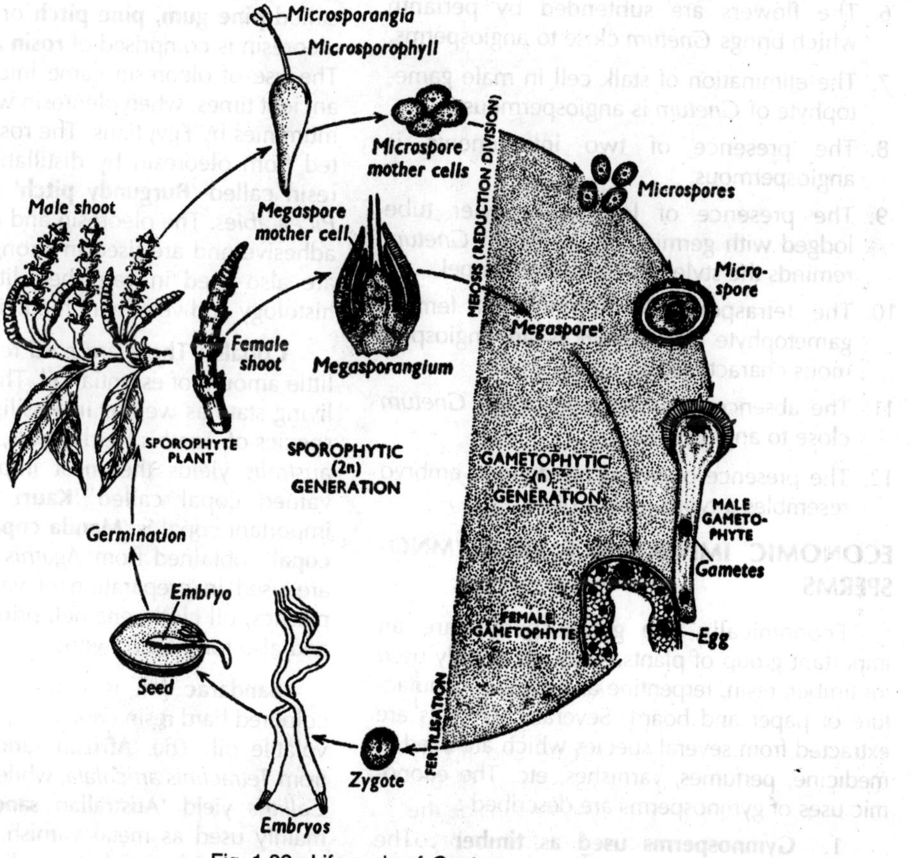

Fig. 1.82 : Life cycle of *Gnetum*

Angiospermous characteristics of *Gnetum*

1. The external appearance of Gnetum resembles a typical dicot plant.
2. The leaves of *Gnetum* are broad, dorsiventral with reticulate venation as in dicot plants.
3. The presence of vessels in xylem of *Gnetum* is angiospermous.
[However, ontogenically the vessels of *Gnetum* are quite distinct from the angiosperm vessels. The vessels of *Gnetum* develop from tracheids having bordered pits on the end wall, while the angiospermous vessels develop from tracheids with narrow scalariform perforations.]
4. The shoot apex of *Gnetum* with a definable tunica-corpus layers is a typical angiospermous characteristics.
5. The male and female strobili of *Gnetum* are compound and represent the inflorescences and flowers of angiosperms.
6. The flowers are subtended by perianth which brings *Gnetum* close to angiosperms.
7. The elimination of stalk cell in male gametophyte of *Gnetum* is angiospermous.
8. The presence of two integuments is angiospermous.
9. The presence of long micropyler tube lodged with germinating pollen in *Gnetum* reminds the style of angiosperm carpel.
10. The tetrasporic development of female gametophyte in *Gnetum* is an angiospermous characteristics.
11. The absence of archegonia brings *Gnetum* close to angiosperms.
12. The presence of two cotyledons in embryo resembles a typical dicot plant.

ECONOMIC IMPORTANCE OF GYMNOSPERMS

Economically, the gymnosperms are an important group of plants. They are mainly used for timber, resin, terpentine and for the manufacture of paper and board. Several chemicals are extracted from several species which are used in medicine, perfumes, varnishes, etc. The economic uses of gymnosperms are described :

1. **Gymnosperms used as timber** : The coniferous wood is very light or hard,, strong or soft, coarse or straight grained and durable with distinct annual rings.

Pinus, Cedrus, Abies, Agathis, Araucaria, Cupressus, etc. yield good quality of timber used in various purposes. These are used for making building materials, furnitures, boxes, paper pulp, wood wool, ply wood, railway sleeper, sound board of musical instruments, match boxes, matchsticks, pencil, poles, etc. etc.

2. **Gymnosperms as sources of resins** : Resins are coniferous plant exudates which are only soluble in organic solvents. Commmercially resins are of two types : superior grade resin and inferior grade resin. The former is used in varnishes, plasters, enamels, paper sizing, medicine and ointments, while the latter shows its use in the manufacture of printing ink, laundry soap, oil and grease, adhesive, insulators, insecticides plastics, shoe polish and oil cloth. The different types of resins are discribed :

* **Oleoresin** and **rosin** : Several species of *Pinus* produce oleoresins which are commonly called **pine gum, pine pitch** or **turpentine**. The oleoresin is comprised of **rosin** and **essential oil.** The use of oleoresin came into existence since ancient times, when oleoresin was used to smear mummies by Egyptians. The rosin can be extracted from oleoresin by distillation. The purified resin called '**Burgundy pitch**' is obtained from *Picea abies.* The oleoresin and rosin are strongly adhesive and are used in strongly adhesive and are also used in varnishes, lithographic work, histology and veterinary medicines.

Copals : These are hard resins containing a little amount of essential oil. The copal occurs in living state as well as in fossilised form. Several species of *Agathis* yield copals, of which *Agathis australis* yields the most important and most valued copal called '**Kauri Copal**'. Another important copal is '**Manila copal**' or '**East Indian copal**', obtained from *Agathis alba.* The copals are used in preparation of varnishes linoleum, plastics, oil cloth, enamel, printing ink, lacquers and also for interior work.

Sandarac : It is a pale yellow or orange coloured hard resin containing about one percent volatile oil. The '**African sandarac**' is obtained from *Tetraclinis articulata,* while several species of *Callitris* yield '**Australian sandarac**'. These are mainly used as metal varnish, paper varnish or leather varnish. Sandarac is also used for cementing glass, porceline and preservation of fine paintings.

Canada balsam : It is a transparent resin which does not granulate or crystallise on drying. *Abies balsamea* is the main source of Canada balsam. Moreover, *Tsuga canadensis and Pseudotsuga taxifolia* also yield a somewhat similar resin.

Canada balsam is widely used as a mounting medium for microscopic studies for its high refractive (1.5) index nearing that of glass.

3. **Gymnosperms as sources of essential oils and fatty oils** : Almost all conifers yield essential oils along with resins. '**Spruce oil**' is obtained from *Picea abies* which is used in room sprays, bath salts and deodorant. The '**Siberian fir needle oil**', obtained from *Abies sibirica* is also used in deodorant, toilet and shaving soaps, room sprays etc. The '**Cedar oil**' is obtained from *Cedrus deodara* and *Juniperus virginiana* and is used in perfumary and also for use with oil immersion lens in microscopic studies. *Juniperus mexicana* yields '**Cedarwood oil**' which is used in deodorant, room sprays scenting soaps, insect proofing. The oil, obtained from various conifers such as *Dacrydium franklinii, Cedrus atlantica, Thuja occidentalis,* is used against many diseases like bronchitis, tuberculosis, skin diseases, gonorrhoea, etc.

Fatty oils obtained from seeds of many gymnosperms like *Pinus cembra, Torreya nucifera, Cephalotaxus drupacea,* etc. are used for food and paint.

4. **Gymnosperms as sources of papers** : Conifers yield good quality timbers which are used for making paper pulp. Thus, conifers are the major source of paper pulp in the newsprint industry. Super quality of papers such as writing and printing papers are manufactured from the wood of *Picea, Abies, Larix* and *Tsuga,* whereas kraft papers are produced from *Pinus* and *Cryptomeria.*

5. **Gymnosperms used as drugs** : The green branches of *Ephedra* (e.g., *E. sinica, E. gerardiana, E. equisetina*) are important sources of alkaloid, ephedrine which has been used as bronchodilator in cough mixture for the treatment of bronchial asthma.

The leaves of *Ginkgo biloba* is an important source of C_{20} trilactone ginkgolide compound. This compound is used to antagonise platelet activating factor (PAF) in vertebrate blood system. The leaves of *G. biloba* are also used in the treatment of cerebral insufficiency and vertigo.

Taxus brevifolia is an important source of an anticancer drug **taxol** which has been used in the treatment of ovarian cancer, breast cancer, melanoma and colon cancer. The leaves of *Taxus baccata* are useful in the treatment of asthma, bronchitis, hiccough, epilepsy and indigestion. *Taxus* also yield a toxic substance called **taxine** which is an active heart poison.

An extract of *Cycas revoluta* leaves is useful in treatment of flatulence and vomiting blood. The powdered stem of *Cycas pectinata* is commonly used by Assamese as hair wash for the treatment of diseased hair roots. The seeds of *Cycas* are used as herbal drug for the cure of boils, wounds and sores.

6. **Gymnosperms as sources of tannins** : The bark of *Tsuga canadensis, Picea alba, Sequoia sempervirens, Larix decidua, Araucaria,* etc., yield tannins which are used in medicine, leather and petroleum industry and also for dyeing purposes.

7. **Gymnosperms used as ornamental plants** : Gymnosperms such as pine, cedar, fir, spruce, juniper or hemlock, are widely planted as ornamental plants in gardens, parks and around tombs and religious buildings for their stately appearance, symmetrical growth and evergreen habit.

8. **Gymnosperms used as food** : The stem (pith) starch called '**sago**' is obtained from the species of *Cycas* (*C. revoluta, C. rumphii,* and *C. circinalis*), *Zamia* and *Macrozamia* (*M. spiralis,* and *M. reidlei*).

The seeds of *Cycas* spp., *Pinus gerardiana, P. pinea, P. edulis, P. pumila, Dioon edule, D. spinulosum, Araucaria araucana, Ginkgo biloba,* are edible for their rich source of starch. In North America, the seeds of *Pinus* spp. are commercially exploited in the manufacture of candies, nut-coffee and caramels.

The young unfolded succulent leaves and tender shoots of *Cycas circinalis, C. pectinata, C. rumphii* and *C. siamensis* are eaten as cooked food in Indonesia, Philippines, Malay and Indian peninsula.

Comparative account of *Cycas*, *Pinus*, *Ginkgo* and *Gnetum*

Characteristics	*Cycas*	*Pinus*	*Ginkgo*	*Gnetum*
1. Systematic position	Class : Cycadopsida Order : Cycadales Family : Cycadaceae	Class : Coniferopsida Order : Coniferales Family : Pinaceae	Class : Ginkgopsida Order : Ginkgoales Family : Ginkgoaceae	Class : Gnetopsida Order : Gnetales Family : Gnetaceae
2. Habit	Tropical, evergreen, palm-like unbranched trees.	Sub-tropical to temperate, evergreen, tall. profusely branched trees.	Sub-tropical to temperate, deciduous, tall, profusely branched trees.	Tropical, evergreen, shrubs or trees, majority are lianes.
3. Root				
(a) Morphology	Two types : (a) short-lived tap roots, (b) coralloid roots.	One type : Tap roots may persist, lateral roots grow extensively and become mycorhizic.	One type : Normal tap roots.	Like *Ginkgo*.
(b) Anatomy	Normal root is differentiated into epidermis, multilayered cortex with tannin cells and mucilage cells and central vascular cylinder. Stele is diarch to tetrach. In coralloid root, there is an algae zone in the middle cortex. Mature root shows secondary growth.	Primary root resembles a typical dicot root, composed of epidermis, cortex with resinous ducts and central vascular cylinder. The stele is diarch to hexarch. Mature root shows secondary growth.	Primary root is differentiated into epidermis, multilayered cortex with mucilage ducts and central vascular cylinder. The stele is diarch. Mature root shows secondary growth.	Primary root is differentiated into epidermis, cortex with starch grains and central vascular cylinder. The stele is diarch or triarch. Mature root shows secondary growth.
4. Stem				
(a) Morphology	Unbranched, columnar, covered with armour of persistent leaf bases.	Branched (excurrent type), regular and dimorphic : (a) longs shoots, (b) dwarf shoots.	Branched (excurrent type), becomes irregular, dimorphic : (a) long shoots, (b) dwarf shoots.	Branched in *G. gnemon*; in other species stems are erect or prostrate with lianing habit, dimorphic : (a) long shoots, (b) dwarf shoots.
(b) Anatomy	Stem shows a thick cortex, a thin vascular cylinder and a thick pith.	Stem shows a thin cortex, a thick vascular cylinder and a thin pith.	Long shoot like those of *Pinus*, but dwarf shoot resembles *Cycas*.	Stem like those of *Pinus*.
	Girdle of leaf traces	Direct leaf traces.	Like *Pinus*.	Like *Pinus*.
	Mucilage canals are present in cortex and pith.	Resinous ducts are present in cortex and pith.	Like *Cycas*.	Mucilage or resinous canals are absent.

Contd.

Contd.

Characteristics	*Cycas*	*Pinus*	*Ginkgo*	*Gnetum*
	Vessels absent.	Like *Cycas*.	Like *Cycas*.	Vessels present in secondary xylem.
	Cambium short-lived, secondary growth polyxylic, medullary rays uniseriate, no annual ring formation.	Cambium persistent, secondary growth normal and extensive, medullary rays uniseriate or multiseriate, annual rings prominent.	Like *Pinus*.	*G. gnemon* is like those of *Pinus*, but climbing species show anomalous secondary growth with successive vascular rings or forming eccentric bundles.
5. Leaf				
(a) Morphology	Dimorphic : (a) scale leaves, (b) foliage leaves.	Like *Cycas*.	Like *Cycas*.	One type
	Foliage leaves large, pinnately compound, arranged spirally on the top of stem forming crown.	Foliage leaves simple, needle-like develop at the apex of dwarf shoots.	Foliage leaves simple, long petiolate fan-shaped, with dichotomous venations, bilobed, borne on both long and dwarf shoots.	Foliage leaves simple, short petiolate, broad with reticulate venations, arranged in opposite decussate pairs. Scale leaves often present.
	Young leaflet shows circinate ptyxis.	No circinate ptyxis	Like *Pinus*.	Like *Pinus*.
(b) Anatomy	Mesophyll tissue differentiates into palisade and spongy parenchyma.	Mesophyll tissue not differentiated, made up of only polygonal cells having peg-like infolding.	Mesophyll tissue of such leaves borne on long shoots differentiates into palisade and spongy parenchyma, but mesophyll tissue of such leaves borne on dwarf shoots remains undifferentiated.	Like *Cycas*.
	With one vascular bundle, xylem mesarch.	With two vascular bundles, xylem endarch.	With many vascular bundles, xylem endarch.	With many vascular bundles, xylem mesarch.
6. Vegetative reproduction	The plants reproduce vegetatively by adventitious buds or bulbils.	No vegetative reproduction.	Like *Pinus*.	Like *Pinus*.
7. Flower	Flowers are naked, dioecious.	Flowers are naked, monoecious.	Flowers are naked, dioecious.	Flowers are subtended by perianth, dioecious.

Contd.

Contd.

Characteristics	*Cycas*	*Pinus*	*Ginkgo*	*Gnetum*
8. Male fructification	Large (50 cm in length), cylindrical, compact, borne terminally at the growing point of the stem.	Small (2–3 cm in length), ovoid, compact, borne singly in the axils of scale leaves of long shoots.	It is a catkin-like inflorescence (4–6 cm in length), borne in the axils of scale leaves of dwarf shoots.	A short stout axis-like structure (up to 6 cm in length), borne solitary or axillary or in cluster at the shoot apex.
9. Microsporophyll	A flattened wood structure, narrow below and broadened above into an expanded sterile apophysis. Numerous microsporangia are borne in sori (group of 3–5) on abaxial surface.	A scaly structure with a short stalk and foliar expanded part. The tip of the foliar part is slightly bent upwards forming apophysis. Two elongated microsporangia are borne on abaxial surface.	A slender structure, turned upwards into a hump containing a large mucilage cavity and bears two pendent microsporangia.	Male flower consists of two coherent bracts (representing perianth) and two anthers on the apex of short stalk.
10. Pollen	Oval, monosulcate, exine smooth, non-winged.	Boat-shaped, monosulcate, exine reticulate, expanded in two lateral sides to form two wings.	Oval, monosulcate, exine granulate, two lateral ears are seen due to incomplete nature of exine, non-winged.	Ovoid, inaperturate, exine spinulate, non-winged.
11. Female fructification	Megasporophylls are not aggregated to form female cone.	Compound comprising of bract-scale complexes, borne in the axils of scale leaves of the long shoots.	Each cone consists of a stalk which bifurcates into two branches. Each branch terminates with a single sessile ovule.	Compact, slender axis-like structure called inflorescence, borne solitary or axillary or in cluster at the shoot apex.
12. Megasporophyll	Large, brown in colour, covered with hairs, divided into upper dissected laminar part and lower stalk bearing 1–5 pairs of large ovules on both sides of the stalk.	A reduced secondary shoot, consists of a hard, woody and dry ovuliferous scale bearing two ovules. The ovuliferous scale is subtended by a small bract scale.	A single ovule is borne on the top of each of the bifurcated branches. Each ovule is surrounded by a collar which represents the sporophyll as per some suggestions.	Female flower is stalked and consists of an ovule surrounded by a third envelop, called perianth.
13. Ovule	Orthotropous, large, short-stalked, integument three-layered, nucellus fused with integument except at the top of the nucellus. The ovule is supplied with 3 vascular strands.	Anatropous, small, short-stalked, integument three-layered, nucellus free from integument except at the basal region. The ovule is not supplied with vascular trace.	Orthotropous, large, long-stalked, integument three-layered, nucellus free from integument except at the base. Two vascular strands supplying the base of the ovule.	Orthotropous, large, short-stalked, protected by 3 envelops : inner envelop (= inner integument), middle envelop (= outer integument), outer envelop (= perianth). Inner integument

Contd.

Contd.

Characteristics	*Cycas*	*Pinus*	*Ginkgo*	*Gnetum*
				elongated, nucellus is free from integument, except at the base. There are separate vascular strands for three envelops.
14. Male gametophyte	Consists of a prothallial cell (persistent), a tube cell, a stalk cell and two spermatozoids.	Consists of two prothallial cells (both degenerate), a tube cell, a stalk cell and two male cells.	Consists of two prothallial cells (one persistent), a tube cell, a stalk cell and two spermatozoids.	Consists of a prothallial nucleus, a tube nucleus, two male cells (other view : a tube nucleus, a stalk cell, two male cells).
15. Sperms	Large, spirally coiled, multiciliated; motile.	Small, non-motile cells.	Like *Cycas*, but cilia are restricted to the apical part.	Like *Pinus*.
16. Female gametophyte	Monosporic.	Like *Cycas*.	Like *Cycas*.	Tetrasporic.
	Female gametophyte develops from a functional megaspore through free nuclear divisions, followed by cell wall formation. The cellular gametophyte (endosperms) differentiates into two regions : the upper region of larger cells near the chalaza and lower region of smaller cells towards the micropylar end. Several archegonia (2–8) develop from the cells of gametophytic tissue at micropylar end.	Like *Cycas*, the gametophytic tissue (endosperm) is composed of one type of cell. Two to three archegonia are formed.	A delicate membrane is formed on the outer surface of the gametophyte. The endosperm tissue is green which contains chloroplasts. Two or rarely three archegonia are formed.	Like *Cycas*, but the gametophyte becomes wider above and gradually becomes elongated to give an appearance of an inverted flask. Archegonia are not formed, a few large free nuclei at the micropylar end function as eggs. The lower part of gametophyte becomes cellular, while the upper part remains free nuclear during fertilization.
17. Archegonia	It consists of two neck cells, a ventral nucleus and an egg.	It consists of neck of eight cells in two tiers, a ventral canal cell and an egg.	It consists of a two-celled neck, a ventral canal cell and an egg.	Archegonia are absent.

Contd.

Contd.

Characteristics	*Cycas*	*Pinus*	*Ginkgo*	*Gnetum*
18. Pollination	Pollen grains are shed at 3-celled stage.	Pollen are shed at 4-celled stage.	Like Pinus.	Like *Cycas*.
19. Embryo	Only one embryo is formed, there is no polyembryony. Mature embryo in seed consists of a radicle, a plumule and two cotyledons.	Polyembryony occurs. Mature embryo consists of a radicle, the hypocotyl, a plumule and 3-many cotyledons.	Like *Cycas*, often contains 3-cotyledons.	Only one embryo is formed, there is often polyembryony. Mature embryo consists of a root clothed by a root cap, two cotyledons, a stem tip and a large feeder.
20. Seed	Seeds are red, orange or dark brown in colour. The nucellus is mostly consumed by the growing embryo. The embryo is protected by three-layered integument. Seeds are non-winged.	Seeds are dark brown in colour. The unused nucellus, endosperms and embryo are protected by integument. The outer fleshy layer of integument disappears, so the middle stony layer develops into seed coat. A thin layer of the ventral surface of ovuliferous scale attached to seed becomes winged. Thus, the seeds are winged.	Seeds are dark brown in colour. The unused nucellus, endosperms and embryo are protected by three-layered integument. Seeds are non-winged.	Seeds are green to red in colour. The unused nucellus, endosperms and embryo are protected by three-layered envelops. Seeds are non-winged.
	Germination is hypogeal.	Germination is epigeal.	Like *Cycas*.	Like *Pinus*.

FOSSILS AND FOSSILISATION

What are fossils ?

The word fossil is derived from the Latin verb *'fodere'* which means to dig up. Thus, a fossil refers to anything that is excavated from the earth and not fashioned by man. Actually, the fossil refers to organic remains taken from the earth. So in common sense, fossils are the traces of the past life forms in the womb of the earth.

According to Arnold, the fossils are "the relic of some former living things, plants or animals, embedded in or dug out of the superficial deposits in past geological periods." Schopf (1975) defines fossil as any specimen that demonstrates physical evidence of occurrence of ancient life.

Types of fossil

The fossils may be categorised in the following types :

1. **Chemical fossils** : These are the remnants of organic compounds preserved in sediments or in parts of fossilised structures without undergoing any or minimal change. These include amino acids, hydrocarbons, fatty acids, lipids, carbohydrates and the derivatives of other organic compounds. The chemical composition of Pre-Cambrian rocks is an important criterion to establish the biogenicity of putative unicellular or multicellular organisms present in Pre-Cambrian rocks. The existence of insoluble **kerogen** is used as proof of biogenicity. Similarly, the occurrence of **pristane** and **phytane**, degradable products of chlorophyll molecule, may be used as proof of photosynthesis.

2. **Trace fossils or Ichnofossils** : Sometimes, indications of prior existence of organisms in the sediments of earth may be regarded as trace fossils or Ichnofossils. These include animal tracks or foot print preserved in rocks, burrows of invertebrates, coprolites (fossil excretes), gastroliths (polished stones in the abdomen of dinosaurs), gnawed bones, etc.

3. **Microfossils** : Microscopic organisms like bacteria, spores and pollen grains, fungal and algal spores, foraminifera, diatoms, epidermal and wood fragments of plants etc. preserved in the sedimentary deposits are referred to as microfossils. Microfossils are visible only after maceration of sediments.

4. **Megafossils** : Large parts of plants like leaf, stem, root, flower, seed, etc. and animal remains as whole organism or in parts, preserved in the sedimentary deposits are called megafossils. These are visible to naked eyes and are the better source of morphological as well as anatomical studies. The megafossils may be categorised into the following five types on the basis of the nature of fossilisation :

(a) **Compressions** : These are plant parts, compressed by the vertical pressure of the sediments (Fig. 1.83). The plant fragments like leaves, stem, seeds get flattened and are retained as thin carbonaceous films with outline of external features. Generally, internal structure is not preserved, however, in rare instances cuticles, stomate, etc. are retained.

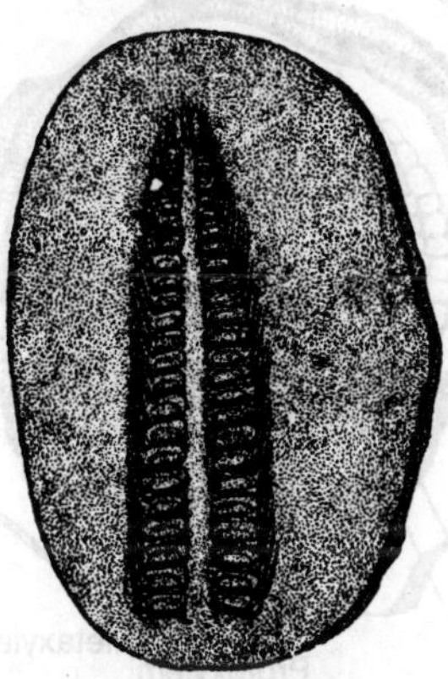

Fig. 1.83 : Compression of *Lepidostrobus* cone

(b) **Impressions** : Impression may be defined as the negative of a compression. These are just impression of plant parts which do not contain organic matters as in compression (Fig. 1.84). The sediments containing the flattened plant parts become hardened and when split open shows the negative imprint i.e. impression.

(c) **Petrifactions** : These are the best, but rarest types of fossil which preserve the external form as well as the internal structures. The cellular details are preserved due to the infiltration of minerals like SiO_2, $CaCO_3$, $MgCO_3$, FeS, etc.

Fig. 1.84 : Impression of *Neuropteris* leaf

into the tissue. The petrified fossils can be cut into small pieces and series of section can be made for anatomical studies (Fig. 1.85).

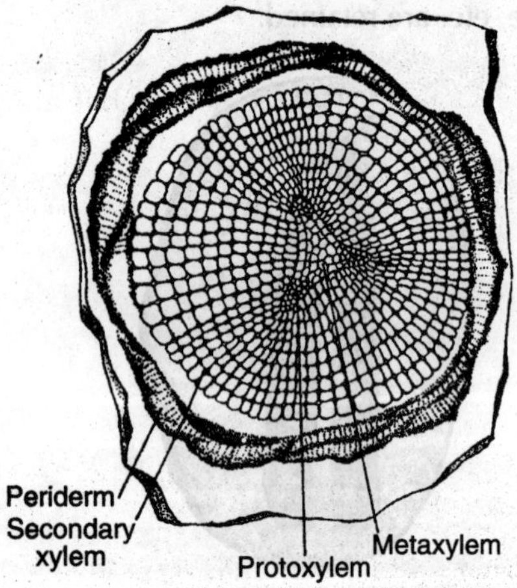

Fig. 1.85 : T.S. of a petrified *Sphenophyllum* stem

(d) **Casts or Incrustations and Molds** : In these types, the deposition of iron and carbonate minerals occurs in the form of a hard cast around the plant parts. The internal structure is degraded to form a cavity which is completely filled up by the surrounding sediments. Thus, the external preserved surface of the plant part is called a mold (Fig. 1.86), while the replaced internal structure of the plant part is called a cast. Stigmarian root system is an example of mold, while the pith cast of a *Calamites* stem is a common example of incrustation (Fig. 1.87).

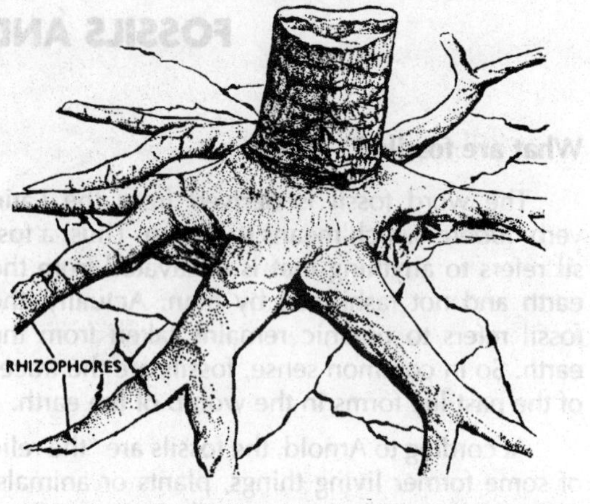

Fig. 1.86 : Mold of *Stigmaria*

In these types, only external forms are preserved, while internal cellular details are not preserved.

Fig. 1.87 : Pith cast of a *Calamites* stem

5. **Sub-fossils** : A certain time period is required for the formation of a fossil. If the plant or animal parts are excavated before they completed their fossilisation process, they are called sub-fossils. Coal is a compressed fossil, while peat, an early stage of coalification is referred to as sub-fossil.

6. **Pseudofossils** : Sometimes inorganic rocks that appear to be fossils are actually mineral deposition. These fake structures are mistaken for plant or animal remains. These are known as pseudofossils.

7. **Derived fossils** : The fossilised organisms that held in a stratum younger or older than the fossil themselves are called derived fossils. These are results of tectonic movement of earth or other geological upheaval.

8. **Coal balls** : The petrified spherical balls containing plant parts are commonly termed coal balls (Fig. 1.88). These spherical balls are formed as a result of infiltration of plant debris in swamps by carbonates of calcium or magnesium, thus restric-ting the conversion of the debris into coal. Coal balls occur in localised regions and they range in few centimeters to several meters and weigh from a few to several pounds. Coal balls are specifically significant in palaeobotanical studies.

9. **Paper coal** : It consists of thin dead leaves, dispersed in organic matrix. The inner tissues of leaves are destroyed, thus the paper coal consists of layer after layer of cuticles, often with decomposed stems. The carbonaceous lime stone horizon at Tovarkovo, in Toula in Russia, is an example of paper coal.

10. **Amber** : The fossilised resin of extinct coniferous trees, *Pinus succinifera* in particular, is called Amber. The resinous exude flowed due to injury caused by boring insects which eventually accumulated on the forest floor and got hardened forming amber. Insects and flowers are often found preserved in amber. Amber has high economic value and used in jewelllery.

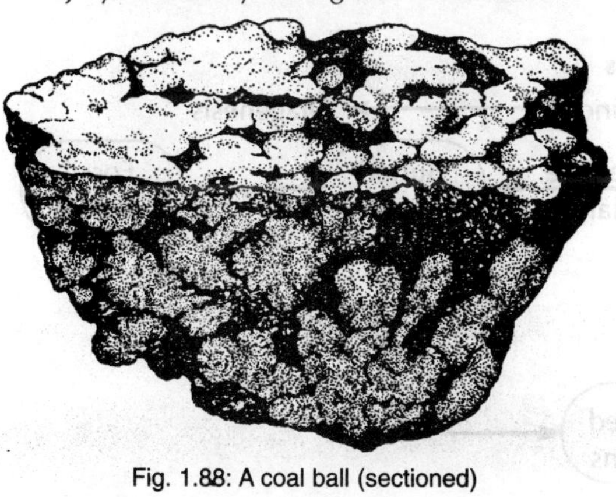

Fig. 1.88: A coal ball (sectioned)

11. **Index fossils** : The organisms that help in dating other fossils found in the same sedimentary layer are called index fossils. Such fossils are found widely distributed geographically, and limited in time span having very distinct characteristic features. *Monograptus* is an index fossil of Lower Devonian, while *Myrepollenites* is a marker of Eocene. Foraminifera, pollen grains, spores etc. are also used as index fossils.

Nomenclature of Fossils

The whole plant is not preserved, but only detached plant parts like stem, root, cone, leaf, etc. are preserved as fossils. These detached plant parts are being discovered in different times by different authors. Thus, these detached plant parts or organs are given a bionomial (generic and specific name) by the same set of rules under the International Rules of Botanical Nomenclature which have been framed for living plants.

The first valid description of *Lepidodendron* came into existence from the publication of **Sternberg** in 1820. Thus, this date has been considered as the starting point of palaeobotanical nomenclature like that of Linnaeus's *'Species Plantarum'* in 1753 for the nomenclature of modern vascular plants.

Each detached organs or fragments is given a different name. Each of these names acquires the status of a genus. The generic name in fossils is applicable for only a plant part like root, stem, leaf, cone or other organ, without indicating to what plant it belongs. Thus, the genus is termed **form genus** or **artificial genus** in contrast to **natural genus** for living plants. A form genus cannot reliably be assigned to a single family, however, it may be assigned to an order or other higher taxonomic rank. For example, *Stigmaria* is a form genus of the order Lepidodendrales which cannot be assigned to any one of the three families : Lepidodendraceae, Sigillariaceae or Bothrodendraceae.

When the relationships among different organs like stem, root, leaf, and reproductive structures are established and can be assigned to the same family, then the genera can be called **organ genera**. For example, stem genus *Bucklandia,* leaf genus *Ptilophyllum,* male fructification *Weltrichia* and female fructification *Williamsonia* are genetically related and assigned to the same family Williamsoniaceae. Thus, all are considered to be organ genera. However, there is no provision in the International Rules of Botanical Nomenclature for the use of organ genera.

During reconstruction, the palaeobotanists should select the earliest (after 1820) validly published generic name applied to any one of its parts as per Rule of Priority. He or she will use any one of the form genera as the generic name for the whole organism. Say, for example, the validly published female fructification, *Williamsonia* has been used for naming the whole plant.

Rules for naming form genera

A particular suffix is used for naming a form genus which signifies the organ it belongs. The suffixes applied to different plant parts are :

Suffix	Applied to organ	Examples
Dendron	Stem	*Lepidodendron, Lyginodendron.*
Xylon	Woody part	*Dadoxylon, Cordaixylon, Mesoxylon.*
Phyllum	Leaf	*Ptilophyllum, Nipaniophyllum, Brachyphyllum*
Pteris	Fern-like stem or frond	*Sphenopteris, Lyginopteris, Etapteris, Archaeopteris*
Spermum	Seed	*Corystospermum, Mitrospermum*
Carpon	Seed or seed like	*Lepidocarpon, Mazocarpon, Calamocarpon*
Carpus	seed	*Trigonocarpus, Cardiocarpus*
Stoma	seed	*Lagenostoma, Stamnostoma*
Theca	Microsporangia	*Codonotheca, Aulacotheca, Crossotheca*
Strobus	Cone	*Lepidostrobus, Androstrobus*

Fossilisation

Fossilisation is a process of preservation of an organism or its parts in the form of fossils. In other way, fossilisation can be defined as what happens to an organism after its death and until its discovery as a fossil. It includes decomposition, postmortem transport, burial, compaction and other physical, chemical or biological activities which affect the remains of the organism. The process that occurs between the death of an organism and its subsequent burial in the sediments are termed **biostratinomy**. This generally includes decomposition, abrasion, bioerosion, etc. On the other hand, **diagenesis** is the physical and/or chemical effects after burial. This includes dissolution, replacement or recrystallisation, etc.

Schematic representation of fossilisation process

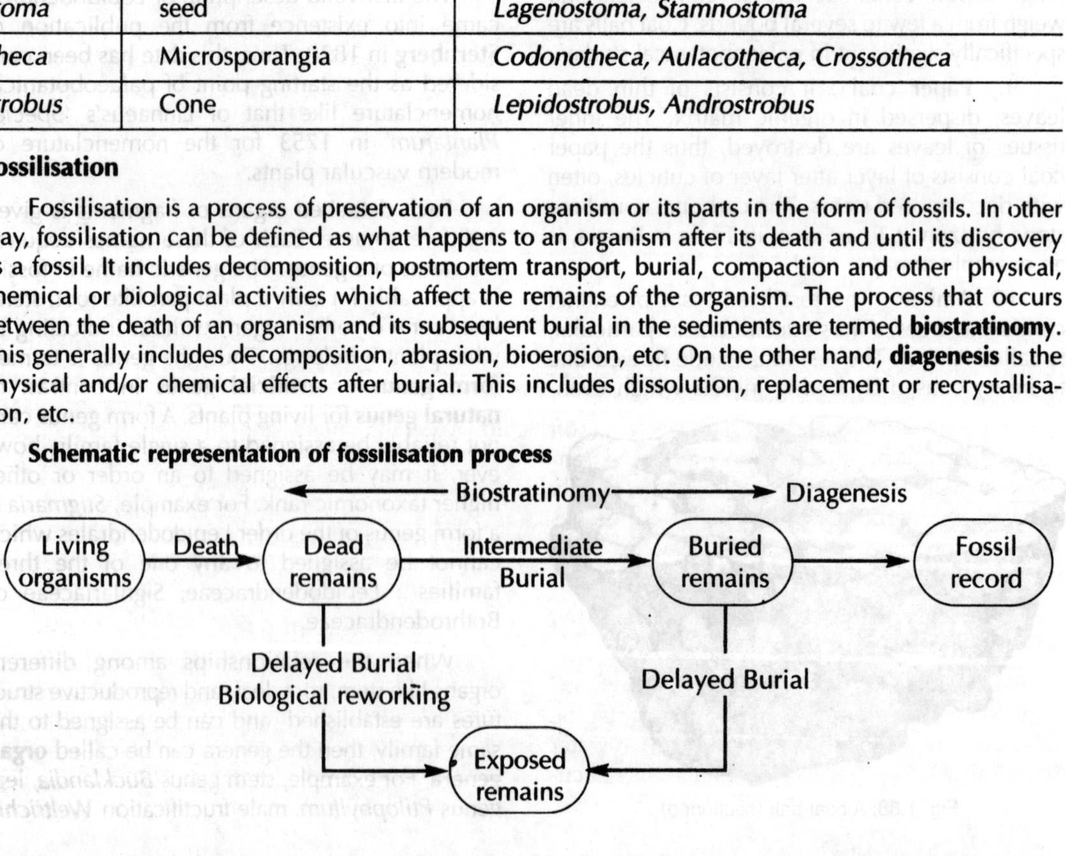

Conditions for fossilisation

It is a rare instance that an organism is preserved intact. Most of the known fossils are imperfect, where only external features are preserved. The perfect permineralised fossils showing cellular details are very rare. The conditions for perfect fossilisation process can be categorised under the following heads :

1. Sites of fossilisation
2. Nature of the tissue undergoing fossilisation
3. Events that occur before, during and after fossilisation.

1. **Sites of fossilisation** : The preservation of plants depends on removing the organic materials from the zone of aerobic decomposition. This can most easily be accomplished by burying the plants in enclosed water reservoir in which sediments accumulate on the surface at a fast rate. Therefore, swamps, deltas, lakes, lowland flood plains are good sites for fossilisation.

2. **Nature of the tissue undergoing fossilisation** : The extent of preservation is directly proportional to the mechanical resistance of the cells, tissues and organs undergoing fossilisation. Thin-walled soft organs like flowers, juicy fruits, thin delicate leaves are unsuitable for preservation, while thick walled hard organs like stem, seeds etc. are preserved as compressions, casts or petrifactions for their resistant nature.

Plant cell walls primarily comprising of cellulose together with hemicellulose, lignin and suberin are far more resistant to decomposition. Cutinised epidermal layers of leaf, sporopolleninous exine of spores and pollen grains, siliceous wall of diatoms, hard skeletal parts of calcareous algae, suberised bark layers are better preserved for their high degree of resistance to destruction. In a similar way, soft tissues like cortex, phloem and xylem parenchyma, sieve tubes, pith, etc., are rapidly decayed by microorganisms, while hard tissues such as tracheary elements, fibres, sclerenchyma, etc. are well preserved.

3. **Events that occur before, during and after fossilisation** :

(a) **Events that occur before fossilisation** : Plant parts undergoing fossilisation must be deposited in an enclosed or protected water reservior in which fine-grained sediments accumulate on the surface with sufficient rapidity to cause quick burial.

In deep water body, the oxygen content is low, thus the intensity of microbial disintegration is also very low. Moreover, environment rich in humic acids and clay minerals can retard decay by blocking the chemical sites onto which decomposers fasten their degrading enzymes.

The plant parts must be preserved *in situ* (**autochthonas deposition**) which provides an environment with minimum destruction of plant parts. On the other hand, transportation of plant parts (**allochthonas deposition**) for long distance results in breakage and decay.

(b) **Events that occur during fossilisation** : A plant part is subjected to many factors during the course of fossilisation which ultimately determines the characteristic of a fossil.

The depth of water in which the plant parts sink is important, because it can avoid two disintegrative forces such as (a) **decay** and **hydrolysis,** (b) **mechanical action** of water, wind, scouring (rubbing action) sand, and rolling stones. The activities of bacteria and saprophytic fungi are greatly reduced, because in deep water the concentration of decomposers are very poor. Moreover, it affords good protection from wave action as intensive wave action or rolling boulders can reduce even the most resistant plant tissues to pulp.

The sediment that accumulates on the surface of plant part is usually of fine textures, such as clay, silt or fine sand favouring good preservation in greater details.

Most decomposers are aerobic and require oxygen for their metabolism. In deep water, oxygen content is low and plant parts are generally preserved in such anaerobic sediments. Thus, low oxygen content, and relatively high concentration of toxic substances also retard decay and hydrolysis.

The cells of the plant parts must be quickly filled with mineral deposits to avoid pilling up of the overlaying sediments.

(c) **Events that occur after fossilisation** : Well-preserved fossils may suffer severe damages during destructive processes of earth, such as seismic activity, earthquake, volcanic eruption, etc., at the sites of their origin. Thus, undisturbed sediments provide a proper fossil record.

Modes of preservation

J. M. Schopf (1975) has recognised four distinctly different modes of fosssil preservaton. These are : (a) Cellular permineralisation, (b) Coalified compression, (c) Authigenic preservation, and (d) Duripartic preservation.

(a) **Cellular permineralisation** : This includes all of the specimens that have previously been called petrifaction fossils. This involves infiltration followed by intracellular and interstitial precipitation of soluble minerals like silicates, carbonates, iron compounds through cell walls. The buried plant part undergoes partial disintegration to release free carbons which interact with the sulphides present in water and lead to the formation of carbonates of Ca, Mg, Fe. Thus, the soluble minerals like carbonates, silicates etc. are deposited within cell walls through infiltration and precipitation. As mineral deposition continues within plant tissues, water is expelled as a result of compaction of the sediments. This causes the sediments and the buried plant part to solidify, thus completing cellular permineralisation. This is analogous to the embedding of plant parts in paraffin blocks for microtome sectioning. Some of the well-known deposits that contain permineralised fossils are Rhynie chert, Gunflint chert, Bitter Spring Formation, petrified forest of Arizona, Deccan Intertrappean bed in India, etc.

Since a cellular permineralisation reveals the cellular details of the plant, such as nature of cortical cells, vascular bundle, pit connections, secondary wall thickenings, nature of ray cells etc. — it has served as a tool to unravel the taxonomic dispute.

(b) **Coalified compression** : Unmineralised parts are deposited in sediment, followed by softening of cell walls and collapse of internal cell spaces. This leads to loss of gas moisture and soluble materials. As a result of pressure exerted by accumulated sediments and water, the residues are altered and consolidated to form a black coaly deposit. The distortion or compression is directional and only in the vertical plane. The splitting of rocks commonly yields the coalified compression on one face and its counterpart i.e. impression on the opposite face. On weathering, coaly part is lost or exfoliated, thus an impression may be revealed on the rock.

The extent of compression varies with the degree of hard nature of plant parts. Leaves are commonly retained in their natural form, but cylindrical or rounded organs such as stem, root and seed become dorsiventrally flattened. Most of the fossils recovered as coalified compression are remains of leaves, occasionally stems, roots, flowers, cones and seeds.

Since a coalified compression reveals leaf form, venation pattern, epidermal characteristics, etc., it has served as a tool in establishing systematic position and afinity of extinct plants.

(c) **Authigenic preservation or cementation** : It involves very early cementation in soft sediments by iron and carbonate compounds. The plant material develops an electric charge as soon as it starts to decay. Thus, it attracts oppositely charged ionised particles of sediments. As a result sediments comprising of iron pyrite sphalerite, chalerite, agate, opal, carbonate along with mud and sand accumulate around the plant part. Later, the sediments become cemented and the plant part is entombed in the sediment. The internal structure is degraded to form a cavity which is completely filled up by the surrounding sediments. After lithification, the external surface of the plant part is preserved as mold and the replaced internal structure of the plant part is called a cast. In this process, the internal cellular details are not preserved. *Stigmaria* stump of *Lepidodendron* stem is a mold, while the pith cast of *Calamites* stem is an example of cast.

(d) **Duripartic (hard part) preservation** : Certain hard parts of both plants and animals are resistant to decay and oxidation and also resistant to physical distortion. Preservation of such hard parts without being changed by chemical or physical factors is referred to as duripartic preservation. Such sites of preservation are abundant in the Phanerozoic epicontinental seas. Skeletal parts of lime-precipitating algae (Dasycladaceae, Characeae), coccoliths, diatom frustules (diatomite) and radiolarian skeletons (radiolarian ooze) are some examples of duripartic preservation.

Techniques for fossil study

There are several techniques for study of different types of fossils. Some of the important techniques are :

1. **Ground Thin Section** : This technique is suitable for the study of petrified fossils which preserve cellular details.

To begin with, the specimen is cut to a convenient size with a lapidiary's saw or similar instrument and is smoothened on one surface with abrasive (300 or 400 carborundum). The smooth surface is attached to a glass slide with melted resin. Then, the specimen is cut as close to the glass as possible, thus a moderately thin section is obtained. The thin section fixed to the glass slide is further ground on a revolving lap with 100 carborundum till sufficiently translucent, so that the section can be viewed under the microscope. Finally, the slice is mounted with a cover glass using suitable mounting media.

Advantage : A thin section of petrified wood can be made that preserves the cellular structure in an unchanged condition.

Disadvantages : (a) The technique needs extensive labour and is time-consuming. (b) A small number of sections can be made from a given specimen, thus a lot of material is wasted.

2. **Film or Peel Technique** : This is a conventional and most suitable technique for study of well-preserved petrifactions with considerable organic materials.

The specimen to be studied is ground and smoothened and is etched with 5% hydrochloric acid (if the material has calcium carbonate) and 10% hydrofluoric acid (if the material has silica) for 5 and 10 mns respectively. The etched surface is gently washed in running water for the removal of acid and is air-dried and then covered with a solution or a thin film of nitrocellulose. The specimen is air-dried for at least 6 hours. Then the film is carefully peeled by loosening one edge and is permanently mounted on a slide with a coverglass using suitable mounting media.

An improvement in the peel technique has been made that makes possible rapid preparation of serial sections.

Advantages

(a) A series of sections can be made from a single specimen.

(b) It is less expensive and quicker to prepare.

(c) The sections (peels) obtained are translucent, thinner and durable.

3. **Transfer Technique** : This technique is most suitable for study of coalified compression which reveals additional details of venation, epidermal pattern and hairs.

The face of the specimen adjoining the rock surface is cleaned either mechanically or by washing in an acid for removal of rock particles. The prepared surface of the rock is coated with a solution of nitrocellulose or with a cellulose acetate film. When the film is dried, it is loosened from the rock surface. Sometimes coalified materials are adhered to the film. Occasionally the film is treated with strong oxidising agent to make the film more transparent. Finally, the film is dried and permanently mounted on the slide with a coverglass using a suitable mounting medium.

Advantage : This technique is very useful for study of coalified compression. It helps to learn about leaf form, venation pattern, stomatal and epidermal characteristics which are important features used in establishing systematics and phylogeny of extinct plants.

4. **Maceration technique** : This technique is most suitable for study of peat, lignite and coal. It is very useful for pollen and spore analysis. The methods of peat, lignite and coal analysis are discussed schematically :

(a) **Peat analysis (Acetolysis technique)**

Sample meterial
↓
Cleaning, disaggregation and weighing
↓
Treatment with 10% KOH for removal of humic substances
↓
Wash and centrifuge
↓
The residue is treated with 40% Hydrofluoric acid for digestion of silica
↓
Wash and centrifuge
↓
The residue is treated with Acetolysis mixture (Acetic anhydride : Conc. H_2SO_4 = 9:1) for removal of unwanted materials and also for colouration of spore/pollen wall
↓
Wash and centrifuge
↓
Samples containing pollen/spores are kept in 50% glycerine
↓
A drop of sample is mounted on a slide and observe under microscope

(b) **Lignite and coal analysis**

Sample meterial
↓
Cleaning, disaggregation and weighing
↓
2-3 gm of coal/lignite pulverised and decalcified
↓
Treatment with 10% HCl
↓
Oxidation with Schulze's solution
↓
Treatment with 5% HCl/dilute NH_4OH/20% K_2CO_3/acetone for removal of humic acid
↓
Stain with Safranin-O
↓
Sample containing pollen/spores are stored in 50% glycerine
↓
A drop of sample is mounted on a slide and observe under microscope

Advantages

1. A small portion of sediment is enough to get considerable amount of pollen and spores.
2. Due to the presence of resistant chemical sporopollenin in exine, pollen and spores are considerably preserved in deposits like peat/ coal/lignite, etc.
3. The maceration process is very simple and reliable.

Significance of studying fossil plants

Palaeobotany is a branch of plant science which deals with the study of plant fossils. The scope of the science of palaeobotany includes both basic and applied aspects.

The various applications of palaeobotany are :

Basic aspect : Such studies help us to understand the past vegetation, climate and palaeoecology including the history of plant evolution, biostratigraphy, etc. Some of the important basic aspects of such application are :

(a) **To decipher palaeovegetation and palaeoclimate** : Like all living organisms, plants ultimately face death and decay. Sometimes, however, they escape from final destruction and survive as fossils. There are remains which allow palaeobotanists to place together the scattered jigsaw of plant life forms. Thus, plant fossils serve as tool for past vegetational studies.

From remote geological past a correlation exists between fossil plants and their surrounding environments which serve as an index for palaeoclimate. The fossil plants are used for interpretation of the past climate, because the present-day living plants are well understood for their integral association with the idea "present is the key to the past". Palaeoclimate factors, therefore, are assessed on the basis of modern-day plants, showing their morphological and anatomical characteristics commensurate with climatic condition including adaptive factors. Such data reveal the character of fossil plants.

(b) **Provide evidence for origin and evolution of plants** : The study of fossil plants has an importance in understanding the origin and evolution of different plant groups through ages.

There are several exampes in support of the origin of land plants. A most dramatic event took place in the history of plant life i.e. the evoluton of land plants during Silurian period. To acquire terrestrial habit leaving their aquatic habitat the plants needed to be self-supportive and they had to be able to withstand the drying effect of the atmosphere leading to a series of adaptations, such as development of cuticle in their outer surface to check desiccation, formation of roots as anchoring and absorbtive organ, stomata for gaseous exchange, vascular tissue for conduction. Thus, during Silurian-Devonian periods more and increasingly complex fossil plants represented by Rhyniopsida appeared with such adaptive features. Gymnosperms became more successful than the pteridophytes by having selective advantage of seed formation. This group of plants first appeared in the Devonian period and reached a climax in the Mesozoic Era which constituted the world's dominant vegetation. Still later, at the end of Mesozoic Era (early Cretaceous) the angiosperms got evolved and became dominant replacing most of the pteridophytes and gymnosperms.

(c) **To ascertain palaeophytogeography** : The important aspect of studying palaeobotany is to ascertain the palaeophytogeographic condition. The most astonishing example of such aspect is the concept of Gondwanaland, which consisted of a single land mass formed by the union of the faraway continents of the Southern Hemisphere, such as South America, South Africa, Madagascar, Australia, New Zealand, Antarctica and India. The fossil evidence shows a clear floral continuity ranging in age from the Upper Carboniferous to the beginning of the Cretaceous.

(d) **To make biostratigraphic correlation** : The study of fossil plants has an applied significance in understanding the biostratigraphical sequence which provides evidence to trace the plant evolution through ages. In this context there are two aspects to be considered :

(i) Correlation of the data showing quantitative and qualitative value of retrieved fossil data, and

(ii) Use of fossils for comparison to get the empirical value in terms of appearance, duration of dominance and then gradual disappearance through migration or extermination due to climatic and other factors establishing a biostratigraphic scale.

For example, in the coal bearing strata of the Middle Carboniferous of West Europe, seven successive vegetational sequences were established using plant megafossils. Thus, an analysis of plant fossils of a given stratigraphic zone, provides the needed data to interpret its floristic composition and evolutionary sequence.

(e) **Calculation of age of the rocks** : Radioactive elements are generally used for dating the rocks. But sometimes, index fossils are used to date a rock of unknown origin. Any index fossil the rock contains must be carefully studied with precise information about its systematics and age. Using of such index fossils the age of an unknown rock can be determined.

Applied aspect

To provide useful information in the exploration of fossil fuel like coal and oil

The plant inhabitants of Palaeozoic-Mesozoic swamps served as a source of coal and formed coal seams. Accumulation of plant materials with a variety of minerals coupled with mud, silt and other organic materials constitute a coal bed. Many plant parts get beautifully preserved as fossils in the stratified sedimentary layers which are closely adjacent to coal layers. A stratified scale based on fossils can be made to establish the age of coal deposits and their position in the succession of rocks. Palaeobotanical studies have served as a tool to ascertain age of coal layers, their lateral extent and quality of coal deposits. Such information are required to specify the suitability of a particular coal for energy production. In India, the palaeobotanical study has helped to demarcate the nature and quality of Raniganj coal (Permian Age) from that of Rajmahal coal (Jurassic Age).

The assemblage of fossil pollen grains and spores has contributed significantly in the field of oil exploration. The problem of oil exploration begins with search of oil reservoirs. Exploration of oil is done by :

(i) **Determining the Thermal Alteration Index** : The sporopollenin present in the walls of pollen and spores undergo post depositional thermal changes in course of the geological ages. These thermal changes brought about carbonisation resulting in changes in exine colour of fossil pollen and spores in transmitted light. The basic idea involved is the usage of the variation in the exine colour as an indicator of the degree of carbonisation in the rocks to predict their changes of bearing reservoir. **Pearson's colour chart** directly relates exine colour to a numerical index called **Thermal Alteration Index** (TAI) which is a measure of the degree of carbonisation. TAI having a range of 2^{+}–3^{+} and exinite flourescence colour white-yellow, dark yellow-brown indicates the possibility of exinite containing rock to possess liquid petroleum and even natural gas.

(ii) **Palynostratigraphy** : To avoid unnecessary and costly drilling the determination of oil zone is made by comparing the biostratigraphic data of one to those of the others.

(iii) **Defining of ancient shorelines** : The sediments parallel to sea shore are rich in oil. The density of pollen and spores decreases in the seaward direction. Sedimentary environment with pollen assemblages are limited to near shore marine or lacustrine waters. Thus by the study of microfossils along with marine microfossils if presents, one can determine the distance and direction of ancient shore lines, possibly bearing oil deposits.

GEOLOGICAL TIME SCALE AND DIVERSITIES OF LIFE FORMS IN DIFFERENT GEOLOGICAL AGES

A bewildering and varied display of plants and animals of all shapes and sizes is encountered on the surface of the earth. Arising out of a very simple form of life an intricate balance has been maintained for over 4,000 million years [one million year (m.y.) = 10^6 years], as slowly more complex plants and animals have evolved. In a normal course of life, plants and animals ultimately die. In some cases, however, they escape from final destruction and leave the sign of their existence as fossils in rocks. Such remains when carefully brought together may indicate the course of evolution of life.

The arrangement of geological time scale has been given in Table 1.2. The main divisions, or **Eras** (viz. Pre-Cambrian, Palaeozoic, Mesozoic and Cenozoic), embrace a number of subdivisions or **Periods** and these periods are further refined into **Epochs**. The sample of past life especially of Pre-Cambrian Age is uneven and incomplete. Innumerable organisms made up of soft and delicate tissues, like algae, protozoan, jelly fishes, flatworms, etc., are not favourable for preservation. In contrast, the hard woody plant parts, the shells of molluscs, the bones of vertebrate are relatively common as fossils. Despite the above facts, the available fossil assemblage contains an extraordinary amount of information.

Pre-Cambrian life forms are not well-known because :

- Rocks are poorly exposed.
- Many rocks have been eroded or metamorphosed.
- Most are deeply buried beneath younger rocks.
- Fossils seldom found.

Fossil records of different geological strata

A. Pre-Cambrian

The Pre-Cambrian includes approximately 90% of the geologic time which covers approximately 4,200 million years (m.y.) of the earth history (Table 1.2). The Pre-Cambrian world was almost certainly as diverse and complex a place as today's world. Most information are from **Cratons** i.e. the large portions of continents which have not been deformed since Pre-Cambrian. The exposed cratons are called **Pre-Cambrian shields** viz. Canadian shield. The Pre-Cambrian divides into three Epochs or Eons, viz. Hadean, Archaean or Archaeozoic and Proterozoic.

(a) **Hadean (4,,700–3,800 m.y. old)** : There is no fossil record in this Epoch. A major times of this epoch was involved on the origin of the earth and solar system, ultimately led to the differentiation of the earth to form crust, mantle and core. The events that took place in the Hadean are : origin of primary and secondary atmosphere; condensation of water vapour, formation of rain; run off leads to lakes, rivers and oceans; origin of continental crust. The oldest dated Hadean rock is 3,960 million years old, located in Canada (e.g. **Canadian shield**).

(b) **Archaen or Archaeozoic (3,800–2,500 m.y. old)** : The earliest known life forms (Prokaryotes) from Archaean rocks are known. They are microscopic rod-shaped bacteria (*Eobacterium isolatum*) and spherical coccoid bacteria (*Archaeospheroides barbartonensis*) recovered from **Onverwacht series** (Fig. 1.89A) and **Fig Tree series** of South Africa (3,700–3,200 m.y. old). A plenty of stromatolites containing the colony of chroococcalean cyanobacteria or blue-green algae has been recovered from the carbonate sediments (3,500–3,400 m.y. old) of **Towers Formation** and **Apex Basalt** of the Warrawoona group in Archaean rocks. The earliest bacterial cells had to form and exist in anoxic (oxygen-free environment) conditions probably chemosynthetic in nature, produced CO_2, or H_2S. They were obviously heterotrophs which consumed simple organic compounds. Eventually, some amount of oxygen was released by the cyanobacteria which might have been used up by minerals through oxidation.

Evidence for a lack of free oxygen in the earth's early atmosphere

(i) Urananite and pyrite are readily oxidised today, but are found unoxidised in early Precambrian sediments.

(ii) There are no early Precambrian iron oxides (no red beds).

(iii) Banded iron formations appear in stratigraphic record in mid or late Precambrian (2,500–1,800 m.y.).

(iv) Evidence from mid or late Precambrian soils shows that oxygen was only about 2% of modern levels.

(v) Chemical building biocks of life (amino acids, DNA) could not have formed in presence of oxygen.

(vi) The simplest living organisms have an anaerobic metabolism. They are killed by oxygen which include some bacteria (such as botulism), some or all Archaebacteria which inhabit unusual conditions.

(c) **Proterozoic (2,500–570 m.y. old)** : The first autotrophic chlorophyllous alga (possibly blue green alga) was reported from **Gunflint rock** of Canada which dated back to 2,000 m.y. The best-known alga was *Animikiea* (Fig. 1.89B) (similar to modern blue-green alga, *Oscillatoria*) which was able to perform photosynthesis to liberate oxygen, eventually giving rise to ozone. It is believed that the stratospheric ozonosphere was formed through this process. The existence of free oxygen was established by the occurrence of chemosynthetic bacterium known as *Gunflintia* (Fig. 1.89C) in the iron banded Gunflint rock. The putative eubacterial spores, namely *Huroniospora* (Fig. 1.89D) and *Eostrion* (Fig. 1.89E) were reported from Gunflint rock. The ozone layer screened the lethal UV rays forming a situation conducive to the growth of other autotrophic primitive (Cyanobacteria) as well as advanced (Green algae) algae, etc. in free-floating condition. The more diversified prokaryotes and eukaryotes were observed from several sediments like **Duck Creek Dolomite** of California (1,300 m.y.), **Bitter Spring Formation** of Australia (900-800 m.y.) and **Svanbergfiellet shale** of Spitsbergen island (800-700 m.y.). Several cyanobacteria were described from Proterozoic age which include *Palaeolyngbaea* (Fig. 1.89F) (comparable to modern *Lyngbaea*),

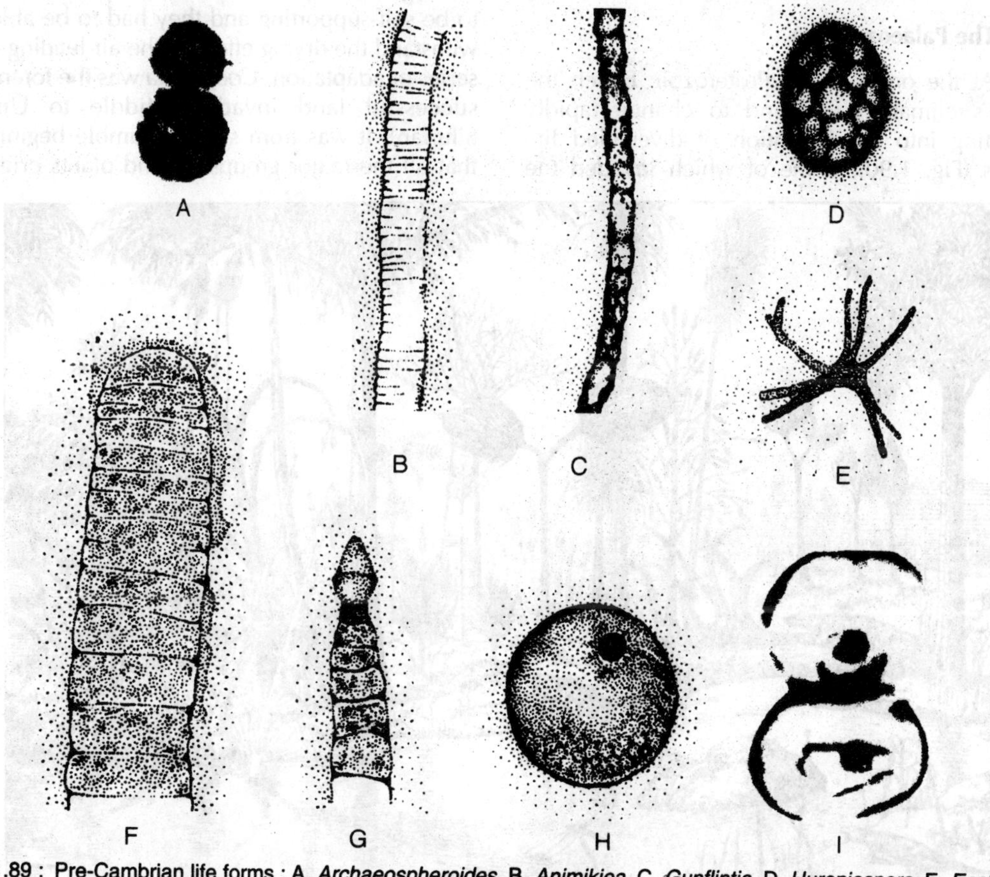

Fig. 1.89 : Pre-Cambrian life forms : A. *Archaeospheroides*, B. *Animikiea*, C. *Gunflintia*, D. *Huroniospora*, E. *Eostrion*, F. *Palaeolyngbya*, G. *Cephalophytarion*, H. *Glenobotrydion*, I. *Caryospheroides*

Cephalophytarion (Fig. 1.89G) (comparable to extant *Microcoleus*) and putative eukaryotes like *Glenobotrydion* (Fig. 1.89H) and *Caryosphe-roides* (Fig. 1.89I).

Stromatolites

Stromatolities, the oldest known fossils, are rock-like structures built by photosynthesised cyanobacteria of more than 3,000 m.y. old. Those cyanobacteria thrived in warm aquatic environment and built reefs much the same way as coral does today. They were likely responsible for the creation of much of the planet's free breathable oxygen, which allowed the development of oxygen-breathing organisms. They were dominant life forms on the earth for over 2,000 million years. Stromatolites were thought to have been extinct until 40 years ago. However, with the discovery of modern stromatolites in Shark Bay, Australia, and elsewhere, it is now known that some living cyanobacteria are still growing in rocks on top of each other at a rate of 1 mm a year.

B. The Palaeozoic Era

At the onset of the Proterozoic Epoch the earth's climate was found to change rapidly resulting into the formation of diversified life forms (Fig. 1.90). Some of which invaded the land escaping from aquatic habit and initially being amphibian in nature. For example, while acquiring land habit from water, the basal part of the Bryophytes (moss-like plants) developed in such a way that it served the purpose of anchorage and absorption, whereas the erect part performed photosynthetic function and some part of the branching system became flattered to form rudimentary leaves and, finally, simple reproductive organs at maturity.

The Cambrian and the Ordovician periods were characterised by a diverse group of algae and bryophytes in the warm ocean and island seas. About 410 m.y. ago, during the Silurian period, a most dramatic event took place in the history of plant life i.e., evolution of land plants. By this time the early plants added oxygen which was 20% of modern level in the atmosphere through photosynthesis. Ozone (a product of oxygen) prevented UV rays from reaching the earth's surface. To acquire terrestrial habit leaving their aquatic habitat, the plants needed to be self-supporting and they had to be able to withstand the drying effect of the air leading to a series of adaptation. *Cooksonia* was the foremost successful land invader (Middle to Upper Silurian). It was from such a humble beginning that all the major groups of land plants origina-

Fig. 1.90 : A Carboniferous forest (Imaginary picture)

ted. In the following geological period, the Devonian (410-345 m.y. old), more and increasingly complex plants appeared with some modifications, through the formation of vascular tissue (for conduction), epidermal cuticle (to check desiccation of water) and stomata (for gaseous exchange), which were visualised in the three early vascular plant groups, viz. Rhyniopsida (primitive land plants). Zosterophyllopsida (ancestor of microphylls) and Trimerophytopsida (ancestor of megaphylls).

The existence of such plants made possible the subsequent emergence of a variety of arborescent (tree-sized) plants that flourished in the swamp of Carboniferous period (345-325 m.y.) in Northern Hemisphere. The most dominant plants were *Lepidodendron* (giant club moss). *Calamities* (giant horsetail), etc. which attained a height of 30-50 metres. During the optimum period of diversification a new plant group i.e. Gymnosperms (naked seeded plants) got evolved. Those plants mostly comprising of Pteridosperms (the primitive seed bearing plants), *Cordaites* (the progenitors of the modern conifers), became more successful land plants because of their selective advantage of seed formation. The embryos of the seeds were well-protected and had the potentiality to overcome the adverse condition.

C. Mesozoic Era

Most of the Palaeozoic flora failed to survive and was largely replaced by the naked seeded gymnosperms. The gymnospermous plants mostly cycads, cycadeoids, conifers, ginkgos along with ferns reached their climax in the Jurassic period which constituted the world's dominant vegetation co-existent with the giant reptiles known as *Dinosaurs.* Glossopterids were totally disappeared and sphenopsids and lycopsids were less conspicuous. Some angiosperm-like leaves and tectate pollen were reported from the Jurassic period. Still later, in the end of Mesozoic era (i.e., in Cretaceous period) the closed seeded plants i.e., angiosperms got evolved which subsequently became dominant in Cenozoic Era replacing cycads, cycadeoids, conifers and ginkgos. There was a rapid diversification of angiosperms in the Cretaceous period where ferns and conifers still continued and cycadeoids got extinct. The evidences of both monocot and dicot plants were established by the occurrence of leaves, flowers with closed carpels, perianths and stamens, tectate collumelate pollen, etc. The Mesozoic Era is called "**The age of gymnosperms**' because of the diverse assemblage of cycads, ginkgos (maiden hair tree) and primitive conifers.

D. Cenozoic Era

In the Cenozoic Era (65 m.y. to present day), the dominance of gymnosperms steadily declined in number and distribution, as in the meanwhile, the angiosperms diversified and occupied most of the land mass surface. Now the gymnosperms are represented by only 730 species while angiosperms are distributed globally by having over 300,000 species. It has now been accepted that the angiosperms evolved from gymnospermic ancestors, and now is considered to be the highest evolved and dominant terrestrial plant life-form. Angiosperms grow in a greater range of environments (tropical, temperate, alpine, coastal etc.) and a variety of habitats (aquatic, terrestrial, epiphytic, etc.) displaying an astonishing array of morphological, anatomical and physiological variations, which include the world's major food-yielding plants and other plants of great economic importance. Hence the Cenozoic Era is often known as "**The age of angiosperms**".

Fruit-forming plants (angiosperms) exerted an influence on the evolution of birds and mammals. Birds, which feed on fruits, seeds, flowers, evolved rapidly in co-association with angiosperms. The emergence of herbivorous mammals coincided with the widespread distribution of grasses and herbs over the plains. In turn, the herbivours furnished the setting for the evolution of carnivorous animals, thus maintaining an intricate ecological balance by making interdependence between plants and animals.

The study of fossils has an applied significance in understanding the biostratigraphical sequence which provides to trace the plant as well as animal evolution through ages. In this context there are two aspects to be considered :

1. Correlation of the data showing quantitative and qualitative value to retrieve fossil taxa, and
2. Use of microfossils for comparison to get the empirical value in terms of appearance, duration of dominance and then gradual disappearance through migration or extermination due to climatic and others establishing a biostratigraphical scale.

For example, in the coal-bearing strata of the Middle Carboniferous of West Europe, seven successive vegetaional sequences were established using plant fossils. It has been demonstrated that climatic episodes were directly or indirectly connected with the change in floral and faunal composition creating sharp boundaries between :

Devonian and Carboniferous

Lower and Middle Carboniferous

Middle and Upper Carboniferous

Carboniferous and Permian

Permian and Triassic

Middle and Upper Jurassic

Createaceous and Early Tertiary

To sum up, stratigraphic zones during a given time interval witnessed evolution, migration or extinction of both plant and animal groups. It is thus implied that the fossil assemblages in these zones correspond a definite span of time and throw light on the evolutionary trend to the plant groups. Thus an analysis of fossils of a given stratigraphic zone provides the needed data to interpret its changing floral composition and consequent evolutionary sequence (Table 1.2) :

Table 1.2 : Geological time scale showing climatic fluctuation as evidenced by fossil plants (modified after Stewart and Rothwell, 1993) [Not shown in original scale]

Age	Era	Period	Epoch	Characteristic Plant Association	Climatic Inference
5,000–10,000 yrs.	CENOZOIC	Quaternary	Holocene	Modern flora	Retreat of ice sheet
2.5 m.y.			Pleistocene	Redistribution of floras	Successive glaciations
7 m.y.		Tertiary	Pilocene	Spread of grasslands, caused local restriction or extinction of some species	Elevation of Andes, climatic changes in temperate latitude
26 m.y.			Miocene	Establishment of present-day forest association	Continental uplift, rise of Alps, Himalayas
38 m.y.			Oligocene	Widespread occurrence of relic taxa like *Metasequoia*	Mild-temperate climate
54 m.y.			Eocene	Many angiosperm extinct and appearance of new types. Distinct forests in Northern and Southern latitudes	Subtropical climate, with heavy rainfall
65 m.y.			Palaeocene	Primitive angiosperms like Magnoliaceae, Lauraceae, Junglandaceae	Trend from temperate to mild-subtropical climate, seasonal variations
141 m.y.	MESOZOIC	Cretaceous	Upper Lower	First appearance of angiosperm	Climate tending to be uniform, temperate
195 m.y.		Jurassic	Upper Middle Lower	Cycads, ginkgos, conifers, ferns, cycadeoids, some pteriodosperms	Uniform, mild climates from North to South Poles
225 m.y.		Triassic	Upper Middle Lower	Rise of cycads and ginkgos. Diversification of conifers and ferns	Arid to semi-arid savanna-type climate

Contd.

Contd.

Age	Era	Period		Epoch	Characteristic Plant Association	Climatic Inference
280 m.y.	PALAEOZOIC	Permian		Upper Lower	Diversification of Southern hemisphere glossopterids, Extinction of arboscent lycopods and sphenopsids	Cooler and drier climates with extensive glaciation in Southern hemisphere
325 m.y.		CARBONIFEROUS	Pennsyl-vanian	Upper Middle Lower	Origin of conifers, formation of great arboreal swamp forest	Uniformly warm, humid climate
345 m.y.			Mississi-ppian	Upper Lower	Primitive ferns, seed ferns, arborescent lycopods, *Calamites*	Warm, equable climate
395 m.y.		Devonian		Upper Middle Lower	Diversification of vascular plants, establishment of heterospory and seed habit	Heavy rainfall and aridity, extensive inundation of continents
435 m.y.		Silurian		Upper Lower	First vascular land plant *Cooksonia*	Mild climate low-lying continents with epicontinental flooding
500 m.y.		Ordovician		Upper Lower	Abundance of red and green algae	Warm, mild climate, epicontinental seas
570 m.y.		Cambrian		Upper Middle Lower	Abundance of blue-green, green and red algae	Warm and equable
2,500 m.y.	PRE-CAMBRIAN	Proterozoic			First cell bacteria, growth of blue-green algae (stromatolite), and possibly green algae	Warm, organic evolution
3,800 m.y.		Archaeozoic				
4,700 m.y.		Hadean			No fossil record	Chemical evolution. Hot origin of earth.

PROGYMNOSPERMOPSIDA

Prior to 1960, the two form genera, *Archaeopteris* and *Callixylon* were described as two different types of plants that occurred abundantly in close association in Late Devonian. *Archaeopteris* was discovered as a fern-like foliage with pteridophytic free sporing characteristic. While, *Callixylon* was discovered as fragments of stem, branch or twig showing pycnoxylic wood with unique type of pitting (circular bordered pits) on radial walls. Thus, *Callixylon* was considered to have been a member of the conifer alliance.

Charles B. Beck in 1960 discovered an organic connection between the fern-like *Archaeopteris* fronds and the conifer-like *Callixylon* from a 80 cm long specimen recovered from the Upper Devonian beds in New York State. This discovery led to the establishment of a new class of vascular plants called Progymnospermopsida that possessed gymnospermous anatomy but reproduced by free-sporing pteridophytic method. Thus, two form-genera were merged and the generic name *Archaeopteris,* first used by Dawson in 1871, was retained by Beck for the whole plant as per the Rule of Priority. Beck (1962) reconstructed *Archaeopteris* as a large tree, 18 m tall, with a large crown of spirally arranged fronds.

The class Progymnospermopsida is divided into three orders — (i) Aneurophytales, (ii) Archaeopteridales, and (iii) Protopityales which extend from the Middle Devonian (early Eifelian) to the Lower Mississipian (Tournaisian).

GENERAL CHARACTERISTICS OF PROGYMNOSPERMS

1. The progymnosperms having gymnospermic secondary wood with free-sporing pteridophytic reproduction existed from the Middle Devonian to the Lower Mississipian i.e. approximately 350 m.y. ago.
2. They were widely distributed in the localities of New York, West Virginia and Pennsylvania in the USA, Eastern Canada, Russia, Yorkshire and Dunbartonshire in Ireland, Great Britain, Belgium, Germany, Ellesmere Island and Bear Island in the Arctic region.
3. The plants possessed a shrubby to arborescent habit.
4. The plants showed a pseudomonopodial branching pattern without forming any axillary bud.
5. The ultimate appendages were either dichotomously branched units or laminate leaves with dichotomous venation.
6. The vascular systems were of different types, ranging from a protostele to a eustele with mesarch primary xylem.
7. A bifacial cambium produced both centrifugal secondary phloem and centripetal secondary xylem.
8. The pycnoxylic wood comprised of tracheids with circular bordered pits and rays.
9. The plants produced fusiform sporangia that were borne along the adaxial or lateral surface of branches or on modified leaves.
10. The plants were either homosporous or heterosporous.

Classification

The class Progymnospermopsida has been divided into three orders : Aneurophytales, Archaeopteridales and Protopityales.

Order : **Aneurophytales**

Age : Middle to Upper Devonian.

Distinguishing Characteristics : There are at least three orders of helically arranged branches, with three-dimensional non-laminate, ultimate appendages; stem protostelic; primary xylem mesarch; pre-fronds showing different degrees in the evolution of leaf-like characteristics; homosporous.

Reconstructed plants : *Aneurophyton, Protopteridium, Tetraxylopteris, Triloboxylon* and *Proteokalon.*

Order : **Archaeopteridales**

Age : Upper Devonian.

Distinguishing characteristics : Stem eustelic with a ring of sympodia around the pith; primary xylem mesarch; flattended, laminated leaves which are opposite and decussate on the ultimate branches, although helically arranged on the penultimate branches; both homosporous and heterosporous.

Reconstructed plant : *Archaeopteris.*

ORDER : PROTOPITYALES

Age : Lower Mississipian.

Distinguishing characteristics : Stem eustelic with a pair of protoxylem strands divides below the node to form two strands that are incorporated in an arc-shaped trace which is emitted from the side of the stele; primary xylem endarch; appendages and pre-fronds borne in an alternate and distichous manner on the stem; heterosporous.

Reconstructed plant : *Protopitys.*

Among the three orders, the Archaeopteridales has extensively been studied which shows its evolutionary significance. Thus, the plant *Archaeopteris* is discussed in this chapter.

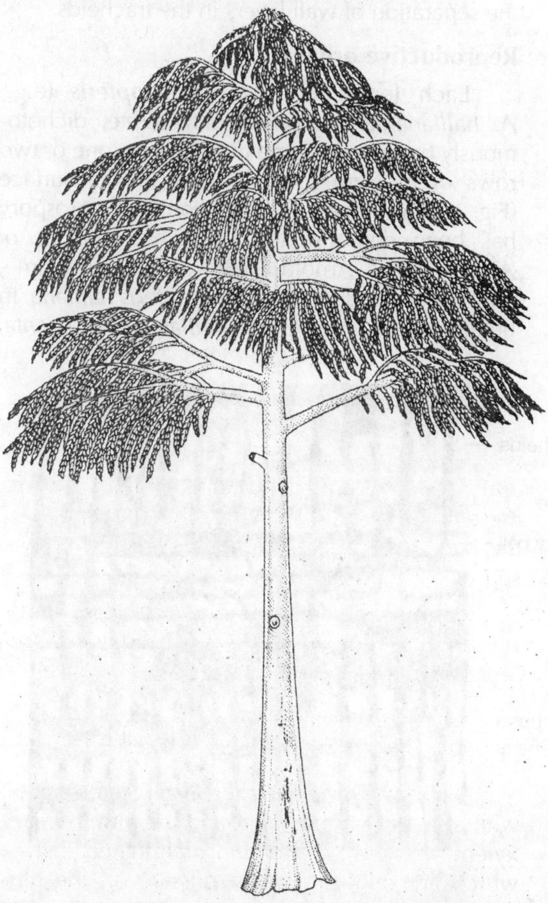

Fig. 1.91 : A reconstructed *Archaeopteris* tree

ARCHAEOPTERIS

Geological age : Upper Devonian.

Habit : Large trees up to 18 m high with a large crown of spirally arranged pseudomonopodial branches (Fig. 1.91) in which the penultimate branches bear both ultimate branches and leaves in a spiral.

Morphology : Both the branches and leaves show determinate growth which are developed in the same organotactic spiral. There is a continuity between the xylem of the branches and the parental axes. This suggests that the branches of *Archaeopteris* were not produced by lateral buds, rather developed from primordia like those that produced the leaves.

The fertile ultimate branches may occur as the most basal units of a branch system. In some species, both fertile branches (strobili) and sterile

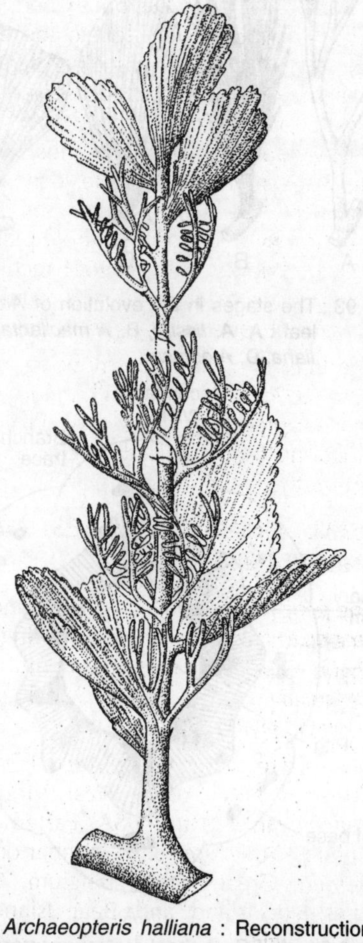

Fig. 1.92 : *Archaeopteris halliana* : Reconstruction of a branch bearing sterile and fertile leaves

branches (leaves) are intermixed and are arranged spirally (Fig.8.92). The fertile branches bear one or two rows of fusiform sporangia on adaxial surface.

Archaeopteris shows considerable variations of ultimate branches in different species, suggesting the origin of megaphyllous leaves from a planated and webbed telome truss (Fig. 1.93A-D). In *A. fissilis* (Fig. 1.93A), the leaves are represented by planated terete telomes that dichotomise two to three times. In *A. macilenta* (Fig. 1.93B), the telomes are planated with various degrees of incomplete webbing so that the distal margins of leaves are deeply incised. The wedge-shaped megaphyllous leaves of *A. halliana* (Fig. 1.93C) and *A. obtusa* (Fig. 1.93D) are formed due to the complete webbing where a single vein enters the decurrent base and forms repeated dichotomies.

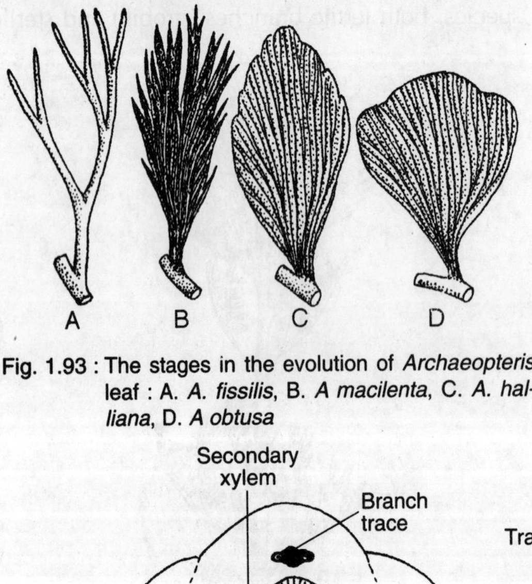

Fig. 1.93 : The stages in the evolution of *Archaeopteris* leaf : A. *A. fissilis*, B. *A macilenta*, C. *A. halliana*, D. *A obtusa*

Stem anatomy

Several logs of *Archaeopteris* (*Callixylon*) were discovered from the Upper Devonian strata which were up to 10 m long and about 1.5 m in diameter. *Archaeopteris* possesses a eustele with a ring of mesarch primary bundles surrounding a pith (Fig. 1.94A). The primary bundles are surrounded by thick-walled, compactly arranged secondary xylem (pycnoxylic wood), composed of radially arranged tracheids and narrow rays. In a few specimens, the wood shows growth rings. The wood rays are generally narrow, but may be broader in the region of the pith. The rays are variable in height, comprising of ray tracheids and ray parenchyma. The secondary tracheids have circular bordered pits arranged on the radial walls in groups (6-20) and are multiseriate in most species. These groups of pits are arranged horizontally and are separated by unpitted tracheid walls (Fig. 1.94B). The crassulae are often present in the tracheids and those are formed as a result of the separation of wall layers in the tracheids.

Reproductive organs

Each fertile leaf of *Archaeopteris* (e.g., *A. halliana*) is comprised of a terete, dichotomously branched telome truss bearing one or two rows of fusiform sporangia on adaxial surface (Fig. 1.95A). Although homosporous, heterospory has been demonstrated in several species of *Archaeopteris*. Arnold (1939), for the first time, reported heterospory in *Archaeopteris halliana*. In heterosporous species (*A. halliana, A macienta,*

Fig. 1.94 : A. T.S. of shoot of *Archaeopteris macilenta*, B. R.L.S. of secondary wood of *Archoeopteris* (*Callixylon*)

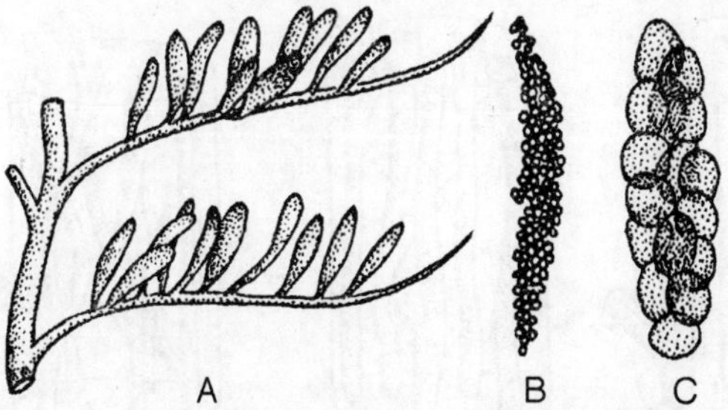

Fig. 1.95 : *Archaeopteris halliana* : A. Fertile appendages, B. Cluster of microspores, C. Cluster of megaspores

A fissilis), the micro- and megasporangia are more or less alike in terms of their sizes. The sporangia are thick-walled with longitudinal dehiscence. Sometimes, stomata are present on the epidermal cells of the sporangia. A single microsporangium contains more than 100 microspores of 33-70 μm in diameter (Fig. 1.95B), while a single megasporangium contains 16-32 megaspores of 110-500 μm in diameter (Fig. 1.95C). The microspore has a long trilete laesura. The proximal surface of the microspore is psilate, while the distal surface is ornamented. The microspores of *Archaeopteris* are similar to the dispersed form genus *Cyclogranisporites*. The megaspores are also trilete with elevated laesura scar. Like microspores, the proximal surface of megaspore is psilate, while the distal surface is ornamented with folds. The megaspores of *Archaeopteris* are included in the dispersed form genus *Biharisporites*.

Origin of the Progymnosperms

The Aneuroplytales is the primitive order among progymnosperms. The basic pattern of the vascular system in Middle to Upper Devonian Aneurophytales is either a three- or four-ribbed protostele. The three-ribbed protostele is similar to the vascular system in the Lower to Middle Devonian Trimerophytales (e.g. *Psilophyton dawsonii*), the group from which the progymnosperms (especially aneurophytes) are believed to have evolved. This idea is based on the common characteristics of progymnosperms (aneurophytes) and the trimerophytes.

Common characteristics of two groups are :

1. The presence of dischotomous and pseudomonopodial branching.
2. The ultimate and penultimate branches are modified into such structures that are interpreted as prefronds.
3. The presence of massive three-ribbed protostele.
4. The primary xylem is mesarch or centrarch.
5. The sporangia are borne terminally in cluster on ultimate branches of fertile telome truss.
6. The sporangia are fusiform and are dehisced through a longitudinal slit.

Role of Progymnosperms in the origin and evolution of Gymnosperms

After the establishment of the group Progymnospermopsida, it has widely been accepted that the progymnosperms are the ancestral plexus from which the gymnosperms evolved. However, opinions vary as to whether the gymnosperms are monophyletic or polyphyletic in origin.

Beck (1971) hypothesised that conifers have evolved from *Archaeopteris*. Because, both the groups possess pycnoxylic wood and simple decurrent leaves on their branches. Thus, the determinate axis of *Archaeopteris* have given rise to long and dwarf shoots of conifers.

It has now been accepted that progymnosperms are the ancestor of both cycadophytes and coniferophytes. The Aneurophytales are the ancestral progymnosperms from which the Archaeopteridales and cycadophytes evolved. The Archaeopteridales, in turn, has given rise to the coniferophytes. Support for this hypothesis includes the presence of primitive three-ribbed protostele in Aneurophytales and eustele in Archaeopteridales, cycadophytes and coniferophytes that evolved from the protostele. Moreover, the simple leaves of the coniferophytes and fronds of cycadophytes evolved from the primitive prefronds of the Aneurophytales.

Namboodiri and Beck (1968) put forth the concept that gymnospermous eustele evolved from a dissected protostele and the leaf traces developed directly from sympodial bundles of the eustele without formation of leaf gaps that are not homologous with that of siphonostelic ferns.

The evolution of gymnosperm stele from the ancestral protostele of aneurophytes as proposed by Namboodiri and Beck (1968) is described below :

1. The three-ribbed protostele of Aneurophytales is the ancestral form which initially formed secondary wood between the arms of the primary xylem (Fig. 1.96A).

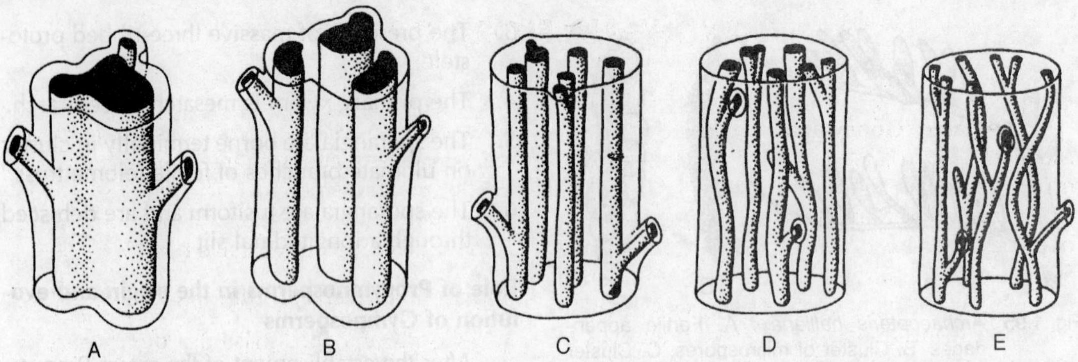

Fig. 1.96 : A–E. The stages in the evolution of primary vascular system in gymnosperms : A. Three-ribbed protostele (e.g. *Aneurophyton*), B. Longitudinal dissection of three-ribbed protostele resulting in the formation of discrete sympodial bundles (e.g. *Stenomyelon*), C. Sympodial bundles dividing tangentially at nodes with radially diverging leaf traces and reparatory strand (e.g. *Archaeopteris*), D. Same to C, but the divergence of leaf trace is tangential (e.g. *Lyginopteris*), E. Same to D, but showing undulating path of the vascular bundles (e.g. Modern conifers)

2. The ancestral protostele is longitudinally dissected into three-stranded vascular system by thin plates of parenchyma and a broad zone of secondary xylem. The condition is found in *Stenomyelon* (Fig. 1.96B).
3. The three-stranded vascular system is further longitudinally dissected into many equal columns, accompanied by the formation of central pith. Each of these pericolumns or sympodia divides longitudinally forming two radially aligned strands, of which one forms a leaf trace (diverging radially) and the other a reparatory strand. This condition is observed in *Archaeopteris* (Fig. 1.96C).
4. In *Lyginopteris, Callistophyton* and *Poroxylon,* the separation of strands is like that of *Archaeopteris,* but the initial divergence of leaf trace is tangential (Fig. 1.96D).
5. In most of the living conifers, the stele is similar to *Lyginopteris,* but showing undulating path of vascular bundle during upward course with helical phyllotaxy (Fig. 1.96E).

Chaloner and Pettitt (1987) proposed the origin of a linear tetrad in living gymnosperm from the *Archaeopteris* ancestor bearing many megaspores (Fig. 1.97A). There is a tendency in the reduction of megaspore number to a single tetrad.

The primitive condition is observed in *Archaeopteris* where a megasporangium contains many megaspores (Fig. 1.97A).

In *Bothrodendrostrobus,* the number of megaspore is reduced to four where all the four megaspores are equal in size (Fig. 1.97B).

The further reduction in number of megaspores is seen in *Bensonites,* where two of the four megaspores are larger and the remaining two are smaller (Fig. 1.97C).

In *Archaeosperma,* the single basal megaspore is large and functional and the remaining three are greatly reduced in size (non-functional) (Fig. 1.97D).

A further reduction is observed in *Cardiocarpus* where the three non-functional megaspores in a tetrahedral tetrad are greatly reduced (Fig. 1.97E).

The final stage is a linear tetrad where the three non-functional megaspores are greatly reduced in size and the functional megaspore becomes much larger (Fig. 1.97F). This condition is seen in most of the living gymnosperms.

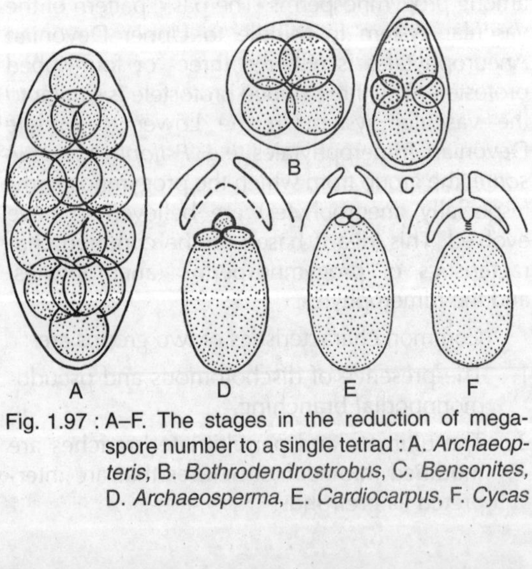

Fig. 1.97 : A–F. The stages in the reduction of megaspore number to a single tetrad : A. *Archaeopteris*, B. *Bothrodendrostrobus*, C. *Bensonites*, D. *Archaeosperma*, E. *Cardiocarpus*, F. *Cycas*

GONDWANA FLORA OF INDIA

The name '**Gondwana**', suggested by H. B. Medlicott (1872), is derived from the ancient tribal Dravidian Gond Kingdom of Madhya Pradesh ruled by Rani Durgabati during the reign of Akbar the Great. Later the term '**Gondwana System**' was published by Ottokar Feistmantel (1876), comprising of continental rock formation with coal seams and fossil plants and animals of the Upper Palaeozoic as well as Mesozoic Eras.

Other scientists discovered similar continental rock formations with coal seams and fossils from faraway continents of the Southern Hemisphere like South America, South Africa, Antarctica, Australia, Madagascar and New Zealand. Thus, the concept '**Gondwanaland**' (Fig. 1.98) was introduced by Edward Suess (1885) which was originally conceived by Blanford (1875).

It is believed that the Gondwana Era began after the second major period of earth revolution, the **Hercynian Revolution** during Upper Carboniferous. The Gondwanaland lasted up to the beginning of the Cretaceous when the third major revolution of earth i.e., the **Alpine-Himalayan Revolution** took place. This gave rise to the concept of two supercontinents by Du Toit (1937), one forming the Northern continental mass, the **Laurasia** and the Southern continent **Gondwanaland**, separated by the great equatorial **Tethys sea**. Thus, the single landmass, Gondwanaland, existed, showing a clear floral continuity ranging in age from the Upper Carboniferous to the beginning of the Cretaceous, from bottom to the top. There was free and easy migration of plants and animals in this land mass, exhibiting similarity in floral composition and climatic conditions. There was

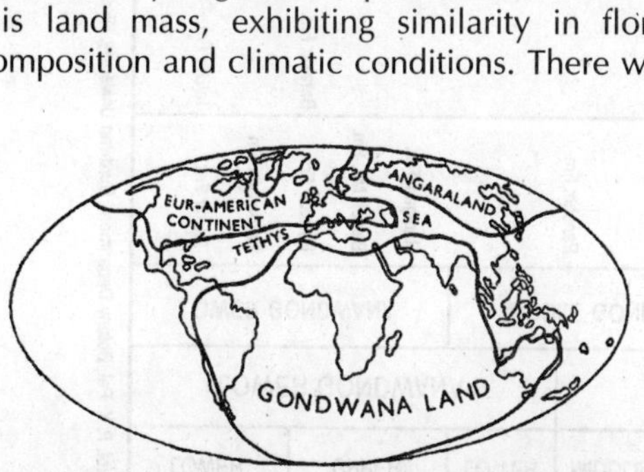

Fig. 1.98 : The continents during Upper Carboniferous to Lower Cretaceous period, shown by thick lines

several local minor revolutions of earth that changed the pattern of flora or fauna of the region in some periods. Though, these minor revolutions did not affect the whole landmass and many of them must have been localised.

Indian Gondwana stratigraphy

Gondwanaland existed from the Lower Permian to the Lower Cretaceous with a span of 225 m.y. The Gondwana system is comprised of continental deposits of conglomerates, sandstones, shales and coal measures of mainly fluviatile (river) and locustrine (lake) origin. Recent works also indicate the presence of some marine intercalations in the Gondwana sediments of central and east coastal India. The sediments were deposited in trough-like basins, bounded on either side by older Archaean rocks which endured faulting. Archaean rocks protected the Gondwana deposits from denudation and thus preserved the sediments in their original horizontal stratification. There is a striking homogeneity of Gondwana sediments with a uniform alteration of sandstones and shales bearing coal seams.

Distribution of Indian Gondwana rocks

The Gondwana rocks are mainly distributed in major river valleys of Peninsular India. These are — (i) Damodar valley, (ii) Sone-Mahanadi valley, and (iii) Wardha-Godavari valley. A few exposures also occur in the Eastern coast (e.g., Cauvery, Palar, Krishna-Godavari and Orissa), Western Rajasthan, Saurashtra and Kutch. The Gondwana Basins have been mentioned in Table 1.3.

In the extra-peninsular, some detached outcrops are found in northern India such as Punjab Salt Range and Kashmir; and also in the sub-Himalayan regions of Sikkim, Bhutan, Arunachal Pradesh and Assam. The extra-peninsular outcrops have been subjected to metamorphism, showing very poor fossil content.

Classification of the Gondwanas

The classification of Indian Gondwanas is a matter of controversy.

G. de Cotton (1917) and Fox (1931) divided the Gondwana system into two major divisions (Table 1.3). Lower Gondwana (Lower Permian to

Table 1.3 : Correlation of Peninsular Indian Gondwana Formations

Standard Scale		Gondwana Divisions		Damodar Koel	Rajmahal	Mahanadi Sone	Upper Narmada	Lower Narmada	Wardha Pranhita Godavari	East Coast: Cauvery	East Coast: Palar	East Coast: Krishna-Godavari	Orissa Coast	Jaisalmer	Kathiawar	Kachchh
L. CRETACEOUS						Bansa beds								Pariwar Fm.		Bhuj Fm.
JURASSIC	UPPER	UPPER GONDWANA	UPPER GONDWANA		Rajmahal Fm.		Jabalpur Fm.	Lower Nimar sst.	Gangapur Fm.	Shivganga Fm.	Satyavadu Fm. Sriperumbudur Fm.	Tirupati sst. Pavalur sst. Ragavpuram shale Vemavaram shale	Athgarh sst.		Than beds	Jhuran Fm.
JURASSIC	MIDDLE	UPPER GONDWANA	UPPER GONDWANA								Golapilli sst.	Budavada sst.		Lathi Fm.		Jumara Fm. Jhurio Fm.
JURASSIC	LOWER	UPPER GONDWANA	UPPER GONDWANA		Dubrajpur Fm.	Hartala Hill beds			Kota Fm.							
TRIASSIC	UPPER	UPPER GONDWANA	MIDDLE GONDWANA	Suprapanchet		Tiki Fm. Parsora Fm.	Bagra Fm. Denwa Fm.		Dharmaram Fm. Maleri Fm.							
TRIASSIC	MIDDLE	UPPER GONDWANA	MIDDLE GONDWANA			Nidpur beds	Pachmari Fm.		Bhimaram Fm. Yerrapalli Fm.							
TRIASSIC	LOWER	LOWER GONDWANA	MIDDLE GONDWANA	Panchet Fm.		Pali Fm.	Almod Fm.		Mangli beds/ Kamthi Fm.							
PERMIAN	UPPER	LOWER GONDWANA	LOWER GONDWANA	Raniganj Fm. Barren Ms. Fm.		Hingir Fm.	Bijori Fm. Motur Fm.		Motur Fm.							
PERMIAN	LOWER	LOWER GONDWANA	LOWER GONDWANA	Barakar Fm. Karharbari Fm. Talchir Fm.	Barakar Fm. Talchir Fm.	Barakar Fm.	Barakar Fm. Karharbari Fm.		Barakar Fm. Talchir Fm.							

[Courtesy : Prof. P. K. Pal, Botany Department, Burdwan University, West Bengal, India]

Lower Triassic) and Upper Gondwana (M. Triassic to L. Cretaceous) which are characterised by *Glossopteris* and *Ptilophyllum* floras, respectively.

Feistmantel (1876) and Vredenburg (1910) proposed a three-fold classification of Gondwana : Lower, Middle and Upper Gondwanas (Table 8.3) which are correspondingly equivalent to the Permian, Triassic and Jurassic. All the localities containing Lower Gondwana fossils, but bearing the lithological characteristics of the Upper Gondwanas such as Panchet, Pachmari, Parsona, Maleri etc. are retained in the Middle Gondwana (Table 1.3).

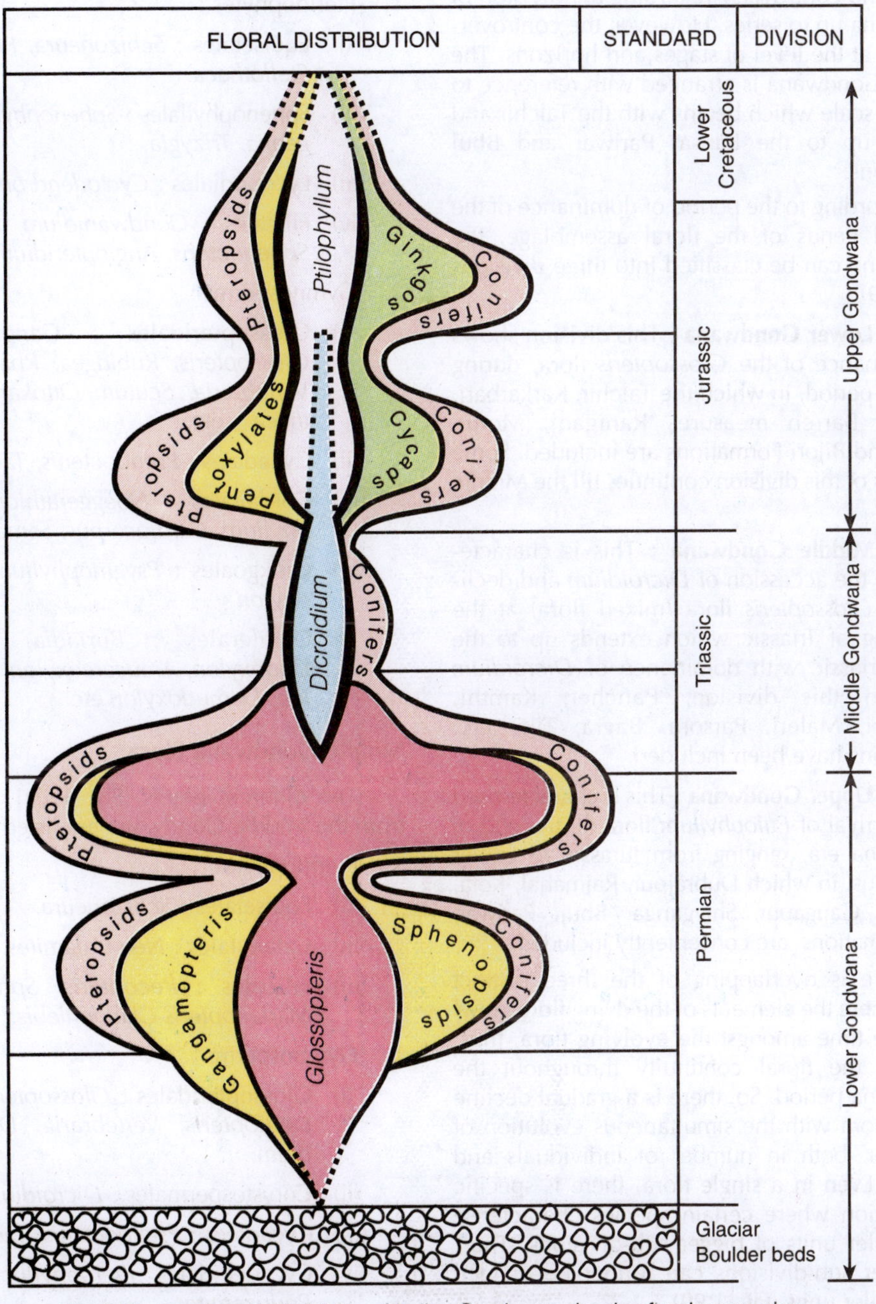

Fig. 1.99 : The three-fold division of Indian Gondwana showing floral succession

The Middle Gondwana is characterised by *Dicroidium* flora, present in between the *Glossopteris* and *Ptilophyllum* floras. Subsequently, most of the palaebotanists supported the three-fold division of Gondwana.

The sub-divisions of Gondwana system are further divided into various series and stages. There is no controversy regarding the division of Gondwana up to series. However, the controversy arises at the level of stages and horizons. The flora of Gondwana is arranged with reference to the time scale which begins with the Talchir and extends up to the Bansa, Pariwar and Bhuj Formations.

According to the period of dominance of the principal genus of the floral assemblage, the Gondwana can be classified into three divisions (Fig. 1.99) :

1. **Lower Gondwana** : This division shows predominance of the *Glossopteris* flora, during Permian period, in which the Talchir, Karkarbari, Barakar, Barren measures Raniganj, Motur, Hingir and Bijori Formations are included. Some elements of this division continues till the Middle Triassic.

2. **Middle Gondwana** : This is characterised by the accession of *Dicroidium* and declining of *Glossopteris* flora (mixed flora) at the beginning of Triassic which extends up to the Upper Triassic with dominance of *Dicroidium* flora. In this division, Panchet, Kamthi, Pachmari, Maleri, Parsora, Bagra, Tiki, etc. Formations have been included.

3. **Upper Gondwana** : This is characte-rised by the arrival of *Ptilophyllum* flora till the end of Gondwana era, ranging from Jurassic to Lower Cretaceous, in which Dubrajpur, Rajmahal, Kota, Jabalpur, Gangapur, Shivganga, Bhuj, Pariwar, etc., Formations are conveniently included.

There is overlapping of the three distinct floras where the elements of the dying flora crawl for some time amongst the evolving flora, thus, showing the floral continuity throughout the Gondwana period. So, there is a gradual decline of one flora with the simultaneous evolution of the other, both in number of individuals and species. Even in a single flora, there is specific delimitation where certain species characterise the smaller units of bigger sub-divisions. Thus, the larger sub-divisions can further be divided into smaller units (Fig. 1.99).

The major floral distribution of Gondwanas is enumerated :

Lower Gondwana flora

The different plants (Fig. 1.100) recovered from the various stages of the Lower Gandwara are listed :

1. Pteridophytes :
   - (i) Equisetales : *Schizoneura, Phyllotheca, Stellotheca.*
   - (ii) Sphenophyllales : *Sphenophyllum, Raniganjia, Trizygia.*
   - (iii) Lycopodiales : *Cyclodendron*
   - (iv) Filicales : *Gondwanidium, Pecopteris, Sphenopteris, Angiopteridium,* etc.
2. Gymnosperms :
   - (i) Glossopteridales : *Gangamopteris, Glossopteris, Rubidgea, Rhabdotaenia, Vertebraria, Scutum, Ottokaria, Glossotheca,* etc.
   - (ii) Cycadales : *Pseudoctenis, Taeniopteris.*
   - (iii) Cordaitales : *Noeggerathiopsis, Euryphyllum, Cardiocarpus, Samaropsis.*
   - (iv) Ginkgoales : *Psygmophyllum, Ginkgophyton.*
   - (v) Coniferales : *Buriadia, Barakaria, Dadoxylon, Araucareoxylon, Agathoxylon, Damudoxylon* etc.

Middle Gondwana Flora

The different plants (Fig. 1.101) recovered from the Middle Gondwana are listed :

1. Pteridophytes
   - (i) Equisetales : *Schizoneura.*
   - (ii) Calamitales : *Neo-Calamites.*
   - (iii) Filicales : *Pecopteris, Sphenopteris, Marianopteris Cladophlebis.*
2. Gymnosperms
   - (i) Glossopteridales : *Glossopteris, Macrotaeniopteris, Vertebraria, Dictyopteridium.*
   - (ii) Corystospermales : *Dicroidium.*
   - (iii) Cycadales : *Pseudoctenis, Taeniopteris.*
   - (iv) Cycadeoidales : *Pterophyllum, Zamites, Podozamites.*

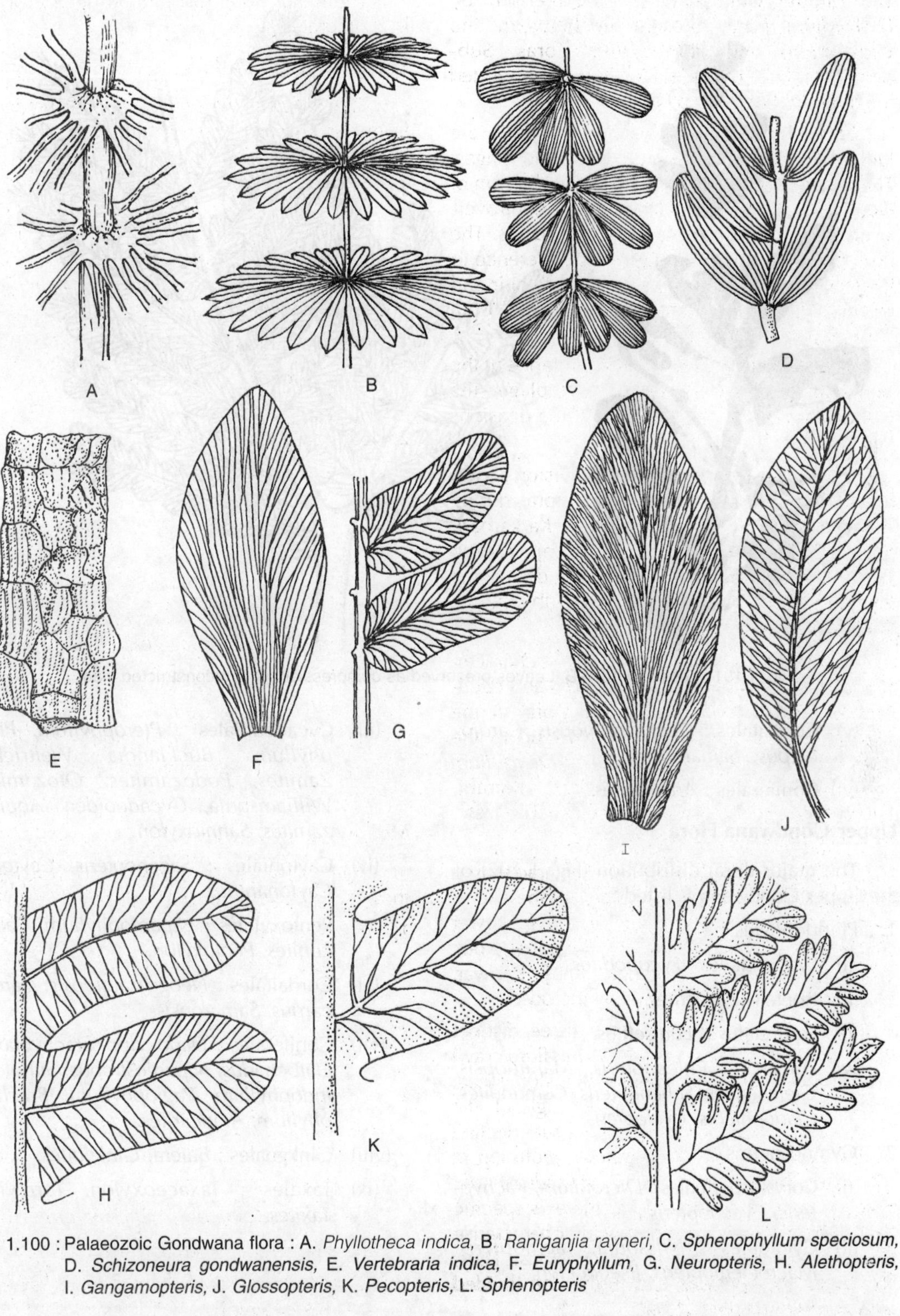

Fig. 1.100 : Palaeozoic Gondwana flora : A. *Phyllotheca indica*, B. *Raniganjia rayneri*, C. *Sphenophyllum speciosum*, D. *Schizoneura gondwanensis*, E. *Vertebraria indica*, F. *Euryphyllum*, G. *Neuropteris*, H. *Alethopteris*, I. *Gangamopteris*, J. *Glossopteris*, K. *Pecopteris*, L. *Sphenopteris*

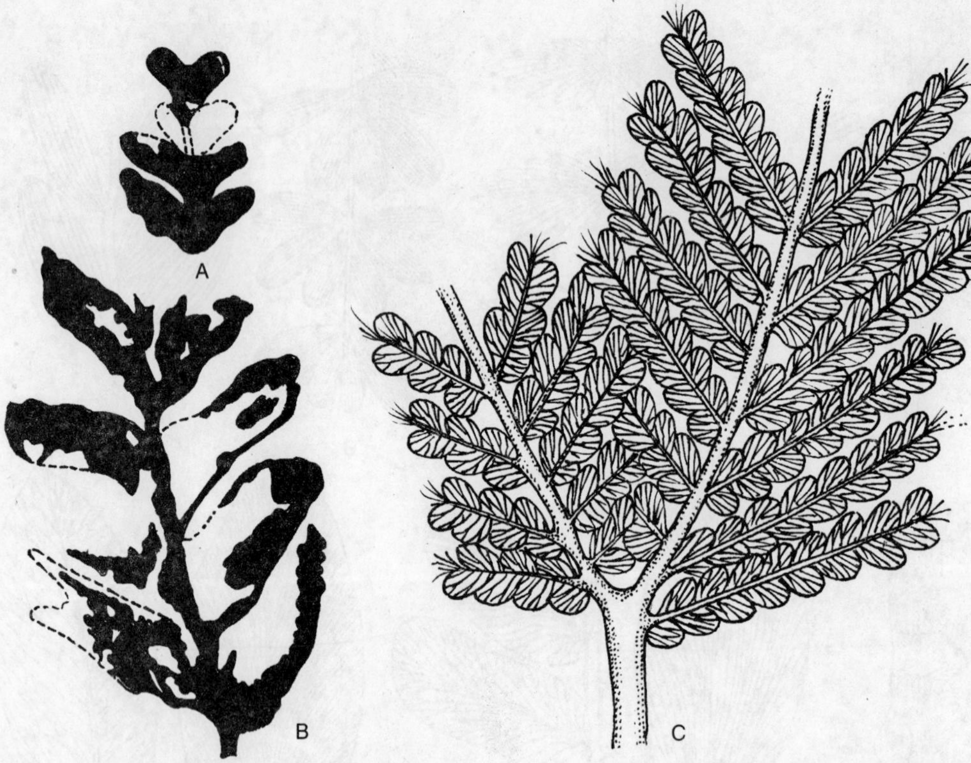

Fig. 1.101 : *Dicroidium* : A–B. Leaves preserved as compression. C. A reconstructed plant

(v) Cordaitales : *Neoggerathiopsis, Cardiocarpus, Samaropsis.*

(vi) Coniferales : *Araucarites.*

Upper Gondwana Flora

The major floral distribution (Fig. 1.102) of the Upper Gondwana is listed :

1. Pteridophytes

   (i) Lycopodiales : *Lycopodites.*

   (ii) Isoetates : *Isoetites.*

   (iii) Equisetales : *Equisetites.*

   (iv) Filicales : *Phlebopteris, Marattiopsis, Gleichenites, Sphenopteris, Osmundites, Cladophlebis, Cyclopteris.*

2. Gymnosperms

   (i) Corystospermales : *Dicroidium, Pachypteris, Cycadopteris.*

   (ii) Cycadales : *Nilssonia, Taeniopteris, Macrotaeniopteris, Pseudoctenis.*

   (iii) Cycadeoidales : *Pterophyllum, Ptilophyllum, Bucklandia Weltrichia, Zamites, Podozamites, Otozamites, Williamsonia, Cycadeoidea, Anomozamites, Sahnioxylon.*

   (iv) Caytoniales : *Sagenopteris, Caytonia, Caytonanthus.*

   (v) Pentoxylales : *Nipaniophyllum, Carnoconites, Pentoxylon.*

   (vi) Cordaitales : *Neoggerathiopsis, Cardiocarpus, Samaropsis.*

   (vii) Coniferales : *Dadoxylon, Stachyotaxus, Elatochadus, Nipaniostrobus, Sitholeya, Indophyllum, Pagiophyllum, Brachyophyllum, Araucarites.*

   (viii) Ginkgoales : *Baiera, Ginkgoites.*

   (ix) Taxales : *Taxaceoxylon, Torreyites, Taxites.*

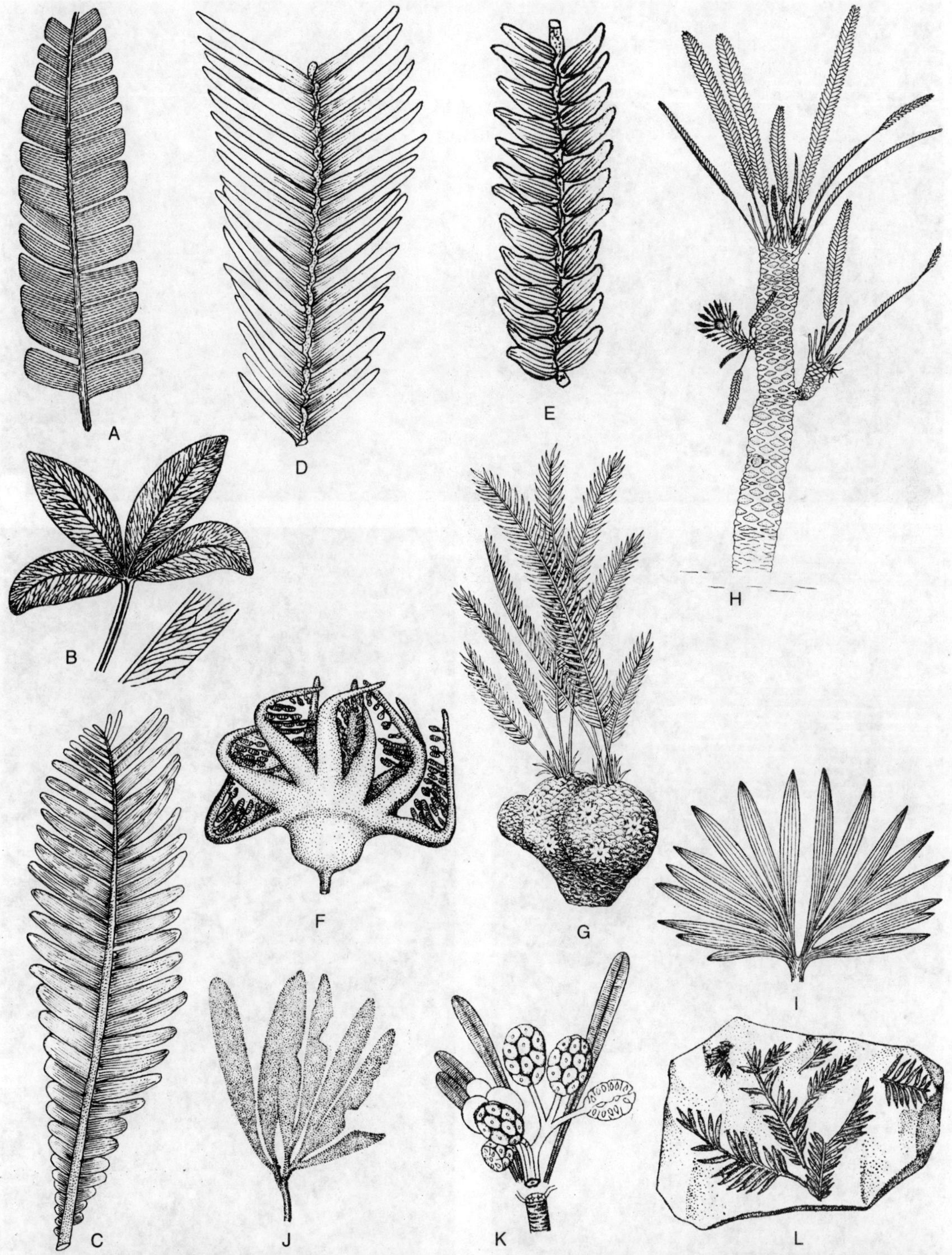

Fig. 1.102 : Mesozoic Gondwana flora : A. *Nilssonia compta,* B. *Sagenopteris,* C. *Pterophyllum,* D. *Ptilophyllum,* E. *Otozamites,* F. *Weltrichia spectabilis,* G. *Cycadeoidea,* H. *Williamsonia sewardiana,* I. *Baiera gracilis,* J. *Ginkgoites hermelini,* K. Shoot of *Pentoxylon* with *Nipaniophyllum* leaves and *Carnoconites* fructification, L. *Elatocladus*

Fig. 1.103 : The oldest known plant, *Pinus aristata*, growing at Inyo National Forests of California, USA (age about 4,600 years). A. Forest showing the plant association, B. A single plant, C. Plant bearing cones

[Photo courtesy : Prof. Debashis Bhattacharya, Visva-Bharati]

2 Angiosperm Morphology

Plants are living organisms containing chlorophyll, this enables them to live wholly on inorganic substances. They have specialised sense organs and the power of voluntary movement.

An angiospermic plant consists of different organs like root, stem, banches, leaves, flowers, fruits and seeds. Seed is the most protective organ that contains a baby plant, the embryo. After germination, it grows to an adult plant.

SEED

Definition

Seed is a fertilised (ripen) ovule, consisting of an embryo enclosed by protective seed coat(s) derived from the integument(s). In brief, the seed is defined as **"a matured, integumented indehiscent megasporangium"**.

Generally, all flowering plants bear seeds. A seed is a miniature plant which encloses an inactive or dormant embryo. With the onset of favourable condition, it becomes active and germinates to produce an adult plant.

General structure of seed

A typical matured seed of angiosperms consists of two parts : 1. **Seed coat,** and 2. **Kernel.**

1. **Seed coat** : Seed coat is the protective covering of seed, developed from integument of the ovule. It is made up of two layers :
   - i. the outer layer is called **testa**, which is usually hard and of various appearances (Fig. 2.1), and
   - ii. the inner layer is called **tegmen**, which is very thin and papery.

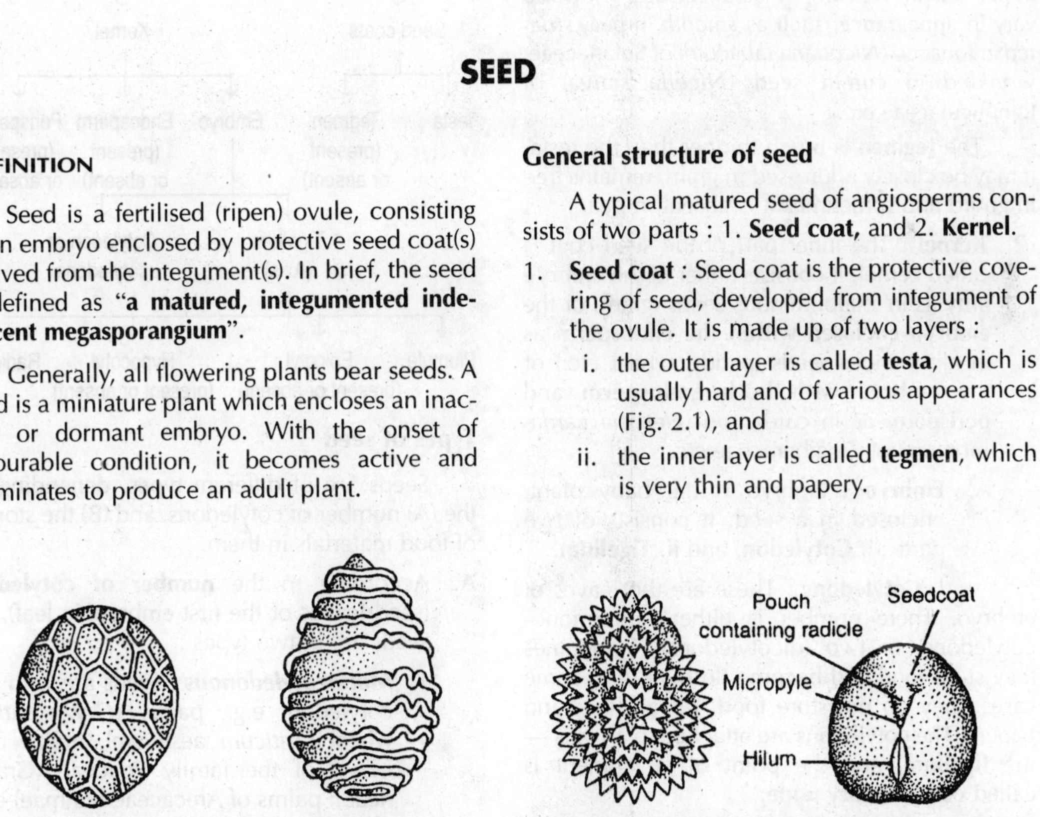

Fig. 2.1 : Seeds showing different nature of testa. A. reticulate (*Nasturtium officinale*), B. wrinkled (*Nigella sativa*), C. ridged and furrowed (*Delphinium majus*) and D. smoth (*Pisum sativum*)

The following **structures** are present **on the testa** :

At one end of the testa, there is a scar, called **hilum**. The hilum is the position of attachment of seed with its stalk, the **funiculus**. Near the hilum a small opening is present, called **micropyle**, through which water enters the seed. Sometimes a ridge is present beyond the hilum, opposite the micropyle, it is called **raphe**. It represents the base of the funiculus which becomes fused with integument.

(The testa are of various **natures** : It may be *tough* and *leathery* in gram (*Cicer arietinum*) and pea (*Pisum sativum*) of Fabaceae; *hard* and *horny* in castor (*Ricinus communis*) of Euphorbiaceae; *soft* and *fleshy* in gourd (*Cucurbita maxima*) of Cucurbitaceae. The **colour** of the testa is generally brown, but it may also be *black* in Indian shot (*Canna indica*) of Cannaceae; *red* in carb's eye (*Abrus precatorius*) of Fabaceae; *whitish* or *yellow* in pea (*Pisum sativum*) of Fabaceae etc. They also vary in **appearance**, such as *smooth* in pea, *striated* in tobacco (*Nicotiana tabacum*) of Solanaceae; *wrinkled* in cumin seed (*Nigella sativa*) of Ranunculaceae etc.)

The **tegmen** is much thinner than the testa. It may be closely adpressed in gram, remains free in gourd and sometimes it is absent in pea.

2. **Kernel** : The inner part of the seed coat is called kernel. It consists either of the **embryo only** as in exalbuminous seeds (gram) or the **embryo enclosed within the endosperm** as in albuminous seeds (paddy, wheat etc.) or the **embryo with both endosperm** and **perisperm** as in cardamon (*Elettaria cardamomum*) of Zingiberaceae etc.

A. **Embryo** : Embryo is the baby plant, enclosed in a seed. It consists of two parts : i. **Cotyledon,** and ii. **Tigellum.**

i. **Cotyledons** : These are the leaves of embryo. There number is either one (monocotyledons) or two (dicotyledons). Sometimes they store food and become fleshy. But in some cases, they do not store food, remain thin and papery. The cotyledons are attached to an axis — the **tigellum,** and the point of attachment is called **cotyledonary node**.

ii. **Tigellum** (axis) : It is the main axis, of which one end is pointed and protrudes out of the cotyledon, it lies next to the micropyle and is called **radicle,** i.e., the rudimentary root. The other end of the tigellum is the **plumule,** the first apical bud of shoot (Fig. 2.3E). In monocotyledonous seeds, both radicle and plumule remain covered by protective sheaths, known as **coleorhiza** and **coleoptile,** respectively.

The portion of axis above the point of attachment of cotyledon is called **epicotyl** and its lower portion is called **hypocotyl** (Fig. 2.3E).

B. **Endosperm** : Endosperm is the food-reservoir of seed. It may be starchy in wheat (*Triticum aestivum*), oily in mustard (*Brassica nigra*) and proteinaceous in pulses (*Cicer arietinum*).

C. **Perisperm** : It is the remnant of nucellus of ovule and is also nutritive in function.

General structure of seeds

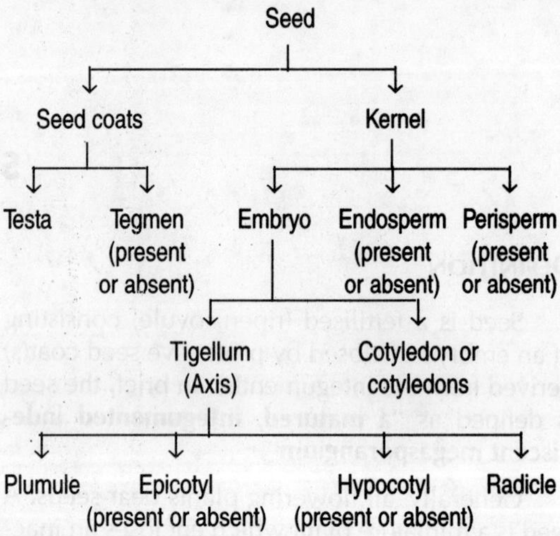

Types of seed

Seeds are of different types, depending on the (A) number of cotyledons, and (B) the storage of food materials in them.

A. According to the **number of cotyledons** (seed leaves or the first embryonic leaf), the seeds are of two types :

1. ***Monocotyledonous*** : Seeds are with *one* cotyledon, e.g., paddy (*Oryza sativa*), wheat (*Triticum aestivum*), maize (*Zea mays*) of the family Poaceae (Gramineae); palms of Arecaceae (Palmae) etc.
2. ***Dicotyledonous*** : Seeds are with two cotyledons, e.g., mango (*Mangifera indica*) of Anacardiaceae, gram (*Cicer arietinum*) and pea (*Pisum sativum*) of Fabaceae

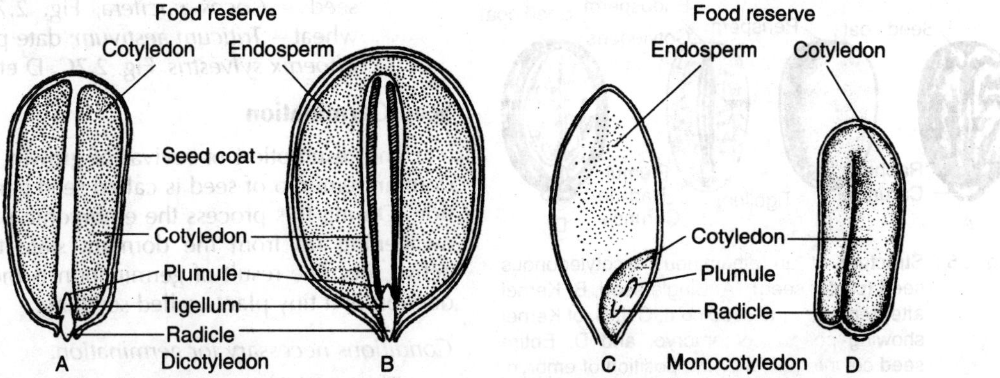

Fig. 2.2 : General structural differences among the four different types of seeds : A. *Dicotyledonous* exalbuminous (non-endospermic), B. *Dicotyledonous* albuminous (endospermic), C. *Monocotyledonous* albuminous (endospermic). D. *Monocotyledonous* exalbuminous (non-endospermic), Embryo = Cotyledon + tigellum (axis) i.e. radicle + plumule

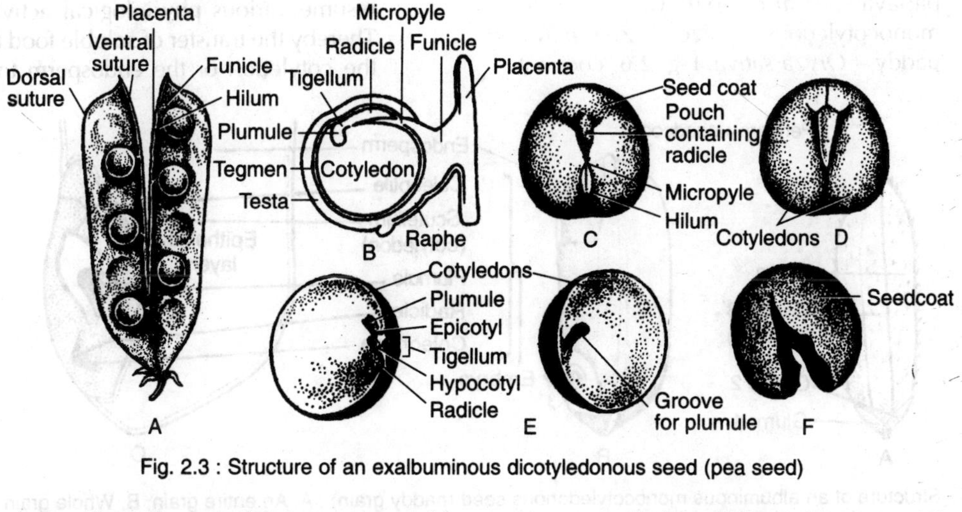

Fig. 2.3 : Structure of an exalbuminous dicotyledonous seed (pea seed)

(Leguminosae), castor (*Ricinus communis*) of Euphorbiaceae, gourd (*Cucurbita maxima*) of Cucurbitaceae etc.

B. According to the **storage of food materials** in them, the seeds are of two types (Fig. 2.2) :

1. ***Exalbuminous (non-endospermic) seeds*** : In such seeds the food is stored in the cotyledon. The endosperm of the seed is consumed in young stages by the developing cotyledons and do not have endosperm at maturity (Fig. 2.2A,D). This type of seeds is found in both dicotyledons (pea – *Pisum sativum,* Fig. 2.3; tamarind – *Tamarindus indicus,* gram – *Cicer arietinum* etc.) and monocotyledons (*Alisma,* Fig. 2.4 – Alismataceae, *Vallisneria* – Hydrocharitaceae etc.).

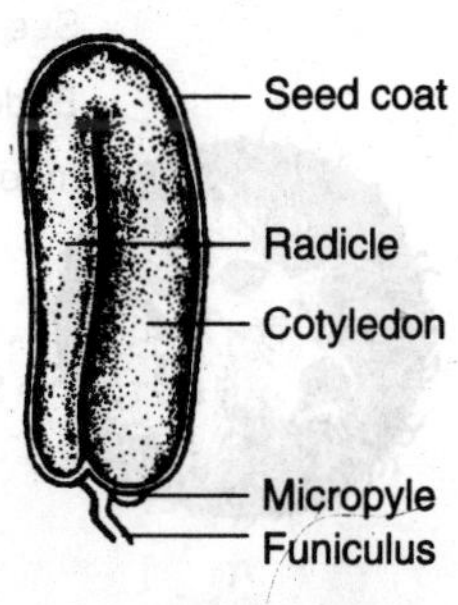

Fig. 2.4 : Structure of an exalbuminous monocotyledonous seed (*Alisma plantago* seed)

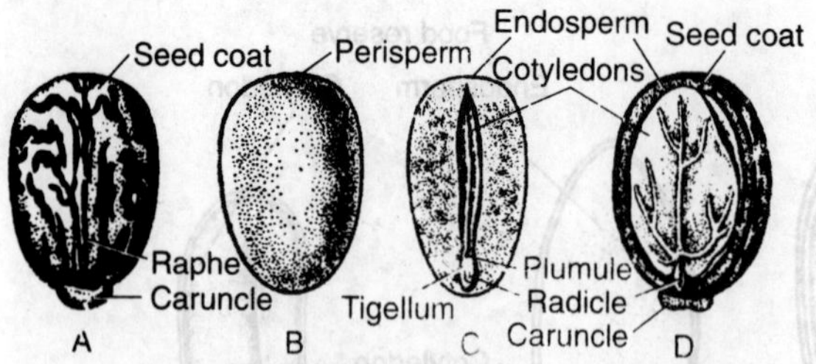

Fig. 2.5 : Structure of an albuminous dicotyledonous seed (castor seed) : A. Single seed, B. Kernel after removing the seed coat, C. L.S. of Kernel showing position of embryo, and D. Entire seed cut into half showing position of embryo

2. ***Albuminous (endospermic) seeds*** : In such seeds the food is stored outside the cotyledon i.e., in a separate tissue, the endosperm (Fig. 2.2B, C). This type of seed is also found in both dicotyledons (caster – *Ricinus communis*, Fig. 2.5; papaya – *Carica papaya* etc.) and monocotyledons (maize – *Zea mays*; paddy – *Oryza sativa*, Fig. 2.6; coconut seed – *Cocos nucifera*, Fig. 2.7A, B; wheat – *Triticum aestivum*; date palm – *Phoenix sylvestris*. Fig. 2.7C, D etc.).

Seed Germination

The resumption or revival of growth of the dormant embryo of seed is called seed germination. During this process the embryonic plant is awakening up from the dormant state to the active life. The result of germination is the production of a tiny plant, called seedling.

Conditions necessary for germination.

The conditions necessary for germination of a seed are :

A. **External factors**

1. **Water**. It plays an important role in the germination of seed. A viable seed, after absorbing some amount of water, will resume various physiological activities. Thereby the transfer of soluble food from the cotyledon or the endosperm to the

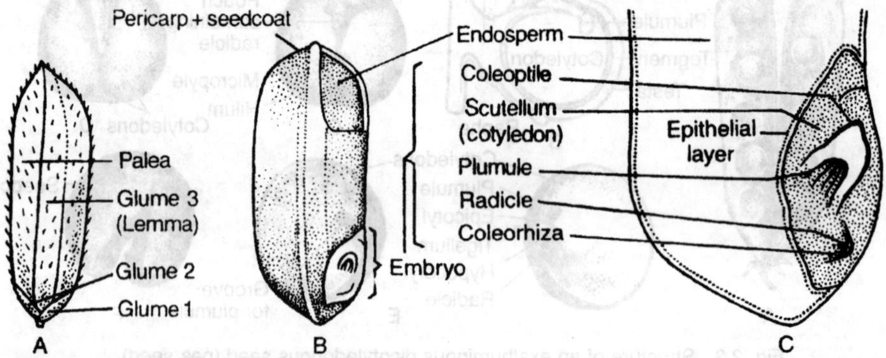

Fig. 2.6 : Structure of an albuminous monocotyledonous seed (paddy grain) : A. An entire grain, B. Whole grain after removing husk, showing sectional view of a portion of embryo and endosperm, C. Magnified lower part of the grain showing embryonic region in detail

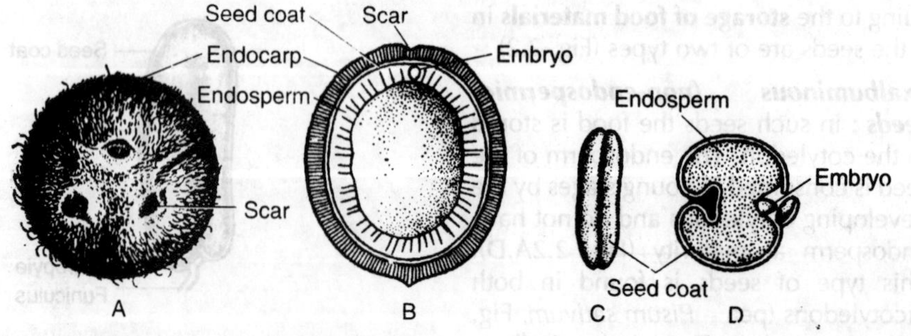

Fig. 2.7 : Structure of an albuminous monocotyledonous seed : A. Coconut seed, B. L.S. of same seed, C. Date palm seed and D. L.S. of same seed showing embryo in a pit covered by a lid

actively growing region of the embryo takes place to build up a new cell. Due to imbibition of water, the seed coat(s) become soften and burst easily during the growth of the developing embryo.

2. **Oxygen**. During germination, the seed requires more energy for its growth, thus the rate of respiration becomes high and oxygen is essentially required.
3. **Temperature**. Most seeds fail to germinate below 5°C and above 45°C. The optimum temperature lies between 25°C-30°C.
4. **Light**. Light has variable effect on the germination of seeds. Based on the sensitivity of light, they are divided into three types :
   i. Light-sensitive i.e., the germination is favoured by light.
   ii. Light-hard i.e., the germination is inhibited by light, and
   iii. Light-indifferent i.e., the germination is not affected by the presence or absence of light.

B. **Internal factors**

1. **Reserve food**. Food is absolutely necessary for the normal growth and development of the growing embryo. This comes from cotyledon or endosperm.
2. **Growth regulators**. Some growth regulators like auxins are developed inside the seed at the time of germination. These are required for the growth of the embryo during germination.
3. **Completion of dormancy**. Dormancy is a state or condition of inactivity exhibited by many seeds before germination. The period of dormancy varies from plant to plant. It is also a means to overcome the period, unsuitable for a successful germination.
4. **Viability** (vitality) and **longevity.** Viability is the germinating capacity of the seeds, but longevity denotes the period the seed can remain dormant and viable.

Changes during germination

The important physico-chemical changes that take place in the seeds during germination are :

1. Swelling of seeds by rapid imbibition and osmosis of water.
2. Digestion of stored food by rapid respiration and copious secretion of enzymes.
3. Insoluble food becomes soluble and complex food becomes simple.
4. Translocation of food to the growing organs (plumule and radicle).
5. Assimilation of the food by the growing organ, which enables the growth of the seedling.

Types of germination

Germination is of three types : **epigeal** or epigeous, **hypogeal** or hypogeous and **viviparous.**

1. *Epigeal* or *epigeous germination.* In this type, the *hypocotyl* grows first and it pushes the cotyledonary node and all other parts of the seed out of the soil.

This type of germination is found in many dicotyledonous and a few monocotyledonous plants. Common examples are :

   a. *Exalbuminous* dicotyledonous, e.g., gourd, bean, tamarind, etc.
   b. *Exalbuminous* monocotyledonous, e.g., *Alisma plantago,* etc.
   c. *Albuminous* dicotyledonous, e.g., castor etc.
   d. *Albuminous* monocotyledonous, e.g., onion, *Typha,* etc.

2. *Hypogeal* or *hypogeous germination.* In this type, the epicotyl grows first and only the plumule is pushed out of the soil, while the cotyledons and all other parts remain at or below the soil surface.

This type of germination is noted in some dicotyledonous and many monocotyledonous plants. Common examples are :

   a. *Exalbuminous* dicotyledonous, e.g., pea, gram, mango, etc.
   b. *Exalbuminous* monocotyledonous, e.g., *Pothos, Monstera,* etc.
   c. *Albuminous* dicotyledonous, e.g., lotus, custard apple, etc.
   d. *Albuminous* monocotyledonous, e.g., maize, wheat, paddy, etc.

3. *Viviparous germination.* In this type, the seed(s), instead of being liberated from the fruit, germinates inside the fruit which is still attached with the mother plant. The embryo comes out of the fruit with the well-developed hypocotyl pointing downwards. Later on, it falls vertically downwards into the mud present around the mother plant by its own weight and fixes itself in the ground.

Vivipary is common in some mangrove plants like *Rhizophora mucronata, Ceriops roxburghiana, Kandelia rheedi* etc. of Rhizophoraceae; *Aegiceras majus* of Myrsinaceae and *Avicenia officinalis* of Verbenaceae. Vivipary has also been reported in some mesophytes like squash (*Sechium edule*) of Cucurbitaceae, jack (*Artocarpus heterophyllus*), etc.

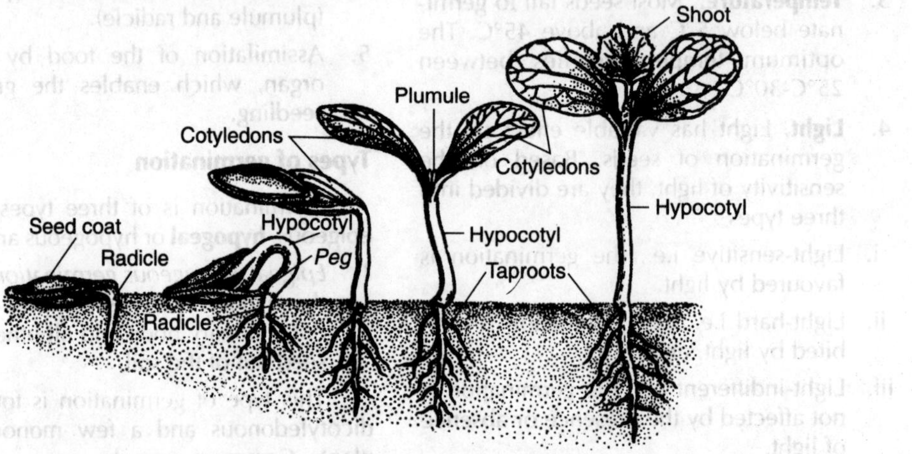

Fig. 2.8 : Epigeal germination of an *exalbuminous* dicotyledonous seed (Gourd)

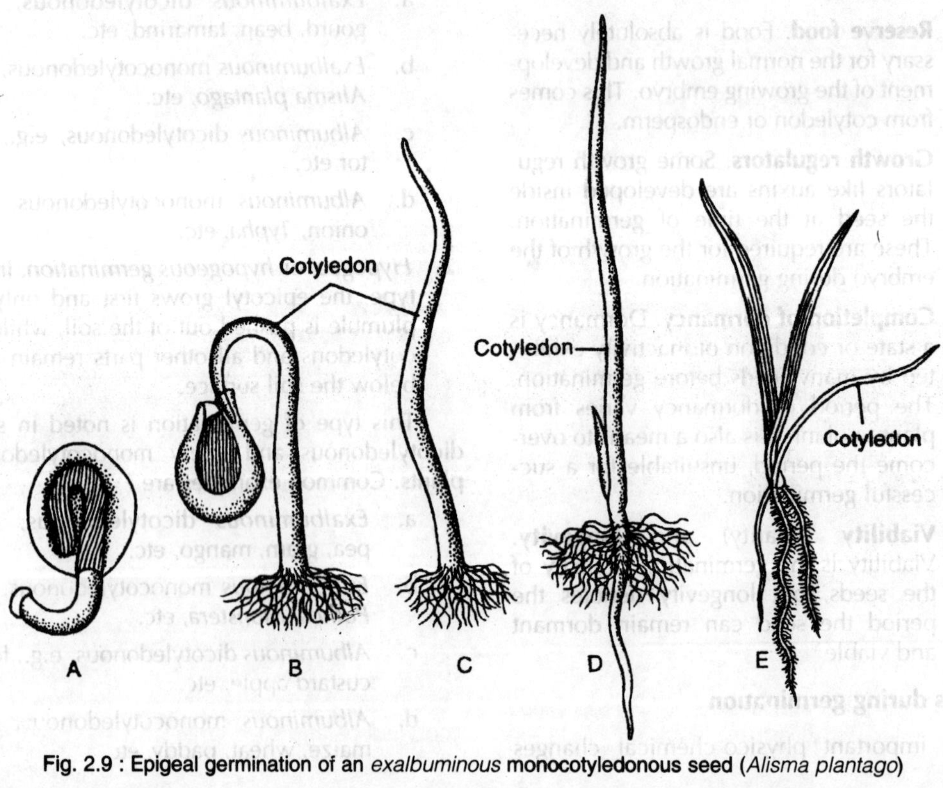

Fig. 2.9 : Epigeal germination of an *exalbuminous* monocotyledonous seed (*Alisma plantago*)

Germination of some common seeds

1. ***Epigeal type***

   a. Germination of an *exalbuminous* dicotyledonous seed.

Gourd (*Cucurbita maxima* of Cucurbitaceae)

Initially, after soaking water, the radicle comes out of the seed, goes straight downward and fixes the seed with the soil. The hypocotyl next grows very quickly and forms a loop-like structure and comes out of the soil. Gradually, it pulls the entire seed on the ground. Similarly the lower half of the seed coat gets attached to the peg-like projection developed at the base of the hypocotyl. Due to this, the seed coats cast off very easily and the cotyledons come out in the air. The cotyledon then opens like leaves. Later on, the plumule (within the cotyledon) becomes exposed and develops into an aerial shoot (Fig. 2.8).

   b. Germination of an exalbuminous monocotyledonous seed.

Alisma plantago of Alismataceae

Initially, after soaking water, the hypocotyl comes out of the seed. From the base of the hypocotyl, a bunch of root hair develops around the insignificant radicle. After that, the cotyledon comes out of the seed as an erect stand by its own development. Meanwhile the radicle elongates to give rise to the primary root. The plumule develops into a shoot (Fig. 2.9).

   c. Germination of an albuminous dicotyledonous seed.

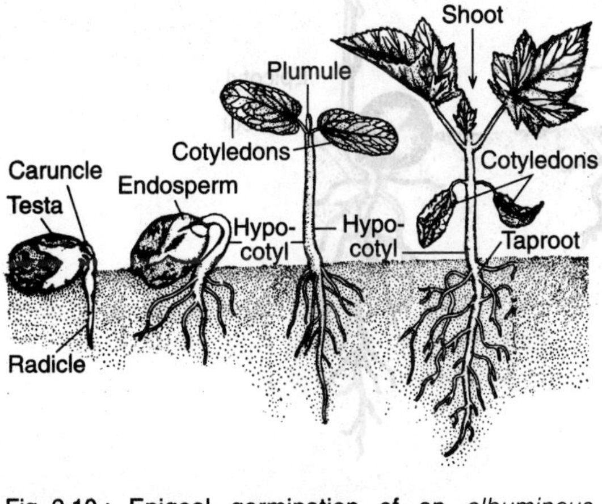

Fig. 2.10 : Epigeal germination of an *albuminous* dicotyledonous seed (Castor)

Caster (*Ricinus communis* of Euphorbiaceae)

During germination, the hard testa first bursts near the *caruncle*. Gradually the radicle comes out and fixes into the soil. With subsequent growth of the hypocotyl, the cotyledons (enclosed by endosperm) come out of the soil and the testa cracks and cast off. The endosperm becomes consumed by the cotyledons. The cotyledons open like leaves and become green in colour. The remnant of the endosperm gradually withers and drops off. The plumule between the cotyledons gradually grows up and develops into a new leafy shoot (Fig. 2.10).

   d. Germination of an albuminous monocotyledonous seed.

Onion (*Allium cepa* of Liliaceae)

During germination, the radicle, along with the base of the curved cotyledon, comes out of the seed. The radicle penetrates the soil. The other end of the cotyledon remains in the endosperm and draws nutrition. The base of the cotyledon further grows and becomes green, which pushes the seed out of the soil. For a long time, the plumule is not visible; as it covered by the cotyledonary base in the form of a sheath above the radicle. The plumule now pierces the sheath and forms the first foliage leaf. By this time, the adventitious root develops from the region above the radicle and forms fibrous root system (Fig. 2.11).

This type of germination is very rare in monocotyledonous plants.

2. ***Hypogeal type***

   a. Germination of an exalbuminous dicotyledonous seed.

   i. **Pea seed** (*Pisum sativum* of Fabaceae)

In presence of all essential requirements at a satisfactory level, the radicle and plumule start to develop. The radicle usually grows first and comes out of the seed coat through the micropyle. It grows vertically downwards and, after entering the soil, produces first primary and then gradually secondary, tertiary branches in acropetal order — thus it forms a root system.

During the radicle development the epicotyl begin to grow, initially by forming a loop. Later on, it straightens and comes out of the cotyledons and seed coat. Then it grows rapidly

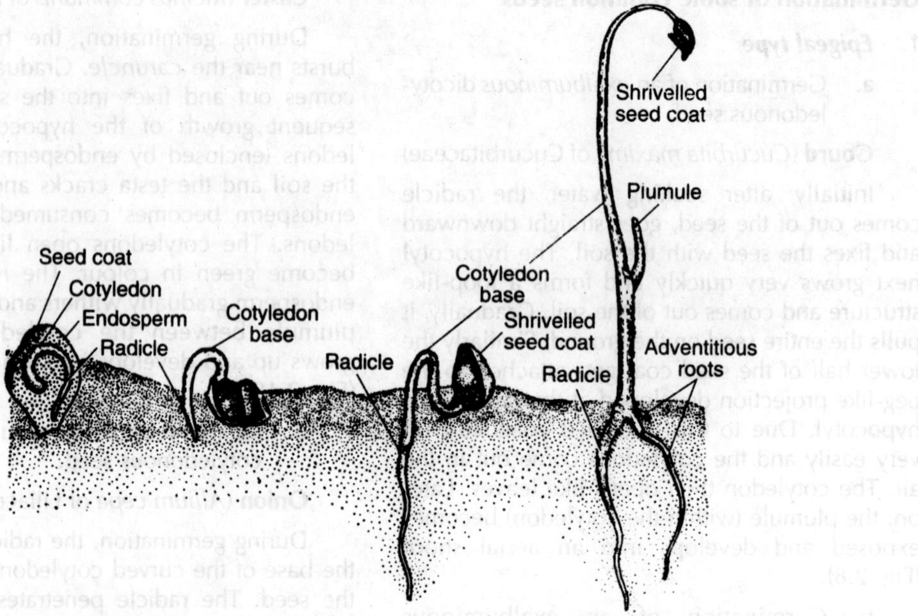

Fig. 2.11 : Epigeal germination of an *albuminous* monocotyledonous seed (Onion)

and gives rise to stem and then leaves towards the apical region.

During the above process of germination, the cotyledon does not change its position and remains inside the seed coat. Hence the germination is of hypogeal type. The cotyledons are generally reduced in size and gradually wither away (Fig. 2.12).

ii. **Gram seed** (*Cicer arietinum* of Fabaceae)

The mode of germination of gram is similar to pea i.e., hypogeal type (Fig. 2.13).

b. Germination of an albuminous monotyledonous seed.

i. **Paddy** (*Oryza sativa* of Poaceae)

After water soaking, the coleorhiza pierces

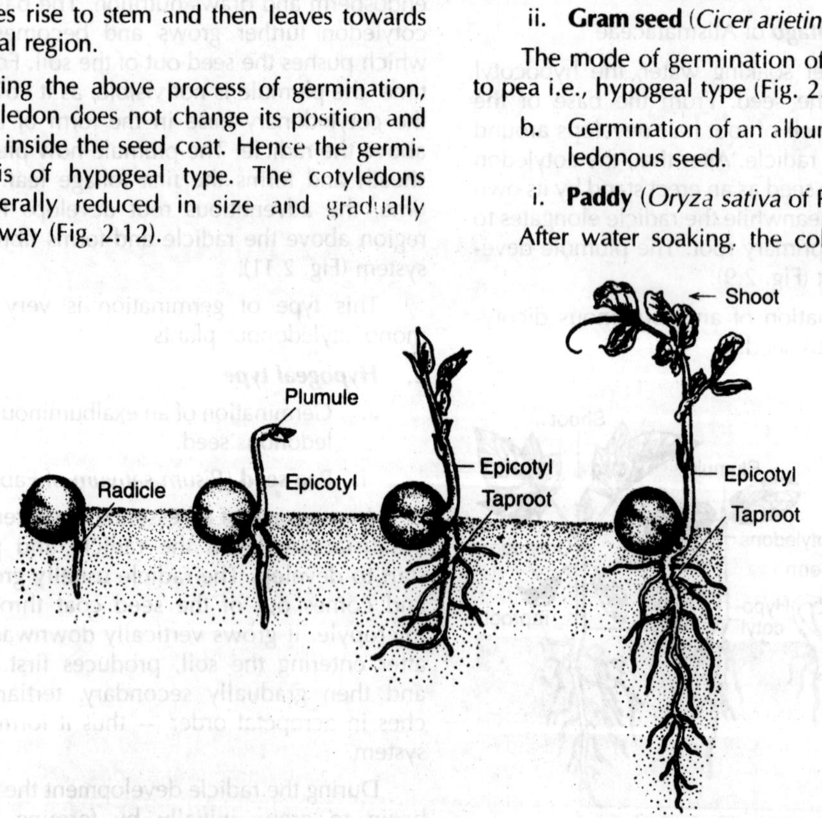

Fig. 2.12 : Hypogeal germination of an *exalbuminous* dicotyledonous seed (Pea)

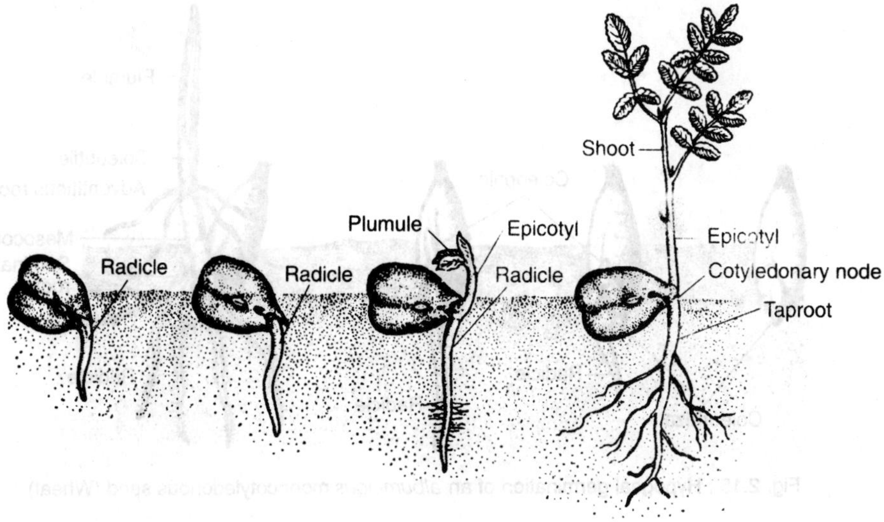

Fig. 2.13 : Hypogeal germination of an *albuminous* dicotyledonous seed (Gram)

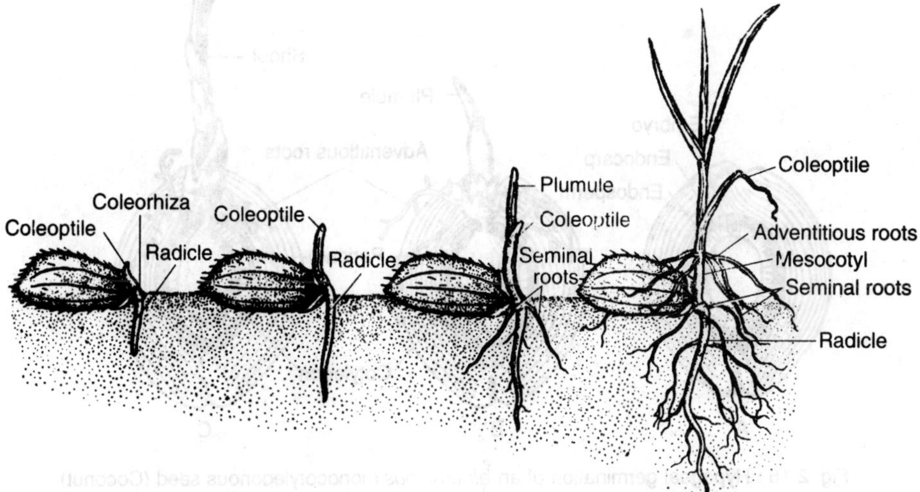

Fig. 2.14 : Hypogeal germination of an *albuminous* monocotyledonous seed (Paddy)

the base of the grain and appears as a knob. The radicle then splits the coleorhiza and penetrates the soil. The coleoptile then comes out. Concomitantly, after the emergence of the radicle, two other roots grow from its base — the *seminal roots.* The radicle and seminal roots give rise to secondary branches. Unlike dicotyledons, the radicle and seminal roots do not form the root system. At the base of the coleoptile, the mesocotyl becomes slightly elongated. The plumule then comes out of the coleoptile. *Adventitious roots* then develop from the top of the mesocotyl i.e., from the base of hypocotyl. These adventitious roots actually form the fibrous root system. Within short time the coleoptile becomes damaged (Fig. 2.14).

ii. **Wheat** (*Triticum aestivum* of Poaceae)

The germination of wheat is almost like paddy. But the number of *seminal roots* is generally 4-5. Adventitious roots also develop above the mesocotyl. Some of the adventitious roots may develop on the coleoptile.

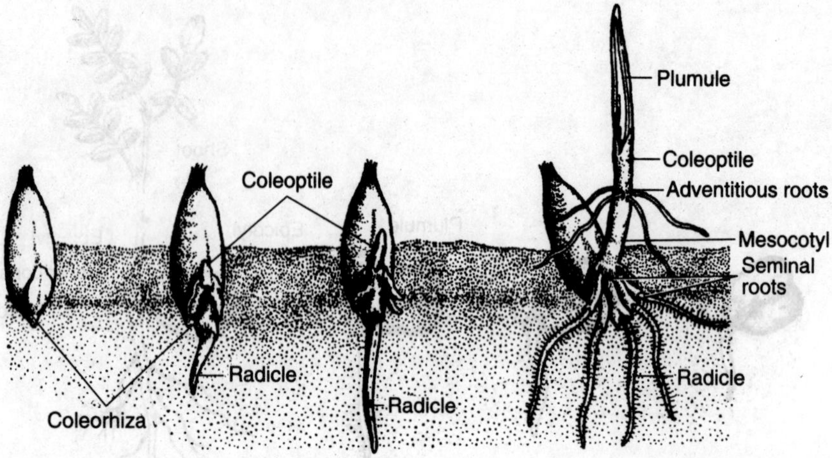

Fig. 2.15 : Hypogeal germination of an *albuminous* monocotyledonous seed (Wheat)

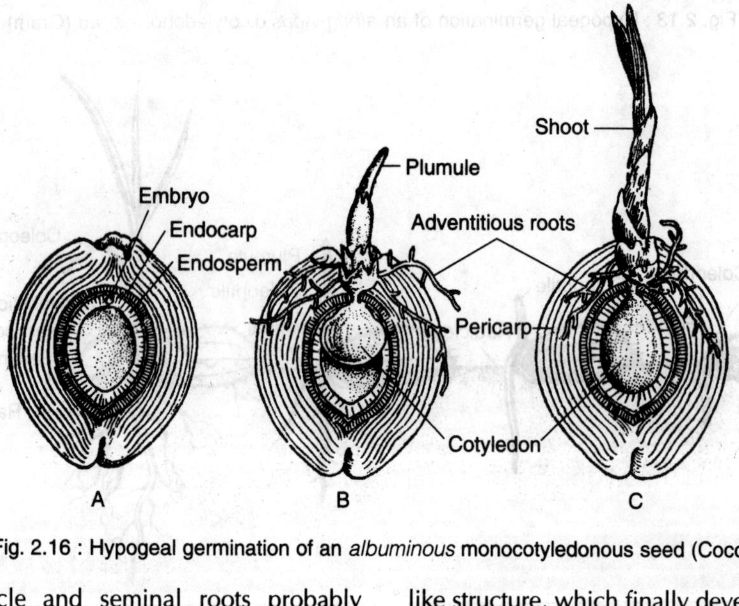

Fig. 2.16 : Hypogeal germination of an *albuminous* monocotyledonous seed (Coconut)

The radicle and seminal roots probably persist along with the adventitious root and function normally (Fig. 2.15).

iii. **Coconut** (*Cocos nucifera* of Arecaceae)

Initially, the small embryo remains undifferentiated below the eye on the endocarp inside the endosperm. During germination the lower end of the embryo develops a more or less round spongy structure — the cotyledon. It absorbs nutrition from the endosperm and becomes enlarged in size. The upper end develops through the eye carrying the radicle and plumule. Later on, the plumule pierces the mesocarp and epicarp and finally comes out in the air in the form of a horn-like structure, which finally develops into an aerial shoot. The radicle does not develop any more. Several adventitious roots grow from the base of the plumule. The roots also penetrate the mesocarp and epicarp. Later on, the roots come in contact with the soil and establish the plant (Fig. 2.16) on the substratum.

3. ***Vivipary*** or ***Viviparous type***

In this type, the seed — instead of being liberated from the fruit — germinates inside the fruit that is still attached to the mother plant.

Vivipary is common in some mangrove plants like *Rhizophora mucronata, Ceriops roxburghiana, Kandelia rheedi* etc. of Rhizophoraceae;

Aegiceras majus of Myrsinaceae and *Avicenia officinalis* of Verbenaceae that grow in the Sundarbans in Gangetic delta and Indo-Burmese rivers. The embryo of this type of seeds has no resting period and continues to grow inside the fruit. The radicle initially comes out of the fruit and then the hypocotyl begins to grow vigorously, thereby it looks like a club, whose length is usually 2-10 inch (5-25 cm). The hypocotyl points downward during its growth. Slight growth of the plumule also takes place, while the cotyledons remain inside the fruit acting as haustoria. The *Rhizophora* (Fig. 2.17) grows in the deepest water and has largest hypocotyl. When the hypocotyl becomes heavy, the entire fruit or the elongated axis i.e., the radicle, hypocotyl and plumule get detached from the plant and fall vertically downwards into the mud present around the mother plant by its own weight and fixes itself in the ground.

It is however, visible in squash, *Sechium edule* of Cucurbitaceae; a vegetable fruit commonly found in Indian hill towns; where the seeds germinate inside the fruit still attached with the mother plant. It is also visible in coconut, *Cocos nucifera* of Palmae. Vivipary is also found in jack, orange, tomato, lemon, *Agave* etc.

Economic importance of seeds

The seeds are the most important source of different types of food for the human being and also for animals like cattles etc.

Uses of different types of seeds are :

1. **Main food** : Seeds of rice, wheat and maize are used as the main staple food for the human being. In addition to the above three, other seeds like millet, barley, oat, sorghum etc., are used as food by both man and cattle.
2. **Pulses** : Seeds of lentil, pea, gram, soya beans, beans and others are used as pulses due to rich source of vegetable proteins.
3. **Other foods** : Different kinds of nuts like walnut, cashew nut, chestnut, coconut, almond etc. are also directly consumed as food.
4. **Oil** :
   a. **As vegetable oil** : Oil extracted from the seeds of groundnut, coconut, mustard, linseed etc. are the vegetable oils, used in cooking.
   b. **As hair oil** : Oil extracted from the dried endosperm of coconut is used as hair oil.

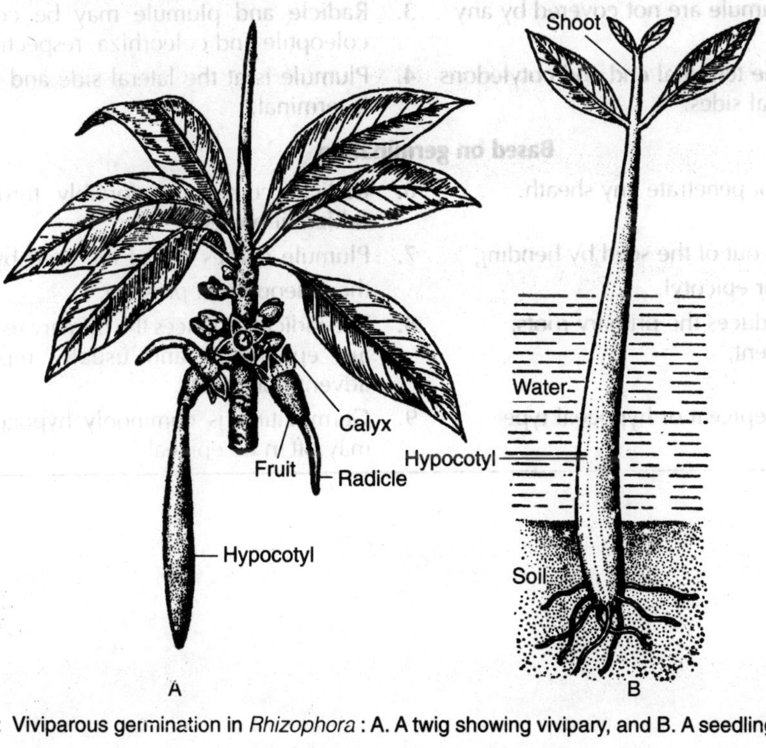

Fig. 2.17 : Viviparous germination in *Rhizophora* : A. A twig showing vivipary, and B. A seedling established in tidal soil

c. **In paint** and **varnish** : Oil extracted from cotton, linseed, and flax, tung oil seed etc., are used extensively in paint and varnish industries.

d. **As engine lubricant** and **fuel** : The castor oil extracted from the seeds of castor is used as a fuel and engine-lubricant.

e. **In medicine** : The castor oil is used as medicine. The *Chaulmoogra oil* extracted from the seeds of *Taraktogenos kurzii* (Bixaceae) is extensively used in the treatment of skin diseases.

f. **As adulterant** : Oil from the seeds of *Argemone mexicana* of Papaveraceae, *Guizotia abyssinica* of Asteraceae, *Madhuca latifolia* of Sapotaceae, etc. are often used in adulterating edible oils.

5. **Coffee** : The commercial coffee obtained from the seeds of *Coffea arabica* of Rubiaceae is used as beverage.

6. **Alcoholic drinks** : Alcoholic drinks are usually prepared from the seeds of some cereals like barley etc.

7. **Silk cotton** : The silk cotton fibre is the hairy outgrowth of the seed. It is not only used in textile industry, medical treatment, but also in the manufacture of tyres, explosives, cellophane etc. It is also used in the preparation of pillows, mattresses etc.

8. **Dye** : Dye like **annatto**, obtained from the seeds of *Bixa orelana* (Bixaceae) is used in dyeing and staining of butter.

9. **In Jewellery** : The red seeds of *Abrus precatorius* of Fabaceae are light in weight and are used in jewellery.

Table 2.1 : Differences between Dicotyledonous and Monocotyledonous seeds

Dicotyledonous Seeds	Monocotyledonous Seeds
Based on the structure	
1. Usually seed coats are free from pericarp.	1. Usually seeds coats are fused with the pericarp in the albuminous seeds.
2. Embryo is large and has two cotyledons.	2. Embryo is very small and has one cotyledon.
3. Radicle and plumule are not covered by any sheath.	3. Radicle and plumule may be covered by coleoptile and coleorhiza, respectively.
4. Plumule is at the terminal end and cotyledons are on the lateral sides.	4. Plumule is at the lateral side and cotyledon is terminal.
Based on germination	
6. Radicle does not penetrate any sheath.	6. Radicle comes out forcibly through cole orhiza (if present).
7. Plumule comes out of the seed by bending the hypocotyl or epicotyl.	7. Plumule comes out of the seed by piercing the coleoptile (if present).
8. The radicle produces the primary roots, which is persistent.	8. The radicle produces the primary roots, which are ephemeral and usually replaced by adventitious roots.
9. Germination is epigeal or hypogeal type.	9. Germination is commonly hypogeal, but it may often be epigeal.

ANGIOSPERM PLANT BODY

A seed contains an embryo (baby plant) which germinates in favourable condition and gives rise to an adult plant.

The body of an angiosperm plant consists of two systems : the **root system** that grows below the soil and the **shoot system** that grows above the soil. A typical root system anchors the plant in the soil and absorbs water along with various dissolved raw materials. On the other hand, the typical shoot system consists of a main stem, along with leaves, branches and sometimes with flowers (Table 2.2). Leaves contain chlorophyll by which they produce food through photosynthesis. The water along with dissolved minerals passes to the leaf through the stem and the produced food is translocated normally to the diffe-rent organs through this stem only. Sometimes the foods are stored in special organs for future use. Later on, the plants develop flowers. Flowers are the organs that concerned with reproduction (Fig. 2.18). Sometimes the stems become so small, they are not visible from outside. Such stems are called acaulescent, e.g., onion (*Allium cepa*), garlic (*A. sativum*), tuberose (*Polianthes tuberosa*), etc.

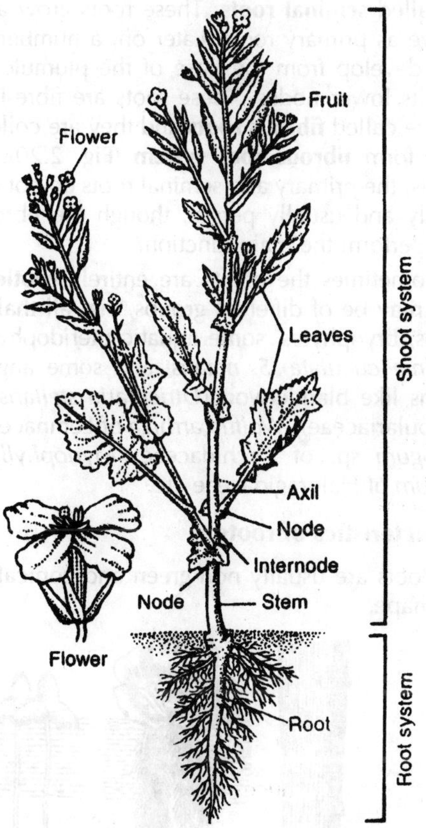

Fig. 2.18 : Different parts of an Angiospermic plant (*Brassica nigra*) of Brassicaceae

Table 2.2 : Parts of a typical angiospermic plant are tabulated :

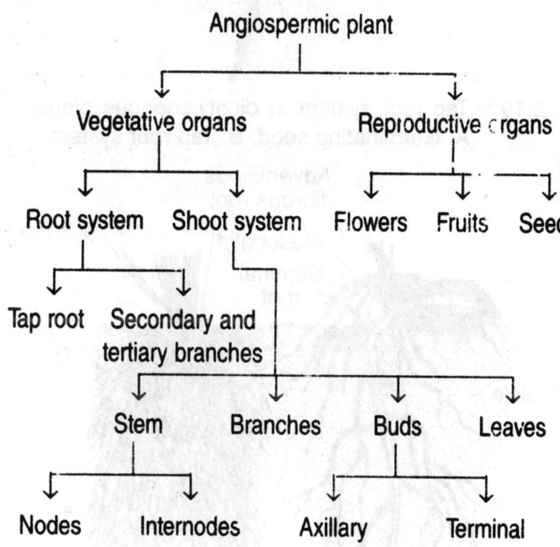

ROOT

Root is the descending part of plant which generally penetrates inside the soil and serves the function of fixation and absorption. The root which develops from the radicle is called **primary root**. The primary roots giving rise to lateral branches are called **secondary branches**, those become further branched and called **tertiary branches**. The primary root along with its branches forms **primary** or **normal** or **tap root system** (Fig. 2.19). The root branches arise from

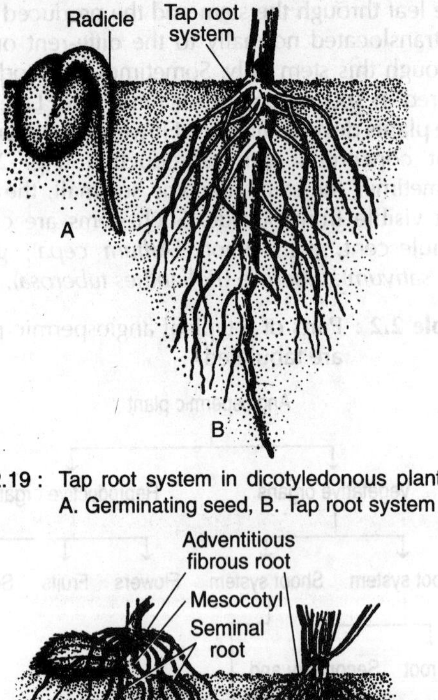

Fig. 2.19 : Tap root system in dicotyledonous plants : A. Germinating seed, B. Tap root system

Fig. 2.20 : Adventitious fibrous root system in monocotyledonous plant : A. Root region of rice seedling showing different regions of root, B. Rice plant with fibrous root system

an inner layer (pericycle) i.e., *endogenous in origin*. [While the shoot branches arise from the superficial layer (cortex) i.e., they are *exogenous in origin*]. Sometimes roots arise from anywhere (node of stem or on margin of leaves) except the radicle are called **adventitious root.** In most of the monocotyledons, the primary root either dies very early or its growth becomes arrested and new roots develop from the base of the radicle are called **seminal roots**. These roots grow and behave as primary roots. Later on, a number of roots develop from the base of the plumule or from its lower nodes. These roots are fibre-like and are called **fibrous roots** and they are collectively form **fibrous root system** (Fig. 2.20). In cereals, the primary and seminal roots do not die quickly and usually persist, though the fibrous roots perform the main function.

Sometimes the plants are entirely **rootless.** They may be of different groups, viz. all thallophytes, bryophytes, some aquatic pteridophytes (*Salvinia cuculata, S. natans*) and some angiosperms like bladderwort (*Utricularia stellaris* of Lentibulariaceae), *Wolffia arrhiza* of Lemnaceae, *Epipogum* sp. of Orchidaceae, *Myriophyllum indicum* of Haloragidaceae etc.

Characteristics of roots

1. Roots are usually non-green and conical in shape.

Fig. 2.21 : Root caps and root pockets : A. Simple root cap of hanging aerial prop root of *Ficus benghalensis*, B. Multiple root cap of *Pandanus foetidus*, C. Plant body of *Pistia stratiotes* showing root pockets and D. Plant body of *Lemna trisulca* showing root pocket

2. They are usually **negatively phototropic** (grow away from light), **positively geotropic** (grow in the direction of gravitational force) and **positively hydrotropic** (grow in water medium).
3. They do not bear leaves, flowers, except a few cases.
4. They develop branches of endogenous origin (develop from pericycle).
5. The growing apex is normally protected by a cap, called **root cap** or **calyptra** (Fig. 2.21a, b). In some water plants, an elongated cap-like structure (like a finger of a glove) develops at the apex, called **root pocket** (Fig. 2.21C, D).
6. The root cap may be **simple** (banyan, *Ficus benghalensis* of Moraceae; *Amaryllis* of Amaryllidaceae) or **multiple** (screw pine, *Pandanus foetidus* of Pandanaceae). [During growth, the root forces its way through the soil. At that time, the outer surface of cap becomes gelatinous and secretes a slimy substance to minimise the friction with the soil particles. With further growth the old root cap withers away and new root cap develops at the apex. The root pocket, unlike root cap, does not arise from epiblema and does not show regular renovation.]
7. Normally the root consists of four regions (Fig. 2.22). These are (from the apex) : (i) root cap, (ii) growing region, (iii) root hair region, and (iv) permanent region. [**Root hairs are the unicellular structure, visible with naked eye.** The root hairs normally live a few days to a few months or rarely even few years in some members of Asteraceae.]
8. Some minor modifications are often available with the roots growing in different environmental conditions. These are :
   a. The aquatic roots do not bear root hairs.
   b. The epiphytic root (in some orchids) develops a special kind of absorbing sheath, the **velamen** (Fig. 2.26A).
   c. The xerophytic plants (plants grow in dry condition like desert) normally have deep root system to get more water and some have an additional superficial root to absorb atmospheric precipitation like rain etc.
   d. The halophytic plants (plants grow in saline soil at or near the sea coast region) have negatively geotropic aerial roots with many pores, called **pneumatophores** (Fig. 2.24) or **respiratory roots** in addition to normal root system.

Classification

According to structure and mode of development, the roots are classified into two types. These are :

1. **Tap root** (Fig. 2.19B). It forms by the continuous growth of radicle and is usually found in dicotyledonous plants, e.g., mango (*Mangifera indica* of Anacardiaceae).
2. **Adventitious root** (Fig. 2.20B). The roots neither develop from radicle nor primary roots and are called adventitious roots. They may develop from any other region of the plant body, viz. (i) from **leaves**, e.g., *Bryophyllum calycinum* and *Kalanchoe laciniata* of Crassulaceae, (ii) from **stems**, e.g., *Zea mays* (maize), *Bambusa tulda* (bamboo) and *Saccharum officinarum* (sugarcane) of Gramineae and (iii) from **lower end of stem cuttings**, e.g., *Hibiscus rosa-sinensis* (china-rose) of Malvaceae, *Rosa centifolia* (rose) of Rosaceae etc.

[The schematic representation of various types of roots has been given in Table 2.3].

Regions of a typical root

A typical root has the following regions (Fig. 2.22) :

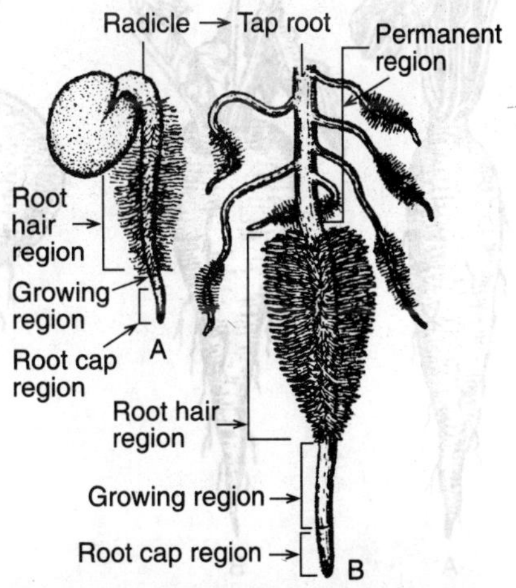

Fig. 2.22 : Roots showing four different regions : A. Single germinating seed, and B. Apical region of root of a mature plant

1. **Root cap zone**. At the apex of the root a cap-shaped structure is present, called root cap. It gives protection to the young apical region during growth of root against soil particles.

2. **Growing region** or **Meristematic zone**. This region lies just above the root cap region. Growth of the roots at this region takes place by mitotic cell division and cell elongation. This zone consists of two regions : (a) region of cell division, and (b) region of cell elongation. The cells of this region are thin walled with dense cytoplasm and with single large nucleus.

3. **Root hair region**. This region is situated just behind the growing region and is covered by unicellular root hairs. The root hairs serve the functions of both anchorage and absorption of water and solutes.

4. **Permanent region**. It is the region situated behind the root hair zone. The growth of this region is seized. It serves the function of anchorage and translocation of substances absorbed by root hairs.

Modification of roots

In addition to normal functions like anchorage and absorption of water with minerals, roots perform some special functions for which they get modified in various ways :

A. **Modified Tap Roots**. The tap roots are completely or partially modified according to their functional need. These are :

(a) For storage of food-matters : Due to accumulation of reserve food, the tap roots become fleshy and swollen and assume different shapes. These are of four different types :

i. *Fusiform*. These roots are thick in the middle and gradually taper at both the ends. On both the sides, secondary and tertiary roots are developed (Fig. 2.23A), e.g., radish (*Raphanus sativus*) of Brassicaceae.

ii. *Conical*. These roots are broadest at the base and gradually taper at the lower end (Fig. 2.23B), e.g., carrot (*Daucus carota* of Apiaceae).

iii. *Napiform* These roots get much swollen above, but abruptly taper towards the lower end (Fig. 2.23C), e.g., beet (*Beta vulgaris* of Chenopodiaceae), turnip (*Brassica rapa* of Brassicaceae).

iv. *Tuberous*. These roots get swollen in any portion, thus they do not have a regular shape (Fig. 2.23D), e.g., *Ruellia tuberosa* of Acanthaceae, 4'o clock plant (*Mirabilis jalapa* of Nyctaginaceae).

(b) For physiological functions (centre for N_2 fixation, respiration etc.)

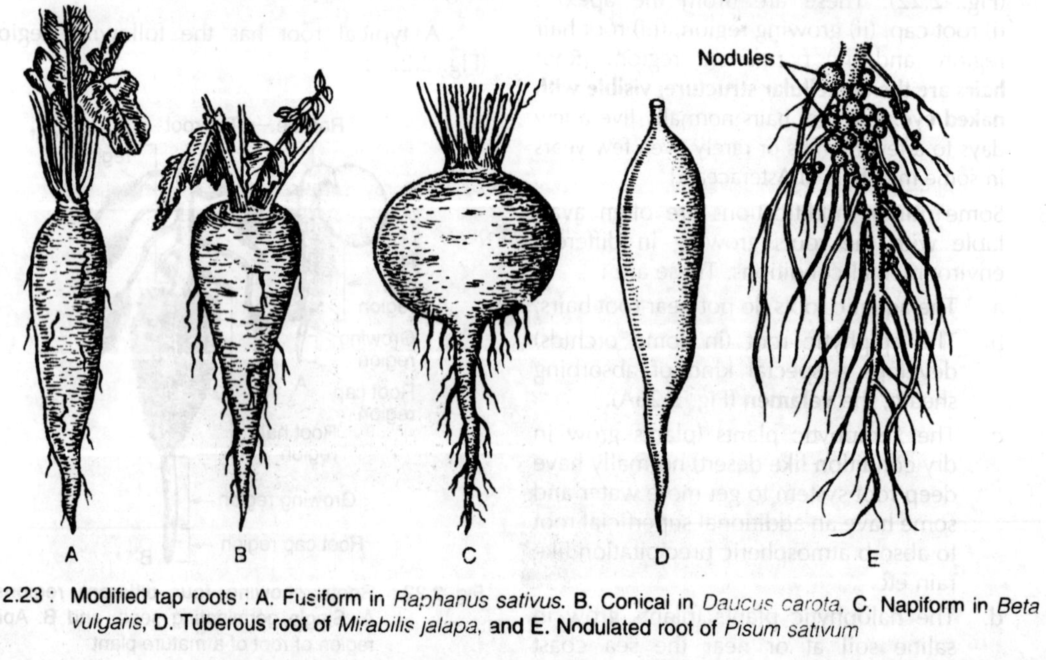

Fig. 2.23 : Modified tap roots : A. Fusiform in *Raphanus sativus*. B. Conical in *Daucus carota*, C. Napiform in *Beta vulgaris*, D. Tuberous root of *Mirabilis jalapa*, and E. Nodulated root of *Pisum sativum*

i. *Nodulated root.* The nodules are formed on the main root as well as on the branches (Fig. 2.23E). These nodules contain nitrogen-fixing bacteria. Viz., different members of Leguminosae (Fabaceae) such as gram (*Cicer arietinum*), garden pea (*Pisum sativum*) etc.

ii. *Respiratory root.* These are the branches of normal tap root system; those grow vertically above the soil or water-level (negatively geotropic) for respiration. The apices of these roots have many pores (**pneumathodes**) on their wall, through which air is conducted down into the subterranean roots. These negatively geotropic roots are called **pneumatophores** or **respiratory roots** (Fig. 2.24A, B). These are found in halophytes (plants grow in saline soil) such as *Rhizophora mucronata, Ceriops roxburghiana* of Rhizophoraceae, *Sonneratia apetala* of Sonnaratiaceae, *Heritiera minor* of Sterculiaceae, etc.

B. **Modified Adventitious Roots.** The adventitious roots are variously modified for different functions : (a) Mechanical and (b) Physiological.

(a) **For Mechanical functions** (fixation, support etc.).

i. *Prop roots.* These roots are developed from the horizontal branches of some trees in vertically downward direction. On reaching the soil each root grows into thick and woody pillar-like structure which gives support to the branch from where it developed. These are found in Indian rubber (*Ficus elastica*) and banyan (*Ficus benghalensis*, Fig. 2.25A) of Moraceae.

ii. *Stilt roots.* These are stout roots developed obliquely from the lower part of the trunk; grows on soft soil. On reaching the soil they help the mother plant to keep erect and also give an additional anchorage support to prevent from being uprooted. These are found in screwpine (*Pandanus foetida*, Fig. 2.25B) of Pandanaceae; maize, sugarcane etc. of Poaceae.

iii. *Root-buttresses* or *plank-buttresses.* These are plank-like structures developed with the contribution of both stem and root, come out as erect projections from the basal region of the trunk. These plank-like structures are called root-buttresses or plank-buttresses (Fig. 2.25C). These are found in *Bombax ceiba* of Bombacaceae; *Terminalia catappa* of Combretaceae, *Sterculia alata* of Sterculiaceae etc.

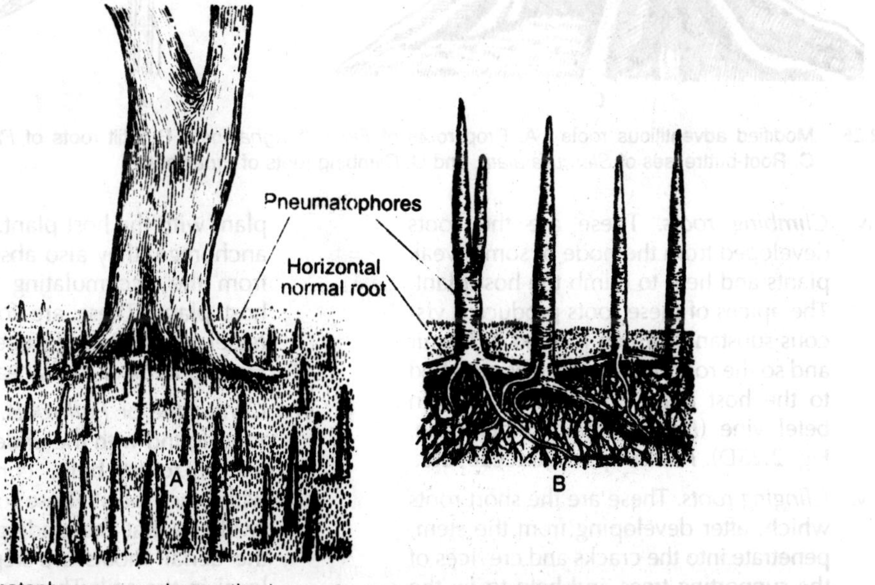

Fig. 2.24 : A. Pneumatophores at and near the base of mangrove tree B. Pneumatophores developed from the horizontal normal root

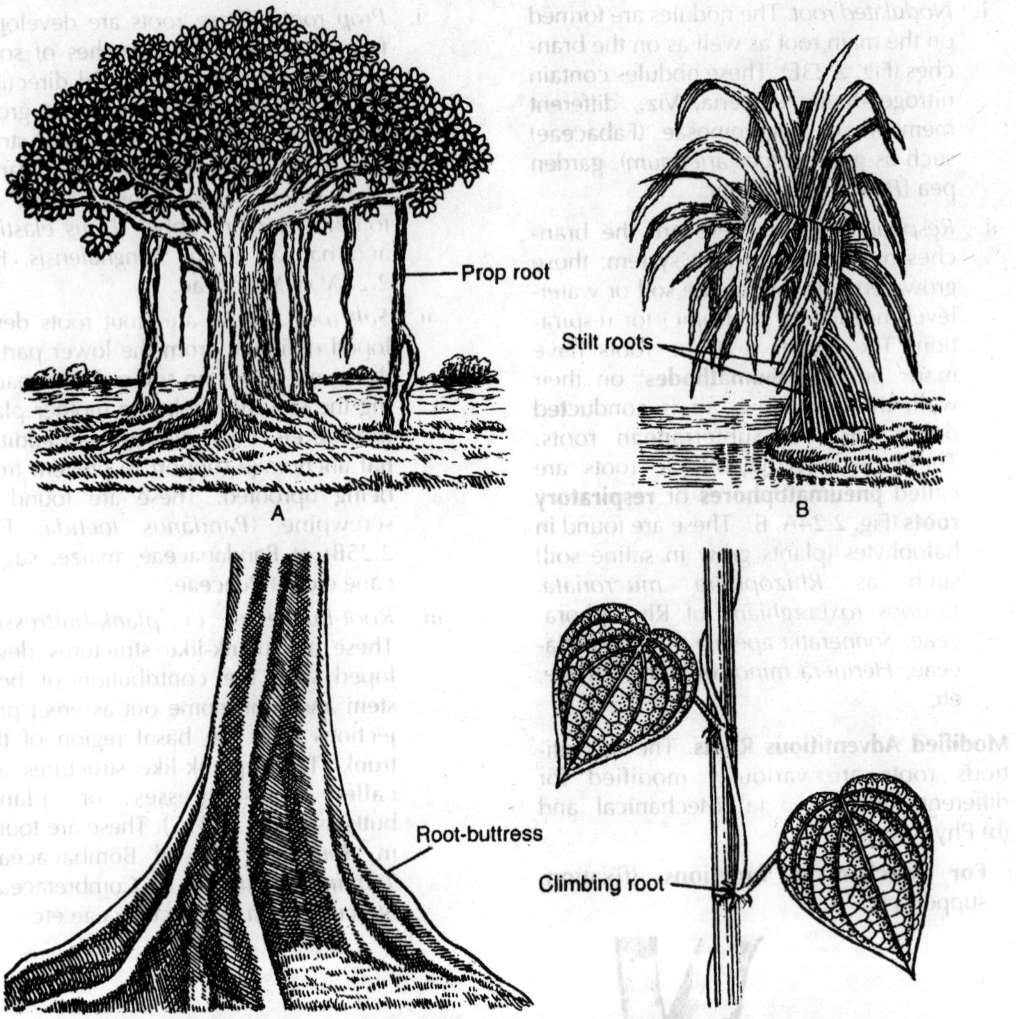

Fig. 2.25 : **Modified adventitious roots : A. Prop roots of *Ficus benghalensis*, B. Stilt roots of *Pandanus foetida*, C. Root-buttresses of *Sterculia alata*, and D. Climbing roots of *Piper betel***

iv. *Climbing roots*. These are the roots developed from the node of some weak plants and help to climb the host plant. The apices of these roots produce a viscous substance which is dried in the air and so the roots get themselves attached to the host plant. These are found in betel vine (*Piper betel* of Piperaceae, Fig. 2.25D), *Pothos* sp. of Araceae, etc.

v. *Clinging roots*. These are the short roots which, after developing from the stem, penetrate into the cracks and crevices of the supporting trees and help to fix the plant with the host plant. In addition to anchorage, they also absorb some food from the accumulating debris on the host plant. These are found in *Vanda tesselata* of Orchidaceae (Fig. 2.26A), *Hedera helix* of Araliaceae.

vi. *Contractile* or *Pull roots*. These are the roots found with the underground stems like rhizome, bulb, corm etc., along with other roots. These roots contract or swell so that the underground part or the aerial shoots are kept at a proper level in the soil. These roots are called

Contractile or *pull roots* (Fig. 2.26B). This type of roots is found in *Allium, Crocus, Canna,* etc.

vii. *Floating roots.* These are small, inflated, spongy and white coloured roots developed adventitiously from the upper side on the node of the stem. These roots contain huge amount of air and help the plant to float on water; and also in respiration. These are found in *Jussiaea repens* of Onagraceae (Fig. 2.26C).

viii. *Root-thorns.* In some plants, adventitious roots developed from the base of the stem, being hard, pointed and thorn-like, are called root-thorns. As an armature they help the plant to protect itself

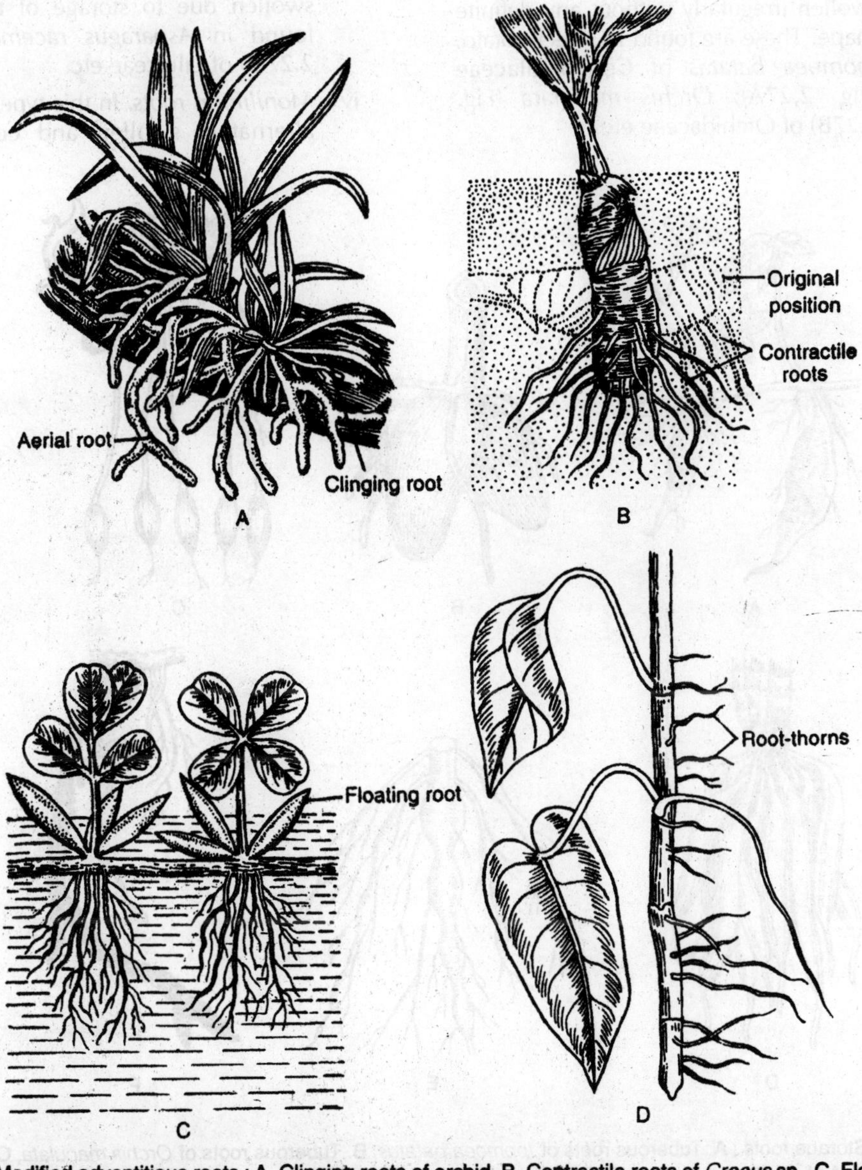

Fig. 2.26 : Modified adventitious roots : A. Clinging roots of orchid, B. Contractile roots of *Crocus* sp., C. Floating roots of *Jussiaea repens*, and D. Root-thorns of *Pothos armatus*

from marauding animals. These are found in *Pothos armatus* (Fig. 2.26D) and in many palms like *Iriartea* and *Acanthorhiza.*

(b) **For physiological functions** (storage of food, absorption of moisture, assimilation etc.).

For storage of food-matters

i. *Tuberous roots.* These roots arise from nodes of the prostrate stem and become swollen irregularly without any definite shape. These are found in sweet potato, *Ipomoea batatas* of Convolvulaceae (Fig. 2.27A); *Orchis maculata* (Fig. 2.27B) of Orchidaceae etc.

ii. *Nodulose roots.* This type of roots, after coming out of the rhizome, becomes swollen at their apex as a nodule-like structure due to storage of food. It is found in mango-ginger (*Curcuma amada*) and *Costus speciosus* (Fig. 2.27C) of Zingiberaceae.

iii. *Fasciculated roots.* In this type, all the fibrous adventitious roots — after originating in cluster from a common point at the base of the stem — become swollen due to storage of food. It is found in *Asparagus racemosus* (Fig. 2.27D) of Liliaceae etc.

iv. *Moniliform roots.* In this type, roots are alternately swollen and constricted,

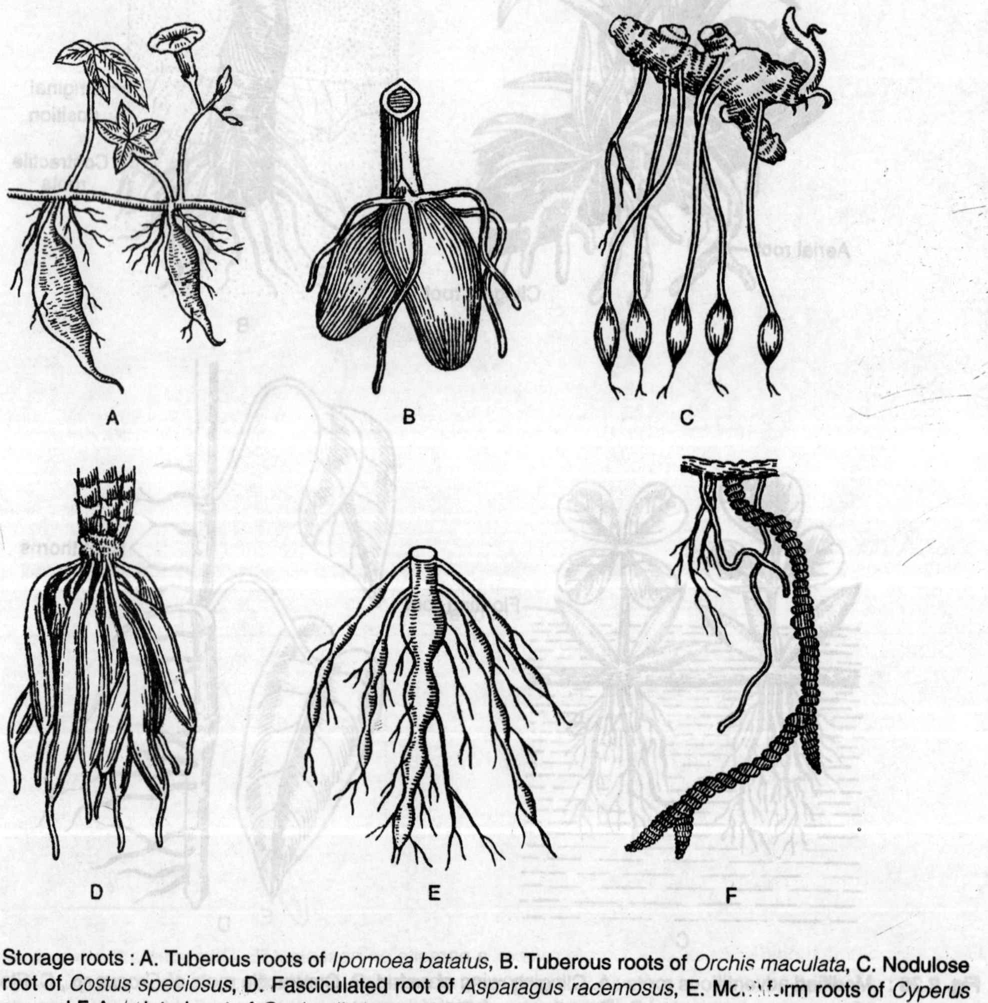

Fig. 2.27 : Storage roots : A. Tuberous roots of *Ipomoea batatus*, B. Tuberous roots of *Orchis maculata*, C. Nodulose root of *Costus speciosus*, D. Fasciculated root of *Asparagus racemosus*, E. Mc..'.f..rm roots of *Cyperus* sp. and F. Annulated root of *Cephaelis ipecacuanha*

thus they form a beaded structure. This is found in *Dioscorea alata* of Dioscoreaceae, *Cyperus* sp. of Cyperaceae (Fig. 2.27E), *Vitis trifolia* of Vitaceae etc.

v. *Annulated roots.* The roots look as if formed by several discs placed side by side. This is found in ipecac (*Cephaelis ipecacuanha* of Rubiaceae, Fig. 2.27F) — a medicinal plant found in the Himalayas.

For absorbing moisture from air

vi. *Epiphytic roots.* These are hanging **aerial roots** of epiphytic plants. The roots are covered by a thin greyish layer (4 -5 cell-layer in thickness) known as **velamen.** The velamen can absorb moisture from the atmosphere. It is found in orchid root (*Vanda tesselata* of Orchidaceae, Fig. 2.26A).

For sucking moisture from the host

vii. *Parasitic roots.* These are also called sucking roots or haustoria. The small root-like structure, after developing from the stem of the parasite, penetrates the tissue of the host plant and absorbs food.

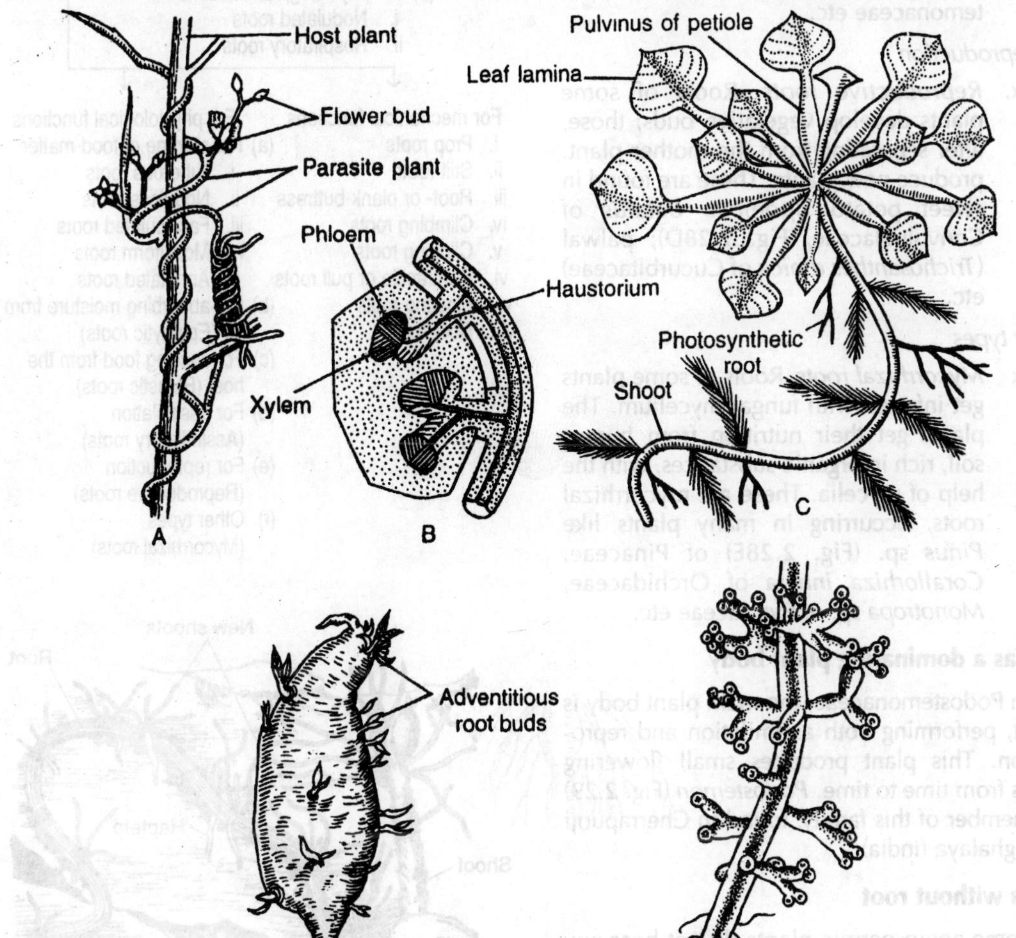

Fig. 2.28 : Some modified roots : A. *Cuscuta reflexa*, a parasite on host plant, B. Diagrammatic representation of T.S. of host along with *Cascuta* parasite showing penetration into the host, C. Assimilatory or photosynthetic root of *Trapa bispinosa*, D. Reproductive roots of *Ipomoea batatus*, and E. Mycorrhizal root of *Pinus* sp.

These roots are found in dodder (*Cuscuta reflexa* of Convolvulaceae Fig. 2.28A, B), sandal wood (*Santalum album* of Santalaceae), *Orobanche racemosa* of Orobanchaceae etc.

For assimilation

viii. *Assimilatory or photosynthetic roots.* Roots of some plants develop chlorophyll and become green. These roots are able to prepare carbohydrate foods through photosynthesis. These roots are found in *Trapa sp.* of Trapaceae (Fig. 2.28C), *Tinospora cordifolia* of Menispermaceae, *Podostemon* sp. of Podostemonaceae etc.

For reproduction

ix. *Reproductive roots.* Roots of some plants develop vegetative buds; those, after separating from the mother plant, produce new plants. These are found in sweet potato (*Ipomoea batatus* of Convolvulaceae, Fig. 2.28D); palwal (*Trichosanthes dioica* of Cucurbitaceae) etc.

Other types

x. *Mycorrhizal roots.* Roots of some plants get infested with fungal mycelium. The plants get their nutrition from humus soil, rich in organic substances, with the help of mycelia. These are mycorrhizal roots, occurring in many plants like *Pinus* sp. (Fig. 2.28E) of Pinaceae, *Corallorhiza innata* of Orchidaceae, *Monotropa* sp. of Pyrolaceae etc.

Root as a dominating plant body

In Podostemonaceae, the main plant body is a root, performing both assimilation and reproduction. This plant produces small flowering shoots from time to time. *Podostemon* (Fig. 2.29) is a member of this family, found in Cherrapunji of Meghalaya (India).

Plants without root

Some angiospermic plants do not bear any root. *Urticularia* (Fig. 2.76), a submerged hydrophyte, has some finely dissected leaves which carry on the function of the root. ***Wolffia arrhiza***, a very small angiosperm, does **not have** any root. Root is also absent in ***Epipogium***.

Table 2.3 : Roots of different types :

Roots
- True roots
  - Normal
  - Modified
    - (a) For storage of food-matter
      - i. Fusiform
      - ii. Conical
      - iii. Napiform
      - iv. Tuberous
    - (b) For physiological functions
      - i. Nodulated roots
      - ii. Respiratory roots
- Adventitious roots
  - Normal (Fibrous root)
  - Modified
    - For mechanical functions
      - i. Prop roots
      - ii. Stilt roots
      - iii. Root- or plank-buttress
      - iv. Climbing roots
      - v. Clinging roots
      - vi. Contractile or pull roots
      - vii. Floating roots
      - viii. Root-thorn
    - For physiological functions
      - (a) For storage of food-matter
        - i. Tuberous roots
        - ii. Nodulose roots
        - iii. Fasciculated roots
        - iv. Moniliform roots
        - v. Annulated roots
      - (b) For absorbing moisture from air (Epiphytic roots)
      - (c) For sucking food from the host (Parastic roots)
      - (d) For assimilation (Assimilatory roots)
      - (e) For reproduction (Reproductive roots)
      - (f) Other types (Mycorrhizal roots)

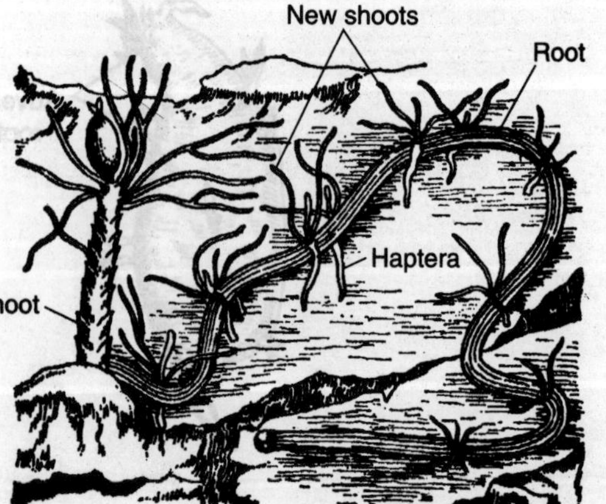

Fig. 2.29 : *Podostemon*, the main plant body is a assimilatory root clinging to the rock surface by haptera. Short-living shoots develop from this root

Functions of roots

The functions of roots are of two types. These are :

A. **Normal functions.**

a. *Mechanical.* Root helps in the fixation i.e., anchoring the plant with the substratum.

b. *Physiological.* Roots serve different physiological functions like :

i. Absorption of water containing various nutrients (sap) from the soil.

ii. Conduction of sap in upward direction.

iii. Storage of some amount of food in the permanent region.

B. **Special functions.** Besides the above normal functions, roots also carry some special functions of both mechanical and physiological types.

a. *Mechanical.*

i. Support to the long horizontal branches, e.g., prop root of banyan (*Ficus benghalensis*).

ii. Clinging and climbing, e.g., orchid (clinging root) and betel vine (climbing root).

iii. Support and additional anchorage, e.g., stilt root of screwpine (*Pandanus foetida*).

iv. Protection, like root thorns of some palms.

b. *Physiological.*

i. Assimilation, e.g., assimilatory root of *Trapa*.

ii. Storage, e.g., storage root of sweet potato (*Ipomoea batatas*).

iii. Absorption of moisture from air, e.g., epiphytic root of orchid (*Vanda tessellata*).

iv. Sucking from host, e.g., haustoria of dodder (*Cuscuta reflexa*).

v. Breathing, e.g., pneumatophores of mangrove plants (*Rhizophora mucronata*).

vi. Reproduction, e.g., reproductive root of sweet potato (*Ipomoea batatas*).

Economic importance

Economically, roots are very important to the mankind. Some of the beneficial importance of the roots are :

1. *Source of food for human being.* Sweet potato, carrot, beet, turnip are the tuberous roots used as human food.

2. *Source of sugar.* Modified root like sugar beet is a source of sugar.

3. Source of medicine. Some roots have high quantity of active substances of great medicinal importance. These are :

   i. Ipecac from *Cephaelis ipecacuanha* of Rubiaceae,

   ii. Aconite from *Aconitum napellus* of Ranunculaceae,

   iii. Sarsaparilla from *Smilax macrophylla* of Liliaceae,

   iv. Gentian from *Gentiana lutea* of Gentianaceae,

   v. Liquorice from *Glycyrrhiza glabra* of Fabaceae,

   vi. Rubarb from *Rheum officinale* of Polygonaceae.

4. Source of rubber. Root of *Taraxacum koksaghyz* of Asteraceae is a source of rubber.

5. Source of insecticide. Pyrethrum powder from *Chrysanthemum cinerariafolium* of Asteraceae is used as insecticide.

SHOOT : THE BUD AND STEM

The shoot consists of stem with branches and leaves, develops from the plumule. The shoot is usually positively phototropic and negatively geotropic.

The stem consists of *long axis* with *terminal bud* and *leaves* at the lateral sides. The bud develops in the axil (i.e., the angle between leaf stalk and stem) and is called axillary bud.

BUD

Bud is a condensed young shoot in which there are shortened internodes and the leaves envelop one another (Fig. 2.30A). During development of bud, the internodes become elongated and the leaves get separated, e.g., cabbage (*Brassica oleracea* var. *capitata* of Cruciferae, Fig. 2.30B, C).

Kinds of bud

Buds are of different types :

A. Depends on the organ it develops.

1. *Leaf bud.* It develops into a leaf, e.g., banyan tree (*Ficus benghalensis* of Moraceae).
2. *Stem bud.* It develops into a leafy branch.
3. *Flower bud.* It develops into flower.
4. *Mixed bud.* It develops into vegetative shoot and reproductive structures (i.e., flower), e.g., apple (*Pyrus malus* of Rosaceae), buckeye (*Aesculus ohioensis* of Hippocastanaceae) etc.

B. Depends on the position on plant.

Depending on the position of the plant, buds are divided into two groups : 1. Normal buds, and 2. Adventitious buds.

1. *Normal bud.* The buds developed either on the apex of the stem or in the axil of the leaves are called normal buds. Normal bud is also divided into two types : a. Terminal or Apical bud, and b. Axillary bud. The length of plant increases by the growth of apical bud, whereas the branching system depends on the activity of axillary buds.

   a. *Terminal* or *Apical bud.* The bud, when situated at the terminal position i.e., at the apex, is called terminal or apical bud (Fig. 2.31A).

   b. *Axillary bud.* The bud, when developed in the axil of leaves, is called axillary bud. Based on the **position** and **arrangement**, the axillary buds are divided into three types :

i. *alternate*, ii. *opposite,* and iii. *whorled*; they give rise to different branching patterns. All axillary buds do not develop into branches. Some may be always **active**, some remain **dormant**, but become active when required (if the main stem is cut off, the dormant lateral bud becomes active); and the buds that fall off instead of developing into shoots are called **deciduous buds.**

In addition to normal bud, two or more buds may develop in the axil. Depending on their

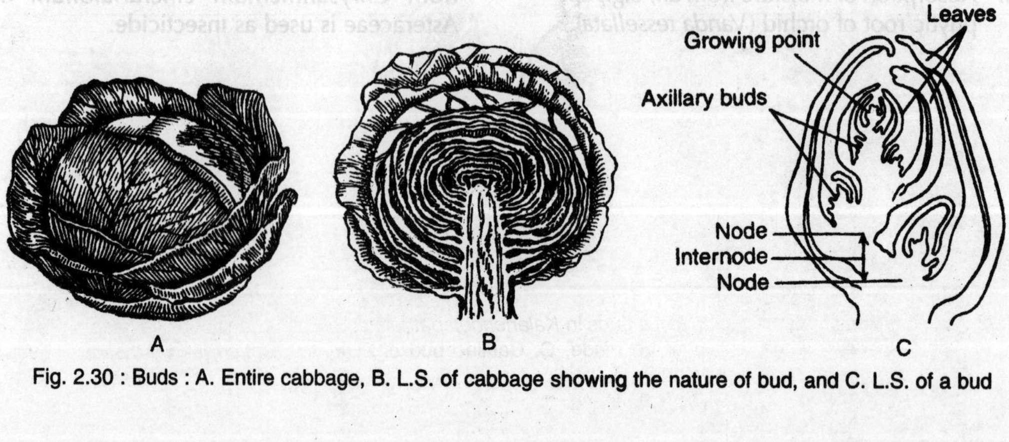

Fig. 2.30 : Buds : A. Entire cabbage, B. L.S. of cabbage showing the nature of bud, and C. L.S. of a bud

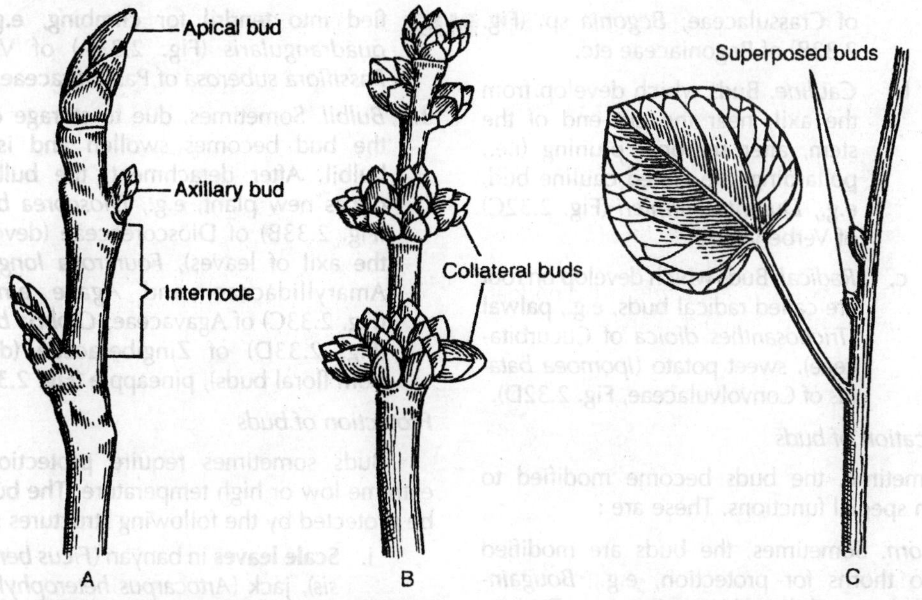

Fig. 2.31 : Different types of bud : A. Apical and axillary buds of *Aesculus* sp., B. Collateral buds of *Acer* sp. and C. Superposed buds of *Aristolochia* sp.

position with respect to one another, they are of two types : i. *Collateral,* and ii. *Superposed.*

i. *Collateral.* When the buds are placed side by side, they are called collateral buds, e.g., brinjal (*Solanum melongena*), *Acer* sp. (Fig. 2.31B) etc.

ii. *Superposed.* When the buds are placed one above the other, they are called superposed buds, e.g., walnut (*Juglans regia*), *Aristolochia altissima* (Fig. 2.31C) etc.

The axillary buds except the normal one are called **supernumerary** or **accessory** buds. Rarely, the superposed buds which develop a little above the axil are called **extra-axillary buds**.

2. Adventitious bud. The buds that develop from a position other than the normal one (i.e., apex or axil) are called adventitious buds. The adventitious buds are of three types : a. *Epiphyllous,* b. *Cauline* and c. *Radical.*

a. *Epiphyllous.* Buds which arise on leaf are called epiphyllous type, e.g., *Bryophyllum calycinum* and *Kalanchoe spathulata* (Fig. 2.32A)

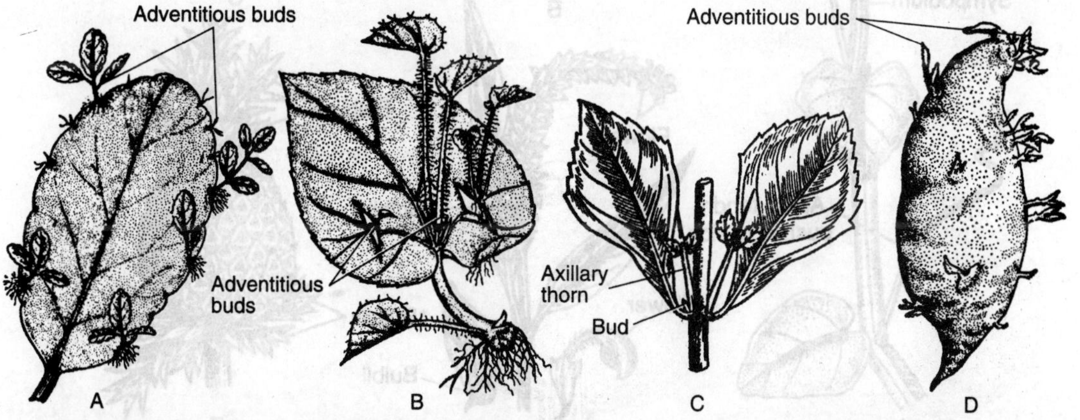

Fig. 2.32 : Adventitious buds : A. Epiphyllous buds in *Kalanchoe spathulata*, B. Epiphyllous buds develop on *Begonia* sp. especially when incisions are made, C. Cauline bud of *Duranta plumieri*, and D. Radical buds of *Ipomoea batatus* (Sweet potato)

of Crassulaceae; *Begonia* sp. (Fig. 2.32B) of Begoniaceae etc.

b. *Cauline*. Buds which develop from the axil, near the cut end of the stem, after seasonal pruning (i.e., pollarding) are called cauline bud, e.g., *Duranta plumieri* (Fig. 2.32C) of Verbenaceae.

c. *Radical*. Buds which develop on root are called radical buds, e.g., palwal (*Trichosanthes dioica* of Cucurbitaceae), sweet potato (*Ipomoea batatas* of Convolvulaceae, Fig. 2.32D).

Modification of buds

Sometime, the buds become modified to perform special functions. These are :

1. *Thorn*. Sometimes, the buds are modified into thorns for protection, e.g., *Bougainvillea spectabilis* of Nyctaginaceae, *Duranta plumieri* (Fig. 2.32C) of Verbenaceae etc.

2. *Tendril*. Sometimes, the bud becomes modified into tendril for climbing, e.g., *Vitis quadrangularis* (Fig. 2.33A) of Vitaceae, *Passiflora suberosa* of Passifloraceae etc.

3. *Bulbil*. Sometimes, due to storage of food, the bud becomes swollen and is called bulbil. After detachment, the bulbil produces new plant, e.g., *Dioscorea bulbifera* (Fig. 2.33B) of Dioscoreaceae (develops in the axil of leaves), *Fourcroya longaeva* of Amaryllidaceae; and *Agave americana* (Fig. 2.33C) of Agavaceae; *Globba bulbifera* (Fig. 2.33D) of Zingiberaceae (develops from floral buds), pineapple (Fig. 2.33E).

Protection of buds

Buds sometimes require protection from extreme low or high temperature. The buds may be protected by the following structures :

i. **Scale leaves** in banyan (*Ficus benghalensis*), jack (*Artocarpus heterophyllus*, Fig. 2.34A), *Michelia champaca* etc.,

ii. **Leaf base** protects as a sheath in *Wormia burbizia* of Dilleniaceae (Fig. 2.34B),

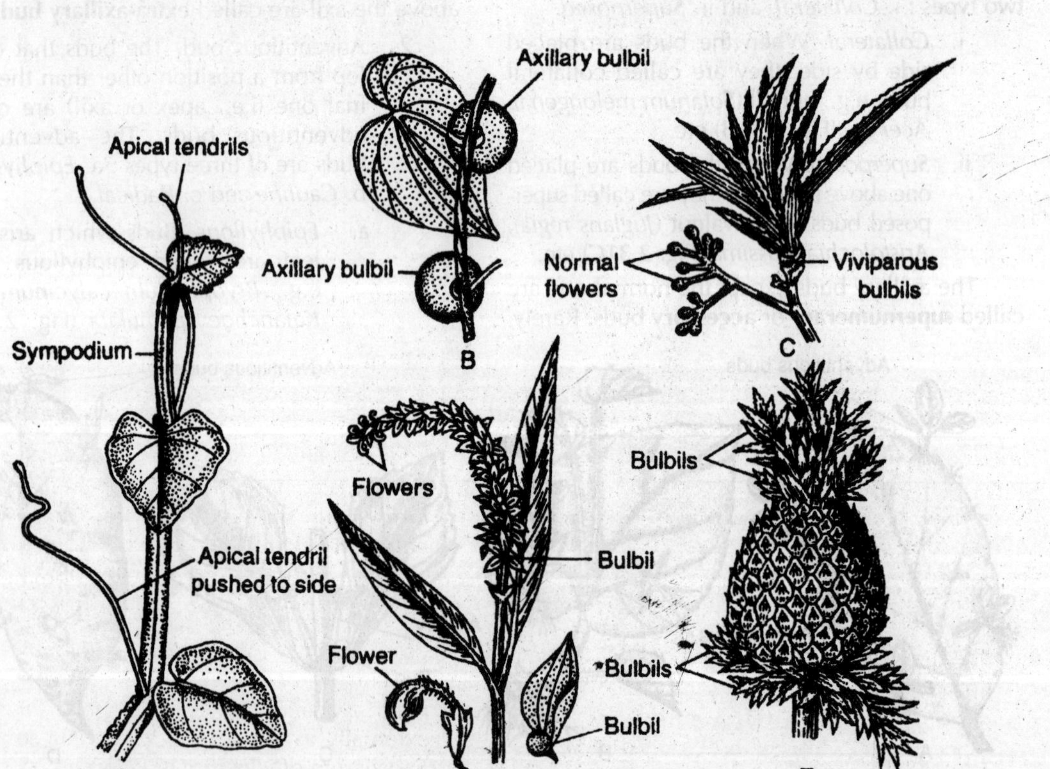

Fig. 2.33 : Modified buds : A. Tendril of *Vitis quadrangularis*, B-E. Bulbils in different plants (B. in *Dioscorea bulbifera*, C. in *Agave americana*, D. in *Globba bulbifera* and E. in Pineapple)

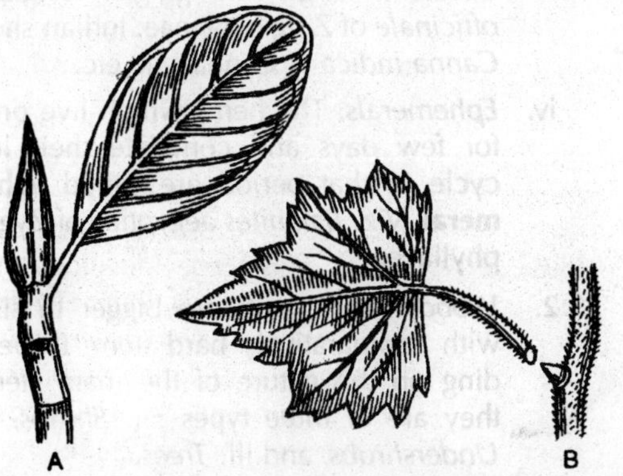

Fig. 2.34 : Protection of buds : A. by scale in *Artocarpus heterophyllus*, B. by sheathing leaf base in *Wormia burbizia*

iii. **Coating of dense hairs** in jujube (*Ziziphus jujuba*),

iv. **Coating of oil, wax** or **resinous matters** to prevent loss of water in *Aesculus indicus* of Hippocastanaceae etc., and

v. **Deeply seated position** in the bark of camphor (*Camphorum aromaticum* of Lauraceae).

THE STEM

Stem is the ascending axis of the plant; develops from plumule of the embryonic axis. The stem may remain simple and unbranched (e.g., many palms) or branched (e.g., mango). The stems are normally erect and subaerial (e.g., Mango, *Mangifera indica*), but they may be subterranean (e.g., Ginger, *Zingiber officinale*) or may creep along the surface of soil (e.g., Indian pennywort, *Centella asiatica*) or may climb up on any object (e.g., passion flower, *Passiflora suberosa*; glory lily, *Gloriosa superba*). In some forms, the stem becomes much condensed and looks like stemless (**acaulescent**), e.g., onion (*Allium cepa*, Fig. 2.35), garlic (*Allium sativum*) etc.

Characteristic features of stem

1. The stem and/or its branches bear leaves and flowers,
2. It is differentiated into nodes and internodes,
3. The branches are sub-epidermal in origin,
4. The apex of the stem is usually terminated by a bud and is never protected by cap,

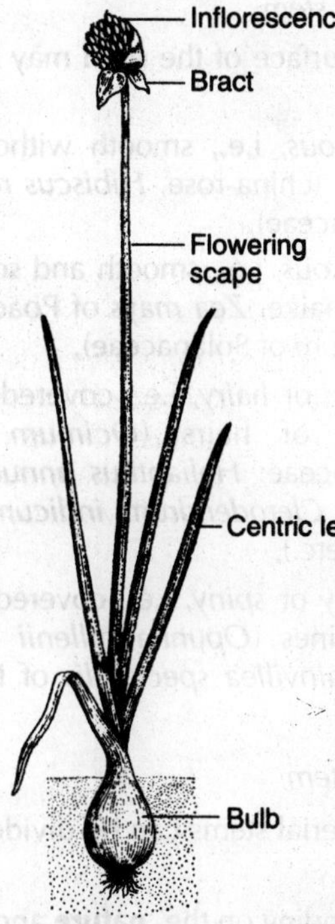

Fig. 2.35 : Flowering scape of *Allium cepa*

5. Hairs, if present, are multicellular,
6. They are usually positively phototropic i.e., grow towards the source of light.

Forms of stem

The forms of stem may vary with plants :

1. *Cylindrical*, i.e., it shows circular outline in transverse section (wheat, *Triticum aestivum* of Poaceae),
2. *Angular*, i.e., it reveals three or four-angles in transverse section (3-angled in sedge, *Cyperus rotundus*; 4-angled in basil, *Ocimum sanctum*),
3. *Ribbed*, i.e., it shows alternate ridges and furrows in transverse section (*Casuarina equisetifolia*),
4. *Fistular*, i.e., solid at the nodes and hollow at the internodes (bamboo, *Bambusa tulda* etc. of Poaceae),
5. *Articulated* or *jointed*, i.e., articulation or joint appears in the node which bulges externally in the form of a ring (sugarcane, *Saccharum officinarum* of Poaceae),
6. *Flattened*, i.e., plate-like structure (prickly pear, *Opuntia dillenii* of Cactaceae).

Surface of stem

The surface of the stem may be of different types :

1. *Glabrous*, i.e., smooth without any emergence (china-rose, *Hibiscus rosa-sinensis* of Malvaceae),
2. *Glaucous*, i.e., smooth and somewhat shining (maize, *Zea mays* of Poaceae; *Solanum glaucum* of Solanaceae),
3. *Hispid* or *hairy*, i.e., covered with different types of hairs (*Ocimum sanctum* of Lamiaceae; *Helianthus annuus* of Compositae, *Clerodendrum indicum* of Verbenaceae etc.),
4. *Prickly* or *spiny*, i.e., covered with prickles or spines (*Opuntia dillenii* of Cactaceae, *Bougainvillea spectabilis* of Nyctaginaceae etc.)

Kinds of stem

The aerial stems can be divided into several categories :

A. Depending on the '**nature** and **size**' they are of two types : 1. Herbaceous, and 2. Woody.

1. *Herbaceous.* Plants are small in size with soft stem. They are called **herbs**. Depending on the '**duration of life**', the herbs can be divided into four groups : i. *Annuals*, ii. *Biennials*, iii. *Perennials*, and iv. *Ephemeral*.

   i. *Annuals.* The herbs which complete their life cycle in one season only are called **annuals**. Viz. mustard, *Brassica nigra* of Brassicaceae; paddy, *Oryza sativa* of Poaceae etc.

   ii. *Biennials.* The herbs which complete their life cycle in two successive seasons are called **biennials**. In the first season they store up food material mainly in the roots and utilise the food in the next season for the development of flowers and fruits. Viz. carrot, *Daucus carota* of Apiaceae; radish, *Raphanus sativa* of Brassicaceae etc.

   iii. *Perennials.* The herbs which live more than two seasons are called **perennials.** This type of herbs develop aerial shoots once in a year for vegetative growth and reproduction. The aerial shoot dies during unfavourable season. Viz. turmeric, *Curcuma longa* and ginger, *Zingiber officinale* of Zingiberaceae; Indian shot, *Canna indica* of Cannaceae etc.

   iv. *Ephemerals.* The herbs which live only for few days and complete their life cycle in that period are called **ephemeral**. Viz. *Balanites aegyptica* of Zygophyllaceae.

2. *Woody.* The plants are bigger in size with comparatively hard stem. Depending on the nature of the main stem, they are of three types : i. *Shrubs*, ii. *Undershrubs*, and iii. *Trees.*

   i. *Shrubs.* These are comparatively small, woody and perennial plants, without clear differentiation of main trunk and lateral branches. Viz. china-rose, *Hibiscus rosa-sinensis* of Malvaceae; rose, *Rosa centifolia* of Rosaceae; custard apple, *Annona squamosa* of Annonaceae etc.

   ii. *Undershrubs.* Plants having characteristics between herbs and shrubs are called undershrubs. Viz. brinjal (*Solanum melongena*) and chilli (*Capsicum frutescens*) of Solanaceae.

   iii. *Trees.* These are taller and more robust than the shrubs. Initially, the main axis remains unbranched for some distance starting from the ground; the remaining part generally becomes profusely branched. The unbranched erect part is called **trunk**. Viz. mango, *Mangifera indica* of Anacardaceae; tamarind, *Tamarindus indicus* of Fabaceae (Leguminosae), banyan, *Ficus benghalensis* of Moraceae etc.

B. Depending on their '**habit**' they are of two types : 1. *Strong* (*Erect*), and 2. *Weak.*

1. *Strong* (*Erect*) *stem.* They are more or less cylindrical, unbranched or branched plant body and strong enough to keep the plant in an upright position. They are of different types :

   i. *Excurrent.* The tree that takes the shape of a pyramid by racemose branching is called **excurrent type**. Viz. *Polyalthia longifolia* (Fig. 2.36A) of Anonaceae, *Pinus* sp. of Pinaceae etc.

   ii. *Deliquescent.* The tree that becomes dome-shaped due to cymose branching

Fig. 2.36 : Plants with erect stem : A. Excurrent habit in *Polyalthia longifolia*, B. Caudex in *Cocos nucifera*, and C. Culm in *Bambusa* sp.

is called **deliquescent type**. Viz. mango, *Mangifera indica* of Anacardaceae; *Ficus benghalensis* of Moraceae.

iii. *Caudex*. The tall tree with erect and unbranched stem having leaf scars and crown of leaves at the apex is called **caudex**. Viz. coconut palm (*Cocos nucifera*, Fig. 2.36B) and palmyra palm (*Borassus flabellifer*) of Arecaceae (Palmae).

iv. *Culm or Haulm*. The erect and close jointed stem is called **Culm** or **Haulm.** Viz. bamboo (*Bambusa tulda*, Fig. 2.36C) and sugarcane (*Saccharum officinarum*) of Poaceae.

2. *Weak stem*. These stems are delicate and are not strong enough to keep them in the upright position. The weak stems are of the following types :

i. *Trailer*. The stem trails over the surface of the soil without rooting at the nodes. It is of two types :

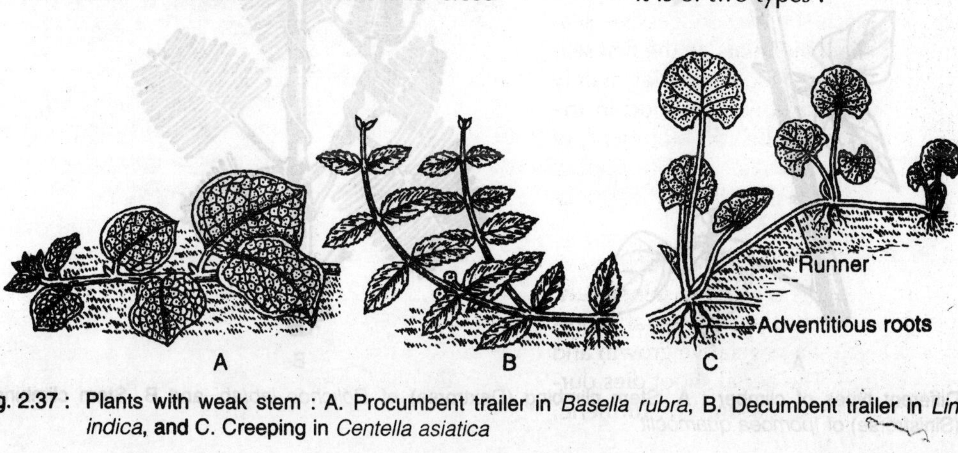

Fig. 2.37 : Plants with weak stem : A. Procumbent trailer in *Basella rubra*, B. Decumbent trailer in *Lindenbergia indica*, and C. Creeping in *Centella asiatica*

a. Procumbent. The stem tip lies flat over the surface. Viz. *Ipomoea reptans* and *Evolvulus nummularius* of Convolvulaceae, *Basella rubra* (Fig. 2.37A) of Basellaceae etc.

b. Decumbent. The stem tip turns upward, leaving the surface of the soil. Viz. *Tridax procumbens* of Asteraceae, *Lindenbergia indica* (Fig. 2.37B) of Scrophulariaceae etc.

ii. *Creeper.* The plant that grows horizontally on the soil surface and develops roots at each node is called creeper. Viz. *Centella asiatica* (Fig. 2.37C) of Apiaceae; *Phyla nodiflora* of Verbenaceae, *Cynodon dactylon* of Poaceae etc.

iii. *Climber.* The plants that climb up on other plants or objects are called climbers. Darwin recognised four groups of climbers : a. Stem climbers or twiners, b. Tendril climbers, c. Root climbers, d. Scramblers. In addition to these, three other types may be added e. Hook climbers, f. Adhesive climbers, and g. Lianes.

a. **Stem climbers** or **twiners**. The plants can climb up around the object by twining their stem. According to the direction of twining they are of two types : **Dextrorse** i.e., the twining takes place in clockwise direction (from right to left), e.g., *Dioscorea alata* of Dioscoreaceae, *Mikania scandens* of Asteraceae, *Dolichos lablab* of Fabaceae (Fig. 2.38A) etc.; and **Sinistrorse** i.e., the twining takes place in anti-clockwise direction (from left to right), e.g., *Clitoria ternatea* of Fabaceae, *Ipomoea quamoclit* (Fig. 2.38B) of Convolvulaceae, *Aganosma caryophyllata* and *Beaumontia grandiflora* of Apocynaceae etc.

b. **Tendril climbers**. In this type, the plants can climb up by means of a very sensitive thread-like structure, the tendril. The tendrils are very sensitive to contact with the object. The tendrils are formed by the modification of different type of organs like *stem* (Grape vine, *Vitis vinifera,* Fig. 2.39A of Vitaceae); Passion flower, (*Passiflora suberosa,* Fig. 2.39B of Passifloraceae); *leaves* (Lathyrus *aphaca,* Fig. 2.39C of Fabaceae); petiole (*Clematis gouriana*; Fig. 2.39D of Ranunculaceae); *stipule* (*Smilax zeylanica,* Fig. 2.39E of Liliaceae); *leaflet* (*Pisum sativum*; Fig. 2.39F of Fabaceae); *leaf apex* (*Gloriosa superba,* Fig. 2.39G of

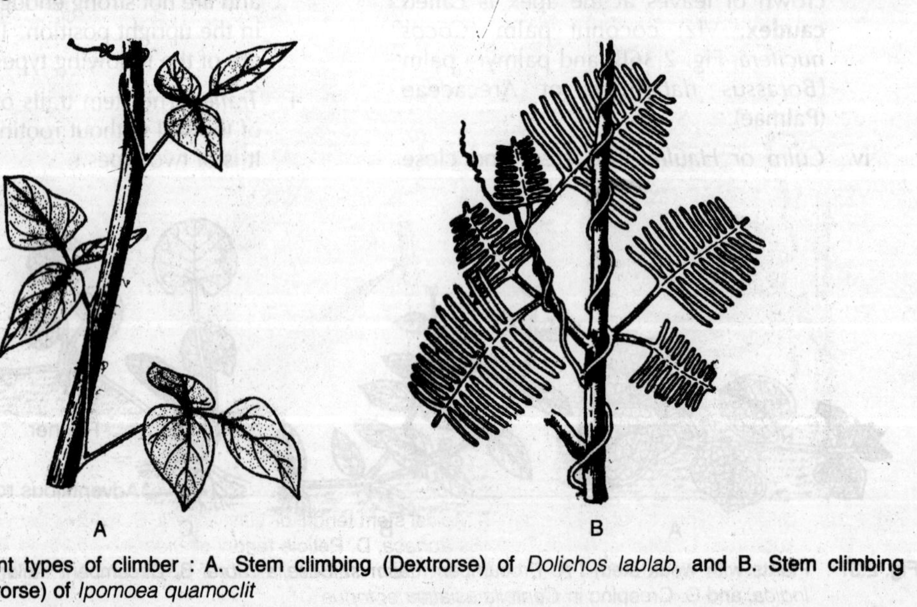

Fig. 2.38 : Different types of climber : A. Stem climbing (Dextrorse) of *Dolichos lablab,* **and B. Stem climbing (Sinistrorse) of** *Ipomoea quamoclit*

Liliaceae): *inflorescence axis* (*Quisqualis malabaricum* of Combretaceae; *Antigonon leptopus*, Fig. 2.39H of Polygonaceae) etc.

c. **Root climbers**. These plants can climb up by means of adventitious roots developed on the surface of stem on contact with the support. In most of the cases, the root becomes profusely branched and helps in climbing. Viz. *Pothos auriens* of Araceae; *Piper betel* (Fig. 2.40A) of Piperace etc.

d. **Scramblers** or **ramblers**. These plants can climb up by means of thorns, spines or prickles through scrambling and stick to support on other plants. Viz. *Calamus rotang* (Fig. 2.40B) of Arecaceae; *Bougainvillea spectabilis* and *Pisonia aculeata* of Nyctaginaceae; *Rosa* sp. (Fig. 2.40C) of Rosaceae; *Capparis sepiaria* and *C. zeylanica* of Capparidaceae; *Zizyphus jujuba* (Fig. 2.40D) of Rhamnaceae etc.

e. **Hook climbers**. Plants of this group can climb up with the help of hooks, developed by the modification of *flower stalk* i.e., pedicel (*Artabotrys uncinatus* of Anonaceae), *terminal leaflet* (*Bignonia unguiscati*, Fig. 2.40E, of Bignoniaceae) and inflorescence axis (*Artabotrys uncinatus*, Fig. 2.40F, of Anonaceae).

f. **Adhesive climber**. These plants can climb up even on flat surface with the help of adhesive discs. The adhesive discs may develop from

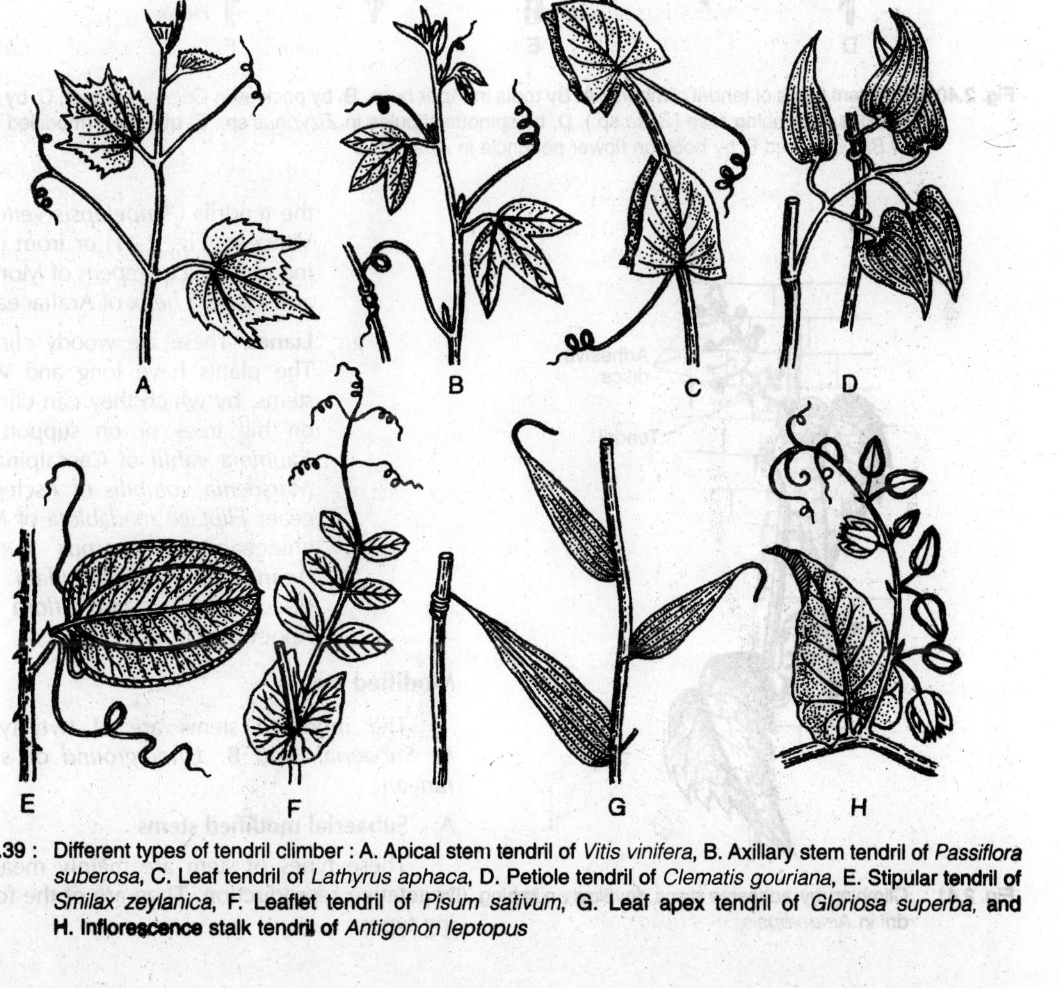

Fig. 2.39 : Different types of tendril climber : A. Apical stem tendril of *Vitis vinifera*, B. Axillary stem tendril of *Passiflora suberosa*, C. Leaf tendril of *Lathyrus aphaca*, D. Petiole tendril of *Clematis gouriana*, E. Stipular tendril of *Smilax zeylanica*, F. Leaflet tendril of *Pisum sativum*, G. Leaf apex tendril of *Gloriosa superba*, and H. Inflorescence stalk tendril of *Antigonon leptopus*

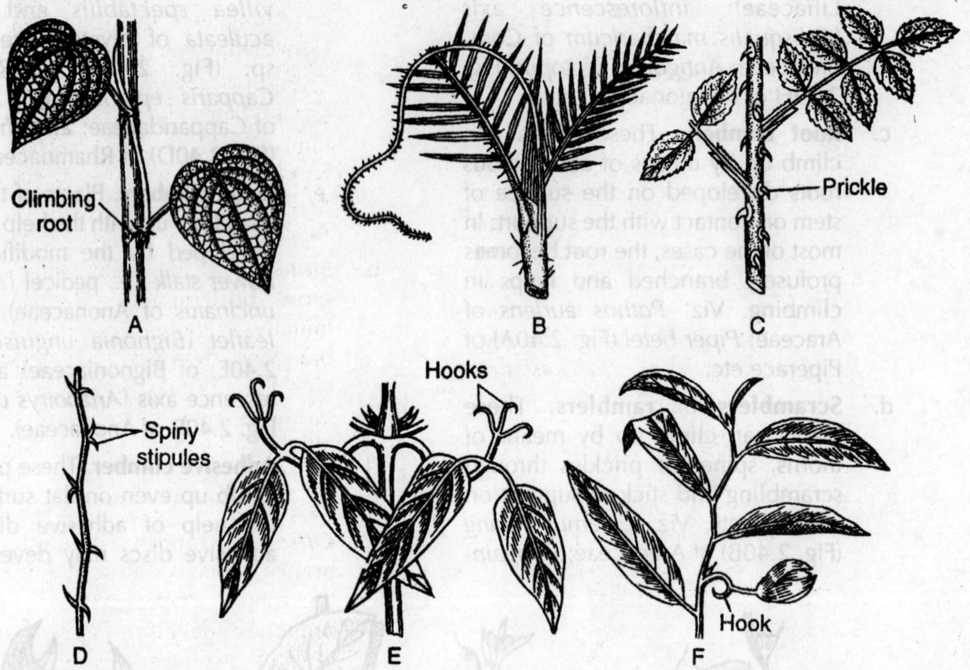

Fig. 2.40 : Different types of tendril climbers : A. By roots in *Piper betel*, B. by prickles in *Calamus* (cane), C. by curved prickles in climbing rose (*Rosa* sp.), D. by spinous stipules in *Zizyphus* sp., E. by hooks (modified leaflet) in *Bignonia* and F. by hook on flower peduncle in *Artabotrys*

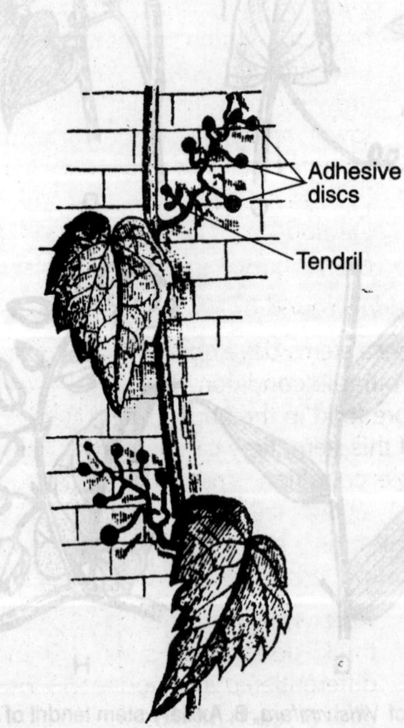

Fig. 2.41 : Climbing by adhesive discs develop on tendril in *Ampelopsis*.

the tendrils (*Ampelopsis veitchii* of Vitaceae, Fig. 2.41) or from climbing roots (*Ficus repens* of Moraceae and *Hedera helix* of Araliaceae).

g. **Lianes**. These are woody climbers. The plants have long and woody stems, by which they can climb up on big trees or on support. Viz. *Bauhinia vahlii* of Caesalpinaceae; *Marsdenia volubilis* of Asclepiadaceae; *Hiptage madablota* of Malpighiaceae; *Ichnocarpus fruticosa*, *Aganosma caryophyllata* and *Beaumontia grandiflora* of Apocynaceae.

Modified stem

The modified stems are of two types : A. *Subaerial*, and B. *Underground* or *subterranean.*

A. Subaerial modified stems

These types of stem are mainly meant for vegetative reproduction. They are of the following types :

a. ***Runner***. In this type, a slender and prostrate branch develops from the axil of a leaf close to the ground, which runs along the surface of the soil. After crossing some distance, it produces new leaves and roots at its apex. A number of such branches may develop from a single point, thereby a single plant can spread a large area within a very short time. Later on, these branches get separated from the mother and form new individuals. Viz. *Oxalis corniculata* (Fig. 2.42A) of Oxalidaceae; *Centella asiatica* of Apiaceae, *Ipomoea aquatica* of Convolvulaceae etc.

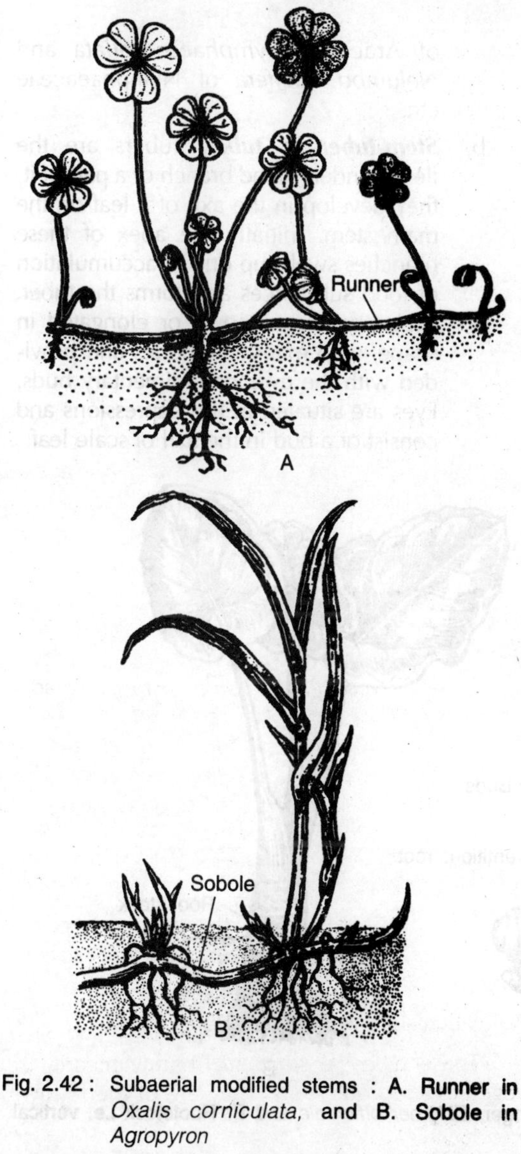

Fig. 2.42 : Subaerial modified stems : A. Runner in *Oxalis corniculata*, and B. Sobole in *Agropyron*

Sometimes the runners may be underground and are called **sobole** (sometimes it may be considered as **thin rhizome**), e.g., grassy weeds like *Imperata cylindrica*; *Saccharum spontaneum*; *Agropyron* sp. (Fig. 2.42B) etc.

b. ***Stolon***. It is like runner, but does not grow horizontally on the ground. Initially, it grows upward like normal branches, then arches down and touches the ground. At the contact point it produces new plant after developing roots, e.g., *Mentha piperita* (Fig. 2.43A) of Lamiaceae; *Fragaria vesca* of Rosaceae etc.

c. ***Offset***. Like runner, it develops prostrate branch which is shorter and thicker than the runner. The apex of the prostrate branches develops a tuft of roots towards the lower side and a rosett of aerial leaves, e.g., *Pistia stratiotes* of Araceae; *Eichhornia crassipes* (Fig. 2.43B) of Pontederiaceae etc.

d. ***Sucker***. It is an underground runner. Like runner, it develops prostrate branch, which originates from the underground portion of the stem. After growing for a short distance, it develops lower adventitious roots and the apex emerges out into the air and produces leafy aerial shoot, e.g., *Mentha viridis* of Lamiaceae, *Chrysanthemum coronarium* (Fig. 2.43C) of Asteraceae etc.

B. Underground or subterranean stems

These stems develop leaves and flowers during favourable condition. The plants manufacture and store food in the underground stem. With the help of this stem, they can perennate during unfavourable condition. Thus this underground stem primarily acts as an organ of perennation and storage, but it also serves the function of vegetative propagation. It is of the following types :

a. ***Rhizome***. Rhizome is an elongated, thick, dorsiventral stem or branch. It is differentiated into nodes and internodes and at each node there are brownish scale leaves, having a bud in its axil. Some of the axillary buds develop aerial shoot annually. Usually it branches

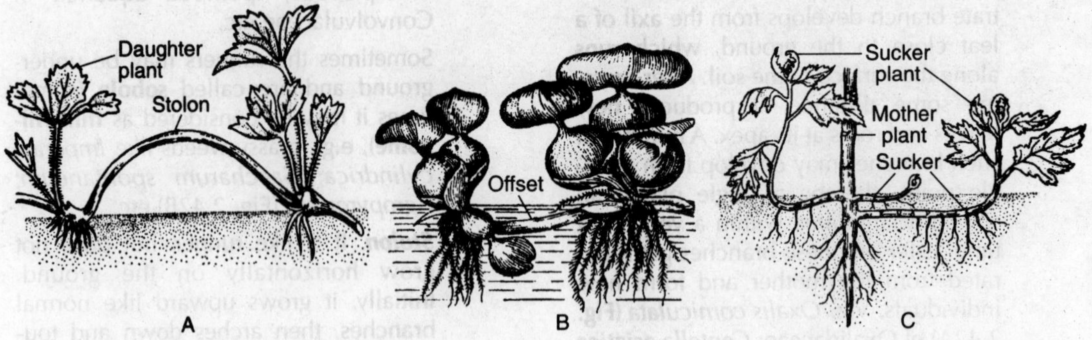

Fig. 2.43 : Subaerial modified stems : A. Stolon in *Mentha piperita*, B. Offset in *Eichhornia crassipes* and C. Sucker in *Chrysanthemum coronarium*

freely. Large numbers of adventitious roots are developed from the lower side of the rhizome. A terminal bud is present at the growing point which either helps in the elongation of rhizome or develops into new aerial shoot. The older portion of the stem gradually dies with age. Viz. *Zingiber officinale* (Fig. 2.44A) and *Curcuma longa* of Zingiberaceae; *Canna indica* of Cannaceae etc.

When the rhizome grows vertically instead of horizontally, it is called **rootstock**, e.g., *Musa paradisiaca* of Musaceae, *Alocasia indica* (Fig. 2.44B) of Araceae; *Nymphaea stellata* and *Nelumbo nucifera* of Nymphaeaceae etc.

b. ***Stem-tuber*** or ***tuber***. Tubers are the fleshy underground branch or a part of it, they develop in the axil of a leaf on the main stem. Initially the apex of these branches swells up due to accumulation of food substances and forms the tuber. Tubers are round, oval or elongated in shape, with many internodes and provided with the eyes or rudimentary buds. Eyes are situated in the depressions and consist of a bud in the axil of scale leaf.

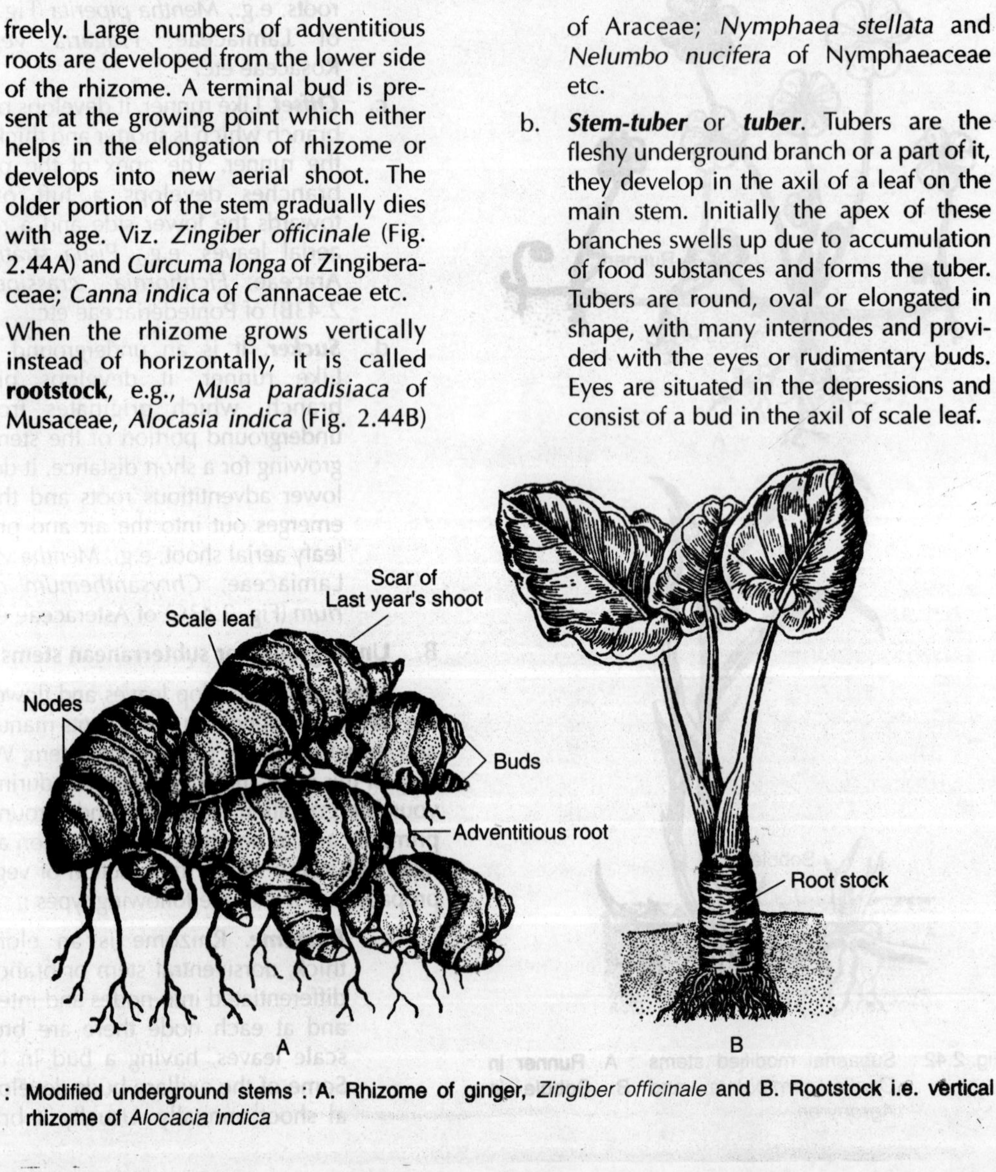

Fig. 2.44 : Modified underground stems : A. Rhizome of ginger, *Zingiber officinale* and B. Rootstock i.e. vertical rhizome of *Alocacia indica*

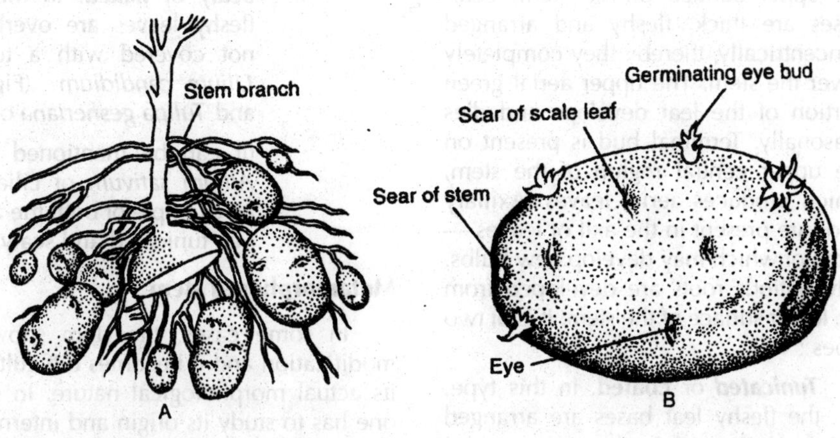

Fig. 2.45 : Modified underground stems : A. Lower part of the potato plant showing the development of tubers i.e., Potato on stem branches and B. Single tuber

Usually the tubers do not develop roots but during favourable condition (during cultivation) axillary buds develop into new shoot and from the base of stem adventitious roots are developed. Viz. *Solanum tuberosum* (Fig. 2.45A) of Solanaceae; *Cyperus rotundus* of Cyperaceae etc.

c. ***Corm*** or ***solid bulb***. The corm is more or less round, solid, stout and vertically fleshy underground stem. Buds are developed at the nodes in the axils of thin and membranous scale leaves, remain scattered around the entire body. Some buds develop new corm and the older corm gradually dies. Adventitious roots are developed from the base or all over its body. The apical bud is comparatively larger in size which grows out and develops into large foliage in early spring and later on flowers. Viz. *Amorphophallus campanulatus* (Fig. 2.46A) and *Colocasia* sp. (Fig. 2.46B) of Araceae; *Crocus sativus* of Liliaceae etc.

d. ***Bulb***. The bulb is a modified underground stem, it looks like a bud. The stem is extremely reduced to a short and **convex disc** with compressed internodes. The leaves are developed from

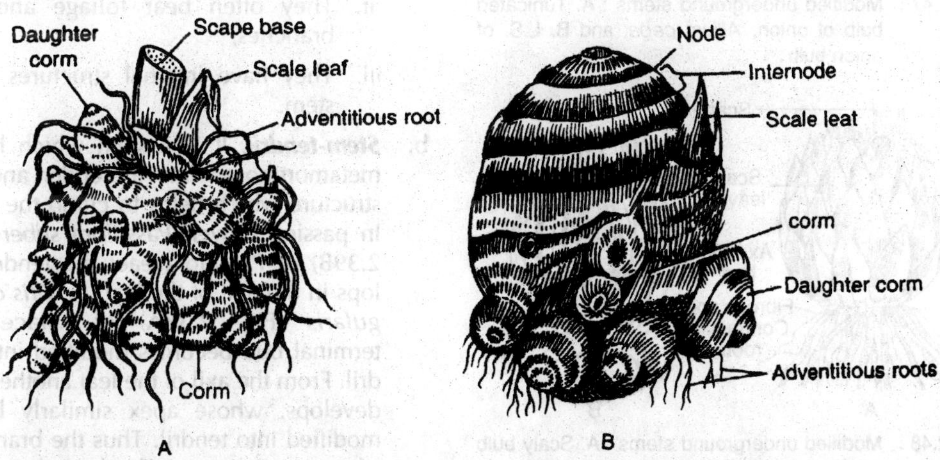

Fig. 2.46 : Modified underground stems : A. Corm of *Amorphophallus campanulatus*, and B. Corm of *Colocasia*

the upper surface of the stem. Leaf-bases are thick, fleshy and arranged concentrically, thereby they completely cover the stem. The upper aerial green portion of the leaf develops and dies seasonally. Terminal bud is present on the upper central region of the stem, which produces aerial leaves. Axillary buds are present in the axil of leaves — some of which may develop new bulbs. Adventitious roots are developed from the lower surface of the stem. It is of two types :

i. ***Tunicated*** or **coated**. In this type, the fleshy leaf bases are arranged regularly and enclosed in a concentric manner. The outer one becomes dry and scaly, which forms the tunica, e.g., *Allium cepa* (Fig. 2.47A, B) of Liliaceae.

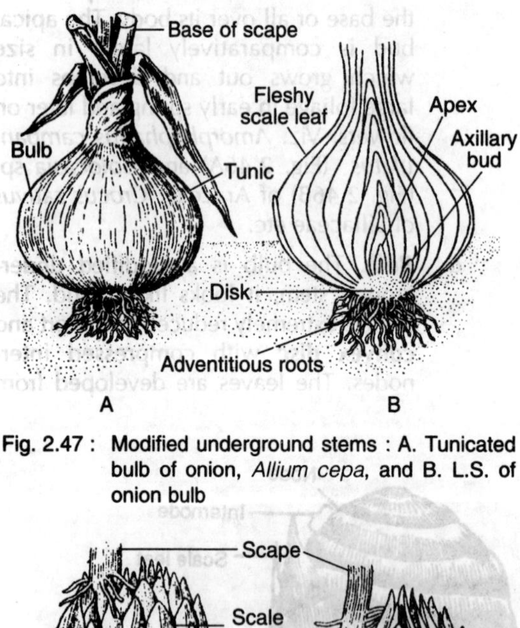

Fig. 2.47 : Modified underground stems : A. Tunicated bulb of onion, *Allium cepa*, and B. L.S. of onion bulb

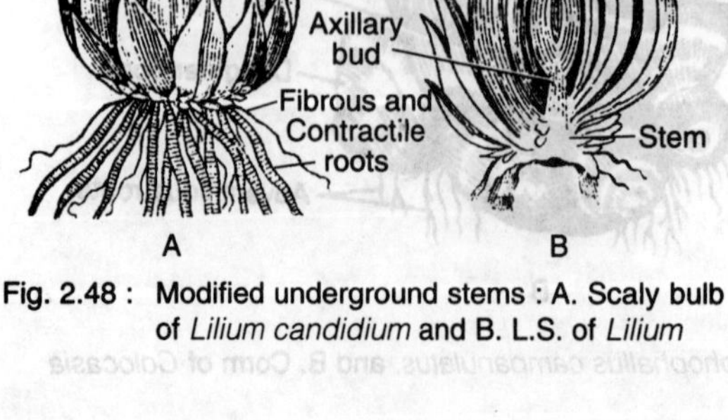

Fig. 2.48 : Modified underground stems : A. Scaly bulb of *Lilium candidium* and B. L.S. of *Lilium*

ii. ***Scaly*** or **naked**. In this type, the fleshy leaves are overlapped and not covered with a tunica, e.g., *Lilium candidium* (Fig. 2.48A,B) and *Tulipa gesneriana* of Liliaceae.

(It can be mentioned that garlic, *Allium sativum* of Liliaceae, cites an example of both the above types i.e., tunicated and scaly.)

Metamorphosed stem

In some cases, the stem shows extreme modification and it becomes difficult to find out its actual morphological nature. In such cases, one has to study its origin and internal structure to confirm it as a stem. These extremely modified stems are said to be metamorphosed stem. The *metamorphosed* stems are of various types :

a. ***Thorn*** or ***stem spine***. These are axillary branches whose growth become arrested and metamorphose into hard and sharply pointed structures, the **thorns**. The thorn acts as a defensive organ. They are sub-epidermal in origin. They may be simple, e.g., *Duranta plumieri* (Fig. 2.49A) of Verbenaceae; *Hygrophyla auriculata* (Fig. 2.49B) of Acanthaceae; *Aegle marmelos* (Fig. 2.49C) of Rutaceae; *Prunus* sp. (Fig. 2.49D) of Rosaceae; or compound, e.g., *Carissa carandas* (Fig. 2.49E) of Apocynaceae; *Flacourtia cataphracta* (Fig. 2.49F) of Flacourtiaceae etc. These are proved to be metamorphosed, because of the following facts :

   i. They develop in the axil of leaves,

   ii. They often bear foliage and lateral branches,

   iii. They have internal structures like the stem.

b. ***Stem-tendril***. It is a stem which becomes metamorphosed into a slender and coiled structure which helps to climb the support. In passion flower, *Passiflora suberosa* (Fig. 2.39B) of Passifloraceae; the tendril develops in the axil of the leaf. In *Vitis quadrangularis* (Fig. 2.33A) of Vitaceae, the terminal bud becomes modified into a tendril. From the axil of the leaf another branch develops, whose apex similarly becomes modified into tendril. Thus the branching is of sympodial type. The branches become

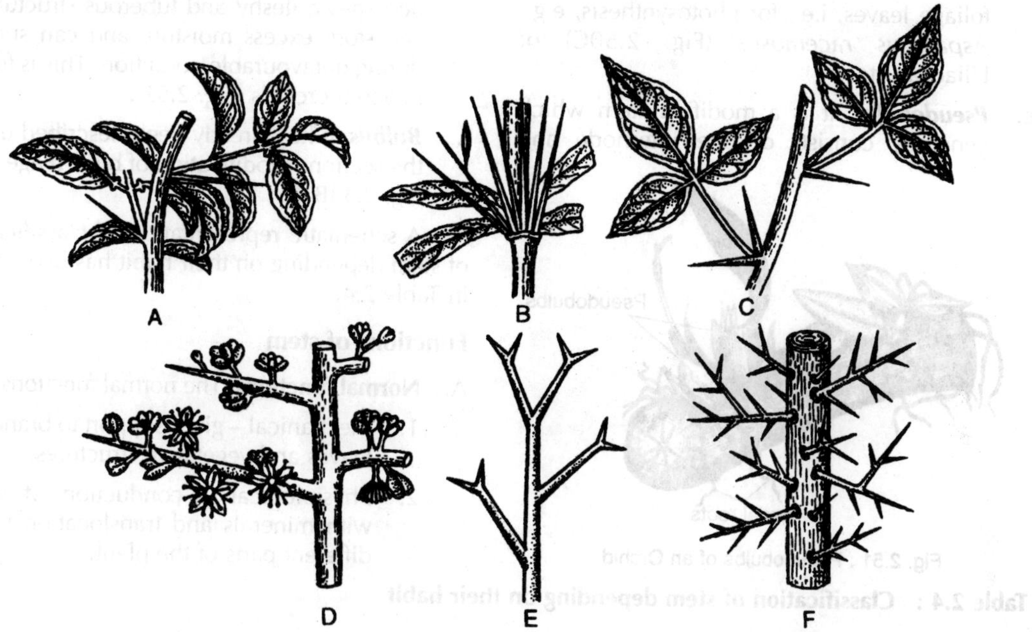

Fig. 2.49 : Different type of thorns : A-D. Simple thorns, A. *Duranta plumieri*, B. *Hygrophila auriculata*, C. *Aegle marmelos*, D. *Prunus* sp., E-F. Compound thorns, E. Branched thorn of *Carissa carandas* (develop axillary), and F. Branched thorns of *Flacourtia cataphracta*

straight and the tendrils are present opposite to the leaves.

c. ***Phylloclade*** or ***Cladophyll***. It is the flat leaf-like metamorphosed stem, consisting of many nodes and internodes and functions as leaf. The leaves may either fall off or become very small or modified into spines, thus the rate of transpiration becomes reduced. Sometimes it acts as an organ of water storage. Viz. prickly pear, *Opuntia dillenii* (Fig. 2.50A) of Cactaceae, *Muehlenbeckia platyclados* (Fig. 2.50B) of Polygonaceae etc.

d. ***Cladode***. The metamorphosed leaf-like branches consisting of one internode are called cladode. They develop in the axil of minute scale leaves and serve the function of

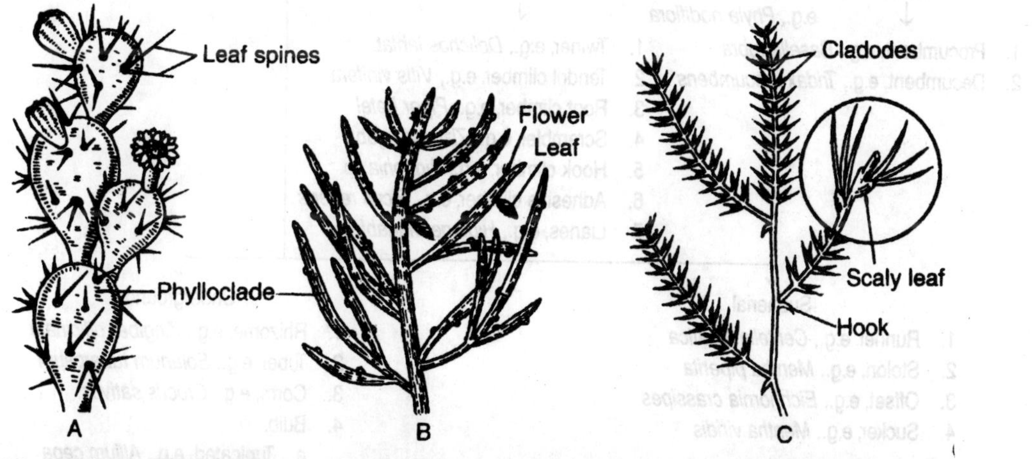

Fig. 2.50 : Phylloclades and cladode : A. Phylloclade or cladophyll of *Opuntia dillenii*, B. Phylloclade of *Muehlenbeckia platyclados* and C. Cladode of *Asparagus racemosus*

foliage leaves, i.e., for photosynthesis, e.g., *Asparagus racemosus* (Fig. 2.50C) of Liliaceae etc.

e. ***Pseudobulbs.*** It is a modified stem which generally consists of one internode and becomes a fleshy and tuberous structure. It can store excess moisture and can survive during unfavourable condition. This is found in aerial orchids (Fig. 2.51).

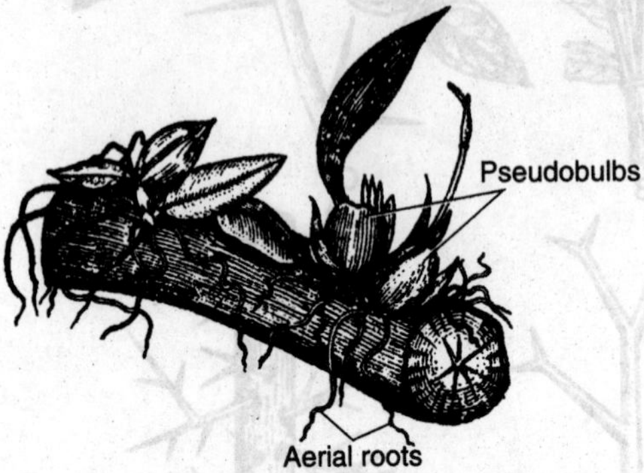

Fig. 2.51 : Pseudobulbs of an Orchid

f. ***Bulbils.*** It has already been described under the section "modification of bud" (Page 152, Fig. 2.33B-E).

A schematic representation of classification of stem depending on their habit has been given in Table 2.4.

Functions of stem

A. **Normal functions.** The normal functions are :

1. Mechanical – gives support to branches, leaves and vegetative structures.
2. Physiological – conduction of water with minerals and translocation to the different parts of the plant.

Table 2.4 : Classification of stem depending on their habit

Stem

- Aerial
  - Strong (Erect)
    1. Excurrent, e.g., *Pinus* sp.
    2. Deliquescent, e.g., *Mangifera indica*
    3. Caudex, e.g., *Cocos nucifera*
    4. Culm, e.g., *Bambusa tulda*
  - Weak
    - Trailer
      1. Procumbent, e.g., *Basella rubra*
      2. Decumbent, e.g., *Tridax procumbens*
    - Creeper e.g., *Phyla nodiflora*
    - Climber
      1. Twiner, e.g., *Dolichos lablab*
      2. Tendril climber, e.g., *Vitis vinifera*
      3. Root climber, e.g., *Piper betel*
      4. Scrambler, e.g., *Zizypus jujuba*
      5. Hook climber, e.g., *Bignonia* sp.
      6. Adhesive climber, e.g., *Ficus repens*
      7. Lianes, e.g., *Hiptage madablota*
- Modified
  - Subaerial
    1. Runner, e.g., *Centella asiatica*
    2. Stolon, e.g., *Mentha piperita*
    3. Offset, e.g., *Eichhornia crassipes*
    4. Sucker, e.g., *Mentha viridis*
  - Underground
    1. Rhizome, e.g., *Zingiber officinale*
    2. Tuber, e.g., *Solanum tuberosum*
    3. Corm, e.g., *Crocus sativus*
    4. Bulb.
       a. Tunicated, e.g., *Aliium cepa*
       b. Scaly, e.g., *Lilium candidium*
- Metamorphosed
  1. Thorn, e.g., *Aegle marmelos*
  2. Stem tendril, e.g., *Passiflora suberosa*
  3. Cladode, e.g., *Asparagus racemosus*
  4. Pseudobulbs, e.g., Orchid
  5. Bulbil, e.g., *Globba bulbifera*

B. **Special functions.** In addition to normal functions, the stems also carry some other functions :

1. Water storage, in many cacti;
2. Food storage, in underground modified stems;
3. Assimilation, in cladodes and phylloclades;
4. Defence structures, as thorns in *Aegle marmelos* of Rutaceae, *Duranta plumieri* of Verbenaceae etc.
5. Climbing, by stem tendril in *Vitis vinifera* of Vitaceae;
6. Vegetative reproduction, in the subterranean and subaerial modified stems.

Branching pattern

Branches are similar but smaller copies of the mother shoot. It arises in the axil of a leaf, develops from axillary bud or the bifurcation of the apex (growing point) of the main axis. These branches further develop lateral buds on them which form secondary and tertiary branches. The branches follow the same pattern of growth. ***The patterns in which the branches are arranged on the main axis are called branching.***

Branches are endogenous in origin. In angiosperm, the branches develop from the leaf axil and have connection of vascular supply with the main stem.

The small undeveloped branches are called *twigs,* while the large branches are called *boughs.* The pattern of branching determines the shape of the plant. Depending upon the presence or absence of branching pattern (Fig. 2.36), the plants may be of ***caudex*** (the main stem is unbranched), ***excurrent*** (apical bud grows indefinitely) or ***deliquescent*** (growth of lateral buds are more than the apical bud).

Types of branching

In cryptogams, the branches are formed by splitting of the apical bud, but in angiosperm, it is always concerned with lateral branches.

It is divided into two types : A. Dichotomous and B. Lateral :

A. **Dichotomous branching.** It takes place due to division of the growing point of the main axis, thus two branches are developed in forked manner. The branches may or may not develop with equal vigour. Thus the main axis remains at the base of the fork and is called *foot* or *podium.* The forked region again becomes branched following the same pattern. It is of two kinds : a. Normal or monopodial or true, and b. Sympodial.

a. *Normal* or *true dichotomy.* In this type, the apex of the main axis bifurcates and gives rise to two daughter axes of equal vigour. The daughter axes further become branched like the main axes. Viz. *Selaginella monospora, Lycopodium clavatum* (Fig. 2.52A, 2.53A), *Psilotum nudum,* all are pteridophytes; *Dictyota dichotoma,* an alga; *Clinogyne dichotoma* of Marantaceae, an angiosperm etc.

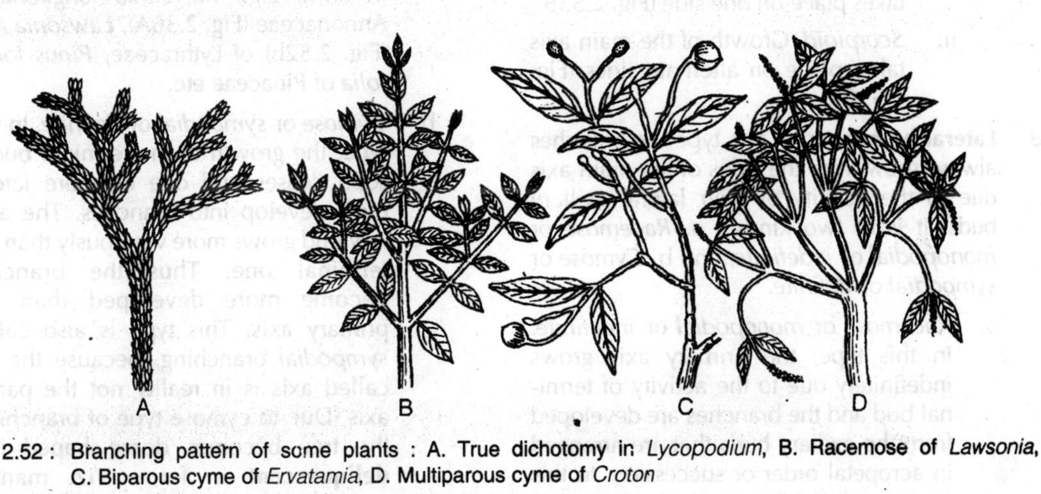

Fig. 2.52 : Branching pattern of some plants : A. True dichotomy in *Lycopodium,* B. Racemose of *Lawsonia,* C. Biparous cyme of *Ervatamia,* D. Multiparous cyme of *Croton*

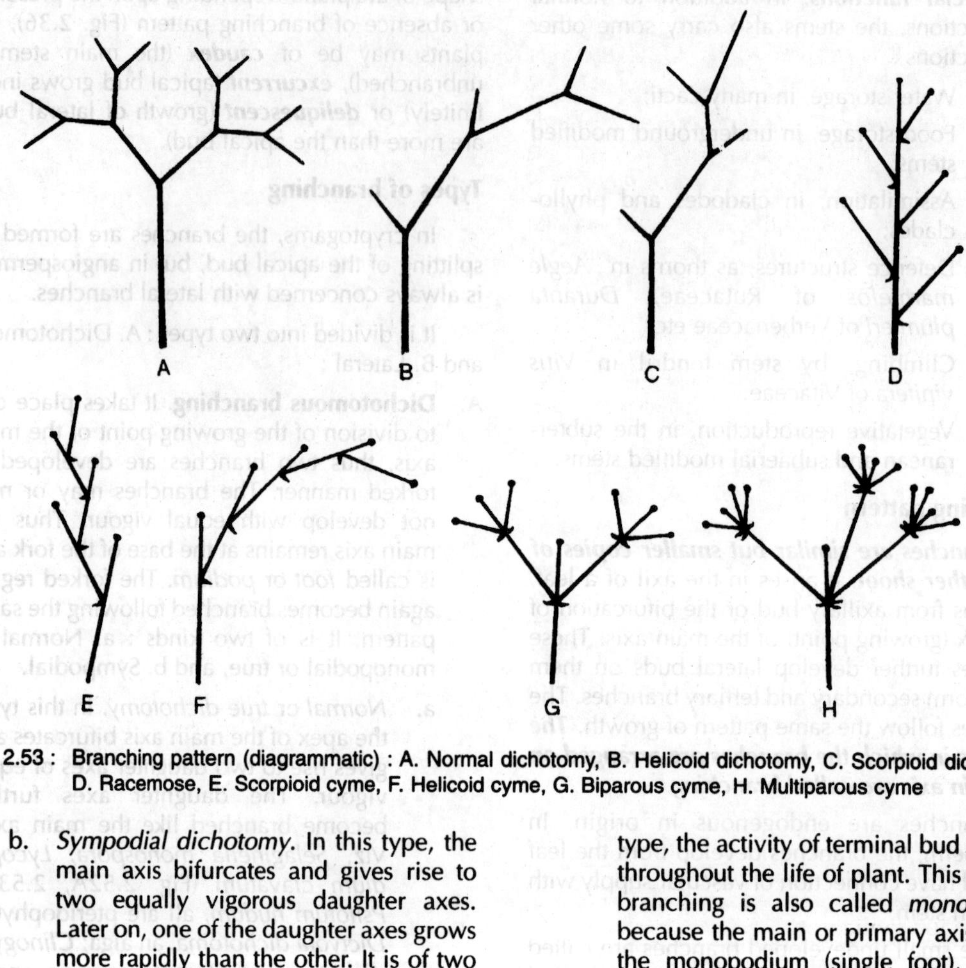

Fig. 2.53 : Branching pattern (diagrammatic) : A. Normal dichotomy, B. Helicoid dichotomy, C. Scorpioid dichotomy, D. Racemose, E. Scorpioid cyme, F. Helicoid cyme, G. Biparous cyme, H. Multiparous cyme

b. *Sympodial dichotomy.* In this type, the main axis bifurcates and gives rise to two equally vigorous daughter axes. Later on, one of the daughter axes grows more rapidly than the other. It is of two kinds : i. *Helicoid,* and ii. *Scorpioid.*

i. *Helicoid.* Growth of the main axis takes place on one side (Fig. 2.53B).

ii. *Scorpioid.* Growth of the main axis takes place on alternate sides (Fig. 2.53C).

B. **Lateral branching.** In this type the branches always grow from the sides of the main axis due to the activity of their lateral axill.ıy buds. It is of two kinds : a. *Racemose* or *monopodial* or *indefinite,* and b. *Cymose* or *sympodial* or *definite.*

a. *Racemose* or *monopodial* or *indefinite.* In this type, the primary axis grows indefinitely due to the activity of terminal bud and the branches are developed from the axillary buds that are arranged in acropetal order or succession. In this type, the activity of terminal bud persists throughout the life of plant. This type of branching is also called *monopodial,* because the main or primary axis forms the monopodium (single foot), which supports the lateral branches. Thus the plant looks like a pyramid i.e., **excurrent** in form, e.g., *Polyalthia longifolia* of Annonaceae (Fig. 2.36A), *Lawsonia alba* (Fig. 2.52b) of Lythraceae, *Pinus longifolia* of Pinaceae etc.

b. *Cymose* or *sympodial* or *definite.* In this type, the growth of the terminal bud is soon ceases and one or more lateral buds develop into branches. The axillary bud grows more vigorously than the terminal one. Thus the branches become more developed than the primary axis. This type is also called *sympodial* branching, because the so-called axis is in reality not the parent axis. Due to cymose type of branching, the tree becomes dome-shaped i.e., **deliquescent** in form. Viz. mango,

Mangifera indica of Anacardiaceae; banyan, *Ficus benghalensis* of Moraceae etc. It is of three kinds : 1. *Uniparous*, 2. *Biparous* and 3. *Multiparous* :

1. *Uniparous*. In this type, only one branch is developed. It is of two types : i. *Helicoid*, and ii. *Scorpioid* type.
   i. *Helicoid*. If the successive daughter branches are developed on one side only, it is called helicoid type (Fig. 2.53F), e.g., *Saraca indica* of Caesalpiniaceae etc.
   ii. *Scorpioid*. If the successive daughter branches are developed alternately from left and right, it is called scorpioid type (Fig. 2.53E), e.g., *Vitis quadrangularis* of Vitaceae etc.
2. *Biparous*. When two branches are developed just below the apical point of the main axis, it is called biparous type (Fig. 2.53G), e.g., *Carissa carandas* and *Ervatamia indica* (Fig. 2.52C) of Apocyanaceae, *Mirabilis jalapa* of Nyctaginaceae
3. *Multiparous*. When more than two branches are developed just below the apical point of the main axis, it is called multiparous type (Fig. 2.53H), e.g. *Croton bonplandianum* (Fig. 2.52D) of Euphorbiaceae.

The schematic representation of branching pattern has been given in Table 2.5.

Table 2.5 : The branching pattern

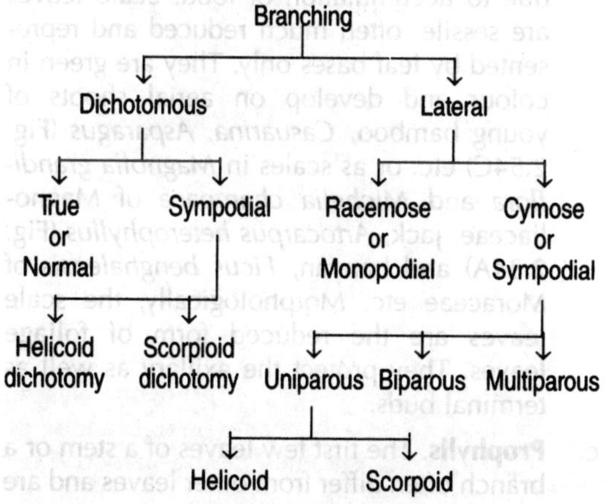

Economic importance of stem

Economically, the stems are very much important and serve many useful purposes :

1. Sugar. Stem containing sap of sugarcane plant is highly rich in sugar. It is used in the preparation of sugar, molasses and also as drinks after extracting by machine or by chewing.
2. Condiments. Rhizomes of ginger (*Zingiber officinale*), mango ginger (*Curcuma amada*) and turmeric (*Curcuma longa*) of Zingiberaceae are used as condiments.
3. Arrowroot. Rhizome of Indian arrowroot (*Curcuma zeodoria*) is used in the preparation of arrowroot, but of inferior quality.
4. Timber and fuel. Stems of the plants like Sal, *Shorea robusta* of Dipterocarpaceae; Teak, *Tectona grandis* of Verbenaceae, *Albizia lebeck*, *Melanoxylon brauna* of Fabaceae and *Pinus wallichiana* of Pinaceae are used as timber and those of *Tamarindus indicus* and *Acacia nilotica* of Fabaceae are used as fuel.
5. Paper. Stems of Bamboo, *Bambusa tulda*, *B. vulgaris* etc. of Poaceae; *Eucalyptus* sp. are used in the preparation of paper. Stem bark of *Broussnetia papyrifera* of Moraceae is used in paper-making. Stem of castor-oil plant (*Ricinus communis* of Euphorbiaceae) yield pulp and cellulose, are used in the preparation of cardboard-containers, newspapers etc.
6. Fibre. The fibres present in the stem of jute (*Corchorus capsularis* and *C. olitorius* of Tiliaceae), flax (*Linum usitatissimum* of Linaceae), *Crotalaria juncea* of Fabaceae, hemp (*Hibiscus cannabinus* of Malvaceae), etc. are utilised for the manufacture of gunny bag, rope, clothings and other commercial commodities.
7. Medicine. Various types of medicines are prepared from the stem of different plants :
   a. Quinine. It is obtained from the bark of *Cinchona calisaya*, *C. succirubra*, *C. ledgerians*, *C. officinalis* of Rubiaceae used to control malaria.
   b. Other medicines. Dry rhizomes of ginger, *Zingiber officinalis* of Zingiberaceae are used to relieve cough; decoction of bark of *Saraca indica* of Fabaceae

is used as a remedy for menstrual disorder of women; stemtuber of *Cyperus rotundus* and *Scirpus grossus* of Cyperaceae are used as medicine.

8. Cinnamon. It is a product of stem bark of *Cinnamomum zeylanica* of Lauraceae, used as spice.
9. Rubber. It is prepared from latex of the stem of different plants like *Ficus elastica* of Moraceae, *Parthenium argentatum* and *Taraxacum koksaghye* of Compositae, *Hevea brasilensis* and *Manihot glaziovii* of Euphorbiaceae.
10. Gums, resins and balsam. These are the secretions of stem of various kinds of plants. Gum from the bark of *Acacia nilotica* of Fabaceae, is used to stick papers etc.; gum from *Acacia catechu*, called 'Khair' or 'Katha' of commerce, is used for dressing betel leaf. Resin, methyl alcohol and turpentine are exctracted from the stem of *Pinus roxburghii*. Balsam, a sticky transparent gum available from *Abies balsamina*, a gymnosperm, is used as mounting medium during mounting of specimen(s) on slide.
11. Laboratory reagent. The laboratory reagent 'haematoxylin' is obtained from the heartwood of *Haematoxylon campechianum* of Fabaceae.

LEAF

Leaves are usually the expanded, dissimilar, lateral members of stem and/or branches with limited growth. They are exogenous in origin and bear buds in their axis. They develop on stem and/or branches in acropetal order. The variation of characteristics that has been noted in response to the variation of environment serves different functions.

Types of Leaves depending on their duration

Depending on duration, the leaves are of three kinds : a. Caducous, b. Deciduous, and c. Persistent.

a. **Caducous** or **fugacious**. The leaves fall off very soon after its development (prematurely shed), e.g., *Acacia recurva* of Mimosaceae.
b. **Deciduous** or **annual**. The leaves fall off after one growing season (maturely shed), e.g., silk cotton tree, *Bombax ceiba* of Bombacaceae; banyan, *Ficus benghalensis* and peepal, *Ficus religiosa* of Moraceae; hog plum, *Spondias mangifera* of Anacardiaceae etc.
c. **Persistent** or **evergreen** or **perennial**. The leaves remain with the plant for more than one season and fall off with age, e.g., jack, *Artocarpus heterophyllus* of Moraceae; mango, *Mangifera indica* of Anacardiaceae; etc.

Kinds of leaf

Leaves show great variation in their form and functions. They are :

a. **Seed leaves** or **cotyledons**. The first embryonal leaf is called seed leaf or cotyledon (Fig. 2.54A). Details about the cotyledons have been discussed under seed (Page no. 128).
b. **Scale leaves** or **cataphylls** or **scales**. These leaves are usually non-green and appear brown in most cases. These are thin, scaly or membranous structures (e.g., rhizome of ginger, Fig. 2.54B) but may become fleshy due to accumulation of food. Scale leaves are sessile, often much reduced and represented by leaf bases only. They are green in colour and develop on aerial shoots of young bamboo, *Casuarina, Asparagus* (Fig. 2.54C) etc. or as scales in *Magnolia grandiflora* and *Michelia champaca* of Magnoliaceae; jack, *Artocarpus heterophyllus* (Fig. 2.34A) and banyan, *Ficus benghalensis* of Moraceae etc. Morphologically, the scale leaves are the reduced form of foliage leaves. They protect the axillary as well as terminal buds.
c. **Prophylls**. The first few leaves of a stem or a branch often differ from other leaves and are

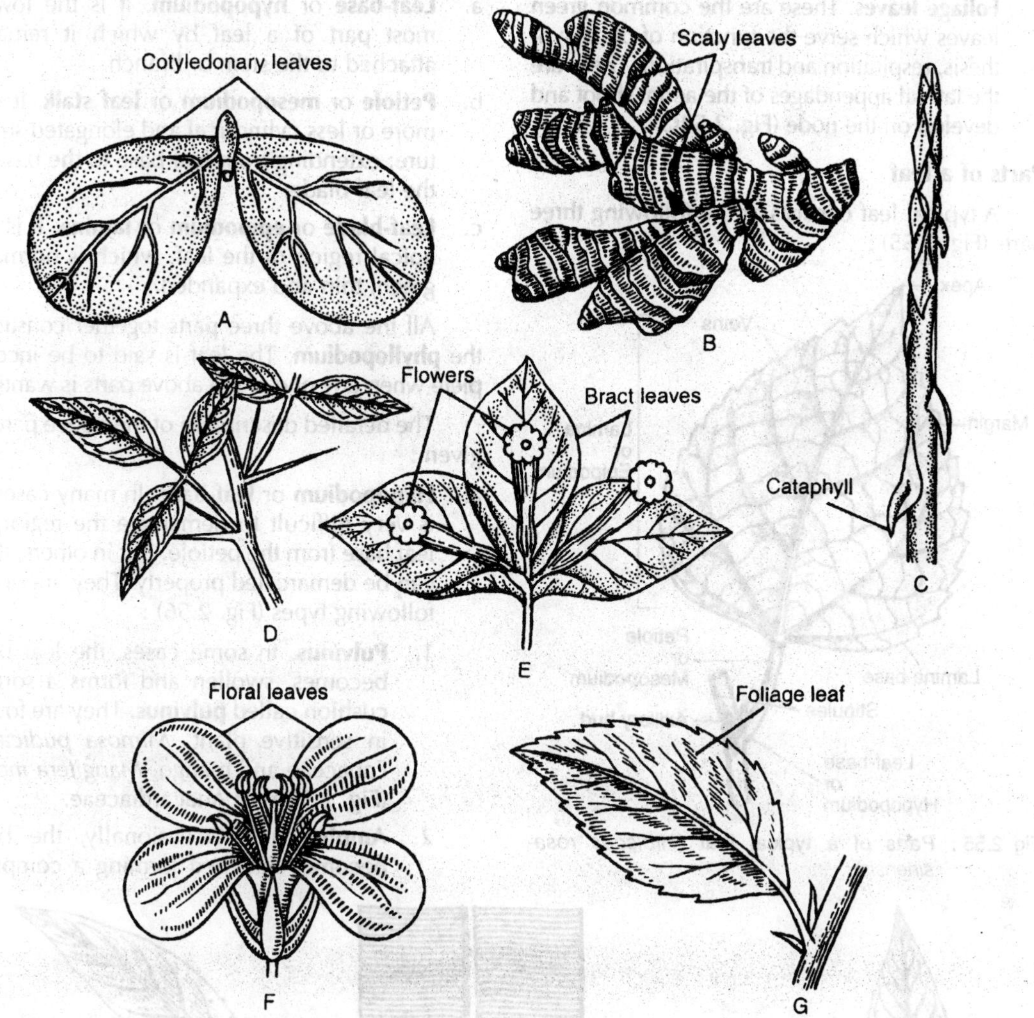

Fig. 2.54 : Kinds of leaves : A. Cotyledonary leaves of *Ricinus*, B. Scale leaves of ginger, C. Cataphyll of *Asparagus*, D. Prophyll of *Aegle*, E. Bract leaves of *Bougainvillea*, F. Floral leaves of a Flower, and G. Foliage leaf

termed as *prophylls*. Commonly, two prophylls are found in dicotyledons and one in monocotyledons. Sometimes prophylls may be modified into one (orange, *Citrus* spp. of Rutaceae) or two spines (in wood apple, *Aegle marmelos*, Fig. 2.54D) or tendril (*Cucurbita*).

d. **Bract leaves** or **Hypsophylls**. These are the special leaves either modified in colour and structure or may be the foliage leaves reduced in form, developed at the base of individual flower or floret or inflorescence. They may be leafy (*Acaiypha indica* of Euphorbiaceae), or petaloid (*Bougainvillea spectabilis* of Nyctaginaceae, Fig. 2.54E; *Poinsettia pulcherrima* of Euphorbiaceae), or spathe (*Colocasia antiquorum* of Araceae; *Musa paradisiaca* of Musaceae), or glume (*Triticum aestivum* and *Oryza sativa* of Poaceae), or epicalyx (china-rose, *Hibiscus rosa-sinensis* of Malvaceae); or involucre (in Asteraceae) or cupule (oak, *Quercus spicata* of Fagaceae; birch, *Betula bhojpatra* of Betulaceae) etc.

e. **Floral leaves**. These are specialised leaves of a typical flower (Fig. 2.54F), which constitute sepals, petals, stamens and carpels. A detailed discussion is available with flower.

f. **Foliage leaves**. These are the common green leaves which serve the function of photosynthesis, respiration and transpiration. These are the lateral appendages of the aerial shoot and develop on the node (Fig. 2.54G).

Parts of a leaf

A typical leaf consists of the following three parts (Fig. 2.55) :

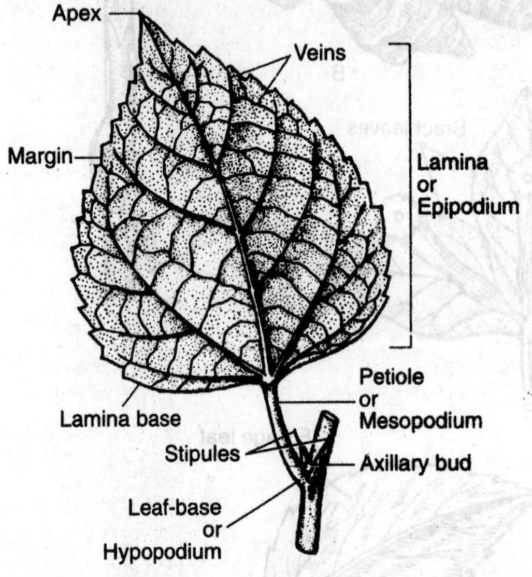

Fig. 2.55 : Parts of a typical leaf (*Hibiscus rosa-sinensis*)

a. **Leaf-base** or **hypopodium**. It is the lowermost part of a leaf by which it remains attached to the stem or branch.

b. **Petiole** or **mesopodium** or **leaf stalk**. It is a more or less cylindrical and elongated structure; extending from leaf base to the base of the leaf blade.

c. **Leaf-blade** or **epipodium** or **lamina**. It is the apical region of the leaf, which is normally green, thin and expanded.

All the above three parts together constitute the **phyllopodium**. The leaf is said to be incomplete when any one of the above parts is wanting.

The detailed description of the above parts is given :

a. **Hypopodium** or **leaf-base**. In many cases, it is very difficult to demarcate the region of leaf base from the petiole. But in others, they can be demarcated properly. They are of the following types (Fig. 2.56) :

1. **Pulvinus**. In some cases, the leaf base becomes, swollen and forms a sort of cushion called **pulvinus.** They are found in sensitive plant, *Mimosa pudica* of Fabaceae and mango, *Mangifera indica* (Fig. 2.56A) of Anacardiaceae.

2. **Amplexicaul**. Occasionally, the base becomes flattened forming a complete

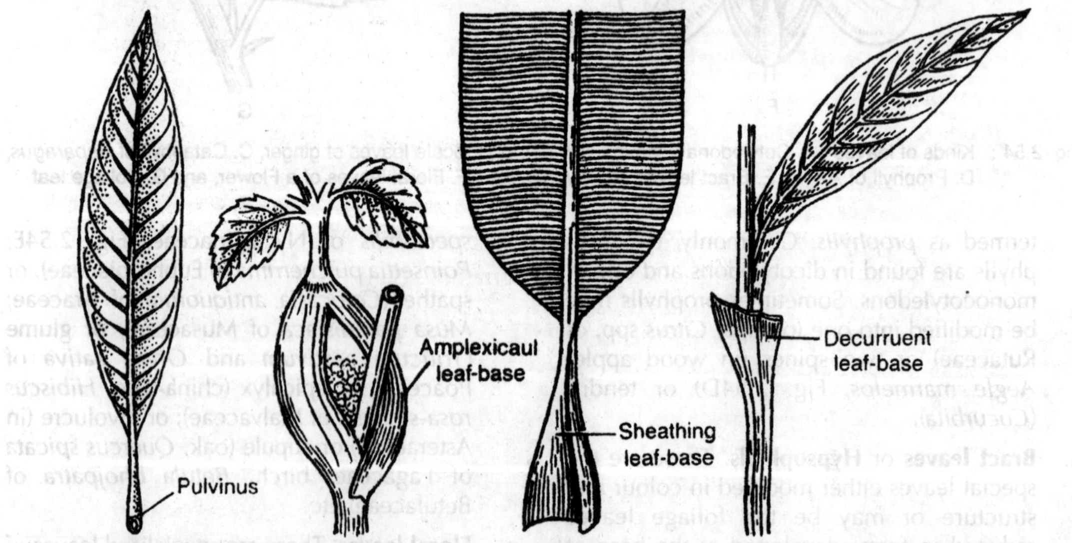

Fig. 2.56 : Different types of hypopodium or leaf-base : A. Pulvinus of Mango (*Mangifera indica*), B. Amplexicaul sheathing leaf base of *Aethusa cynapium*, C. Semi-amplexicaul (sheathing leaf base) of *Musa paradisiaca*, and D. Decurrent of *Symphytum officinale*

sheath around the stem, called **amplexicaul**, e.g., *Polygonum* sp. of Polygonaceae, *Aethusa cynapium* (Fig. 2.56B) :

3. **Semi-amplexicaul**. When the leaf base becomes flattened and sheathed around the stem partially, it is called **semi-amplexicaul** as in palms. These are also called **sheathing leaf bases**. They are common in monocotyledons, but also found in dicotyledonous families like Apiaceae, Araliaceae, and Ranunculaceae, etc. Sometimes the leaf bases become much extended and together form a stem-like structure i.e., **false stem** as found in banana, *Musa balbisiana* and plantain, *Musa paradisiaca* (Fig. 2.56C) of Musaceae.
4. **Decurrent**. In some plants, leaf base and petiole both become flat, broad and winged. These ensheath the upper region of the node and are called decurrent leaf-bases, e.g., *Symphytum officinale* (Fig. 2.56D) of Boraginaceae, *Laggera* of Asteraceae.

Ligule and Auricle

Ligule is the membranous outgrowth located at the junction of the petiole and lamina (Fig. 2.57). The ligules may be scaly, hairy etc. They are found in the members of the family Zingiberaceae and Poaceae (Gramineae). On the other hand, the **auricle** is the winged expansion of leaf base or hypopodium, which is continuous with the lamina (Fig. 2.57). Thus the axis appears to be overlapped or clasped by such expanded leaf base. It is found in the members like *Oryza sativa* of Poaceae; *Aegialitis rotundifolia* of Plumbaginaceae, *Lonicera caprifolium* of Caprifoliaceae etc.

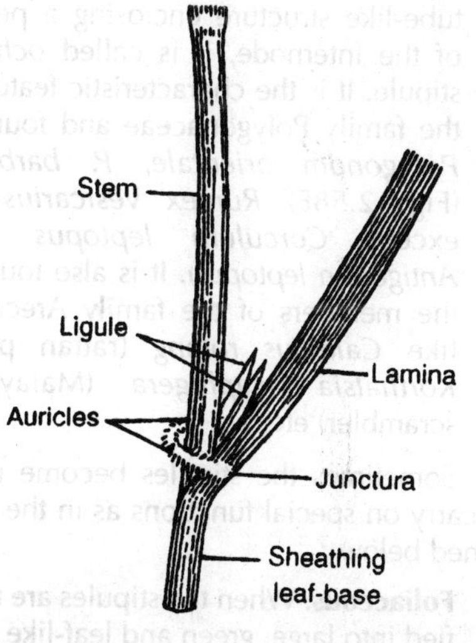

Fig. 2.57 : Ligule and auricles of rice (*Oryza sativa*)

The stipules

The lateral outgrowths from the leaf bases are called stipules. They usually occur in pairs. They are present in many dicotyledonous plants, but are rare in monocotyledonous plants.

Leaves with stipules are called **stipulate** (e.g., China rose, *Hibiscus rosa-sinensis* of Malvaceae etc.) and the leaves without stipule are called **exstipulate** (e.g., mango, *Mangifera indica* of Anacardiaceae; guava, *Psidium guajava* of Myrtaceae etc.).

Stipules are usually simple structures but they may be branched as in *Ipomoea quamoclit* of Convolvulaceae.

Depending on their longevity, they are of three kinds :

i. ***Caducous***. When the stipules fall off before the leaf unfolds, they are called *caducous*, e.g., banyan, *Ficus benghalensis* of Moraceae, *Michelia champaca* of Magnoliaceae etc.
ii. ***Deciduous***. When the stipules fall off soon after the unfolding of the leaf, they are called *deciduous*, e.g., *Dillenia indica* of Dilleniaceae etc.
iii. ***Persistent***. When the stipules remain throughout the life of leaf. They are called *persistent*. e.g., rose, *Rosa centifolia* of Rosaceae etc.

Forms of stipule

The stipules are of ten (10) different types. These are (Fig. 2.58) :

1. Free-lateral, 2. Adnate, 3. Intrapetiolar, 4. Interpetiolar, 5. Ochreate 6. Foliaceous, 7. Tendrillar, 8. Spiny, 9. Convolute or bud scales, and 10. Winged. The last five types (6-10) are the modified forms of the stipule.

1. **Free-lateral**. When two distinct tiny stipules are present on two sides of the leaf base, it is called *free-lateral* type, e.g., china-rose, *Hibiscus rosa-sinensis* of Malvaceae (Fig. 2.58A), They are also

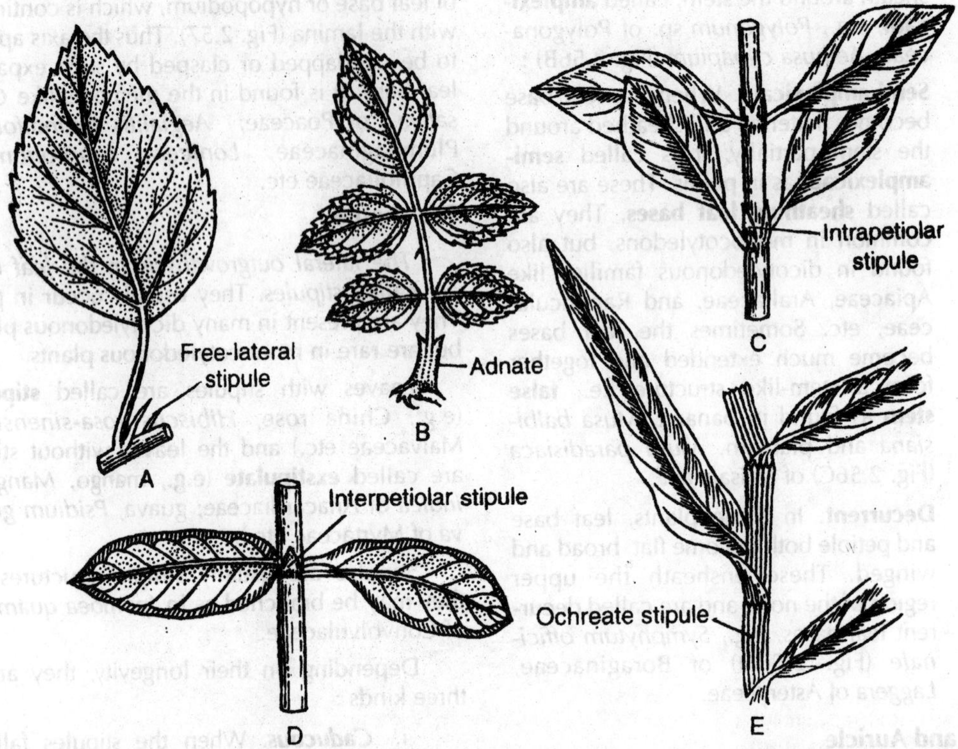

Fig. 2.58 : Different types of stipule : A. Free lateral stipule of *Hibiscus rosa-sinensis*, B. Adnate of *Rosa* sp., C. Intrapetiolar stipule in *Gardenia* sp., D. Interpetiolar stipule in *Ixora coccinia*, and E. Ochreate stipule in *Polygonum*

found in other families like Tiliaceae, Sterculiaceae, Fabaceae etc.

2. **Adnate**. When the stipules are attached on both the sides of the petiole up to a certain height it is called *adnate* type, e.g., rose, *Rosa centifolia* of Rosaceae (Fig. 2.58B).

3. **Intrapetiolar**. When the stipules occur on both the sides of opposite leaves, unite together by their inner margins and lie in the axil of leaves, it is called *intrapetiolar* type. It is also called **axillary** type, e.g., *Paederia foetida* and *Gardenia jasminoides* (Fig. 2.58C) of Rubiaceae etc.

4. **Interpetiolar**. When the union takes place between stipules of opposite leaves by their outer margins, thereby the four stipules are transformed into two fused stipules, lies on the node between (inter) the petiole of the opposite leaves, it is called *interpetiolar* type, e.g., kadam, *Anthocephalus indicus*; rangan, *Ixora coccinia* (Fig. 2.58D) and some others of Rubiaceae; *Strychnos nux-vomica* of Loganiaceae etc.

5. **Ochreate**. When the two stipules develop at the leaf base and fuse to form a tube-like structure enclosing a portion of the internode, it is called *ochreate* stipule. It is the characteristic feature of the family Polygonaceae and found in *Polygonum orientale*, *P. barbatum* (Fig. 2.58E) *Rumex vesicarius* etc. except *Corculum leptopus* (Syn. *Antigonon leptopus*). It is also found in the members of the family Arecaceae like *Calamus rotang* (rattan palm), *Korthalsia scaphigera* (Malayasian scrambler) etc.

Sometimes the stipules become modified to carry on special functions as in the cases mentioned below :

6. **Foliaceous**. When the stipules are modified into large, green and leaf-like structures which serve all the functions of a

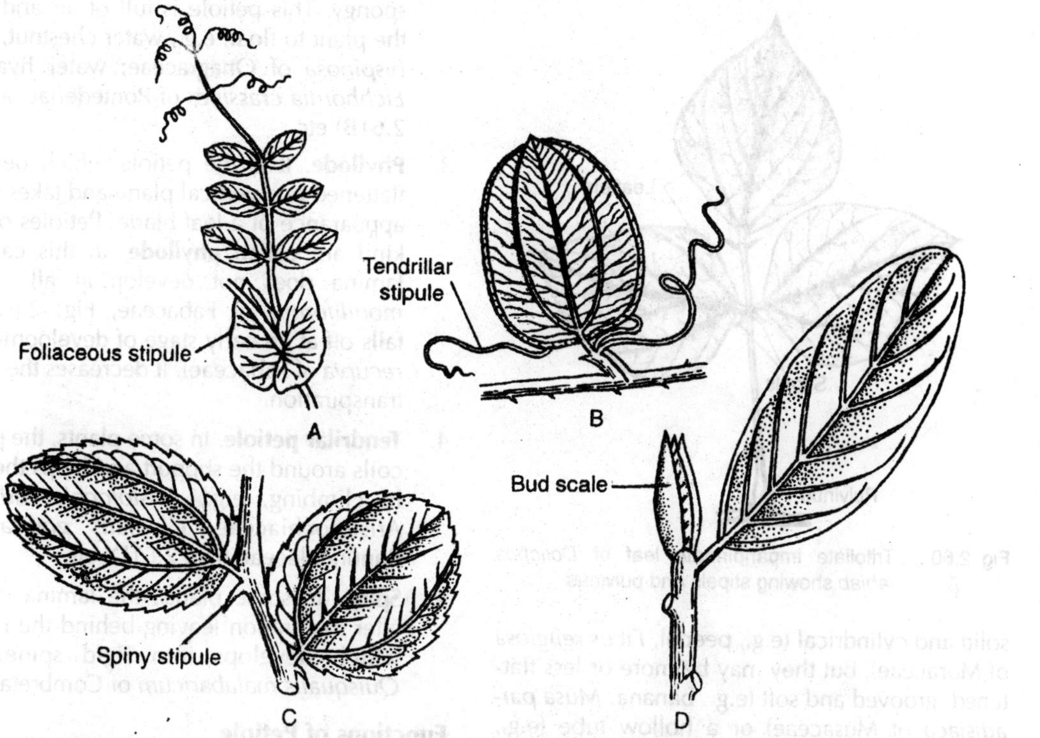

Fig. 2.59 : Modified stipules : A. Foliaceous stipule in *Pisum sativum*, B. Tendrillar stipule in *Smilax macrophylla*, C. Spinous stipule in *Zizyphus mauritiana*, and D. Bud scale in *Artocarpus* sp.

foliage leaf, it is called *foliaceous* type, e.g., wild pea, *Lathyrus aphaca* and pea, *Pisum sativum* (Fig. 2.59A) of Fabaceae.

7. **Tendrillar**. When the stipule is modified into tendril it is called tendrillar type, e.g., sarsaparilla, *Smilax macrophylla* (Fig. 2.59B) of Liliaceae.
8. **Spiny**. When the stipules are modified into spines it is called *spiny* type, e.g., jujube, *Ziziphus mauritiana* (Fig. 2.59C) of Rhamnaceae, *Acacia arabica* of Fabaceae. Such stipules serve as defensive organ.
9. **Convolute** or **Bud scales**. When the stipules form membranous protective structures covering the buds and fall off during unfolding of leaves, they are called *bud scales*, e.g., peepal, *Ficus religiosa* and banyan, *F. benghalensis* and *Artocarpus* sp. (Fig. 2.59D) of Moraceae etc.
10. **Winged**. When the stipules are modified into an expanded wing-like structure it is called *winged* type, e.g., *Crotalaria alata* of Fabaceae.

Function of stipule

Stipule serves the following functions :

1. It prevents the axillary bud to fall-off,
2. It protects the next higher leaves as bud-scales,
3. Foliaceous stipule acts as photosynthetic organ,
4. Spiny stipule acts as defensive organ,
5. Tendrillar stipule acts as climbing organ,
6. Some stipules secrete mucilage and protect the buds.

Stipels

When leafy outgrowth develops at the base of each leaflet of a compound leaf, it is called **stipels**, e.g., *Vicia faba; Dolichos lablab* (Fig. 2.60) and *Desmodium gangeticum* of Fabaceae.

b. **Petiole or mesopodium or leaf stalk.**

It is the part of a leaf which connects leaf blade with the leaf base. The petioles are usually

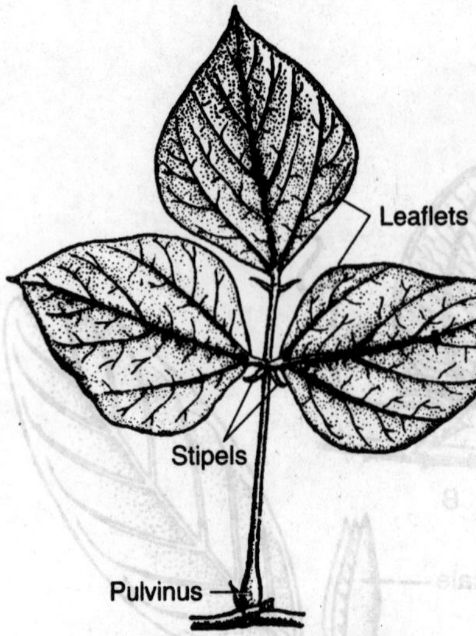

Fig. 2.60 : Trifoliate imparipinnate leaf of *Dolichos lablab* showing stipels and pulvinus

solid and cylindrical (e.g., peepal, *Ficus religiosa* of Moraceae), but they may be more or less flattened, grooved and soft (e.g., banana, *Musa paradisiaca* of Musaceae) or a hollow tube (e.g., papaw, *Carica papaya*). The leaf without petiole is called *sessile* (glory lily, *Gloriosa superba* of Liliaceae; Mexican poppy, *Argemone mexicana* of Papaveraceae), while a leaf with a petiole is called *petiolate* (mango, *Mangifera indica* of Anacardiaceae).

The petioles get modified in different ways :

Modified Petioles

1. **Winged petiole**. In this type, the petiole becomes flattened and winged, which looks like lamina and carries out photosynthesis, e.g., *Citrus* sp. of Rutaceae (Fig. 2.61A). In *Nepenthes* the petiole becomes partly winged and partly tendrillar.
2. **Swollen** or **bulbous petiole**. In some aquatic plants, the petiole becomes swollen and spongy. This petiole is full of air and helps the plant to float, e.g., water chestnut, *Trapa bispinosa* of Onagraceae; water hyacinth, *Eichhornia crassipes* of Pontederiaceae (Fig. 2.61B) etc.
3. **Phyllode**. It is the petiole which becomes flattened in a vertical plane and takes up the appearance of a leaf blade. Petioles of such kind are called **phyllode**. In this case the lamina does not develop at all (*Acacia moniliformis* of Fabaceae, Fig. 2.61C) or falls off at an early stage of development (*A. recurva* of Fabaceae). It decreases the rate of transpiration.
4. **Tendrilar petiole**. In some plants, the petiole coils around the support and helps the plant in climbing, e.g., *Aristolochia indica* of Aristolochiaceae, *Clematis gouriana* of Ranunculaceae (Fig. 2.61D) etc.
5. **Spine**. In some plants, the lamina falls off after maturation leaving behind the petiole, which develops into rigid spine, e.g., *Quisqualis malabaricum* of Combretaceae.

Functions of Petiole

Petiole serves the following functions :

1. It helps to conduct material in both the direction,
2. Due to bending and twisting action, the petiole brings the leaf blade in a position to get sufficient sunlight for photosynthesis,
3. The swollen or bulbous petiole helps in floating,
4. The tendrilar petiole helps in climbing.

c. **Leaf-blade** or **epipodium** or **lamina**.

The leaf blade or lamina is the most important green and flattened part of a leaf. It is generally thin and usually dorsiventrally differentiated. Sometimes, it becomes reduced, modified or

Table 2.6 : Differences between phylloclade and phyllode

Phylloclade	Phyllode
1. It forms by the modification of stem.	1. It forms by the modification of petiole.
2. It is differentiated into nodes and internodes.	2. It does not have nodes and internodes.
3. It bears flowers.	3. It does not bear any flower.
4. Modified leaves of different forms like spines, scales etc., are present on the phylloclade.	4. It does not have spines and scales.

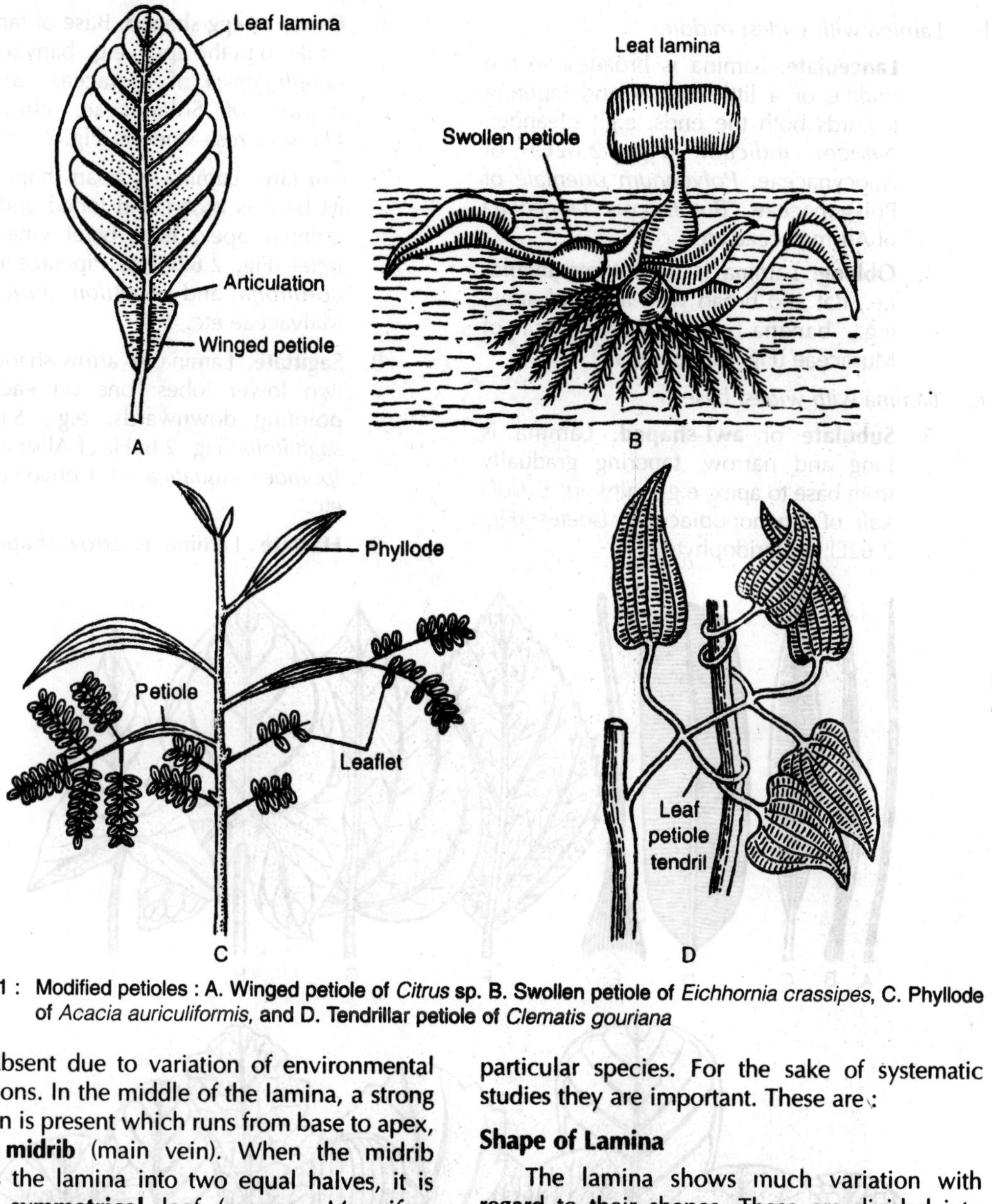

Fig. 2.61 : Modified petioles : A. Winged petiole of *Citrus* sp. B. Swollen petiole of *Eichhornia crassipes*, C. Phyllode of *Acacia auriculiformis*, and D. Tendrillar petiole of *Clematis gouriana*

even absent due to variation of environmental conditions. In the middle of the lamina, a strong midvein is present which runs from base to apex, called **midrib** (main vein). When the midrib divides the lamina into two equal halves, it is called **symmetrical** leaf (mango, *Mangifera indica*). On the other hand, in some cases, the lateral halves are unequal and the leaf is said to be **asymmetrical** (*Mirabilis jalapa* of Nyctaginaceae and in some members of Bignoniaceae etc.). The midrib or the main vein branches out in the lateral halves and arrange either in parallel or in reticulate fashion.

Lamina shows much variation as to their general outline, surface, apex and other characteristics. But the characteristics are constant for a particular species. For the sake of systematic studies they are important. These are :

Shape of Lamina

The lamina shows much variation with regard to their shapes. These are divided into three main groups :

a. *Lamina with almost same width throughout.*

1. **Acicular**. Lamina is needle-shaped, long, thin and pointed, e.g., *Pinus* sp. (Fig. 2.62A).
2. **Linear**. Lamina is long and very narrow with nearly parallel margins, e.g., tube-rose, *Polyanthes tuberosa*, rice, wheat and in many grasses (Fig. 2.62B).

b. *Lamina with widest middle.*

3. **Lanceolate**. Lamina is broadest in the middle or a little below and tapering towards both the ends, e.g., oleander, *Nerium indicum* (Fig. 2.62C) of Apocynaceae, *Polygonum orientale* of Polygonaceae, *Butomopsis lanceolata* of Alismataceae etc.
4. **Oblong**. Lamina is almost rectangular i.e., flat and broad with a round apex, e.g., banana, *Musa paradisiaca* of Musaceae (Fig. 2.62D) etc.

c. *Lamina with widest base.*

5. **Subulate** or **awl-shaped**. Lamina is long and narrow, tapering gradually from base to apex, e.g., saltwort, *Salsola kali* of Chenopodiaceae; *Isoetes* (Fig. 2.62E), a pteridophyte.
6. **Ovate** or **egg-shaped**. Base of lamina is wider than the apex, e.g., banyan, *Ficus benghalensis* of Moraceae; *Solanum nigrum* of Solanaceae; china-rose, *Hibiscus rosa-sinensis* (Fig. 2.62F) etc.
7. **Cordate**. Lamina is heart-shaped, i.e., its base is broad and lobed and has a pointed apex, e.g., betel vine, *Piper betel* (Fig. 2.62G) of Piperaceae; *Sida cordifolia* and *Abutilon indicum* of Malvaceae etc.
8. **Sagittate**. Lamina is arrow-shaped i.e., two lower lobes, one on each side pointing downwards, e.g., *Sagittaria sagittifolia* (Fig. 2.62H) of Alismataceae; *Ipomoea aquatica* of Convolvulaceae etc.
9. **Hastate**. Lamina is arrow-shaped, but

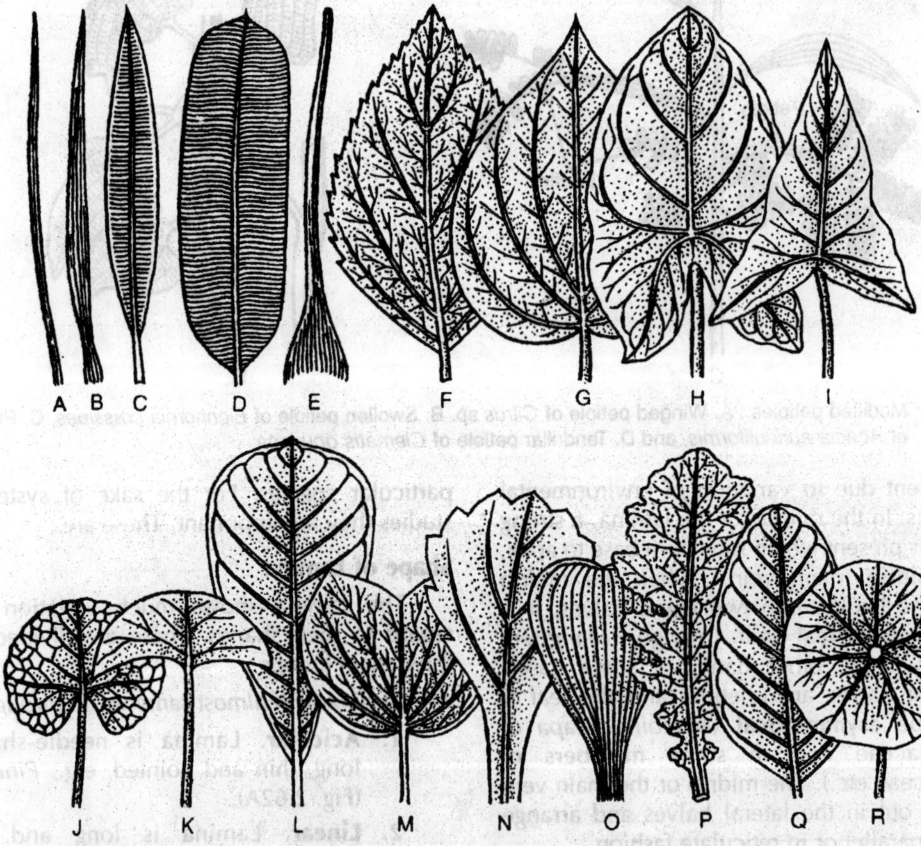

Fig. 2.62 : Lamina Shapes : A. *Acicular* in pine, B. *Linear* in grass, C. *Lanceolate* in *Nerium*, D. *Oblong* in banana, E. *Subulate* in *Isoetes*. F. *Ovate* in china-rose, G. *Cordate* in betel vine, H. *Sagittate* in *Sagittaria*, J. *Hastate* in *Ipomoea*. J. *Reniform* in *Centella*, K. *Lunate* in *Passiflora lunata*, L. *Obovate* in jack-fruit, M. *Obcordate* in *Bauhinia*, N. *Spathulate* in *Phyla*, O. *Cuneate* in *Pistia*, P. Lyrate in *Brassica*, Q. *Elliptical* in guava, and R. *Rotund* in lotus

the two lower lobes are directed outwards, e.g., *Ipomoea* sp. (Fig. 2.62I) of Convolvulaceae; *Typhonium trilobatum* of Araceae.

10. **Reniform** or **kidney-shaped**. Lamina is rounded above with a deep notch at its base, e.g., Indian pennywort, *Centella asiatica* (Fig. 2.62J) of Apiaceae.
11. **Lunate**. Lamina is like a half-moon with pointed basal lobes i.e., semicircular in outline, e.g., *Adiantum lunatum*, a pteridophyte; *Passiflora lunata* (Fig. 2.62K) of Passifloraceae etc.

d. *Lamina with widest apex.*

12. **Obovate**. Lamina is like an inverted egg (reverse of ovate), e.g., jack, *Artocarpus heterophyllus* (Fig. 2.62L) of Moraceae; *Cassia obovata* of Fabaceae etc.
13. **Obcordate**. Lamina is like an inverted heart i.e., its apical region is broad and bilobed, e.g., *Bauhinia variegata* (Fig. 2.62M) of Fabaceae; *Oxalis corniculata* of Oxalidaceae etc.
14. **Spathulate** or **spathula-shaped**. Lamina is broad and rounded at the apex and gradually tapers towards the base, e.g., *Phyla nodiflora* (Fig. 2.62N) and *Duranta repens* of Verbenaceae; *Drosera burmanii* of Droseraceae etc.
15. **Cuneate** or **wedge-shaped**. Lamina looks like the hood of a snake i.e., its breadth increases towards the apex, e.g., water lettuce, *Pistia stratiotes* (Fig. 2.62O) of Araceae.
16. **Lyrate** or **lyre-shaped**. Lamina looks like a lyre having a large oval terminal lobe and two or more smaller lobes, e.g., radish, *Raphanus sativus* and mustard, *Brassica nigra* (Fig. 2.62P) of Brassicaceae.

c. *Lamina symmetrical.*

17. **Elliptical** or **Oval**. Lamina is like an ellipse, e.g., Indian rubber, *Ficus elastica* of Moraceae; guava, *Psidium guajava* of Myrtaceae (Fig. 2.62Q) etc.
18. **Orbicular (circular)** or **rotund** or **peltiform**. Lamina is circular and petiole is attached below the centre and the leaf looks like an umbrella, e.g., lotus, *Nelumbo nucifera* (Fig. 2.62R) of Nymphaeaceae etc.

Base of Lamina

The base of lamina is of different types :

1. **Auriculate**. Leaf-bases of the sessile leaves form two lobes like that of wings, which partially surround the stem, e.g., Mexican poppy, *Argemone mexicana* of Papaveraceae, *Calotropis procera* (Fig. 2.63A) of Asclepiadaceae etc.
2. **Perfoliate**. Lobes of sessile leaf become fused together, thereby the leaf completely surrounds the stem, e.g., *Canscora perfoliata* (Fig. 2.63B) of Gentianaceae, *Aloe perfoliata* of Liliaceae etc.
3. **Connate**. Bases of two sessile leaves with opposite phyllotaxy are completely fused together, e.g., *Canscora diffusa* (Fig. 2.63C) and *Swertia chirata* of Gentianaceae etc.

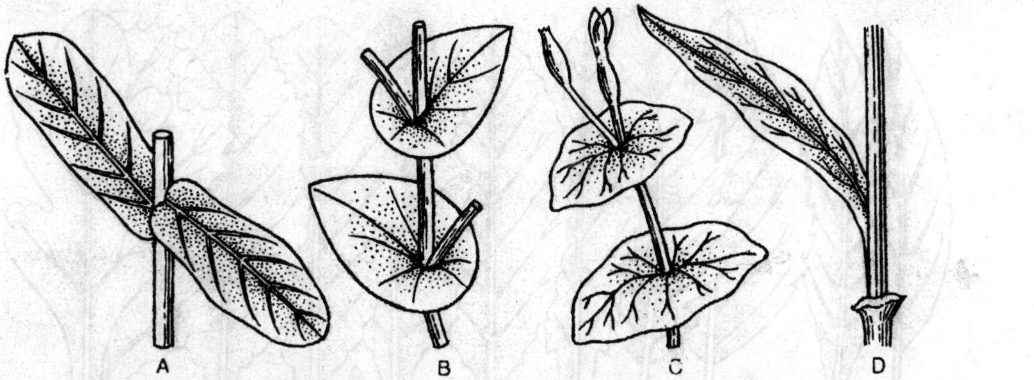

Fig. 2.63 : Base of lamina : A. Auriculate in *Calotropis*, B. Perfoliate in *Canscora perfoliata*, C. Connate in *C. diffusa*, and D. Decurrent in *Sphaeranthus*

4. **Decurrent**. Leaf base is winged and fused with the stem, e.g., *Laggera alata* and *Sphaeranthus indicus* (Fig. 2.63D) of Asteraceae.

Surface of Lamina

The surface of lamina is of different types :

1. **Glabrous**. Surface of lamina is smooth and free from hairs, e.g., *Pongamia glabra* of Fabaceae; pink, *Dianthus chinensis* of Caryophyllaceae.
2. **Glaucus**. Surface of lamina has a coating of wax and appears to be shiny, e.g., *Solanum glaucum* of Solanaceae.
3. **Viscose**. Surface of lamina becomes sticky due to some sugary exudation, e.g., *Cleome viscosa* and *Polanisia icosandra* of Capparidaceae.
4. **Scabrous**. Surface of lamina is rough due to some short rigid points, e.g., *Ficus hispida* of Moraceae.
5. **Rugose**. Surface of lamina is somewhat wrinkled, e.g., *Rubus rugosus* of Rosaceae.
6. **Gland-dotted**. Surface of lamina becomes covered with glands, e.g., lemon, *C. aurantifolia* of Rutaceae.
7. **Hairy**. Surface of lamina becomes covered with hairs. These are of the following types :
   a. *Pilose*. Covered with long, soft and scattered hairs, e.g., *Grewia flavescens* of Tiliaceae.
   b. *Hispid*. Covered with long, rigid scattered hairs, e.g., Cucurbits.
   c. *Hirsute*. Covered with stiff, fine and scattered hairs, e.g., *Eclipta alba* of Asteraceae.
   d. *Tomentose*. Covered with short, dense and cottony hairs, e.g., *Calotropis procera* of Asclepiadaceae, *Guazuma tomentosa* of Sterculiaceae etc.
   e. *Woolly*. Covered with long and interwoven hairs like wool, e.g., *Solanum verbescifolium* of Solanaceae.
8. **Spinose**. Surface of lamina is overed with prickles. e.g., *Hibiscus esculentus* of Malvaceae.

Margin of Lamina

The margin of the lamina may be of the following types :

1. **Entire**. Margin is smooth i.e., without indentations (Fig. 2.64A), e.g., mango, *Mangifera indica*; banyan, *Ficus benghalensis.*
2. **Repand**. Margin is wavy with indentations $< {}^{1}/_{16}$ distance to midrib (Fig. 2.64B), e.g., *Polyalthia longifolia* of Annonaceae.
3. **Serrate**. Margin is like the teeth of a saw, pointed upwards (Fig. 2.64C), e.g., china-rose, *Hibiscus rosa-sinensis.*
4. **Biserrate**. Margin is toothed but the teeth again become serrated (Fig. 2.64D), e.g., elm tree.
5. **Retroserrate**. Margin is toothed and pointed downwards (Fig. 2.64E).
6. **Dentate**. Margin is toothed and the teeth are pointed outward at right angles to midrib (Fig. 2.64F), e.g., water-lily.

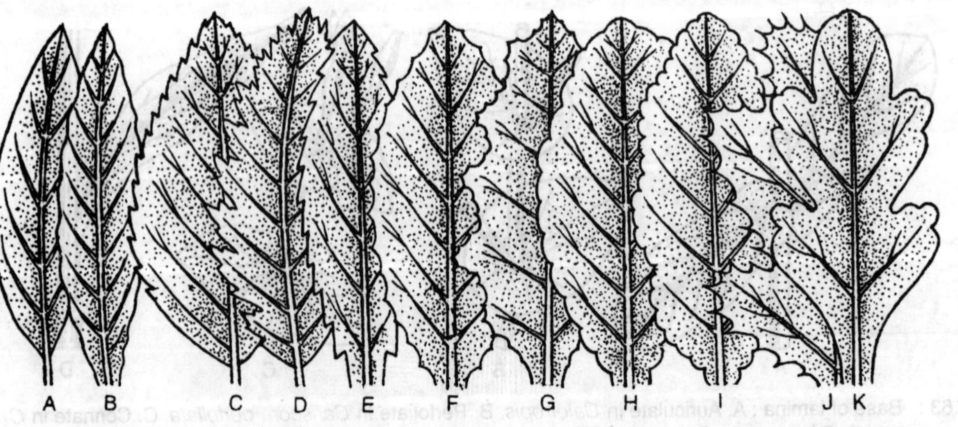

Fig. 2.64 : Margin of lamina : A. Entire, B. Wavy or repand, C. Serrate, D. Biserrate, E. Retroserrate, F. Dentate, G. Bidentate, H. Crenate, I. Bicrenate, J. Spiny, and K. Lobed (Incised)

7. **Bidentate**. Margin is toothed and the teeth are again dentate (Fig. 2.64G).
8. **Crenate**. Margin is toothed and the teeth are rounded (Fig. 2.64H), e.g., *Centella asiatica* of Apiaceae.
9. **Bicrenate**. Margin is toothed and the teeth are crenate (Fig. 2.64I).
10. **Spiny**. Margin is toothed and the teeth are pointed to form spines (Fig. 2.64J), e.g., Mexican poppy, *Argemone mexicana* of Papaveraceae; *Solanum xanthocarpum* of Solanaceae.
11. **Incised** or **lobed**. Margin is incised into various depths and divided into small lobes (Fig. 2.64K), e.g., mustard, *Brassica nigra*; radish, *Raphanus sativa* etc.

Apex of Lamina

The apex of lamina is of the following types :

1. **Acute**. Apex is pointed forming a terminal angle of 45°–90°, e.g., mango, *Mangifera indicia* (Fig. 2.65A); china-rose, *Hibiscus rosa-sinensis* etc.
2. **Acuminate**. Apex is slender and much prolonged like a tapering tail forming a terminal angle of < 45°, e.g., peepul tree, *Ficus religiosa* (Fig. 2.65B) of Moraceae, *Bauhinia acuminata* of Fabaceae (Caesalpiniaceae).
3. **Obtuse**. Apex is broad angled (>90°) and blunt, e.g., banyan, *Ficus benghalensis* (Fig. 2.65C).
4. **Mucronate**. Apex is as broad as or broader than long and forms a sharp point, e.g., *Catharanthus roseus* (Fig. 2.65D) of Apocynaceae.
5. **Cuspidate** or **spiny**. Apex ends in a hard and pointed structure, e.g., date palm, *Phoenix sylvestris* (Fig. 2.65E) of Arecaceae, *Agave cantula* of Agavaceae etc.
6. **Tendrillar**. Apex becomes narrowed and forms a tendril, e.g., glory lily, *Gloriosa superba* (Fig. 2.65F) of Liliaceae.
7. **Cirrhose**. Apex ends in a fine thread-like coiled or flexuous structure, e.g., banana, *Musa paradisiaca* (Fig. 2.65G) of Musaceae.
8. **Truncate**. Apex is cut across abruptly, almost at right angle to midrib, e.g., *Indigofera linifolia* of Fabaceae, *Paris polyphylla* (Fig. 2.65H) of Liliaceae etc.
9. **Retuse**. Obtuse apex is slightly notched, e.g., *Clitorea ternatea* of Fabaceae, water lettuce, *Pistia stratiotes* (Fig. 2.65I) of Araceae etc.
10. **Emarginate**. Obtuse apex is deeply notched, e.g., *Bauhinia variegata* (Fig. 2.65J) of Fabaceae

Texture of Lamina

1. **Coriaceous**. Lamina is thick and leathery, e.g., mango, *Mangifera indica*; Indian rubber, *Ficus elastica; Vanda roxburghii* and *Vanilla planifolia* of Orchidaceae.
2. **Herbaceous**. Lamina is thin and membranous, e.g., china-rose, *Hibiscus rosa-sinensis*; rose, *Rosa centifolia* etc.
3. **Succulent**. Lamina is fleshy and more or less brittle, e.g., *Aloe indica* of Liliaceae, *Bryophyllum calycinum* of Crassulaceae.
4. **Gland-dotted** or **glandular**. Lamina is dotted with glands and the glands are filled with essential oil, e.g., lemon, *Citrus limon* of Rutaceae; custard apple, *Annona squamosa* of Annonaceae.

Fig. 2.65 : Apex of lamina : A. Acute, B. Acuminate, C. Obtuse, D. Mucronate, E. Cuspidate, F. Tendrillar, G. Cirrhose, H. Truncate, I. Retuse, and J. Emarginate

VENATION

Laminae always have veins which are more prominent on the lower surface. Veins are arranged variously in different leaves. ***The mode of arrangement of veins in leaves is known as Venation.*** The prominent vein which runs from base to apex of a lamina is called *midrib*. The ultimate branches, by which they anastomose between the veins, are called *veinlets*.

Venation is of two types : 1. **Reticulate**, and 2. **Parallel** (striate).

[The reticulate venation is common in dicotyledons, with the exception of members like *Garcinia cowa, Calophyllum inophyllum* and some other plants of the family Clusiaceae, those have parallel venation. On the other hand, parallel venation is common in monocotyledons, with the exception of members like *Smilax macrophylla* of Liliaceae, *Colocasia antiquorum* of Araceae, *Dioscorea alata* of Dioscoreaceae etc. — those have reticulate venation.]

1. **Reticulate venation**. When the main vein divides into various branches and forms a net-like structure in the lamina, it is called *reticulate venation*. Based on the number of mid-veins, it is divided into two types : A. **Unicostate** and B. **Multicostate**.

   A. **Unicostate** or **pinnate**. There is only one mid-vein in the centre of lamina. The mid-vein gives out branches on two lateral sides like feather and those ultimately anastomose by veinlets forming a network. Viz. jack, *Artocarpus heterophyllus* and pipal tree, *Ficus religiosa* (Fig. 2.66A) of Moraceae; mango, *Mangifera indica* of Anacardaceae etc.

   B. **Multicostate** or **palmate**. In this type, many mid-veins develop from the tip of the petiole and progress towards the upper side of lamina. It is of two kinds : a. Convergent, and b. Divergent.

      a. *Convergent* type. In this type, the mid-veins after developing from the basal region meet or appear to meet at the apex. Viz. jujube, *Ziziphus mauritiana* of Rhamnaceae; *Strychnos nux-vomica* and *S. potatorum* of Loganiaceae; cinnamon, *Cinnamomum tamala* (Fig. 2.66B) of Lauraceae etc.

      b. *Divergent* type. In this type, the mid-veins, after developing from the basal region, spread at different directions. Viz. gourd, *Cucurbita maxima* (Fig. 2.66C); cotton, *Gossypium herbaceum* etc.

2. **Parallel** or **striate venation**. Here all veins run parallel to each other, either from the base to the apex of the lamina or at right angles to the midrib. They do not form any network. Based on the number of mid-veins, it is also divided into two types : A. **Unicostate**, and B. **Multicostate**.

   A. **Unicostate** or **pinnate**. In this type, there is only one mid-vein in the centre of lamina. The mid-vein gives out branches on two lateral sides perpendicular to the midrib. The branches run parallel to each

Table 2.7 : The venation is tabulated :

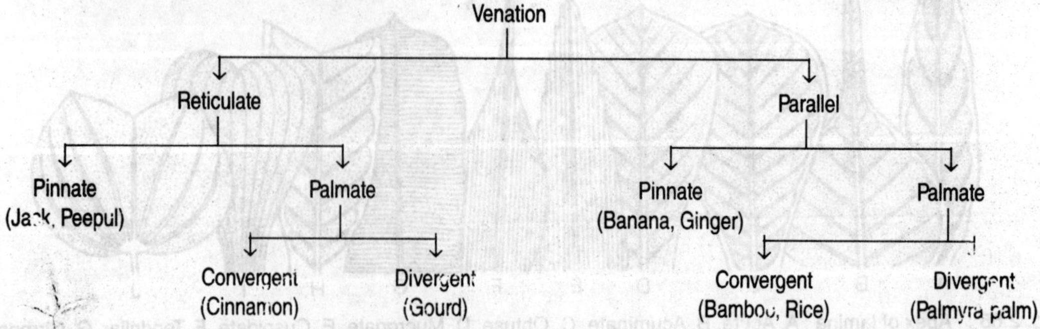

other towards the margin or apex of lamina. Viz. ginger, *Zingiber officinale; Canna indica*; banana, *Musa bulbisiana* (Fig. 2.66D) etc.

B. **Multicostate** or **palmate**. Here many mid-veins develop from the apex of the petiole. It is of two kinds : a. Convergent and b. Divergent.

a. *Convergent* type. Here all the mid-veins run parallel to each other from base and unite at the apex, e.g., rice, *Oryza sativa;* bamboo, *Bambusa tulda* (Fig. 2.66E) etc.

b. *Divergent* type. Here all the mid-veins, after originating from the apex of the petiole, spread towards the periphery away from each other. Viz. palmyra palm, *Borassus flabellifer* (Fig. 2.66F) of Arecaceae.

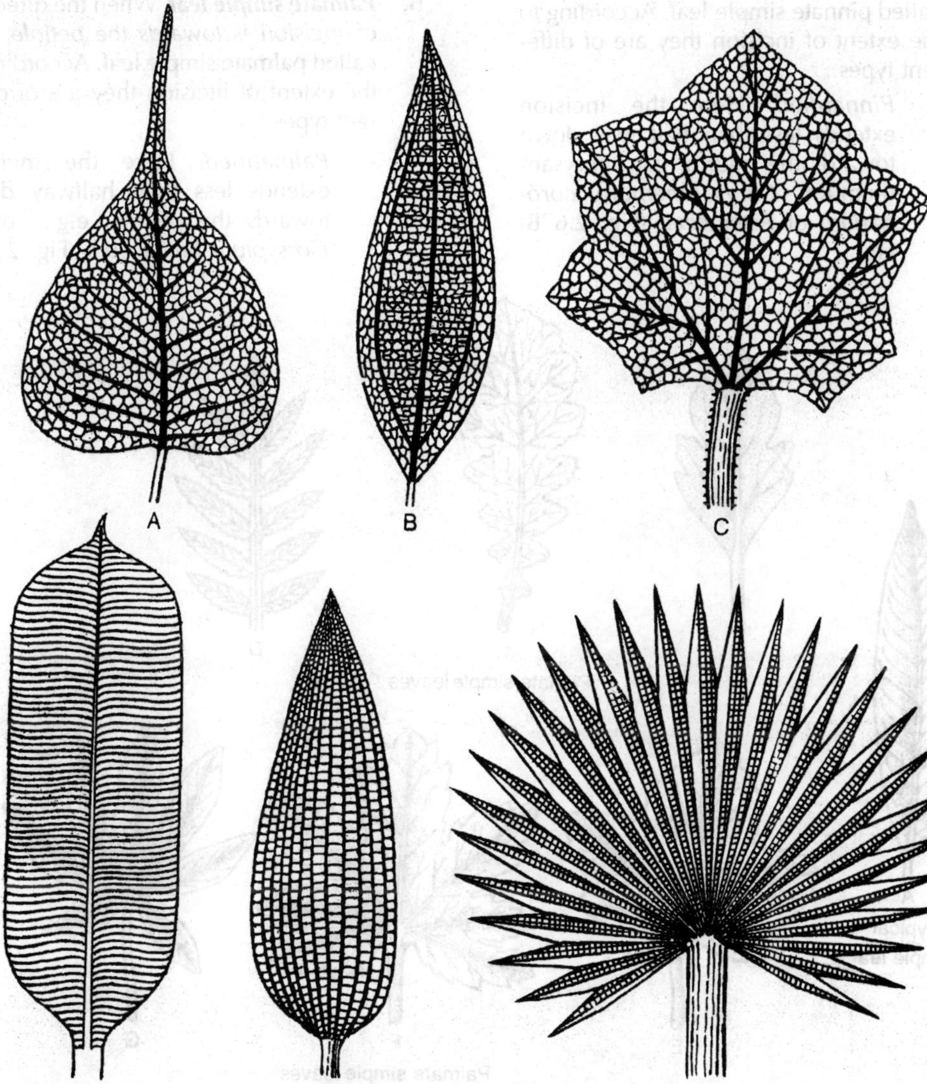

Fig. 2.66 : Venation of leaves : A. Reticulate unicostate (pinnate), B. Reticulate multicostate (palmate) convergent, C. Reticulate multicostate (palmate) divergent, D. Parallel (unicostate) pinnate, E. Parallel multicostate (palmate) convergent and F. Parallel multicostate (palmate) divergent

Type of Leaves

Based on the incision of lamina, the leaves are divided into two groups : Simple leaf and Compound leaf :

1. **Simple leaf**. When the leaf consists of a single lamina with usually entire (Fig. 2.67A) or incised margin, but the incision never touches the mid-vein, it is called simple leaf. The simple leaves with incised margins are of two types :

   A. ***Pinnate simple leaf***. When the direction of *incision is towards the midrib,* it is called pinnate simple leaf. According to the extent of incision they are of different types :

      a. *Pinnatified.* Here the incision extends less than half way down towards the midrib, e.g., chrysanthemum, *Chrysanthemum coronarium* of Asteraceae (Fig. 2.67B) etc.

      b. *Pinnatipartite.* Here the incision extends more than halfway down towards the midrib, e.g., Mexican poppy, *Argemone mexican* of Papaveraceae; radish, *Raphanus sativus* and mustard, *Brassica nigra* (Fig. 2.67C) of Brassicaceae.

      c. *Pinnatisect.* Here the incisions almost reach the midrib, e.g., marigold, *Tagetes patula* (Fig. 2.67D) of Asteraceae; *Ipomoea quammoclit* of Convolvulaceae etc.

   B. ***Palmate simple leaf***. When the direction of *incision is towards the petiole,* it is called palmate simple leaf. According to the extent of incision they are of different types :

      a. *Palmatified.* Here the incision extends less than halfway down towards the petiole, e.g., cotton, *Gossypium herbaceum* (Fig. 2.67E)

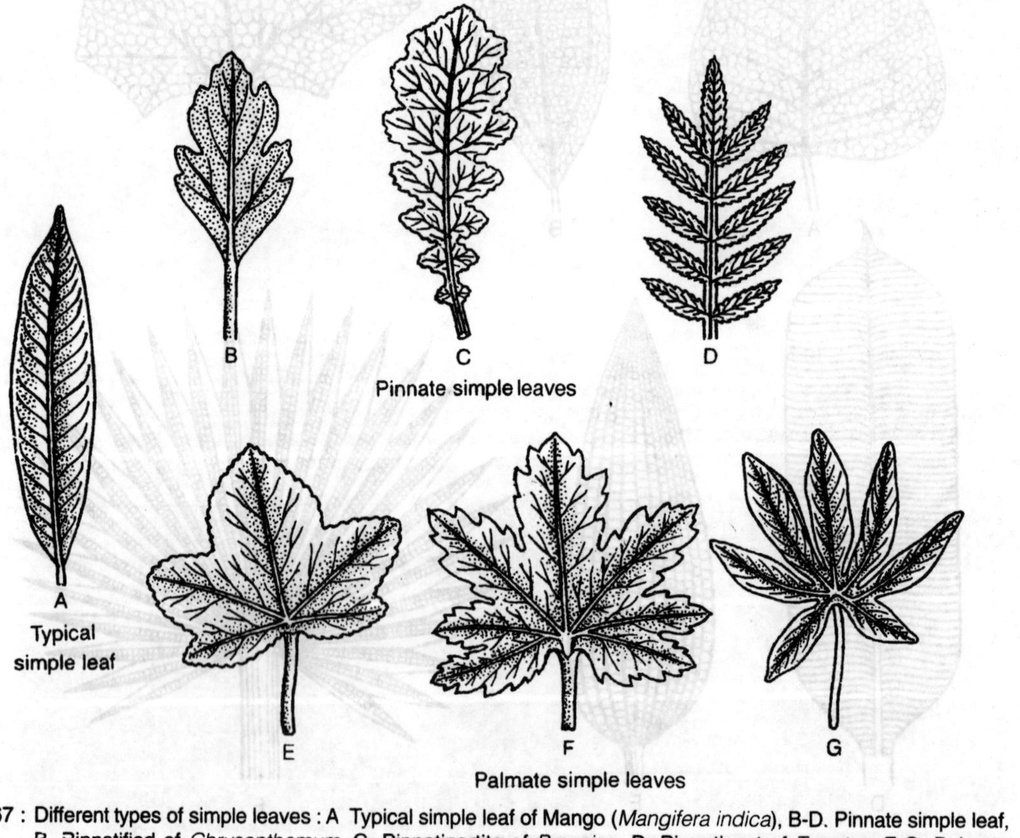

Fig. 2.67 : Different types of simple leaves : A Typical simple leaf of Mango (*Mangifera indica*), B-D. Pinnate simple leaf, B. Pinnatified of *Chrysanthemum*, C. Pinnatipartite of *Brassica*, D. Pinnatisect of *Tagetes*, E-G. Palmate simple leaf, E. Palmatified of *Gossypium*, F. Palmatipartite of *Ricinus*, and G. Palmatisect of *Ipomoea*

and lady's finger, *Abelmoschus esculentus* of Malvaceae etc.

b. *Palmatipartite.* Here the incision extends up to halfway down towards the petiole, e.g., *Ricinus communis* (Fig. 2.67F) of Euphorbiaceae; bitter gourd, *Momordica charantia* of Curcurbitaceae etc.

c. *Palmatisect.* Here the incision extends more than halfway down towards the petiole e.g., *Ipomoea paniculata* and *I. pulchella* (Fig. 2.67G) of Convolvulaceae etc.

2. **Compound leaf**. In this type, the lamina becomes incised and the incision reaches up to the midrib or petiole. Thus, the lamina is divided into several small parts (leaflets) forming compound leaf. It is of two types :

A. ***Pinnate compound leaf.*** When the incision of lamina is directed towards the midrib (rachis) and the leaflets are arranged on both sides of it or on its branches, it is called pinnate compound leaf. Based on the division of rachis it is of four types :

a. *Unipinnate.* When the leaflets are directly attached on the midrib i.e., rachis, it is called unipinnate. It is of two types :

i. *Paripinnate.* When leaflets are arranged in pairs (even numbers) on the rachis, the leaf is called paripinnate, e.g., tamarind, *Tamarindus indicus* (Fig. 2.68A), *Cassia fistula*, gold mohur, *Delonix regia* etc.

ii. *Imparipinnate.* When leaflets are arranged in pairs (even numbers) on the rachis, except the apical region which is terminated by an unpaired or odd leaflet, it is called

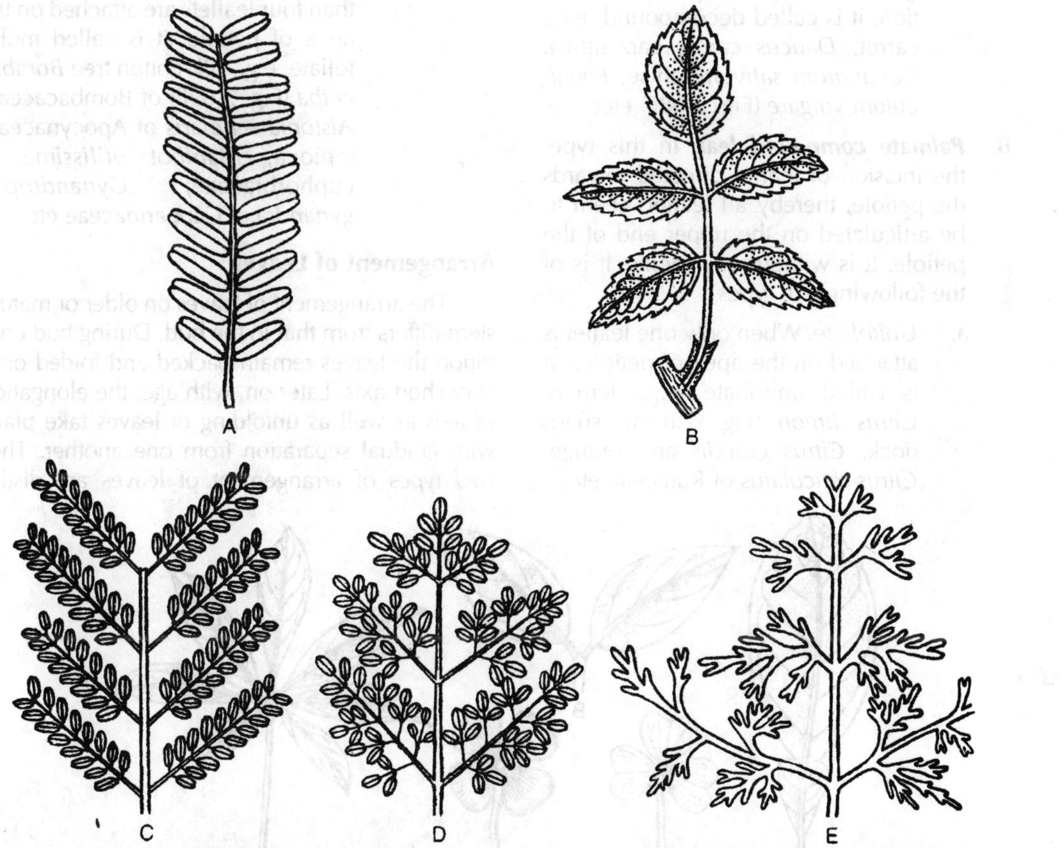

Fig. 2.68 : Pinnate compound leaves : A. Paripinnate of *Tamarindus*, B. Imparipinnate of *Rosa*, C. Bipinnate of *Acacia*, D. Tripinnate of *Moringa*, and E. Decompound of *Foeniculum*

imparipinnate, e.g., rose, *Rosa centifolia* (Fig. 2.68B) of Rosaceae; *Clitorea ternatea* of Fabaceae; *Azadirachta indica* of Meliaceae etc.

b. *Bipinnate.* When rachis divides and develops secondary axis, where leaflets are arranged on both the sides, it is called bipinnate, e.g., *Mimosa pudica, Caesalpinia pulcherrima,* and *Acacia nilotica* (Fig. 2.68C) of Fabaceae.

c. *Tripinnate.* When the secondary axis divides and gives rise to tertiary axis, where leaflets are arranged on both the sides, it is called tripinnate. Viz. horse radish, *Moringa oleifera* (Fig. 2.68D) of Moringaceae, *Oroxylon* sp. of Bignoniaceae etc.

d. *Decompound.* When the incision goes beyond the tripinnate condition, it is called decompound, e.g., carrot, *Daucus carota* var. *sativa*; *Coriandrum sativum,* anise, *Foeniculum vulgare* (Fig. 2.68E) etc.

B. ***Palmate compound leaf.*** In this type, the incision of leaf is directed towards the petiole, thereby all leaflets seem to be articulated on the upper end of the petiole. It is without any rachis. It is of the following five types :

a. *Unifoliate.* When only one leaflet is attached on the apex of petiole, it is called unifoliate, e.g., lemon, *Citrus limon* (Fig. 2.69A); shaddock, *Citrus grandis* and orange, *Citrus reticulatus* of Rutaceae etc.

b. *Bifoliate* or *binnate.* When two leaflets are attached on the apex of petiole, it is called bifoliate, e.g., *Bignonia grandiflora* of Bignoniaceae; *Balanites aegyptiaca* of Simaroubaceae; *Inga dulcis* and *Hardwickia binata* (Fig. 2.69B) of Fabaceae etc.

c. *Trifoliate* or *ternate.* When three leaflets are attached on the apex of petiole, it is called trifoliate, e.g., *Vitex negundo* of Verbenaceae; *Oxalis corniculata* (Fig. 2.69C) of Oxalidaceae etc.

d. *Quadrifoliate* or *quadrinate.* When four leaflets are attached on the apex of petiole, it is called quadrifoliate, e.g., *Marsilea quadrifolia,* a Pteridophyte (Fig. 2.69D) etc.

e. *Multifoliate* or *digitate.* When more than four leaflets are attached on the apex of petiole, it is called multifoliate, e.g., silk-cotton tree *Bombax ceiba* (Fig. 2.69E) of Bombacaceae; *Alstonia scholaris* of Apocynaceae; tapioca, *Manihot utilissima* of Euphorbiaceae; *Gynandropsis gynandra* of Capparidaceae etc.

Arrangement of Leaves

The arrangement of leaves on older or mature stem differs from that in the bud. During bud condition the leaves remain packed and folded on a very short axis. Later on, with age, the elongation of axis as well as unfolding of leaves take place with gradual separation from one another. Thus *two types* of arrangement of leaves are distin-

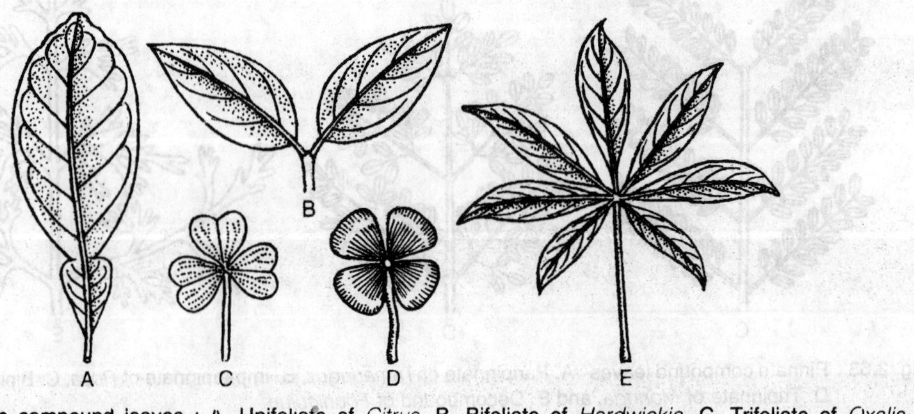

Fig. 2.69 : Palmate compound leaves : A. Unifoliate of *Citrus*, B. Bifoliate of *Hardwickia*, C. Trifoliate of *Oxalis*, D. Quadrifoliate of *Marsilea*, and E. Digitate of *Bombax*

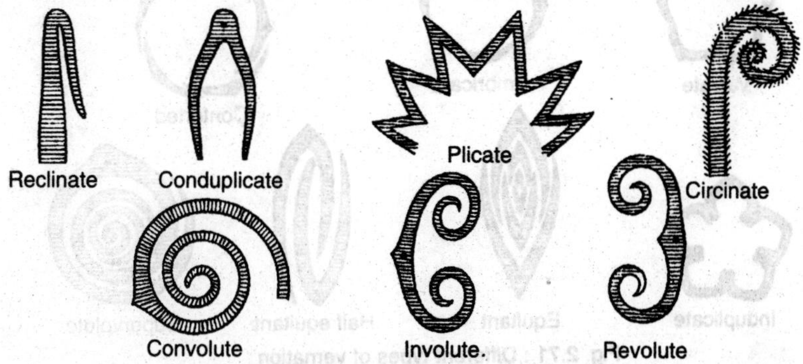

Fig. 2.70 : Different types of ptyxis

guished : A. **Prefoliation**. It is the disposition of leaves in the bud; and B. **Phyllotaxy**. It is the arrangement of leaves on stem or branches.

Prefoliation

The leaves in the bud are arranged in a definite sequence. It can be discussed under two heads : a. **Ptyxis**. It is the pattern in which the individual leaf (consider independently of others) remains rolled, folded or bent in the bud, and b. **Vernation**. It is the relative arrangement of leaves of individual bud.

a. **Ptyxis**. The ptyxis (Fig. 2.70) can be discussed under *two* heads : i. Based on leaf *bend* or *fold*, and ii. Based on leaf roll.

i. Based on leaf *bend* or *fold :*

Reclinate or **inflexed**. In this type, the upper part of lamina is inwardly half bent towards the base of lamina, e.g., loquat, *Eriobotrya japonica* of Rosaceae.

Conduplicate. In this type, the leaf is folded lengthwise along the midrib, so in T.S. it is almost V-shaped in appearance, e.g., *Magnolia grandiflora* of Magnoliaceae; rose, *Rosa centifolia* of Rosaceae etc.

Plicate. In this type, the leaf is folded several times along the main veins like a close fan, e.g., *Geodorum* sp. of Orchidaceae, palmyra palm, *Borassus flabellifer* of Arecaceae etc.

Crumpled. In this type, the leaf becomes folded irregularly in all directions, e.g., cabbage, *Brassica oleracea* of Brassicaceae.

ii. Based on leaf *roll* :

Circinate. In this type, the leaf is rolled up from apex to base on its axis like spring of a watch, e.g., ferns, *Cycas* sp (a gymnosperm).

Convolute. In this type, the leaf is rolled up from one margin to the other, e.g., Indian shot, *Canna indica* of Cannaceae; banana, *Musa paradisiaca* of Musaceae etc.

Involute. In this type, both the margins rolled upward towards the midrib, e.g., banyan, *Ficus beghalensis* and jack, *Artocarpus heterophyllus* of Moraceae; water lily, *Nymphaea stellata* of Nymphaeaceae etc.

Revolute. In this type, both the margins rolled downward towards the midrib, e.g., *Nerium indicum* of Apocynaceae etc.

b. **Vernation**. As observed by transverse section of bud, the vernation is of the following types (Fig. 2.71) :

Valvate. In this type, the margins of leaves are touched one another but do not overlap.

Twisted or **contorted**. In this type, the margins of leaves are overlapped one another in a regular pattern.

Imbricate. In this type, the margins of leaves are arranged in such a way that one becomes completely internal, another one completely external, and the rest are like the twisted one.

Induplicate. This type is similar to valvate, but the margins of leaves are slightly bent towards the inner side.

Equitant. In this type, the margins of leaves are overlapped the next opposite one in a regular pattern, e.g., corn flag, *Gladiolus communis* of Iridaceae, *Vanda roxburghii* of Orchidaceae etc.

Half-equitant or **Obvolute**. This type is like that of the equitant type, but only half of the leaf blade remains inside the other, e.g., *Salvia officinalis* of Lamiaceae (Labiatae).

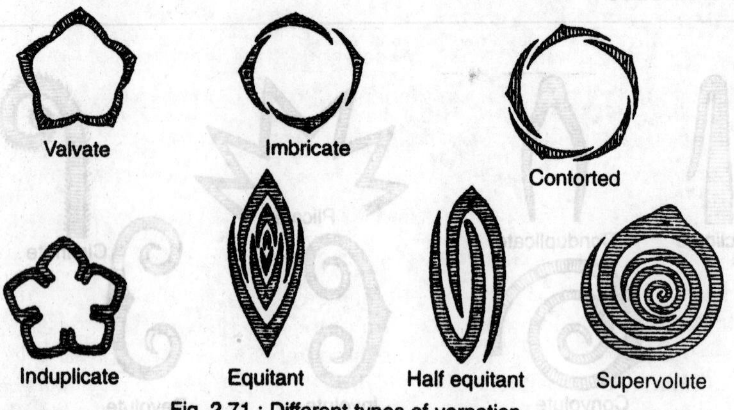

Fig. 2.71 : Different types of vernation

Supervolute. In this type, the convolute leaves completely enclose other leaves forming a single compact bud. e.g., *Prunus persica* of Rosaceae.

Phyllotaxy

The leaves develop on the stem or branches in a definite sequence. When the leaves develop as a cluster or in a rosette from the very short stem just on the top of the root, they are called *radical* leaves (the leaves do not actually develop from the root but from a condensed stem). The leaves develop on the stem with short or long internodes between them are called *cauline* leaves. The number of these leaves on nodes may be one or more and they arrange in a definite sequence. According to Eames and MacDaniels (1947) "the leaf arrangement on the stem or phyllotaxy and variations in size and shape are doubtless in part correlated with the exposure of the photosynthetic surface to light". *The mode of arrangement of leaves on the stem or branches is called* **Phyllotaxy**. It is of three types : 1. **Alternate** or *Acylic* or Spiral, 2. **Opposite,** and 3. **Whorled** or Verticillate. The last two types are called *Cyclic* type.

1. **Alternate** or **Acyclic** or **Spiral**. When one leaf develops from each node, it is called alternate type. The leaves are arranged in a spiral manner all round the stem. This type is very common in plants like china-rose, *Hibiscus rosa-sinensis* (Fig. 2.72A) of Malvaceae; mango, *Mangifera indica* of Anacardiaceae; mustard, *Brassica nigra* of Cruciferae etc.

Table 2.8 : Differences between simple and compound leaf

Simple leaf	Compound leaf
1. The leaves have single lamina.	1. The lamina is divided into many small leaflets.
2. Simple leaf has one axillary bud.	2. Bud is present in the axil of rachis and not in the axil of leaflets.
3. Stipule, if present, is at the base of leaf.	3. Stipule, if present, is at the base of the rachis, but not at the base of leaflets.

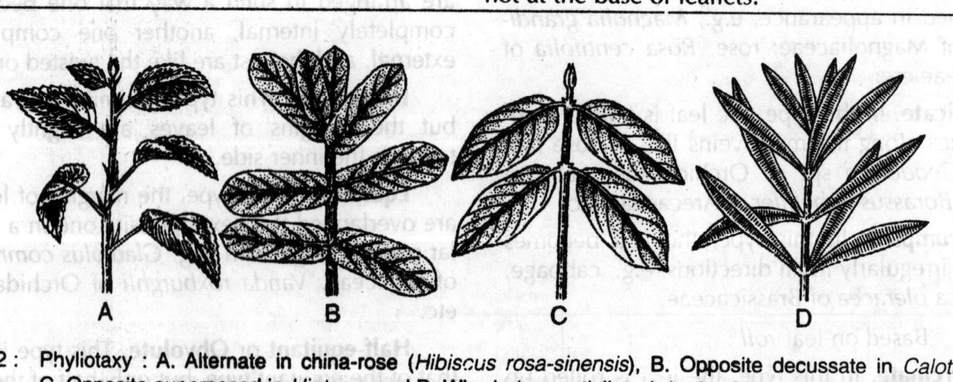

Fig. 2.72 : Phyllotaxy : A. Alternate in china-rose (*Hibiscus rosa-sinensis*), B. Opposite decussate in *Calotropis*, C. Opposite superposed in *Hiptage*, and D. Whorled or verticilate (whorls of three) in *Nerium indicum*

The alternate or spiral phyllotaxy may be categorised into different types :

i) *Distichous* ($^1/_2$ phyllotaxy). viz. members of Poaceae, Orchidaceae

ii) *Tristichous* ($^1/_3$ phyllotaxy). viz. members of Cyperaceae

iii) *Pentastichous* ($^2/_5$ phyllotaxy). viz. *Hibiscus rosa-sinensis, Ficus benghalensis*

iv) *Octastichous* ($^3/_8$ phyllotaxy). viz. *Carica papaya, Thevetia peruviana*

v) *Parastichous* ($^5/_3$, $^8/_{21}$, etc., phyllotaxy). viz. *Phoenix* spp.

The fundamental type of leaf arrangement may be expressed as a fraction ($^1/_2$, $^1/_3$, $^2/_5$, $^3/_8$, $^5/_{13}$, $^8/_{12}$, etc.) where each succeeding fraction is the sum of the two previous numerators and the sum of the two previous denominators. The **numerator** is the number of turns or spirals (e.g., 1 for $^1/_2$, 2 for $^2/_5$, 5 for $^5/_{13}$, etc.) around a stem before one leaf is directly above another. The **denominator** is the number of leaves in the turns or spirals (e.g., 2 for $^1/_2$, 3 for $^1/_3$, 5 for $^2/_5$, 13 for $^5/_{13}$, etc.) before one is directly above the other. Thus, we would get the series in the following manner (i.e., sum of two preceeding numerators/sum of two preceding denominators) :

$$\underset{\text{(i)}}{\frac{1}{2}};\ \underset{\text{(ii)}}{\frac{1}{3}};\ \frac{1+1}{2+3}=\underset{\text{(iii)}}{\frac{2}{5}};\ \frac{1+2}{3+5}=\underset{\text{(iv)}}{\frac{3}{8}};\ \frac{2+3}{5+8}=\underset{\text{(v)}}{\frac{5}{13}};\ \frac{3+5}{8+13}=\underset{\text{(vi)}}{\frac{8}{21}}\ \ldots\ldots$$

Thus, the spiral arrangement of leaves can be interpreted in terms of **divergence** (distance formed upon the spiral of the stem between two successive leaves) using mathematical calculation. This is known as **Fabonacci series** and the mean angle of divergence of successive leaves is called **Fabonacci angle.** The Fabonacci series is also named as **Schimpes-Brown series** after its discoverer.

In districhous ($^1/_2$) phyllotaxy, there are one twist and two leaves before one leaf is directly above the other. Thus, the angle of divergence would be 180° ($^1/_2$ of 360°) between succeeding leaves in the stem.

Tritichous ($^1/_3$) phylotaxy would mean one twist and three leaves in a complete turn or an angle of divergence of 120° ($^1/_3$ of 360°) between succeeding leaves in the centre of the stem.

Pentastichous ($^2/_5$) phyllotaxy means two twists and five leaves in a complete turn. The angle of divergence would be 144° ($^2/_5$ of 360°) between succeeding leaves in the stem.

Similarly, octastichous ($^3/_8$) phyllotaxy would mean three twists and eight leaves before one leaf is directly above the other or an angle of divergence of 135° ($^3/_8$ of 360°) between succeeding leaves in the stem.

2. **Opposite.** When two leaves are attached at each node opposite to each other (i.e., it makes an angular distance of 180°), it is called opposite type. It is of two types : i. Decussate, and ii. Superposed.

   i. **Decussate.** When consecutive pair of opposite leaves are arranged at right angle to each other, it is called opposite decussate type, e.g., basil, *Ocimum sanctum* of Lamiaceae; *Calotropis procera* (Fig. 2.72B) of Asclepiadaceae etc.

   ii. **Superposed.** When consecutive pair of opposite leaves are arranged in one plane i.e., one set of leaves above the other, it is called opposite superposed type, e.g., *Quisqualis* sp. of Combretaceae, *Hiptage madablota* (Fig. 2.72C) of Malpighiaceae, *Aganosma caryophyllata* of Apocynaeae; *Psidium guajava* of Myrtaceae etc.

3. **Whorled** or **verticilate.** When more than two leaves develop from each node forming a whorl or vertical, it is called whorled type, e.g., oleander, *Nerium indicum* (Fig. 2.72D); *Alstonia scholaris* of Apocynaceae.

Modification of leaves

The lamina or leaf blade becomes modified either wholly or in part to perform different functions. A brief account of such modifications are :

1. **Leaf tendrils.** In some plants, the lamina becomes modified either partially or wholly into tendril to climb up on the support. These are :

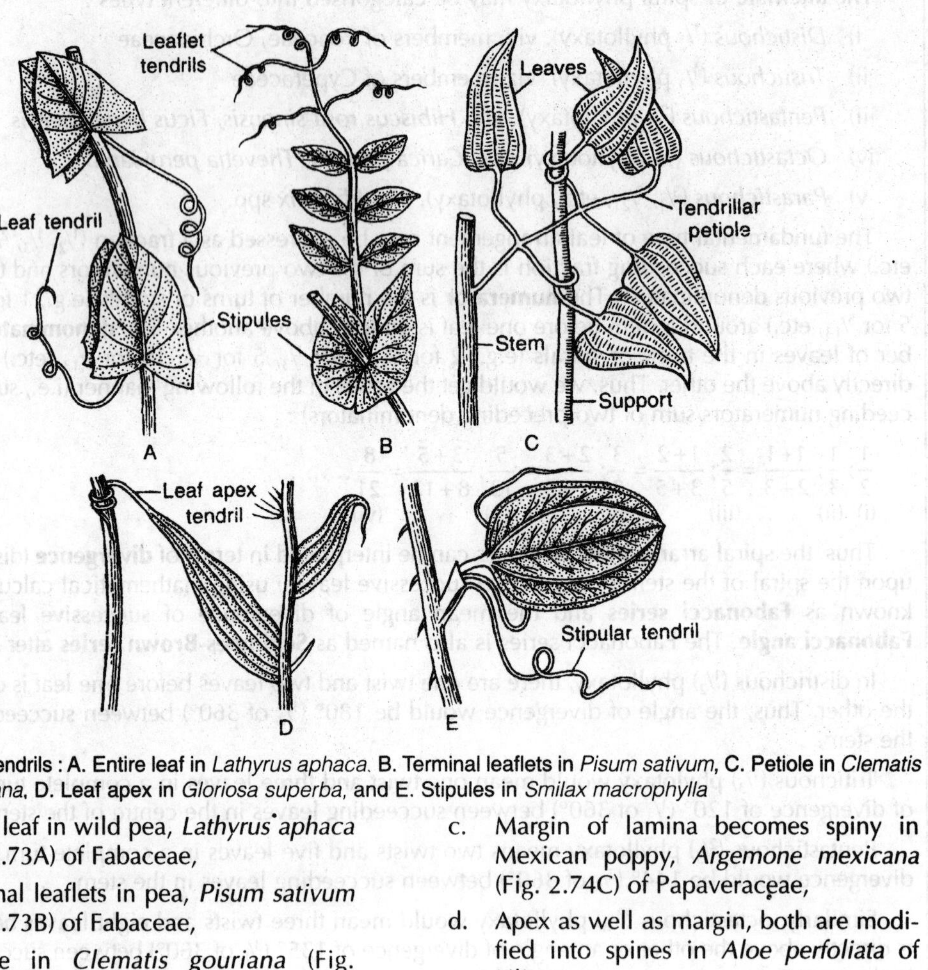

Fig. 2.73 : Leaf tendrils : A. Entire leaf in *Lathyrus aphaca*. B. Terminal leaflets in *Pisum sativum*, C. Petiole in *Clematis gouriana*, D. Leaf apex in *Gloriosa superba*, and E. Stipules in *Smilax macrophylla*

a. Entire leaf in wild pea, *Lathyrus aphaca* (Fig. 2.73A) of Fabaceae,

b. Terminal leaflets in pea, *Pisum sativum* (Fig. 2.73B) of Fabaceae,

c. Petiole in *Clematis gouriana* (Fig. 2.73C) of Ranunculaceae,

d. Leaf apex in glory lily, *Gloriosa superba* (Fig. 2.73D) of Liliaceae,

e. Stipules in *Smilax macrophylla* (Fig. 2.73E) of Liliaceae.

2. **Leaf spines**. The leaf spines are the modification of either whole or part of lamina and perform as defensive organ, to protect the plants from herbivorous animals. In some xerophytic plants, it also reduces the rate of transpiration in addition to defense.

a. Stipules are modified into spines in jujube, *Zizyphus mauritiana* (Fig. 2.74A) of Rhamnaceae; *Acacia arabica* of Fabaceae etc.

b. Apex of lamina is modified into spine in date palm, *Phoenix sylvestris* of Arecaceae (Palmae); *Yucca* sp. (Fig. 2.74B).

c. Margin of lamina becomes spiny in Mexican poppy, *Argemone mexicana* (Fig. 2.74C) of Papaveraceae,

d. Apex as well as margin, both are modified into spines in *Aloe perfoliata* of Liliaceae,

e. Leaf of axillary bud becomes modified into spines (normal leaves are ill-developed and fall down very soon) in prickly pear, *Opuntia dillenii* (Fig. 2.74D) of Cactaceae,

f. Leaves are modified into spines (but the leaves of axillary buds remain normal) in barberry, *Berberis vulgaris* (Fig. 2.74E) of Berberidaceae,

g. Leaves are modified into spines (branches present in the axis of these plants are modified into large and thick thorns) in *Ulex*, (Fig. 2.74F).

3. **Fleshy leaves**. Sometimes the leaves become fleshy due to storage of water, mucilage and food materials and serve as storage tissue. This is common in xerophytes and also in halophytes, e.g., *Basella rubra* (Fig. 2.37A)

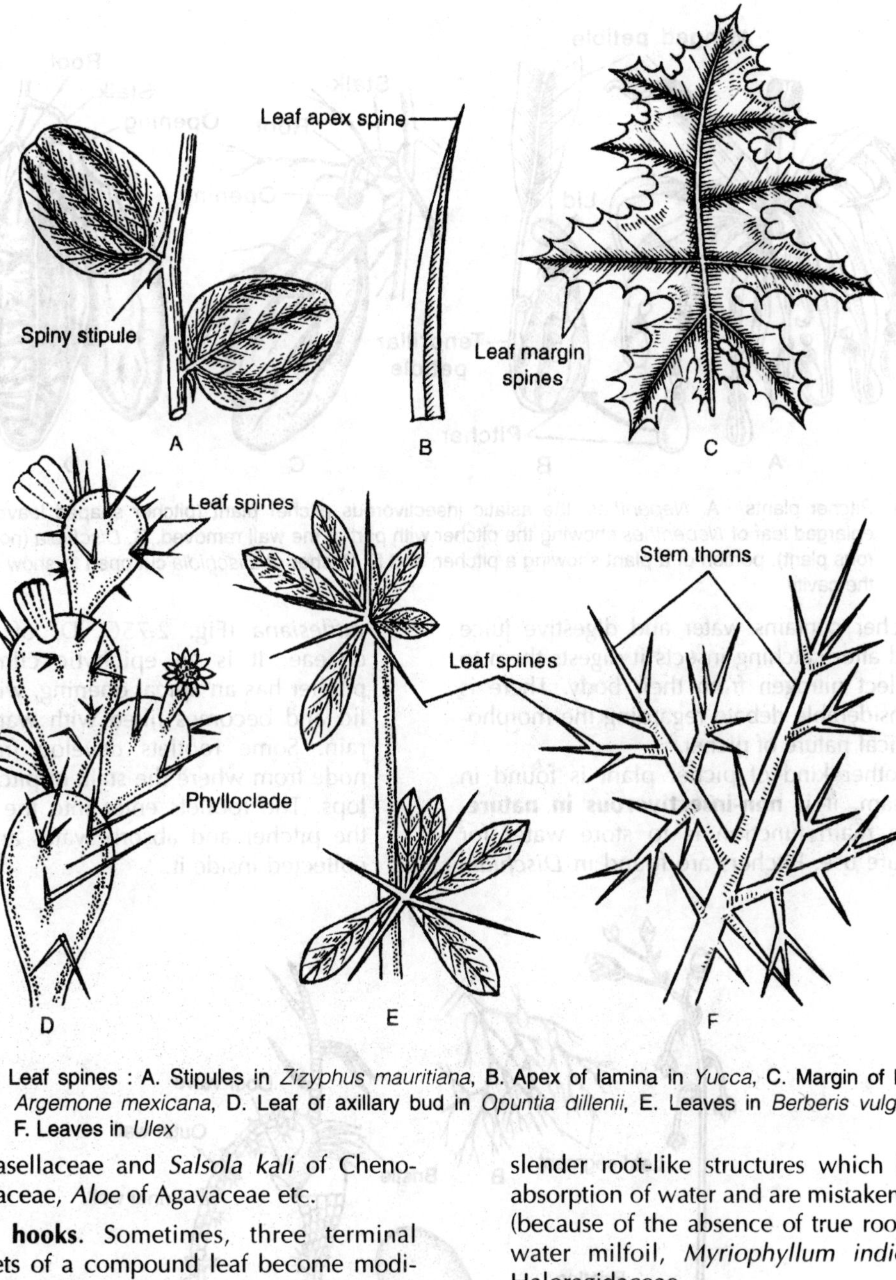

Fig. 2.74 : Leaf spines : A. Stipules in *Zizyphus mauritiana*, B. Apex of lamina in *Yucca*, C. Margin of lamina in *Argemone mexicana*, D. Leaf of axillary bud in *Opuntia dillenii*, E. Leaves in *Berberis vulgaris*, and F. Leaves in *Ulex*

of Basellaceae and *Salsola kali* of Chenopodiaceae, *Aloe* of Agavaceae etc.

4. **Leaf hooks**. Sometimes, three terminal leaflets of a compound leaf become modified into sharp, curved and pointed hooks like the palm of cat and entangle the bark of trees for climbing, e.g., *Bignonia anguiscati* (Fig. 2.40E) of Bignoniaceae.

5. **Phyllode**. It is modified petiole (see page no. 175, Fig. 2.61C).

6. **Root-like structure**. In some aquatic plants, the submerged leaves are modified into slender root-like structures which help in absorption of water and are mistaken as root (because of the absence of true roots). Viz. water milfoil, *Myriophyllum indicum* of Haloragidaceae.

7. **Pitcher shaped leaves**. In the pitcher plants like *Nepenthes khasiana* (Fig. 2.75A, B) of Nepanthaceae (**an insectivorous plant**), the leaf is modified into a **pitcher** with a lid develops from the apex. The anterior part of the petiole coils like tendril and helps the pitcher to remain erect while the posterior part becomes flattened like a leaf. The

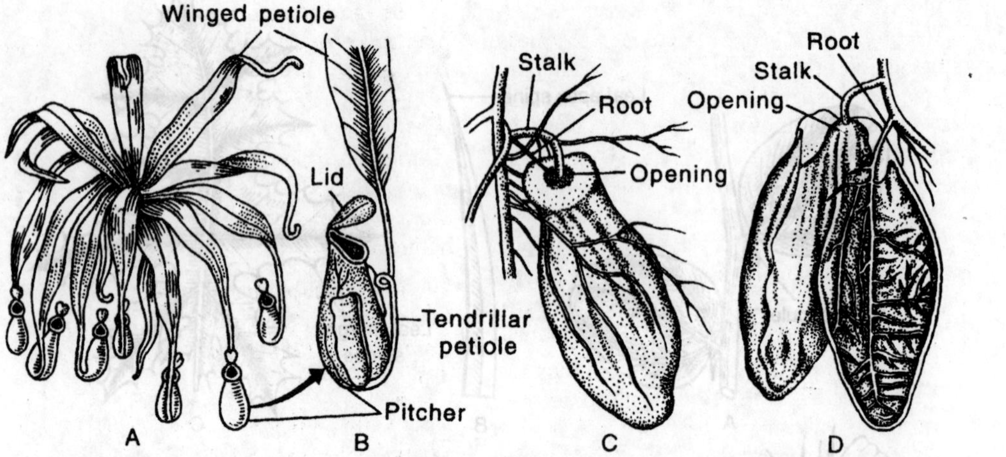

Fig. 2.75 : Pitcher plants : A. *Nepenthes*, the asiatic insectivorous pitcher plant (pitcher shaped leaves), B. One enlarged leaf of *Nepenthes* showing the pitcher with part of the wall removed, C. *Dischidia* (non-insectivorous plant), portion of a plant showing a pitcher, and D. Pitcher of *Dischidia* cut open to show roots inside the cavity

pitcher contains water and digestive juice and after catching insects it digests them to collect nitrogen from their body. There is considerable debate regarding the morphological nature of picher.

8. Another kind of picher plant is found in Assam, it is **non-insectivorous in nature**. The main function is to store water for future use. Pitchers are found in *Dischidia rafflesiana* (Fig. 2.75C, D) of Asclepiadaceae. It is an epiphytic climber. The pitcher has an apical opening, without any lid and becomes filled with water during rain. Some rootlets develop from same node from where the stalk of pitcher develops. The rootlets enter into the cavity of the pitcher and absorb water and humus collected inside it.

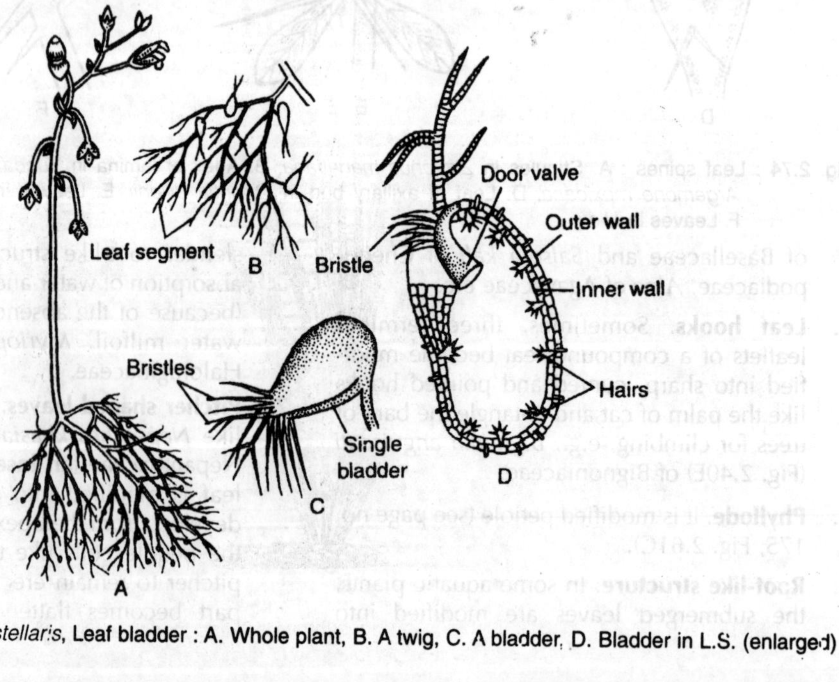

Fig. 2.76 : *Utricularia stellaris*, Leaf bladder : A. Whole plant, B. A twig, C. A bladder, D. Bladder in L.S. (enlarged)

9. **Bladder**. The bladderwort, *Utricularia stellaris* (Fig. 2.76) of Lentibulariaceae, a rootless herb, is commonly found in ponds. The submerged leaves are much segmented. Some of the segments become modified to form bladder-like structures. They can trap insects to obtain nitrogen from their body.

Heterophylly

A number of plants have foliage leaves of different forms. This condition is called **heterophylly**. It is found in many aquatic and land plants.

In **aquatic plants**, there are two kinds of leaves : *submerged* and *floating*. The submerged leaves are narrow or finely dissected, while the floating leaves are larger in size, undivided or lobed. Heterophylly is found in the plants like water chestnut, *Trapa bispinosa* of Onagraceae; water crowfoot, *Ranunculus aquatilis* (Fig. 2.77A) of Ranunculaceae; the arrowhead, *Sagittaria sagittifolia* (Fig. 2.77B) and *Alisma plantago* of Alismataceae, *Limnophila heterophylla* (Fig. 2.77C) of Scrophulariaceae etc.

In **land plants**, the heterophylly is found in plants like *Artocarpus chaplasa* and *Ficus heterophyllus* of Moraceae, *Coriandrum sativum* of Apiaceae, *Ranunculus sceleratus* of Ranunculaceae, *Coriandrum sativum* of Apiaceae, *Goldfussia glomerata* of Acanthaceae etc. Some plants growing in marshy places, like *Cardenthera difformis* of Acanthaceae, frequently show heterophylly.

In **Pteridophytes**, like *Dymoglossum piloselloides* and *Pteris ludens* of Polypodiaceae; *Ceratopteris thalictroides* of Parkeriaceae show distinct heterophylly.

In **Gymnosperms**, like *Ginkgo biloba* of Ginkgoaceae also shows heterophylly.

Anisophylly

It is the condition where the plants having opposite leaves are of unequal in size. This is known as **anisophylly**. It is found in *Pellionia daveanana* of Urticaceae, *Goldfussia glomerata* (Fig. 2.78) of Acanthaceae, etc.

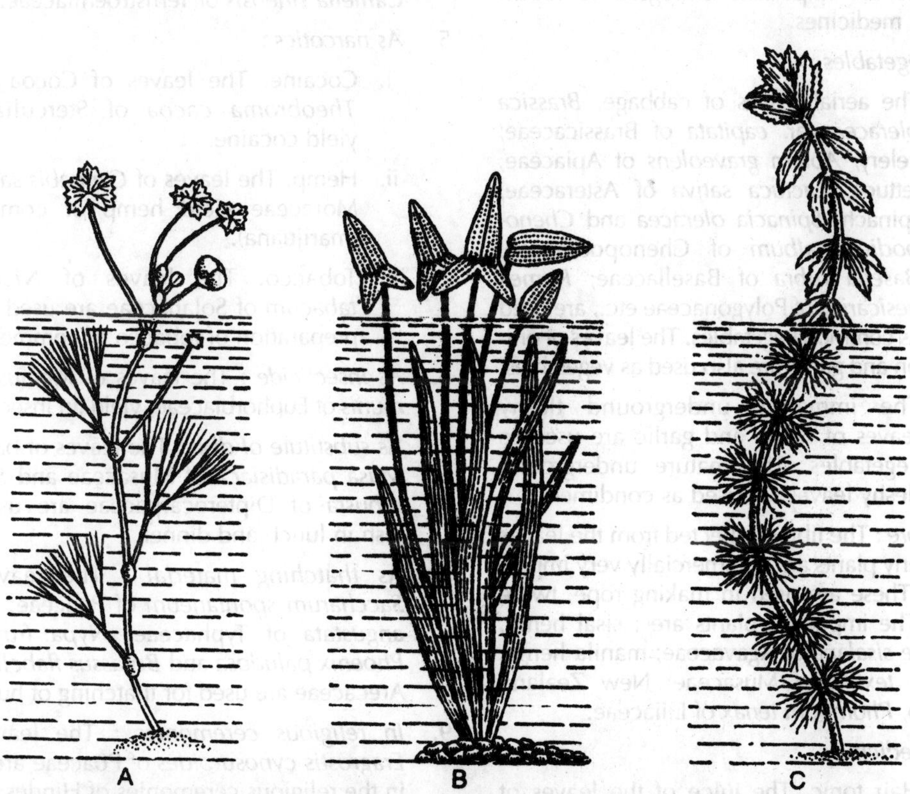

Fig. 2.77 : Heterophylly in the aquatic plants showing submerged and aerial leaves in different appearances : A. *Ranunculus aquatilis*, B. *Sagittaria sagittifolia* and C. *Limnophila heterophylla*

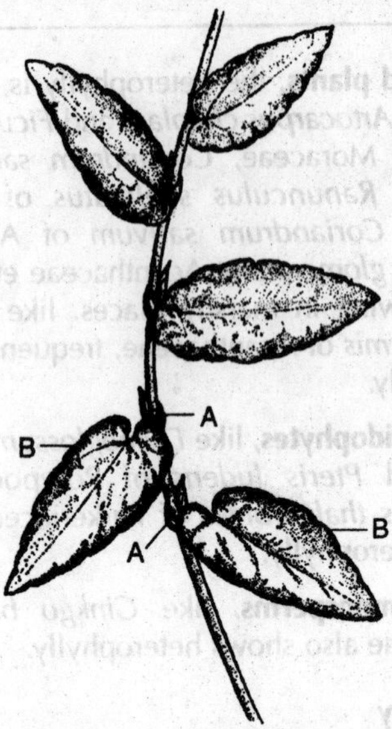

Fig. 2.78 : Anisophylly in *Goldfussia glomerata*, with opposite leaves (A and B) those are very unequal in size.

Economic importance of leaf

Leaves are important in regard to foods, fibres and medicines.

1. *As vegetables* :

   i. The aerial leaves of cabbage, *Brassica oleracea* var. *capitata* of Brassicaceae; celery, *Apium graveolens* of Apiaceae; lettuce, *Lactuca sativa* of Asteraceae; spinach, *Spinacia oleracea* and *Chenopodium album* of Chenopodiaceae; *Basella rubra* of Basellaceae; *Rumex vesicarius* of Polygonaceae etc., are used as common vegetables. The leaves of onion and garlic are also used as vegetables.

   ii. The immature underground fleshy leaves of onion and garlic are used as vegetables. The mature underground fleshy leaves are used as condiments.

2. *As fibre* : The fibres collected from the leaves of many plants are commercially very important. These are used in making rope, twine etc. The important plants are : sisal hemp, *Agave sisalana* of Agavaceae; manila hemp, *Musa textilis* of Musaceae; New Zealand hemp, *Phormium tenax* of Liliaceae.

3. *As medicine* :

   i. Hair tonic. The juice of the leaves of *Calendula officinalis, Eclipta alba* and *Wedelia calendulacea* of Asteraceae are used as hair tonic.

   ii. Digitalin. This drug is extracted from the leaves of *Digitalis purpurea* of Scrophulariaceae.

   iii. Stramonium. This drug is extracted from the leaves of *Datura stramonium* of Solanaceae.

   iv. Atropine and belladonna. These drugs are extracted from the leaves of *Atropa belladona* of Solanaceae.

   v. Leaves of other plants like *Andrographis paniculata* and *Adhatoda vasica* of Acanthaceae, *Paederia foetida* of Rubiaceae, *Bacopa monnieri* of Scrophulariaceae, *Ocimum sanctum* of Lamiaceae, *Centella asiatica* of Apiaceae, *Swertia chirata* of Gentianaceae, *Eupatorium ayapana* of Asteraceae are also popularly used as medicine.

4. *As beverage* : The very common beverage tea is obtained from the leaves of tea plant, *Camelia sinensis* of Ternstroemiaceae.

5. *As narcotics* :

   i. Cocaine. The leaves of Cocoa plant, *Theobroma cacoa* of Sterculiaceae, yield cocaine.

   ii. Hemp. The leaves of *Cannabis sativa* of Moraceae yield hemp of commerce (marijuana).

   iii. Tobacco. The leaves of *Nicotiana tabacum* of Solanaceae are used in the preparation of tobacco of commerce.

6. *As insecticide* : The leaves of *Ricinus communis* of Euphorbiaceae, yield an insecticide.

7. *As substitute of dish* : The leaves of banana, *Musa paradisiaca* of Musaceae and *Shorea robusta* of Dipterocarpaceae are used as dish in lunch and dinner.

8. *As thatching material* : The leaves of *Saccharum spontaneum* of Poaceae; *Typha angustata* of Typhaceae; *Nypa fruticans, Phoenix paludosa* and *Borassus flabellifer* of Arecaceae are used for thatching of huts etc.

9. *In religious ceremonies* : The leaves of *Eragrostis cynosuroides* of Poaceae are used in the religious ceremonies of Hindus.

INFLORESCENCE

Flowers are arranged in various ways in different plants. *The arrangement of flowers on the floral axis (shoot system) of a plant is called* ***inflorescence.***

The character of inflorescence (Fig. 2.79) has profused taxonomic significance. This characteristic is found to be constant in a number of families and helps in taxonomic identifications, e.g., capitulum in Asteraceae, verticellaster in Lamiaceae (Labiatae), umbel in Apiaceae (Umbelliferae) etc. In some cases it helps to identify the genera like *Hamelia* of Rubiaceae with helicoid cyme and *Heliotropium* of Boraginaceae with scorpioid cyme etc.

The main axis or stalk of a solitary inflorescence is called the **peduncle** (Fig. 2.79A, D). In many cases, the main axis branches out and bears flower at the branch apex, the main axis is called **peduncle** and the stalk of individual flower is called **pedicel**. A long, simple or branched peduncle is called a **rachis** (Fig. 2.79D). The small central axis of a grass or sedge spikelet is called **rachilla**. The unbranched naked peduncle developing from the underground stem is called **scape** or **radical peduncle** (Fig. 2.79C). The flower with pedicel is called **pedicellate flower** and those without them are called **sessile flower**. The dilated or flattened peduncle is called **receptacle** (Fig. 7.79E). The stage or platform on which the floral parts situated is called **thalamus**. The conical receptacle is called **torus**. Inflorescence with unbranched peduncle is called **simple inflorescence** and if branched, it is called **compound inflorescence**.

Sometimes, the flowers as well as inflorescences are subtended by an expanded leafy organ, called **bract** and flowers having bract is called **bracteate flower** and without bract it is called **ebracteate flower**. Sometimes, very small thin bract-like (leafy or scaly) structures are developed on flower stalk in between flower and bracts, called **bracteoles** or **secondary bracts.**

Types of Bracts

Bracts are of different types. These are :

a. *Leafy or foliaceous.* These bracts are typically green leaf-like, e.g., *Acalypha indica* (Fig. 2.80A, B) of Euphorbiaceae etc.

b. *Scaly.* These bracts are thin and scale-like, e.g., disc florets of capitulum inflorescence in *Helianthus annuus* of Asteraceae (Fig. 2.80C).

c. *Spathy.* These bracts are large, fleshy and boat-shaped, which partially or completely covered the inflorescence, e.g., spathe (of spadix inflorescence) of coconut, *Cocos nucifera* of Palmae; spathe of *Colocasia antiquorum* (Fig. 2.80D) of Araceae etc.

d. *Petaloid.* These bracts are coloured and showy like petals, e.g., *Euphorbia pulcherrima* of Euphorbiaceae, *Bougainvillea spectabilis* (Fig. 2.80E) of Nyctaginaceae etc.

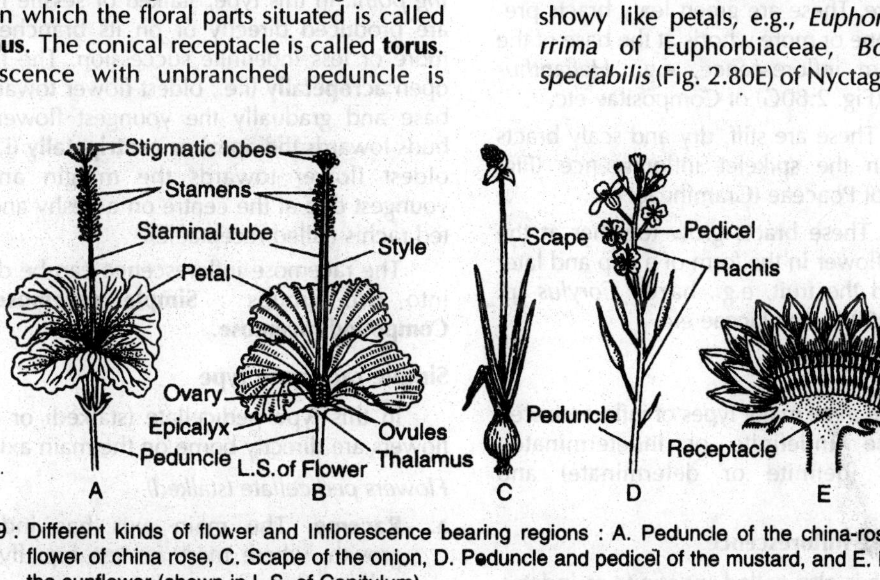

Fig. 2.79 : Different kinds of flower and Inflorescence bearing regions : A. Peduncle of the china-rose, B. L.S. of flower of china rose, C. Scape of the onion, D. Peduncle and pedicel of the mustard, and E. Receptacle of the sunflower (shown in L.S. of Capitulum)

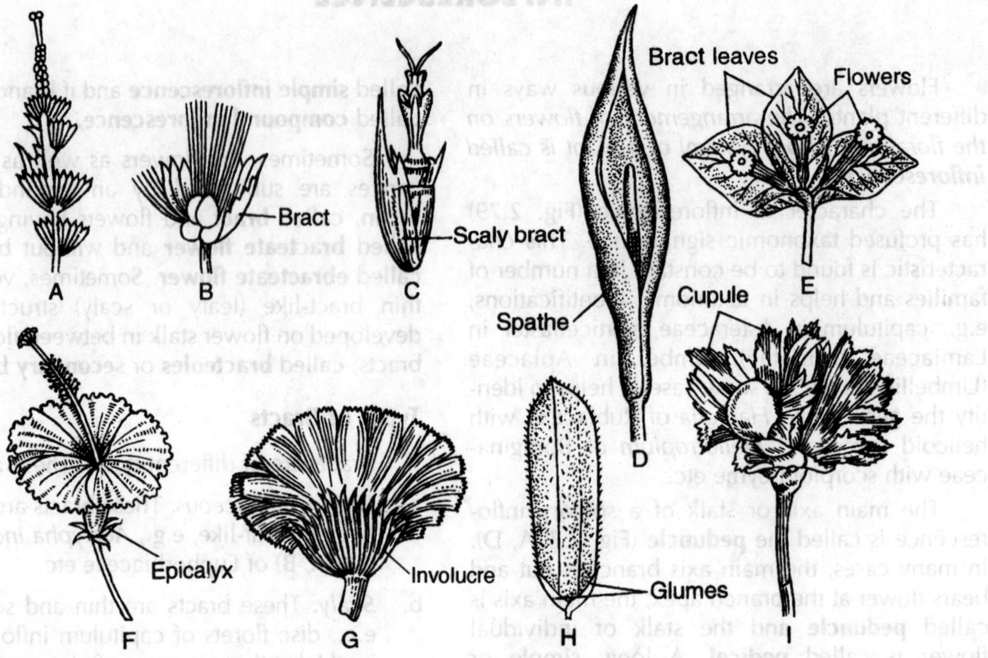

Fig. 2.80 : Different type of bracts : A & B. Inflorescence (A) and female flower (B) of *Acalypha indica* showing leafy bract, C. Scaly bract of disc floret of *Helianthus*, D. Spathe of *Colocasia*, E. Petaloid bracts of *Bougainvillea*, F. Epicalyx of *Hibiscus*, G. Involucre of *Helianthus* (Compositae), H. Glumes of *Oryza sativa*, and I. Cupule of hazel (*Corylus* sp.) formed by husky bracts

e. *Epicalyx*. These are present at the base of the calyx and are bracteole in nature, e.g., *Hibiscus rosa-sinensis* (Fig. 2.80F) and many other members of Malvaceae.

f. *Involucre*. These are green leafy bracts present in one or more whorls at the base of the capitulum inflorescence, e.g. *Helianthus annuus* (Fig. 2.80G) of Compositae etc.

g. *Glume*. These are stiff, dry and scaly bracts found in the spikelet inflorescence (Fig. 2.80H) of Poaceae (Gramineae) etc.

h. *Cupule*. These bracts grow together at the base of flower in the form of a cup and later surround the fruit, e.g., hazel, *Corylus* sp. (Fig. 2.80I) of Betulaceae etc.

Types of Inflorescence

There are *three* major types of inflorescence :

I. Racemose (indefinite or indeterminate),
II. Cymose (definite or determinate) and
III. Mixed.

I. Racemose Inflorescence

This type is also called *indefinite* or *indeterminate* or *botryose* inflorescence. A ***racemose* inflorescence** *is one whose rachis (simple or branched) never ends in a flower and it continues to elongate by means of a persistent growing point*. In this type, stalked or sessile flowers are produced directly or on its branches in a more or less indefinite succession. The flowers open **acropetally** (i.e., oldest flower towards the base and gradually the youngest flowers and buds towards the apex) or **centripetally** (i.e., the oldest flower towards the margin and the youngest one at the centre on a fleshy and dilated rachis called receptacle).

The racemose inflorescence can be divided into two groups : **Simple racemose** and **Compound racemose.**

Simple racemose type

In this type pedicellate (stalked) or sessile flowers are directly borne on the main axis.

Flowers pedicellate (stalked).

a. **Raceme**. The main axis has indefinite growth, where more or less equally pedicellate flowers are borne, e.g., radish,

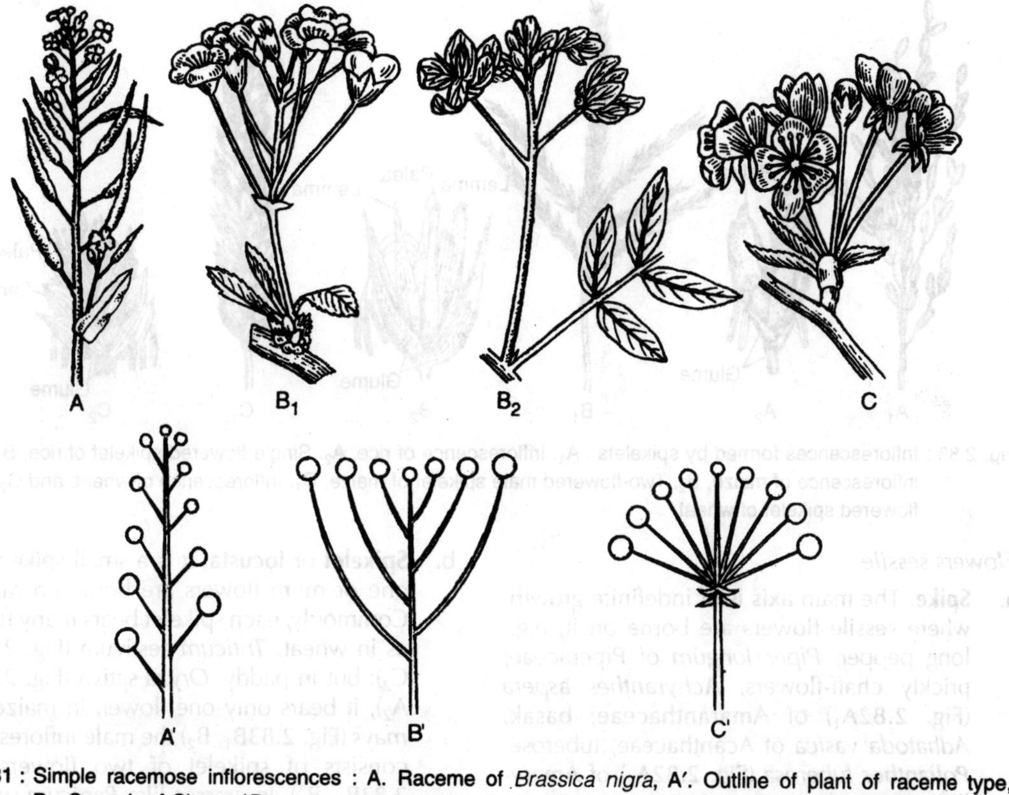

Fig. 2.81 : Simple racemose inflorescences : A. Raceme of *Brassica nigra*, A'. Outline of plant of raceme type, B_1. Corymb of Cherry (*Prunus cerasus*) at maturity, B_2. Corymb of *Cassia sophera*, B'. Outline of plant of corymb type, C. Umbel of Cherry (*P. cerasus*) when young, and C'. Outline of plant of Umbel type

Raphanus sativus and mustard, *Brassica nigra* (Fig. 2.81A, A') of Brassicaceae; *Gynandropsis gynandra* and *Polanisia icosandra* of Capparidaceae etc.

b. **Corymb.** The main axis is comparatively shorter and the lower flowers have much larger pedicels than the upper ones, so that all the flowers are brought more or less at the same level, e.g., cherry, *Prunus cerasus* of Rosaceae (Fig. 2.81B_1, B'); *Cassia sophera* (Fig. 2.81B_2) of Fabaceae; candytuft, *Iberis amara* of Brassicaceae etc.

c. **Umbel.** The main axis is much shortened and the flowers appear to develop from the same point. The older flowers are towards the periphery and the younger flowers towards the centre. Thus, it looks like an open umbrella. Flowers are usually bracteate and the bracts collectively form an involucre at the base of the pedicellate flowers, e.g., Indian pennywort, *Centella asiatica* and *Coriandrum sativum* of Apiaceae (Umbelliferae); cherry, *Prunus cerasus* during young stage (Fig. 2.81C, C').

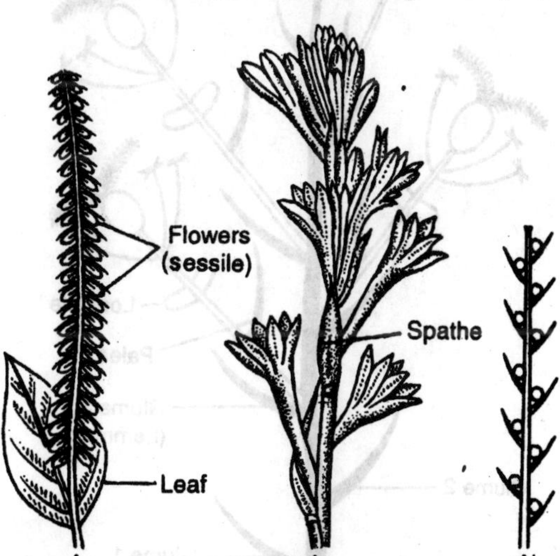

Fig. 2.82 : A_1. Spike of *Achyranthes aspera*, A_2. Spike of *Polyanthes tuberosa*, A'. Outline plan of spike

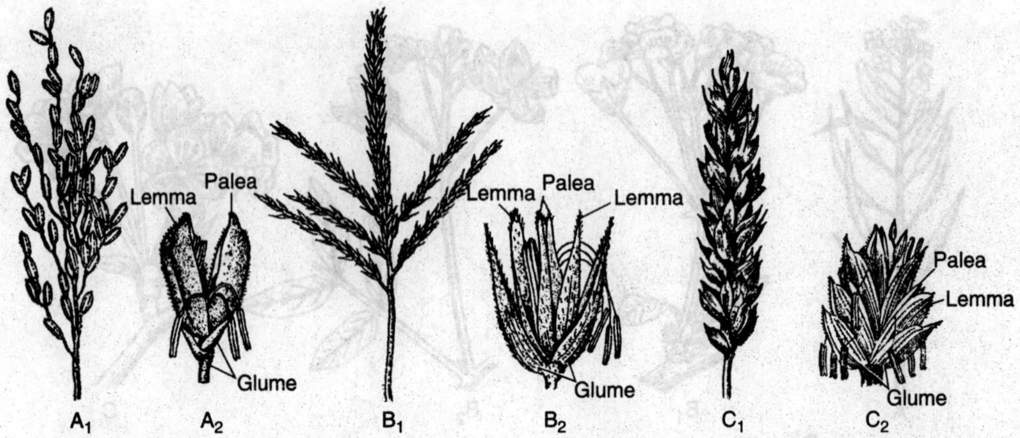

Fig. 2.83 : Inflorescences formed by spikelets : A_1. Inflorescence of rice, A_2. Single flowered spikelet of rice, B_1. Male inflorescence of maize, B_2. Two-flowered male spikelet of maize, C_1. Inflorescence of wheat, and C_2. Many flowered spikelet of wheat

Flowers sessile

a. **Spike**. The main axis is of indefinite growth, where sessile flowers are borne on it, e.g., long pepper, *Piper longum* of Piperaceae; prickly chaff-flowers, *Achyranthes aspera* (Fig. 2.82A_1) of Amaranthaceae; basak, *Adhatoda vasica* of Acanthaceae; tuberose, *Polianthes tuberosa* (Fig. 2.82A_2) of Amaryllidaceae; etc.

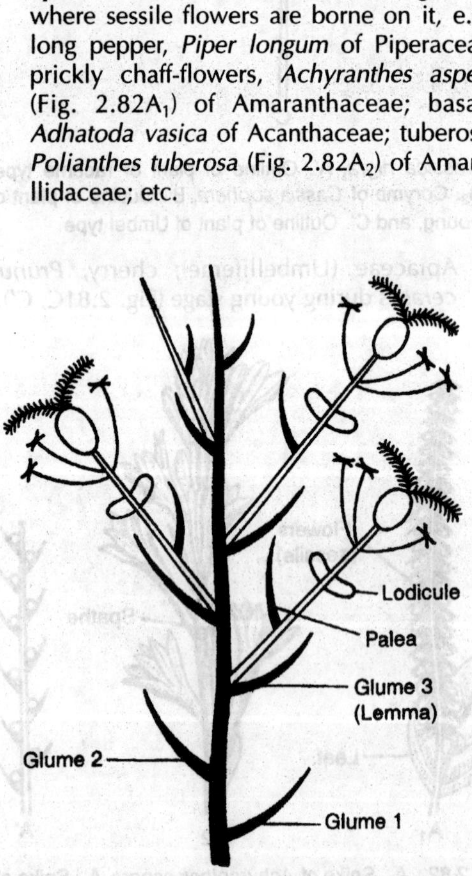

Fig. 2.84 : Spikelet inflorescence (Diagrammatic)

b. **Spikelet** or **locusta**. It is a small spike where one or more flowers are borne on rachilla. Commonly, each spikelet bears many flowers as in wheat, *Triticum* aestivum (Fig. 2.83C_1, C_2); but in paddy, *Oryza* sativa (Fig. 2.83A_1, A_2), it bears only one flower. In maize, *Zea mays* (Fig. 2.83B_1, B_2) the male inflorescence consists of spikelet of two flowers (Fig. 2.83B_1, B_2). In grasses like *Panicum* sp. (Fig. 2.84) the entire inflorescence bears at its base two sterile bracts, the **empty glumes**. Above the empty glumes, there are one or more fertile glumes, the **flowering glumes** or **lemmas**. Each lemma, bears single sessile flower in its axil. Opposite the lemma a small glume is present, called **palea**. In wheat, *Triticum aestivum* (Fig. 2.83C_1, C_2) of Poaceae; the spikelets are sessile and develop on an elongated axis, but in oat, *Avena sativa* of Poaceae, they are produced on a more or less branched axis.

c. **Spadix**. It is a spike with fleshy axis having both male and female flowers. Entire structure is surrounded by a large bract called **spathe**. In some aroids like *Acorus calamus,* the spathe is absent.

The female flowers are always found towards the base of the axis and male flowers towards the apex, whereas the sterile flowers are situated between these two. The terminal portion is barren and called as **appendix**, e.g., *Colocasia antiquorum* (Fig. 2.85A,B) of Araceae.

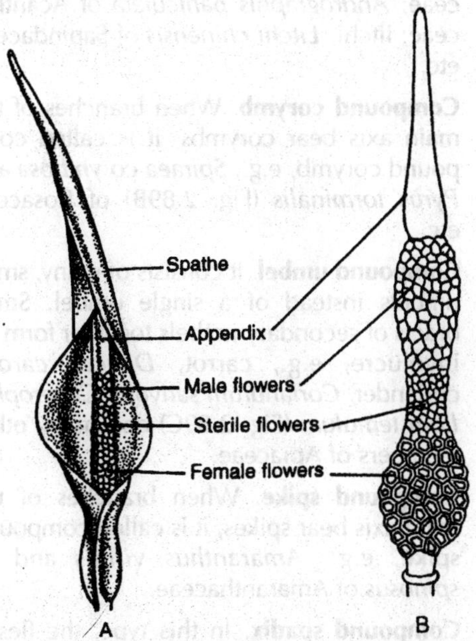

Fig. 2.85 : A. Spadix of *Colocasia antiquorum*, B. Spadix of *C. antiquorum* after removing the spathe

d. **Catkin** or **amentum.** It is the pendulous spike with fleshy and delicate axis which bears naked unisexual flower (Fig. 2.86A′) that falls as a unit at maturity. Viz. hazel, *Corylus* sp. of Betulaceae (Fig. 2.86A); mulberry, *Morus indica* of Moraceae; *Trewia nudiflora* and *Acalypha hispida* of Euphorbiaceae; Oak, *Quercus* sp. of Fagaceae etc.

e. **Strobile.** It is a modified spike, where the pistillate flowers are borne singly in the axil of a persistent membranous bract. e.g., hops, *Humulus lupulus* (Fig. 2.86B) of Cannabinaceae etc.

f. **Capitulum** or **anthodium** or **head**. In this type (Fig. 2.87A), the main axis is much shortened and broadened out to form a flat or more or less convex **receptacle** on which numerous sessile and small florets are arranged in a **centripetal** manner i.e., youngest at the centre and oldest towards the periphery. Individual florets are bracteate (Fig. 2.87B). The cluster of florets is surrounded by a whorl of bracts collectively called **involucre.** Two kinds of florets are distinguished : **ray florets** (Fig. 2.87C) those at the periphery with strap-shaped corolla. These florets are female and are always *zygomorphic,* arrange in one or two whorls. **Disc florets** (Fig. 2.87D) are grouped at the centre and are bisexual and *actinomorphic.* This inflorescence is the characteristic feature of the family Asteraceae (Compositae), e.g., sunflower, *Helianthus annuus; Tridax procumbens, Eclipta alba* etc. of Asteraceae.

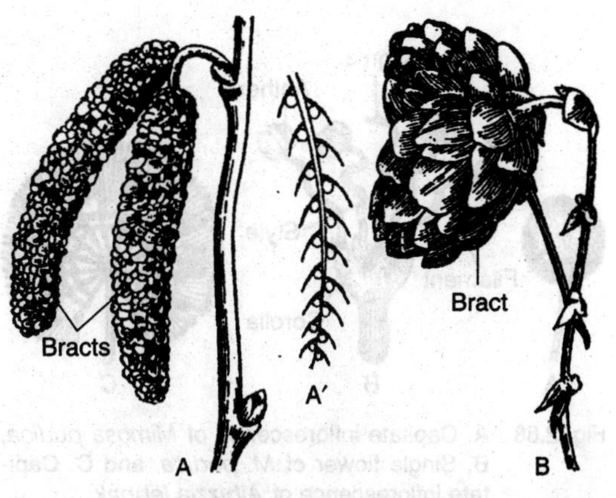

Fig. 2.86 : A. Catkin of *Corylus* sp. (Hazel), A′. Outline plan of Catkin, and B. Strobile of *Humulus* sp. (Hops)

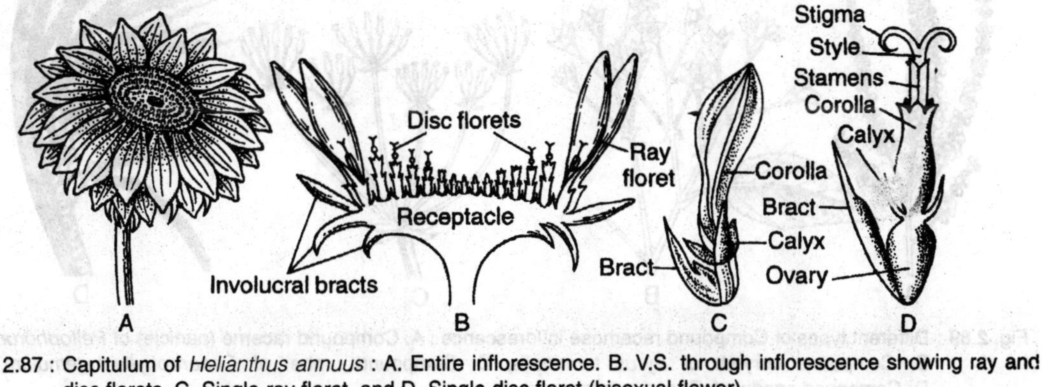

Fig. 2.87 : Capitulum of *Helianthus annuus* : A. Entire inflorescence. B. V.S. through inflorescence showing ray and disc florets, C. Single ray floret, and D. Single disc floret (bisexual flower)

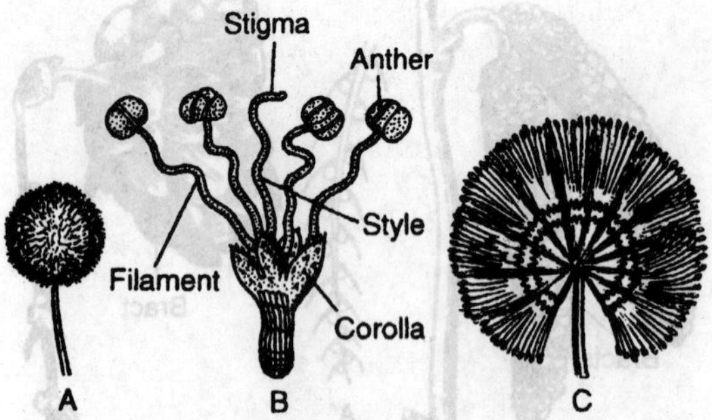

Fig. 2.88 : A. Capitate inflorescence of *Mimosa pudica*, B. Single flower of *M. pudica*, and C. Capitate inflorescence of *Albizzia lebbek*

g. **Capitate**. In this type, a dense cluster of sessile flowers arise upon a compressed rachis; thereby they give rise to a somewhat globose structure (Fig. 2.88A, C). e.g., *Acacia nilotica, Mimosa pudica* (Fig. 2.88A, B), *Albizzia lebbek* (Fig. 2.88C) *Trifolium* sp. etc. of Fabaceae.

Compound racemose type

In this type, pedicellate (stalked) or sessile flowers are borne on the branches of the main axis.

a. **Compound raceme** or **panicle**. In this type, each branch of the main axis develop a cluster of stalked flowers like the raceme. Viz. *Peltophorum ferrugineum* (Fig. 2.89A), mango, *Mangifera indica* of Anacardiaceae; mahagony, *Swietenia mahagoni* of Meliaceae; *Andrographis paniculata* of Acanthaceae; litchi, *Litchi chinensis* of Sapindaceae etc.

b. **Compound corymb**. When branches of the main axis bear corymbs, it is called compound corymb, e.g., *Spiraea corymbosa* and *Pyrus torminalis* (Fig. 2.89B) of Rosaceae etc.

c. **Compound umbel**. It consists of many, small umbels instead of a single umbel. Small bracts of secondary umbels together form an involucre, e.g., carrot, *Daucus carota*; coriander, *Coriandrum sativum; Chaerophyllum temulum* (Fig. 2.89C) and many other members of Apiaceae.

d. **Compound spike**. When branches of the main axis bear spikes, it is called compound spike, e.g., *Amaranthus viridis* and *A. spinosus* of Amaranthaceae.

e. **Compound spadix**. In this type, the fleshy axis is repeatedly branched and each branch bears sessile unisexual flowers. When young, the entire inflorescence or each branch separately enclosed in a spathe, e.g., coconut, *Cocos nucifera* (Fig. 2.89D); betel nut, *Areca catechu*; date palm, *Phoenix sylvestris* and many other members of Arecaceae.

f. **Compound capitulum**. In this type, the capitulum is composed of many small capitula, surrounded by involucre of bracts, e.g., *Echinops echinatus* of Asteraceae.

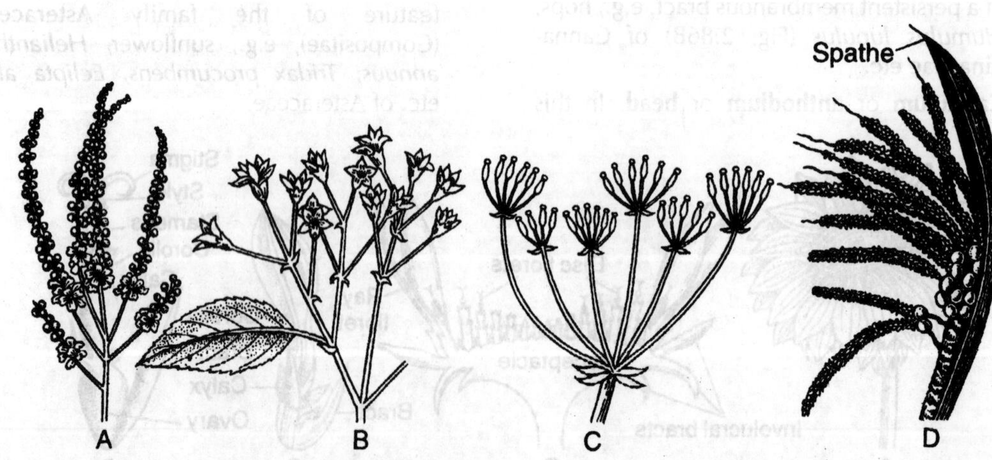

Fig. 2.89 : Different types of Compound racemose inflorescence : A. Compound raceme (panicle) of *Peltophorum* sp., B. Compound corymb of *Pyrus torminalis*, C. Compound umbel of *Chaerophyllum temulum*, and D. Compound spadix of *Cocos nucifera*

II. Cymose Inflorescence

This type is also called *definite* or *determinate* inflorescence. *A* **cymose inflorescence** *is one whose rachis (simple or branched) becomes terminated by a flower bud at an early stage and subsequent buds are developed gradually towards the lower side of the axis.* The flowers open **basipetally** i.e., oldest flower at the apex and gradually the youngest flowers and buds towards the base or **centrifugally** i.e., the oldest flower towards the centre and the youngest one at the periphery on a fleshy and dilated rachis, called receptacle.

The cymose inflorescences are divided into the following four types :

a. ***Solitary***. It is the simplest type of cymose. Here the rachis is unbranched and always terminated by a flower, e.g., *Magnolia grandiflora* and *Michelia champaca* of Magnoliaceae; china-rose, *Hibiscus rosa-sinensis* (Fig. 2.90) of Malvaceae etc.

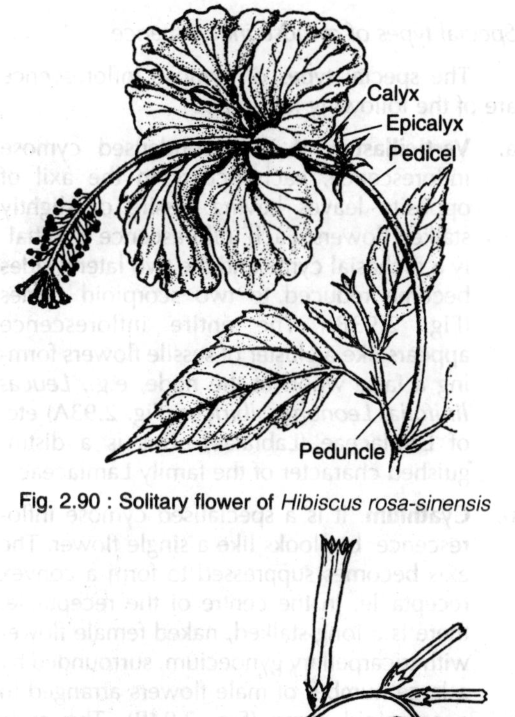

Fig. 2.90 : Solitary flower of *Hibiscus rosa-sinensis*

b. ***Uniparous cyme*** or **monochasial cyme** *or* **monochasium**. In this type, the primary axis ends in a flower and gives rise to only one daughter axis, which behaves as the mother. It is of two types :

i. ***Helicoid cyme*** or **bostryx**. In this type, the flowers are developed on one side, either clockwise or anti-clockwise of the subsequent daughter axes, e.g., *Hamelia patens* (Fig. 2.91A, A') of Rubiaceae; forget-me-not, *Myosotis palustris* of Boraginaceae; day lily, *Hemerocallis fulva* of Liliaceae etc.

[A tight, modified helicoid cyme, in which pedicels are short on the developed side, is often called **cincinnus**.]

ii. ***Scorpioid cyme***. In this type, the flowers are developed alternately on either side of the successive daughter axes, thereby it appears as a zigzag structure, e.g., *Heliotropium ovalifolium* (Fig. 2.91B', B) of Boraginaceae, *Ranunculus bulbosus* of Ranunculaceae etc.

c. ***Biparous cyme*** or **dichasial cyme** or **dichasium**. In this type, the primary axis ends in a flower and develops two daughter axes with apical flower bud from a single node, a little distance behind the apex, e.g., jasmine, *Jasminum* sp. and *Nyctanthus arbor-tristis* of Oleaceae; *Clerodendrum viscosum* of Verbe-

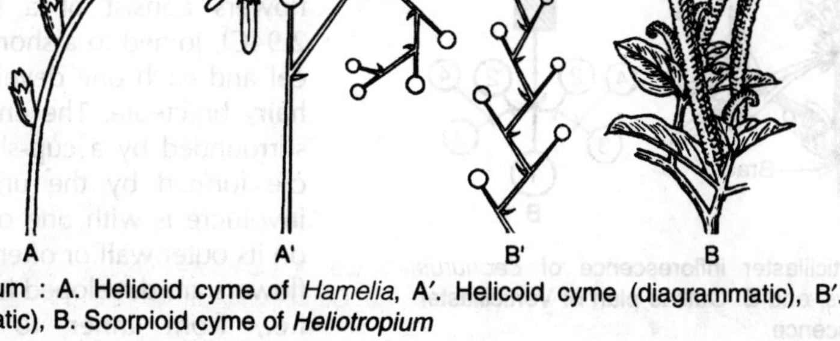

Fig. 2.91 : Monochasium : A. Helicoid cyme of *Hamelia*, A'. Helicoid cyme (diagrammatic), B'. Scorpioid cyme (diagrammatic), B. Scorpioid cyme of *Heliotropium*

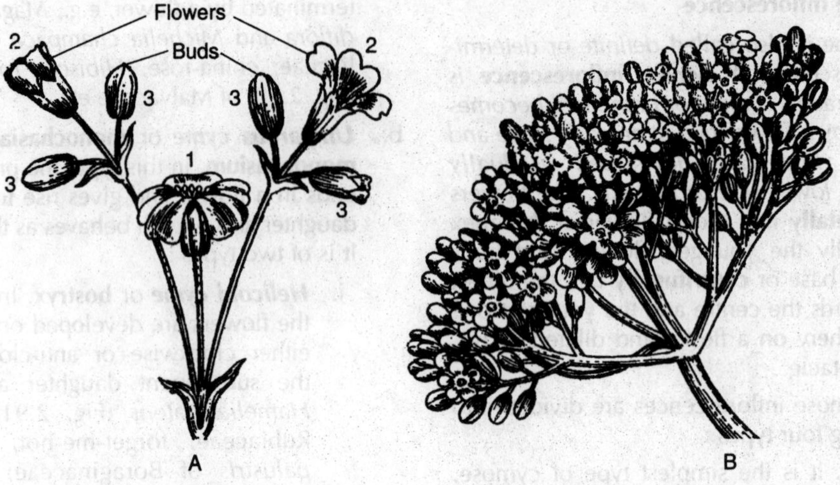

Fig. 2.92 : A. Dichasium of *Dianthus chinensis*, B. Polychasium of *Viburnum tinus*.

naceae ; *Dianthus chinensis* (Fig. 2.92A) of Caryophylaceae etc.

d. ***Multiparous cyme*** or **polychasial cyme** or **polychasium** or **pleiochasium.** In this type, the primary axis ends in a flower and develops more than two daughter axes with apical flower bud from a single node, a little distance behind the apex. The daughter axes, in their turn, also behave like mother, e.g., *Kleinhovia hospita* and *Dombeya mastersii* of Sterculiaceae; *Calotropis procera* of Asclepiadaceae; *Viburnum tinus* (Fig. 2.92B) of Caprifoliaceae etc. (The multiparous cyme often looks alike to racemose umbellate inflorescence, but is distinguished by the presence of oldest flower at the centre).

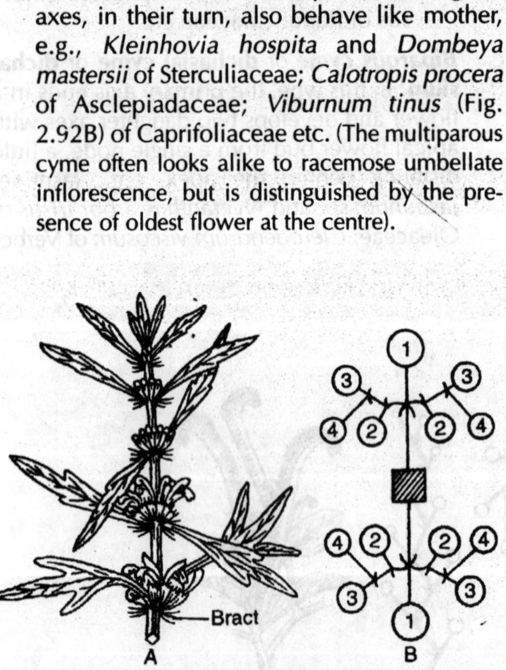

Fig. 2.93 : A. Verticillaster inflorescence of *Leonurus sibiricus*, and B. Outline plan of Verticillaster inflorescence

Special types of cymose inflorescence

The special types of cymose inflorescence are of the following four types :

a. **Verticillaster.** It is a condensed cymose inflorescence, each occurs in the axil of opposite leaves having sessile or slightly stalked flowers. Each inflorescence is initially a dichasial cyme and the two lateral sides become reduced to two scorpioid cymes (Fig. 2.93B). The entire inflorescence appears like a cluster of sessile flowers forming a false whorl at the node, e.g., *Leucas linifolia, Leonurus sibiricus* (Fig. 2.93A) etc. of Lamiaceae (Labiatae). This is a distinguished character of the family Lamiaceae.

b. **Cyathium.** It is a specialised cymose inflorescence, but looks like a single flower. The axis becomes suppressed to form a convex receptacle. In the centre of the receptacle, there is a long-stalked, naked female flower with tricarpellary gynoecium, surrounded by a large number of male flowers arranged in a scorpioid cyme (Fig. 2.94B). The male flowers consist of a single stamen (Fig. 2.94C), joined to a short stalk i.e., the pedicel and each one develops in the axil of a hairy bracteole. The entire inflorescence is surrounded by a cup-shaped green involucre formed by the union of bracts. The involucre is with one or two nectar glands on its outer wall or often without gland. The flowers are developed in *centrifugal* manner i.e., from inner to outer side, e.g.,

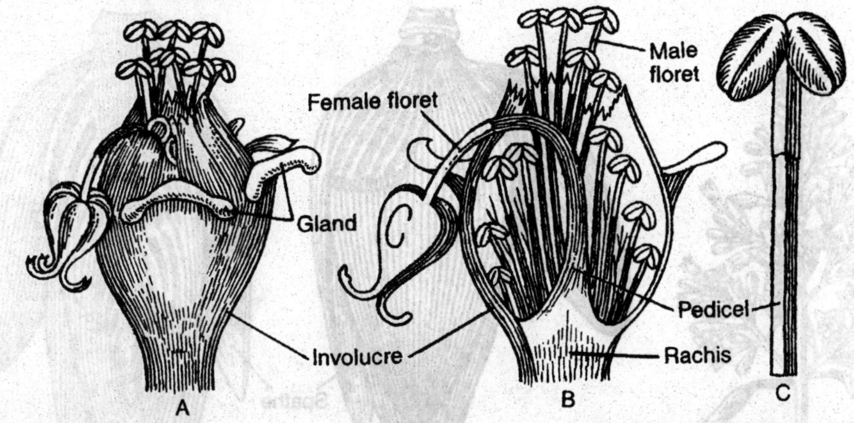

Fig. 2.94 : Cyathium of *Poinsettia pulcherrima* : A. External view, B. L.S. showing male and female florets with bracteoles in between, and C. A male floret

Pedilanthus tithymaloides, Euphorbia microphylla, Poinsettia pulcherrima (Fig. 2.94A) and some other members of Euphorbiaceae.

c. **Hypanthodium**. In this type, a hollow sphere-like receptacle (**syconium**) is formed by the fusion of the rachis of three closely placed cymes. The spherical receptacle is like a closed fleshy vessel with a small opening at the apex. Three types of unisexual flowers (male, fertile female and sterile female) are arranged on the inner surface of the receptacle in cymose groups, e.g., fig., *Ficus cunia* (Fig. 2.95A,B) and banyan, *F. benghalensis* of Moraceae.

d. **Coenanthium**. It is like hypanthodium, but the receptacle is somewhat saucer-shaped with margins curved upwardly, e.g., pick-aback plant, *Dorstenia cordifolia* (Fig. 2.95C) of Moraceae.

III. Mixed Inflorescence

In addition to typical racemose and cymose inflorescences, many inflorescences are found to show the combination of characteristics of both racemose and cymose types or of two types of racemose i.e., combination of raceme and umbel, raceme and spike, etc. The mixed inflorescences are of the following types :

a. **Mixed panicle**. The general pattern of the inflorescence is like a panicle, but the apical flower opens first and in all the branches the flowers open in a basipetal order i.e., cymose type, e.g., *Ligustrum vulgare* (Fig. 2.96A) of Oleaceae.

b. **Mixed spadix**. In this type, cymose group of flowers are covered by spathe and are arranged racemosely on a fleshy stalk. The entire structure is also covered by spathe, when young, e.g., Banana, *Musa para-*

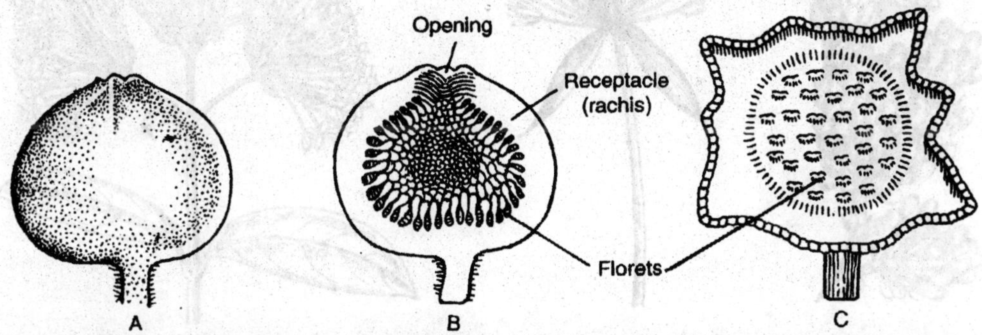

Fig. 2.95 : A. Hypanthodium inflorescene (external view) of fig. (*Ficus cunia*), B. L.S. of hypanthodium showing opening and florets inside, and C. Coenanthium of *Dorstenia cordifolia* showing florets on saucer-shaped receptacle

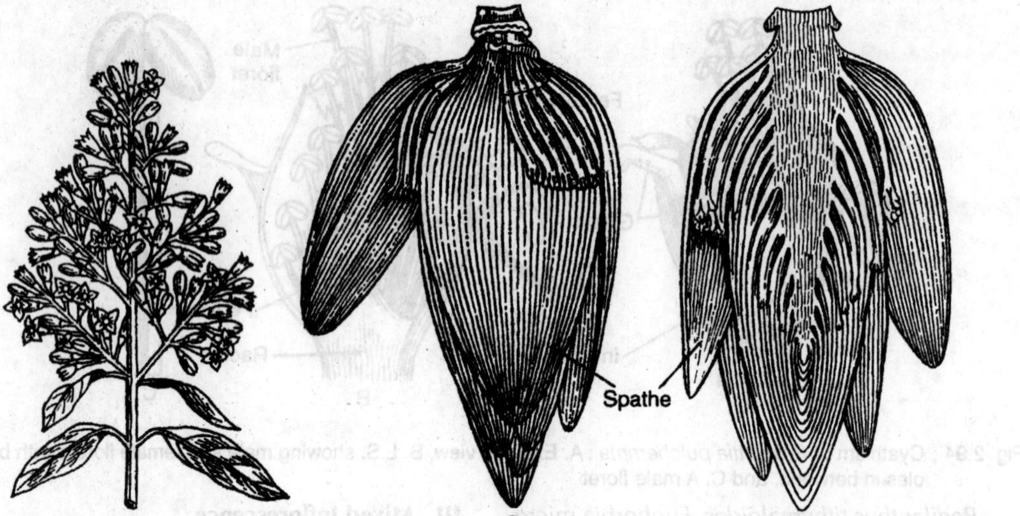

Fig. 2.96 : A. Mixed panicle of *Ligustrum vulgare*, B. Mixed spadix of banana (external view), and C. L.S. of spadix

disiaca (Fig. 2.96B,C), *M. sapientum* etc. of Musaceae.

c. **Thyrsus**. The inflorescence is very much condensed mixed type of panicle, where the entire inflorescence is of racemose type, but later on, the flowers are developed in cymose pattern. Each flower has a separate whorl of bracts and the flowers open centrifugally, in both directions, e.g., grape vine, *Vitis vinifera* (Fig. 2.97A) of Vitaceae; *Syringa* sp. of Oleaceae etc.

d. **Umbellate cyme** or **cymose umbel**. In this type, the group of flowers looks like an umbel, but the oldest flower is at the centre, e.g., *Butomopsis lanceolata* of Alismataceae; *Phyla nodiflora* and *Lippia alba* of Verbenaceae; Onion, *Allium cepa* (Fig. 2.97B) of Liliaceae etc.

e. **Corymbose cyme** or **cymose corymb**. In this type, the cluster of flowers looks like a corymb, but the oldest one is at the centre, e.g., *Ixora coccinia* (Fig. 2.97C),

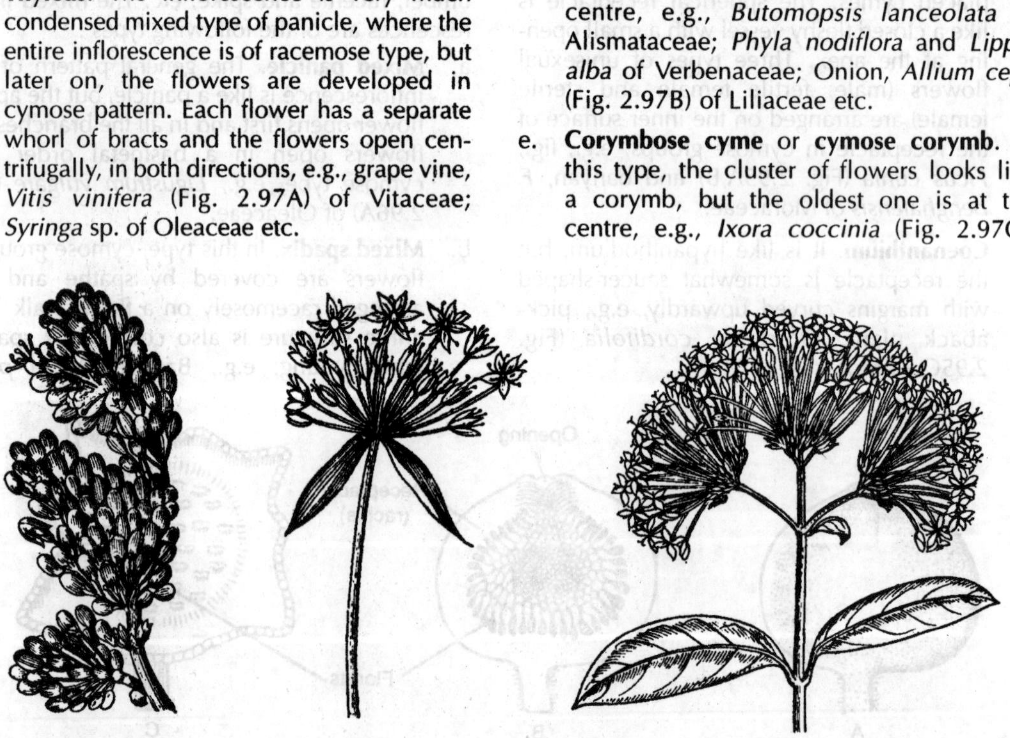

Fig. 2.97 : Mixed inflorescences : A. Thyrsus of *Vitis vinifera*, B. Umbellate cyme of *Allium cepa*, and C. Corymbose cyme of *Ixora coccinia*

Oldenlandia corymbosa of Rubiaceae; *Holarrhena antidysenterica, Alstonia scholaris* and many other members of Apocynaceae etc.

f. **Fascicle.** It is a modified form of corymbose cyme, where the successive axes are much shortened and the flowers remain at the same level forming a flat-topped cyme which resembles a corymb, e.g., *Dianthus barbatus* of Caryophyllaceae.

Diagramatic representation of some inflorescence is given in Fig. 2.98.

Table 2.9 : The *inflorescence* may be tabulated as :

Inflorescence

- **Racemose**
  - **Simple**
    - **Flower pedicellate**
      1. Raceme e.g. *Brassica nigra*
      2. Corymb e.g. *Cassia sophera*
      3. Umbel e.g. *Coriandrum sativum*
    - **Flower sessile**
      1. Spike e.g. *Polyanthes tuberosa*
      2. Spikelet e.g. *Oryza sativa*
      3. Spadix e.g. *Colocasia antiquorum*
      4. Catkin e.g. *Morus indica*
      5. Strobile e.g. *Humulus lupulus*
      6. Capitulum e.g. *Helianthus annuus*
      7. Capitate e.g. *Acacia nilotica*
  - **Compound**
    1. Panicle e.g. *Mangifera indica*
    2. Compound corymb e.g. *Spiraea corymbosa*
    3. Compound umbel e.g. *Daucus carota*
    4. Compound spike e.g. *Amaranthus viridis*
    5. Compound spadix e.g. *Cocos nucifera*
    6. Compound capitulum e.g. *Echinops echinatus*
- **Cymose**
  1. Solitary e.g. *Michelia champaca*
  2. Uniparous cyme
     a. Helicoid e.g. *Hamelia patens*
     b. Scorpoid e.g. *Heliotropium viscosum*
  3. Biparous cyme e.g. *Dianthus chinensis*
  4. Multiparous cyme e.g. *Calotropis procera*

  SPECIAL TYPE OF CYMOSE
  1. Verticillaster e.g. *Leonurus sibiricus*
  2. Cyathium e.g. *Poinsettia pulcherrima*
  3. Hypanthodium e.g. *Ficus cunia*
  4. Coenanthium e.g. *Dorstenia cordifolia*
- **Mixed**
  1. Mixed panicle e.g. *Ligustrum vulgare*
  2. Mixed spadix e.g. *Musa paradisiaca*
  3. Thyrsus e.g. *Vitis vinifera*
  4. Umbellate cyme e.g. *Allium cepa*
  5. Corymbose cyme e.g. *Ixora coccinia*
  6. Fascicle e.g. *Dianthus barbatus*

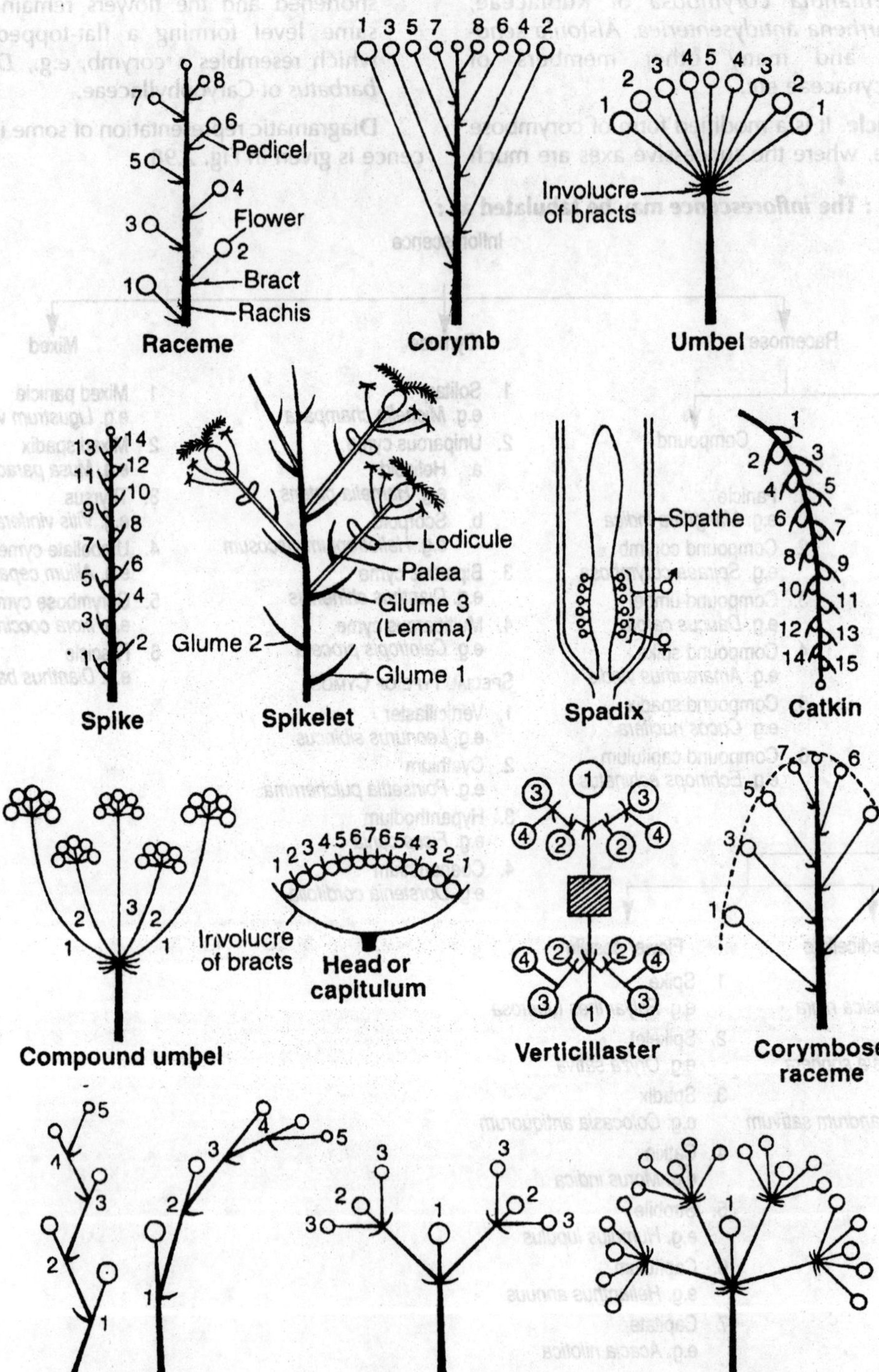

Fig. 2.98 : Diagrammatic representation of different types of inflorescence

FLOWER

Flower is an important organ which draws our attention mainly for the attractive colour of petals. Sometimes other parts of a flower become modified and may help in beautification. Later on, seeds are developed after fertilisation and help the plant to continue the life process.

Definition

1. A flower is a reproductive structure, which bears either stamens or carpels or both, along with or without calyx and corolla.
2. Based on the morphological point of view, **flower is a highly condensed and modified shoot**.
3. It can also be defined as "**a specialised shoot in which the leaves are modified into floral structures**".

Parts of a typical flower

A typical flower (Fig. 2.99A) has a stalk situated at the lower side, called **pedicel** and the swollen upper part of the pedicel, called **thalamus**. Like ordinary vegetative shoot, the thalamus has both nodes and internodes, but the internodes are generally suppressed, thereby the nodes become very much closely placed. On the thalamus, four types of structures, i.e., the floral leaves are arranged in four whorls.

The four floral whorls taken from outside are : 1. Calyx, 2. Corolla, 3. Androecium, and 4. Gynoecium.

The outer two whorls, i.e., calyx and corolla, are the **accessory whorls** and inner two whorls, i.e., androecium and gynoecium, are the **essential whorls** (Fig. 2.99B, C).

1. **Calyx**. It is the lowermost, i.e., the first accessory, whorl. The unit member of calyx is called *sepal*.
2. **Corolla**. It is the second accessory whorl, situated above the calyx. The unit member of corolla is called *petal*.
3. **Androecium**. It is the first essential whorl, situated above the corolla. The unit member of androecium is called *stamen*.
4. **Gynoecium** (also spelt as **gynaecium**) or **pistil**. It develops above the androecium whorl, i.e., present at the top of the thalamus. The unit member of gynoecium is called *carpel*.

Functions of the different floral whorls :

Calyx

i. It protects the other whorls when young,
ii. When green, it performs photosynthesis,
iii. When coloured, it helps in beautification and to attract insects for pollination.

Corolla

i. It attracts insects for pollination and helps in beautification,

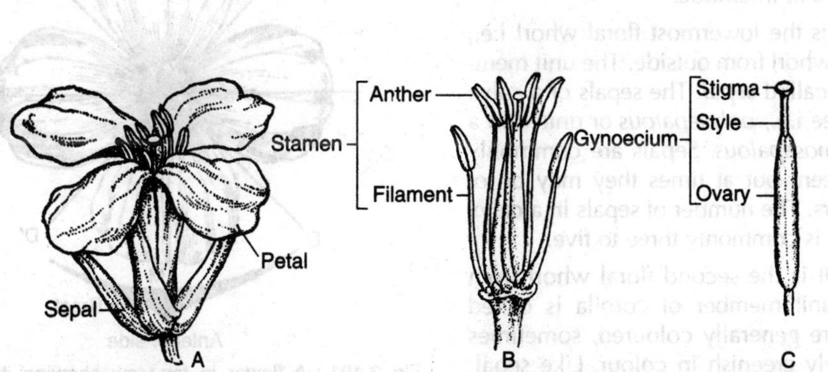

Fig. 2.99 : A typical flower (*Brassica nigra*) and its parts : A. An entire flower, B. Single flower after removing sepals and petals, and C. Gynoecium

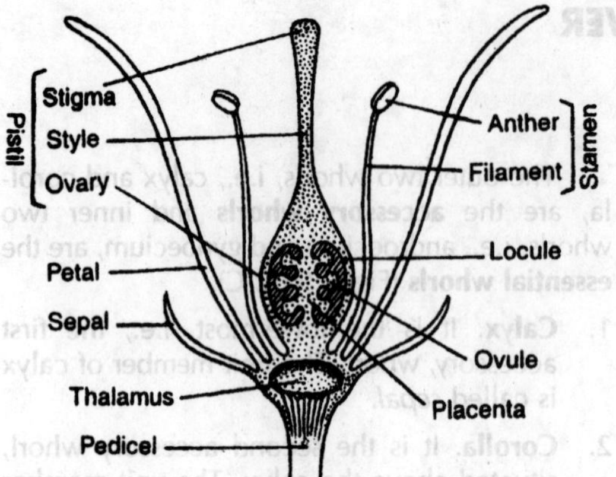

Fig. 2.100 : L.S. of a typical flower showing different whorls on thalamus

ii. It protects the essential whorls i.e., the androecium and gynoecium.

iii. At times, the base of petals has glands which secrete nectar.

Androecium.

i. Production of pollen grains (male gametophytes) inside the anther lobes.

Gynoecium.

i. Production of megaspore (female gametophyte) and ultimately the fruits and seeds.

Brief description of the different parts of a flower is given (Fig. 2.100) :

Thalamus : It is the flattened disc on which floral members are borne. It is also called torus, axis and sometimes *receptacle*. Usually it is convex or concave and often it is very short and suppressed. Sometimes it becomes elongated (*axis*) and shows distinct internodes.

Calyx : It is the lowermost floral whorl i.e., first accessory whorl from outside. The unit member of calyx is called *sepal*. The sepals of a calyx may remain free i.e., *polysepalous* or united in a group i.e., *gamosepalous*. Sepals are commonly sessile and green, but at times they may be of different colours. The number of sepals in a calyx is variable, but is commonly three to five.

Corolla : It is the second floral whorl from outside. The unit member of corolla is called *petal*. Petals are generally coloured, sometimes white and rarely greenish in colour. Like sepal, the petals of a corolla may remain free i.e., *polypetalous* or united into a group i.e., *gamopetalous*. Sometimes the lower part of the petal becomes very thin and looks like a stalk, called *claw* (*Brassica campestris*) and the upper expanded portion is called *limb*. The number of petals is variable i.e., from one to many.

Perianth : When both the accessory whorls of a flower (i.e., sepals and petals) are not differentiated from one another, they are called *perianth*, e.g., tuberose, *Polyanthes tuberosa* of Amaryllidaceae etc. The unit member of the perianth is called *tepal*. Members of the perianth may be sepaloid (palmyra palm, *Borassus flabellifer* and coconut, *Cocos nucifera* of Palmae etc.) or petaloid (*Michelia champaca* of Magnoliaceae etc.)

Androecium : It is the third floral whorl and the first essential whorl from outside. The unit member of androecium is called *stamen*. A stamen consists of a long slender stalk, the *filament*, bears at its tip the sac-like structure, the *anther*. The anther may be one or two celled and contains many microspores or pollen grains. The number of stamens in an androecium is variable. The stamens may remain free or united or remain attached with its inner (gynoecium) or outer (petal) floral whorls.

Gynoecium or **pistil** : It is the fourth floral whorl and the second essential whorl from outside, situated at the top. Gynoecium consists of three parts : i. the basal swollen structure i.e., *ovary*, ii. the middle elongated structure i.e., *style*

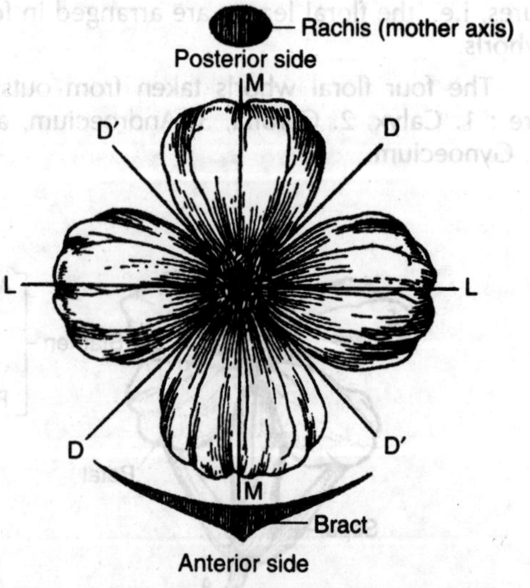

Fig. 2.101 : A flower in top-view showing its different plane. MM = median plane; LL = lateral plane; DD and D′D′ = diagonal planes

and iii. the uppermost knob-like structure i.e., *stigma*. The unit member of gynoecium is the *carpel*. The number of carpels in a gynoecium is variable. The carpels may remain united i.e., *syncarpous* pistil (*syncarpous gynoecium*) or free i.e., *apocarpous pistil* (*apocarpous gynoecium*).

Flowers are arranged on the inflorescence axis in a definite way. The side of a flower which faces the stem or the rachis is called the **posterior** (back) **side** and its opposite side is called **anterior** (front) **side**. The imaginary plane which passes through the axis of the stem, the back and the front of the flower is called *median plane* or *antero-posterior plane* and at right angles to it is the *lateral* or *transverse plane*. The planes which pass between median and transverse plane are called *diagonal planes* (Fig. 2.101).

Different types of flower

1. ***Complete flower***. When all the four floral whorls i.e., calyx, corolla, androecium and gynoecium are present in a flower, it is called *complete flower*. Viz. china-rose, *Hibiscus rosa-sinensis* of Malvaceae; mustard, *Brassica nigra* of Brassicaceae (Cruciferae) etc.
2. ***Incomplete flower***. Out of four floral whorls of a complete flower when any one or more whorls become absent, it is called *incomplete flower*. In *Cucurbita maxima* of Cucurbitaceae, flower is incomplete (unisexual) due to the absence of either androecium or gynoecium.
3. ***Bisexual flower*** (**perfect** or **hermaphrodite** or **monoclinous**). When both the essential whorls (i.e., androecium and gynoecium) are present in a flower, it is called *bisexual flower*. (All complete flowers are bisexual flowers, but all bisexual flowers are not complete flowers). Viz. china-rose, *Hibiscus rosa-sinensis* of Malvaceae; mustard, *Brassica nigra* of Brassicaceae (Cruciferae) etc.
4. ***Unisexual flower*** (**imperfect** or **diclinous**). When out of two essential whorls (androecium and gynoecium) any one becomes absent, it is called *unisexual flower*. In *Cucurbita maxima* (Fig. 2.102) of Cucurbitaceae, flower is incomplete (unisexual) due to the absence of either androecium or gynoecium. The unisexual flowers are of two types :

   Pistillate or ***female flower***. The unisexual flowers that only bear pistil or gynoecium.

   Staminate or ***male flower***. The unisexual flowers that only bear stamen or androecium.
5. ***Neuter*** or ***sterile flower***. When both the essential whorls (androecium and gynoecium) become absent or sterile, it is called *neuter flower*, e.g., some ray florets of capitulum inflorescence of Asteraceae i.e., Compositae.
6. ***Nude*** or ***naked*** or ***achlamydeous flower***. When either androecium or gynoecium or both are present, but *calyx* and *corolla* are absent, it is called **nude flower**. In *Beta* sp. of Chenopodiaceae, the flower is nude as corolla whorl is absent. (Flower having only one accessory whorl i.e., either calyx or corolla or perianth is called *monochlamydeous* or *haplochlamydeous*, e.g., tuberose, *Polyanthes tuberosa* of Amaryllidaceae; ordinary flower having both calyx and corolla is called *dichlamydeous*, e.g., pea, *Pisum sativum* of Fabaceae etc.)

   (*All unisexual, neutral and nude flowers are incomplete flowers*)
7. ***Actinomorphic (regular) flower***. When flower can be divided into two equal and symmetrical halves after cutting it **through any vertical plane** passing through the axis, it is called *actinomorphic flower*. It is also called radially symmetrical. This happens due to proportionate arrangement of floral members around the thalamus, e.g., china-rose, *Hibiscus rosa-sinensis* of Malvaceae.
8. ***Zygomorphic (irregular) flower***. When flower can be divided into two equal and symmetrical halves after cutting it **through**

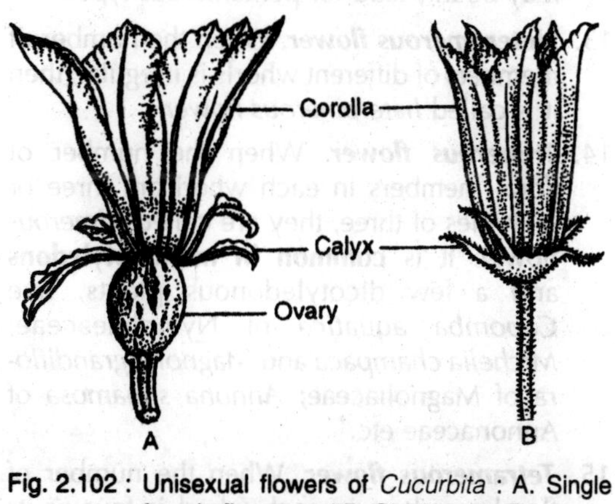

Fig. 2.102 : Unisexual flowers of *Cucurbita* : A. Single female flower and B. Single male flower

only one vertical plane passing through the axis, it is called ***symmetrically* zygomorphic** or *medianly zygomorphic* or *bilaterally symmetrical flower*, e.g., pea, *Pisum sativum* of Fabaceae; *Leonurus sibiricus* and *Leucas linifolia* of Lamiaceae; etc. But in some cases, the flower **may not even be cut** into two equal halves **through any vertical plane**, is called *asymmetrically zygomorphic flower*, e.g., Indian shot, *Canna indica* of Cannaceae etc. Zygomorphic flowers are also called *irregular flowers*.

9. ***Cyclic flower***. When the members of the floral leaves, i.e., sepals, petals, stamens and carpels are arranged on the thalamus in separate whorls, they are called *cyclic flowers* (Fig. 2.99A). Most of the members of angiosperms are cyclic, e.g., *Gynandropsis gynandra* of Capparidaceae; mustard, *Brassica nigra* of Brassicaceae (Cruciferae) etc.
10. ***Acyclic flower***. When all the floral leaves are arranged spirally on the thalamus, they are called *acyclic flowers*, e.g., lotus, *Nelumbo nucifera* of Nymphaeaceae; *Ranunculus* sp. of Ranunculaceae etc.
11. ***Spirocyclic* (hemicyclic) *flower***. When out of four sets of floral leaves some sets are arranged in whorls, while others are in spiral, they are called *spirocyclic flower*, e.g., water lily, *Nymphaea stellata* of Nymphaeaceae; *Michelia grandiflora* and *Michelia champaca* of Magnoliaceae etc.
12. ***Isomerous flower***. When the number of floral leaves i.e., the number of sepals, petals, stamens and carpels are same or of the multiples, it is called *isomerous flower*. It may be tri-, tetra- or pentamerous type.
13. ***Heteromerous flower***. When the number of members of different whorls is irregular, then it is called *heteromerous flower*.
14. ***Trimerous flower***. When the number of floral members in each whorl are three or multiples of three, they are called *trimerous flower*. It is **common in monocotyledons** and a few dicotyledonous plants, like *Cabomba aquatica* of Nymphaeaceae; *Michelia champaca* and *Magnolia grandiflora* of Magnoliaceae; *Annona squamosa* of Annonaceae etc.
15. ***Tetramerous flower***. When the number of floral members in each whorl is four or its multiple, it is called *tetramerous flower*, e.g., radish, *Raphanus sativa* and mustard, *Brassica nigra* of Brassicaceae; *Gynandropsis gynandra* of Capparidaceae; *Scoparia dulcis* of Scrophulariaceae; sensitive plant *Mimosa pudica* of Fabaceae etc.
16. ***Pentamerous flower***. When the number of floral members in each whorl is five or any multiple of it, it is called *pentamerous flower*. It is **common in dicotyledons.**

Terminologies related with flower

1. ***Monoecious plant***. When both staminate (male) or pistilate (female) flowers are borne on the same plant (e.g. palwal, *Trichosanthes dioica* and red pumpkin, *Cucurbita maxima* of Cucurbitaceae etc.), it is called *monoecious plant*.
2. ***Dioecious plant***. If male and female plants are different from one another, such plants are called *dioecious plants*, e.g., palmyra palm, *Borassus flabellifer* of Arecaceae etc.
3. ***Trioecious plant***. When male, female and bisexual flowers are borne on the three different plants, they are called *trioecious plants*, e.g., *Silene* sp. of Caryophyllaceae etc.
4. ***Polygamous plant.*** When unisexual, bisexual and neutral flowers are borne on the same plant, it is called *polygamous plant*, e.g., mango, *Mangifera indica* of Anacardiaceae etc.

FLOWER IS A MODIFIED SHOOT

The vegetative shoot has an elongated stem, differentiated into nodes and internodes, which bears leaves at their nodes in different orientation. The flower is exactly like a shoot, but it becomes condensed due to shortening of internodes and modification of leaves. But at times, in all the parts it does not follow the strict principle of a typical flower. Some of the flowers show variation, those proved that flowers are the modified shoots. The modifications of various parts are discussed below : A. *Axis nature of thalamus*, B. *Leaf nature of the floral members*, and C. *Homology of floral buds* :

A. *Axis nature of thalamus* : Flower consists of a condensed axis, the thalamus. In some cases, it becomes modified and forms an axis like elongated structure. The modifications are discussed :

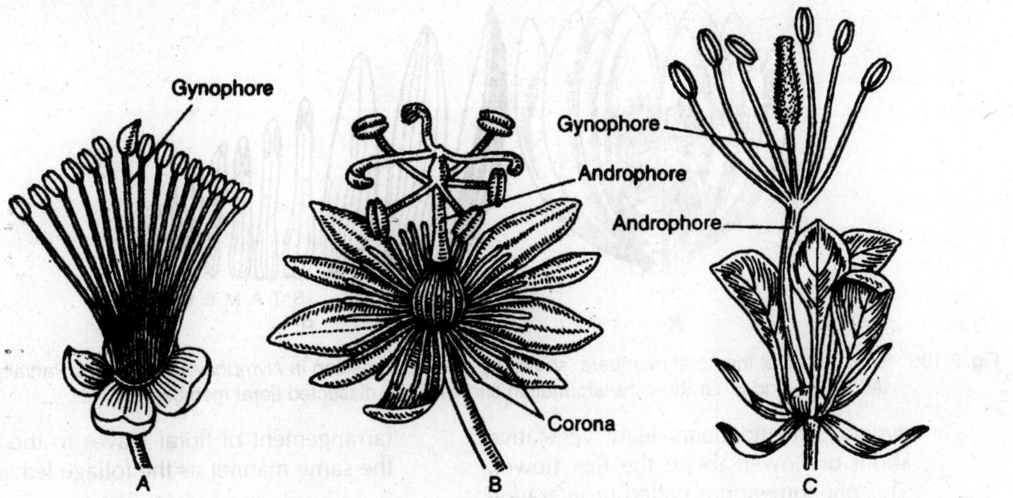

Fig. 2.103 : Flowers showing axis nature of thalamus : A. Gynophore of *Capparis sepiaria*, B. Androphore of *Passiflora suberosa*, and C. Androphore and gynophore of *Gynandropsis gynandra*

1. Flower consists of a condensed axis, the thalamus, formed due to shortening of internodes. But in some cases, it shows variation by having internodes like the vegetative shoot. The floral leaves developed at the nodes remain as usual like a typical flower. The region of elongated internode between petal and androecium is called *androphore* and between androecium and gynoecium is called *gynophore*. Only gynophore is present in *Capparis sepiaria* (Fig. 2.103A) of Capparidaceae, androphore in *Passiflora suberosa* (Fig. 2.103B) of Passifloraceae, and both androphore and gynophore, together called as *androgynophore* in spider flower, *Gynandropsis gynandra* (Fig. 2.103C) of Capparidaceae.
2. The growth of the thalamus is normally checked by the carpels, but sometimes the thalamus grows further beyond the

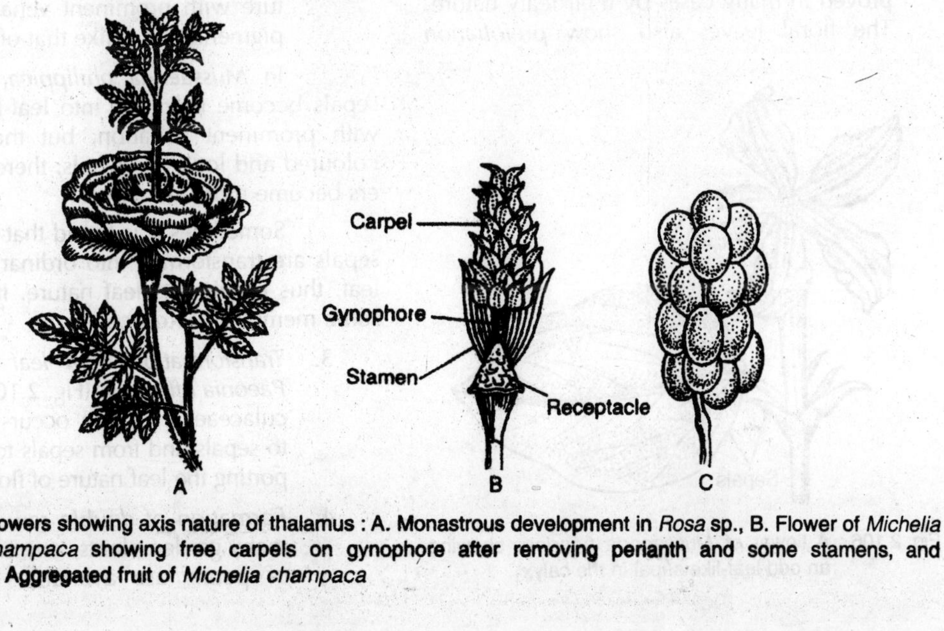

Fig. 2.104 : Flowers showing axis nature of thalamus : A. Monastrous development in *Rosa* sp., B. Flower of *Michelia champaca* showing free carpels on gynophore after removing perianth and some stamens, and C. Aggregated fruit of *Michelia champaca*

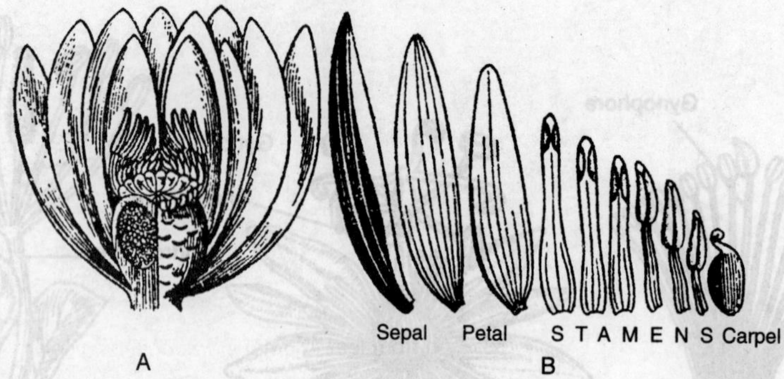

Fig. 2.105 : Leaf nature of the floral members, shown by gradual transition in *Nymphaea stellata* : A. Partially dissected flower, and B. Gradual transformation shown in the dissected floral members

gynoecium and bears leafy vegetative shoot or flower above the first flower. This phenomenon is called *proliferation* or *monstrous development*, e.g., rose, *Rosa* sp. (Fig. 2.104A); pear, *Pyrus communis* of Rosaceae.

3. In some cases, the free carpels are arranged spirally on the thalamus (Fig. 2.104B) and after fertilisation it elongates like an ordinary stem giving rise to an aggregated fruit, e.g., *Michelia champaca* (Fig. 2.104B,C) of Magnoliaceae and *Polyalthia longifolia* of Annonaceae.

B. *Leaf nature of the floral members* : The floral leaves i.e., the sepals, petals, stamens and carpels are the modified leaves. This can be proved in many cases by their leafy nature. The floral leaves also show *prefoliation* (arrangement of floral leaves in the bud) in the same manner as the foliage leaves.

Fig. 2.106 : A flower of *Mussaenda frondosa* showing an odd leaf-like sepal in the calyx

1. *Gradual transition of floral members.* In this type, the floral leaves are spirally arranged on the thalamus. The outermost sepal is green with distinct venation, and there is gradual transition of sepal to petal. The petals gradually become narrowed bearing anther at the tip, then to typical stamen with filament and anther. It is found in white water lily, *Nymphaea stellata* (Fig. 2.105A,B) of Nymphaeaceae.

2. *Modification of sepal into leaf.* In *Mussaenda frondosa* (Fig. 2.106) of Rubiaceae, out of five sepals *one* becomes modified into leaf-like structure with prominent venation, but the pigmentation is like that of petals.

In *Mussaenda philippica,* all the five sepals become modified into leaf-like structure with prominent venation, but the sepals are coloured and look like petals, thereby the flowers become more attractive.

Sometimes it is found that one or more sepals are transformed into ordinary compound leaf, thus proving its leaf nature. It is found in some members of Rosaceae.

3. *Transformation from leaf to petal.* In *Paeonia officinalis* (Fig. 2.107) of Ranunculaceae, transition occurs from leaves to sepals and from sepals to petals, supporting the leaf nature of floral parts.

4. *Formation of double row of petals.* In many garden plants, *one* or *double* row of petals are available in a single

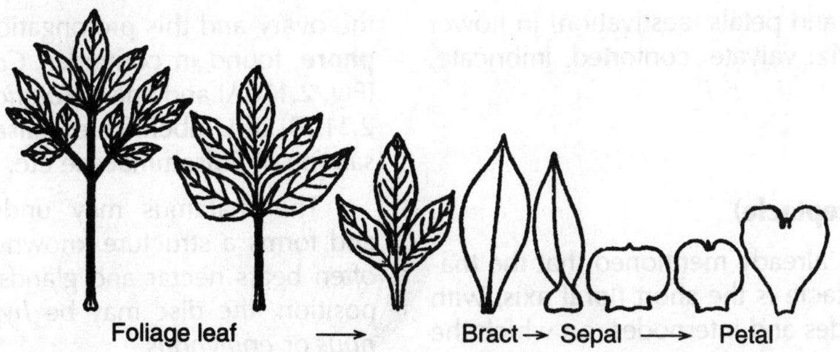

Fig. 2.107 : Transformation from leaf to petal in *Paeonia officinalis*

variety. The doubling of petals has been raised by horticultural technique, where the stamens are remodified into petals. e.g., china-rose, *Hibiscus rosa-sinensis* of Malvaceae; *Rosa* sp. of Rosaceae etc.

5. *Leafy petals.* In some cases, sepals look like foliage leaves and the petals are also green and leafy in nature. It is found in Green rose.

6. *Petaloid stamen.* In Indian shot, *Canna indica* (Fig. 2.108) of Cannaceae; the stamen becomes sterile and forms petaloid staminode, but in one stamen, $^1/_2$ anther lobe becomes petaloid and the other $^1/_2$ remains fertile.

7. *Petaloid and sepaloid carpel.* The carpels of *Zinnia* sp. of Asteraceae become petaloid or sepaloid.

8. *Leafy nature of carpel.* The carpel is also leafy in nature and this can be proved on examination of the single carpel of pea, *Pisum sativum* (Fig. 2.109) and other members of Fabaceae. The gynoecium is formed by the folding of a single leaf along its midrib to form one chambered ovary containing seeds. The upper elongated part of the leaf forms style and its apex forms stigma.

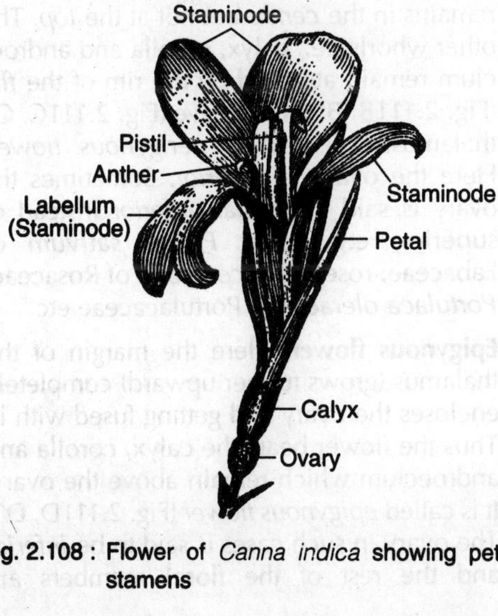

Fig. 2.108 : Flower of *Canna indica* showing petaloid stamens

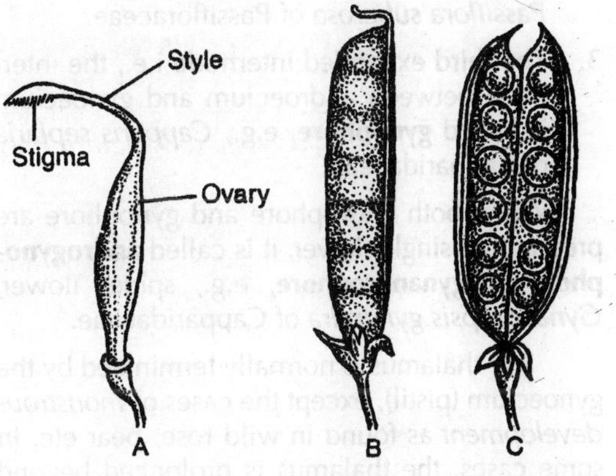

Fig. 2.109 : Leaf nature of carpel of *Pisum sativum* : A. Young gynoecium, B. Single mature pod, and C. Fruit split open

C. *Homology of floral buds.*

1. Flower buds sometimes become transformed into vegetative buds or bulbils, e.g., *Agave* sp. of Agavaceae, *Allium sativum* of Liliaceae.

2. Floral buds occupy the same terminal or axillary position as that of vegetative (shoot) bud.

3. The nature of arrangement of leaves (vernation) in vegetative bud is more or less identical with the arrangement of

sepals and petals (aestivation) in flower bud. Viz. valvate, contorted, imbricate, etc.

FLORAL PARTS

Thalamus (Receptacle)

It has been already mentioned that the thalamus or receptacle is the short floral axis, with compressed nodes and internodes, on which the floral leaves are inserted. In a complete flower, the thalamus consists of four nodes and three compressed internodes. In a few cases, however, the internodes become elongated. The elongated or expanded internodes are named according to the position of the internode. These are :

1. The first expanded internode i.e., the internode between the calyx and corolla is called **anthophore**, e.g., *Silene* sp. and a few plants of Caryophylaceae.
2. The second expanded internode i.e., the internode between corolla and androecium is called **androphore**, e.g., passion flower, *Passiflora suberosa* of Passifloraceae.
3. The third expanded internode i.e., the internode between androecium and gynoecium is called **gynophore**, e.g., *Capparis sepiaria* of Capparidaceae.

When both androphore and gynophore are present in a single flower, it is called **androgynophore** or **gynandrophore**, e.g., spider flower, *Gynandropsis gynandra* of Capparidaceae.

The thalamus is normally terminated by the gynoecium (pistil), except the cases of *monstrous development* as found in wild rose, pear etc. In some cases, the thalamus is prolonged beyond

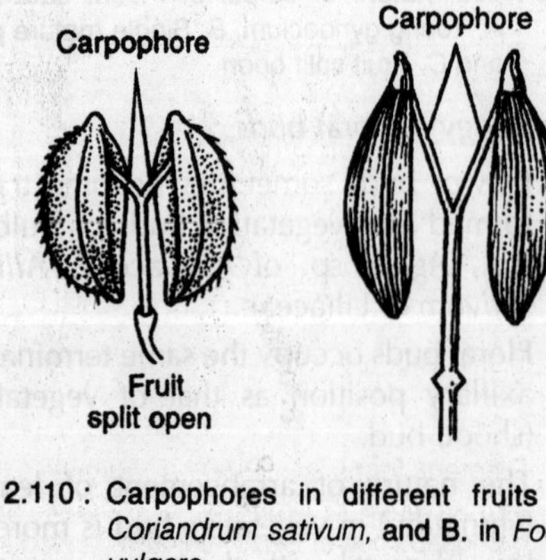

Fig. 2.110 : Carpophores in different fruits : A. in *Coriandrum sativum*, and B. in *Foeniculum vulgare*

the ovary and this prolongation is called **carpophore**, found in coriander, *Coriandrum sativum* (Fig. 2.110A) and anise, *Foeniculum vulgare* (Fig. 2.110B) of Umbelliferae; balsam, *Impatiens balsamina* of Balsaminaceae etc.

The thalamus may undergo modification and forms a structure known as **disc**. This disc often bears nectar and glands. Depends on the position, the disc may be *hypogynous, perigynous* or *epigynous*.

Insertion of Floral Leaves on the Thalamus

The calyx, corolla, androecium and gynoecium develop on the thalamus in different orientation. Depending on the position of three lower whorls with respect to the position of ovary, the flowers are of three types :

A. **Hypogynous flower**. When the ovary occupies the highest position on the thalamus, and the stamens, petals and sepals are separately and successively arranged below the ovary, the flower is called *hypogynous* (Fig. 2.111A, A'). Here the ovary is *superior* and all other members are *inferior*, e.g., chinarose, *Hibiscus rosa-sinensis* of Malvaceae. This is very common in many dicotyledonous (Brassicaceae, Solanaceae, Ranunculaceae etc.) as well as in monocotyledonous (Arecaceae, Araceae, Commelinaceae, Liliaceae etc) families.

B. **Perigynous flower**. Here the thalamus becomes flattened to form a **flat** or **cup-shaped** concave structure, where the ovary remains in the *centre* but not at the *top*. The other whorls i.e., calyx, corolla and androecium remain attached on the rim of the *flat* (Fig. 2.111B, B') or *concave* (Fig. 2.111C, C') thalamus. It is called *perigynous flower*. Here the ovary is *superior*. Sometimes the ovary is said to be *half inferior* instead of superior, e.g., pea, *Pisum sativum* of Fabaceae; rose, *Rosa centifolia* of Rosaceae; *Portulaca oleracea* of Portulacaceae etc.

C. **Epigynous flower**. Here the margin of the thalamus (grows further upward) completely encloses the ovary and getting fused with it. Thus the flower bears the calyx, corolla and androecium which remain above the ovary. It is called *epigynous flower* (Fig. 2.111D, D'). The ovary in such cases is said to be *inferior* and the rest of the floral members are

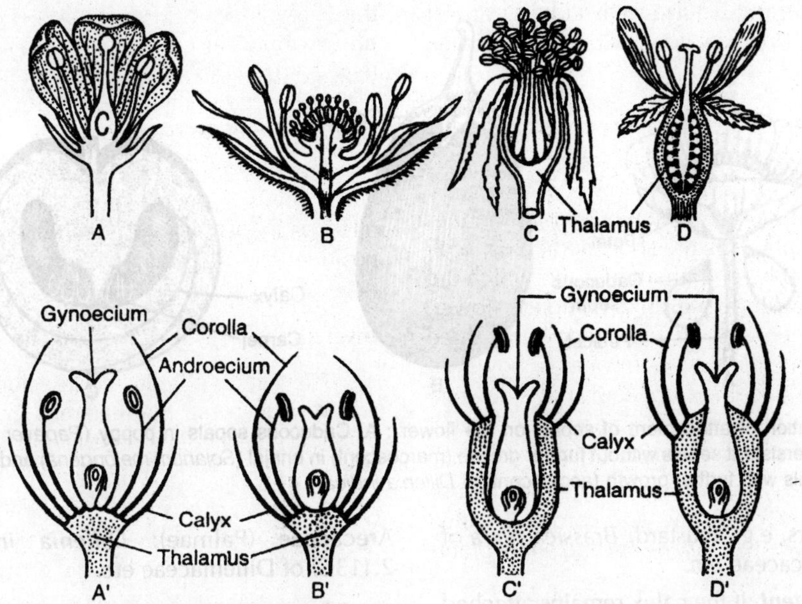

Fig. 2.111 : Insertion of floral leaves : A. Hypogynous, B. Perigynous with flat receptacle in strawberry, C. Perigynous with concave receptacle in rose, D. Epignous and A'–D' Diagrammatic representation of the figures of respective members

superior. Viz. sunflower, *Helianthus annuus* of Compositae; red pumpkin, *Cucurbita maxima* of Cucurbitaceae etc.

Calyx

Calyx is the outermost (lowermost) whorl, made up of sepals. The sepals are usually green in colour, but sometimes coloured i.e., like petals, called **petaloid** or **petaline**, e.g., cumin, *Nigella sativa* of Ranunculaceae; *Saraca indica* of Fabaceae (Leguminosae). The number of sepa-ls is generally three in monocotyledons and four to five in dicotyledons.

Calyx may be regular or irregular in shape. Depending on the various nature of calyx, they are grouped variously.

A. Based on the **arrangement of the unit member of calyx,** i.e., sepals, they are of two types: a. *Polysepalous,* and b. *Gamosepalous.*

a. ***Polysepalous.*** When the sepals remain free, they are called polysepalous, e.g., mustard, *Brassica nigra* (Fig. 2.112A) of Brassicaceae etc.

b. ***Gamosepalous.*** When the sepals remain united, they are called gamosepalous, e.g., *Leonurus sibiricus* (Fig. 2.112B) of Lamiaceae, *Datura metel* (Fig. 2.112C) of Solanaceae, *Hibiscus rosa-sinensis* of Malvaceae etc.

B. Based on the duration of **attachment of sepals upon the flower**, they are divided into three types :

a. ***Caducous*** or **Fagacious.** If the calyx falls just at the time of opening of flower from bud, it is called caducous, e.g., poppy, *Papaver somniferum* (Fig. 2.113A) of Papaveraceae etc.

b. ***Deciduous.*** Here the calyx remains attached on the flower until the flower

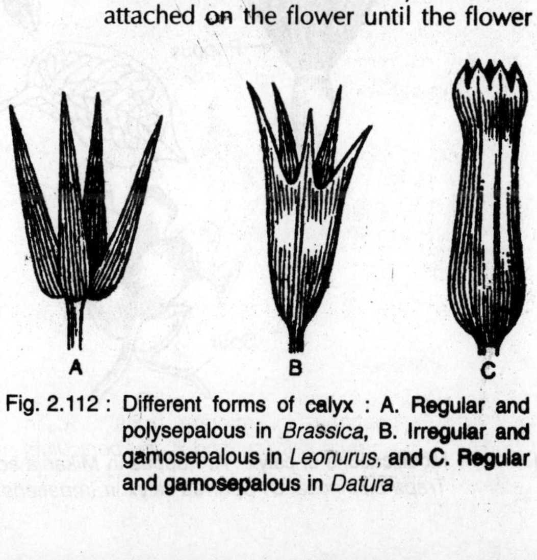

Fig. 2.112 : Different forms of calyx : A. Regular and polysepalous in *Brassica*, B. Irregular and gamosepalous in *Leonurus*, and C. Regular and gamosepalous in *Datura*

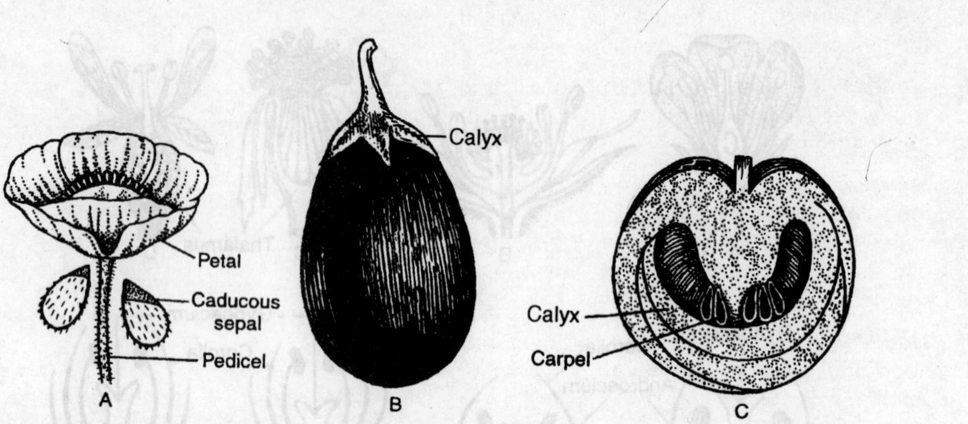

Fig. 2.113 : Duration of attachment of sepals on the flower : A. Caducous sepals in poppy (*Papaver somniferum*), B. Persistent sepals without further growth (marcescent) in brinjal (*Solanum melongena*) and C. Persistent sepals with further growth (accrescent) in *Dillenia indica*

withers, e.g., mustard, *Brassica nigra* of Brassicaceae etc.

c. ***Persistent***. If the calyx remains attached even on fruit, it is called persistent. It is of two types :

Marcescent. The calyx remains persistent but ceases to grows further, e.g., tomato, *Lycopersicum esculentum* and brinjal, *Solanun melongena* (Fig. 2.113B) of Solanaceae.

Accrescent. The calyx remains persistent and grows further along with the development of the fruit, e.g., coconut, *Cocos nucifera* of Arecaceae (Palmae); *Dillenia indica* (Fig. 2.113C) of Dilleniaceae etc.

Modification of Calyx

Sometimes sepals are modified into various forms, such as :

a. ***Pappus***. When the sepals are modified into hairy or bristly structure, it is called pappus. e.g., *Mikania scandens* (Fig. 2.114A) and other members of Asteraceae, etc.

b. ***Leafy***. Here sepal(s) are modified into leaf-like structure, e.g., *Mussaenda frondosa* –

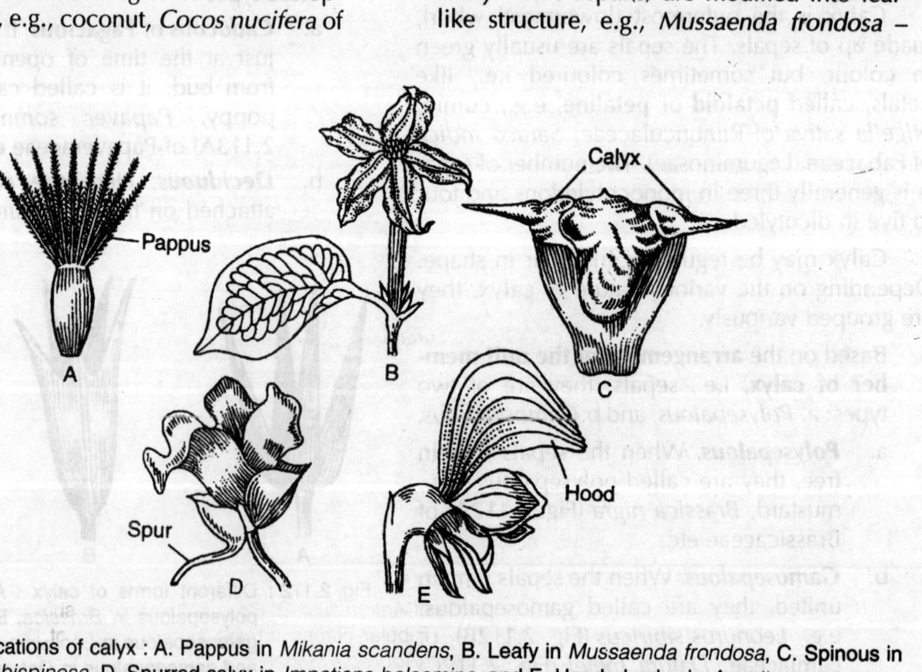

Fig. 2.114 : Modifications of calyx : A. Pappus in *Mikania scandens*, B. Leafy in *Mussaenda frondosa*, C. Spinous in *Trapa bispinosa*, D. Spurred calyx in *Impatiens balsamina*, and E. Hooded calyx in *Aconitum* sp.

one sepal (Fig. 2.114B) or *M. philippica* (all the five sepals) of Rubiaceae etc.

c. ***Spinous***. Here sepals are modified into spines, e.g. *Trapa bispinousa* (Fig. 2.114C) of Onagraceae.

d. ***Spurred***. When the calyx is modified into long tubular or beak-like structure, it is called spurred, e.g., *Impatiens balsamina* (Fig. 2.114D) of Balsaminaceae, *Delphinium* sp. of Ranunculaceae etc.

e. ***Hooded***. Here the calyx is modified into hood, thus covering the whole flower, e.g., *Aconitum* sp. (Fig. 2.114E) of Ranunculaceae etc.

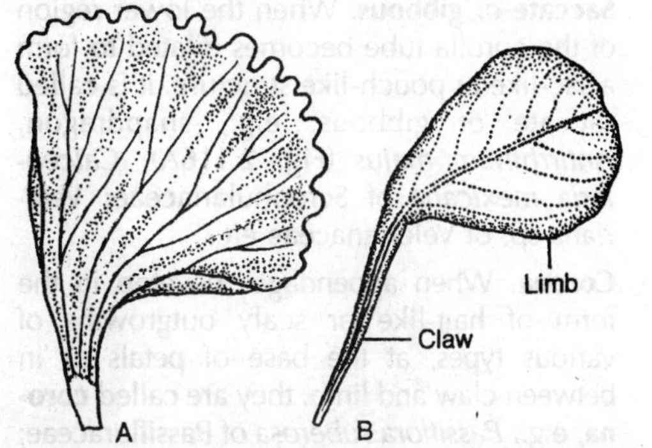

Fig. 2.115 : Petals : A. Single petal of *Hibiscus rosa-sinensis*, B. Single petal of *Brassica* with limb and claw

Corolla

Corolla is the second floral whorl from outside, consisting of **petals**. Petals are generally broad and bright-coloured; those are inserted on the thalamus by a comparatively narrow base, e.g., *Hibiscus rosa-sinensis* (Fig. 2.115A). Sometimes the petals are green in colour, called **sepaloid** or **sepaline**, e.g., *Polyalthia longifolia, Annona squamosa* of Annonaceae; *Pergularia daemia* of Asclepiadaceae.

Sometimes, the lower part of the petal becomes very thin and looks like a stalk, called *claw* and the upper expanded portion is called *limb* (*Brassica campestris*, Fig. 2.115B). The number of petals is variable i.e., from one to many.

Corolla may be regular or irregular in shape. Depending on the nature of sepals, they are grouped variously.

A. Based on the **arrangement of the unit member of corolla** i.e., petals, they are of two types : a. *Polypetalous* and b. *Gamopetalous* or *Sympetalous*.

a. ***Polypetalous***. When the petals remain free, they are called polypetalous, e.g., mustard, *Brassica nigra* of Brassicaceae etc.

b. ***Gamopetalous*** or ***Sympetalous***. When the petals remain united, they are called gamopetalous, e.g., *Lantana camara* of Verbenaceae; *Catharanthus roseus* of Apocynaceae etc. Sometimes, the gamopetalous corolla forms a lower tubular structure, called **corolla tube** and the free upper region is called **corolla lobes.** The junction point between the tube and the lobes is called **throat**.

Appendages of Corolla

Sometimes different types of appendages develop on the corolla. These are :

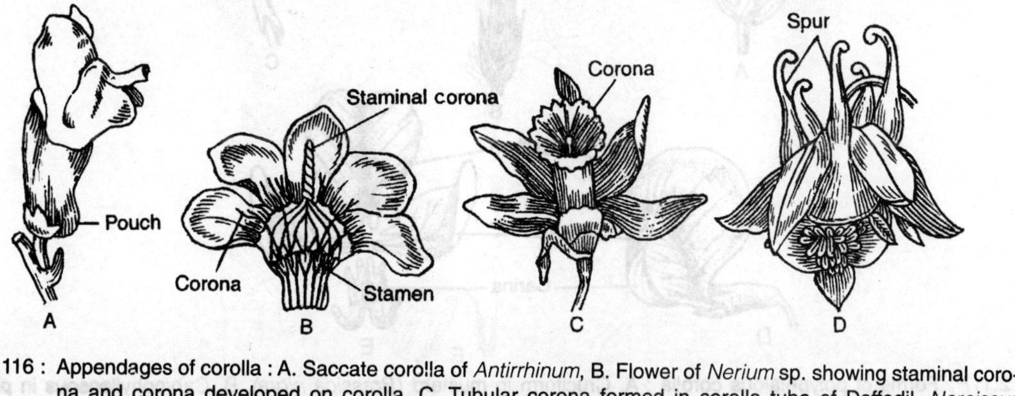

Fig. 2.116 : Appendages of corolla : A. Saccate corolla of *Antirrhinum*, B. Flower of *Nerium* sp. showing staminal corona and corona developed on corolla, C. Tubular corona formed in corolla tube of Daffodil, *Narcissus pseudonarcissus*, and D. Spurred petals of *Aquilegia vulgaris*

1. **Saccate** or **gibbous**. When the lower region of the corolla tube becomes dilated to form a sac-like or pouch-like structure, it is called saccate or gibbous, e.g., snapdragon, *Antirrhinum majus* (Fig. 2.116A), *Calceolaria mexicana* of Scrophulariaceae; *Veleriana* sp. of Velerianaceae etc.
2. **Corona**. When appendages develop in the form of hair-like or scaly outgrowths of various types, at the base of petals or in between claw and limb, they are called **corona**, e.g., *Passiflora suberosa* of Passifloraceae; *Nerium indicum* (Fig. 2.116B) of Apocynaceae; *Scoparia dulcis* of Scrophulariaceae; Daffodil, *Narcissus pseudonarcissus* of Amaryllidaceae (Fig. 2.116C) etc.
3. **Spurred** or **calcarate**. Here petal or petals are prolonged to form a tube-like structure, called **spurred** or calcarate, e.g., *Delphinum majus* and *Aquilegia* sp. (Fig. 2.116D) of Ranunculaceae; *Viola tricolor* of Violaceae etc.

Forms of Corolla

The corolla is of various types. The principal types are grouped variously and described :

1. Polypetalous and Regular
   a. **Cruciform**. Here four clawed petals are arranged crosswise in one whorl, e.g., mustard, *Brassica nigra* (Fig. 2.117A) and radish, *Raphanus sativa* of Brassicaceae.
   b. **Caryophyllaceous**. Here corolla consists of five-clawed petals with spreading limbs i.e., the limbs and the claws are at right angles to each other, e.g., *Dianthus chinensis* (Fig. 2.117B) and *Saponaria vaccaria* of Caryophyllaceae.
   c. **Rosaceous**. Here corolla consists of five or many petals without or with very short claws having outwardly spreading very wide limbs e.g., wild rose, *Rosa lucida* (Fig. 2.117C) of Rosaceae; tea, *Cammelia sinensis* of Ternstroemiaceae etc.
2. Polypetalous and Irregular

Papilionaceous. Here, out of five irregular free petals, the posterior odd petal is the largest and outermost (i.e., the *vexillum* or *standard*), which overlaps the two lateral petals (i.e., the *wings* or *alae*) and the lateral petals in their turn partially overlap the two smallest and innermost petals (i.e., the *keel* or *carina*). Viz. Corolla of all members belonging to the sub-family Papilionaceae under Family Fabaceae, such as pea, *Pisum sativum* (Fig. 2.117D, E); *Sesbania grandriflora*, *Clitoria ternatea* etc.

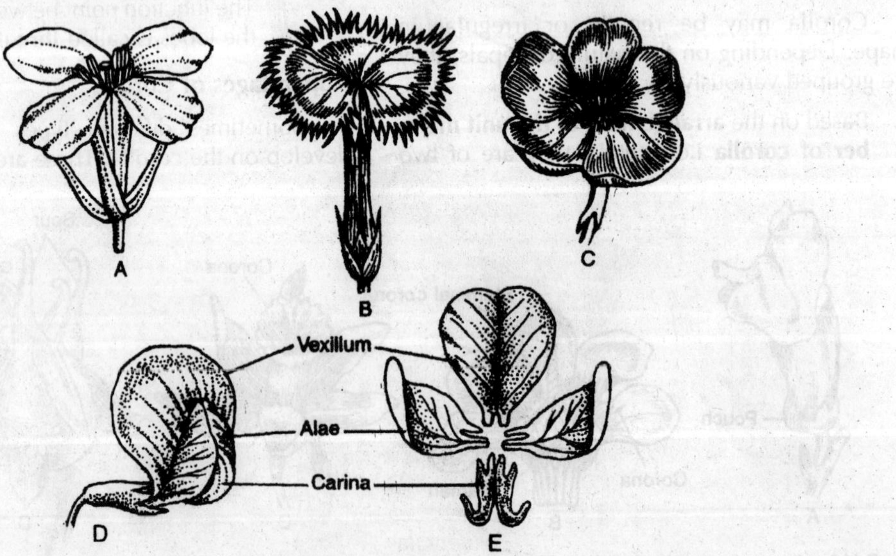

Fig. 2.117 : Forms of polypetalous corolia : A. Cruciform in mustard (*Brassica nigra*), B. Caryophyllaceous in pink (*Dianthus chinensis*), C. Rosaceous in wild rose, D–E. Papilionaceous in pea, D. Complete flower, and E. Corolla dissected to show arrangement and forms of petals

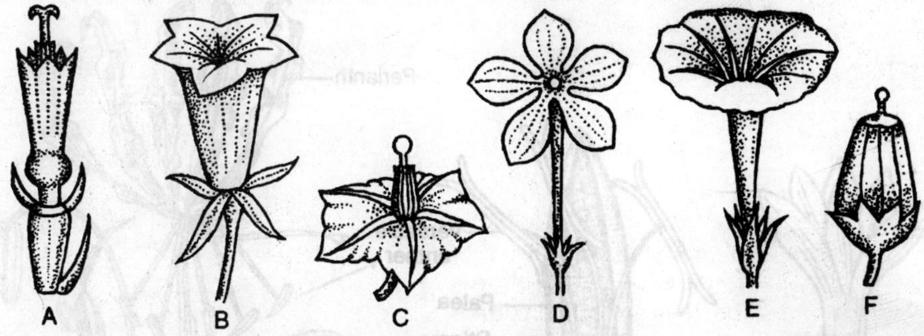

Fig. 2.118 : Gamopetalous regular corolla : A. Tubular in *Helianthus*, B. Campanulate in *Physalis*, C. Rotate in *Solanum*, D. Hypocrateriform in *Catharanthus*, E. Infundibuliform in *Ipomoea*, and F. Urceolate in *Bryophyllum*

3. Gamopetalous and Regular

a. **Tubular**. Here all the petals are united to form a corolla tube, which is more or less cylindrical, e.g., disc florets of sunflower, *Helianthus annuus* (Fig. 2.118A) and many other members of Asteraceae.

b. **Campanulate** or **bell-shaped**. Here corolla forms bell-shaped structure, e.g., gourd, *Cucurbita maxima* of Cucurbitaceae; yellow oleander, *Thevetia peruviana* of Apocynaceae; *Physalis peruviana* (Fig. 2.118B) of Solanaceae etc.

c. **Rotate** or **wheel-shaped**. Here the corolla tube is short with spreading wheel-shaped limbs develop at right angles to the tube, e.g., *Scoparia dulcis* of Scrophulariaceae; brinjal, *Solanum melongena* (Fig. 2.118C) of Solanaceae; *Calotropis procera* of Asclepiadaceae etc.

d. **Hypocrateriform** or **salver-shaped.** Here the corolla tube is long and slender with spreading limbs develop at right angles to the tube (like the rotate type), e.g., *Ixora coccinea* of Rubiaceae; *Ipomoea quammoclit* of Convolvulaceae; *Catharanthus roseus* (Fig. 2.118D) of Apocynaceae etc.

e. **Infundibuliform** or **funnel-shaped.** Here the corolla is tubular below and gradually becomes widen at the top like that of a funnel, e.g., morning glory, *Ipomoea pulchella* (Fig. 2.118E) of Convolvulaceae; *Ruellia tuberosa* of Acanthaceae; thorn apple, *Datura fastuosa* of Solanaceae etc.

f. **Urceolate** or **urn-shaped**. Here the corolla tube becomes inflated in the middle and comparatively narrow at both the ends, e.g., *Rhododendron* sp. of Ericaceae; *Bryophyllum calycinum* (Fig. 2.118F) of Crassulaceae etc.

4. Gamopetalous and Irregular

a. **Ligulate** or **strap-shaped.** Here the upper part of corolla is flattened like a strap and the lower part forms a short tube, e.g., ray florets of sunflower, *Helianthus annuus* (Fig. 2.119A) and many other members of Asteraceae.

b. **Bilabiate** or **two-lipped**. Here the upper part of corolla tube is divided into two lips with a wide mouth which remains open at the throat, e.g., *Ocimum sanctum, Anisomeles indica, Leucas linifolia, Leonurus sibiricus* (Fig. 2.119B) of Lamiaceae; *Andrographis paniculata, Adhatoda vasica* of Acanthaceae etc.

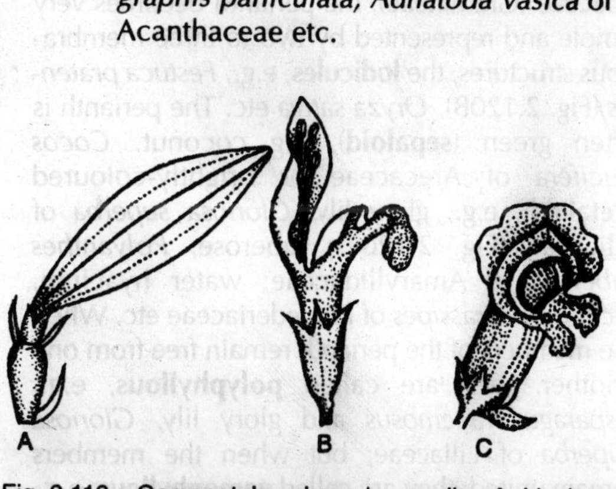

Fig. 2.119 : Gamopetalous irregular corolla : A. Ligulate in *Helianthus*, B. Bilabiate in *Leonurus* and C. Personate in *Antirrhinum*

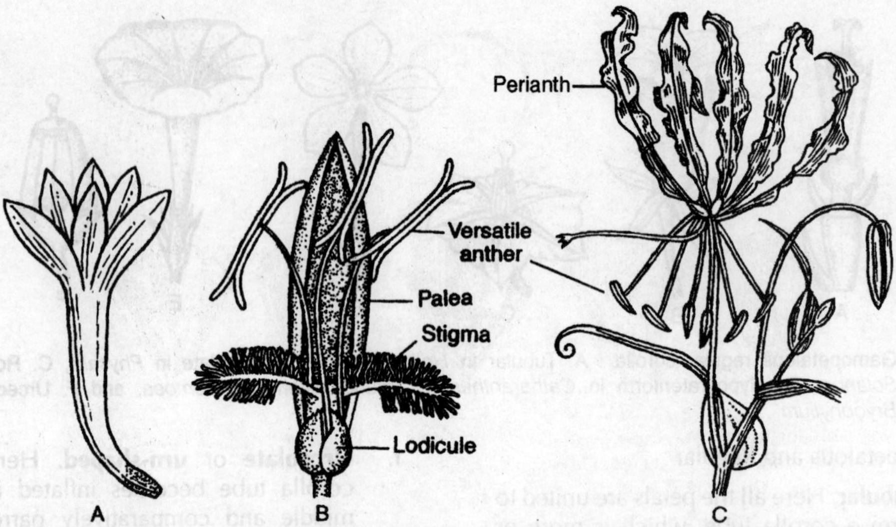

Fig. 2.120 : Perianths of different kinds : A. Petaloid perianth of tuberose, *Polyanthes tuberosa*, B. Flower of *Festuca pratensis* showing lodicules on removal of lemma, and C. Flower of *Gloriosa superba* showing petaloid polyphyllous perianth

c. **Personate** or **masked**. Here the limb of corolla is two-lipped like bilabiate, but the mouth remains closed by a projection (**palate**) of the lower lip, e.g., snapdragon, *Antirrhinum majus* (Fig. 2.119C), *Lindenbergia indica* of Scrophulariaceae etc.

Perianth

In majority of the monocotyledonous flower, the calyx and corolla are not differentiated and form a single accessory whorl, the **perianth** as found in *Crinum, Polyanthes* (Fig. 2.120A) etc. The unit member of the perianth is called **tepal**. In Poaceae (Gramineae), the perianth becomes very simple and represented by two to three membranous structures, the **lodicules**, e.g., *Festuca pratensis* (Fig. 2.120B), *Oryza sativa* etc. The perianth is often green (**sepaloid**), e.g coconut, *Cocos nucifera* of Arecaceae or brightly-coloured (**Petaloid**), e.g., glory lily, *Gloriosa superba* of Liliaceae (Fig. 2.120C); tuberose, *Polyanthes tuberosa* of Amaryllidaceae; water hyacinth, *Eichhornia crassipes* of Pontederiaceae etc. When the members of the perianth remain free from one another, they are called **polyphyllous**, e.g., *Asparagus racemosus* and glory lily, *Gloriosa superba* of Liliaceae; but when the members remain united, they are called **gamophyllous**, e.g., *Polyanthes tuberosa* of Amaryllidaceae etc.

Aestivation or Prefoliation

The aestivation is the pattern of arrangement of sepals, petals and tepals with respect to one another in budding stage. It is of the following five types (Fig. 2.121) :

1. **Valvate**. Here the margins of sepals, petals or tepals either touch or lie in close association, but they do not overlap (Fig. 2.121A, A′), e.g., calyx of china-rose, *Hibiscus rosa-sinensis* of Malvaceae; both calyx and corolla of sensitive plant, *Mimosa pudica* and gum tree, *Acacia nilotica* of Fabaceae etc.

2. **Twisted** or **Contorted**. Here one margin of each sepal, petal or tepal overlaps the margin of neighbouring one (Fig. 2.121B, B′), e.g., corolla (petals) of china-rose, *Hibiscus rosa-sinensis* of Malvaceae; corolla of karabi, *Nerium odoratum* of Apocynaceae etc.

3. **Imbricate**. Here the margins of sepals, petals or tepals are arranged in such a manner that one becomes completely external, another completely internal, and the rest members are arranged like that of twisted one (Fig. 2.121C, C′), e.g., petals of *Solanum torvum,* and *Nicotiana tabacum* of Solanaceae; petals of *Clerodendrum viscosum* of Verbenaceae etc.

4. **Quincuncil**. When the margins of sepals, petals or tepals are arranged in such a man-

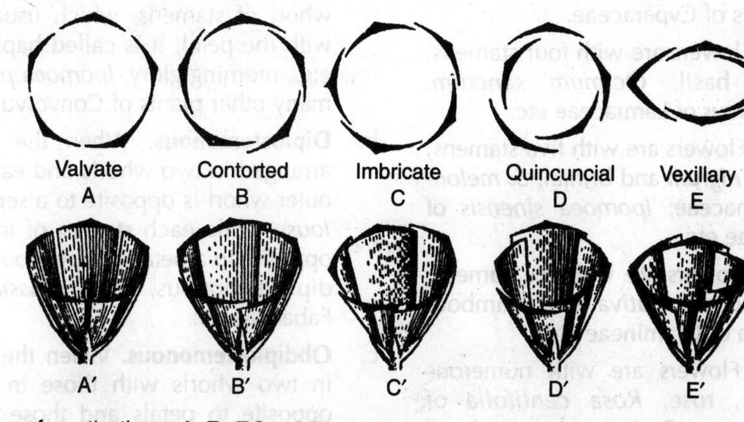

Fig. 2.121 : Types of aestivation : A–E. T.S. of flower bud and A′–E′. Lower half of flower bud, A–A′. Valvate, B-B′. Contorted, C-C′. Imbricate, D-D′. Quincuncial, and E-E′. Vexillary

ner that the two become completely external, other two completely internal and the rest members are like that of twisted (Fig. 2.121D, D′), e.g., petals of *Lantana camara* of Verbenaceae; sepals of *Holarrhena antidysenterica* of Apocynaceae etc.

5. **Vexillary**. In this type, out of five members the posterior odd petal is the largest and outermost (i.e., the *vexillum*), which overlaps the two lateral petals (i.e., the *wings*) and the lateral petals in their turn partially overlap the two smallest and innermost petals (i.e., the *keel*) (Fig. 2.121E, E′). Viz. corolla of all members belonging to the subfamily Papilionaceae under Family Fabaceae (Leguminosae).

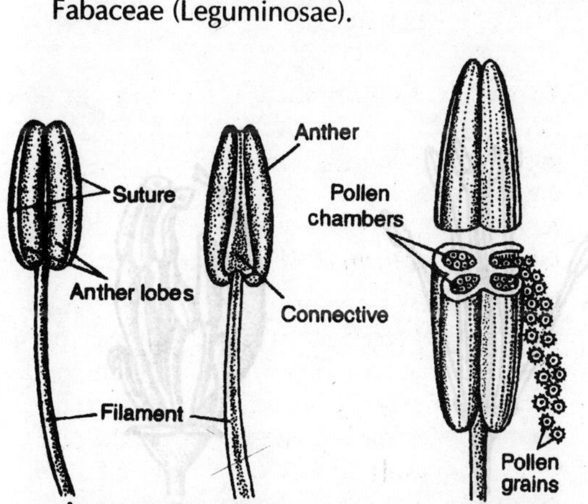

Fig. 2.122 : Stamen (dithecous) showing different parts : A. Ventral view, B. Dorsal view and C. Enlarged with T.S. showing chambers and pollen grains

Androecium

Androecium is the first set of essential whorl vis-a-vis the third set of the floral whorl. The unit member of the androecium is **stamen**. They look quite different from sepals and petals.

Stamen generally consists of an elongated stalk, the **filament** and bears at its apex a sac-like structure, the **anther** (Fig. 2.122). Generally, each anther consists of two **anther lobes** which are connected by a tissue formed by the extension of the filament, called the **connective**. The connective is visible towards the dorsal side of the anther lobes. There are longitudinal grooves or **sutures** along the ventral face of the anther demarcating the **pollen chambers**. Each anther lobe has **two** chambers, called **pollen sacs** or **microsporangia**. Each stamen, therefore, consists of four microsporangia and each microsporangium contains innumerable number of **pollen grains** or **microspores**. The anthers generally have two lobes, thus called **dithecous**, but in Malvaceae, it is with one lobe and is called **monothecous**.

The number of stamens varies from one to many. Depending on the number of stamens, the flowers are grouped as :

a. **Monandrous**. Flowers are with one stamen, e.g., turmeric, *Curcuma longa* and ginger, *Zingiber officinale* of Zingiberaceae etc.

b. **Diandrous**. Flowers are with two stamens, e.g., *Adhatoda vasica* of Acanthaceae etc.

c. **Triandrous**. Flowers are with three stamens, e.g., wheat, *Triticum aestivum* of Poaceae;

Scirpus articulatus, Cyperus rotundus and many members of Cyperaceae.

d. **Tetrandrous**. Flowers are with four stamens, e.g., sacred basil, *Ocimum sanctum, Leonurus sibiricus* of Lamiaceae etc.

e. **Pentandrous**. Flowers are with five stamens, e.g., *Solanum nigrum* and brinjal, *S. melongena* of Solanaceae; *Ipomoea sinensis* of Convolvulaceae etc.

f. **Hexandrous**. Flowers are with six stamens, e.g., paddy, *Oryza sativa* and bamboo, *Bambusa tulda* of Gramineae etc.

g. **Polyandrous**. Flowers are with numerous stamens, e.g., rose, *Rosa centifolia* of Rosaceae; castor, *Ricinus communis* of Euphorbiaceae, *Acacia nilotica* of Fabaceae etc.

On the thalamus, the floral members may arrange in one or in many whorls, or they may arrange spirally as in *Magnolia grandiflora* of Magnoliaceae.

Usually, there is only one whorl of stamens, which alternate with the petals. Sometimes, two or more whorls of stamens are present and these may alternate with the petals or lie opposite to them. When the stamens lie alternate with the petals (opposite to the sepal), they are called **antisepalous** and on the other hand, if they alternate with the sepals (opposite to the petals), they are called **antipetalous**. Depending on the number of whorl and their arrangement, they are of the following types :

a. **Haplostemonous**. When there is one whorl of stamens, which usually alternate with the petal, it is called haplostemonous, e.g., morning glory, *Ipomoea pulchella* and many other plants of Convolvulaceae.

b. **Diplostemonous**. When the stamens are arranged in two whorls and each stamen of outer whorl is opposite to a sepal (*antisepalous*), while each stamen of inner whorl is opposite to a petal (antipetalous), it is called diplostemonous, e.g., *Cassia fistula* of Fabaceae etc.

c. **Obdiplostemonous.** When the stamens are in two whorls with those in outer whorl opposite to petals and those in the inner whorl opposite to sepals, it is called obdiplostemonous. e.g., many members of Rutaceae, Caryophyllaceae, *Geranium* sp. of Geraniaceae etc.

d. **Polystemonous**. When stamens occur in more than two whorls, it is called polystemonous. e.g., *Delphinium* sp. of Ranunculaceae etc.

In a flower, when the number of stamens is equal to the number of sepals, petals and carpels, then the androecium is called **isostemonous**.

Depending on the position of stamens on the thalamus with reference to ovary, they may also be *hypogynous, perigynous* and *epigynous*.

The stamens in an androecium vary in length and accordingly they are of the following types :

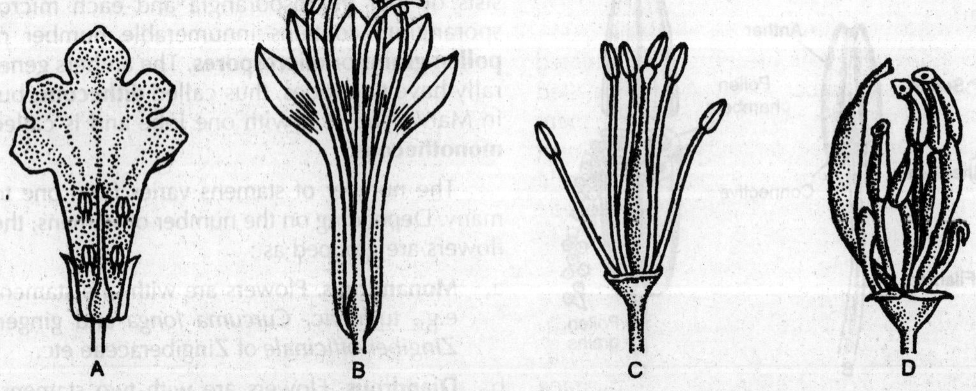

Fig. 2.123 : Stamens of different lengths : A. Epipetalous and didynamous stamen of *Lantana camara* of Verbenaceae, B. Didynamous stamen of *Clerodendrum viscosum* of Verbenaceae, C. Tetradynamous stamen of *Brassica nigra* of Cruciferae, and D. Heterodynamous stamens of *Cassia tora* of Leguminosae

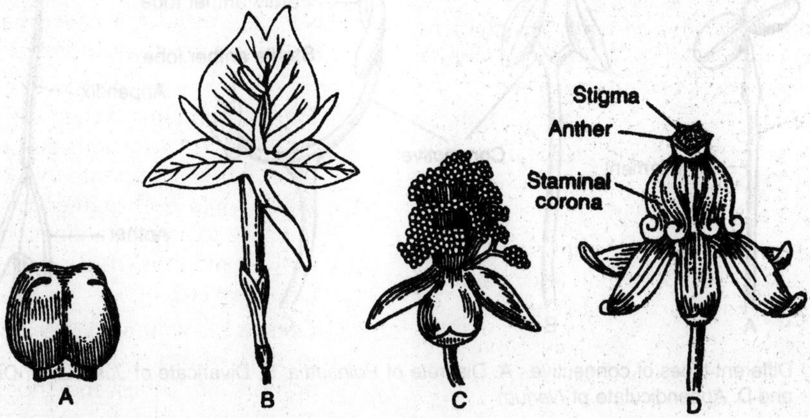

Fig. 2.124 : Filaments : A. Sessile anther of *Arum*, B. Petaloid staminode of *Zingiber officinale*, C. Branched filaments of *Ricinus communis* and D. Staminal corona of *Calotropis procera*

a. **Didynamous**. There are four stamens in one whorl in which two are longer than the other two, it is termed as *didynamous type*, e.g., *Leonurus* and other members of Lamiaceae; also common in Acanthaceae, Scrophulariaceae and Verbenaceae (Fig. 2.123A,B) etc.

b. **Tetradynamous**. When six stamens are in two whorls in which two of the outer whorl are shorter than the four of the inner whorl, it is termed as *tetradynamous type*, e.g., mustard, *Brassica nigra* (Fig. 2.123C) and other members of Brassicaceae, etc.

c. **Heterodynamous**. When stamens of different lengths occur in one whorl, it is termed as *heterodynamous* type, e.g., *Cassia tora* (Fig. 2.123D) of Fabaceae.

Filament

The elongated stalk-like structure, situated at the base of an anther of a stamen, is called **filament.** The stamens may be without filament i.e., *sessile* (*Arum maculatum*, Fig. 2.124A) of Araceae. The filament may not bear fertile anther i.e., *sterile* and is called **staminode** (*Cassia sophera*, Fig. 2.123D of Fabaceae). In *Canna* of Cannaceae, *Zingiber officinale* (Fig. 2.124B) of Zingiberaceae, the sterile stamens form petal like structure, the **petaloid staminode.** The gradual transition from petal to stamen is observed in Indian blue lotus, *Nymphaea stellata* and white water lily, *Nymphaea alba* (Fig. 2.105B) of Nymphaeaceae. Normally the filaments are unbranched, but it may be branched as in castor, *Ricinus communis* (Fig. 2.124C) of Euphorbiaceae. Sometimes it bears horny appendages, called **staminal corona,** found in *Calotropis procera* (Fig. 2.124D) of Asclepiadaceae.

Connective

The connective is a tissue which connects the two parallel anther lobes. It is of different types :

a. **Discrete**. The connective is very small or wanting, where the anther lobes remain very close to each other, e.g., *Poinsettia pulcherrima* (Fig. 2.125A) of Euphorbiaceae.

b. **Divaricate**. The connective develops in such a way that the anther lobes get separated from each other, e.g., *Lindenbergia indica* of Scrophulariaceae, *Justicia gendarussa* (Fig. 2.125B) of Acanthaceae etc.

c. **Distractile**. The connective is long stalk-like body which separates the two anther lobes, of which one is sterile and the other one is fertile, e.g., *Salvia plebeja* (Fig. 2.125C) of Lamiaceae.

d. **Appendiculate**. The connective is prolonged beyond the anther into a feathery appendix, e.g., oleander, *Nerium indicum* (Fig. 2.125D) of Apocynaceae.

Anther

Generally the anthers are bilobed and *quadrilocular,* but rarely the anthers become

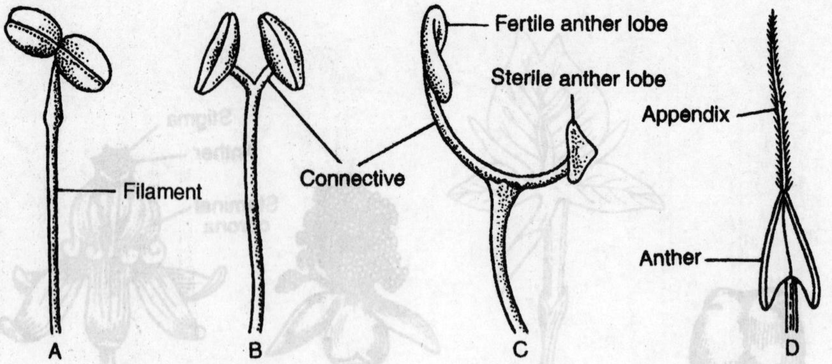

Fig. 2 125 : Different types of connective : A. Discrete of *Poinsettia*, B. Divaricate of *Justicia*, C. Distractile of *Salvia*, and D. Appendiculate of *Nerium*

unilocular (Fig. 2.126C) either due to abortion of one lobe and destruction of partition wall between the chamber or destruction of all partition tissue. One-chambered i.e., unilocular anther is found in the family Malvaceae.

The anthers may be of different shapes. It may be ***rounded*** (*Mercurialis annua* of Euphorbiaceae, Fig. 2.126A), ***linear*** (*Acalypha indica* of Euphorbiaceae, Fig. 2.126B), ***reniform*** (*Hibiscus rosa-sinensis* of Malvaceae, Fig. 2.126C), ***sagittate*** (*Catharanthus roseus* of Apocynaceae, Fig. 2.126D), **sinuous** i.e., rolled-up like an inverted 'S' (gourd, *Cucurbita maxima* of Cucurbitaceae, Fig. 2.126E), ***appendiculate*** (*Erica cinerea* of Ericaceae, Fig. 2.126F) etc.

The ventral side of the anther generally faces inwardly towards the centre of the flower, it is called **introrse** (*Polyanthes tuberosa* of Amaryllidaceae, *Nymphaea stellata* of Nymphaeaceae etc.) and when it faces outwardly towards the petal, it is called **extrorse** (*Colchicum luteum* and *Gloriosa superba* of Liliaceae etc.).

Attachment of Anther with the Filament

The anther becomes attached with the filament in different ways. These are :

a. **Adnate**. The filament or the connective remains attached to the back of the anther throughout the whole length (Fig. 2.127A), e.g., *Magnolia* sp. of Magnoliaceae, *Nymphaea* sp. of Nymphaeaceae etc.

b. **Basifixed** or **innate**. The apex of the filament becomes firmly attached to the base of the anther (Fig. 2.127B), e.g., *Brassica* sp. of Brassicaceae, *Solanum* sp. of Solanaceae, *Carex* sp. of Cyperaceae etc.

c. **Dorsifixed**. The apex of the filament becomes firmly attached to the back of the

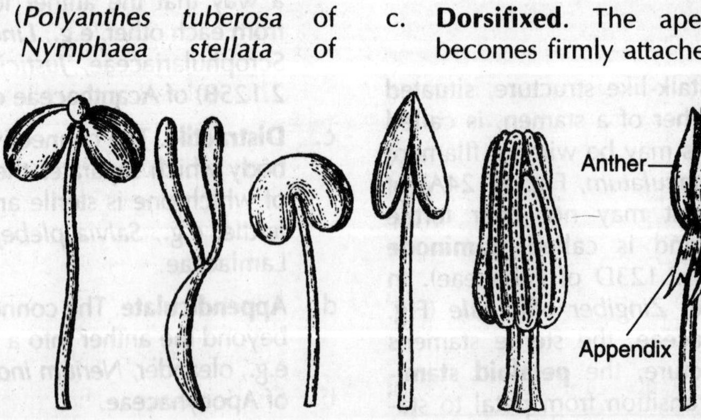

Fig. 2.126 : Anther of different shapes : A. Rounded in *Mercurialis*, B. Linear in *Acalypha*, C. Unilocular and reniform in *Hibiscus rosa-sinensis*, D. Sagittate in *Catharanthus*, E. Sinuous in *Cucurbita maxima*, and F. Appendiculate in *Erica cinerea*

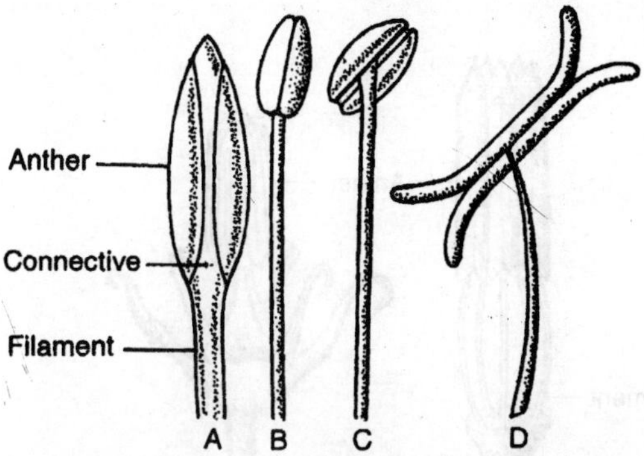

Fig. 2.127 : Attachment of anther with the filament : A. Adnate, B. Basifixed, C. Dorsifixed, and D. Versatile

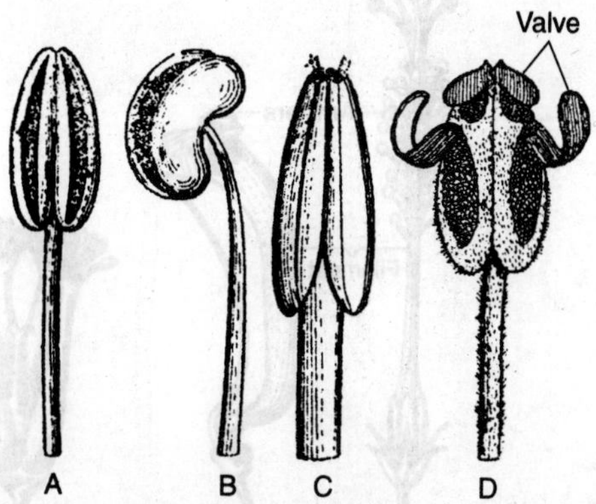

Fig. 2.128 : Dehiscence of anthers : A. Longitudinal, B. Transverse, C. Apical, and D. Valvular

anther (Fig. 2.127C), e.g., *Sesbania* sp. of Fabaceae, *Passiflora* sp. of Passifloraceae, *Polianthes tuberosa* of Amaryllidaceae etc.

d. **Versatile.** The apex of the filament becomes attached at a point near the middle of the back of the connective, so that the anther can move freely (Fig. 2.127D), e.g., wheat, *Triticum aestivum* and rice, *Oryza sativa* of Poaceae; glory lily, *Gloriosa superba* of Liliaceae etc.

Dehiscence of Anthers

The pollen grains inside the mature anther exert some pressure on the outer wall, thereby the anther lobes burst open and the pollen grains are set free. The dehiscence may be of different types :

a. **Longitudinal**. The anther lobes burst lengthwise along the line of *suture* from base to apex (Fig. 2.128A), e.g., *Datura* sp. of Solanaceae; *Vitis vinifera* of Vitaceae etc.

b. **Transverse.** The anther splits up transversely (as the *suture* is present crosswise) (Fig. 2.128B), e.g., china-rose, *Hibiscus rosa-sinensis* and other members of Malvaceae.

c. **Apical** or **porous**. Pollens are discharged through apical pores (Fig. 2.128C), e.g., potato, *Solanum tuberosum*; brinjal, *S. melongena* and many other members of Solanaceae.

d. **Valvular**. Pollens are discharged through a trap-door-like valve formed on the anther wall, e.g., *Berberis vulgaris* of Berberidaceae, *Cinnamomum zeylanicum* and *Laurus nobilis* (Fig. 2.128D) of Lauraceae etc.

Union of Stamens

Union of stamens among themselves is known as **cohesion**, and the union of stamens with other members i.e., *perianth, petal* or *gynoecium* is known as **adhesion**.

Cohesion of Stamens

When the stamens are united by their filaments only (anthers remain free), it is called **adelphous condition**, and when the stamens are united by their anthers only (filaments remain free), it is called **syngenesious condition.** If, however, the filaments as well as the anthers of different stamens become united, it is called **synandrous condition**.

The cohesion may be of the following types :

a. **Monadelphous**. All stamens are united by their filaments (anthers remain free), so forming a tube around style, e.g., china-rose, *Hibiscus rosa-sinensis* (Fig. 2.129A); lady's finger, *Abelmoschus esculentus* and other members of Malvaceae; few members of Euphorbiaceae, Fabaceae etc.

b. **Diadelphous**. The filaments are united (anthers remain free) and form *two bundles*, e.g., pea, *Pisum sativum* (Fig. 2.129B); gram, *Cicer arietinum* and some other members of Fabaceae.

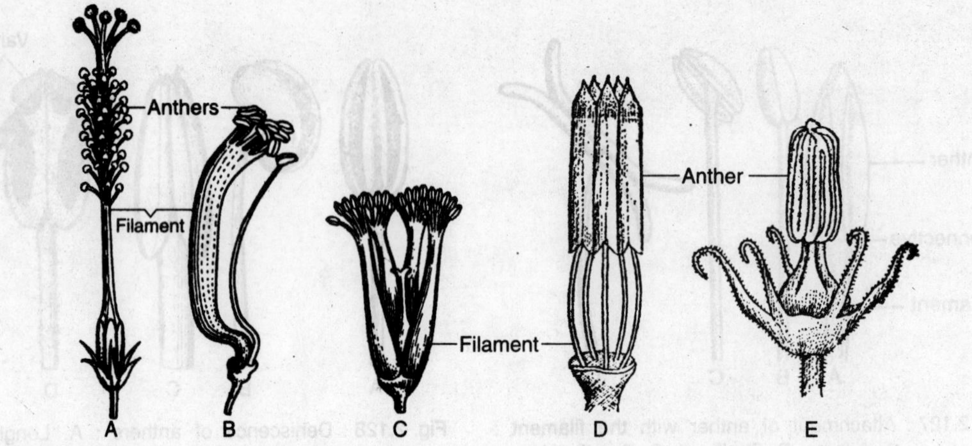

Fig. 2.129 : Different types of cohesion of stamens : A. Monadelphous of *Hibiscus*, B. Diadelphous of *Pisum*, C. Polydelphous of *Bombax*, D. Syngenesious of *Helianthus*, and E. Synandrous in *Cucurbita*

c. **Polydelphous**. The filaments are united in *more than two bundles* and anthers remain free, e.g., *Bombax ceiba* (Fig. 2.129C) of Bombacaceae; lemon, *Citrus* sp. of Rutaceae etc.

d. **Syngenesious**. The anthers are united into a bundle, but filaments remain free, e.g., sunflower, *Helianthus annuus* (Fig. 2.129D) of Asteraceae etc.

e. **Synandrous**. The anthers as well as filaments are united throughout their length to form a compact body, e.g. gourd, *Cucurbita maxima* (Fig. 2.129E) and many members of Cucurbitaceae.

Adhesion of Stamens

Adhesion of stamens is of the following types :

a. **Epiphyllous**. The stamens are attached to or arise from perianths, e.g., tuberose, *Polyanthes tuberosa* (Fig. 2.130A); spider lily, *Crinum asiaticum* and other plants of Amaryllidaceae etc.

b. **Epipetalous**. The stamens are attached to or arise from petals, e.g., *Solanum nigram* (Fig. 2.130B) of Solanaceae, *Lantana camara* of Verbenaceae, *Catharanthus roseus* of Apocynaceae etc.

c. **Gynandrous**. The stamens are attached with the gynoecium, either wholly or by anthers only. In the members of Asclepiadaceae (*Calotropis procera*, (Fig. 2.130C), the stamens adnate to the stigma i.e., *gynostegium*, but in Orchidaceae the adnation of stamens takes place with the style and stigma forming *gynostemium*.

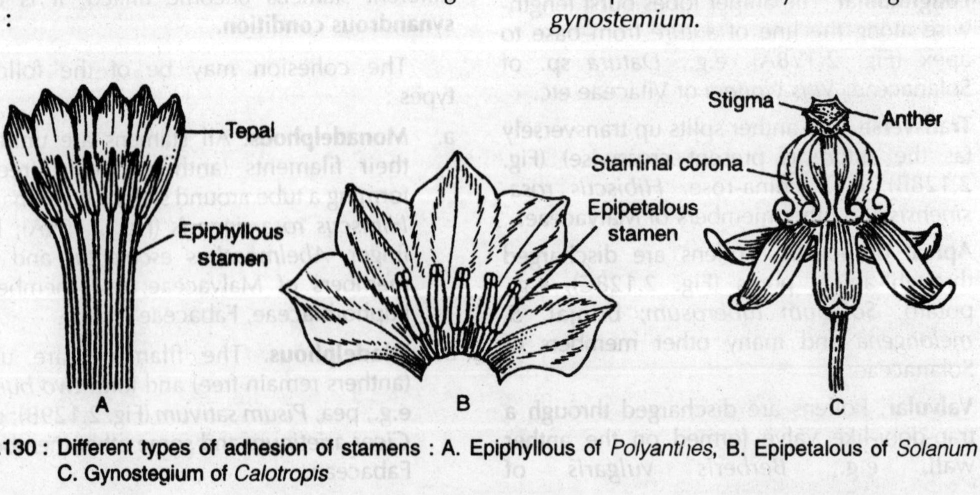

Fig. 2.130 : Different types of adhesion of stamens : A. Epiphyllous of *Polyanthes*, B. Epipetalous of *Solanum* and C. Gynostegium of *Calotropis*

Gynoecium

The gynoecium, also called **pistil** (Fig. 2.131A, B), is the innermost or the topmost whorl on the thalamus. It is differentiated into three regions :

a. **Ovary**. It is the lowermost swollen part containing ovules.

b. **Style**. It is the elongated stalk or thread-like part attached to the apex of the ovary.

c. **Stigma**. It is the apex of the style, the receptive region on which pollen grains are deposited during pollination.

The sterile pistil or gynoecium (devoid of functional ovule) is called **pistillode**.

The gynoecium consists of one or more **carpels**, the unit of pistil. It is of two types :

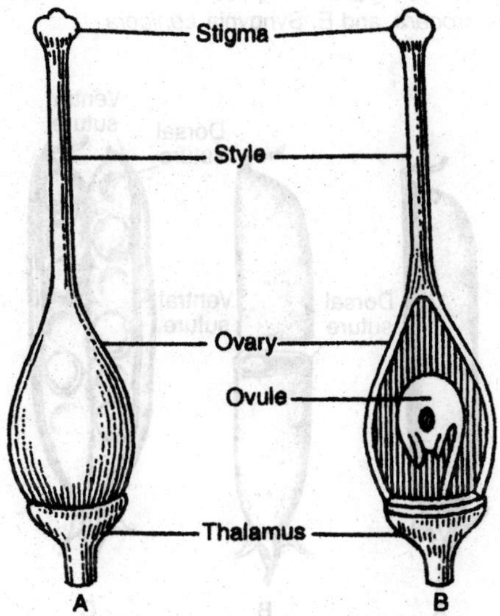

Fig. 2.131 : Gynoecium (=Pistil) : A. External morphology, and B. Gynoecium, showing internal ovule

A. **Simple** or **monocarpellary**. When the gynoecium is made up of one carpel, it is called *simple* or *monocarpellary gynoecium*, e.g., members of Fabaceae (Fig. 2.132A) and Poaceae (Gramineae).

B. **Compound** or **polycarpellary**. When the gynoecium is made up of two or more carpels, it is called *compound or polycarpellary gynoecium*, e.g., members of Solanaceae, Lamiaceae, Malvaceae etc. The

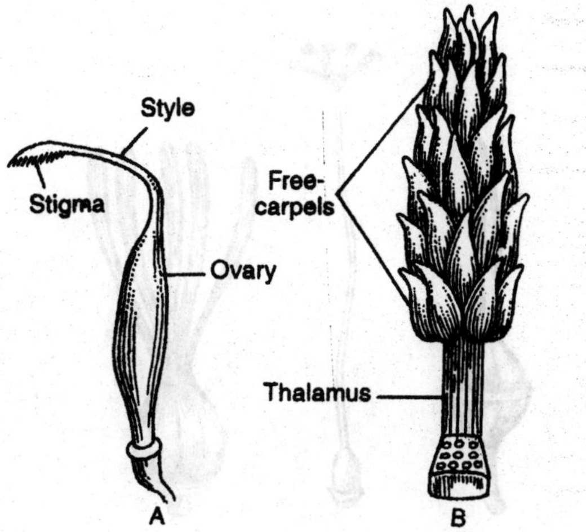

Fig. 2.132 : Types of gynoecium : A. Monocarpellary of *Pisum sativum* of Fabaceae, and B. Polycarpellary and apocarpous of *Michelia champaca* of Magnoliaceae, showing elongated thalamus

polycarpellary gynoecium is of two types :

a. **Apocarpous**. The carpels remain completely free from one another, e.g., Members of Magnoliaceae (Fig. 2.132B), Alismaceae, Ranunculaceae, Nymphaeaceae etc.

b. **Syncarpous**. The carpels are united with each other. This type is very common in angiospermic families like Brassicaceae, Malvaceae, Liliaceae, Asteraceae (Compositae) etc.

In syncarpous gynoecium, the ***cohesion of carpels*** are of different types :

1. Carpels are united throughout their length (Fig. 2.133A), e.g., Acanthaceae, Solanaceae, Boraginaceae etc.
2. Carpels are united at the ovary and style, but the stigmas are free (Fig. 2.133B), e.g., Malvaceae etc.
3. Carpels are united at the ovary but the style and stigma remain free (Fig. 2.133C), e.g., *Linum usitatissimum* of Linaceae.
4. Carpels are free, except at the upper part of style and stigma (Fig. 2.133D), e.g., Asclepiadaceae, Apocynaceae etc.

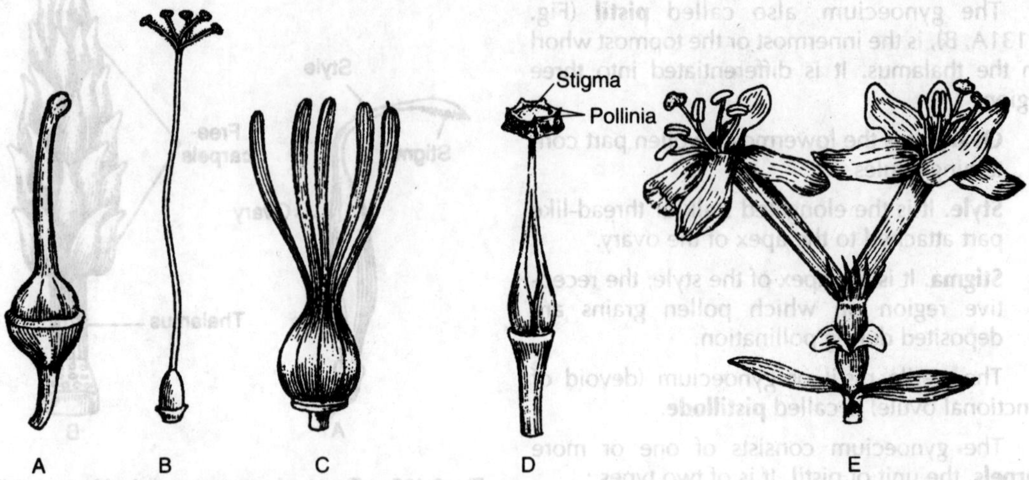

Fig. 2.133 : Different types of polycarpellary and syncarpous gynoecium : A. Completely united of *Solanum melongena*, B. With free stigma of *Hibiscus rosa-sinensis*, C. With free style and stigmas of *Linum usitatissimum*, D. With free ovary and lower part of style of *Calotropis procera*, and E. Syngynia *Lonicera*

5. Ovaries of two adjoining flowers may fuse, while other parts of the flowers remain free (i.e., **syngynia**), e.g., *Lonicera* sp. (Fig. 2.133E).

Ovary

The ovary is the swollen region formed by the folding up of the lamina (i.e., carpel), containing one or many ovules. The ovules develop into seeds. The foliar origin of ovary is evidenced in the simple ovary of pea, *Pisum sativum* of Fabaceae (Fig. 2.134A, B and C). One can easily understand that a leaf-like carpel i.e., the megasporophyll becomes folded towards the midrib and forms one-chambered ovary by the fusion of the margins. The marginal line through which the carpel fuses is called **ventral suture** and the midrib being the **dorsal suture**. The ovules or megasporangia usually develop on **placenta** produced on the inner surface of the united margins of the carpel.

Depending on the *position of ovary on the thalamus* with respect to other floral whorls, it is of the following types :

a. ***Superior***. The ovary is situated at the top on the thalamus with respect to other floral whorls, e.g., members of Solanaceae, Lamiaceae, Malvaceae; Fig. 2.135A) etc.

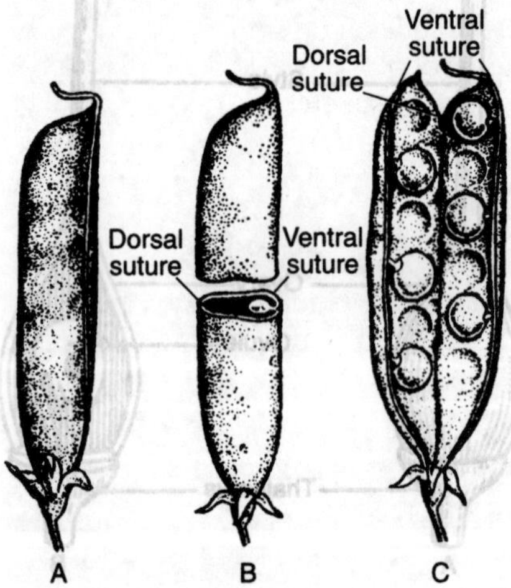

Fig. 2.134 : Pod of Pea (*Pisum sativum*) : A. Single pod, B. Single pod, cut transversely into two, showing the placenta and ovule on ventral suture, and C. Pod split along the ventral suture

b. ***Inferior***. The ovary is situated on the thalamus at a level lower than the other floral whorls, e.g., Members, of Cucurbitaceae (Fig. 2.135B), Rubiaceae, Asteraceae etc.

c. ***Half*** or ***semi-inferior***. The ovary is partly adherent to the receptacle i.e., an intermediate position of ovary between the inferior and superior types (Fig. 2.135C), e.g., pea,

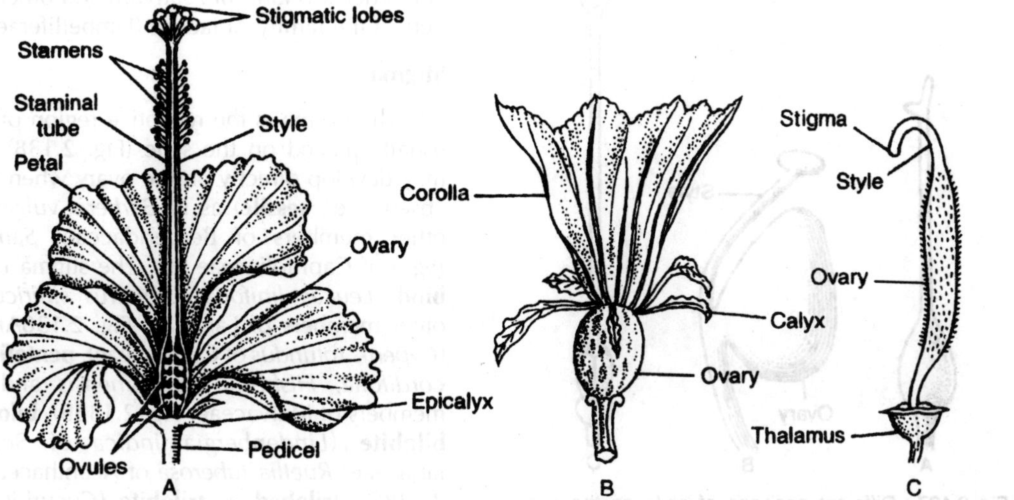

Fig. 2.135 : Position of ovary on the thalamus : A. L.S. of China-rose (*Hibiscus rosa-sinensis*) showing superior ovary, B. Female flower of *Cucurbita* showing inferior ovary, and C. Female flower of *Cassia sophera* showing half inferior ovary

Pisum sativum of Fabaceae; rose, *Rosa centifolia* of Rosaceae; *Portulaca oleracea* of Portulacaceae etc.

The ovary of a **monocarpellary** gynoecium has a **single locule** or chamber (Fig. 2.136A). The ovary with **polycarpellary** gynoecium may be of the following types :

a. **Unilocular** (1-*chambered*), (Fig. 2.136B), e.g., poppy, *Papaver somniferum* of Papaveraceae; *Carica papaya* of Caricaceae etc.

b. **Bilocular** (2-*chambered*), (Fig. 2.136C), e.g., Members of Acanthaceae, Scrophulariaceae etc.

c. **Trilocular** (3-*chambered*), (Fig. 2.136D), e,g, Members of Euphorbiaceae, Amaryllidaceae, Liliaceae etc., and

d. **Plurilocular** or **multilocular** (*many-chambered*), (Fig. 2.136E and F), e.g., Members of Malvaceae, Rutaceae, Sterculaceae etc.

Style

The stalk-like extension of ovary is the style. Depending on their position on the ovary, they are of the following types :

a. **Apical** or **terminal**. The style develops on the top of the ovary (Fig. 2.137A), e.g., brinjal, *Solanum melongena* and other members of Solanaceae; Members of the other families like Euphorbiaceae, Rubiaceae, Fabaceae, Cucurbitaceae Verbenaceae etc.

b. **Lateral**. The style develops from one side of the ovary (Fig. 2.137B), e.g., mango, *Mangifera indica* of Anacardiaceae; some members of Rosaceae etc.

c. **Gynobasic**. The style appears to arise from the base of the invaginated ovary (Fig.

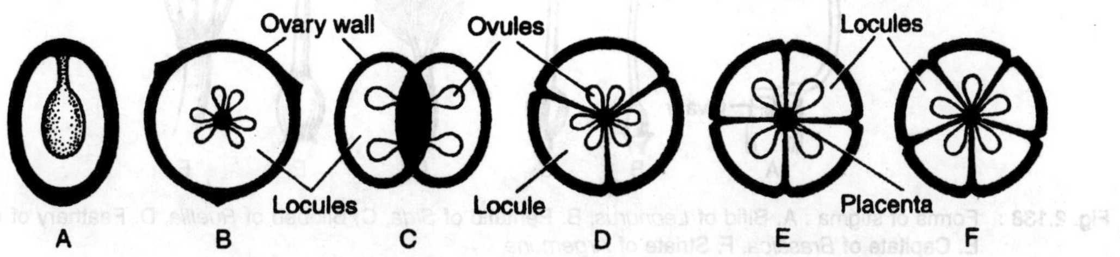

Fig. 2.136 : A. T.S. of monocarpellary gynoecium with 1-locule, B–F. T.S. of polycarpellary ovary showing number of locules. (B. Unilocular, C. Bilocular, D. Trilocular and E–F. Multilocular)

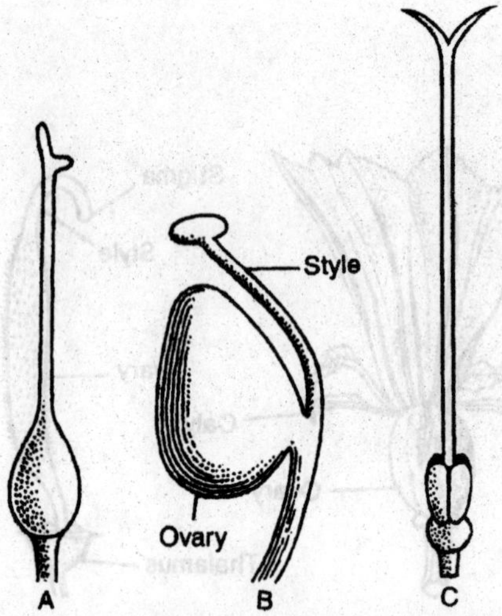

Fig. 2.137 : Different positions of style on the ovary : A. Terminal style of *Lantana camara* (Verbenaceae), B. Lateral style of *Mangifera indica* (Anacardiaceae), and C. Gynobasic style of *Leucas linifolia* of (Lamiaceae)

2.137C), e.g., members of the family Lamiaceae (Labiatae) and Boraginaceae.

Styles are usually **deciduous** i.e., they fall off or dried soon after fertilisation. In some cases, it may remain attached during development of fruit and is called **persistent**, e.g., *Clematis gouriana* and *Naravelia zeylanica* of Ranunculaceae. The style may be flat and petal-like i.e., **petaloid**, e.g., *Iris nepalensis* of Iridaceae, Indian shot, *Canna indica* of Cannaceae etc. Sometimes the base of the style is swollen and is known as **stylopodium**, e.g., coriander, *Coriandrum sativum* and other members of the family Apiaceae (Umbelliferae).

Stigma

The stigma is the receptive region of ovary, usually placed on the style (Fig. 2.138). But it may develop directly on the ovary when style is absent (i.e., **sessile**) as in *Berberis vulgaris* and other members of Berberidaceae; *Sambucus nigra* of Caprifoliaceae etc. The stigma may be **bifid** (*Leucas linifolia, Leonurus sibiricus* and other members of Labiatae, Fig. 2.138A), **trifid** (*Cyperus rotundus* of Cyperaceae), **pentafid** (*Sida cordifolia, Hisbiscus rosa-sinensis* and other members of Malvaceae, Fig. 2.138B), **bilobed** or **bilobite** (*Lindenbergia indica* of Scrophulariaceae, *Ruellia tuberosa* of Acanthaceae, Fig. 2.138C), **trilobed** or **trilobite** (*Cucurbita pepo* and other members of Cucurbitaceae), **capitate** i.e., head-like (*Ipomoea sinensis* and other members of Convolvulaceae, Fig. 2.138E), **laciniate** i.e., cut into narrow lobes of equal length (*Rumex maritimus* of Polygonaceae), **feathery** (*Oryza sativa* and other members of Poaceae, Fig. 2.138D), **striate** or **radiate** i.e., there are distinct rays of stigma in the form of a star (*Argemone mexicana* of Papaveraceae, Fig. 2.138F), **dumb-bell** shaped (*Catharanthus roseus, Alstonia scholaris* and other members of Apocynaceae).

In Orchidaceae, out of three stigmas, one becomes projected and is functionless (sterile). It bends over and conceals the receptive portion of the stigma. It is called **rostellum**.

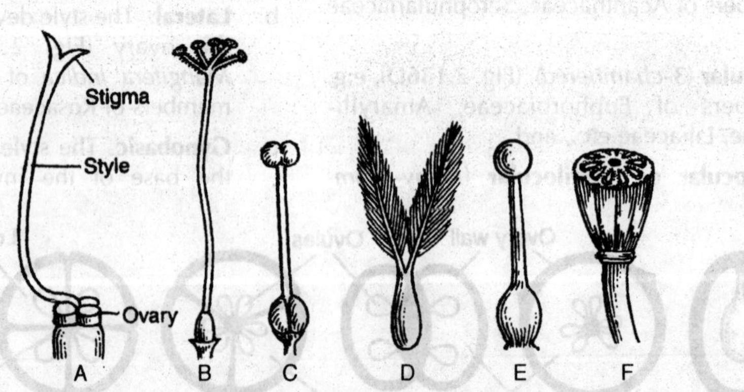

Fig. 2.138 : Forms of stigma : A. Bifid of *Leonurus*, B. Pentafid of *Sida*, C. Bilobed of *Ruellia*, D. Feathery of *Oryza*, E. Capitate of *Brassica*, F. Striate of *Argemone*

PLACENTATION

The ovules are attached on the inner surface of the ovarian wall on one or more cushions, called **placenta**. The distribution of placenta (pl. placentae) on the ovary wall and consequently the arrangement of ovules are called **placentation**.

Types of placentation

The placentation is of the following types :

1. **Marginal**. When single placenta develops along the junction of two fused margins of a single-chambered ovary, it is called *marginal*, e.g., *Mimosa pudica* (Fig. 2.139A); pea, *Pisum sativum* of Fabaceae (monocarpellary) and *Magnolia grandiflora* of Magnoliaceae (in each carpel of multicarpellary apocarpous gynoecium).
2. **Parietal**. The gynoecium is formed by the fusion of two or more carpels by their neighbouring margins forming unilocular compound ovary, where two or more placentae are developed on the ventral sutures, e.g., mustard, *Brassica campestris* (Fig. 2.139B_1) of Brassicaceae; *Argemone mexicana* (Fig. 2.139B_2) of Papaveraceae, *Cucumis sativus* of Cucurbitaceae (Fig. 2.139B_3) etc.
3. **Axile**. The gynoecium is formed by two or more carpels, the adjacent carpellary margins fuse and grow inwards to meet in the centre of the ovary forming an axis, thus making the ovary multichambered. The ovules are deve-

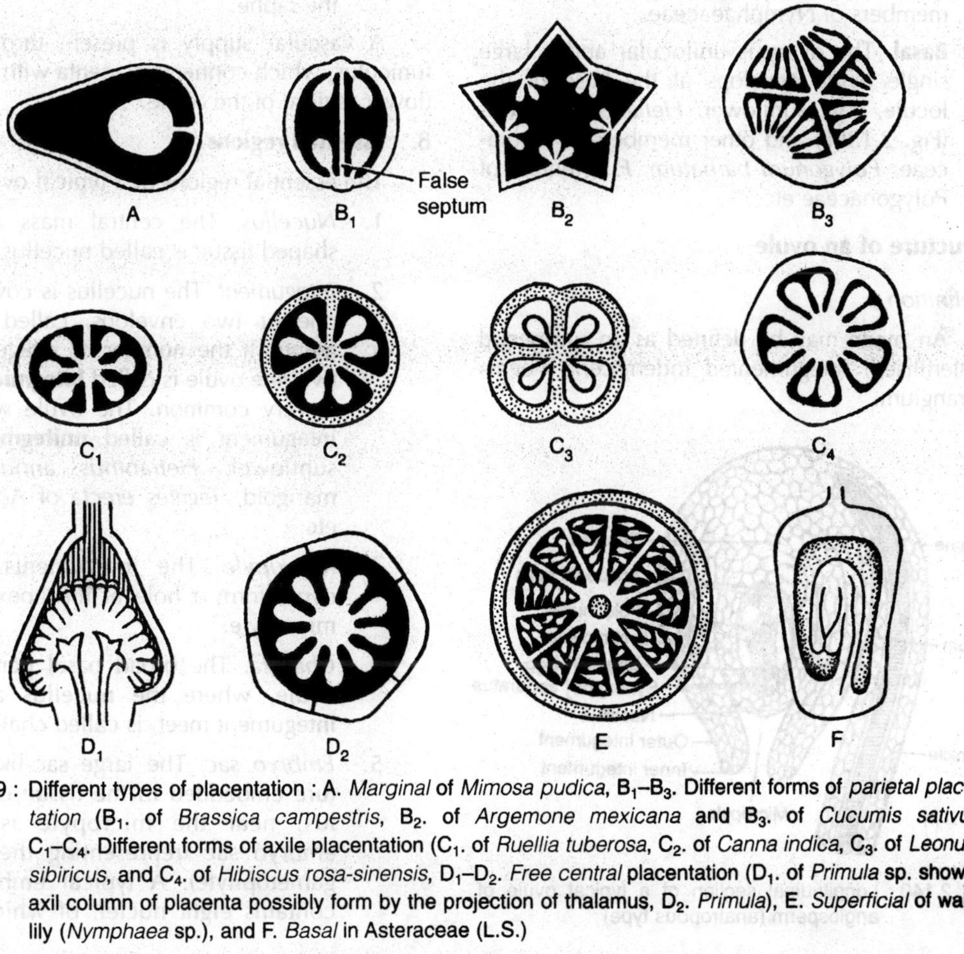

Fig. 2.139 : Different types of placentation : A. *Marginal* of *Mimosa pudica*, B_1–B_3. Different forms of *parietal placentation* (B_1. of *Brassica campestris*, B_2. of *Argemone mexicana* and B_3. of *Cucumis sativus*). C_1–C_4. Different forms of axile placentation (C_1. of *Ruellia tuberosa*, C_2. of *Canna indica*, C_3. of *Leonurus sibiricus*, and C_4. of *Hibiscus rosa-sinensis*, D_1–D_2. *Free central* placentation (D_1. of *Primula* sp. showing axil column of placenta possibly form by the projection of thalamus, D_2. *Primula*), E. *Superficial* of waler-lily (*Nymphaea* sp.), and F. *Basal* in Asteraceae (L.S.)

loped on the central axis, e.g., *Ruellia tuberosa* (Fig. 2.139C_1) of Acanthaceae; *Canna indica* (Fig. 2.139C_2) of Canaceae; *Leonurus sibiricus* (Fig. 2.139C_3) of Lamiaceae; *Solanum nigrum* of Solanaceae; *Hibiscus rosa-sinensis* (Fig. 2.139C_4) of Malvaceae etc.

4. **Free central.** The development of gynoecium is like that of axile type, but the partition walls are broken and thereby the ovary becomes unilocular and the ovules are borne on the axis in the centre of the ovary, e.g., *Primula* sp. (Fig. 2.139D_1, D_2), *Dianthus caryophyllus, Stellaria media* and many members of Caryophyllaceae.
5. **Superficial** or **Laminal.** The gynoecium is multicarpellary and syncarpous with whole inner wall of ovary lined with placentae and a large numbers of ovules are borne without any specific order, e.g., water-lily, *Nymphaea* sp. (Fig. 2.139E) and many other members of Nymphaeaceae.
6. **Basal.** The ovary is unilocular and a large single ovule develops at the base of the locule, e.g., sunflower, *Helianthus annuus* (Fig. 2.139F) and other members of Asteraceae; *Polygonum barbatum, P. orientale* of Polygonaceae etc.

Structure of an ovule

Definition

An ovule may be defined as an unripened (unfertilised) integumented indehiscent megasporangium.

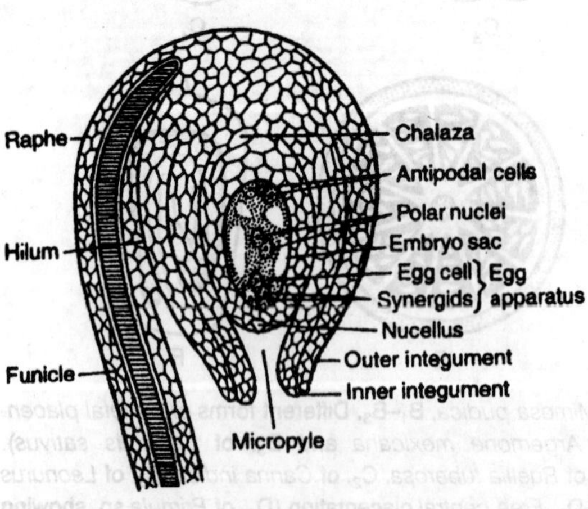

Fig. 2.140 : Longitudinal section of a typical ovule of angiosperm (anatropous type)

Structure

The structure of commonly available ovule i.e., **anatropous** (Fig. 2.140) or inverted type is described below. It is differentiated into two regions : A. **Accessory regions**, and B. **Essential regions.**

A. **Accessory regions.**

The accessory regions are :

1. *Funicle* or *funiculus.* It is a stalk-like structure by which the ovules remain attached with the placenta.
2. *Hilum.* It is the point of attachment of the funiculus to the body of the ovule.

An ovule may be devoid of funicle and remains attached directly with the placenta, is called *sessile.*

3. *Raphe.* The extended region of the funicle beyond the hilum, which remains closely attached with the ovule wall, is the raphe.

A vascular supply is present through the funiculus, which connects placenta with chalaza (lower region) of the ovule.

B. **Essential regions.**

The essential regions of a typical ovule are :

1. *Nucellus.* The central mass of oval-shaped tissue is called nucellus.
2. *Integument.* The nucellus is covered by one or two envelops, called integuments. If the number of integument is two, the ovule is called **bitegmic**, which is very common. The ovule with one integument is called **unitegmic**, e.g., sunflower, *Helianthus annuus* and marigold, *Tagetes erecta* of Asteraceae etc.
3. *Micropyle.* The integuments of the ovule form a hole at the apex, called micropyle.
4. *Chalaza.* The broad basal part of the ovule, where the nucellus and the integument meet, is called chalaza.
5. *Embryo sac.* The large sac-like structure embedded in the tissue of nucellus, near the micropyle is called embryo sac (representing the female gametophyte). A typical embryo sac contains eight nuclei, of which three

towards the micropyle, three towards the chalaza and the rest two are in the centre. The micropylar three nuclei consist of a larger **egg** or **ovum** or **oosphere** and two **synergids** or **auxiliary cell**, one on each side of the egg. The egg and synergids together constitute the **egg apparatus**. At the chalazal end, the three cells are called **antipodals** or **antipodal cells**. The central two nuclei are called polar nuclei; those subsequently fuse together and form a **secondary nucleus** or **definitive nucleus** or **polar fusion nucleus**.

After fertilisation, the ovule develops into seed. In **angiosperm**, the ovule remains protected by the ovary, thus seeds are covered, while in **gymnosperm**, ovules are not covered, thus the seeds remain naked.

Types of ovules

Usually, there are six types of ovules (Fig. 2.141) found in angiosperms. These are :

1. **Orthotropous** or **straight** or **atropous.** The ovule is erect so that the funiculus, chalaza and micropyle lie on a straight line. The chalaza lies at the base and the *micropyle at the apex* of the ovule, e.g., *Polygonum orientale, Rumex vesicarius* and other members of Polygonaceae; members of Piperaceae, Nyctaginaceae, Urticaceae etc.
2. **Anatropous** or **inverted**. The body of the ovule becomes completely inverted (*micropyle is directed towards the base*), so that the micropyle and hilum come very close to each other and a part of the funiculus is fused with the body of the ovule forming the **raphe**, e.g., common in the members of Liliaceae, Amaryllidaceae, Lamiaceae, Boraginaceae, Euphorbiaceae etc.
3. **Campylotropous**. When the ovule is curved in such a way that the micropyle is directed towards chalaza and the chalaza is situated at right angles to funicle, e.g., members of Brassicaceae, Fabaceae etc.
4. **Amphitropous**. The ovule is placed transversely at right angles to the funiculus, e.g., *Lemna* of Lemnaceae, *Ranunculus* of Ranunculaceae etc.
5. **Hemianatropous** or **hemitropous**. The ovule body is half-inverted, so that the embryo sac becomes curved like a horse shoe, e.g., members of Butomaceae, Portulaccaceae, Alismaceae etc.

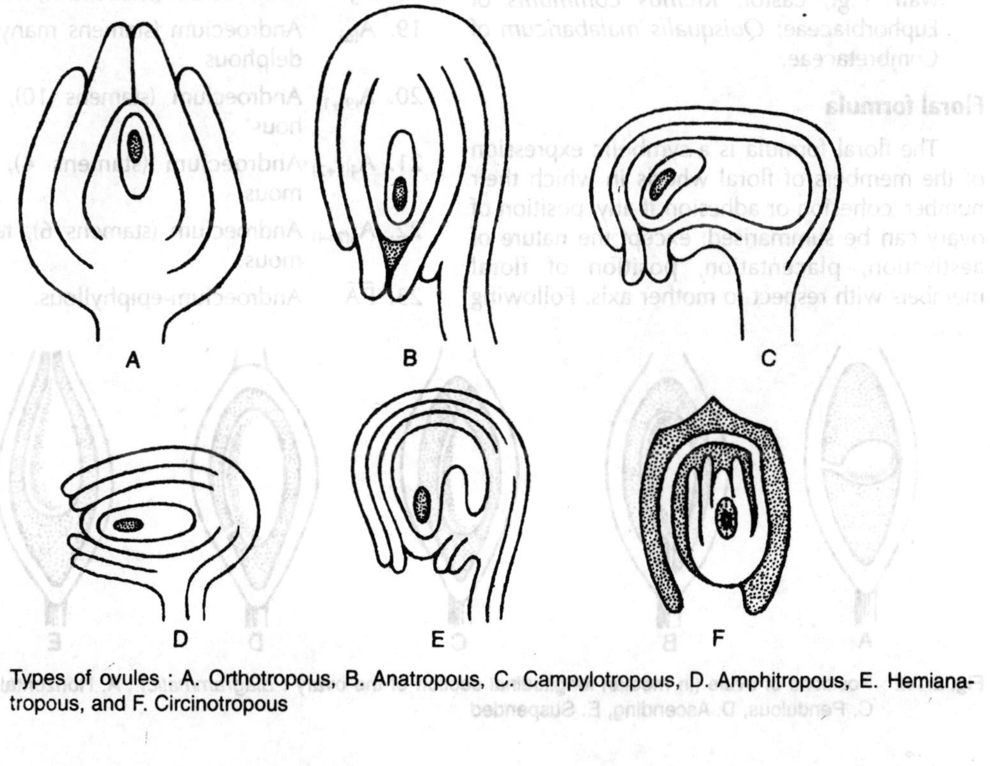

Fig. 2.141 : Types of ovules : A. Orthotropous, B. Anatropous, C. Campylotropous, D. Amphitropous, E. Hemianatropous, and F. Circinotropous

6. **Circinotropous**. The funicle is very much long and forms a complete circle around the ovule, which is free from it except for a small area at the end of the funicle, e.g., members of Plumbaginaceae and Cactaceae (*Opuntia* sp.)

Position of ovule

The position of ovule in the ovarian chamber may be of various types (Fig. 2.142) :

1. *Horizontal*. The ovule grows horizontally from one side of the ovarian wall, e.g., *Podophyllum emodi* of Berberidaceae.
2. *Erect*. The ovule grows from the base of the ovary and remains erect, e.g., members of Asteraceae (compositae), Polygonaceae etc.
3. *Pendulous*. The ovule grows from the side or from the top of ovarian wall and hangs downwardly, e.g., *Linum usitatissimum* of Linaceae, members of Euphorbiaceae, Annonaceae etc.
4. *Ascending*. The ovule grows from the side of the ovarian wall and directed upward, e.g., some members of Ranunculaceae.
5. *Suspended*. The ovule grows perpendicularly downward from the top of the ovarian wall, e.g., castor, *Ricinus communis* of Euphorbiaceae; *Quisqualis malabaricum* of Combretaceae.

Floral formula

The floral formula is a symbolic expression of the members of floral whorls in which their number, cohesion or adhesion if any, position of ovary can be summarised, except the nature of aestivation, placentation, position of floral members with respect to mother axis. Following symbols are used in constructing the floral formula :

1. ¢ Male flower.
2. ″ Female flower.
3. \` or ⚥ Bisexual flower.
4. ⊕ Actinomorphic flower.
5. ·|· or ↓ Zygomorphic flower.
6. Ep Epicalyx.
7. K Calyx.
8. Kp Calyx represented by pappus.
9. K_5 Calyx (sepals 5), free i.e., polysepalous.
10. $K_{(5)}$ Calyx (sepals 5), united i.e., gamosepalous.
11. C Corolla.
12. C_5 Corolla (petals 5), free i.e., polypetalous.
13. $C_{(5)}$ Corolla (petals 5), united i.e., gamopetalous.
14. P Perianth.
15. P_5 Perianth (tepals 5), free.
16. $P_{(5)}$ Perianth (tepals 5), united.
17. A Androecium.
18. A_5 Androecium (stamens 5), free.
19. $A_{(\alpha)}$ Androecium (stamens many), monadelphous.
20. $A_{(9)+1}$ Androecium (stamens 10), diadelphous
21. $A_{4(2+2)}$ Androecium (stamens 4), didynamous.
22. $A_{6(2+4)}$ Androecium (stamens 6), tetradynamous.
23. $\overline{PA}$ - Androecium-epiphyllous.

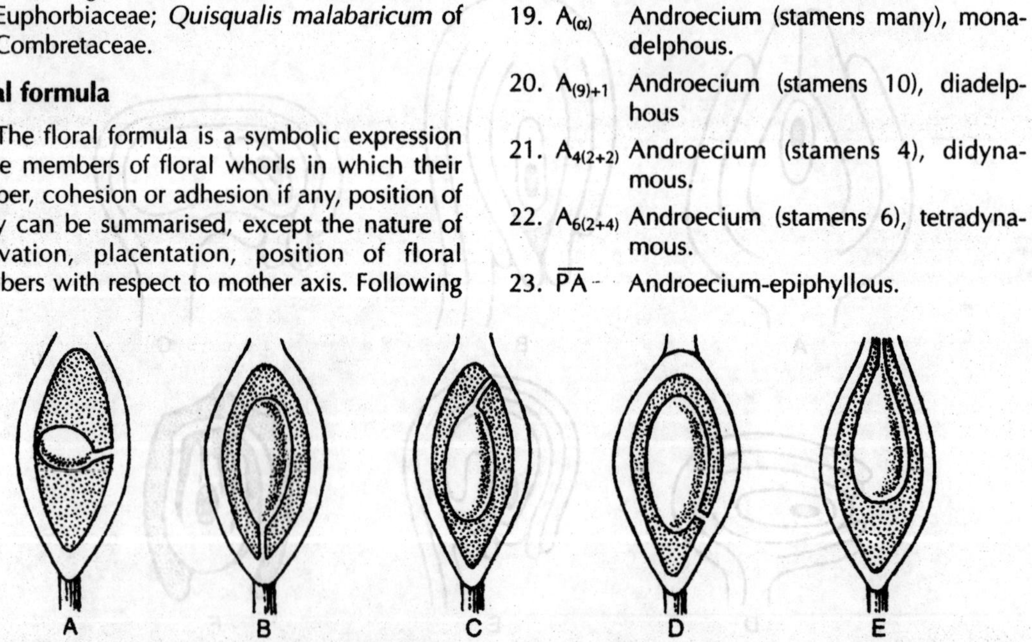

Fig. 2.142 : Positions of ovule (in median longitudinal section of the ovary : *diagrammatic*) : A. Horizontal, B. Erect, C. Pendulous, D. Ascending, E. Suspended

24. $\overline{CA}$ Androecium-epipetalous.
25. G Gynoecium.
26. G_2 Gynoecium (carpels 2), free i.e., apocarpous.
27. $G_{(2)}$ Gynoecium (carpels 2), united i.e., syncarpous.
28. $\underline{G}_{(2)}$ Gynoecium (carpels 2), united i.e., syncarpous; ovary-superior.
29. $\overline{G}_{(2)}$ Gynoecium (carpels 2), united i.e., syncarpous; ovary-inferior.
30. $\overline{AG}$ Gynostegium.

The floral formula of some common plants of different families are given :

A. Monocotyledons :

1. Onion, *Allium cepa* (Liliaceae) :
   $\oplus$ ⚥ $P_{6(3+3)}A_{6(3+3)}\underline{G}_{(3)}$
2. Tuberose, *Polyanthes tuberosa* (Amaryllidaceae) : $\oplus$ ⚥ $P_{6(3+3)}A_{6(3+3)}\overline{G}_{(3)}$
3. Banana, *Musa balbisiana* (Musaceae) :
   ·|· ⚥ $P_{6(3+3)}A_5\overline{G}_{(3)}$
4. Paddy, *Oryza sativa* (Poaceae) :
   $\oplus$ ⚥ $P_0A_{3+3}\underline{G}_1$
5. Rasna, *Vanda tessellata* (Orchidaceae)
   ·|· ⚥ $P_{3+3}A_1\overline{G}_{(3)}$

B. Dicotyledons :

1. Black mustard, *Brassica nigra* (Brassicaceae) :
   $\oplus$ ⚥ $K_{2+2}\,C_4\,A_{6(2+4)}\,\underline{G}_{(2)}$
2. Sun-hemp, *Crotalaria juncea* (Fabaceae) :
   ·|· ⚥ $K_{(5)}\,C_{(5)}\,A_{(10)}\,\underline{G}_1$
3. Motherwort, *Leonurus sibiricus* (Lamiaceae) :
   ·|· ⚥ $K_{(5)}\,\widehat{C_{(5)}\,A_4}\,\underline{G}_{(2)}$
4. Nighshade, *Solanum nigrum* (Solanaceae) :
   $\oplus$ ⚥ $K_5\,\widehat{C_{(5)}\,A_5}\,\underline{G}_{(2)}$
5. Gourd, *Cucurbita maxima* (Cucurbitaceae) :
   Male flower. $\oplus$ ¢ $K_{(5)}\,C_{(5)}\,A_{\underline{(5)}}$
   Female flower. $\oplus$ " $K_{(5)}\,C_{(5)}\,\overline{G}_{(3)}$
6. Golden champa, *Michelia champaca* (Magnoliaceae) : $\oplus$ ⚥ $P_\alpha A_\alpha \underline{G}_\alpha$

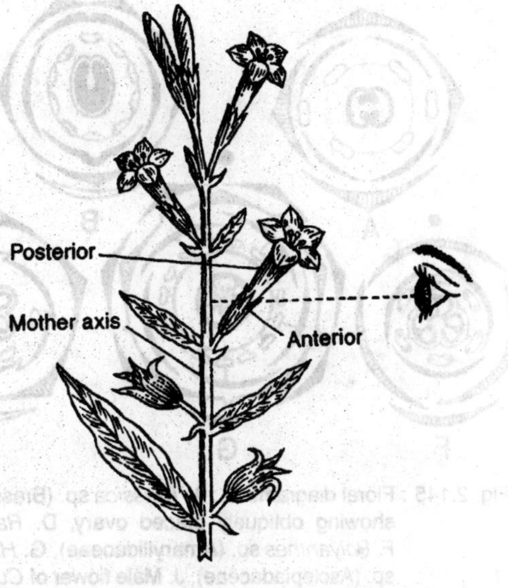

Fig. 2.143 : Flowering twig of *Nicotiana plumbaginifolia*, showing mother axis, anterior and posterior side of flower

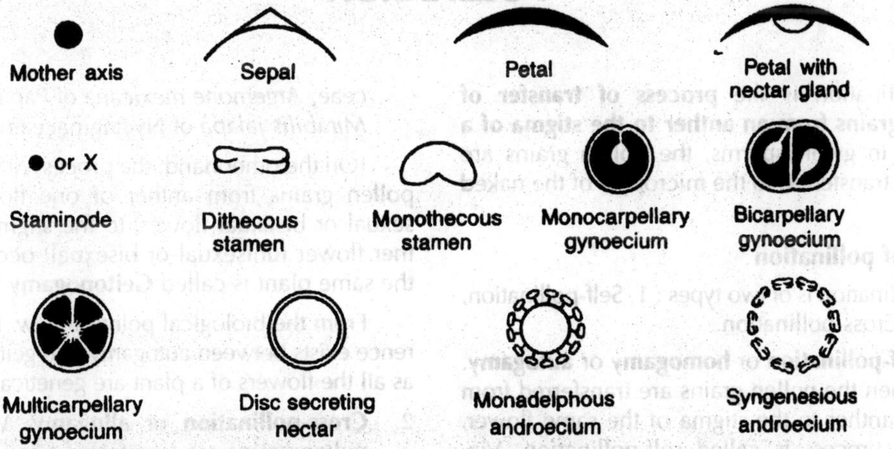

Fig. 2.144 : Signs used in floral diagram

7. Swallow-wort, *Calotropis gigantea* (Asclepiadaceae) : $\oplus$ ⚥ $K_{(5)}C_{(5)}\widehat{A_5G}_{(2)}$
8. Climbing tempweed, *Mikania scandens* (Asteraceae) Disc floret : $\oplus$ ⚥ $K_p\overline{C_{(5)}A_{(5)}}\,\overline{G}_{(2)}$

Floral diagram

Floral diagram illustrates the number and relative position of floral whorls with reference to mother axis (Fig. 2.143).

The flowers can be imaginarily dissected by different planes (see Fig. 2.101, page 206). All the characteristics of a flower can be depicted in a floral diagram, except the nature of insertion of floral members with respect to thalamus (i.e., hypogyny, perigyny or epigyny).

The various signs (Fig. 2.144) are used to construct a floral diagram.

Floral diagram of some plants are given below (Fig. 2.145) :

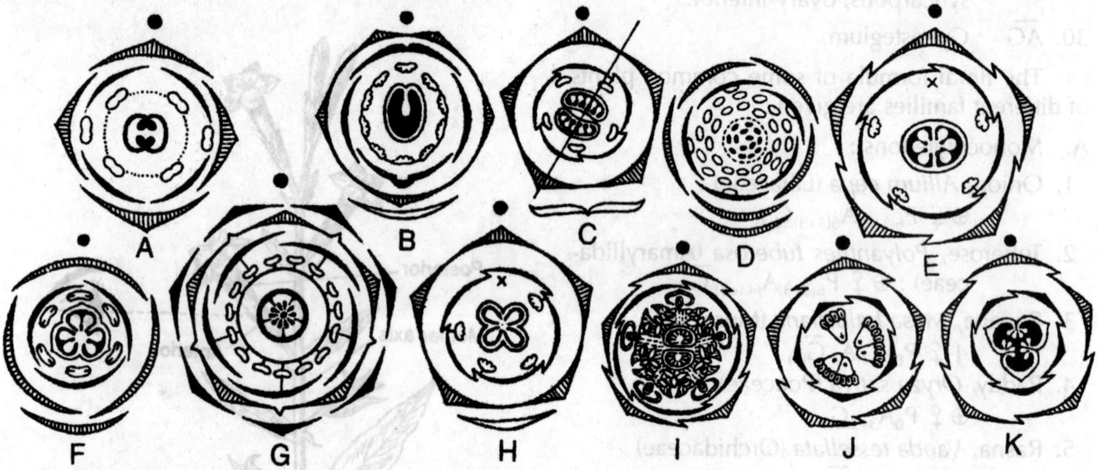

Fig. 2.145 : Floral diagrams of : A. *Brassica* sp. (Brassicaceae), B. *Pisum* sp. (Fabaceae), C. *Solanum* sp. (Solanaceae) showing obliquely placed ovary, D. *Ranunculus* sp. (Ranunculaceae), E. *Ruellia* sp. (Acanthaceae), F. *Polyanthes* sp. (Amaryllidaceae), G. *Hibiscus* sp. (Malvaceae), H. *Leonurus* sp. (Labiatae), I. *Calotropis* sp. (Asclepiadaceae), J. Male flower of *Cucurbita* sp. (Cucurbitaceae), and K. Female flower of *Cucurbita* sp.

POLLINATION

Pollination is the **process of transfer of pollen grains from an anther to the stigma of a flower**. In gymnosperms, the pollen grains are directly transferred to the micropyle of the naked ovule.

Types of pollination

Pollination is of two types : 1. Self-pollination, and 2. Cross-pollination.

1. **Self-pollination** or **homogamy** or **autogamy**. When the pollen grains are transferred from an anther to the stigma of the same flower, the process is called self-pollination. Viz. *Commelina benghalensis* of Commelinaceae, *Argemone mexicana* of Papaveraceae, *Mirabilis jalapa* of Nyctaginaceae etc.

[On the other hand, the process of transfer of pollen grains from anther of one flower (unisexual or bisexual flower) to the stigma of another flower (unisexual or bisexual) occurring on the same plant is called **Geitonogamy**.]

From the biological point of view, little difference exists between autogamy and geitonogamy, as all the flowers of a plant are genetically same.

2. **Cross-pollination** or **allogamy**. When the pollen grains are transferred from an anther of one flower to the stigma of another flower

borne on different plants of same or different species, the process is called cross-pollination. Viz. *Michelia champaca* of Magnoliaceae, *Leonurus sibiricus* of Lamiaceae etc.

[The cross-pollination taking place between two different plants of the same species is called **Xenogamy** and between two different species is called **Hybridism.**]

Contrivances for self-pollination

Self-pollination occurs only in bisexual flowers and never in unisexual ones.

There are several **devices** which ensure self-pollination :

1. **Cleistogamy**. Some plants produce small and closed bisexual flowers, in addition to normal flowers. These small flowers remain either on or under the ground and never open and so self-pollination is obvious. These flowers are said to be **cleistogamous** or **cleistogenes**, e.g., *Commelina benghalensis* (Fig. 2.146) of Commelinaceae, sundew, *Drosera burmanii* of Droseraceae, *Oxalis corniculata* of Oxalidaceae, *Impatiens balsamina* of Balsaminaceae, *Polygala polygama* of Polygalaceae etc. It is of **two** types :
   a. **Obligate** or **habitual cleistogamy**. The flowers are typically cleistogamous and never open, e.g., small underground flowers of *Commelina benghalensis* of Commelinaceae etc.
   b. **Facultative** or **pseudo-cleistogamy**. The flowers remain open for a short period of time and then they become closed permanently, e.g., sundew, *Drosera burmanii* of Droseraceae, *Oxalis corniculata* of Oxalidaceae, *Portulaca oleracea* of Portulacaceae etc.

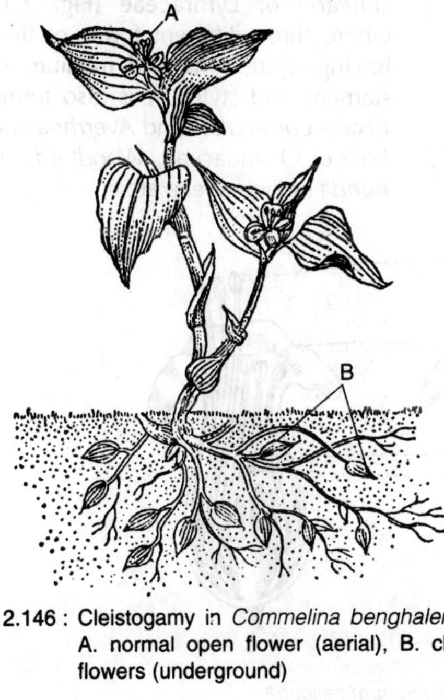

Fig. 2.146 : Cleistogamy in *Commelina benghalensis* : A. normal open flower (aerial), B. closed flowers (underground)

2. **Homogamy**. When anthers and stigmas become matured almost at the same time, the self-pollination takes place. This condition is called homogamy.

The **chasmogamous** flowers (i.e., the flowers normally open during anthesis) are commonly cross-pollinated, but due to failure of cross-pollination they may undergo self-pollination by one of the following processes :

a. The style and filament get spirally coiled, e.g., *Clerodendrum viscosum* of Verbenaceae, *Mirabilis jalapa* of Nyctaginaceae etc.
b. The style elongates to receive the pollen, e.g., *Gardenia florida* of Rubiaceae.
c. The stigma either contracts or bends to come in close contact with the anthers, e.g., *Grewia subinaequalis* of Tiliaceae.
d. The stigma curls back to reach the anthers, e.g., sunflower, *Helianthus annuus* of Asteraceae etc.

Contrivances for cross-pollination

There are several devices which ensure cross-pollination :

1. **Dicliny** or **unisexuality**. When unisexual (diclinous) flowers i.e., staminate or male flowers and pistillate or female flowers are borne on the different plants of the same species (**dioecious plant**), then the cross-pollination is a must, e.g., *Bryonia dioica* and *Trichosanthes dioica* of Cucurbitaceae; *Carica papaya* of Caricaceae; *Borassus flabellifer* of Arecaceae etc.

   In **monoecious plants**, however, *geitonogamy* may take place, e.g., maize, *Zea mays* of Poaceae; gourd, *Cucurbita maxima* of Cucurbitaceae etc.

2. **Dichogamy**. In some bisexual flowers, the stamens and carpels do not mature at the

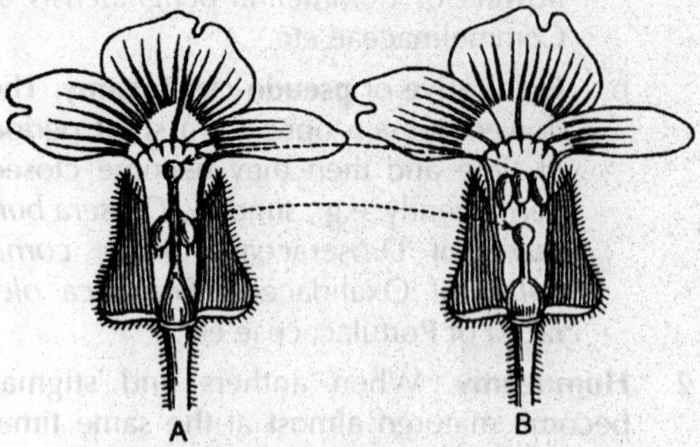

Fig. 2.147 : Dimorphic flowers of *Primula sinensis*

same time, thereby the self-pollination becomes hindered. It is of two types :

a. *Protandry* or *proterandry*. The anthers of a flower mature earlier than the carpels, e.g., *Leonurus* sp. of Lamiaceae (Labiatae), *Helianthus annuus* of Asteraceae (Compositae), *Hibiscus* sp. of Malvaceae etc.

b. *Protogyny* or *proterogyny*. The carpels of a flower mature earlier than the anthers, e.g., *Magnolia grandriflora* and *Michelia champaca* of Magnoliaceae, *Pedilanthus tithymaloides* and *Poinsettia pulcherrima* of Euphorbiaceae etc.

3. **Self-sterility** or **incompatibility**. In this condition, the pollen grains are not able to fertilise the ovule of the same flower as they failed to germinate on the stigma, e.g., *Reseda odorata* of Resedaceae, a few fruit trees of Rosaceae, *Solanum tuberosum* of Solanaceae, *Passiflora* sp. of Passifloraceae etc.

4. **Herkogamy**. In some flowers, due to some physical barriers between anther and style, the pollination becomes impossible. In orchids and members of Asclepiadaceae like *Calotropis procera*, where the *pollens are aggregated in pollinia*, the pollination is completely depends on the mercy of the insects. Due to the *extrorse* anthers in *Gloriosa superba* of Liliaceae (Fig. 2.120C), the pollens become out of reach of its own stigma.

5. **Heteromorphism**. In some plants, the flowers are of two (**dimorphic**) or three (**trimorphic**) different forms having anthers and stigmas developed at different levels. The di- or trimorphism usually involves **heteroanthy** (i.e., anthers of different types) and **heterostyly** (i.e., styles of different lengths). Thus, the cross-pollination takes place between stamens and stigmas of same length.

a. The ***dimorphism*** is found in *Primula sinensis* (Fig. 2.147) of Primulaceae, where one type having long stamen and a short style and the other having short stamen and a long style. It is also found in *Polygonum orientale* and *Fagopyrum esculentum* of Polygonaceae, *Biophytum sensitivum* of Oxalidaceae etc.

b. The ***trimorphism*** is found in *Lythrum salicaria* of Lythraceae (Fig. 2.148), where three different forms of flowers having short, long and medium sized stamens and styles. It is also found in *Oxalis corniculata* and *Averrhoa carambola* of Oxalidaceae, *Woodfordia floribunda* of Lythraceae etc.

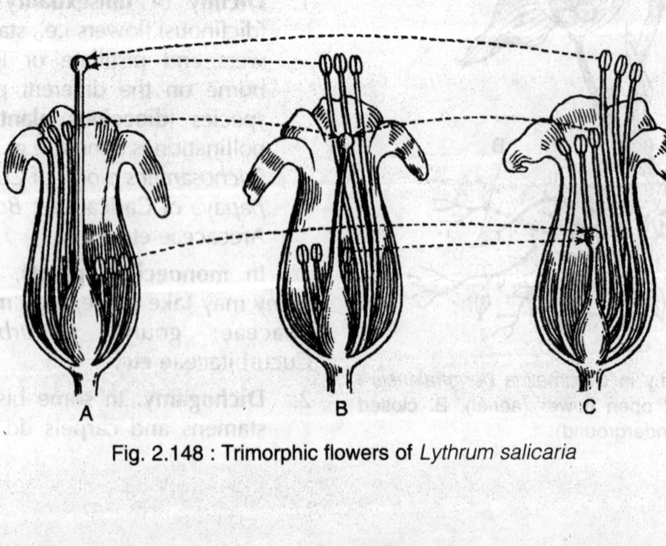

Fig. 2.148 : Trimorphic flowers of *Lythrum salicaria*

Agents for cross-pollination

Different agents can transfer pollen grains from anther of one flower to the stigma of other flowers. The agents are : **Wind** (anemophily); **water** (hydrophily); **insects** (entomophily); **slugs, snails** and **squirrels** (malacophily); **birds** (ornithophily); **bat** (chiropteriphily). There are different types of pollination based on the pollinating agents :

1. **Anemophily** or **anemogamy**. The pollination with the help of **wind** is called *anemophily* or *anemogamy*. The anemophilous plants produce enormous amount of pollen grains. The pollen grains are *small, smooth, dry* and *light in weight*. Pollens of such plants are blown off at a distance more than 1,000 km. A single plant of *Mercurialis annua* has been estimated to produce more than 135 crores of pollen grains. The pollen grains of *Pinus* (Gymnosperm) are *winged* and help in wind pollination. In grasses, the flowers may be *borne on long axis situated much above the leaves* and the *anthers are versatile,* thereby the anther can oscillate in all directions at the tip of the filament. On the other hand, the flowers have adequate device to catch the air-borne pollen grains. In grasses, the stigma is usually large and feathery, which helps to catch the pollen grains. In *Typha,* the stigma is brush-like which helps to catch more pollen grains.

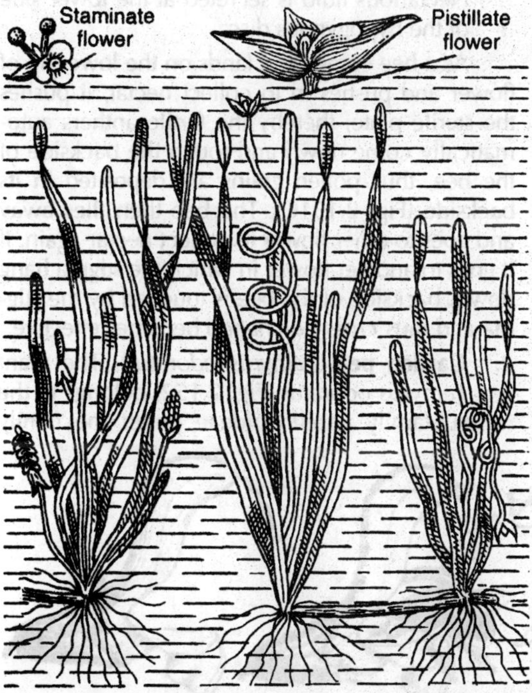

Fig. 2.149 : Pollination (epihydrogamy) in *Vallisneria spiralis*

2. **Hydrophily** or **hydrogamy**. The pollination with the help of water is called *hydrophily* or *hydrogamy*. It is of two types : a. Hypohydrogamy and b. Epihydrogamy.

   a. **Hypohydrogamy**. When the pollination takes place **inside the water**, it is called *hypohydrogamy,* e.g., *Najas* sp., *Ceratophyllum* sp. etc.

   b. **Epihydrogamy**. When pollination takes place **on the water surface**, it is called *Epihydrogamy,* e.g., *Vallisneria spiralis* (ribbon weed). In *Vallisneria* (Fig. 2.149), *the* flowers are borne under water. After maturation, the pistillate flowers are brought to the surface by their long stalk and form a cup-like depression. The male flowers get detached from the parent plant and float on the surface of water. If male flower gets lodged into the depression, pollination occurs. After pollination, the elongated stalk of the pistillate flower undergoes spiral coiling, thereby it again comes under the water.

3. **Zoophily** or **zoogamy** or **zoidiophily**. When pollination takes place with the help of

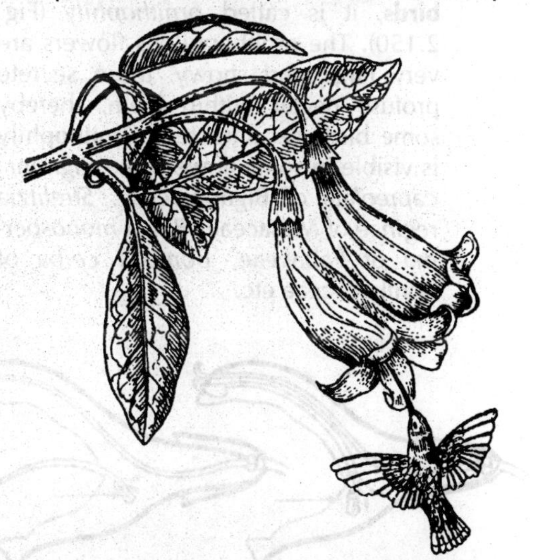
Fig. 2.150 : Pollination by humming bud during collection of nectar from *Bignonia capreolata* (After Bates)

animals, it is called *Zoophily*. It is of the following types :

a. **Entomophily** or **entomogamy**. When pollination takes place with the help of **insects**, it is called *entomophily*. The insects those help in pollination are bees, flies, beetles etc. The flowers which open during sunrise are generally brightly coloured, acting as flag apparatus for attracting insects. But the flowers those open after sunset are white in colour, thus become visible in night. In addition to colour, other devises to attract the insects are smell and/or nectar. The pollen grains are larger in size, the exine is pitted, spiny etc., so they can be adhered firmly on the sticky stigma. Approximately, 80% of the pollination done by the insects is carried by bees.

b. **Malacophily** or **malacogamy**. When pollination takes place with the help of **other animals like slugs, snails, squirrels etc.**, it is called *malacophily*. During their visit from one tree to other, the slugs help in pollination. Like wise, snails and squirrels also pollinate some aroid flowers.

c. **Ornithophily** or **ornithogamy**. When pollination takes place with the help of **birds**, it is called *ornithophily* (Fig. 2.150). The ornithophilous flowers are very large and showy, those secrete profuse nectar within them, thereby some birds get attracted. Ornithophily is visible in some flowers like *Bignonia capreolata* of Bignoniaceae, *Strelitzia reginae* of Musaceae, *Butea monosperma* of Fabaceae, *Bombax ceiba* of Bombacaceae etc.

d. **Chiropteriphily** or **chiropterigamy**. When pollination takes place with the help of **bats**, it is called *chiropteriphily*. The flowers of *Bombax ceiba* of Bombacaceae, *Anthocephalous cadamba* of Rubiaceae etc., are pollinated by bat.

Mechanism of pollination in some common plants

1. **Garden sage** (*Salvia plebeja*) : The plant *Salvia plebeja* (Fig. 2.151) belongs to the family Lamiaceae (Labiatae) having bisexual flower with bilabiate corolla (with 2-lobed upper lip and 3-lobed lower lip). Two stamens are attached to the corolla tube and out of two anther lobes of each anther, one half is sterile and other one is fertile (Fig. 2.151A,B). The sterile halves of both the anthers jointly form a sterile plate of tissue which is placed above the lower lip near the mouth of the flower. The fertile anther lobe and the sterile anther knob are connected to one another by a curved connective. The fertile anther lobes (one from each anther) are situated just under the upper corolla lobe. The lower lip of corolla forms a landing stage for the insects (bees). Nectarious fluid is secreted at the lower side of the hypogynous disc.

As a bee comes and lands on the lower lip of flower and pushes in to collect nectar, it pushes the sterile plate; thereby the fertile anthers automatically come down and touch the backside of the bee, thus pollen grains are deposited in its backside (Fig. 2.151C). The bee then flies away and goes to other flower to collect nectar again. If it goes to another flower in which the stigma hang down, backside of the bee is touched by the stigma and thus *cross-pollination* becomes effective.

2. **Garden pea** (*Lathyrus odoratus*). The plant *Lathyrus odoratus* (Fig. 2.152) belongs to the subfamily Papilionoideae under the family

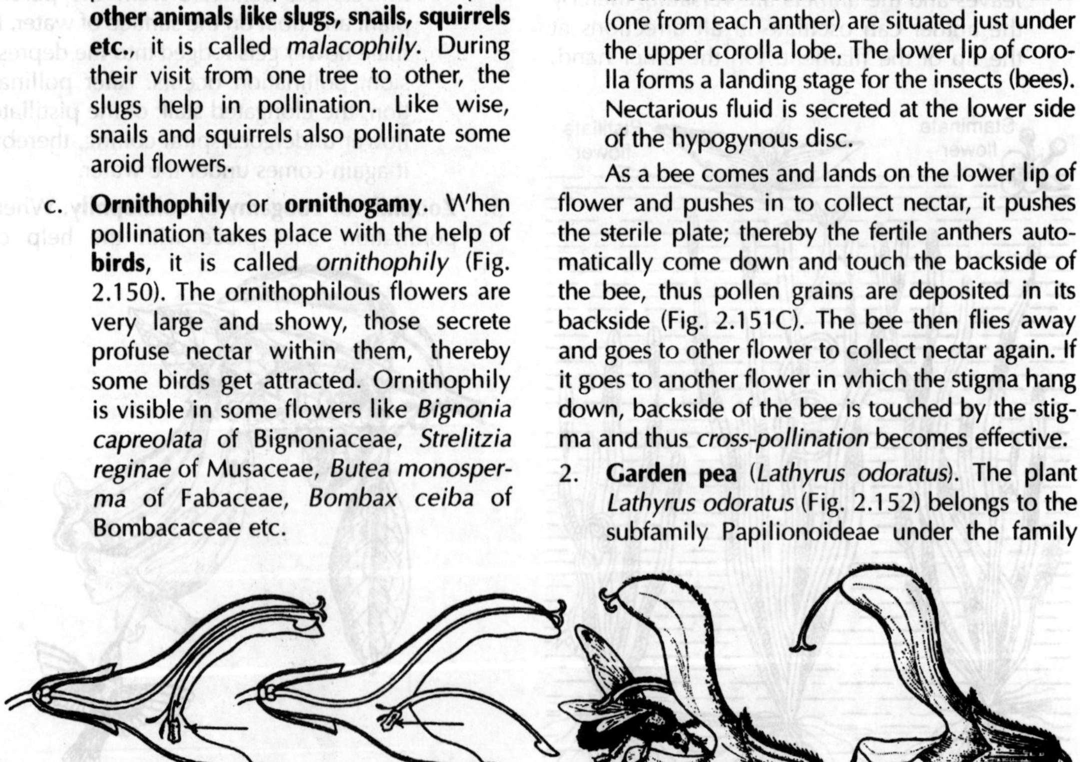

Fig. 2.151 : Pollination in *Salvia plebeja* : A & B. L.S. of flower showing immature pistil and movement of stamen when pressed in the direction of arrow, C. A bee becoming dusted with pollen, and D. A flower with mature stigma

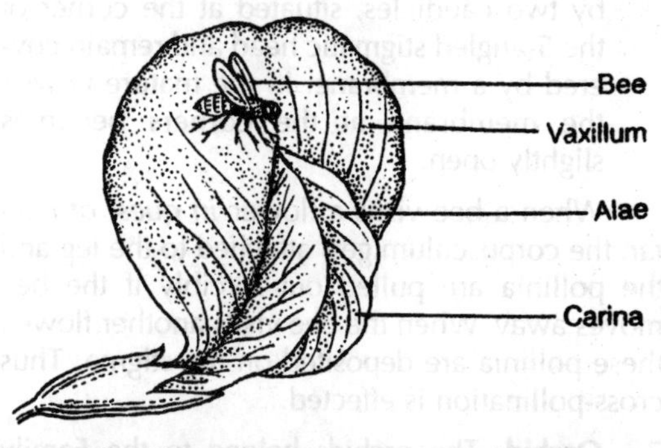

Fig. 2.152 : Pollination in garden pea (*Lathyrus* sp.), a bee is going to land on the alae (wings)

Fabaceae, having papilionaceous corolla. The flowers are variously coloured and sweet-scented. The flowers are adapted for *cross-pollination* by insects, commonly by bee. The *vexillum* is the flag apparatus, the *alae* (wings) are the landing stage for the insects. The flower is constructed in such a way that the weight of the bee on wings causes the *keel* (carina) to come down and thereby the hidden anthers and stigma are jerked out and come in contact with the body of the insect, where pollen becomes dusted. The bee then flies away and both the sex organs go back to their previous position.

When the bee visits another flower, some pollen from its body gets attached to the stigma and pollination occurs. The visitor bee may also receive fresh pollen from the second plant.

3. **Fig** (*Ficus carica*). The plant *Ficus carica* (Fig. 2.153) belongs to the family Moraceae, having hypanthodium inflorescence. The flowers are adapted for *cross-pollination* by insect, the gall wasp (*Blastophaga*). There are three types of flowers : male, female (long-styled) and gall female (short-styled). There is a narrow orifice at the top of the hypanthodium through which insects can enter. The male flowers are situated near the orifice, while long-styled female flowers and short-styled gall flowers are towards the base. The gall wasp crawls into the receptacle and lays eggs inside the ovule of gall flowers. The eggs develop larvae which feed on ovule and form galls. The larva then develops into a mature wasp which comes out of the hypanthodium.

During coming out of the hypanthodium, the newly developed wasp brushes against male flowers and carries the pollen grains. When it visits another inflorescence, it pollinates the long-styled female flowers and lays egg within the gall flower.

4. **Birthwort** (*Aristolochia indica*). The plant *Aristolochia indica* (Fig. 2.154) belongs to the family Aristolochiaceae having flowers with a long perianth tube, swollen at the

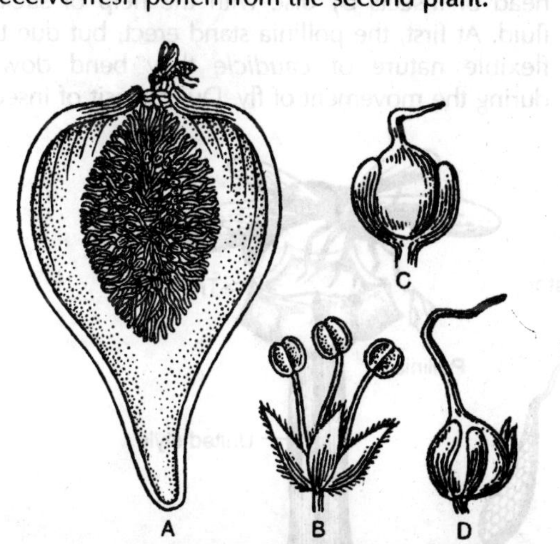

Fig. 2.153 : A. L.S. of fig (*Ficus carica*) showing the gall wasp (*Blastophaga*) entering through orifice, B. Male flower, C. Gall flower, and D. Female flower

Fig. 2.154 : Young and upright flower of *Aristolochia* sp. with mature stigmas. Small insects can crawl down but cannot come out because of the hairs pointing downward

base. The swollen part encloses the sex organs. The inner side of the tube is slippery and with deflected hairs. The flowers are protogynous (i.e., gynoecium matures earlier than anthers).

Small insects (flies) are attracted by the foul odour and bright colour of the flowers. The deflected hairs allow the entry of the flies into the base, but prevent their escape. During its attempt to come out from the flower, it rubs against the stigma and thereby pollination is effected, if the fly carries pollen of other flower. Later on, the hairs wither, enabling the fly to escape.

5. **Tape grass** (*Vallisneria spiralis*). In *Vallisneria* (Fig. 2.149), the flowers are borne under water. After maturation, the pistillate flowers are brought to the surface by their long stalk and form a cup-like depression. The male flowers get detached from the male plant and float on the surface of water. If the male flower gets lodged into the depression, pollination occurs. After pollination, the elongated stalk of the pistillate flower undergoes spiral coiling, thereby it again goes under the water.

6. **Swallow-wort** (*Calotropis gigantea*). The plant *Calotropis gigantea* (Fig. 2.155) belongs to the family Asclepiadaceae; it shows an elaborate mechanism for *cross-pollination* by insects. In flowers, there are corona-like appendages which are associated with the secretion of nectar. Pollens are enclosed in pollinia. The two pollinia of different anthers get attached to a common gland, the **corpusculum** or **retinaculum** (composed of glandular adhesive discs connected by translator), by two caudicles, situated at the corner of the 5-angled stigmatic head and remain covered by a membrane. In the mature flower, the membrane at the corners becomes slightly open.

When a bee visits a flower *in quest of nectar*, the corpusculum gets attached to the leg and the pollinia are pulled out forcibly if the bee moves away. When the bee visits another flower, these pollinia are deposited on the stigma. Thus cross-pollination is effected.

7. **Orchid.** The orchids belong to the Family Orchidaceae. The highest level of adaptation for cross-pollination by insects is developed in orchid flowers (Fig. 2.156). The **labellum** (i.e., the posterior median member of the inner whorl of petal which is larger in size) acts as the flag and landing stage for the flies. The **rostellum** (i.e., the sterile stigma) secretes a sticky fluid, by which the pollinia are adhered to the body of the insects. The **gynostemium** is formed (by the union of stamens and carpels) bearing two naked pollinia at its top, and below them lies the **rostellum** concealing the stigmatic surface (formed by the fusion of two fertile stigmas).

Usually the long-tongued bees come and land on labellum to get nectar concealed in the spur. During its attempt, the insect pushes back the rostellum and even breaks it. Thus the pollinia are released and get stuck to leg or forehead of insects by disc with the help of sticky fluid. At first, the pollinia stand erect, but due to flexible nature of *caudicle* they bend down during the movement of fly. During visit of insect

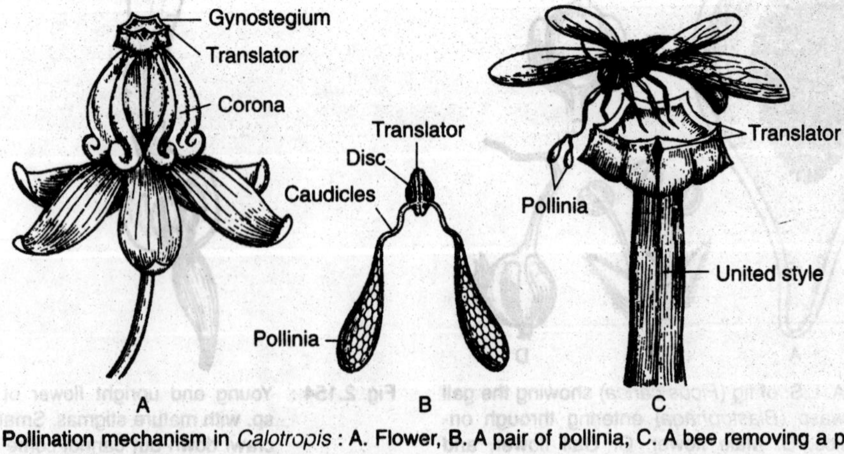

Fig. 2.155 : Pollination mechanism in *Calotropis* : A. Flower, B. A pair of pollinia, C. A bee removing a pair of pollinia from the gynostegium

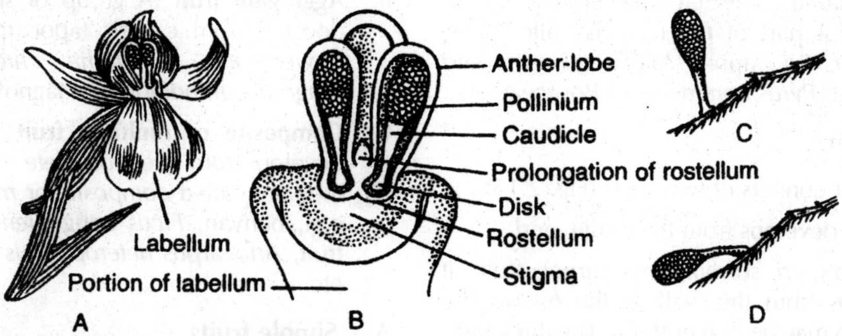

Fig. 2.156 : Pollination of Orchid : A. An entire flower, B. V.S. of the same after removal of the accessory whorls, C–D. Different positions of pollinia, while attached to the leg of a bee

to other flower the pollinia get caught by the sticky fluid of the stigma and gradually the pollens are released. Thus the cross-pollination is effected.

Advantages and Disadvantages of self and cross-pollination

Several advantages and disadvantages are there in nature in both self and cross-pollination :

A. **Self-pollination**

Advantages

1. The purity of the race is maintained through self-pollination.
2. The wastage of pollen grain is negligible.
3. Chances of failure of pollination are very less.

Disadvantages

1. The production of new species and varieties is inhibited.
2. Due to continued self-pollination for several generations, the progeny becomes weak.

B. **Cross-pollination**

Advantages.

1. Cross-pollination is useful in improving the crop and also for production of new varieties of different crops like fruits, vegetables etc.
2. The offsprings become more healthy and strong.
3. The production of more viable seeds is increased.
4. The plants are better adapted in the nature.

Disadvantages

1. Pollination is uncertain, because it depends on external agencies like wind, insects, water etc.
2. The wastage of pollen is much more as it depends on external agents. Wastage of pollen is more in anemophily (wind pollination) than other types.
3. The wastage of energy is much more.

FRUIT

Fruit may be defined as the fertilised mature ovary, where fruit wall develop from ovary wall and seeds develop from ovules.

The fruit developed from unfertilised ovule is said to be **parthenocarpic fruit**. This type of fruit is also called seedless fruit, e.g., banana, *Musa sapientum* of Musaceae; papaya, *Carica papaya* of Caricaceae etc. Sometimes abortive seeds may remain inside the fruit, e.g., grapes etc.

The fruits are of two types :

a. **True fruit**. The fruit which develops from ovary is called *true fruit.* It takes place normally and is very common in angiospermic plants, e.g., mango, *Mangifera indica* of Anacardiaceae.

b. **False fruit.** The fruit in which, along with ovule, some other floral parts like thalamus, calyx etc. may grow and form major part of the fruit, is called *false fruit*, e.g., apple, *Malus sylvestris* and pear, *Pyrus communis* of Rosaceae etc.

Parts of fruit

The fruit consists of two parts (Fig. 2.157) :

a. *Seed*. It develops from the ovule, and

b. *Pericarp* (*peri*, surrounding; *carpos*, fruit). It develops from the wall of the ovary. The pericarp may be thin or thick. The thick pericarps are differentiated into three regions :

1. Epicarp or exocarp. It is the outermost thin layer which forms the skin of fruit.
2. Mesocarp. The middle thick and fleshy layer which forms the pulp, and
3. Endocarp. The innermost layer which may be membranous or hard.

Function of fruit

The main functions of the fruit are :

1. It stores the food material for future use.
2. It protects the seeds, and
3. It helps in seed dispersal.

Classification of fruits

There are three major types of fruits based on their origin, texture and dehiscence : A. Simple fruit, B. Aggregate fruit, and C. Composite or multiple fruit.

A. **Simple fruit**. The fruits develop from the ovary of a solitary pistil in a single flower, e.g., pea, *Pisum sativum* of Fabaceae; rice, *Oryza sativa* of Poaceae etc.

B. **Aggregate fruit**. A group of separate fruits develop from carpels (apocarpous) of one flower, e.g., *Michelia champaca* and *Magnolia grandiflora* of Magnoliaceae etc.

C. **Composite** or **multiple fruit**. When fruits develop from the *complete inflorescence*, they are called *composite* or *multiple fruits*, e.g., banyan, *Ficus benghalensis* and jack-fruit, *Artocarpus heterophyllus* of Moraceae etc.

A. Simple fruits

These fruits are divided into two types : a. **Dry fruits**, and b. **Succulent fruits** (**Fleshy fruits**).

a. **Dry fruits**. In this type, the pericarp of fruit is simple, dry, membranous, leathery or woody and is not differentiated into three layers. The dry fruits are divided into the following **three** types : 1. **Dehiscent**, 2. **Indehiscent**, and 3. **Schizocarpic fruits.**

1. **Dry dehiscent fruits** (split open at maturity). The pericarp of the fruits rupture after ripening and then the seeds are dispersed. These fruits are divided into the following **five** types :

   i. **Legume** or **pod**. When the fruits develop from *monocarpellary superior ovary*, they are called legumes. Ovary has one locule and many ovules are arranged on marginal placentation. Mature fruits dehisce by splitting up along two sutures, one on each side, e.g., pea, *Pisum sativum* (Fig. 2.158A); French bean, *Phaseolus vulgaris*; hyacinth bean, *Dolichos lablab* of Fabaceae etc.

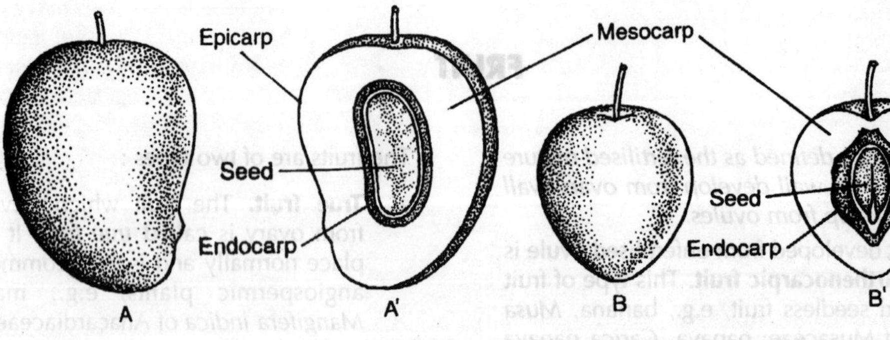

Fig. 2.157 : Parts of fruits : A. Entire mango (*Mangifera indica*). A'. L.S. of Mango showing different parts, B. Entire Indian plum (*Zizyphus*) and B'. L.S. of Indian plum showing different parts

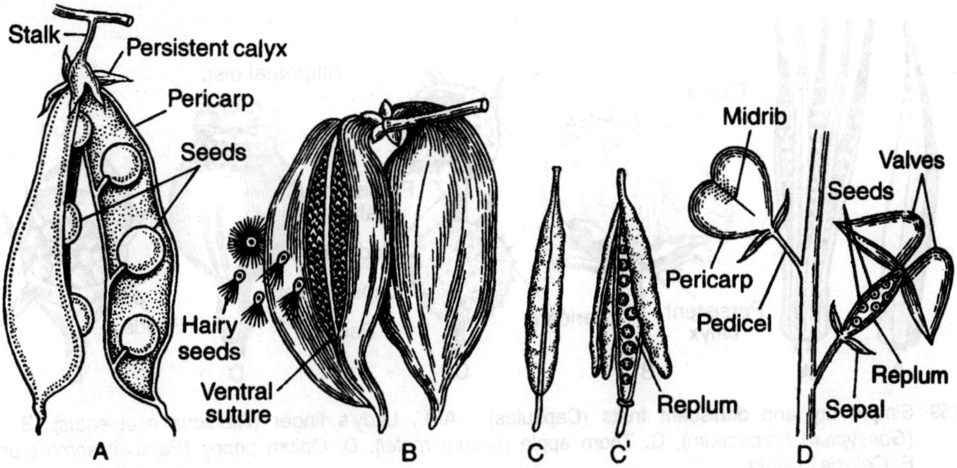

Fig. 2.158 : Different types of simple dry and dehiscent fruits : A. Legume of pea (*Pisum sativum*), B. Paired follicle of *Calotropis*, C–C′. Siliqua of mustard (*Brassica*) and D. Silicula of *Capsella*

ii. **Follicle**. The fruits develop like legume, but dehisce by splitting up along one suture. These are usually found in pairs or more than two in a group, e.g., *Calotropis procera* (Fig. 2.158B) of Asclepiadaceae, *Catharanthus roseus* of Apocynaceae etc.

iii. **Siliqua**. The fruits develop from *bicarpellary, syncarpous, superior ovary*. Placentation is parietal. Initially ovary is unilocular, but later on it becomes bilocular by a persistent false septum, the **replum**. On maturity, the pericarp ruptures in two valves starting from base to apex, leaving the seeds attached to the replum. The fruit is more in length than breadth, e.g., mustard, *Brassica nigra* (Fig. 2.158C-C′) and many other plants of Brassicaceae.

iv. **Silicula**. It resembles siliqua, but is short, flat and wide. Its length is same as, or shorter than breadth, e.g., shepherd's purse, *Capsella bursa-pastoris* (Fig. 2.158D) and candytuft, *Iberis odorata* of Brassicaceae.

v. **Capsule**. When the fruit develops from *bi-* or *multicarpellary, syncarpous, usually superior* (sometimes inferior — *Canna indica* of Cannaceae etc.) *ovary,* it is called *capsule*. The capsule generally opens by valves or pores. The capsules may open in different ways (Fig. 2.160) :

- *Septicidal capsule*. In this type, the fruit splits into component carpels along with their line of union and thus the placentae become separated in the middle, e.g., *Linum usitatissimum* of Linaceae, *Ricinus communis* of Euphorbiaceae, *Hibiscus esculentus* (Fig. 2.159A) of Malvaceae etc.
- *Loculicidal capsule*. In this type, the fruit splits along the middle of each carpel and thus the placentae are found along the united margins, e.g., *Andrographis paniculata, Adhatoda vasica* and many other members of Acanthaceae; *Gossypium herbaceum* (Fig. 2.159B) of Malvaceae etc.
- *Septifragal capsule*. In this type, the fruit splits either in the septicidal or loculicidal manner, but the septa break and the placenta along with seeds remain in the middle, e.g., thorn apple, *Datura metel* (Fig. 2.159C) of Solanaceae; *Oroxylum indicum* of Bignoniaceae etc.
- *Porocidal capsule*. In this type, the opening takes place by means of pores at the apices of the ripened carpels, e.g., opium poppy, *Papaver somniferum* (Fig. 2.159D) of Papaveraceae.
- *Pyxis* or *Circumscissile capsule*. In this type, the dehiscence takes place

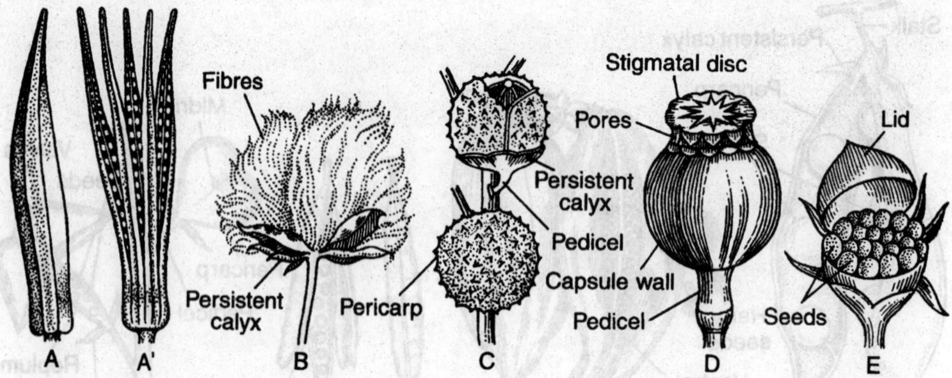

Fig. 2.159 : Simple dry and dehiscent fruits (Capsules) : A-A′. Lady's finger (*Hibiscus esculentus*), B. Cotton (*Gossypium herbaceum*), C. Thorn apple (*Datura metel*), D. Opium poppy (*Papaver somniferum*) and E. *Celosia cristata*

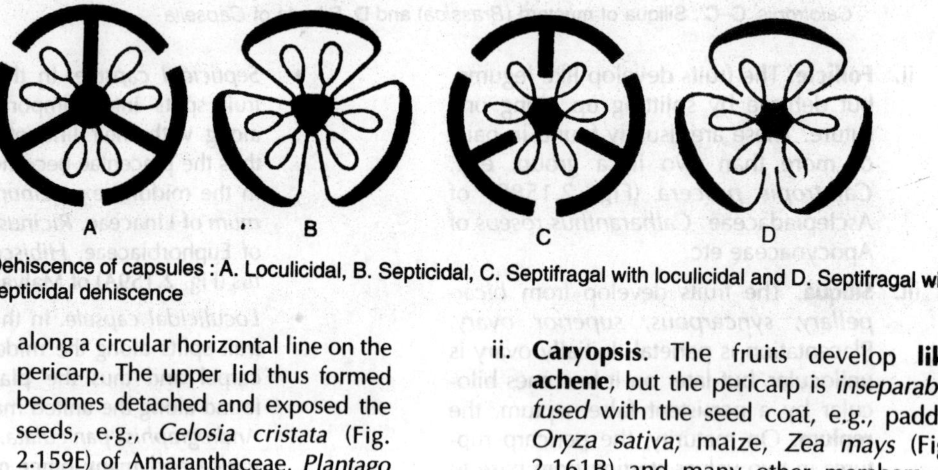

Fig. 2.160 : Dehiscence of capsules : A. Loculicidal, B. Septicidal, C. Septifragal with loculicidal and D. Septifragal with septicidal dehiscence

along a circular horizontal line on the pericarp. The upper lid thus formed becomes detached and exposed the seeds, e.g., *Celosia cristata* (Fig. 2.159E) of Amaranthaceae, *Plantago* sp. of Plantaginaceae.

2. **Dry indehiscent fruits** (achenial fruits). The *fruits do not dehisce* even after ripening and the seeds remain inside the fruits. These fruits are divided into the following **five** types :

i. **Achene**. The fruits develop from *monocarpellary superior ovary*. The fruits are unilocular and one-seeded. *Pericarp is free from the seed coat* and the seed is attached to fruit wall at one point only, e.g., *Mirabilis jalapa* of Nyctaginaceae, *Rumex vesicarius* of Polygonaceae, *Nigella sativa* (Fig. 2.161A) of Ranunculaceae etc.

Usually achenes are found in groups, because they develop from multicarpellary apocarpous ovary, e.g., *Nigella sativa, Clematis gouriana* of Ranunculaceae.

ii. **Caryopsis**. The fruits develop **like achene**, but the pericarp is *inseparably fused* with the seed coat, e.g., paddy, *Oryza sativa*; maize, *Zea mays* (Fig. 2.161B) and many other members of Poaceae (Gramineae).

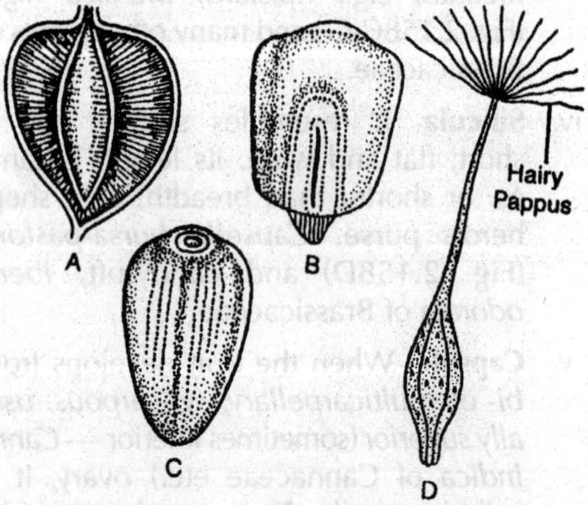

Fig. 2.161 : Simple dry indehiscent fruits : A. Achene of *Nigella* (a fruitlet), B. Caryopsis of *Zea*, C. Cypsella of *Helianthus* and D. Cypsella of *Taraxacum*

iii. **Cypsella**. The fruits develop from *bicarpellary, syncarpous, inferior one-chambered ovary*. Pericarp is free from the seed coat, e.g., Marigold, *Tagetes patula*; sunflower, *Helianthus annuus* (Fig. 2.161C) and many other members of Asteraceae.

In many cases the fruits are attached with hairy **pappus** formed by the modification of calyx, e.g., *Blumea lacera, Vernonia cinerea, Taraxacum* sp. (Fig. 2.161D) of Asteraceae.

iv. **Nut**. The fruits develop from *bi-* or *multi-carpellary, syncarpous, superior ovary* and the pericarp is leathery or woody, e.g., oak, *Quercus* sp. (Fig. 2.162A,B) of Fagaceae; *Sapindus trifoliatus* of Sapindaceae; cashewnut, *Anacardium occidentale* (Fig. 2.162C,D); water chestnut, *Trapa bispinosa* (Fig. 2.162E) of Onagraceae etc.

The fruit of litchi, *Litchi chinensis* of Sapindaceae is also a nut, where a fleshy aril (funiculus) is present (Fig. 2.162F).

v. **Samara**. The fruits are two- or more chambered and a few seeds develop from *superior multicarpellary ovary* where the pericarp is flattened into wings, e.g., chilbil, *Holoptelea indica* (Fig. 2.163A,B); maple, *Acer* sp. (Fig. 2.163C) of Aceraceae; lumber, *Fraxinus americana* of Oleaceae; *Hiptage benghalensis* of Malpighiaceae etc.

In many cases, the persistent calyx develops into wings. Thus the calyx helps in dispersal. These fruits are called **samaroid**. In this case, the ovary may be *inferior* (e.g., *Combretum* sp. of Combretaceae) or *superior*. e.g., *Shorea robusta*

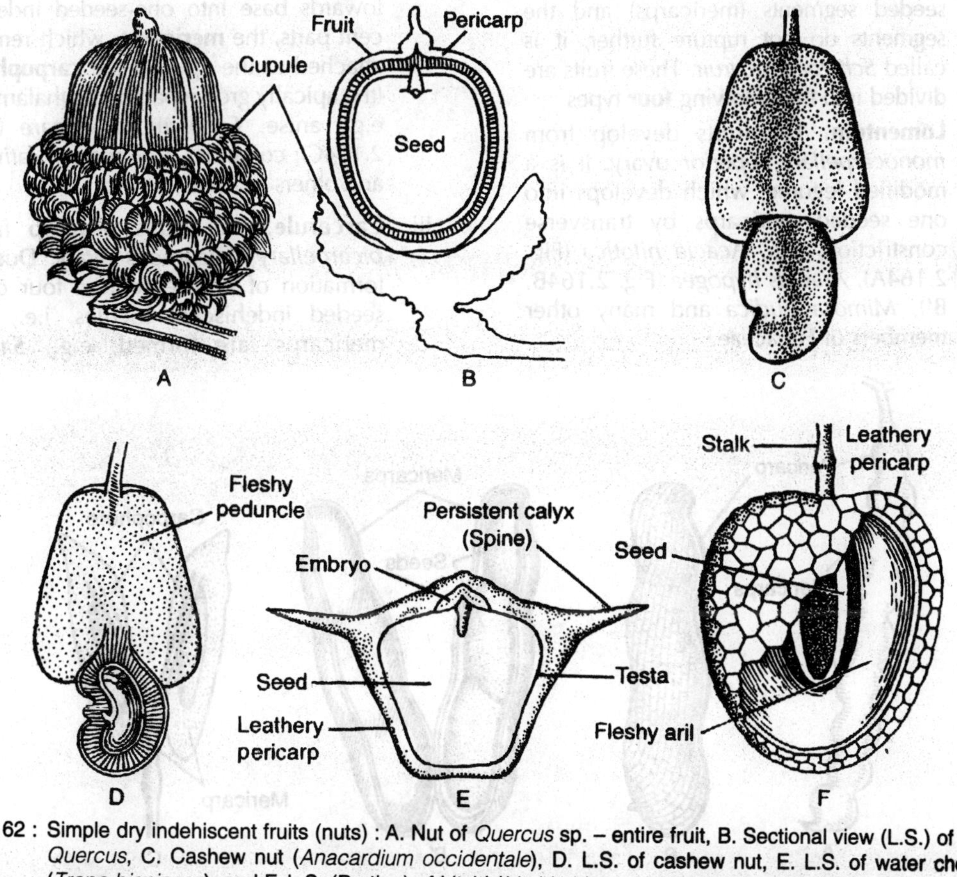

Fig. 2.162 : Simple dry indehiscent fruits (nuts) : A. Nut of *Quercus* sp. – entire fruit, B. Sectional view (L.S.) of nut of *Quercus*, C. Cashew nut (*Anacardium occidentale*), D. L.S. of cashew nut, E. L.S. of water chestnut (*Trapa bispinosa*), and F. L.S. (Portion) of Litchi (*Litchi chinensis*) showing inner parts

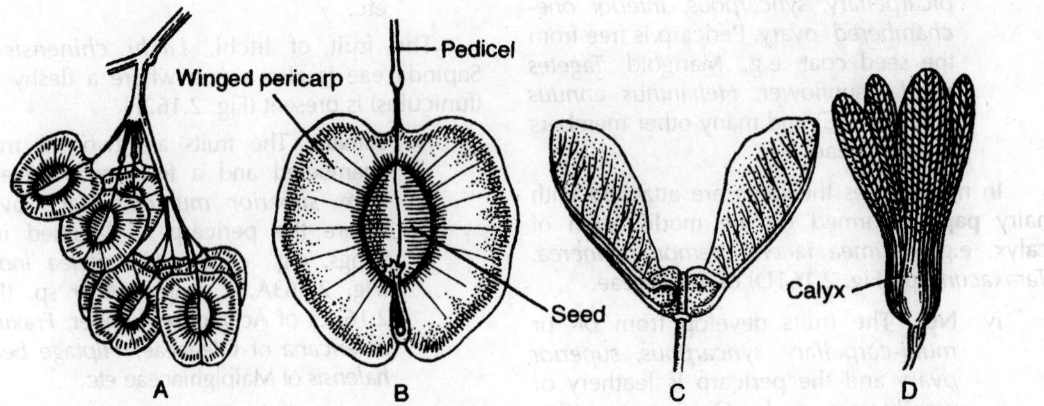

Fig. 2.163 : Simple dry indehiscent fruits : A. Samara of *Holoptelea* in bunch, B. Single fruit of *Holoptelea*, C. Samara of *Acer*, and D. Samaroid fruit of *Shorea robusta*

(Fig. 2.163D) of Dipterocarpaceae, *Ulma* sp. of Ulmaceae etc.

3. **Schizocarpic fruits** (splitting fruits). When the ripe fruits are divided into one-seeded segments (mericarps) and the segments do not rupture further, it is called *Schizocarpic fruit.* These fruits are divided into the following **four** types :

i. **Lomentum**. The fruits develop from *monocarpellary superior ovary.* It is a modified legume, which develops into one seeded mericarps by transverse constriction, e.g., *Acacia nilotica* (Fig. 2.164A), *Arachis hypogea* (Fig. 2.164B, B′), *Mimosa pudica* and many other members of Fabaceae.

ii. **Cremocarp**. The fruits are bilocular and two-seeded; developed from a *bicarpellary, syncarpous* and *inferior ovary*. After maturation, it splits open from apex towards base into one-seeded indehiscent parts, the **mericarps**; which remain attached to the apex of the **carpophore** (the apically grown part of the thalamus), e.g., anise, *Foeniculum vulgare* (Fig. 2.164C); coriander, *Coriandrum sativum* and others of Apiaceae etc.

iii. **Carcerule**. The fruits develop from *bicarpellary syncarpous ovary*. Due to formation of a false septum four one-seeded indehiscent locules, i.e., the mericarps, are formed, e.g., *Salvia*

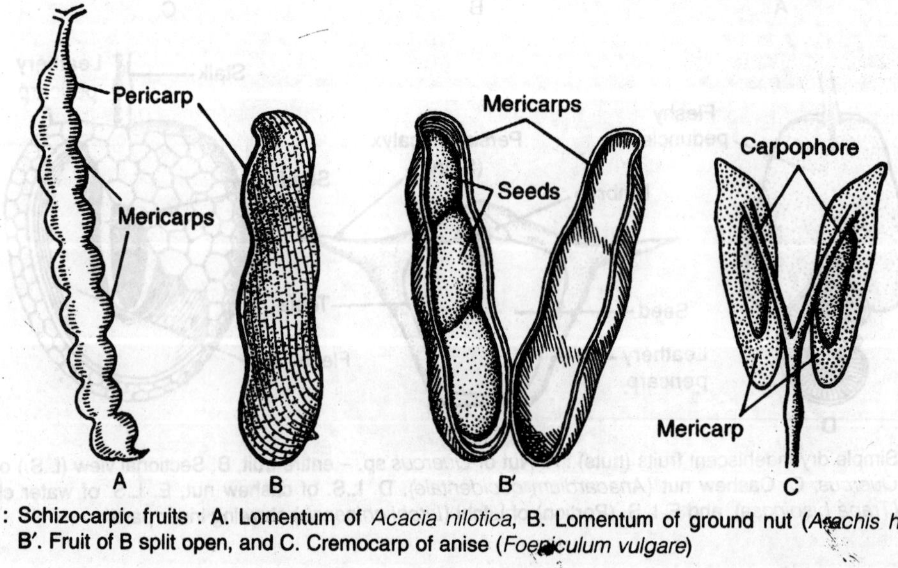

Fig. 2.164 : Schizocarpic fruits : A. Lomentum of *Acacia nilotica*, B. Lomentum of ground nut (*Arachis hypogea*), B′. Fruit of B split open, and C. Cremocarp of anise (*Foeniculum vulgare*)

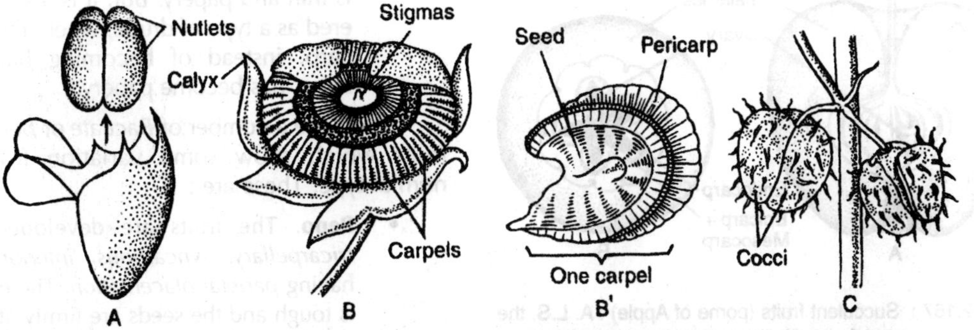

Fig. 2.165 : Schizocarpic fruits : A. Carcerule of *Ocimum*, B. Carcerule of Hollyhock (*Althea rosea*), B'. Single mericarp of B, and C. Regma of castor (*Ricinus communis*)

plebeja, Ocimum sanctum (Fig. 2.165A) of Lamiaceae.

In Hollyhock, *Althea rosea* (Fig. 2. 165B, B') of Malvaceae; the fruits are developed from multicarpellary ovary and, hence, the number of mericarps are much more.

iv. **Regma**. The fruits develop from multicarpellary ovary and after maturation they divide into parts — as many as carpels. Each one-seeded part is called **coccus**. Thus the number of cocci is three in castor, *Ricinus communis* (Fig. 2.165C) of Euphorbiaceae and five in *Geranium* sp. of Geraniaceae.

b. Succulent fruits (Fleshy fruits). The fruits usually remain succulent and juicy even at maturity. The pericarp of the fruit is differentiated into three layers : epicarp, mesocarp and endocarp. The mesocarp is thick, fleshy or fibrous. The fruits are indehiscent, thus the seeds are separated after the decay of fleshy tissue enclosing them. These fruits are divided into the following **seven** types :

i. **Drupe (stone fruit)**. The fruits are *generally one-seeded* and develop from *monocarpellary superior ovary*. The pericarp is differentiated into an outer **exocarp**, a middle fleshy **mesocarp**, and inner hard (stony) **endocarp**, e.g., mango, *Mangifera indica* (Fig. 2.166A) of Anacardiaceae; peach, *Prunus persica* of Rosaceae etc.

Sometimes the fruit contains two seeds (developed in multicarpellary ovary) as in almond, *Prunus amygdalus* of Rosaceae, where the mesocarp is dry and tough. In palmyra palm, *Borassus flabellifer* (Fig. 2.166B) of Arecaceae, the drup contains three seeds, each has got a separate endocarp.

Sometimes the fruits are developed from multicarpellary syncarpous ovary, where the mesocarp is *fibrous*, hence it is called **fibrous drupe**, e.g., coconut, *Cocos nucifera* (Fig. 2.166C) of Arecaceae (Palmae) etc.

ii. **Pome**. The fruits developed from the *fleshy thalamus of syncarpous inferior*

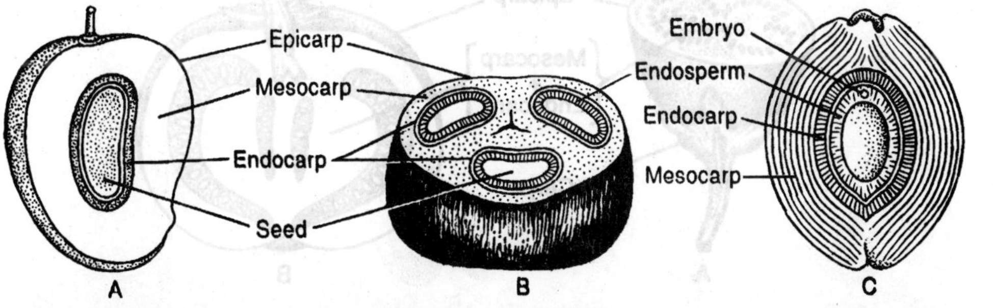

Fig. 2.166 : Succulent fruits (Drupe) : A. L.S. of mango (*Mangifera indica*), B. Palmyra palm (*Borassus flabellifer*) with three seeds in T.S., and C. L.S. of coconut, *Cocos nucifera*

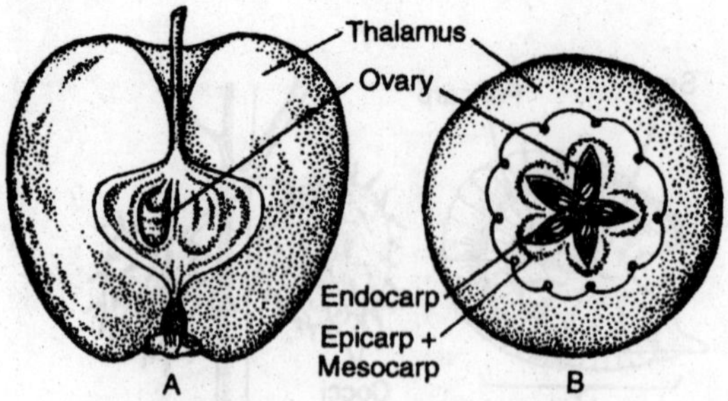

Fig. 2.167 : Succulent fruits (pome of Apple) : A. L.S. the real fruit with its pericarp, surrounded by a fleshy edible thalamus. Of the pericarp, the epicarp and mesocarp are fleshy while the endocarp is cartilaginous, and B. T.S. of fruit

ovary are called **false fruit**. Outer skin and flesh develop from the thalamus. True fruit is present inside the fleshy thalamus. Seeds are enclosed in thin ovarian wall. The fleshy thalamus is the edible part of the fruit, e.g., pear, *Pyrus communis* and apple, *P. malus* (Fig. 2.167) of Rosaceae, etc.

iii. **Berry** or **Bacca**. The fruits are developed from the *multicarpellary, syncarpous, superior* or *inferior ovary*. Placentation is axile or parietal. The seeds lie embedded freely in the massive pulp developed from mesocarp and endocarp. The epicarp remains as the outer thin covering of the fruit, e.g., brinjal, *Solanum melongena* (Fig. 2.168A); tomato, *Lycopersicon esculentum* (Fig. 2.168B) and many others of Solanaceae. One-seeded berry is found in date palm, *Phoenix sylvestris* etc. [The date palm is usually considered as a one-seeded berry in which endocarp is thin and papery. But, it is also considered as a type of drupe, where the endocarp, instead of becoming hard and stony, has become papery.]

There are a number of *Baccate* or *berry-type* fruits which show some variation from the normal type. These are :

- **Pepo**. The fruits are developed from *tricarpellary, syncarpous, inferior ovary* having *parietal placentation*. The exocarp is tough and the seeds are firmly attached to the placentae, e.g., cucumber, *Cucumis sativa* (Fig. 2.169A); gourd, *Cucurbita maxima* and many others of the family Cucurbitaceae.
- **Hesperidium**. The fruits are many-chambered, developed from the *multicarpellary, syncarpous, superior ovary* having *axile placentation*. The epicarp and mesocarp are fused together forming the skin, while the endocarp projects inwards forming distinct chambers. Many unicellular juicy hairs are present on the inner side of the endocarp, e.g., sweet orange, *Citrus sinensis*; lemon, *Citrus aurantium* (Fig. 2.169B) and many others of Rutaceae.
- **Balausta**. The fruits are many-chambered, many-seeded and develop from the *multicarpellary, syncarpous, inferior ovary*. The seeds have seed coat consisting of outer fleshy *testa, forming the edible part,* and inner hard tegmen. Seeds are irregularly arranged inside the fruit. The pericarp is rough and leathery with persistent calyx, e.g., Pomegranate, *Punica granatum* (Fig. 2.169C, C′) of Punicaceae.

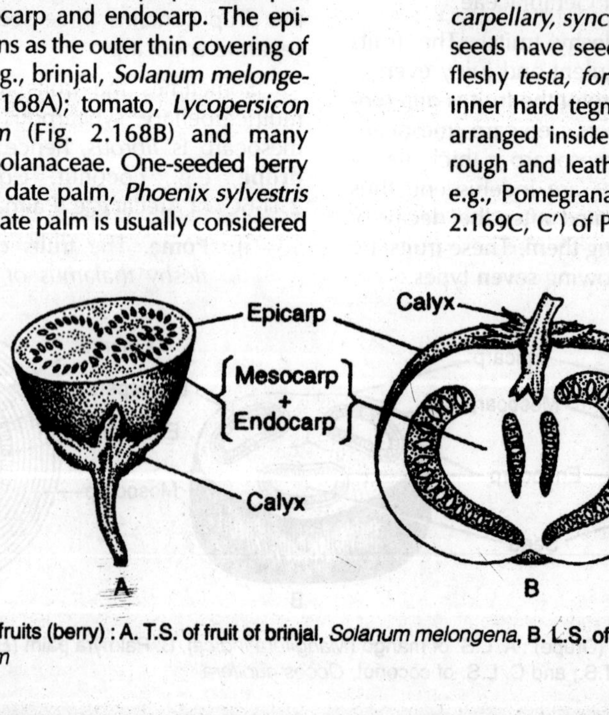

Fig. 2.168 : Succulent fruits (berry) : A. T.S. of fruit of brinjal, *Solanum melongena*, B. L.S. of fruit of tomato, *Lycopersicon esculentum*

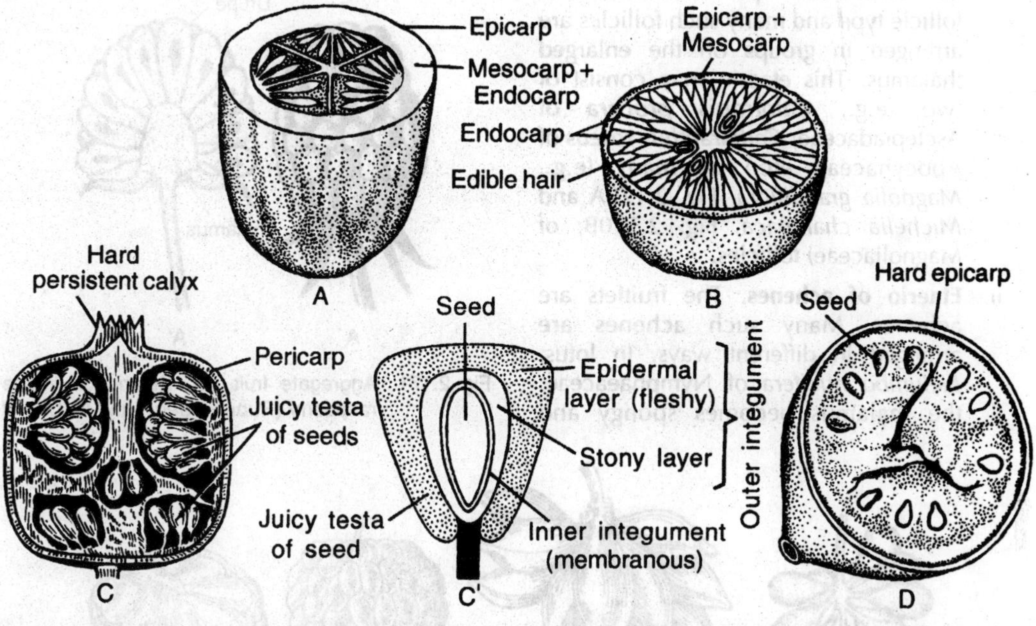

Fig. 2.169 : Succulent fruits : A. Pepo (T.S. of Cucumber, *Cucumis sativa*), B. Hesperidium (T.S. of Lemon, *Citrus aurantinum*), C. Balausta (L.S. of pomegranate, *Punica granatum*) C'. L.S. of a single seed of pomegranate, and D. Amphisarca (T.S. of wood apple, *Aegle marmelos*)

- **Amphisarca**. The fruits are many-chambered, many-seeded and develop from the *multicarpellary, syncarpous, superior ovary*. Epicarp is hard; mesocarp and endocarp are fleshy — inside these seeds are scatteredly arranged. Mesocarp, endocarp and the swollen placenta are the edible parts, e.g., wood apple, *Aegle marmelos* (Fig. 2.169D) and elephant apple, *Feronia limonia* of Rutaceae.

B. Aggregate fruit

Actually the aggregate fruit is the collection of small fruits, known as **etaerio** of fruitlets. This collection of small fruitlets develops from a *single flower with multicarpellary apocarpous ovary*.

These fruits are divided into the following **four** types :

i. **Etaerio of follicles**. In this type, each

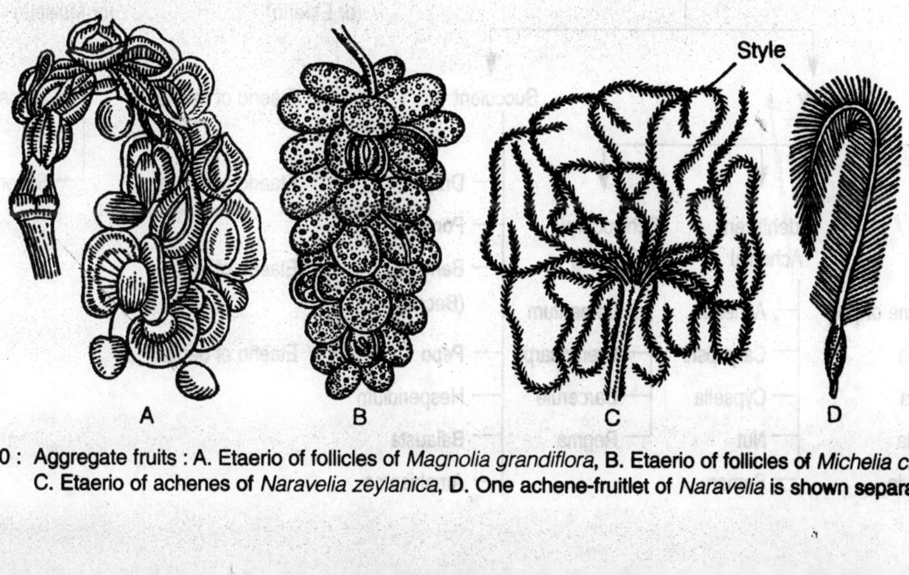

Fig. 2.170 : Aggregate fruits : A. Etaerio of follicles of *Magnolia grandiflora*, B. Etaerio of follicles of *Michelia champaca*, C. Etaerio of achenes of *Naravelia zeylanica*, D. One achene-fruitlet of *Naravelia* is shown separately

free carpel develops into a fruitlet of follicle type and many such follicles are arranged in groups on the enlarged thalamus. This etaerio may consist of two (e.g., *Calotropis procera* of Asclepiadaceae; *Catharanthus roseus* of Apocynaceae etc.) or *many* (e.g., *Magnolia grandiflora*, Fig. 2.170A and *Michelia champaca*, Fig. 2.170B; of Magnoliaceae) follicles.

ii. **Etaerio of achenes**. The fruitlets are achenes. Many such achenes are arranged in different ways. In lotus, *Nelumbo nucifera* of Nymphaeaceae; the thalamus becomes spongy and

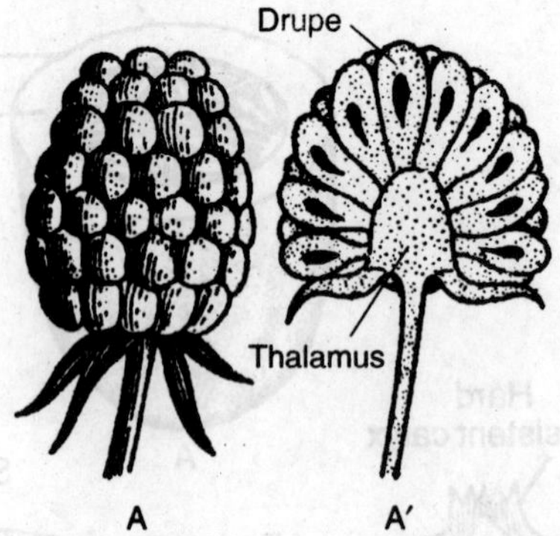

Fig. 2.171 : Aggregate fruit : A. Etaerio of drupes in raspberry (*Rubus idaeus*) and A′. L.S. of A

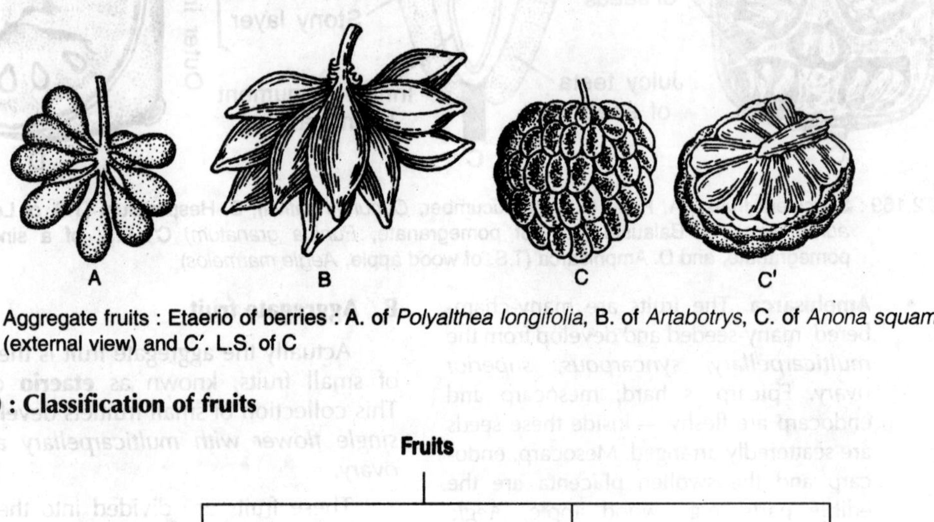

Fig. 2.172 : Aggregate fruits : Etaerio of berries : A. of *Polyalthea longifolia*, B. of *Artabotrys*, C. of *Anona squamosa* (external view) and C′. L.S. of C

Table 2.10 : Classification of fruits

Fruits

- **I. Simple fruits**
  - **Dry**
    - **Dehiscent Capsular**
      - Legume or pod
      - Follicle
      - Siliqua
      - Silicula
      - Capsule
    - **Indehiscent Achenial**
      - Achene
      - Caryopsis
      - Cypsella
      - Nut
      - Samara
    - **Schizocarpic Splitting**
      - Lomentum
      - Cremocarp
      - Carcerule
      - Regma
  - **Succulent**
    - Drupe
    - Pome
    - Berry or (Bacca)
    - Pepo
    - Hesperidium
    - Balausta
    - Amphisarca
- **II. Aggregate fruits (or Etaerio)**
  - Etaerio of follicles
  - Etaerio of achenes
  - Etaerio of drupes
  - Etaerio of berries
- **III. Composite fruits (or Multiple)**
  - Sorosis
  - Syconus

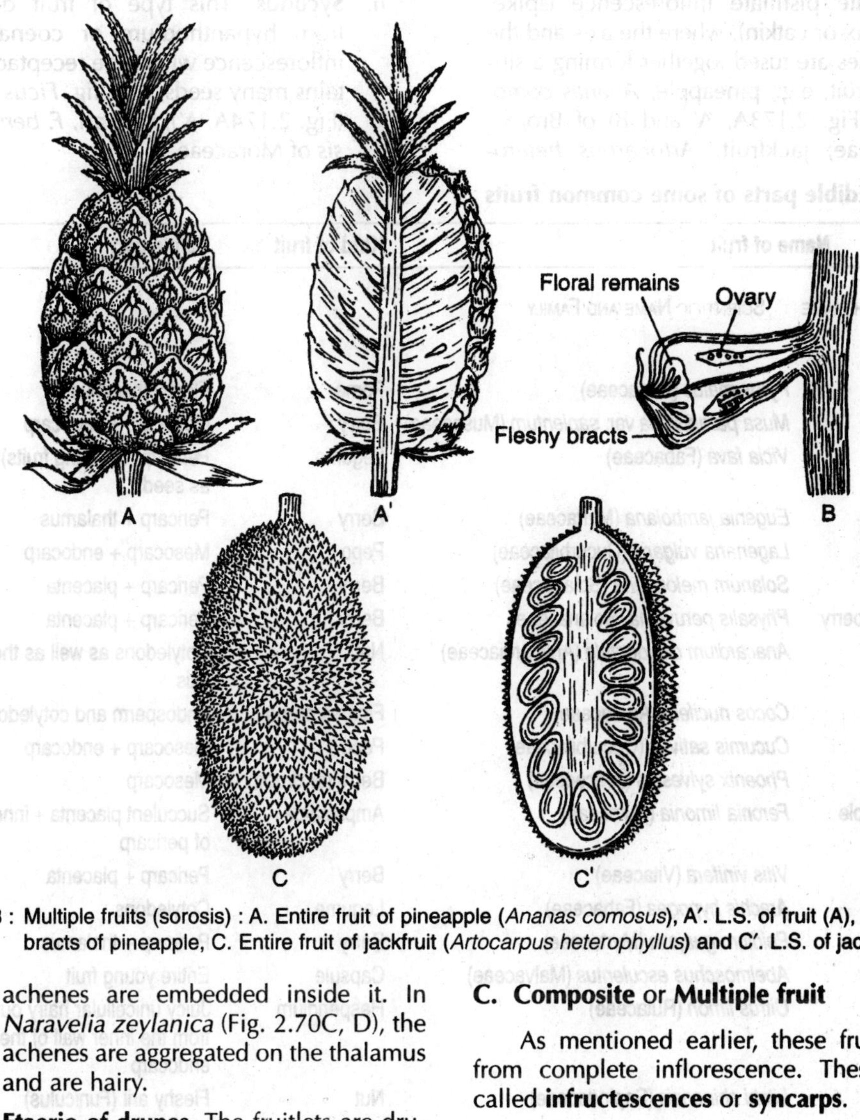

Fig. 2.173 : Multiple fruits (sorosis) : A. Entire fruit of pineapple (*Ananas comosus*), A'. L.S. of fruit (A), and B. Fleshy bracts of pineapple, C. Entire fruit of jackfruit (*Artocarpus heterophyllus*) and C'. L.S. of jackfruit

achenes are embedded inside it. In *Naravelia zeylanica* (Fig. 2.70C, D), the achenes are aggregated on the thalamus and are hairy.

iii. **Etaerio of drupes**. The fruitlets are drupes. Many small drupes are arranged in groups on the fleshy thalamus, e.g., strawberry, *Fragaria vesca* and raspberry, *Rubus idaeus* (Fig. 2.171A, A') of Rosaceae.

iv. **Etaerio of berries**. The fruitlets are berries. Many such small berries are arranged on all sides of fleshy thalamus, where the apical part is fused with each other forming a common rind, e.g., mast tree, *Polyalthia longifolia* (Fig. 2.172A); *Artabotrys odoratissima* (Fig. 2.172B) and *Anona squamosa* (Fig. 2.172C, C') of Annonaceae.

C. Composite or Multiple fruit

As mentioned earlier, these fruits develop from complete inflorescence. These are also called **infructescences** or **syncarps**.

These fruits are of the following **two types** :

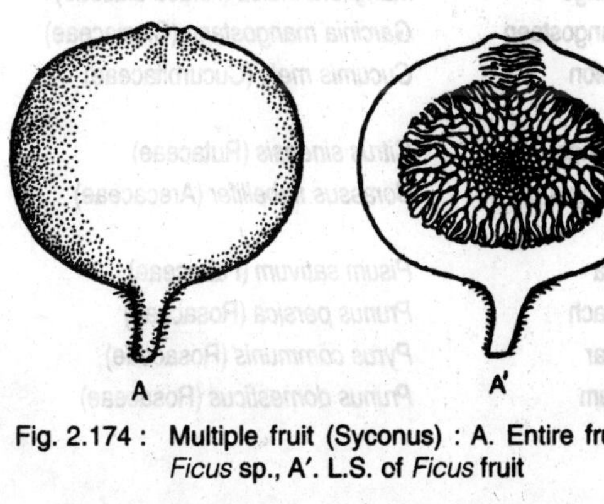

Fig. 2.174 : Multiple fruit (Syconus) : A. Entire fruit of *Ficus* sp., A'. L.S. of *Ficus* fruit

i. **Sorosis**. This type of fruit develops from a spicate pistillate inflorescence (spike, spadix or catkin), where the axis and the ovaries are fused together forming a single fruit, e.g., pineapple, *Ananas comosus* (Fig. 2.173A, A′ and B) of Bromeliaceae; jackfruit, *Artocarpus heterophyllus* (Fig. 2.173C, C′) of Moraceae.

ii. **Syconus**. This type of fruit develops from hypanthodium or coenanthium inflorescence where the receptacle contains many seeds, e.g., fig, *Ficus hispida* (Fig. 2.174A, A′); banyan, *F. benghalensis* of Moraceae.

Table 2.11 : Edible parts of some common fruits

Name of fruit		Kind of fruit	Edible part
COMMON ENGLISH NAME	SCIENTIFIC NAME AND FAMILY		
A. Simple fruit			
Apple	*Pyrus malus* (Rosaceae)	Pome	Fleshy thalamus
Banana	*Musa paradisiaca* var. *sapientum* (Musaceae)	Berry	Mesocarp + endocarp
Bean	*Vicia fava* (Fabaceae)	Legume	Pericarp (of young fruits) as well as seeds
Black berry	*Eugenia jambolana* (Myrtaceae)	Berry	Pericarp + thalamus
Bottle gourd	*Lagenaria vulgaris* (Cucurbitaceae)	Pepo	Mesocarp + endocarp
Brinjal	*Solanum melongena* (Solanaceae)	Berry	Pericarp + placenta
Cape gooseberry	*Physalis peruviana* (Solanaceae)	Berry	Pericarp + placenta
Cashewnut	*Anacardium occidentale* (Anacardiaceae)	Nut	Cotyledons as well as the fleshy axis
Coconut	*Cocos nucifera* (Arecaceae)	Fibrous drupe	Endosperm and cotyledon
Cucumber	*Cucumis sativus* (Cucurbitaceae)	Pepo	Mesocarp + endocarp
Date palm	*Phoenix sylvestris* (Arecaceae)	Berry (or drupe)	Mesocarp
Elephant apple	*Feronia limonia* (Rutaceae)	Amphisarca	Succulent placenta + inner portion of pericarp
Grape	*Vitis vinifera* (Vitaceae)	Berry	Pericarp + placenta
Groundnut	*Arachis hypogea* (Fabaceae)	Legume	Cotyledons
Guava	*Psidium guajava* (Myrtaceae)	Berry	Pericarp + thalamus
Lady's finger	*Abelmoschus esculentus* (Malvaceae)	Capsule	Entire young fruit
Lemon	*Citrus limon* (Rutaceae)	Hesperidium	Juicy unicellular hairy outgrowth from the inner wall of the endocarp
Litchi	*Litchi chinensis* (Sapindaceae)	Nut	Fleshy aril (Funiculus)
Maize	*Zea mays* (Poaceae)	Caryopsis	Endosperm + embryo
Mango	*Mangifera indica* (Anacardiaceae)	Drupe	Mesocarp
Mangosteen	*Garcinia mangostana* (Clusiaceae)	Berry	Fleshy aril
Melon	*Cucumis melo* (Cucurbitaceae)	Pepo	Mesocarp + endocarp as well as a portion of the receptacle
Orange	*Citrus sinensis* (Rutaceae)	Hesperidium	As in lemon
Palmyra palm	*Borassus flabellifer* (Arecaceae)	Drupe	Mesocarpic juice as well as cotyledon and endosperm
Pea	*Pisum sativum* (Fabaceae)	Legume	Mainly cotyledons
Peach	*Prunus persica* (Rosaceae)	Drupe	Epicarp + mesocarp
Pear	*Pyrus communis* (Rosaceae)	Pome	Fleshy thalamus
Plum	*Prunus domesticus* (Rosaceae)	Drupe	Epicarp + mesocarp

Contd.

Name of fruit		Kind of fruit	Edible part
COMMON ENGLISH NAME	SCIENTIFIC NAME AND FAMILY		
Pomegranate	*Punica granatum* (Punicaceae)	Balausta	Succulent testa
Rice	*Oryza sativa* (Poaceae)	Caryopsis	Mainly endosperm
Roseapple	*Eugenia jambos* (Myrtaceae)	Berry	Fused pericarp and thalamus
Sweet gourd	*Cucurbita maxima* (Cucurbitaceae)	Pepo	As in cucumber, as well as seeds
Tamarind	*Tamarindus indicus* (Caesalpiniaceae)	Lomentum	Mesocarp + endocarp
Tomato	*Lycopersicon esculentum* (Solanaceae)	Berry	Pericarp + placenta
Walnut	*Juglans regia* (Juglandaceae)	Drupe	Cotyledons
Wheat	*Triticum aestivum* (Poaceae)	Caryopsis	Endosperm + embryo
Wood apple	*Aegle marmelos* (Rutaceae)	Amphisarca	As in elephant apple
B. Aggregate fruits			
Bullock's heart	*Annona reticulata* (Annonaceae)	Etaerio of berries	Pericarp
Custard apple	*Annona squamosa* (Annonaceae)	Etaerio of berries	Pericarp
Lotus	*Nelumbium speciosum* (Nymphaeaceae)	Etaerio of achenes	Thalamus and seeds
Strawberry	*Fragaria* sp. (Rosaceae)	Etaerio of achenes	Fleshy thalamus
C. Composite fruits			
Fig	*Ficus* sp. (Moraceae)	Syconus	Receptacle
Jackfruit	*Artocarpus heterophyllus* (Moraceae)	Sorosis	Bract, perianth and seed
Mulberry	*Morus indica* (Moraceae)	Sorosis	Perianth
Pineapple	*Ananas comosus* (Bromeliaceae)	Sorosis	Fused fleshy rachis, bract, perianth and pericarp

DISPERSAL OF FRUITS AND SEEDS

Dispersal is the phenomenon of transferring of plant parts to distant place. Dispersal of fruits and seeds are common in angiosperms. It has great importance in their life process. The dispersal is essential because of the following facts :

1. If all the fruits and seeds fall under the parent plant, all may not be able to germinate to develop seedling.
2. Seedlings will not get sufficient light and air for normal living.
3. Soil nutrition will be insufficient for them as these have been mostly used up by the parent plant, and there will be sharing of the existing nutrition present in the soil.
4. A group of same type of seedling will be developed under the parent plant.

For the above facts, there will be **struggle for existence** amongst themselves, thereby most of the seedlings will become very weak and liable to perish under such unfavourable condition. Still, the **Natural hazards** (floods, drought etc.) may destroy the total plant population.

To avoid such problem, nature has arranged wonderful provisions to scatter them in different localities far away from the parent plant. Thereby some or most of them may find favourable conditions for their growth.

Most of the plants have no power to disperse their seeds to distant place, thereby they depend on some agents for their dispersal. In some plants, the fruits are provided with **some explosive mechanism** to gain the goal. The dispersal of fruits and seeds on the basis of

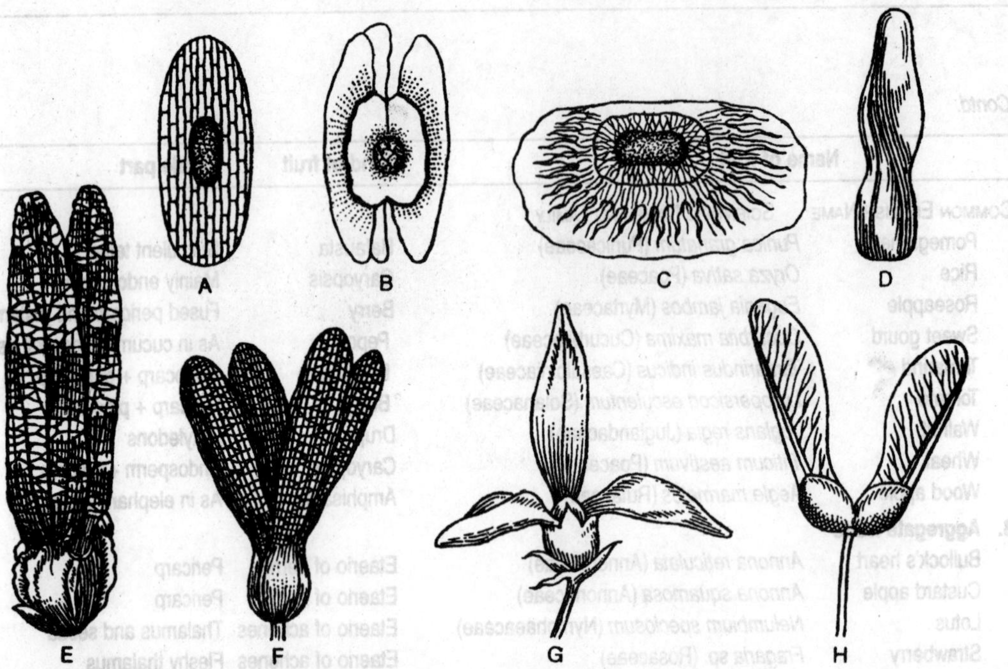

Fig. 2.175 : Fruits and seeds adopted for wind dispersal : A. Minute and light seed of orchid, B–H. Winged fruits and seeds. Wings develop from seed coat of *Moringa* (B), *Oroxylum* (C), *Swietenia* (D), Wings develop from persistant sepal of *Dipterocarpus* (E), *Shorea* (F), Pericarp expands into wings of *Hiptage* (G) and *Acer* (H)

these agents can be divided into the following groups :

A. **Dispersal by wind**. Wind plays an important role in the dispersal of some fruits and seeds. Plants which require wind as an agent are called **anemochorous.** For wind dispersal, the seeds should be light in weight, so that their buoyancy may help them to fly for long distance. These fruits and seeds have the following adaptations which help them to disperse long distance away from the mother plant.

1. **Minute and light seeds**. Seeds of some plants are very minute, dry and light in weight and are easily dispersed by wind, e.g., orchids (one seed of orchid weigh about 0.004 mg), Fig. 2.175A; grains (fruits and seeds) of grasses etc.

2. **Winged fruits and seeds**. Wing-like structures of different types are developed on fruits and seeds of some plants — these help them to float in air.

   i. Wings may develop *from seed coat* (testa), e.g., horse radish, *Moringa oli-fera* of Moringaceae (Fig. 2.175B); *Oroxylun indicum* of Bignoniaceae (Fig. 2.175C); mahogany, *Swietenia mahogani* of Meliaceae (Fig. 2.175D) etc.

   ii. Wings may develop *from persistent sepals,* e.g., *Dipterocarpus* sp. (Fig. 2.175E), *Shorea robusta* (Fig. 2.175F) and many other plants of Dipterocarpaceae.

   iii. Wings may develop *from pericarp* of fruit, e.g., *Hiptage benghalensis* (Fig. 2.175G) of Malpighiaceae; country almond, *Terminalia catappa* of Combretaceae; *Dioscorea alata* of Dioscoreaceae; maple, *Acer pseudoplatanum* (Fig. 2.175H) of Aceraceae etc.

3. **Parachute mechanism.** Appendages develop on some fruits and seeds, act as parachutes by which they remain in the air for long time and can disperse for long distance. The appendages are :

   i. **Pappus**. Hair-like persistent structure develops on the fruit by the modification of calyx and called *pappus.* The pappus helps in the dispersal of fruits of many

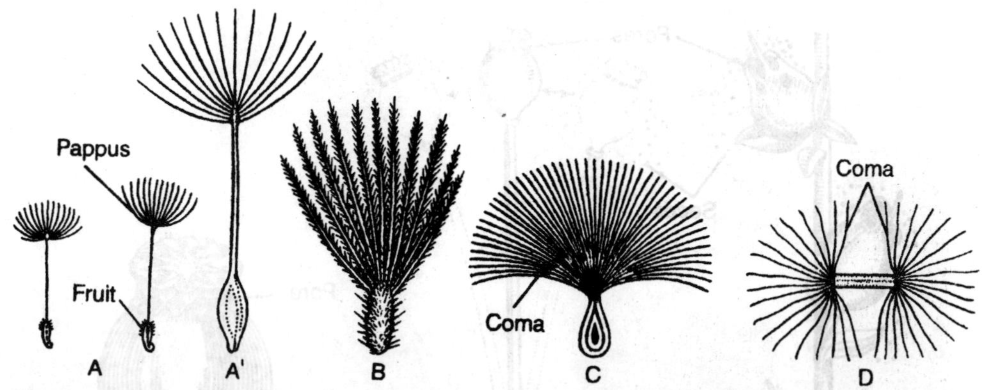

Fig. 2.176 : Seeds adopted for wind dispersal : A–A'. Pappus (looks like parachute) of *Taraxacum*, B. Pappus of *Tridax procumbens*, C. Coma of *Calotropis*, and D. Double coma of *Alstonia*

members of Asteraceae (Compositae), e.g., *Vernonia cinerea, Blumea lacera, Ageratum conyzoides, Taraxacum officinale* (Fig. 2. 176A, A') *Tridax procumbens* (Fig. 2.176B) of Asteraceae etc.

ii. **Coma**. These are tufts of hairy outgrowth of *testa* developed at one or both ends of the seeds and help to disperse for long distance, e.g., seeds with **one comma**, *Calotropis procera* (Fig. 2.176C) of Asclepiadaceae; seeds with **double comma** (one each at both the ends), *Alstonia scholaris* (Fig. 2.176D) of Apocynaceae.

iii. **Hairy outgrowth on seeds**. This type of structure is present all over the seeds as an outgrowth of testa, e.g., Silk cotton, *Bombax ceiba* (Fig. 2.177A) of Bombaceae.

iv. **Persistent hairy style**. The persistent hairy styles are seen in the **achene fruits**, by means of which fruits can float in the air and disperse for long distance, e.g., *Clematis gouriana* (Fig. 2.177B), *Naravelia zeylanica* of Ranunculaceae.

v. **Balloon-shaped appendages**. The balloon-shaped appendages are formed by the modification of different parts of the plant; they help the fruits and seeds to remain in air for longer period. These are formed by inflated calyx in *Physalis minima* (Fig. 2.177C) of Solanaceae, inflated fruits (capsule) in *Cardiospermum helicacabum* (Fig. 2.177D) of Sapindaceae, inflated ovaries of legume in *Colutea arborescens* (Fig. 2.177E).

4. **Censer mechanism.** It is the mechanism by which very small seeds are liberated

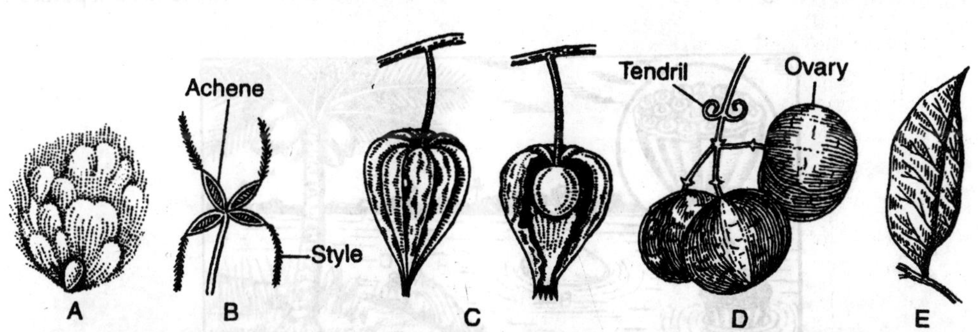

Fig. 2.177 : Fruits and seeds adopted for wind dispersal : A. Hairy outgrowth on seeds of *Bombax ceiba*, B. Persistent hairy style of *Clematis gouriana*, C–E. Balloon-shaped appendages [by inflated calyx of *Physalis* (C), by inflated fruits of *Cardiospermum* (D) and by inflated ovary of *Colutea* (E)]

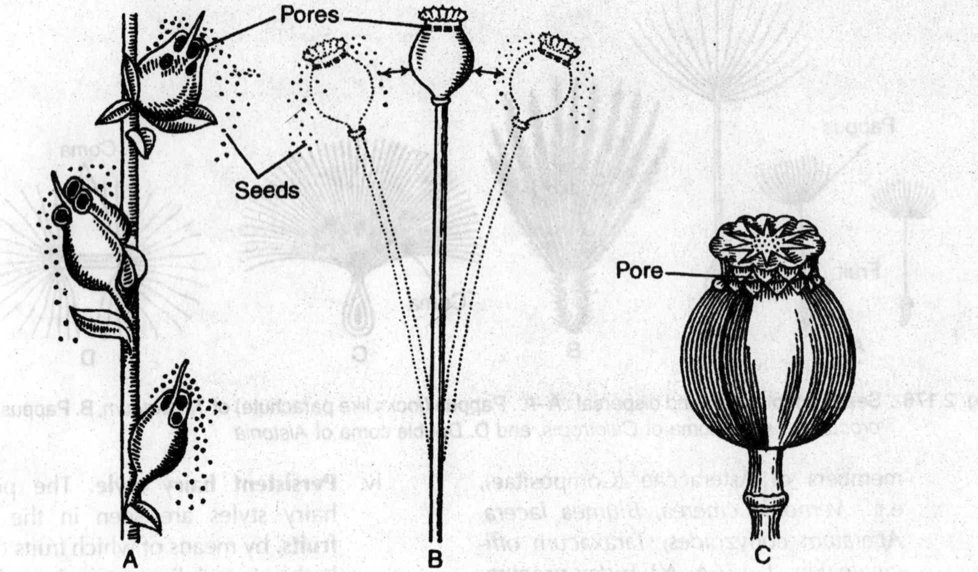

Fig. 2.178 : Seed dispersal by wind through censer mechanism from porous capsules of *Antirrhinum majus* (A) and opium poppy, *Papaver somniferum* (B); C. Single enlarged capsule of *P. somniferum*

slowly from the fruit through minute pores (apertures). These pores are so minute that when the fruits are shaken by wind only some seeds can come out through the pore at a time and disperse, e.g., snapdragon, *Antirrhinum majus* (Fig. 2.178A) of Scrophulariaceae; *Papaver somniferum* (Fig. 2.178B,C) of Papaveraceae.

B. **Dispersal by water**. The aquatic plants and the plants growing on the bank of river or sea-shore disperse their seeds and fruits by water current for a long distance. The seeds and the fruits have the floating devices in the form of fibrous or spongy water-proof coats. The plants requiring water as an agency for dispersal are called **hydrochorous**.

The **fruitlets** may lie embedded in the spongy and watertight thalamus and can be carried away by the water current, e.g., lotus, *Nelumbo nucifera* (Fig. 2.179A) of Nymphaeaceae etc.

The **fruits** may have watertight coverings of sufficient air spaces, by which they can easily float on water and drift away to distant places, by

Fig. 2.179 : Dispersal by water : A. Lotus, *Nelumbo nucifera*, and B. Coconut, *Cocos nucifera*, and C. Germinating coconut

water current. e.g., coconut, *Cocos nucifera* (Fig. 2.179B,C); betel nut, *Areca catechu* of Arecaceae. *Cerbera odollam* of Apocynaceae etc.

In some cases, **seeds** have air spaces in their coverings, which enable them to float on water and can be dispersed easily by water current, e.g., *Sagittaria sagittifolia, Alisma plantago* and many others of Alismataceae.

C. **Dispersal by animals**. The fruits and seeds with their adhering structures are dispersed by animals. Plants having such devices are called **zoochorous**. The nature of dispersal is different in **dry** and **fleshy fruits** and seeds.

Dry fruits and seeds

Many dry fruits and seeds have spines, barbs, bars, hooks or sticky glands by means of which they can attach themselves to the body of animals or human beings and thus be transported to a distant place. Fruits may provide with :

i. Spines. The fruits provided with spines become attached to the body of animals including human beings and dispersed at distant places, e.g., *Achyranthes aspera* (Fig. 2.180A) of Amaranthaceae, *Andropogon aciculatus* of Poaceae etc.

ii. Hooked bars. The fruits provided with many hooked bars become attached to the body (tail etc.) of animals including the clothes of human beings when passing through the plant population. The fruits are removed from their adhering surfaces naturally or artificially, thereby dispersal takes place, e.g., spiny cocklebur, *Xanthium strumarium* (Fig. 2.180B) of Asteraceae (Compositae); *Urena lobata* (Fig. 2.180C) of Malvaceae; *Martynia diandra* (have two hooks) (Fig. 2.180D), of Martyniaceae.

iii. Sticky glands. The fruits of some plants provided with sticky glands on pericarp, get attached with the body of animals and are dispersed to distant place, e.g., *Polanisia icosandra* of Capparidaceae, *Boerhaavia repens* (Fig. 2.180E) of Nyctaginaceae etc.

iv. Curved hairs. The fruits provided with large curved hairs, get attached with the animal body and are dispersed, e.g., *Chrysopogon* sp., *Aristida* sp. (Fig. 2.180F) of Poaceae.

Fleshy fruits

Fleshy fruits possess beautiful colour or tastes or both; invite the animals specially the birds. These animals and birds disperse fruits and seeds in various ways :

v. The red colour of the fruits of banyan, *Ficus benghalensis* of Moraceae, attracts the birds like crow etc. After eating the fruits they flow away to distant places. When the birds rub their beak to wipe of these seeds, the seeds are dispersed. Due to this, the plants are visible on the roof, cornices of old buildings, temples, on other places etc.

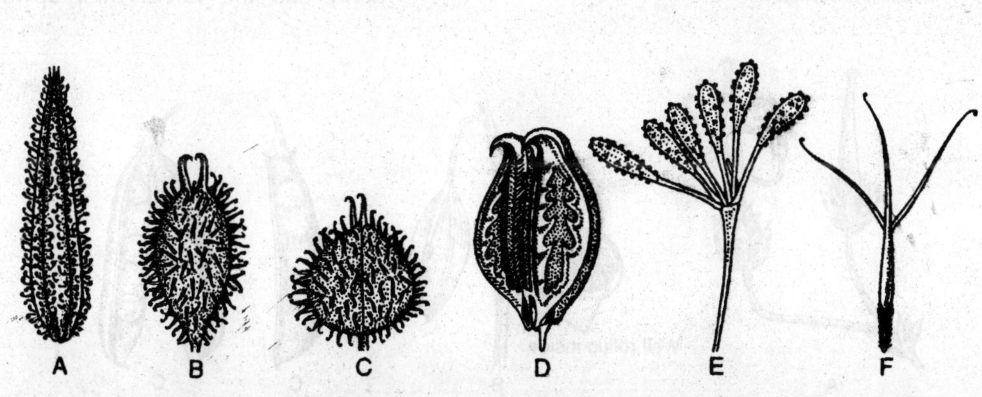

Fig. 2.180 : Dispersal by animals : A. By spines in *Achyranthes*, B–D. By hooked bars (B. in *Xanthium*, C. in *Urena* and D. in *Martynia*), E. By sticky glands in *Boerhaavia* and F. By curved hairs in *Aristida*

vi. Fruits like *Litchi chinensis* of Sapindaceae; mango, *Mangifera indica* of Anacardiaceae are collected by birds from the trees and then they fly away to distant places. After eating the sweet pulp, the seeds are left, thereby the seeds are dispersed.

vii. The tasty fruits of date palm, *Phoenix sylvestris* of Arecaceae, the jujube, *Ziziphus jujuba* of Rhamnaceae etc., are delicacies to dogs, jackal etc. When they devour the fruits, the well-protected seeds are not digested and are ejected along with droppings at distant places. These seeds develop into new plants on germination.

viii. Seeds and fruits of some aquatic plants are dispersed from one pond to the other, getting attached with the leg of aquatic birds like snipes, duck etc.

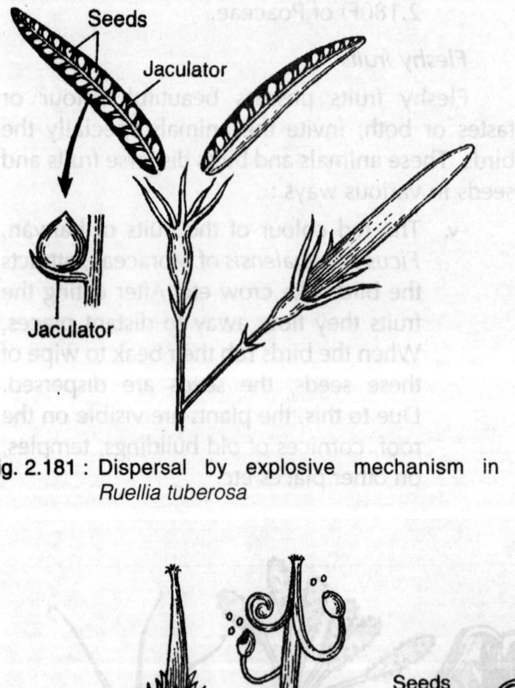

Fig. 2.181 : Dispersal by explosive mechanism in *Ruellia tuberosa*

ix. The fruits and seeds are also dispersed by human beings according to their wish and need to distant place.

x. The dehisced pods of *Abrus precatorius* of Fabaceae (Leguminosae) look like caterpillar and are mistaken by birds. When they are able to understand their mistake, the pods are thrown away and thus the seeds are dispersed.

D. **Dispersal by explosive mechanism**. In some cases, the pericarp of fruit — after ripening and drying — becomes stretched due to some external pressure in contact with air, water drops, animals or change of humidity. In this method, the fruits dehisce suddenly and the seeds are dispersed, but only for a short distance.

i. The mature fruits of *Andrographis paniculata, Ruellia tuberosa* (Fig. 2.181A) of Acanthaceae, when in contact with moisture, suddenly burst with an explosive sound and the seeds are dispersed in the nearby regions. Sometimes the seeds are provided with curved hooks, the jaculators — these help in their dispersal.

ii. The mature fruits of *Geranium palustre* (Fig. 2.182A) of Geraniaceae and balsam, *Impatiens balsamina* (Fig. 2.182B) of Balsaminaceae burst suddenly and the valves twist upwards,

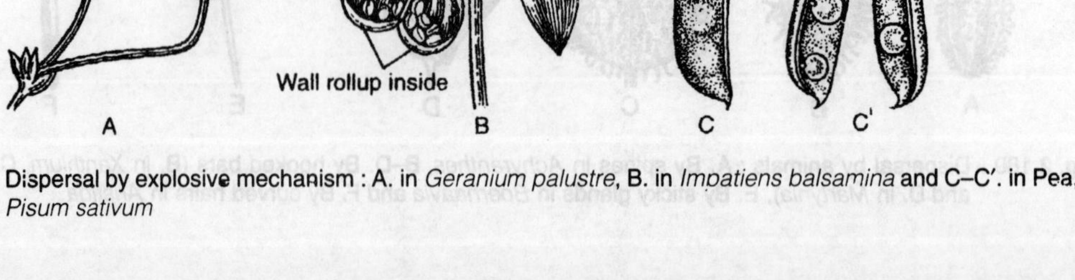

Fig. 2.182 : Dispersal by explosive mechanism : A. in *Geranium palustre*, B. in *Impatiens balsamina* and C–C′. in Pea, *Pisum sativum*

Fig. 2.183 : Dispersal by explosive mechanism in *Ecballium elaterium* showing dispersal of seed like a fountain

thereby the seeds are dispersed in the adjacent region.

iii. The mature fruits of pea, *Pisum sativum* (Fig. 2.182C) and many other plants of Fabaceae, etc., suddenly burst into valves and the valves twist upwards. The seeds are then ejected with some force.

iv. The mature fruits of castor, *Ricinus communis* of Euphorbiaceae are normally broken up elastically and the seeds are dispersed.

v. The mature fruits of squirting cucumber, *Ecballium elaterium* (Fig. 2.183) of Cucurbitaceae (grows in the Mediterranean region); contain a mucilaginous substance. The fruits drop off from their stalk even by a slight touch and the mucilaginous substance comes out with seeds as a fountain through the holes left at one end. Similar situation is also found in *Luffa aegyptiaca* of Cucurbitaceae.

3 Embryology

INTRODUCTION

The **Phanerogams** (the flowering-plants) are also called **spermatophytes** (the seed bearing plants). These plants propagate mainly through seeds. The seed is a structure in which the embryo is enclosed. Adjacent to the embryo, foods are stored either inside the endosperm (albuminous) or in cotyledon (exalbuminous) for future use.

Embryology is the study of structure and development of embryo, including the structure and development of male and female reproductive organs, fertilisation and similar other processes.

The Greek philosopher **Aristotle** (384-322 B.C.) is the **'Father of the Study of Natural Sciences'**. The greater part of his work is not available. With the available data, it seems that he did not recognise the presence of sex in plants. He believed that *male* and *female* plants were so blended that they generated on their own and the offspring developed from the superfluous (above what is enough) food of the plant. Later **Theophrastus** (370-287 B.C), the **'Father of Botany' in third century B.C.** referred the pollination of the date palm. After this, the nature of sexuality of plants was forgotten for about 2,000 years and his report was considered as useless. After a long gap, it was **Nehemiah Grew** who first clearly stated, in 1682, that the stamens are nothing but the male sex organs of a plant. Later, in 1694, **R. J. Camerarius** suggested that the anthers are the male sex organs of a plant, while the ovaries with the styles were the female sex organs. After many years, **J. G. Kölreuter** (1761) fully confirmed the idea of Camerarius.

Hanstein (1870) first studied in detail the nature of embryogeny in *Alisma* and *Capsella*, though he had no clear conception on **syngamy** (fusion between male and female gametes). **Strasburger** (1884) was the first to discover the phenomenon of **syngamy** in *Monotropa hypopitys*. Later, **Nawaschin** (1898) had established the phenomenon of **double fertilisation** in *Fritillaria tenella* and *Lilium martagon*. Later on, the different aspects of embryology were studied by different workers. Most of the facts on development of gametophytes and embryo were discovered later.

During the 20th century, particular interest was developed on *experimental embryology* which includes the problems on storage and viability of pollen grains, effect of environmental factors on the growth of pollen tube, production of seedless fruits, embryo culture, induction of parthenogenesis etc. Thus embryology maintains an intimate association with genetics and physiology and opens many promising lines of research for the useful and economic development of crops. The pioneer workers like **Schnarf, Soueges, Finn, Juel, Gustafsson, Coulterb, Johonson** and many others have made many important contributions in this field. In **India, Maheshwari, Swamy, Banerjee, Johri, Vasil, Shivanna,** and many others also contributed a lot in this field.

Embryology played a significant role in consideration of systemic botany. Although **Hofmeister** and **Strasburger** pointed out the possibility of utilising embryological characteristics in taxonomy, it was **Schnarf** (1931) who first prominently brought into focus the **role of embryology in taxonomy**. It has been proved that many embryological characteristics have taxonomic significance. **Maheshwari** (1950, 1963) had listed 12 characteristics to be of special interest in consideration of systemic botany. The valuable contributions are available with the genera like *Paeonia, Trapa, Butomus, Peganum, Parassia* etc., and families like Crassulaceae, Liliaceae, Loranthaceae, Lemnaceae, Cyperaceae etc.

Some important embryologists and their contributions :

1. Camerarius, R. J. (1694) : Discovered plant sex organs and pollination.
2. Amici, C. B. (1824) : Discovered pollen tube.
3. Hofmeister (1848) : Discovered the nature and development of male and female gametophytes.
4. Strasburger, E. (1884) : Discovered the development of male and female gametophytes and syngamy.
5. Schlliden (1897) : Described the origin and development of embryology.
6. Nawaschin, S. (1898) : Discovered double-fertilisation.
7. Schnarf, K. (1929) : Wrote two important reference books on Embryology. Later, in 1931, he first prominently brought into focus the **role of embryology in taxonomy**.
8. Maheswari, P. (1950-1966) : Wrote an important book on "Embryology of Angiosperm". He made several contributions to embryology.
9. Johri, B. M. (1950-2006) : Made several contributions to experimental embryology.

MICROSPOROGENESIS AND MICROGAMETOGENESIS

Development of Microspore mother cell and Microsporogenesis (Development of microspores i.e., pollen grain)

Microspores i.e., the pollen grains, are developed inside microsporangia. The microsporangia are developed inside the corners of the 4-lobed anther.

Young anthers are more or less oblong in shape in section and made up of homogeneous mass of meristematic cells without intercellular space (Fig. 3.1A). With further development, the anther becomes 4-lobed. The outer layer of anther is called epidermis. Below the epidermis, at each corner, some cells become differentiated from others by their dense protoplasm — **archesporium** or **archesporial cells** (Fig. 3.1B). Each archesporial cell then divides mitotically and

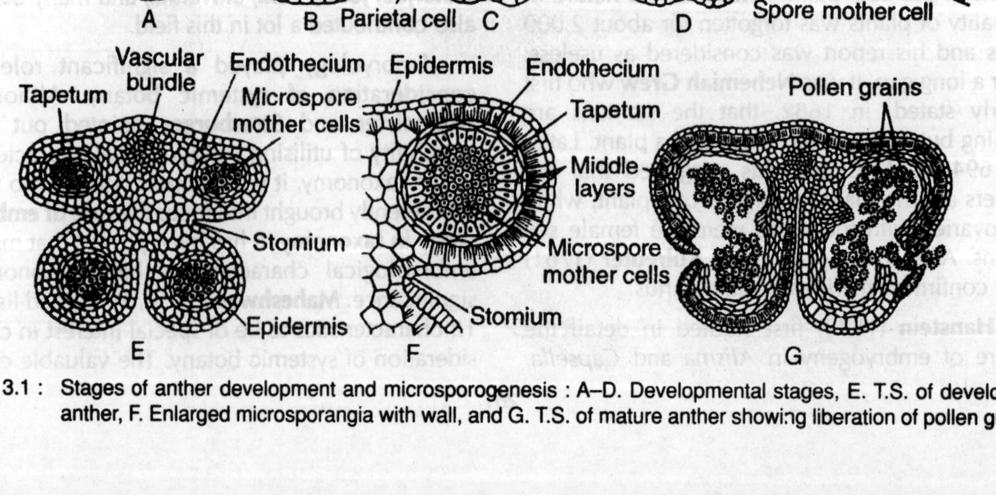

Fig. 3.1 : Stages of anther development and microsporogenesis : A–D. Developmental stages, E. T.S. of developing anther, F. Enlarged microsporangia with wall, and G. T.S. of mature anther showing liberation of pollen grains

forms an outer **primary parietal cell** and an inner **primary sporogenous cell**. The outer primary parietal cells form primary parietal cell layer at each corner (Fig. 3.1C). Below the parietal cell layer, the primary sporogenous cells remain in groups i.e., the *sporogenous tissue*. The cells of primary parietal layer then divide both periclinally and anticlinally and form multilayered **antheridial wall** (Fig. 3.1D). The innermost layer of antheridial wall, which remains in close contact with the sporogenous tissue, functions as nutritive layer, called **tapetum** (Fig. 3.1D). The primary sporogenous cells either *directly* function as spore mother cells or divide *mitotically* into a number of cells which function as **spore mother cells** (Fig. 3.1D). The spore mother cell undergoes *meiotic* division and gives rise to 4 microspores arranged tetrahedrally (Fig. 3.2).

Structure of Microspores i.e., Pollen grains

Microspore i.e., the pollen grain, is the first cell of the male gametophyte, which contains only one haploid nucleus. These are of various shapes — polyhedral (milk thistle, *Sonchus palustris* of Asteraceae), cubical (*Basella alba* of Basellaceae), trigonal (common in Onagraceae), cylindrical (*Rheo discolor* of Commelinaceae) etc. The size of the pollen grains generally varies from 10–80μm, but the size may be even 100μm in diameter. The pollen grains have two walls — outer **exine** (*the exine is further differentiated into two regions, outer* ***sexine*** *and inner* ***nexine***) and inner **intine**. The exine is cutinised and tough with different ornamentations. It may be warty, spiny etc. It can protect the pollen from external injury. The intine is very thin, elastic and delicate. The structure of different pollen grains is given in Fig. 3.3.

Usually the mature pollens are not attached in tetrad and they get separated from one another. In some plants like *Typha angustata* of Typhaceae etc., they do not get separated from

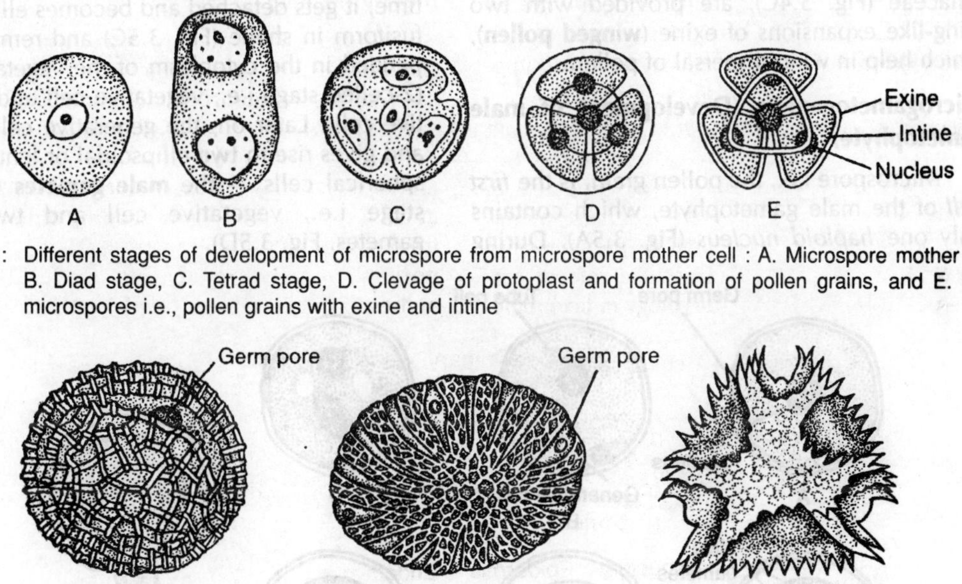

Fig. 3.2 : Different stages of development of microspore from microspore mother cell : A. Microspore mother cell, B. Diad stage, C. Tetrad stage, D. Clevage of protoplast and formation of pollen grains, and E. Four microspores i.e., pollen grains with exine and intine

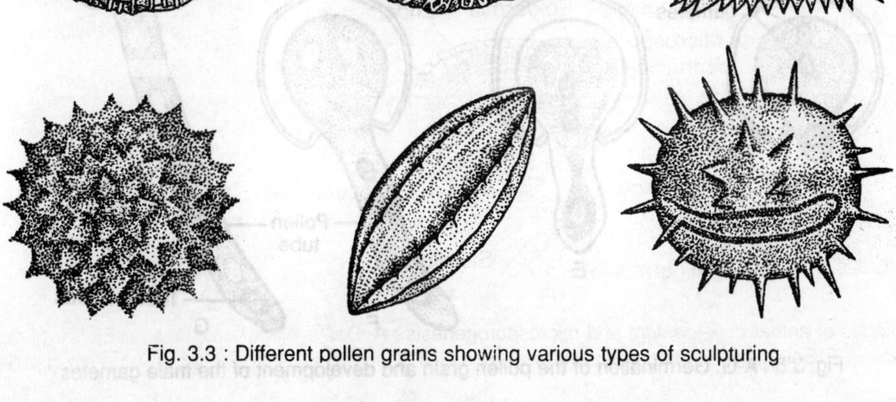

Fig. 3.3 : Different pollen grains showing various types of sculpturing

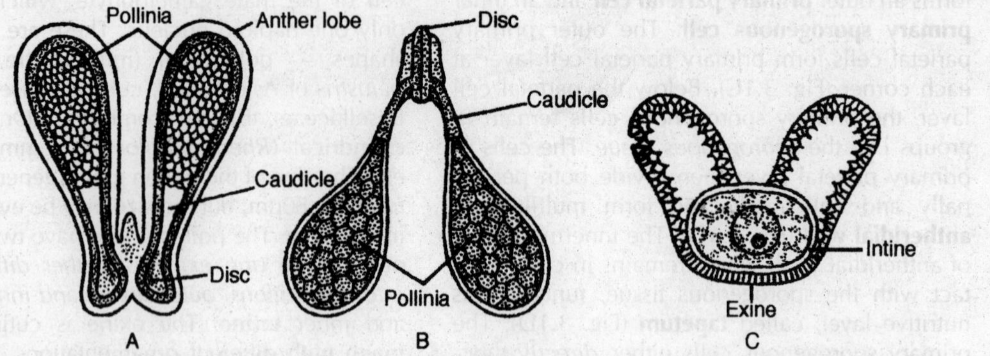

Fig. 3.4 : Pollen grains and Pollinia : A. Pollinia of *Orchis*, B. Pollinia of *Calotropis*, and C. Pollen of *Pinus*

one another (**compound pollen grain**). In orchids (Fig. 3.4A) and members (*Calotropis procera* etc.) of Asclepiadaceae (Fig. 3.4B); all the pollen grains within each pollen sac remain united forming the structure called **pollinium** (pl. pollinia). The pollen grains of *Pinus* spp. of Pinaceae (Fig. 3.4C), are provided with two wing-like expansions of exine (**winged pollen**), which help in wind dispersal of pollen.

Microgametogenesis (Development of male gametophyte)

Microspore i.e., the pollen grain, is the *first cell* of the male gametophyte, which contains only one *haploid nucleus* (Fig. 3.5A). During early stage of development, it remains within the microsporangium. The cell undergoes unequal division and forms a small **generative cell** and a large **vegetative** or **tube cell** (Fig. 3.5B). Initially the *generative cell* remains lying at one corner of the spore wall. Within short time, it gets detached and becomes ellipsoid or fusiform in shape (Fig. 3.5C) and remains suspended in the cytoplasm of the vegetative cell (2-celled stage i.e., vegetative cell and generative cell). Later on, the generative cell divides and gives rise to **two** ellipsoidal or lenticular or spherical cells — the **male gametes** (3-celled stage i.e., vegetative cell and two male gametes, Fig. 3.5D).

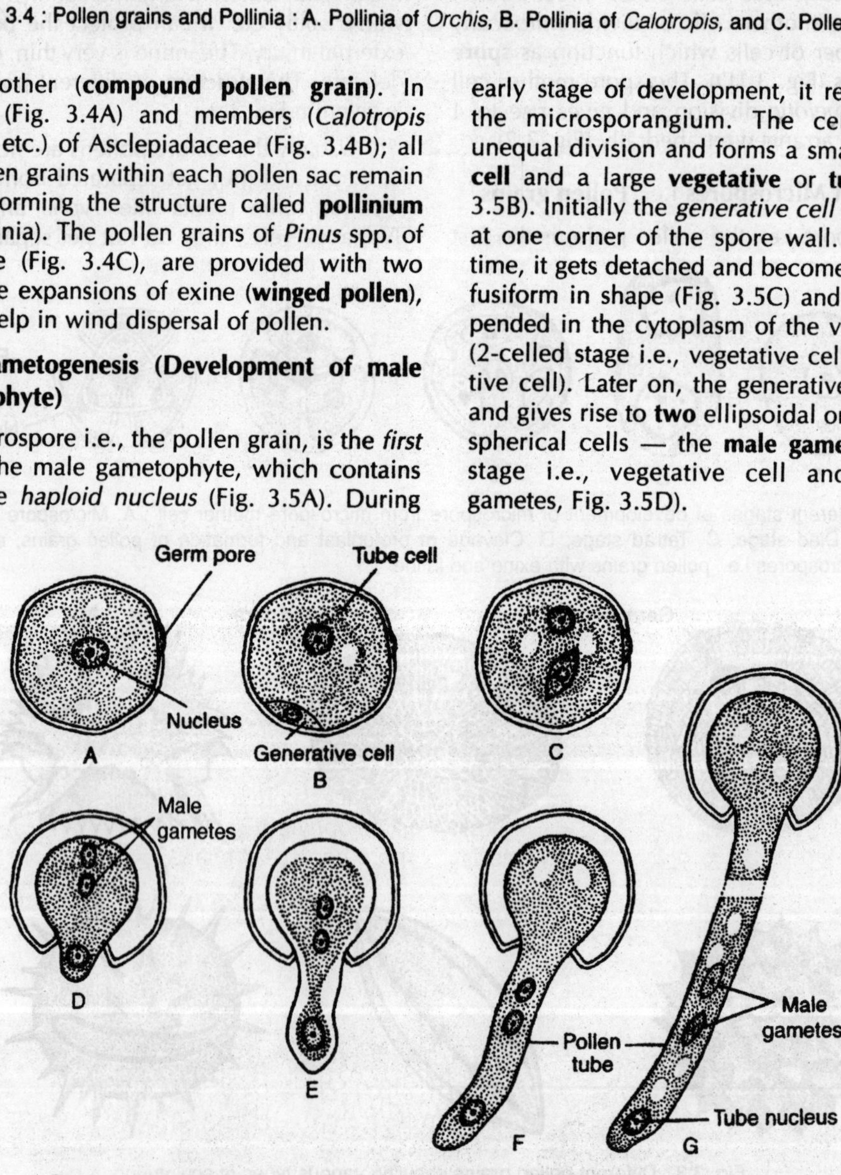

Fig. 3.5 : A-G. Germination of the pollen grain and development of the male gametes

The second division i.e., the division of generative cell, may take place either in the pollen grain or in the pollen tube which develops through germ pore after pollination. The nucleus of the vegetative cell is commonly known as **tube nucleus** (Fig. 3.5D). It usually shows sign of degeneration with the maturation of generative cell. Finally the tube nucleus remains within spore or may enter the pollen tube (Fig. 3.5E, F and G). Sooner or later it may be degenerated completely.

Significance of **tube nucleus** : Earlier workers thought that the tube nucleus had great significance in the direction of growth of the pollen tube, as it is usually present just behind the growing point within the pollen tube. However, recent workers differ with the above opinion and consider it as a purely non-functional vestigial structure, based on the following facts :

1. In branched pollen tube, the tube nucleus remains in one tube, but all the tubes grow normally.
2. It does not always occupy the position behind the growing point within the pollen tube, but in many cases it lies behind the male gametes.
3. In some cases, the growing pollen tube does not have any tube nucleus as it degenerates prior to the development of pollen tube.

MEGASPOROGENESIS AND MEGAGAMETOGENESIS

Development of Megaspore mother cell

The ovule develops as multicellular placental outgrowth including the epidermal and a number of hypodermal cells. With further development, this gives rise to nucellus and one or two integuments from its basal region. In ovules, with two integuments, usually the inner one is formed first than the outer one. The inner one is more delicate and inconspicuously developed than the outer one.

One hypodermal cell of the nucellus becomes differentiated from the other by its bigger size, dense cytoplasm and conspicuous nucleus, called **archesporial cell** (Fig. 3.6A). The archesporial cell divides transversely and forms an inner **primary sporogenous cell** and an outer **primary parietal cell** (Fig. 3.6B). The primary sporogenous cell functions as **megaspore mother cell** (Fig. 3.6C) and the primary parietal cell undergoes repeated vertical divisions and forms layers of **parietal cells** (Fig. 3.6C). Sometimes, the archesporial cell does not divide and directly functions as megaspore mother cell.

Megasporogenesis (Development of megaspores)

The megaspore mother cell is diploid (2n), which undergoes meiosis and forms four haploid (n) megaspores (Fig. 3.6D). The first division of megaspore mother cell is transverse, forming two cells. Both the cells again divide transversely and thus four (4) haploid megaspores are formed. The megaspores are then arranged in an axial row, called **linear tetrad** (Fig. 3.6D). Out of four megaspores, only one which remains towards the chalazal end behaves as functional megaspore and the other three which remain towards the micropylar end, gradually degenerate (Fig. 3.6E). The functional megaspore forms the **female gametophyte** i.e., the **embryo sac.**

Megagametogenesis (Formation of female gametophyte i.e., Embryo sac)

Megaspore (n) is the first cell of the female gametophyte (Fig. 3.7A). The functional megaspore becomes enlarged at the expense of tape-

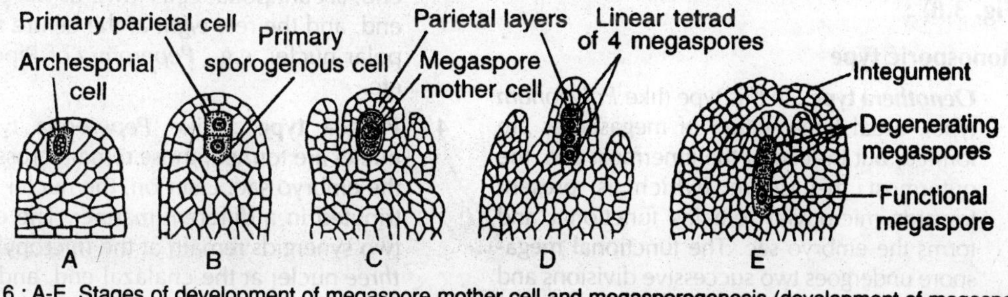

Fig. 3.6 : A-E. Stages of development of megaspore mother cell and megasporogenesis (development of megaspore)

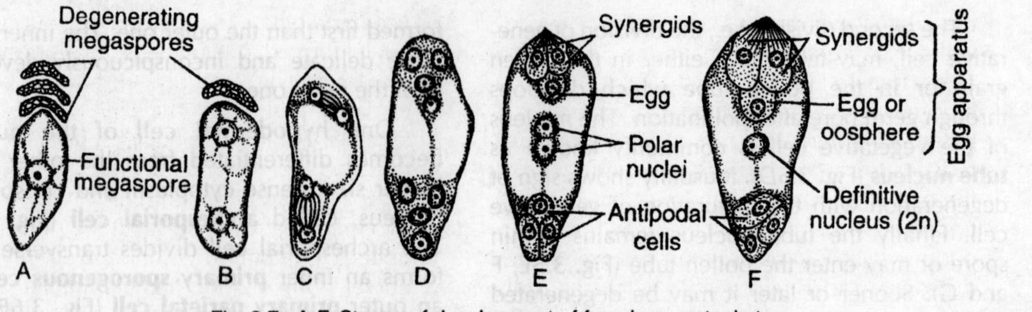

Fig. 3.7 : A-F. Stages of development of female gametophyte

tum and the nucellus and thus forms the **female gametophyte** i.e., the **embryo sac**. Initially, the embryo sac is uninucleate and with further growth its nucleus divides by three successive divisions and forms eight nuclei (Fig. 3.7B, C and D). Out of eight nuclei, initially four remain towards the micropyle end and the other four towards the chalazal end. One nucleus from each pole then moves towards the centre and forms a pair of **polar nuclei** (Fig. 3.7E). These nuclei fuse together and form 2n nucleus, the **definitive nucleus**. It is also known as **polar fusion nucleus** or **secondary nucleus**. The three nuclei of the micropylar end form the **egg apparatus** and the rest three at the chalazal end are called **antipodal cells**. In the egg apparatus, each nucleus is surrounded by viscous mass of cytoplasm without any wall, of which the middle one is the largest and called **egg, ovum** or **oosphere** and the rest two (one on each side of the egg) are the **synergids** or helping cells. The antipodal cells have viscous mass of cytoplasm, covered by cellulosic wall (Fig. 3.7F).

This type of embryo sac development is very common in angiosperms and is known as **ordinary type** or **normal type** or ***Polygonum* type**. This type is also known as **monosporic type**, because, out of four megaspores, only one remains functional and forms the embryo sac.

***Other types of embryo sac* development (Fig. 3.8) :**

Monosporic type

1. ***Oenothera* type** : In this type (like *Polygonum* type), usual linear tetrad of megaspores are formed, but instead of the innermost one, the outermost megaspore (which is present towards micropyle) remains functional and forms the embryo sac. The functional megaspore undergoes two successive divisions and forms 4 nuclei. All the nuclei remain towards the micropyle. Out of **four** nuclei, *three* nuclei form the egg apparatus (egg and two synergids) and the remaining one forms a *single* polar nucleus. **Second polar nucleus and antipodal cells are absent**, e.g., *Oenothera* and other members of Onagraceae.

Bisporic type

2. ***Allium* type** : The megaspore mother cell divides to form two cells, the upper one quickly degenerates. The lower one then divides and forms two nuclei, distributed in the two poles. Later on, both the nuclei undergo two successive divisions and form usual octant type of embryo sac, i.e., *polygonum* type. Here two megaspore nuclei take part in the development of embryo sac i.e., **bisporic type**, e.g., *Allium, Scilla, Trillium* etc., of Liliaceae.

Tetrasporic type

3. ***Peperomia* type** : The megaspore mother nucleus undergoes meiotic division and forms *four* nuclei which remain crosswise in the embryo sac without any wall. All the nuclei undergo two successive divisions and form 16 nuclei which remain dispersed inside the sac. Later on, out of 16 nuclei, egg and one synergid remain at the micropylar end, *six* antipodal cells towards the chalazal end, and the rest *eight* at the centre forming polar nuclei, e.g., *Peperomia* of Piperaceae etc.

4. ***Penaea* type** : Like *Peperomia* type, 16 nuclei are formed, those remain crosswise in the embryo sac. Later on, the nuclei are distributed in a different manner. The egg and two synergids remain at the micropylar end, *three* nuclei at the chalazal end, and *four* at

the centre and *three* each on the two side walls, e.g., *Penaea* of Penaeaceae.

5. ***Drusa* type** : Like *Peperomia* type, initially four megaspores are formed, these are distributed in different ways. One megaspore remains towards the micropyle, and the rest three at the chalazal end. All the nuclei undergo two divisions and form 16 nuclei, out of which *four* nuclei remain towards the micropyle and the rest *twelve* at the chalazal end. In the mature embryo sac, egg and two synergids remain towards the micropyle, two (one from each pole) at the centre and the rest *eleven* at the chalazal end, e.g., *Drusa oppositifolia* of Apiaceae.

6. ***Fritillaria* type** : Like *Drusa type*, out of four nuclei formed, one nucleus remains towards the micropyle, and the rest three at the chalazal end. The chalazal nuclei fused together and form 3n nucleus. Both the cells thus undergo one mitotic division and again form a tetrasporic stage. Out of four nuclei, two remain at each pole. All the nuclei then undergo mitotic division and form eight nuclei. Out of four haploid nuclei at the micropyle, one egg and two synergids are formed, those remain at the micropylal end; three triploid nuclei at the chalazal end and one from each pole remain at the centre (one haploid and the other one triploid),

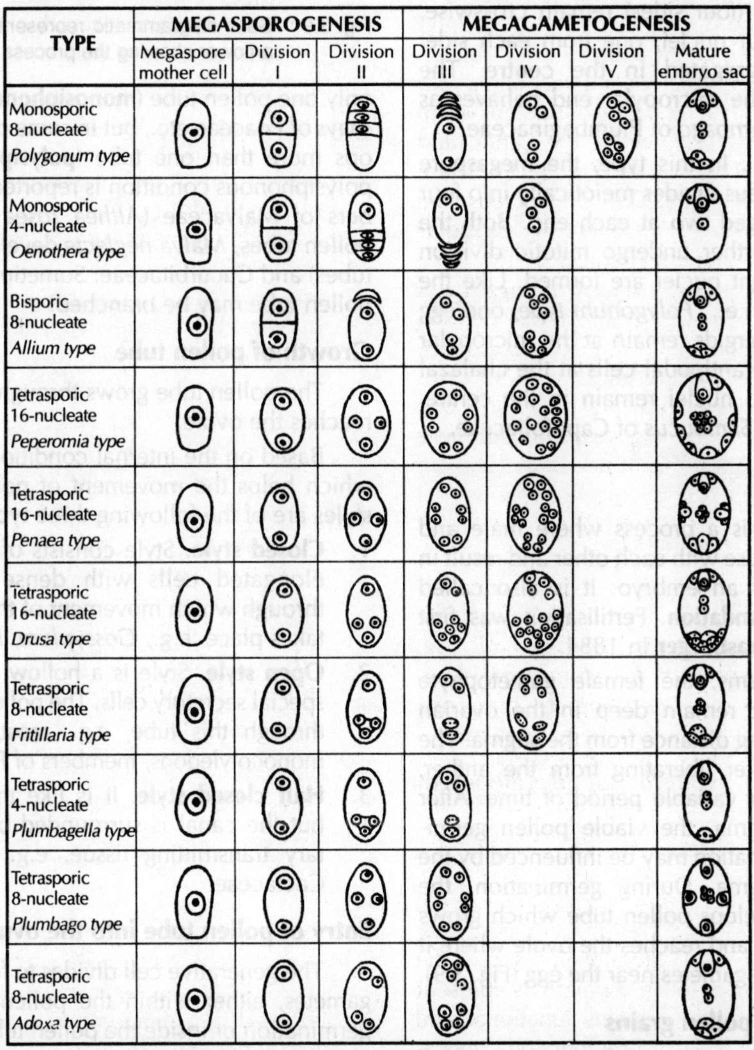

Fig. 3.8 : Development of different types of embryo sac in angiosperms (*after* Maheshwari)
[Micropyle above in all illustrations]

e.g., *Fritillaria, Tulipa* and some other members of Liliaceae.

7. ***Plumbagella* type** : It is like *Fritillaria* type which forms 1st and 2nd tetrasporic stage with two haploid nuclei at the micropyle and two triploid nuclei at the chalazal end of the embryo sac. Later on, the nuclei are distributed in such a way that the egg is at the micropyle, one triploid nucleus at the chalazal end and one triploid plus one haploid nuclei at the centre, e.g., *Plumbagella* of Plumbagellaceae.

8. ***Plumbago* type** : It is like *Penaea* type where firstly four nuclei are formed followed by eight nucleated embryo sac. The two nuclei at each side (four sides) remain crosswise. Later on, four nuclei, one from each side, become aggregated in the centre. The nucleus at the micropylar end behaves as egg, e.g., *Plumbago* of Plumbaginaceae.

9. ***Adoxa* type** : In this type, the megaspore mother nucleus divides meiotically into four nuclei arranged two at each end. Both the nuclei — further undergo mitotic division and thus eight nuclei are formed. Like the normal type i.e., *Polygonum type*, one egg and two synergids remain at the micropylar region, three antipodal cells at the chalazal end and two nuclei remain in the centre, e.g., *Adoxa, Sambucus* of Caprifoliaceae.

FERTILISATION

Fertilisation is a process where male and female gametes fuse with each other and result in the formation of an embryo. It is also called **syngamy** or **fecundation**. Fertilisation was first discovered by **Strasburger** in 1884.

In angiosperms, the female gametophyte i.e., embryo sac remain deep in the ovarian cavity, quite a long distance from the stigma. The pollen grains, after liberating from the anther, remain viable for variable period of time. After reaching the stigma, the viable pollen germinates. The germination may be influenced by the fluid of the stigma. During germination, the pollen grain develops pollen tube which grows through the style and reaches the ovule where it releases the male gametes near the egg (Fig. 3.9).

Germination of pollen grains

During germination, pollen grain develops pollen tube. The pollen grain usually develops

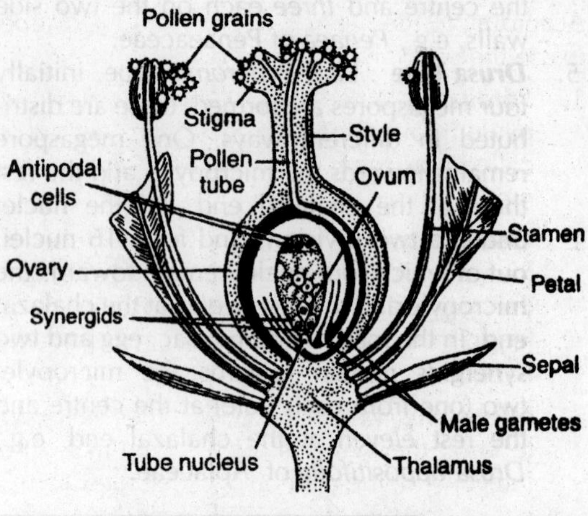

Fig. 3.9 : Semi-diagrammatic representation of L.S. of a flower showing the process of fertilisation

only one pollen tube (**monosiphonous**), e.g., *Zea mays* of Poaceae etc., but in some plants it develops more than one tube (**polysiphonous**). The polysiphonous condition is reported in the members of Malvaceae (*Althea rosea* develops 10 pollen tubes, *Malva neglecta* develops 14 pollen tubes) and Cucurbitaceae. Sometimes, the single pollen tube may be branched.

Growth of pollen tube

The pollen tube grows through the style and reaches the ovule.

Based on the internal condition of the style, which helps the movement of pollen tube, the styles are of the following three types :

1. **Closed style.** Style consists of solid core of elongated cells with dense protoplasm, through which movement of the pollen tube takes place, e.g., *Gossypium, Datura* etc.
2. **Open style**. Style is a hollow tube lined by special secretory cells. The pollen tube creeps through this tube, e.g., most members of monocotyledons, members of Papaveraceae.
3. **Half closed style**. It is like the open type, but the canal is surrounded by a rudimentary transmitting tissue. e.g., members of Cactaceae.

Entry of pollen tube into the ovule

The generative cell divides to form two male gametes, either within the pollen grain before germination or inside the pollen tube during germination. After reaching the top of the ovary, pollen tube may enter through different routes.

According to the method of entry of pollen tube inside the ovary, it is of three types :

1. **Porogamy**. When the pollen tube enters the ovule through the micropyle, the method is called porogamy. This process is very common.
2. **Chalazogamy**. When the pollen tube is found to enter the ovule through chalaza, the method is called chalazogamy. It is found in *Casuarina* of Casuarinaceae, members of Betulaceae, Fagaceae etc.
3. **Mesogamy**. When the pollen tube is found to enter the ovule through funiculus or the integument, the method is called mesogamy. e.g., *Cucurbita* of Cucurbitaceae etc.

Entry of pollen tube into the embryo sac

After the entry of pollen tube inside the ovary, it proceeds through the nucellus and reaches the embryo sac. The pollen tube is then progressed inside, either :

1. between one synergid and the egg, or
2. between one synergid and the wall of embryo sac, or
3. directly into one synergid, thereby it is being destroyed.

Release of male gametes from pollen tube

After the entry of pollen tube inside the embryo sac, the gametes are released following either of the three processes :

1. Two openings are formed just behind the apex of the pollen tube and one gamete is discharged through each opening.
2. Bursting takes place at the apex of the pollen tube and the gametes are discharged, or
3. The apex of the pollen tube divides into two unequal short branches, of which the shorter one reaches the egg and the longer one towards the secondary nucleus. Later on, the apex of each branch bursts and the gametes are released.

Fusion

The fusion between male and female gametes is called **syngamy** and the fusion between male gamete and definitive nucleus i.e., secondary nucleus (formed by the fusion of two polar nuclei) is called **triple fusion** (Fig. 3.10A-D).

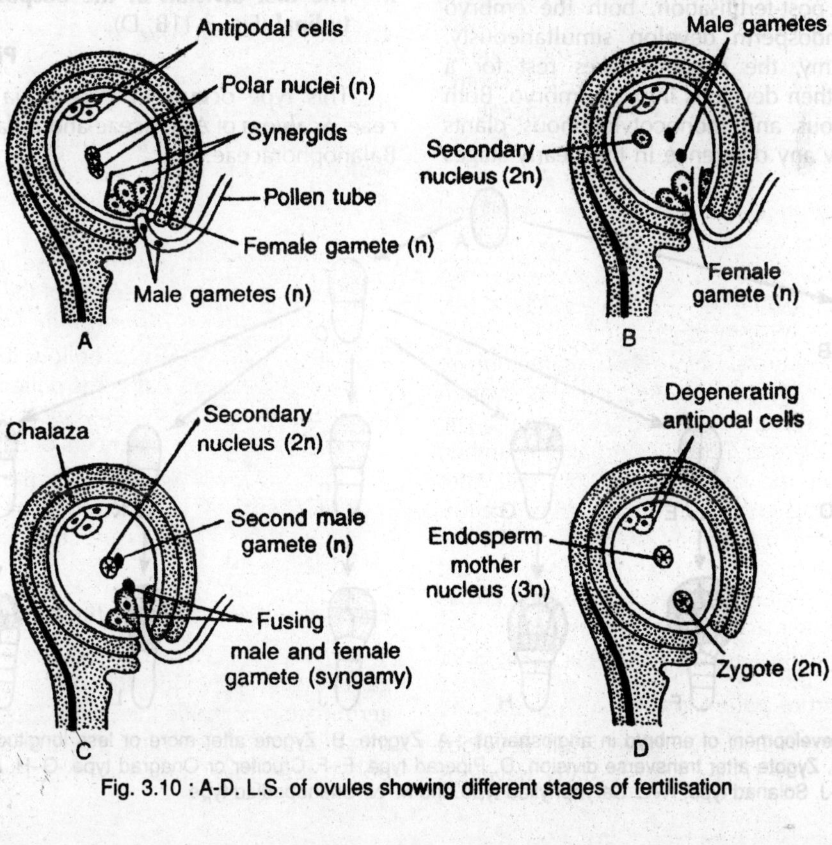

Fig. 3.10 : A-D. L.S. of ovules showing different stages of fertilisation

After releasing from the pollen tube, one male gamete (n) fuses with the egg (n) and forms the oospore or zygote (2n). This is called **syngamy** or **true fertilisation**. The other male gamete (n) fuses with the definitive nucleus i.e secondary nucleus (2n) and forms a triploid nucleus (3n), called **endosperm mother nucleus** i.e., the first cell of the endosperm. This is called **triple fusion**.

Both the male gametes take part in fertilisation i.e., the fertilisation takes place twice. This is called **double fertilisation**. So the *double fertilisation includes syngamy or true fertilisation and triple fusion*. The double fertilisation was first observed by **Nawaschin** (1898) in species of *Fritillaria* and *Lilium*.

The oospore or zygote develops into embryo and the endosperm mother nucleus develops the endosperm. The endosperm is used for nutrition by the developing embryo.

The other components i.e., synergids and antipodal cells, gradually degenerate after fertilisation.

DEVELOPMENT OF EMBRYO (i.e., embryogeny)

During post-fertilisation, both the embryo and the endosperm develop simultaneously. After syngamy, the oospore takes rest for a period and then develops into an embryo. Both dicotyledonous and monocotyledonous plants do not show any difference in their early stages of development, but difference is visible in the latter stages.

Johansen (1950) recognised *six types of embryo development* in angiosperms based on the patterns of differentiation of the terminal and basal cells. These are :

1. *Piperad type*, e.g., *Peperomia, Balanophora* etc.
2. *Crucifer* or *Onagrad type*, e.g., commonly found in Brassicaceae (Cruciferae), Ranunculaceae, Onagraceae, Rutaceae etc.
3. *Asterad type*, e.g., members of Compositae, *Polygonum* of Polygonaceae etc.
4. *Caryophyllad type*, e.g., members of Caryophyllaceae, *Saxifraga* of Saxifragaceae etc.
5. *Solanad type*, e.g., members of Solanaceae, *Linum* of Linaceae, *Papaver* of Papaveraceae etc.
6. *Chenopodiad type*, e.g., members of Chenopodiaceae etc.

The above types of embryo development can be represented schematically (Fig. 3.11) :

I. The first division of the oospore is longitudinal (Fig. 3.11B, D).

Piperad type

This type occurs in *Peperomia* of Piperaceae, *Scabiosa* of Asteraceae and *Balanophora* of Balanophoraceae.

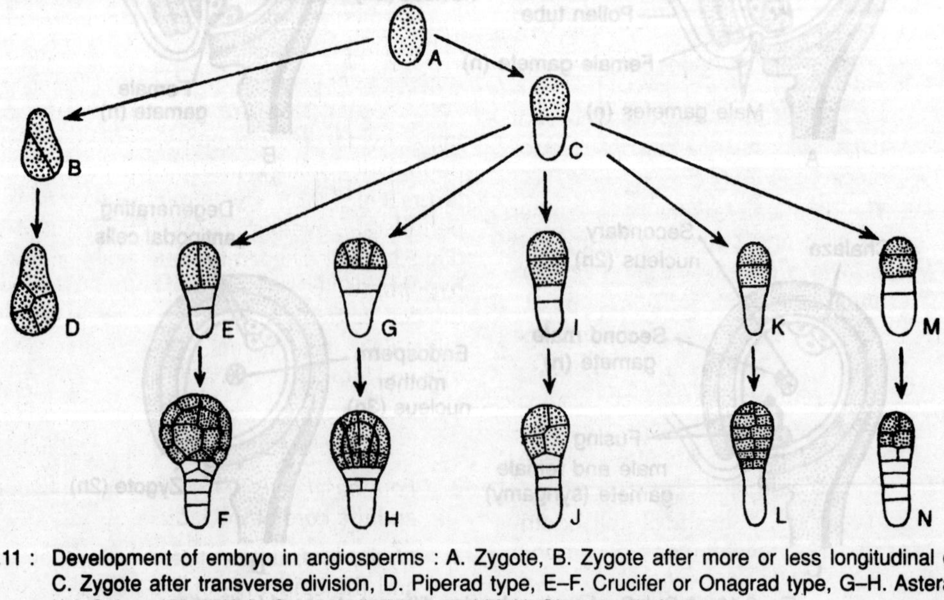

Fig. 3.11 : Development of embryo in angiosperms : A. Zygote, B. Zygote after more or less longitudinal division, C. Zygote after transverse division, D. Piperad type, E–F. Crucifer or Onagrad type, G–H. Asterad type, I–J. Solanad type, K–L. Caryophyllad type and M–N. Chenopodiad type

II. The first division of the oospore is transverse.

A. **The division of the apical cell of the two celled proembryo is longitudinal :**

1. The basal cell plays no part or only a minor part in the subsequent development of embryo (Fig. 3.11E, F).

Crucifer or **Onagrad type**

This type occurs in *Capsella* of Brassicaceae, *Veronica* of Verbenaceae, *Euphorbia* of Euphorbiaceae, *Mentha* of Lamiaceae etc.

2. The basal as well as apical cells contribute to the development of embryo (Fig. 3.11G, H).

Asterad type

This type occurs in almost all members of Compositae (Asteraceae), *Polygonum* of Polygonaceae, *Oxalis* of Oxalidaceae, *Poa* of Poaceae, *Urtica* of Urticaceae, *Lamium* of Lamiaceae etc.

B. **The division of the apical cell of the two celled proembryo is transverse** :

a. The basal cell plays no or only a minor part in the subsequent development of embryo.

1. The basal cell commonly forms 2 or more celled suspensor (Fig.3.11I, J)

Solanad type

Physalis, Nicotiana, Datura and *Hyocyamus* of Solanaceae; *Sherardia* of Rubiaceae, *Papaver* of Papaveraceae, *Hydnora* of Hydnoraceae, *Linum* of Linaceae etc.

2. The basal cell undergoes no further division and suspensor (if present) is always developing from the terminal cell (Fig. 3.11K, L).

Caryophyllad type

This type is found in *Sagina* of Caryophyllaceae, *Medicago* of Fabaceae, *Sagittaria* of Alismataceae, *Drosera* of Droseraceae, *Ruppia* of Potamogetonaceae etc.

b. Both basal cell and apical cell contribute to the development of embryo (Fig. 3.11M, N).

Chenopodiad type

This type occurs in *Myosotis* of Boraginaceae, *Beta* and *Chenopodium* of Chenopodiaceae and *Polemonium* of Polemoniaceae etc.

a. Typical dicotyledonous type (Crucifer type)

The typical dicotyledonous type of embryo is found in *Capsella bursa-pastoris* of Brassicaceae (Cruciferae). The ovule is *campylotropous type*, thus the embryo sac and the endosperm as well as embryo become horseshoe-shaped and the embryo looks upside down.

The oospore (Fig. 3.12A), first divides transversely into a large **basal cell** (b) and a small **terminal cell** (a), (Fig. 3.12B). The basal cell divides transversely and forms two cells, c and d, whereas the terminal cell (a) divides longitudinally to form two cells (a_1 and a_2). Thus, a 4-celled **proembryo** is formed (Fig. 3.12C).

Both the cells (c and d) of basal region, undergo repeated transverse division and form 5-10 celled suspensor (Fig. 3.12D, E). The lowermost cell of the suspensor becomes swollen disproportionately and forms a vesicle, called the **vesicular cell** or the **haustorial cell**. The suspensor pushes the developing embryo into the endosperm and the basal cell functions as **haustorium**. The uppermost cell of the suspensor is called **hypophysis** (Fig. 3.12F).

Meanwhile, the two terminal cells further divide longitudinally at right angle to the first longitudinal division and form a quadrant i.e., 4 celled structure (Fig. 3.12D). The quadrant then undergoes transverse division and forms octant i.e., 8 celled structure (Fig. 3.12E). The octant then divides periclinally (i.e., parallel to the peripheral surface) and forms outer and inner layers (Fig. 3.12F). The terminal octant mass with peripheral covering is called **embryonal mass** (Fig. 3.12G). The **embryonal mass** along with **hypophysis** i.e., the terminal cell of suspensor (h) divides further and forms a mass of tissue. Ultimately the different sets of tissue form the different structures of a seed which are :

1. *Four terminal cells of the octant* differentiate to form plumule and two cotyledons.
2. *Four basal cells of the octant* form hypocotyl and the core of the radicle.
3. *Hypophysis* differentiates to form the cortex and the epidermis of the radicle as well as the root-cap.

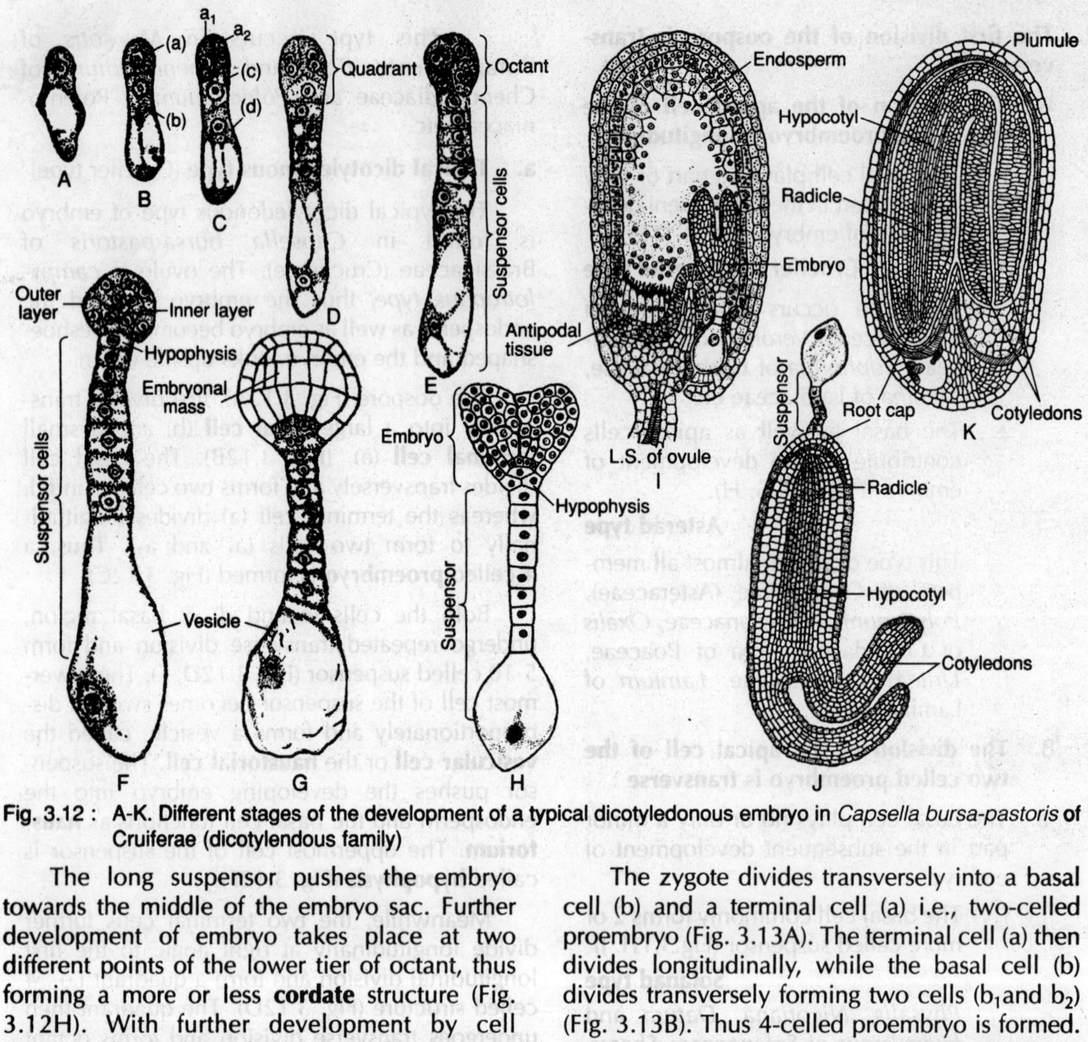

Fig. 3.12 : A-K. Different stages of the development of a typical dicotyledonous embryo in *Capsella bursa-pastoris* of Cruiferae (dicotylendous family)

The long suspensor pushes the embryo towards the middle of the embryo sac. Further development of embryo takes place at two different points of the lower tier of octant, thus forming a more or less **cordate** structure (Fig. 3.12H). With further development by cell division, the hypocotyl and cotyledon become horse shoe-shaped (Fig. 3.12I, J). This shape appears as per the shape of the embryo sac. During further maturation, the suspensor cells wither gradually.

This typical type of embryo development is called **Crucifer type** or **Onagrad type**.

The endosperm formed in the embryo sac is mostly consumed by the embryo. In mature seeds the remnant also disappears and thus the seed becomes **exalbuminous** (Fig. 3.12K).

b. Typical monocotyledonous type

It is found in *Luzula forsteri* of Juncaceae. The basic development of embryo up to octant stage is found to be similar in both dicotyledonous and monocotyledonous plants.

The zygote divides transversely into a basal cell (b) and a terminal cell (a) i.e., two-celled proembryo (Fig. 3.13A). The terminal cell (a) then divides longitudinally, while the basal cell (b) divides transversely forming two cells (b_1 and b_2) (Fig. 3.13B). Thus 4-celled proembryo is formed. Two terminal cells then divide longitudinally at right angle to the first one and develop a *quadrant* (Fig. 3.13C). A little later, at the basal region the cell (b_1) divides longitudinally and the other cell (b_2) divides transversely into b_2' and b_2'' cells (Fig. 3.13C). The cells of b_1 further divide and give rise to a tissue m (Fig. 3.13D, E). *The* b_2'' *cell acts as* ***suspensor*** and the rest cells take part in the formation of embryo. The quadrant of apical region undergoes divisions and forms a globose tissue. The tissue then undergoes *periclinal divisions* and forms outer and inner layers. The globose tissue of the terminal region becomes differentiated into two regions, the extreme terminal region (a_1) and behind it is the a_2 region (Fig. 3.13D). With further division both a_1 and a_2 regions enlarge in volume and designated as l_1 and l_2 region, respectively (Fig. 3.13E).

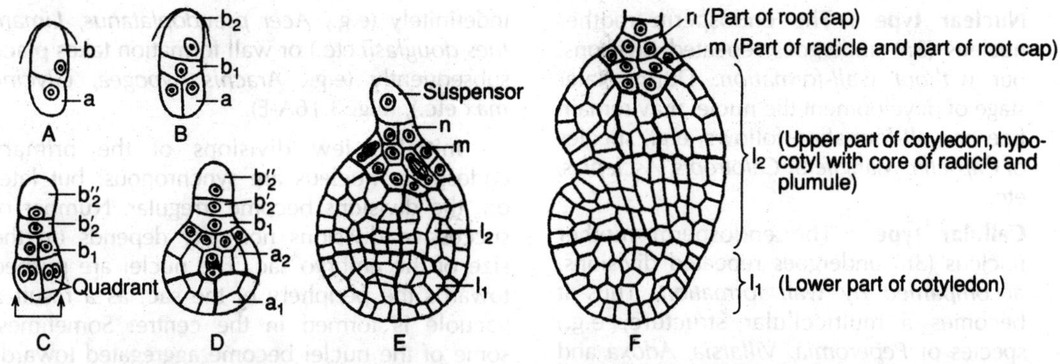

Fig. 3.13 : A-F. Different stages of the development of a typical monocotyledonous embryo in *Luzula forsteri* of Juncaceae

Ultimately the different sets of tissue (Fig. 3.13F) form the different structures of a seed :

1. The l_1 **region** gives rise to *lower part of the cotyledon,*
2. The l_2 **region** gives rise to *upper part of the cotyledon, the hypocotyl with core of radicle and later the plumule,*
3. The cells of **m** becomes differentiated into *upper part of the radicle* and part of root cap,
4. The **n** develops the part of root cap,
5. The b_2' develops the tip of root-cap, and
6. The b_2'' cell acts as **suspensor**.

ENDOSPERM

The endosperm plays a very important role, because it is the main source of food for the growing embryo (Fig. 3.14). In **angiosperms**, the endosperm forms *after fertilisation* and is **triploid** (3n) in nature as it is formed by triple fusion i.e., fusion of two polar nuclei (n + n) and further fusion with one of the male gametes (n) during double fertilisation. But in **gymnosperms**, the endosperm forms *before fertilisation* and is **haploid** (n) in nature, which is formed by the repeated divisions of megaspore nucleus. Actually, it represents the female gametophyte.

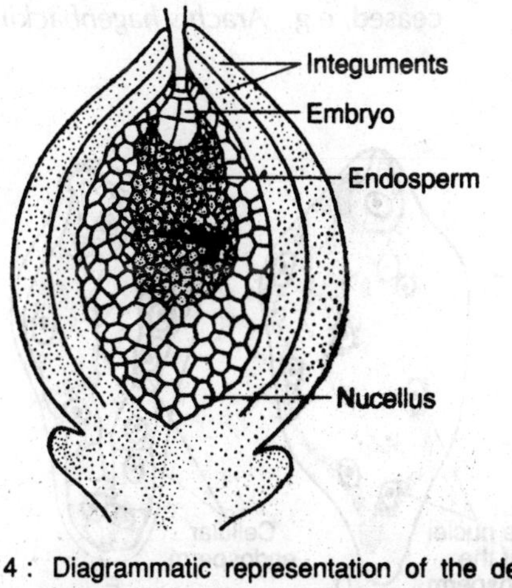

Fig. 3.14 : Diagrammatic representation of the development of embryo and endosperm inside ovule

Endosperm formation is suppressed in the families **Orchidaceae, Podostemonaceae** and **Trapaceae**. In those families double fertilisation takes place, but the triploid nucleus undergoes one or two divisions or degenerates immediately. Other angiosperms show **three general types of endosperm development**. These are (Fig. 3.15) : 1. Nuclear type, 2. Cellular type and 3. Helobial type.

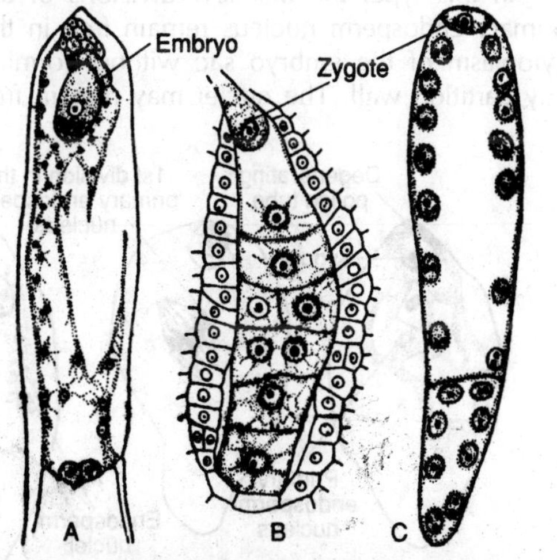

Fig. 3.15 : Different types of endosperm development. A. Nuclear type in *Capsella*. B. Cellular type in *Villarsia* (after Stolt), and C. Helobial type in *Eremurus* (*after* Stenar)

1. **Nuclear type** : The endosperm mother nucleus (3n) undergoes repeated divisions, but *without wall-formation.* During later stage of development the nuclei may remain free or wall-formation follows, e.g., species of *Capsella, Mangifera, Calotropis, Asclepias,* etc.
2. **Cellular type** : The endosperm mother nucleus (3n) undergoes repeated divisions, *accompanied by wall formation.* Thus it becomes a multicellular structure, e.g., species of *Peperomia, Villarsia, Adoxa* and many other members of Boraginaceae, Annonaceae, Aristolochiaceae, Gentianaceae etc.
3. **Helobial type** : It is an *intermediate between nuclear and cellular type* and is common in the order Helobiales (e.g., *Vallisneria, Limnophyton, Eremurus* etc.). A partition wall develops between the two nuclei results from the first division of the endosperm mother nucleus (3n) and thus a very big micropylar cell and a small chalazal cell are formed. *The micropylar cell divides by free nuclear division, later followed by wall formation and the nucleus of the chalazal cell undergoes a few or no division at all.*

The development of different types of endosperm can be described briefly as :

Nuclear type

In this type, the first few divisions of the primary endosperm nucleus remain free in the cytoplasm of the embryo sac without forming any partition wall. The nuclei may remain free indefinitely (e.g., *Acer pseudoplatanus, Limanthes douglasii* etc.) or wall formation takes place subsequently (e.g., *Arachis hypogea, Glycine max* etc.), (Fig. 3.16A-E).

Initially, few divisions of the primary endosperm nucleus are synchronous, but later on, the divisions become irregular. Number of successive divisions normally depends on the size of the embryo sac. The nuclei are pushed towards the periphery of the sac; as a result a vacuole is formed in the centre. Sometimes, some of the nuclei become aggregated towards both the chalazal and micropylar ends and the rest remains suspended in the periphery of cytoplasm, lining the wall of the embryo sac.

The nature of further growth and their behavior vary with plants. It is generally of 4 types :

(a) Primary endosperm nucleus forms hundreds of free nuclei which remain along the periphery of the embryo sac and the wall-formation takes place at a very late stage, e.g., *Mangifera, Citrus, Arachis, Malva* etc.

(b) Wall-formation takes place only at 8- or 16-nuclear stage, e.g., *Calotropis, Xeranthenum* etc.

(c) Wall-formation does not take place and the nuclei remain free throughout, e.g., *Cardiospermum, Lopezia, Tropaeolum,* etc.

(d) Primary endosperm nucleus undergoes first or second division, but the nuclei degenerate — thus the further growth is ceased, e.g., *Arachis hagenbackii.*

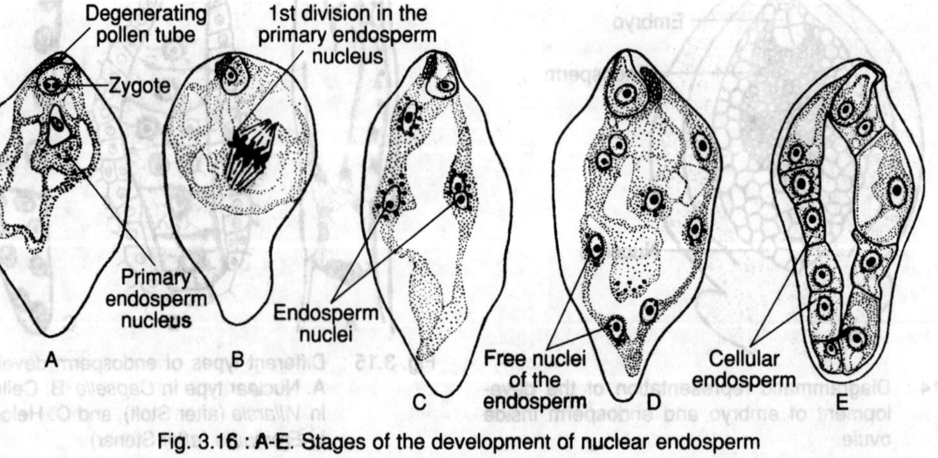

Fig. 3.16 : A-E. Stages of the development of nuclear endosperm

All nuclei of endosperm are not of the same size. Commonly the chalazal nuclei are larger than those of the micropylar nuclei. But the micropylar nuclei are larger than chalazal nuclei in *Fagraea*.

Usually wall formation takes place centripetally i.e., from periphery towards the centre (e.g., *Euphorbia, Bocconia* etc.) or it may start at the micropylar end (e.g., *Rumex, Polygonum, Phaseolus* etc.). But the extent of cell formation also varies :

(a) Entire endosperm may be cellular, e.g., *Gagea stipitala, Allium fistulosum, Nyssa sylvatica* etc., or

(b) Entire endosperm does not become cellular, e.g., *Arachis, Cassia, Crotalaria* (Fig. 3.17) etc.

Cocos nucifera is an unique example of nuclear endosperm. The watery liquid endosperm contains many free nuclei. It is known as **liquid syncitium**. Later on, cell formation takes place centripetally. The tissue thus formed is called **coconut meal**. The liquid endosperm contains growth-promoting **coconut milk factor**, used as nutrient medium in laboratory cultures.

Endosperm haustoria. The endosperm haustoria absorb the nutrition from the nucellar tissue. In members of Cucurbitaceae, Leguminosae

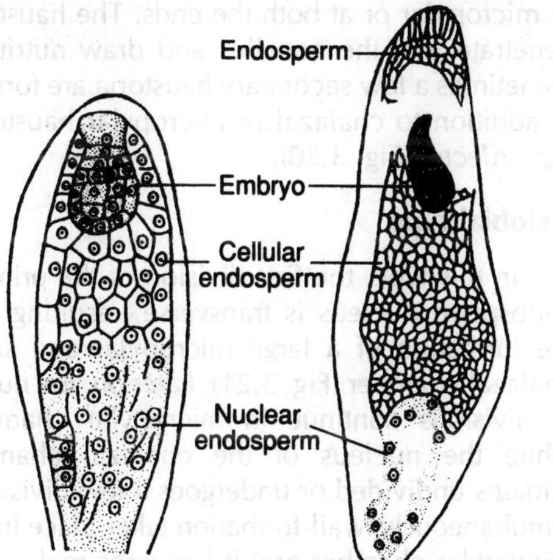

Fig. 3.17 : A & B. Nuclear endosperm : A. Endosperm of *Cassia articulata* after cellularisation; note a small portion at chalazal end that continues to be free nuclear, B. Cellularisation of endosperm of *Crotalaria* around embryo and a part remains free nuclear

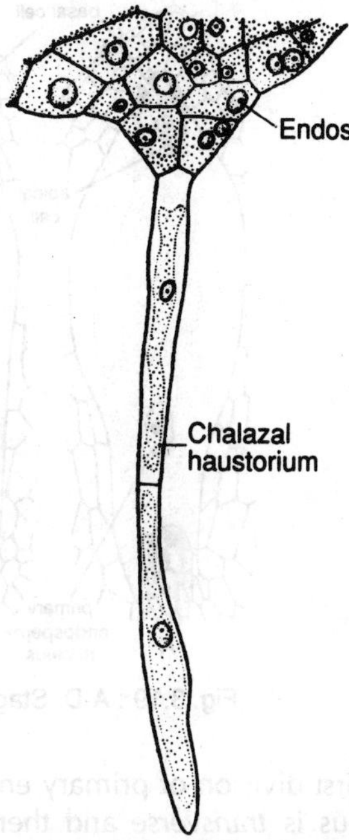

Fig. 3.18 : Endosperm haustorium of *Macadamia ternifolia*

(Fabaceae) etc., the cell formation is restricted towards the micropylar end and nuclei at the chalazal end remain free. The coenocytic part of this chalazal end develops tubular haustorium, e.g., *Macadamia ternifolia* (Fig. 3.18).

Cellular type

There is no free nuclear phase in this type, because the wall-formation starts with the first division of the primary endosperm nucleus (Fig. 3.19).

The primary endosperm nucleus may remain at the centre or at the chalazal end (e.g., *Scrophularia* etc.). The first division is usually transverse and divides the embryo sac into chalazal and micropylar chambers. But in some cases, it may be vertical or oblique. Based on the pattern of wall-formation of a few subsequent divisions of primary endosperm nucleus, it is divided into 5 types :

(a) The first division of primary endosperm nucleus is *vertical* (i.e., longitudinal to the embryo sac) and the second one is also vertical, but at right angle to the first one — thus four elongated cells are formed. The subsequent wall-formations take place at the micropylar end only, e.g., *Cetranthus, Adoxa* etc.

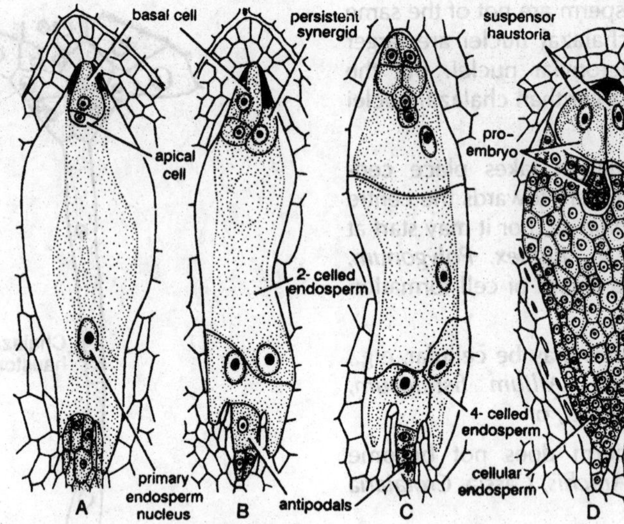

Fig. 3.19 : A-D. Stages of the development of cellular endosperm

(b) The first division of primary endosperm nucleus is *transverse* and then one or both the daughter cells divide vertically, e.g., *Verbascum* of Verbenaceae etc.

(c) The first division of primary endosperm nucleus is transverse and then one or both the daughter cells again divide transversely, e.g., members of Annonaceae and Palmae (Arecaceae).

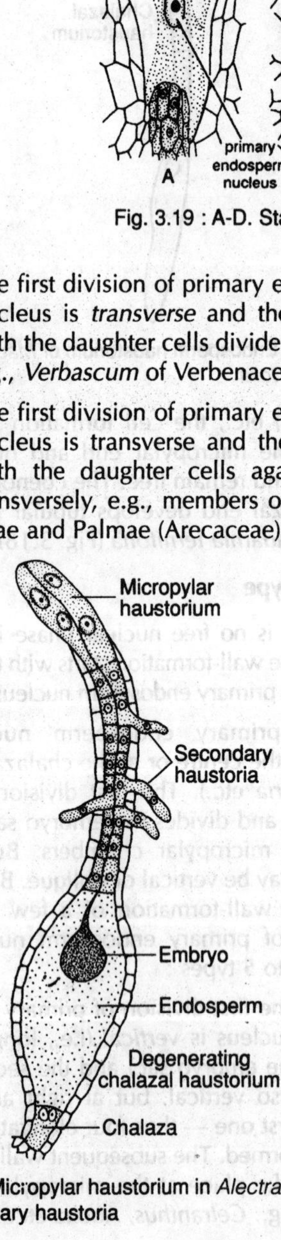

Fig. 3.20 : Micropylar haustorium in *Alectra* with secondary haustoria

(d) The first division of primary endosperm nucleus is *oblique* and the resultant two cells may be equal or unequal in size, e.g., *Myosotis arvensis* of Boraginaceae.

(e) The orientation of the first wall is indefinite, e.g., *Gunnera, Senecio* etc.

Endosperm haustoria. In this type, one or more cells become specialised and function as haustoria. The haustoria are formed at chalazal or micropylar or at both the ends. The haustoria penetrate into the nucellus and draw nutrition. Sometimes a few secondary haustoria are formed in addition to chalazal or micropylar haustoria, e.g., *Alectra* (Fig. 3.20).

Helobial type

In this type, the first division of the primary endosperm nucleus is transverse, resulting into the formation of a large micropylar and small chalazal chamber (Fig. 3.21). Later on, the nuclear divisions continue in micropylar chamber, while the nucleus of the chalazal chamber remains undivided or undergoes a few divisions. Simultaneously, wall-formation takes place in the micropylar chamber and it becomes multicellular, whereas wall formation does not take place in the chalazal chamber. The chalazal chamber contains one or a few disorganised nuclei.

Endosperm haustoria. In this type, the haustoria normally develop from the micropylar tissue. They are unicellular, tube-like and pene-

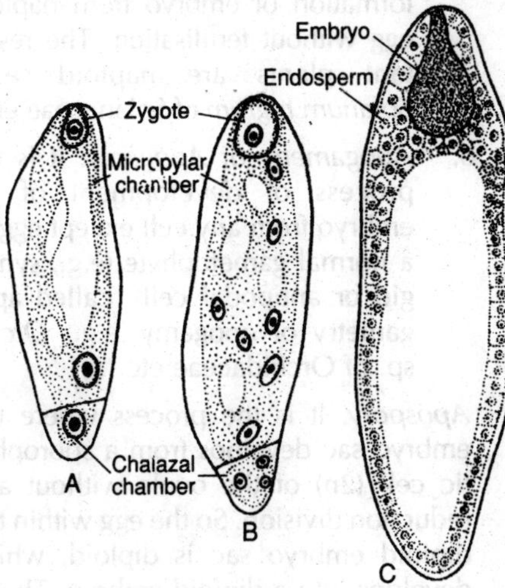

Fig. 3.21 : A-C. Stages of development of the Helobial endosperm.

trate the nucellar tissue at the chalazal end. It is the unique character of endosperm in the order Helobiae under monocotyledons.

Histological nature of endosperm

The endosperm cells are usually isodiametric in shape and can store food materials. The nature of endosperm tissue varies from plant to plant. Generally, they are thin-walled and without pits. The cells of endosperm tissue are devoid of intercellular space. Endosperm may be oily in castor, starchy in rice etc. In cereals like rice etc., the outermost layer of endosperm is made of aleurone tissue (protein).

Fate of endosperm

Depending on the presence or absence of endosperm in the mature seed, the seeds are classified into two types : 1. Endospermic seeds and 2. Non-endospermic seeds :

1. **Endospermic seeds** (*albuminous seeds*). The seeds where the endosperm forms a permanent tissue which persists till the germination of seeds are called *endospermic seeds.* They are also called *albuminous seeds*, e.g., date palm, *Phoenix sylvestris* and coconut, *Cocos nucifera* of Arecaceae (Palmae); wheat, *Triticum aestivum* and maize, *Zea mays* of Poaceae (Gramineae), etc.
2. **Non-endospermic seeds** (*exalbuminous seeds*). The seeds where the endosperm becomes used up by the growing embryo and is no longer seen in mature seeds, are called *non-endospermic seeds.* They are also called *exalbuminous seeds,* e.g., pea, *Pisum sativum* and ground nut, *Arachis hypogea* and many others of Fabaceae; pumpkin, *Cucurbita pepo* and many others of Cucurbitaceae.

Morphological nature of endosperm

In *gymnosperms,* the endosperm forms *before fertilisation* from the megaspore and is gametophyte in nature i.e., *haploid.* However, in *angiosperms,* it develops *after fertilisation* and is *triploid.* However, its morphological nature is a debatable question. The considerations of the different workers regarding its nature are :

1. *As gametophyte.* Strasburger (1900), Coulter and Chamberlain (1911) considered the endosperm as gametophyte.
2. *As sporophyte.* Monnier (1980) considered the endosperm as sporophyte.
3. *As an entirely new structure* (i.e., *Tissue sui generic*). Brink and Copper (1940) stated that double fertilisation causes the entry of extra amount of chromosome, thereby it becomes more active in collecting food from the integument and nucellus which compensates the extremely reduced gametophyte of angiospermic plants.

Table 3.1 : Brief representation of the events that occur in an ovary during seed and fruit formation

Parts of Ovary	Parts of Fruit
1. Ovarian wall	– Pericarp
2. Ovule	– Seed
3. Nucellus	– Perisperm
4. Outer integument	– Testa i.e., outer seed coat
5. Inner integument	– Tegmen i.e., inner seed coat
6. Micropyle	– Micropyle
7. Funicle	– Stalk of seed
8. Hilum	– Hilum
Inside embryo sac	
9. Egg cell	– Embryo
10. Definitive nucleus	– Endosperm (may or may not be left in mature seed)
11. Synergids and antipodal cells	– Degenerate.

FORMATION OF FRUIT AND SEED

1. *Formation of fruit* : Fruit is a mature ovary enclosing the seeds. The fruit wall develops from the ovarian wall (Table 3.1).
2. *Formation of seed* : After double fertilisation, many changes takes place in ovule and finally the ovule becomes converted into seeds (Table 3.1). Outer integument becomes hard and forms the outer seed coat, the **testa**, and the inner integument becomes thin and papery, forming the inner coat, the **tegmen.** The scar is the place where it is attached with the funicle.

APOMIXIS

During normal sexual cycle, fertilisation (*amphimixis* i.e., fusion of male and female gametes), leads to the formation of embryo and, later on, seed. The seed germinates to produce new plant. Later on, male and female gametophytes are formed on the plant after meiosis. This normal sexual cycle involves both fertilisation and meiosis. The haploid and diploid phase regularly alternate each other and is known as **alternation of generations.**

However, in many plants, the normal process of sexual reproduction is substituted by an asexual process. Thus the phenomenon of substitution of sexual process by asexual methods is known as **apomixis**. The plant which shows the phenomenon is called **apomictic plant**.

The term 'apomixis' was coined by **Winkler** in 1908. According to him, the term apomixis (i.e., away from mixing) refers to the substitution of sexual reproduction by any such method which does not involve *syngamy* and *meiosis.* It is of the following types :

A. *Vegetative reproduction.* In this type new plants are developed from stem, leaf, buds etc. Here seeds are not required.

B. *Agamospermy.* In this type, seeds and embryo are formed without meiosis and fertilisation. It is of the following types :

1. *Adventive embryony.* Embryo is formed from nucellar tissue or integuments.
2. *Diplospory.* In this type, the megaspore mother cell develops an unreduced embryo sac. It is of two types :
   a. *Parthenogenesis.* It is the process of formation of embryo from haploid egg without fertilisation. The resultant plants are haploid, e.g., *Solanum nigrum* of Solanaceae etc.
   b. *Apogametry* or *Apogamy.* It is the process of development of an embryo from any cell except egg of a normal gametophyte (e.g., synergid or antipodal cell), called apogametry or apogamy, e.g., *Orchis* sp. of Orchidaceae etc.
3. *Apospory.* It is the process where the embryo sac develops from a sporophytic cell (2n) of the ovule without any reduction division. So the egg within the diploid embryo sac is diploid, which develops into a diploid embryo. This is also called diploid parthenogenesis. Development of a gametophyte on the sporophyte without reduction division is called apospory, e.g., *Eupatorium glandolusum, Taraxacum albidium,* several species of *Hieracium* etc.

POLYEMBRYONY

It is the phenomenon of development of more than one embryo inside a seed. Leeuwenhoek (1719) first discovered polyembryony in orange (Fig. 3.22). It is of two types :

A. *True polyembryony.* When many embryos are developed in the same embryo sac it is called *true polyembryony.*

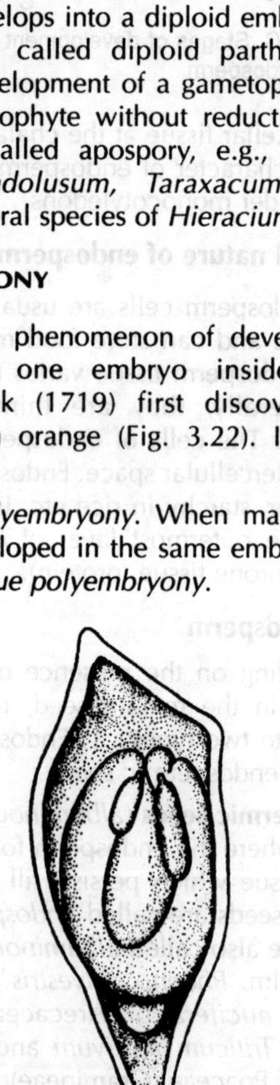

Fig. 3.22 : Many embryos inside a seed of orange (*after* Penzig)

B. *False polyembryony.* When more than one embryo develop in different embryo sacs inside the ovule it is called false *polyembryony*.

Polyembryony may develop in seeds by one of the following methods :

1. *Cleavage polyembryony.* When more than one embryo is developed by cleavage of zygote it is called **cleavage polyembryony**. This is commonly found in gymnosperm. In angiosperms, this is found in *Nymphea advena, Nicotiana rustica, Erythronium americanum*, etc.

2. *Embryo from the other embryo sac in the ovule*, e.g., *Citrus* sp. of Rutaceae.

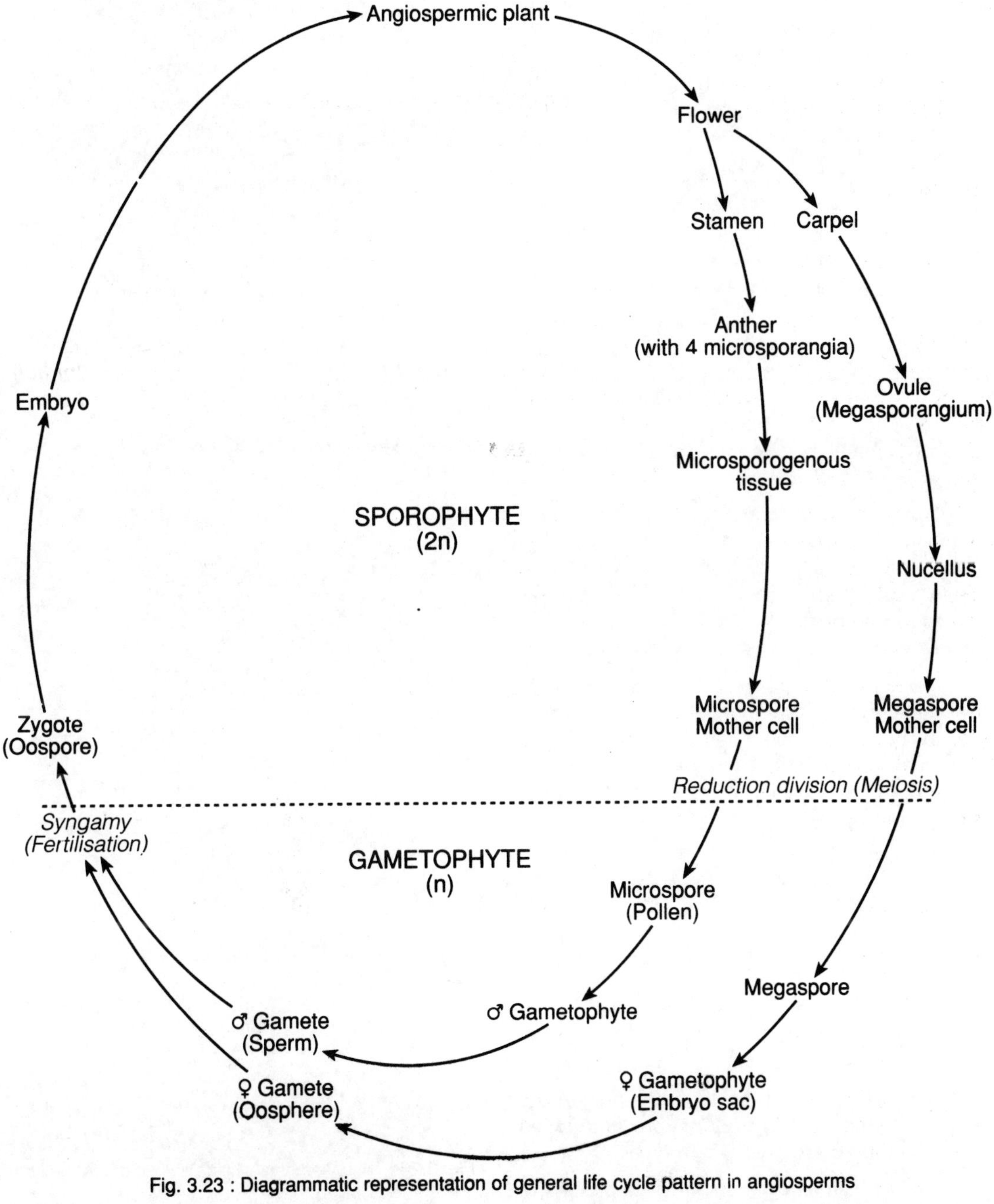

Fig. 3.23 : Diagrammatic representation of general life cycle pattern in angiosperms

3. *Embryo from nucellus* i.e., *Adventive embryony*. Embryos develop from the cells of nucellus which penetrate the embryo sac. e.g., *Mangifera indica* of Anacardiaceae.
4. *Embryo from endosperm*, e.g., *Balanophora* sp. of Balanophoraceae.
5. *Embryo from synergids*, e.g., *Sagittaria* sp. of Allismaceae.
6. *Embryo from antipodal cell*, e.g., *Ulmus americana* of Ulmaceae.

4 Taxonomy of Angiosperms

Taxonomy (Greek, *taxis* = arrangement, *nomous* = law, rule) means the *"arrangement by rules"* or *"lawful arrangement"*. The botanists agreed on a nomenclature — arrangement of plants into convenient groups for the proper and easy handling of vast number of plants in a suitable process following certain principles or rules. The term **Taxonomy** was first introduced in plant science by A. P. de Candolle (1813), a French botanist, as *the theory of plant classification*. Later on and till date, it is considered as a part of plant science which includes *identification, nomenclature* and *classification*. According to G. H. M. Lawrence (1955) "Taxonomy is a science which includes *identification, nomenclature* and *classification* of objects and is usually restricted to objects of biological origin". When the taxonomy is concerned with plants, it is referred to as **systematic botany**.

In respect to plant, both the terms are considered as synonym but are not accepted by all. Both the terms are often used variously, like *chemotaxonomy* (based on chemical content), *cytotaxonomy* (based on chromosome structure and numbers), *biosystematics* (systematics of living organisms), etc.

Initially, the taxonomy was based on a few macromorphological informations like habit, sex organ, etc. i.e., the *artificial* system. Later on, the classification was developed after considering many morphological characteristics, the *natural* systems. During the post-Darwinian period the taxonomy was based on evolutionary relationships i.e., the *phylogenetic* systems.

However, modern taxonomy is not restricted on morphology only. It depends on other branches of botany for good information like anatomy, cytology, physiology, phytochemistry, genetics, embryology, ecology etc.

The term 'taxon' was first introduced by Adolf Meyer (1926), a German biologist, for the animal groups. Later, in 1948, it was Herman J. Lam who proposed the term in plant science and it was accepted in the Seventh International Congress (1950). The term *"taxon"* indicates a taxonomic group like a variety, species, genus or any higher group.

Taxonomy and Systematics

There is no harmonic opinion in the meaning and use of both the terms *taxonomy* and *systematics*. Mason (1950) considered taxonomy as a vast field in biological science including four main streams. These are *systematics* i.e., comparative study of the organisms, *taxonomic systems, nomenclature,* and *documentation*.

On the other hand, authors like Simpson (1961), Heywood (1967), Ross (1974) and many others have treated *systematics* as a field which covers the study of diversity, differentiation and the relationship that exists among the organisms. According to them, taxonomy is a part of systematics. According to Solsbrig (1966), taxonomy includes nomenclature and classification but tilt massively on systematics for its concept. Later, Small (1989) reviewed the above and defined systematics as 'The science of organisation and pattern of heritable relationships among the kinds and diversity of organisms' and, on the other hand, taxonomy as 'a very substantial but imprecisely separated part of systematics, that is especially concerned with the production of formal classifications of living things on the basis of genetic relationships'.

Finally, it can be concluded that due to loose and interchangeable use in past, a proper explanation at present is very difficult. The literal meaning of Taxonomy in Greek is *putting in order* or *lawful arrangement* and Systematics means *putting together*. Lam (1959) and Turrill (1964) boldly expressed their opinion to treat these terms synonymously and later it was followed by many others.

Aims of Taxonomy

The aim of taxonomy includes three aspects I. Identification, II. Nomenclature, and III. Classification of plants. But to develop the construction of the above, the taxonomists approach in different ways. These are of two types : 1. Empirical approach, and 2. Interpretative approach.

1. **Empirical approach** : This type of approach is based on the thorough observation and characterisation of the organism which finally leads to the construction of a classification. Due to thorough observation of many characteristics of the specimens and development of a classification, this type becomes popularly acceptable. This type of approach was developed during pre-Darwinian period.
2. **Interpretative approach** : This type of approach of classification is based on the interpretation of evolution of a taxon. This is called phylogenetic classification, which developed during post-Darwinian period (i.e., after the publication of *'The Origin of Species by Means of Natural Selection'* (1859). This type of classification needs the data from the past history of a taxon.

The modern taxonomy made an attempt to merge the above two approaches. The aims of modern taxonomy are :

1. To supply an appropriate method of identification;
2. To contribute classification based on natural affinities of the specimens;
3. To contribute a catalogue of taxa by studying the different flora;
4. To trace the evolution after proper observation and interpretation;
5. To supply an integrating and synthetic role to maintain the relationship among the different biological fields.

Principles of Taxonomy

Taxonomy is the oldest branch of Botany and was practiced in many countries like India, Greece, Rome, China, England from long back. During early part, taxonomy was mainly aimed to develop some convenient methods of classification. One of the earliest known Indian works dealing with plants in a scientific manner is *Vrikshayurveda*. Later, it was completed by Parasar before Christ. This book contained a classification based on comparative morphology of plants. The **artificial system** reached its climax in Carolus Linnaeus (1707-1778), the **Father of modern taxonomy**. Later, Michel Adanson (1727-1806) was perhaps the first person to reject all artificial systems in favour of **natural system**. He proposed the idea that all characteristics are important in grouping of plants, which in the recent years came as **Numerical Taxonomy**. At that time, it was very difficult to consider all the characteristics of a plant in grouping of plants. For the above problem, importance was given on flower characteristics. In this system, a group of characteristics are considered in grouping of plants.

The natural system reached its climax in George Bentham and Joseph Dalton Hooker in their book *'Genera Plantarum'* in July 1862 and the last part in April 1883. Before the publication of their book, the **phylogenetic concept** came in focus after the publication of Darwin's concept of *'Origin of Species'* (1859). This system is still continuing with eminent personalities like Engler (1886-1892), Hutchinson (1926-1973), Takhtajan (1969, 1980), Cronquist (1968, 1981) and many others. Thus, the principles have evolved through time.

However, **Cronquist** (1968) has **formulated certain basic principles of Taxonomy** in *"Evolution and Classification of Flowering Plants"* :

1. Taxa are properly established on the basis of multiple correlations of characters.
2. Taxonomic importance of a character is determined by how well it correlates with other characters. This means that the taxonomic importance of a character is determined by *a posteriori* rather than *a priori*.
3. An important feature of taxonomy is its predative value.

The most important principle of taxonomy is the *multiple correlations of characters.* To find out correlation of taxa with others, single character should not be considered, it should always be considered along with other characters.

The selected characters should show maximum correlation with other characters. The significance of a selected characters depends on the degree of correlation. Characters with no distinct correlation with other characters is often accepted as an anomalous one and usually it is not important from the taxonomic point of view. Thus, the regular flower (*Scoparia dulcis*) of Scrophulariaceae, exstipulate leaves (*Corculum leptopus*) of Polygonaceae, zygomorphic flower (*Delphinium ajacis*) of Ranunculaceae and many such exceptional characters are taken as anomalous, because they are not correlated with common characters of taxa. The above examples clearly indicate that they are not significant taxonomically, but are important from their identification. But the anomalous characters often help the taxa with other groups of plants where these same characters are normally available.

Phases of Taxonomy

The taxonomy includes *identification, nomenclature* and *classification.* According to Davis and Heywood (1963) the classification is achieved in four (4) different consecutive phases :

1. **Pioneer phase** : This is also called *exploratory phase.* In this phase, different plants of a taxon have to be collected throughout the world and identified.
2. **Consolidation phase** : In this phase, plant is studied both in field as well in the herbarium and a range of variations are evaluated. The new group or groups, if discovered, are fully described. Thus a monograph is obtained.
3. **Biosystematic or experimental phase** : This phase deals with much more detailed knowledge of a taxon based on the above two phases along with the geographical variation, physiological characteristics, cytological characteristics, etc. The progress of this phase is very slow and it requires a team work rather than individual effort. The success in this phase is remarkable in some countries (U.K., U.S.A. etc.) but at and below the generic level.
4. **Encyclopaedic or holotaxonomic phase** : This phase is a coordination of the above three phases.

According to Turrill (1938) the classification is of two types:

1. **Alpha classification** : The first two phases i.e., pioneer and consolidation phases are based on gross morphological characteristics and the classifications of these phases are called Alpha classification or *Alpha taxonomy.*
2. **Omega classification** : The last two phases, i.e., biosystematic and encyclopaedic phase, are based on the data collected from fields, herbarium, laboratory and library and then properly analysed with the help of computer. The classifications of plants of these phases are called Omega classification or *Omega taxonomy.*

Identification

A taxon is determined after studying the overall similarity and dissimilarity with other taxa. Correct decision is essential for identification, because further studies are based on it. It is generally done directly after comparing with the plants of known identity or indirectly with the help of literature. Sometimes, the specimen may not tally with the existing identified specimens. In that case it is considered as a new one. During this situation, the person concerned feels the excitation of discovery of a *new plant.*

Nomenclature

The nomenclature may be defined simply as the system of naming of plants. After determination of a new taxon based on some important characteristics, it becomes essential to give a scientific name. The name of the taxon is given following the rules of International Code of Botanical Nomenclature (ICBN). With the help of this procedure one can communicate his discovery to the world. So, many new plants may be known to the scientific world by the exploration of new areas. These plants are then included in the list of the existing flora. The rules of plant naming are constructed by the International Botanical Congress held at different intervals in different places.

Before the mid 18th century, botanical names were in the form of short descriptive phases i.e., **polynomials**. Thus the presently known plant *Sida acuta* Burm. f. was named by Plukenet as *'Althaea coromandelina angustis prolongis foliis semine bicorneo'*. Similarly, *Strychnos nux-vomica* L. was known as *'Strychnos follies ovalis caule enerme'*. Similar type of naming is available in the earlier works of Linnaeus (1707-1778). This polynomial system serves as a label and the diagnosis of taxa. These two aspects later on proved to be contradictory and, therefore, Linnaeus separated them.

According to the review work of Choate (1912) on *"Origin and evolution of the binomial system of nomenclature in Botany"*, **Caspar Bauhin** (1560-1624) **was the first to implement the binomial system**, though he was not consistent in its use. Later on, *this track was chosen by* **Carolus Linnaeus** who used binomials for the plants described in his *'Species Plantarum'* (1753) and later works.

Latin is used in biological nomenclature. But the ICBN gives one liberty of selecting names of taxa from any language or source but insists that they are deemed to be Latin.

Binomial nomenclature

The binomial nomenclature is the system where naming of plants consists of two words — a **generic name** and a **specific name**. The first one is the generic name and the second one is the specific name. Both the names together form a binary or binomial name. In a binomial, only the generic name should start with capital letter and all others in small letters. After selecting the name of a particular plant, it must be added with the name of the author. If the author's name is too long, it should be mentioned in abbreviated form. The names of the plants are written in Latin. The scientific name i.e., the Latin name of Paddy is *Oryza sativa*. The first name, *Oryza* is the generic name and the second name *sativa* is the specific name. To complete the name, the author's name in abbreviated form should be added at the end. So the complete scientific name of Paddy is *Oryza sativa* L. The L. indicates the name of Linnaeus, who has given the name.

Binomials should be typed in Italic type face or in case of handwriting both generic and specific epithet should be underlined separately.

Nomenclatural rules

1. Scientific names and its purpose

Primitive man became interested in plants long back. The plants were useful to them for different aspects — medicines, food and so many other purposes. Same plants were available in different localities, different regions of a country and even in different countries. But the same plants were called by different names in different places or the same name had been applied for several different plants. Sadly it can be said that no uniformity could be achieved regarding the naming of the plant. Thus the common names were different and restricted in some particular regions — either small or larger one.

For the above, botanists felt the requirement or need of *uniform naming* and to assign a single scientific name to a particular type of plant. Thus different types of plants have different scientific names used uniformly throughout the scientific world, e.g., paddy is called by different names in different regions of India, such as **Dhan** in West Bengal, **Dhana** in Orissa, **Nellu** in Tamil Nadu, **Beeum** in Andhra Pradesh, **Chaul** in Uttar Pradesh and Bihar etc. But the scientific name *Oryza sativa* is uniformly used throughout the scientific world.

2. Elementary knowledge of "International Code of Botanical Nomenclature" (ICBN)

Caspar Bauhin (1560-1624) was the first to construct or design the binomial system. But this system was not harmonious in its use. After many years, it was **Carolus Linnaeus** who picked up the idea of Caspar Bauhin and used the binomial for all plants described in his *Species Plantarum* (1753) and later works. Later the *Species Plantarum* (1753) achieved much importance in Plant Taxonomy and the year of its publication is considered as the starting point of botanical nomenclature. The simple nomenclatural rules he considered were discussed earlier in his publications *Fundamenta Botanica* (1736), *Critica Botanica* (1737) and *Philosophia Botanica* (1750).

Due to the subsequent activities of the botanists after Linnaeus, it was found that the total number of species of flowering plants is about 300,000, which is much more than 10,000 as reported by Linnaeus. It was found that the Linnaean rule is insufficient to govern plant nomenclature. Later on, **Augustin de Candolle** published detail rules on plant nomenclature in *Theorie elemaintaire de la botanique* (1813). The confusion also arose among the taxonomists by the insufficient flow of message to the different corners of the world. Thereby same species was described by different names by different taxonomists from different corners of the world. Due to this the number of plants recorded in *Index Kewensis* during 1753-1965 is 830,000 — much more than the actual existing specimens! The above happened due to multiple publication of about $^1/_3$rd of the above specimens.

For the above, taxonomists felt the necessity of formulation of internationally accepted rules and regulations for the naming of plants.

As an organised effort, the first 'International Botanical Congress' was held in London in 1866, but nothing was settled there. In that conference **Alphonse de Candolle** (son of Augustin de Candolle) was entrusted to prepare a draft of the law of nomenclature. The second 'International Botanical Congress' was held in Paris in 1867. In that conference **Alphonse de Candolle** proposed the laws of Botanical Nomenclature (*Lois de la nomenclature botanique*) which was accepted and adopted with some modifications as the guideline for nomenclature of vegetable kingdom. These rules are now known as *Paris Code* or *de Candolle rules* (1867) published in 1868. Till date the Code has been periodically reviewed in different congress and subjected to addition, deletion and alteration. Subsequent meetings of the International Botanical Congress were held periodically in 1892 (Rochester Code), 1905 (Vienna Code), 1907 (American Code) and 1912 (Brussels Code), while in true sense the first ICBN was framed in the 6th Botanical Congress held in Cambridge (Cambridge Code) in 1930.

Saint Louis Code

International Code of Botanical Nomenclature (ICBN) is a book of LAW, containing the Rules and Recommendations prescribed for naming plants in an international context and comprises of three divisions. Division I deals with six PRINCIPLES containing mandatory provisions which are inviolable and form the basis of the System of Nomenclature. Division II deals with Rules and Recommendations the system of Nomenclature. Division III, on the other hand, outlines the procedures to be adopted if any of the Articles and Recommendations in Division II are proposed to be modified.

ICBN is a living document, revised and updated every 5 or 6 years by the Nomenclatural Session of the International Association for Plant Taxonomy (IAPT). The 16th International Code of Botanical Nomenclature (ICBN) was held at Saint Louis, Germany in 2000 under the Chairmanship of Prof. Werner Greuter. This 16th ICBN is called **Saint Louis Code**. Although the Saint Louis Code, 2000 replaces the Tokyo Code (1994), it does not alter the basic structure of the hitherto published Codes. The St. Louis Code has merely added some minor changes, clarified certain ambiguous terms like Epitype, and deleted some highly controversial provisions like REGISTRATION of new names, etc., added some new NOTES and EXAMPLES and in certain cases re-numbered the Articles, by bringing together Rules, such as those involving Typification, or deleted some duplicating examples, and clarified some ambiguous words here and there. The BASIC STRUCTURE of the CODE remains unaltered and thereby such steps ensure continuance of the ICBN as a universally acceptable document.

The ICBN of the **Saint Louis Code** (2000) comprises three Divisions : Principles, Rules and Recommendations and Provisions for the governance of the code.

A. **Division I. Principles** :

This division provides 6 Principles which form the basis of the system of Botanical Nomenclature. These are :

I. Botanical nomenclature is independent of zoological and bacteriological nomenclature. The Code applies equally to names of taxonomic groups treated as plants whether or not these groups were originally so treated.

II. The application of names of taxonomic group is determined by means of nomenclatural type.

III. The nomenclature of taxonomic group is based on the priority of publication.

IV. Each taxonomic group with a particular circumscription, position and rank can bear only one correct name, the earliest that is in accordance with the Rules, except in specified cases.

V. Scientific names of taxonomic groups are treated as **Latin**, regardless of their derivation.

VI. The Rules of Nomenclature are retroactive unless expressly limited.

B. **Division II. Rules and Recommendations** :

This division contains the Rules distributed in 62 Articles and Recommendations. The names which contravene (contradict) any one or more of the above Rules are illegitimate. Recommendations deal with subsidiary points. The object is to bring about greater uniformity in Nomenclature. Names contrary (contradictory) to Recommendations need not be rejected, but are not examples of the following :

C. **Division III. Provisions for the Governance of the Code** :

The provisions for modifications are mentioned in 7 Appendices. These are :

I. Names of hybrids and some special categories.

IIA. *Nomina familiarum algorum, fungorum, pteridophytorum et fossilium conservanda et rejieienda.*

IIB. *Nomina familiarum bryophytorum et spermatopytorum conservenda.*

IIIA. *Nomina generica conservanda et rejicienda.*

IIIB. *Nomina specifica conservanda et rejicienda.*

IV. *Nomina utique rejicienda.*

V. *Opera utique oppressa.*

There are FIVE Appendices split up into several subappendices and also INDEXES to Appendices II–IV, as also subject Index and, finally, Index of scientific names. Thus, while the CODE proper comprising of all the three Divisions, is covered by 104 pages, the APPENDICES and INDEXES spread over 370 pages form the bulk of the CODE.

Of the 62 Articles, Arts. 3, 4 and 5 deal with the sequence of the RANKS of taxa, whose relative sequence cannot be altered. Effective and valid publication of a name are covered by Arts 29–31 and Arts. 32–45 and Art. 61 or H. 9 respectively. Arts. 7–10 deal with problems of TYPIFICATION; Arts. 11–12, with the question of PRIORITY of a validly published name, whereas Arts. 13–15 outline the limitations of the Principle of Priority. Art. 14 provides guidelines for CONSERVATION OF NAMES, to by-pass disadvantages that may accrue because of strict application of the RULES OF PRIORITY.

The single largest of the changes made in the St. Louis Code concerns TYPIFICATION. Art. 8.2 defines that a specimen is a gathering, or part of a gathering of a single species or infraspecific taxon, made at one time, mounted on a Herbarium sheet or is preserved in a box, packet, jar or as microscopic slide and therefore, none of such a part-mounting can be selected as a lectotype.

In the St. Louis Code, entries of all conserved generic names, within each major group, are printed alphabetically, abandoning the former system of numerical classification of the obsolete Dalla Torre and Harms system. Instructions to review/revise the list of conserved family names of Spermatophyta, abandoning A. L. Jussieu (*Genera Plantarum*, 1789) and accepting Adanson (*Familles des plantes,* Vol. 2, 1763, Vol. 1, 1764) as the starting point date of family-names and introducing numerous other changes of authorship and date of publication, based on J. E. Reveal's list of suggested changes, a less dependable compilation, have been accepted in the St. Louis Code (2000). Citation of Berchfold and J. Presl, as authors of family names, were modified, because the word **rad** they used refers to *orders,* not to *families* (**celed**).

The Nomenclatural Session at St. Louis denied registration of newly proposed names and refused to discuss even the question of protection of NAMES IN CURRENT USE (NCU) and all suggestions for harmonised terminology in biological nomenclature (**Biocode**), so painstaking laboured upon over two decades, by the host of nomenclaturalists under the leadership of Prof. Werner Greuter, were rejected. And, all traces of registration of Botanical names found mentioned in the TOKYO Code were deleted and the entire procedure adopted in the Nomenclature Session at St. Louis to conduct the Proceedings of the session, is considered, by Greuter, as "unprecedented". A series of proposals to bring harmony between the RULES governing ICBN, ICZN, Bacterial Code and the Viral Nomenclature, were all sidestepped. All these developments led Dr. Werner Greuter (Chairman) and Dr. David L. Hawksworth (Secretary), to lament and nickname the Saint Louis Code, 2000 AD (printed with a black jacket cover) as the **BLACK CODE,** after naming the Tokyo Code as the **PURPLE CODE**.

The 17th International Botanical Congress was held in Vienna in 2005.

Table 4.1 : Ranks and endings

	Category	Ending of names	Specific examples
1.	Division	- phyta	Magnolio**phyta**
2.	Subdivision	- icae	Pin**icae** (Conifers)
3.	Class	- opsida	Magnoli**opsida**, Lili**opsida**
4.	Subclass	- idae	Magnoli**idae**, Lili**idae**
5.	Order	- ales	Magnoli**ales**, Lili**ales**
6.	Suborder	- ineae	Magnoli**ineae**, Lili**ineae**
7.	Family	- aceae	Magnoli**aceae**, Lili**aceae**
8.	Subfamily	- oideae	Magnoli**oideae**
9.	Tribe	- eae	Magnoli**eae**
10.	Subtribe	- ineae	**Rutineae.**

Phylocode

The **Phylocode** is an unofficial code of biological nomenclature in relation to phylogenetic relationship (C. Langer, *Taxon* 2001, Vol. 50 : 1091–1096). Phylocode advocates abolition of ranks, but this would have undesirable consequences for biodiversity studies. Because classifications are utilitarian, they should be worked out by the systematists and users of classifications. In defence, one may assume that the Phylocode is defined "by Law" rather than "by people". But what role should it play in taxonomy is still to be worked out.

3. Ranks of taxa

The taxa (sing. taxon) indicate the 'taxonomic groups of any rank' as mention in Art. 1 of the code. The rank of species is the basic unit and the successive higher taxa are genus, family, order, class, division and kingdom. Thus species is included under a genus, genus under a family, family under an order and so on. The code also provides that the name of each taxon of different rank should end with specific manner. Name of the different taxa with their specific ending along with examples are given in Table 4.1.

The botanists are authorised to develop additional categories with the addition of a prefix 'sub' to the next higher categories. This can be exemplified as suborder, subfamily etc.

The name of specimens is mentioned in binomials but in intraspecific categories trinomials are used with their own epithets added with the respective species to which they belong, e.g., *Brassica campestris* var. *dichotoma* Watt (Kalisarson); *B. campestris* var. *toria* Duth. (Indian rape or Toria); *B. oleracea* var. *capitata* L. (Cabbage); *B. oleracea* var. *botrytis* (Cauliflower); *Anisomeles indica* var. *albiflora* (Hassk.) Backer.

4. Rules of Priority

Many more names are published for plant taxon than the actual number available in the world flora. The above fact indicates that several names are given for the same taxon. To avoid such confusion it becomes essential to eliminate the duplication and to decide which one will be selected for a taxon out of several published names.

To avoid such problem, certain provisions are there in the code. According to the **Principles of Priority** (Arts.11 and 12), each family or taxon of lower rank with particular circumscription, position and rank can bear only one correct name. The first validly published name of a taxon becomes its valid name. If other names are subsequently published for a particular taxon, those will be considered as synonyms. The botanists have agreed with the opinion that for the flowering plants the starting point shall be from *Species Plantarum* of Carolus Linnaeus, published in 1st May, 1753.

(i) Thus there should be only one valid name for a particular plant, but it may have many synonyms. For the above, date of publication have to be mentioned for each name.

(ii) This rule is not applicable to the names of taxa above the rank of family.

Some examples of the validity of the genus is given below :

A. Valid generic name – *Xyris*.

1. The name '*Xyris*' was published by

Carolus Linnaeus in *Species Plantarum* (1753).

2. The same plant was named as *'Schismaxow'* published by Steudel in Bot. Zeitung 14, 1856.

So the priority should be given to *Xyris* as it was published earlier than *Schismaxow*.

B. Valid generic name – *Mondo*.

1. The name *Mondo* was given by Adanson in 1763.
2. The same plant was named as *Ophiopogon* by Kerl Gawl in 1807.

So the name *Mondo* was retained as a valid name.

C. Valid generic name – *Polygala*.

1. The name *Polygala* was given by Linnaeus in 1753.
2. Later on, the same plant was named as *Poligala* Neck (1768), *Polygaloides* Agosti (1770) and *Phylax* Nor. (1790) are the synonyms.

5. Alternative names

The name of the family should end with the suffix **'aceae'** added with the legitimate name of a genus included in the family, e.g., Magnoliaceae from *Magnolia*, Ranunculaceae from *Ranunculus*, etc.

Some exceptions are also permitted at the family level. The alternative names are permitted to be used for eight (8) families, those which do not end with 'aceae'. The new names are given according to the name of type genus of the particular family. The old names are also treated as valid, as such, because of long usages. The old name, new names (mentioned in parenthesis) and the type genus is given below :

	Name (Alternative name)	**Type genus**
1.	Palmae (Arecaceae)	*Areca*
2.	Gramineae (Poaceae)	*Poa*
3.	Cruciferae (Brassicaceae)	*Brassica*
4.	Leguminosae (Fabaceae)	*Faba*
5.	Guttiferae (Clusiaceae)	*Clusia*
6.	Umbelliferae (Apiaceae)	*Apium*
7.	Labiatae (Lamiaceae)	*Lamium*
8.	Compositae (Asteraceae)	*Aster*

When Papilionaceae (Fabaceae; Type, *Faba* Mill.) are regarded as a family distinct from the remainder of the Leguminosae, the name Papilionaceae is conserved against Leguminosae.

When the Papilionaceae are included in the family Leguminosae as a subfamily, the name Papilionoideae may be used as an alternative to Faboideae.

6. Effective and valid publication

The condition for effective and valid publication have been mentioned in Articles 29-45 of the Code.

Publication regarding new name of a plant with taxon is **effective** only when the printed matter is distributed (through gift, exchange or sale) to the general public (particularly botanist) or at least to botanical institution with libraries accessible to botanists generally with the consequent said matter published in the recognised periodicals or journals concerning botany. The publication is not effected by communication of new names in local newspaper, gardens open to the public, public meeting etc. The date of effective publication is the date of availability of printed matters.

The publication of name of new taxa is considered as **valid** only when it will fulfil the following conditions :

1. Publication must be effective.
2. It must be accompanied by a description of a taxon or by a reference to a previously and effectively published description.
3. The new name, if published after 1.1.1953, must have clear indication of the rank of the taxon.
4. It must be accompanied by a Latin diagnosis or reference to a previously and effectively published Latin diagnosis.
5. The name of a new taxon of the rank of family or below must be indicated by nomenclatural type if published on or after 1.1.1958.

7. How names are given

Botanical names are given by the following procedures :

Name of genera :

i. *Latinisation of common plant names* from any other language. e.g., *Ravenala* from Madagascan, *Nelumbo* from Ceylonese.

ii. *Commemoration of some botanist or patron of Botany.* e.g., *Hookerea, Jacobinia, Gerardiana, Caesalpinia,* etc.

iii. *Combination of two or more Greek or Latin words.* e.g., *Hygrophyla* (marsh loving), *Acanthospermum* (spiny fruits), *Asterocantha* (spreading spine), *Polygala* (more milk), etc.

Name of species :

i. *Based on habitat of plant,* e.g., *aquatica* or *aquatilis* (grows in water), *montanus* (grows on mountains), *arvensis* (grows on cultivated land) etc.

ii. *Based on the names of locality or country,* e.g., *indica* (India), *mexicana* (Mexico), *americana* (America), *zeylanica* (Ceylon), *sinensis* (China) etc.

iii. *Formed from a noun with a suffix indicating relationship,* e.g., *portulacastrum* (like *Portulaca*), *boerhavifolia* (leaves like *Boerhavia*), etc.

iv. *Based on the name of some honourable persons.* e.g., *hookeriana, griffithii, roxburghii, wallichii,* etc.

v. *The specific epithet should agree with the gender of the generic name.* In Latin, there are three genders : masculine, feminine and neuter gender. These are :

a. Masculine gender. e.g., *japonicus, albus,* etc.

b. Feminine gender. e.g., *japonica, alba,* etc.

c. Neuter gender. e.g., *japonicum, album* etc.

Similarly, above procedure is followed in generic name with some exceptions : e.g., *Fagus, Pyrus, Quercus* etc. ends with *us* but are feminine.

8. Author's citation

A. The name of a taxon should append the name of the author or authors who first published validly. This helps the taxonomists to verify the dates of publication. Commonly the names are used in abbreviated forms, e.g.,

i. *Vitex* Linn.

ii. *Vitex trifolia* Linn.

iii. *Vitex trifolia* var. *simplicifolia* Cham.

B. When a taxon of a lower rank is altered in a rank by retaining its older name, the name of the earlier author must be mentioned in *parenthesis,* followed by the name of the author who altered it. This is called *double citation,* e.g.,

i. *Citrus aurantium* var. *grandis* L., when promoted to the rank of species by Osbeck ; the new name of the plant becomes *Citrus grandis* (L.) Osbeck.

C. Similarly, the above rule is applied when a subdivision of a genus or a species is transferred or placed to another genus, e.g.,

i. When *Oxalis sensitivum* L. was transferred to the genus *Biophytum* by A. de Candolle, it becomes *Biophytum sensitivum* (L.) DC.

ii. When *Antirrhinum spurium* L. was transferred to the genus *Linaria* by Miller, it becomes *Linaria spuria* (L.) Mill.

D. Similar rules are applied during intraspecific transfer or change of rank, e.g.,

i. When *Sesbania aegyptiaca* var. *picta* Prain was transferred to *Sesbania sesban* by Santapau, it becomes *Sesbania sesban* var *picta* (Prain) Santapau.

E. When the name is published by two authors, their names must be linked with *et* or by an ampersand (&), e.g.,

i. *Didymopanax gleasonii* Britton *et* Wilson (or Britton & Wilson).

F. When the name is published by more than three authors, the name must be appended by the first author and is followed by *et al.,* e.g.,

i. *Streptomyces albo-niger* Hesseltine *et al.*

G. When an author who suggested the name but published validly by other author or to an author who published the name before the starting point of the group; the name of the latter author should be connected with the name of the author who validly published the name by an *ex,* e.g.,

i. *Allmannia nodiflora* R. Br. *ex* Hook. *f.*

ii. *Acalypha racemosa* Wall. *ex* Baill.

iii. *Arenaria uligenosa* schleich. *ex* Schlechtend.

H. During citing the name of a taxon of garden origin, the name must be ascribed to *hort* (*hortulanorum*) and *ex.* should be added before the name of the author who published it. e.g.,

i. *Gresneria donklarii* hort. *ex.* Hook.

9. Retention and Choice of Names

Some rules adopted from time to time by different congresses are mentioned below :

A. When a genus is divided into two or more genera, the original name is to retain for the one which includes the type species. e.g.,

   i. The genus *Dicera* J. R. et. G. Forster (1776) was divided by Rafinesque (1838) into two genera *Misipus* and *Skidanthera* and did not follow the rule and consequently the name *Dicera* was retained as one of the genera and kept for the part of *Dicera* based on lectotype *Dicera dentata.*

B. When a species or intraspecific taxa is divided, it will also follow the above procedure, viz.

   i. The *Heliotropium indicum* Linn. is divided into two species. The part which includes the type retains the name *Heliotropium indicum* L. and the other one as *H. keralense* Manilal *et* Sivarajan.

C. When a genus or a species is transferred to another genus or another generic name of the same species, the legitimate name have to be retained.

   i. The *Melochia cordata* Burm. f. when transferred to the genus *Sida* becomes *Sida cordata* (Burm. f.) Borssum Waalkes.

   ii. When *Antirrhinum spurium* L. was transferred to the genus *Linaria* by Miller, it becomes *Linaria spuria* (L.) Mill.

D. When two or more taxa of the same rank are combined, the oldest legitimate epithet should be retained, unless it contravenes other provisions of the Code.

   i. K. Schumann combined 3 genera *Sloanea* L. (1753), *Echinocarpus* Bl. (1825) and *Phoenicosperma* Miq. (1865), adopted the oldest one *Sloanea* L.

   ii. When *Sida rhombifolia* Linn. (1753) and *S. rhomboidea* Roxb. *ex* Flem. (1810) are combined, it becomes accepted as *Sida rhombifolia* Linn.

E. The autonym will get the priority over names of the same date that automatically established it (Art. 57.3 of Sydney Code), e.g.,

   i. When *Heracleum sibiricum* Linn. (1753) included *H. sibiricum* subsp. *lecokii* (Godron & Gren.) Nyman (1879), the subsp. *sibiricum* (1879) automatically established at the same time.

The abbreviated names of the different authors and their full names are :

BSP. = Britton, Sterns and Poggenberg;
Carr. = Carriere;
DC. = A. P. de Candolle;
A. DC. = Alphonse de Candolle;
HBK. = Humboldt, Bonpland and Kunth;
Hook. = William Hooker;
Hook. f. = Sir J. D. Hooker (son of William Hooker), f. represents the Latin word *filius,* son;
Koch. = Karl Koch;
L. or Linn. = Carolus Linnaeus;
Marsh. = Humphrey Marshall;
Mill. = Miller;
R. Br. = Robert Brown;
R&P = Ruiz Lopez and Pavon;
T&G = Torrey and Gray;
Willd. = Carl Willdenow;
Wm. Br. = William Brown;
W & K = Waldtein and Kitaibel etc.

10. Rejection of Names

A *legitimate name* is one in accordance with the Rules of Botanical Nomenclature, while those which do not satisfy the rules are known as *illegitimate names.* A **synonym** is a rejected name due to misapplication or difference in taxonomic judgement. Some rules have been framed for the rejection of names. These are :

i. **Superfluous name**. The name which is given to a taxon when already some name is present in existence is called superfluous name. The superfluous name must be rejected. The *Chrysophyllum sericeum* Salisb. (1796) is superfluous, because this taxon already has a name *C. cainito* L.

ii. **Basionym**. It is a specific or infraspecific epithet which has priority and is retained when transferred to a new or different taxon. viz. *Desmodium ochroleucum* Curtis includes *Meibomia ochroleuca* (Curtis) Kuntze, where the latter is a legitimate name based on the same type in which basionym *ochroleuca* is correctly retained.

iii. **Homonym**. The homonym must be rejected. (It is the same name given to other taxon). The *Astragalus rhizanthus* Boiss (1843) is a homonym of *Astragalus rhizanthus* Royle (1835) and it should be rejected.

iv. **Tautonyms**. The tautonym must be rejected. (When the specific epithet repeats exactly the generic name is called tautonyms). *Malus malus, Linaria linaria* are the tautonyms.

v. **Nomen nudum**. The names, those are published without any description, called *nomina nuda*. They are rejected unless and until published with proper description.

vi. If the name is based on discordant elements and monstrosities should be rejected.

vii. If the name is used in different senses and has become a persistent source of error should be rejected.

viii. If the naming does not follow the rules of nomenclature should be rejected.

11. Typification

The names of taxa in the rank of family or a lower rank i.e., genus and species are determined by means of *nomenclatural types.*

A nomenclatural type is that element of a taxon to which the name of a taxon is permanently attached (either as correct name or as a synonym). The nomenclatural type of a species is an individual herbarium specimen mounted on a sheet. It is that element with which the name is permanently attached. For small herbs, the type consists of more than one individual mounted on a sheet. Careful preservation is essential as they are the records of the present and the past. If a specimen cannot be preserved in dried form it should be recorded by diagnostic drawing, photographs and proper description. The type of a family is a genus and that of a genus is a species.

The following terms are commonly used in the nomenclature of types :

A. **Holotype** : It is the one specimen or other element used by the author or designated by him as the nomenclatural type.

B. **Isotype** : It is a duplicate copy of the holotype. It is always a specimen. During collection, several plants or several branches of a plant are collected. At that time one is selected as a holotype and the rest are considered as isotypes

C. **Lectotype** : A lectotype is a specimen or other element selected from the original material to serve as nomenclatural type when no holotype was designated at the time of publication or as long as it is missing.

D. **Syntype** : If the author does not designate a single holotype from the multiple collections and has used all or more than one specimen then the specimens are called syntypes.

E. **Neotype** : A neotype is a specimen or other element selected to serve as nomenclatural type as long as all of the material on which the name of the taxon was based is missing.

F. **Paratype** : It is a specimen other than an isotype or a syntype cited by the author while describing a taxon. (Where the author has cited two or more specimens as types, the remaining specimens are designated as paratype and not syntype. However, when no holotype was designated there will also be no paratype because all the cited specimens will be syntypes.)

After the publication of Sydney Code (1983), the above concept of nomenclatural type has now been extended to the names of taxa of higher ranks when the names are ultimately based on generic names. The genera which typify the alternate specific names should be taken as types of the respective family names which are not based on generic names, e.g., the genus *Faba* is the type genus of the family Leguminosae (alternate name Fabaceae), *Aster* that of Compositae (alternate name Asteraceae) and similarly with other six families having alternate names. The type can be defined as *an element on which the original valid description of a taxon is based.* The type is simply a nomenclatural concept. The types are the materials to which the respective names always remain attached.

12. Key

Key is a device or tool useful to identify an unknown taxon.

The aims of key are :

1. To provide an ease and certainty of identification (Davis and Heywood, 1963).

2. To make artificial separation of groups using one or two easily observable characteristics.

Types of key :

The keys are of two types : 1. Punched card key, and 2. Dichotomous key. The latter is again divided into two types : a. Indented key, and b. Bracketed key.

1. **Punched card key** : It consists of cards of appropriate size where the names of all taxa like families, genera or species for which the key have to be prepared should be printed on each. Each card has to be numbered serially. Each card should be printed with any one characteristic at any corner. All the taxa having the characteristics should be indicated by a hole in front of their name and others remain without it.

Many set of cards have to be developed, one set for each characteristics. During identification of a specimen, select the cards showing characteristics possessed by the specimens. The characteristic combination shown by the specimen will permit only one hole in the set of cards chosen. The particular sample of specimen is then assigned to that family to which the cards indicate the hole.

2. **Dichotomous key** : The dichotomous key consists of the following characteristics :
   - i. It consists of a pair of contrasting characteristics i.e., *cuplets* and each statement of a couplet is called a *lead.*
   - ii. Lead should be numbered.
   - iii. Both the leads of a couplet begins with same word as far as possible.
   - iv. The characteristics used in the key should be easily observable and contrast.
   - v. The quantitive characteristics are usually preferred than qualitative characteristics.
   - vi. More than one contrasting characteristics should be selected to differentiate the closely related taxa with overlapping characteristics.

The dichotomous key is of two types :

a. **Indented key**. In this type the collateral leads of a couplet are arranged in yokes and leads are identified by a figure or letter. Successive yokes are arranged one above the other. An example of dichotomous key with indented or yoked leads is given :

b. **Bracketed keys**. In this type both the leads of each *couplet* always remain together. An example of dichotomous key with bracketed or parallel leads is given :

Key to some genera of Cruciferae

Indented Key

1. Fruit dehiscent
   1. Pod short, cylindrical; seeds minute, biseriate; sepals spreading; flowers yellow. .... *Nasturtium.*
   2. Pod short, flattened, acute; seeds uniseriate; sepals erect; flowers white. .... *Cardamine.*
   3. Pod with indehiscent beak; cotyledons incumbent or folded.
      - a. Beak cylindrical or conical; seeds uniseriate; sepals yellow, pouched at the base and marked with green veins.......................... *Brassica.*
      - b. Beak flattened; seeds biseriate; sepals lilac, pouched at the base. ... . *Eruca.*
2. Fruit indehiscent
   4. Pod long, yellow or separate, breaking into 1-seeded fragments; seeds globose; sepals pouched at the base. .... *Raphanus.*
   5. Pod short, globose; sepals spreading, not pouched at the base. ...... *Senebiera.*

Bracketed key

1. Fruit dehiscent. 2
   Fruit indehiscent. 4
2. Pod short, cylindrical; seeds minute, biseriate; sepals spreading; flowers yellow. ....*Nasturtium.*
   Pod short, flattened, acute; seeds uniseriate; sepals erect; flowers white. ...... *Cardamine.*
   Pod with indehiscent beak; cotyledons incumbent or folded. 3
3. Beak cylindrical or conical; seeds uniseriate; sepals yellow, pouched at the base and marked with green veins. ....*Brassica.*

Beak flattened; seeds biseriate; sepals lilac, pouched at the base. .... *Eruca.*

4. Pod long, yellow or separate, breaking into 1-seeded fragments; seeds globose; sepals pouched at the base. .... *Raphanus.*

Pod short, globose; sepals spreading, not pouched at the base. .... *Senebiera.*

Advantages and disadvantages

1. **Punched card key** : The punched card key is tackled by the beginners like college students, who find more interest in taxonomy. But the system is very costly for the cards, printing of names etc.
2. **Dichotomous key** : This key is very suitable for the taxonomists and is far less costly. The indented key gives a visual representation of the group, thereby the user can readily obtain a picture about the taxon. But if the key is too large then the *leads* of a *couplet* get widely separated from each other, on the other hand, in bracketed or parallel *leads,* the advantage lies that the *couplets* occur side by side.

Good knowledge of characteristic differences and sound knowledge of the flora is essential in the construction of a key.

Herbarium

Herbarium is a place of depository of plant specimens in dry condition. According to H. M. Lawrence (1967) "a Herbarium is a collection of plant specimens that usually have been dried and pressed, are arranged in the sequence of an accepted classification, and are available for reference or other scientific study". In some cases, such as succulents and the specimens unsuitable for pressing are preserved in some preservatives like formaldehyde solution 4-5% or in F.F.A (Formyl acetic acid).

Composition of some common preservatives

A. Formaldehyde solution (5%):

i. Commercial formaldehyde (40% pure) .... 12.5ml.

ii. Water .... 87.5ml.

[Calculate the amount of formaldehyde following the formula $V_1S_1 = V_2S_2$. For the preparation of 100 ml of 5% solution from 40% concentration : 100 ml x 5% = V_2 x 40%. So V_2 = 500ml/40 = 12.5 ml].

B. *F.A.A. (Formyl acetic acid)* :

i.	Commercial formaldehyde (36-40% pure)	5 ml.
ii.	Glacial acetic acid	5 ml.
iii.	70% Ethanol	90 ml.

(The F.A.A. is used for long term preservation).

C. *Solution to preserve green colour of plants* :

i.	50% Ethanol	90.00 ml.
ii.	Commercial formaldehyde	5.00 ml.
iii.	Glycerine	2.50 ml.
iv.	Glacial acetic acid	2.50 ml.
v.	Cupric chloride	20.00 gm.
vi.	Uranium nitrate	2.50 gm.

Preparation of Herbarium

To prepare a Herbarium following equipments and techniques are generally required :

Equipments

1. **Secateur**. This is useful for cutting the twigs of shrubs and trees for Herbarium specimens.
2. **Digger**. The digger (Fig. 4.1A) is used to take out the important underground part of the plants like rhizome, bulb, corm etc. which may serve as an important diagnostic characteristic.
3. **Pruning shears**. The pruning shears (Fig. 4.1B) is useful to collect the twig from tall trees.
4. **Vasculum**. The vasculum (Fig. 4.1C) is an elongated (45-60 cm.) container with oval ends generally made of aluminium or tin. It has a hinged tight lid. It is used to keep the collected plant material fresh for long time. The vasculum is coated with white paint to reflect light and also to locate easily due to its contrast with the green vegetation. A belt is fitted at both the ends, thereby one can carry it like a bag.
5. **Magnifying lens**. Sometimes a hand lens is required to observe some characteristics of the plant in the field.
6. **Plant press**. It is made up of two wireguage frames (Fig. 4.2A, B) of 30 cm. × 45 cm. Three (3) iron plates with two holes, one at

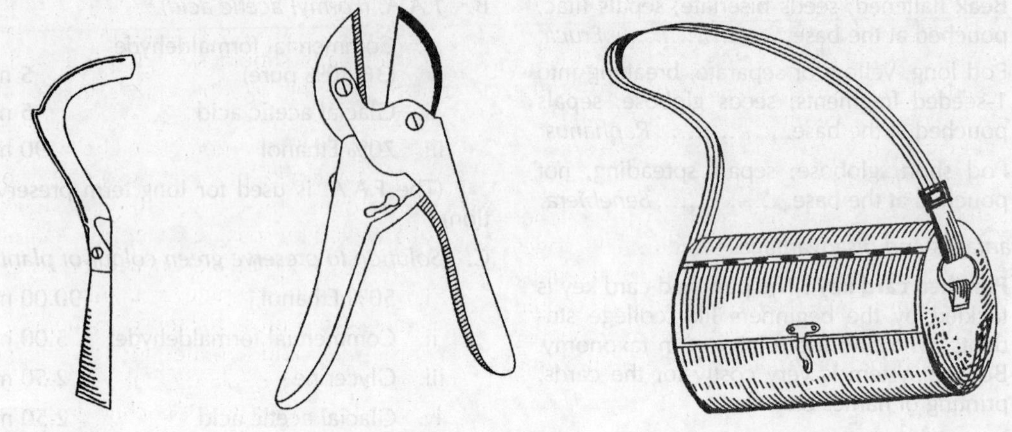

Fig. 4.1 : A. Digger, B. Pruning shears, C. Vasculum

each end are fitted at one side of the wire-gauge frame at equal distance. Similarly the other is also fitted with another three plates. Six thin ropes or chains (three on each side) should tighten at one end with the holes present on both sides of the plates. After placing the plants in-between the blotting papers or newspapers, it should be kept between the wiregauge and then tightened properly after passing the rope through the holes of plates of opposite wiregauge frame.

7. **Photographic equipment**. It is necessary to take photographs mainly of the plants where preparation of Herbarium makes difficult.

Techniques

1. **Selection and collection of plant species**. The plant species selected for collection must have flowers and fruits (if present). The herbaceous plants should be collected with underground parts but for shrubs and trees a twig have to be collected. The size of the

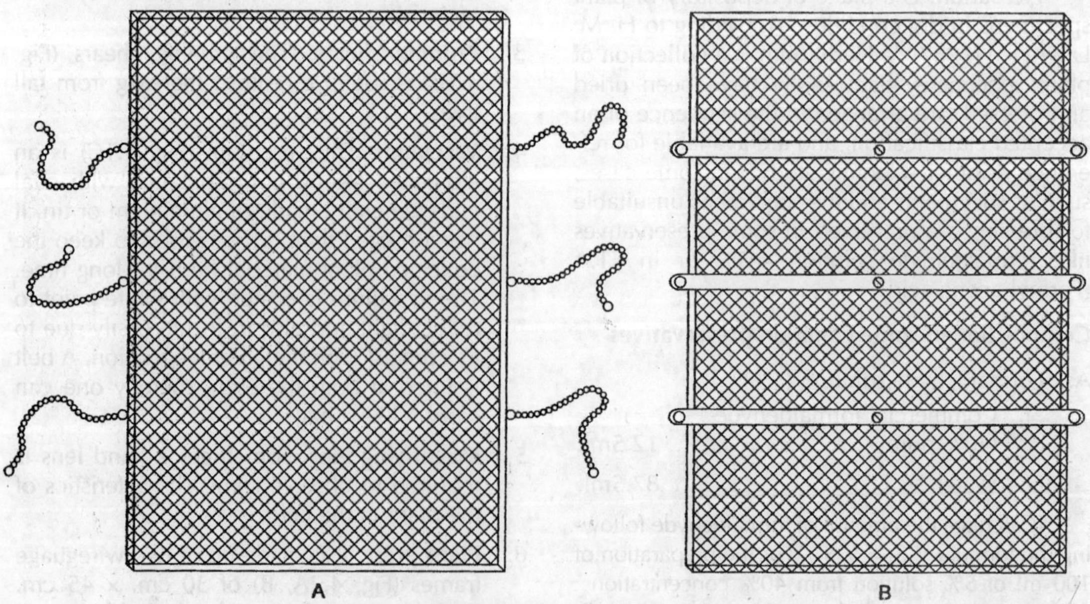

Fig. 4.2 : Plant press : A. Lower frame, B. Upper frame

collected specimen should be about 25 cm. There is no fixed rule for the number of specimens to be collected, but normally 4-6 sample plants are collected during routine collection. (Gymnosperms are collected along with cones and Pteridophytes along with cones or sori).

2. **Records on Voucher book**. The voucher book (Fig. 4.3) usually contains 100 pages each. The numbers are maintained serially i.e., from 1–100 and the second book should start with 101 and ends in 200. The voucher book should be printed with the following things :

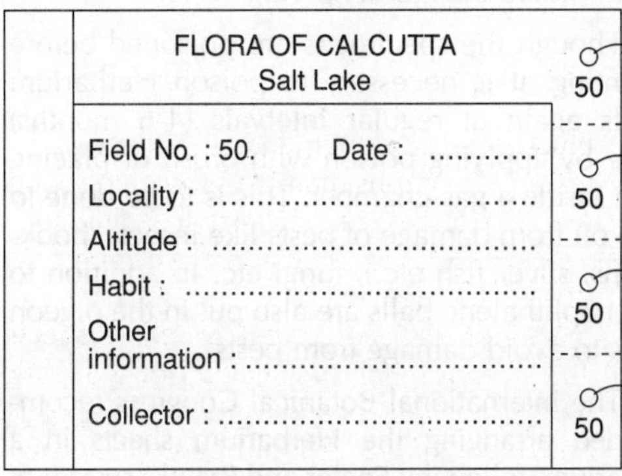

Fig. 4.3 : Single page of voucher book with 4 tickets attached with it. These numbering tickets are to be torn off through perforation and attached it by thread with the collected specimens

i. Name of the flora (where collection is conducted should be printed at the top)
ii. Field No.
iii. Date of collection
iv. Locality
v. Altitude
vi. Habit
vii. Any other information that can't be preserved, and
viii. Name of collector.

3. **Collection in vasculum** or **in polythene bag**. After proper record the plants are collected in vasculum or in moist polythene bag, thereby the collected plants or plant parts would remain fresh.

4. **Pressing of specimens for drying.** Before pressing, the plant material should be cleaned from mud or dust, if any. Plant or plant parts should spread nicely and kept between the blotting papers or old newspapers. Many such papers with collected plants are kept in-between the wire gauge and tightened properly with the rope as mentioned above (Fig. 4.4). The papers will be wetted by soaking water from the plants. The press should be opened on following days and after keeping the plants in other dry papers, the press tightened as before. This process should continue for 15 days for the proper drying of plants. During rainy season or in humid weather the changing of paper should be done twice a day to obtain good result.

Plants which can hold more water are very difficult to dry. Sometimes growth of the plant may take place during drying inside the press. So these plants require to be killed before drying. The killing can be done by dipping the plants in boiling water containing a little amount of formaldehyde solution.

5. **Poisoning**. The dried plants are then poisoned by dipping in saturated solution of mercuric chloride ($HgCl_2$) in ethanol or in rectified spirit. After some time, the ethanol

Fig. 4.4 : Specimens are between drying papers in press

or rectified spirit evaporates and the specimens become dry.

6. **Mounting**. The dry poisoned plants are then mounted on a comparatively hard sheet of paper — 42 cm × 29 cm (Fig. 4.5). The mounting is done by *adhesive coated cellophane tape, gum* or *thread*. The gum is applied with brush.

7. **Labelling**. Label is a rectangular piece of paper of usually 12 cm × 10 cm size, pasted at the lower right of the Herbarium sheet. The label should contain :

   i. HERBARIUM SPECIMEN (Printed at the top),

   ii. No.

   iii. Date

   iv. Name

   v. Family or Natural order

   vi. Locality

   vii. Altitude

   viii. Collected by

   ix. Identified by.

   A model of the label is given below (Fig. 4.5).

Fig. 4.5 : Single Herbarium sheet with label
[Photo courtesy : Dr. Subrata Mondal, Visva-Bharati]

Thus the Herbarium sheet may be defined as "a sheet of thick and tough paper having a measurement of 42 cm × 29 cm (international size), upon which perfectly dried and well displayed plant specimen is mounted and properly labelled (Fig. 4.5) for permanent retention at any Herbarium".

After the preparation of Herbarium sheet, it should be covered with paper. The species are covered with a thin paper, called *species cover,* and all the species under a genus should be covered with comparatively thick paper called *genus cover.* The Herbarium sheets with their proper cover are kept in special cupboards or pigeon-holed cabinet (Fig. 4.6).

Though the specimens are poisoned before mounting, it is necessary to poison Herbarium sheets again at regular intervals (4-6 months) either by applying poison with brush or placing them inside a gas-chamber. This is to be done to keep off from damage of pests like insects (book-worms, silver fish etc.), fungi etc. In addition to that, naphthalene balls are also put in the pigeon holes to avoid damage from pests.

The International Botanical Congress recommended arranging the Herbarium sheets in a herbarium, following Engler and Prantl's system of classification. But Bentham and Hooker's system is followed by many Herbaria including Royal Botanic Garden, Kew, London; Central National Herbarium, Sibpur, Howrah, W. Bengal, India, and many others Indian Herbaria.

Important Herbaria

Lanjouw and Staffeu (1952, 1964) made an index of world Herbaria — about 850 Institutions including 50 most important Herbaria. The name of some important Herbaria along with acronyms (given in bracket) and number of collections are given :

A. **Important Herbaria of the world**

1. Museum of Natural History, Paris (P) established in 1635 9,500,000
2. Royal Botanic Garden, Kew, England (K) established in 1841 .... More than 7,000,000
3. New York Botanic Garden (NY), established in 1891..................... 7,000,000
4. Conservatoire at Jardin Botaniques de Geneve (G) founded in 1824 ... 6,000,000

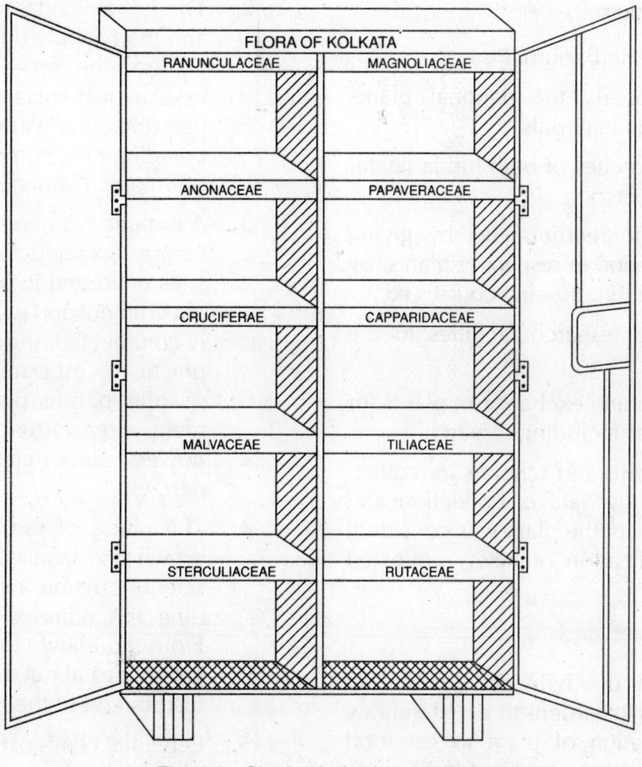

Fig. 4.6 : Single Herbarium almirah

5. V. L. Komarov Botanical Institute, St. Petersburg (Leningrad), Russia (LE), established in 1823 .... 5,770.000
6. Herbarium of Missouri Botanic Garden, St. Louis(MO), established in 1859...5,400,000
7. British Museum of Natural History (BM) established in 1753 5,200,000
8. Harvard University, Cambridge, Mass. (GH), established in 1864 5,005,000
9. Nuturhistorisches Museum, Win, Austria (W), established in 1807 5,000,000
10. Swedish Museum of Natural History, Stockholm (S), established in 1739 4,400,000
11. Smithsonian Institute, Washington D.C. (US), established in 1848 4,368,000
12. Central National Herbarium, Sibpur, Howrah, W. Bengal, India (CAL), established in 1793 2,500,000

B. **Important Herbaria of India**

1. Central National Herbarium, Sibpur, Howrah, W. Bengal (CAL), established in 1793....................................... 2,500,000
2. Herbarium of the Forest Research Institute, Dehradun (DD) 300,000
3. Madras Herbarium, Coimbatore (MH) 150,000
4. Botanical Survey of India, Pune ... 120,000
5. Regional Herbarium of the Botanical Survey of India, Shillong (ASSAM) 86,000
6. Herbarium of the National Botanical Research Institute, Lucknow (LWG) 80,000
7. Industrial Section, Botanical Survey of India, Indian Museum, Kolkata 40,000
8. Herbarium of the Botanical Survey of India, Allahabad (BSA) 26,440

* One important Herbarium — Berlin-Dahlem, Botanischer Garten (Germany) — established in 1646 having 4,000,000 collection was destroyed completely during Second World War.

Functions of Herbarium

The functions of Herbarium are :

1. To preserve properly the national plants including the type materials.
2. To maintain researches of both fundamental and applied aspects.
3. To fill the public requirements by giving scientific information in respect of plants, by organising exhibition, training courses etc.
4. To give job and research facilities to the juniors.
5. To assist the loan and exchange of plants for different purposes including research.
6. To help the students and scholars for collection (by seeing the date of collection and locality, from where the plant was collected) and easy identification of newly collected plant materials.

Botanic Gardens

From early part of civilisation, man had begun to grow plants in gardens to get useful raw material. The cultivation of plant to get food and medicine was known in ancient India since 4000 BC. The available record indicates that it is from the time of **Theophrastus** (370-287 BC) — the Father of Botany — that the gardens have contributed a lot to Botany. During the period of ancient Mediterranean civilisation, it was found that people develop garden in front of palaces and temples for beautification. The '**Hanging Gardens**' of Babylon was one of the wonders of ancient world. After that, so many plants have grown and died in different gardens and so much water has passed through different rivers — till we are able to see the modern picture, with different functions and importances of Botanic Gardens.

A garden is a place for growing plants to get flowers, fruits and/or vegetables. But a botanica garden can be defined as a place to grow and maintain plants of different countries, according to their habitat and importance, for the academic and economic benefit of mankind.

A. Functions of Botanic Garden

1. Botanic Garden performs as an institute for different types of botanical research, particularly the flora of that region.
2. The entire fundamental i.e., taxonomy and many applied researches e.g., silviculture (the growing and tending of trees as a branch of forestry); horticulture (the art of garden cultivation), plant breeding etc., come within the territory of Botanic Garden.
3. A Botanic Garden may act as a training centre to conduct training for the persons interested in tree plantation. Many government and public gardens regularly conduct training (both theoretical and practical), especially on technique of making bonsai plant, flower arrangement, preparation and display of dry flowers, development of kitchen garden, etc.
4. The plants of the Botanic Garden are grown and labelled properly with their scientific name and with a note including the name of its native country. From the above, the students of botany get an idea about the phytogeographical distribution of the plants.
5. Perennial plants of different countries are planted in different plots according to their habitat, thereby students and interested persons are able to get an idea about the vegetation of that country.
6. Botanic Gardens act as an educational centre for students (School, College and University) and also the interested persons.
7. The beauty of the Botanic Garden and its charming environment gives pleasure and enjoyment to the general public.

B. Importance of Botanic Garden

1. Due to overgrowing population it seems that the tropical ecosystems are on the margin or edge of extinction. For the above, the Botanic Gardens throughout the country have taken special role in the *conservation of tropical plants.*
2. Botanic Gardens also play a very important role in maintaining *germ-plasm bank* of different plants and act as a depository of 'extinct' and 'endangered' plants.
3. Botanic Gardens play an important role in the *ex situ* conservation of plant species.

4. Botanic Gardens act as 'outdoor laboratory' for the students and also for research workers.
5. Recently the Gardens are considered as a very useful centres for acclimatisation (accustoming to a new climate) studies, store house of ornamental plants of different countries.
6. Many Botanic Gardens organise popular courses for interested persons about :
   i. Making of Bonsai plants.
   ii. Tapiary (pruning of trees or shrubs to get different shapes like different animals and birds).
   iii. Flower arrangement.
   iv. Preparation and display of dry flowers.
   v. Gardening.
   vi. Landscape gardening (art or practice of laying out ornamental grounds in a miniature form intimating natural scenery), etc.

With the above training people can learn about and can earn a lot through commercial implementation.

Role of Botanic Garden in conservation strategies

Biodiversity can be defined as "the variety and variability among living organisms and the ecological complexes in which they occur" The environment is under a constant pressure to change, mainly due to urbanisation and of modern high technology farming. Thus the protection of plants, either in a restricted place or to protect them in their natural habitat through restriction, is essential to preserve genetic diversity i.e., conservation. Thus, conservation may be defined as "the planning and management of resources in a way so as to secure their wider use and continuous supply, maintaining their quality, value and diversity". Through conservation different types of diversity, like species diversity — genetic diversity and community or ecological diversity — can be preserved.

Conservation can be done in two ways :

1. ***Ex-situ*** **or** ***off-site*** **conservation** : In this process the species of a plant can be preserved (prevented from extinction) in artificial condition under the supervision of human beings.
2. ***In-situ*** **or** ***on-site*** **conservation** : In this process the species of a plant can be preserved in its area of natural distribution.

Botanic Garden is a place where the plants of different countries are grown and maintained according to their habitat and importance. Here the plants of different phytogeographical regions, ornamental plants, economically important plants and also the plants facing danger of extinction are grown and maintained under strict scientific monitoring. Hence, Botanic Garden plays an active role not only in academic and economical benefit, but also in *ex-situ* conservation of biodiversity.

Conference on Botanic Gardens and the World Conservation Strategies (W.C.S) was held in Las Palnas Gran Canaria in 1985, to review the role of Botanic Gardens in implementing the conservation strategies and to discuss to make gardens more effective in conservation of plants. The conservation strategy of Botanic Garden outlines the contributions that it can make to achieve what the W.C.S wants. These are :

1. to maintain essential ecological processes and life support system;
2. to preserve genetic diversity, and
3. to ensure that the utilisation of species and ecosystem is sustainable.

The Botanic gardens will grow more plants for conservation :

1. to cultivate as many as threatened species as practicable for conservation;
2. to cultivate with proper care the artificially threatened species to prevent genetic erosion;
3. to make available for research work, educational programme and to display for the general visitors;
4. to make available more plants and their propagating stocks like seeds etc., to reintroduce into the other parts of the world and also to reinforce in wild population in the previous or/ and other new areas.

Role of Botanic Gardens in *ex-situ* conservation

During *ex-situ* conservation the entire plant or any of its parts (i.e., the propagating stock) can be conserved. The long term alternative of *in-situ* conservation is the *ex-situ* conservation in Botanic Garden, where the plants as a whole or its propagating stocks are conserved properly. Thus *ex-situ* conservation of plants in Botanic Gardens can provide stimulus to research workers and also to serve the future management of plants available in *in-situ* condition. Simultaneously, the Botanic Garden can increase public awareness about the plant and/or plants through publicity.

Role of Botanic Gardens in seed and gene banking

It is necessary to build up seed banks in as large amount as possible with the seeds collected from the species in the field where they are distributed. The gene bank conserves propagating stock — commonly seeds, pollens of a threatened species — at the temperature of liquid nitrogen. Botanic Gardens are the best place to set up seed banks of wild plants with the aim of preserving rare or threatened plants which will be available for future research and also for applied purposes. Some Botanic Gardens play an active role in exchange of germplasm and/or plant as a whole of economically important species from one part to other part of the world. Seed banking of crop plants is usually co-ordinated by the *International Bureau of Plant Genetic Resources* (IBPGR). It is also involved in collecting and storing the wild species of different genera like *Phaseolus, Glycine, Brassica, Arachis,* etc. In Wakehuest Place, the satellite garden of Royal Botanic Garden, Kew, England have been conserved the seeds of Brassicaceae from West Mediterenial region. National Botanic Garden Research Institute, Uttar Pradesh, India, is very active in seed banking.

Cryopreservation through gene bank and seed bank and also through *ex-situ* conservation are the important ways to restore the threatened and endangered taxa. The Botanic Gardens of the different corners of the world have functioned as medicinal gardens, herbal gardens, economic species garden etc. — they conserve huge number of germplasm of the world flora. Gardens attached with Botany and Horticulture Departments of different universities can play a historic role in the preservation and exchange programme.

The names of some Botanic Gardens are :

A. **Some Botanic Gardens of the World** (year of establishment is given in parenthesis) :

1. Botanic Garden of Padua, Padua, Italy (1545).
2. Botanic Garden of Heidelberg, Germany (1595).
3. National Museum of Natural History, Paris, France (1635).
4. Botanic Garden and Museum, Berlin, Germany (1646).
5. Royal Botanic Garden, Edinburg, Scotland (1670).
6. Botanic Garden of the Academy of Sciences, Leningrad, Russia (1712).
7. Jardin Botanico do Rio de Janeiro, Brazil (1808).
8. Nikitisky Botanic Garden, Yalta (1812).
9. Kebun Raya (Botanic Garden) Bogor, Java (1817).
10. Royal Botanic Garden, Kew, London (1841).
11. Missouri Botanic Garden, U.S.A. (1859).
12. Arnold Arboretum, Boston, U.S.A. (1872).
13. New York Botanic Garden, U.S.A. (1891).
14. Montreal Botanic Garden, Montreal, Canada (1936).

B. **Botanic Gardens of India** (year of establishment is given in parenthesis) :

1. Indian Botanic Garden, Sibpur, Kolkata (1787).
2. Lloyd Botanic Garden, Darjeeling, West Bengal (1878).
3. National Botanic Garden, Lucknow (1946).
4. The Mysore State Botanic Garden, Bangalore, Karnataka (1856).
5. Garden of the Forest Research Institute, Dehra Doon (1934).

Indian Botanic Garden, Sibpur, Howrah (1787)

It is one of the greatest Botanic Gardens of the world. Col. Robert Kyd, a Military Engineer of the East India Company, was a lover of gardening. He planed to develop the nursery of different plants of economic importance, such as medicinal plants, timber trees, spices etc., and grow them in the garden. The seedling could be distributed to cultivate in different parts of the country and the Company would get the profit directly from the increase of trade and commerce. With the prior permission of the East India Company, **Col. Robert Kyd established the garden in July 1787** at Sibpur, Howrah, Bengal (now West Bengal). The garden covers an area of about 273 acres of land by the side of the Ganges. At that time Col. Kyd was appointed the Hony. Superintendent. Initially the garden was named as **Hon'ble East India Company's Botanical Garden**. Later, when the East India Company went under the British crown, it was named as **Royal Botanical Garden**. The Garden and its Herbarium (Central National Herbarium) were handed over to the Government of West Bengal after independence in 1947. Now it is called **Indian Botanic Garden** as a section of the Botanical Survey of India. Now the garden contains many plants of tropical regions of the world.

Col. Robert Kyd arranged the Garden according to his own idea, but he died after 6 years. Later, **William Roxburgh** was appointed as Superintendent of the garden and worked up to 1813. He introduced many interesting plants and gave the Garden a botanical shape. He prepared and published a catalogue of the plants cultivated in the Garden at that time. He had many important contributions on Indian flora and is known as the **'Father of Indian Botany'**.

Later, Roxburgh was succeeded by **Dr. Nathanial Wallich**. The Herbarium, which is now known as Central National Herbarium, was started by him. He also made many important contributions on Indian flora.

After that, the Garden was headed by many renowned botanists such as W. Griffith, Voigt, T. Anderson, George King, David Prain (wrote the book '*Bengal Plants*'), Gage, D. Chatterjee and many others.

The Herbarium of the Garden contains good number of collection which includes more than 1,000 **Type material** of Indian species of flowering plants.

The interesting and attractive feature of the Garden is the **Great Banyan Tree** (*Ficus benghalensis*), It forms the **world's largest (spreading) tree crown** which has enormous trunks up to 100 feet (30 m) or more in circumference of more than 200 years old, that covers more than four acres and stands with more than 1,000 pillar-like legs formed by aerial roots i.e., prop roots (Fig. 4.7). Another plant, the **Amazon lily** (*Victoria amazonica* of Nymphaeaceae) with large dish-like floating leaves having raised margin (Fig. 4.8) attracts the general public. In addition to the above, so many important places like large palm house, small palm house, *Charac uddyan* (housing economically important plants), orchid house, fern house, etc., and various shrubs and trees are present — they attract the students and also the general public.

Fig. 4.7 : *World's largest tree crown.* The Indian banyan tree (*Ficus benghalensis*)
Photograph (portion) taken from Indian Botanic Garden. Sibpur, Kolkata

Fig. 4.8 : **Amazon lily** (*Victoria amazonica* of Nymphaeaceae) with large dish-like floating leaves having raised margin

CLASSIFICATION

It is impossible for any person to study the entire flora of the world. So, it becomes necessary to make the grouping of plants based on similarity and differences among themselves. The similar plants are grouped under a *species*; single or many species under a genus; single or many genera under a *family*; single or different families under an *order* and so on. These groups are then arranged following the rules of nomenclature. The grouping at the lower level is very easy, but complexity arises towards the upper taxa.

Basically, there are three types of classification : Artificial, Natural and Phylogenetic.

Definitions :

1. **Artificial system** : The system where the organisms are classified into different taxa based on one or a few superficial characteristics only, e.g., Classification of Theophrastus, John Ray, Sexual system of Carolus Linnaeus, etc.
2. **Natural system** : The system where the plants are grouped on the basis of common natural characteristics and placed into different taxa like classes, orders, families and genera, etc. This type of system does not consider the *evolutionary sequence* and hence the *affinity* and *relationship* between the taxa are not considered. viz., Classifications of Michel Adanson, Robert Brown, G. Bentham and J. D. Hooker, etc.
3. **Phylogenetic system** : The system where the construction of classification is based on the course of evolution and tries to arrange the plants based on genetic and phylogenetic relationships, e.g., Classifications of A. Engler and Karl A. E. Prantl, John Hutchinson, A. Takhtajan, A. Cronquist etc.

Artificial system of classification

It is the earliest system of classification dominated from 285 BC to about 1830 AD.

In this system, the organisms are classified into different taxa merely on the basis of one or a few superficial characteristics only.

These systems were proposed by many like Theophrastus (370-287BC), Caius Plinius Secundus (23-79AD), Pedanios Dioscorides (62-128AD), Andrea Cesalpino (1519-1603AD), Gaspard Bauhin (1560-1624AD), John Ray (1627-1705AD), Joseph Pitton de Tournefort (1656-1708AD), Rudolf Camerarius (1665-1721AD) and Carolus Linnaeus (1707-1778AD) (Figs. 4.9-4.13). Out of the above classifications, the classification of Carolus Linnaeus gained much popularity. Brief knowledge of some of the above classifications is needed :

Theophrastus (370-287 BC) : He was a Greek naturalist and pupil of **Plato** and **Aristotle**. Based on the habit, he classified the plants into four (4) groups : **Herbs, Undershrub, Shrubs** and **Trees**. He described and named about 480 plants and published his work in *Historia Plantarum*, known as the oldest botanical work in existence. Some of his important contributions are mentioned :

Fig. 4.9 : Theophrastus (370-287 B.C.)

Fig. 4.10 : Andrea Cesalpino (1519-1603)

Fig. 4.11 : Gaspard Bauhin (1560-1624)

1. Fundamental differences between Dicotyledons and Monocotyledons;
2. Difference between centripetal (racemose) and centrifugal (cymose) type of inflorescences;
3. Difference between polypetalous and gamopetalous corolla;

Fig. 4.12 : John Ray (1627-1705)

Fig. 4.13 : Carolus Linnaeus (1707-1778)

4. Difference between superior and inferior ovary, etc.

Some of the names given by him received general recognition and some are also used as generic names in modern classification, e.g., Aspharagos (*Asparagus*), Daukan (*Daucas*) etc.

For the above immense contributions, Theophrastus is called the **"Father of Botany"**.

Andrea Cesalpino (1519–1603) classified the plants first on the basis of habit and secondarily on the characteristics of fruits and seeds. He is usually regarded as the **"First Plant Taxonomist"**.

Later. **Gaspard** (Casper or Kaspar) **Bauhin** (1560–1624) in his work frequently referred the plants with a binary nomenclature and made distinction between genus and species.

John Ray (1627-1705) : He was a British botanist. Initially, he proposed his classification in *Methodus Plantarum Nova* (1682). Later on, he represented his classification in *Historia Plantarum* (1686-1704) in three volumes. Like other old systems, he classified the plants into **herbs** and **trees** and further into Dicotyledons and Monocotyledons. His system was much advanced than the earlier classifications and approached the natural system.

The outline of his system is :

A. **Herbae** (herbs)
   - i. Imperfectae (Non-flowering plants i.e., cryptogams).
   - ii. Perfectae (Flowering plants).
     - a. Dicotyledons.
     - b. Monocotyledons.

B. **Arborae** (trees and shrubs).
   - a. Dicotyledons.
   - b. Monocotyledons.

Carolus Linnaeus (1707-1778) : He was a Swedish naturalist. For his outstanding contribution he is called the "**Father of Modern Botany**". He published his work in different times since 1730 to final form in 1753. The publications are :

1. 1730 — *Hortus Uplandicus* (contains the plants of Uppsala Botanic Garden at that time).
2. 1735 — *Systema Naturae.*
3. 1737 — *Hortus Cliffortianus* (contains the plants in the Garden of George Clifford at Hartecamp).
4. 1737 — *Genera Plantarum.*
5. 1738 — *Classes Plantarum.*
6. 1751 — *Philosophia Botanica* (contains the revised revision of his earlier publications *Systema Naturae* and *Classes Plantarum.*
7. 1753 — *Species Plantarum* (contains some 7,300 species described and arranged according to his system of classification). In this book, he constantly used the **binomial system** in plant names. The Binomial system consists of two names of a specimen, where the first one is the generic epithet and the second is the specific epithet. This binomial system was used subsequently, even by the modern botanists till date.

The classification of Linnaeus is an **artificial** one. The significance of flower and fruit structures was first recognised by him. He emphasised the basic numerical characteristics of sexual parts i.e., stamens and carpels. Thus the Linnaeus system is also known as **sexual system.** Based on the number, size, length and union of stamens, he classified the plant kingdom into 24 classes, such as *Monandria* (flower with one stamen), *Diandria* (flower with two stamens), etc. The classes were further divided into orders, based on the number of styles in each flower such as Monogynia (flower with one style), *Digynia* (flower with two styles) etc. The outline of **Linnaeus System** of classification with twenty four (24) classes is given :

Classes

1. Monandria (flowers with 1 stamen).
2. Diandria (,, ,, 2 stamens).
3. Triandria (,, ,, 3 stamens).
4. Tetrandria (,, ,, 4 stamens).
5. Pentrandria (,, ,, 5 stamens).
6. Hexandria (,, ,, 6 stamens).
7. Heptandria (,, ,, 7 stamens).
8. Octandria (,, ,, 8 stamens).
9. Enneandria (,, ,, 9 stamens).
10. Decandria (,, ,, 10 stamens).
11. Dodecandria (,, ,, 12 stamens).
12. Icosandria (,, ,, more than 20 stamens are attached to the calyx).
13. Polyandria (,, ,, more than 20 stamens are attached to the receptacle).
14. Didynamia (stamens didynamous).
15. Tetradynamia (stamens tetradynamous).
16. Monadelphia (stamens monadelphous).
17. Diadelphia (stamens diadelphous).
18. Polyadelphia (stamens polyadelphous).
19. Syngenesia (stamens syngenesious).
20. Gynandria (stamens adnate to the pistil).
21. Monoecia (plants monoecious).
22. Dioecia (plants dioecious).
23. Polygamia (plants polygamous).
24. Cryptogamia (non-flowering plants i.e., cryptogams which include algae, fungi, mosses and ferns).

Due to simple and convenient approach, the Sexual System became very popular for more than 75 years before the publication of natural system by A. L. de Jussieu and A. P. de Candolle.

The merits and demerits of the Linnaeus system are :

Merits and Demerits

Merit :

The only merit of this system is the quick and easy identification of plants based on one or a few characteristics.

Demerits :

1. The system is not at all sexual in proper sense, but based only on *numerical relationship of sex organs.* Thus, it can be said that this system was developed on differences rather than on similarities of sex organs.
2. The closely related members go apart and the distantly related members become very close to each other for the numerical relation of the sex organs.
3. The Gymnosperms were placed in the 14th class *Didynamia* along with Labiatae, an angiosperm family.
4. The Monocotyledons, Dicotyledons and Gymnosperms i.e., phanerogamic plants are not considered separately. Thus, the members of Dicotyledons and Monocotyledons become very close. e.g.,
   i. In the 1st class *Monandria* (flower having 1 stamen) – *Globba* (Zingiberaceae of Monocotyledons) and *Mangifera* (Anacardiaceae of Dicotyledons).
   ii. In the 6th class *Hexandria* (flowers having 6 stamens) –*Alisma* (Alismaceae of Monocotyledons) and *Rumex* (Polygonaceae of Dicotyledons) come very close. Many other examples are also available.

Benson (1957) made the comment that "certainly it did not contribute anything directly to knowledge of plant relationships, but it did give a tremendous impetus to study of plant kingdom".

5. Thus it can be said that this classification is just like a dictionary where the words are arranged alphabetically without maintaining any relationship.

Natural system of classification

This type of classification is based on the proper utilisation of all facts and figures available in nature. Based on common natural characteristics, the plants are grouped and placed into different taxa like classes, orders, families and genera. This type of system does not consider the *evolutionary sequence* and, hence, the *affinity* and *relationship* between the taxa are not considered.

The natural system of classification serves the practical need of easy identification.

Perhaps **Michel Adanson** (1727-1806) is the first person to reject all the artificial systems and propose his classification in favour of natural system. He was a French botanist and published his work in *Families de Plantes* (1763) in two volumes. His grouping of the plants is equivalent

Fig. 4.14 : Antoine Laurent de Jussieu (1748-1836)

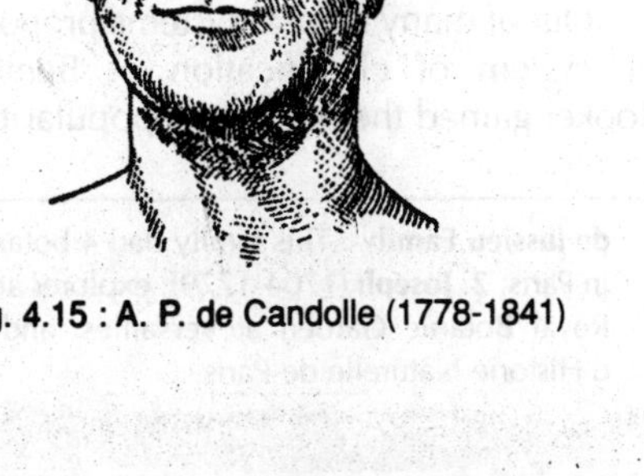

Fig. 4.15 : A. P. de Candolle (1778-1841)

Fig. 4.16 : Robert Brown (1773-1858)

to orders and families of the modern system. Later on, many botanists published (name of the publication with year is given in parentheses) their system and some of them are :

1. Antoine Laurent de Jussieu (*Genera Plantarum* in 1789);
2. Augustin Pyramus de Candolle (*Theorie Elementaire de la Botanique* in 1813). In this book he proposed his approach to classify the plants. Later on, he published his work in *Prodromus Systematis Naturalis Regni Vegetabilis.* The first volume appeared in 1824 and was able to complete only 7 volumes during his life time. Later on his son Alphonse de Candolle published the rest in 10 volumes and the last one appeared in 1844.
3. Robert Brown (*Prodromus Florae Novae Hollandiae* in 1827);
4. John Lindley (*Introduction to the Natural System of Botany* in 1830);
5. Stephan Endlicher (*Genera Plantarum* from 1838-1841);
6. George Bentham and Joseph Dalton Hooker (*Genera Plantarum* from 1862-1883).

Out of many natural systems proposed so far the system of classification of Bentham and Hooker gained the maximum popularity.

George Bentham (1800-1884) and Joseph Dalton Hooker's (1817-1911) system

Both the British botanists (Fig. 4.17, 4.18) were associated with the Royal Botanic Garden, Kew, England. They worked together for about 25 years and jointly published their work in ***Genera Plantarum*** (1862-1883) in 3 volumes. The 1st volume was published in July 1862 and the 3rd volume in April 1883. They described every species available at that time.

Robert Brown held that "Their *Genera Plantarum* was at that time a near approach to finality. Hitherto its supremacy has not been challenged".

Their system was preferred by Britain and Commonwealth countries but not in Europe and North America, where Engler's system is preferred.

They divided the seed plants into two **Subkingdoms** : *Cryptogamia* and *Phanerogamia.* The Phanerogamia was divided into two **Divisions** : *Gymnospermae* and *Angiospermae.* The Angiospermae was further divided into two **Classes** *Dicotyledons* and *Monocotyledons.* The Dicotyledons was further divided into three subclasses Polypetalae, Gamopetalae and Monochlamydeae. Polypetalae was divided into three **Series** : *Thalamiflorae, Disciflorae* and *Calyciflorae;* Gamopetalae was divided into

Fig. 4.17 : George Bentham (1800-1884)

* **de Jussieu Family** : This family had 4 botanists : the brothers were 1. **Antoine** (1688-1758), Director of the Jardin des in Paris; 2. **Joseph** (1704-1779), explorer and collector in South America, and 3. **Bernard** (1699-1777), Founder of the Royal Botanic Garden at Versailles, and 4. their nephew **Antoine-Laurent** (1748-1836), Founder of the Musse d'Historie Naturelle de Paris

Fig. 4.18 : Sir Joseph Dalton Hooker (1817-1911)

three **Series** : *Inferae, Heteromerae* and *Bicarpellatae;* Monochlamydeae into eight **Series** and the **Class** Monocotyledons was divided into seven **Series**.

The Series were further divided into **Cohorts** (= **Orders**), Cohorts into **Natural Orders** (= **Families**), Natural orders into **Genera** and, finally, Genera into **Species**.

They have placed *Gymnospermae,* in between *Dicotyledons* and *Monocotyledons.* According to their system, *Dicotyledons* starts with the Natural Order Ranunculaceae and ends in Ceratophyllaceae. On the other hand, *Monocotyledons* starts with the Natural Order Hydrocharitaceae and ends in Gramineae.

The summary of the system of classification of Bentham and Hooker is given :

Class. DICOTYLEDONS
(*Seed with two cotyledons*).

Subclass a. **POLYPETALAE**
(*Petals are free*).

Series I. THALAMIFLORAE
(*Flowers are hypogynous*).

Cohort 1. Ranales
2. Parietales
3. Polygalineae
4. Caryophyllineae
5. Guttiferales
6. Malvales

Series II. DISCIFLORAE
(*Receptacle expanded into a disc; stamens are inserted on the disc; ovary is superior*).

Cohort 1. Geraniales
2. Olacales
3. Celastrales
4. Sapindales

Series III. CALYCIFLORAE
(*Flowers perigynous or sometimes epigynous*)

Cohort 1. Rosales.
2. Myrtales.
3. Passiflorales.
4. Ficoidales.
5. Umbellales.

Subclass b. **GAMOPETALAE**
(*Petals are united*).

Series I. INFERAE (*Ovary inferior*)

Cohort 1. Rubiales.
2. Asterales.
3. Campanulales.

Series II. HETEROMERAE
(*Ovary superior, stamens as many or twice as many as corolla lobes, carpels more than two*)

Cohort 1. Ericales.
2. Primulales.
3. Ebenales.

Series III. BICARPELLATAE
(*Ovary superior, stamens as many as corolla lobes or fewer, carpels usually two*)

Cohort 1. Gentianales
2. Polemoniales.
3. Personales.
4. Lamiales.

One order 'Anomalus Plantagineae'

Subclass c. **MONOCHLAMYDEAE**
(*Flowers incomplete, either unisexual or perianth undifferentiated i.e., haplochlamydous*).

Series I. CURVEMBRYEAE. (*Embryo curved, ovules usually one*).

II. MULTIOVULATAE AQUATICAE. (*Ovules numerous, plants immersed aquatics*).

III. MULTIOVULATAE TERRESTRIS. (*Ovules numerous, plants terrestrial*).

IV. MICROEMBRYEAE. (*Embryo minute, ovule usually one*).

V. DAPHNALES. (*Ovary monocarpellary, ovule is usually one*).

VI. ACHLAMYDOSPOREAE (*Ovary unilocular, ovules one to three, seeds without testa*).

VII. UNISEXUALES. (*Flowers unisexual, ovules one to three*).

VIII. ORDINES ANOMALI. (*Anomalous orders*).

Division. GYMNOSPERMAE (*Naked seeded plants*)

It includes three natural orders: Gnetaceae, Coniferae and Cycadaceae.

Class. MONOCOTYLEDONS (*Seed with one cotyledon*)

Series I. MICROSPERMAE. (*Ovary inferior, seeds minute and exalbuminous*).

II. EPIGYNAE. (*Ovary inferior, seeds large*).

III. CORONARIEAE. (*Perianth petaloid, ovary superior*).

IV. CALYCINAE. (*Perianth sepaloid, ovary superior*).

V. NUDIFLORAE. (*Perianth mostly lacking or reduced to scale*).

VI. APOCARPAE. (*Carpels free or only one, ovary superior, seeds without albumin*).

VII. GLUMIFLORAE. (*Flowers in spikelet or in heads, perianth small, scale-like or chaffy, seeds with albumin*).

Due to busy official work of J. D. Hooker as the Curator of the Royal Botanic Garden, England, two-third of the work was done by G. Bentham. Charles Darwin had published the *Origin of Species* (1859) earlier to the publication of their system. They had also the knowledge of theory of evolution, but it was Bentham who did not accept the essentiality of Darwin's work in their system, as was proposed by Hooker to recognise the entire system based on this theory.

Merits and Demerits

Merits :

1. This is a great natural system of its own kind.
2. This one is suitable for all practical purposes especially for identification and description of plants.
3. Beginning of the Subclass Polypetalae with the Cohort (= Order) Ranales, comprised of Ranunculaceae and its allied Natural Orders (= Families).
4. The Dicotyledons are primitive than Monocotyledons.
5. They have studied the actual specimens available at that time and represented them in the classification after careful comparative examination.

Demerits :

1. The classification is based on the assumption of constancy of species. Hence, closely related Cohorts (= Orders) are often widely separated.
2. Authors were silent about the phylogeny of Dicotyledons and Monocotyledons and the anomalous position of Gymnosperms in between Dicotyledons and Monocotyledons.
3. The authors did not consider the phylogenetic relationship. Thus closely related Natural Orders (= Families) were placed apart and unrelated Natural Orders were placed very near. In Dicotyledons, Euphorbiaceae was placed in Monochlamydeae though it is related to Malvaceae of Polypetalae. The Natural Orders like Polygonaceae, Nyctaginaceae, Amaranthaceae, Chenopodiaceae etc. are grouped together and placed under the Subclass Monochlamydeae which is also unnatural, because the above Natural Orders are

related with the orders having differentiated perianth.

4. In Monocotyledons, they placed Scitamineae and Orchidaceae at the very beginning. But these should not be regarded as primitive with so many advanced characteristics.
5. In the Class Monocotyledons proper emphasis has not been given on :
   i. The relative position of ovary, and
   ii. Characteristics of the perianth.

Based on the above, it was considered that Iridaceae and Amaryllidaceae are more nearly related to Liliaceae than Scitamineae or Bromeliaceae which are associated for epigynous flowers. Hydrocharideae, though having inferior ovary, were placed nearer to aquatic cohorts with non-endospermic seeds.

Phylogenetic system of classification

After the publication of Darwin's theory of the *'Origin of Species'* in 1859, the taxonomists switched over in their thinking towards the construction of classification based on the course of evolution and tried to arrange the plants based on genetic and phylogenetic relationships. Out of many, the most widely known systems are of Engler (1886-1892), Hutchinson (1926, 34, 48, 59, 69 and 73) and more recently by Takhtajan (1969, 80, 97) and Cronquist (1968, 81) (Figs. 4.19-4.23). The above systems are detailed :

Adolf Engler (1844-1930)

The best known and widely accepted phylogenetic system is that by **Adolf Engler**, Professor of Botany, University of Berlin. In 1892, he published a system of classification mainly based on **August Wilhelm Eichler** in the book *'Syllabus der Vorlesungen'* as a guide to study the plants available in the Breslau Botanic Garden. During 1887-1915, Engler and his associate **Karl Prantl** made a monographic work, the *"Die naturlichen Pflanzenfamilien"* in 20 volumes, including all the known genera of plants from algae to the phanerogams along with key to identify the plants. Engler, in collaboration with Gilg, and later with Diels, published the works in a single volume *'Syllabus der Pflanzenfamilien'*. After his death, the book was revised by followers in several editions and the latest (12th) one in 2 volumes in 1954 and 1964.

The system of Engler has been widely used in the American and Europian continents.

Engler divided the plant kingdom into thirteen (13) **Divisions**. The brief outline of his system is given below :

Divisions

I. Schizophyta

Class. 1. Schizomycetes

2. Schizophyceae.

II. Myxothallophyta, Myxomycetes, etc.

III. Flagellatae

IV. Dinoflagellatae

V. Bacillariophyta

Fig. 4.19 : Adolf Engler (1844-1930)

Fig. 4.20 : Karl Prantl (1849-1893)

VI. Conjugatae

VII. Chlorophyceae

Class.1. Protococcales

2. Ulotrichales

3. Siphonocladales

4. Siphonales

VIII. Charophyta

IX. Phaeophyceae

X. Rhodophyceae

Class.1. Bangiales

2. Florideae

XI. Eumycetes

Class.1. Phycomycetes

2. Ascomycetes

3. Protomycetes

4. Basidiomycetes

XII. Embryophyta Asiphonogama (Archegoniatae)

A. *Subdivision Bryophyta*

Class.1. Hepaticae

2. Musci

B. *Subdivision Pteridophyta*

Class.1. Filicales

2. Sphenophyllales

3. Equisetales

4. Lycopodiales

5. Psilotales

6. Isoetales

XIII. Embryophyta Siphonogama

A. *Subdivision Gymnospermae*

Class.1. Cycadofilicales

2. Cycadales

3. Bennettitales

4. Ginkgoales

5. Coniferae

6. Cordaitales

7. Gnetales

B. *Subdivision Angiospermae*

Class.1. Monocotyledoneae

2. Dicotyledoneae

Subclass. 1. Archichlamydeae

2. Metachlamydeae (Sympetalae)

The thirteenth (13) **Division** is the **Embryophyta Siphonogama** (the seed-bearing plants i.e., Spermatophyta). It is divided into two **Subdivisions**, *Gymnospermae* and *Angiospermae*. The *Angiospermae* is divided into two **Classes** — *Monocotyledonae* and *Dicotyledonae*. The Class *Monocotyledonae* is divided directly into 11 **Orders**. On the other hand, the Class *Dicotyledonae* is divided into two **Subclasses** — *Archichlamydeae* i.e., lower dicotyledons, and *Metachlamydeae* or *Sympetalae* i.e., higher dicotyledons. The *Archichlamydeae* is further divided into 33 Orders, and *Metachlamydeae* into 11 Orders. The Orders are divided into **Suborders, Families, Genera** and finally into **Species**.

In this system, the **Plant Kingdom** contains 309 families. The Class *Monocotyledonae* starts with the family *Typhaceae* and ends in *Orchidaceae,* while the class *Dicotyledonae* starts with the family *Casuarinaceae* and ends in *Compositae*.

In this system, Engler considered that in Embryophyta Siphonogama the *flower without perianth is the primitive one.* Thus, plants like Oak, Willow etc., with woody stem and unisexual apetalous flowers (Amentiferae), are treated as primitive Dicotyledons.

The main distinctive features of Engler's system that separate it from that of Bentham and Hooker's system are :

1. The Polypetalae and Monochlamydeae of Bentham and Hooker are amalgamated and placed into a single group (Subclasses) *Archichlamydeae.*
2. The families of the flowering plants are arranged in ascending order with the increasing complexity of the flowers (mainly on floral envelope).
3. Monocotyledons are placed before Dicotyledons.
4. The term **Natural order** has been replaced by **Family**.
5. The term **Series** or **Cohort** has been replaced by **Order**.

The summary of the system of classification of Engler and Prantl is given :

13th Division: Embryophyta Siphonogama

Subdivision A. Gymnospermae.

Class 1. Cycadofilicales.
2. Cycadales.
3. Bennettitales.
4. Ginkgoales.
5. Coniferae.
6. Cordaitales.
7. Gnetales.

Subdivision B. Angiospermae.

Class 1. Monocotyledoneae.

Order 1. Pandanales.
2. Helobiae.
3. Triuridales.
4. Glumiflorae.
5. Principes.
6. Synanthae.
7. Spathiflorae.
8. Farinosae.
9. Liliiflorae.
10. Scitamineae.
11. Microspermae.

Class 2. Dicotyledoneae.

Subclass 1. Archichlamydeae (Choripetalae and Apetalae).

Order 1. Verticillatae.
2. Piperales.
3. Hydrostachyales.
4. Salicales.
5. Garryales.
6. Myricales.
7. Balanopsidales.
8. Leitneriales.
9. Juglandales.
10. Julianiales.
11. Batidales.
12. Fagales.
13. Urticales.
14. Podostemonales.
15. Proteales.
16. Santalales.
17. Aristolochiales.
18. Balanophorales.
19. Polygonales.
20. Centrospermae.
21. Ranales.
22. Rhoeadales.
23. Sarraceniales.
24. Rosales.
25. Pandales.
26. Geraniales.
27. Sapindales.
28. Rhamnales.
29. Malvales.
30. Parietales.
31. Opuntiales.
32. Myrtiflorae.
33. Umbelliflorae.

Subclass 2. Metachlamydeae (Sympetalae).

Order 1. Diapensiales.
2. Ericales.
3. Primulales.
4. Plumbaginales.
5. Ebenales.
6. Contortae.
7. Tubiflorae.
8. Plantaginales.
9. Rubiales.
10. Cucurbitales.
11. Campanulatae.

Merits and Demerits

Merits :

1. The entire Plant Kingdom was broadly treated with excellent illustrations, and phylogenetic arrangement of many groups of plants was made.
2. The amalgamation of Polypetalae and Monochlamydeae into Archichlamydeae is justified.
3. Consideration and placing of Orchidaceae at the end of Monocotyledons and Compositae at the end of Dicotyledons are justified — since they are most highly evolved.

4. Juncaceae, Amaryllidaceae and Iridaceae are placed judiciously nearer to Liliaceae.

Demerits :

1. The placement of Amentiferae and Centrospermae almost at the beginning of Dicotyledones, even before Ranales, are not justified.
2. The assemblage of all sympetalous members under Metachlamydeae increased the distance of closely related orders.
3. The placing of Monocotyledons before Dicotyledons is not appropriate, because it is generally agreed that monocots have arisen from dicotyledons by reduction.
4. The placing of the order Helobiae between the advanced orders Pandanales and Glumiflorae is questionable. Araceae was placed much earlier than Liliaceae, from which it has been derived.
5. Fossil evidences gave little support to this system.

John Hutchinson (1884-1972)

John Hutchinson was a British botanist associated with Royal Botanic Gardens, Kew, England. He developed and proposed his system based on Bentham and Hooker and also on Bessey. His phylogenetic system first appeared as *"The Families of Flowering Plants"* in two volumes. The first volume contains Dicotyledons (published in 1926) and second volume contains Monocotyledons (published in 1934). He made several revisions in different years. The final revision of *"The Families of Flowering Plants"* was made just before his death on 2nd September 1972 and the 3rd i.e., the final edition, was published in 1973.

The following principles were adopted by Hutchinson to classify the flowering plants :

1. Evolution takes place in both upward and downward direction.
2. During evolution all organs do not evolve at the same time.
3. Generally, evolution has been consistent.
4. Trees and shrubs are more primitive than herbs in a group like genus or family.
5. Trees and shrubs are primitive than climbers.
6. Perennials are older than annuals and biennials.
7. Terrestrial angiosperms are primitive than aquatic angiosperms.
8. Dicotyledonous plants are primitive than monocotyledonous plants.
9. Spiral arrangement of vegetative and floral members are primitive than cyclic arrangements.
10. Normally, simple leaves are more primitive than compound leaves.
11. Bisexual plants are primitive than unisexual plants and monoecious plants are primitive than dioecious plants.
12. Solitary flowers are primitive than flowers on inflorescence.
13. Types of aestivation gradually evolved from contorted to imbricate to valvate.
14. Polymerous flowers precede oligomerous flowers.
15. Polypetalous flowers are more primitive than gamopetalous flowers.
16. Flowers with petals are more primitive than apetalous flowers.
17. Actinomorphic flowers are more primitive than zygomorphic flowers.
18. Hypogyny is considered as more primitive from which perigyny and epigyny gradually evolved.

Fig. 4.21 : John Hutchinson (1884-1972)

19. Apocarpous pistil is more primitive than syncarpous pistil.
20. Polycarpy is more primitive than gynoecium with few carpels.
21. Flowers with many stamens are primitive than flowers with few stamens.
22. Flowers with separate anthers are primitive than flowers with fused anthers and/filaments.
23. Endospermic seeds with small embryo is primitive than non-endospermic one with a large embryo.
24. Single fruits are primitive than aggregate fruits.

He divided the Phylum *Angiospermae* into two Subphyla *Dicotyledones* and *Monocotyledones.* The Dicotyledones are further divided into two divisions — *Lignosae* (arboreal) and *Herbaceae* (herbaceous). The Lignosae includes, fundamentally, the woody representatives derived from Magnoliales and Herbaceae includes most of the predominantly herbaceous families derived from Ranales. The subphylum Monocotyledones are divided into three divisions — *Calyciferae, Corolliferae* and *Glumiflorae.*

1. The division *Lignosae* was further divided into 54 orders beginning with Magnoliales and ending in Verbenales.
2. The division *Herbaceae* was divided into 28 orders beginning with Ranales and ending in Lamiales.
3. The division *Calyciferae* was divded into 12 orders beginning with Butamales and ending in Zingiberales.
4. The division *Corolliferae* was divided into 14 orders beginning with Liliales and ending in Orchidales.
5. The division *Glumiflorae* was divided into 3 orders beginning with Juncales and ending in Graminales.

So in the latest system of Hutchinson, the Dicotyledones consists of 83 orders and 349 families and Monocotyledones consists of 29 orders and 69 families.

The summary of the system of classification of Hutchinson is given :

Phyllum ANGIOSPERMAE

Subphylum I. Dicotyledones

Division I. Lignosae.

Order 1. Magnoliales.
2. Annonales.
3. Laurales.
4. Dilleniales.
5. Coriariales.
6. Rosales.
7. Leguminales.
8. Cunoniales.
9. Styracales.
10. Araliales.
11. Hamamelidales.
12. Salicales.
13. Leitneriales.
14. Myricales.
15. Balanopales.
16. Fagales.
17. Juglandales.
18. Casuarinales.
19. Urticales.
20. Bixales.
21. Thymelaeales.
22. Proteales.
24. Pittosporales.
25. Tamaricales.
26. Violales.
27. Polygalales.
28. Loasales.
29. Passiflorales.
30. Cucurbitales.
31. Cactales.
32. Tiliales.
33. Malvales.
34. Malpighiales.
35. Euphobiales.
36. Theales.
37. Ochnales.
38. Ericales.
39. Guttiferales.

40. Myrtales.
41. Celastrales.
42. Olacales.
43. Santalales.
44. Rhamnales.
45. Myrsinales.
46. Ebenales.
47. Rutales.
48. Meliales.
49. Sapindales.
50. Loganiales.
51. Apocynales.
52. Rubiales.
53. Bignoniales.
54. Verbenales.

Division.II. Herbaceae.

Order 1. Ranales.
2. Berberidales.
3. Aristolochiales.
4. Piperales.
5. Rhoeadales.
6. Brassicales.
7. Resedales.
8. Caryophyllales.
9. Polygonales.
10. Chenopodiales.
11. Onagrales.
12. Gentianales.
13. Primulales.
14. Plantaginales.
15. Saxifragales.
16. Sarraceniales.
17. Podostemales.
18. Umbellales.
19. Valerianales.
20. Campanales.
21. Goodeniales.
22. Asterales.
23. Solanales.
24. Personales.
25. Geraniales.
26. Polemoniales.
27. Boraginales.
28. Lamiales.

Subphylum II. Monocotyledones

Division I. Calyciferae.

Order 1. Butomales.
2. Alismatales.
3. Triuridales.
4. Juncaginales.
5. Aponogetonales.
6. Potamogetonales.
7. Najadales.
8. Commelinales.
9. Xyridales.
10. Eriocaulales.
11. Bromeliales.
12. Zingiberales.

Division II. Corolliferae.

Order 1. Liliales.
2. Alstroemeriales.
3. Arales.
4. Typhales.
5. Amaryllidales.
6. Iridales.
7. Dioscoreales.
8. Agavales.
9. Palmales.
10. Pandanales.
11. Cyclanthales.
12. Haemodorales.
13. Burmanniales.
14. Orchidales.

Division III. Glumiflorae.

Order 1. Juncales.
2. Cyperales.
3. Graminales.

Merits and Demerits

Merits :

1. Hutchinson proposed the monophyletic origin of angiosperms from some hypothetical Proangiosperms having Bennettitalean characteristics.
2. He made a valuable contribution in

phylogenetic classification by his careful and critical studies.

3. Monocots have been derived from Dicots.
4. According to him, the definitions of orders and families are mostly precise, particularly in case of subphylum Monocotyledones.

Demerits :

1. There is undue fragmentation of families.
2. Too much emphasis is laid on habit and habitat. Thus, creation of Lignosae and Herbaceae is thought to be a defect reflecting the Aristotelean view.
3. The origin of angiosperms from Bennettitalean-like ancestor is criticised by many, because the anatomical structures of the early dicotyledons are not tenable with such ancestry.

Armen Takhtajan (1911-)

Takhtajan was a reputed palaeobotanist of Komarov Botanical Institute of Leningrad, U.S.S.R. (now in Russia). He also made great contributions in the field of angiosperm taxonomy. In 1942, he proposed preliminary phylogenetic arrangement of the orders of higher plants, based on the structural types of gynoecium and placentation. After 12 years i.e., in 1954, the actual system of classification was published in *"The Origin of Angiospermous Plants"* in Russian language. It was translated in English in 1958. Later on, in 1964, he propsed a new system in Russian language. To trace the evolution of angiosperm, he was particularly inspired by Hallier's attempt to develop a synthetic evolutionary classification of flowering plants based on Darwinian philosophy. The classification was published in *'Flowering Plants: Origin and Dispersal'* (1969) in English language. Later on, in 1980, a new revision of his system was published in *"Botanical Review"*.

Takhtajan (1980) included the angiospermic plants under the Division Magnoliophyta. The Magnoliophyta is divided into two classes Magnoliopsida (Dicotyledons) and Liliopsida (Monocotyledons). The class Magnoliopsida consists of 7 subclasses, 20 superorders, 71 orders and 333 families. On the other hand,

Fig. 4.22 : Armen Takhtajan (1911-)

Liliopsida comprises of 3 subclasses, 8 superorders, 21 orders and 77 families. The class Magnoliopsida starts with the order Magnoliales and ends in Asterales and the class Liliopsida begins with Alismatales and ends in Arales.

The class Magnoliopsida is considered to be monophyletic in origin, probably derived from Bennettitales-like ancestors or stocks ancestral to them. On the other hand, the Liliopsida have been considered to be originated from the stocks ancestral to Nymphaeales. He considered Magnoliopsida more primitive than the Liliopsida. The principles as adopted by Takhtajan (1980) for interpreting the evolutionary lineages in higher plants are mentioned in Table 4.2.

The outline of the classification up to order is given :

Division. MAGNOLIOPHYTA (Angiospermae)

Class. Magnoliopsida (Dicotyledons)

Subclass Magnoliidae

Superorder. Magnolianae

Order. 1. Magnoliales (Annonales)

2. Illiciales
3. Laurales
4. Piperales
5. Aristolochiales

Superorder. Rafflesianae

Order. 1. Rafflesiales

Table 4.2 : Principles adopted by Takhtajan (1980) to classify flowering plants

Primitive	Advanced
A. Growth habit.	A. Growth habit.
1. Woody habit.	1. Herbaceous habit.
2. Small woody angiosperms.	2. Large trees of tropical rain forests.
3. Sparingly branched trees.	3. Trees with numerous slender branches.
4. Evergreen plants.	4. Deciduous woody plants.
B. Leaves.	B. Leaves.
1. Simple leaves.	1. Compound leaves.
2. Reticulate venation.	2. Parallel venation.
3. Alternate leaves.	3. Opposite leaves.
C. Stomata.	C. Stomata.
1. Mesogenous type.	1. Perigenous type.
2. With subsidiary cells.	2. Lacking subsidiary cells.
D. Nodal structure.	D. Nodal structure.
1. Tri to pentalacunar type.	1. Unilacunar type.
E. Inflorescence.	E. Inflorescence.
1. Cymose.	1. Racemose.
F. Floral structure.	F. Floral structure.
1. Indefinite and variable number of floral parts arranged spirally on an elongated axis.	1. Fixed number of floral parts arranged in cyclic pattern on a short axis.
G. Pollen grains.	G. Pollen grains.
1. Exine devoid of external sculpturing.	1. Sculpturing is of various types.
2. Monocolpate (dicots).	2. Tricolpate (dicots).
H. Gynoecium.	H. Gynoecium.
1. Apocarpous.	1. Syncarpous.
I. Ovules.	I. Ovules.
1. Anatropous type.	1. Other types.
2. Crassinucellate ovules.	2. Tenuinucellate ovules.
J. Pollination.	J. Pollination.
1. Entomophily.	1. Anemophily.
K. Gametophyte and fertilization.	K. Gametophyte and fertilization.
1. Monosporic *Polygonum* type.	1. Tetrasporic type.
L. Seeds.	L. Seeds.
1. Abundant endosperm with a minute and undifferentiated embryo.	1. Endosperm is reduced or wanting and the embryo is large and well-differentiated.
M. Fruits.	M. Fruits.
1. Many seeded follicular fruit develop from multicarpellary apocarpous gynoecia.	1. Coenocarpous fruits.

Superorder. Nymphaeanae

Order. 1. Nymphaeales

2. Nelumbonales

Subclass. Ranunculidae

Superorder. Ranunculanae

Order. 1. Ranunculales

2. Papaverales
3. Sarraceniales

Subclass. Hamamelididae
Superorder. Hamamelidanae
Order. 1. Trochodendrales
2. Circidiphyllales
3. Eupteleales
4. Didymelales
5. Hamamelidales
6. Eucommiales
7. Utricales
8. Barbeyales
9. Casuarinales
10. Fagales
11. Balanopales
12. Leitneriales
Superorder. Juglandanae
Order. 1. Myricales
2. Juglandales

Subclass. Caryophyllidae
Superorder. Caryophyllanae
Order. 1. Caryophyllales
2. Polygonales
Superorder. Plumbaginanae
Order. 1. Plumbaginales

Subclass. Dilleniidae
Superorder. Dillenianae
Order. 1. Dilleniales
2. Paeoniales
3. Theales
4. Violales
5. Begoniales
6. Capparales
7. Tamaricales
8. Salicales
Superorder. Ericanae
Order. 1. Ericales
2. Ebenales
3. Primulales
Superorder. Malvanae
Order.1. Malvales
2. Euphorbiales
3. Thymeleales

Subclass. Rosidae.
Superorder. Rosanae
Order. 1. Saxifragales
2. Rosales
3. Fabales
4. Connarales
5. Podostemales
6. Nepenthales
Superorder. Myrtanae
Order. 1. Myrtales
Superorder. Rutanae
Order. 1. Rutales
2. Sapindales
3. Geraniales
4. Polygalales
Superorder. Aralianae
Order. 1. Cornales
2. Araliales (Apiales)
Superorder. Celastranae
Order. 1. Celastrales
2. Santalales
3. Balanophorales
4. Rhamnales
5. Elaeagnales
Superorder. Proteanae
Order. 1. Proteales

Subclass. Asteridae
Superorder. Gentiananae
Order. 1. Gentianales
2. Oleales
3. Dipsacales
4. Loasales
Superorder. Lamianae
Order. 1. Polemoniales
2. Lamiales
3. Scrophulariales
Superorder. Asteranae
Order. 1. Campanulales (including Goodeniales)
2. Calycerales
3. Asterales

Class. Liliopsida (Monocotyledons)

Subclass. Alismatidae

Superorder. Alismatanae

Order. 1. Alismatales

2. Najadales (including Aponogetonales, Potamogetonales and Zosterales)

Subclass. Liliidae (including Commelinidae and Zingiberidae)

Superorder. Triuridanae

Order. 1. Triuridales

Superorder. Lilianae

Order. 1. Liliales

2. Smilacales

3. Burmanniales

4. Orchidales

5. Bromeliales

Superorder. Juncanae

Order. 1. Juncales

2. Cyperales

Superorder. Commelinanae

Order. 1. Commelinales

2. Eriocaulales

3. Restionales

4. Hydatellales

5. Poales

Superorder. Zingiberanae

Order. 1. Zingiberales

Subclass. Arecidae

Superorder. Arecanae

Order. 1. Arecales

2. Cyclanthales

3. Pandanales

4. Typhales

Superorder. Aranae

5. Arales

Later, **Armen Takhtajan** (1997) revised his classification in the book *"Diversity and Classification of Flowering Plants"*. He treated the flowering plants as a division or phylum (**Magnoliophyta**) with two classes **Magnoliopsida** (dicots) and **Liliopsida** (monocots). The class Magnoliopsida has been divided into 11 subclasses, 56 superorders and 225 orders and the class Liliopsida has been divided into 6 subclasses, 16 superorders and 58 orders. The class Magnoliopsida starts with the family Degeneriaceae and ends in Hippuridaceae. On the other hand, the class Liliopsida starts with Tofieldiaceae and ends in Typhaceae.

The outline of the classification (1997) up to order is given :

Class Magnoliopsida (Dicots)

Subclass. Magnoliidae

1. Superorder : Magnolianae

Order. 1. Magnoliales

2. Winterales.

3. Canellales.

4. Illiciales.

5. Austrobaileyales.

6. Eupomatiales.

7. Annonales.

8. Myristicales.

9. Aristolochiales.

2. Superorder : Lactoridanae

Order. 1. Lactoridales

3. Superorder : Piperanae

Order. 1. Piperales.

4. Superorder : Lauranae

Order. 1. Laurales.

2. Calycanthales.

3. Chloranthales.

5. Superorder : Rafflesianae

Order. 1. Hydnorales.

2. Rafflesiales.

6. Superorder : Balanophoranae

Order.1. Cynomoriales.

2. Balanophorales.

Subclass. Nymphaeidae

1. Superorder : Nymphaeanae

Order. 1. Hydropeltidales.

2. Nymphaeales.

2. Superorder : Ceratophyllanae

Order. 1. Ceratophyllales.

Subclass. Nelumbonidae

1. Superorder. Nelumbonanae

Order. 1. Nelumbonales.

Subclass. Ranunculidae

1. Super order : Ranunculanae
   Order. 1. Lardizabales.
   2. Menispermales.
   3. Berberidales.
   4. Ranunculales.
   5. Circaeasterales.
   6. Hydrastidales.
   7. Glaucidiales.
   8. Paeoniales.
   9. Papaverales.

Subclass. Caryophyllidae

1. Superorder : Caryophyllanae
   Order. 1. Caryophyllales.
2. Super order : Gyrostemonanae
   Order. 1. Gyrostemonales.
3. Superorder : Polygonanae
   Order. 1. Polygonales.
4. Superorder : Plumbaginanae
   Order. 1. Plumbaginales.

Subclass. Hamamelididae

1. Superorder : Trochodendranae
   Order. 1. Trochodendrales.
   2. Cercidiphyllales.
   3. Eupteleales.
2. Superorder : Myrothamnanae
   Order. 1. Myrothamnales.
3. Superorder : Hamamelidanae
   Order. 1. Hamamelidales.
4. Superorder : Barbeyanae
   Order. 1. Barbeyales.
5. Superorder : Daphniphyllanae
   Order. 1. Daphniphllales.
   2. Balanopales.
6. Superorder : Buxanae
   Order. 1. Didymelales.
   2. Buxales.
   3. Simmondsiales.
7. Superorder : Faganae
   Order. 1. Fagales.
   2. Corylales.
8. Superorder : Casuarinanae
   Order. 1. Casuarinales.
9. Superorder : Juglandanae
   Order. 1. Myricales.
   2. Rhoipteleales.
   3. Juglandales.

Subclass. Dilleniidae

1. Superorder : Dillenianae
   Order. 1. Dilleniales.
2. Superorder : Theanae
   Order. 1. Paracryphiales.
   2. Theales.
   3. Hypericales.
   4. Physenales.
   5. Medusagynales.
   6. Ochnales.
   7. Elatinales.
   8. Ancistrocladales.
   9. Dioncophyllales.
   10. Lecythidales.
3. Superorder : Sarracenianae
   Order. 1. Sarraceniales.
4. Superorder : Nepenthanae
   Order. 1. Nepenthales.
   2. Droserales.
5. Superorder : Ericanae
   Order. 1. Actinidiales.
   2. Ericales.
   3. Diapensiales.
   4. Bruniales.
   5. Geissolomatales.
   6. Fouquieriales.
6. Superorder : Primulanae
   Order. 1. Styracales.
   2. Sapotales.
   3. Myrsinales.
   4. Primulales.
7. Superorder : Violanae
   Order. 1. Violales.
   2. Passiflorales.
   3. Caricales.
   4. Salicales.
   5. Tamaricales.

6. Cucurbitales.
7. Begoniales.
8. Capparales.
9. Moringales.
10. Batales.

8. Superorder : Malvanae
Order. 1. Cistales.
2. Elaeocarpales.
3. Malvales.

9. Superorder : Urticanae
Order. 1. Urticales.

10. Superorder : Euphorbianae
Order. 1. Euphorbiales.
2. Thymelaeales.

Subclass. Rosidae

1. Superorder : Saxifraganae
Order. 1. Cunoniales.
2. Saxifragales.
3. Cephalotales.
4. Greyiales.
5. Francoales.
6. Haloragales.
7. Podostemales.
8. Gunnerales.

2. Superorder : Rosanae
Order. 1. Rosales.
2. Crossosomatales.
3. Chrysobalanales.

3. Superorder : Rhizophoranae
Order. 1. Anisophylleales.
2. Rhizophorales.

4. Superorder : Myrtanae
Order. 1. Myrtales.

5. Superorder : Fabanae
Order. 1. Fabales.

6. Superorder : Rutanae
Order. 1. Sapindales.
2. Tropaeolales.
3. Sabiales.
4. Connarales.
5. Rutales.
6. Leitneriales.
7. Coriariales.
8. Burserales.

7. Superorder : Geranianae
Order. 1. Linales.
2. Oxalidales.
3. Geraniales.
4. Biebersteiniales.
5. Balsaminales.
6. Zygophyllales.
7. Vochysiales.
8. Polygalales.

8. Superorder : Corynocarpanae
Order. 1. Corynocarpales.

9. Superorder : Celastranae
Order. 1. Brexiales.
2. Parnassiales.
3. Celastrales.
4. Salvadorales.
5. Icacinales.
6. Metteniusales.
7. Cardiopteridales.

10. Superorder : Santalanae
Order. 1. Medusandrales.
2. Santalales.

11. Superorder : Rhamnanae
Order. 1. Rhamnales.
2. Elaeagnales.

12. Superorder : Proteanae
Order. 1. Proteales.

13. Superorder : Vitanae
Order. 1. Vitales.

Subclass. Cornidae

1. Superorder : Cornanae
Order. 1. Hydrangeales.
2. Desfontainiales.
3. Roridulales
4. Cornales.
5. Garryales.
6. Aucubales.
7. Griseliniales.
8. Eucommiales.
9. Aralidiales.
10. Torricelliales.

2. Superorder : Aralianae
   Order. 1. Helwingiales.
   2. Araliales.
   3. Pittosporales.
   4. Byblidales.

3. Superorder : Dipsacanae
   Order. 1. Viburnales.
   2. Adoxales.
   3. Dipsacales.

Subclass. Asteridae

1. Superorder : Campanulanae
   Order. 1. Campanulales.
   2. Goodeniales.
   3. Stylidales.
   4. Menyanthales.

2. Superorder : Asteranae
   Order. 1. Calycerales.
   2. Asterales.

Subclass. Lamiidae

1. Superorder : Gentiananae
   Order. 1. Gentianales.
   2. Rubiales.
   3. Apocynales.

2. Superorder : Solananae
   Order. 1. Solanales.
   2. Convolvulales.
   3. Polemoniales.
   4. Boraginales.
   5. Limnanthales.

3. Superorder : Loasanae
   Order. 1. Loasales.

4. Superorder : Oleanae
   Order. 1. Oleales.

5. Superorder : Lamianae
   Order. 1. Scrophulariales.
   2. Lamiales.
   3. Callitrichales.
   4. Hydrostachyales.
   5. Hippuridales.

Class. LILIOPSIDA (Monocots)

Subclass. Liliidae

1. Superorder : Lilianae
   Order. 1. Melanthiales.
   2. Colchicales.
   3. Trilliales.
   4. Liliales.
   5. Alstroemeriales.
   6. Iridales.
   7. Tecophilaeales.
   8. Burmanniales.
   9. Hypoxidales.
   10. Orchidales.
   11. Amaryllidales.
   12. Asparagales.
   13. Xanthorrhoeales.
   14. Hanguanales.

2. Superorder : Dioscoreanae
   Order. 1. Stemonales.
   2. Smilacales.
   3. Dioscoreales.
   4. Taccales.

Subclass. Commelinidae

1. Superorder : Bromelianae
   Order. 1. Bromeliales.
   2. Velloziales.

2. Superorder : Pontederianae
   Order. 1. Philydrales.
   2. Pontederiales.
   3. Haemodorales.

3. Superorder : Zingiberanae
   Order. 1. Musales.
   2. Lowiales.
   3. Zingiberales.
   4. Cannales.

4. Superorder : Commelinanae
   Order. 1. Commelinales.
   2. Mayacales.
   3. Xyridales.
   4. Rapateales.
   5. Eriocaulales.

5. Superorder : Hydatellanae
   Order. 1. Hydatellales.

6. Superorder : Juncanae
   Order. 1. Juncales.
   2. Cyperales.

7. Superorder : Poanae
   Order. 1. Flagellariales.
   2. Restionales.
   3. Centrolepidales.
   4. Poales.

Subclass. Arecidae

1. Superorder : Arecanae
   Order. 1. Arecales.

Subclass. Alismatidae

2. Superorder : Alismatane
   Order. 1. Butomales.
   2. Hydrocharitales.
   3. Najadales.
   4. Alismatales.
   5. Aponogetonales.
   6. Jucaginales.
   7. Juncaginales.
   8. Potamogetonales.
   9. Posidoniales.
   10. Cymodoceales.
   11. Zosterales.

Subclass. Triurididae

1. Superorder : Triuridanae
   Order. 1. Petrosaviales.
   2. Triuridales.

Subclass. Aridae

1. Superorder : Aranae
   Order. 1. Arales.
   2. Acorales.
2. Superorder : Cyclanthanae
   Order. 1. Cyclanthales.
3. Superorder : Pandananae
   Order. 1. Pandanales.
4. Superorder : Typhanae
   Order. 1. Typhales.

Merits and Demerits

Merits :

1. The classification of Takhtajan is more phylogenetic than that of earlier systems.
2. This classification is in a general agreement with the major contemporary systems of Cronquist, Dahlgren, Thorne, and others. Both phylogenetic and phenetic informations were adopted for delimination of orders and families.
3. Due to the abolition of several artificial groups like Polypetalae, Gamopetalae, Lignosae, Herbaceae, many natural taxa came close together, viz. Lamiaceae (earlier placed under Herbaceae) and Verbenaceae (placed under Lignosae) are brought together under the order Lamiales.
4. Nomenclature adopted in this system is in accordance with the ICBN, even at the level of division.
5. The treatment of Magnoliidae as a primitive group and the placement of Dicotyledons before Monocotyledons are in agreement with the other contemporary systems.
6. The derivation of monocots from the extinct terrestrial hypothetical group of Magnoliidae is found to be logical.

Demerits :

1. In this system, more weightage is given to cladistic information in comparison to phenetic information.
2. This system provides classification only up to the family level, thus it is not suitable for identification and for adoption in Herbaria. In addition, no key has been provided for identification of taxa.
3. Takhtajan reognised angiosperms as division which actually deserve a class rank like that of the systems of Dahlgren (1983) and Throne (2003).
4. Numerous monotypic families have been created in 1997 due to the further splitting and increase in the number of families to 592 (533 in 1987), resulting into a very narrow circumscription.
5. Takhtajan incorrectly suggested that smaller families are more "natural".
6. Although the families Winteraceae and Canellaceae showed their 99-100% relationship by multigene analyses, Takhtajan placed these two families in two separate orders.

Arthur Cronquist (1919-1992)

Fig. 4.23 : Arthur Cronquist (1919-1992)

Arthur Cronquist was the Senior Curator of New York Botanic Garden and Adjunct Professor of Columbia University. He presented an elaborate interpretation of his concept of classification in *"The Evolution and Classification of Flowering Plants"*(1968). The further edition of his classification was published in *"An Integrated System of Classification of Flowering Plants"* (1981). The latest revision was published in the 2nd edition in 1988 in *"The Evolution and Classfication of Flowering Plants"*. He discussed a wide range of characteristics important to phylogenetic system. He also provided synoptic keys designed to bring the taxa in an appropriate alignment. He also represented his classification in charts to show the relationships of the orders within the various subclasses. His system is more or less parallel to Takhtajan's system, but differs in details.

He considered that the Pteridosperms i.e., the seed ferns as probable ancestors of angiosperm.

The following principles were adopted by Cronquist (1981) to classify the flowering plants :

1. The earliest angiosperms were shrubs rather than trees.
2. The simple leaf is primitive than compound leaf.
3. Reticulate venation is primitive than parallel venation.
4. Paracytic stomata is primitive than the other types.
5. Slender, elongated, long tracheids with numerous scalariform pits are primitive. Further specialisation leads to shorter broad vessels with somewhat thinner walls and transverse end walls with few larger perforations. Later on, the perforation becomes single and large.
6. Long and slender sieve elements with very oblique end walls where the sieve areas scattered along the longitudinal wall with groups of minute pores are primitive. Whereas, the phloem with short sieve tube elements where end walls having a single transverse sieve plate with large openings is a derived condition.
7. The area and activity of cambium and also the length of fusiform initial is more in primitive form which gradually becomes reduced in advanced one.
8. Plants with vascular bundles arranged in a ring are primitive rather than scattered vascular bundle as found in monocots.
9. Plants with large and terminal flowers are primitive, those may arrange in monochasia or dichasia and the other type of inflorescences have been derived from these types.
10. Flowers with many large, free and spirally arranged petals; many linear and spirally arranged stamens and free carpels as found in Magnoliaceae are primitive, and other types got evolved through gradual reduction, aggregation, elaboration and differentiation of floral members.
11. Plants with unisexual flowers are evolved from bisexual floral ancestors.
12. The large and indefinite number of floral members are primitive than the small and definite numbers.
13. Androecium with many stamens is primitive than the reduced numbers.
14. Linear stamens with embedded pollen sacs as found in some Magnolian genera are considered more primitive than the others.
15. Uniaperturate pollen grains are considered as primitive and the triaperturate type are derived from it.
16. Insect pollinated plants are considered as primitive from which wind pollinated plants got evolved.

17. The gynoecium comprising of many carpels arranged spirally on a more or less elongated receptacle is considered as primitive. Further evolution leads to the reduction of the number of carpels which are arranged in a single whorl and then undergo further fusion.
18. Axial placentation is primitive from which other types have been evolved.
19. Anatropous ovule is primitive from which other types have been evolved.
20. Ovule with two integuments (bitegmic) is primitive and, either by fusion or abortion, unitegmic condition has been evolved.
21. Embryo-sac with 8-nuclei (*Polygonum*-type) is primitive from which embryo-sac with 4-nuclei (*Oenothera*-type) has been derived through reduction.
22. Monocotyledons have been developed from dicotyledons through abortion of one cotyledon.
23. The follicle (fruit) is considered as primitive. Further, dry and dehiscent fruit is more primitive than fleshy and indehiscent fruit.

According to him *"many of the evolutionary trends bear little apparent relation to survival value and that there are some reversals"*.

In 1981, he divided the Division Magnoliophyta (Angiosperms) into two classes Magnoliatae (Dicotyledons) and Liliatae (Monocotyledons). He divided Magnoliatae into 6 subclasses and 55 orders, of which magnoliales is the primitive and Asterales is the advanced taxa. On the other hand, the class Liliatae has been divided into 4 subclasses and 18 orders, of which Alismatales is the primitive and Orchidales is the advanced taxa. The class Magnoliatae consists of 291 families and Liliatae with 61 families.

The summary of the system of classification of Cronquist (1981) is given :

Division. MAGNOLIOPHYTA

Class. Magnoliatae

1. Subclass : Magnoliidae

Order. 1. Magnoliales
2. Piperales
3. Aristolochiales
4. Nymphaeales
5. Ranunculales

2. Subclass. Hamamelidae

Order. 1. Trochodendrales
2. Hamamelidales
3. Eucomiales
4. Urticales
5. Leitneriales
6. Juglandales
7. Myricales
8. Filgales
9. Casuarinales

3. Subclass. Caryophyllidae

Order. 1. Caryophylales
2. Batales
3. Polygonales
4. Plumbaginales

4. Subclass. Dilleniidae

Order. 1. Dilleniales
2. Theales
3. Malvales
4. Lecythidales
5. Serraceniales
6. Violales
7. Salicales
8. Capparales
9. Ericales
10. Diapensiales
11. Ebenales
12. Primulales

5. Subclass. Rosidae

Order. 1. Rosales
2. Podostemales
3. Haloragales
4. Myrtales
5. Proteales
6. Cornales
7. Santalales
8. Rafflesiales
9. Celastrales
10. Euphorbiales
11. Rhamnales
12. Sapindales
13. Geraniales

14. Linales
15. Polygalales
16. Umbellales

6. Subclass. Asteridae
Order. 1. Gentianales
2. Polemoniales
3. Lamiales
4. Plantaginales
5. Scrophulariales
6. Campanulales
7. Rubiales
8. Dipsacales
9. Asterales

Class. Liliatae

1. Subclass. Alismatidae
Order. 1. Alismatales
2. Hydrocharitales
3. Najadales
4. Triuridales

2. Subclass. Commelinidae
Order. 1. Commelinales
2. Eriocaulales
3. Restionales
4. Juncales
5. Cyperales
6. Typhales
7. Bromeliales
8. Zingiberales

3. Subclass. Arecidae
Order. 1. Arecales
2. Cyclanthales
3. Pandanales
4. Arales

4. Subclass. Liliidae
Order. 1. Liliales
2. Orchidales.

In 1988, Cronquist broadly classified angiosperms (Division Magnoliphyta) into two classes : Magnoliopsida (Dicots) and Liliopsida (Monocots), unlike Magnoliatae and Liliate of the previous classifications (1968, 1981). Cronquist divided Magnoliopsida into 6 subclasses, 64 orders (55 orders in 1981) and 320 families (291 in 1981), comprising of 169,400 species. The class Liliopsida has been divided into 5 subclasses (4 subclasses in 1981), 19 orders (18 orders in 1981) and 66 families (61 in 1981), comprising of 49,900 species.

The broad outline of Cronquist's classification (1988) is given :

Division. MAGNOLIOPHYTA

(2 classes, 11 subclasses, 83 orders, 386 families and 219,300 species)

Class. 1. Magnoliopsida (Dicots)

Subclass. 1. Magnoliidae (8 Orders : Magnoliales, Laurales, Piperales, Aristolochiales, Illiciales, Nymphaeales, Ranunculales and Papaverales).

2. Hamamelidae (11 orders : Trochodendrales, Hamamelidales, Daphniphyllales, Didymelales, Eucommiales, Urticales, Leitneriales, Juglandales, Myricales, Fagales and Casuarinales).

3. Caryophyllidae (3 orders : Caryophyllales, Polygonales and Plumbaginales).

4. Dilleniidae (13 orders : Dilleniales, Theales, Malvales, Lecythidales, Nepenthales, Violales, Salicales, Capparales, Batales, Ericales, Diapensiales, Ebenales and Primulales).

5. Rosidae (18 orders : Rosales, Fabales, Proteales, Podostemales, Haloragales, Myrtales, Rhizophorales, Cornales, Santalales, Rafflesiales, Celastrales, Euphorbiales, Rhamnales, Linales, Polygalales, Sapindales, Geraniales and Apiales).

6. Asteridae (11 orders : Gentianales, Solanales, Lamiales, Callitricales, Plantaginales, Scrophulariales, Campanulales, Rubiales, Dipsacales, Calycerales and Asterales).

Class 2. Liliopsida (Monocots)

Subclass. 1. Alismatidae (4 orders : Alismatales, Hydrocharitales, Najadales and Triuridales).

2. Arecidae (4 orders : Arecales, Cyclanthales, Pandanales and Arales).

3. Commelinidae (7 orders : Commelinales, Eriocaulales, Restionales, Juncales, Cyperales, Hydatellales and Typhales).

4. Zingiberidae (2 orders : Bromeliales and Zingiberales).

5. Lillidae (2 orders : Liliales and Orchidales).

Merits and Demerits

Merits :

1. There is general agreement of Cronquist's system with that of other contemporary systems like Takhtajan, Dahlgren and Thorne.
2. Detailed information on anatomy, ultrastructure phytochemistry and chromosome — besides morphology — was presented in the revision of the classification in 1981 and 1988.
3. The system is highly phylogenetic.
4. Nomenclature is in accordance with the ICBN.
5. The family Asteraceae in Dicotyledons and Orchidaceae in Monocotyledons are generally regarded as advanced and are rightly placed towards the end of respective groups.
6. The relationships of different groups have been described with diagrams which provide valuable information on relative advancement and size of the various subclasses.
7. The family Winteraceae (vessel-less wood present similar to Pteridosperms) placed at the beginning of dicotyledons is favoured by many authors.
8. The subclass Magnoliidae is considered as the most primitive group of Dicotyledons. The placement of Dicotyledons before Monocotyledons finds general agreements with modern authors.
9. As the text is in English, the system has been readily adopted in different books.

Demerits :

1. Though highly phylogenetic and popular in U.S.A., this system is not very useful for identification and adoption in Herbaria since Indented keys for genera are not provided.
2. Dahlgren (1983, 89) and Thorne (1980, 83) treated angiosperms in the rank of a class and not that of a division.

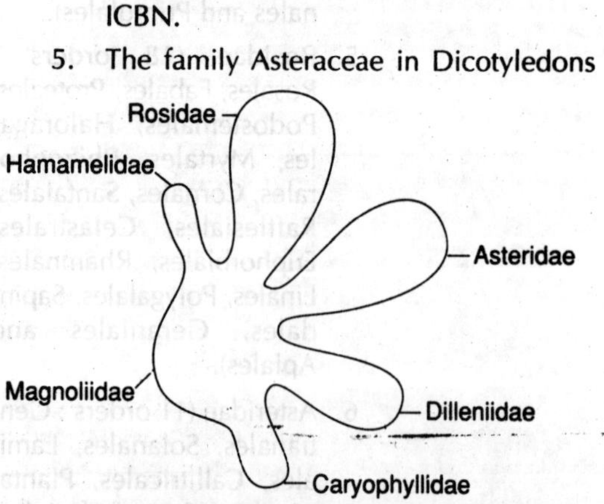

Fig. 4.24 : Probable relationship among subclasses of Magnoliopsida (Dicots). Sizes of balloons are proportional to the number of species under each subclass (Cronquist, 1988)

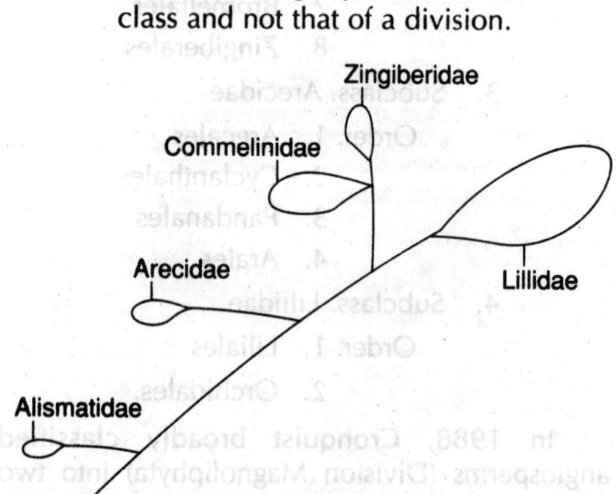

Fig. 4.25 : Probable relationship among subclasses of Liliopsida (Monocots). Sizes of balloons are proportional to the number of species under each subclass (Cronquist, 1988)

3. Superorder as a rank above order has not been recognised here, though it is present in other contemporary classifications like Takhtajan, Thorne and Dahlgren.
4. The subclass Asteridae represents a loose assemblage of several diverse sympetalous families.
5. Ehrendorfer (1983) pointed out that the subclass Hamamelidae does not represent an ancient side branch of the subclass Magnoliidae, but is remnant of a transition from Magnoliidae to Dilleniidae, Rosidae, and Asteridae.
6. There is a difference in opinion with other authors regarding the systematic position of some orders like Typhales, Arales, Urticales etc.

Rolf Dahlgren (1932-1987)

Rolf Dahlgren, working at the Botanical Museum in the University of Copenhagen, Denmark, published a new method in Danish in 1974 to illustrate an angiosperm system in a text book of angiosperm taxonomy (Dahlgren *et al.*, 1974). Later on in 1975 he published *"A System of Classification of Angiosperms to be Used to Demonstrate the Distribution of Characteristics"* in *Botanische Notiser* in English.

The revised and improved version of his system gradually appeared in the subsequent years :

Fig. 4.26 : Rolf Dahlgren (1932-1987)

i. In 1980 in *"Botanical Journal of the Linnean Society"*.
ii. In 1981, in *"Phytochemistry and angiosperm Phylogeny"* (Edited by Young and Siegier).
iii. In 1983, in *"Nordiac Journal of Botany"*.

In his system he included the information at different levels as much as possible. In his classification he extensively used the chemical characteristics. He considered the following morphological and chemical characteristics in his classification.

1. Morphological characteristics :
   i. Chloripetalae and Sympetalae.
   ii. Apocarpous, syncarpous and monocarpellate condition.
   iii. Types of microsporogenesis.
   iv. Bi- and trinucleate pollen grain.
   v. Tenuinucellate, pseudocrassinucellate and crassinucellate.
   vi. Bi- or unitegmic ovules etc.
2. Chemical characteristics :
   i. Benzylisoquinoline alkaloids.
   ii. Pyrrolisidine alkaloids.
   iii. Tropane alkaloids.
   iv. Silica bodies.
   v. Ellagic acid and ellagitannins
   vi. Various groups of flavonoids etc.

He did not consider angiosperms to be originated polyphyletically from different gymnosperms, but believed that the combination of different characteristics like 8-nucleate embryo sac, secondary endosperm etc. would hardly have evolved independently from different groups of gymnosperms.

According to Dahlgren (1980), the class Magnoliopsida (Angiosperms) has been divided into two subclasses, Magnoliidae and Liliidae. The Magnoliidae includes 24 Superorders, those start with Magnoliiflorae and end in Lamiflorae; 80 orders, those start with Annonales and end in Lamiales and 346 families. On the other hand, the subclass Liliidae includes 7 Superorders, those start with Alismatiflorae and end in Areciflorae; 26 orders, those start with Hydrocharitales and end in Pandanales and 92 families.

The summary of the system of classification of Dahlgren (1980) is given :

Class. MAGNOLIOPSIDA

Subclass. Magnoliidae (Dicotyledons)

1. Superorder. Magnoliiflorae
   - Order. 1. Annonales
   - 2. Illiciales
   - 3. Laurales
   - 4. Nelumbonales
   - 5. Aristolochiales
   - 6. Refflesiales
   - 7. Magnoliales
2. Superorder. Nymphaeaflorae
   - Order. 1. Nymphaeales
   - 2. Piperales
3. Superorder. Ranunculiflorae
   - Order. 1. Papaverales
   - 2. Ranunculales
4. Superorder. Caryophylliflorae
   - Order. 1. Caryophyllales
5. Superorder. Polygoniflorae
   - Order. 1. Polygonales
6. Superorder. Malviflorae
   - Order. 1. Paeoniales
   - 2. Elaeagnales
   - 3. Dilleniales
   - 4. Malvales
   - 5. Plumbaginales
   - 6. Thymelaeales
   - 7. Urticales
   - 8. Euphorbiales
   - 9. Rhamnales
7. Superorder. Violiflorae
   - Order. 1. Violales
   - 2. Salicales
   - 3. Tamaricales
   - 4. Capparales
   - 5. Salvadorales
8. Superorder. Theiflorae
   - Order. 1 Theales
   - 2. Droserales
9. Superorder. Primuliflorae
   - Order. 1. Ebenales
   - 2. Primulales
10. Superorder. Rosiflorae
    - Order. 1. Trochodendrales
    - 2. Hamamelidales
    - 3. Fagales
    - 4. Balanopales
    - 5. Juglandales
    - 6. Myricales
    - 7. Casuarinales
    - 8. Buxales
    - 9. Cunoniales
    - 10. Saxifragales
    - 11. Gunnerales
    - 12. Rosales
    - 13. Malales
11. Superorder. Podostemiflorae
    - Order. 1. Podostemales
12. Superorder. Fabiflorae
    - Order. 1. Fabales
13. Superorder. Proteiflorae
    - Order. 1. Proteales
14. Superorder. Myrtiflorae
    - Order. 1. Myrtales
    - 2. Haloragales
    - 3. Rhizophorales
15. Superorder. Rutiflorae
    - Order. 1. Rutales
    - 2. Sapindales
    - 3. Balsaminales
    - 4. Polygalales
    - 5. Geraniales
    - 6. Tropaeolales
16. Superorder. Santaliflorae
    - Order. 1. Celastrales
    - 2. Vitidales
    - 3. Santalales
17. Superorder. Balanoflorae
    - Order. 1. Balanophorales
18. Superorder. Araliiflorae
    - Order. 1. Pittosporales
    - 2. Araliales
19. Superorder. Asteriflorae
    - Order. 1. Campanulales
    - 2. Asterales
20. Superorder. Solaniflorae
    - Order. 1. Solanales
    - 2. Boraginales

21. Superorder. Corniflorae
    Order. 1. Fouquieriales
    2. Serraceniales
    3. Ericales
    4. Eucomiales
    5. Cornales
    6. Dipsacales
22. Superorder. Loasiflorae
    Order. 1. Loasales
23. Superorder. Gentianiflorae
    Order. 1. Goodeniales
    2. Gentianales
    3. Oleales
24. Superorder. Lamiflorae
    Order. 1. Scrophulariales
    2. Hydrostachyales
    3. Hippuridales
    4. Lamiales

Subclass. Liliidae (Monocotyledons)

1. Superorder. Alismatiflorae
    Order. 1. Hydrocharitales
    2. Zosterales
    3. Alismatales
2. Superorder. Triuridiflorae
    Order. 1. Triuridales
3. Superorder. Ariflorae
    Order. 1. Arales
4. Superorder. Liliiflorae
    Order. 1. Dioscoreales
    2. Asparagales
    3. Hydatellales
    4. Liliales
    5. Burmanniales
    6. Philydrales
    7. Orchidales
    8. Velloziales
    9. Bromeliales
    10. Haemodorales
    11. Pontederiales
    12. Typhales.
5. Superorder. Zingiberiflorae
    Order. 1. Zingiberales
6. Superorder. Commeliniflorae
    Order. 1. Commelinales
    2. Eriocaulales
    3. Juncales
    4. Cyperales
    5. Poales
7. Superorder. Areciflorae
    Order. 1. Arecales
    2. Cyclanthales
    3. Pandanales

The recent revision of both the Takhtajan's and Dahlgren's systems of classification is similar in many respects, which happens due to handling of same data.

After the demise of Rolf Dahlgren in 1987, his wife Gertrud Dahlgren continued the work of her husband on the angiosperm classification. She published "*The last Dahlgrenogram*", system of classification of dicots in 1989, followed by a classification of monocots in the same year based on the hypothesis of Rolf Dahlgren. She changed the suffix for the superorders from **-florae** to **-anae** for keeping the nomenclatural uniformity. The system is comprised of 35 superorders, 111 orders and 447 families, of which dicots consist of 25 superorder and monocots 10 superorders. The relationship of various families are also shown in a bubble diagram.

An outline of updated Dahlgren's classification as presented by Gertrud Dahlgren is given :

Dicotyledons

(25 superorders, 87 orders, 343 families)

Superorders

1. Magnolianae (10 orders)
2. Nymphaeanae (2 orders)
3. Ranunculanae (2 orders)
4. Caryophyllanae (1 order)
5. Polygonanae (1 order)
6. Plumbaginanae (1 order)
7. Malvanae (7 orders)
8. Violanae (7 orders)
9. Theanae (4 orders)
10. Primulanae (2 orders)
11. Rosanae (15 orders)
12. Proteanae (2 orders)
13. Myrtanae (2 orders)
14. Rutanae (9 orders)
15. Vitanae (1 order)
16. Santalanae (1 order)
17. Balanophoranae (1 order)

18. Aralianae (2 orders)
19. Asteranae (2 orders)
20. Solananae (2 orders)
21. Ericanae (5 orders)
22. Cornanae (3 orders)
23. Loasanae (1 order)
24. Gentiananae (3 orders)
25. Lamianae (3 orders)

Monocotyledons
(10 superorders, 24 orders, 104 families)

Superorders

1. Alismatanae (2 orders)
2. Triuridanae (1 order)
3. Aranae (1 order)
4. Lilianae (6 orders)
5. Bromelianae (6 orders)
6. Zingiberanae (1 order)
7. Commelinanae (3 orders)
8. Arecanae (2 orders)
9. Cyclanthanae (1 order)
10. Pandananae (1 order)

Merits and Demerits

Merits :

1. Detailed information on morphology, phytochemistry and embryology was presented in the classification of Dahlgren.
2. This system is highly phylogenetic where angiosperms are ranked as a class like other recent systems.
3. The arrangement of taxa in the form of a bubble diagram gives an idea about the relationship of superorders, orders and even families.
4. The use of superorders and the suffix **–anae** are in accordance with the other modern systems like those of Takhtajan, Thorne, etc.

Demerits :

1. This system provides classification of angiosperms only up to the family level, thus it is not suitable for identification and for adoption in Herbaria.
2. Dahlgren classified angiosperms into dicots and monocots, which shows inconformity with the recent classification of APG II (2003) and Throne (2003).
3. Dahlgren placed monocots in-between dicots, while in modern classification monocots are placed in-between primitive angiosperms and the eudicots.
4. Although the families Winteraceae and Canellaceae showed their 99–100% relationship by multigene analysis, yet Dahlgren placed these two families in two separate orders.

Robert F. Thorne (1920-)

Robert F. Thorne, an American taxonomist, associated with the Rancho Santa Ana Botanic Garden, California, U.S.A., initially published the principles of his classification in 1958 and 1963. Later, in 1968, he published *"Synopsis of a putatively phylogenetic classification of the flowering plants"* in Aliso. The subsequent revisions were published in 1974, 1976, 1981, 1983, 1992 and 2000. The electronic version of his classification was published in 1999 which was finally revised in 2003.

Thorne gave much emphasis on phytochemical approach. In addition to the above, many other different aspects were also considered by him. These are :

1. Seed morphology.
2. Pollen morphology.
3. Comparative anatomy.
4. Embryology.
5. Cytology.
6. Host-parasite relationship.
7. Plant geography.
8. Paleobotany etc.

He believed that the Angiospermae are monophyletic.

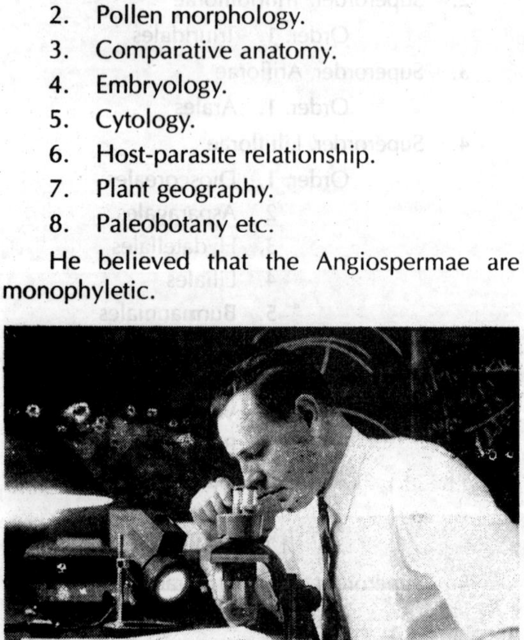

Fig. 4.27 : Robert F. Thorne (1920-)

In 1983, he divided the Class Angiospermae (Annonopsida) into two subclasses Dicotyledoneae (Annonidae) and Monocotyledoneae (Liliidae). The Dicotyledoneae is further divided into 19 superorders which start with Annoniflorae and ends in Asteriflorae; 41 orders, which start with Annonales and ends in Asterales; and 297 families. On the other hand, Monocotyledoneae is further divided into 9 superorders, which start with Liliiflorae and end in Commeliniflorae; 12 orders, which start with Liliales and ends in Zingiberales. Thus, he preferred the name Annonopsida for angiosperms, Annonidae for dicots, replacing Magnoliflorae by Annoniflorae and Magnoliales by Annonales. He, however, abandoned these nomenclatures and adopted the conventional names Magnoliopsida, Magnoliidae and Magnoliales since 1992.

Robert F. Thorne (1992) revised his classification ("*Classification and Geography of Flowering Plants*") and published it in Botanical Review. He followed the arrangements of different taxa in descending order such as subclasses (*-idae*), superorders (*-anae*), orders (*-ales*), suborders (*-inae*), families (*-aceae*), subfamilies (*-oideae*), and tribes (*-ineae*). He treated the flowering plants as Class with an initial bifurcation into two Subclasses, **Magnoliidae** (Dicots) and **Liliidae** (Monocots). The subclass Magnoliidae has been divided into 19 superorders, 52 orders and the subclass Liliidae has been divided into 9 superorders, 24 orders. The Magnoliidae starts with the order Magnoliales and ends in Lamiales, whereas the Liliidae starts with Triuridales and ends in Restionales.

The summary of the system of classification of Thorne (1992) is given :

Subclass. MAGNOLIIDAE (Dicots)

1. Superorder. Magnolianae
   - Order. 1. Magnoliales
   - 2. Ceratophyllales
   - 3. Nelumbonales
   - 4. Paeoniales
   - 5. Berberidales
2. Superorder. Nymphaeanae
   - Order. 1. Nymphaeales
3. Superorder Rafflesianae
   - Order. 1. Rafflesiales
4. Superorder. Caryophyllanae
   - Order. 1. Caryophyllales
5. Superorder. Dillenianae
   - Order. 1. Dilleniales
   - 2. Ericales
   - 3. Fouquieriales
   - 4. Styracales
   - 5. Primulales
   - 6. Polygonales
6. Superorder. Celastranae
   - Order. 1. Celastrales
7. Superorder. Malvanae
   - Order. 1. Malvales
   - 2. Urticales
   - 3. Rhamnales
   - 4. Euphorbiales
8. Superorder. Violanae
   - Order. 1. Violales
   - 2. Brassicales
   - 3. Batales
9. Superorder. Santalanae
   - Order. 1. Santalales
   - 2. Balanophorales
10. Superorder. Geranianae
    - Order. 1. Linales
    - 2. Rhizophorales
    - 3. Geraniales
    - 4. Polygalales
11. Superorder. Rutanae
    - Order. 1. Rutales
12. Superorder. Proteanae
    - Order. 1. Proteales
13. Superorder. Rosanae
    - Order. 1. Hamamelidales
    - 2. Casuarinales
    - 3. Balanopales
    - 4. Bruniales
    - 5. Juglandales
    - 6. Betulales
    - 7. Rosales
    - 8. Saxifragales
    - 9. Podostemales
    - 10. Cunoniales
14. Superorder. Aralianae
    - Order. 1. Brexiales
    - 2. Cornales
    - 3. Pittosporales
    - 4. Araliales
    - 5. Dipsacales
15. Superorder. Asteranae
    - Order. 1. Asterales
    - 2. Campanulales

16. Superorder. Solananae
    Order. 1. Solanales
17. Superorder. Loasanae
    Order. 1. Loasales
18. Superorder. Myrtanae
    Order. 1. Myrtales
19. Superorder. Lamianae
    Order. 1. Rubiales
    2. Lamiales

Subclass. Liliidae (Monocots)

1. Superorder. Triuridanae
    Order. 1. Triuridales
2. Superorder. Acoranae
    Order. 1. Nartheciales
    2. Acorales
3. Superorder. Aranae
    Order. 1. Arales
4. Superorder. Alismatanae
    Order. 1. Alismatales
    2. Potamogetonales
5. Superorder. Lilianae
    Order. 1. Liliales
    2. Taccales
    3. Burmanniales
    4. Orchidales
6. Superorder. Pandananae
    Order. 1. Pandanales
    2. Cyclanthales
    3. Velloziales
7. Superorder. Arecanae
    Order. 1. Arecales
    2. Dasypogonales
    3. Hanguanales
8. Superorder. Hydatellanae
    Order. 1. Hydatellales
9. Superorder. Commelinanae
    Order. 1. Commelinales
    2. Cannales
    3. Bromeliales
    4. Xyridales
    5. Juncales
    6. Typhales
    7. Restionales.

Since 1992, Thorne revised his classification in electronic version in the website : **http:/ www.inform.umd.edu/PBIO/fam/thorneangiosp99.html** put up in 1999. This was further published in '*Botanical Review* (2000)'. In this system, he abandoned the traditional division of angiosperms into dicots and monocots. He treated angiosperms as class **Angiospermae** which has been classified into 10 subclasses wherein the traditional dicot taxa are distributed in 7 subclasses and traditional monocot taxa are distributed in 3 subclasses. He made a major revision of his classification in 2003 after the publication of *Angiosperm Phylogeny Group (APG) II*, put up on the Website of Rancho Santa Ana Botanic Garden as **http:/www.rsabg.org/ publications/angiosp.htm.**

In the latest classification (2003), Thorne retained the same number of subclasses with some significant changes from the earlier versions. The classification recognised 33 superorders and 90 orders comprising of 489 families with an estimated 13,260 genera and 2,54,990 species. Many of the changes are in accordance with APG II. An outline of Thorne's classification (2003) is presented :

Class : Angiospermae

Subclass

1. Magnoliidae (3 superorders, 7 orders, 32 families, 307 genera, 9,095 species)
    Superorder. 1. Magnolianae
    2. Nymphaeanae
    3. Rafflesianeae
2. Alismatidae (3 superorders, 6 orders, 19 families, 177 genera, 3,530 species)
    Superorder. 1. Acoranae
    2. Aranae
    3. Alismatanae
3. Liliidae (3 superorders, 9 orders, 54 families, 1,257 genera, 28,540 species)
    Superorder. 1. Pandananae
    2. Taccanae
    3. Lilianae
4. Commelinidae (2 superorders, 9 orders, 35 families, 1,293 genera, 25,730 species)
    Superorder. 1. Arecanae
    2. Commelinanae
5. Ranunculidae (2 superorders, 9 orders, 18 families, 299 genera, 6,330 species)
    Superorder. 1. Proteanae
    2. Ranunculanae
6. Hamamelididae (1 superorder, 6 orders, 26 families, 162 genera, 4,695 species)
    Superorder. 1. Hamamelidanae
7. Caryophyllidae (1 superorder, 5 orders, 34 families, 688 genera, 11,680 species)
    Superorder. 1. Caryophyllanae
8. Rosidae (12 superorders, 20 orders, 146 families, 3,758 genera, 72,360 species)
    Superorder. 1. Santalanae

2. Balanophoranae
3. Dillenianae
4. Rosanae
5. Celastranae
6. Oxalidanae
7. Brassicanae
8. Rutanae
9. Myrtanae
10. Violanae
11. Hypericanae
12. Geranianae

9. Asteridae (4 superorders, 15 orders, 81 families, 2,733 genera, 44,680 species)

Superorder. 1. Cornanae
2. Ericanae
3. Aralianae
4. Asteranae

10. Lamiidae (2 superorders, 4 orders, 44 families, 2,635 genera, 48,340 species)

Superorder. 1. Solananae
2. Lamianae

Merits and Demerits

Merits :

1. Detailed information on molecular systematics and chemotaxonomy was presented in the classification of Throne.
2. This system is highly phylogenetic where angiosperms are ranked as a class like those of other recent systems.
3. The traditional groups, dicots and monocots have been abolished and angiosperms are divided into 10 subclasses which are in conformity with the recent phylogenetic thinking.
4. Several closely related taxa are placed nearer to one another, viz. the orders Malvales, Urticales, Rhamnales and Euphorbiales have been included under the superorder Malviflorae.

Demerits :

1. This system has no practical utility for identification and for adoption in Herbaria, because identification keys for genera are not provided.
2. The systematic position of the five genera — namely *Emblingia, Guametela, Haptanthus, Heteranthia* and *Pteleocarpa* — has not been mentioned.
3. The placement of Asteridae before Lamiidae is not justified.
4. Segregation of Grewiaceae from Tiliaceae is highly questionable, as in the recent APG classification all the members of Tiliaceae, Bombacaceae and Sterculiaceae are placed under Malvaceae.

Angiosperm Phylogeny Group (APG) Classification

In true sense, the first cladistic approach in angiosperm classification was expounded by Bremer and Wanntorp in 1979. They treated angiosperms as subclass Magnoliidae under the class Pinatae (seed plants), where the traditional division of angiosperms into monocots and dicots was abandoned. They argued that the treatment of dicots and monocots as independent entities would make the group paraphyletic. Such cladistic concept was revived later, because of the availability of sufficient molecular data on plant characteristics. A group of scientists (headed by A. Bremer, M.W. Chase, P.F. Stevens and others) proposed a classification of 462 families of angiosperms in a paper entitled, *"An Ordinal Classification for the families of Flowering Plants"* in 1998 in the *"Annals of the Missouri Botanic Garden"* (Vol. 85, pp. 531-553). This classification is known as **APG** (Angiosperm Phylogeny Group) **Classification**, where all the families were grouped into 40 putative monophyletic orders under a small number of informal monophyletic groups. The families were divided into 11 informal groups : 1. Monocots, 2. Commelinids, 3. Magniliids, 4. Eudicots, 5. Core Eudicots, 6. Rosids, 7. Eurosids I, 8. Eurosids II, 9. Asterids, 10. Euasterids I, 11. Euasterids II. Subsequently, the APG classification was modified from time to time. Judd *et al.* (1999) documented a total of 51 orders excluding out some families from informal groups. A revised 2nd edition of APG II classification was presented by Judd *et al.* in 2002. A revision of APG (APG II; version 4, May 2003) by P. F. Stevens came in the website (**htpp://www.mobot.org/MOBOT/research/AP web/**).

Stevens, in the periodic updated AP web (version 7, May 2006), has encounted 61 families and has reduced the number of unplaced families to 20 with one unplaced genus, *Brachynema*. This classification is very useful, because it reflects more or less the true phylogenetic picture of angiosperms based on the concept of monophyly. The classification is based on the all available information, subtantiated by molecular phylogeny.

MODERN TAXONOMY : SUPPORTIVE EVIDENCES

The morphological characteristics are very useful to taxonomists for diagnosis and classification of plants. With the progress of scientific research it becomes clear that only the morphological characteristics are not sufficient enough and the data of studies of other branches of botany have considerable value to assess the proper systematic status and phylogeny of taxa. Findings through the recent scientific studies indicate that data from many different branches of botany are equally useful in systematic studies. Such branches are anatomy, cytology, embryology, palynology and chemical compositions etc.

The importance of the different branches of botany is discussed :

1. External Morphology

The taxonomists still rely to a great extent on the morphological characteristics, because they are easily visible and can conventionally use in the classification of plants. In addition to the conventional morphological characteristics that are now popular to the students, other characteristics like *habit, underground organs, leaves, seedling morphology, stipules, non-conventional characteristics of floral parts, seeds* etc. are used in the identification of various groups of plants. In recent years, due to thorough and careful studies some characteristics neglected in the past came in focus and help in taxonomic practices :

A. *Habit* : Size, branching pattern, spread, density etc. of a plant are treated commonly in taxonomic description. The shape of a tree i.e., bushy, umbrella-shaped, flat topped, cylindric, oblong etc. are used for the recognition of a tree. It is *Excurrent in Polyalthea longifolia, Caudex* in *Cocos nucifera, Deliquescent in Mangifera indica, Culm in Bambusa tulda*, etc. The characteristics of bark (colour, thickness, fissuring, texture etc.) are used to distinguish different species of *Betula* and *Pinus.*

B. *Underground organs* : Generally the underground parts are not collected during the preparation of a Herbarium. In some cases, they have been shown to be very useful in the identification of different taxa. Chouard (1936) have used the characteristics of bulbs and underground parts in the classification of Scilleae of Liliaceae. In Liliaceae two species of *Chlorophytum* viz. *C. glaucoides* Blatt. and *C. glaucum* Dalz. are morphologically alike except the characteristics of underground parts. In *C. glaucoides* the root-fibres are slender and end in ellipsoid tubers, while in *C. glaucum* the root fibres are thick and without terminal tubers. The structure and morphology of root-tubers are used in the differentiation of large number of species in *Dioscorea* of Dioscoreaceae (Burkill and Prain, 1936).

C. *Leaves* : Leaf characteristics are extensively used in the differentiation of species of *Betula* (Natho, 1959), *Ulmus* (Melville, 1955) and in many other genera. In Papilionaceae, different species of *Dalbergia* viz. *D. sissoo* Roxb., *D. latifolia* Roxb. and *D. sympathetica* Nimmo are distinguished by shape, size and arrangement of the leaflet on the rachis. Venation is also useful in differentiation of species. Numerous venation patterns in fossil genus *Glossopteris* was described by Pant (1958).

D. *Stipules* : The *stipule* characteristic is also very useful in the differentiation of species. In Rosaceae the characteristics of stipule is variable, but almost constant within a species. The morphology of stipule is useful to differentiate the different species of *Viola* and *Trifolium.*

E. *Seedling morphology* : The characteristics of seedling have also been used in tracing phylogenetic relationships (Ladiges *et al.* 1884). Sampathkumar (1982) found that characteristics of cotyledons are of taxonomic importance in some Convolvulaceae. In recent years, the seedling morphology is also useful for keying out different genera and species.

F. *Non-conventional characteristics of floral parts* : The characteristics of flowers and inflorescences are commonly used in the classification of plants. But more careful observations on floral parts show some non-conventional characteristics which become very useful in the identification of different taxa and their differentiation with others. Some of them are :

a. The degree of branching of inflorescence is useful in the differentiation of various species of *Pinus*.

b. The configuration of floral disc and also nectaries have great value in the diagnosis of Cruciferae (Brassicaceae) and *Tamarix* of Tamaricaceae.

c. Staminal appendages contribute useful data to diagnose the species of *Alyssum* of Cruciferae.

d. Degree of pubescence on young ovaries and the type of papillae on stylar branches are useful in the classification of Compositae (Asteraceae).

G. *Seeds* : The characteristics of seed are also useful in the differentiation of taxa at specific level. Duke (1961) reported that the characteristics of seed shape, colour and sculpturing contribute a critical indication of the systematic positions of species in *Drymaria* of Caryophyllaceae. In Liliaceae, the two genera *Antherium* and *Chlorophytum* can be differentiated only by the shape and number of seeds (Obermeyer, 1962). Similarly, other characteristics like hairy growth on testa, their colour and length contribute useful data to differentiate the genera and species of different families like Acanthaceae, Malvaceae, Convolvulaceae, etc.

2. **Micromorphology**

A. **Epidermis** : The characteristics of epidermis, such as shape, thickness, nature of sculpturing and inclusions of epidermal cells provide important data useful in taxonomic consideration. Sometimes the divisions of epidermal cell also consider in taxonomic analysis. Narrow epidermal cells are the characteristics of the family Stylidiaceae (Mildbread, 1908) and sclerification of the epidermal cell wall is of taxonomic importance in the tribe Mutisieae of Compositae (Carlquist, 1958).

In Cyperaceae, the distribution of silica-body on the surface of epidermis provide an important data of taxonomic importance (Metcalfe,1968) :

a. Spherical and hemispherical warty bodies found in *Scleria, Acriulars* and *Bisboeckelera*,

b. Spherical and hemispherical echinulate bodies found in *Rhynchospora*,

c. Hemispherical smooth bodies in *Lophoschoenus*,

d. Wedge-shaped silica bodies are restricted in *Ptilianthelium, Scirpodendron,* and *Neesenbeckia* etc.

e. Bridge-shaped bodies in *Mapania* etc.

Based on epidermal characteristics like outline of epidermal cells in surface view, type of hairs, structure of stomata; Jain and Singh (1974) and Singh and Jain (1975) differentiated the species of *Pyrus* and *Prunus*. Jain and Singh (1974) also distinguished Himalayan species of oaks, based on epidermal characteristics.

B. **Trichomes**. The trichomes show great variation in the structure and are used in taxonomy and phylogeny of angiosperms. It has been successfully used in the classification of genera and species. The plant parts may bear one or more types of trichomes. The anatomical characteristics of trichomes are also used in this respect.

Some important observations are :

a. Cruciferae : Rollins (1941) observed that in *Arabis* the trichomes are commonly unicellular and eglandular, but in some species they are simple and in some others they are branched, either star-shaped or dendritic type (tree-like). In some species, they are terete, flattened or with swollen base.

b. Compositae : Rollins (1944, 1949) observed that the two species of *Parthenium* viz. *P. argentatum* and *P. incanum* can be distinguished by trichomes. The former is T-shaped and the latter whip-like with a long thread. The hybrids between the species show intermediate type of trichomes. Ramayya (1969) develop the generic key of Indian members of Compositae, based on trichome characteristics.

c. Ericaceae : Cowan (1950) observed the trichomes of *Rhododendron* that are useful in the taxonomic separation in infrageneric and specific levels.

d. Oleaceae : Imandar (1967) studied the structure and ontogeny of trichome and confirm the position of *Nyctanthes* in Oleaceae.

Ramayya (1972) thoroughly observed the trichomes of flowering plants and explained their phylogeny. According to him, unicellular stigmatic papillae are the most primitive among angiosperms. Thus the order Magnoliales is a heterogenous group and the angiosperms are a polyphyletic group. Based on the morphology and ontogeny of trichomes in the tribe Heliantheae of Compositae (Asteraceae), Ramayya (1962, 1969) supported the view that Heliantheae is the most primitive tribe of Asteraceae. Characteristics of trichomes have also been found valuable in the classification of Indian species of *Malvastrum* (Shanmukha Rao and Ramayya, 1977), in some members of Capparidaceae (Aleykutty and Imander, 1978) and in Caesalpiniaceae (Leelavathi and Ramayya, 1977). Gornall (1986) utilised the anatomical characteristics in the systematics of the genus, *Saxifraga*.

C. **Stomata**. The development and morphology of stomata may be helpful in assigning taxa of uncertain affinity to their true position and also to find out the evolutionary trends among the families of Angiosperms (Paliwal and Anand, 1978).

Some evidences are :

a. *Structure and distribution* : These play important roles in the study of taxonomic and phylogenetic significance as discussed by Shah (1968) and Kothari and Shah (1974, 1975) in Papilionaceae. The anomoperigenous and aniso-mesoperigenous types i.e., the primitive types are more common in the tribes Sophoreae, Podalyrieae and Gensiteae, whereas the advanced types occur more frequently in Phaseoleae, Dalbergieae and also in some members of Hedysareae.

b. *Ontogeny* : Both Caricaceae and Cucurbitaceae possess aperigenous and aniso-mesoperigenous types indicating close affinity between them.

i. Bentham, Hooker, and Engler placed the families Umbelliferae, Araliaceae and Cornaceae under the order Umbellales. Later, Cronquist (1965) segregated them and placed the first two families under the order Umbellales and the last one Cornaceae under a separate order, Cornales, along with Alangiaceae. The ontogenic study of stomata support the placing of the above families. The former two families have anisocytic and paramesogenous types, while the latter two possess aperigenous stomata (Kannabiran and Krishnamurthy, 1979).

ii. Bremekamp (1953) suggested the transfer of the genus *Elytraria* from Acanthaceae to Scrophulariaceae. Paliwal (1963) studied ontogeny of stomata and found that the genus *Elytraria* has diacytic stomata and syndetochelic mode of development, clearly indicating its close relationship with Acanthaceae than Scrophulariaceae.

iii. The transfer of *Nyctanthes* is suggested from Oleaceae to Verbenaceae. The ontogenic studies of stomata (Inamdar, 1968) indicate that, like other members of Oleaceae, it possesses tetramesoperigenous and occasionally aperigenous stomata which differ from the diamesogenous stomata of Verbenaceae.

3. **Anatomy**

A. **Leaf anatomy**. Carlquist (1961) has asserted that "the leaf is perhaps anatomically most varied organ of angiosperms and its anatomical variations often concur closely with generic and specific and occasionally familial lines". The important structures of leaves considered useful in taxonomy are :

- Structure of mesophyll.

- Structure of vein sheath.
- Stomatal crypt and time of formation.
- Suberisation and location of abscission layer.
- Presence and absence of hydathodes, etc.

Duval-Jouve (1875) realised that the anatomy of leaf is one of the most reliable characteristics useful in grass systematics. After that, many workers have contributed a lot in this line :

a. Brown (1958) has recognised 6 different types of tissue arrangement in grass leaf and correlated it with the taxonomic groups. Based on leaf anatomy, he also discussed the evolutionary tendencies in different groups of grasses.

b. Culter (1965) studied on the genus *Thurnia* and found **inverted bundles** unique in its leaf. This gives support to the view that *Thurnia* should be placed in a separate family Thurniaceae separating from Rapateaceae and Juncaceae — as was placed previously.

c. Govindarajalu (1969) made a key based on leaf anatomy to separate the different species of *Cyperus.*

d. Ayensu (1974) reported that the genera *Barbacenia* and *Vellozia* of Velloziaceae can be differentiated on the basis of form of sclerenchyma in leaf.

e. Many botanists have also worked on different taxa and made important contributions, e.g., Morly (1953) on *Mouriri*, Hagerup (1953) on Ericaceae, Tomlinson (1956, 59) on Musaceae and Zingiberaceae etc.

B. **Petiole anatomy**. The petiole shows various types of vascular structure as illustrated by Metcalfe and Chalk (1950). Vascular structure of dicotyledonous petioles also shows taxonomic importance.

Azizian and Culter (1982) reported the importance of vascular structure of petiole in the diagnostic significance of *Phlomis* and *Eremostachys* of Labiatae (Lamiaceae). The petioles of *Phlomis* show two distinct groups. In one type, 1-2 median arcs are noted in *Phlomis* and species of *Eremostachys* and, in other type, numerous separate bundles are present and often form a complete ring in some *Phlomis.* The above observation establishes the distinction of the two groups in *Phlomis* and its close relation with *Eremostachys*.

C. **Nodal anatomy**. Sinnott (1914) distinguished three fundamental types of nodal anatomy in dicotyledons. These are unilacunar, trilacunar and multilacunar. According to him, the trilacunar node is primitive and the other two types are more advanced. Many years later, Marsden and Bailey (1955) reported the 4th type as unilacunar two trace, which is now considered as the basic type for angiosperms. Sinnott and Bailey (1914) made a comparative study of nodal anatomy of angiosperms and showed the relationship and distinction among different genera and species.

a. Canright (1955) recognised the apparent phylogenetic sequence in the mature unilacunar node of the family Monimiaceae. The different genera of the family show the number of traces i.e., 2 in *Trimenia, Piptocalyx* etc., followed by 3, 5 or 7 in *Anthobembix, Hortonia* etc. and finally 1- broad arc-shaped leaf trace in Siparuneae.

b. Bailey and Howard (1941) divided the subfamily Icacinoideae of Icacinaceae into two different sections. One section shows unilacunar nodes and the other trilacunar nodes.

D. **Wood anatomy**. Wood anatomy is found to be very much useful in solving some taxonomic and phylogenetic dispute. Carlquist (1961) observed that "the well-established trends of xylem evolution should be taken into account by any systematist interested in formulating relationships and evolutionary status of angiosperm taxa".

Some important characteristics are :

a. *Vessel elements.* Characteristics like distribution, diameter and frequency of vessels, perforation and thickening and presence of tyloses have been found useful in taxonomic and phylogenetic syudies. The non-porous wood without vessels is more primitive, as in Gymnosperms; compared to porous wood found in angiosperms. Plants with diffused porous wood are considered more primitive than ring porous wood.

b. *Vascular rays.* The different characteristics of vascular rays offer important taxonomic significance. These include cellular composition and width of rays, dimensions of rays (in tangential section), degree of wall-thickness of ray cells etc. *Heterogeneous rays* i.e., rays with both radially and vertically elongated cells are considered more primitive, than the *homogenous rays* — rays with only radially elongated cells.

c. *Axial parenchyma.* The characteristics and distribution of axial parenchyma cells are the distinctive features of secondary wood. These characteristics are used in systematics and phylogeny of angiosperms. Based on distribution, they are divided into two main groups : *apotracheal type* i.e., parenchyma distributed without any specific relation to the vessels, and *paratracheal type* i.e., parenchyma distributed in close association with the vessels. Absence of parenchyma is considered primitive, as found in some members of Winteraceae. The diffuse arrangement of cells of parenchyma are considered more primitive than the diffuse-in-aggregates. The various aggregate arrangements (apotracheal banded with many paratracheal types) are considered as advanced types.

d. *Storied wood.* It is the pattern of arrangement of cells or tissues in horizontal series as seen in tangential section. Storied woods are considered as advanced over the non-storied woods.

Wood anatomy has played an important role to solve certain taxonomic problems :

i. Parietales : The anatomical findings indicate that the order Parietales of Engler and Prantl (1889) is a heterogeneous assemblage and the segregation of the order into Parietales and Guttiferales as done by Wettstein is justified.

ii. *Degeneria* : The genus *Degeneria* was placed under Magnoliaceae by Hutchinson (1959). Wood anatomy supported the segregation, placing it in a separate family Degeneriaceae. Thus the Degeneriaceae, Magnoliaceae and Eupomatiaceae are the closely related families.

iii. *Illicium* : As *Illicium* contains vessels, so it should be removed from Winteraceae (a vessel less family).

iv. Annonaceae : The family Annonaceae is placed under the order Annonales by Hutchinson (1959) and under Magnoliales by Takhtajan (1969). Studies on wood anatomy of the family by Purkayastha (1980) suggested that the Annonaceae have much advanced type of wood — this supported the view of Hutchinson.

v. Rhoipteleaceae : The family has been placed either in Urticales or in Juglandales (Takhtajan). The anatomical studies of Withner (1941) supported the systematics of Rhoipteleaceae as done by Takhtajan. The scalariform perforation plates are found both in Juglandales and Rhoipteleaceae, but are absent in Urticales.

E. **Floral anatomy**. The floral anatomy also proved useful to solve the problems related to taxonomic and phylogenetic studies. This has been discussed in detail by different scientists like Eames, 1935; Puri, 1952, 62 and Murthy and Puri, 1980. A flower has typical arrangements i.e., sepals with 3, petals, stamens with 1 and carpels with 3 traces. During the course of evolution, the vascular plane undergoes various modifications through amplification, reduction, cohesion and adhesion of vascular bundles. Such modifications are used as important tools in systematic studies.

Some of the examples are :

a. *Paeonia* : The genus *Paeonia* has been separated from the family Ranunculaceae and placed in a separate family, Paeoniaceae, by the modern taxonomists based on floral anatomy.

The anatomical study shows :

i. Other genera of Ranunculaceae have sepals with 3 and petals with 1 trace, but *Paeonia* have few to several traces.

ii. In other genera, stamen traces are derived independently, but in *Paeonia* stamens are supplied by few large trunk bundles.

iii. In other genera the development of stamen is centripetal, but it is centrifugal in *Paeonia*.

iv. In other genera the carpels have 3-traces, but *Paeonia* receives several traces.

The above differences justify the separation of *Paeonia* from Ranunculaceae and placed in a separate family of its own.

b. *Trapa* : The genus *Trapa* has been included in Trapaceae, after removing it from Onagraceae. This has been justified by the following data based on vascular anatomy :

i. The vascular plan of flowers in *Trapa* differs from the other genera of Onagraceae.

ii. The inferior ovary of *Trapa* has been considered to be receptacular due to down turning of the receptacular bundles. On the other hand, in other genera of Onagraceae, the ovary has been regarded as appendicular due to fusion of the bundles present in the same radii.

c. *Hydrocotyle asiatica* : The *Hydrocotyle asiatica* L. has been transferred to the genus *Centella* by Urban and named it as *Centella asiatica* (L.) Urban. This was confirmed by Mittal (1955) based on floral anatomy

The observations are :

i. The inflorescence of *Hydrocotyle* is umbellose raceme, but in *H. asiatica* it is a cyme like *Centella*.

ii. In the other species of *Hydrocotyle*, the ovular traces are derived from the placental strands, but in *H. asiatica* they are derived from the alternate bundles in each carpel as in the genus *Centella*.

4. **Cytology**. Cytology plays an important role in the study of evolution, phylogeny and classification of plants. Classification of angiosperms in general and monocotyledons in particular has been a subject of considerable debate. Contradictory views have been held not only with broader subdivisions but also with the position of minor taxa. So the subject cytotaxonomy develops which utilises the cytological characteristics and phenomenon for the elucidation of taxonomic problems regarding the position of taxa at different level.

The *Chromosome number, morphology* and *behaviour at meiosis* are the parameteristics considered in this respect. In addition, *cytochemistry* and *banding patterns* are also used.

A. *Chromosome number* : Chromosome number is constant in many taxa (species) which is considered as an important taxonomic characteristic. Phanerogams with lowest chromosome number (n=2) is found in *Haplopappus gracilis* of Compositae and the highest number (n= 263-265) is found in *Poa litorosa* of Poaceae (Davis and Heywood, 1963). In higher groups of plants, three main cytological types are found — diploid, aneuploid, and polyploids. In addition, haploid type is also reported in some species like *Cocos nucifera, Prunus persica* and *Theobroma cacao*. The chromosome number may or may not help identify the different taxa.

a. Not help to identify : Due to uniformity in different species, the chromosome number does not make any help to identify the taxa, but it is useful to differentiate them from other taxa. All the species of *Pinus* have n=12. Similarly *Quercus* and most other members of Fagaceae also have n=12. In the above cases chromosome number does not make any help to differentiate the species. But in both the cases, the species are easily distinguishable by morphological characteristics. The species differentiation has come through gene mutation without differentiation of their karyotype.

b. Help to identify : The polyploid plants are not recognised as species unless they are morphologically distinct. Only when the diploid and polyploids are clearly distinct, they can be recognised as different species, such as in *Saxifraga hyperborea* (2n=26) and *S. rivularis* (2n=52). Thus it is not accepted in *Galium aparine* of Rubiaceae, having the numbers 2n= 22, 44, 66 and even 88. In *Salix,* different species are considered to the different polyploidy members.

The basic number in *Salix* is 19 and the different diploid and polyploidy species that have been recognised are :

i. *Salix viminalis* is diploid (2n=38),
ii. *S. atrocinerea* is tetraploid (2n=76),
iii. *S. phylicifolia* is hexaploid (2n=114) and
iv. *S. myrsinites* is octaploid (2n=152).

In aneuploidy, the diploid monosomics are inviable normally and in trisomics (2n+1), the extra chromosomes produce certain minor variations, thus limiting their taxonomic significance. Several disomic (2n) species have been recognized in *Taraxacum.*

Chromosome number may help to maintain the actual position of taxon at generic level.

In *Tephrosia* of Fabaceae all the species have 2n=22, except *T. constricta* where 2n=16, justifying its separation as a new genus *Sphinctospermum.*

Both the plants *Cicendia filiformis* and *Mixcrocala pusilla* of Gentianaceae were placed under *Cicendia.* The former have the basic chromosome number 13 and the latter with 10 justifying their separation into different genera.

B. *Chromosome morphology* : Delaunay (1926) was the first to formulate the concept of karyotype as 'a group of species resembling each other in the number, size and form of their chromosomes'. Karyotype studies mainly concerned with morphological aspects and are represented by figures termed as 'Idiograms' or 'Karyograms'.

The principle parameters of such study to get the differences are :

a. The number of chromosome,
b. The relative size and form of chromosomes of the same set,
c. The size and the number of secondary constrictions and satellites,
d. Absolute size of the chromosomes, and
e. The distribution of eu- and heterochromatin.

The absolute size of the chromosomes in a karyotype is fairly constant and is considered as species-specific characteristic (Davis and Heywood, 1963).

It has been noted in *Chrysanthemum* and *Dianthus* that the diploid have longer chromosomes than the polyploids. In *Crepis,* the annual species have smaller chromosomes than the perennials. The length of chromosomes alone always does not show taxonomic significance.

Another type of supernumery chromosome found in plants of many families (at least 250) is the **B-chromosomes**. These chromosomes are small and heterochromatic in nature. Variation occurs in their number. They do not pair with A-chromosomes but can pair with each other.

C. *Behaviour of chromosome at meiosis* : Behaviour *of chromosome* at meiosis helps to find out the relationship of different species and the population as a whole. Two important aspects of chromosome behaviour are considered in respect of phylogenetic significance. These are : i. Synopsis during meiotic prophase, and ii. Frequency of chiasma.

Cytological studies in the above aspects help to solve the different taxonomic problems. Before dealing with the actual problem one should know the characteristics considered as criteria for deciding the status of a taxon.

a. *Symmetrical karyotype* (when all the chromosomes are of equal size with the centromere at the middle) indicates primitive characteristic and *asymmetrical karyotype* (chromosomes differ in size and are acrocentric with the centromere almost near the tip) indicates advanced characteristic.

b. *Large chromosomes* represent a primitive characteristic from which through gradual deletion *small chromosomes* have been evolved. (The reduction of size may involve in all or in some chromosomes of a genome. The condition further leads to asymmetry in the nature of karyotype).

c. Plants with low *chromosome number* indicate primitive and high *chromosome number* indicates as an advanced condition.

Briefly, it can be considered that the *symmetrical karyotype* with *low chromosomes numbers* and *long metacentric chromosomes* indicate **primitive characteristics,** whereas *asymmetrical karyotype* with *high chromosome number* and *short acrocentric chromosomes* are the **advanced characteristics.**

Some taxonomic problems can be explained at the different level of taxa. Some examples are :

a. Pandanales : Engler considered Pandanales, including 3 principal genera *Pandanus, Typha* and *Sparganium* as the ancient stock from which the monocotyledons have been evolved. On the other hand, Hutchinson considered Pandanales as an advanced group and the aquatic family Alismataceae as the progenitor of Monocots.

Cytological studies show that the basic chromosome number of all the three genera, *Pandanus* (n=30), *Typha* (n=15) and *Sparganium* (n=15) of Pandanales is x=15 and the chromosomes are small and acrocentric. Thus it shows that such a high number of small-sized chromosome undoubtedly represent an advanced level in evolution. From the above data, it can be concluded that the consideration of Hutchinson as an advanced group for Pandanales is justified and the Engler's idea about the primitiveness is not acceptable.

Based on habitat, Hutchinson separated Engler's Pandanales into two orders, Pandanales (includes *Pandanus,* nearly terrestrial) and Typhales (includes *Typha* and *Sparganium,* both are aquatic).

Cytological studies show that in all the three genera the basic number is x=15 and they show chromosome homogeneity. All the genera are characterised by very small chromosomes with nearly identical types of constriction. Based on karyotype, they look very similar to each other indicating the homogeneity of Pandanales. Thus the splitting of the order as done by Hutchinson is not acceptable and the idea of placing of all the three genera in a single order Pandanales as done by Engler is justified.

b. Alismatales and Butomales complex : John Hutchinson considered the Alismatales and Butomales complex as the primitive group of Monocotyledons because of the aquatic habit, and thought that all other members of Monocotyledons have been evolved from it.

The genus *Alisma* of Alismataceae has the characteristic feature of an ancient genus *Alisma plantago-aquatica.* Because, it has :

i. Low chromosome number starting from x=5.

ii. Long chromosomes.

iii. Symmetrical karyotype.

The above characteristics indicate the potentiality of this complex to be the ancestral stock. But this is not true for all the members of Alismatales and Butomales complex. The genera like *Hydrocharis* of Hydrocharitaceae and *Butomus* and *Limnocharis* of Butomaceae and also others show asymmetrical karyotype with high chromosome number and are shorter in size. So the above genera must not be recognised as primitive and the homogeneity of the complex group cannot be considered. Lastly, it can be concluded that the group should be broken up into several groups and should not be considered as the progenitors of all other groups of Monocotyledons. Only *Alisma* might have given rise to some of the groups.

c. Liliaceae and Amaryllidaceae : Liliaceate have superior ovary that can be distinguished from Amaryllidaceae having inferior ovary. Engler placed both the families under the order Liliiflorae. Similarly, Hutchinson also placed the families Liliaceae and Amaryllidaceae into two separate orders : Liliales and Amaryllidales based mainly on the characteristics of inflorescence, but he constructed a new order Agavales. In his Agavales, he included several genera like *Agave, Yucca, Dracaena* etc. from Liliaceae and *Polianthes* from Amaryllidaceae.

The three tribes like Agapantheae, Allieae and Gillesieae were transferred to Amaryllidaceae by Hutchinson from Engler's Liliaceae due to the presence of umbellate scapose cyme like Amaryllidaceae.

Cytological studies with the members of Agapantheae, Allieae and Gillesieae show that the chromosomes are long and look almost alike with the members of Amaryllidaceae. Though the karyotypes do not fully match, even in that manner they do not show drastic difference to make any obstacle in the inclusion of the above three tribes under the family Amaryllidaceae.

The five genera — *Agave, Yucca, Dracaena, Sauservia* and *Cordyline* — of Liliaceae and *Polianthes* of Amaryllidaceae were placed under Agavales. *Funkia* was not taken out by Hutchinson as was done by the modern cytotaxonomists.

Cytological studies on the three genera *Agave, Yucca* and *Funkia* show that they have chromosome no. x =30, with 25S (short) and 5L (long). The genus *Polianthes* also shows same result. The above result proves the justification to include *Polianthes* along with the other genera under Agavales. On the other hand, the genera like *Dracaena, Sauservia* and *Cordyline* show x=18-21 and are of graded type. The above three genera do not fit with *Agave* and others and should go back to the family Liliaceae. Thus it can be said that the creation of Agavaceae with genera being taken out from two different families by Hutchinson was no doubt a bold approach.

It can be said that chromosomal characteristics alone are not always a reliable taxonomic criteria, nor they are considered as good phylogenetic markers.

5. **Embryology** : Embryology plays a very significant role in systematic consideration and phylogeny. The role of embryology in the taxonomy of flowering plants has been discussed by Maheswari (1958, 1964), Johri (1963), Kapil and Bhatnagar (1980) and Cocucci (1983).

The embryology includes not only the characteristics of embryogeny and mature embryo, but also the characteristics of micro- and megasporogenesis. Maheswari (1950, 1963) considered the following characteristics to discuss the taxonomic problems :

a. Arrangement and number of anther chambers, shape and thickening of endothecium and nature of tapetum.
b. Development and arrangement of the pollen grain.
c. Types of quadripartition in pollen mother cell.
d. Structure and development of the ovule.
e. Nature of nucellus.
f. Origin and extent of sporogenous tissue inside the ovule.
g. Arrangement of megaspores and mode of development of embryo sac.
h. Types of mature embryo sac.
i. Path of pollen tube growth and the time taken between pollination and fertilisation.
j. Presence and absence of endosperm, endosperm haustoria and also the types of endosperm.
k. Structure of mature embryo, etc.

Some taxonomic problems can be explained at the different level of taxa. Some examples are :

a. *Paeonia* : In earlier classifications, the genus was included in Ranunculaceae. Later, Hutchinson (1959) placed it in a separate family Paeoniaceae. The embryological studies on *Paeonia* and other members of the Ranunculaceae show :

   i. In *Paeonia,* the generative cell is large and elongated, the embryo sac is narrow and long, the seed coat is massive and seeds germinate by hypogeal means. The above characteristics contrast with the other genera of Ranunculaceae.

   ii. In *Paeonia,* the embryogeny is unique which is not available in other members of Ranunculaceae.

The above embryological data support the thinking of Hutchinson (1959) in the separation (from Ranunculaceae) and placing of *Paeonia* in a separate family Paeoniaceae.

b. *Butomus* : Bentham and Hooker divided the family Alismataceae into two tribes, Alismeae and Butomeae. Engler and Prantl treated the above two tribes in the rank of family, Alismataceae and Butomaceae (includes *Butomus* and other 5 genera *Ostenia, Hydrocleis, Limnocharis, Tenagocharis* and *Elattosis*). Hutchinson placed the families under separate orders Alismatales and Butomales, respectively. The embryological studies indicate :

i. The genus *Butomus* having anatropous ovules and straight embryo can be separated from the other five genera showing campylotropous ovules and 'horse-shoe'-shaped embryo.

ii. *Butomus* has a *Polygonum* type of embryo sac, but the other genera have *Oenothera* type of embryo sac.

Therefore, Pichon (1946) suggested that all the five genera should be transferred to Alismaceae, although the placentation of the carpel is laminar.

Later, both Cronquist (1968, 1981) and Takhtajan (1969, 1980) included only *Butomus* under the family Butomaceae and the rest five genera under a separate family Limnocharitaceae and placed the families Butomaceae (unigenic), Limnocharitaceae and Alismaceae under the order Alismales (Alismatales).

c. *Trapa* : The genus *Trapa* was placed in the family Onagraceae by Bentham and Hooker. Later, Engler segregated out the genus *Trapa* and placed it in a separate family Trapaceae for its perigynous flower, bilocular ovary and spiny fruit which has been accepted by most of the recent taxonomists. The embryological study shows :

i. In *Trapa*, the embryo sac is *Polygonum* type, but other genera show *Oenothera* type of embryo sac.

ii. In *Trapa*, the endosperm is lacking, but others have nuclear endosperm.

iii. In *Trapa*, the development of embryo is Solanad type, but it is Onagrad type in Onagraceae.

iv. In *Trapa*, the suspensor is well developed showing haustorial type, but it is short and inconspicuous in the family Onagraceae.

v. In *Trapa*, the cotyledons are unequal, but they are equal in other members of Onagraceae.

The above embryological studies support the Engler's separation of *Trapa* from Onagraceae to a separate family Trapaceae.

The embryological studies are also useful to solve some problems at family level :

a. Saxifragaceae and Crassulaceae : Both the families are placed very close under the order Rosales by the earlier taxonomists. Presence of the following embryological characteristics of both the families justifies their consideration :

i. Multilayered anther wall.

ii. Bitegmic ovules.

iii. *Polygonum* type of embryo sac development.

iv. Endosperm cellular.

v. Embryo development Caryophyllid type.

b. Gramineae and Cyperaceae : The earlier taxonomists like Bentham, Hooker and Engler and others placed the two families very close to each other. The embryological studies show that the following characteristics are common in both the families justifying their consideration :

i. In Gramineae, the microsporogenesis is of successive type and the tetrad members are separated, whereas, in Cyperaceae, it is of simultaneous type and the tetrad members do not separate, where three degenerate and only one remains functional.

ii. In Gramineae, the archesporium is without parietal cell, but it is present in Cyperaceae.

iii. In Gramineae, the antipodals form a complex or become coenocytic, but it is ephemeral in Cyperaceae.

iv. In Gramineae, the testa and pericarp are fused and either only the inner integument form the testa or both are obliterated, but in Cyperaceae they are distinctly free and the testa consists of both the integuments.

v. In Gramineae, the embryogeny is of variable type, but it is of Onagrad type in Cyperaceae.

c. Lemnaceae : The family is considered to have originated from *Pistia* of Araceae or from Helobiales. S. C. Maheshwari (1954, 1956. 1958) studied the embryological aspects and clearly indicates that the Lemnaceae are evolved from Araceae and are not related with the order Helobiales.

The Lemnaceae resemble with Araceae by the following characteristics :

i. Presence of bitegmic ovules with a micropylar cap.
ii. Cellular endosperm with a chalazal haustorium.
iii. Irregular sequence in the division of embryo development.
iv. Short and stocky suspensor.
v. Integumentary operculum in the seed.

The Helobiales having the following characteristics by which it differs :

i. The ovules do not have any micropylar cap.
ii. The endosperm is Helobial or nuclear type.
iii. Regular sequence in the division of embryo development.
iv. Prominent basal cell in the embryo.

6. **Palynology** : G. Erdtman, published *"Pollen Morphology and Plant Taxonomy of Angiosperms"* in 1952, which opens a new phase in the role of palynology in taxonomy. Since then, it aggravated the use of pollen characteristics in taxonomic studies. Later, Erdtman (1969) published an excellent review on the application of palynology in taxonomy. Investigations in this line become more aggravated by the use of Electron Microscope in pollen studies. The pollen characteristics are then being extensively used in taxonomic and evolutionary studies as mentioned in the publication of Blackmore, (1984). The following pollen characteristics are used in taxonomic studies :

a. **Size and shape**. The sizes are very small, small, medium, large, very large and gigantic. The shape also varies with different views : in *polar view* it may be circular, triangular etc.; in *equatorial view* it may be prolate, perprolate, peroblate, spheroidal etc. and in *lateral view* the bilateral pollens are planoconvex, biconvex or concavoconvex.

b. **Exine sculpturing**. The surface of the pollen may be smooth or they may be variously sculptured. It may be of two types : Excrescences type and Depression type.

Excrescences type :

i. Spinulose (having spinules).
ii. Granulose (having minute granules).
iii. Gemmate (having rounded warts with concentric base).
iv. Tuberculate (having tubercule like excrescences).
v. Clavate (having club-shaped excrescences).
vi. Baculate (having rod-shaped excrescences), etc.

Depression type :

i. Striate (lumina parallel).
ii. Rugulate (lumina anastomosing).
iii. Reticulate (lumina forms a network).
iv. Foveolate (lumina circular and closely placed).
v. Scrobiculate (lumina circular and distantly placed).
vi. Fossulate (lumina elongated), etc.

c. **Apertures**. Based on the number, position and character (NPC) of aperture, the pollens are of different types.

i. Number : 1-many.
ii. Position : Proximal, distal, zonal or global.
iii. Characteristics : Colpate (furrow), porate (circular) and inaperturate.

d. **Ultrafine structures**. The ultrafine structures of sporoderm have been studied by Transmission Electron Microscopy which show considerable taxonomic value.

e. **Pollen association**. Pollens are generally associated in tetrads in angiosperms. Such association is common in dicots (Annonaceae, Ericaceae, Mimosaceae, Droseraceae etc.), less commom in monocots (Cyperaceae, Eriocaulaceae and Juncaceae) and the extreme cohesion is found in Asclepiadaceae (dicot) and Orchidaceae (monocot), where the pollens remain in pollinia.

f. **Nuclear number in pollen**. The nuclear condition of pollen in angiosperms has taxonomic importance (Dufou, 1961). At the time of anthesis, pollens are 2-nucleate in monocotyledons, 2-nucleate in polypetalous dicotyledons and apetalous taxa, 3-nucleate in gamopetalous dicotyledons.

The characteristics of pollen are useful in the differentiation of various taxa and also to find out the phylogenetic relationship.

a. **Family level** :

Araceae and Lemnaceae. Hutchinson in his Arales included both Araceae and Lemnaceae. The family Araceae is eurypalynous with 1-2-4-colpate, 3-porate or inaperturate pollen with exine sculpturing and, on the other hand, Lemnaceae is stenopalynous with 1-porate and spinous pollen.

Bombacaceae. Bentham and Hooker divided the family into four subfamilies, Malveae, Ureneae, Hibisceae and Bombaceae. Later, Engler treated Bombaceae as a separate family Bombacaceae. The exine of most of the members of Malvaceae is *spinous,* but the exine of Bombaceae is reticulate. The above study supporting the separation of Bombaceae as a family is justified.

Berberidaceae. The family Berberidaceae consists of 12 genera. Modern taxonomists removed the genus *Podophyllum* from Berberidaceae and placed it in a separate family Podophyllaceae. The pollen grains in *Podophyllum* remain united, but they are free in other genera, supporting the removal of *Podophyllum.*

Orchidaceae. Vij and Kashyap (1978) studied pollens of 50 orchid species and found three groups :

i. Single pollen grain (monad) (*Cypripedium cordigerum, Paphiopedilum,* etc.)

ii. Tetrads (in tribes Epidendrieae and Neottieae), and

iii. Perfect massulae (in Orchideae and some members of Neottieae).

b. **Genus level** :

Salicaceae. The family Salicaceae consists of 2 genera, *Salix* and *Populus* that can be distinguished on the basis of pollen characteristics. The genus *Populus* has spherical pollen without distinct aperture, whereas there is long and narrowed 3-furrowed pollen in *Salix.*

Phytolaccaceae. The family consists of 22 genera. The pollen studies on two genera *Rivinia* and *Phytolacca* indicate that they can be distinguished very easily on pollen characteristics. The pollen of *Rivinia* is pantocolpate, whereas it is 3-zonocolpate in *Phytolacca.*

c. **Species level** :

The pollen characteristics help in differenciating the species within a single genus :

Anemone. Based on germinal aperture, the different species of *Anemone* (Ranunculaceae) can be distinguished. The pollens are 3-zonocolpate in *Anemone obtusiloba,* pantocolpate in *A. rivularis,* pantoporate in *A. alchemillaefolia* and spiraperturate in *A. fulgens.*

Malva. Based on pollen size, the different species of *Malva* (Malvaceae) can be distinguished. The size of pollen ranges from 105-126μm in *M. sylvestris,* but in *M. rotundifolia* it is only 74-84 μm.

Bauhinia. Based on the exine patterns, the different species of *Bauhinia* (Fabaceae) can be differentiated (Nair, 1974) :

a. Pilate in *B. acuminata.*

b. Reticulate in *B.racemosa.*

c. Striate in *B. krugii.*

d. Verrucate in *B. retusa.*

e. Spinulate in *B. malabarica* and

f. Reticulate-tuberculate in *B. purpurea.*

The different species of *Ranunculus* can easily be identified by studying the pollens. Pollen characteristics also help to distinguish the cultivers of *Canna, Bougainvillea* etc. Different sizes of the areoles on the exine surface help to differentiate the cultivers and hybrids of *Cajanus caian* (Fabaceae).

7. **Phytochemistry**. The chemical characteristics of plants attract the plant taxonomists since the early part of 1960s and since then it demanded as promising subdiscipline of taxonomy. This subdiscipline is called as *Chemotaxonomy* or *Biochemical systematics.*

The phytochemical characteristics of taxonomic significance have been classified into three main groups (Mentzer, 1966) :

i. *Primary constituents* : These are proteins, nucleic acids, chlorophylls and polysaccharides.

ii. *Secondary constituents* : These are simple phenolic compounds (benzoic, caffeic and nicotinic acid) and polyphenolic compounds (terpenes, flavonoids, coumarines and pigments).

iii. *Miscellaneous substances* : This group includes ellagitannins, iridoid compounds, etc.

The following content deals with some of the above compounds in relation to taxonomy :

a. **Amino acids** : In relation to the distribution of non-protein amino acids in seeds, Bell (1962) recognised 7 infrageneric groups in *Lathyrus.* Similarly 4 groups have been recognized in *Vicia.* The differentiation based only on protein characteristc is very difficult. Thus, Cronquist (1976) made the comment that "It is the phenotype, not individual genes, that characterises the organism and is subject to selection. This whole is greater than the sum of its parts, and it cannot be reconstructed by separate analysis of all those parts".

The serological studies help in finding the relation among the different taxa. The genus *Liriodendron* had been found to be quite distinct from other members of the family Magnoliaceae and both the genera *Michelia* and *Magnolia* displayed closest affinity within the family (Johnston and Fairbrothers, 1965). Later, Pickering and Fairbrothers (1970), based on their finding, supported the classification of Umbelliferae into Hydrocotyloideae, Saniculoideae and Apioideae and concluded that the Apioideae is more closely related to Saniculoideae than the Hydrocotyloideae.

b. **Phenolic compounds** : The phenolic compounds are *ellagic*, caffeic, benzoic and nicotinic acids. The ellagic acid is present only in the tribe Kerrieae under the subfamily Rosoideae of Rosaceae, but it is absent in other tribes of the Rosoideae.

c. **Polyphenolic compounds** : The polyphenolic compounds are flavonoids, coumarins, terpenes, alkaloids etc. The *flavonoides* have been widely exploited in chemotaxonomic studies. Their role has been reviewed by Harborne *et al.* (1975). Swain (1975) reported that the structural complexity of flavonoides increases along with evolution, thus the *Chara* has glycoflavones, whereas proanthocyanidines are present in higher pteridophytes (ferns), gymnosperms and in some angiosperm progenitors. Flavonoides have been used to trace the species relationships in *Chenopodium* (Crawford & Evans, 1978) and to predict generic affinities among Ulmaceae (Giannasi, 1978). Giannasi and Chuang (1976) studied the flavonoides in *Perideridia* of Umbelliferae. Though the 16 species of *Perideridia* have same flavonoids, they are divided into three groups based on certain specific combinations. These are :

i. Producing only flavonoides,

ii. Producing mainly flavonols along with a flavones, and

iii. Producing mainly flavones.

d. **Alkaloids** : These are heterogeneous groups of organic nitrogen containing bases, often with a heterocyclic ring. Alkaloid content has been considered as a useful source of taxonomic consideration. The members of Fabaceae have *lupin* alkaloides, Papaveraceae have *isoquinoline* alkaloids and Solanaceae have *tropane* derivatives.

In some cases, their distribution is restricted to a particular species, such as *conine* to a few members of Apiaceae, *strychnine* to a few species of *Strychnos* and *morphine* is found only in *Papaver somniferum.*

In the subfamily Lotoideae of Fabaceae, three tribes Podalyrieae, Genisteae and

Sophoreae are characterised by the presence of lupin alkaloids suggesting their origin from common ancestor.

Presence of isoquinoline alkaloids in both the family Papaveraceae and Fumariaceae indicate their close relationship.

e. **Betalains** : Betalains differ from rest of the phenolic compounds owing to the presence of nitrogen in them. These are red and yellow (betacyanins and betaxanthins) pigments and found in only 10 families of angiosperms that have been traditionally included under a single order Centrospermae. The Cactaceae was placed in an order of their own, the Cactales or Opuntiales. But according to phytochemists for the presence of betalains, they should be kept under Centrospermae instead of Cactales.

f. **Terpenoids** : These are volatile compounds and are almost universal in their occurrence. Earlier, plants were characterised by smell due to the presence of various terpenes e.g., species of *Cymbopogon* of Poaceae, *Ocimum* of Lamiaceae, the tribe Tagitineae of Asteraceae and so on.

Mirov (1961) done the classical work in plant taxonomy on gum terpentines of *Pinus*. The terpenes such as monoterpenes, diterpenes and sesquiterpenes have been studied in relation to taxonomy. Irwin (1968) distinguished 3 sympatric taxa in *Hedeoma* under the family Lamiaceae by using their terpenoides. Flake and Turner (1973) studied the terpenoids in various populations of *Juniperus virginiana* and tried to explain their affinity.

g. **Other substances** : Some visible chemical substances such as starch grains and raphides also provide taxonomic significance. Form of starch grains along with other characteristics have been used in the classification of Gramineae (Poaceae). Forms of the crystals of calcium oxalate on ovarian wall in Compositae (Asteraceae) have considerable taxonomic value. Similarly presence or absence of raphides provides an useful characteristic in the classification of Rubiaceae.

8. **Ecology**. Ecological studies also play a very significant role in systematic consideration. The ecological variation in plants is the result of response to environment involving different factors like climatic, edaphic and biotic. These factors have direct effect on phenotype and indirect effect on the genotype. Thus the appearance of the individual plant is the result of their genetic constitution and environmental influences.

Turesson collected a series of population samples from different parts and grown in uniform condition in his experimental garden at Akarp in Sweden. He observed that in some cases the differences observed in the field disappeared in respect of height, habit flowering time, etc. These plants are called ecotypes. Others, which show persistent differences, appeared to be due to genetic differences.

Later, Clausen, Keck and Hiesey have researched on this between 1940 and 1950. They carried out the geneological studies with population of various species of *Achillea millefolium* of Asteraceae and *Potentilla glandulosa* of Rosaceae. Experimental gardens were established at varying altitudes at 30 to 500 metres and also at 3,300 metres and were maintained free from any other vegetation including weeds. They observed the response of the above plants in different climatic conditions. Crossing between the ecotypes resulted into fully fertile hybrid and produced new genotype with new adaptive features. So, it indicated that the interaction between habitat of plant and its genotype could produce ecotype.

George *et al.* (1936) grown different races of *Plantago maritima* of Plantaginaceae in experimental garden under uniform environmental conditions and observed that the pattern of ecotype variation was more frequently continuous than discontinuous, corresponding to gradients shown in the natural habitat. The continuity of ecotype variation was also supported by Bocher (1943) with experiment on *Plantago lanceolata.*

PHENETICS AND CLADISTICS

Systematics is the study of the nature, cau-ses, patterns and trends in variation among taxa. The **plant systematics** encompasses plant species which can be recognised, classified, circumscribed, and named with a logical relationship based on evolution exist among these units.

These are two basic methods used in establishing relationship between the organisms. These are (a) Phenetic method, and (b) Phylogenetic or Cladistic method.

(a) **Phenetic method** : In this method, the classification is based on the overall similarity of taxa from all available sources such as morphology, anatomy, palynology, embryology, phytochemistry, cytology, ultrastructure, etc., along with all other field data. Thus, in phenetic classification, the taxa are grouped by shared features, regardless of whether their similarities are **symplesiomorphies** (shared primitive characters) or **synapomorphies** (shared derived characters) in a phylogenetic sense. This classification is denoted by a **phenogram** (graphic presentation).

(b) **Phylogenetic or Cladistic method** : This classification is based on the phylogenetic studies, indicating the presumed ancestor/descendent relationship. Thus, only shared derived, characters (synapomorphies) are used to assess phylogenetic relationship, while the shared ancestral characters (symplesiomorphies) are of no value for such classification. The grouping of taxa is made by a genealogical descent, as denoted by a **genealogical tree** or **phylogenetic tree** or **cladogram.** The approach of such study is called **cladistics** and the scientists using this method are known as **cladists.**

PHENETICS VERSUS CLADISTICS

The character states are of two basic types : homologous characters and analogous characters. The **homologous characters** are those which have been inherited from a common ancestor. The **analogous characters** are the derived characters which are the results of independent evolutionary events such as paralled or convergent evolution. The ancestral characters are termed **plesiomorphies** and derived characters are called **apomorphies**. Thus, shared ancestral characters are called **symplesimorphies**, similarly shared derived characters are known as **synapomorphies**. Only, synapomorphies are used to assess phylogenetic relationship (cladistics) and symplesiomorphies are used to interprete the relationship among the groups under investigations. Thus, synapomorphies alone are of no value for interpreting relationships among the groups (phenetics) to be studied, vis-a-vis symplesiomorphies are not used in cladistic analysis.

In **phenogram** and **cladogram** construction, the resultant groupings are quite different, even though the same data matrix is used. For example, in the data from the character x taxon matrix (Fig. 4.28), taxa B and C would be grouped together in a phenogram of phenetic classification (Fig. 4.28). Because, taxon B and taxon C share the states of more character (state 0 of characters 2 and 3) than either does with taxon A.

In cladistic approach, with the addition of a hypothetical ancestor XYZ (Fig. 4.29), it is apparent that taxa A and B share a derived common feature (state 1 of character 1) indicating their recent common ancestor than either does to taxon C. Therefore, taxa A and B would be grouped together in cladogram. The similarity between taxon A and taxon C are all symplesiomorphic (shared ancestral character states) and do not bear any information with regard to phylogeny.

The phenetics may be preferable in classifying taxa of low rank such as species. The cladistics rely on the principle of parsimony and assume the existence of past evolutionary events which can be detected today as a synapomorphy in defining a particular monophyletic group.

	1	2	3
A	1	1	1
B	1	0	0
C	0	0	0

(a)

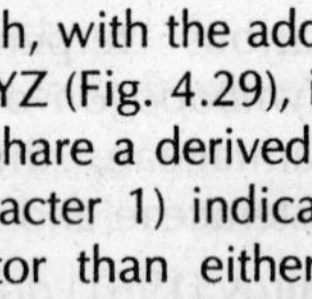

(b)

Fig. 4.28 : (a) Character x taxon matrix of taxa A, B, C; (b) Dendrogram for taxa A, B and C. (*after* Radford 1984)

	1	2	3
A	1	1	1
B	1	0	0
C	0	0	0
XYZ	0	0	0

(a)

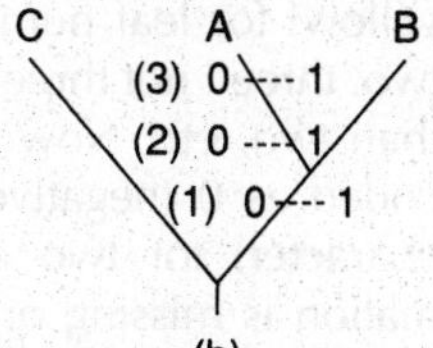

(b)

Fig. 4.29 : (a) Character x taxon matrix of taxa A, B, C, and ancestor XYZ; (b) Cladogram for taxa A, B and C (*after*, Radford 1984)

Numerical Taxonomy and determination of phenetic structure

Phenetic classifications are based on overall similarities among the organisms including those which were not used in the original classification. Phenetic classification is often called numerical taxonomy (coined by Sokal and Sneath, 1963). However, these two are not synonymous. **Numerical taxonomy** or **mathematical taxonomy** is the numerical expression of data in systematics, even though it is entirely predicted on the phenetic approach. However, phenetics need not always be based on numerical analysis. Similarly, cladistic data are also expressed numerically like that of phenetic data.

Actually, the basis of numerical taxonomy begins with the publication of Michel Adanson (1763). He used all the available characters giving equal weightage, and classified the taxa based on the concept of similarity. This classification in popularly known as '**Adansonian classification**', which received much attention by the modern scientists. Thus, a new branch of taxonomy called numerical taxonomy had emerged with the initial contributions from Sneath (1957), Michener and Sokal (1957), and Sokal and Michener (1958). Those classifications are the extensions of the Adansonian classification following mathematical procedures. Actually, the principles of modern numerical taxonomy developed after the epoch-making publication of Sokal and Sneath (1963) in their book "**Principles of Numerical Taxonomy**".

Numerical taxonomy determines the phenetic relationship between organisms or taxa. The phenetic relationship may be defined as arrangement by over-all similarity, based on as much evidence as is available without any weighting (Harrison, 1960). Sneath and Sokal (1973) defined phenetic relationship as "similarity (resemblance) based on a set of phenotypic characters of the objects or organisms under study". The phenetic relationship is represented by a graphical diagram called **phenogram.**

There are several operations involved in phenetic classification. These include :

(a) Selection of taxa which are called Operational Taxonomic Units (OTUs).
(b) Selection of characters, generally more than 60 unit characters.
(c) Description and measurement of character states.
(d) Comparison of states to determine phenetic relationship between each pair of OTUs and to detect the possible groups and subgroups among all OTUs.
(e) Ranking of all OTUs in the hierarchy of classification.

Operational Taxonomic Units (OTUs)

The OTUs are the fundamental units from which the data is collected for numerical analysis. As OTUs are expressed in numerical method, so these are not comparable to formal taxonomic units. Generally, the taxa at lower rank are selected as OTUs. Thus, for analysis of a species, OTUs would be of various population. Similarly, OTUs for the analysis of a genus would be of different species, and for the study of a family they would be of different genera. In the higher ranks there would be much variation, so it is customary to employ species as an OTU for better interpretation. A list of OTUs is to be prepared after the final selection of OTUs.

Taxonomic characters

The taxonomic character may be defined as a character which distinguishes one taxon from another. Heywood (1967) defined a character as "any attribute of form, structure, physiology or behaviour which is considered separately from the whole organism for a particular purpose, such as comparison, identification or interpretation". But, from the point of numerical value, Michener and Sokal (1957) defined a character as a feature that varies from one orga-nism to another. Thus, position of ovary is a character, and the inferior ovary and superior ovary are its two character states. For numerical analysis, it is customary to select a unit character. Sokal and Sneath (1963) defined a unit character as **a taxonomic character as one of two or more states,**

which within the group of study cannot be subdivided logically.

The characters can be divided into four categories : (i) General characters, (ii) Phyletic or phylogenetic or evolutionary characters, (iii) Cladistic characters, and (iv) Phenetic characters. Among the general categories, the characters which can be measured or counted are called **quantitative characters**, (e.g., leaf size, seed number, number of stamen). The characters which cannot be measured are **qualitative characters** (e.g., flower colour, leaf arrangement). Only phenotypic characters are chosen to make a list of unit character. Say for example, trichomes may be unicellular or multicellular. The unicellular trichomes may be clavate or falcate. Similarly, multicellular trichomes may be of uncinate or moniliform types. In such case, unicellular trichome is one unit character and multicellular trichome is another unit character. If all the unicellular trichomes are of similar type and all the multicellular trichomes are of same nature, the trichomes may be selected as a unit character. It is important to select such characters which show variations within the entire sample of organisms. Leaf length, flower diameter, seed number are unit characters, while leaf shape is not a unit character, because leaf shapes can again be subdivided according to length and breadth ratio.

All taxonomic characters are systematic characters, but all systematic characters are not taxonomic characters. The conditions or expressions of characters are called **character states**. A large number of characters, not less than 60, are desirable for numerical studies, and all the characters should be weighted equally.

Coding of characters

As a large number of data are to be handled, so the numerical analysis is done by computer programme. The analysis of data is done by **two-state** (binary) coding method. The characters should be suitably coded (0 or 1) or are given some symbol or mark (– or +). All characters may not be two-state, often may be qualitative multistate (e.g., flower colours : yellow, white, red, etc.) or quantitative multistate (leaves one, two, three, four, five at each node). Such multistate characters have to be converted to two-state. Say for example, for flower colour they may be white : coloured, red : non-red, yellow : non-yellow; for leaf numbers they may be two : not two, three : not three, two or less than two : more than two, etc. Now the two-state characters are coded as 0 (negative character) and 1 (positive character) for two alternate state. If the information is missing or is irrelevant in a OTU, the character must be scored as NC (no comparison).

Measurement of resemblance

To analyse the data statistically, all characteristics must be numerical based on binary data coded as 1 and 0. The data that have been collected can be presented in a character-taxon (OTUs) matrix. The characters are the rows and the OTUs are the column. There are two major aspects of this comparison : (a) The similarity or dissimilarity is calculated on the basis of some clearly defined coefficient of similarity, (b) The phenetic structure (generally hierarchial) is determined based on obtained similarity coefficients by clustering or ordination of the OTUs. As there is no general mathematical theory for the assessment of phenetic relationships, so it is important to select a specific statistics for measurement of similarity as well as an algorithm for determination of phenetic relationships among all the OTUs.

The statistics which are generally employed in calculation of affinity between pairs of OTUs are of four basic types, (a) assocation coefficients, (b) distance coefficients, (c) correlation coefficients, and (d) probabilistic similarity-coefficients (Sokal and Sneath, 1973). Then, the values obtained between each pair of OTUs are placed in a data matrix of similarity coefficients prior to clustering or ordination.

Association coefficients :

For calculation of similarity or dissimilarity based on binary data, the following chart shows all the types of matches possible between two OTUs (A and B). This data could be used by the scientists for assessing association coefficients :

		OTU A	
		1	0
OTU B	1	a (1, 1)	b (1, 0)
	0	c (0, 1)	d (0, 0)

Jaccard (Sneath) Coefficient (S_J) was introduced by Jaccard (1908) and Sneath (1957) and is easiest to comprehend for binary data. The coefficient is represented as :

$$S_J = \frac{a}{a+b+c}$$

where, a (1, 1) is a positive match, b (1, 0) and c (0,1) are mismatches. The mismatch d (0, 0) is a negative match and is not considered meaningful.

The **Simple Matching Coefficient** (S_{SM}) is another commonly used association coefficient (Sokal and Michener, 1958) using binary data. The coefficient is represented as :

$$S_{SM} = \frac{a+d}{a+b+c+d}$$

where the sum of the positive match, a(1, 1) and negative match, d (0, 0) is divided by all the possible matches (a, b, c, d).

The **Gower's General Similarity Coefficient** (S_G) introduced by Gower (1971), is used for binary, multistate and quantitative data. It is represented as :

$$S_G = \frac{\sum_{i=1}^{n} W_{ijk} S_{ijk}}{\sum_{i=1}^{n} W_{ijk}}$$

The similarity is assessed in character i between OTUs k and j. The W (weights) and S (scores) for each character in each OTU are related to the nature of the data (whether binary, multistate or quantitative).

Distance coefficients

These measure the distance or dissimilarity between OTUs in a space which are actually the converse of similarity coefficient. The most commonly used distance coefficient is the **Euclidean Distance** which is represented as :

$$D_E = \left[\sum_{i=1}^{n} \left(X_{ij} - X_{ik}\right)^2\right]^{\frac{1}{2}}$$

where, X_{ij} is the character state value for character i in the taxon J, and similarity, X_{ik} is the character state value for character i in the taxon k.

Some other distance measures are also commonly used, of which **Manhattan** (or **City-block**) **distance** measures the absolute number of character state differences between two OTUs :

$$D_M = \sum_{i=1}^{n} \left|X_{ij} - X_{ik}\right|$$

where, X_{ij} and X_{ik} are the values of the ith character for every pair of OTUs.

Among others, the **Mean Character Difference** (D_{MC}) and **Coefficient of Divergens** (D_{CP}) are two distance measures commonly applied in many plant groups.

Correlation coefficient :

The **Pearson Product-Moment correlation coefficient** is the most widely used correlation in phenetic studies as well as in phytosociological studies. This is represented as :

$$\gamma_{jk} = \frac{\sum_{i=1}^{n} (X_{ij} - \overline{X}_j)(X_{ik} - \overline{X}_k)}{\sqrt{\sum_{i=1}^{n} (X_{ij} - \overline{X}_j)^2 \sum_{i=1}^{n} (X_{ik} - \overline{X}_k)^2}}$$

where, X_{ij} is the character state value of character i in OTU_j, X_j is mean of all state values for OTU_j, n is the number of characters sampled.

Probabilistic similarity coeffcients :

These are the complex coefficients which measure the distribution of the frequencies of the character states over the set of OTUs. The **Goodall's Similarity Index** is a commonly used statistics which determines the overall probability that a pair of OTUs will be as similar or more similar than can be observed for each character on the basis of the distribution of states in all the OTUs.

Determination of phenetic structure

The last step is the detemination of the phenetic structure among all the OTUs. To map the position of each of the taxa as a point in space on the basis of the similarity matrix, it would form n – 1 dimensions for placing each point accurately. Thus, two taxa M and N could be separated on a single line to represent distance from each other. Similarly, a two-dimensional plane is required for three taxa, M, N, O to be separated. With the addition of another taxon P, a three-dimensional space is required. With addition of more and more taxa, it becomes impossible to visualise the relative position of taxa in space. Thus, it becomes necessary to reduce the space for understanding the relationships among taxa. There are two basic approaches involved in reducing dimensions. These are : (a) Clustering, and (b) Ordination.

Cluster analysis : In cluster analysis, character states are compared from one OTU to the next. There are several statistical methods for making such comparison, but the **simple matching coefficient** (Sokal and Michener, 1958) is the simplest method. This is done by counting the number of character states for each character in common between two OTUs. These are the positive matches. The number of positive matches is then divided by the total number of characters used (positive + negative matches). The higher values signify the more closely relations of the OTUS.

The **agglomerative method** is also useful in hierarchial clustering which starts with n clusters, each containing just one object. A proximity matrix for the n object is n × n. The correlation coefficient between two OTUs is measured. A measure of distance between two OTUs is also measured based on correlation coefficients. Thus, the similarity measure is transferred to a distance. Hence, each OTU forms a cluster. The first step is to fuse the two or more nearest OTUs into a single cluster and the rest containing one OTU each. The second step is to fuse the two nearest of these clusters into lesser number of clusters and continues on this manner until at the last step (at most [n – 1] step).

A phenogram is constructed which graphically expresses the relationships among all the OTUs in a conventional form (Fig. 4.30). Thus, taxonomic groups or sub-groups among OTUs can be constructed.

The Table 4.3 shows that minimum distance value in the matrix is zero (0) between OTUs **D** and **F**, **D** and **G**, **F** and **G**, and, as such, they (**D, F** and **G**) are linked at that level (Fig. 4.30) . The next minimum dissimilarity value of 0·70 is between OTUs **A** and **H**, hence these two forms a cluster linked at 0·70 (Fig. 4.30). Similarly, **E** and **C** are in a cluster with a distance value of 0·86. The process is repeated until all OTUs are agglomerated into a single cluster at a maximum distance value of 0·98.

Cladistics

Phylogeny is defined as the study of genealogy and evolutionary history of groups of organisms, where **genealogy** is the study of ancestral relationship and lineages. The phylogenetic studies make genealogical branching pattern indicating the presumed ancestor/descendant relationship. Thus, the purpose of phylogenetic studies is to develop the most probable pathways for the evolution of selected taxa based on analysis of a finite number of useful characteristics.

Phylogenetic or **cladistic analysis** helps in identification of evolutionary lineages or clades

Table 4.3 : Distance values of the different paired taxa (A – H)
[*after* Bhattacharya & Bishayee, 1998]

	Taxon A	Taxon B	Taxon C	Taxon D	Taxon E	Taxon F	Taxon G	Taxon H
Taxon A	0·00	0·94	0·94	0·98	0·94	0·98	0·98	0·70
Taxon B		0·00	0·94	0·98	0·94	0·98	0·98	0·94
Taxon C			0·00	0·98	0·86	0·98	0·98	0·94
Taxon D				0·00	0·98	0·00	0·00	0·98
Taxon E					0·00	0·98	0·98	0·94
Taxon F						0·00	0·00	0·98
Taxon G							0·00	0·98
Taxon H								0·00

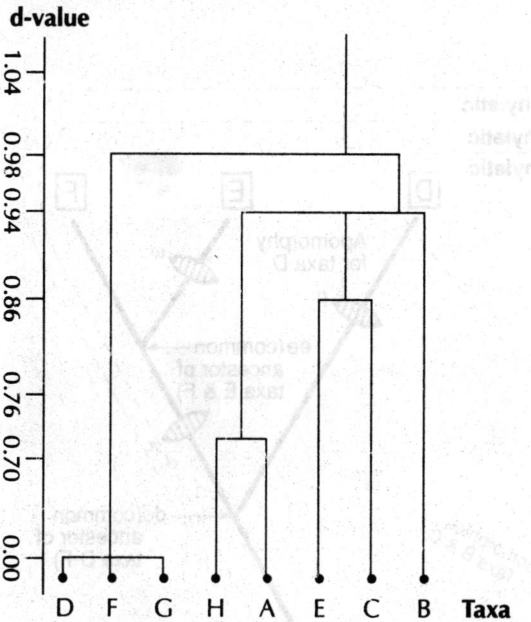

Fig. 4.30 : Dendrogram showing the correlation among 8 taxa (*after* Bhattacharya & Bishayee, 1998)

and also gives emphasis to separate the clades from groups or organisms which have independently reached the similar level of evolutionary advancement.

Concept of cladogram

Cladistic analysis is defined as the concept and methods for the determination branching patterns of evolution (Stuessy, 1980). Like phenetic analysis, cladistic data are also analysed by numerical methods (i.e., selection of OTUs, coding of characteristics, measurement of resemblance and graphical representation).

A branching diagram (dendrogram) constructed by cladistic method is called a **cladogram** or **phylogenetic tree**, representing the evolutionary patterns of descent. The vertical axis of a cladogram is an implied (relative) time scale (Fig. 4.31), where extant or terminal taxa are placed at the top of the time scale. Each node of the cladogram represents a hypothesised ancestral taxon. Cladogram internodes are called **lineages** or **clades** which represent ancestral-descendant sequences of taxa (Fig. 4.31). Any branching of the cladogram represents lineage divergens and the diversification of lineages are from one common ancestor. Hence, each of the bifurcation finally represents a past speciation in two separate lineages.

Plesiomorphy and apomorphy

The evolution of an individual or taxon can be recognised as a change from an ancestral character state to a new derived character state. The ancestral characters are called **plesiomorphies**, while derived characters are termed **apomorphies**. Plesiomorphies shared by two or more taxa are called **symplesiomorphies**, while shared derived characters are known as **synapomorphies**. In cladistics, only synapomorphous characters are used for inferring the pattern of evolutionary history of a group of taxa. Thus, symplesiomorphies are of no value in cladistic analysis. The apomorphy restricted to any taxon, is called **autapomorphy**. Two taxa which share the most recent common ancestor are called **sister groups.**

Concept of monophyletic, paraphyletic and polyphyletic groups

A cladogram serves as the basis of phylogenetic classification. The most important concept in cladistics is the **monophyletic groups**. A monophyletic group includes a common ancestor and all and only all descendants of that ancestor. In Fig. 4.31, the group including **A, B, C, D, E** and **F** (and common ancestor **aa**) is monophyletic taxon, identified by synapomorphy of character α. The monophyletic taxon of **B, C, D, E** and **F** (and common ancestor **bb**) is identified by the synapomorphy of character ε, similarly the taxon **D** is grouped with **E** and **F** by the synapomorphy of character π. The derived state of character ρ representing an evolutionary event unique for taxon **D** is **autamorphy**. Thus, taxon **E** is the sister group of **F** with the most recent common ancestor **ee**, similarly **B** is the sister group of **C** with the most recent common ancestor **cc**.

In contrast, **paraphyletic group** is one consisting of a common ancestor and some, but not all known descendants of that common ancestor. For example, a group including ancestor **aa** and the descendants **B, C, D, E** and **F** is paraphyletic, because it has excluded taxon **A** which is also a descendant of common ancestor **aa** (Fig. 4.31).

A **polyphyletic group** is one which has two or more ancestral sources. Thus, there are two or more separate groups, each having a separate common ancestor. For example, a group containing **C** and **D** alone could be designated as

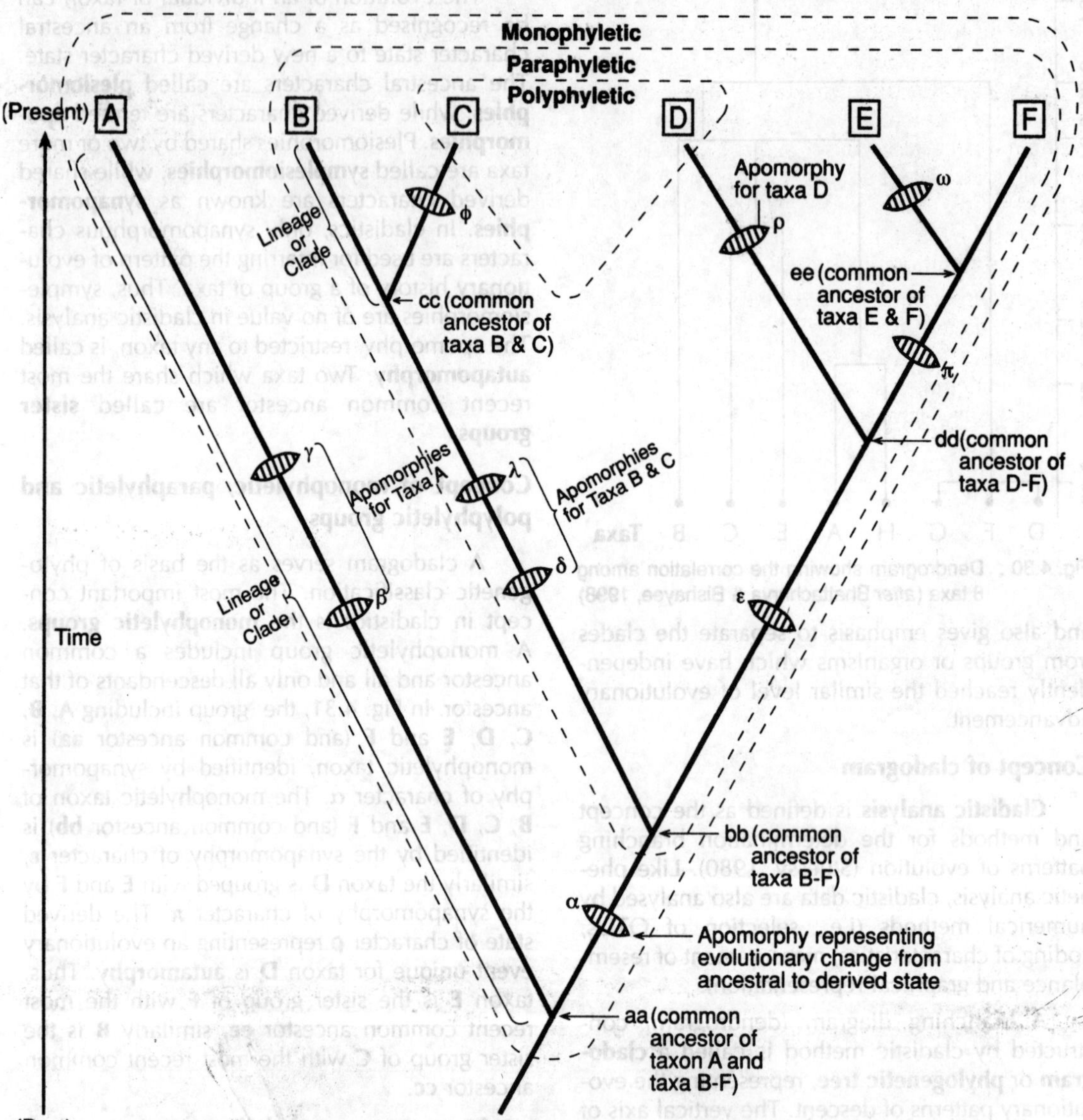

Fig. 4.31 : Cladogram of taxa A-F including common ancestor (aa-ee) plus hypothetic ancestor XYZ, showing lineages or clades and apomorphies (α, β, γ, δ, λ, φ, ε, ρ, π, and ω). [*after*, Simpson, 2006]

polyphyletic as these 2 taxa do not have a single common ancestor that is part of the group (Fig. 4.31).

Monophyletic and paraphyletic groups are not considered as natural evolutionary units and are thus abandoned in formal classification system.

Concept of Parsimony

Homology is defined as similarity between organisms resulting from common ancestry. **Homoplasy** (also termed **analogy**) is defined as similarity between organisms resulting from independent evolutionary origin, but not by common ancestry. Homoplasy can occur either by **convergence** or by a **reversal**. Convergence is the independent erolution of a similar feature in two or more different lineages. For example, sporophytic leaves of *Lycopodium* and gametophytic leaves of *Porella* evolved independently as photosynthetic appendages. Reversal is the loss of a derived feature with the replacement of an original ancestral condition. For example, the reduced flowers of *Lemna* lack the perianth by secondary loss, thus reverting to a condition prior to the evolution of a reproductive shoot having a perianth-like structure.

In constructing a cladogram, a single branching pattern is selected from among many possibilities. The number of possible dichotomously branching cladograms increases with a corresponding increase in the number of taxa. For two taxa, there is only one cladogram (Fig. 4.32A), for three taxa, three dichotomously branched cladograms can be constructed (Fig. 4.32B), for four taxa, 15 dichotomously branched cladograms are possible (Fig. 4.32C). There would be many possible trees for any given data set, and one of the major methods for reconstruction of phylogenetic relationship is the **Parsimony Analysis** or the **Principle of Parsimony**. The parsimony analysis states that of all possible cladograms for a given group of taxa, the one (or more) exhibiting the least number of convergence or reversals is accepted as being the best estimates of phylogeny. The principle of parsimony is actually the **Ockham's Razor** of phylogenetics which states that given more or more competing hypothesis, each of which can explain the facts, the simplest one is accepted. Thus, the parsimony analysis is the simplest explanation which minimises the requirement for **ad hoc hypothesis** (i.e., the hypothesis for which there is no direct evidence). Thus, of all possible cladograms for a given group of taxa, the one (or more) that minimises the total number of character state changes is accepted.

Say for example, for three taxa **a**, **b** and **c**, three dichotomously cladograms can be constructed (Fig. 4.33) From the results of the character x taxon matrix, the Fig. 4.33B with a minimum of four character state changes, is more parsimonious than that at Cladogram **C** or Cladogram **D**, each with a six apomorphies, two of which are either reversals (second cladogram of **C** and **D**) or pairs of convergences (first Cladogram of **C** and **D**) (Fig. 4.33).

The principle of parsimony is a valid working hypothesis and various computer algorithm (programme) are being used to determine the most parsimonious cladogram for a given character x taxon matrix.

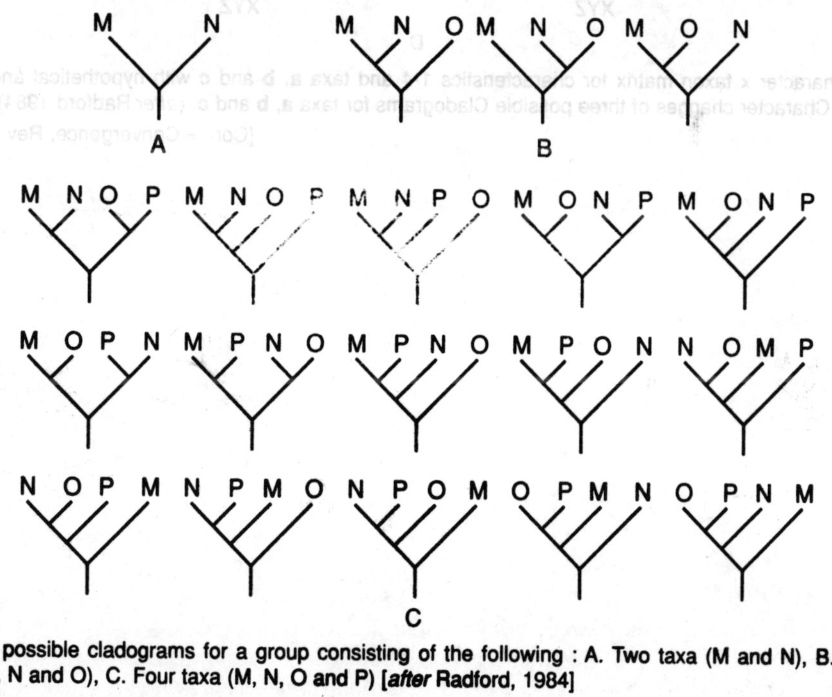

Fig. 4.32 : All possible cladograms for a group consisting of the following : A. Two taxa (M and N), B. Three taxa (M, N and O), C. Four taxa (M, N, O and P) [*after* Radford, 1984]

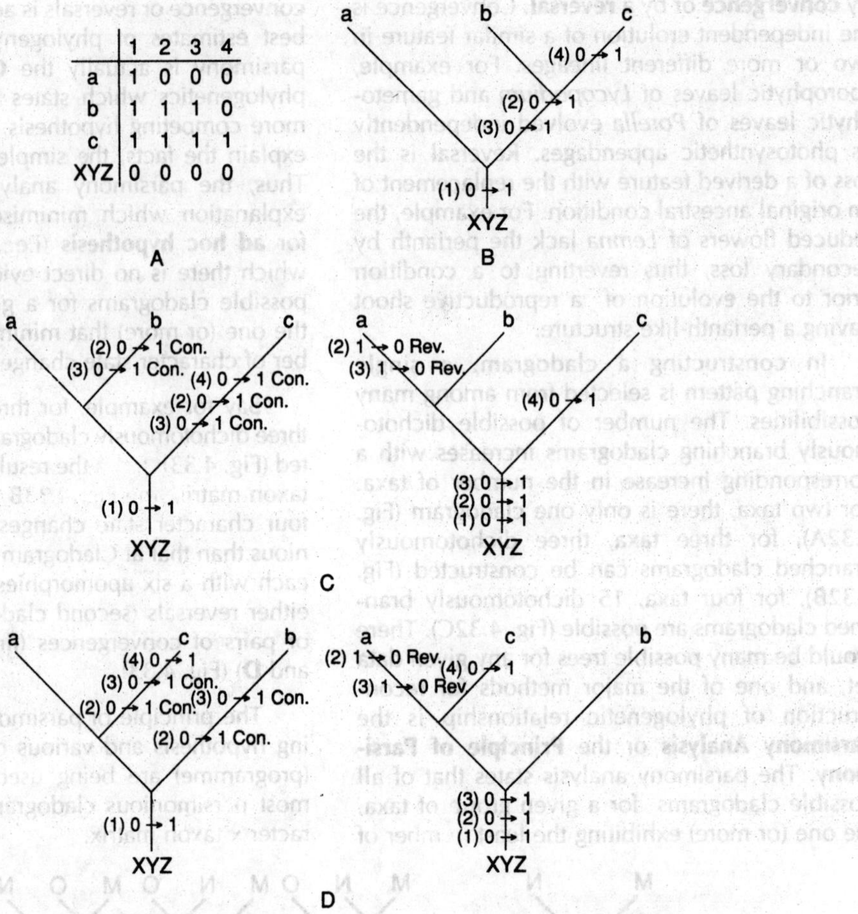

Fg. 4.33 : A. Character x taxon matrix for characteristics 1-4 and taxa **a**, **b** and **c** with hypothetical ancestor XYZ. B-D. Character changes of three possible Cladograms for taxa **a**, **b** and **c**. (*after* Radford 1984)

[Con. = Convergence, Rev. = Reversal]

ORIGIN OF ANGIOSPERMS

The angiosperms may be defined as a group of flowering plants whose seeds are enclosed within the carpel(s), with an upper extended region, the style and its apex, the stigma.

To consider the origin, the three parameters, namely *time, place* and the *nature of ancestor,* have to be taken into account. There are several opinions proposed regarding the time of origin :

a. Some morphologists considered their origin during Carboniferous or Permian period;
b. Some considered their origin during the 200-250 million years ago;
c. Some others considered them to be developed about 150 million years ago i.e., in Jurassic period;
d. Still some others considered them to be originated about 100 million years ago i.e., at the late Cretaceous period.

Due to lack of the fossil evidences it becomes very difficult to make valid comments on their origin. Eames (1961) and Takhtajan (1969) are of opinion that "angiosperms undoubtedly originated long before the Cretaceous period". Some pollen retrieved from the pre-Cretaceous deposits considered at one time being the pollen grains of angiosperms, which have been suggested later as unicellular green alga, *Tetrahedron.*

Many theories have been proposed to explain the ancestry of angiosperms :

1. ***Isoetes*-Monocotyledon theory**

Isoetales, Lycopodales, ferns etc. have shown to possess many common characteristics with monocotyledons. Engler and his associates postulated that the monocotyledons have originated from various groups of Pteridophyte through a hypothetical herbaceous intermediate, the *Proangiosperms.* According to Campbell (1928) the herbaceous angiosperms are primitive, those inherited from filicinean ancestor. He also pointed out that the *Isoetes,* a living heterosporous genus, shows possible relationship with monocotyledons. Like most of the members of Filicineae, the *Isoetes* grows in aquatic or amphibious environment indicating relationship with monocotyledons. Campbell suggested that though *Isoetes* shows similarity with monocotyledons like *Najas flexilis* with regard to habitat, anatomy of old sporophyte and resemblances in embryo, the differences in respect of simplest angiosperm flower and sporophylls of *Isoetes* need more evidence to support the hypothesis.

The Lycopodiales and Equisetales are now considered as completely separate and independent evolutionary lines.

The monocots are now considered to be more advanced than dicotyledons and their resemblance with *Isoetes* is superficial in nature.

2. **Conifer-Amentiferae theory**

According to this theory, the higher gymnosperms like Conifers, Cordaites and others are considered as the probable ancestors of angiosperms. Engler (1882, 1892) and Rendle (1904, 1930) found the similarities of angiosperms with the conifers and considered the amentiferous group (Casuarinaceae, Fagaceae, Salicaceae etc.) as the most primitive dicotyledons. The amentiferous members show the following feature resemblances with conifers :

i. Flowers are simple and naked like conifers;
ii. Stamens of amentiferae are like that of bisporangiate microsporophylls of conifers;
iii. Seeds are covered in angiosperms which have also been found in *Araucaria* and *Agathis* of gymnosperm;
iv. During fertilisation in *Agathis* and *Araucaria* and in other conifer, the pollens are deposited on the scale and only pollen tubes enter into the micropyle. This feature is very similar to angiosperm (Doyle, 1945).

It has now been proved that the above similarities are probably due to *parallelism* between the two groups. The simple flower present in the amentiferous taxa are not considered as primitive, but developed due to *reduction at higher level.*

The vessel-less angiosperms like *Tetracentron, Trochodendron* etc. have more *primitive secondary xylem* than the conifers and amentifers. The amentifers are now considered to be more specialised and not considered as primitive one.

It has now been established that the primitive angiosperm flower is bisexual, whereas the cones of conifers are unisexual.

Lastly, it can be said that this theory does not gain proper rooting, because of the modern concept that amentifers are proved definitely as advanced members of the angiosperms.

3. Gnetales-Angiosperm theory

Richard von Wettstein (1901) emphasised the close relationship between Gnetales and angiosperms, after modifying the Engler's system as mentioned in *Handbuch der Systematische Botanik*. Later on, Markgraf (1930) and Fagerlind (1947) boldly supported the above view. Fagerlind demonstrated the homology among the three genera of Gnetales (*Ephedra, Welwitschia* and *Gnetum*) and proposed that Gnetales and Pro-angiosperms evolved from a common ancestor.

Similarities between Gnetales and angiosperms :

i. Presence of two cotyledons;
ii. Vessels in the secondary wood;
iii. Venation – reticulate;
iv. Unisexual inflorescence (similar to catkin of many Amentifers);
v. In *Welwitschia*, the male flowers are actually bisexual formed by reduction of female organs;
vi. Stamens of *Ephedra, Welwitschia* and *Gnetum* are apparently similar to angiosperms;
vii. The female flowers of all the above three genera have 2 or more envelops around the nucellus like the perianth of angiosperms;
viii. The well-developed micropylar tube (formed by the elongation of the inner envelop of the ovule) of female flowers looks like the style of angiosperm flower;
ix. Female gametophyte of *Gnetum* is tetrasporic and without archegonia like angiosperms;
x. First division of zygote in *Gnetum* is accompanied by cell wall formation like angiosperms.

Differences between Gnetales and Angiosperms :

i. *Gnetum* have vessels in secondary wood, while it is absent in some primitive dicots like *Tetracentron, Trochodendron,* etc.;
ii. The vessels of Gnetales originate in an entirely different way from those of angiosperms (Thompson, 1918);
iii. The presence of scalariform pitted tracheids, a characteristic of primitive angiosperms, is absent in both primary and secondary xylem of *Gnetum*. While circular bordered pits appear in *Gnetum* during the ontogeny of secondary xylem;
iv. *Gnetum* also differs from angiosperms in respect of development of vessels (Cheadle, 1953).

Based on the above discussions, it can be said that the similarity between Gnetales and certain angiosperms is now considered as *superficial* and appears due to tilting towards convergence during evolution. Similarities in the number of cotyledons (two) also appear due to convergence. The angiosperms might have been evolved from polycotyledonous ancestors rather than dicotyledonous one. The polycotyledon (3-4 cotyledones) is reported in an extant genus *Degeneria*, a primitive group of angiosperm. The absence of archegonia in *Gnetum* indicates its affinity with angiosperms due to *parallelism*. But the vascular anatomy strongly opposes any suggestion regarding the relationship between angiosperms and Gnetales.

4. Bennettitalean theory

According to Saporta and Marion (1885), and Arber and Parkin (1907), the Bennettitales of gymnosperm is the possible ancestor of angiosperms. They considered Benettitales as the possible ancestor of angiosperms due to similarities between the **strobili** of *Cycadeoidea*, a Mesozoic genus and the **flower** of *Magnolia*.

Similarities :

i. Both the structures (strobili and flower) are bisexual having an elongated central axis;

ii. In *Cycadeoidea* the different parts like **bracts, microsporophylls** and **megasporophylls** are arranged from the bottom towards the apex, while in *Magnolia*, perianth, stamens and carpels are arranged in a similar way.

In spite of the above superficial similarities, both the groups show much differences by the detail investigation. The differences are :

Differences :

i. **Stem** : Benettitalean stems have a thick cortex, a comparatively thin vascular cylinder and a large pith, but the angiospermic plants have thin cortex, a thick vascular cylinder and a small pith.

ii. **Microsporophylls** (Stamens) : In Benettitales, the microsporophylls are arranged in whorled and mostly connate, but in *Magnolia* the stamens are free and spirally arranged on the axis.

iii. **Megasporophylls** (Carpels) : In Benettitales, the megasporophylls are greatly reduced having a simple stalk on which a single terminal ovule is present. Interseminal scales (sterile scales) are present in between megasporophylls for protection. The flower of *Magnolia* does not show such structures.

iv. **Micropylar tube** : The micropylar tube is present in Benettitales, but such structure is unknown in angiosperms.

v. **Seed** : The seeds of Benettitales are exalbuminous i.e., non-endospermic having a large embryo, but the seeds of primitive angiosperms are albuminous i.e., endospermic and with a small embryo.

The above facts show that Benettitales is not in a position to be considered as the ancestor of angiosperms. Both the groups show much difference than the similarities. The above similarities may result due to common ancestry or parallel evolution. According to Arber and Parkin (1907), both the groups have a common origin from seed ferns and they might have diversed very early. Later, Takhtajan (1980) mentioned in his classification that Magnoliophyta are regarded to be monophyletic in origin, most probably derived from Bennettitale-like ancestors or stocks ancestral to them.

5. Caytonialean theory

The Caytonialean theory was proposed by H. Hamshaw Thomas (1925, 1936). His proposition was based on comparative studies of angiosperms with the certain member of Caytoniales (the fossil members of middle Jurassic period).

Some fossil members of Caytoniales had anthers developed singly or in groups on pinnately branched structures that are described as sporophylls (Fig. 4.34). The above structures can be compared with the branched stamens as found in plants like *Ricinus* (Euphorbiaceae), *Calothamnus* (Myrtaceae) and *Hypericum* (Clusiaceae). He also explained how angiosperms originate from Caytoniales. According to him, the carpel wall of angiospermic plants represents a pair of concrescent cupules and the probable origin of stigma should be judged in the light of these primitive forms.

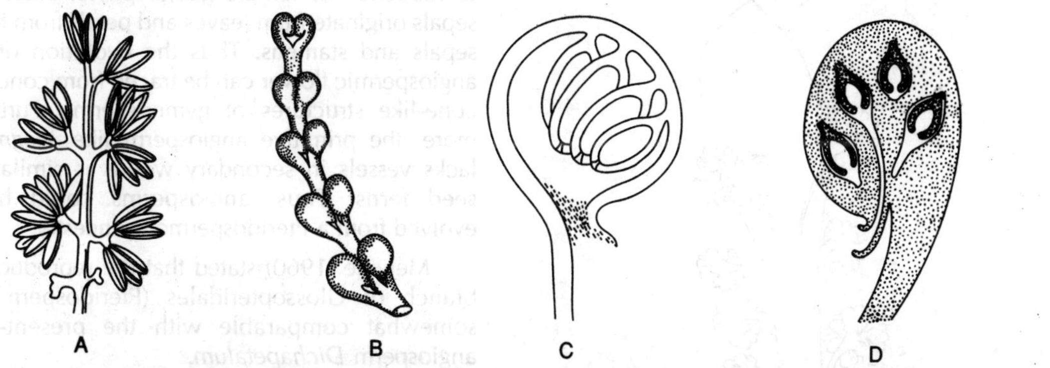

Fig. 4.34 : *Caytonia* : A. Microsporophyll of *Caytonia sewardi* called *Caytonanthus* (*Antholithus*) *arberi*, B. *Caytonia nathorsti* (megasporophyll with two rows of cupules, C. Vertical L. S. of *Caytonia sewardi* cupule showing seeds inside, D. Vertical L. S. of fleshy cupule fruit of *Caytonia thomasi*

The Caytoniales show affinity with the Pteridosperms and are now considered as Mesozoic remnants of that group. Based on the morphological studies of the nature of carpel, it can be said that their ancestor should have open megasporophyll with exposed ovule, thus the ovary-like pouches of the form-order Caytoniales cannot be considered as the ancestor of the angiosperm carpel.

Later on, Stebbins (1974) supported the theory based on homology of ovules of the two groups : angiosperms and Caytoniales. According to him, ovule of angiosperm is homologous with the cupule of Caytoniales including both Caytoniaceae and Corystospermaceae (Thomas, 1933, 1955). The anatropous ovule with well-developed funiculus can be compared with the cupule with the stalk. In Caytoniaceae, the cupules are multiovulate and dehiscent, but in Corystospermaceae the cupules are uniovulate and indehiscent. If one considers the cupule of Corystospermaceae to be homologous with the angiosperm ovule, then the outer integument becomes homologous with the cupule wall and the inner one with the single integument of *Caytonia* ovule as well as with the ovule of other gymnosperms.

The lateral ovules on flattened sporophylls of angiosperms show a different situation in contrast to the terminal position of cupule on branched axes in Caytoniales. According to Stebbins, the above situation can be explained on the basis of structure of *Ligettonia* (Fig. 4.35), a fossil member (Thomas, 1958) of Permian age reported from South Africa in association with *Glossopteris* leaves. It was proposed that a branched cupule-bearing rachis became adnate to its subtending bract. Thus the carpel is considered as a compound structure alike to the ovule bearing cone-scale of conifers (Florin, 1944). In *Ligettonia*, the flattened and elliptical sporophyll likely bears two rows of dehiscent cupules on its surface.

Fig. 4.35 : *Ligettonia* : Fertile branch with cupules

Thus, according to this theory, the predecessor (forefather) of angiosperms were comparatively recent pteridosperms belonging to Glossopteridalean-Caytonialean alliance available during Jurassic to Permian Age.

6. **Pteridosperm theory**

The Pteridosperms are also known as Cycadofilicales or seed ferns. They are so called because the plants show fern-like leaves; some of them bears leaves associated with cycad-like stem showing difference in stem anatomy with ferns. They bear true seeds. They were reported abundantly from Upper Devonian to Permian of Paleozoic Age. Plants are often monoecious, but micro- and megasporophylls are not arranged in definite strobili. Based on the above characteristics of stem, leaf and seed, they are considered as the probable ancestors of Bennettitales.

The various evidences and interpretations discussed the possibility of Cycadales, Bennettitales or any other gymnosperms to be the ancestor of angiosperms. Therefore, emphasis was given to the Pteridosperms as the possible ancestor. Cronquist (1968) stated that "it is a long way, morphologically, from any known seed fern to an angiosperm, but each of the differences could logically be bridged in the course of evolution". The angiosperms show double fertilisation with triploid (3n) endosperm, possibly due to reduction of female gametophyte. Likewise, sepals originate from leaves and petals from both sepals and stamens. Thus the evolution of an angiospermic flower can be traced from cones or cone-like structures of gymnosperms. Furthermore, the primitive angiosperm like *Magnolia* lacks vessels in secondary woods is similar to seed ferns. Thus, angiosperms could have evolved from a Pteridospermous ancestry.

Melville (1960) stated that the reproductive branch of Glossopteridales (Pteridosperm) is somewhat comparable with the present-day angiosperm *Dichapetalum*.

The external similarity of Bennettitales with angiosperms can also be explained in terms of common ancestry of both these groups in the

Pteridosperms. Takhtajan (1969) stated that the "angiosperms arose from some very ancient group of gymnosperms, which must have had primitive secondary xylem of scalariform tracheids at least in the early wood, and primitive bisexual strobili. The strobili must have been of a type that could have diversed to give rise to the primitive Bennettitalean strobilus and the primitive angiosperm flower".

7. **Pentoxylales theory**

The Pentoxylales is a group of fossil under Pteridosperms. Meeuse (1961) proposed that angiosperms have originated from Pentoxylales. He found several similarities between Pentoxylales and the living order Pandanales.

The characteristics common with the genus *Pandanus* are :

a. Plants are erect or suberect and dioecious;
b. Stem slender and cylindrical;
c. Stem bears a terminal tufts of strap-shaped leaves arranged spirally;
d. Inflorescences develop either on short lateral branches in the axil of leafy bract or among the leaves;
e. The peduncle have tracheids with spiral thickening;
f. In Pentoxylales, the seeds are enclosed in sarcotesta, a fleshy layer; seed coat consists of two layers, endosperm copious and embryo minute. All the characteristics are found in the seeds of *Pandanus*.

The above characteristics indicate the possible origin of angiosperms from Pentoxylales (Fig. 4.36).

Later, Pant and Kidwai (1971) stated that the above similarity appears as a result of parallel evolution.

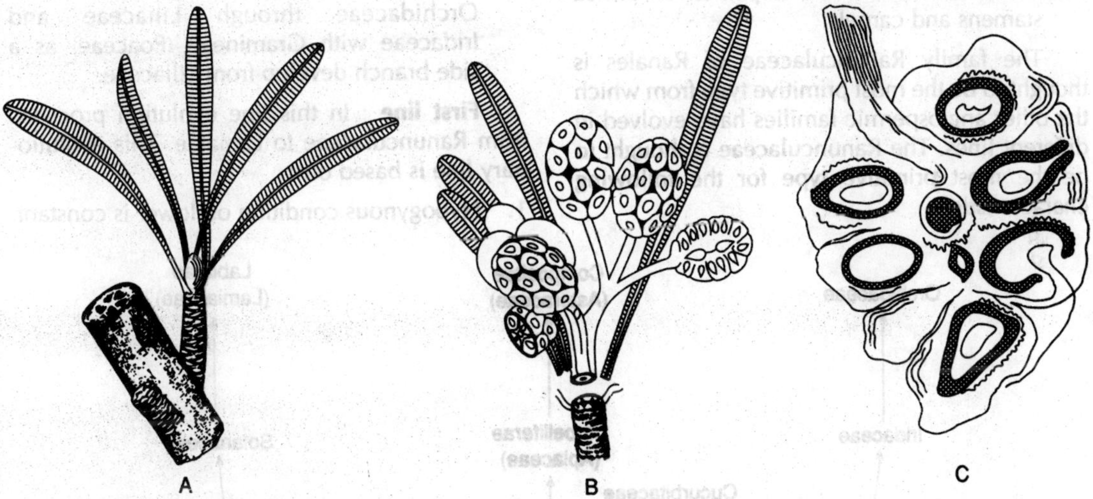

Fig. 4.36 : *Pentoxylon* : A. Leafy shoot (after Sahni), B. Female shoot (*Carnoconites*), C. L.S. of female cone

EVOLUTION OF ANGIOSPERMS

The angiosperms are thought to represent the climax of an extensive evolution and it would appear that within this group evolution has been still a continuous process.

The principal evolutionary tendencies of the flowers are :

1. Elongated floral axis to shortened floral axis.
2. Spiral arrangement of floral parts to whorled arrangement.
3. Complete and perfect flowers to incomplete and imperfect flowers.
4. Regular flowers to irregular flowers.
5. Hypogynous flowers to epigynous flowers.
6. Many free sepals and petals to few united sepals and petals.
7. Many free stamens and carpels to few united stamens and carpels.

The family Ranunculaceae of Ranales is thought to be the most primitive type from which the other angiospermic families have evolved in different lines. The Ranunculaceae is thought to be the most primitive type for the following characteristics :

1. The floral parts i.e., calyx, corolla, stamens and carpels, are numerous and arranged in close spirals on the receptacle.
2. In most cases the floral parts are not at all fused together.

From the Ranalian type of flower, the evolution is thought to precede in at least 3 main lines (Fig. 4.37) :

1. *First line* : This line culminates in Labiatae (Lamiaceae) through Malvaceae and Solanaceae with Cruciferae (Brassicaceae) and Salicaceae as side branches.
2. *Second line* : This line culminates in Compositae (Asteraceae) through Rosaceae and Umbelliferae (Apiaceae) with Leguminosae (Fabaceae) and Cucurbitaceae as side branches.
3. *Third line* : This line culminates in Orchidaceae through Liliaceae and Iridaceae with Gramineae (Poaceae) as a side branch develop from Liliaceae.

First line : In this line evolution proceeds from Ranunculaceae to Labiatae. This evolutionary line is based on :

1. Hypogynous condition of flower is constant.

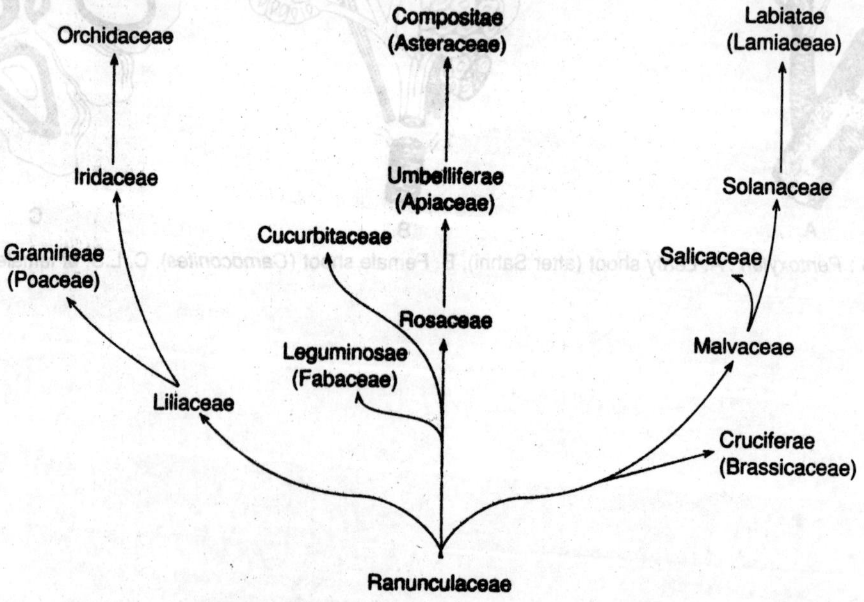

Fig. 4.37 : Evolutionary tendencies in the angiosperms

2. Presence of syncarpous ovary as well as the number of floral parts are the early phenomena.
3. Other type of union and the shift to irregular flowers have appeared somewhat later, but still early in this line.

The line was traced through different families based on these characteristics :

Malvaceae :

i. This family has comparatively primitive characteristics like actinomorphic, complete and perfect flowers.
ii. Stamens α and monadelphous, carpels 5-more and syncarpous.

Cruciferae (Brassicaceae) :

i. The members show the reduction in the number of floral parts and only the carpels are united.

Solanaceae :

i. Flowers are regular and perfect which show union of sepals, petals and carpels.
ii. Stamens epipetalous.

Salicaceae : It is a comparatively advanced family and shows the following characteristics :

i. Reduction in both the size and number of floral parts.
ii. Calyx and corolla are absent and the flowers are arranged in catkins.
iii. Flowers are unisexual.
iv. Male flower : Number of stamens has been reduced to 1 or 2.
v. Female flower : Hypogynous condition is still continuing.
vi. Pollination takes place by wind rather than insect.

Based on the above the Salicaceae is considered as a side branch much advanced than Malvaceae. Finally the Labiatae is considered as the end point in this line based on the following characteristics :

i. Flowers are irregular but remain hypogynous.
ii. Flowers show union in sepals, petals and carpels.
iii. Epipetalous condition of stamens.
iv. Flowers are specially adopted for insect pollination.

Second line : In this line, evolution proceeds from Ranunculaceae to Compositae (Asteraceae). Hypogyny to epigyny is noticed during the early part of evolution. The main line is divided into two branches. One branch shows actinomorphy without coalescence and other one shows zygomorphy with coalescence.

Rosaceae : A primitive family in this line shows much variation in floral structure. These are :

i. Members of the comparatively primitive group shows, regular, perfect and slightly perigynous flower with numerous parts.
ii. Other members show gradual coalescence, reduction in the number of floral parts and also from slightly perigynous condition to typical perigynous condition.

Sidelines are considered to be developed from the nearby region of Rosaceae. These are Leguminosae (Fabaceae) and Cucurbitaceae. This consideration is based on the following characteristics of the two families :

Leguminosae :

i. Reduction in the number of floral parts.
ii. Union of some petals and corolla parts, thus became irregular.
iii. Union of stamens.
iv. Gynoecium with solitary carpel is commonly hypogynous, but perigyny is also visible.

Cucurbitaceae :

i. Flowers imperfect.
ii. Epigyny in female flower.
iii. Union of different floral parts.

The line of evolution to be progressed further from Rosaceae to Umbelliferae is based on the following characteristics :

Umbelliferae :

i. Flowers epigynous.
ii. Union of carpels i.e., syncarpy.

Finally, the Compositae (Asteraceae) is considered as the culminating point for the following characteristics :

Compositae :

i. Reduction in the number of floral parts.
ii. Flowers imperfect.
iii. Modification of calyx into pappus in most cases.
iv. Union of each whorl and adnation.

Third line : In this line, evolution proceeds from Ranunculaceae to Orchidaceae. The primitive monocotyledons, like the members of Alismataceae, are very much common with Ranunculaceae. The line of evolution to be progressed further to Liliaceae is based on the following characteristics :

Liliaceae :

i. Reduction in the number of floral parts.
ii. Union of carpels.

The line of evolution that progressed further to Iridaceae is based on the following characteristics :

Iridaceae :

i. Flowers show epigyny.

Finally the Orchidaceae is considered as the culminating point for the following characteristics :

Orchidaceae :

i. Flowers are irregular.
ii. Flowers are showy and highly specialised for insect pollination.

The family Gramineae that evolved from Liliaceae as a side line is based on the following characteristics :

Gramineae :

i. Loss of perianth members.
ii. Reduction in the number of stamens and carpels.

MAJOR FAMILIES OF ANGIOSPERMS

DICOTYLEDONS

1. RANUNCULACEAE

(Buttercup, Crowfoot family)

The family is represented by 50 genera and more than 1,900 species found to occur mainly in the North Temperate Zone. In India, 20 genera and 154 species are found to grow mostly in the hilly regions.

The type genus of the family is ***Ranunculus***.

Systematic position

Bentham & Hooker

Division. Angiospermia
Class. Dicotyledons
Subclass. Polypetalae
Series. Thalamiflorae
Cohort. Ranales

Engler & Prantl

Subdivision. Angiospermae
Class. Dicotyledoneae
Subclass. Archichlamydeae
Order. Ranales

Hutchinson

Phylum. Angiospermae
Subphylum. Dicotyledones
Division. Herbaceae
Order. Ranales

Cronquist

Division. Magnoliophyta
Class. Magnoliopsida
Subclass. Magnoliidae
Order. Ranunculales

Takhtajan

Division. Magnoliophyta
Class. Magnoliopsida
Subclass. Ranunculidae
Superorder. Ranunculanae
Order. Ranunculales.

Thorne

Subclass. Magnoliidae
Superorder. Magnolianae
Order. Barberidales

Identifying characteristics

1. Plants are **generally herbs**.
2. **Leaves radicle** or **cauline**, alternate, exstipulate, **with sheathing leaf base and incised blades.**
3. Flowers **solitary** or in **racemose** or **cymose** inflorescence; bisexual, hypogynous and commonly actinomorphic.
4. **Stamens usually many, free, spirally arranged, anthers extrorse.**
5. Carpels few to many, **apocarpous** (carpels free), placentation **parietal**.
6. Fruit aggregate of **follicles** or **achene.**

Floral formula

$\oplus$ ⚥ $K_5C_5A_{5 \text{ or } \alpha}\ G_{1\text{-}5 \text{ or } \alpha}$

Ranunculaceae : Distinctive features

Common characteristics	Exceptions
Vegetative characteristics	
Plant. Generally annual or perennial herbs, occasionally shrubs (*Paeonia*), often growing in marshy places.	Woody climber – *Naravelia, Clematis, Atragene.* Aquatic – *Ranunculus aquatilis.*
Leaves. Simple and palmately lobed, radical or alternate, exstipulate, often with twisted sheathing leaf base. (Aquatic species shows heterophylly with submerged leaves dissected very finely.)	Pinnately compound – *Clematis.* Decompound – *Thalictrum.*
Reproductive characteristics	
Inflorescence. Cymose or racemose.	Solitary and terminal – *Nigella, Anemone.* Branched panicle – *Clematis, Thalictrum.*
Flower. Bisexual, regular (actinomorphic), hypogynous, hemicyclic or acyclic.	Unisexual – Species of *Thalictrum.* Zygomorphic – *Delphinium, Aconitum.*

Contd.

Distinctive features (Contd.)

Common characteristics	Exceptions
Calyx. Sepals 5-8, free, usually deciduous, imbricate or valvate.	Sepals petaloid, irregular (posterior sepal is hood-like or sepaloid) – *Aconitum, Delphinium.*
Corolla. Petals usually 5, free, may be spurred and store honey (*Aquilegia*).	Petals absent – *Clematis, Caltha* and *Anemone.*
Androecium. Stamens — many, free, spirally arranged, anther extrorse (Nectaries usually present, formed by the modification of petals).	Stamens arranged in a definite ring – *Nigella, Aquilegia* and *Helleborus.*
Gynoecium. Carpel many, free, ovary — superior, unilocular and has one to many anatropous ovules on parietal placentation along the ventral suture; style-1, stigma-1.	Carpel – 1 in *Cimicifuga,* Carpel – 3 in *Delphinium,* Carpel – 5 in *Aquilegia.* Syncarpous – *Nigella,* Carpels connate at the base – *Helleborus.* Placentation : Basal – *Ranunculus,* Apical – *Clematis,* Marginal – *Delphinium.* Stigma 5 to 12 – *Nigella.*
Fruit. Few to many seeded follicle (*Caltha, Delphinium* etc.) or 1-seeded achene (*Ranunculus* etc.).	Capsule – *Nigella,* Berry – *Actaea.*
Seed. Seeds with copious oily endosperm and single small embryo.	

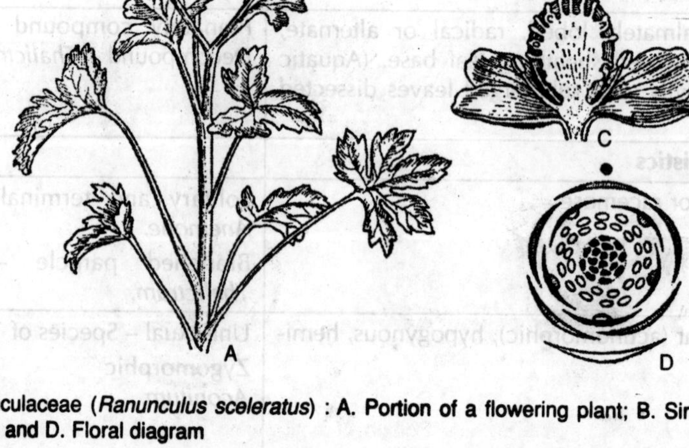

Fig. 4.38 : Ranunculaceae (*Ranunculus sceleratus*) : A. Portion of a flowering plant; B. Single flower; C. V.S. of a flower, and D. Floral diagram

Phylogeny and systematics

The family shows affinity with both monocotyledons (Alismataceae, Hydrocharitaceae etc.) and dicotyledons (Magnoliaceae, Berberidaceae etc.).

Many taxonomists consider the family as primitive one of Dicotyledons. They are considered as primitive family because of the following characteristics :

1. Herbaceous habit,
2. Bisexual and actinomorphic flower,
3. Stamens numerous, free and spirally arranged,
4. Carpels numerous, free and spirally arranged,
5. Ovule anatropous,
6. Fruit follicular.

The advanced characteristics of this family are :

1. Presence of unisexuality,
2. Few and whorl appendages,
3. Zygomorphic flower,
4. Connation in gynoecium, and
5. Fruit-achene.

Bentham and Hooker placed the Natural order Ranunculaceae in the cohort (= order) Ranales of series I Thalamiflorae under subclass Polypetalae and class Dicotyledons.

Engler did not consider the family as primitive over the amentiferous monochlamydous families and placed it in more advanced order after the 17th order Centrospermae under subclass Archichlamydeae of class Dicotyledoneae.

Hutchinson placed the family in the order Ranales under the division II, Herbaceae of subphylum Dicotyledones.

Bentham and Hooker divided the Natural order Ranunculaceae into 5 tribes, Clematideae, Anemoneae, Ranunculeae, Helleboreae and Paeonieae. Engler and Prantl and many others have considered only 3 tribes, Helleboreae, Anemoneae and Paeonieae. The tribe Helleboreae have *more than one ovule in a carpel* and have *follicular fruits*, thereby Hutchinson (1959, 69) separated it from Ranunculaceae and raised as a family Helleboraceae.

The genus *Paeonia* differs from other members of Ranunculaceae because of the following characteristics :

1. Numerous centrifugal stamens,
2. Presence of disc around the carpels and
3. Presence of arillate seeds.

For the above, Worsdell (1908) separated *Paeonia* and created a separate unigeneric family

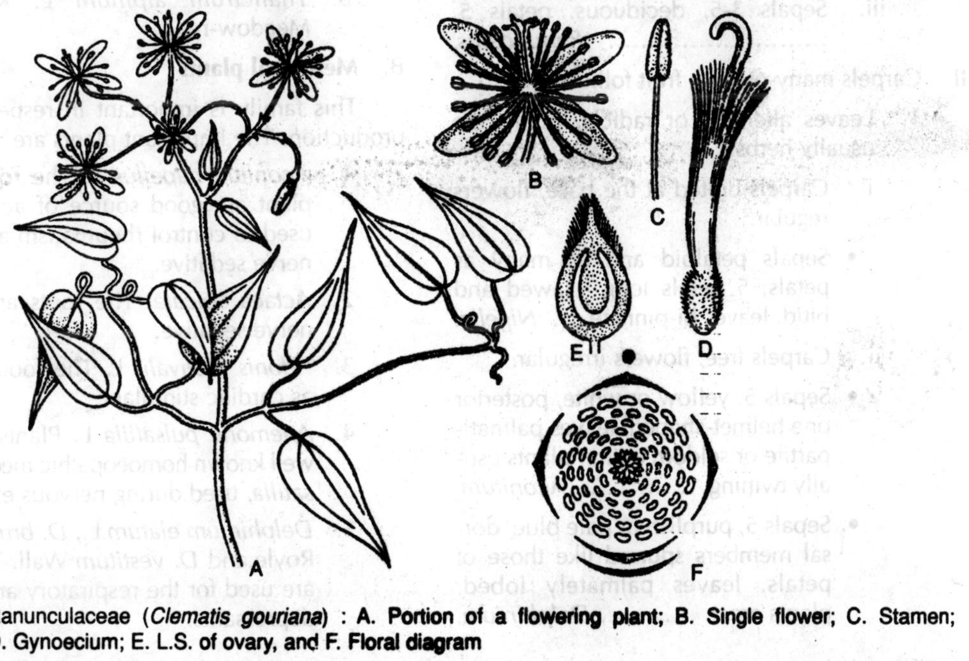

Fig. 4.39 : Ranunculaceae (*Clematis gouriana*) : A. Portion of a flowering plant; B. Single flower; C. Stamen; D. Gynoecium; E. L.S. of ovary, and F. Floral diagram

Paeoniaceae. Later, this was recognised by Corner (1946) and many other recent workers like Hutchinson (1959, 69), Takhtajan (1969, 80) and Cronquist (1968, 81).

Common plants

1. *Clematis gouriana* Roxb.;
2. *Naravelia zeylanica* DC.;
3. *Nigella sativa* L.;
4. *Ranunculus sceleratus* L.;
5. *Thalictrum javanica* Bl.

Key to genera

I. Carpel 1–ovuled, fruits achene;
   1. Leaves opposite, compound; plants are usually climbing.
      i. Sepals 4, petaloid; petals and tendril absent. .... *Clematis.*
      ii. Sepals as many as petals–4 or 5, third or terminal leaflet modified into tendril. .... *Naravelia.*
   2. Leaves alternate or radicle; plants are usually herbs.
      i. Sepals petaloid, petals absent, flowers involucrate. .... *Anemone.*
      ii. Sepals petaloid, petals absent, flowers non-involucrate. .... *Thalictrum.*
      iii. Sepals 3-5, deciduous, petals 5. .... *Ranunculus.*

II. Carpels many-ovuled, fruit follicles;
   1. Leaves alternate or radicle, plants are usually herbs.
      i. Carpels united at the base, flowers regular.
         - Sepals petaloid and as many as petals; 5 petals long, clawed and bifid, leaves bi-pinnate. ... *Nigella.*
      ii. Carpels free, flowers irregular.
         - Sepals 5, yellow or white, posterior one helmet-shaped, leaves palmatipartite or seldom entire, plants usually twining. .... *Aconitum.*
         - Sepals 5, purplish or pale blue, dorsal members spurred like those of petals, leaves palmately lobed, plants erect. .... *Delphinium.*

Plants of economic importance

The family has a number of ornamental plants, medicinal plants. Seeds of some are used as condiments. These are :

A. **Ornamental plants**

The following are the plants, used in gardening because of their attractive flowers :

1. *Aconitun bicolor* Schult. and *A. napellus* L. known as Aconite.
2. *Anemone sylvestris* L. and *A. parviflora* Michx. These are known as Wind flower.
3. *Aquilegia vulgare* L. is known as Columbine.
4. *Caltha palustris* L. Commonly known as Marsh marigold.
5. *Clematis vitaba* L. and *C. paniculata* Thunb. These are commonly known as Virgin-bowers.
6. *Delphinium cashmerianum* Royle and *D. ajacis* L. Commonly known as Larkspur.
7. *Nigella damascene* L. is known as Love-in-a-mist.
8. *Paeonia emodi* Wall. Commonly called Peony.
9. *Ranunculus asiaticus* L. and *R. repens* L. Commonly known as Buttercups.
10. *Thalictrum alpinum* L. Known as Meadow-rue.

B. **Medicinal plants**

This family is important in respect of drug production. The important plants are :

1. *Aconitun napellus* L. The roots of this plant are good source of aconite. It is used to control rheumatism and also as nerve sedative.
2. *Actaea spicata.* The roots are used as nerve sedative.
3. *Adonis aestivalis* L. The roots are used as cardiac stimulant.
4. *Anemone pulsatilla* L. Plants yield the well known homoeopathic medicine *Pulsatilla*, used during nervous exhaustion.
5. *Delphinium elatum* L., *D. brunonianum* Royle.and *D. vestitum* Wall. The plants are used for the respiratory and cardiac depressants.

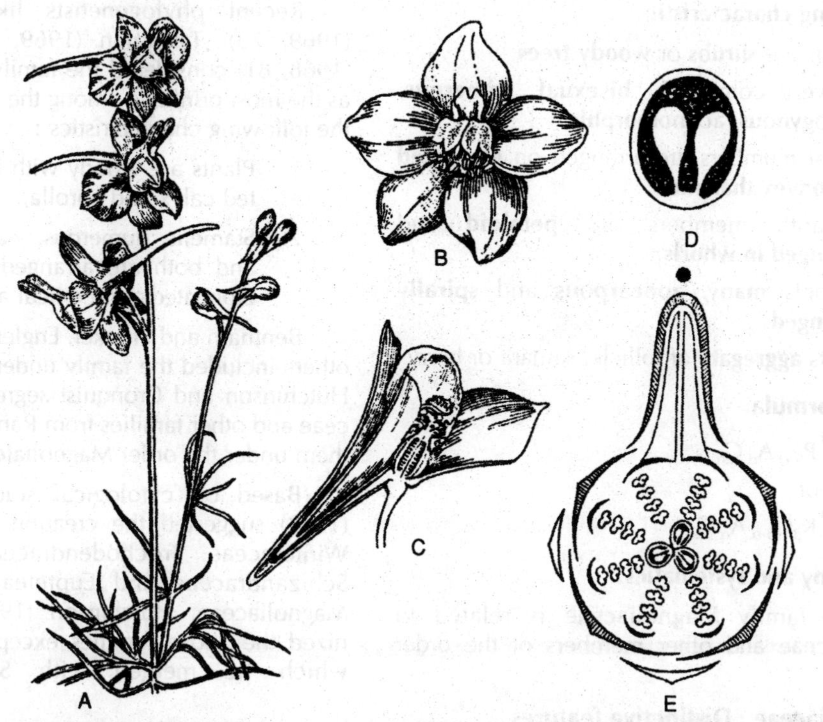

Fig. 4.40 : Ranunculaceae (*Delphinium ajacis*) : A. Portion of a flowering plant; B. Single flower; C. V.S. of a flower; D. T.S. of ovary and E. Floral diagram

6. *Helleborus niger* L. The plant is used to control worms and skin infections.
7. *Hydrastis canadensis* L. (Orange root). Roots of the plant are used as an antidote for snake bite.
8. *Nigella sativa* L. (Kala jeera, Black fennel). The plants are used in the treatment of bronchial asthma and cough.
9. *Ranunculus muricatus* L. and *R. arvensis* L. The juice of the plants is used to control intermittent fever.

C. **Condiments**

1. *Nigella sativa* L. (Kala jeera, Black fennel). The seeds are used as condiments in cooking.

2. MAGNOLIACEAE
(Magnolia Family)

The family Magnoliaceae is represented by 7 genera and 230 species found to grow in tropical region of both Asia and America and also extends to North Temperate Zone. The family is represented in India by 8 genera and about 27 species found to grow mainly in the north temperate regions of Himalayas.

The type genus of the family is ***Magnolia***.

Systematic position

Bentham & Hooker		**Engler & Prantl**	
Division.	Angiospermia	Subdivision.	Angiospermae
Class.	Dicotyledons	Class.	Dicotyledoneae
Subclass.	Polypetalae	Subclass.	Archichlamydeae
Series.	Thalamiflorae	Order.	Ranales
Cohort.	Ranales.		

Hutchinson		**Cronquist**	
Phylum	Angiospermae	Division.	Magnoliophyta
Subphylum	Dicotyledones	Class.	Magnoliopsida
Division.	Lignosae	Subclass.	Magnoliidae
Order.	Magnoliales	Order.	Magnoliales.

Takhtajan		**Thorne**	
Division.	Magnoliophyta	Class.	Angiospermae
Class.	Magnoliopsida	Subclass.	Magnoliidae
Subclass.	Magnoliidae	Superorder.	Magnolianae
Superorder.	Magnolianae	Order.	Magnoliales
Order.	Magnoliales.		

Identifying characteristics

1. Plants are **shrubs** or **woody trees**.
2. Flowers commonly **bisexual, trimerous, hypogynous, actinomorphic**.
3. Floral members are arranged on **elongated** or **convex thalamus**.
4. Perianth members are **petaloid** and **arranged in whorls**.
5. Carpels **many, apocarpous** and **spirally arranged**.
6. Fruits **aggregate** of follicle, samara or berry.

Floral formula

$\oplus$ ⚥ $P_{9-\alpha} A_{\alpha} \underline{G}_{\alpha}$

or

$\oplus$ ⚥ $K_3 C_{6-\alpha} A_{\alpha} \underline{G}_{\alpha}$

Phylogeny and systematics

The family Magnoliaceae is related to Annonaceae and other members of the order Ranales.

Recent phylogenetists like Hutchinson (1969, 73), Takhtajan (1969, 80), Cronquist (1968, 81) considered the family Magnoliaceae as the most primitive among the angiosperms for the following characteristics :

1. Plants are woody with little differentiated calyx and corolla,
2. Stamens numerous, carpel numerous and both are arranged spirally on an elongated and conical axis.

Bentham and Hooker, Engler and Prantl and others included the family under Ranales. Later, Hutchinson and Cronquist segregate Magnoliaceae and other families from Ranales and placed them under the order Magnoliales.

Based on cytological studies, Whittaker (1933) suggested the creation of the families Winteraceae, Trochodendraceae, Illiciaceae, Schizandraceae and Eupteleaceae from the Magnoliaceae. Hutchinson (1959, 69) recognized the above families except Eupteleaceae, which he merged with Schizandraceae.

Magnoliaceae : Distinctive features

Common characteristics	Exceptions
Vegetative characteristics	
Plant. Trees or shrubs, usually evergreen.	Woody climber – *Schizandra*.
Leaf. Simple, petiolate, alternate, mostly entire with bud scales or stipules covering the bud, which is early deciduous.	Stipules absent – *Drimys, Schizandra, Illicium* etc.
Reproductive characteristics	
Inflorescence. Usually solitary, axillary or terminal.	Flowers crowded near the apex – *Illicium*.
Flowers. Showy, aromatic, actinomorphic, bisexual, hypogynous with an elongated floral axis on which floral members are arranged spirally.	Flowers unisexual – *Schizandra, Drimys*.
Perianth. Tepals 9 – many, petaloid or differentiated into calyx and corolla, deciduous, arranged spirally or in whorls.	Perianth absent – *Euptelea*.
Androecium. Numerous, free, spirally arranged on the lower part of floral axis called androphore, connective of anther not projecting. Anthers are adnate or basifixed, introrse.	Connective extends beyond the anther lobe – *Illicium*. Anther extrorse – *Liriodendron, Michelia*.
Gynoecium. Carpels many, free, arranged spirally on floral axis above the stamens. Ovary superior, 1-chambered with 2 to many anatropous ovules on marginal placentation.	Carpel fused (syncarpous) – *Zygogynum, Pachylarnax*. Carpels (5-20) arranged in a single whorl – *Drimys* and *Illicium*.
Fruit. Mostly etaerio of follicle.	Etaerio of berries – *Schizandra*.
Seed. Embryo small with copious fleshy and oily endosperm.	Samaras – *Liriodendron*.

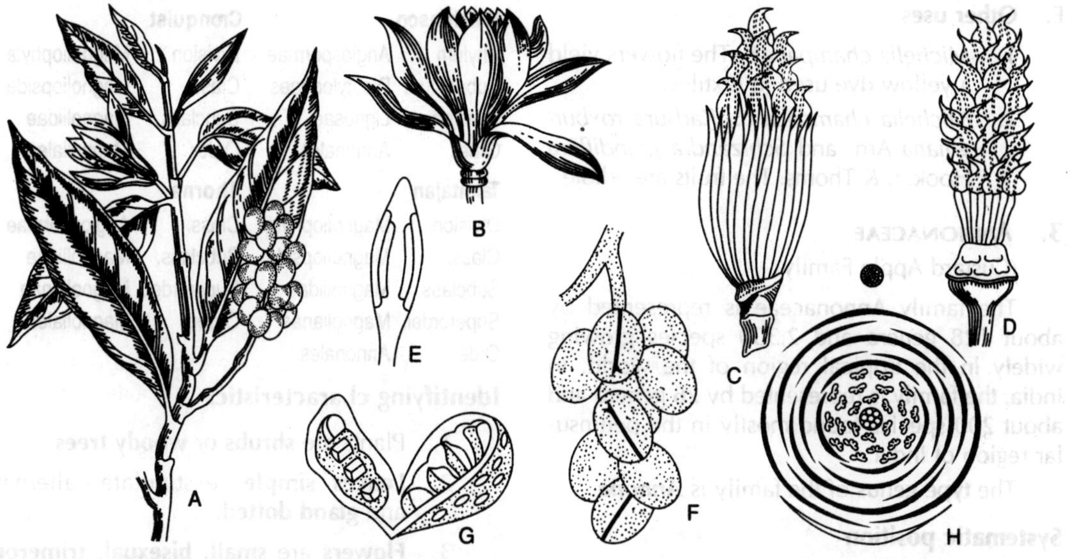

Fig. 4.41 : Magnoliaceae (*Michelia champaca*) : A. Portion of a flowering plant; B. A single flower; C. Flower after removing the perianth; D. Carpel on floral axis; E. A single stamen with broad filament; F. Aggregate fruit; G. Split follicle and H. Floral diagram

Cronquiest (1968) and Takhtajan (1969) included Winteraceae and Magnoliaceae under Magnoliales and Illiciaceae and Schizandraceae in a new order Illiciales derived from the Magnoliales.

Common plants

1. *Magnolia grandiflora* L.;
2. *M. montana* Bl.;
3. *Michelia champaca* L.
4. *M. doltsopa* Buch. - Ham.
5. *M. cathcartii* Hk. f. & T.

Key to genera

I. Erect trees or shrubs, conspicuous;
   1. Gynophore absent –
      i. Carpels closely aggregated, flowers dehiscing dorsally, ovules 2 or few, flowers terminal .... *Magnolia.*
   2. Gynophore present –
      i. Carpels in loose spike, gynophore stalked, persistent, flowers axillary and solitary .... *Michelia.*

Plants of economic importance

The plants of this family are not so economically important. Some are used as ornamental plants, timbers, condiments, volatile oils etc.

A. **Ornamental plants**

Magnolia denudata Desv., M. grandiflora L., *M. pterocarpa* Roxb.; *Michelia champaca* L., *M. fuscata* Blume and *Liriodendron tulipifera* L. (Tulip tree). The plants are cultivated as ornamental plants for their beautiful fragrant flowers and also for the glossy leaves.

B. **Timber yielding plants**

Liriodendron tulipifera L. (Tulip tree), *Michelia doltsopa* Buch.-Ham. (Syn. *M. excelsa* DC. ex Wall) and different species of *Magnolia* and *Talauma* are very much useful for their timber. These timbers are used for tea boxes, furniture, planking, etc.

C. **Condiments**

The fruits of *Illicium verum* Hook. f. and *I. griffithii* Hook. f. are used as condiment for flavouring confectionary, curries, etc.

D. **Volatile oil**

1. The flowers of *Magnolia grandiflora* L. and *Michelia champaca* L. are the source of a volatile oil used in cosmetics and perfumery.
2. The fruits of *Illicium verum* Hook. f. and *I. griffithii* Hook. f. yield an essential oil which relieves colic pain and cough. It is also used in stomachic and carminative.

E. **Other uses**

1. *Michelia champaca* L. The flowers yield a yellow dye used in textiles.
2. *Michelia champaca* L., *Kadsura roxburghiana* Arn. and *Schizandra grandiflora* Hook. f. & Thoms. The fruits are edible.

3. Annonaceae

(Custard Apple Family)

The family Annonaceae is represented by about 128 genera and 2,300 species growing widely in the tropical region of the world. In India, the family is represented by 26 genera and about 200 species found mostly in the peninsular region of India.

The type genus of the family is ***Annona***.

Systematic position

Bentham & Hooker		**Engler & Prantl**	
Division.	Angiospermia	Subdivision.	Angiospermae
Class.	Dicotyledons	Class.	Dicotyledoneae
Subclass.	Polypetalae	Subclass.	Archichlamydeae
Series.	Thalamiflorae	Order.	Ranales
Cohort.	Ranales.		

Hutchinson		**Cronquist**	
Phylum	Angiospermae	Division.	Magnoliophyta
Subphylum	Dicotyledones	Class.	Magnoliopsida
Division.	Lignosae	Subclass.	Magnoliidae
Order.	Annonales	Order.	Magnoliales.

Takhtajan		**Thorne**	
Division.	Magnoliophyta	Class.	Angiospermae
Class.	Magnoliopsida	Subclass.	Magnoliidae
Subclass.	Magnoliidae	Superorder.	Magnolianae
Superorder.	Magnolianae	Order.	Magnoliales
Order.	Annonales.		

Identifying characteristics

1. Plants are **shrubs** or **woody trees.**
2. **Leaves simple**, exstipulate, alternate and **gland dotted.**
3. **Flowers** are **small, bisexual, trimerous** and **spirocyclic.**
4. **Perianth** in **three whorls**, outer sepaloid and inner two petaloid.
5. **Stamens numerous, spirally arranged;** anthers long with tunicate connective, extrorse.

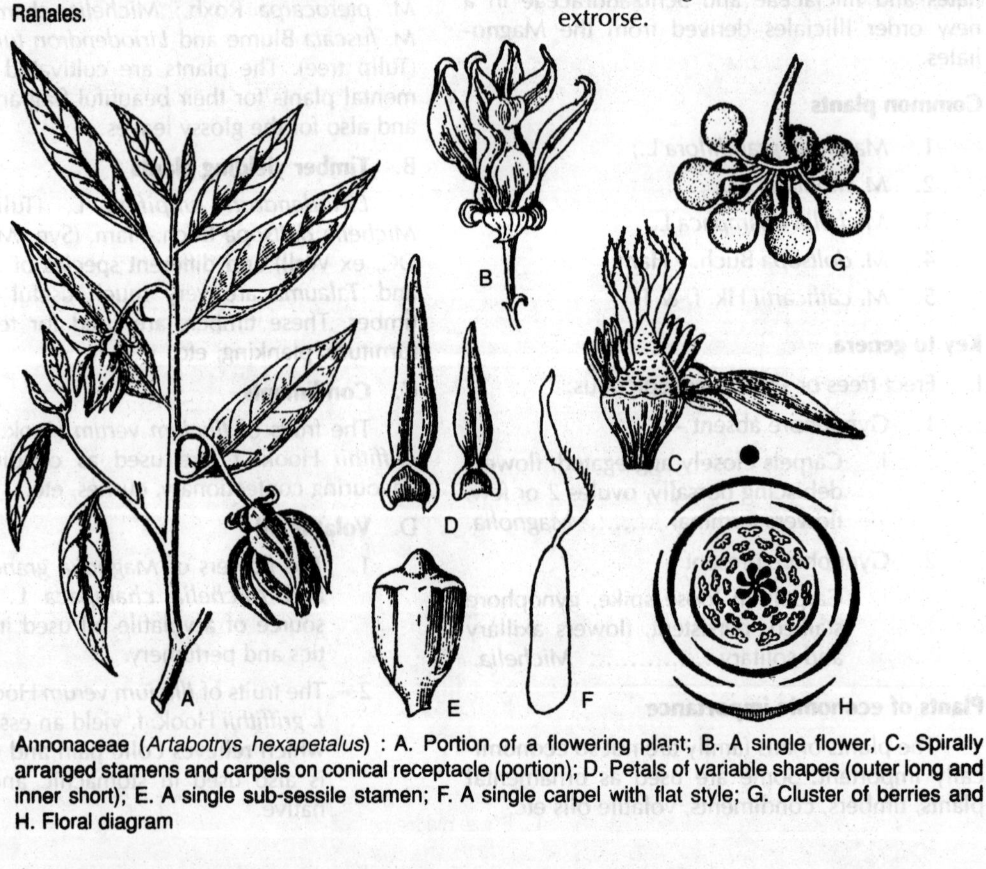

Fig. 4.42 : Annonaceae (*Artabotrys hexapetalus*) : A. Portion of a flowering plant; B. A single flower, C. Spirally arranged stamens and carpels on conical receptacle (portion); D. Petals of variable shapes (outer long and inner short); E. A single sub-sessile stamen; F. A single carpel with flat style; G. Cluster of berries and H. Floral diagram

6. **Carpels numerous** or few **on slightly elongate convex thalamus.**
7. Fruits **etaerio of drups** or **berries.**

Floral formula

$$\oplus ⚥ K_{3 \text{ or } (3)} C_{3\text{-}6} A_{\alpha} \underline{G}_{\alpha}$$

Annonaceae : Distinctive features

Common characteristics	Exceptions
Vegetative characteristics	
Plant. Plants are trees or shrubs.	Woody climbers – *Melodorum* and *Oxymitra*.
Leaves. Alternate, simple, exstipulate, entire and gland dotted.	
Reproductive characteristics	
Flowers. Solitary, regular, bisexual, hypogynous, spirocyclic, trimerous, sometimes aromatic.	Unisexual – *Stelechocarpus*. Zygomorphic – *Monodora*.
Calyx. Sepals 3, free or united at the base, persistent, valvate.	Form 3-lobed cup – *Cyathocalyx*.
Corolla. Petal 6 in two whorls, equal in size, free or connate at the base, valvate or imbricate.	Inner whorl absent – *Unona*. Petals slightly unequal – *Monodora*.
Androecium. Stamens numerous, free, spirally arranged on convex receptacle, filaments short and thick, anthers are adnate and extrorse, anther lobes linear with prolonged connective.	
Gynoecium. Carpels many, free or *united*, usually stipitate, ovary superior, 1-celled, ovules 1-many, anatropous, arranged on *basal* or *marginal* placentation.	
Fruit. Aggregate of berries or drupe.	Fruit follicular – *Anaxagorea*.
Seed. Large, often arillate, embryo small, endosperm ruminated or marbled.	

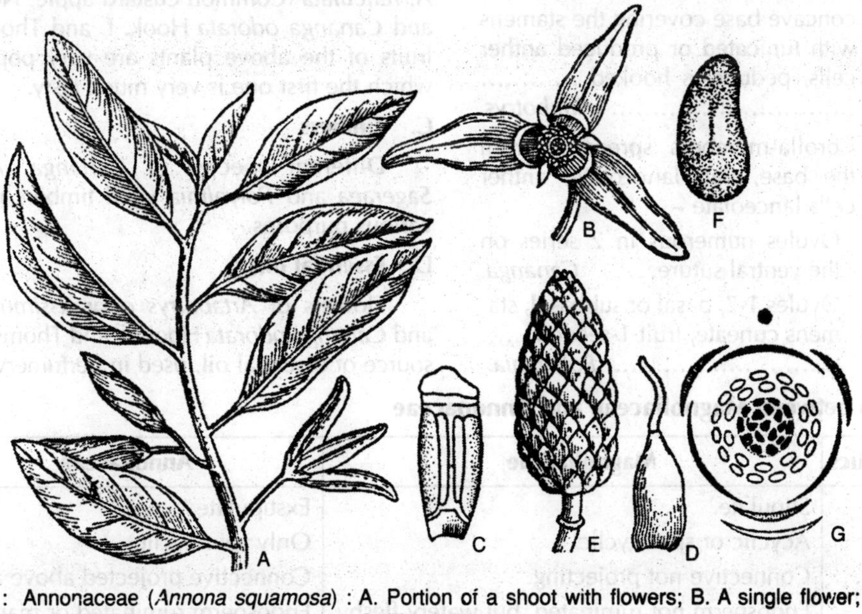

Fig. 4.43 : Annonaceae (*Annona squamosa*) : A. Portion of a shoot with flowers; B. A single flower; C. A Stamen; D. A Pistil; E. A Fruit; F. A Seed and G. Floral diagram

Phylogeny and systematics

The family Annonaceae is nearly related and advanced than Magnoliaceae. The Annonaceae and Magnoliaceae are very close and possibly derived from the same ancestral stock.

Bentham and Hooker, Engler and Prantl and others included the family under Ranales. But, Hutchinson (1959, 69) placed the family Annonaceae under the order Annonales. Later, Takhtajan (1969, 80) and Cronquist (1968, 81) included Annonaceae under the order Magnoliales.

Common plants

1. *Annona reticulata* L.;
2. *A. squamosa* L.;
3. *Cananga odorata* Hook. f. & Thoms.);
4. *Polyalthia cerasoides* Benth. & Hook.;
5. *P. longifolia* L. etc.

Key to genera

I. Carpels numerous, distinct, indehiscent :
   1. Petals imbricate in bud –
      i. Corolla-members biseriate, sepals valvate, connective truncate, torous flat. .... *Uvaria.*
   2. Petals valvate in bud –
      i. Corolla-members conniving at the concave base covering the stamens with tunicated or produced anther cells, peduncles hooked .... *Artabotrys.*
      ii. Corolla-members spreading from the base, flat, lanceolate, anther cells lanceolate –
         Ovules numerous in 2 series on the ventral suture. ...... *Cananga.*
         Ovules 1-2, basal or subbasal, stamens cuneate, fruit 1-seeded ....*Polyalthia.*
         Ovules 2-6, uniseriate on the ventral suture. .... *Unona.*

II. Carpels numerous, united into a fleshy fruit :
   1. Petals 3-6 in 2 series, valvate in bud, stamens numerous, ovaries numerous, partially united, ovule 1, erect...... *Annona.*

Plants of economic importance

The members of this family are useful as ornamental plants, edible fruits, timbers, essential oils and they also have other uses.

A. Ornamental plants

The following plants are grown in the garden for their fragrant flowers and attractive glossy leaves :

1. *Artabotrys odoratissimus* R. Br. (Climbing ylang-ylang). The plants are shrubs with glossy leaves and yellowish very fragrant flowers.
2. *Cananga odorata* Hook. f. & Thoms. (Ylang-ylang). It is a small tree with very fragrant yellowish flowers.
3. *Polyalthia longifolia* Benth. & Hook. It is a lofty tree with drooping branches and leaves. The plants are selected to grow in the fencing of the gardens.

B. Edible fruits

Annona squamosa L. (Custard apple, Ata), *A. reticulata* (Common custard apple, Nona ata) and *Cananga odorata* Hook. f. and Thoms. The fruits of the above plants are very popular, of which the first one is very much tasty.

C. Timbers

Different species of *Cananga, Miliusa, Sageraea* and *Polyalthia* yield timbers used for various purposes.

D. Essential oils

Flowers of *Artabotrys odoratissimus* R. Br. and *Cananga odorata* Hook. f. and Thoms are the source of essential oil, used in perfumery, etc.

Comparison between Magnoliaceae and Annonaceae

Characteristics	Magnoliaceae	Annonaceae
1. Leaves	Stipulate.	Exstipulate.
2. Flowers.	Acyclic or spirocyclic	Only spirocyclic.
3. Stamens.	Connective not projecting.	Connective projected above anthers.
4. Fruits.	Endosperm not ruminated, but watery-fleshy.	Endosperm ruminated or marbled.

E. **Other uses**

Annona squamosa L. The seeds have insecticidal properties and are used in killing lice and insects.

Unona pannosa Dalz. The inner bark has a strong fibre, used for paper-making and cordage (ropre in mass).

4. Nymphaeaceae.

(Water Lily Family)

The family Nymphaeaceae having 8 genera and about 70 species grows at different habitats (cosmopolitan), but are very common in tropical region. In India, there are 4 genera and 7 species growing almost everywhere including some hilly regions.

The type genus of the family is ***Nymphaea***.

Systematic position

Bentham & Hooker		**Engler & Prantl**	
Division.	Angiospermia	Subdivision.	Angiospermae
Class.	Dicotyledons	Class.	Dicotyledoneae
Subclass.	Polypetalae	Subclass.	Archichlamydeae
Series.	Thalamiflorae	Order.	Ranales
Cohort.	Ranales.		

Hutchinson		**Cronquist**	
Phylum.	Angiospermae	Division.	Magnoliophyta
Subphylum.	Dicotyledones	Class.	Magnoliopsida
Division.	Herbaceae	Subclass.	Magnoliidae
Order.	Ranales	Order.	Nymphaeales

Takhtajan		**Thorne**	
Division.	Magnoliophyta	Class.	Angiospermae
Class.	Magnoliopsida	Subclass.	Magnoliidae
Subclass.	Nymphaeidae	Superorder.	Nymphaeanae
Superorder.	Nymphaeanae	Order.	Nymphaeales
Order.	Nymphaeales.		

Identifying characteristics

1. Plants are **aquatic herbs with rhizomatous stem.**
2. Leaves **simple with long petiole** and **peltate** or **pseudopeltate.**
3. Flower **solitary, bisexual**, hypogynous **develop on long peduncle**.
4. Perianth members 6-many, differentiated into calyx and corolla.
5. Stamens 6-many, spirally arranged on thalamus.
6. **Gradual changes from sepal to petal, petal to stamen are clearly visible.**

Nymphaeaceae : Distinctive features

Common characteristics	Exceptions
Vegetative characteristics	
Plant. Plants are aquatic perennial herbs.	Annual – *Euryale.*
Stem. Usually rhizomatous, erect (*Brasenia*) or creeping.	Cauline – *Cabomba.*
Leaves. Simple, long and spongy, long petioled, usually large (up to 2 meters diameter in *Victoria*), peltate or pseudopeltate, usually floating.	Sessile – *Cabomba.* Leaves emerge from water – *Nelumbo.*
Reproductive characteristics	
Flower. Solitary, showy, long peduncled, regular, bisexual, hypogynous, or epigynous, spirocyclic but rarely cyclic, thalamus cup-shaped and fleshy, float on the surface of water.	Flowers project out from water surface – *Nelumbo.* Flowers partially submerged – *Euryale.*
Perianth. Calyx and corolla are poorly differentiated.	
Calyx. Sepals 3-5 (3 in *Brasenia,* 4 in *Nelumbo* and 5 in *Barclaya*), free, green, imbricate.	Sepals petaloid – *Nuphar.*
Corolla. Petals 3 (*Brasenia*) to indefinite (*Nymphaea*), free, spirally arranged, coloured or white and showy, there is gradual transition from sepals to petals and petals to stamens.	Petals are united into a tube – *Barclaya.*

Contd.

Distinctive features (Contd.)

Common characteristics	Exceptions
Androecium. Stamens 3-6 (*Cabomba*) or indefinite (*Nelumbo*), free, adnate to cup-shaped thalamus (torus), filament-foliaceous, introrse.	
Gynoecium. Carpels 3-many, free or loosely connate or sunken in spongy torus, ovary superior, ovary 1-many celled with ovules disposed on ovary wall on superficial placentation.	Inferior ovary – *Euryale*.
Fruit. Spongy berry.	Follicle – *Cabomba*. Aggregate of achene – *Nelumbo*.
Seed. Often arillate with endosperm and perisperm. Embryo straight.	

7. Carpels 3-many, free or connate, **placentation lamellate** or **parietal**.
8. **Fruits** are generally **spongy berry**.

Floral formula

$\oplus$ ⚥ $K_{3-5}C_{3-\alpha}A_{6-\alpha}\underline{G}_{3-\alpha}$ or $\overline{G}_{3-\alpha}$

The family divided into 3 subfamilies, having different floral formula :

Subfamily 1. Cabomboideae

$\oplus$ ⚥ $P_{3+3}A_6\underline{G}_3$

Subfamily 2. Nelumboideae

$\oplus$ ⚥ $P_\alpha A_\alpha \underline{G}_\alpha$

Subfamily 3. Nymphaeoideae

$\oplus$ ⚥ $K_{4-5}C_\alpha A_\alpha \underline{G}_{(\alpha)}$ or $\overline{G}_{(\alpha)}$

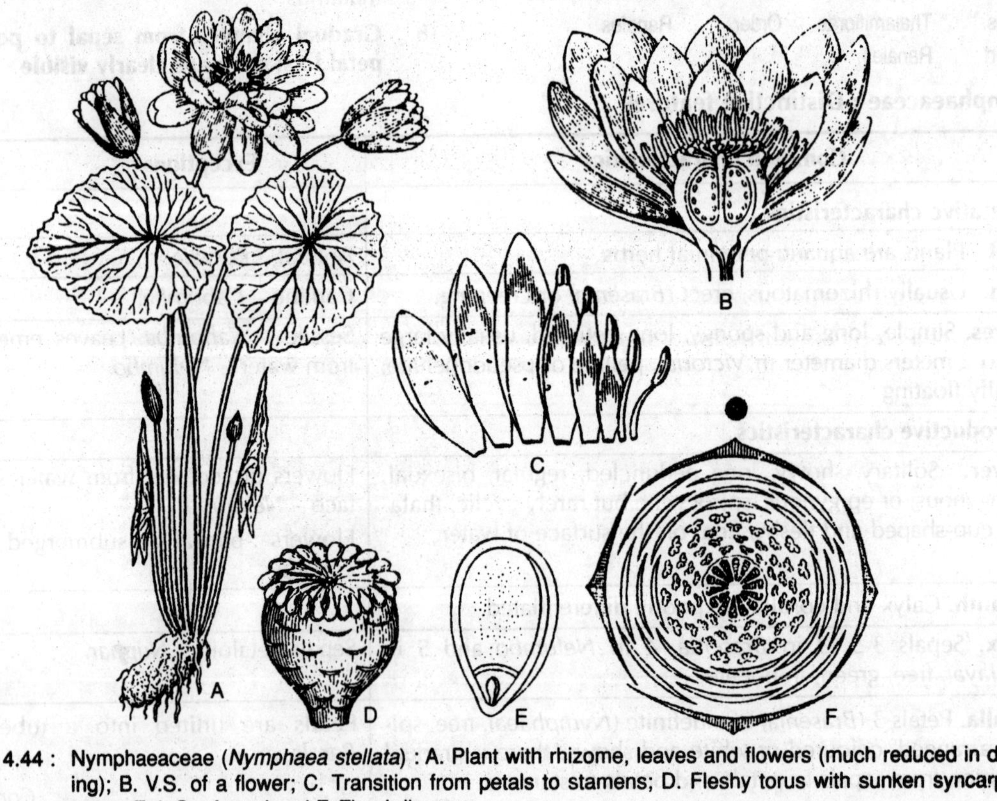

Fig. 4.44 : Nymphaeaceae (*Nymphaea stellata*) : A. Plant with rhizome, leaves and flowers (much reduced in drawing); B. V.S. of a flower; C. Transition from petals to stamens; D. Fleshy torus with sunken syncarpous ovary; E. L.S. of seed and F. Floral diagram

Phylogeny and systematics

The family Nymphaeaceae is very close to the family Ranunculaceae.

The family is generally considered to consist of three separate subfamilies : Cabomboideae, Nelumboideae, and Nymphaeoideae. The first two have free carpels and the last one has fused carpels. Due to their floral variation, many authors further divided the family into 2-5 families. Bessey (1915) changed the status of subfamilies to separate families such as Cabombaceae, Nelumbaceae, and Nymphaeaceae. This was followed by Small (1931), Willis (1977) etc. Takhtajan (1969) considered 3 families such as Cabombaceae, Nelumbonaceae, and Barclayaceae. Hutchinson (1959, 69 and 73) recognised two families i.e., Cabombaceae (includes subclass Cabomboideae) and Nymphaeaceae (including the last two subfamilies Nelumboideae and Nymphaeoideae).

Bentham and Hooker placed the natural order Nymphaeaceae under a distinct cohort (= order) Ranales of subclass Polypetalae under the class Dicotyledons.

Engler and Prantl placed it in the order Ranales of subclass Archichlamydeae under the class Dicotyledoneae.

Hutchinson placed the family in the order Ranales of division Herbaceae under subphylum Dicotyledones.

Most of the researchers, placed the family under Ranales, but Takhtajan (1969, 80) placed both Cabombaceae and Barclayaceae under Nymphaeales and the last one Nelumbonaceae under the order Nelumbonales.

Common plants

1. *Nelumbo nucifera* Gaertn. (East Indian lotus).
2. *Nymphaea alba* L. (European white waterlily).
3. *N. esculenta* Roxb. (Indian white waterlily).
4. *N. nouchali* Burm. f. (Indian red waterlily).
5. *N. stellata* Willd (Indian blue lotus)
6. *Nuphar luteum* L.
7. *Euryale ferox* Salisb.

Key to genera

I. Carpels embedded irregularly in the pits of obscure thalamus :
   Sepals 4-5, petals and stamens numerous, seeds exalbuminous. .... *Nelumbo.*

II. Carpel confluent with one another or with ovary :
   1. Sepals, petals and stamens half-inferior, inserted on a disc, confluent with the carpels. .... *Nymphaea.*
   2. Sepals, petals and stamens inferior, fruit prickly berry. .... *Euryale.*

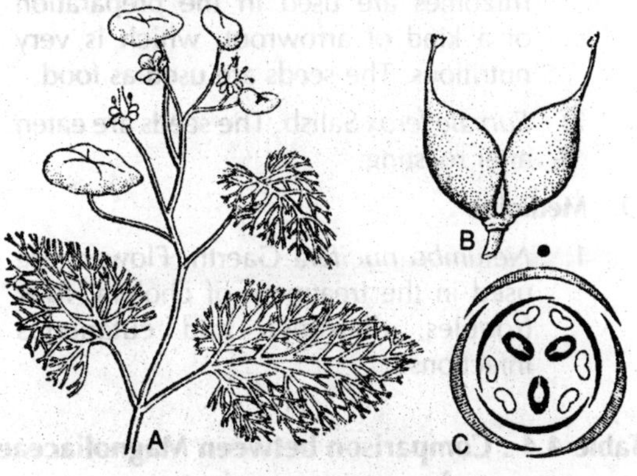

Fig. 4.45 : Nymphaeaceae (*Cabomba aquatica*) : A. Portion of a plant with flowers; B. Fruit, and C. Floral diagram

Plants of economic importance

Members of this family have great ornamental value, because of bright colour and sweet scent. In addition to the above, the members have some medicinal and food value.

A. Ornamental plants

The following plants are used in water gardening :

1. *Nelumbo nucifera* Gaertn. (Syn. *Nelumbium speciosum* Gaertn.). The lotus of India used for offering in temple in addition to beautification.
2. Beside the above one, following are the plants used in water gardening :
   i. *Nymphaea nouchali* Burm. f. (Syn. *N. lotus* L.). Indian red waterlily;
   ii. *N. alba* L.-European white waterlily.

3. *Victoria amazonica* Mart. & Zucc. (Syn. *Victoria regia* Lindl.). The Royal water-lily, was discovered in 1801 in the back water (stagnant water fed from a stream) of Amazon and the first name given in honour of Queen Victoria of Britain.

B. **Vegetable**

Nelumbo nucifera Gaertn. The tender leaves, petioles, rhizomes, flowers and fruiting torus are used as vegetable.

C. **Food**

1. *Nelumbo nucifera* Gaertn. The fleshy rhizomes are used in the preparation of a kind of arrowroot, which is very nutritious. The seeds are used as food.
2. *Euryale ferox* Salisb. The seeds are eaten after roasting.

D. **Medicine**

1. *Nelumbo nucifera* Gaertn. Flowers are used in the treatment of cholera, liver troubles, diarrhoea and cutaneous infections.
2. *Nymphaea stellata* Willd. The decoction of flowers is used in curing heart diseases.
3. *N. nouchali* Burm. f. The underground stem is used in the treatment of dysentery.

5. PAPAVERACEAE
(Poppy Family)

The family is represented by 40 genera and 770 species, found mainly in the north temperate regions. There are 5 genera and about 20 species found in India, which are mostly confined to the Himalayan zone.

The type genus of the family is ***Papaver***.

Systematic position

Bentham & Hooker		**Engler & Prantl**	
Division.	Angiospermia	Subdivision.	Angiospermae
Class.	Dicotyledons	Class.	Dicotyledoneae
Subclass.	Polypetalae	Subclass.	Archichlamydeae
Series.	Thalamiflorae	Order.	Rhoeadales
Cohort.	Parietales.		

Table 4.4 : Comparison between Magnoliaceae and Nymphaeaceae

Characteristics	Magnoliaceae	Nymphaeaceae
1. Plant	Trees or shrubs.	Aquatic herbs.
2. Leaves	Simple, short-petioled, ovate to elongate, entire, often stipulate.	Simple, long and spongy petioled, peltate or pseudopeltate.
3. Flower	Solitary on short peduncle, bisexual.	Solitary on long spongy peduncle, bisexual but rarely unisexual.
4. Perianth	Tepals are often many in several whorls or arranged spirally on long torus, petaloid or sometimes differentiated into calyx and corolla.	Perianth members 6-many, differentiated into calyx and corolla; sepals 3-6, imbricate; petals 3-many, spirally arranged.
5. Androecium	Stamens many, free, spirally arranged on thalamus.	Stamens 6-many, spirally arranged on elongated cup-shaped thalamus.
6. Gynoecium	Carpel many, free, generally spiral on torus but sometimes whorled. Ovary superior, 1-chambered with 1 ovule on marginal placentation.	Carpel 3-many, free or loosely connate or sunken on thalamus. Ovary superior or inferior, many chambered, ovules 1-many on the inner wall of carpel. Placentation lamellate or parietal.
7. Fruit	Fruit aggregate of follicles or berries.	Fruit spongy berry or aggregate of achenes.
8. Seed	With abundant fleshy and oily endosperm.	Arillate with endosperm and perisperm.

Hutchinson		**Cronquist**	
Phylum.	Angiospermae	Division.	Magnoliophyta
Subphylum.	Dicotyledones	Class.	Magnoliopsida
Division.	Herbaceae	Subclass.	Magnoliidae
Order.	Rhoeadales	Order.	Papaverales

Takhtajan		**Thorne**	
Division.	Magnoliophyta	Class.	Angiospermae
Class.	Magnoliopsida	Subclass.	Magnoliidae
Subclass.	Ranunculidae	Superorder.	Magnolianae
Superorder.	Ranunculanae	Order.	Berberidales
Order.	Papaverales.		

Identifying characteristics

1. Plants are generally **herbs** with **latex**.
2. Leaves alternate, **simple** and **exstipulate**.
3. Flowers are generally **solitary, regular, bisexual** and **hypogynous**.
4. Calyx-sepals 2, free, imbricate.
5. Corolla-petals 4 in 2 whorls, imbricate.
6. Stamens **many, free** and **present in several whorls**.
7. Carpel 2-**many, syncarpous, ovary superior**.
8. Fruit **capsule** with small oily or mealy endosperm.

Floral formula

$\oplus$ ⚥ $K_{2\text{-}3 \text{ or } 3+3}\ C_{4\text{-}6}\ A_{\alpha}\ \underline{G}_{2\text{-}\alpha}$

Phylogeny and systematics

Bentham and Hooker included the natural order (= family) under Parietales. Later, both Engler and Prantl, and Hutchinson placed the family under the order Rhoeadales — an order derived from Ranales. Both Takhtajan and Cronquist placed the family in a separate order, the Papaverales, and placed it after Ranunculales.

Common plants

1. *Argemone mexicana* L. (Prickly poppy or Mexican poppy),

Papaveraceae : Distinctive features

Common characteristics	Exceptions
Vegetative characteristics	
Plant. Herbs with coloured or milky latex.	Shrubs – *Dactylicapnos.* Creeper – *Hypecoum procumbens.*
Leaves. Alternate, simple, radicle or cauline, lobed, exstipulate; rarely crowded at the base.	Leaves compound or decompound – *Dactylicapnos, Corydalis.*
Reproductive characteristics	
Flower. Generally solitary, showy, regular, bisexual, hypogynous.	Raceme or panicle – *Macleya.* Umbel – *Chelidonium.*
Calyx. Sepals 2, free, caducous, imbricate.	Sepals 3 – *Argemone.* Sepals nited - *Eschscholzia.*
Corolla. Petals 4 in 2 whorls, deciduous, crumpled, imbricate.	Petals 6 – *Argemone.* Corolla absent – *Macleya.*
Androecium. Stamens many, free, arranged in several whorls. Anther 2-celled, basifixed, extrorse.	
Gynoecium. Carpel 2-many, syncarpous, ovary superior, 1 or more celled. Placentation parietal, ovule-anatropous or campylotropous. Style short or absent, stigmas many as carpel.	
Fruit. Capsule, dehiscing by pores or valves (*Papaver*) under the persistent stigma.	
Seed. Small, with copious oily or mealy endosperm and small embryo.	

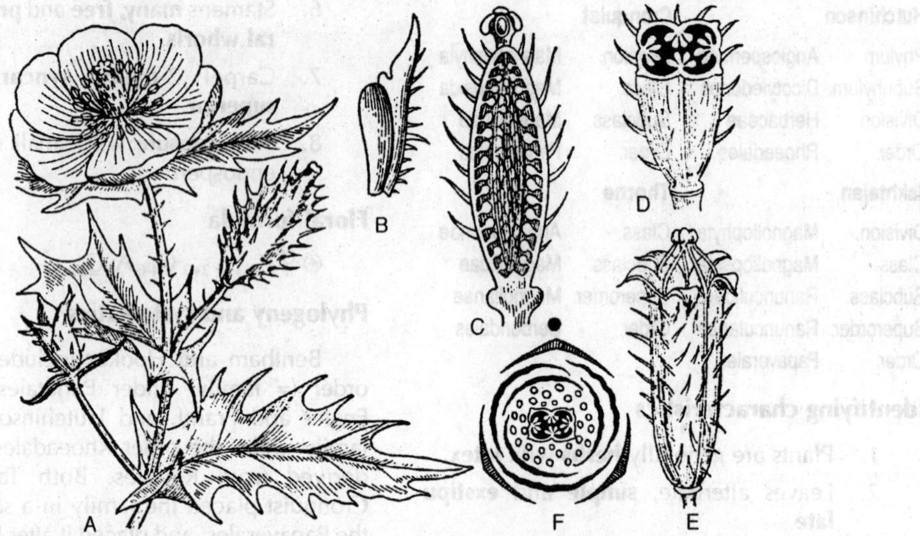

Fig. 4.46 : Papaveraceae (*Argemone mexicana*) : A. Portion of a plant with flowers and fruit; B. Sepal; C. L.S. of ovary; D. Ovary in T.S. along with lower part; E. Mature fruit, and F. Floral diagram

2. *Meconopsis nepaulensis* DC. (Blue poppy),
3. *Papaver argemone* L. (Garden poppy),
4. *P. orientale* L. (Oriental poppy),
5. *P. somniferum* L. (Opium poppy).

Key to genera

I. Capsules opening by apical pore or valve :

i. Stigma sessile, radiating from disc; capsule globose, floral symmetry 2 or its multiple, latex white............ *Papaver.*

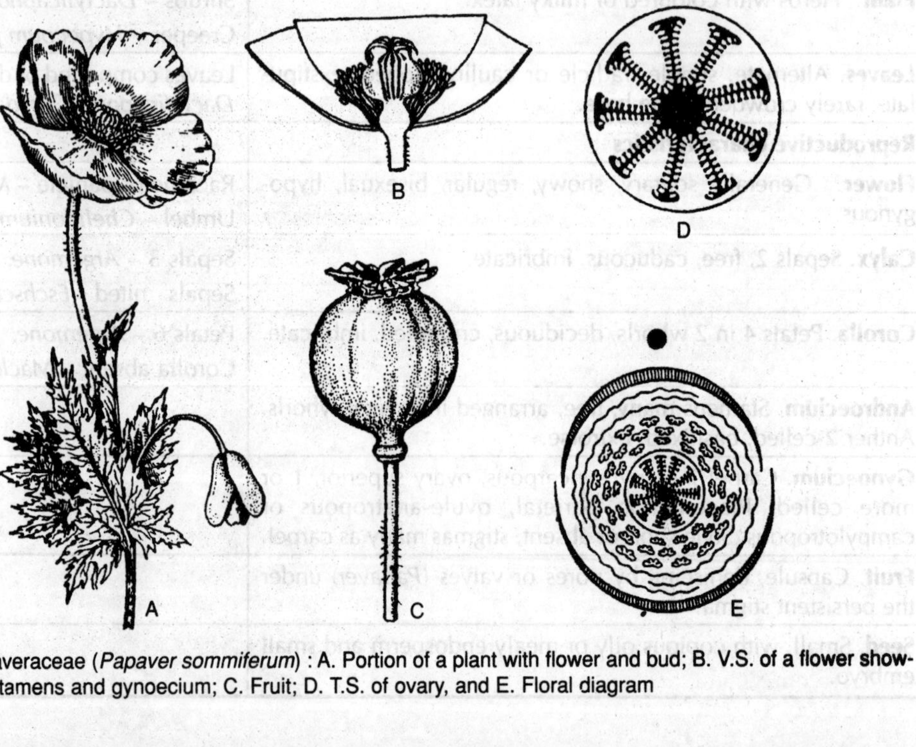

Fig. 4.47 : Papaveraceae (*Papaver sommiferum*) : A. Portion of a plant with flower and bud; B. V.S. of a flower showing stamens and gynoecium; C. Fruit; D. T.S. of ovary, and E. Floral diagram

ii. Stigma arising from depressed style; capsule oblong; floral symmetry 3 or its multiple; latex yellow. ...... *Argemone.*

iii. Stigma decurrent with style; capsule elongated; floral symmetry 2 or 3; latex yellow. .... *Meconopsis.*

Plants of economic importance

Members of these family have medicinal, food, ornamental and other uses. Out of many, the opium poppy (*Papaver somniferum* L.) is very important from both useful and harmful uses. Details about the plants are :

A. **Medicinal**

Papaver somniferum L. (Opium poppy). The plants are cultivated in the warmer regions in different countries including India. The '**opium**' is prepared from the milky juice of wall of the unripe capsules. The opium contains several alkaloids like codeine, morphine, narcotine and papaverine. They are used as medicines to relieve pain, anxiety, sleeplessness and also to control stomach troubles.

B. **Food**

Papaver somniferum L. (Opium poppy). The seeds of the plant do not contain any narcotic substances. It is used as food after forming a paste water — either directly or added with some vegetables during preparation.

C. **Ornamental plants**

Many plants are cultivated in garden for their attractive flowers. These are :

1. *Eschscholzia californica* (Californian poppy),
2. *Meconopsis nepaulensis* DC. (Blue poppy),
3. *Papaver orientale* (Oriental poppy),
4. *P. nudicaule* (Iceland poppy),
5. *P. rhoeas* (Corn poppy), etc.

D. **Other uses**

1. *Papaver somniferum* L. (Opium poppy). The **opium** (prepared from the milky juice of the wall of unripe fruits) contains several alkaloids like codeine, morphine, narcotine and papaverine. The toxic dose of opium causes respiratory depression, paralysis, more sleep, coma, and, finally, death.
2. *Argemone mexicana* L. The bitter and non-edible oil extracted from the seeds is used in illumination, and applied externally to cure skin diseases and also used as lubricant.

The oil is also used as **adulterant** with mustard oil and possibly causes the **severe outbreak of dropsy**.

6. CRUCIFERAE (BRASSICACEAE)

(Mustard Family)

The family Brassicaceae consists of 419 genera and 4,130 species, distributed widely throughout the world and is found to grow abundantly in North Temperate Zone. About 51 genara and 138 species are found in India, mostly in the alpine and temperate regions of Himalayas and some are also grows in the plains. Some species are also grown in the plains of South India.

The type genus of the family is ***Brassica***.

Systematic position

Bentham & Hooker		**Engler & Prantl**	
Division.	Angiospermia	Subdivision.	Angiospermae
Class.	Dicotyledons	Class.	Dicotyledoneae
Subclass.	Polypetalae	Subclass.	Archichlamydeae
Series.	Thalamiflorae	Order.	Rhoeadales
Cohort.	Parietales.		

Hutchinson		**Cronquist**	
Phylum.	Angiospermae	Division.	Magnoliophyta
Subphylum.	Dicotyledones	Class.	Magnoliopsida
Division.	Herbaceae	Subclass.	Dilleniidae
Order.	Brassicales	Order.	Capparales

Takhtajan		**Thorne**	
Division.	Magnoliophyta	Class.	Angiospermae
Class.	Magnoliopsida	Subclass.	Magnoliidae
Subclass.	Ranunculidae	Superorder.	Violanae
Superorder.	Ranunculanae	Order.	Brassicales
Order.	Papaverales.		

Identifying characteristics

1. **Herbs** with **watery sap**.
2. **Leaves simple** exstipulate, **lyrate in shape**.
3. Inflorescence **raceme**.
4. Flower **bisexual, actinomorphic** and *without bract* i.e., *ebracteate*.
5. Calyx-Sepals 4 (2+2).
6. **Corolla**-Petals 4, **cruciform**.

7. **Stamens** 6 (2+4) i.e., **tetradynamous.**
8. **Carpels 2, syncarpous,** ovary 1-chambered, falsely two-chambered by **replum, placentation parietal.**
9. **Fruit siliqua** or **silicula.**

Floral formula

$\oplus$ ⚥ $K_{4(2+2)}C_4A_{6(2+4)}\underline{G}_{(2)}$

Phylogeny and systematics

The family Cruciferae is very close to the family Capparidaceae. This family is thought to be derived either from the papaveraceous ancestors (Bentham and Hooker, 1862-83) or capparidaceous ancestors (Puri, 1941; Takhtajan, 1969, 80 and Cronquist, 1968, 81).

Bentham and Hooker placed the family in the cohort (= order) Parietales under series Thalamiflorae, subclass Polypetalae and class Dicotyledons.

Engler and Prantl and many others placed the family under the order Rhoeadales next to Ranales. Hutchinson placed the family in the order Brassicales under the division Herbaceae. Takhtajan and Cronquist placed the family in Capparales.

Common plants

1. *Brassica campestris* var. *dichotoma* watt. (Mustard);
2. *B. campestris* var. *toria* Duth. (Indian rape or Toria);
3. *B. juncea* (L.) Czern and Coss. (Indian mustard, Rai);
4. *B. nigra* Koch (Black mustard);
5. *Eruca sativa* Lamk.;
6. *Iberis amara* L.;
7. *Raphanus sativus* L. (Radish);
8. *Rorippa indica* (L.) Hochr. (Syn. *Nastartium indicum* (L.) DC.
9. *Cardamine debilis* Don.
10. *Capsella bursa-pastoris* Moench. (Shephard's purse).

Key to genera

I. Fruit dehiscent.
   1. Pod short, cylindrical; seeds minute, biseriate; sepals spreading; stigma shortly divided; flowers yellow. .... *Rorippa.*

Cruciferae : Distinctive features

Common characteristics	Exceptions
Vegetative characteristics	
Plant. Annual or perennial herbs with pungent juice.	
Leaves. Simple, radicle or cauline (usually alternate), exstipulate, auriculate, often lyrate.	Leaves pinnately compound – *Cardamine, Loxostemon.*
Reproductive characteristics	
Inflorescence. Racemose (raceme), ebracteate.	Corymb – *Iberis.*
Flower. Bisexual, regular (actinomorphic), hypogynous.	Irregular – *Iberis.*
Calyx. Sepals 4, in two whorls, polysepalous, imbricate.	
Corolla. Petals 4, polypetalous, cruciform, each petal with distinct limb and claw, imbricate or twisted.	
Androecium. Stamens 6, of which outer 2 short and inner 4 long i.e., tetradynamous (2+4), nectaries present at the base of long stamens.	Stamens 16 in number – *Megacarpaea.*
Gynoecium. Carpel 2, syncarpous, 1-celled with parietal placentation but subsequently becomes 2-celled by false septum (replum), ovules many, campylotropous.	Carpel 4 or more – *Tetrapoma* and *Holargidium.* Carpel 3 – *Lepidium sativum.*
Fruit. Siliqua or Silicula.	
Seed. Round, oily, exalbuminous, embryo large and curved.	

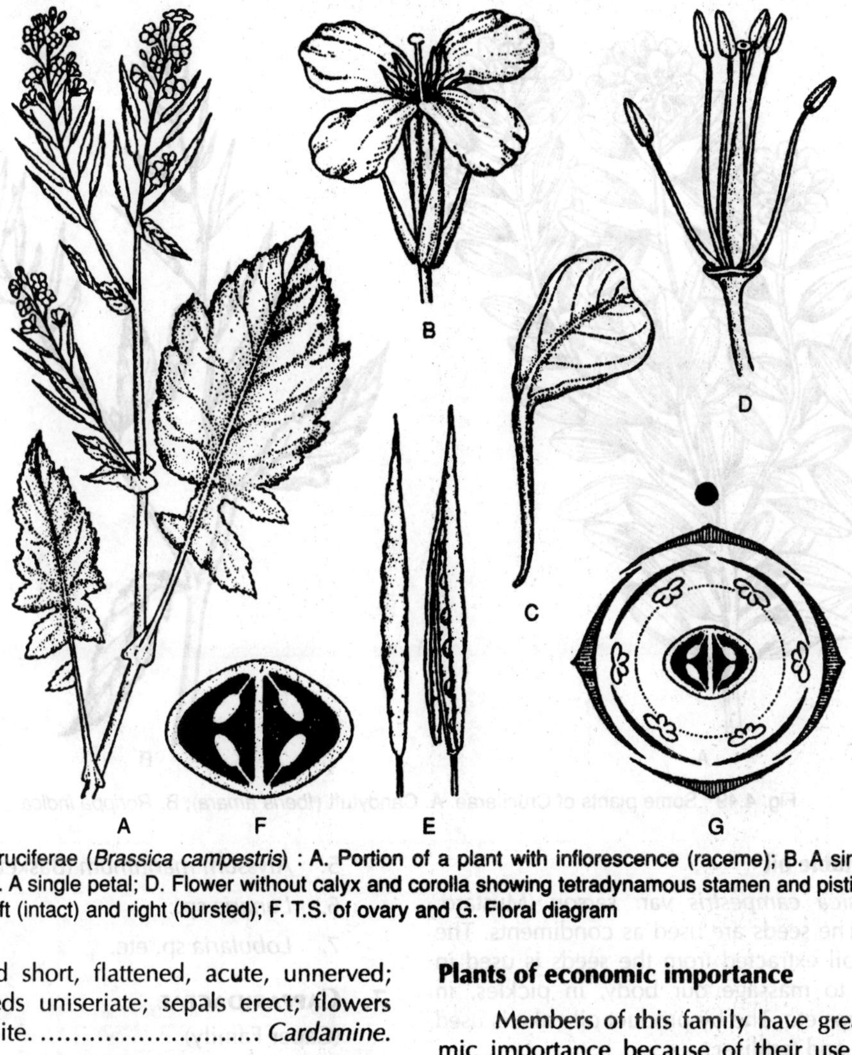

Fig. 4.48 : Cruciferae (*Brassica campestris*) : A. Portion of a plant with inflorescence (raceme); B. A single flower; C. A single petal; D. Flower without calyx and corolla showing tetradynamous stamen and pistil; E. Fruits, left (intact) and right (bursted); F. T.S. of ovary and G. Floral diagram

2. Pod short, flattened, acute, unnerved; seeds uniseriate; sepals erect; flowers white. .... *Cardamine.*
3. Pod with indehiscent beak; cotyledons incumbent or folded.
   i. Beak cylindrical or conical; seeds uniseriate; sepals yellow, pouched at the base and marked with green veins. .... *Brassica.*
   ii. Beak flattened; seeds biseriate; sepals lilac, pouched at the base. ... . *Eruca.*

II. Fruit indehiscent.

1. Pod long, hollow or septate, breaking into 1-seeded fragments; seeds globose; sepals pouched at the base.... *Raphanus.*
2. Pod short, globose; sepals spreading, not pouched at the base....... *Senebiera.*

Plants of economic importance

Members of this family have great economic importance because of their use as vegetable, for cooking oil and for ornamental value.

A. **Vegetable**

1. *Brassica oleracea* var. *capitata* (Cabbage). The leaves are used as vegetable.
2. *B. oleracea* var. *botrytis* (Cauliflower). The inflorescence is used as vegetable.
3. *B. oleracea* var. *gongylodes* (Knolkhol). The swollen stem is used as vegetable.
4. *B. rapa* (Turnip). The swollen root is used as vegetable.
5. *B. campestris* var. *sarson* (Sarson). The leaves are used as vegetables.
6. *Raphanus sativus* (Radish). The swollen root is used as vegetable.

Fig. 4.49 : Some plants of Cruciferae. A. Candytuft (*Iberis amara*); B. *Rorippa indica*

B. **Vegetable oil**

Brassica campestris var. *sarson* (Mustard, Sarson). The seeds are used as condiments. The mustard oil extracted from the seeds is used in cooking, to massage our body, in pickles, in small lamps etc. The by-product oil cake is used as fodder and fertiliser.

Other oil-yielding species are :

1. *B. nigra* Koch. (Black mustard, Kali sarson);
2. *B. juncea* hook. f. & Thoms. (Indian mustard, Rai);
3. *B. napus* L. (Rape seed, Toria);
4. *B. alba* Boiss. (White mustard, Safed sarson) etc.

C. **Ornamental value**

The following plants are used in gardening :

1. *Iberis amara* (Candytuft);
2. *Cheiranthus cheiri* (Wall flowers);
3. *Mathiola* (Stocks);
4. *Arabis* (Rock cress);
5. *Alyssum maritimum* (Basket of gold);
6. *Lunaria* sp.;
7. *Lobularia* sp. etc.

7. Capparidaceae

(Caper Family)

The family Capparidaceae is represented by 42 genera and more than 900 species found to grow in warm temperate and tropical region of the world. In India, 7 genera and about 53 species are available in southern and western part and also in some region of Himalayas.

The type genus of the family is ***Capparis***.

Systematic position

Bentham & Hooker		**Engler & Prantl**	
Division.	Angiospermia	Subdivision.	Angiospermae
Class.	Dicotyledons	Class.	Dicotyledoneae
Subclass.	Polypetalae	Subclass.	Archichlamydeae
Series.	Thalamiflorae	Order.	Rhoeadales
Cohort.	Parietales.		

Hutchinson		**Cronquist**	
Phylum.	Angiospermae	Division.	Magnoliophyta
Subphylum.	Dicotyledones	Class.	Magnoliopsida
Division.	Lignosae	Subclass.	Dilleniidae
Order.	Capparidales	Order.	Capparales
Takhtajan		**Thorne**	
Division.	Magnoliophyta	Class.	Angiospermae
Class.	Magnoliopsida	Subclass.	Magnoliidae
Subclass.	Ranunculidae	Superorder.	Violanae
Superorder.	Ranunculanae	Order.	Brassicales
Order.	Papaverales.		

Identifying characteristics

1. Plants are generally **herbs**.
2. Inflorescence **racemose**.
3. Flowers **bisexual, hypogynous, regular** (actinomorphic), **tetramerous, bracteate, gynophore** or **gynandrophore** often present.
4. **Petals 4, free** (*corolla never cruciform*).
5. Stamens **4** or **indefinite, free**.
6. Carpel usually 2, placentation parietal.

Floral formula

$$\oplus ⚥ K_{4(2+2)} C_4 A_{4-\alpha} \underline{G}_{(2)}$$

Phylogeny and systematics

The family Capparidaceae is closely related to Cruciferae (Brassicaceae) and Papaveraceae. Like Cruciferae, Bentham and Hooker included the family Capparidaceae under the order Parietales; Engler and Prantl included under Rhoeadales and Hutchinson included under Capparales. Recent phylogenetists like Takhtajan, Cronquist and others also placed the family under the Capparales.

Capparidaceae : Distinctive features

Common characteristics	Exceptions
Vegetative characteristics	
Plant. Herbs or shrubs, without latex.	Small tree – *Crataeva*. Climber – *Capparis*.
Leaves. Alternate, simple or palmately compound, with or without stipule.	
Reproductive characteristics	
Inflorescence. Cymose or racemose.	
Flower. Bisexual, bracteate, hypogynous. usually tetramerous, actinomorphic, often with gynophore, sometimes with androphore.	Unisexual – *Podandrogyne*.
Calyx. Sepals generally 4, polysepalous or sometimes connate at the base, valvate (*Maerua*) or imbricate.	
Corolla. Petal 4, polypetalous, valvate or imbricate.	Petal 2 – *Cadaba*. Petal absent – *Roydsia*.
Androecium. Stamens usually many (*Capparis crataeva*), free, filament long.	Stamens 4 – *Cleome tetrandra*. Stamens 6, but not tetradynamous – *Cleome gynandra*.
Gynoecium. Carpel 2 or more, syncarpous, ovary superior (supported by long or short gynophore), 1-celled with parietal placentation, ovules many and campylotropous.	2 chambered by replum – *Capparis*. Gynophore absent – *Cleome viscosa*.
Fruit. Capsule (*Cleome*) or berry (*Capparis*) or drupaceous (*Roydsia*).	
Seed. Reniform or angled, albuminous with an incurved embryo.	

Bentham and Hooker (1862-67) divided the family into two tribes, Cleomeae and Cappareae. But Hutchinson (1967, 69) shifted Cleomeae from Capparidaceae and raised a new family Cleomaceae.

Common plants

1. *Capparis spinosa* L.;
2. *Cleome gynandra* L. (Syn. *Gynandropsis gynandra* Briq);
3. *C. rutidospermum* L.;
4. *Crataeva nurvala* Buch.-Ham. (Syn. *C. religiosa* Rorst. f.);
5. *Cleome viscosa* L. (Syn. *Polanisia viscosa* DC.) etc.

Key to genera

I. Fruit capsule; plants usually herbs.

1. Stamens 12 to 20, sessile; corolla imbricate in bud; leaves digitately compound. .... *Cleome*.
2. Stamens 6 on androphore, corolla open in bud; leaves compound, 5-7 foliate. .... *Gynandropsis*.

II. Fruit berry or capsule.

1. Disc present below sepaline tube; open in bud; plants usually small trees. .... *Crataeva*.
2. Disc hemispheric; sepals 4, 2-seriate; stamens 8 to numerous, long purple, inserted at the base of gynophore; plants climbing with spinous stipule. .... *Capparis*.

Plants of economic importance

The plants of this family are not so economically important. Some are used as ornamental plants, medicines and other purposes.

A. Ornamental plants

Capparis spinosa L. (Caper plants), *Cleome spinosa* (Giant spider) and *Crataeva adansonii* DC. These are grown in the garden as ornamental plants.

B. Medicinal plants

1. The *roots of Maerua arenaria* HK. f. & Th. are used as tonic and also as stimulant.
2. The *leaves* and *barks* of *Crataeva nurvala* Buch.-Ham. (Syn. *C. religiosa*

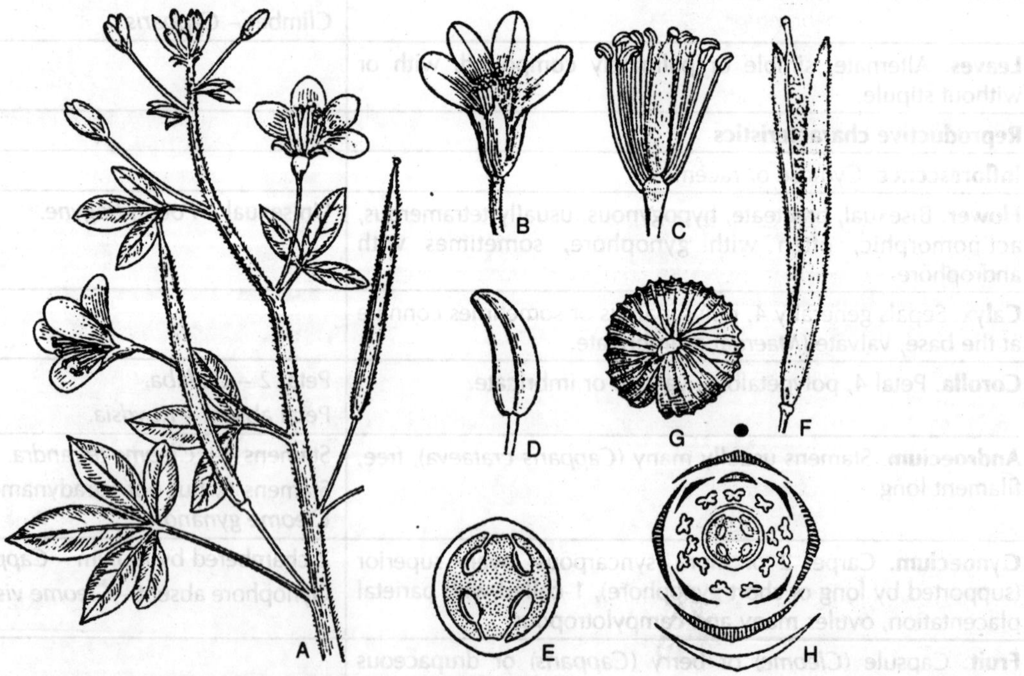

Fig. 4.50 : Capparidaceae (*Cleome viscosa*) : A. Portion of a plant with flower and fruit; B. A single flower; C. A flower after removal of sepals and petals; D. Stamen; E. T.S. of ovary; F. Single fruit; G. Seed and H. Floral diagram

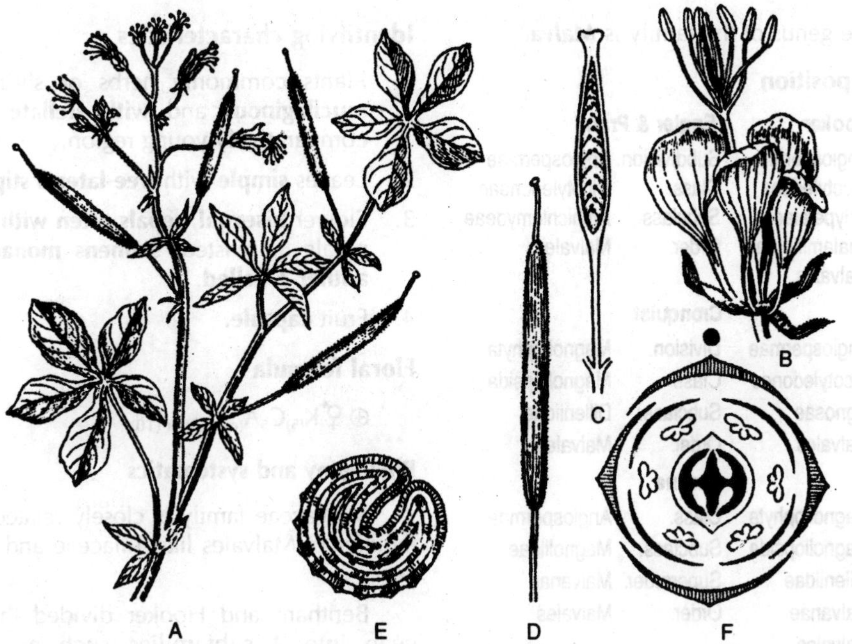

Fig. 4.51 : Capparidaceae (*Cleome gynandra*) : A. Portion of a plant with flowers and fruits; B. Flower with androgynophore; C. V.S. of ovary with stalk; D. Single fruit; E. L.S. of seed and F. Floral diagram

Forst. f. var. *nurvala* Buch.-Ham.) — are diuretic.

3. *Cleome gynandra* DC. (Syn. *Gynandropsis pentaphylla* L.). The *seeds* are internally employed to control round worms.

C. **Other plants**

1. The flower buds of *Capparis spinosa* L. are used in flavouring of foods.
2. *Capparis decidua* (Forsk.) Pax. (Syn. *Capparis aphylla* Roth.) and *C. zeylanica* L. (Syn. *Capparis horrida* L.) The fruits of these are used in pickles.

8. MALVACEAE
(Mallow Family)

The family Malvaceae is represented by 197 genera and 2,865 species, found to grow in the tropical and subtropical regions of the world. In India, there are 22 genera and about 110 species available in tropical regions.

Table 4.5 : Comparison between Cruciferae and Capparidaceae

Characteristics	Cruciferae	Capparidaceae
1. Plant	Herb.	Herbs, shrubs or trees.
2. Leaves	Simple, exstipulate, radicle or cauline, lyrate in shape.	Simple or palmately compound, stipulate or exstipulate.
3. Flowers	Ebracteate (i.e. without bract), without androphore and gynophore.	Bracteate, with gynophore, sometimes with both androphore and gynophore.
4. Corolla	Cruciform.	Never Cruciform.
5. Androecium	Stamens 6, tetradynamous (2 + 4).	Stamens few or many, never tetradynamous.
6. Fruit	Siliqua or silicula.	Various types : Siliqua, Berry, Capsule, Samara, Nut or Drupe.
7. Endosperm	Absent.	Absent or scanty.

The type genus of the family is ***Malva***.

Systematic position

Bentham & Hooker		**Engler & Prantl**	
Division.	Angiospermia	Subdivision.	Angiospermae
Class.	Dicotyledons	Class.	Dicotyledoneae
Subclass.	Polypetalae	Subclass.	Archichlamydeae
Series.	Thalamiflorae	Order.	Malvales
Cohort.	Malvales.		

Hutchinson		**Cronquist**	
Phylum.	Angiospermae	Division.	Magnoliophyta
Subphylum.	Dicotyledones	Class.	Magnoliopsida
Division.	Lignosae	Subclass.	Dilleniidae
Order.	Malvales	Order.	Malvales.

Takhtajan		**Thorne**	
Division.	Magnoliophyta	Class.	Angiospermae
Class.	Magnoliopsida	Subclass.	Magnoliidae
Subclass.	Dilleniidae	Superorder.	Malvanae
Superorder.	Malvanae	Order.	Malvales
Order.	Malvales.		

Identifying characteristics

1. Plants commonly herbs or shrubs, **often mucilaginous** and **with stellate hairs** on comparatively young region.
2. Leaves **simple** with **free-lateral stipule**.
3. Flower **bisexual, sepals often with epicalyx;** petals 5, twisted; **stamens monadelphous, anther 1-celled**.
4. Fruit **capsule**.

Floral formula

$\oplus$ ⚥ $K_{(5)} C_5 A_{(\alpha)} \underline{G}_{(5) \text{ or } (10)}$

Phylogeny and systematics

Malvaceae family is closely related to other families of Malvales like Tiliaceae and Sterculiaceae.

Bentham and Hooker divided the Malvaceae into 4 subfamilies such as Malveae,

Malvaceae : Distinctive features

Common characteristics	Exceptions
Vegetative characteristics	
Plant. Usually herbs or shrubs, often with mucilaginous juice and young regions are covered with stellate hairs.	Tree – *Kydia, Thespesia.*
Leaves. Alternate, simple, entire or variously lobed, mostly palminerved, stipulate (free-lateral).	
Reproductive characteristics	
Inflorescence. Cymose.	
Flower. Bisexual, regular, hypogynous, often subtended by epicalyx consists of 3-many bracteoles (3 in *Malva*, 5 in *Urena* and many in *Hibiscus*), pentamerous and cyclic.	Unisexual – *Napacea.* Epicalyx absent – *Abutilon* and *Sida.*
Calyx. Sepals 5, gamosepalous, valvate.	
Corolla. Petals 5, free but adnate at the base to the staminal column, twisted.	
Androecium. Stamens many, usually monadelphous, anther 1-celled, reniform, dehiscence-transverse, pollen grains large and spiny.	
Gynoecium. Carpel 2-many, syncarpous, superior, 1 or more anatropous ovules in each locules on axile placentation. Style 1, passing through the hollow staminal column; stigmas as many or twice as many as carpels.	Carpel 1 or 2 – *Plagianthus.* Carpel 15-20 – *Malva.* Carpel numerous – *Malope trifida.*
Fruit. Capsule or schizocarp.	Berry – *Malvasviscus.*
Seed. Reniform with scanty endosperm, often mucilaginous and a curved embryo with folded cotyledons.	Seeds are clothed with wooly hairs – *Gossypium herbaceum.*

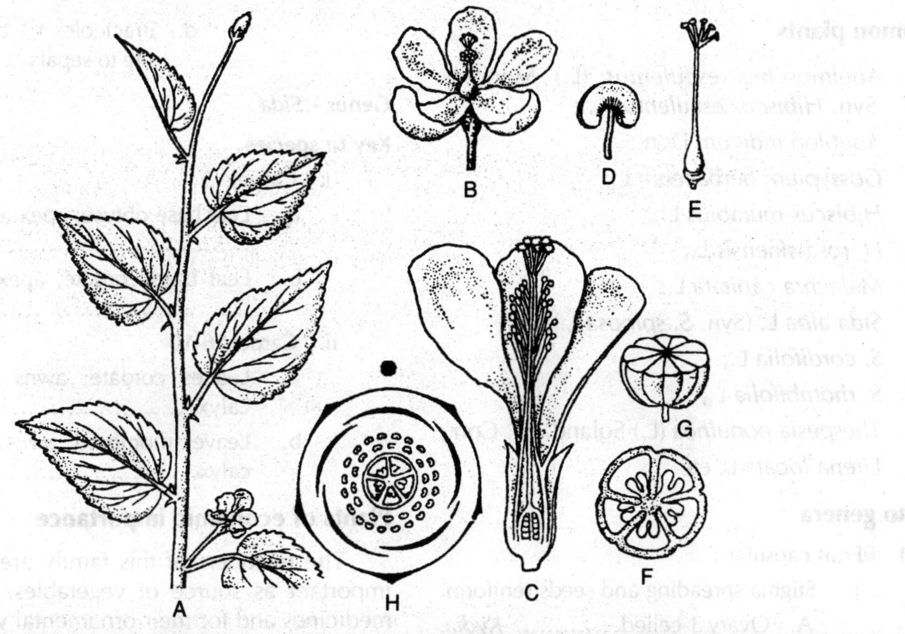

Fig. 4.52 : Malvaceae (*Sida cordifolia*) : A. Portion of a plant with flower; B. Single flower; C. Vertical section of flower; D. Stamen; E. Gynoecium; F. T.S. of ovary; G. Fruit and H. Floral diagram

Ureneae, Hibisceae and Bombaceae. The subfamily Bombaceae include the members with large trees, flowers bisexual — often appear before the leaves, leathery sepals, monadelphous or polydelphous stamens, hairy outgrowth on pericarp etc. Engler treated this tribe Bombaceae as a separate family.

Malvaceae is related to Tiliaceae and Sterculiaceae but differ in monothecous anther and staminal tube. Most of the taxonomists included it under Malvales along with Tiliaceae and Sterculiaceae. Hutchinson (1959, 69, 1973), however, retained only Malvaceae under the order Malvales and Sterculiaceae and Tiliaceae along with Bombacaceae, Gonystylaceae and Scytopetalaceae under a single order Tiliales.

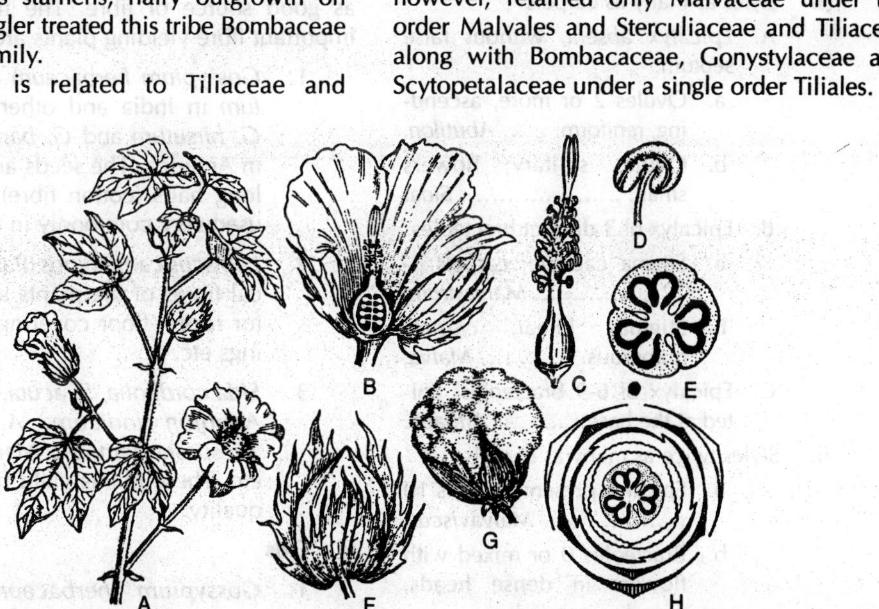

Fig. 4.53 : Malvaceae (*Gossypium herbaceum*) : A. Part of a plant with flower; B. V.S. of a flower; C. Monadelphous stamen with gynoecium; D. Stamen; E. T.S. of ovary; F. Fruit encircled by calyx; G. Dehiscent fruit and H. Floral diagram

Common plants

1. *Abelmoschus esculentus* (L.) Moench. (Syn. *Hibiscus esculentus* L.);
2. *Abutilon indicum* Don.;
3. *Gossypium herbaceum* L.;
4. *Hibiscus mutabilis* L.;
5. *H. rosa-sinensis* L.;
6. *Malachra capitata* L.;
7. *Sida alba* L. (Syn. *S. spinosa* L.);
8. *S. cordifolia* L.;
9. *S. rhombifolia* L.;
10. *Thespesia populnea* (L.) Soland. ex. Corr.;
11. *Urena lobata* L. etc.

Key to genera

1. Fruit capsular :
   - i. Stigma spreading and seeds reniform.
     - A. Ovary 3-celled. .... *Kydia*.
     - B. Ovary 5- celled. ...... *Hibiscus*.
   - ii. Stigma cohering in club-shaped mass.

     Bracteoles (epicalyx) 3, large, cordate; seeds cottony. .... *Thespesia*.
2. Fruit of ripe carpels separating from the axis.
   - i. Styles as many as carpels
     - A. *Epicalyx absent,* without false septum.
       - a. Ovules 2 or more, ascending, reniform. ...... *Abutilon*.
       - b. Ovules solitary; flowers small. .... *Sida*.
     - B. Epicalyx of 3 distinct bracteoles.
       - a. Stigma capitate; carpels 8-12. .... *Malvastrum*.
       - b. Stigmas linear, carpels numerous. .... *Malva*.
     - C. Epicalyx of 6-9 bracteoles, united at the base. .... *Althaea*.
   - ii. Styles twice as many as carpels
     - a. Epicalyx present; carpels 10 .... *Malvaviscus*.
     - b. Bracteoles 0 or mixed with flowers in dense heads; carpels unarmed. .... *Malachra*.
     - c. Bracteoles 5; carpels opposite to petals. .... *Urena*.
     - d. Bracteoles 10; carpels opposite to sepals. ...... *Pavonia*.

Genus - *Sida*

Key to species

- i. Carpels 5
  - a. Leaf base obtuse, apex acute. .... *acuta*.
  - b. Leaf base cordate, apex acuminate. .... *Cordata*.
- ii. Carpels 6-10
  - a. Leaves cordate; awns longer than calyx. .... *cordifolia*.
  - b. Leaves rhomboid; awns shorter than calyx. .... *rhombifolia*.

Plants of economic importance

The members of this family are very much important as source of vegetables, fibres, oils, medicines and for their ornamental value.

A. Vegetables

1. *Abelmoschus esculentus* (L.) Moench. (Ladies finger). The young fruits are used as vegetable.

B. Fibres

Many members of this family are important as good source of fibre. The names of some important fibre yielding plants are given :

1. *Gossypium herbaceum* and *G. arboretum* in India and other countries and *G. hirsutum* and *G. barbadense* grown in America. The seeds are covered with long hairs (cotton fibre). The cotton is used very commonly in textiles.
2. *Hibiscus cannabinus* (Patsan). The cortical fibres of the plants are widely used for ropes, floor coverings, coarse sackings etc.
3. *Sida cordifolia, S. acuta, S. rhombifolia; Abutilon indicum, A. polyandrum; Thespesia lampas, Malachra capitata* etc. These plants yield fibres of inferior quality.

C. Oils

1. *Gossypium herbaceum* and other species. The seeds are the source of fatty oil which is edible. The oil is also used as lubricant, manufacture of soaps, etc. The oil cake is used as cattle feed.

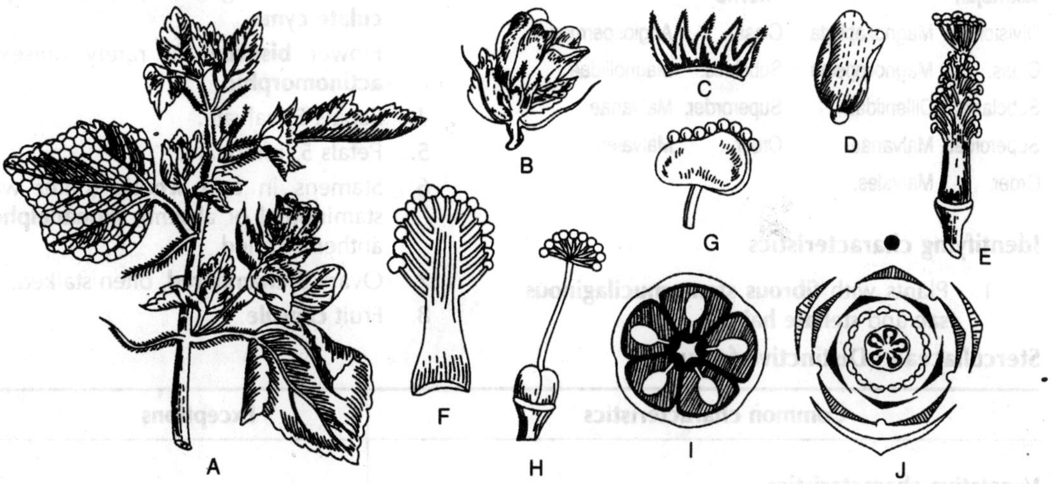

Fig. 4.54 : Malvaceae (*Malachra capitata*) : A. Part of a flowering plant; B. Single flower; C. Calyx; D. Petal; E. Flower after removal of calyx and corolla; F. Monadelphous stamens; G. Anther; H. Pistil (Gynoecium); I. T.S. of ovary and J. Floral diagram

2. *Hibiscus cannabinus* L. (Patsan). The seeds are also the source of fatty oil. The oil is used as vegetable oil and also as a lubricant in the preparation of varnishes and linoleum paints.

D. **Medicines**

1. *Urena lobata* L. The barks and roots are used as a cure of hydrophobia.
2. *Malachra capitata* L. The roots are used in the treatment of rheumatism.
3. *Malva verticillata* L. The roots are used for whooping cough. The ash of dried leaf is used on scabies.

E. **Ornamental plants**

Different species of Malvaceae are grown in the garden for their large and showy flowers. Some of them are :

1. *Althea rosea* (Hollyhock).
2. *Hibiscus mutabilis* (Cotton rose).
3. *H. rosa-sinensis* (China-rose).
4. *H. sabdariffa* (Jamaica sorell).
5. *H. syriacus* (Rose-of-Sharon).
6. *Malva sylvestris* (Mallow).
7. *Thespesia populnea* (Indian tulip tree).

F. **Other uses**

1. *Pavonia odorata*. The roots are used in perfumery.
2. *Thespesia populnea* (Indian tulip tree), and *Kydia calycina* (Pula). The timber of these plants is used in making toys, pencils, match factory and in agricultural implements.

9. STERCULIACEAE

(Sterculia Family)

The family Sterculiaceae is represented by 60 genera and 700 species, found to grow mainly in the tropical regions of both northern and southern hemisphere of the world. In India, there are 18 genera and more than 90 species available mostly in tropical regions. A few species are also available in Himalayas.

The type genus of the family is ***Sterculia***.

Systematic position

Bentham & Hooker		**Engler & Prantl**	
Division.	Angiospermia	Subdivision.	Angiospermae
Class.	Dicotyledons	Class.	Dicotyledoneae
Subclass.	Polypetalae	Subclass.	Archichlamydeae
Series.	Thalamiflorae	Order.	Malvales
Cohort.	Malvales.		

Hutchinson		**Cronquist**	
Phylum.	Angiospermae	Division.	Magnoliophyta
Subphylum.	Dicotyledones	Class.	Magnoliopsida
Division.	Lignosae	Subclass.	Dilleniidae
Order.	Tiliales	Order.	Malvales.

Takhtajan		**Thorne**	
Division.	Magnoliophyta	Class.	Angiospermae
Class.	Magnoliopsida	Subclass.	Magnoliidae
Subclass.	Dilleniidae	Superorder.	Malvanae
Superorder.	Malvanae	Order.	Malvales
Order.	Malvales.		

Identifying characteristics

1. Plants with **fibrous stem, mucilaginous sac** and **stellate hairs.**
2. Inflorescence **generally cymose** or **paniculate cyme.**
3. Flower **bisexual** or rarely unisexual, **actinomorphic.**
4. Sepals 5, valvate.
5. Petals 5, imbricate.
6. Stamens in **two whorls** (outer whorl staminodial or absent), **monadelphous,** anther 2-celled.
7. Ovary **5-chambered**, often stalked.
8. Fruit **capsule.**

Sterculiaceae : Distinctive features

Common characteristics	Exceptions
Vegetative characteristics	
Plant. Generally soft wooded tree or shrubs with mucilage and stellate hairs.	Herb – *Pentapetes* and *Melochia.* Climber – *Ayenia.* Mangrove – *Heritiera.*
Leaves. Simple, alternate, entire, palmately lobed or compound, stipulate, stipules free-lateral and caducous.	
Reproductive characteristics	
Inflorescence. Usually cymose, either axillary or terminal.	
Flower. Bisexual, regular or rarely zygomorphic, hypogynous and pentamerous.	Unisexual – *Sterculia, Heritiera* and *Cola.* Zygomorphic – Some species of *Helicteres.*
Calyx. Sepals 3-5, more or less united at the base, valvate.	
Corolla. Petals 5, free, sometimes adnate to the base of the staminal column, deciduous, contorted or imbricate in bud.	Petals absent – *Heritiera, Sterculia* and *Tetradia.* Corolla persistent – *Dombeya.*
Androecium. Stamens many in two whorls (inner whorl usually fertile and outer whorl form staminode or 0) or monadelphous, anther 2 celled.	Stamens 5 – *Melochia, Waltheria.* Stamens 10 – *Melhania.* Stamens 20 – *Pterospermum* and *Pentapetes.* Staminode absent – *Eriolaena, Chiranthodendron* and *Waltheria.*
Gynoecium. Carpel 2-5, syncarpous, ovary superior, often on gynophore, styles as many as carpels, free or united, locules as many as carpels, 2-many anatropous ovules are present in each locule on axile placentation.	Carpels free except style and stigma – *Sterculia.* Carpe! 1 – *Waltheria.* Gynophore absent – *Melochia, Guazuma, Thomasia.*
Fruit. Capsule or follicle.	
Seed. With fleshy endosperm, embryo flat, rolled or folded, cotyledon leafy.	Seed arillate – Some species of *Sterculia.* Seed winged – *Pterospermum.*

Floral formula

$\oplus$ ⚥ $K_{(3-5)}\, C_5\, A_{(\alpha)\ \text{or}\ \alpha}\, \underline{G}_{(5)}$

Phylogeny and systematics

The family Sterculiaceae is very close to Malvaceae and Euphorbiaceae.

Sterculiaceae is related to Tiliaceae and Malvaceae. The Malvaceae makes difference with others by having monothecous anther and staminal tube. Most of the taxonomists included it under Malvales along with Tiliaceae and Malvaceae. Hutchinson (1959, 69, 73), however, retained only Malvaceae under the order Malvales and Sterculiaceae and Tiliaceae along with Bombacaceae, Gonystylaceae and Scytopetalaceae under a single order Tiliales.

Common plants

1. *Abroma augusta* L. f.;
2. *Dombeya mastersii* Gaertn.;
3. *Guazuma ulmifolia* Lamk. var. *tomentosa* K. schum. (Syn. *Guazuma tomentosa* Kunth.);
4. *Heritiera fomes* Buch. – Ham. (Syn. *H. minor* Roxb.);
5. *Melochia corchorifolia* L.;
6. *Pentapetes phoenicea* L.;
7. *Sterculia alata* Roxb.;
8. *S. foetida* L.;
9. *Kleinhovia hospita* L. etc.

Key to genera

I. Flowers bisexual :
   1. Petals flat, deciduous; anthers alternating with staminodes ...
      i. Leaves oblique below; ovary at the base of the staminal column; anther-cells parallel; seeds winged. .... *Pterospermum*.
      ii. Leaves simple; ovary at the top of the staminal column; anther-cells divaricate; capsule woody with wingless seed. .... *Helicteres*.
   2. Petals flat, persistent; anthers alternating with staminodes ...
      i. Ovary 5-celled; styles 5. ... *Melochia*.
      ii. Ovary 1-celled, monocarpellary. .... *Waltheria*.
      iii. 15 anthers in 5 groups of 3, alternating with staminodes. ... *Pentapetes*.
   3. Petals deciduous, unequal; capsule, inflated pyriform. .... *Kleinhovia*.
   4. Petals concave; anthers in groups between each pair of staminodes...
      i. Petals clawed; fruit membranous; 5-angled 5-winged capsule, truncated above; ovules many in each loculus. .... *Abroma*.
      ii. Fruits woody, tubercled capsule, globose; ovules many in each loculus. .... *Guazuma*.
      iii. Petals ligulate; fruit armed, prickly or bristly capsule; ovules 2 in each loculus. .... *Buettneria*.

II. Flower unisexual or polygamous; petals 0 :
   1. Staminal column with ring of sessile anthers...
      i. Anthers 5; ripe carpels indehiscent. .... *Heritiera*.
      ii. Anthers 5-25; ripe carpels dehiscent. .... *Sterculia*.

Plants of economic importance

Economically, the family is very important at different aspects :

A. Medicines

1. *Eriolaena quinquelocularis*. The thick extract of the roots used as poultice (a soft composition applied in a cloth to sores) on the wounded area.
2. *Guazuma ulmifolia*. The roasted seeds have the property of astringency (to bind together) and used in diarrhoea.
3. *Helicteres isora* (Marorphali). The juice extracted from the roots is used to control diabetes.
4. *Heritiera littoralis*. The extract from seeds is used for the treatment of dysentery and diarrhoea.
5. *Sterculia balanghas*. The fruits are used as laxative.

B. Timber

1. *Eriolaena candollei, Heritiera littoralis, Pterospermum acerifolium, P. diversifolium, Sterculia villosa, S. foetida* and *S. urens*. The timber of the above plants are used in the preparation of toys, furniture, match box, tea box, construction works, etc.

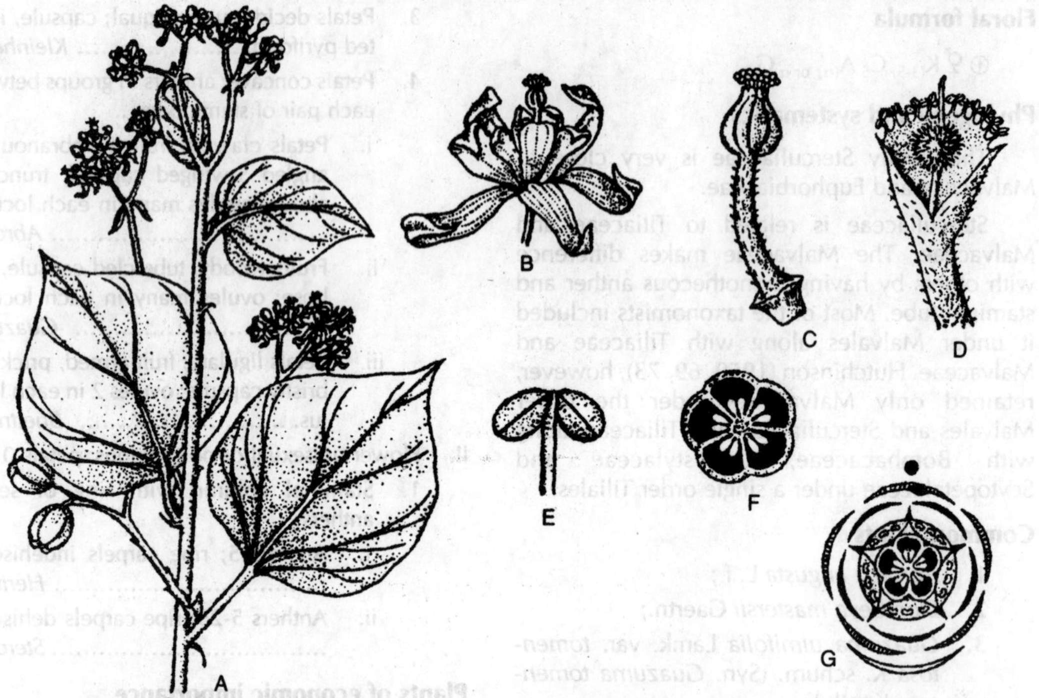

Fig. 4.55 : Sterculiaceae (*Kleinhovia hospita*) : A. Portion of a plant (tree) with flowers and fruits; B. Single flower; C. Staminal tube; D. Same split open to show ovary on gynophore; E. Divergent anthers; F. T.S. of ovary and G. Floral diagram

C. **Fibre**

1. *Abroma augusta, Eriolaena hookeriana, Guazuma ulmifolia, Helicteres isora, Pterygota alata, Sterculia villosa, S. guttata* and *S. balanghas.* The fibres extracted from the bark of the above plants are used in the preparation of cord (a short thin rope).

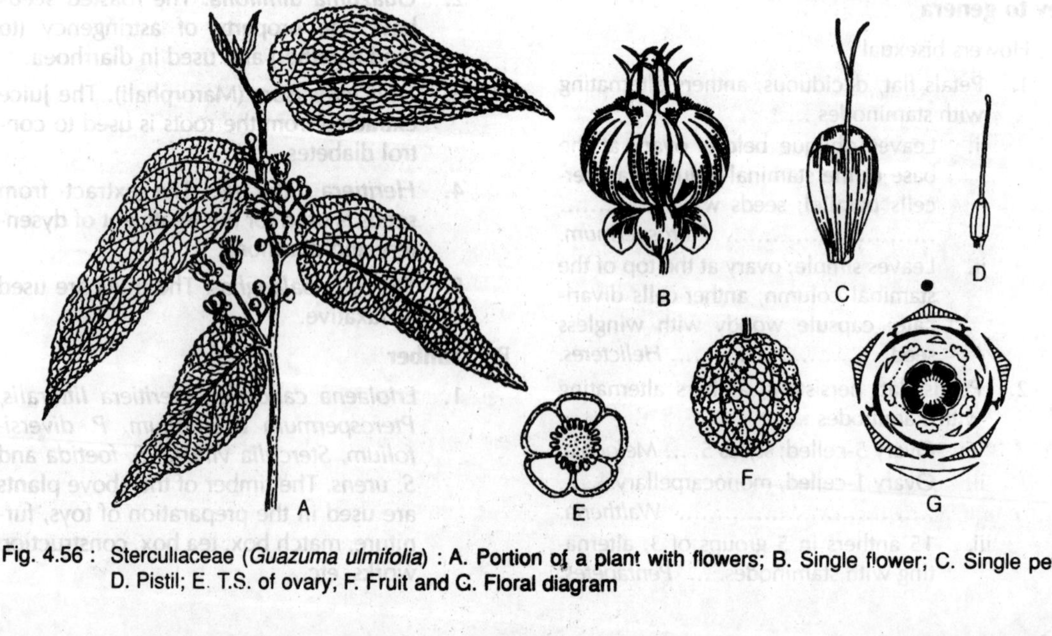

Fig. 4.56 : Sterculiaceae (*Guazuma ulmifolia*) : A. Portion of a plant with flowers; B. Single flower; C. Single petal; D. Pistil; E. T.S. of ovary; F. Fruit and G. Floral diagram

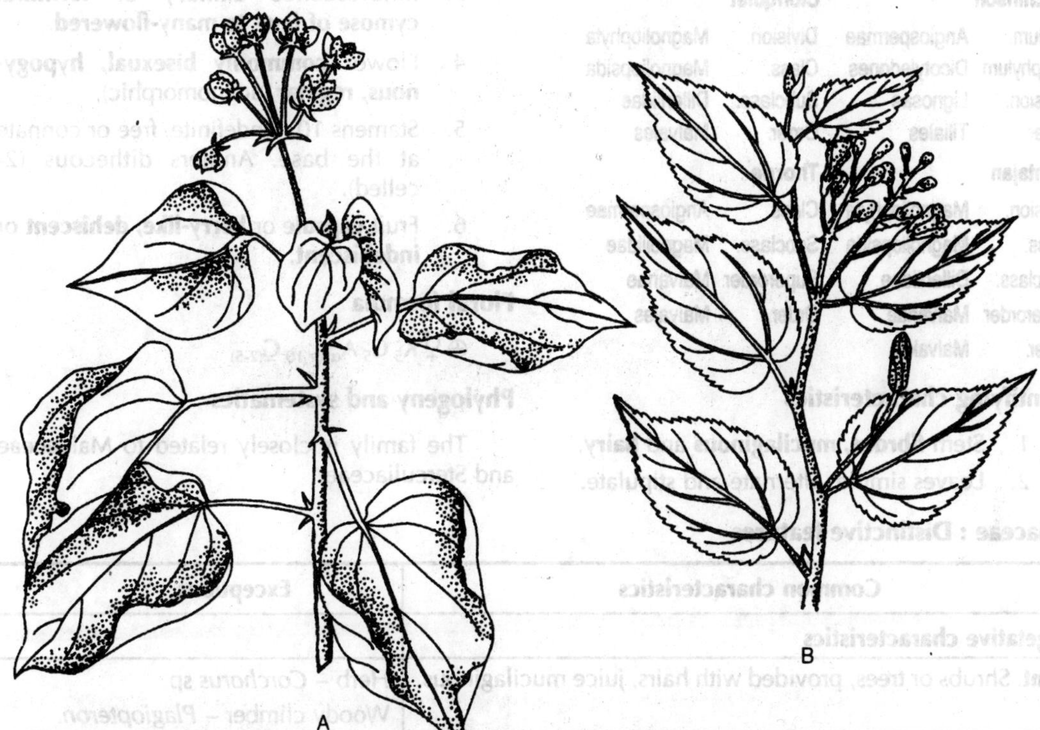

Fig. 4.57 : Some plants of Sterculiaceae : A. *Dombeya mastersii* and B. *Helicteres isora*

D. **Beverages**

1. *Cola acuminata* (Kola nut). The seeds contain significant amount of **theobromine** and **caffeine**, used in soft drinks and beverages.
2. *Theobroma cacao* (Cocoa tree). **Cocoa powder** is prepared from roasted seeds and is used in chocolate and beverage.

E. **Ornamental plants**

The following plants are cultivated for ornamentation and also as avenue trees :

1. *Dombeya natalensis* and *D. spectabilis.*
2. *Guazuma ulmifolia*
3. *Pentapetes phoenicea* (Dophariya). Herb with red flowers which open at noon.
4. *Pterospermum acerifolium* (Kanak champa)
5. *Sterculia balanghas, S. alata, S. foetida* and *S. parviflora.*

F. **Other uses**

1. *Guazuma ulmifolia* and *Sterculia villosa.* The leaves of the plants are used as fodder.
2. *Sterculia urens.* Karaya gum is collected from this plant. It is used in leather, paper and textile industry. It is also used in pharmaceuticals.

10. TILIACEAE

(Linden Family)

The family Tiliaceae is represented by 62 genera and 800 species, found to grow in the tropical and temperate regions of the world. The plants mainly grow in Brazil and South East region of Asia. In India, 14 genera and 110 species of this family is found mostly in the warmer regions.

The type genus of the family is ***Tilia***.

Systematic position

Bentham & Hooker		**Engler & Prantl**	
Division.	Angiospermia	Subdivision.	Angiospermae
Class.	Dicotyledons	Class.	Dicotyledoneae
Subclass.	Polypetalae	Subclass.	Archichlamydeae
Series.	Thalamiflorae	Order.	Malvales
Cohort.	Malvales		

Hutchinson		**Cronquist**	
Phylum.	Angiospermae	Division.	Magnoliophyta
Subphylum.	Dicotyledones	Class.	Magnoliopsida
Division.	Lignosae	Subclass.	Dilleniidae
Order.	Tiliales	Order.	Malvales
Takhtajan		**Thorne**	
Division.	Magnoliophyta	Class.	Angiospermae
Class.	Magnoliopsida	Subclass.	Magnoliidae
Subclass.	Dilleniidae	Superorder.	Malvanae
Superorder.	Malvanae	Order.	Malvales
Order.	Malvales.		

Identifying characteristics

1. Stem **fibrous, mucilaginous** and **hairy**.
2. Leaves simple, alternate and stipulate.
3. Inflorescence **axillary** or **terminal cymose of few- to many-flowered**.
4. Flower **commonly bisexual, hypogynous, regular** (actinomorphic),
5. Stamens 10 – indefinite, free or connate at the base. Anthers dithecous (2-celled).
6. Fruit **capsule** or **berry-like, dehiscent** or **indehiscent**.

Floral formula

$\oplus$ ⚥ $K_5\ C_5\ A_{\alpha \text{ or } 10}\ \underline{G}_{(2\text{-}5)}$

Phylogeny and systematics

The family is closely related to Malvaceae and Sterculiaceae.

Tiliaceae : Distinctive features

Common characteristics	Exceptions
Vegetative characteristics	
Plant. Shrubs or trees, provided with hairs, juice mucilaginous.	Herb – *Corchorus* sp. Woody climber – *Plagiopteron*.
Leaves. Simple, alternate, stipulate (caducous), entire or toothed.	Opposite – *Plagiopteron*.
Reproductive characteristics	
Inflorescence. Cymose (either axillary or terminal), sometimes provided with bracts (*Tilia*).	Axillary racemes – *Elaeocarpus*.
Flower. Regular, bisexual, hypogynous and pentamerous, sometimes with an androphore(*Grewia*).	Unisexual – *Vasivaea* and *Carpodiptera*.
Calyx. Sepals 5, free or basally connate, caducous, valvate.	Sepals 4 – *Sparmannia*.
Corolla. Petals 5 or fewer or 0, free, imbricate or valvate.	Petals 4 – *Sparmannia*. Petals may be absent – *Chartacalyx*.
Androecium. Stamens 10-many, free or slightly connate at the base or in 5-10 bundles (polyadelphous), usually inserted at the base of the petals, anther 2-celled.	Stamens form 5 bundles – *Pentace*. Stamens modified into petaloid staminode – *Brownlowia* and *Pentace*. Androgynophore present – *Grewia*.
Gynoecium. Carpel 2-10, syncarpous, ovary superior, sessile, 2-10 celled with 1-many anatropous ovule on axile placentation, style 1, stigma as many as the number of carpel.	Carpels many – *Apeiba*. Ovary unilocular – *Elaeocarpus*. Placentation parietal – *Goethalsia* and *Mollia*.
Fruit. Fleshy or dry and dehiscent or indehiscent. It is of different types : Drupaceous (*Grewia*) or loculicidal capsule (*Corchorus*).	Berry – *Grewia subinaequalis*.
Seed. Endospermic, embryo large, straight or curved.	

Bentham and Hooker placed it in Malvales of series Thalamiflorae, subclass Polypetalae under the class Dicotyledons. Engler and Prantl placed the family under Malvales, subclass

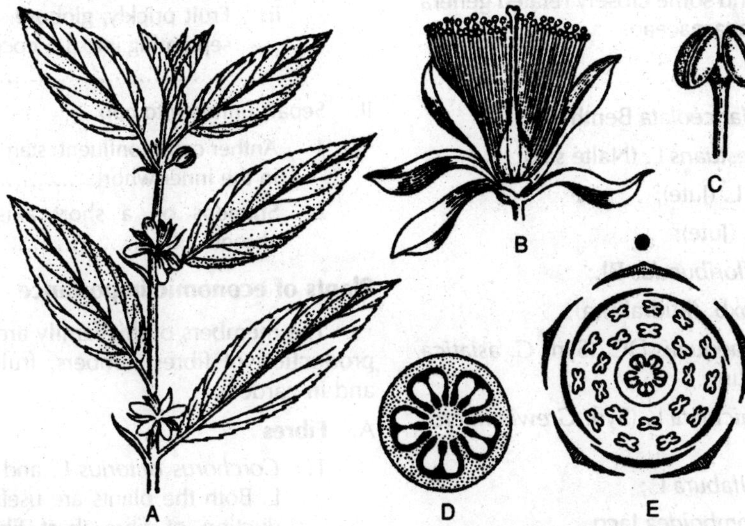

Fig. 4.58 : Tiliaceae (*Corchorus capsularis*) : A. Portion of a flowering shoot; B. Single flower; C. Stamen; D. T.S. of ovary and E. Floral diagram

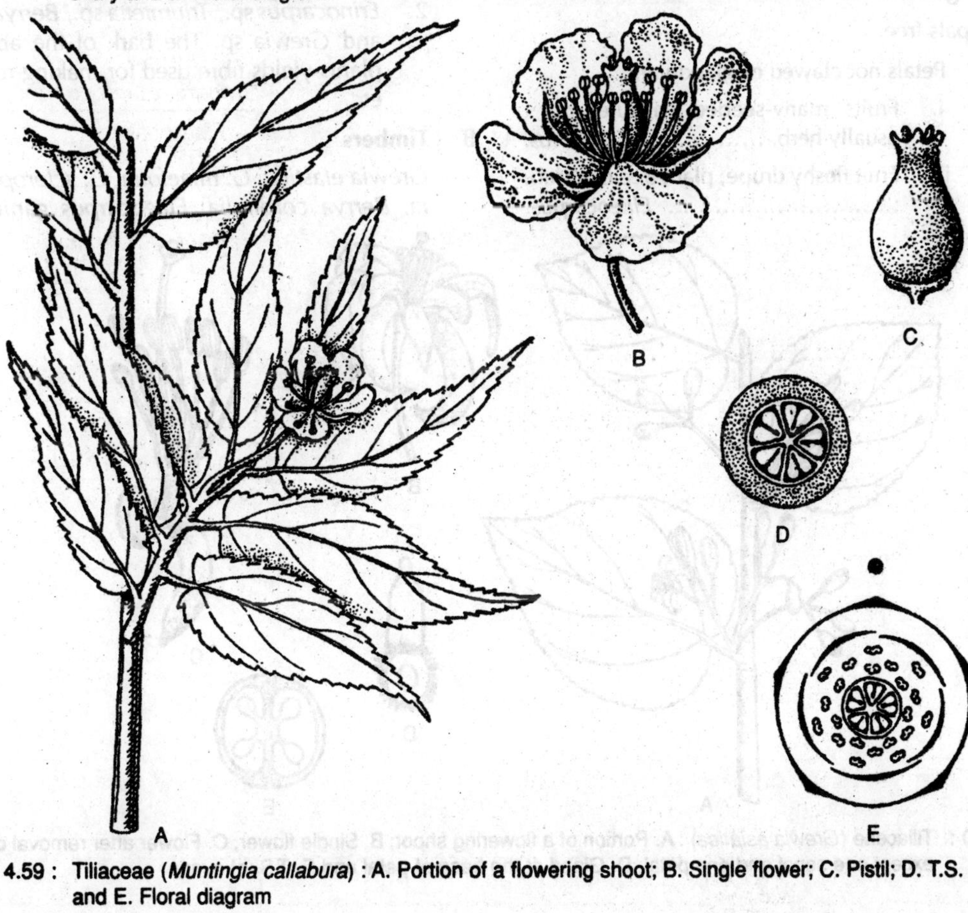

Fig. 4.59 : Tiliaceae (*Muntingia callabura*) : A. Portion of a flowering shoot; B. Single flower; C. Pistil; D. T.S. of ovary and E. Floral diagram

Archichlamydeae and class Dicotyledoneae. Recent taxonomists like Takhtajan (1969, 80) and Cronquist (1968, 81) followed Engler and included *Elaeocarpus* and some closely related genera in the family Elaeocarpaceae.

Common plants

1. *Brownlowia lanceolata* Benth.;
2. *Corchorus aestuans* L. (Nalte sak);
3. *C. capsularis* L. (Jute);
4. *C. olitorius* L. (Jute);
5. *Elaeocarpus floribundus* Bl.;
6. *E. ganitrus* Roxb. (Rudraksha);
7. *Grewia subinaequalis* DC. (Syn. *G. asiatica* Masters non Linn.);
8. *Microcos paniculata* L. (Syn. *Grewia microcos* L.);
9. *Muntingia callabura* L.;
10. *Triumfetta rhomboidea* Jacq.

Key to genera

I. Sepals free.
   1. Petals not clawed or glandular.
      i. Fruits many-seeded capsule; plant usually herb. .... *Corchorus.*
      ii. Fruit fleshy drupe; plant tree or shrub. .... *Elaeocarpus.*
   2. Petals with thickened or glandular claw.
      i. Fruit smooth, drupaceous; leaves 3-9 nerved. .... *Grewia.*
      ii. Fruit prickly, globose, indehiscent or separating into 3-5 cocci. .... *Triumfetta.*

II. Sepals united into cup.
   1. Anther cells confluent; staminodes present in the inner whorl. .... *Brownlowia.*
   2. Stamens on a short axis; staminodes absent. .... *Berrya.*

Plants of economic importance

The members of this family are useful in the production of fibres, timbers, fruits, medicines and in gardens.

A. **Fibres**

1. *Corchorus olitorius* L. and *C. capsularis* L. Both the plants are useful in the production of fibre (bast fibre), used for making carpets, bunny bags, twins etc.
2. *Erinocarpus* sp., *Triumfetta* sp., *Berrya* sp. and *Grewia* sp. The bark of the above plants yields fibre used for making ropes etc.

B. **Timbers**

Grewia elastica, G. tiliaefolia, G. sclerophylla, Berrya cordifolia, Elaeocarpus ganitrus,

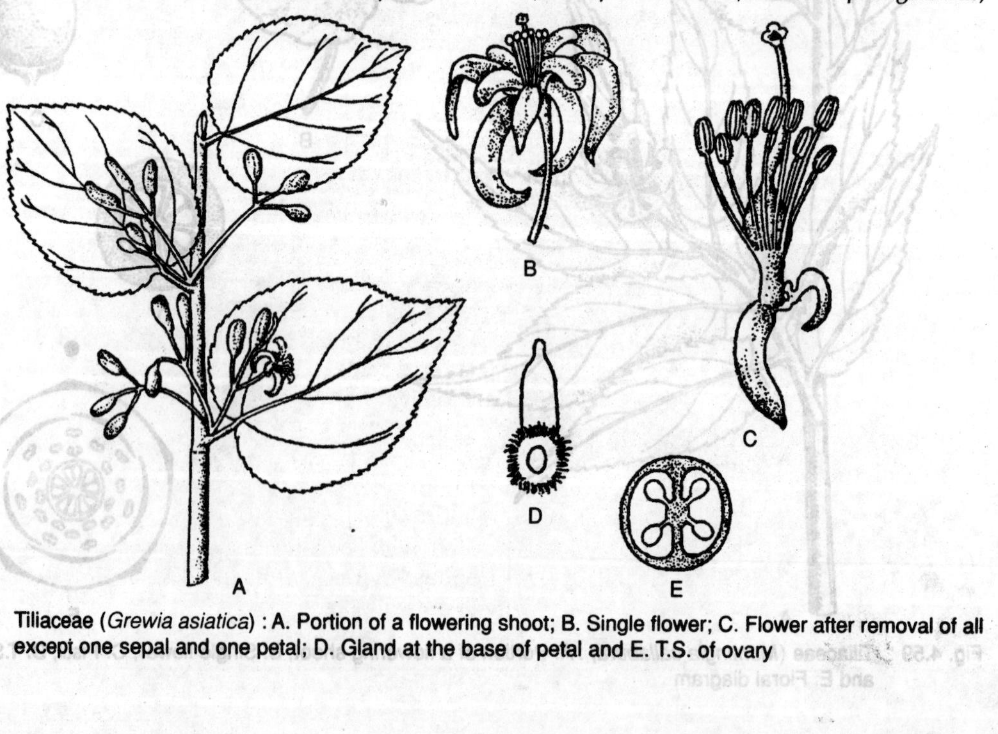

Fig. 4.60 : Tiliaceae (*Grewia asiatica*) : A. Portion of a flowering shoot; B. Single flower; C. Flower after removal of all except one sepal and one petal; D. Gland at the base of petal and E. T.S. of ovary

Tilia vulgaris etc. The timber of the above plants is used for various purposes.

C. **Ornamental plants**

Muntingia calabura, Elaeocarpus ganitrus, Grewia elastica, G. tiliaefolia, G. serrulata, Tilia vulgaris are cultivated as ornamental plants.

D. **Fruits**

The fruits of *Grewia subinaequalis* (Phalsa) and its other species are edible.

E. **Medicines**

1. Plants of *Triumfetta rhomboidea* is used to control gonorrhoea.
2. Roots of *Grewia vallosa* is used to control diarrhoea.
3. Roots of *Grewia sclerophylla* is used to control intestine irritation and cough.
4. Plants of *Microcos paniculata* are used to control itch and eczema.
5. Bark of *Grewia tiliaefolia* is used to control dysentery.

F. **Other uses**

1. As fodder : The young twigs and also the leaves of *Grewia glabra, G. tenax* and some other species are used as food for goats, camels etc.
2. In lac culture : The young leaves and twigs of *Grewia glabra* and *G. serrulata* are used to feed insect in lac culture.

11. RUTACEAE

(Citrus Family)

The family Rutaceae is represented by 150 genera and 900 species, distributed widely in temperate and tropical regions especially in Australia and South Africa. In India, there are 23

Table 4.6 : Comparison among Malvaceae, Sterculiaceae and Tiliaceae

Characteristics	Malvaceae	Sterculiaceae	Tiliaceae
1. Plant	Usually herbs or shrubs, often with mucilage.	Usually trees or shrubs, sometimes herbs.	Usually shrubs or trees, sometimes herbs.
2. Leaves	Alternate, simple, stipulate, margin entire or variously lobed.	Alternate, simple or palmately compound, stipulate, margin lobed or entire.	Alternate, simple, stipulate, margin entire, lobed or dentate.
3. Flowers	Regular, bisexual.	Regular or irregular, bisexual or unisexual.	Regular, bisexual or unisexual.
4. Calyx	Sepals-5, usually surrounded by epicalyx.	Sepals 3-5, free or slightly united at the base.	Sepals-5, free or united.
5. Corolla	Petals-5, almost free; twisted aestivation.	Petals-5, sometimes 0; twisted aestivation.	Petals-5 or less or sometimes 0; imbricate aestivation.
6. Androecium	Stamens many, usually monadelphous.	Stamens many in 2 whorls, outer whorl staminodal or 0, but the inner whorl is fertile or monadelphous.	Stamens 10-many, free or slightly united or polydelphous; rarely situated on an androgynophore.
7. Anthers	1-celled, reniform.	2-celled.	2-celled.
8. Gynoecium	Carpels- 2 to many.	Carpels-4 to 5.	Carpel-1 to 5.
	Ovary 2-many celled with one to many anatropous ovules in each cell.	Ovary often stalked, 5-celled with 2-many anatropous ovules in each cell on axile placentation.	Ovary 2-10 celled with 1-many pendulous or ascending ovule in each cell.
9. Fruit	Capsule or schizocarp.	Capsule or follicle.	Capsule.

genera and about 80 species available in subtropical and tropical regions of Himalayas and Western Peninsular region of India. Economically the family is very important.

The type genus of the family is ***Ruta***.

Systematic position

Bentham & Hooker		**Engler & Prantl**	
Division.	Angiospermia	Subdivision.	Angiospermae
Class.	Dicotyledons	Class.	Dicotyledoneae
Subclass.	Polypetalae	Subclass.	Archichlamydeae
Series.	Disciflorae	Order.	Geraniales
Cohort	Geraniales.		

Hutchinson		**Cronquist**	
Phylum.	Angiospermae	Division.	Magnoliophyta
Subphylum.	Dicotyledones	Class.	Magnoliopsida
Division.	Lignosae	Subclass.	Rosidae
Order.	Rutales	Order.	Sapindales

Takhtajan		**Thorne**	
Division.	Magnoliophyta	Class.	Angiospermae
Class.	Magnoliopsida	Subclass.	Magnoliidae
Subclass.	Rosidae	Superorder.	Rutanae
Superorder.	Rutanae	Order.	Rutales
Order.	Rutales.		

Rutaceae : Distinctive features

Common characteristics	Exceptions
Vegetative characteristics	
Plant. Usually shrubs or trees, often provided with spines (*Aegle* and *Citrus*).	Herb – *Boenninghausenia*.
Leaves. Alternate or opposite, simple or compound (pinnately, or *palmately*), smoth, gland-dotted, exstipulate.	
Reproductive characteristics	
Inflorescence. Usually cymose.	
Flower. Regular (actinomorphic), bisexual, hypogynous with a cushion-like or annular disc generally present above or between the stamens. Generally pentamerous.	Unisexual – *Zanthoxylum*. Trimerous – *Lunasia*. Tetramerous – *Acronychia, Ruta*. Slightly zygomorphic – *Dictamnus, Galipea* and *Almeidea*.
Calyx. Sepals 4-5, free or united, glandular, imbricate.	
Corolla. Petals 4-5, free, valvate or imbricate in bud.	Petals united – *Correa*.
Androecium. Stamens as many as or twice the number of petals and then they become obdiplostemonous, usually free, unequal in size, anther 2-celled, introrse.	Numerous – *Citrus* and *Aegle*. Diplostemonous – *Fagara*. Polydelphous – *Citrus*. Monadelphous – *Atalantia*.

Contd.

Identifying characteristics

1. Generally **shrubs** or **trees**, rarely herbs.
2. Leaves **simple** or **compound, exstipulate** and commonly **gland dotted**.
3. Inflorescence usually **cymose**.
4. Flower **bisexual, regular** and **hypogynous**.
5. Perianth **tetra-** or **pentamerous**.
6. Stamens **isomerous** or **double the number of perianth**, free or variously united.
7. Ovary superior, carpels are commonly **free at the apex and united below**.
8. Fruit **capsule, baccate** or a **drupe**.

Floral formula

$\oplus$ ⚥ $K_{4-5}C_{4-5}A_{8-10}\underline{G}_{(4-5)}$

Phylogeny and systematics

The family Rutaceae is related to Meliaceae and Anacardiaceae. It is also allied to Geraniaceae.

Bentham and Hooker placed Rutaceae under Geraniales, series Disciflorae, subclass Polypetalae and class Dicotyledons. Engler and Prantl also placed it under Geraniales. Both the group of taxonomists considered it to be related

Common characteristics	Exceptions
Gynoecium. Carpel 4-5, syncarpous, ovary superior, elevated on disc, multilocular with 1-2 or more ovules in each chamber on axile placentation, style short, stigma capitate.	Ovary 1 – chambered with parietal placentation – *Feronia*. Carpels are semi-free – *Xanthoxylum*.
Fruit. Berry (Hesperidium in *Citrus* or Amphisarca in *Aegle* and *Feronia*).	Capsule – *Ruta*. Drupe – *Skimmia*.
Seed. May or may not contain endosperm, embryo large, straight or curved.	Polyembryony – *Citrus*.

with Geraniaceae. But Hutchinson (1969) did not consider to be related to Geraniaceae and included Rutaceae together with Burseraceae and Simaroubaceae under Rutales.

Common plants

1. *Aegle marmelos* Corr. (Wood apple);
2. *Atalantia monophylla* Corr.;
3. *Citrus sinensis* Osbeck. (Sweet orange);
4. *Clausena heptaphylla* W. & A.;
5. *Feronia elephantum* Correa (Elephant apple);
6. *Glycosmis pentaphylla* Corr.;
7. *Murraya exotica* L. (Orange jasmine);
8. *M. koenigii* Spreng (Curry-leaf plant);
9. *Zanthoxylum budranga* L., etc.

Key to genera

I. Fruit syncarpous; ovary entire; flowers polygamous :

1. Stem unarmed

   i. Flowers usually 5-merous, leaves simple .... *Skimmia*.

   ii. Flowers 4-merous, leaves 1-3-foliate ....*Acronychia*.

2. Stem armed

   Flowers 2-5 merous, 1-3-foliate .... *Toddalia*.

II. Fruit capsule; ovary deeply 2-5-lobed; flowers polygamous; leaves 3-foliate or imparipinnate .... *Zanthoxylum*.

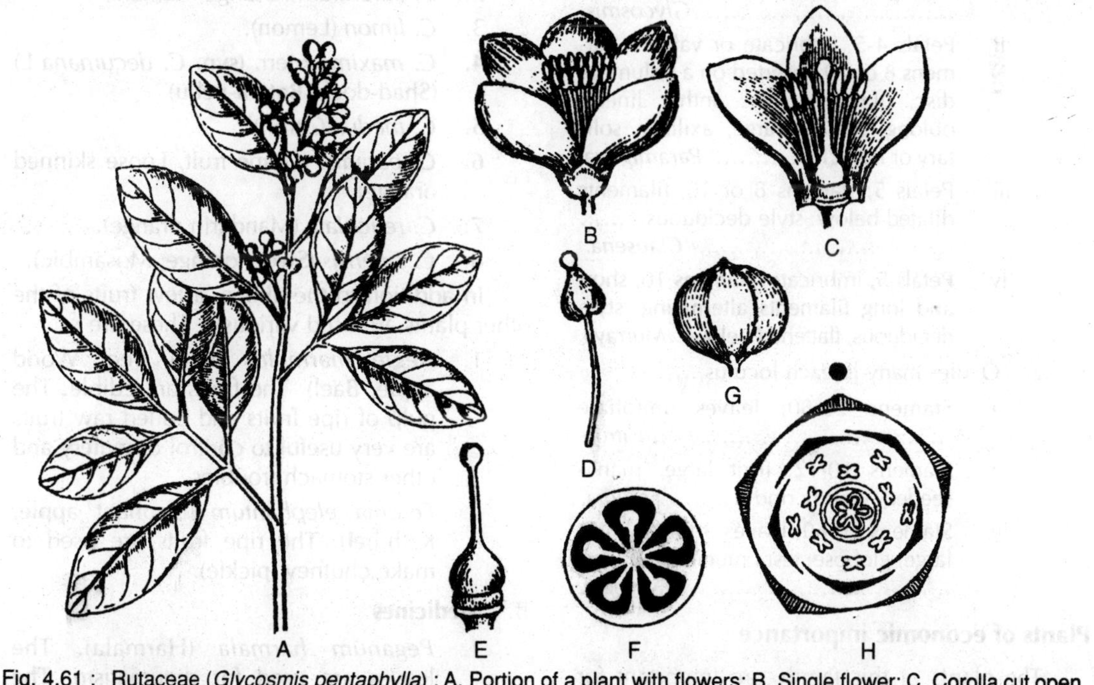

Fig. 4.61 : Rutaceae (*Glycosmis pentaphylla*) : A. Portion of a plant with flowers; B. Single flower; C. Corolla cut open showing stamens and gynoecium; D. Single stamen; E. Pistil (Gynoecium); F. T.S. of ovary; G. Fruit and H. Floral diagram

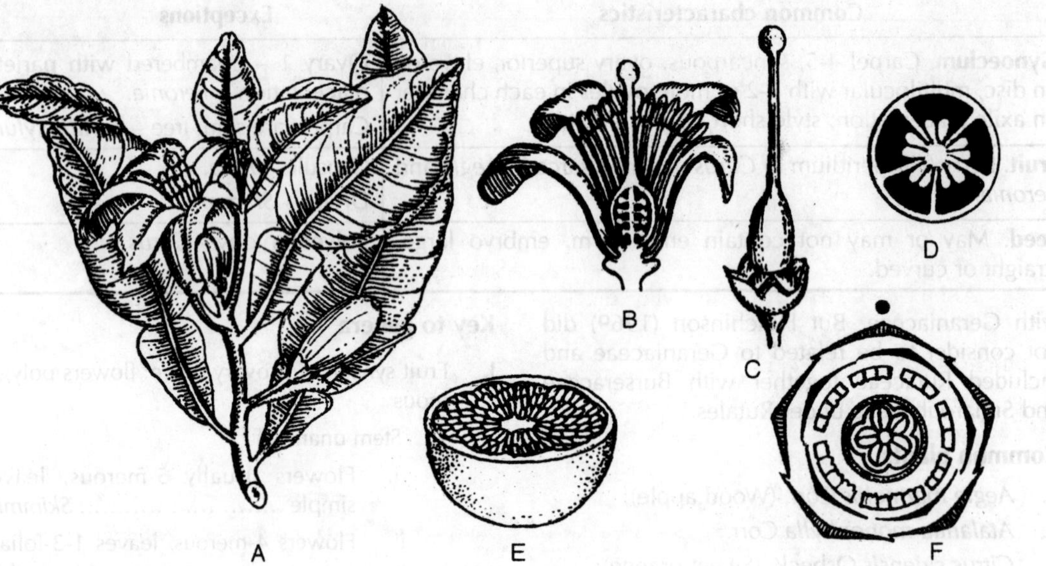

Fig. 4.62 : Rutaceae (*Citrus aurantinum*) : A. Portion of a plant with flowers; B. V.S. of a flower; C. Pistil; D. T.S. of ovary; E. T.S. of fruit and F. Floral diagram

III. Fruit berry; ovary entire; flowers bisexual; leaves compound :

1. Ovules solitary or 2 in a loculus
   i. Petals 4-5, imbricate; ovary 2-5-celled; style very short; persistent. .... *Glycosmis.*
   ii. Petals 4-5, imbricate or valvate; stamens 8 or 10, inserted on a columnar disc, filaments free; anther linear-oblong; flowers large, axillary, solitary or fascicled. .... *Paramignya.*
   iii. Petals 5; stamens 8 or 10; filaments dilated below; style deciduous. .... *Clausena.*
   iv. Petals 5, imbricate; stamens 10, short and long filaments alternating; style deciduous, flattened below...*Murraya.*
2. Ovules many in each loculus ...
   i. Stamens 20-60; leaves unifoliate .... *Citrus.*
   ii. Stamens 10-12; fruit large, many-seeded, woody rind .... *Feronia.*
   iii. Stamens 30-60; leaves 3-foliate; fruit large, globose; testa mucilagenous ... . *Aegle.*

Plants of economic importance

The plants of this family are important for fruits and medicines. A few also have ornamental value.

A. **Fruits**

The family is very much important for its fruits. Name of some important fruit plants are :

1. *Citrus aurantifolia* (Lime).
2. *C. aurantium* (Orange, Kamala)
3. *C. limon* (Lemon).
4. *C. maxima* Merr. (syn. *C. decumana* L) (Shad-dock, Batabi-nebu).
5. *C. media* (Citron).
6. *C. paradisi* (Grape fruit, Loose skinned orange),
7. *C. reticulata* (Mandarin orange).
8. *C. sinensis* (Sweet orange, Mosambic).

In addition to the above, a few fruits of the other plants are used variously. These are :

1. *Aegle marmelos* (Bael tree, Wood apple, Bael). The fruits are edible. The pulp of ripe fruits and boiled raw fruits are very useful to control dysentery and other stomach troubles.
2. *Feronia elephantum* (Elephant apple, Kath-bel). The ripe fruits are used to make chutney (pickle).

B. **Medicines**

1. *Peganum harmala* (Harmala). The leaves are used for rheumatism. The extract of the seeds are used in fevers, asthma and rheumatism.

2. *Toddalia asiatica*. The extract of the root bark is used as antimalarial drug.
3. *Zanthoxylum armatum* (Tejpat). The plant is an insect repellant. Oil extracted from the fruits has an antiseptic properties and is used in tooth-paste and soap industries.
4. *Z. nitidum*. The roots are used in toothache.
5. *Z. limonella*. The fruits are very digestive.

C. **Spices**

1. *Murraya koenigii* (curry-leaf plant). The aromatic leaves are used for flavouring curries and other items mainly in southern part of India.
2. *Zanthoxylum limonella*. The leaves are used as condiments.

D. **Ornamental plants**

The following plants are grown as ornamentals :

1. *Dictamnus albus*.
2. *Glycosmis pentaphylla*.
3. *Murraya paniculata* (syn. *M. exotica*) (Orange jasmine, Kamini). Plants with white fragrant flower.
4. *Ruta graveolens* (Common rue).
5. *Skimmia arborescens*.

12. ANACARDIACEAE

(Cashew or Sumac Family)

The family Anacardiaceae is represented by 70 genera and about 615 species, distributed mostly in tropical regions and some members are found to grow in temperate regions of Asia, America and also in Europe. About 22 genera and 110 species are commonly found in Himalayan range of India.

The type genus of the family is ***Anacardium***.

Systematic position

Bentham & Hooker		**Engler & Prantl**	
Division.	Angiospermia	Subdivision.	Angiospermae
Class.	Dicotyledons	Class.	Dicotyledoneae
Subclass.	Polypetalae	Subclass.	Archichlamydeae
Series.	Disciflorae	Order.	Sapindales
Cohort.	Sapindales.		

Hutchinson		**Cronquist**	
Phylum.	Angiospermae	Division.	Magnoliophyta
Subphylum.	Dicotyledones	Class.	Magnoliopsida
Division.	Lignosae	Subclass.	Rosidae
Order.	Sapindales	Order.	Sapindales

Takhtajan		**Thorne**	
Division.	Magnoliophyta	Class.	Angiospermae
Class.	Magnoliopsida	Subclass.	Magnoliidae
Subclass.	Rosidae	Superorder.	Rutanae
Superorder.	Rutanae	Order.	Rutales
Order.	Burserales.		

Identifying characteristics

1. Plant shrub or woody trees.
2. Leaves generally **simple, exstipulate** and **alternate.**
3. Inflorescence **usually axillary** or **terminal panicle.**
4. Stamens **commonly 10**, often reduced to **1, develop from** or **under the intrastaminal disc.**
5. Ovary **superior, carpel 1-5, 1-5 chambered**, with **marginal** or **axial placentation.**
6. Fruit **drupe.**

Floral formula

$\oplus$ ⚥ $K_{3-5} C_{3-5} A_{10} \underline{G}_{(1-5)}$

Phylogeny and systematics

The family is closely related with Sapindaceae and Rutaceae.

Bentham and Hooker placed Anacardaceae in the Cohort (= Order) Sapindales under the Series Disciflorae, Subclass Polypetalae of Class Dicotyledons. They divided the family into two tribes :

1. Anacardieae. Ovary commonly 1-celled, e.g., *Mangifera, Anacardium, Pistacia* etc.
2. Spondieae. Ovary 2-5 celled. e.g., *Dracontomelum* and *Spondias*.

Airy Shaw (1973) separated the genus *Pistacia* and placed it in a separate family Pistaciaceae for the following characteristics :

1. Plants are dioecious,
2. Absence of petals, and
3. Structure of pollen grains.

Anacardiaceae : Distinctive features

Common characteristics	Exceptions
Vegetative characteristics	
Plant. Trees or shrubs, often with resinous or milky juice.	
Leaves. Alternate, usually simple, exstipulate.	Opposite – *Dobinea*. Compound – *Rhus, Spondias* and *Odina*.
Reproductive characteristics	
Inflorescence. Panicle, axillary or terminal.	
Flower. Small, regular, bisexual, hypogynous, more or less cup-shaped, glandular disc is present between the stamens.	Unisexual – *Odina, Semecarpus, Rhus, Pistacia*. Perigynous – *Holigarna*.
Calyx. Sepals 3-5, connate.	
Corolla. Petals 3-5, free.	Flower apetalous – *Pistacia*. Pentamerous – *Buchanania*.
Androecium. Stamens 10, free or connate at the base, anther 2-celled.	Stamens (10) all fertile – *Buchanania*. Stamens (7–10) but 1 fertile – *Anacardium*. Stamens 5, all fertile – *Rhus*. Stamens (5) but 1 fertile – *Mangifera*.
Gynoecium. Carpel usually 3 but functionally 1, ovary 1-celled, ovule solitary, anatropous, ascending or pendulous with dorsal raphe. Style usually 1, stigma capitate.	Carpel 1 – *Anacardium, Mangifera*. Ovary 5 loculed – *Spondias* and *Buchanania*. Gynophore present – *Melanorrhoea*.
Fruit. Commonly a drupe, rarely nut.	
Seed. Exalbuminous, embryo large with fleshy and curved cotyledons.	

Engler and Prantl placed the family under the Order Sapindales, Subclass Archichlamydeae of Class Dicotyledoneae. Hutchinson also placed the family under the Order Sapindales and placed it near Rutales.

Most of the taxonomists placed the family under Sapindales, but Takhtajan (1960, 80) placed it under Rutales instead of Sapindales.

Common plants

1. *Anacardium occidentale* L. (Cashew nut);
2. *Buchanania lanzan* Spreng. (Syn. *B. latifolia* Roxb.) (Piyal);
3. *Mangifera indica* L. (Mango);
4. *Semecarpus anacardium* L. f.;
5. *Spondias pinnata* (L. f.) Kurz. (Syn. *S. mangifera* Willd);
6. *Odina* woodier Roxb (Syn. *Lannea coromandelina* Merr.).

Key to genera

I. Ovary 1- or 2-celled :
   1. Leaves alternate; perianth not fleshy ...
      i. Calyx 4-5; corolla 4-6; leaf pinnate. .... *Rhus*.
      ii. Calyx 4-5; stamens 5 (usually 1 fertile); style 1, filiform; leaf simple. ... . *Mangifera*.
      iii. Calyx 5; stamens 8-10 (usually with staminodes) : style 1, filiform; torus striated; leaf simple ...... *Anacardium*.

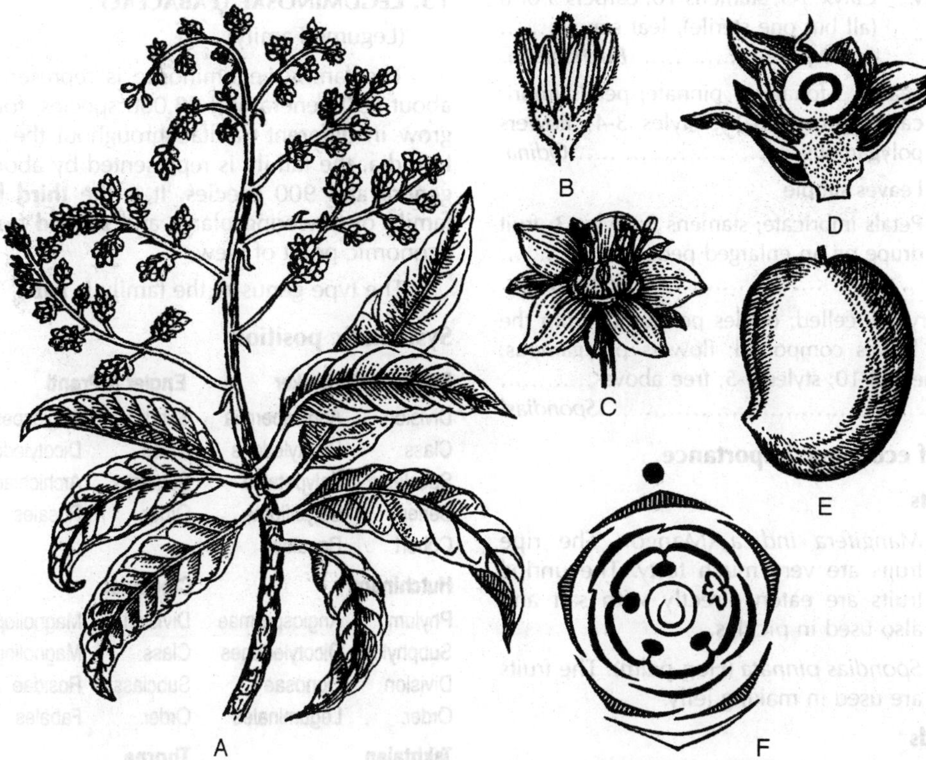

Fig. 4.63 : Anacardiaceae (*Mangifera indica*) : A. Portion of a flowering twig with terminal panicle; B. Single flower; C. Male flower; D. V.S. of a bisexual flower; E. Fruit and F. Floral diagram

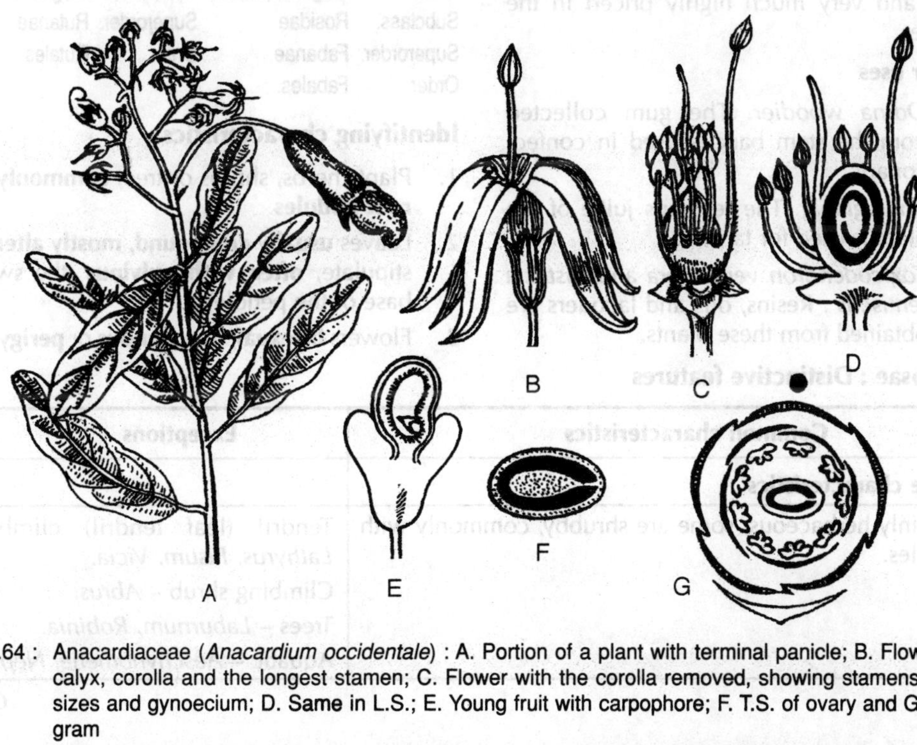

Fig. 4.64 : Anacardiaceae (*Anacardium occidentale*) : A. Portion of a plant with terminal panicle; B. Flower showing calyx, corolla and the longest stamen; C. Flower with the corolla removed, showing stamens of unequal sizes and gynoecium; D. Same in L.S.; E. Young fruit with carpophore; F. T.S. of ovary and G. Floral diagram

iv. Calyx 3-5; stamens 10; carpels 5 or 6 (all but one sterile), leaf simple .... *Buchanania.*

2. Leaves 3-foliate or pinnate; petals imbricate; stamens 8-10; styles 3-4; flowers polygamous............................ *Odina.*
3. Leaves simple

   Petals imbricate; stamens 5; styles 3; fruit drupe on an enlarged peduncle .... *Semecarpus.*

II. Ovary 2-5-celled; ovules pendulous from the top; leaves compound; flowers polygamous; stamens 8-10; styles 4-5, free above ....*Spondias.*

Plants of economic importance

A. **Fruits**

1. *Mangifera indica* (Mango). The ripe fruits are very much tasty. The unripe fruits are eaten directly with salt and also used in pickles.
2. *Spondias pinnata* (Hog-palm). The fruits are used in making jelly.

B. **Seeds**

Anacardium occidentale (Cashew nut) and *Pistacia vera* (Pistachio nut). The seeds are tasty and very much highly priced in the market.

C. **Other uses**

1. *Odina woodier.* The gum collected from the stem bark is used in confectionary.
2. *Rhus glabar.* The resinous juice of the plant is used for tanning.
3. *Toxicodendron vernicifera* and *Pistacia lentiscus* : Resins, oils and lacquers are obtained from these plants.

13. Leguminosae (Fabaceae)

(Legume Family)

The family Leguminosae is represented by about 630 genera and 18,000 species, found to grow in different habitat throughout the world. In India, the family is represented by about 138 genera and 900 species. It is the **third largest** family of flowering plants and **second** from the economic point of view.

The type genus of the family is ***Faba*.**

Systematic position

Bentham & Hooker		**Engler & Prantl**	
Division.	Angiospermia	Subdivision.	Angiospermae
Class.	Dicotyledons	Class.	Dicotyledoneae
Subclass.	Polypetalae	Subclass.	Archichlamydeae
Series.	Calyciflorae	Order.	Rosales
Cohort.	Rosales.		
Hutchinson		**Cronquist**	
Phylum.	Angiospermae	Division.	Magnoliophyta
Subphylum.	Dicotyledones	Class.	Magnoliopsida
Division	Lignosae	Subclass.	Rosidae
Order.	Leguminales	Order.	Fabales
Takhtajan		**Thorne**	
Division.	Magnoliophyta	Class.	Angiospermae
Class.	Magnoliopsida	Subclass.	Magnoliidae
Subclass.	Rosidae	Superorder.	Rutanae
Superorder.	Fabanae	Order.	Rutales
Order.	Fabales.		

Identifying characteristics

1. Plants herbs, shrubs or tree; **commonly with root nodules.**
2. Leaves **usually compound, mostly alternate,** stipulate, **often with pulvinus** (the swollen base of the petiole).
3. Flowers **bisexual, hypogynous** or **perigynous.**

Leguminosae : Distinctive features

Common characteristics	Exceptions
Vegetative characteristics	
Plant. Mainly herbaceous, some are shrubby, commonly with root nodules.	Tendril (leaf tendril) climber – *Lathyrus, Pisum, Vicia.* Climbing shrub – *Abrus.* Trees – *Laburnum, Robinia.* Aquatic – *Aeschynomene, Neptunia.*

Contd.

Contd.

Common characteristics	Exceptions
Leaves. Alternate, usually compound (pinnately – *Sesbania, Dalbergia, Abrus* or digitately – *Parochetus* and *Trifolium*), stipulate, leaf base frequently swollen forming pulvinus.	Leaves simple – *Crotalaria juncea*. Stipules are adnate – *Trifolium*.
Reproductive characteristics	
Inflorescence. Racemose.	
Flower. Bisexual, regular or irregular, lateral in position, slightly perigynous.	
Calyx. Sepals 5, united or free, odd sepals anterior, imbricate.	Calyx petaloid – *Saraca asoka, Dialium, Apalatoa*. Sepals 3 – *Mimosa juniperina*. Sepals 4 – *Mimosa pudica*, some species of *Acacia*.
Corolla. Petals 5, free, imbricate or valvate.	Petals absent – *Saraca asoka, Dialium, Apalatoa*. Petals 3 – *Mimosa juniperina*. Petals 4 – *Mimosa pudica*, some species of *Acacia*.
Androecium. Stamens 10 or many, definite or indefinite, free (*Ormosia, Sophora*) or united at the base, usually diadelphous (9+1), anther 2-celled.	Monadelphous – *Crotalaria, Inga*.
Gynoecium. Carpel 1, ovary superior, 1 celled with 1-many ovules on marginal placentation, style 1, long, stigma 1.	Carpels 2 – *Affonsea, Tounatea*. Carpel 10 to 15 – *Archidendron*.
Fruit. Usually Legume, sometimes Lomentum.	
Seed. Nonendospermic, embryo large and curved, cotyledons fleshy and plano-convex.	

4. Sepals 5, **odd sepals anterior**.
5. Petals 5, free.
6. **Carpel-1**, 1-chambered, **placentation marginal**.
7. **Fruits legume** or **lomentum**.

Subdivisions of the family

This family is the second largest family of the members of phanerogams. The family has been divided into three subfamilies based mainly on the characteristics of flower : A. **Papilionoideae**, B. **Caesalpinioideae** and C. **Mimosoideae**.

A. **Papilionoideae.** Leaves are imparipinnate; Inflorescence raceme; Flower solitary or in raceme, irregular (papilionaceous); Petals 5, vaxillary in aestivation, posterior petal biggest and outermost; Stamens usually 10, commonly diadelphous (9) +1, sometimes monadelphous (*Crotalaria, Arachis* etc.) rarely 9 (*Abrus*); Embryo curved. viz., *Desmodium, Crotalaria,* etc.

B. **Caesalpinoideae.** Leaves are usually paripinnate or bipinnate; Inflorescence raceme or corymb; Flowers slightly irregular (but not papilionaceous); Petals 5, imbricate, posterior petal innermost and smallest; Stamens 10 or less, some are sterile. Embryo straight. viz., *Cassia, Bauhinia, Caesalpinia* etc. [Some exceptions : Herb-*Cassia absus, C. tora*; Woody climber-*Caesalpinia bonducella*; Climbing by stem tendril-*Bauhinia vahlii*.]

C. **Mimosoideae.** Leaves are usually bipinnate; Inflorescence head or spike; Flowers regular (actinomorphic); Petals 4, united or free, valvate; Stamens all fertile-numerous, sometimes definite and exerted; Embryo straight. viz., *Mimosa, Acacia,* etc.

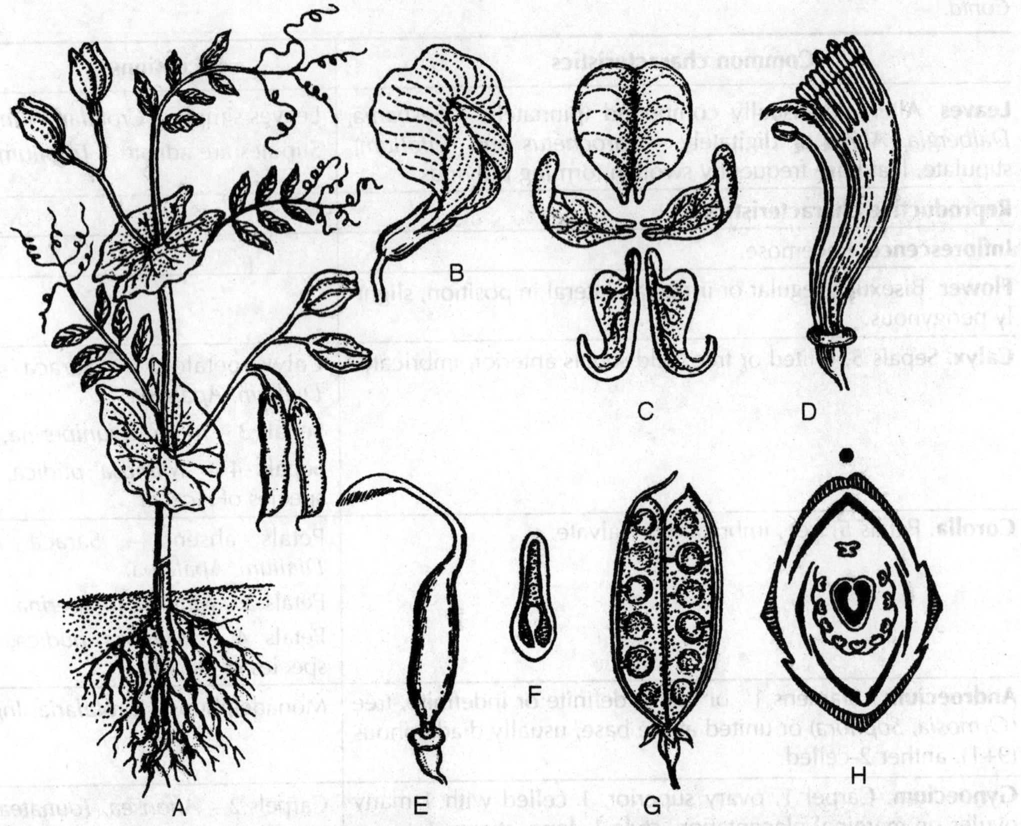

Fig. 4.65 : Leguminosae : Papilionoideae (*Pisum sativum*) : A. Plant with flowers and fruits; B. Single flower; C. Arrangement of petals; D. Flower with calyx and corolla removed, showing diadelphous androecium and pistil; E. Pistil; F. T.S. ovary; G. Splitted fruit, showing seeds and H. Floral diagram

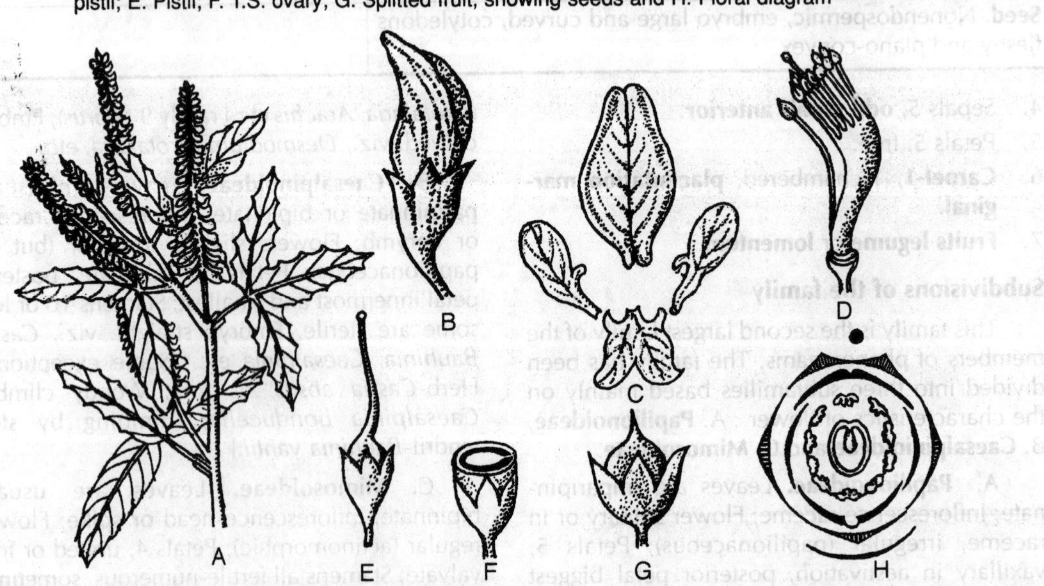

Fig. 4.66 : Leguminosae : Papilionoideae (*Melilolus alba*) : A. Portion of a flowering plant; B. Single flower; C. Arrangement of petals; D. Flower with calyx and corolla removed; E. Pistil; F. T.S. of ovary along with lower part; G. Fruit and H. Floral diagram

Distribution of the subfamilies

A. **Papilionoideae.** The subfamily Papilionoideae is represented by about 440 genera and 12,800 species, grows abundantly in the tropical and subtropical regions of the world. About 100 genera and more than 748 species are found to grow in the Himalayas and Peninsular India. Economically the subfamily is very much important than the other two subfamilies.

B. **Caesalpinoideae.** The subfamily Caesalpinoideae is represented by about 150 genera and 2,700 species, grows in both tropical and subtropical regions, but are very common in America. In India, the subfamily is represented by 23 genera and 80 species, found to grow abundantly in tropical region of Eastern Himalayas and also in the western part of Peninsular India.

C. **Mimosoideae.** The subfamily Mimosoideae consists of about 40 genera and 2,500 species, found to grow in tropical and subtropical regions, but profusely grows in Southern Hemisphere. In India, it is represented by 15 genera and 72 species, grow commonly in western part of Peninsular India and also in tropical and subtropical Himalayas.

Floral formula

Subfamily. Papilionoideae

$\circ|\circ$ ⚥ $K_{(5)}C_5A_{(9)+1 \text{ or } (10)}\underline{G}_1$

Subfamily. Caesalpiniodeae

$\circ|\circ$ ⚥ $K_{(5) \text{ or } 5}\, C_5A_{10}\underline{G}_1$

Subfamily. Mimosoideae

$\oplus$ ⚥ $K_{(4\text{-}5)}C_{4\text{-}5}A_{\alpha \text{ or } 10 \text{ or } 4}\,\underline{G}_1$

Phylogeny and systematics

The family Leguminosae is closely related to Rosaceae.

Bentham and Hooker placed the Natural order (= family) Leguminosae in the cohort Rosales of series Calyciferae, subclass Polypetalae and class Dicotyledons. The family is divided into three subfamilies as Papilionoideae, Caesalpiniodeae and Mimosoideae.

Engler and Prantl placed the family Leguminosae under the order Rosales of subclass Archichlamydeae and class Dicotyledoneae.

Hutchinson treated the family Leguminosae as a separate order Leguminales which contains

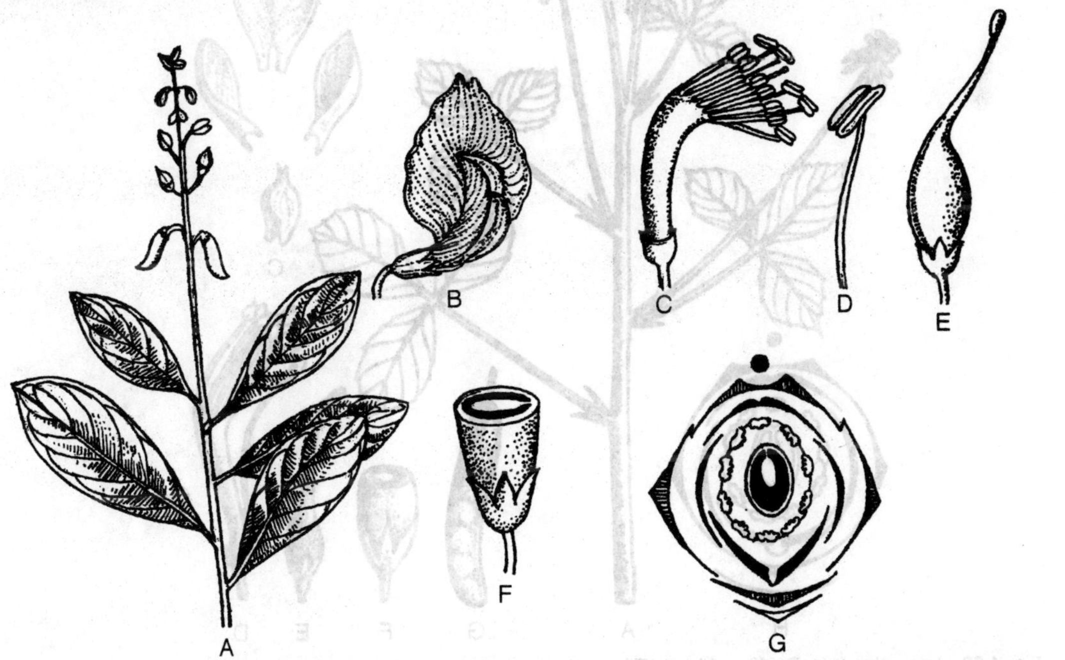

Fig. 4.67 : Leguminosae : Papilionoideae (*Crotalaria juncea*) : A. Portion of a flowering plant; B. Single flower; C. Monadelphous stamen around the gynoecium; D. Stamen; E. Pistil; F. T.S. of ovary along with lower part and G. Floral diagram

three independant families : Mimosaceae, Caesalpiniaceae and Papilionaceae. Takhtajan and Thorne treated Leguminosae as a family Fabaceae after the Type genus *Faba*. Takhtajan placed this family under the order Fabales of the superorder Fabonae while Thorne placed it under the order Rutales of the superorder Rutanae.

Common plants

Subfamily. **Papilionoideae**

1. *Abrus precatorius* L.
2. *Aeschynomene aspera* L.;
3. *Butea monosperma* (Lam.) Taub.;
4. *Clitoria ternatea* L.;
5. *Crotalaria pallida* Aiton. (Syn. *C. striata* DC.);
6. *Desmodium gyrans* DC.;
7. *Dalbergia sissoo* Roxb.;
8. *Erythrina indica* Lamk;
9. *Indigofera tinctoria* L.;
10. *Pongamia pinnata* (L.) Pierre.;
11. *Pterocarpus marsupium* Roxb.;
12. *Tephrosia purpurea* Pers. etc.

Subfamily. **Caesalpinioideae**

1. *Bauhinia purpurea* L.;
2. *B. acuminata* L.
3. *Caesalpinia pulcherrima* Sw.;
4. *Cassia sophera* L.;
5. *C. tora* L.;
6. *C. fistula* L.;
7. *Delonix regia* Raf.;
8. *Peltophorum ferrugineum* Berth.;
9. *Tamarindus indicus* L.

Subfamily. **Mimosoideae**

1. *Acacia auriculiformis* A. Cunn.;
2. *A. arabica* Willd;

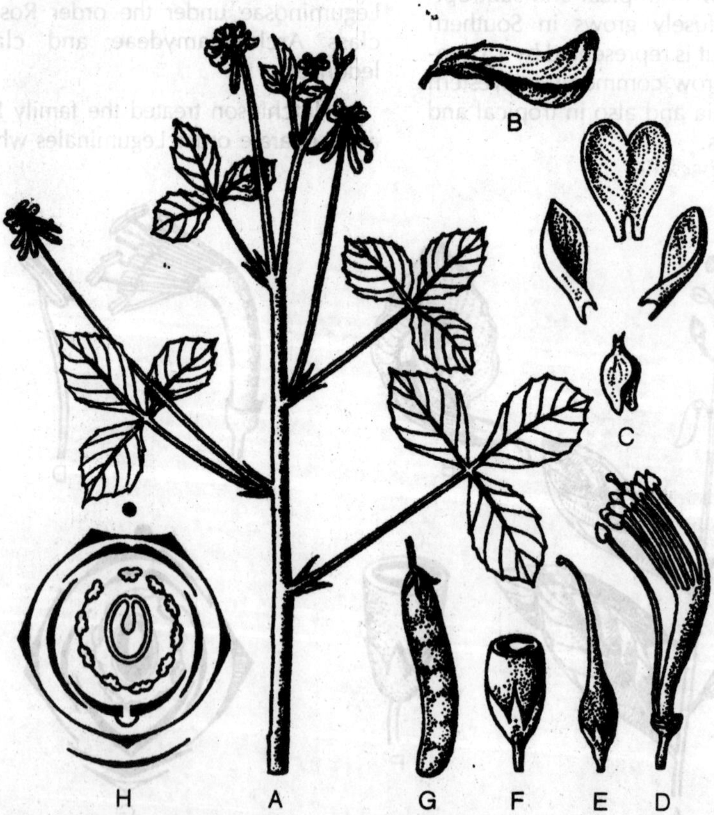

Fig. 4.68 : Leguminosae : Papilionoideae (*Trigonella corniculata*) : A. Portion of a shoot with flowers; B. Single flower; C. Arrangement of petals; D. Diadelphous stamen around the gynoecium; E. Gynoecium; F. T.S. of ovary with lower part of gynoecium; G. Fruit; and H. Floral diagram

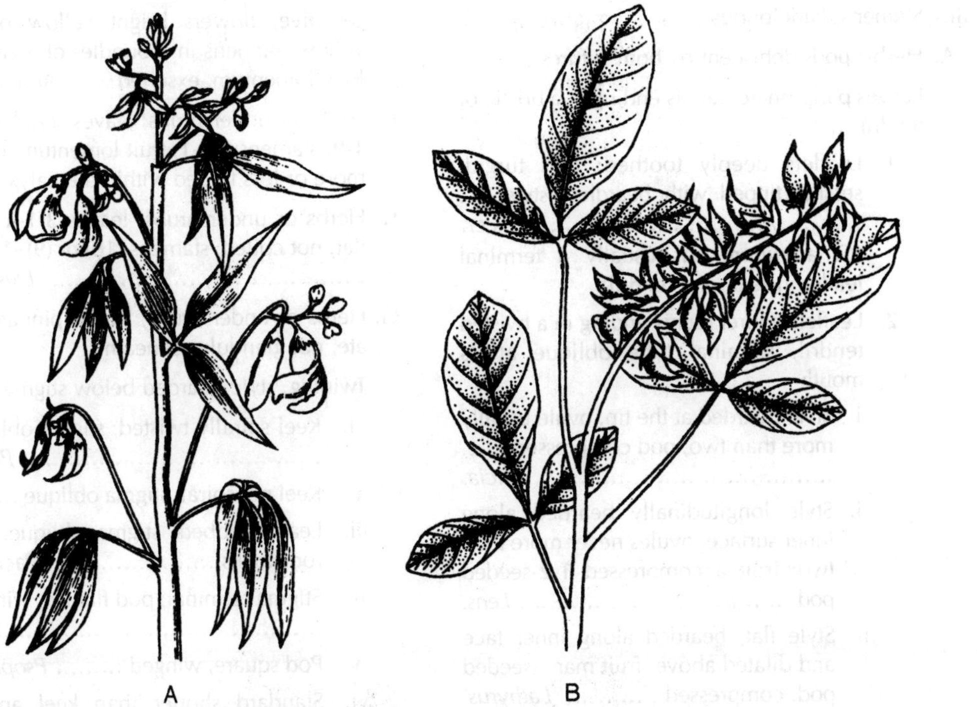

Fig. 4.69 : Leguminosae : Papilionoideae. Portion of plants with flowers : A. *Cajanus cajan* and B. *Crotalaria pallida*

3. *Albizzia lebbeck* Benth;
4. *Entata pursaetha* DC.;
5. *Inga dulcis* Willd. (Syn. *Pithecolobium dulce* Benth.);
6. *Mimosa pudica* L.;
7. *Samanea saman* (Jacq.) Merr. etc.

Key to sub-families

I. Flowers irregular.

1. Stamens diadelphous, very often monadelphous; posterior petal outermost. .... *Papilionoideae.*
2. Stamens free; posterior petal innermost. ... . *Caesalpinioideae.*

II. Flowers regular. .... *Mimosoideae.*

Key to genera

Papilionoideae

I. Stamen free :

1. Straggling or scandent shrubs; leaves 1-foliate; bracteoles large, hiding flowers. .... *Dalhausiea.*
2. Shrubs; leaves imparipinnate; bracteoles 0 .... *Sophora.*
3. Erect trees; leaves imparipinnate; bracteoles minute .... *Ormosia.*

II. Stamens monadelphous :

1. Prostrate herbs; leaves pinnately compound with 2 pairs of leaflets; fruit hypogeal burying itself to ripen underground.............................. *Arachis.*
2. Annual herbs; leaves digitately 3- or 4-foliate, gland-dotted .... *Zornia.*
3. Herbs or shrubs; leaves simple, sessile, or digitately 3-7-foliate; fruit turgid, non-septate, dehiscent .... *Crotalaria.*
4. Stem woody and twining; leaves paripinnate; stamens (9), petals vexillary, filament absent............................. *Abrus.*
5. Large twining or prostrate herbs ...
   i. Leaves pinnately 3-foliate .... *Canavalia.*
   ii. Leaves pinnately 3-foliate, leaflets lobed............................... *Pueraria.*
   iii. Leaves pinnately 3-foliate, 5 alternate stamens sterile .... *Teramnus.*

III. Stamens diadelphous :

A. Herbs; pods dehiscent by both sutures ...

Leaves paripinnate, rachis ending in a bristic or tendril ...

1. Leaflets deeply toothed; pod turgid, sessile, tipped with beardless style...... ....*Cicer*. (*C. arietinum* has usually a terminal leaflet instead of a tendril).
2. Leaflets entire, rachis ending in a twisted tendril, staminal tube oblique at its mouth
   - i. Style bearded at the tip; ovule usually more than two; pod compressed .... *Vicia*.
   - ii. Style longitudinally bearded along inner surface; ovules never more than two; fruit a compressed 1-2-seeded pod .... *Lens*.
   - iii. Style flat, bearded along inner face and dilated above; fruit many-seeded pod, compressed .... *Lathyrus*.
   - iv. Style 3-cornered, dilated upwards; pod turgid .... *Pisum*.

B. Herbs; pods indehiscent or dehiscent along the ventral suture ...

Leaves pinnately 3-foliate, veins produced as marginal teeth in leaflets, stipules adnate ...

- i. Flowers yellow or white; pod nearly round, 1- or 3-seeded ...... *Melilotus*.
- ii. Flowers usually purple, violet or blue, sometimes yellow; pods larger than calyx, downy and loosely spiral ....*Medicago*.
- iii. Flowers whilte or yellow; pod long, thin and jointed .... *Trigonella*.
- iv. Flowers usually reddish or purple; pod slender, 1- or more seeded .... *Indigofera*.
- v. Flowers blue; pod 1-seeded, indehiscent .... *Psoralea*.

C. Herbs or undershrubs; leaves pinnate, sensitive, 3-5 pairs, exstipulate, the rachis ending in a bristle or leaflet; flowers bright yellow; stamens in 2-bundles of 5 each; fruit a lomentum, folded inside the calyx .... *Smithea*.

D. Undershrubs growing in jheels and ditches, soft-wooded; leaves imparipinnate, numerous, sensitive; flowers bright yellow or orange-yellow; stamens in 2 bundles of 5 each; fruit a long lomentum, exserted ...... *Aeschynomene*.

E. Herbs or undershrubs; leaves stipulate, leaflets 1-9; stamens (9)+1; fruit lomentum, coiled and more or less folded within the calyx ... *Uraria*.

F. Herbs or undershrubs; leaves 3-1-foliate; pod flat, not coiled; stamens (10) or (9)+1 .... *Desmodium*.

G. Herbs or undershrubs; leaves pinnately 3-foliate, not glandular beneath

Twining, style bearded below stigma ...

- i. Keel spirally twisted; stigma oblique .... *Phaseolus*.
- ii. Keel not spiral; stigma oblique ...... *Vigna*.
- iii. Leaflets lobed, stigma oblique, tuberous rootstocks .... *Pachyrrhizus*.
- iv. Stigma terminal; pod flattish, winged ....*Dolichos*.
- v. Pod square, winged .... *Psophocarpus*.
- vi. Standard shorter than keel and wings; pods usually beset with stinging hairs .... *Mucuna*.

H. Trees with prickly branches; keel and wings shorter than standard; leaflets more or less cordate .... *Erythrina*.

I. Trees or woody climbers; flowers large showy, orange-red; calyx silky, wings free from keel ... . *Butea*.

J. Woody twiners; flowers small; corolla not much larger than calyx; wings adnate to the Keel .... *Spalholobus*.

K. Trees undershrubs, or climbers, pod dehiscent or indehiscent.

Leaves pinnately 5 or more foliate with terminal leaflet ...

(a) Pod flat, indehiscent

- i. Woody climbers or trees

  Leaflets alternate; stamens (9)+1 or (10) or (5) + (5) ....*Dalbergia*.
- ii. Timbers trees; flowers yellow

  Leaflets alternate; stamens as above; pod thin, orbicular ...... *Pterocarpus*.
- iii. Trees; leaflets opposite; flowers pink or white; ovary 2-ovuled; pod almost woody, wingless .... *Pongamia*.

(b) Pod dehiscent

i. Herbs or shrubs, usually climbing Flowers blue, white or purple; standard spoon-shaped; stamens (9)+1 or (10); anthers uniform .... *Clitoria*.

ii. Erect herbs or undershrubs, more or less hairy with an offensive smell; leaflets imparipinnate, with 6-10 pairs; flowers red or purple .... *Tephrosia*.

iii. Herbs, shrubs or soft-wooded trees, sometimes prickly; leaves paripinnate; leaflets may be 40 pairs; flowers large, yellow, dotted with purple or white or red; roots with copious tubercles .... *Sesbania*.

Caesalpinioideae

I. Leaves simple, 2-lobed .... *Bauhinia*.

II. Leaflets paripinnate ...

A. Calyx 5, Petals 5, Stamen 10, anthers basifixed ....*Cassia*.

B. Calyx petaloid, scarlet, 4-lobed; corolla 0; Stamens 8 ....*Saraca*.

C. Petals 0; calyx 5-lobed, green; stamens 10; leaflets 1-3, jugate .... *Hardwickia*.

D. Petals 3; stamens 3, monadelphous ....*Tamarindus*.

III. Leaflets bipinnate ...

A. Trees, rusty tomentose shoots and rusty panicles or showy yellow flowers .... *Peltophorum*.

B. Trees, shrubs or prickly climbers Calyx deeply cleft, the lowest lobe largest; petals orbicular, usually clawed ....*Caesalpinia*.

C. Unarmed trees; Calyx-lobes valvate and almost equal, flowers large, bright scarlet and variegated with yellow ... *Poinciana*.

D. Small armed trees; main leaf-rachis spinescent, pinnae with much-flattened rachis, small .... *Parkinsonia*.

Mimosoideae

I. Stamens definite, 4, 5, 8 or generally 10 ...

A. Marshy plants with flattened or angular stems; heads of flowers on long stalks; flowers hermaphrodite, male or neuter ... .*Neptunia*.

B. Flowers in round heads, all flowers similar

(a) Woody tree, leaf with one pair of pinnae .... *Xylia*.

(b) Woody climber with tendrils...*Entada*.

(c) Erect unarmed tree with bipinnate leaves, fruit linear, falcate, seeds scarlet .... *Adenanthera*.

(d) Prickly trees or shrubs, pod turgid and spongy inside .... *Prosopis*.

(e) Herbs or shrubs with sensitive leaflets, pod flat and jointed...... *Mimosa*.

II. Stamens indefinite, much exerted ...

A. Erect or climbing shrubs or trees armed with spiny stipules or recurved prickles; flowers white, yellow or purple in round heads or cylindrical spikes .... *Acacia*.

B. Large unarmed trees, rarely climber; leaves bi-pinnate, pinnae and leaflets several; flowers in round heads; stamens monadelphous; pods thin, strap-shaped, continuous within .... *Albizzia*.

C. Tall trees; leaves bipinnate with pinnae one pair and leaflet one pair only, sometimes spinescent; stamens monadelphous; pod twisted .... *Inga*.

D. Big trees; leaves bipinnate, leaflets many; stamens monadelphous; pod septate between seeds, valves spongy or fleshy ... .*Enterolobium*.

Genus *Crotalaria*

Key to species

1. Small herb; leaves simple. .... *albida*.
2. Undershrub; leaves compound, 3-foliate .... *striata*.

Genus *Desmodium*

Key to species

1. Leaves trifoliate; flowers solitary. ...... *triflorum*.
2. Leaves unifoliate; flowers solitary. .... *gangeticum*.

Genus *Cassia*

Key to species

1. Pods impressed between the seeds. .... *occidentales*.
2. Pods not impressed between the seeds.
   i. Leaflets usually 3 pairs, oblong-ovate. ... . *tora*.
   ii. Leaflets 7-10 pairs, lanceolate. ... *sophera*.

Plants of economic importance

Subfamily. **Papilionoideae**

A. Vegetables

The fruits of the following species are used as vegetables :

1. *Canavalia ensiformis* (Jack bean).
2. *Cyamopsis tetragonolobus* (Cluster bean, Gwar).
3. *Dolichos lablab* (Sim).
4. *Glycine max* (Soybean, Soya).
5. *Phaseolus vulgaris* (Kidney bean).
6. *P. lunatus* (Lima bean).
7. *P. coccineus* (Multiflora bean).

B. Pulses

Seeds of different species have high content of protein and starch used widely as food. The commonly used plants are :

1. *Cajanus cajan* (Pigeon pea, Arhar).
2. *Cicer arietinum* (Chick pea, Gram, Chola).
3. *Glycine max* (Soyabean, Soya).
4. *Lens culinaris* (Lentil, Masur).
5. *Pisum sativum* (Garden pea, Matar).
6. *Phaseolus mungo* (Black gram, Mush-kalai).
7. *P. aureus* (Green gram, Mung)
8. *Vigna aconitifolia* (Moth).

C. Oils

Arachis hypogea (Groundnut). The oil extracted from the seeds is used as vegetable oil. The oil is used in the manufacture of vanaspati. It is also used in cosmetics and soap industry. The oil cake is used as fodder and fertiliser.

D. Fibres

1. *Crotalaria juncea* (Sunhemp). The bast fibre extracted from the stem is very strong and is used in making ropes, gunny bags, hair of goddess, etc.
2. *Sesbania cannabina*. The plant also yields good quality of fibre.

E. Timbers

The following are the valuable timber yielding plants found in this subfamily :

1. *Dalbergia latifolia* (Rose wood).
2. *D. sissoo* (Sissoo).
3. *Pterocarpus dalbergioides* (Red wood).
4. *P. indicus* (Malay padauk).
5. *P. marsupium* (Indian kino tree).
6. *P. santalinus* (Red sandal wood).

F. Fodder

Medicago sativa and different species of *Trifolium* (clover) i.e., *Trifolium alexandrinum, T. hybridum, T. pratense, T. repens* are used as fodder.

G. Dyes

1. *Indigofera tinctoria, I. sumatrana* and *I. suffruticosa*. The plants are the source of 'indigo' dye used in printing, dyeing and paints.
2. *Pterocarpus santalinus*. The red resinous substance 'santalin' extracted from wood is used for dyeing silk and cotton cloths.

H. Gum

Different species of plants are the source of gum used in paper and textile industry and also in food :

1. *Astragalus gummiferus*. The 'Tragacanth' gum found from this plant is used in textile industry as a sizing material and in cosmetic and confectionary as emulsifier.
2. *Cyamopsis tetragonolobus* (Cluster bean). The 'Gwar gum' is extracted from the seeds used in paper and textile industry, etc.

I. Medicines

Many plants are used as the source of drugs. These are :

1. *Abrus precatorius*. The roots and leaves are used to control cough and cold.
2. *Glycyrrhiza glabra*. The medicine Liquorice, extracted from the root and stem, used in cough syrups.
3. *Krameria triandra*. The drug "Krameria" extracted from the roots is used to control chronic inuate.
4. *Moghania strobilifera*. The decoction of the roots is used to relieve pain and induce sleep.

5. *Psoralea corylifolia*. The seeds are used in the treatment of leprosy, leucoderma and also as laxative.
6. *Teramnus labialis*. The plants are used in the treatment of tuberculosis, rheumatism, etc.

J. **Ornamental plants**

Various members of this subfamily are grown as ornamental plant.

Some of them are :

1. *Clitoria ternatea* (Butterfly pea).
2. *Erythrina indica* and *E. variegata* (Mother plant).
3. *Pongamia pinnata* (Pongam oil tree).
4. *Sesbania sesban* (Common sesban).
5. *Sophora japonica* (Japanese pagoda tree).
6. *Wisteria sinensis* (Chinese wisteria).

K. **Other uses**

1. *Aeschynomene aspera*. The dried soft stem is the '**shola**' of commerce used in various decoration purposes.
2. *Abrus precatorius*. The seeds are light in weight and used by goldsmiths for necklace and also as jewellers' weight.

Subfamily. **Caesalpinioideae**

A. **Vegetables**

Bauhinia purpurea (Deva kanchan) and *B. variegata* (Camelís foot tree, Rakto kanchan). The flower buds of both the plants are used as vegetables and also used in pickles.

B. **Medicines**

1. *Cassia fistula* (Indian laburnum, Bandar lathi). The fruits are used as laxative.
2. *Cassia sophera* (Kalkashunda, Kasunda). The decoction of leaves is used in the treatment of bronchitis and ringworm.
3. *Cassia surattense* (*syn. C. glauca*). The bark of stem and leaves are used in the treatment of gonorrhoea and diabetes.
4. *Saraca asoka*. The decoction of bark is used in the treatment of menstrual disorder of women.

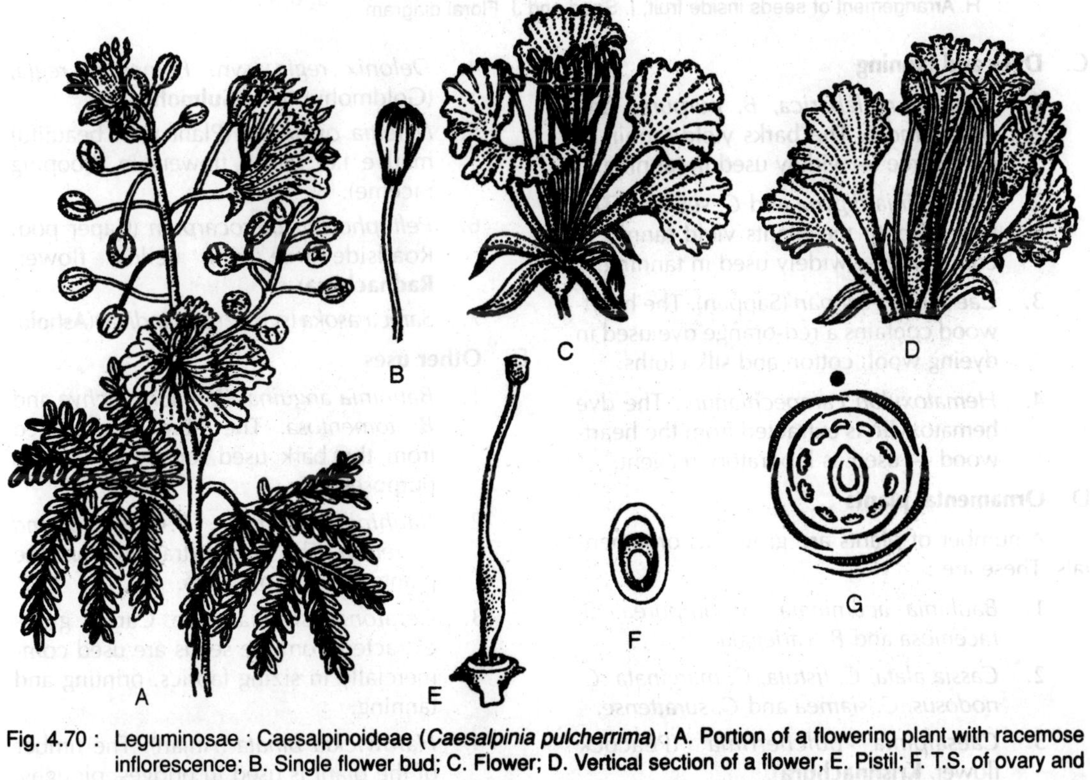

Fig. 4.70 : Leguminosae : Caesalpinoideae (*Caesalpinia pulcherrima*) : A. Portion of a flowering plant with racemose inflorescence; B. Single flower bud; C. Flower; D. Vertical section of a flower; E. Pistil; F. T.S. of ovary and G. Floral diagram

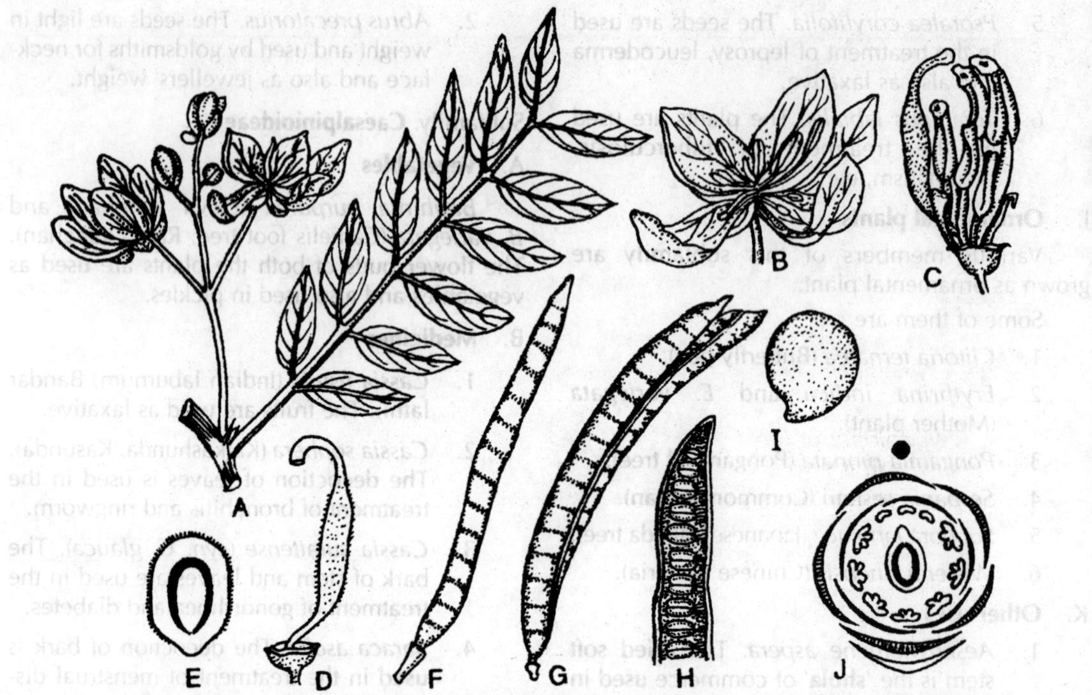

Fig. 4.71 : Leguminosae : Caesalpinoideae (*Cassia sophera*) : A. Portion of a flowering plant; B. Single flower; C. Flower after removal of Calyx and Corolla; D. Pistil; E. T.S. of ovary; F. Fruit; G. Bursted fruit; H. Arrangement of seeds inside fruit; I. Seed and J. Floral diagram

C. **Dyes and tanning**

1. *Bauhnia malabarica, B. purpurea* and *B. racemosa*. The barks yield tannin of commerce — widely used in tanning.
2. *Caesalpinia digyna* and *C. sepiaria*. The bark of both the plants yield tannin of commerce — widely used in tanning.
3. *Caesalpinia sappan* (Sappan). The heartwood contains a red-orange dye used in dyeing wool, cotton and silk cloths.
4. *Hematoxylon campechianum*. The dye hematoxylin is extracted from the heartwood — used as laboratory reagent.

D. **Ornamental plants**

A number of plants are grown as ornamentals. These are :

1. *Bauhinia acuminata, B. purpurea, B. racemosa* and *B. variegata*.
2. *Cassia alata, C. fistula, C. marginata, C. nodosus, C. siamea* and *C. surattense*.
3. *Caesalpinia pulcherrima* (Peacock flower, **Krishnachura**).
4. *Delonix regia* (syn. *Poinciana regia*) (Goldmohur tree, **Gulmohur**).
5. *Milettia ovalifolia* (Plant with beautiful mauve i.e. bluish flowers in drooping raceme).
6. *Peltophorum pterocarpum* (Caper pod. Roadside tree with yellow flower, **Radhachura**).
7. *Saraca asoka* (syn. *Saraca indica*)(**Ashok**).

E. **Other uses**

1. *Bauhinia anguina, B. macrostachya* and *B. tomentosa*. The fibre is extracted from the bark used locally in various purposes.
2. *Bauhinia racemosa, B.variegata and B. retusa*. The gum extracted from the plants is of various uses.
3. *Ceratonia siliqua*. The Caroh gums extracted from the seeds are used commercially in sizing fabrics, printing and tanning.
4. *Hardwickia binata* (Anjan). The timber of the plant is used in bridges, ploughs,

cart wheels, etc. The fibres from the bark are used in rope making.

5. *Phanera vahlii* (Camelís foot climber). The gum extracted from the plants is of commercial importance.
6. *Tamarandus indicus* (Tamarind tree, Tentul). The pulp of the fruit is acidic and used in sauces, curries, chutney, etc.

Subfamily. **Mimosoideae**

A. **Edibles**

1. *Neptunia prostrata*. Apical region of the stem and the pods are used as vegetable.
2. *Pithecolobium dulce*. The juicy and sweet aril is edible. (Aril : an outgrowth from the funiculus or hilum of a seed, which is often spongy and gelatinous, and sometimes covered the seed.)

B. **Timbers**

1. *Acacia catechu, A. nilotica, A. modesa*. The stems of the plants are used as source of fuel, cart wheels, tool handles, etc.
2. *Albizzia amara, A. lucida, A. chinensis, A. lebbeck*. The timbers of the above plants are valuable and used for furniture, agricultural implements, etc.
3. *Xylia xylocarpa* (syn. *X. dolabriformis*). The stem is used in the construction of boat, etc.

C. **Tannins**

Acacia nilotica, A. catechu, A. pinnata. The tannins extracted from the bark are used for various purposes.

D. **Gums**

1. *Acacia catechu*. The gum (Khair or Katha) extracted from the bark is used for dressing of betel leaf during preparation of 'paan'.
2. *A. nilotica*. The gum is found to accumulate on injured surface of the stem, used in binding of books, fixing of papers, etc.
3. *A. senegal*. The gum (*gum* arabic) obtained from this plant is used in confectionary, textile and medicine industry.

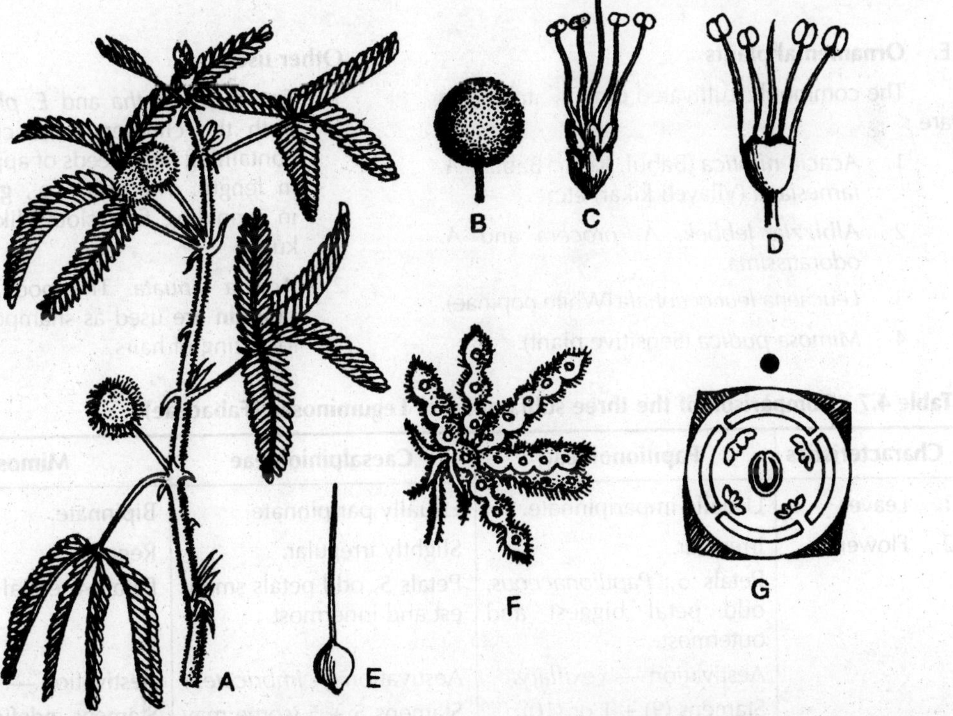

Fig. 4.72 : Leguminosae : Mimosoideae (*Mimosa pudica*) : A. Portion of a flowering plant; B. Inflorescence; C. Single flower; D. Flower after removal of calyx and corolla; E. Pistil; F. One bunch of pods, and G. Floral diagram

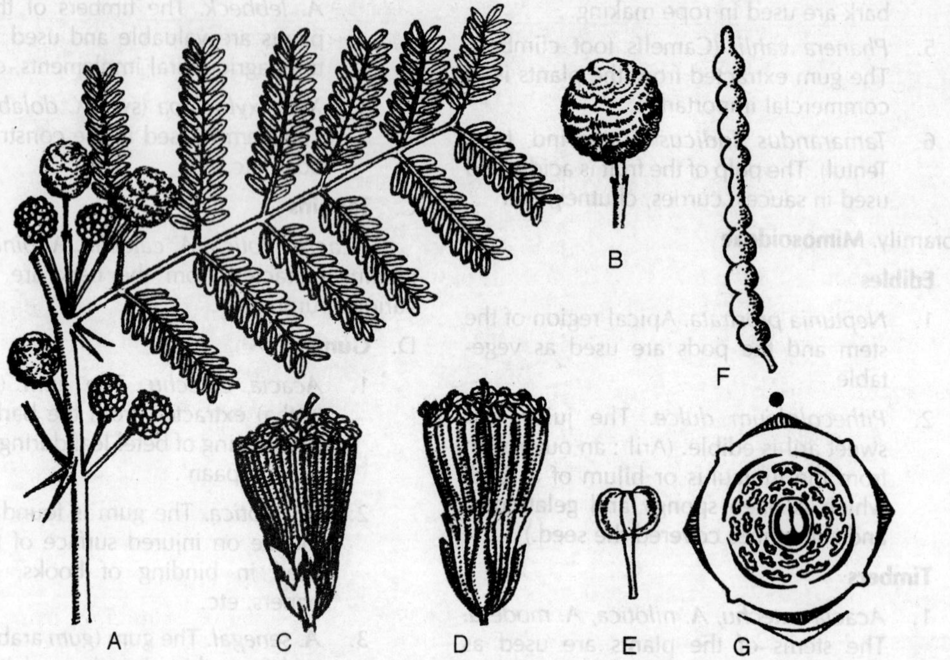

Fig. 4.73 : Leguminosae : Mimosoideae (*Acacia nilotica*) : A. Portion of a plant with flowers; B. Inflorescence; C. Flower; D. V.S. of a flower; E. Single stamen; F. Fruit and G. Floral diagram

E. **Ornamental plants**

The commonly cultivated ornamental plants are :

1. *Acacia nilotica* (Babul, Kikar, Babla), *A. farnesiana* (Vilayeti kikar) etc.
2. *Albizzia lebbek, A. procera* and *A. odoratissima.*
3. *Leucaena leucocephala* (White popinae).
4. *Mimosa pudica* (Sensitive plant).

F. **Other uses**

1. *Entada pursaetha* and *E. phaseoloides.* Both the climbers have curved fruits containing large seeds of approx. 50 cm in length. The seed i.e., **gilla** is used in wrinkling fine cloths like hands of kurta.
2. *Acacia sinuata.* The pods containing **saponin** are used as shampoo and also for sizing of hairs.

Table 4.7 : Comparison of the three subfamilies of Leguminosae (Fabaceae)

Characteristics	Papilionoideae	Caesalpinioideae	Mimosoideae
1. Leaves	Usually imparipinnate.	Usually paripinnate.	Bipinnate.
2. Flowers	Irregular.	Slightly irregular.	Regular.
	Petals 5, *Papilionaceous,* odd petal biggest and outermost.	Petals 5, odd petals smallest and innermost.	Petals 4, equal in size.
	Aestivation — *vexillary.*	Aestivation — *imbricate.*	Aestivation — *valvate.*
	Stamens (9) + 1 or (10).	Stamens 5 + 5 (some may be sterile) or less (due to abortion), free.	Stamens indefinite or definite.

3. *A. franesiana*. Cassie perfume is extracted from the flower of this plant.

14. ROSACEAE

(Rose Family)

The family is represented by 110 genera and 3,100 species, is cosmopolitan in distribution. More than 25 genera and 215 species are growing widely in the temperate region of Himalayas up to an about 1,980 m (6,500 ft). The plants of this family are important as cash crops.

The type genus of the family is ***Rosa***.

Systematic position

Bentham & Hooker		Engler & Prantl	
Division.	Angiospermia	Subdivision.	Angiospermae
Class.	Dicotyledons	Class.	Dicotyledoneae
Subclass.	Polypetalae	Subclass.	Archichlamydeae
Series.	Calyciflorae	Order.	Rosales
Cohort.	Rosales.		

Hutchinson		Cronquist	
Phylum.	Angiospermae	Division.	Magnoliophyta
Subphylum.	Dicotyledones	Class.	Magnoliopsida
Division.	Lignosae	Subclass.	Rosidae
Order.	Rosales	Order.	Rosales

Takhtajan		Thorne	
Division.	Magnoliophyta	Class.	Angiospermae
Class.	Magnoliopsida	Subclass.	Magnoliidae
Subclass:	Rosidae	Superorder.	Rosanae
Superorder.	Rosanae	Order.	Rosales
Order.	Rosales.		

Identifying characteristics

1. Plants herbs, shrubs or trees.
2. Leaves **alternate, simple** or **compound. Usually stipulate** and more commonly of **adnate type.**
3. Flowers **usually bisexual** and **regular, pentamerous.**
4. Sepals 5 and are basally connate.
5. **Petals 5** or **more, free, inserted on receptacular cup** or **hypanthium.**
6. Stamens **numerous.**
7. Ovary **superior** or **inferior, ovules on axile placentation.**
8. Carpel **1-many, free** or **connate.**

Floral formula

⊕ ⚥ $K_{(5)}C_{5 \text{ or } \alpha}A_{\alpha}\underline{G}_{1-(\alpha)} \text{ or } \overline{G}_{1-\alpha \text{ or } (\alpha)}$

Rosaceae : Distinctive features

Common characteristics	Exceptions
Vegetative characteristics	
Plant. Herbs (*Fragaria, Agrimonia*), shrubs (*Prinsepia* and *Spiraea*) or tree (*Prunus, Eriobotrya, Pyrus* etc.). Stems are covered with prickles (species of *Rubus* and *Rosa*) or sharp spines (*Crataegus, Prinsepia*).	
Leaves. Alternate, usually pinnately compound, stipulate (stipule adnate).	Leaves simple – *Prunus* and *Pyrus*. Palmately compound – *Potentilla* and *Fragaria*. Stipules absent – *Spiraea*.
Reproductive characteristics	
Inflorescence. Various types, such as cymes (*Fragaria*), racemes (*Pygeum*), corymb (*Rubus*) or panicle (*Eriobotrya*).	Rarely solitary – Certain species of *Rubus* and *Rosa*.
Flower. Regular, usually bisexual, pentamerous, perigynous with disc, odd sepals posterior.	Zygomorphic – *Parinarium, Parastemon* and *Chrysobalanus*. Unisexual – *Aruncus*. Epigynous – *Pyrus* and *Eriobotrya*. Hypogynous – *Prunus*.

Contd.

Contd.

Common characteristics	Exceptions
Calyx. Sepals 4-5, connate to receptacle, persistent, imbricate, odd 5th sepal posterior.	Epicalyx present – *Potentiella* and *Fragaria*. Sepals foliaceous – *Rosa*.
Corolla. Petals 4-5, free, imbricate, deciduous.	Petals absent – *Alchemilla*. Petals many – Cultivated species of *Rosa*. Petals very small – *Neurada*.
Androecium. Stamens many, 1-many whorls of 5 stamens, free, bent inward in bud, anthers small, 2 celled.	Stamens 1 to 4 – *Alchemilla*. Stamens 2 and unilateral – *Parastemon*. Monadelphous – *Chrysobalanus*.
Gynoecium. Carpels 1-many, free or united, ovary usually superior, sometimes inferior, 1 celled or 2-many celled with 1 or more anatropous ovules in each cell.	Carpel 1 – *Prunus*. Ovary inferior – *Pyrus*, *Rosa*, *Eriobotrya*.
Fruit. Achene, drupe, berry, pome or aggregate of follicle.	
Seed. Generally without endosperm (exalbuminous), embryo with fleshy cotyledon.	

Phylogeny and systematics

The family Rosaceae is related to the family Leguminosae (Fabaceae). This family is an assemblage of fairly primitive and advanced characteristics. Most of the taxonomists included the family under the order Rosales. Bentham and Hooker placed the Natural order Rosaceae under the cohort Rosales, subclass Polypetalae and class Dicotyledons. Engler and Prantl also placed it in Rosales after Sarraceniales, under the subclass Archichlamydeae and class Dicotyledonae. Hutchinson placed Rosaceae under Rosales and given a comparatively primitive position and placed after the 5th order Coriariales under division Lignosae and sub-

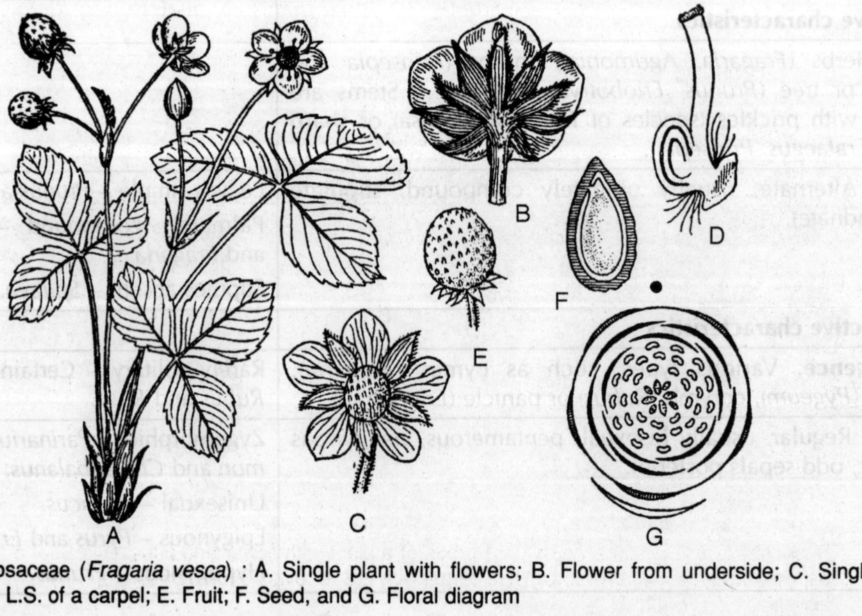

Fig. 4.74 : Rosaceae (*Fragaria vesca*) : A. Single plant with flowers; B. Flower from underside; C. Single flower; D. L.S. of a carpel; E. Fruit; F. Seed, and G. Floral diagram

phylum Dicotyledones. Recently, Takhtajan and Cronquist placed Rosaceae under Rosales of sub-class Rosidae, class Magnoliopsida. Although the family Rosaceae is a more natural assemblage, still many taxonomists treated most of the tribes (= subfamilies) as independent families.

Common plants

1. *Eriobotrya japonica* Lindl.;
2. *Fragaria vesca* L.;
3. *Malus sylvestris* (L.) Mill. (Syn. *Pyrus malus* L.);
4. *Potentilla nepalensis* Hook;
5. *Prunus ceresus* L.;
6. *Pyrus communis* L.;
7. *Rosa alba* L. (Indian white rose);
8. *R. centifolia* L. (Cabbage rose);
9. *R. chinensis* Jacq. (Syn. *R. indica* L.)
10. *R. damascena* Mill.;
11. *Rubus hexagynous* Roxb.

Key to genera

I. Ovary superior; ripe carpels not enclosed within the calyx-tube
   1. Carpel solitary
      i. Leaves simple, quite entire, fruit 2-seeded .... *Pygeum.*
      ii. Leaves simple, serrate, fruit 1-seeded .... *Prunus.*
   2. Carpels many
      A. Unarmed herbs; ripe carpels dry; ovules solitary
         i. Leaves compound, fruit achenes on fleshy receptacle...... *Fragaria.*
         ii. Leaves compound fruit achenes on dry receptacle ...... *Potentilia.*
      B. Armed shrubs; leaves simple or compound, with stipules adnate or free; carpels many or few on a convex receptacle; fruit drupe .... *Rubus.*

II. Ovary inferior; ripe carpels enclosed within the calyx-tube :
   Carpels many, not confluent when ripe :
   A. Armed shrubs; leaves compound with adnate stipules; fruit a fleshy calyx-tube enclosing bony achenes ... *Rosa.*
   B. Shrubs or trees; leaves simple, coriaceous, usually serrate; flowers white, panicled; fruit 1-seeded dry or succulent berry, endocarp membranous .... *Eriobotrya.*
   C. Shrubs or trees; leaves simple or compound; flowers white, red or pink, in terminal corymbs.................*Pyrus.*

Plants of economic importance

The members are the good source of large number of economically important fruits, some are ornamental and some other medicinal plants. Some plants are also used for different purposes.

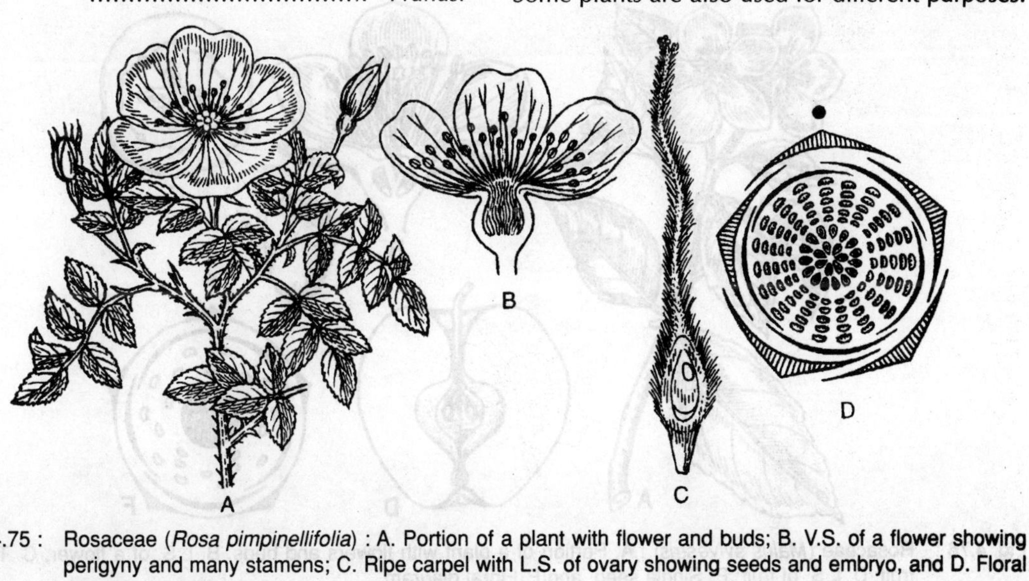

Fig. 4.75 : Rosaceae (*Rosa pimpinellifolia*) : A. Portion of a plant with flower and buds; B. V.S. of a flower showing perigyny and many stamens; C. Ripe carpel with L.S. of ovary showing seeds and embryo, and D. Floral diagram

A. **Edible fruits**

Some members are good source of edible fruits and are economically very important. These are :

1. *Eriobotrya japonica* (Loquat),
2. *Fragaria chiloensis* (Garden strawberry),
3. *F. vesca* (Alpine strawberry),
4. *Malus sylvestris* (Apple),
5. *Prunus amygdalus* (Almond),
6. *P. armeniaca* (Apricot),
7. *P. avium* (Sweet cherry),
8. *P. cerasus* (Sour cherry),
9. *P. domestica* var. *institia* (Plum),
10. *P. persica* (Peach),
11. *Pyrus communis* (Pear),
12. *P. pyrifolia* var. *culta* (syn. *P. sinensis*) (Chinese pear).

B. **Ornamental plants**

Many plants with beautiful flowers are extensively cultivated in the garden. These are :

1. *Rosa* (Garden rose). Different species of *Rosa* are commonly cultivated in the garden. These are *R. banksiae, R. centrifolia, R. damascena, R. indica, R. multiflora,* etc.
2. *Spiraea* (Spirea). Different species are cultivated in the garden in fences, lawn and also in borders. These are *S. canescens, S. cantoniensis, S. japonica,* etc.
3. *Potentilla, Rubus* and *Sorbus*. Different species are grown in the garden as ornamental plants.

C. **Medicinal plants**

1. *Genum urbanum*. Root stocks are used in chronic dysentery, diarrhoea etc.
2. *Hagenia abyssinica*. The panicles of female flowers are used as the source of anthelmintic drug etc.
3. *Potentilla anserina* (Silver weed). The plant is used to control arthritis, kidney stone, etc.

D. **Other uses**

1. Perfume. The flowers of *Rosa damascena* are good source of valuable perfume.
2. Edible oil. The seeds of *Prinsepia utilis* are the source of edible oil used in cooking.
3. Emulsifier. The bark of *Quillaja saponaria* (Soap bark) is used as an emulsifying and foaming agent in hair tonic and in shampoo.
4. Handle of tools, walking sticks, etc. : The woody branches of some plants are used for various purposes like handle of tools, walking sticks, etc. These are *Crataegus oxyacantha, Pyrus communis, Rosa brunonii,* etc.

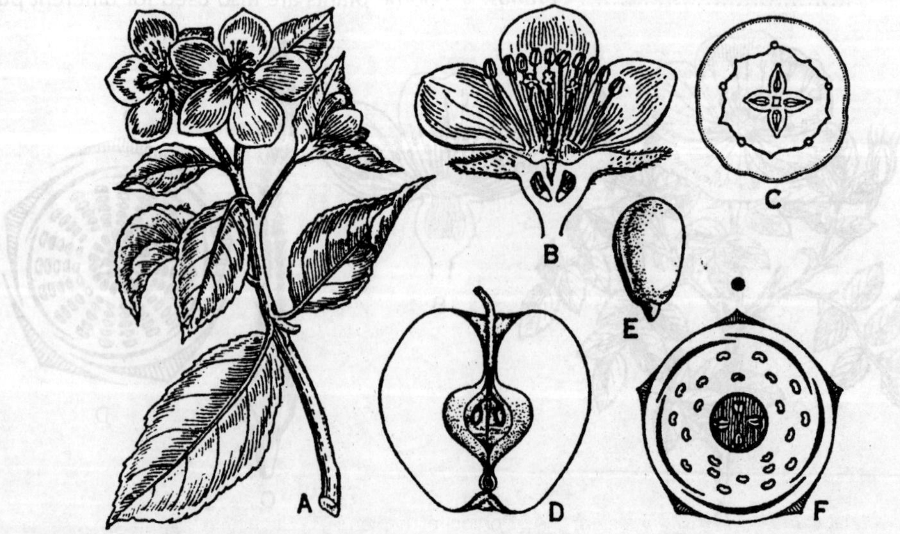

Fig. 4.76 : Rosaceae (*Malus sylvestris*) : A. Portion of a plant with flowers and buds; B. L.S. of a flower; C. T.S. of fruit; D. L.S. of fruit; E. Single seed, and F. Floral diagram

15. CUCURBITACEAE

(Gourd Family)

The family is represented by 118 genera and 760 species, out of which 37 genera and about 97 species are reported from India, mainly from tropical regions. The family is important as cash crop mainly for vegetables.

The type genus of the family is ***Cucurbita***.

Systematic position

Bentham & Hooker		**Engler & Prantl**	
Division.	Angiospermia	Subdivision.	Angiospermae
Class.	Dicotyledons	Class.	Dicotyledoneae
Subclass.	Polypetalae	Subclass.	Metachlamydeae
Series.	Calyciflorae	Order.	Cucurbitales
Cohort.	Passiflorales.		

Hutchinson		**Cronquist**	
Phylum.	Angiospermae	Division.	Magnoliophyta
Subphylum.	Dicotyledones	Class.	Magnoliopsida
Division.	Lignosae	Subclass.	Dilleniidae
Order.	Cucurbitales	Order.	Violales

Takhtajan		**Thorne**	
Division.	Magnoliophyta	Class.	Angiospermae
Class.	Magnoliopsida	Subclass.	Magnoliidae
Subclass.	Dilleniidae	Superorder.	Violanae
Superorder.	Violanae	Order.	Violales
Order.	Cucurbitales		

Identifying characteristics

1. Succulent **herbs with extra axillary tendril**.
2. Flower **unisexual**, actinomorphic, **pentamerous** and **epigynous**,
3. Stamens-5, generally **synandrous** (both anthers and filaments are united),
4. Carpel-3, ovary — **inferior**, ovules many in each cell on **parietal placentation**,
5. Fruit — fleshy **berry-like** or **pepo**.

Floral formula

Male flower :

$\oplus$ ♂ $K_{(5)}C_{(5)}A_{(5)}$

Female flower :

$\oplus$ ♀ $K_{(5)}C_{(5)}\ \overline{G}_{(3)}$

Types of androecium

The androecium shows much variation in its arrangement of stamen. These are :

1. All the five stamens remain free, they alternate with the petals. Anthers dithecous. e.g., *Fevillea*.
2. All the five stamens remain free but anthers monothecous. e.g. *Luffa cylindrica*.
3. Stamens-5, the filaments of 4 stamens are united at the base into two pairs, the 5th

Fig. 4.77 : Cucurbitaceae (*Cucurbita maxima*) : A. Portion of a plant; B. Staminate flower; C. Pistillate flower; D. Androecium showing syngenesious condition; E. Pistil with attached calyx; F. T.S. of ovary; G. Floral diagram of a staminate flower, and H. Floral diagram of a pistillate flower

Cucurbitaceae : Distinctive features

Common characteristics	Exceptions
Vegetative characteristics	
Plants. Generally succulent, annual or perennial, decumbent or trailing herbs usually climb by means of tendril. Stems are green, generally 5-angled, branched.	Shrub – *Acanthosicyos.* Tree (small size) – *Dendrosicyos.* Without tendril – *Ecballium.*
Leaf. Simple but often palmately or pinnately lobed, arranged alternately; exstipulate, petioles are often hollow.	Leafless – *Acanthosicyos.* Thorns in leaf bases – *Acanthosicyos, Kedrostis spinosa.*
Reproductive characteristics	
Nature. Plants are generally monoecious.	Plant dioecious – *Trichosanthes dioica, Momordica dioica, Bryonia dioica.*
Inflorescence. Usually solitary or various cymose.	
Flower. White or yellow; unisexual, ebracteate, pedicellate, actinomorphic, pentamerous.	
Male flower :	
Calyx. Sepals-5, gamosepalous, valvate or imbricate in aestivation.	
Corolla. Petals-5, gamopetalous, campanulate or rotate, imbricate or valvate in aestivation.	Petals are nearly free – *Trichosanthes.* Polypetalous – *Fevillea.*
Androecium. Stamens-5, alternate to petals, sometimes epipetalous, anther dithecous (2-celled), dehiscence longitudinal, generally synandrous. [Androecium shows much variations which are discussed below]	
Gynoecium. absent.	
Female flower : Epigynous.	
Calyx. As in male flower.	
Corolla. As in male flower.	
Androecium. Absent but sometimes 2,3 or 5 staminodes are present	
Gynoecium. Carpel 3, syncarpous, ovary inferior, unilocular with many anatropous ovules in each cell on parietal placentation, style 1, stigma 3, each one bifid.	
Fruit. Pepo and are usually indehiscent.	Dehiscent in *Cyclanthera, Ecballium, Herpetospermum.*
Seed. Many, non-endospermic.Germination is epigeal.	Seed single in *Sechium.* Germination vivipary in *Sechium.*

stamen remains free, anthers monothecous. e.g., *Thladiantha.*

4. Stamens apparently 3 (actually 5), two with dithecous anthers and one with monothecous anther. This condition seems to appear due to the complete union of the filaments and anthers of two pair of stamens, e.g., *Cucumis, Citrullus* and *Lagenaria.*

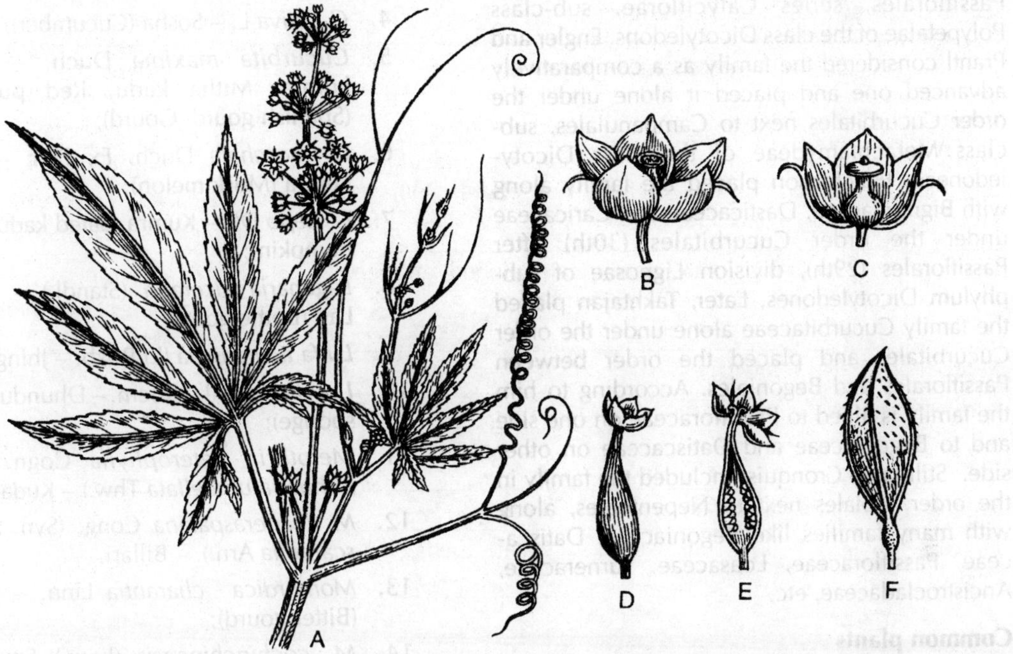

Fig. 4.78 : Cucurbitaceae (*Cyclanthera pedata*) : A. Portion of a plant; B. Staminate flower; C. Staminate flower after partial removal of calyx and corolla; D. Pistillate flower; E. Pistillate flower in L.S. and F. Fruit

5. Stamens apparently 3 (actually 5), two with dithecous anthers and one with monothecous anther. Here the filaments become connate and the anthers become very much twisted. e.g., *Cucurbita*.

6. All the 5 stamens are united into a column and they are remarkably curved. e.g., *Sicyos*.

7. The stamens are strictly *monadelphous*, where all the 5 stamens are united into a column with two ring-like pollen chambers situated around the top. The anthers dehisce by a single transverse slit. e.g., *Cyclanthera*.

Phylogeny and systematics

The family is closely related with Campanulaceae and also with Solanaceae.

The position of Cucurbitaceae shows great variation according to the different taxonomists. Bentham and Hooker placed the natural order

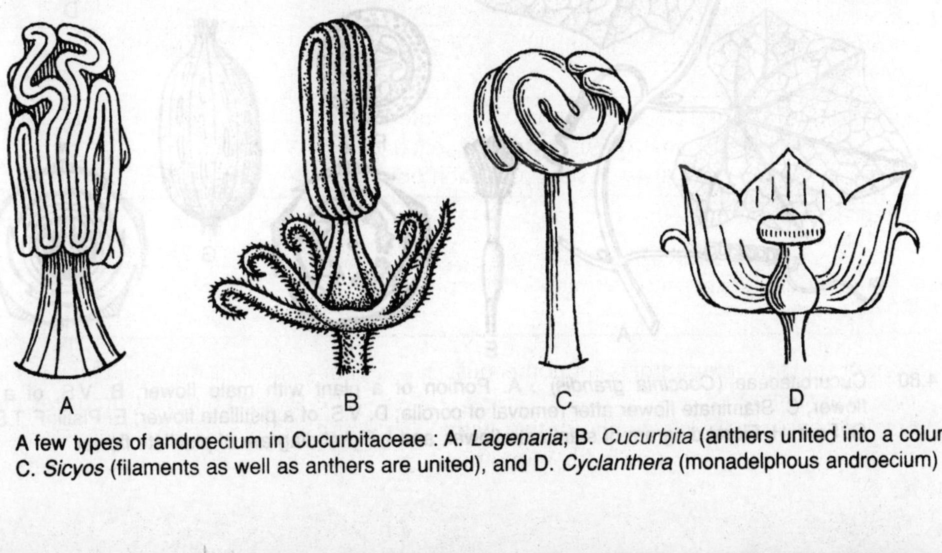

Fig. 4.79 : A few types of androecium in Cucurbitaceae : A. *Lagenaria*; B. *Cucurbita* (anthers united into a column); C. *Sicyos* (filaments as well as anthers are united), and D. *Cyclanthera* (monadelphous androecium)

Cucurbitaceae under the cohort (= order) Passiflorales, series Calyciflorae, sub-class Polypetalae of the class Dicotyledons. Engler and Prantl considered the family as a comparatively advanced one and placed it alone under the order Cucurbitales next to Campanulales, subclass Metachlamydeae of the class Dicotyledoneae. Hutchinson placed the family along with Bignoniaceae, Dasticaceae and Caricaceae under the order Cucurbitales (30th) after Passiflorales (29th), division Lignosae of subphylum Dicotyledones. Later, Takhtajan placed the family Cucurbitaceae alone under the order Cucurbitales and placed the order between Passiflorales and Begoniales. According to him the family is allied to Passifloraceae on one side and to Begoniaceae and Datiscaceae on other side. Still later, Cronquist included the family in the order Violales next to Nepenthales, along with many families like Begoniaceae, Datiscaceae, Passifloraceae, Loasaceae, Turneraceae, Ancistrocladaceae, etc.

Common plants

1. *Citrullus lanatus* (Thunb.) Mak. (Syn. *Citrullus vulgaris* Schard.) Tarbuz (Water melon);
2. *Coccinia grandis* (L.) Voigt. – Telakucha;
3. *Cucumis melo* L. – Kharbuz;
4. *C. sativa* L. – Sosha (Cucumber);
5. *Cucurbita maxima* Duch. – Mitha kumra, Mitha kadu, Red pumpkin (Spanish gourd, Gourd);
6. *C. moschata* Duch. Ex Poir. – Safra kumra (Musk melon);
7. *C. pepo* DC – Kumra, Safed kadu (Field pumpkin);
8. *Lagenaria siceraria* (Standl.) – Louki, Lau (Bottle gourd);
9. *Luffa acutangula* (L) Roxb. – Jhinga;
10. *L. cylindrica* (L.) Roem. – Dhundul (Bath sponge);
11. *Melothria heterophylla* Cogn. (Syn. *Zehneria umbellata* Thw.) – Kudari;
12. *M. maderaspatana* Cong. (Syn. *Mukia scabrella* Arn.). – Billari.
13. *Momordica charantia* Linn. – Karela (Bitter gourd);
14. *M. cochinchinensis* (Lour.) Spreng – Kakrol;
15. *Trichosanthus anguina* L. (Snake gourd, Chichinga);
16. *T. cucumereina* Linn. – Banchichinga;

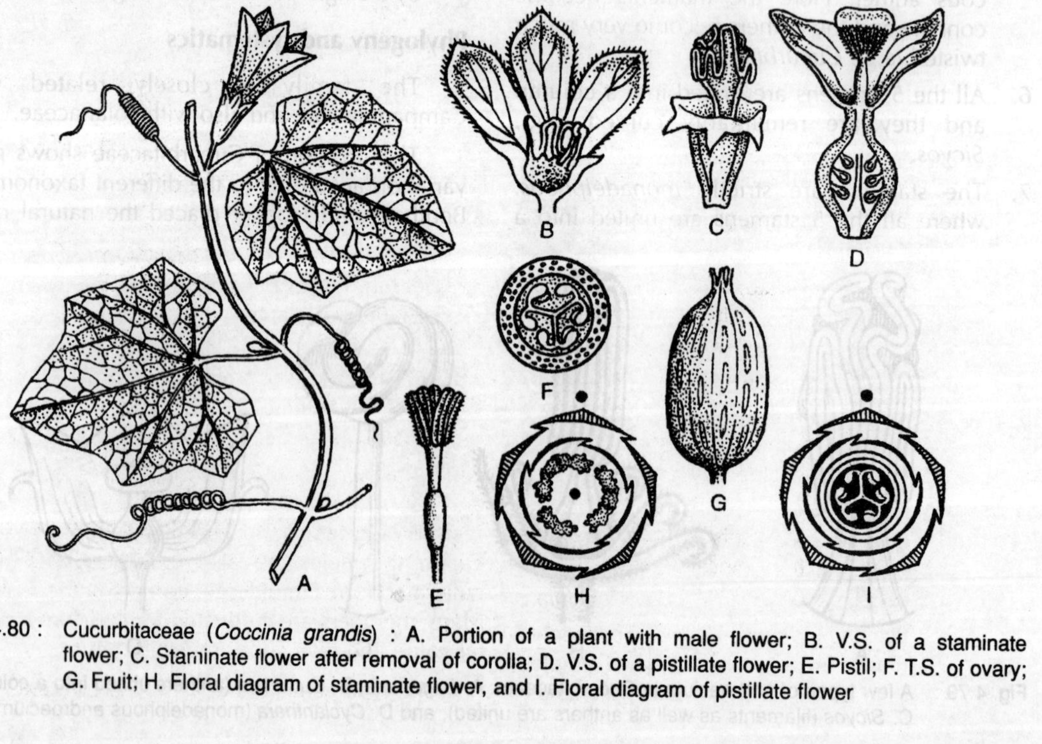

Fig. 4.80 : Cucurbitaceae (*Coccinia grandis*) : A. Portion of a plant with male flower; B. V.S. of a staminate flower; C. Staminate flower after removal of corolla; D. V.S. of a pistillate flower; E. Pistil; F. T.S. of ovary; G. Fruit; H. Floral diagram of staminate flower, and I. Floral diagram of pistillate flower

17. *T. dioica* Roxb. – Palwal, Patol;
18. *T. palmata* Roxb. (Syn. *T. bracteata* (Lamk.) Voigt.) – Mahakala, Makal.

Key to genera

I. Anther-cells 'S'-shaped or folded, Corolla rotate or campanulate, divided or nearly so into 5 petals above :

1. Tendril divided ...
   - i. Flowers white; petals fimbriated at the margin; seeds many ...*Trichosanthes.*
   - ii. Flowers large, white; petals entire; anthers included; petiole with 2 glands .... *Lagenaria.*
   - iii. Flowers pale yellowish-white; petals fimbriated; seeds many ... *Hodgsonia.*

   Stamens inserted at the mouth of the calyx-tube; anthers free; flowers yellow :
   - a. Fruit dry, angled or ribbed, rarely smooth; endocarp fibrous, opening by special lid; male flowers borne in raceme .... *Luffa.*
   - b. Fruit fleshy, soft, indehiscent; male and female flowers both solitary .... *Benincasa.*

   Stamens inserted below the mouth of the calyx-tube :
   - a. Male and female flowers large, yellow, solitary; corolla divided half way down, ovate-obtuse; anthers scarcely connate .... *Citrullus.*
   - b. ♂ and ♀ flowers solitary, yellow, large, corolla lobes short, triangular; anthers connate .... *Cucurbita.*
2. Tendril simple ...
   - i. Leaves cordate; male flowers subraceme; calyx with 2-3 scales; fruit tuberculated, spinous ... *Momordica.*
   - ii. Leaves lobed; male flowers clustered; calyx without scales; fruit large, smooth, edible; connective protruding beyond the anther-cells .... *Cucumis.*
   - iii. Leaves palmately lobed; flowers white; fruit a smooth cylindric berry .... *Coccinia.*
   - iv. Leaves toothed, angled or deeply lobed; flowers yellow; plants simple .... *Zehneria.*

II. Anther-cells straight or somewhat curved. Male and female flowers clustered, small, yellow :

1. Tendril simple........................ *Mukia.*
2. Tendril divided.................... *Bryonia.*

Plants of economic importance

A. **Fruit-vegetables**

1. *Benincasa hispida* (White squash or Chalkumra),
2. *Cucurbita pepo* (Pumpkin),
3. *C. maxima* (Red pumphin),
4. *C. moschata* (musk melon),
5. *Lagenaria siceraria* (Bottle gourd or lau),
6. *Trichosanthus dioica* (Palwal),
7. *T. anguina* (Snake gourd, Chachinga),
8. *Sechium edule* (Squash).

B. **Edible fruits**

1. *Citrullus lanatus* (Tarbuz, Water melon),
2. *Cucumis sativus* (Sosa, Cucumber),
3. *C. melo* (Kharbuz).

C. **Making Musical instruments**

Benincasa hispida and *Lagenaria siceraria* – Mature fruits with hard pericarp are used for making musical instruments and several decorative articles.

D. **Medicinal plants**

1. *Bryonia calliosa* – Seeds are used as vermifuse,
2. *B. rostrata* – Roots are used in the treatment of piles and asthma.

E. **Ornamental plants**

Different species of *Ecballium, Coccinia, Sechium* and *Cyclanthera* are grown as ornamental plants.

16. CACTACEAE

(Cactus Family)

The family Cactaceae consists of about 93 genera (according to someone it is less than 93) and about 1,400 species are restricted in the dry regions of America and also abundant in Mexico. Different species are also found as wild and cultivated forms in different corners of the world. Many members are cultivated for their ornamental value and give good monetary return.

The type genus of this family is ***Cactus.***

Systematic position

Bentham & Hooker		**Engler & Prantl**	
Division.	Angiospermia	Subdivision.	Angiospermae
Class.	Dicotyledons	Class.	Dicotyledoneae
Subclass.	Polypetalae	Subclass.	Archichlamydeae
Series.	Calyciflorae	Order.	Opuntiales
Cohort.	Ficoidales.		
Hutchinson		**Cronquist**	
Phylum.	Angiospermae	Division.	Magnoliophyta
Subphylum.	Dicotyledones	Class.	Magnoliopsida
Division.	Lignosae	Subclass.	Caryophyllidae
Order.	Cactales	Order.	Caryophyllales
Takhtajan		**Thorne**	
Division.	Magnoliophyta	Class.	Angiospermae
Class.	Magnoliopsida	Subclass.	Magnoliidae
Subclass.	Caryophyllidae	Superorder.	Caryophyllanae
Superorder.	Caryophyllnae	Order.	Caryophyllales
Order.	Caryophyllales.		

Identifying characteristics

1. **Xerophytic shrubs** or rarely trees.
2. **Stem fleshy of various shapes (phylloclade).**
3. **Flowers** are usually **solitary, sessile, epigynous, actinomorphic** or **zygomorphic**.
4. Perianth — **tepals many** and **spirally arranged.**
5. **Stamens many, epipetalous.**
6. **Carpels 3-many, unilocular** with many anatropous ovules on **parietal placentation.**
7. Fruit usually **berry**.

Floral formula

$\oplus$ ⚥ $P_{\alpha} A_{\alpha} \overline{G}_{(3-\alpha)}$

Phylogeny and systematics

The family is related with Rhamnaceae based on :

1. Reduction of calyx-lobes,
2. One staminal whorls, and
3. Single ovule.

It also considers to be related with Compositae based on :

1. Reduction of calyx-lobes,
2. Flowers zygomorphic with inferior ovary.

Cactaceae : Distinctive features

Common characteristics	Exceptions
Vegetative characteristics	
Plant. Succulent herbs or shrubs with fleshy, cylindrical, angular or flattened stem (phylloclade), simple or branched and jointed segments.	Epiphyte – species of *Cereus*. Shrubs or small tree – *Pereskiopsis, Quiabentia* and *Pereskia*. Tree (17-18 mtr.) – *Carnegiea gigantea*.
Leaves. Usually modified into tufts of spines or glochids arranged in areoles.	Non-succulent habit with foliage leaves – *Pereskia*.
Reproductive characteristics	
Flower. Solitary, actnomorphic or zygomorphic, bisexual and epigynous.	
Perianth. Numerous in several series, free or united to form tube.	Perianth 2-lipped – *Epiphyllum*.
Androecium. Stamens many, arranged spirally or in clusters, epiphyllous, rarely arise from the thalamus.	
Gynoecium. Carpel 3-many, ovary inferior, unilocular with many anatropous ovules on parietal placentation.	
Fruit. Berry.	
Seed. Usually exalbuminous, embryo straight or curved.	

Bentham and Hooker placed the family under the Cohort Ficoidales along with Ficoideae and placed next to Passiflorales in the series Calyciflorae of subclass Polypetalae. Engler placed the family under the order Opuntiales between the orders Parietales and Myrtiflorae. Hutchinson placed the family under Cactales in between the orders Cucurbitales and Theales. Later, Takhtajan (1980) and Cronquist (1981) placed the family under the order Caryophyllales before the order Polygonales.

Common plant

1. *Opuntia* stricta (Haw.) Haworth, 2. *Cereus hexagonus* (L.) Mill.

Key to genera

Branches 4-6-ribbed .... *Cereus.*

Branches flat .... *Opuntia.*

Plants of economic importance

Different plants of this family are used as fruits, hedges, timbers, medicine, fodder and also as ornamental plants.

A. **Fruits**

Cereus variabilis, Opuntia srricta and in many others. The fruits are edible.

B. **Hedges**

Different species of *Opuntia, Nopalea* are used for hedges along the fencing of cultivated field.

C. **Timbers**

A few species of *Pereskia*. The woody portion of the plants is used as timbers.

D. **Medicines**

Lophophora sp. contains a variety of alkaloids used for therapeutic purposes in Mexico.

E. **Fodder**

The spineless species of *Opuntia* are used as fodder in some dry places.

F. **Ornamental plants**

Different species of *Cereus, Epiphyllum, Echinocereus, Echinocactus* and *Mammillaria* are grown in gardens and also in greenhouses for their peculiar forms and large attractive flowers.

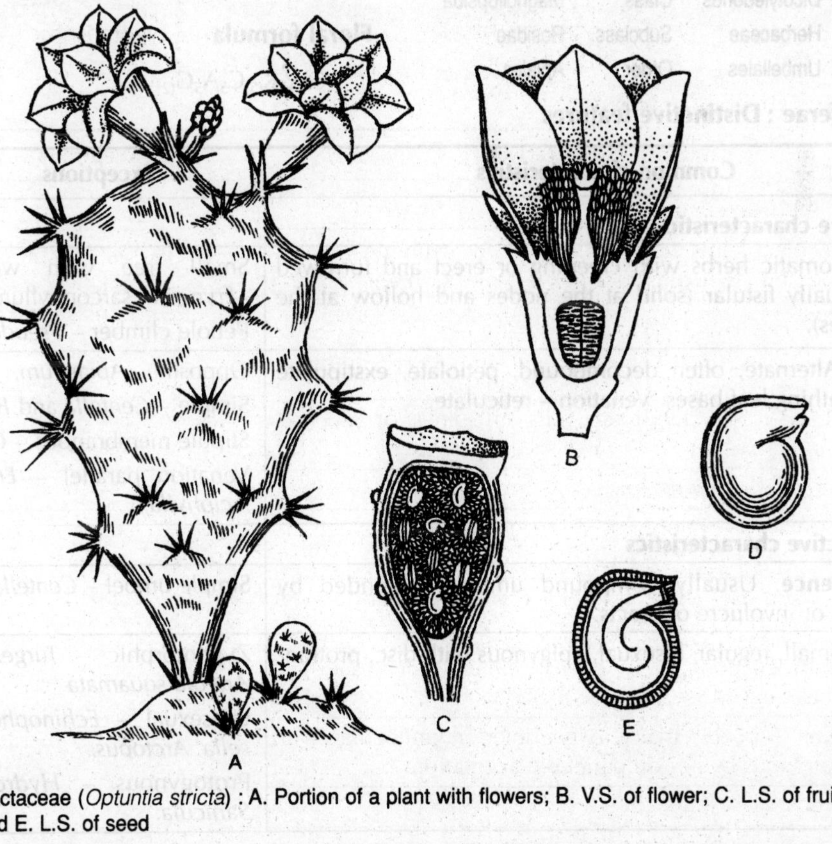

Fig. 4.81 : Cactaceae (*Optuntia stricta*) : A. Portion of a plant with flowers; B. V.S. of flower; C. L.S. of fruit; D. Seed and E. L.S. of seed

17. UMBELLIFERAE (APIACEAE)

(Carrot Family)

The family Umbelliferae consists of 421 genera and 3,220 species distributed throughout the world, but they grow abundantly in north temperate zone. In India, the family is represented by 53 genera and about 200 species found mainly in the temperate and alpine Himalayas, but some species are grown commercially in the warmer parts.

The type genus of the family is ***Apium***.

Systematic position

Bentham & Hooker		**Engler & Prantl**	
Division.	Angiospermia	Subdivision.	Angiospermae
Class.	Dicotyledons	Class.	Dicotyledoneae
Subclass.	Polypetalae	Subclass.	Archichlamydeae
Series.	Calyciflorae	Order.	Umbelliflorae
Cohort.	Umbellales.		

Hutchinson		**Cronquist**	
Phylum.	Angiospermae	Division.	Magnoliophyta
Subphylum.	Dicotyledones	Class.	Magnoliopsida
Division.	Herbaceae	Subclass.	Rosidae
Order.	Umbellales	Order.	Apiales

Takhtajan		**Thorne**	
Division.	Magnoliophyta	Class.	Angiospermae
Class.	Magnoliopsida	Subclass.	Magnoliidae
Subclass.	Rosidae	Superorder.	Aralianae
Superorder.	Aralianae	Order.	Araliales
Order.	Araliales.		

Identifying characteristics

1. Plant **aromatic herb, usually with hollow internodes.**
2. Leaves alternate, **simple or decompound with sheathing leaf bases.**
3. Inflorescence **umbel** (either simple or compound).
4. Flowers small, **bracteate, bisexual** and **epigynous,**
5. Perianth pentamerous,
6. **Ovary inferior**, 2-chambered; styles 2, **free-stylopodic**.
7. Fruit **schizocarp** (consists of two **mericarps**) or **nut-like.**

Floral formula

$\oplus$ ⚥ $K_{(5)}C_5A_5\overline{G}_{(2)}$

Umbelliferae : Distinctive features

Common characteristics	Exceptions
Vegetative characteristics	
Plant. Aromatic herbs with creeping or erect and furrowed stem, usually fistular (solid at the nodes and hollow at the internodes).	Small tree with woody trunk –*Eryngium* sarcophyllum Petiole climber – *Pseudocarum.*
Leaves. Alternate, often decompound, petiolate, exstipulate with sheathing leaf bases. Venation – reticulate.	Opposite – *Apiastrum.* Simple – *Centella* and *Bupleurum.* Stipule membranous – *Centella.* Venation parallel – *Eryngium* and *Aciphylla.*
Reproductive characteristics	
Inflorescence. Usually compound umbel, subtended by involucel or involucre of bracts.	Simple umbel – *Centella.*
Flower. Small, regular, bisexual, epigynous with disc, protrandous.	Zygomorphic – *Turgenia latifolia, Artedia squamata.* Unisexual – *Echinophora, Mathiasella, Arctopus.* Protogynous – *Hydrocotyle* and *Sanicula.*

Contd.

Contd.

Common characteristics	Exceptions
Calyx. Sepals 5, connate at the base, often scarcely indicated by the lobes, valvate.	
Corolla. Petals 5, free, imbricate, apically inflexed.	Pistillate flowers are apetalous – *Mathiasella bupleurioides*. Peripheral flowers having unequal petals – *Coriandrum*.
Androecium. Stamens 5, free, alternating with petals, basifixed or dorsifixed, introrse, dithecous.	
Gynoecium. Carpels 2, syncarpous, ovary inferior, 2 celled with one pendulous and anatropous ovule in each cell on axile placentation. Style 2, commonly with stylopodium (the disc-like enlargement at the base), stigma 2 (bifid).	
Fruit. Cremocarp or nut-like, presence of oil duct in the fruit wall, called *vittae*; cremocarp of two mericarps which are held together by thread-like connection called *Carpophore*.	
Seed. Albuminous (oily endosperm) with a very small embryo.	

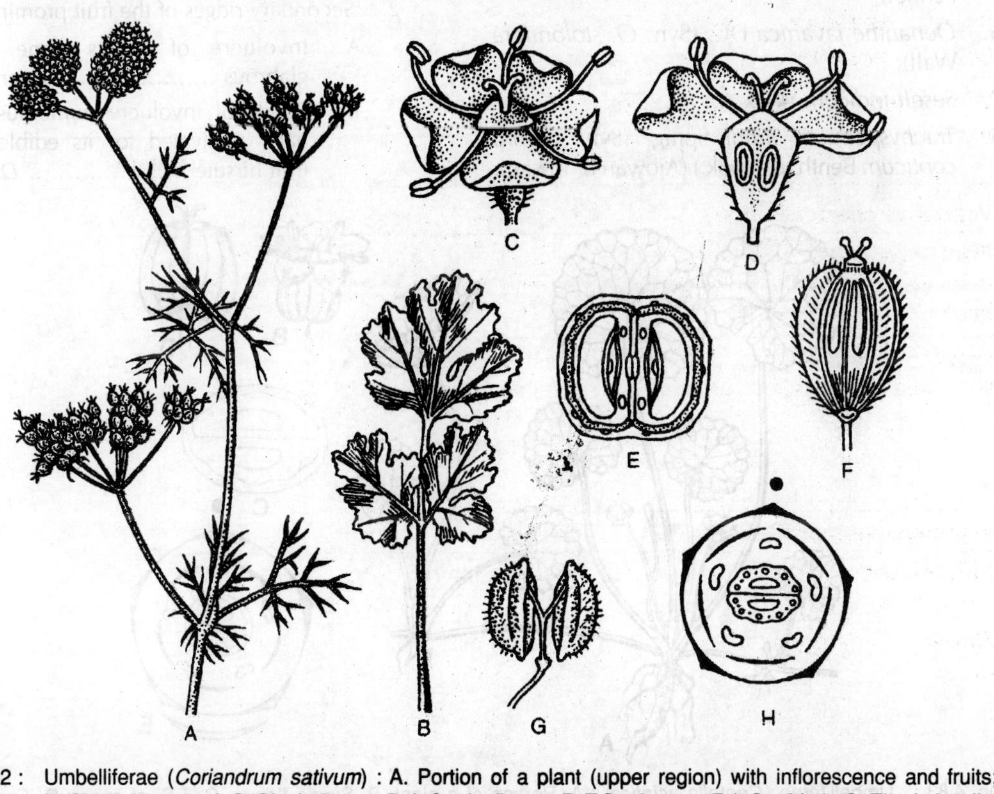

Fig. 4.82 : Umbelliferae (*Coriandrum sativum*) : A. Portion of a plant (upper region) with inflorescence and fruits; B. Single lower leaf; C. Single flower; D. L.S. of a flower; E. T.S. of ovary; F. L.S. of fruit; G. The fruit splits open and H. Floral diagram

Phylogeny and systematics

The family Umbelliferae (Apiaceae) is closely related to the family Compositae and also with the members of Rhamnales.

Bentham and Hooker placed Umbelliferae in the cohort (= order) Umbellales, series Calyciflorae, subclass Polypetalae and class Dicotyledons. Engler and Prantl placed Umbelliferae in the order Umbelliflorae, subclass Archichlamydeae and class Dicotyledoneae. But Hutchinson treated this family as Apiaceae and placed it under the order Umbellales of division Herbaceae and subphylum Dicotyledones.

Common plants

1. *Apium graveolens* L. (Celery);
2. *Carum carvi* L. (Caraway);
3. *Centella asiatica* Urb. (Syn. *Hydrocotyle asiatica* L.) (Pennywort);
4. *Coriandrum sativum* L. (Coriander);
5. *Foeniculum vulgare* Gaertn. (Anise or Fennel);
6. *Oenanthe javanica* DC. (Syn. *O. stolonifera* Wall);
7. *Seseli indicum* W. & A.;
8. *Trachyspermum amni* Sprag. (Syn. *Carum copticum* Benth. & Hook.) (Ajowan or Jowan).

Key to genera

I. Leaves simple :

Umbels simple .... *Hydrocotyle.*

Umbels compound .... *Bupleurum.*

II. Leaves compound :

1. Secondary ridges of the fruit not prominent

A. Primary ridges not winged :

i. Flowers yellow, furrows of the fruit vittate .... *Foeniculum.*

ii. Flowers white, Furrows of the fruit 1-vittate :

a. Lobes of leaves linear, fruit ovoid, laterally compressed ... *Carum.*

b. Each leaflet pinnate/partite, fruit not laterally compressed ... *Seseli.*

c. Marshy herbs, leaves 1-3-pinnate, fruit longer than broad, carpophore 0 .... *Oenanthe.*

B. Primary lateral ridges of the fruit excurrent, winged ...... *Peucedanum.*

2. Secondary ridges of the fruit prominent.

A. Involucre of bracts none, fruits glabrous .... *Coriandrum.*

B. Bracts of involucre numerous, pinnate, cultivated for its edible root, fruit hirsute .... *Daucus.*

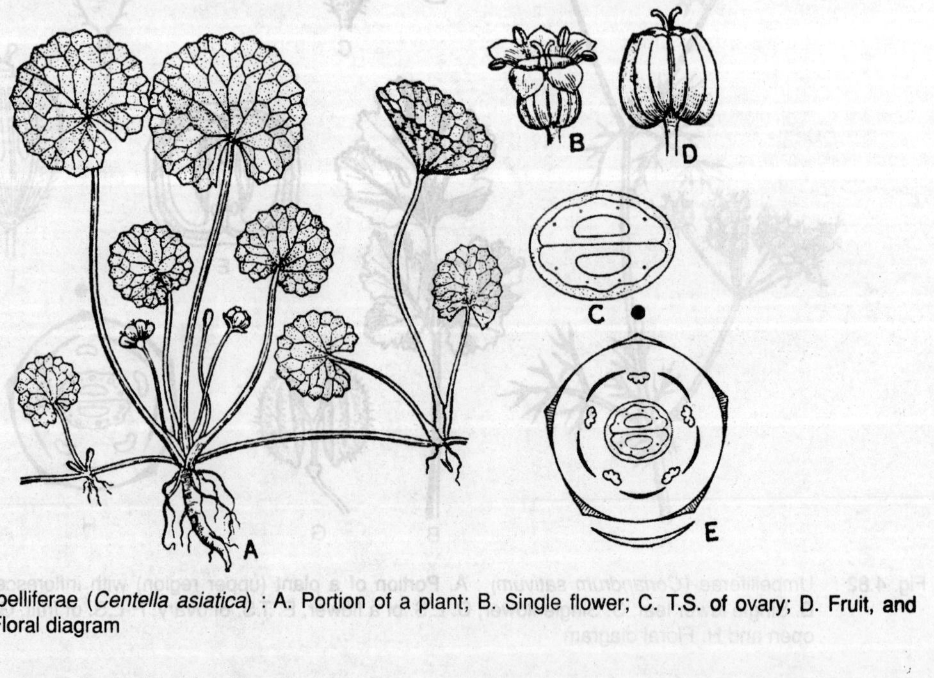

Fig. 4.83 : Umbelliferae (*Centella asiatica*) : A. Portion of a plant; B. Single flower; C. T.S. of ovary; D. Fruit, and E. Floral diagram

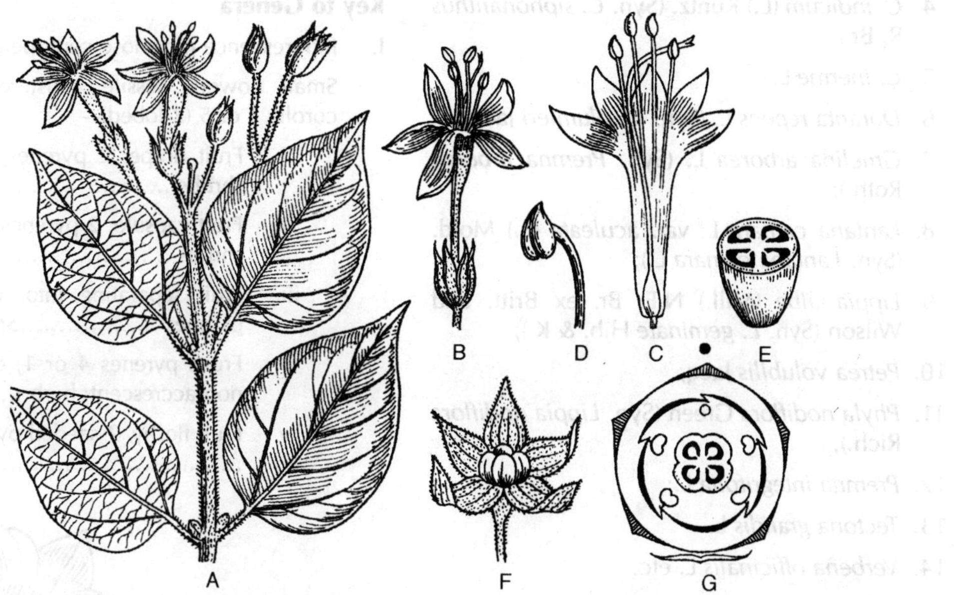

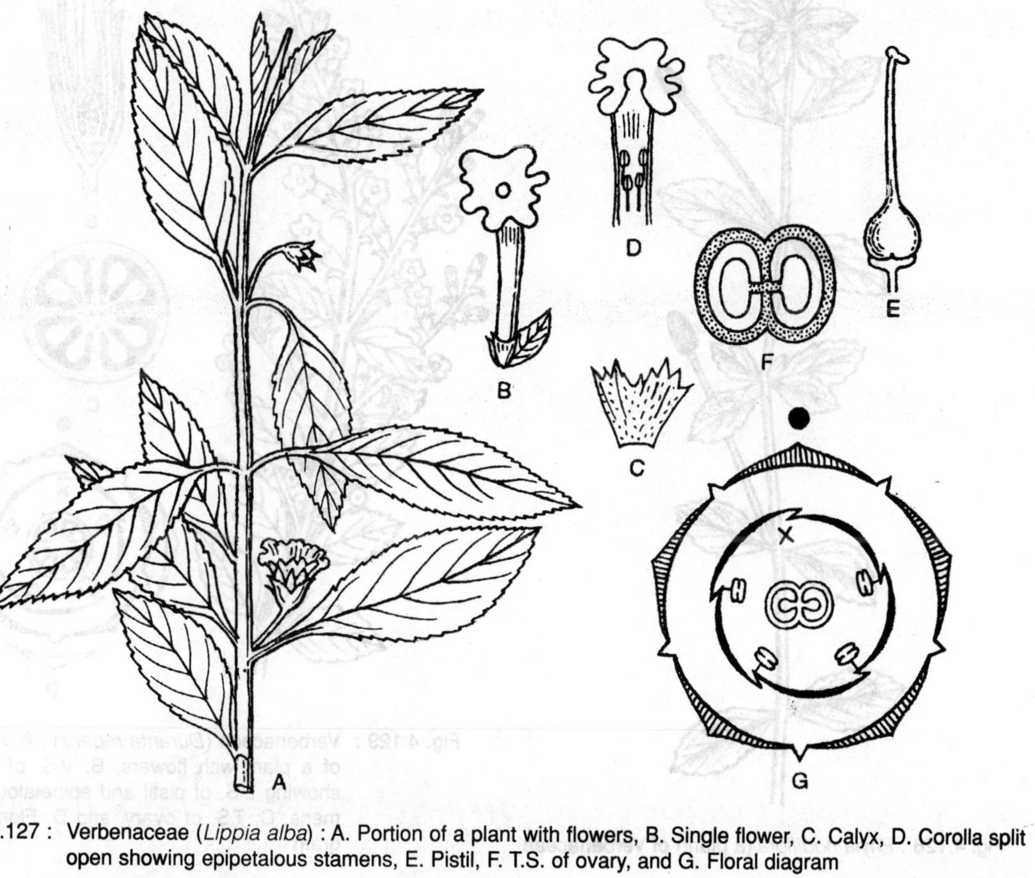

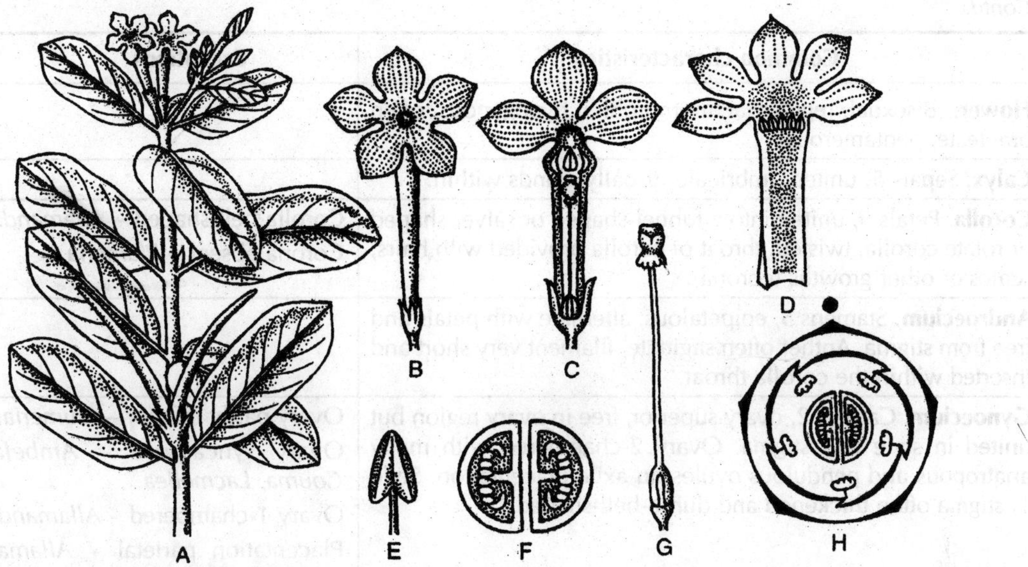

Fig. 4.94 : Apocynaceae (*Catharanthus roseus*) : A. Portion of a branch with fruit and flowers; B. Single flower; C. Corolla split open to show androecium and pistil; D. Corolla with epipetalous stamens; E. Single stamen; F. T.S. of ovary; G. Pistil, and H. Floral diagram

Identifying characteristics

1. **Plants often with milky latex.**
2. Leaves simple and exstipulate.
3. Inflorescence racemose or cymose, some times solitary.
4. Flower **bisexual, hypogynous, regular** (actinomorphic).
5. **Corolla funnel**- or **salver-shaped, throat with corona.**
6. **Stamens 5, epipetalous; anther sagittate.**
7. **Carpels 2, connate** or **free below and connate above** (by styles). **Stigma dumb-bell shaped.**
8. Fruit **commonly a pair of follicle.**

Floral formula

$\oplus$ ♂ $K_{(5)}\overline{C_{(5)}A_5}\underline{G}_{(2)}$

Apocynaceae : Distinctive features

Common characteristics	Exceptions
Vegetative characteristics	
Plant. Generally shrubs, either erect or woody twinging i.e. lianes (*Beaumontia, Aganosma, Landolphia*) or tree (*Alstonia, Holarrhena*) with milky latex.	Herb – *Catharanthus roseus.* Climbers – *Vallaris, Aganosma.* Tendril climber – *Landolphia.* Plants xerophyte – *Adenium.* Mangrove tree – *Cerbera odollam.*
Leaves. Simple, entire, exstipulate, opposite or whorled (*Nerium, Alstonia*).	Alternate – *Landolphia.* Small stipule present – *Tabernaemontana.*
Reproductive characteristics	
Inflorescence. Racemose (panicle) or cymose (dichotomous cyme or cincinnus).	Solitary flower – *Catharanthus major.*

Contd.

Contd.

Common characteristics	Exceptions
Flower. Bisexual, regular (actinomorphic), hypogynous, often bracteate, pentamerous.	
Calyx. Sepals 5, united, imbricate, usually glands within.	
Corolla. Petals 5, united into a funnel-shaped or salver-shaped or rotate corolla, twisted, throat of corolla provided with hairs, scales or other growths (corona).	Corolla bell-shaped – *Allamanda*. Corona absent – *Plumeria.*
Androecium. Stamens 5, epipetalous, alternate with petals and free from stigma. Anther often sagittate, filament very short and inserted within the corolla throat.	
Gynoecium. Carpels 2, ovary superior, free in ovary region but united in style and stigma. Ovary 2-chambered with many anatropous and pendulous ovules on axile placentation. Style 1, stigma often thickened and dumb-bell-shaped.	Ovary partly inferior – *Plumeria*. Ovary syncarpous – *Ambelania, Couma, Lacmellea*. Ovary 1-chambered – *Allamanda*. Placentation parietal – *Allamanda, Willughbeia*.
Fruits. Various types: a pair of follicles, berries (*Carissa*), drupes (*Thevetia, Rauvolfia*) or capsules.	
Seeds. Flat, winged or comose (bunch of hairs at tips) with fleshy or cartilaginous endosperm or non-endospermic. Embryo straight.	

Phylogeny and systematics

The family Apocynaceae is very much related to Asclepiadaceae (but there is difference mainly due to the absence of gynostegium and pollinia) and also bear relationship with Oleaceae and Gentianaceae.

Different taxonomists placed the family in different orders : under Geraniales (Bentham and Hooker), or Contortae (Engler and Prantl) or Apocynales (Hutchinson).

Bentham and Hooker placed the Geraniales under the series Bicarpellatae, subclass

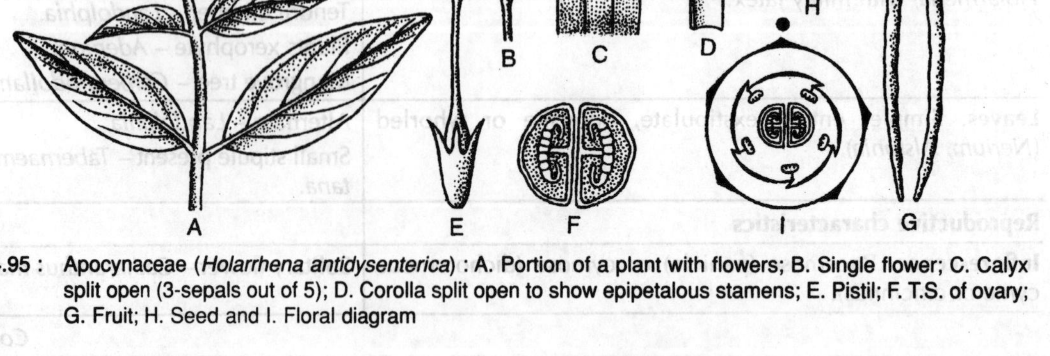

Fig. 4.95 : Apocynaceae (*Holarrhena antidysenterica*) : A. Portion of a plant with flowers; B. Single flower; C. Calyx split open (3-sepals out of 5); D. Corolla split open to show epipetalous stamens; E. Pistil; F. T.S. of ovary; G. Fruit; H. Seed and I. Floral diagram

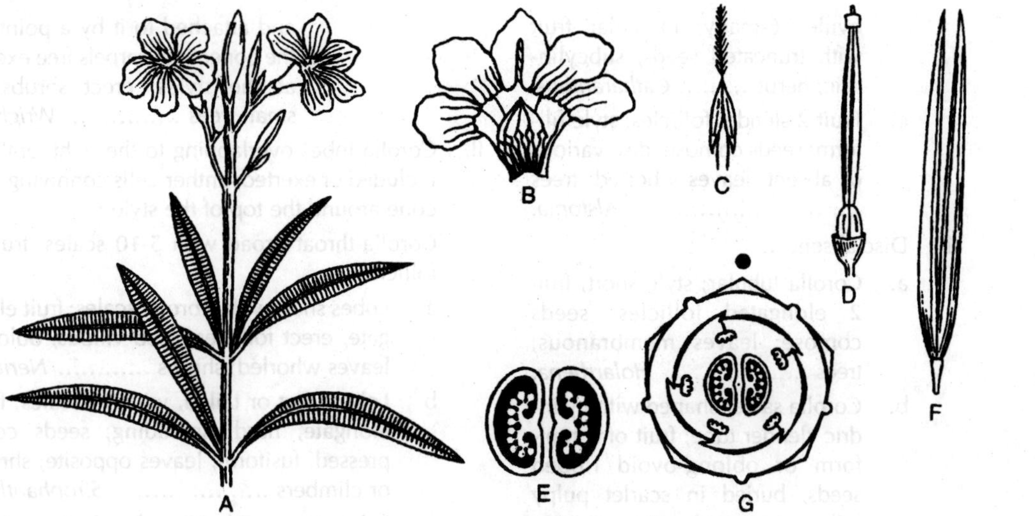

Fig. 4.96 : Apocynaceae (*Nerium indicum*) : A. Portion of a plant with flowers; B. Corolla split open; C. Stamen; D. Pistil; E. T.S. of ovary; F. Fruit, and G. Floral diagram

Gamopetalae and class Dicotyledons. Engler and Prantl placed the order Contortae under the sub-class Metachlamydeae (Sympetalae) and Class Dicotyledoneae. Hutchinson placed Apocyanales under Division Lignosae of sub-phylum Dicotyledones.

Common plants

1. *Aganosma dichotoma* schum. (Syn. *Echites dichotoma* Roth);
2. *Alstonia scholaris* R. Br.;
3. *Catharanthus roseus* (L.) Don (Syn. *Lochnera rosea* Reichb., *Vinca rosea* L.);
4. *Nerium indicum* Mill. (Syn. *N. odorum* Soland);
5. *Plumeria alba* L.; *P. rubra* L.; and *P. acutifolia* Poir.);
6. *Rauvolfia tetraphylla* L. (Syn. *R. canescens* L.);
7. *R. serpentina* (L.) Benth.;
8. *Tabernaemontana divaricata* (L.) R. Br. (Syn. *Ervatamia coronaria* Stapf.);
9. *Thevetia peruviana* (Pers.) Schum. (Syn. *T. neriifolia* Juss.);
10. *Wrightia tomentosa* Roem and Schult. etc.

Key to genera

I. Corolla-lobes overlapping to the left; anthers included, free from the stigma; anther-cells round at the base.

1. Calyx glandular :
   - a. Carpels-2, syncarpous; ovary 2-celled with axile ovules; fruit large, fleshy or pulpy within; seeds wingless or with pencil of hairs; spiny shrubs...*Carissa.*
   - b. Carpels-2, wholly united, corolla salver-shaped; ovary 1-celled with parietal ovules; indehiscent flower; seeds without albumen; climbing .... *Willughbeia.*
   - c. Carpels-2, united by the style; fruit various, rarely a pair of follicles; seeds peltate, winged; trees with alternate leaves ....*Plumeria.*
   - d. Carpels-2, thoroughly distinct or united, style calyptriform at the base, bifid at the top; fruit usually a double berry; disc present; undershrubs .... *Rauvolfia.*
   - e. Corolla-lobes long, yellow or white; fruit a drupe; leaves whorled ....*Thevetia.*
2. Calyx glandular, carpels-2, distinct :
   - i. Disc present ...
     - a. Fruit a product of 2 ripe carpels, usually 1 abortive; leaves opposite; corolla salver-shaped; shrubs .... *Kopsia.*
     - b. Stigma annulate, viscid, sometimes with reflected membrane;

ovules 6-many; follicular fruit with truncated seeds, subcylindric; herbs .... *Catharanthus.*

c. Fruit 2 slender follicles; style filiform; seeds comose; disc various or absent; leaves whorled; trees .... *Alstonia.*

ii. Disc absent ...

a. Corolla tubular; style short; fruit 2 elongated follicles; seeds comose; leaves membranous; trees .... *Holarrhena.*

b. Corolla salver-shaped with cylindric slender tube; fruit of 2 reniform or oblong-ovoid ribbed seeds, buried in scarlet pulpy aril, compressed; stamens sagittate; leaves opposite; shrubs ... . *Tabernaemontana.*

c. Corolla rotate or salver-shaped; mouth of corolla with a ring of scales; anthers conniving in a cone around the top of the style and attached to it by a point on the conective; carpels free except at the styles; erect shrubs or small trees .... *Wrightia.*

II. Corolla-lobes overlapping to the right; anthers included or exerted; anther-cells conniving in a cone around the top of the style :

Corolla-throat broad with 5-10 scales, fruit 2 follicles ...

a. Lobes short with 5 broad scales; fruit elongate, erect follicles; seed villous, oblong; leaves whorled; shrubs .... *Nerium.*

b. Lobes long or tailed, with 10 scales; fruit elongate, hard, spreading; seeds compressed. fusiform; leaves opposite; shrubs or climbers .... *Strophanthus.*

c. Lobes long, with 10 scales; fruit a pair of hard follicles; flowers showy, white or purple; leaves opposite; climbers...*Roupellia.*

d. Lobes large, mouth naked, disc deeply 5-lobed; fruit long, woody, dividing into 2 horizontally spreading follicles; climbers .... *Beaumontia.*

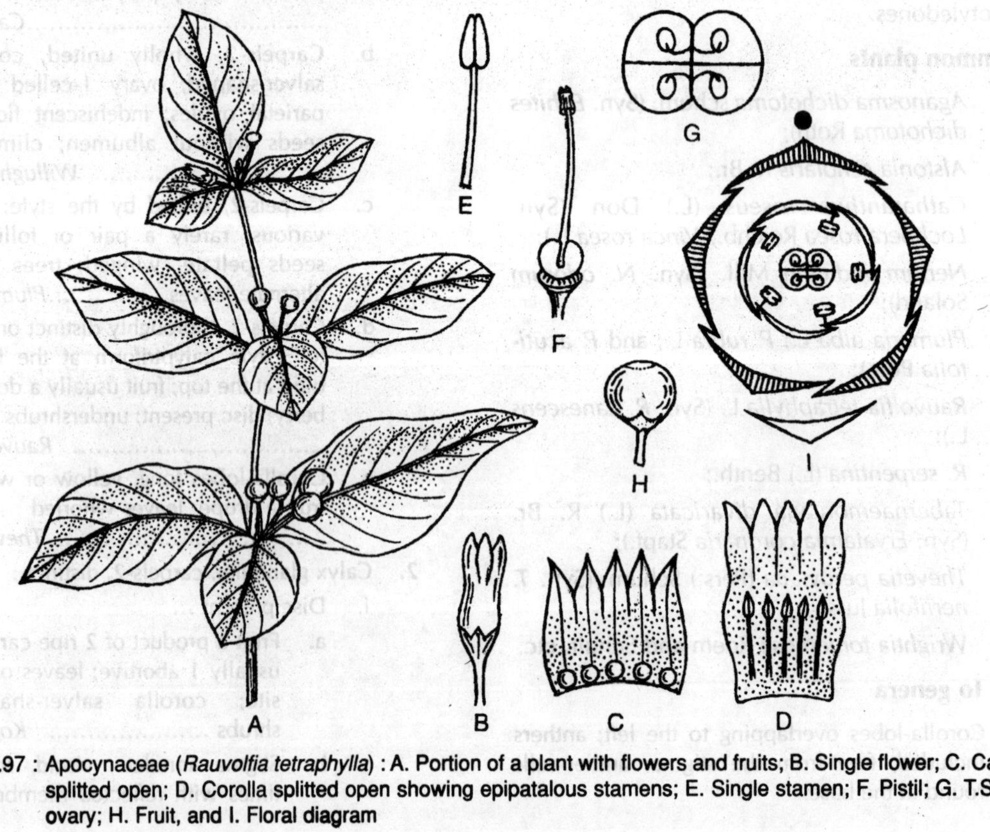

Fig. 4.97 : Apocynaceae (*Rauvolfia tetraphylla*) : A. Portion of a plant with flowers and fruits; B. Single flower; C. Calyx splitted open; D. Corolla splitted open showing epipatalous stamens; E. Single stamen; F. Pistil; G. T.S. of ovary; H. Fruit, and I. Floral diagram

e. Lobes long, mouth naked, yellow or violet; fruit capsule; leaves whorled or opposite; climbers .... *Allamanda.*

f. Lobes straight or slightly twisted to the left in the bud; disc cupular, 5-lobed; fruit follicular, straight or curved; seeds with deciduous coma; leaves opposite; climbers .... *Aganosma.*

g. Lobes short, twisted to the left in the bud; ovary hidden within the disc; seeds beaked, narrowed at the apex; leaves opposite; climbers .... *Ichnocarpus.*

Plants of economic importance

The plants of this family are mostly poisonous. Some have little importance based on economic point of view. A few plants have medicinal and ornamental value. A few others also have different uses.

A. **Medicinal plants**

1. *Alstonia scholaris* (Devil tree, Chatim). The bark of plant is used in the treatment of diarrhoea, dysentery and malaria.
2. *Cerbera manghas.* The drug, *Cerberin*; prepared from the latex is used as purgative and emetic (causing vomiting).
3. *Holarrhena antidysenterica* (Kurchi). The stem bark and roots are used in the treatment of amoebic dysentery.
4. *Rauvolfia serpentina* (Sarpagandha). The well-known drug 'rauvolfia', obtained from the dried roots. It is used for lowering blood pressure and also to relieve the disorder of central nervous system.

B. **Ornamental plants**

The following plants are cultivated for their ornamental value :

1. *Allamanda cathartica* and *A. neriifolia.*
2. *Beaumontia grandiflora* (Heralds trumpet).
3. *Catharanthus roseus* (Syn. *Vinca rosea*) (Periwinkle, Nayantara).
4. *Nerium indicum* (Syn. *N. odoratum*) (Oleander, Karabi).
5. *Plumeria alba* (White champa), *P. acuminata* (Temple tree) and *P. rubra.*
6. *Tabernaemontana divaricata* (Syn. *Ervatamia coronaria*) (Neroís crown, Tagar).
7. *Thevetia peruviana* (Yellow oleander, Kolkey phul). It is a deadly poisonous

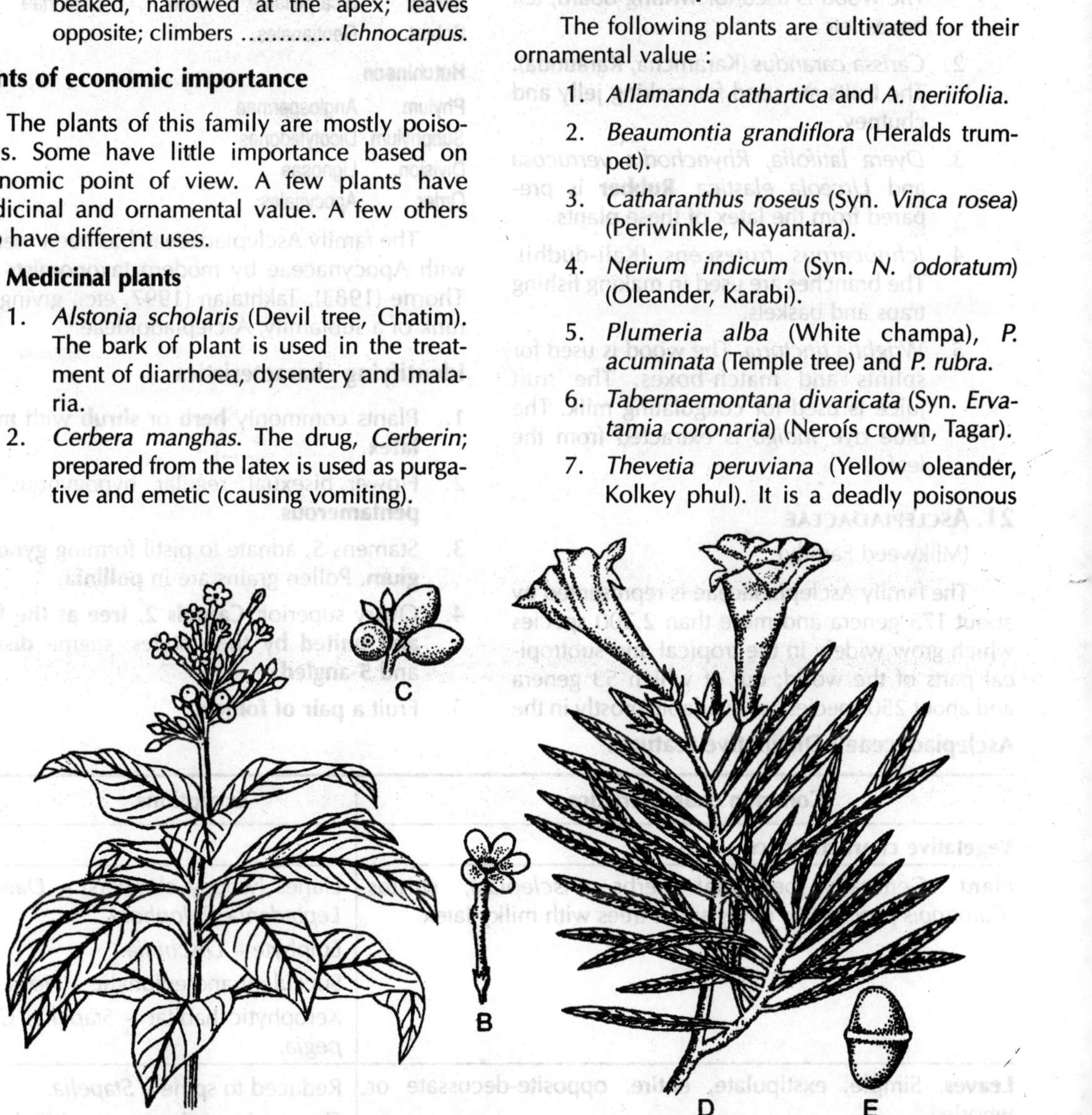

Fig. 4.98 : Plants of Apocynaceae : A. *Rauvolfia serpentina* : Portion of a plant with inflorescence; B. Flower of *R. serpentina*; C. Fruits of *R. serpentina*; D. *Thevetia peruviana* : Portion of a plant with flowers; E. Single fruit of *T. peruviana*

plant. The seeds contain a glycoside, ***'Thevetin'*** which is highly poisonous.

8. *Trachelospermum jasminoides.*
9. *Vallaris solanacea* (Syn. *V. heynei*) (Dudhi bel).
10. *Wrightia tomentosa.*

C. **Other uses**

1. *Alstonia scholaris* (Devil tree, Chatim). The wood is used for writing-board, tea boxes, etc.
2. *Carissa carandus* (Karamcha, Karaunda). The fruits are used for making jelly and chutney.
3. *Dyera latifolia, Rhynchodia verrucosa* and *Urceola elastica.* **Rubber** is prepared from the latex of these plants.
4. *Ichnocarpus frutescens* (Kali-dudhi). The branches are used in making fishing traps and baskets.
5. *Wrightia tinctoria.* The wood is used for splints and match-boxes. The fruit juice is used for coagulating milk. The blue dye *Indigo* is extracted from the leaf.

21. Asclepiadaceae

(Milkweed Family)

The family Asclepiadaceae is represented by about 175 genera and more than 2,200 species which grow widely in the tropical and subtropical parts of the world; out of which 53 genera and about 250 species are available mostly in the tropical Himalayas and Western and Southern parts of India.

The type genus of the family is ***Asclepias.***

Systematic position

Bentham & Hooker		**Engler & Prantl**	
Division.	Angiospermia	Subdivision.	Angiospermae
Class.	Dicotyledons	Class.	Dicotyledoneae
Subclass.	Gamopetalae	Subclass.	Metachlamydeae
Series.	Bicarpellatae	Order.	Contortae
Cohort.	Gentianales.		

Hutchinson

Phylum.	Angiospermae
Subphylum.	Dicotyledones
Division.	Lignosae
Order.	Apocynales

The family Asclepiadaceae has been merged with Apocynaceae by modern taxonomists like Thorne (1983), Takhtajan (1997, etc., giving the rank of a subfamily, Asclepiadoideae.

Identifying characteristics

1. Plants commonly **herb** or **shrub** with **milky latex.**
2. Flower bisexual, regular, hypogynous and **pentamerous.**
3. Stamens 5, adnate to pistil forming **gynostegium**. Pollen grains are in **pollinia**.
4. Ovary superior, **Carpels 2, free at the base and united by their styles**; stigma **discoid** and **5-angled.**
5. Fruit **a pair of follicle.**

Asclepiadaceae : Distinctive features

Common characteristics	**Exceptions**
Vegetative characteristics	
Plant. Generally perennial herbs (*Asclepias*), shrubs (*Calotropis procera*) or rarely small trees with milky latex.	Lianes (woody climbers) – *Daemia, Leptadenia, Tylophora.* Epiphyte – *Dischidia.* Succulent and epiphytic – *Hoya.* Xerophytic habitat – *Stapelia, Ceropegia.*
Leaves. Simple, exstipulate, entire, opposite-decussate or whorled.	Reduced to spine – *Stapelia.* One pair of leaves modified into pitcher to collect rain water – *Dischidia raffesiana.*

Contd.

Contd.

Common characteristics	Exceptions
Reproductive characteristics	
Inflorescence. Cymose (monochasial or dichasial cyme) or racemose (umbel).	
Flower. Bisexual, actinomorphic, hypogynous, *gynandrous*, with a single or double *corona*, borne by the corolla or stamens or by both.	Slightly zygomorphic – *Ceropegia*.
Calyx. Sepals 5, deeply divided or basally connate, quincuncial or imbricate.	
Corolla. Petals 5, gamopetalous, usually rotate, either contorted (twisted) or rarely valvate (*Dregea*).	Campanulate – *Gymnema*. Funnel form – *Cryptostegia*. Pitcher-like corolla – *Ceropegia*.
Anrdoecium. Stamens 5, commonly anthers are attached with the pistil to give rise to a *gynostegium*, anthers usually 2-celled, pollens are present in pollinium (the pollen bag) united in pairs. Pollinia are attached by their stalks (*caudicle* or *translator*) to a black horny structure, the *corpusculum*.	Anther 4 celled – *Secamone*. Pollinia may be in fours (i.e., 20 in all) – *Genianthus*. Pollen not in pollinia – *Hemidesmus*.
Gynoecium. Carpel 2, free at the base and united only by their styles and stigmatic surface; stigma 5-angled to which pollinia are attached; ovary superior with numerous, anatropous and pendulous ovules situated on the ventral suture.	
Fruit. Fruit a pair of follicles, one often aborting.	
Seed. Numerous, comose with large embryo and bony endosperm.	

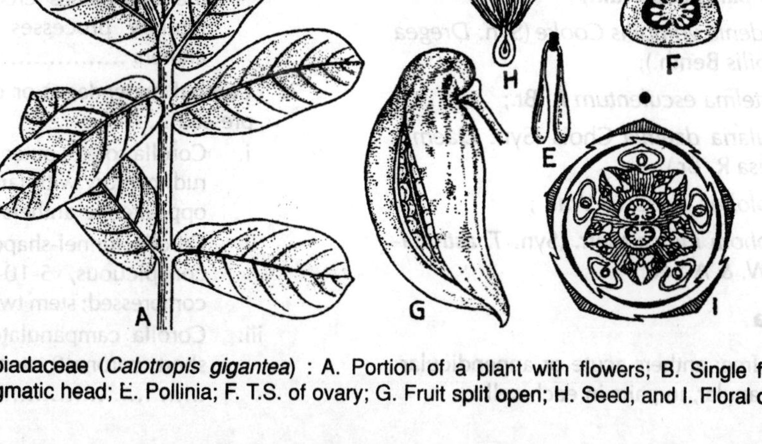

Fig. 4.99 : Asclepiadaceae (*Calotropis gigantea*) : A. Portion of a plant with flowers; B. Single flower; C. Pistil; D. Stigmatic head; E. Pollinia; F. T.S. of ovary; G. Fruit split open; H. Seed, and I. Floral diagram

Floral formula

$\oplus$ ⚥ $K_{(5)}C_{(5)}\overline{A_5\underline{G}_{(2)}}$

Phylogeny and systematics

The family Asclepiadaceae is very much related to and advanced than Apocynaceae, makes the difference mainly in presence of gynostegium and pollinia.

Bentham and Hooker placed the natural order Asclepiadaceae under Geraniales, series Bicarpellatae, subclass Gamopetalae and class Dicotyledons. He divided it into two subfamilies Periploceae (e.g., *Periploca, Cryptolepis* etc.) and Euasclepiadaceae (e.g., *Calotropis, Asclepias* etc.). Engler placed the family under the order Contortae under the subclass Metachlamydeae. But Hutchinson (1969) placed it under Apocynales order of subclass Metachalmydeae. Thorne (1983), Takhtajan (1987, 1997), and others abolished the status of Asclepiadaceae family. They merged this family with Apocynaceae giving a status of subfamily, Asclepiadoideae.

Common plants

1. *Asclepias curassavica* L.;
2. *Boucerosia aucheriana* Decne. ;
3. *Calotropis gigantea* R. Br.;
4. *C. procera* R. Br.;
5. *Cryptolepis buchanani* R. & S.;
6. *Cynanchum callialata* Ham.;
7. *Decalepsis hamiltoni* W. & A.;
8. *Dischidia raffesiana* Wall.;
9. *Finlaysonia maritima* Backer;
10. *Hemidesmus indicus* R. Br.;
11. *Hoya parasitica* Wall.;
12. *Marsdenia volubilis Cooke* (Syn. *Dregea volubilis* Benth.);
13. *Oxystelma esculentum* R. Br.;
14. *Pergularia daemia* Chois (Syn. *Daemia extensa* R. Br.);
15. *Sarcolobus globosus* Wall ;
16. *Tylophora indica* Merr. (Syn. *T. asthmatica* W. & A.) etc.

Key to genera

I. Filaments free; anthers acute or appendicular; pollinia granular, in pairs in each cell :

1. Coronal scales corolline, free, short, thick; seeds with coma ...
   - i. Corolla small, rotate, lobes valvate ... . *Hemidesmus.*
   - ii. Corolla small, rotate, lobes overlapping .... *Cryptolepis.*
   - iii. Corolla large, funnel-shaped, lobes overlapping .... *Cryptostegia.*
2. Coronal scales filiform, 5, free, close to or adnate to the filaments ...
   - i. Corolla-lobes short and broad; filaments without intervening glands; seeds without coma; flowers in pubescent stout cyme ... *Finlaysonia.*

II. Filaments connate; anthers with membranous tips inflexed; pollinia waxy, single in each cell, sessile or pedicelled in pairs, usually pendulous :

1. Coronal appendages double, outer corolline and inner staminal, with ligulate scales
   - i. Corolla wide rotate-campanulate, angularly 5-lobed; plant usually a twiner .... *Oxystelma.*
   - ii. Corolla campanulate; processes of inner corona free from smooth staminal tube .... *Cynanchum.*
2. Coronal appendages single (or while double both are staminal) ...
   - i. Corolla wide campanulate; staminal corona radiating and laterally compressed, spurred on the back; plant erect .... *Calotropis.*
   - ii. Corolla rotate; coronal processes staminal, spoon-shaped; ligulate at the face .... *Asclepias.*
   - iii. Corolla twisted, rotate; outer coronary scales cupular, membranous and the inner ones erect with a dorsal ligulary processes connecting the anthers .... *Sarcostemma.*
3. Coronal appendages or only corona suppressed ...
   - i. Corolla rotate, outer corona absent or rudimentary with larger or only lobes opposite the anthers ...... *Pentatropis*
   - ii. Corolla funnel-shaped; outer corona conspicuous, 5-10-lobed, laterally compressed; stem twining ... *Daemia.*
   - iii. Corolla campanulate, twisted, lobes short; corona 0; stem twining .... *Sarcolobus.*

iv. Corolla funnel- or salver-shaped; coronary scales erect or rarely absent; climbing shrub .... *Stephanotis.*

v. Corolla campanulate or salver-shaped; coronal scales on the back of the anthers, simple .... *Marsdenia.*

vi. Corolla rotate or salver-shaped; coronal scales on the back of the anthers, notched; flowers yellow or orange ... . *Pergularia.*

vii. Corolla rotate; coronal processes fleshy; stamens originating from the throat of the corolla-tube... *Tylophora.*

viii. Corolla rotate, green; coronal scales spreading in stellate processes, fleshy and thick; flowers in pendulous umbel .... *Dregea.*

ix. Corolla rotate; coronal processes large; stellate anthers occasionally without membraneous tip ...... *Hoya.*

x. Corolla urceolate; scales of corona erect, membraneous .... *Dischidia.*

Plants of economic importance

The plants of this family are important as ornamental and drug plants; a few are also used for other purposes.

A. **Ornamental plants**

Asclepias curassavica, Ceropegia woodii, Cryptostegia grandiflora (Rubber-vine), *Hoya parasitica* and *Stapelia gigantea* are grown as ornamental plants.

B. **Medicinal plants**

1. *Cynanchum arnotianum.* The plant is used as insecticide.
2. *Gymnema sylvestris.* It is used to control stomachi, as laxative and diuretic. The plants are used to control cough, sore eyes and biliousness.
3. *Hemidesmus indicus* R. Br. (E.Indian sarpaparilla, B. **Anantamool**). The dried roots are used to control skin and urinary diseases, chronic rheumatism and as blood purifier.
4. *Sarcostema acidum.* Infusion of the roots is used as antidote for snake bite.
5. *Tylophora indica* (Indian Ipecacuanha). The roots are used for the treatment of whooping cough, bronchitis and asthma.

C. **Other uses**

1. *Asclepias curassavica, Calotropis gigantea, Cosmostigma racemosa, Marsdenia*

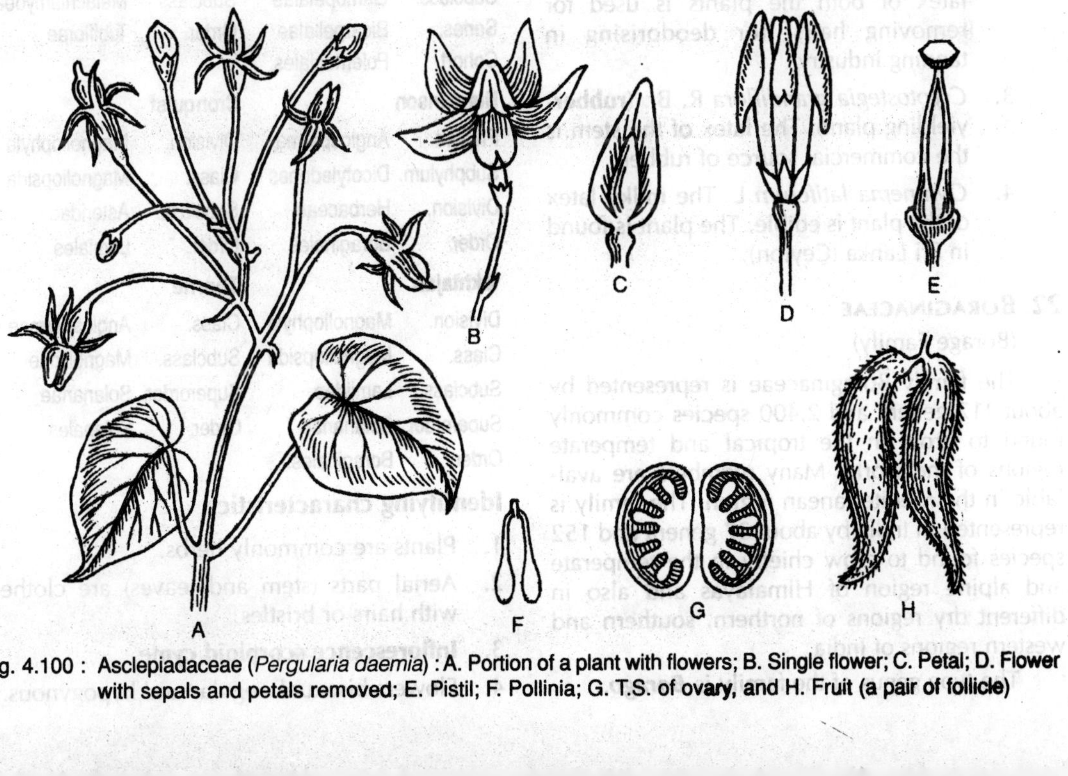

Fig. 4.100 : Asclepiadaceae (*Pergularia daemia*) : A. Portion of a plant with flowers; B. Single flower; C. Petal; D. Flower with sepals and petals removed; E. Pistil; F. Pollinia; G. T.S. of ovary, and H. Fruit (a pair of follicle)

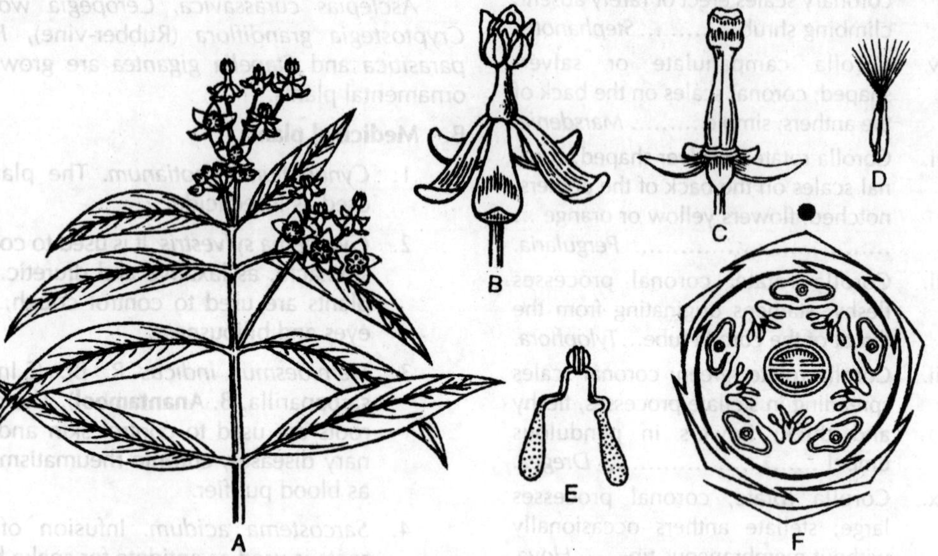

Fig. 4.101 : Asclepiadaceae (*Asclepias curassavica*) : A. Portion of a plant with flowers; B. Single flower; C. Flower after removal of petals; D. Seed; E. Pollinia, and F. Floral diagram

roylei, Leptadenia pyrotechnica. The phloem fibre (bast fibre) obtained from the stem is used in making fishing net, twines, cordage etc.

2. *Calotropis procera* and *C. gigantea*. The latex of both the plants is used for removing hairs, for deodorising in tanning industry.
3. *Cryptostegia grandiflora* R. Br. (**rubber-yielding plant**). The latex of the stem is the commercial source of rubber.
4. *Gymnema latiferum* L. The milky latex of the plant is edible. The plant is found in Sri Lanka (Ceylon).

22. Boraginaceae

(Borage Family)

The family Boraginaceae is represented by about 117 genera and 2,400 species commonly found to grow in the tropical and temperate regions of the world. Many members are available in the Mediterranean region. The family is represented in India by about 37 genera and 152 species found to grow chiefly in the temperate and alpine region of Himalayas and also in different dry regions of northern, southern and western regions of India.

The type genus of the family is ***Borago***.

Systematic position

Bentham & Hooker		**Engler & Prantl**	
Division.	Angiospermia	Subdivision.	Angiospermae
Class.	Dicotyledons	Class.	Dicotyledoneae
Subclass.	Gamopetalae	Subclass.	Metachlamydeae
Series.	Bicarpellatae	Order.	Tubiflorae
Cohort.	Polemoniales.		

Hutchinson		**Cronquist**	
Phylum.	Angiospermae	Division.	Magnoliophyta
Subphylum.	Dicotyledones	Class.	Magnoliopsida
Division.	Herbaceae	Subclass.	Asteridae
Order.	Boraginales	Order.	Lamiales

Takhtajan		**Thorne**	
Division.	Magnoliophyta	Class.	Angiospermae
Class.	Magnoliopsida	Subclass.	Magnoliidae
Subclass.	Lamiidae	Superorder.	Solananae
Superorder.	Solananae	Order.	Solanales
Order.	Boraginales.		

Identifying characteristics

1. Plants are commonly herbs.
2. Aerial parts (stem and leaves) are clothed with hairs or bristles.
3. **Inflorescence scorpioid cyme.**
4. Flowers bisexual, regular and hypogynous.

Boraginaceae : Distinctive features

Common characteristics	Exceptions
Vegetative characteristics	
Plant. Usually pubescent annual or perennial herbs, sometimes shrubs.	Tree – *Cordia dichotoma, Ehretia serrata.*
Leaves. *Alternate*, simple, entire, often hispid and exstipulate.	Opposite – *Trichodesma.*
Reproductive characteristics	
Inflorescence. Simple or double *scorpioid cyme.*	
Flower. Bisexual, hypogynous and actinomorphic (regular).	Slightly zygomorphic – *Lycopsis, Echium.*
Calyx. Sepals 5, free or united below into a tubular or bell-shaped calyx, imbricate.	
Corolla. Petals 5, rotate, funnel or salver-shaped with spreading limb; imbricate. Corolla throat often closed by scale-like outgrowth.	
Androecium. Stamens 5, epipetalous, filaments equal or unequal; anthers 2 celled, introrse.	
Gynoecium. Carpel 2, syncarpous; ovary superior, 2-celled with 2 ovules in each cell or 4-celled with 1 ovule in each cell, placentation axile; style terminal (*Heliotropium, Cordia* etc.) or gynobasic (*Borago, Myosotis* etc.); stigma simple.	Stigma bilobed – *Anchusa, Heliotropium indicum.* Style bipartite – *Echium.*
Fruit. Usually 4 nutlets.	Drupe – *Ehretia, Cordia.*
Seeds. Usually without endosperm but rarely scanty endosperm is present. Embryo curved or straight.	

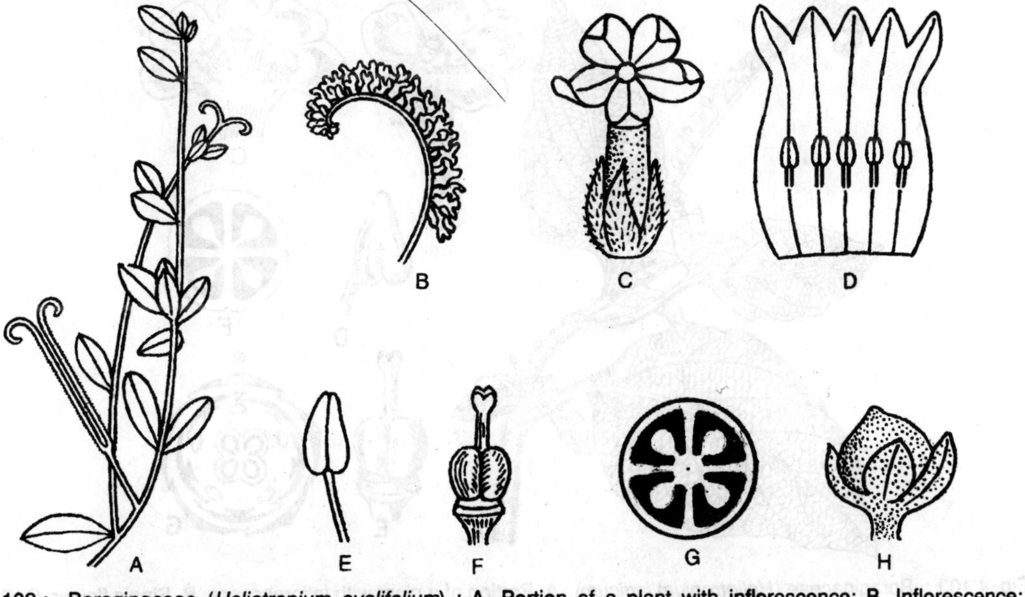

Fig. 4.102 : Boraginaceae (*Heliotropium ovalifolium*) : A. Portion of a plant with inflorescence; B. Inflorescence; C. Single flower; D. Corolla split open; E. Stamen; F. Pistil; G. T.s. of ovary, and H. Single fruit

5. Leaves simple, extipulate, arranged alternately.
6. **Stamens** 5, equal or unequal, **epipetalous**.
7. **Carpel 2, syncarpous**, ovules on axile placentation. Style **terminal** or **gynobasic**.
8. Fruit usually of 4 nutlets.

Floral formula

$\oplus$ ⚥ $K_{5 \text{ or } (5)} \widehat{C_{(5)} A_5} \underline{G}_{(2)}$

Phylogeny and systematics

The family Boraginaceae is closely related to Labiatae (Lamiaceae) and to some extent to Solanaceae. The family Boraginaceae has gynobasic style and deeply–cut nutlets. It shares these characteristics with Lamiacaeae and Verbenaceae. These characteristics may have been developed independently in all the 3 families from similar ancestors.

The family has been divided into 4 subfamilies based on style and fruits. These are Cardioideae, Ehretiodeae, Heliotropiodeae and Boraginoideae. Hutchinson (1969) kept the last two subfamilies in Boraginaceae and first two with woody genera were placed in a separate family Ehretiaceae. However, Takhtajan (1980) accept the family by retaining the components of all the 4 subfamilies. Cronquist (1981) also followed the same principle. Bentham and Hooker and also Takhtajan placed the Boraginaceae in the order Polemoniales but Engler placed it under Tubiflorae.

Common plants

1. *Cordia dichotoma* forst. f. – a small tree;
2. *Coldenia procumbens* L.;
3. *Cynoglossum procumbens* Forst.;
4. *Ehretia serrata* Roxb. (Syn. *E. acuminata* R. Br.) — a large tree found in North Bengal;
5. *Heliotropium indicum* L.;
6. *H. ovalifolium* Forsk ;
7. *Trichodesma indicum* R. Br. etc.

Key to genera

I. Style terminal :
   1. Style 2-partite on ovary; fruit a drupe with four 1-seeded stone; calyx enlarging in fruit; usually trees or shrubs .... *Cordia.*
   2. Style 2-fid at the apex; pyrenes 2 or 4 or rarely 1-seeded; trees or shrubs ... *Ehretia.*
   3. Style more or less separate; pyrenes 4, 1-seeded; prostrate herbs .... *Coldenia.*
   4. Style filiform; stamens 5, anthers lanceolate conniving in a cone; calyx enlarged in fruit; ovary deeply 4-lobed and 4-celled; fruit pyramidal; erect herbs... *Trichodesma.*

Fig. 4.103 : Boraginaceae (*Heliotropium indicum*) : A. Portion of a plant with inflorescence; B. Single flower; C. Corolla split open showing epipetalous stamens; D. Stamen; E. Pistil; F. T.S. of ovary, and G. Floral diagram

5. Style on entire ovary, dilated at the apex or above the base; plants herbaceous .... *Heliotropium*.

II. Style gynobasic :

Flowers all sessile or only the lower ones perdicelled; stamens included; nutlets obovoid, glochidiate, depressed .... *Cynoglossum*.

Plants of economic importance

The plants of this family are of various uses. They are used as edible, ornamentals, medicines and also have other uses.

A. **Edible fruits**

Ehretia acuminata. The ripe fruits are edible.

B. **Vegetable**

Cordia dichotoma. The unripe fruits are used as vegetable.

C. **Ornamental plants**

Cordia sebestena L. (tree with orange red flower), *Heliotropium arborescens* L. (a herb), *Myosotis palustris* (Forget-me-not plant of Europe, is available in Himalayas), *Borago* (Borage), *Echium* (Viperís bugloss), *Cynoglossum* (Houndís tongue) etc. are grown as ornamental plants.

D. **Medicinal plants**

1. *Ehretia macrophylla*. The roots are used as an antidote for vegetable poisoning.
2. *Heliotropium indicum*. The decoction of leaf is used in fever and root in cough.
3. *H. strigosum*. The plants have both diuretic and laxative properties.
4. *Lithospermum arvense* (Corn gromwell). The extract of the plant paralyses the central nervous system due to the presence of an alkaloid *cynoglossine*.
5. *L. officinale* (Gromwell). The decoction of seeds is used in gout and diseases of bladder.
6. *Onosma bracteatum*. The dried leaves and flowers contain the drug *gaozaban*. *Gaozaban* is used as tonic, diuretic, refrigerant etc.

E. **Plants of other uses**

1. *Alkanna tinctoria*. A red dye "*alkanet*" is extracted from the roots.
2. *Arnebia hispidissima* and *Onosma hispida*. The roots are the source a red dye, "*ratanjot*"; used for colouring foods and also used in the preparation of medicines.
3. *Ehretia laevis*. The wood is used for match boxes, brush-backs etc.
4. *E. wallichiana*. The wood is used for making tea boxes and also for building purposes.

23. Convolvulaceae

(Morning-glory Family)

The family Convolvulaceae is represented by about 55 genera and 1,930 species found to grow chiefly in the tropical and subtropical regions of the world. The family is represented in India by 20 genera and about 158 species found to grow chiefly in the western and southern regions.

The type genus of the family is ***Convolvulus***.

Systematic position

Bentham & Hooker		**Engler & Prantl**	
Division.	Angiospermia	Subdivision.	Angiospermae
Class.	Dicotyledons	Class.	Dicotyledoneae
Subclass.	Gamopetalae	Subclass.	Metachlamydeae
Series.	Bicarpellatae	Order.	Tubiflorae
Cohort.	Polemoniales.		
Hutchinson		**Cronquist**	
Phylum.	Angiospermae	Division.	Magnoliophyta
Subphylum.	Dicotyledones	Class.	Magnoliopsida
Division.	Herbaceae	Subclass.	Asteridae
Order.	Solanales	Order.	Polemoniales
Takhtajan		**Thorne**	
Division.	Magnoliophyta	Class.	Angiospermae
Class.	Magnoliopsida	Subclass.	Magnoliidae
Subclass.	Lamiidae	Superorder.	Solananae
Superorder.	Solananae	Order.	Solanales
Order.	Convolvulales.		

Identifying characteristics

1. Plants are **usually twining herbs**, sometimes shrubs or undershrubs.
2. Flower bisexual, regular, hypogynous and **pentamerous**.
3. Corolla **funnel-shaped** or **salver-shaped**.
4. Stamens 5, **unequal in size**.
5. Ovary superior, **Carpels 2**, 2-celled, 1 or 2 ovule on axile placentation.
6. Fruit **usually capsule**.

Convolvulaceae : Distinctive features

Common characteristics	Exceptions
Vegetative characteristics	
Plant. Commonly annual or perennial twining herbs or shrubs, often with watery latex.	Plants erect – *Cressa cretica.* Tree – *Ipomoea wolcottiana* (10m tall).
Leaves. Alternate, simple, exstipulate, entire or variously lobed.	Plant rootless and leafless – *Cuscuta reflexa.*
Reproductive characteristics	
Inflorescence. Usually an axillary biparious cyme (dichasium) or cymose or racemose panicle, bracteate.	Flowers solitary – *Calystegia.* Cymes form large panicle – *Porana.* Monochasial cyme – *Ipomoea pestigridis.*
Flower. Bisexual, actinomorphic (regular), hypogynous, usually showy.	Unisexual, dioecious and tetramerous – *Hildebrandtia.* Slightly zygomorphic – *Mina biloba.*
Calyx. Sepals 5, usually free, quincuncial, often persistent.	
Corolla. Petals 5, united to form a salver-shaped or funnel-shaped corolla, induplicate valvate or twisted in bud.	
Androecium. Stamens 5, *unequal*, epipetalous, inserted at the base of the corolla tube and alternating with petals; anther 2-celled, usually introrse, pollens ectinate globose or ellipsoid, cupular or annular intrastaminal disc usually *present*.	
Gynoecium. Carpel 2, syncarpous; ovary superior, 2-chambered with 2 ovule in each chamber or 4-chambered with 1 ovule in each chamber on axile placentation; style 1, filiform, bipartite or not, stigma capitate or lobed.	Style 2 – *Cuscuta, Argyreia.* Style absent – *Erycibe.*
Fruit. Usually *capsule.*	Berry – *Cuscuta, Argyreia.*
Seeds. Smooth or hairy, exalbuminous, embryo curved with folded cotyledons.	

Floral formula

$\oplus ⚥ K_{5 \text{ or } (5)} \widehat{C_{(5)} A_5} G_{\underline{(2)}}$

Phylogeny and systematics

The family is closely related to Solanaceae and placed it under different orders by different taxonomists. It was placed under Polemoniales by Bentham and Hooker, Takhtajan; Tubiflorae by Engler and Prantl and Solanales by Hutchinson.

Bentham and Hooker placed the order Polemoniales under series Bicarpellatae, subclass Gamopetalae of the class Dicotyledons.

Takhtajan placed the order Polemoniales under the subclass Asteridae of class Magnoliopsida.

Engler and Prantl placed Tubiflorae under the subclass Metachlamydae of class Dicotyledoneae.

Bentham and Hooker divided the family into 5 tribes. These are :

1. Convolvuleae.
2. Dichondreae.
3. Nolaneae.
4. Cresseae.
5. Cuscuteae.

Some taxonomists raised the tribe Dichondreae e.g. *Dichondra,* to the status of a family Dichondraceae and Cuscuteae e.g. *Cuscuta,* to Cuscutaceae. Recent taxonomists

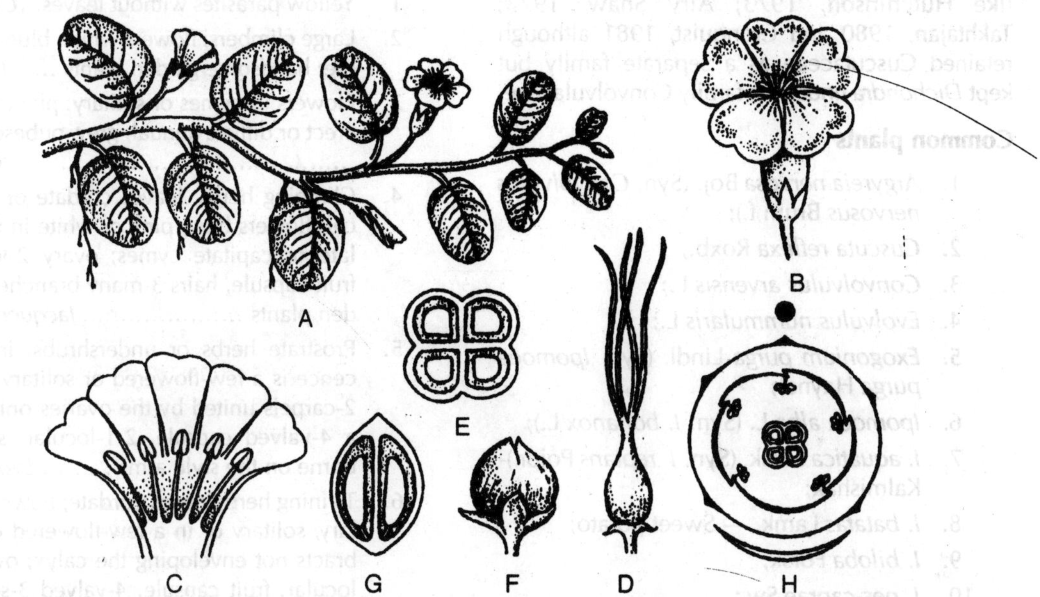

Fig. 4.104 : Convolvulaceae (*Evolvulus nummularius*) : A. Portion of a plant with flowers; B. Single flower; C. Corolla split open showing epipetalous stamens; D. Pistil; E. T.S. of ovary; F. Fruit; G. L.S. of ovary, and H. Floral diagram

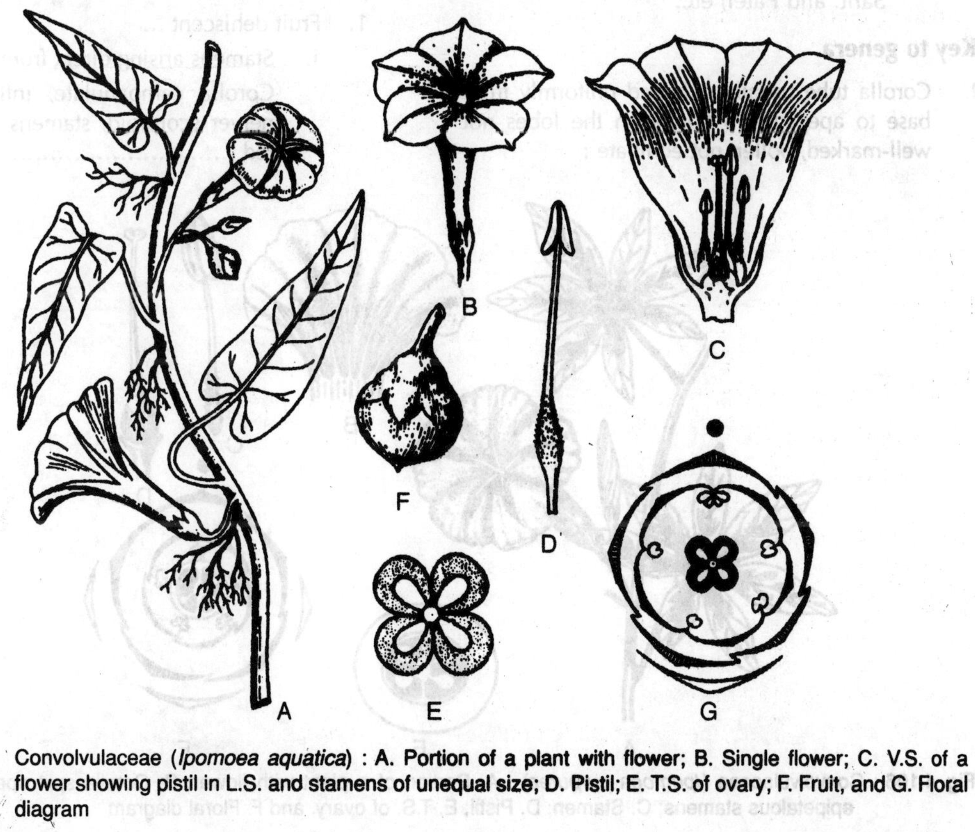

Fig. 4.105 : Convolvulaceae (*Ipomoea aquatica*) : A. Portion of a plant with flower; B. Single flower; C. V.S. of a flower showing pistil in L.S. and stamens of unequal size; D. Pistil; E. T.S. of ovary; F. Fruit, and G. Floral diagram

like Hutchinson, 1973; Airy Shaw, 1973; Takhtajan, 1980 and Cronquist, 1981 although retained Cuscutaceae as a separate family but kept *Dichondra* under the family Convolvulaceae.

Common plants

1. *Argyreia nervosa* Boj. (Syn. *Convolvulus nervosus* Brum f.);
2. *Cuscuta reflexa* Roxb.;
3. *Convolvulus arvensis* L.;
4. *Evolvulus nummularis* L.;
5. *Exogonium purga* Lindl. (Syn. *Ipomoea purga* Hayne);
6. *Ipomoea alba* L. (Syn. *I. bonanox* L.);
7. *I. aquatica* Forsk. (Syn. *I. reptans* Poior.)-Kalmishak;
8. *I. batatas* Lamk. — Sweet potato;
9. *I. biloba* Forsk;
10. *I. pes-caprae* Sw.;
11. *I. quamoclit* L. (Syn. *Quamoclit pinnata* Boj.);
12. *I. sinensis* (Desv.) Choisy (Syn. *I. tropica* Sant. and Patel) etc.

Key to genera

I. Corolla tube usually enlarged uniformly from base to apex, the 5 bands on the lobes not well-marked, pollen not echinate :

1. Yellow parasites without leaves...*Cuscuta.*
2. Large climbers; flowers white, blue or purple; handsome garden plants ... *Porana.*
3. Flowers in cymes or solitary; plants small, erect or diffuse annual, grey, pubescent ... . *Cressa.*
4. Climbing herbs; leaves cordate or elliptical; flowers blue, pink or white in umbellate or capitate cymes; ovary 2-locular; fruit capsule, hairs 3-many branched; garden plants .... *Jacquemontia.*
5. Prostrate herbs or undershrubs; inflorescence is a few-flowered or solitary cyme; 2-carpels united by the ovaries only; fruit a 4-valved capsule, 2-1-locular; stigmas borne on the style arms .... *Evolvulus.*
6. Twining herb; leaves cordate; flowers axillary, solitary or in a few-flowered cymes; bracts not enveloping the calyx; ovary 1-locular; fruit capsule, 4-valved 3-seeded; pollen polyhedral .... *Hewittia.*

II. Corolla tube not uniformly enlarged from base to apex, bands on corolla-lobes prominent, pollen echinate :

1. Fruit dehiscent ...
   i. Stamens arising direct from corolla
   Corolla campanulate, inflorescence nerver scorpioid, stamens not exserted .... *Ipomoea.*

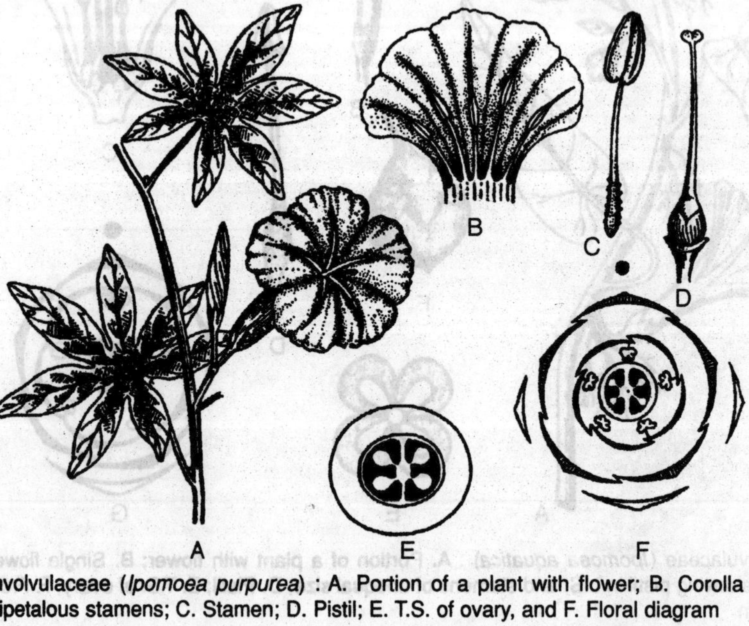

Fig. 4.106 : Convolvulaceae (*Ipomoea purpurea*) : A. Portion of a plant with flower; B. Corolla split open showing epipetalous stamens; C. Stamen; D. Pistil; E. T.S. of ovary, and F. Floral diagram

ii. Stamens exserted ...

a. Flowers small, pink or white, slightly irregular ...... *Quamoclit.*

b. Flowers large, white or purple, quite regular .... *Calonyction.*

2. Fruit indehiscent, woody or mealy or fleshy ...

i. Fruit woody; corolla salver-shaped; stigma elliptic-oblong .... *Rivea.*

ii. Fruit mealy or fleshy, corolla rarely or not salver-shaped

a. Sepals large, orbicular, accrescent, ultimately completely enveloping the fruit ... *Stictocardia.*

b. Sepals small, ovate, not enlarging to envelop the fruit :

Ovary 2-celled .... *Lettsomia.*

Ovary 4-celled .... *Argyreia.*

c. Sepals unequal, outer 3 not larger than the inner 2; carpels – 2; style entire; stigma globose; ovary 2-celled; bands of corolla demarcated by purple lines .... *Merremia.*

Plants of economic importance

Members of this family are not so important. Some are edible; some used as ornamental plants and still some also have other uses.

A. **Edible plants**

1. *Ipomoea aquatica* (Water-binweed, Kalmishak). The young shoots and leaves are used as herbage-vegetable. It *increases the lactation of mother at a very significant amount.* The plants are also used as fodder. In addition to the above the plant juice is also used *as an emetic during arsenic and opium poisoning.*

2. *Ipomoea batatas* (Sweet potato). The fleshy roots develop from the node of the stem, they are rich in sugar and starch. The roots are used as food after boiling, roasting etc. It is also used in industry for alcohol production.

3. *Rivea hypocrateriformis.* The young shoots and leaves are used as vegetable.

B. **Ornamental plants**

Many plants of this family are used as ornamental one. These are :

1. *Argyreia bella* and *A. nervosa.* Commonly known as woolly morning-glory, are grown as ornamental plants.

2. *Calonyction muricatum* and *C. aculeatum.* Commonly known as moon flower, are grown as ornamental plants.

3. Several species of *Ipomoea* are grown in the garden for their attractive campanulate flowers. These are :

i. *Ipomea carica* (A common railway creeper).

ii. *I. coccinea* (Star *Ipomoea*).

iii. *I. nil* (Syn. *I. hederacea*).

iv. *I. purpurea* (Morning-glory).

v. *I. tuberosa* (wood rose).

vi. *I. quamoclit* (Syn. *Quamoclit pinnata*) (Labangalata).

vii. *I. violacea* (heavenly blue).

4. *Porana racemosa* and *p. paniculata.* Commonly known as Christmas vine, are grown as ornamental plants.

C. **Medicinal plants**

Exogonium purga. The roots are used as purgative.

D. **Plants of other uses**

1. *Ipomoea biloba* (Goatís foot, Chagalkaru). The plants are cultivated as soil-binder.

2. *Calystegia soldanella.* The plants are found in sea beaches, they also act as soil binder.

24. SOLANACEAE

(Potato Family)

The family Solanaceae is represented by 83 genera and more than 2,925 species, they grow in the temperate and tropical regions of the world. Their availability is very common in South and Central regions of America. The family is represented in India by 15 genera and 88 species, chiefly found to grow in Eastern and Southern parts and also in Himalayan range.

The type genus of the family is ***Solanum.***

Systematic position

Bentham & Hooker		**Engler & Prantl**	
Division.	Angiospermia	Subdivision.	Angiospermae
Class.	Dicotyledons	Class.	Dicotyledoneae
Subclass.	Gamopetalae	Subclass.	Metachlamydeae
Series.	Bicarpellatae	Order.	Tubiflorae
Cohort.	Polemoniales.		

Hutchinson		**Cronquist**	
Phylum.	Angiospermae	Division.	Magnoliophyta
Subphylum.	Dicotyledones	Class.	Magnoliopsida
Division.	Herbaceae	Subclass.	Asteridae
Order.	Solanales	Order.	Polemoniales

Takhtajan		**Thorne**	
Division.	Magnoliophyta	Class.	Angiospermae
Class.	Magnoliopsida	Subclass.	Magnoliidae
Subclass.	Lamiidae	Superorder.	Solananae
Superorder.	Solananae	Order.	Solanales
Order.	Solanales.		

Identifying characteristics

1. Plants are herbs, or shrubs.
2. Leaves simple, exstipulate and arranged alternately.
3. Inflorescence extra-axillary cymose.
4. Flowers bisexual, regular and hypogynous.

Solanaceae : Distinctive features

Common characteristics	Exceptions
Vegetative characteristics	
Plant. Herbs or shrubs, often with poisonous juice.	Climber – *Solanum jasminoides* and *S. trilobatum*. Prostrate herb – *S. xanthocarpum*. Small tree – *S. verbascifolium*.
Leaves. Alternate, simple, entire or lobed, exstipulate.	
Reproductive characteristics	
Inflorescence. Extra-axillary cymose, cymose panicle or flowers may be solitary.	
Flower. Actinomorphic (regular), bisexual, hypogynous.	Zygomorphic – *Hyoscyamus, Browallia, Brunfelsia.*
Calyx. Sepals 5, persistent, often enlarged in fruits, imbricate.	
Corolla. Petals 5, usually rotate or funnel-shaped, imbricate or plicate;	Corolla more or less bilabiate – *Hyoscyamus, Browallia, Brunfelsia.* Valvate – *Cestrum.*
Androecium. Stamens 5, epipetalous, alternate with petal, anther 2 celled, dehisce longitudinally or by apical pores.	Stamens 2 – *Schizanthus.* Stamens 4 (didynamous) – *Browallia, Brunfelsia, Salpiglossis* and *Schizanthus.*
Gynoecium. Carpel 2, ovary superior, obliquely placed, 2-celled or falsely 3-5 chambered, placentation axile, ovules anatropous; stigma bilobed.	1-celled and 1-ovuled ovary – *Henoonia.* False partition in ovary (4-more chambers) – species of *Datura* and *Solanum.*
Fruit. Usually many-seeded *berry.*	Capsule – *Datura.*
Seeds. Many, albuminous (with endosperm), embryo curved or semi-straight.	

5. **Calyx persistent** on fruit.
6. **Corolla rotate** or **funnel-shaped**.
7. **Carpel** – 2, syncarpous, **ovary obliquely placed**.
8. **Fruit – Usually** many seeded **berry**.

Floral formula

⊕ or ∘|∘ ⚥ $K_{(5)}\overline{C_{(5)}A_5}\underline{G}_{(2)}$

Phylogeny and systematics

The family is closely related to Convolvulaceae and Scrophulariaceae. Many taxonomists placed both the families Solanaceae and Convolvulaceae under Solanales for the similarity in floral characteristics but differ in habit. The family shows similarity with Scrophulariaceae, but differs in the presence of bicollateral vascular bundle, commonly actinomorphic corolla and 4 or 5 stamens.

Bentham and Hooker placed the natural order Solanaceae under cohort (= order) Polemoniales, series Bicarpellatae, subclass Gamopetalae of class Dicotyledons. Engler and Prantl placed the family under the order Tubiflorae, subclass Metachlamydae of class Dicotyledoneae. Later, Hutchinson placed the family under the order Solanales, division Herbaceae in the subphylum Dicotyledones. Takhtajan placed the family under the order Scrophulariales, but Cronquist — following Bentham and Hooker — placed it under Polemoniales.

Bentham and Hooker divided the family into 5 tribes. These are :

1. Solaneae.
2. Atropeae.
3. Hyoscyameae.
4. Cestrineae.
5. Salpiglossidae.

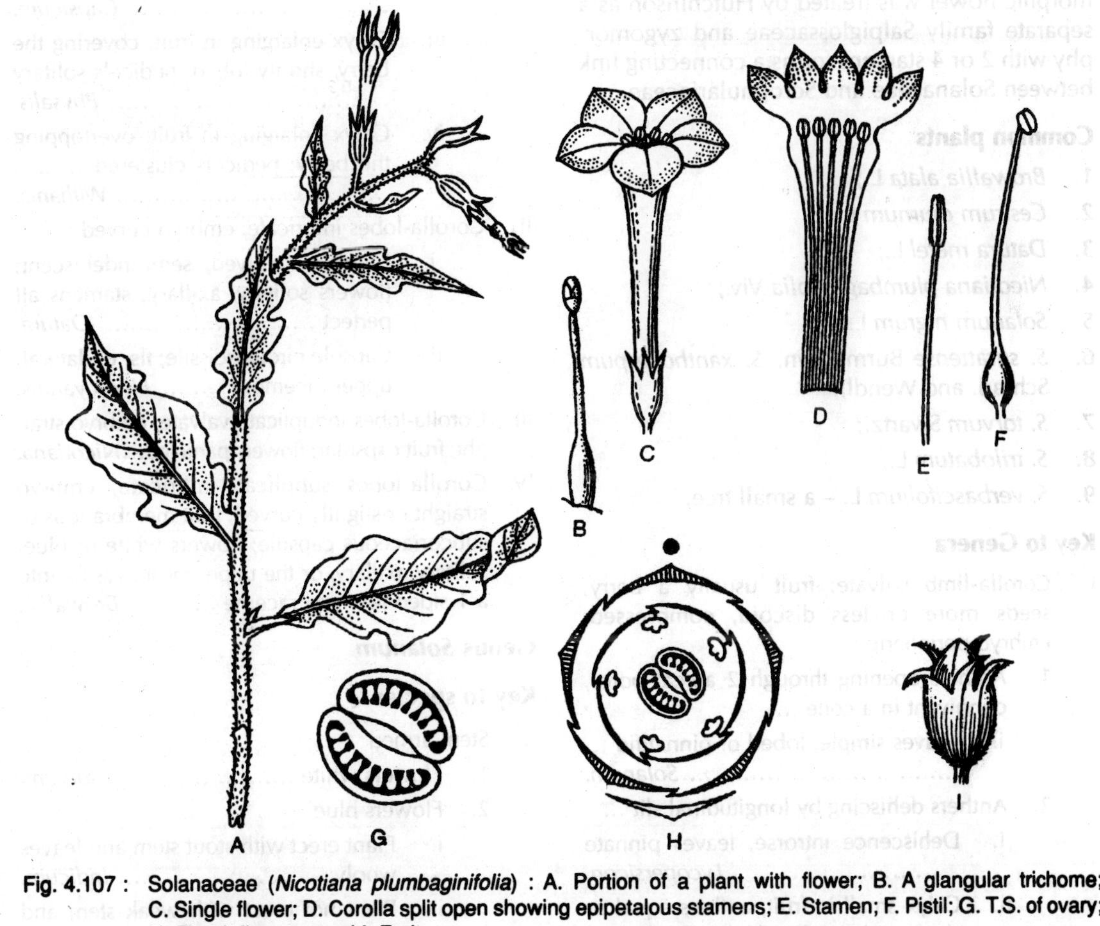

Fig. 4.107 : Solanaceae (*Nicotiana plumbaginifolia*) : A. Portion of a plant with flower; B. A glangular trichome; C. Single flower; D. Corolla split open showing epipetalous stamens; E. Stamen; F. Pistil; G. T.S. of ovary; H. Floral diagram, and I. Fruit

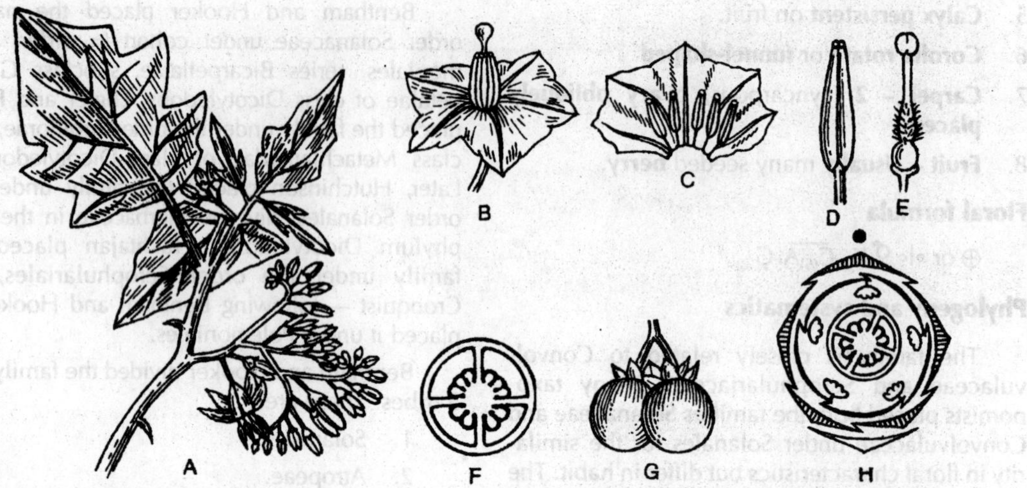

Fig. 4.108 : Solanaceae (*Solanum torvum*) : A. Portion of a plant with flowers; B. Single flower; C. Corolla split open showing epipetalous stamens; D. Stamen; E. Pistil; F. T.S. of ovary; G. Fruits, and H. Floral diagram

The last tribe Salpiglossidae having zygomorphic flower was treated by Hutchinson as a separate family Salpiglossaceae and zygomorphy with 2 or 4 stamens forms a connecting link between Solanaceae and Scrophulariaceae.

Common plants

1. *Browallia alata* L.;
2. *Cestrum diurnum* L.;
3. *Datura metel* L.;
4. *Nicotiana plumbaginifolia* Viv.;
5. *Solanum nigrum* L.;
6. *S. surattense* Burm. (Syn. *S. xanthocarpum* Schrad. and Wendl);
7. *S. torvum* Swartz.;
8. *S. trilobatum* L.;
9. *S. verbascifolium* L. – a small tree.

Key to Genera

I. Corolla-limb valvate; fruit usually a berry; seeds more or less discoid, compressed; embryo peripheric :
   1. Anthers opening through 2 apical pores, connivent in a cone ...
      i. Leaves simple, lobed or pinnatifid ... . *Solanum.*
   2. Anthers dehiscing by longitudinal slit ...
      i. Dehiscence introrse, leaves pinnate .... *Lycopersicon.*
      ii. Calyx small in fruit; anthers equal or shorter than the filament; flowers small, in paired pedicles or simple ... . *Capsicum.*
      iii. Calyx enlarging in fruit, covering the berry, shortly lobed; pedicels solitary .... *Physalis.*
      iv. Calyx enlarging in fruit, overtopping the berry; pedicels clustered .... *Withania.*

II. Corolla-lobes imbricate; embryo curved :
   i. Capsule 4-valved, semi-indehiscent; flowers solitary, axillary; stamens all perfect .... *Datura.*
   ii. Capsule circumscissile; fls. unilateral, upper racemose .... *Hyoscyamus.*

III. Corolla-lobes induplicate-valvate; embryo straight; fruit capsular; flowers panicled... *Nicotiana.*

IV. Corolla-lobes subplicate-imbricate; embryo straight or slightly curved; fruit membranous or subcoriaceous capsule; flowers white or blue, solitary axillary, or the upper-most passing into a 1-sided terminal raceme ...... *Browallia.*

Genus *Solanum*

Key to species

I. Stem armed
   1. Flowers white .... *torvum.*
   2. Flowers blue
      i. Plant erect with stout stem and leaves wooly .... *indicum.*
      ii. Plant not erect with weak stem and leaves glabrescent.

a. Diffuse herb .... *surattense.*
b. Climbing shrub ...... *trilobatum.*

II. Stem unarmed
1. Herb with glabrous stem and leaves .... *nigrum.*
2. Small tree with tomentose stem and leaves .... *verbascifolium.*

Plants of economic importance

The family is very much useful to the mankind for vegetables, fruits, drugs, tobacco and also for ornamental plants.

A. **Vegetables**

1. *Capsicum annum* (Chillies, Mirch). The fruits are used for chilli powder of commerce, a common spice.
2. *C. frutescens.* The fruits are used as vegetable and also used in making hot sauces.
3. *Lycopersicon esculentum* [Syn. *Solanum lycopersicum* (Tomato)]. Both unripe and ripe fruits are used as vegetable. The ripe fruits are used in the preparation of ketchup (a sauce made from mushroom) and sauce.
4. Solanum *tuberosum* (Potato, Alu). The underground stem i.e., the tuber of the plant is rich in carbohydrates and sugars. It is the main food of the people of Ireland like rice in Bengal (India). It is commonly used as an additional food as boiled, fry and in curries along with other vegetables. Tubers are also used in different industries in the production of dextrin, starch and alcohols.
5. *S. melongena* (Brinjal, Begun). The fruits are available in different sizes and are used as vegetables.

B. **Fruits**

Physalis peruviana (Raspberry). The orange red fruits are edible.

C. **Drugs**

1. *Atropa belladonna* (Belladonna). The roots of the plant are the source of *atropine*, an alkaloid. It is used to control spasm of involuntary muscles, relieving pains and dilating pupils. It can act against opium poisoning.
2. *Datura stramonium* (Datura), *D. metel* and *D. inoxia*. The leaves and flowering tops contain the drug, *stramonium*; used in the treatment of asthma and Parkinson diseases. The seeds are deadly poisonous. The fruit juice is used on head to check falling of hairs and also dandruff.
3. *Hyoscyamus niger* (Henbane). The leaves and the flowering tops contain the '*henbane*' drug of commerce. It is used to control whooping cough and asthma.
4. *Mandragora autumnalis.* The roots of the plant have hypnotic and sedative properties.
5. *Nicotiana tabacum* and *N. rustica.* The

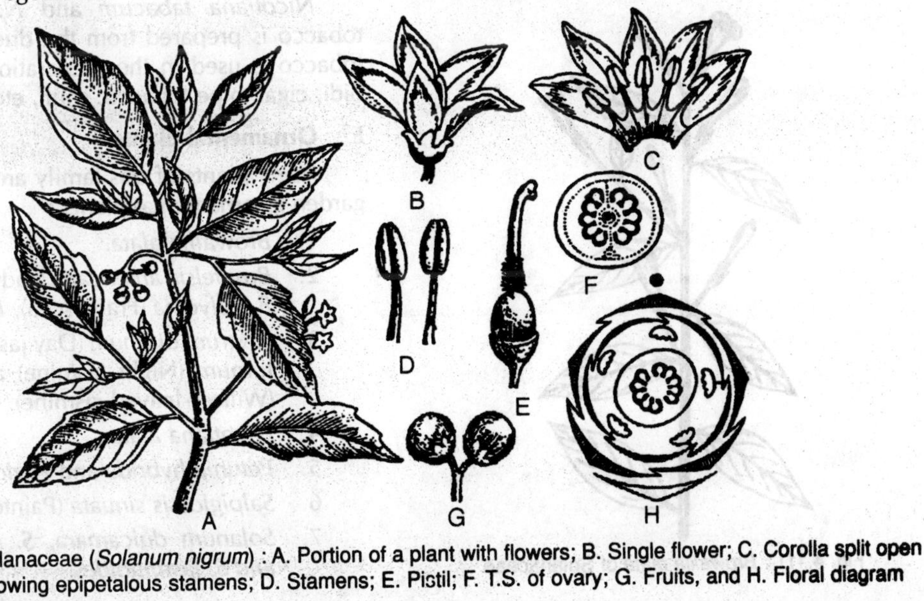

Fig. 4.109 : Solanaceae (*Solanum nigrum*) : A. Portion of a plant with flowers; B. Single flower; C. Corolla split open showing epipetalous stamens; D. Stamens; E. Pistil; F. T.S. of ovary; G. Fruits, and H. Floral diagram

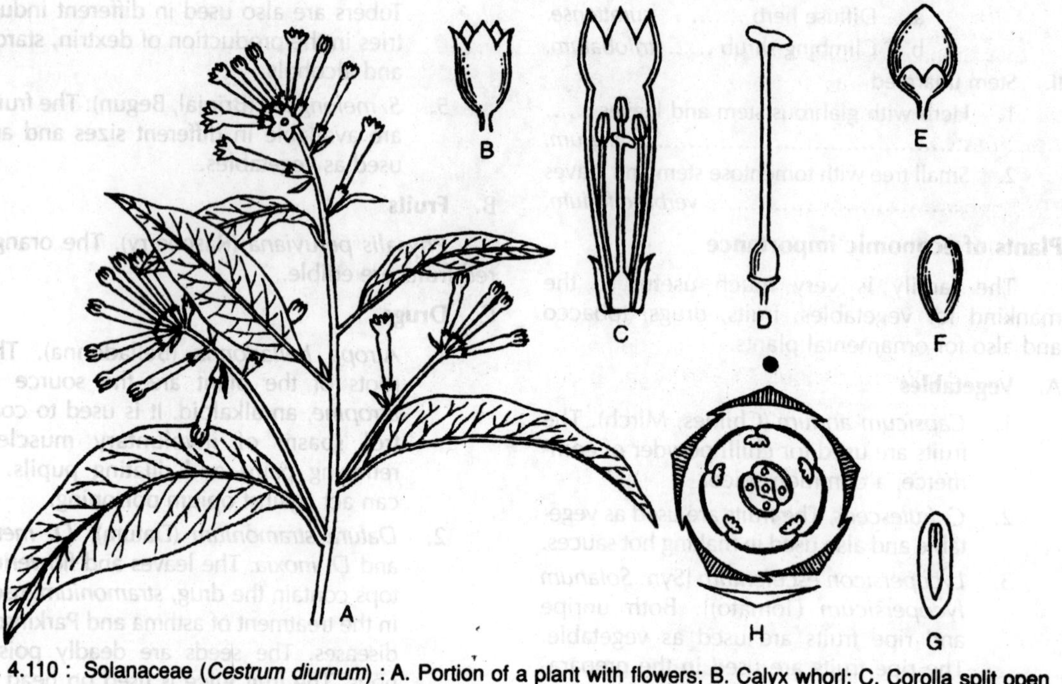

Fig. 4.110 : Solanaceae (*Cestrum diurnum*) : A. Portion of a plant with flowers; B. Calyx whorl; C. Corolla split open showing epipetalous stamens and pistil; D. Pistil; E. Fruit; F. Seed; G. L.S. of seed, and H. Floral diagram

tobacco prepared from the dried and curved leaves are employed in medicine as antispasmodic, sedative and vermifuge.

6. *Withania coagulans* (Indian rennet). The fruits of the plant are used to control liver and colic problem and also for asthma. It is also used as blood purifier.

7. *W. somnifera* (Aswagandha). The roots contain the drug *'Asgandh'* used to control female disorders, rheumatism, hiccup and also for cough.

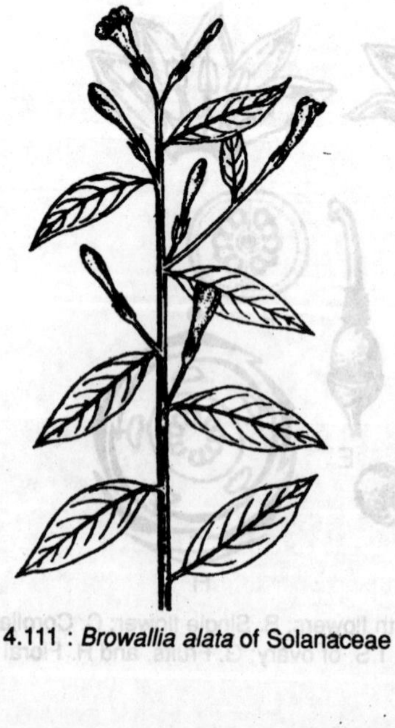

Fig. 4.111 : *Browallia alata* of Solanaceae

D. **Tobacco**

Nicotiana tabacum and *N. rustica*. The tobacco is prepared from the dried leaves. The tobacco is used in the preparation of cigarette, bidi, cigar, pipe tobacco, snuff, etc.

E. **Ornamental plants**

Many plants of this family are grown in the garden as ornamental plants :

1. *Browallia alata*.
2. *Brunfelsia americana* (Lady-of-the night), *B. calycina* (Franciscea), *B.latifolia*.
3. *Cestrum diurnum* (Day jasmine), *C. nocturnum* (Night jasmine) and *C. parqui* (Willow-leaved jasmine).
4. *Nicotiana alata*.
5. *Petunia hybrida* and *P. violacea*.
6. *Salpiglossis sinuata* (Painted tongue).
7. *Solanum dulcamara, S. seaforthianum* and *S. jasminoides*.

F. Other uses

Nicotiana tabacum. The oil extracted from seeds is used in paints and varnishes. The powered leaves are used as insecticide to control insect pests.

25. Scrophulariaceae

(Snapdragon Family)

The family is represented by about 220 genera and 3,000 species found in different climatic conditions and grows abundantly in temperate regions of the world. About 57 genera and more than 350 species occur mainly in the Himalayan range of India.

The type genus of the family is ***Scrophularia***.

Systematic position

Bentham & Hooker		**Engler & Prantl**	
Division.	Angiospermia	Subdivision.	Angiospermae
Class.	Dicotyledons	Class.	Dicotyledoneae
Subclass.	Gamopetalae	Subclass.	Metachlamydeae
Series.	Bicarpellatae	Order.	Tubiflorae
Cohort.	Personales.		

Hutchinson		**Cronquist**	
Phylum.	Angiospermae	Division.	Magnoliophyta
Subphylum.	Dicotyledones	Class.	Magnoliopsida
Division.	Herbaceae	Subclass.	Asteridae
Order.	Personales	Order.	Scrophulariales

Takhtajan		**Thorne**	
Division.	Magnoliophyta	Class.	Angiospermae
Class.	Magnoliopsida	Subclass.	Magnoliidae
Subclass.	Lamiidae	Superorder.	Lamianae
Superorder.	Lamianae.	Order.	Lamiales
Order.	Scrophulariales.		

Identifying characteristics

1. Plants are usually herbs or under-shrubs.
2. Leaves simple, exstipulate, opposite to alternate.
3. Inflorescence racemose or cymose.
4. Flowers are commonly zygomorphic.
5. Corolla – usually **personate.**
6. **Stamens-4, didynamous** (rarely 5 or 2).
7. **Carpel-2, syncarpous** and medianly placed, 2-chambered with many ovules on axile placentation.
8. Fruit usually capsule.

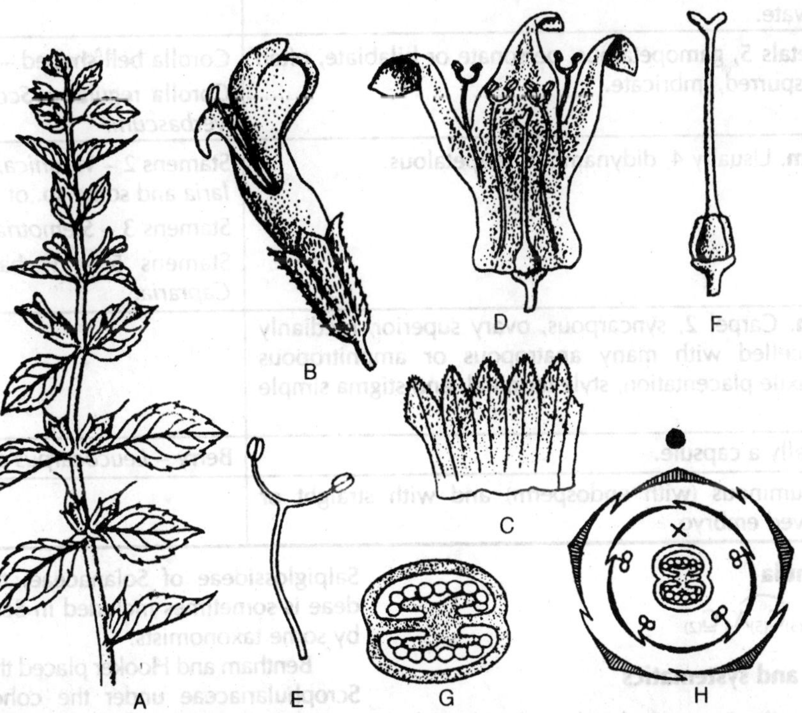

Fig. 4.112 : Scrophulariaceae (*Lindenbergia indica*) : A. Portion of a plant with flowers; B. Single flower; C. Calyx; D. Corolla split open showing epipetalous stamens and pistil; E. Stamen with divaricate anther; F. Pistil; G. T.S. of ovary, and H. Floral diagram

Scrophulariaceae : Distinctive features

Common characteristics	Exceptions
Vegetative characteristics	
Plant. Herbs or undershrubs, often with simple or glandular hairs.	Tree – *Paulownia, Wightia*. Climbers (by means of sensitive petioles) – *Antirrhinum cirrhosum, Rhodochiton* and *Maurandya*. Chlorophyll-less parasite – *Harveya* and *Hyobanche*. Root parasite – some sp. of *Striga, Sopubia, Pedicularis*.
Leaves. Simple; alternate (*Verbascum*), opposite (*Torenia*) or rarely whorled; exstipulate.	Whorled – *Limnophila*. Leaves heterophyllous – species of *Limnophila*.
Reproductive characteristics	
Inflorescence. Raceme or spike.	Flowers are solitary – *Linaria, Bacopa*. Cymose in *Scrophularia*.
Flower. Usually *zygomorphic*, bisexual, hypogynous and generally with bracts.	Actinomorphic (tetramerous) – *Scoparia dulcis*.
Calyx. Sepals 5, gamosepalous, 4-5 partite, persistent, imbricate or valvate.	
Corolla. Petals 5, gamopetalous, personate or bilabiate, often saccate or spurred, imbricate.	Corolla bell-shaped – *Digitalis*. Corolla regular – *Scoparia, Bacopa, Verbascum*.
Androecium. Usually 4, didynamous, epipetalous.	Stamens 2 – *Veronica, Hebe, Calceolaria* and some sp. of *Dopatrium*. Stamens 3 – *Stemotria*. Stamens 5 – *Verbascum nigrum, Capraria*.
Gynoecium. Carpel 2, syncarpous, ovary superior, medianly placed, 2-celled with many anatropous or amphitropous ovules on axile placentation, style terminal, and stigma simple or bilobed.	
Fruit. Usually a capsule.	Berry – *Leucocarpus* and *Halleria*.
Seeds. Albuminous (with endosperm) and with straight or slightly curved embryo.	

Floral formula

$\%$ ⚥ $K_{(5)}\overset{\frown}{C_{(5)}A_4}\underline{G}_{(2)}$

Phylogeny and systematics

The family is very closely related with Solanaceae. Due to close similarity between tribe Verbasceae of Scrophulariaceae and Salpiglossideae of Solanaceae, the Salpiglossideae is sometimes included in Scrophulariaceae by some taxonomists.

Bentham and Hooker placed the natural order Scrophulariaceae under the cohort Personales, series Bicarpellatae, subclass Gamopetalae of class Dicotyledons. Engler placed the family under the order Tubiflorae, subclass Metachla-

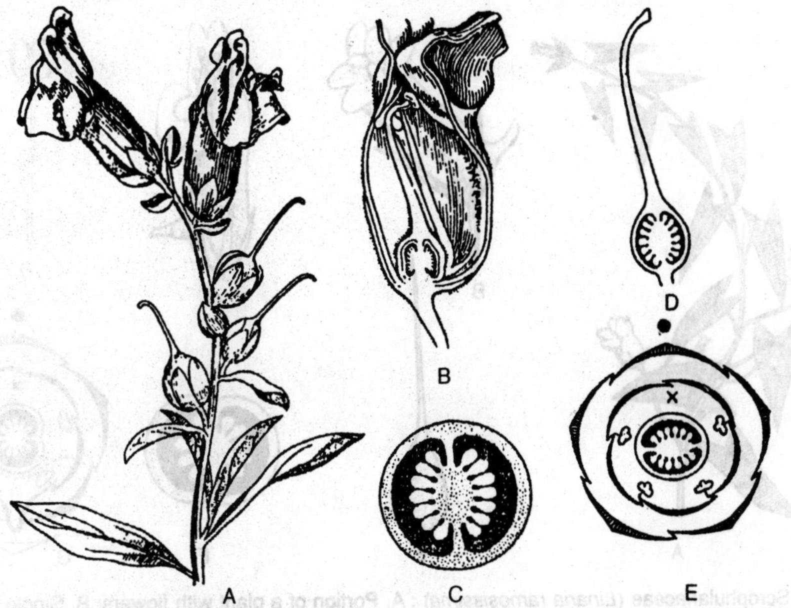

Fig. 4.113 : Scrophulariaceae (*Antirrhinum majus*) : A. Portion of a plant with flowers; B. V.S. of a flower; C. T.S. of ovary; D. L.S. of pistil, and E. Floral diagram

mydeae of class Dicotyledoneae. Later, Hutchinson placed it under the order Personales, division Herbaceae and subphylum Dicotyledones.

Common plants

1. *Bacopa monnieri* (L.) Wetts. (Syn. *Herpestis monnieria* H.B. & K.);
2. *Dopatrium junceum* Ham.;
3. *Lindenbergia indica* (L.) O. Kuntze. (Syn. *L. urticaefolia* Link & Otto.);
4. *L. polyantha* Royle;
5. *Limnophyla heterophylla* (Roxb.) Benth. – an aquatic herb showing heterophylly;

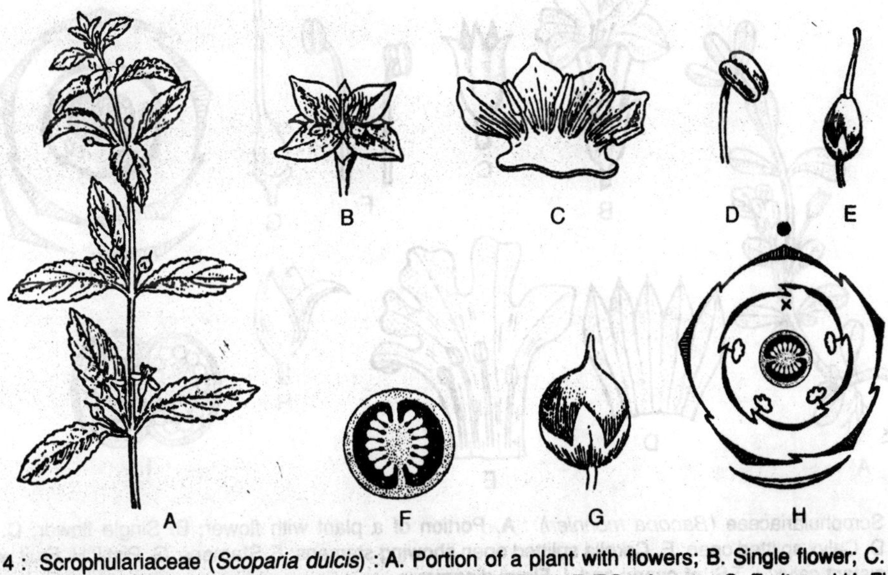

Fig. 4.114 : Scrophulariaceae (*Scoparia dulcis*) : A. Portion of a plant with flowers; B. Single flower; C. Corolla split open showing epipetalous stamens; D. Stamen; E. Pistil; F. T.S. of ovary; G. Fruit, and H. Floral diagram

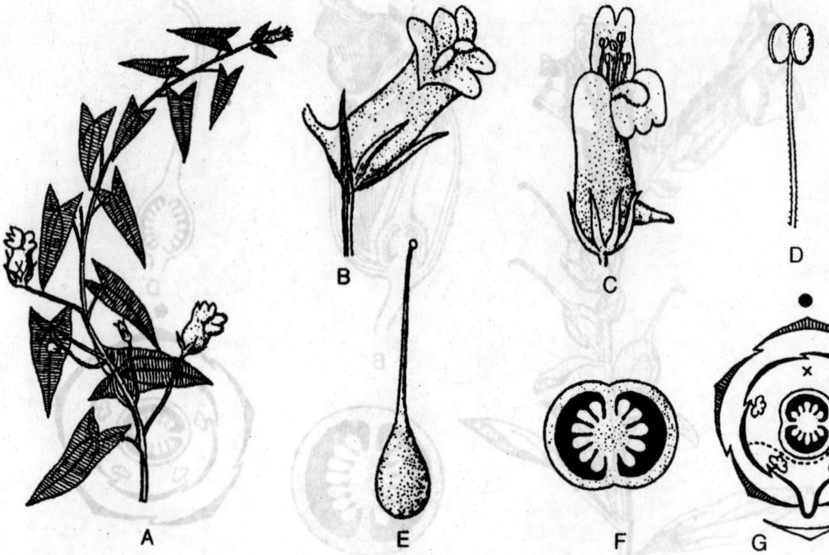

Fig. 4.115 : Scrophulariaceae (*Linaria ramosissima*) : A. Portion of a plant with flowers; B. Single flower; C. Corolla slightly torn open showing stamens; D. Stamen; E. Pistil; F. T.S. of ovary, and G. Floral diagram

6. *Linaria ramosissima* Wall.;
7. *Lindernia ciliata* (Colsm.) Pennel (Syn. *Bonnaya brachiata* Link and Otto.);
8. *L. crustacea* (Linn.) F. Muell. (Syn. *Vandellia crustacea* Benth.);
9. *Scoparia dulcis* L.;
10. *Torenia cordata* (Griff.) Dutta;
11. *Verbascum coromandelianum* O. Ktze. (Syn. *Celsia coromandeliana* Vahl.)

Key to Genera

I. Corolla tubular, tube saccate to spurred; capsule opening by pores; inflorescence uniform :

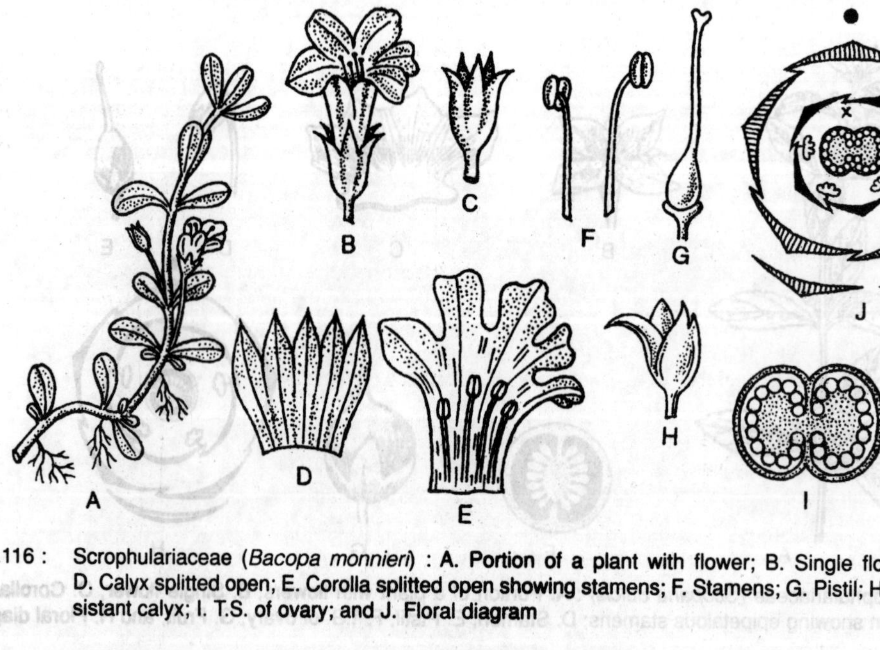

Fig. 4.116 : Scrophulariaceae (*Bacopa monnieri*) : A. Portion of a plant with flower; B. Single flower; C. Calyx; D. Calyx splitted open; E. Corolla splitted open showing stamens; F. Stamens; G. Pistil; H. Fruit with persistant calyx; I. T.S. of ovary; and J. Floral diagram

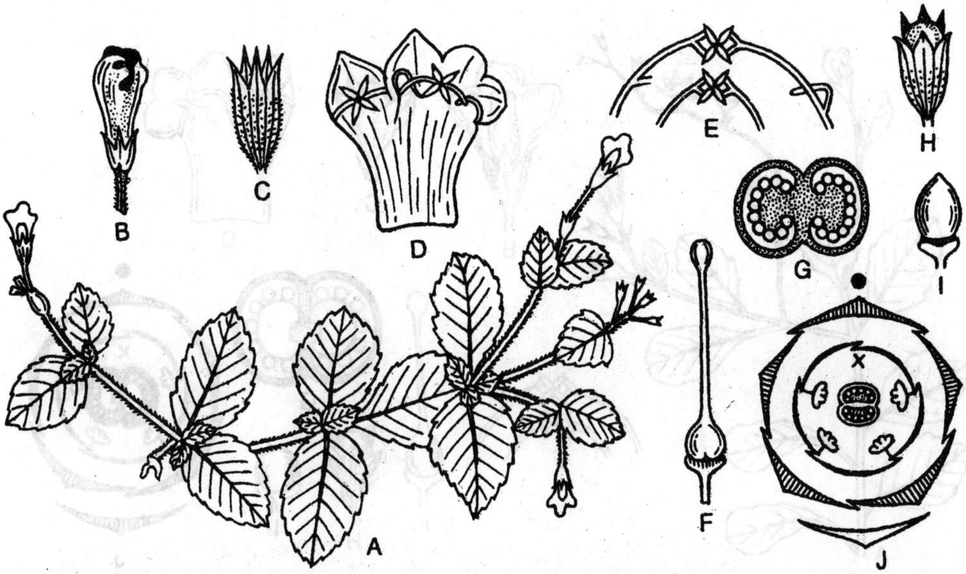

Fig. 4.117 : Scrophulariaceae (*Lindernia multiflora*) : A. Portion of a plant with flowers; B. Single flower; C. Calyx; D. Corolla splitted open showing stamens; E. Arrangement of stamens with appendages; F. Pistil; G. T.S. of ovary; H. Fruit with calyx; I. Single fruit, and J. Floral diagram

i. Corolla spurred; anther-cells distinct .... *Linaria.*

ii. Corolla large, saccate below .... *Antirrhinum.*

II. Corolla personate, with 2 parallel ridges :

i. Calyx campanulate, 5-partite, equal; stamens all perfect; seeds angular; plants usually aquatic with submerged leaves multifid ... *Limnophila.*

ii. Calyx campanulate, 5-fid; anther-cells disjoined; capsule loculicidal ... . *Lindenbergia.*

III. Corolla distinctly bilabiate; the anther-cells contiguous but distinct; calyx wide campanulate, 2-fid .... *Mazus.*

IV. Corolla 5-lobed, without parallel ridges :

i. Calyx 5-partite, imbricate; stamens inserted; anther-cells contiguous; corolla white, tinged with purple and yellow; creeping or erect ... *Herpestis.*

ii. Calyx deeply 5-fid; corolla-throat much widened; stamens 2 perfect, 2 staminodes; flowers violet; annual marshy herbs .... *Dopatrium.*

iii. Calyx 5-toothed; 2 stamens attached to upper lobes, the other 2 stamens to lower lobes, all anthers conniving or cohering in pair .... *Lindernia.*

iv. Calyx tubular, sometimes winged or obliquely 2-lipped; stamens 4, didynamous; flowers showy; fruit a linear capsule .... *Torenia.*

v. Calyx segments 5; stamens 2, both perfect; capsule long ...... *Bonnaya.*

V. Corolla-lobes 4, rotate, white, almost regular; stamens 4, perfect with staminal hairs; fruit a small globular capsule with free central placenta .... *Scoparia.*

VI. Corolla-lobes 5, rotate; calyx partite; stamens equal; anthers sagittate or horseshoe-shaped .... *Hemiphragma.*

VII. Corolla-lobes 5, subovate, with two upper lobes exterior in bud; stamens 4; leaves all alternate .... *Celsia.*

VIII. Corolla tubular, with spreading limbs, abruptly incurved; calyx subcampanulate, 5-toothed; ovary 1-celled, tip mucronate .... *Striga.*

Genus *Lindernia*

Key to species

1. Perfect stamens 2 .... *ciliata.*
2. Perfect stamens 4

i. Flowering calyx cleft down to the middle or less .... *crustacea.*

ii. Flowering calyx cleft down to the base or nearly so .... *viscosa.*

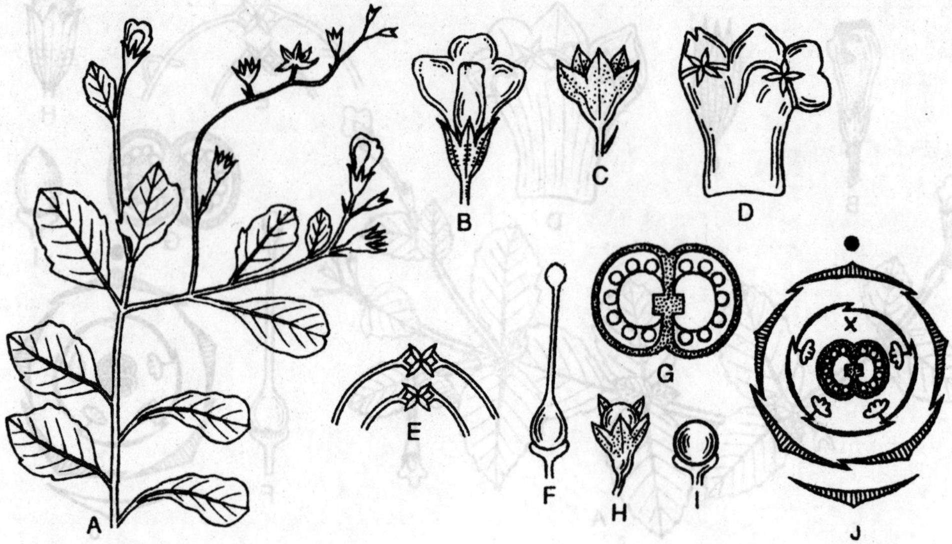

Fig. 4.118 : Scrophulariaceae (*Mazus rugosus*) : A. A portion of a plant with flowers; B. Single flower; C. Calyx; D. Corolla slightly torn open showing stamens; E. Arrangement of stamens; F. Pistil; G. T.S. of ovary; H. Fruit with calyx; I. Fruit, and J. Floral diagram

Table 4.9 : Comparison of Convolvulaceae, Solanaceae and Scrophulariaceae

Characteristics	Convolvulaceae	Solanaceae	Scrophulariaceae
1. Plant	Herbs (annual or perennial).	Herbs, shrubs or rarely small tree.	Herbs or under shrubs.
2. Leaves	Simple (entire or palmately lobed), alternate.	Commonly simple (entire or lobed), alternate.	Simple, alternate or upper alternate and lower opposite or opposite or whorled.
3. Inflorescence	Cymose.	Cymose.	Racemose (spike or raceme) or cymose (dichasia).
4. Flower	Actinomorphic.	Actinomorphic but sometimes zygomorphic.	Zygomorphic.
Calyx	Sepals 5, free or united, often persistent.	Sepals 5, united, persistent, often enlarging on fruit.	Sepals 5, united, more or less deeply 4-5 partite.
Corolla	Petals 5, gamopetalous, funnel or bell-shaped, twisted.	Petals 5, gamopetalous, funnel-shaped or rotate, imbricate.	Petals 5, gamopetalous, usually personate imbricate.
Androecium	Stamens 5, unequal in size.	Stamens 5, didynamous (2+2) in zygomorphic flowers.	Commonly 2+2, rarely 2 or 5.
Gynoecium	Carpel 2, syncarpous.	Carpel 2, syncarpous.	Carpel 2, syncarpous.
	Ovary *medianly placed*, 2-celled with 2 ovules in each cell or 4-celled with 1 ovule in each cell.	Ovary *obliquely placed*, 2-celled or falsely 3-5 celled with many ovules in each cell.	Ovary *medianly placed*, 2-celled with many ovules in each cell.
5. Fruit	Usually capsule, sometimes berry.	Usually berry, sometimes capsule.	Usually capsule, sometimes berry.

Plants of economic importance

The plants of this family are of limited importance based on economic point of view. A few plants have medicinal and ornamental value.

A. **Medicinal plants**

1. *Bacopa monnieri*. The extract of the plants are used to control insanity and epilepsy. It is also used as nerve tonic.
2. *Digitalis purpurea*. The alkaloid '*digitalin*' is extracted from the fully mature leaves and is used in the treatment of heart troubles. It has also some toxic effects like drowsiness and headache.
3. *Picrorhiza kurroa*. The medicine *picrorhiza* is extracted from the dried rhizomes. It acts as cathartic, laxative and stomachic.
4. *Scoparia dulcis* (Sweet Broom-weed). The extract of the leaves is used in the treatment of bronchitis, cough and fever.
5. *Torenia polygonoides*. The extract of the plant is used as a poultice on ulcers and also on sores.

B. **Ornamental plants**

1. *Antirrhinum majus* (Snapdragon).
2. *Calceolaria scabiosaefolia* (Slipperwort).
3. *Digitalis purpurea* (Common foxglove).
4. *Linaria vulgaris*.
5. *Mimulus luteus* (Monkey flower).
6. *Nemesia versicolor* (Nemesia).
7. *Russelia equisetiformis* (Coral blow).
8. *Torenia asiatica*.

26. Acanthaceae

(Acanthus Family)

The family is represented by 256 genera and 2,765 species, mostly in the tropical parts and some are also found in the temperate regions of the world. The family is represented in India by about 68 genera and 337 species found to grow mostly in the hilly regions of tropical and subtropical Himalayas and also in western and southern regions.

The type genus of the family is ***Acanthus***.

Systematic position

Bentham & Hooker		**Engler & Prantl**	
Division.	Angiospermia	Subdivision.	Angiospermae
Class.	Dicotyledons	Class.	Dicotyledoneae
Subclass.	Gamopetalae	Subclass.	Metachlamydeae
Series.	Bicarpellatae	Order.	Tubiflorae
Cohort.	Personales.		
Hutchinson		**Cronquist**	
Phylum	Angiospermae	Division.	Magnoliophyta
Subphylum	Dicotyledones	Class.	Magnoliopsida
Division.	Herbaceae	Subclass.	Asteridae
Order.	Personales	Order.	Scrophulariales
Takhtajan		**Thorne**	
Division.	Magnoliophyta	Class.	Angiospermae
Class.	Magnoliopsida	Subclass.	Magnoliidae
Subclass.	Lamiidae	Superorder.	Lamianae
Superorder.	Lamianae	Order.	Lamiales
Order.	Scrophulariales.		

Identifying characteristics

1. Plants herbs or shrubs.
2. Stem with **swollen nodes**.
3. Leaves simple, exstipulate.
4. Flowers **bisexual, zygomorphic**, hypogynous with **copious bracts** and **bracteoles**.
5. Corolla usually bilabiate or semi-bilabiate.
6. Stamens usually 4, **didynamous**, epipetalous.
7. **Carpels–2, syncarpous**, style terminal.
8. Fruit bilocular capsule.
9. Seeds are non-endospermic and are situated on hook-like outgrowths — the **jaculator** or **retinacula**.

Floral formula

$\circ|\circ$ ⚥ $K_{(5)}\overparen{C_{(5)}A_{4 \text{ or } 2}}\underline{G}_{(2)}$

Phylogeny and systematics

The family Acanthaceae is closely related to Boraginaceae, Scrophulariaceae, Labiatae and Verbenaceae.

Bentham and Hooker placed the natural order Acanthaceae under the cohort (=order)

Acanthaceae : Distinctive features

Common characteristics	Exceptions
Vegetative characteristics	
Plant. Perennial herbs or shrubs, *stem and its branches are swollen at the nodes.*	Climber – *Thunbergia, Mendoncia.* Tree – *Strobilanthus.*
Leaves. Opposite or decussate, simple, entire, exstipulate, *often with cystolith.*	
Reproductive characteristics	
Inflorescence. Racemose or cymose, *usually with copious bracts and bracteoles.*	
Flower. Zygomorphic, bisexual, hypogynous, bracteate.	
Calyx. Sepals 5, gamosepalous, 4- or 5-partite, usually imbricate.	Calyx is reduced to a narrow ring – *Thunbergia.* Calyx 4-lobed – *Barleria, Nelsonia.*
Corolla. Petals 5, *often bilabiate,* sometimes oblique, imbricate or twisted.	Corolla tube is long slender – *Ecbolium.* Lower leaf is rolled – *Symplectochilus.*
Androecium. Stamens 4, didynamous, epipetalous, anther 2-celled, often spurred or hairy. Anthers dehisce either longitudinally or by apical pore.	Stamens 2 – *Justicia, Adhatoda.* Anther I celled (vestigial second anther cell) – *Chaetothylax tocantinus* and *Acanthus.*
Gynoecium. Carpel 2, syncarpous, superior, 2-chambered with 2 or more anatropous ovules in each cell on axile placentation. Style simple and terminal, stigma bilobed, lobes often unequal. **Fruit**. Usually a loculicidal capsule. **Seeds**. Many, exalbuminous, attached to hook-like supports called jaculators or retinacula.	

Personales, series Bicarpellatae, subclass Gamopetalae of class Dicotyledons. Engler placed it under the order Tubiflorae, subclass Metachlamydeae of the class Dicotyledoneae. On the other hand, Hutchinson placed the family under the order Boraginales, division Herbaceae of subphylum Dicotyledones. Takhtajan and Cronquist placed the family under Scrophulariales.

Airy shaw (1973) included the genera *Thunbergia, Meyenia, Pseudocalyx* and *Pounguia* in a separate family Thunbergiaceae and *Mendoncia* and *Gilletiella* in Mendonciaceae.

Common plants

1. *Adhatoda vasica* Nees;
2. *Andrographis paniculata* Nees;
3. *Asterocanthus longifolius* Nees (Syn. *Hygrophyla spinosa* T. And.);
4. *Asystasia gangetica* T. And.;
5. *Barleria cristata* L.;
6. *B. prionitis* L.;
7. *Cardanthera triflora* Ham.;
8. *Daedalacanthus platiferus* Gamble (Syn. *Eranthemum platiferum* Nees);
9. *Dipterocanthus suffruticosus* Voigt non Torr. (Syn. *Ruellia suffruticosa* Roxb.);
10. *D. tuberosus* Nees (Syn. *Ruellia tuberosa* L.);
11. *Ecbolium viride* (Fork) Alston. (Syn. *E. linneanum Kurz.*);
12. *Hemigraphis hirta* T. And.;

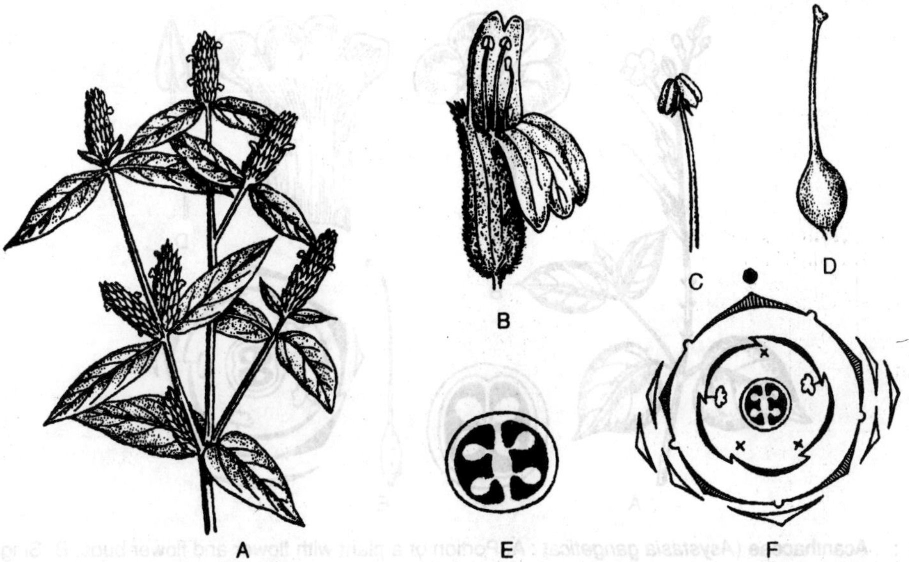

Fig. 4.119 : Acanthaceae (*Justicia simplex*) : A. Portion of a plant with many inflorescence; B. Single flower; C. Stamen with appendage; D. Pistil; E. T.S. of ovary, and F. Floral diagram

13. *Justicia diffusa* Willd.;
14. *J. simplex* Don;
15. *Micranthus oppositifolius* Wendl. (Syn. *Phaylopsis parviflora* Willd);
16. *Nelsonia campestris* R. Br.;
17. *Peristrophe bicalyculata* Nees;
18. *Rostellularia procumbens* Nees (Syn. *Justicia procumbens* Nees);
19. *Rungia parviflora* Nees.

Key to genera

I. Corolla-lobes imbricate in bud; ovules numerous in two long rows; seeds mounted in papilla:

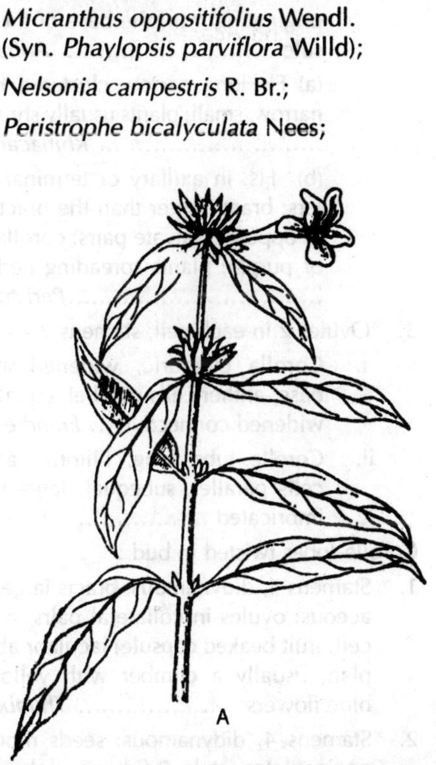

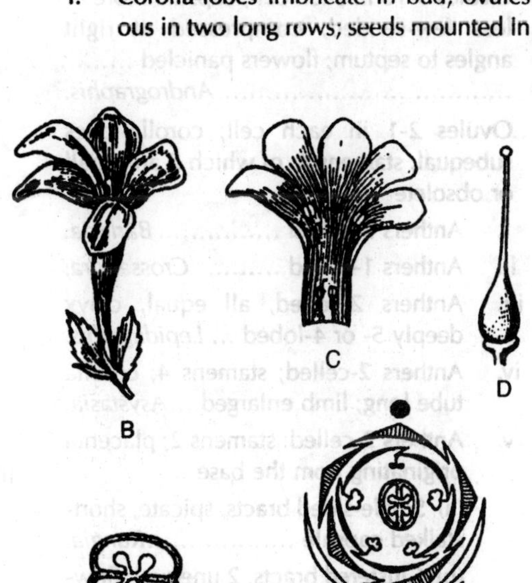

Fig. 4.120 : Acanthaceae (*Barleria cristata*) : A. Portion of a plant with flower; B. Single flower; C. Corolla split open showing epipetalous stamens; D. Pistil; E. T.S. of ovary, and F. Floral diagram

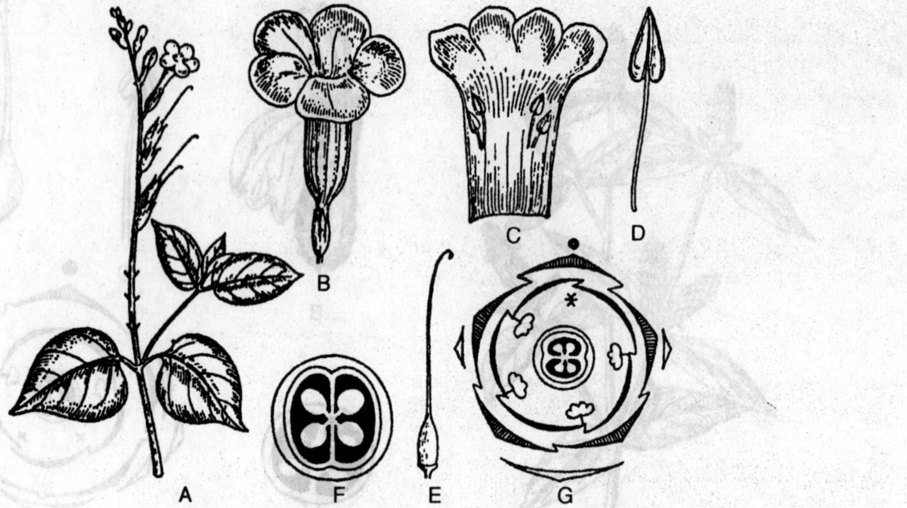

Fig. 4.121 : Acanthaceae (*Asystasia gangetica*) : A. Portion of a plant with flower and flower buds; B. Single flower; C. Corolla split open showing epipetalous stamens; D. Stamen; E. Pistil; F. T.S. of ovary, and G. Floral diagram

1. Stamens 2; calyx 4-sect; herbaceous bracts .... *Nelsonia.*
2. Stamens 4; calyx 5-partite ... *Ebermaiera.*

II. Çorolla-lobes imbricate; jaculators curved; anthers 2-1-celled, often spurred at the base :

1. Ovules 3-10 in each cell; capsule more or less few-seeded, compressed at right angles to septum; flowers panicled .... *Andrographis.*
2. Ovules 2-1 in each cell; corolla-lobes subequal; stamens 4, of which 2 are small or obsolete
   - i. Anthers 2-celled .... *Barleria.*
   - ii. Anthers 1-celled .... *Crossandra.*
   - iii. Anthers 2-celled, all equal, calyx deeply 5- or 4-lobed ... *Lepidagathis.*
   - iv. Anthers 2-celled; stamens 4; corolla tube long; limb enlarged ...*Asystasia.*
   - v. Anthers 2-celled; stamens 2; placenta originating from the base :

     (a) Single-sided bracts, spicate; short-stalked capsule .... *Rungia.*

     (b) Clustered bracts, 2 unequal; flowers shortly pedicellate; club-shaped capsule; calyx 5-partite ... *Dicliptera.*
   - vi. Anthers 2-celled; one placed above the other, lower minutely spurred at the base; herbs or shrubs ... *Justicia.*
   - vii. Anthers 2-celled, apiculate, scarcely spurred; corolla 2-lipped, short-tubed; bracts leafy; placentae not originating from the base of the capsule .... *Adhatoda.*

     (a) Fls. long, narrow, clustered; bracts narrow, small; plants usually shrubs ... .*Rhinacanthus.*

     (b) Fls. in axillary or terminal clusters; bracts larger than the bracteoles in opposite valvate pairs; corolla rose or purple; plants spreading herbs ... .*Peristrophe.*
3. Ovules 2 in each cell; stamens 2 —
   - i. Corolla cylindric, widened at the base; anther-cells parallel, equal with widened connective ... *Eranthemum.*
   - ii. Corolla tube long, filiform; anther-cells parallel, subequal; large bracts imbricated .... *Ecbolium.*

III. Corolla-lobes twisted in bud :

1. Stamens 4, didynamous; bracts large, foliaceous; ovules in collateral pairs in each cell; fruit beaked capsule; jaculaor absent; plant usually a climber with yellow or blue flowers .... *Thunbergia.*
2. Stamens 4, didynamous; seeds mounted on jaculator; style 2-fid, one lobe often suppressed

i. Ovules 3-12 in each cell :

(a) Corolla 2-lipped; flowers sessile, axillary, lower leaves often toothed; jaculator not hooked ... *Cardenthera.*

(b) Corolla 2-lipped; flowers subsessile, axillary; leaves entire; jaculator hooked; capsule narrow; spinescent herb .... *Hygrophila.*

(c) Corolla regular, 5-lobed; large bracteole; stamens muticous; clavate capsule with solid base ...... *Ruellia.*

(d) Corolla 5-lobed; bracteole small or absent; flowers capitate; capsule with seeds from the base .... *Hemigraphis.*

ii. Ovules 2 in each cell :

(a) Corolla 5-lobed; stamens 4 or 2; anthers muticous; placenta from the base of the capsule ... *Strobilanthes.*

(b) Corolla 5-lobed, slightly unequal; stamens 4, didynamous; anthers 2-celled, parallel, base mucronate; placenta from the base of the capsule ... . *Phaylopsis.*

IV. Corolla-lobes short, upper lip reduced, lower one 3-lobed; ovules 2 in each cell; jaculator curved and hardened; anterior filaments with process .... *Acanthus.*

Plants of economic importance

The plants of this family are of limited importance from the economic point of view. A few plants have medicinal and ornamental value.

A. **Medicinal plants**

1. *Adhatoda vasica* (Vasaka). The extract of leaves is used for the remedy of asthma, cough and bronchitis.
2. *Andrographis paniculata* (Kalmegh). The extract of the leaves is used to control liver troubles. The leaves are pasted and made into small tablets that are used in the morning to control small threadworms of human intestine.
3. *Barleria prionitis* (Kala bansa), *B. cristata, B. buxifolia* and *B. longifolia.* The extract of leaves and roots are used for the remedy of cough and also inflammations.
4. *Ecbolium viride.* The roots are used to control jaundice.
5. *Asterocanthus longifolius* (Kuleykhara). The extract of the leaves and roots are used to control rheumatism and jaundice. The extract of young leaves along with stem prepared after boiling with

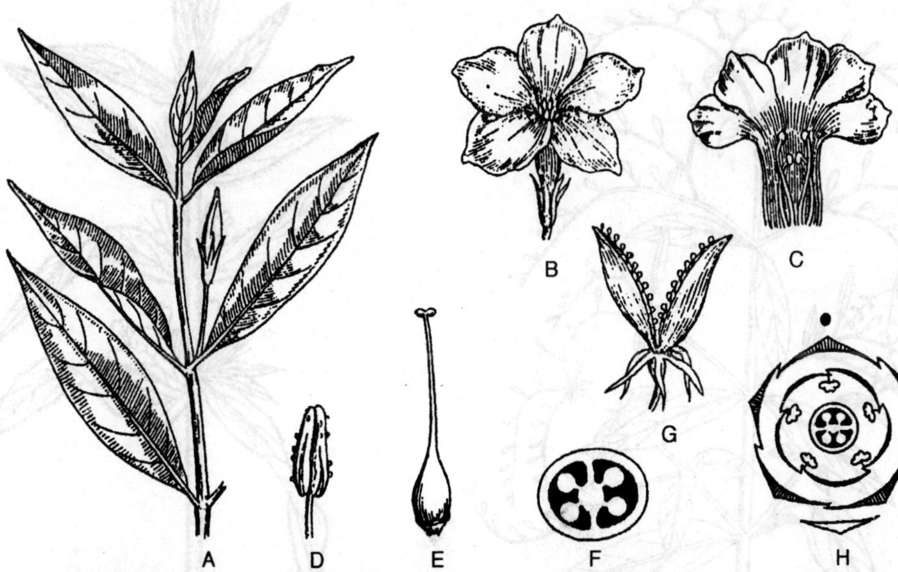

Fig. 4.122 : Acanthaceae (*Dipterocanthus tuberosus*) : A. Portion of a plant with flower bud; B. Single flower; C. Corolla split open showing epipetalous stamens; D. Stamen; E. Pistil; F. T.S. of ovary and G. Dehiscing fruit, and H. Floral diagram

Fig. 4.123 : Acanthaceae (*Adhatoda vasica*) : A. Portion of a plant with flowers; B. Single flower; C. Stamen; D. Pistil, and F. T.S. of ovary

water and taken in the morning that caused increase in hemoglobin content of the blood.

6. *Phlogacanthus thyrsiflorus*. The fruits and leaves are used as febrifuge (any medicine that cures fever).

B. **Ornamental plants**

Several species are cultivated in the garden as ornamental plants. These are :

1. *Beloperone guttata* (Shrimp plant).
2. *Barleria cristata, B. gibsoni, B. lupulina,*

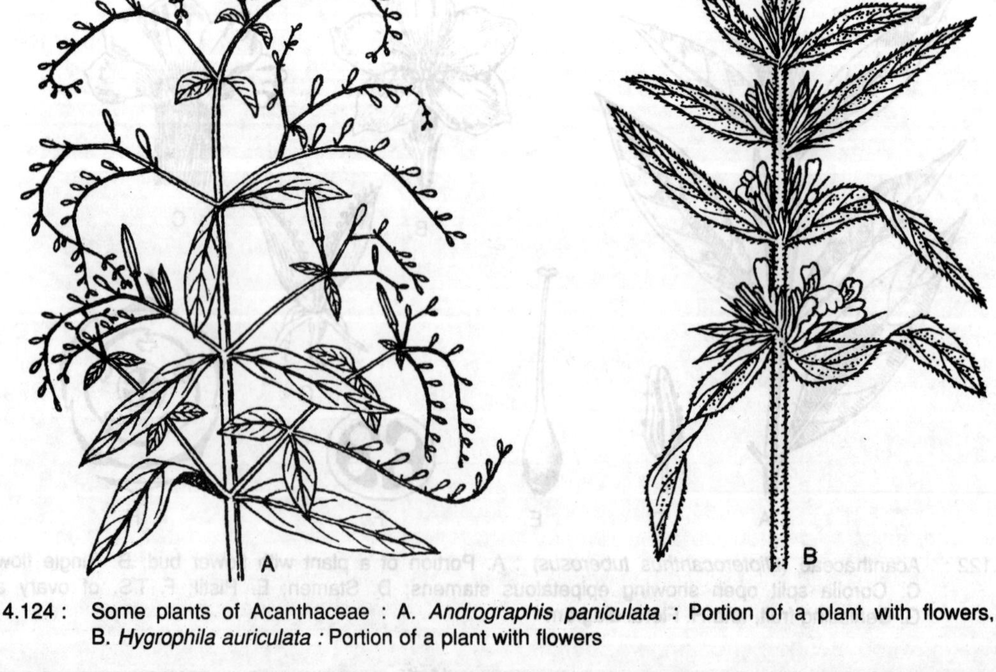

Fig. 4.124 : Some plants of Acanthaceae : A. *Andrographis paniculata* : Portion of a plant with flowers, and B. *Hygrophila auriculata* : Portion of a plant with flowers

B. montana, B. noctiflora, B. polytricha and *B. prionitis.*
3. *Eranthemum bicolor, E. nervosum* and *E. reticulatum.*
4. *Jacobinia tinctoria.*
5. *Justicia gendarussa.*
6. *Ruellia prostata* and *R. brittoniana.*
7. *Strobilanthes isophyllus.*
8. *Thunbergia alata, T. coccinea, T. grandiflora* and *T. laurifolia.*

27. VERBENACEAE

(Verbain Family)

The family is represented by 36 genera and 1,035 species grows commonly in the tropical and subtropical regions of the world. About 21 genera and 125 species are found to grow in the tropical and subtropical Himalayas and also in the Western and Southern regions of India.

The type genus of the family is ***Verbena***.

Systematic position

Bentham & Hooker		**Engler & Prantl**	
Division.	Angiospermia	Subdivision.	Angiospermae
Class.	Dicotyledons	Class.	Dicotyledoneae
Subclass.	Gamopetalae	Subclass.	Metachlamydeae
Series.	Bicarpellatae	Order.	Tubiflorae
Cohort.	Lamiales		
Hutchinson		**Cronquist**	
Phylum.	Angiospermae	Division.	Magnoliophyta
Subphylum.	Dicotyledones	Class.	Magnoliopsida
Division.	Lignosae	Subclass.	Asteridae
Order.	Verbenales	Order.	Lamiales
Takhtajan		**Thorne**	
Division.	Magnoliophyta	Class.	Angiospermae
Class.	Magnoliopsida	Subclass.	Magnoliidae
Subclass.	Lamiidae	Superorder.	Lamianae
Superorder.	Lamianae	Order.	Lamiales
Order.	Lamiales.		

Verbenaceae : Distinctive features

Common characteristics	**Exceptions**
Vegetative characteristics	
Plant. Herbs (*Phyla nodiflora*), shrubs (*Lantana camara*) or trees (*Tectona grandis*), sometimes with square stem, aromatic.	Climber – *Petrea volubilis, Holmskioldia.* Halophyte – *Avicennia.*
Leaves. Opposite or whorled, simple or sometimes compound, exstipulate.	Pinnately compound – *Vitex.*
Reproductive characteristics	
Inflorescence. Racemose (raceme or spike) or cymose, usually bracteate.	
Flower. Zygomorphic, bisexual or polygamous (by abortion), hypogynous, pentamerous, often with bract.	Actinomorphic – *Physopsis, Callicarpa.* Unisexual – *Aegiphila.* Tetramerous – *Physopsis, Callicarpa.*
Calyx. Sepals 5, gamosepalous, bell-shaped or tubular and persistent, valvate.	Calyx petaloid – *Holmskioldia.*
Corolla. Petals 5, gamopetalous, bilabiate or bell-shaped and oblique with spreading limb, imbricate.	Corolla lobes 6-16 in *Symphorema.*
Androecium. Stamens 4, didynamous, epipetalous, staminodes often present in diandrous flowers, anthers 2-celled, dorsifixed, dehisces longitudinally.	Stamens 5 – *Sphenodesma* and *Genusia*; 5-6 in *Tectona* and 6-16 in *Symphorema.*
Gynoecium. Carpel 2, syncarpous, ovary superior, 2- or 4- or rarely 1-celled with 1-2 ovules in each cell on axile placentation. Style 1, terminal; stigma lobes usually as many as carpels.	Carpels 5 – *Geunsia.*
Fruit. Usually drupe, often enclosed in persistent calyx.	4 nutlets – *Verbena.* Capsule – *Avicennia, Caryopteris.*
Seeds. Usually without endosperm, embryo straight.	

Identifying characteristics

1. Plants herbs, shrubs or tree.
2. Leaves **simple** (sometimes compound), **exstipulate**, arranged in opposite or whorled phyllotaxy.
3. Inflorescence-racemose or cymose or cymose panicle.
4. Flowers **bisexual, zygomorphic, hypogynous**, bracteate.
5. Sepals 5, persistent.
6. Corolla **tubular** or bilabiate, **often oblique**.
7. Stamens **commonly 4, didynamous**.
8. **Carpel 2, syncarpous**, ovules 1 in each chamber on axile placentation.
9. Fruit **drupe**.

Floral formula

∘|∘ ⚥ $K_{(5)}\widehat{C_{(5)}A_4}\underline{G}_{(2)}$

Phylogeny and systematics

The family is closely related with Labiatae. Bentham and Hooker placed the Natural Order (= Family) under the Cohort (=order) Lamiales series Bicarpellatae, subclass Gamopetalae of class Dicotyledons. Engler and Prantl placed the family under order Tubiflorae, subclass Metachlamydeae of class Dicotyledoneae. On the other hand, Hutchinson placed the family under the order Verbenales, division Lignosae of sub-phylum Dicotyledones.

The genus *Avicennia* that was included by Bentham and Hooker in Verbenaceae was also retained by Hutchinson, Takhtajan, Cronquist and many others. But taxonomists like Eichler, Airy Shaw and some others placed it in a separate family Avicenniaceae.

The family is considered to be nearly related to Lamiaceae but taxonomists like Hutchinson considered it to be unrelated to Lamiaceae and have been derived from Rubiaceous stock.

Common plants

1. *Avicennia officinalis* L.;
2. *Callicarpa macrophylla* Vahl ;
3. *Clerodendrum viscosum* Vent. (Syn. *Clerodendron infortunatum* Gaertn.);

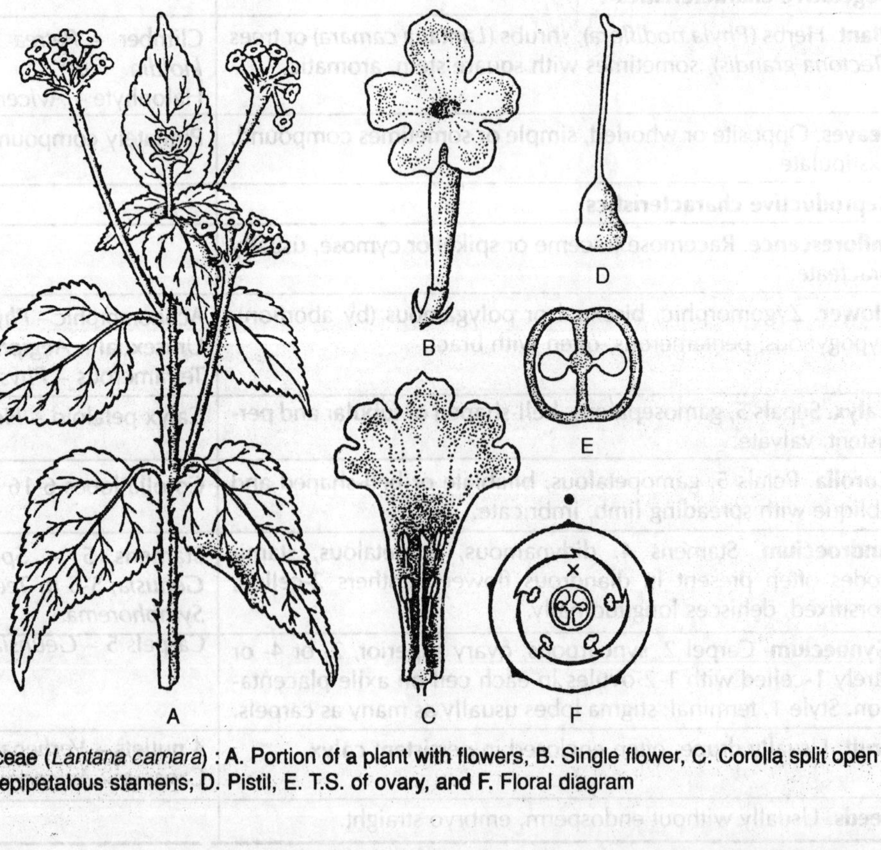

Fig. 4.125 : Verbenaceae (*Lantana camara*) : A. Portion of a plant with flowers, B. Single flower, C. Corolla split open showing epipetalous stamens; D. Pistil, E. T.S. of ovary, and F. Floral diagram

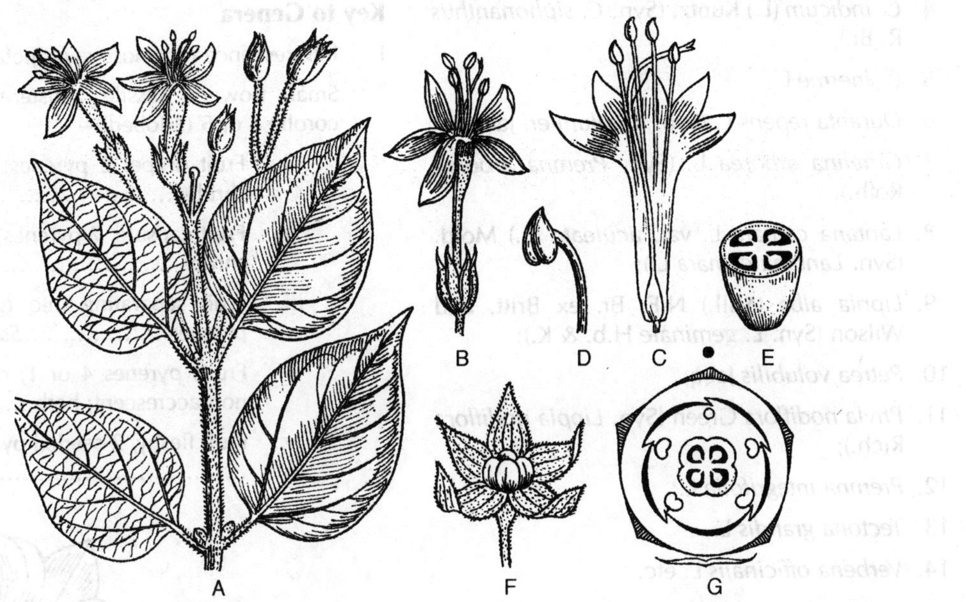

Fig. 4.126 : Verbenaceae (*Clerodendrum viscosum*) : A. Portion of a plant with flowers, B. Single flower, C. Corolla split open showing epipetalous stamens and Pistil, D. Stamen, E. T.S. of ovary, F. Fruit, and F. Floral diagram

Fig. 4.127 : Verbenaceae (*Lippia alba*) : A. Portion of a plant with flowers, B. Single flower, C. Calyx, D. Corolla split open showing epipetalous stamens, E. Pistil, F. T.S. of ovary, and G. Floral diagram

4. *C. indicum* (L.) Kuntz. (Syn. *C. siphonanthus* R. Br.);
5. *C. inerme* L.;
6. *Duranta repens* L. (Syn. *D. plumieri* Jacq.);
7. *Gmelina arborea* L. (Syn. *Premna arborea* Roth.);
8. *Lantana camara* L. var. *aculeata* (L.) Mold. (Syn. *Lantana camara* L.);
9. *Lippia alba* (Mill.) N.E. Br. ex Britt. and Wilson (Syn. *L. geminate* H.b. & K.);
10. *Petrea volubilis* Jacq;
11. *Phyla nodiflora* Green (Syn. *Lippia nodiflora* Rich.);
12. *Premna integrifolia* L.;
13. *Tectona grandis* L.
14. *Verbena officinalis* L. etc.

Key to Genera

I. Inflorescence racemose centripetal spicate :

Small flowers, sessile, clustered in heads, corolla 4 or 5 (2-lobed)—

i. Fruit drupe; 2 pyrenes; small calyx; shrubs .... *Lantana.*

ii. Fruit capsule, 2 pyrenes; small calyx; shrubs .... *Lippia.*

iii. Fruit separating into two 1-seeded pyrenes ....*Stachytarpheta.*

iv. Fruits pyrenes 4 or 1; calyx tubular, non-accrescent; herb ...... *Verbena.*

v. Fruit fleshy, 2-seeded pyrenes .... *Duranta.*

Fig. 4.128 : *Phyla nodiflora* (a plant) of Verbenaceae

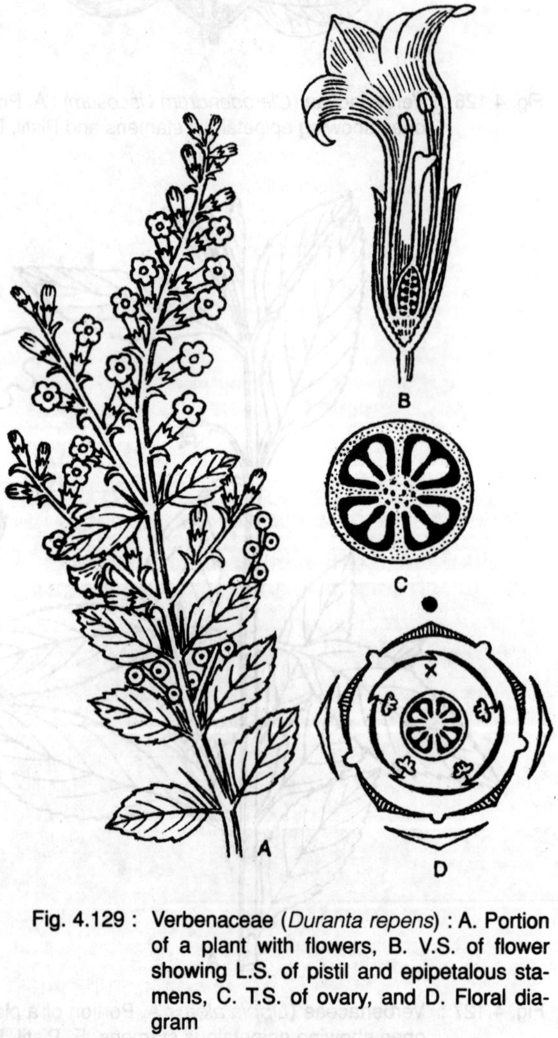

Fig. 4.129 : Verbenaceae (*Duranta repens*) : A. Portion of a plant with flowers, B. V.S. of flower showing L.S. of pistil and epipetalous stamens, C. T.S. of ovary, and D. Floral diagram

II. Inflorescence cymose, centrifugal or panicled :
  a. Small flowers; corolla regular; stamens isomerous; leaves entire –
    i. Flowers 4–6-merous; anthers exserted; fruit a drupe with 1, 4-celled pyrenes, enclosed in fleshy calyx ... . *Tectona*.
    ii. Flowers 4-merous; anthers exserted; fruit a round drupe with 4 pyrenes ... . *Callicarpa*.
  b. Small flowers; corolla 2-lipped; stamens didynamous –
  1. Succulent drupe with single pyrene
    i. Leaves entire or toothed ...... *Premna*.
    ii. Leaves entire; flowers large 1 inch long ....*Gmelina*.
    iii. Leaves digitate; 3–5 folded ...... *Vitex*.
  2. Fruit drupe with 4 pyrenes; leaves simple
    i. Calyx bell-shaped; corolla tube long slender; stamens equal or didynamous; drupe round ... *Clerodendrum*.
    ii. Calyx large, petaloid (red), rotate; usually a climbing or straggling shrub ... .*Holmskioldia*.

III. Inflorescence capitate cyme :
    i. Bracts (involucral) 6 or 3–4; large; fruit usually dry; corolla 2-lipped; large climber .... *Congea*
    ii. Calyx 5-partite, concave, imbricate; fruit fleshy capsule, 1-seeded; flowers small; shrubs with leathery leaves ... . *Avicennia*.

Plants of economic importance

This family is important for its timbers, and also for ornamental plants. A few are used for medicinal and other purposes.

A. **Timber yielding plants**

1. *Gmelina arborea* (Candahar tree, Gambar). The timber of this plant is used for window panels, furniture, paper industry, match industry, shafts, bridges, boats etc.
2. *Premna bengalensis*. The wood of this plant is used in posts, bridges, and also for carving.
3. *Tectona grandis* (Teak, Segun). It is an important hard wood used in ship building, frame of doors and windows, cabinet work, bridge and other heavy constructions.
4. *Vitex altissima*. The wood of this plant is used in boats and other constructional works.

B. **Ornamental plants**

The following plants are cultivated in the garden as ornamental plants :

1. *Caryopteris incana* (blue-beard).
2. *Clerodendrum fragrans, C. speciosum, C. inerme* and *C. indicum*. These plants are commonly known as *glory bower*.
3. *Duranta repens* (Syn. *D. plumeri*). The plant is commonly known as *golden dewdrop*.
4. *Holmskioldia sanguinea* (Chinese hat-plant).
5. *Lantana camara*.
6. *Petrea volubilis* (Queen's wreath).
7. *Verbena bipinnatifida, V. bonariensis* and *V. hybrida*.

C. **Medicinal plant**

1. *Vitex negundo*. The extract of the roots and leaves are used as tonic, febrifuge, and also as diuretic.

D. **Other uses**

1. *Avicennia officinalis*. The bark is used for tanning.

28. Labiatae (Lamiaceae)

(Mint Family)

The family Labiatae consists of 264 genera and 6,990 species and is found to grow almost throughout the world and especially in the Mediteranean region. About 64 genera and 380 species are distributed in India at different habitat mainly in North West and Southern part of India.

The type genus of the family is ***Lamium***.

Systematic position

Bentham & Hooker		**Engler & Prantl**	
Division.	Angiospermia	Subdivision.	Angiospermae
Class.	Dicotyledons	Class.	Dicotyledoneae
Subclass.	Gamopetalae	Subclass.	Metachlamydeae
Series.	Bicarpellatae	Order.	Tubiflorae
Cohort.	Lamiales.		

Hutchinson		**Cronquist**	
Phylum.	Angiospermae	Division.	Magnoliophyta
Subphylum.	Dicotyledones	Class.	Magnoliopsida
Division.	Herbaceae	Subclass.	Asteridae.
Order.	Lamiales	Order.	Lamiales
Takhtajan		**Thorne**	
Division.	Magnoliophyta	Class.	Angiospermae
Class.	Magnoliopsida	Subclass.	Magnoliidae
Subclass.	Lamiidae	Superorder.	Lamianae
Superorder.	Lamianae	Order.	Lamiales
Order.	Lamiales.		

Identifying characteristics

1. Plants are **aromatic** herbs or shrubs.
2. **Stem square** and covered with glandular hairs.
3. Leaves simple, exstipulate, arranged in opposite-decussate phyllotaxy.
4. **Inflorescence — Verticillaster.**
5. Flowers bisexual, zygomorphic.
6. **Corolla bilabiate.**
7. Stamens 4, **didynamous**, epipetalous.
8. **Carpels 2, syncarpous**, 2-chambered (which

Labiatae : Distinctive features

Common characteristics	Exceptions
Vegetative characteristics	
Plant. Usually annual or perennial aromatic herbs or shrubs with square *stem*.	Tree – *Leucosceptrum canum.*
Leaves. Opposite-decussate, simple, hairy, exstipulate.	Leaves whorled – *Leucas ternifolia, Plectranthus ternifolius.* Pinnately compound – *Salvia hydrangea, S. plectranthoidés.* More or less lyrate – *Eremostachys.*
Reproductive characteristics	
Inflorescence. Commonly verticillaster or fascicles of axillary cymes or dense spike.	Spike – *Teucrium.* Axillary cyme – *Leucas prostata.* Racemose – *Scutellaria.*
Flower. Zygomorphic, bisexual, hypogynous, bracteate.	Flowers solitary – *Scutellaria.* Actinomorphic – *Mentha.*
Calyx. Sepals 5, gamosepalous, tubular or bell-shaped, persistent.	Calyx 2-lipped – *Plectranthus.*
Corolla. Petals 5, gamopetalous, bilabiate with 2/3 arrangement (i.e., upper lip with 2 and lower lip with 3 petals), imbricate.	Corolla regular – *Mentha.* Bilabiate 4/1 i.e., 4 upper lip/1 lower lip – *Ocimum, Hyptis* and 0/5 arrangement – *Teucrium.*
Androecium. Stamens 4, didynamous, epipetalous.	Stamens 2 – *Salvia, Meriandra.* Stamens monadelphous – *Coleus.* Anther unilocular and reniform – *Leucosceptrum.*
Gynoecium. Carpel 2, syncarpous, ovary superior, 4-celled with single anatropous ovule in each cell on axile placentation, usually style gynobasic; stigma bifid.	Style terminal – *Teucrium, Ajuga.*
Fruit. Carcerule (consists of 1-seeded 4 nutlets surrounded by persistent calyx).	
Seeds. Endosperm absent or very scanty and fleshy, embryo with flat cotyledons.	

ultimately become 4-chambered). Placentation axile. **Style gynobasic.**

9. Fruit- **one seeded 4 nutlets (carcerule).**

Floral formula

$$\% \ ⚥ \ K_{(5)} \ \widehat{C_{(5)} A_4} \ \underline{G}_{(2)}$$

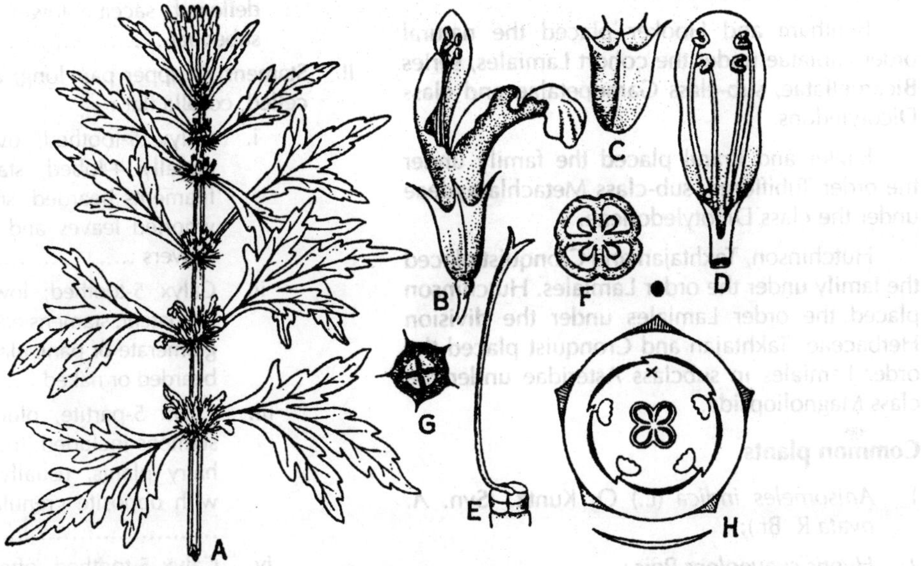

Fig. 4.130 : Labiatae (*Leonurus sibiricus*) : A. Portion of a plant showing verticillaster inflorescences; B. Single flower; C. Calyx; D. Corolla split open showing epipetalous stamens; E. Pistil showing gynobasic style; F. T.S. of ovary; G. Fruit, and H. Floral diagram

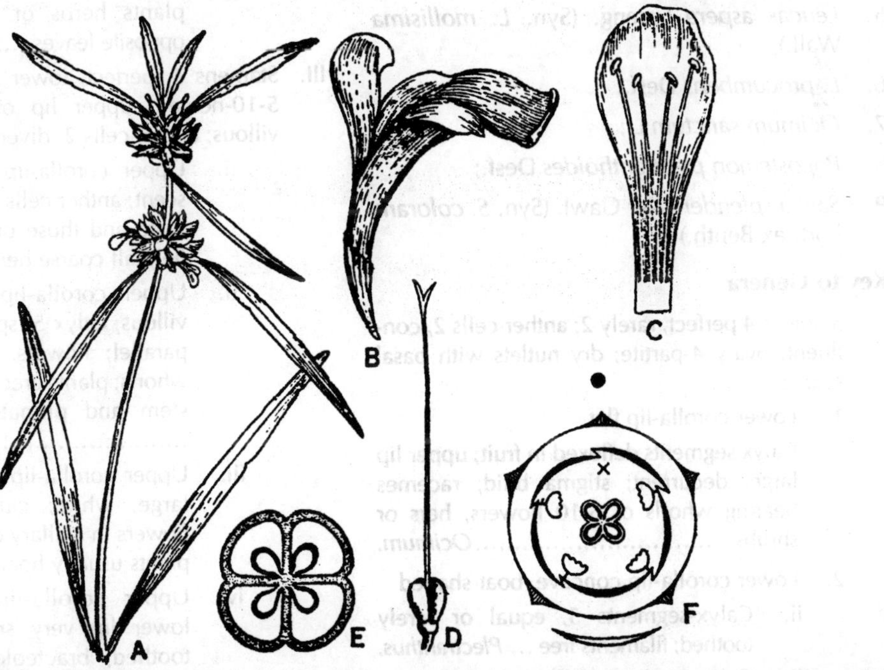

Fig. 4.131 : Labiatae (*Leucas linifolia*) : A. Portion of a plant showing verticillaster inflorescences; B. Single flower; C. Corolla split open showing epipetalous stamens; D. Pistil showing gynobasic style; E. T.S. of ovary, and F. Floral diagram

Phylogeny and systematics

The family is very closely related with Verbenaceae. It resembles Boraginaceae in fruit characteristics but is distinguished them sharply by the position of radicle.

Bentham and Hooker placed the natural order Labiatae under the cohort Lamiales, series Bicarpellatae, sub-class Gamopetalae and class Dicotyledons.

Engler and Prantl placed the family under the order Tubiflorae, sub-class Metachlamydeae under the class Dicotyledoneae.

Hutchinson, Takhtajan and Cronquist placed the family under the order Lamiales. Hutchinson placed the order Lamiales under the division Herbaceae. Takhtajan and Cronquist placed the order Lamiales in subclass Asteridae under the class Magnoliopsida.

Common plants

1. *Anisomeles indica* (L.) O. Kuntz. (Syn. *A. ovata* R. Br.);
2. *Hyptis suaveolens* Poir.;
3. *Leonotis nepetifolia* (L.) R. Br.;
4. *Leonurus sibiricus* L.;
5. *Leucas aspera* Spreng. (Syn. *L. mollisima* Wall.);
6. *L. procumbens* Desf.;
7. *Ocimum sanctum* L.;
8. *Pogostemon plectranthoides* Desf.;
9. *Salvia splendens* Ker-Gawl. (Syn. *S. colorans* hort. ex Benth.)

Key to Genera

I. Stamens 4 perfect, rarely 2; anther-cells 2, confluent; ovary 4-partite; dry nutlets with basal scars :

1. Lower corolla-lip flat
   Calyx segments deflexed in fruit; upper lip large, decurrent; stigma bifid; racemes bearing whorls of 6-10 flowers; hers or shrubs .... *Ocimum*.
2. Lower corolla-lip concave, boat-shaped
   ii. Calyx-segments 5, equal or rarely toothed; filaments free ... *Plectranthus*.
   iii. Calyx 5, always 2-lipped; stamens 4; filaments connate below; herbs or shrubs with fleshy stem and leaves ... . *Coleus*.
3. Corolla obscurely 2-lipped with 2 upper lobes flat, similar and subequal to two latersls, all about as long as the abruptly deflexed, saccate lower lip; calyx teeth subequal .... *Hyptis*.

II. Stamens 4, upper pair long; anthers 1- or 2-celled; corolla flat :

i. Calyx 5-toothed, ovoid; core subequally 4-lobed; stamens exserted; filaments bearded; small plants with whorled leaves and densely spicate flowers .... *Dysophylla*.
ii. Calyx 5-toothed; lower corolla-lobe subpatent; stamens exserted; whorls in glomerate or panicular spike; filament bearded or naked ....*Pogostemon*.
iii. Calyx 5-partite, plumose segments; stamens included; fruit-nutlets ovoid; hairy plants, usually hoary shrubs, with opposite crenulate leaves .... *Colebrookia*.
iv. Calyx 5-toothed, often enlarged and inflated in fruits; corolla tubular, incurved or straight; stamens diverging with divaricate anther-cells; plants herbs or undershrubs with opposite leaves .... *Elsholtzia*.

III. Stamens 4, perfect, lower pair longest; calyx 5-10-nerved; upper lip of corolla hooded, villous; anther-cells 2, diverging

i. Upper corolla-lip short, flat, pubescent; anther-cells of upper pair dimidiate and those of lower ones parallel; tall coarse herbs ... *Anisomeles*.
ii. Upper corolla-lip hooded, purple, villous; calyx 5, spinous; anther-cells parallel; flowers. in dense axillary whorls; plants erect herbs with square stem and pinnately lobed leaves ....*Leonurus*.
iii. Upper corolla-lip hooded; lower lip large, white; calyx 8-10-toothed; flowers in axillary or terminal whorls; plants usually hoary herbs ... *Leucas*.
iv. Upper corolla-lip large, hooded; lower lip very small; calyx 8-10-toothed; bracteoles many; flowers scarlet or yellow, borne in axillary whorls .... *Leonotis*.

IV. Stamens 2, parallel; anther-cells linear :

i. Anther-cells equally, distantly pendulous from a long connective; calyx ovoid, upper lip concave, lower 2-fid; style bilamellate; hoary shrubs ... . *Meriandra.*

ii. Anther-cells unequal or single on prolonged connective; flowers large showy with 2-lipped calyx; upper lip of corolla arched; herbs or shrubs .... *Salvia.*

iii. Floral whorls many-fid; calyx 10-nerved; anther-cells parallel; flowers sometimes borne in terminal spike; corolla 4-lobed, subequal; a strongly scented perennial herb with creeping root-stock .... *Mentha.*

Plants of economic importance

The plants are useful mainly for the source of essential volatile oils. The oils are used for perfumery, flavouring and medicine. A few plants are grown in the garden as ornamental plants and still a few have other uses.

A. **In perfumery**

1. *Lavendula latifolia, L. hybrida* and *L. angustifolia*. The plants are commonly called as *Lavender*. Plants are the great source of lavender oil used in soaps, cosmetics and in perfumery. The oil also used as insect repellent and curminative.
2. *Mentha pulegium*. The plant is the source of *pennyroyal oil*. The oil is used in cosmetic industry and also in the production of synthetic menthol.
3. *Majorana hortensis* (Sweet majoram). The oil is extracted from leaves and inflorescence used in soap, liquor and perfumery industry.
4. *Ocimum gratissimum* (Ram tulsi). The extracted oil is very useful in perfumery.

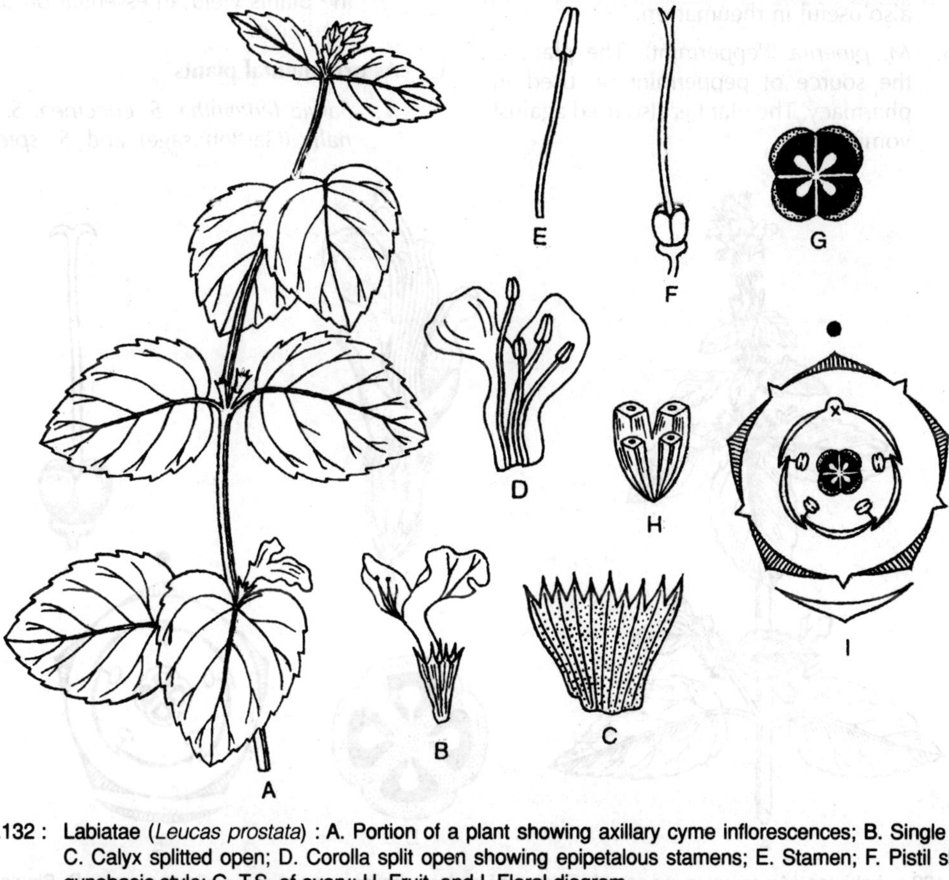

Fig. 4.132 : Labiatae (*Leucas prostata*) : A. Portion of a plant showing axillary cyme inflorescences; B. Single flower; C. Calyx splitted open; D. Corolla split open showing epipetalous stamens; E. Stamen; F. Pistil showing gynobasic style; G. T.S. of ovary; H. Fruit, and I. Floral diagram

5. *Pogostemon benghalense* and *P. cabin.* The leaves of both the plants yield *patchouli oil.* The oil is commercially used in perfumery.

B. **In medicine**

1. *Lallemantia royleana, Salvia aegyptiaca* and *Ocimum basilicum.* The seeds are known as the "**Topmari**" commercially used as poultice for curing boils.
2. *Majorana hortensis* (Sweet majoram). The oil is extracted from leaves and inflorescence used externally during toothache, stiffness of muscles and paralysis.
3. *Mentha arvensis* (Corn mint). The oil is a stimulant and has carminative properties.
4. *M. longifolia* (Horse mint). The decoction of the plant is used for apoplexy of heart and also in fever. The leaves are also useful in rheumatism.
5. *M. piperita* (Peppermint). The plant is the source of peppermint oil used in pharmacy. The plant is also used against vomiting.
6. *Ocimum basilicum* (Sweet basil, Babui tulsi). The plants are used as carminative, stimulant and stomachic.
7. *O. gratissimum* (Ram tulsi). The extracted oil is very useful in medicine.
8. *O. sanctum* (Holy basil, Tulsi). The leaf juice is used to control cold, cough and fever.
9. *Plectranthus mollis.* The oil extracted from the leaves and flowering tops is used as respiratory stimulant and cardiac depressant. The leaf juice is directly used at the cut end to stop bleeding.
10. *Pogostemon benghalense* and *P. cabin.* The leaves of both the plants yield *patchouli oil.* The oil is commercially used in medicine. The leaf extract has antirheumatic activity.
11. *Thymus vulgaris* (Common thyme) and *T. serphyllum* (Mother-of-thyme). Both the plants yield an essential oil used as medicine.

C. **As ornamental plants**

1. *Salvia leucantha, S. coccinea, S. officinalis* (Garden sage) and *S. splendens*

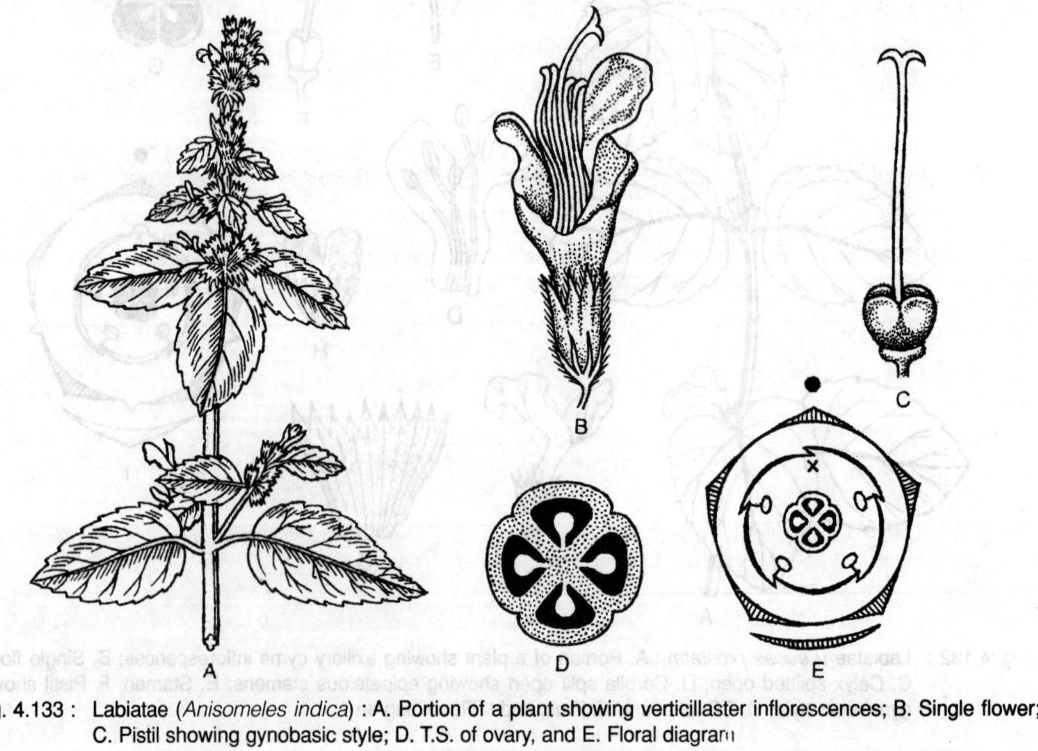

Fig. 4.133 : Labiatae (*Anisomeles indica*) : A. Portion of a plant showing verticillaster inflorescences; B. Single flower; C. Pistil showing gynobasic style; D. T.S. of ovary, and E. Floral diagram

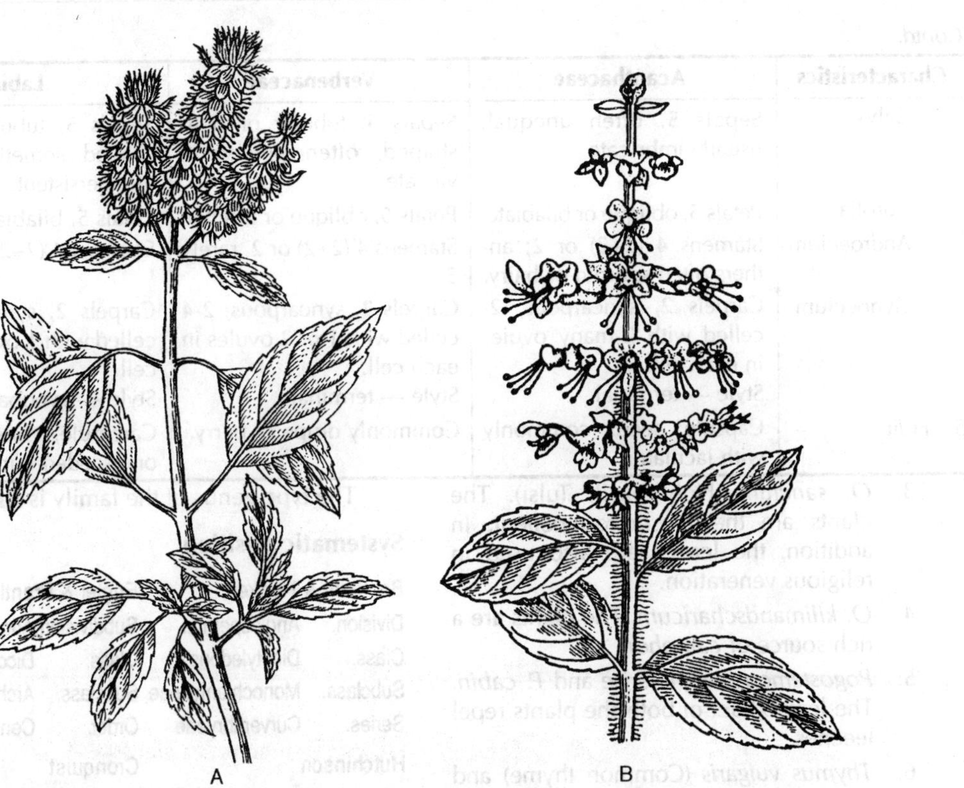

Fig. 4.134 : Some plants of Labiatae : Portion of plants with verticillaster inflorescences : A. *Ocimum basilicum* (Sweet basil), and B. *Ocimum sanctum* (Holy basil)

(Scarlet sage). They are grown in the garden as ornamental plants.

2. *Thymus vulgaris* (Common thyme), *T. serphyllum* (Mother-of-thyme), *Meriandra benghalensis* and *Coleus blumei*. The plants are cultivated as ornamental.

D. **Other uses**

1. *Coleus amboinicus.* The leaves of the plants are used for flavouring food products.
2. *Ocimum basilicum* (Sweet basil, Babui tulsi). The oil extracted from the plants has insecticidal properties.

Table 4.10 : Comparison among Acanthaceae, Verbenaceae and Labiatae

Characteristics	Acanthaceae	Verbenaceae	Labiatae
1. Plant	Commonly herbs or shrubs rarely with square stem; stem swollen at the nodes.	Herbs, shrubs or trees, often with square stem.	Commonly herbs or shrubs with square stem.
2. Leaves	Simple, opposite, sometimes decussate.	Simple, sometimes compound, opposite or whorled.	Simple, opposite-decussate.
3. Inflorescence	Condensed spike (sometimes cymose), usually with many bracts and bractioles.	Racemose or cymose, often with bracts.	Commonly verticillaster.
4. Flowers	Bisexual, hypogynous and zygomorphic.	Bisexual, hypogynous and zygomorphic (rarely actinomorphic).	Bisexual, hypogynous and zygomorphic.

Contd.

Contd.

Characteristics	Acanthaceae	Verbenaceae	Labiatae
Calyx	Sepals 5, often unequal, usually imbricate.	Sepals 5, tubular or bell-shaped, often persistent, valvate.	Sepals 5, tubular or bell-shaped, sometimes bilabiate, persistent.
Corolla	Petals 5, oblique or bilabiate.	Petals 5, oblique or tubular.	Petals 5, bilabiate.
Androecium	Stamens 4 (2+2) or 2; anthers often spurred or hairy.	Stamens 4 (2+2) or 2, rarely 5.	Stamens 4 (2+2) or 2.
Gynoecium	Carpels 2, syncarpous; 2-celled with 2-many ovules in each cell. Style — terminal.	Carpels 2, syncarpous; 2-4 celled with 1 or 2 ovules in each cell. Style — terminal.	Carpels 2, syncarpous; 4-celled with 1 ovule in each cell. Style — gynobasic.
5. Fruit	Capsule; seeds commonly with jaculator.	Commonly drupe or berry.	Carcerule, consisting of 4 one seeded nutlet.

3. *O. sanctum* (Holy basil, Tulsi). The plants are the *mosquito-repellent*. In addition, the leaves are also used in religious veneration.
4. *O. kilimandscharicum*. The plants are a rich source of *camphor*.
5. *Pogostemon benghalense* and *P. cabin*. The leaf extract of both the plants repel leeches.
6. *Thymus vulgaris* (Common thyme) and *T. serphyllum* (Mother-of-thyme). Both the plants are used for culinary purposes.

29. AMARANTHACEAE

(Amaranthus Family)

The family Amaranthaceae is represented by 69 genera and 1,000 species; grows widely in the tropical and temperate regions of the world. Most of the plants are commonly found to grow in India and tropical America. The family is represented in India by 17 genera and more than 50 species.

The type genus of the family is ***Amaranthus***.

Systematic position

Bentham & Hooker		Engler & Prantl	
Division.	Angiospermia	Subdivision.	Angiospermae
Class.	Dicotyledons	Class.	Dicotyledoneae
Subclass.	Monochlamydeae	Subclass.	Archichlamydeae
Series.	Curvembryeae	Order.	Centrospermae
Hutchinson		**Cronquist**	
Phylum.	Angiospermae	Division.	Magnoliophyta
Subphylum.	Dicotyledones	Class.	Magnoliopsida
Division.	Herbaceae	Subclass.	Caryophyllidae
Order.	Chenopodiales	Order.	Caryophyllales
Takhtajan		**Thorne**	
Division.	Magnoliophyta	Class.	Angiospermae
Class.	Magnoliopsida	Subclass.	Magnoliidae
Subclass.	Caryophyllidae	Superorder.	Caryophyllanae
Superorder.	Caryophyllanae	Order.	Caryophyllales
Order.	Caryophyllales.		

Amaranthaceae : Distinctive features

Common characteristics	Exceptions
Vegetative characteristics	
Plant. Usually annual or perennial herbs.	Climbing shrub – *Deeringia*.
Leaves. Alternate or opposite, simple, generally entire, exstipulate.	
Reproductive characteristics	
Inflorescence. Spike (either simple or compound) or raceme or dichasia.	
Flower. Actinomorphic, bisexual, hypogynous and with a pair of membranous bracteoles.	Unisexual – *Amaranthus*. Polygamo-dioecious – *Deeringia amaranthoides*.

Contd.

Contd.

Common characteristics	Exceptions
Perianth. Uniseriate, tepals 5, dry and membranous, often white and shining or coloured, persistent, free or more or less united, imbricate.	
Androecium. Stamens 5, opposite to perianth segments; filaments often united into a membranous tube which may bear simple, fringed or lobed, petaloid outgrowth between two stamens. Anther 2-celled.	Anther lobe one – *Alternanthera*.
Gynoecium. Carpels usually 2, ovary superior, unilocular, frequently with one campylotropous or basal ovule; styles 1-3.	Carpel 3 – Several ovules in each loculus – *Celosia*.
Fruit. One seeded nutlet.	Utricle – *Achyranthes*. Pyxis – *Celosia*.
Seeds. Black, lenticular with bony *shining testa* and mealy endosperm. Embryo annular or horseshoe-shaped.	

Identifying characteristics

1. Plants herbs or shrubs.
2. Inflorescence **spike**.
3. **Flowers bisexual, bracteate.**
4. **Perianth segments 5, sepaloid, dry, hard** or **scarious, pointed** and **persistent.**
5. **Stamens 5**, free or connate.
6. **Ovary 1-chambered with one campylotropous ovule.**
7. Fruit **nutlet** or **utricle.**

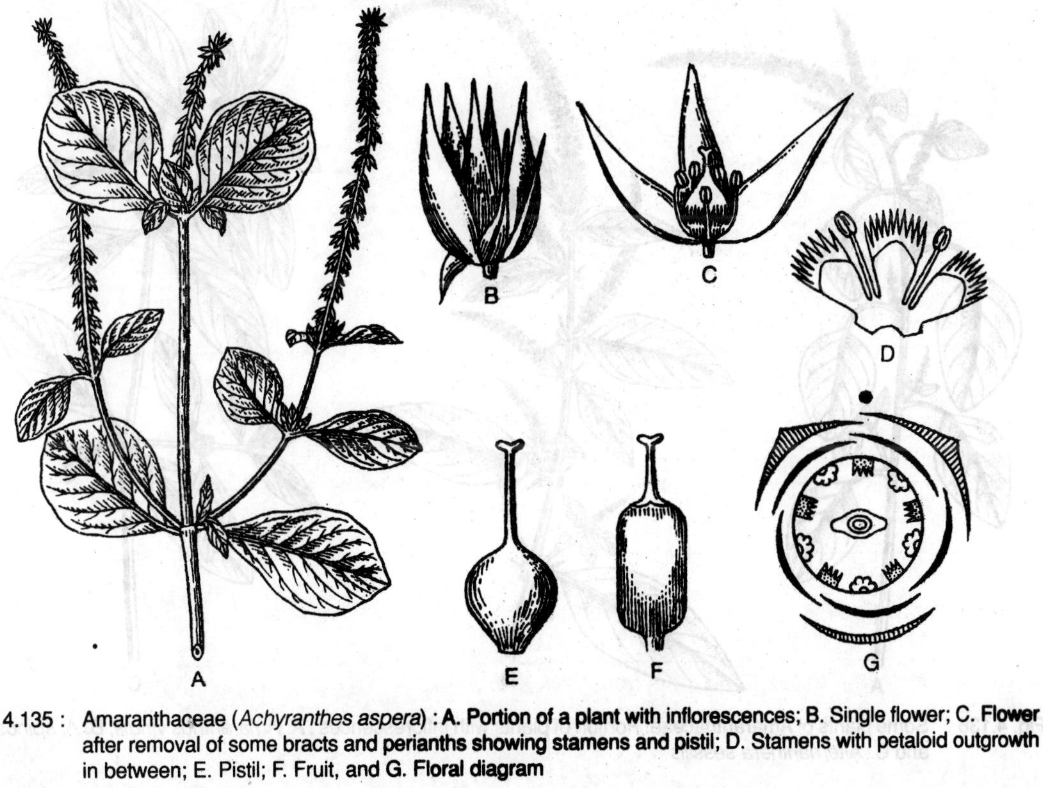

Fig. 4.135 : Amaranthaceae (*Achyranthes aspera*) : **A. Portion of a plant with inflorescences; B. Single flower; C. Flower after removal of some bracts and perianths showing stamens and pistil; D. Stamens with petaloid outgrowth in between; E. Pistil; F. Fruit, and G. Floral diagram**

Floral formula

$\oplus$ ⚥ $P_5A_5\underline{G}_{(2\text{-}3)}$

Phylogeny and systematics

The family is allied to Chenopodiaceae, Nyctaginaceae etc. Many recent phylogenetists agreed that the family Amaranthaceae is not a primitive family which has evolved from Caryophyllaceous stock.

Bentham and Hooker placed Amarantaceae (=Amaranthaceae) under the series Curvembryeae of subclass Monochlamydeae. Engler placed the family in the order Centrospermae under the subclass Archichlamydeae of the class Dicotyledoneae. Later, Hutchinson considered the family as more or less advanced, and placed it under the order Chenopodiales along with Chenopodiaceae, Basellaceae, Phytolaccaceae etc. after Polygoniales.

Common plants

1. *Achyranthes aspera* L.,
2. *Aerva lanata* (L.) Juss.,
3. *Alternanthera sessilis* R. Br.,
4. *Amaranthus tricolor* L. (Syn. *A. gangeticus* L.),
5. *A. spinosus* L.,
6. *A. viridis* l.,
7. *Celosia argentea* L.,
8. *Digera muricata* (L.) Mart.,
9. *Gomphrena globosa* L.

Key to Genera

I. Leaves usually alternate :
  1. Ovary 2-many-ovuled
    i. Plant climbing, undershrub; fruit a berry .... *Deeringia*.
    ii. Annual herb, flowers white or pink, fruit a circumsessile capsule... *Celosia*.
  2. Ovary 1-ovuled
    i. Ovules erect
    Flowers 1-sexual .... *Amaranthus*.
    Flowers bisexual :
      a. Flowers spicate in three, the outer two reduced to crested scales; fruit a nut .... *Digera*.
      b. Flowers capitate, fruit a circumsessile capsule .... *Allmania*.

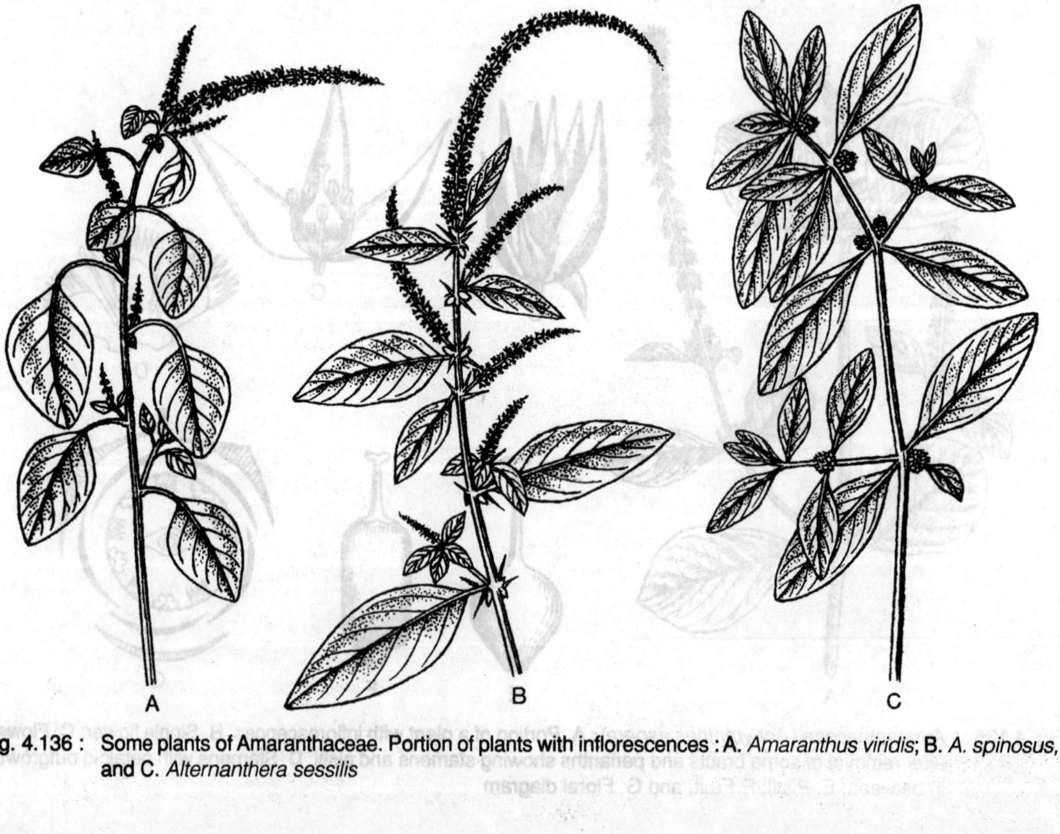

Fig. 4.136 : Some plants of Amaranthaceae. Portion of plants with inflorescences : A. *Amaranthus viridis*; B. *A. spinosus*, and C. *Alternanthera sessilis*

ii. Ovules suspended; flowers perfect; stamens with interposed staminodes .... *Aerva.*

II. Leaves opposite, ovary 1-celled :

1. Anthers 2-celled
   - i. Flowers all perfect
     - a. Stamens with interposed staminodes, bracts spinous ... *Achyranthes.*
     - b. Staminodes absent ... *Psilotrichum.*
   - ii. Flowers perfect and deformed, clustered together
     - a. Stamens with interposed staminodes .... *Cyathula.*
     - b. Staminodes none .... *Pupalia.*
2. Anthers 1-celled
   - i. Flowers small, white, in heads; staminal tube short; stigma capitate, subsessile .... *Alternanthera.*
   - ii. Flowers short, white or coloured, usually capitate; staminal tube long; stigma 2-fid; style long ... *Gomphrena.*

Plants of economic importance

Economically the family has little importance. Some are used as vegetables, a few as medicine and still some others have ornamental value.

Some plants along with their importance are :

A. **Vegetables**

1. *Amaranthus caudatus* L., *A. tricolor* L. are used as leafy vegetable.

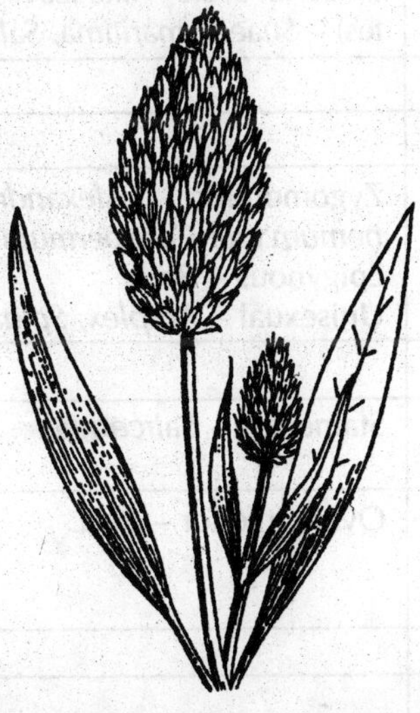

Fig. 4.137 : Amaranthaceae (*Celosia argentea*)

B. **Medicinal**

1. *Achyranthes aspera* (Chirchita, Apang). The flowering spikes ground into paste and are used externally to control insect bites.
2. *Deeringia amaranthoides.* The leaves are applied on sores.

C. **Ornamental**

1. *Amaranthus* spp. Different species (e.g., *A. tricolor, A. salicifolius,* etc.) are cultivated in the garden for their variegated foliage and brightly coloured inflorescence (spike).
2. *Celosia argentea* (Cockscomb), *Deeringia amaranthoides, Gomphrena globosa* (Globe amarnath) and *Iresine lindenii* (Blood leaf) are cultivated as ornamental plants.

30. CHENOPODIACEAE

(Beet Family)

The family Chenopodiaceae consists of 106 genera and 1,425 species in different parts of the world. They are commonly found in Central and Eastern region of Asia, Australia, South Africa (Karroo) and also in the South West Caspian coast. In India, the family is represented by 20 genera and more than 50 species distributed widely from tropical to temperate Himalayas even up to height of 1,750 metres (5,750 ft).

The type genus of the family is ***Chenopodium.***

Systematic position

Bentham & Hooker		**Engler & Prantl**	
Division.	Angiospermia	Subdivision.	Angiospermae
Class.	Dicotyledons	Class.	Dicotyledoneae
Subclass.	Monochlamydeae	Subclass.	Archichlamydeae
Series.	Curvembryeae	Order.	Centrospermae
Hutchinson		**Cronquist**	
Phylum.	Angiospermae	Division.	Magnoliophyta
Subphylum.	Dicotyledones	Class.	Magnoliopsida
Division.	Herbaceae	Subclass.	Caryophyllidae
Order.	Chenopodiales	Order.	Caryophyllales
Takhtajan		**Thorne**	
Division.	Magnoliophyta	Class.	Angiospermae
Class.	Magnoliopsida	Subclass.	Magnoliidae
Subclass.	Caryophyllidae	Superorder.	Caryophyllanae
Superorder.	Caryophyllanae	Order.	Caryophyllales
Order.	Caryophyllales.		

Identifying characteristics

1. **Succulent herbs with swollen nodes.**
2. Inflorescence. **Cyme** or **dense spike** or **panicle.**
3. Flower. **Regular**, hypogynous, **bisexual** and **often with scarious bracts and bracteoles.**
4. Perianth. **Tepals 5**, small, free or united, **sepaloid, persistent.**
5. **Stamens 5 free** or **united** at the base.
6. **Carpels 2** to 3, ovary **superior, 1-celled.**
7. Fruit. **Nutlet** or **Urticle; embryo curved** or **coiled.**

Floral formula

$\oplus$ ⚥ $P_{5 \text{ or } (5)} A_{(5)} \underline{G}_{(2-3)}$

Phylogeny and systematics

The family is closely related to Amaranthaceae which differs in having green and fleshy tepals.

Bentham and Hooker placed the Chenopodiaceae under the Series Curvembryeae, Subclass Monochlamydeae of Class Dicotyledons. Engler and Prantl included the family under the order Centrospermae, Subclass Archichlamydeae of Class Dicotyledones. Hutchinson, on the other hand, placed the family under Order Chenopodiales, Division Archi-chlamydeae of Subphyla Dicotyledones. Later, both Takhtajan and Cronquist placed the family under the Order Caryophyllales.

The two subfamilies Chenopodieae and Baselleae as segregated by Bentham and Hooker, are considered as separate families, Chenopodiaceae and Basellaceae by several taxonomists like Hutchinson, Takhtajan and Cronquist.

Common plants

1. *Arthrocnemum indicum* Moq. Woody and prostrate shrubs commonly grow in Sunderbans.
2. *Beta vulgaris* L. (Sugar-beet),

Chenopodiaceae : Distinctive features

Common characteristics	Exceptions
Vegetative characteristics	
Plant. Plants are mostly annual or perennial succulent herbs with swollen nodes, usually covered with hairs of various forms.	Shrubs – *Suaeda* and *Salicornia.* Tree – *Haloxylon ammodendron.*
Leaf. Simple, alternate, fleshy or membranous, exstipulate, entire, glabrous or mealy (*Salsola, Chenopodium* and *Artiplex*).	Opposite – *Salicornia.* Linear or scale – like leaves (halophytes) – *Suaeda maritima, Salsola.*
Reproductive characteristics	
Inflorescence. Cyme or dense spike or panicle.	
Flower. Regular, hypogynous, bisexual and often with scarious bracts and bracteoles, monochlamydous.	Zygomorphic – *Alexandra, Holocnemum* and *Corispermum.* Epigynous – *Beta.* Unisexual – *Artiplex, Spinacia.*
Perianth. Tepals 5, small, free or united, sepaloid, persistent.	
Androecium. Stamens 5, opposite to the perianth lobes, free or united at the base.	Stamens 2 – *Salicornia.*
Gynoecium. Carpels 2 to 3, ovary superior, 1-celled with one campylotropous pendulous or erect ovule, style terminal short or long, stigma usually 2 to 3 fid.	Ovary inferior – *Beta.*
Fruit. Nutlet or Urticle.	
Seeds. Exalbuminous or albuminous, embryo curved (*Chenopodium, Artiplex, Beta* etc.) or coiled (*Salsola, Suaeda* etc.).	

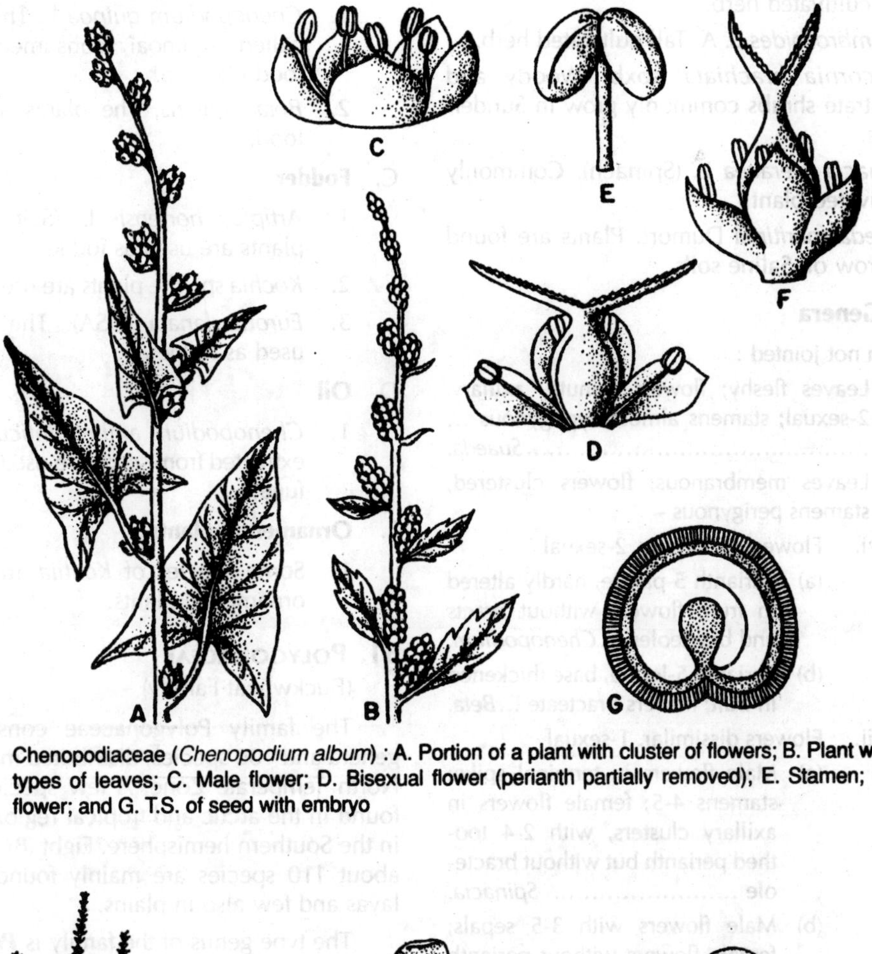

Fig. 4.138 : Chenopodiaceae (*Chenopodium album*) : A. Portion of a plant with cluster of flowers; B. Plant with different types of leaves; C. Male flower; D. Bisexual flower (perianth partially removed); E. Stamen; F. Bisexual flower; and G. T.S. of seed with embryo

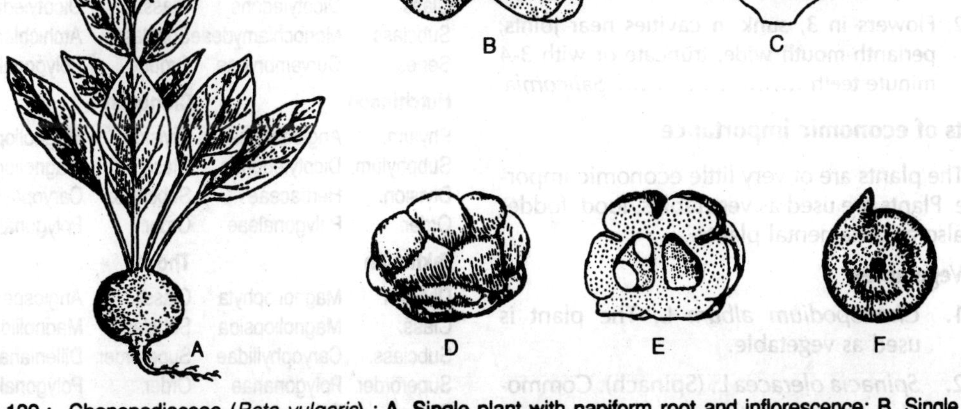

Fig. 4.139 : Chenopodiaceae (*Beta vulgaris*) : A. Single plant with napiform root and inflorescence; B. Single flower; C. V.S. of flower; D. Fruit; E. V.S. of fruit, and F. Seed

3. *Chenopodium album* L. (Lambís quarters). Tall cultivated herb.
4. *C. ambrosoides* L. A Tall cultivated herb.
5. *Salicornia brachiata* Roxb. Woody and prostrate shrubs commonly grow in Sunderbans.
6. *Spinacia oleracea* L. (Spinach). Commonly cultivated plant.
7. *Suaeda maritima* Dumort. Plants are found to grow on Saline soils.

Key to Genera

I. Stem not jointed :
   1. Leaves fleshy; flowers minute, axillary, 2-sexual; stamens almost hypogynous ... . *Suaeda.*
   2. Leaves membranous; flowers clustered; stamens perigynous –
      i. Flowers all similar, 2-sexual
         (a) Perianth 5-partite, hardly altered in fruit; flowers without bracts and bracteoles ... *Chenopodium.*
         (b) Perianth 5-lobed, base thickened in fruit; flowers bracteate ... *Beta.*
      ii. Flowers dissimilar, 1-sexual
         (a) Male flowers in terminal spike; stamens 4-5; female flowers in axillary clusters, with 2-4 toothed perianth but without bracteole .... *Spinacia.*
         (b) Male flowers with 3-5 sepals; female flowers without perianth .... *Artiplex.*

II. Stem joined; leaves 0; flowers bisexual :
   1. Flowers 2-6 in spike; perianth-mouth narrow, 3-4-fid .... *Arthrocnemum.*
   2. Flowers in 3, sunk in cavities near joints; perianth-mouth wide, truncate or with 3-4 minute teeth .... *Salicornia.*

Plants of economic importance

The plants are of very little economic importance. Plants are used as vegetables, food, fodder and also as ornamental plants.

A. **Vegetables**

1. *Chenopodium album* L. The plant is used as vegetable.
2. *Spinacia oleracea* L. (Spinach). Commonly used as vegetables.

B. **Food**

1. *Chenopodium quinoa* L. The seeds are called ëquinoaí, consumed as staple food like rice.
2. *Beta vulgaris.* The plants are used as food.

C. **Fodder**

1. *Artiplex hortensis* L. (Salt bush). The plants are used as fodder.
2. *Kochia* sp. The plants are used as fodder.
3. *Eurotia lanata* (USA). The plants are used as fodder.

D. **Oil**

1. *Chenopodium anthelminticum.* The oil extracted from seeds are used as a vermifuge.

E. **Ornamental plants**

1. Some species of *Kochia* are grown as ornamental plants.

31. POLYGONACEAE

(Buckwheat Family)

The family Polygonaceae consists of 43 genera and 100 species, distributed mainly in the North Temperate Zone. A few species are also found in the arctic and tropical regions and also in the Southern hemisphere. Eight (8) genera and about 110 species are mainly found in Himalayas and few also in plains.

The type genus of the family is ***Polygonum.***

Systematic position

Bentham & Hooker		**Engler & Prantl**	
Division.	Angiospermia	Subdivision.	Angiospermae
Class.	Dicotyledons	Class.	Dicotyledoneae
Subclass.	Monochlamydeae	Subclass.	Archichlamydeae
Series.	Curvembryeae	Order.	Polygonales
Hutchinson		**Cronquist**	
Phylum.	Angiospermae	Division.	Magnoliophyta
Subphylum.	Dicotyledones	Class.	Magnoliopsida
Division.	Herbaceae	Subclass.	Caryophyllidae
Order.	Polygonaleae	Order.	Polygonales
Takhtajan		**Thorne**	
Division.	Magnoliophyta	Class.	Angiospermae
Class.	Magnoliopsida	Subclass.	Magnoliidae
Subclass.	Caryophyllidae	Superorder.	Dillenianae
Superorder.	Polygonanae	Order.	Polygonales
Order.	Polygònales		

Identifying characteristics

1. Generally herbs or shrubs, rarely climber or trees.
2. Stem with **swollen nodes**.
3. Leaves with **ochreate stipule**.
4. Flower usually **bisexual**, small and regular, monochlamydous.
5. Perianth : Tepals undifferentiated.
6. Ovary : **Carpels 2-3, syncarpous**, 1-chambered. Ovary superior, **triangular**.
7. Fruits : **Trigonous nutlets**.

Floral formula

$\oplus ⚥ P_{6(3+3) \text{ or } 5} A_{6\text{-}9} \underline{G}_{(2) \text{ or } (3)}$

Phylogeny and systematics

The family Polygonaceae is closely related with the members of Urticaceae and Caryophyllales. It is generally regarded as a much reduced descendent of the Caryophyllales.

According to Bentham and Hooker, the natural order Polygonaceae is placed in the series Curvembryeae of sub-class Monochlamydeae under the class Dicotyledons.

Engler and Prantl placed the family Polygonaceae in the order Polygonales of sub-class Archichlamydeae under Dicotyledoneae.

Hutchinson, Takhtajan and Cronquist placed the family Polygonaceae in the order Polygonales. Hutchinson placed the order under the

Polygonaceae : Distinctive features

Common characteristics	Exceptions
Vegetative characteristics	
Plant. Plants are mostly herbs.	Shrub – *Calligonum, Pteropyrum, Fagopyrum.* Climber – *Corculum leptopus* Tree – *Coccoloba uvifera.*
Stem. Stems are branched, angular and often with swollen nodes.	Stem and branches are flattened forming **phylloclade** – *Muehlenbeckia platyclados.*
Leaves. Usually alternate, simple and entire or toothed. Leaves with ochreate stipule or ochrea.	Margin lobed – *Rheum,* Stipule absent – *Corculum* and *Koenigia islandica.*
Reproductive characteristics	
Inflorescence. Primarily a compound raceme whose branches are usually cymose.	
Flower. Small, usually bisexual, hypogynous, actinomorphic, trimerous, cyclic or acyclic, protandrous, monochlamydous.	Flower unisexual – *Triplaris* and some species of *Rumex.* Dimorphic flower (long stamen and short style and short stamen and long style) – *Fagopyrum.*
Perianth. Tepals 6 in two whorls (3+3), usually sepaloid, sometimes petaloid or differentiated into outer calyx and inner corolla, polyphyllous, persistent, imbricate.	Dimerous – *Rumex* and *Oxyria digyna.* Pentamerous (pseudo) – *Polygonum.* Tetramerous – *Oxyria.*
Androecium. Stamens 5-9, opposite to perianth segments, anther 2-celled.	Stamens more than 9 – *Calligonum.*
Gynoecium. Carpel-2 to 3, united, ovary superior, triangular, ovule solitary, basal, orthotropous, style–1, stigma–2 to 4.	
Fruit. Biconvex or triangular nut.	
Seeds. Endosperm copious and mealy, embryo curved.	

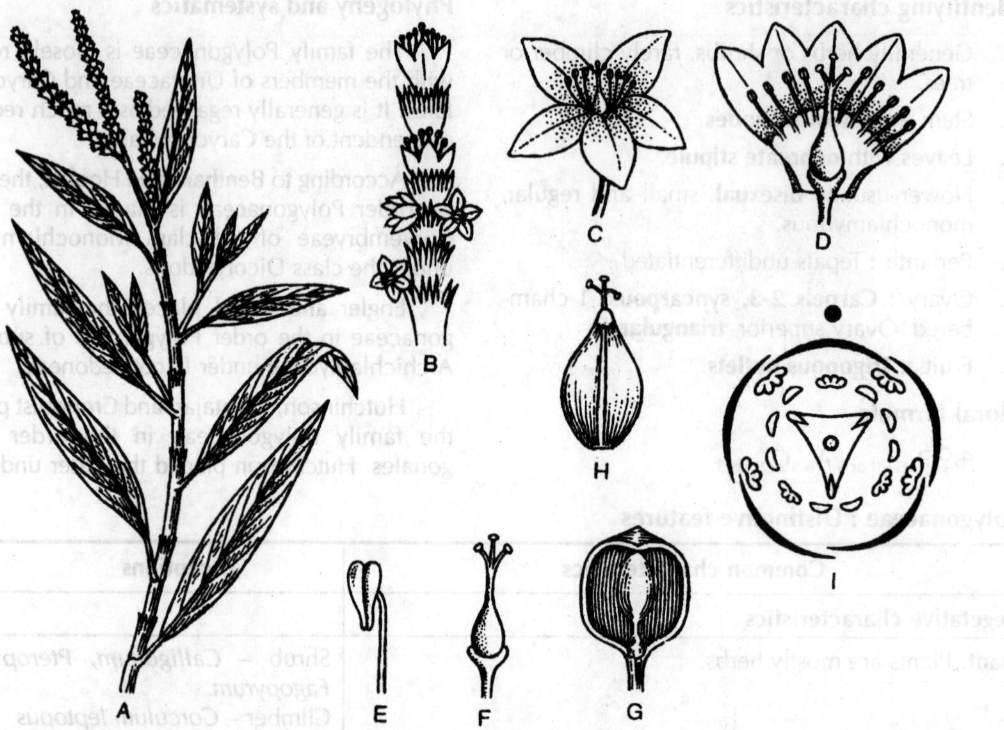

Fig. 4.140 : **Polygonaceae (*Polygonum barbatum*) : A. Portion of a plant with inflorescence; B. Portion of a spike; C. Single flower; D. Flower with perianth split open; E. Stamen; F. Pistil; G. L.S. of ovary; H. Fruit, and I. Floral diagram**

division Herbaceae of subphylum Dicotyledones. But Takhtajan and Cronquist placed the order under the subclass Rosoideae and Caryophyllidae respectively under the Class Magnoliopsida.

Common plants

1. *Corculum leptopus* Stuntz (Syn. *Antigonon leptopus* HK. and Arn.);
2. *Polygonum barbatum* L.;
3. *P. hydropiper* L.;
4. *P. orientale* L.;
5. *Rumex maritimus* L.

Key to Genera

I. Inflorescence ending in a tendril, stipule 0 Perianth 5-cleft; climbing plant ... *Corculum*.

II. Inflorescence not ending in a tendril, stipulate

  i. Perianth 4-5; capitate stigmas .... *Polygonum*.

  ii. Perianth 6-fid, inner 3 usually enlarged in fruit; stigmas horse-shoe-shaped or peltate, fimbriated .... *Rumex*.

Plants of economic importance

The family has little economic importance. Some of the members are useful in respect of vegetables, food, fodder, medicine and also ornamental value.

B. Vegetables

1. *Oxyria digyna* Hill. and *Rumex acetosella* (Garden sorrel). The leaves of the plant are used as salad.
2. *Polygonum alpinum* All., *Rheum nobile* Hook. f. & Th., *R. raponticum* L. and *R. emodi* Wall. The leaf stalks are either eaten as raw or boiled.
3. *Rumex acetosella* L. (Sheep sorrel). The apical regions of the young shoots are used as salad.

C. Food

1. *Fagopyrum esculentum* Moench. (Buckwheat, Kuttu) and *F. tataricum* Gaertn. (Duckwheat). The seeds are used as cereal because of floury endosperm.

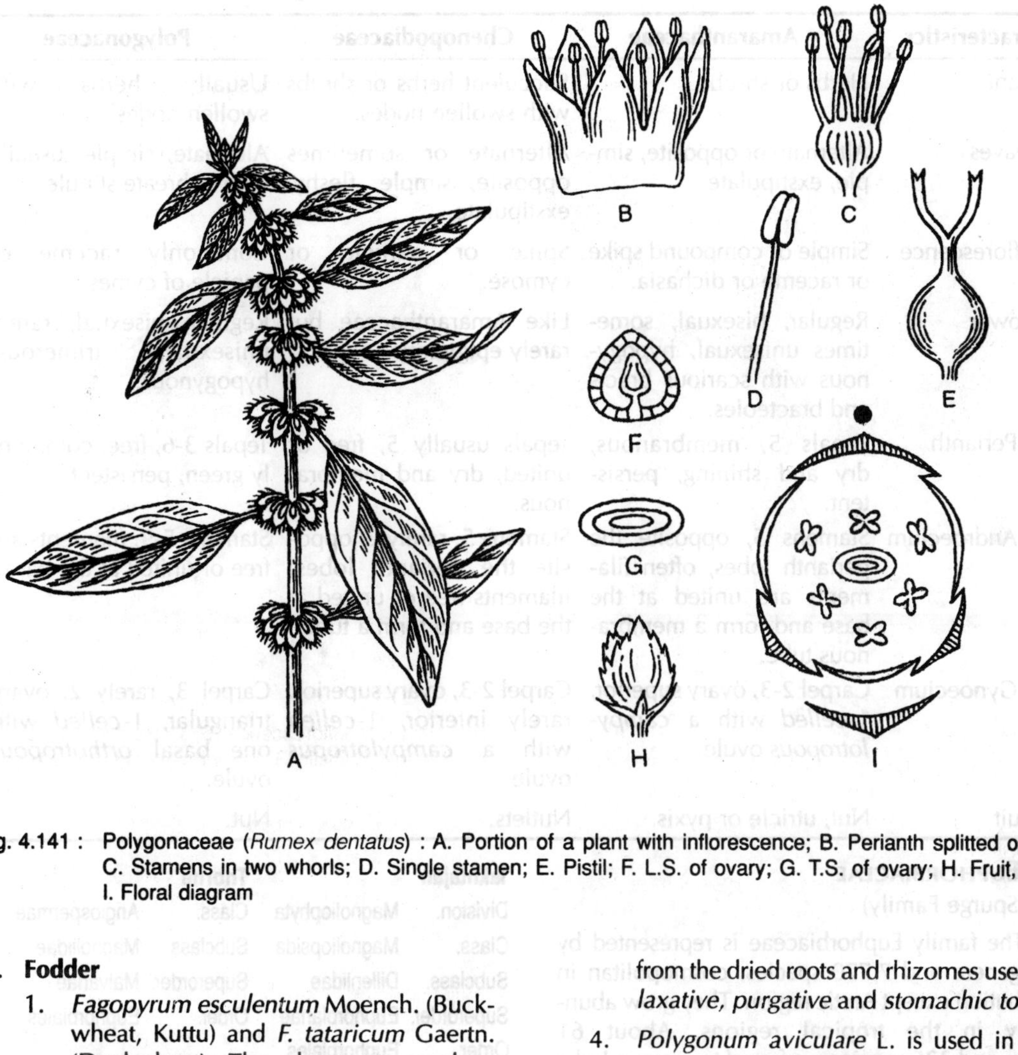

Fig. 4.141 : Polygonaceae (*Rumex dentatus*) : A. Portion of a plant with inflorescence; B. Perianth splitted open; C. Stamens in two whorls; D. Single stamen; E. Pistil; F. L.S. of ovary; G. T.S. of ovary; H. Fruit, and I. Floral diagram

D. **Fodder**

1. *Fagopyrum esculentum* Moench. (Buck-wheat, Kuttu) and *F. tataricum* Gaertn. (Duckwheat). The plants are used as green fodder.

E. **Medicine**

1. *Fagopyrum esculentum* Moench. (Buck-wheat, Kuttu) and *F. tataricum* Gaertn. (Duckwheat). *'Rutin'*, a flavonol glycoside, obtained from the *leaves, flowers,* is used in the treatment of incised capillary fragility associated with hypertension.
2. *Rumex acetosella* L. (Sheep sorrel). The leaves are said to be useful in the treatment of tumours, and also cancer.
3. *Rheum emodi* Wall., *R.officinale* Baill., *R. palmatum* L. and *R. webbiana* Royle. *'Rhubarb'* a well-known drug, is prepared from the dried roots and rhizomes used as *laxative, purgative* and *stomachic tonic.*
4. *Polygonum aviculare* L. is used in the treatment of rheumatism, diabetes and ulcers.

F. **Ornamental plants**

Some plants are cultivated in the garden for beautification. These are :

1. *Muehlenbeckia platyclados* Meissn. (Syn. *Cocoloba platyclada* F. Muell.). A small shrubby climber with ribbon-like green stem and branches grown in the garden as ornamental plant.
2. *Corculum leptopus* Stuntz. (Syn. *Antigonon leptopus* HK. and Arn.). A very much attractive climber with tendril and pink-coloured flowers.

Table 4.11 : Comparison of Amaranthaceae, Chenopodiaceae and Polygonaceae

Characteristics	Amaranthaceae	Chenopodiaceae	Polygonaceae
1. Plant	Herbs or shrubs.	Succulent herbs or shrubs with swollen nodes.	Usually herbs with swollen nodes.
2. Leaves	Alternate or opposite, simple, exstipulate.	Alternate or sometimes opposite, simple, fleshy, exstipulate.	Alternate, simple, usually with ochreate stipule.
3. Inflorescence	Simple or compound spike or raceme or dichasia.	Spike or panicle or cymose.	Commonly raceme or panicle of cymes.
4. Flower	Regular, bisexual, sometimes unisexual, hypogynous with scarious bracts and bracteoles.	Like Amaranthaceae but rarely epigynous.	Regular, bisexual, rarely unisexual, trimerous, hypogynous.
Perianth	Tepals 5, membranous, dry and shining, persistent.	Tepals usually 5, free or united, dry and membranous.	Tepals 3-6, free, commonly green, persistent.
Androecium	Stamens 5, opposite the perianth lobes, often filaments are united at the base and form a membranous tube.	Stamens 5, rarely 2, opposite the perianth lobes; filaments free or united at the base and form a tube.	Stamens 5-9, filaments are free or united at the base.
Gynoecium	Carpel 2-3, ovary superior, 1-*celled* with a *campylotropous* ovule.	Carpel 2-3, ovary superior, rarely inferior, 1-*celled* with a *campylotropus* ovule.	Carpel 3, rarely 2, ovary triangular, 1-*celled* with one basal *orthotropous* ovule.
5. Fruit	Nut, utricle or pyxis.	Nutlets.	Nut.

32. EUPHORBIACEAE

(Spurge Family)

The family Euphorbiaceae is represented by 321 genera and 7,770 species, cosmopolitan in distribution except Arctic region. They grow abundantly in the tropical regions. About 61 genera and 336 species are found to grow in India commonly in the hilly regions of tropical and subtropical Himalayas and also of South India.

The type genus of the family is ***Euphorbia***.

Systematic position

Bentham & Hooker		**Engler & Prantl**	
Division.	Angiospermia	Subdivision.	Angiospermae
Class.	Dicotyledons	Class.	Dicotyledoneae
Subclass.	Monochlamydeae	Subclass.	Archichlamydeae
Series.	Unisexuales	Order.	Geraniales

Hutchinson		**Cronquist**	
Phylum.	Angiospermae	Division.	Magnoliophyta
Subphylum.	Dicotyledones	Class.	Magnoliopsida
Division.	Lignosae	Subclass.	Rosidae
Order.	Euphorbiales	Order.	Euphorbiales.

Takhtajan		**Thorne**	
Division.	Magnoliophyta	Class.	Angiospermae
Class.	Magnoliopsida	Subclass	Magnoliidae
Subclass.	Dilleniidae	Superorder.	Malvanae
Superorder.	Euphorbianae	Order.	Euphorbiales
Order.	Euphorbiales.		

Identifying characteristics

1. Plants with **milky** or **watery latex.**
2. Inflorescence **cyathium, raceme, spike** or **cyme.**
3. **Flower unisexual.**
4. Perianth. Tepals 5, usually in one whorl.
5. **Carpel 3, ovary superior**, 3-chambered with 1 or 2 ovules in each chamber on **axile placentation.**
6. Fruit usually **regma**, sometimes drupe or berry.
7. Seeds **caruncluate**.

Euphorbiaceae : Distinctive features

Common characteristics	Exceptions
Vegetative characteristics	
Plant. Herbs, shrubs or trees, monoecious or dioecious, *often with milky* or *watery latex*.	Climber – *Cnesmone, Tragia*. Aquatic submerged plant – *Phyllanthus fluitans*.
Leaves. Usually alternate, simple, entire or palmately lobed with caducous or persistent stipules.	Opposite – *Choriophyllum* Whorled – *Mischodon* Leaves compound – species of *Bischofia, Hevea*.
Reproductive characteristics	
Inflorescence. Racemose, cymose, mixed or cyathium, subtended by involucral or brightly coloured bracts.	
Flower. Regular (actinomorphic) or slightly irregular, small, hypogynous, unisexual, bracteate.	Perigynous – *Bridelia*
Perianth. Simple, often small, sepaloid, pentamerous, valvate or imbricate.	Differentiated into calyx and corolla – *Jatropha, Wielandia*. Perianth wanting – *Pedilanthus* and *Jatropha*.
Androecium (in male flower). 1-many, filaments free or united, anthers 2-celled, rudiments of ovary present or absent.	
Gynoecium (in female flower). Carpel 3, syncarpous, ovary superior, generally 3-celled with 1 or 2 pendulous anatropous ovules in each cell on axile placentation, micropyle carunculate, staminodes present or absent; styles 3, free or united at the base; stigmas 3 or 6.	Carpel 2 – *Mercurialis*. Carpels 2-15 in *Glochidion*. Semi-inferior – *Bridelia*. Sometimes gynophore present-*Bridelia*. Ovary 1-celled – *Cometia, Antidesma, Drypetes*. Ovary 2-5 celled – *Baccaurea*.
Fruit. Usually 3 chambered schizocarpic-capsule (*regma*) splitting into 3 one-seeded cocci.	Drupe – *Drypetes*. Berry – *Bischofia*.
Seeds. Albuminous, *carunculate*, with fleshy endosperm and straight embryo.	

Variation of stamens :

A. Variation in number (ranges from 1-numerous) :

1. Stamens 1 in *Euphorbia, Pedilanthus,*
2. Stamens 2 or 3 in *Sapium,*
3. Stamens 3 in *Breynia,*
4. Stamens 3-5 in *Phyllanthus,*
5. Stamens 4-8 in *Baccaurea,*
6. Stamens 8 in *Acalypha,*
7. Stamens 10 in 2 alternate whorls in *Jatropha,*
8. Stamens 15 in *Crozophora,*
9. Stamens 8-20 in *Mercurialis,*
10. Stamens 15-30 in *Codiaeum,*
11. Stamens 80-100 in *Croton.*
12. Stamens several hundreds in *Lasiococca, Trewia,* etc.

B. Other variations :

1. Filaments are free or united into a column in *Jatropha* and *Breynia*.
2. Polyadelphy is found in *Lasiococca, Ricinus,* etc.
3. Anthers are united into a ring-like common anther in *Cyclanthera*.
4. Sometimes androphore present in *Bridelia*.

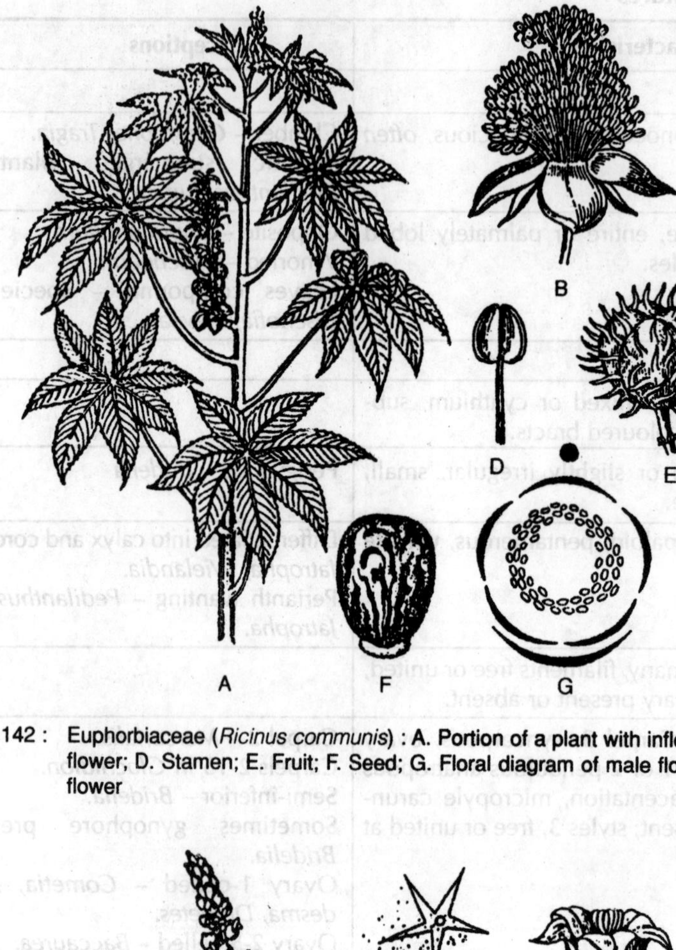

Fig. 4.142 : Euphorbiaceae (*Ricinus communis*) : A. Portion of a plant with inflorescence; B. Male flower; C. Female flower; D. Stamen; E. Fruit; F. Seed; G. Floral diagram of male flower, and H. Floral diagram of female flower

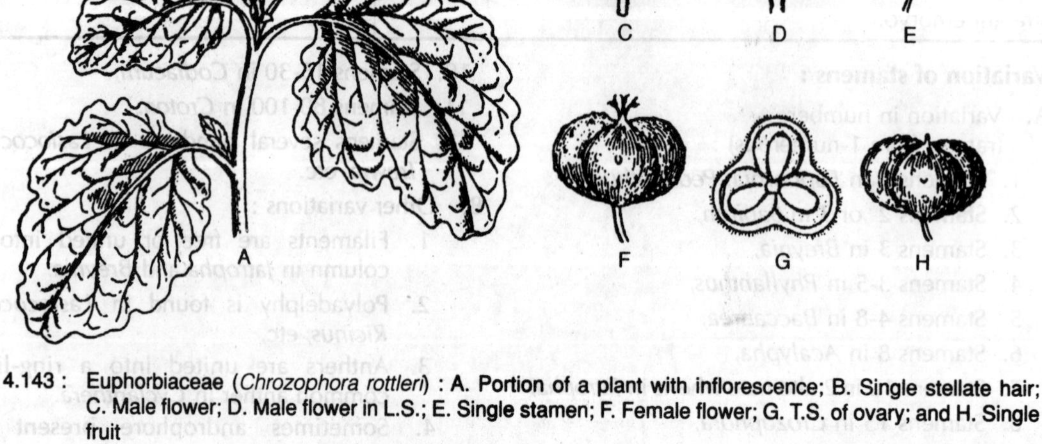

Fig. 4.143 : Euphorbiaceae (*Chrozophora rottleri*) : A. Portion of a plant with inflorescence; B. Single stellate hair; C. Male flower; D. Male flower in L.S.; E. Single stamen; F. Female flower; G. T.S. of ovary; and H. Single fruit

Floral formula

Male flower : $\oplus ♂ P_{3-5} A_{1-\alpha}$

Female flower : $\oplus ♀ P_{3-5} \underline{G}_{(3)}$

Phylogeny and systematics

The family Euphorbiaceae is closely related to Malvales and also related to the members of Sapindales.

Bentham and Hooker placed the family in Unisexuales of subclass Monochlamydeae and class Dicotyledons. Engler and Prantl placed in Geraniales near Malvales of subclass Archichlamydeae under the class Dicotyledoneae. But Hutchinson considers the family as more or less advanced and placed it in the order Euphorbiales near Malpighiales under the division Lignosae.

Takhtajan and Cronquiest followed Hutchinson by placing it under Euphorbiales, but Takhtajan placed it close to Malvales and Cronquist near to Rhamnales.

Common plants

1. *Acalypha indica* L.;
2. *A. hispida* Burm. f.;
3. *Bischofia javanica* Bl.;
4. *Chrozophora rottleri* (Geis) Klotz. (Syn. *C. plicata* Hook. f.);
5. *Croton bonplandianum* Baill. (Syn. *C. sperciflorum* Morr.);
6. *Emblica officinalis* Gaertn. (Syn. *Phyllanthus emblica* L.);
7. *Euphorbia bombaiensis* Santapau (Syn. *E. microphylla* Heyne ex Roth);
8. *E. hirta* L.;
9. *E. nivulia* Buch-Ham;
10. *E. pulcherrima* willd (Syn. *Poinsettia pulcherrima* R.Grah.);
11. *Gelonium multiflorum* A. Juss.;
12. *Jatropha gossypifolia* L.;
13. *Kirganelia reticulata* Bail. (Syn. *Phyllanthus reticulatus* Poir.);
14. *Mallotus philippinensis* Muell.-Arg.;
15. *Pedilanthus tithymaloides* (L.) Poit;
16. *Phyllanthus fraternus Webster*;
17. *P. simplex Retz.*;

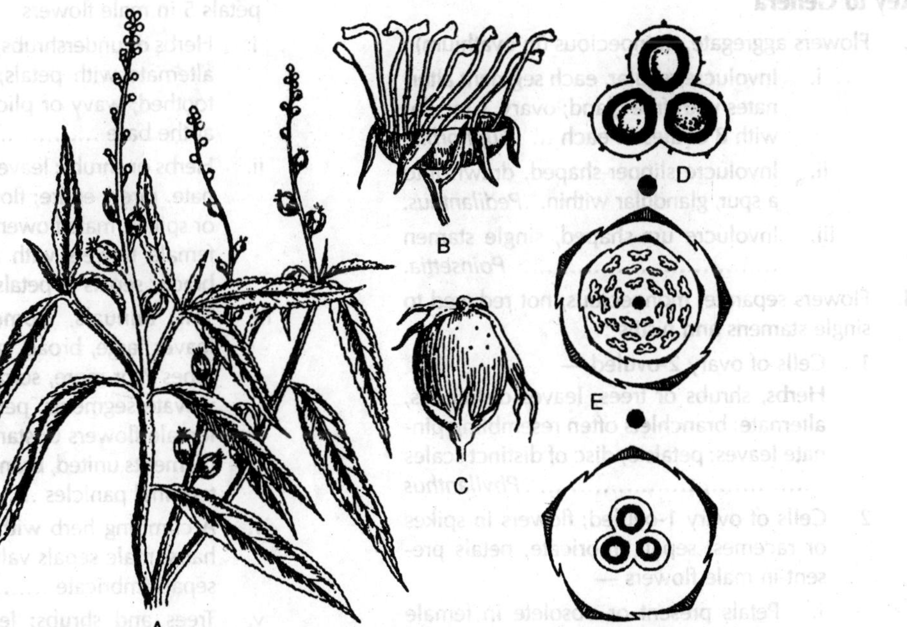

Fig. 4.144 : Euphorbiaceae (*Croton bonplandianum*) : A. Portion of a plant with inflorescence; B. Male flower; C. Female flower; D. T.S. of ovary; E. Floral diagram of male flower, and F. Floral diagram of female flower

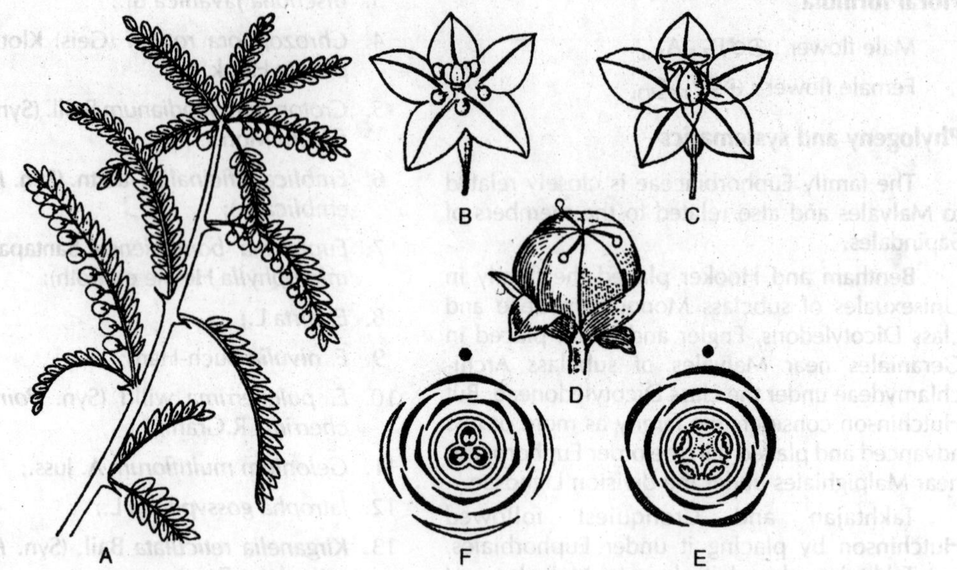

Fig. 4.145 : Euphorbiaceae (*Phyllanthus fraternus*) : A. Portion of a plant; B. Male flower; C. Female flower; D. Fruit; E. Floral diagram of male flower, and F. Floral diagram of female flower

18. *Ricinus communis* L.;
19. *Sapium sebiferum* Roxb.;
20. *Trewia nudiflora* L.

Key to Genera

I. Flowers aggregate, monoecious (in cyathium) :
   i. Involucre regular, each segment alternates with free gland; ovary 3-locular with 8 ovules in each ... *Euphorbia.*
   ii. Involucre slipper-shaped, drawn into a spur, glandular within...*Pedilanthus.*
   iii. Involucre urn-shaped, single stamen .... *Poinsettia.*

II. Flowers separate, monoecious, not reduced to single stamens and pistils :
   1. Cells of ovary 2-ovuled —
      Herbs, shrubs or trees; leaves distichous, alternate; branchlets often resembling pinnate leaves; petals 0; disc of distinct scales ....*Phyllanthus*
   2. Cells of ovary 1-ovuled; flowers in spikes or racemes; sepals imbricate; petals present in male flowers —
      i. Petals present or obsolete in female flowers; disc of 4-6 glands opposite the sepals .... *Croton.*
      ii. Petals absent in female flowers; disc of 5-6 glands alternate with petals; garden shrubs with ornamental leaves .... *Codiaeum.*
   3. Cells of ovary 1-ovuled; flowers in spikes or racemes; calyx segments 5, valvate; petals 5 in male flowers
      i. Herbs or undershrubs; disc of 5 glands alternate with petals; leaves sinuate, toothed; wavy or plicate, 2-glandular at the base .... *Chrozophora.*
      ii. Herbs or shrubs; leaves toothed or crenate, rarely entire; flowers in racemes or spikes, male flowers without bracts, female flowers with accrescent leafy bracts; sepals 4; petals 0 ... *Acalypha.*
      iii. Tall annuals, sometimes shrubs; leaves large, broad, palmately lobed, lobes 7 or more, serrate; calyx of 3-5 valvate segments; petals of male and female flowers 0; stamens numerous, filaments united, branched; flowers in terminal panicles .... *Ricinus.*
      iv. A climbing herb with hispid stinging hairs; male sepals valvate and female sepals imbricate .... *Tragia.*
      v. Trees and shrubs; leaves serrate or toothed; flowers in terminal spikes or racemes; calyx of 2-3 valvate segments; stamens 2 or 3 petals 0 in male and female flowers ... *Sapium.*

vi. Herbs, shrubs or trees, often glandular or prickly; leaves often lobed; sepals and petals 5; stamens many, filaments of inner series united; central flowers in the cyme or in forks usually female, flowers scarlet, purple or yellow .... *Jatropha.*

III. Flowers dioecious :

1. Trees with compound leaves ...... *Bischofia.*
2. Plants with simple leaves –

i. Large shrubs or trees; flowers axillary or leaf-opposed clusters; petals 0; stamens 10-60; pistillode absent; disc obsolete; fruit 3-4 gonous or 2-4 lobed fleshy capsule; leaves alternate .... *Gelonium.*

ii. Trees; male flowers clustered, female subsolitary; calyx 3-6-toothed; stamens 2-5; fruit a small drupe crowned with terminal stigmas .... *Drypetes.*

iii. Shrubs or trees flowers in spikes or racemes; calyx 3-8-toothed; stamens 2-5; fruit a small drupe crowned with terminal stigmas .... *Antidesma*

iv. Trees; flowers in racemes; stamens 4-8, filaments short; ovary 2-3-celled; fruit fleshy capsule; testa with a thick fleshy covering .... *Baccaurea.*

v. Trees or shrubs; leaves opposite or alternate; flowers minute, apetalous in spike or panicle; stamens numerous, anther cells 2, globose, adnate to the broad connective; fruit capsule; female flowers often with disc ...... *Mallotus.*

vi. Trees, with opposite leaves; male flowers in raceme, female flowers few, disc never present; fruit drupe; gall structures commonly occur in leaves ....*Trewia.*

vii. Shrubs or small trees; leaves alternate or opposite; flowers-rarely monoecious; petals in male and female flowers 0, disc 0; fruit a capsule of 3 cocci separating from the columella with twisted valves; common in salt areas of Sunderbans ...... *Excoecaria.*

Plants of economic importance

Plants of this family have various uses. These are :

A. **Rubber**

1. *Hevea brasiliensis.* The latex from the stem yield good quality rubber (**Para rubber**), used in the manufacture of different types of rubber goods.
2. *Manihot glaziovii.* The rubber (**Manicoba** or **Ceara rubber**) prepared from the latex of this plant is also used in different rubber works.

B. **Fruits**

1. *Baccauria sapida* and *Cicca acida*. The fruits (lotka) of the plant are edible.

C. **Medicinal plants**

1. *Emblica officinalis* (Emblic myrobalan,

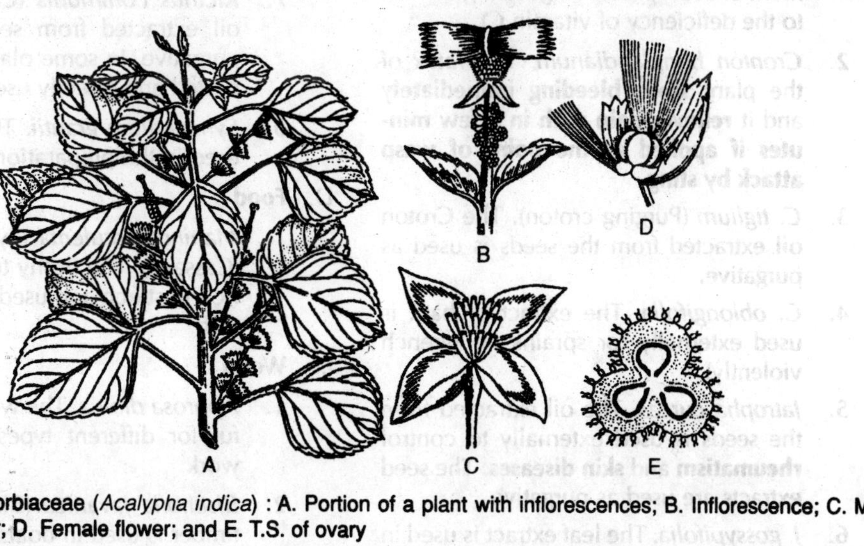

Fig. 4.146 : Euphorbiaceae (*Acalypha indica*) : A. Portion of a plant with inflorescences; B. Inflorescence; C. Male flower; D. Female flower; and E. T.S. of ovary

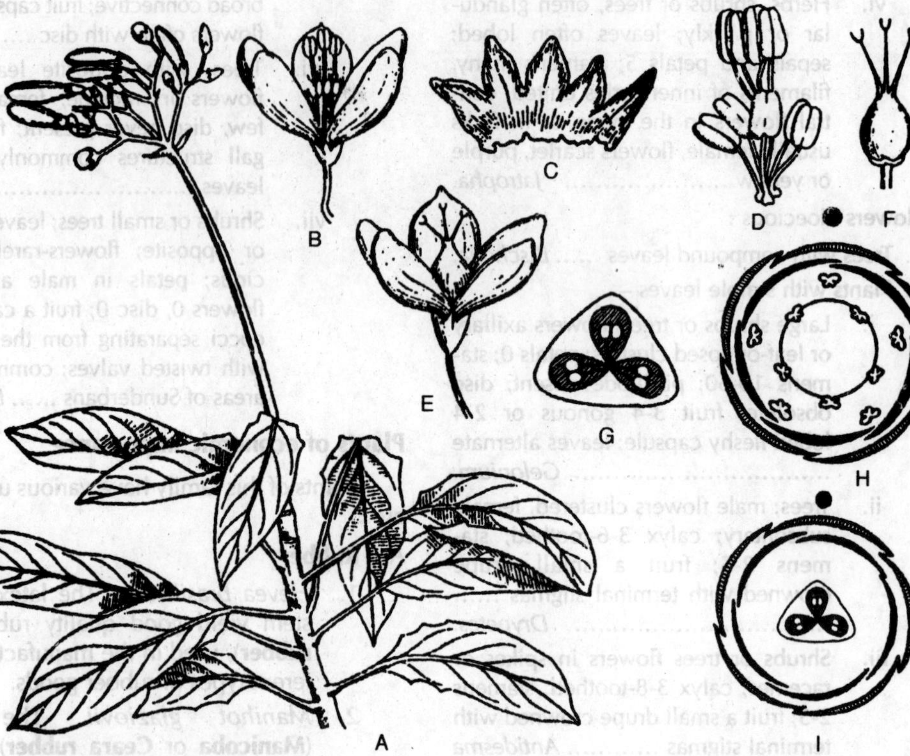

Fig. 4.147 : Euphorbiaceae (*Jatropha gossypifolia*) : A. Portion of a plant with flowers; B. Male flower; C. Calyx; D. Androecium; E. Female flower; F. Pistil; G. T.S. of ovary; H. Floral diagram of male flower, and I. Floral diagram of female flower

Amloki). The fruits are the rich source of **vitamin C** and are used for deodorisation of mouth by chewing the fruits. It is also used in pickles and in the treatment of scurvy (bleeding from gum due to the deficiency of vitamin C).

2. *Cronton bonplandianum.* The *latex* of the plant **stops bleeding immediately** and it **relieves pain with in a few minutes if applied at the point of wasp attack by sting.**
3. *C. tiglium* (Purging croton). The Croton oil extracted from the seeds is used as purgative.
4. *C. oblongifolia.* The extract of bark is used externally for sprains (to wrench violently).
5. *Jatropha curcas.* The oil extracted from the seeds is used externally to control **rheumatism** and **skin diseases**. The seed extracts are used as purgative.
6. *J. gossypifolia.* The leaf extract is used in the eczema and curbuncles. The roots are used to control **leprosy** and as a therapeutic medicine **against snake-bite.**
7. *Ricinus communis* (Castor plant). The oil extracted from seeds are used as purgative. In some places the seeds are pasted and directly used as spice.
8. *Synadenium grantii.* The root extract is used in the preparation of ear drop.

D. **Food**

1. *Manihot esculenta* (Syn. *M. utilissima*) (Cassava). The fleshy tuberous roots are rich in starch and used as valuable food stuff.

E. **Wood**

1. *Aporosa dioica.* The woody stem is useful for different types of construction work.
2. *Bischofia javanica* (Bishop wood). The timber is used in boats, bridges, etc.

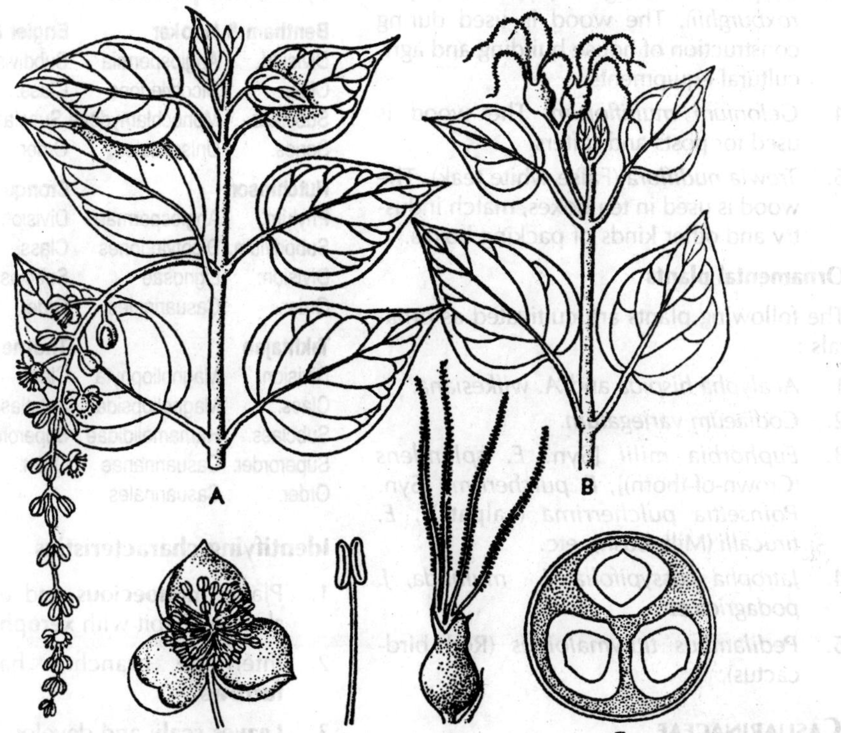

Fig. 4.148 : Euphorbiaceae (*Trewia nudiflora*) : A. Portion of a male plant; B. Portion of a female plant; C. Male flower; D. Stamen; E. Female flower; and F. T.S. of ovary

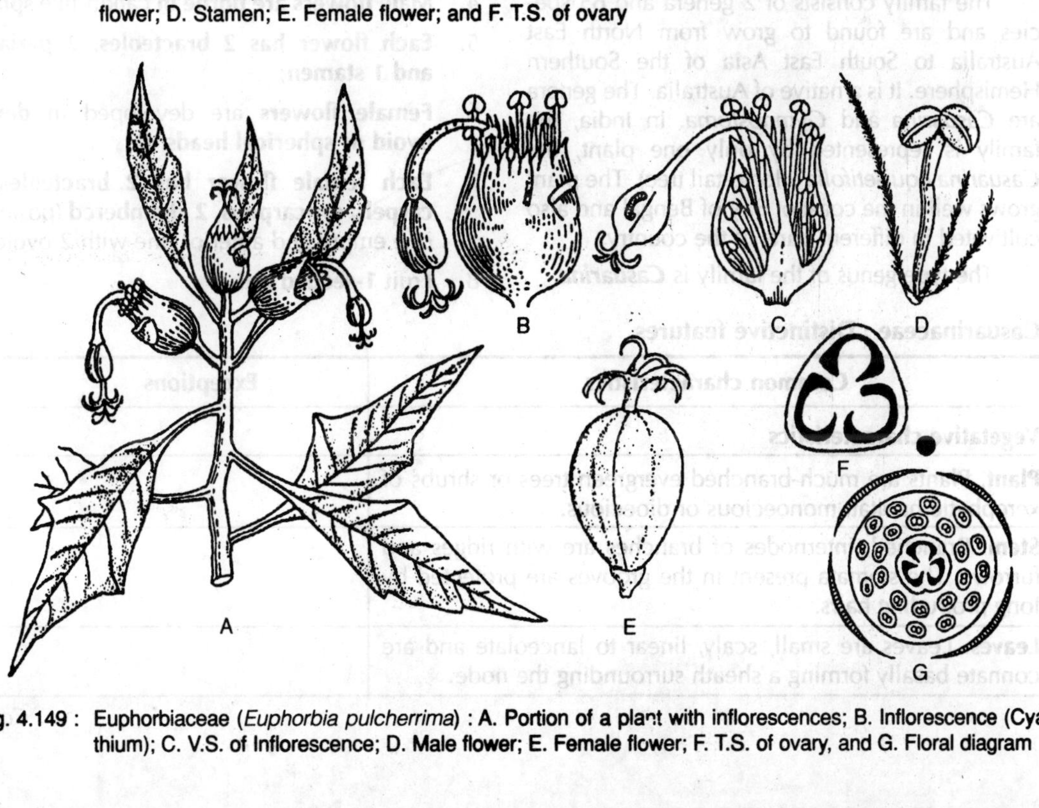

Fig. 4.149 : Euphorbiaceae (*Euphorbia pulcherrima*) : A. Portion of a plant with inflorescences; B. Inflorescence (Cyathium); C. V.S. of Inflorescence; D. Male flower; E. Female flower; F. T.S. of ovary, and G. Floral diagram

3. *Drypetes roxburghii* (Syn. *Putranjiva roxburghii*). The wood is used during construction of house building and agricultural equipments.
4. *Gelonium multiflorum.* The wood is used for posts and rafters.
5. *Trewia nudiflora* (False white teak). The wood is used in tea boxes, match industry and other kinds of packing boxes.

F. **Ornamental plants**

The following plants are cultivated as ornamentals :

1. *Acalypha hispida* and *A. wilkesiana.*
2. *Codiaeum variegatum.*
3. *Euphorbia milii* [Syn. *E. splendens* (Crown-of-thorn)], *E. pulcherrima* [Syn. *Poinsettia pulcherrima* (Lalpata)], *E. tirucalli* (Milk-bush), etc.
4. *Jatropha gossypifolia, J. multifida, J. podagrica* etc.
5. *Pedilanthus tithymaloides* (Red bird-cactus).

33. Casuarinaceae

(She-Oak Family)

The family consists of 2 genera and 65 species and are found to grow from North East Australia to South East Asia of the Southern Hemisphere. It is a native of Australia. The genera are *Casuarina* and *Gymnostoma*. In India, the family is represented by only one plant, the *Casuarina equisetifolia* (Horse-tail tree). The plant grows well in the coast of Bay of Bengal and also cultivated in different parts of the country.

The type genus of the family is ***Casuarina.***

Systematic position

Bentham & Hooker		**Engler & Prantl**	
Division.	Angiospermia	Subdivision.	Angiospermae
Class.	Dicotyledons	Class.	Dicotyledoneae
Subclass.	Monochlamydeae	Subclass.	Archichlamydeae
Series.	Unisexuales	Order.	Verticilatae
Hutchinson		**Cronquist**	
Phylum.	Angiospermae	Division.	Magnoliophyta
Subphylum.	Dicotyledones	Class.	Magnoliopsida
Division.	Lignosae	Subclass.	Hamamelidae
Order.	Casuarinales	Order.	Casuarinales
Takhtajan		**Thorne**	
Division.	Magnoliophyta	Class.	Angiospermae
Class.	Magnoliopsida	Subclass.	Magnoliidae
Subclass.	Hamamelididae	Superorder.	Rosanae
Superorder.	Casuarinanae	Order.	Casuarinales
Order.	Casuarinales.		

Identifying characteristics

1. Plants **monoecious** and evergreen trees or shrubby habit with **xerophytic** habitat;
2. Internode branches have **ridges** and **furrows.**
3. **Leaves scaly** and develop in whorls at each node;
4. **Male flowers are borne in** catkin-like **spikes**;
5. **Each flower has 2 bracteoles, 2 perianth and 1 stamen**;
6. **Female flowers** are developed in **dense, ovoid or spherical heads**;
7. **Each female flower has 2 bracteoles, 2 carpels, syncarpous, 2 chambered** (posterior one empty and anterior one with 2 ovules);
8. Fruit **1-seeded nut.**

Casuarinaceae : Distinctive features

Common characteristics	Exceptions
Vegetative characteristics	
Plant. Plants are much-branched evergreen trees or shrubs of xerophytic habitat, monoecious or dioecious.	
Stem. Branched, internodes of branches are with ridges and furrows. The stomata present in the grooves are protected by long protruding hairs.	
Leaves. Leaves are small, scaly, linear to lanceolate and are connate basally forming a sheath surrounding the node.	

Contd.

Contd.

Common characteristics	Exceptions
Reproductive characteristics	
Inflorescence. Male flowers are developed in catkin-like spikes and female flowers are borne in dense, spherical or ovoid heads those develop at the end of short lateral branches.	
Male flowers. Male flowers are subtended by a pair of bracteoles. Each flower having 2 small perianth leaves and 1 stamen. Anther basifixed, dithecous and pollens are released by longitudinal splitting. Initially filaments are short but elongated during flowering.	
Female flowers. Female flowers are subtended by 1 bract and 2 bracteoles. Perianth absent and thus the female flowers are naked. Ovary superior, carpels 2, syncarpous, 2 chambered of which posterior 1 aborts and anterior one with 2 ovules, of which only one develop into seed. Style short and stigmas 2.	
Fruit. 1-seeded nuts covered by persistent woody bracteoles which separate on maturity. The fruits become aggregated into a dry woody multiple fruit.	
Seeds. Non-endospermic with flat cotyledons and straight embryo.	

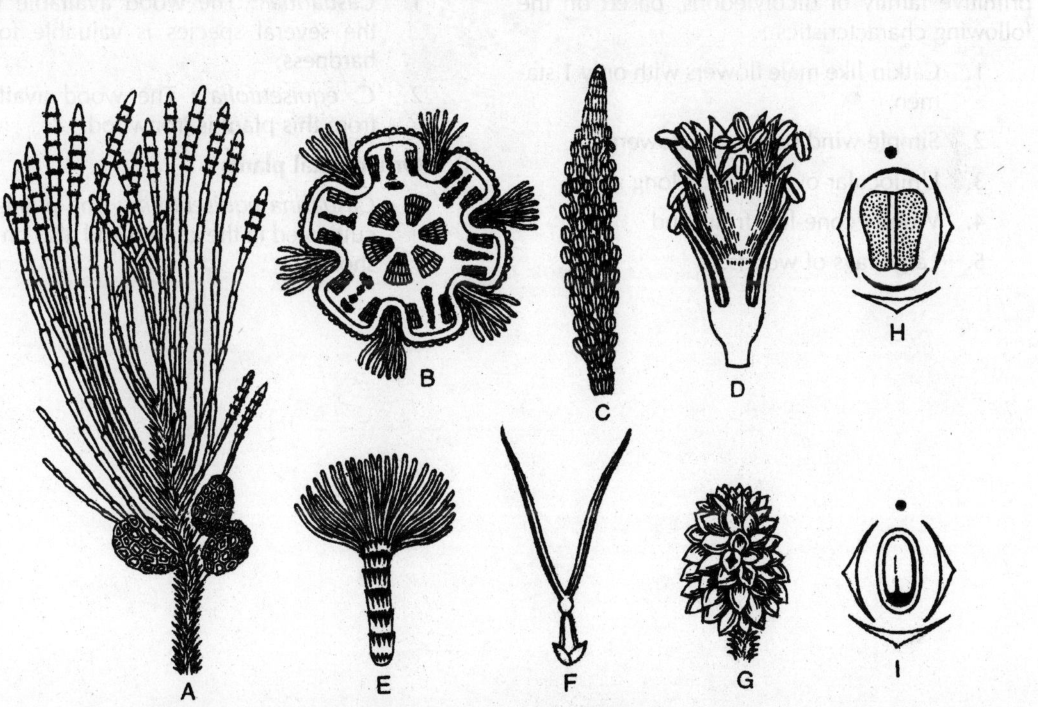

Fig. 4.150 : Casuarinaceae (*Casuarina equisetifolia*) : A. Portion of a flowering plant; B. T.S. of stem; C. Single spike of male flowers; D. Partly excised male flower; E. Female inflorescence; F. Female flower; G. Fruit; H. Floral diagram of male flower, and I. Floral diagram of female flower

Floral formula

Male flower : $Brl_2 \oplus ♂ P_2 A_1$

Female flower : $Br_1 Brl_2$ ○|○ ♀ $P_0 \underline{G}_{(2)}$

Systematics and phylogeny

The family Casuarinaceae resembles with Betulaceae in the structure and development of the pistil and ovary. The leaf and stem are highly adaptive but it is difficult to trace any affinity of this family with the Fagales. The habit of the plants under this family is related with *Ephedra* (a gymnosperm), suggesting the origin of angiospermous flower from *Ephedra*-like ancestor (Wettstein).

The family can readily be distinguished from others by its *Equisetum*-like jointed branches arranged in whorls, minute leaves and its woody cone-like fruits.

The family is placed under the order Unisexuales by Bentham and Hooker; Verticillatae by Engler and Prantl and Casuarinales by Hutchinson, Takhtajan, Cronquist, and Thorne.

Engler and Prantl considered it the most primitive family of dicotyledons, based on the following characteristics :

1. Catkin-like male flowers with only 1 stamen,
2. Simple wind pollinated flowers,
3. Unilocular ovary with 2 long styles,
4. Woody cone-like fruits and
5. Large rays of woods.

Rendle and Wettstein also considered Casuarinaceae to be the most primitive among the angiosperms and considered their derivation from *Ephedra*-like ancestor.

The other group of taxonomists like Bessey, Hutchinson, Cronquist, Takhtajan considered the family Casuarinaceae to be an advanced one derived from Hamamelidales. Based on comparative anatomy, Tippo (1938) considered the family Casuarinaceae as the highly advanced one derived from Hamamelidaceae. Later, Moseley, in 1948, (based on anatomy) also concluded that the Casuarinaceae are a specialised family of the angiosperms and not a primitive group. The morphological studies on flower suggest that it derived from Hamamelidaceae-like ancestors. The structural simplicity of flowers of this family is due to extreme reduction.

Common plants

1. *Casuarina equisetifolia* Forst.

Plants of economic importance

A. **Wood**

1. *Casuarina* : The wood available from the several species is valuable for its hardness;
2. *C. equisetifolia* : The wood available from this plant is ironwood.

B. **Ornamental plants**

1. *Casuarina equisetifolia* : The plants are cultivated in the garden and also on seashores.

MONOCOTYLEDONS

1. Orchidaceae

(Orchid Family)

It is a large family consisting of 788 genera and more than 18,500 species and grows abundantly in the tropical regions. It is less common in the Arctic regions. In India, the family is represented by 128 genera and 881 species, found to grow commonly in Eastern Himalayas, Khasia hills and also in Western Ghats.

The type genus of the family is ***Orchis***.

Systematic position

Bentham & Hooker		**Engler & Prantl**	
Division.	Angiospermia	Subdivision.	Angiospermae
Class.	Monocotyledons	Class.	Monocotyledoneae
Series.	Microspermae	Order.	Microspermae
Hutchinson		**Cronquist**	
Phylum.	Angiospermae	Division.	Magnoliophyta
Subphylum.	Monocotyledones	Class.	Liliopsida
Division.	Corolliferae	Subclass.	Lilidae
Order.	Orchidales	Order.	Orchidales
Takhtajan		**Thorne**	
Division.	Magnoliophyta	Class.	Angiospermae
Class.	Liliopsida	Subclass.	Liliidae
Subclass.	Liliidae	Superorder.	Lilianae
Superorder.	Lilianae	Order.	Orchidales
Order.	Orchidales.		

Identifying characteristics

1. Plants are **perennial herbs**; epiphyte, saprophyte or terrestrial.
2. Flowers are **bisexual, epigynous** and **medianly zygomorphic**, flowers resupinate during anthesis.
3. Perianth 6 in two whorls, of which the posterior median member of the inner whorl is larger in size forming **labellum**.
4. Stamens 1 or 2, pollen grains are present in **pollinia**.
5. Filament is united with the style and forms a column, the **gynandrium**.
6. **Carpel** 3, syncarpous, **ovary inferior**, one-

Orchidaceae : Distinctive features

Common characters	Exceptions
Vegetative characters	
Plant. Perennial herbs, terrestrial or a few are saprophyte associated with mycorrhizal fungus.	Climber – *Vanilla*
Root. The main root is always absent and its place is occupied by adventitious roots developing from the nodes. Roots are of 3 types : aerial, tuberous, and normal cylindrical earth roots. Aerial roots are with velamen.	
Leaves. Usually alternate with closed sheath, thick.	
Reproductive characters	
Inflorescence. Raceme or spike.	
Flower. Zygomorphic, bisexual, epigynous and usually very showy bracteate.	
Perianth. Free, 6 members present in two whorls i.e., 3+3; outer 3 consists of more or less similar sepals and out of inner 3, 2 lateral petals are alike and the posterior median petal is larger and forms the labellum or lip. The labellum is provided with a spur, various in shape and posterior in position. Later on, the position of labellum becomes anterior by the twisting of the ovary or pedicel through 180°.	

Contd.

Contd.

Common characters	Exceptions
Androecium. Stamens 1 or 2; filament adheres with the style (gynandrous) and form a column or gynandrium; pollens remain in pollinium; 2-8 in number. Each pollinium bears a code (caudicle or translator), which ends in a sticky discs or glands (retinacula or viscidia).	Caudicle absent – *Malaxis.* Pollen granular – *Cephalanthera.*
Gynoecium. Carpel 3, syncarpous, ovary inferior, twisted, 1-chambered with many ovules remain on three biseriate placentas, placentation parietal; stigma 3, of which 2 are fertile and the other 1 is sterile and forms *rostellum.*	All 3 stigmas are fertile – *Paphiopedilum, Cypripedium* etc.
Fruits. Capsule.	
Seeds. Minute, numerous, light in weight (apx. 4 μg), exalbuminous, embryo undifferentiated.	

chambered with many ovules on parietal placentation.

7. Stigma 3, of which 2 are fertile and the other 1 is sterile forming **rostellum**.
8. Fruits **capsule** and are **without endosperm**.
9. **Seeds numerous, minute.**

Floral formula

$\circ|\circ$ ⚥ $P_{(3+3)}\ A_{1\ or\ 2}\ \overline{G}_{(3)}$

Phylogeny and systematics

The family shows close relationship with Musaceae, Zingiberaceae and also with Liliaceae.

Bentham and Hooker placed the natural order Orchidaceae at the beginning as series 1 Microspermae along with Hydrocharitaeae and Burmanniaceae under the class Monocotyledons due to the presence of small seeds. Engler and Prantl placed the family Orchidaceae at the end in their 11th series Microspermae under the class Monocotyledoneae. Hutchinson placed the family under a more or less advanced order Orchidales (a monotypic group) next to Burmanniales, division Corolliferae of subphylum Monocotyledones. Takhtajan (1980) placed it under Orchidales next to Burmanniales, subclass Liliidae of class Liliopsida (Monocotyledons).

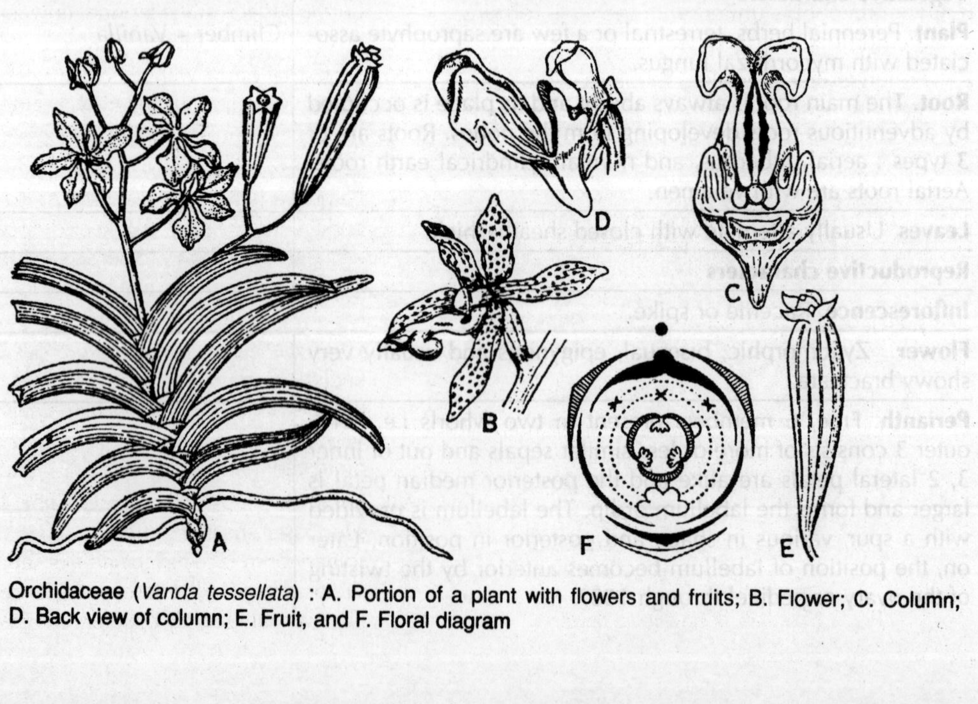

Fig. 4.151 : Orchidaceae (*Vanda tessellata*) : A. Portion of a plant with flowers and fruits; B. Flower; C. Column; D. Back view of column; E. Fruit, and F. Floral diagram

Cronquist (1981) placed the family under the last order Orchidales, class Liliopsida.

The flowers of Orchidaceae are highly specialised. It was considered to have been derived from liliaceous or amaryllidaceous stock. Hutchinson (1959) considered the genera like *Neuwiedia* and *Apostasia* as the connecting link between Liliaceae and orchids.

Consideration of the family Orchidaceae as the highest evolved one among Monocotyledons

The family Orchidaceae is considered by some taxonomists to be the highly advanced family among the Monocotyledons for the following reasons :

1. The family consists of about 788 genera and 18,500 species, are mostly found in the damp tropical and temperate regions of the world;
2. The plants are strictly herbaceous and many of them are epiphytic in nature;
3. Flowers are zygomorphic and epigynous;
4. The inner posterior member of the perianth forming the *labellum,* is one of the highest morphological adaptations for insect pollination. The genus *Ophrys* found in Algeria shows the labellum like a *female bee,* is no doubt a highly adaptive feature for insect pollination;
5. Number of stamens is reduced to one or two;
6. The filament is united with the style forming a column known as *gynandrium*;
7. Pollen grains are united into a mass and present in a special structure the, *pollinium* which checks the wastage of pollens during cross pollination.
8. Out of three stigmas two are receptive and the third one is sterile and forms a structure called *rostellum,* an adaptation for cross-pollination by insects.
9. The seeds are very minute with undifferentiated embryo.

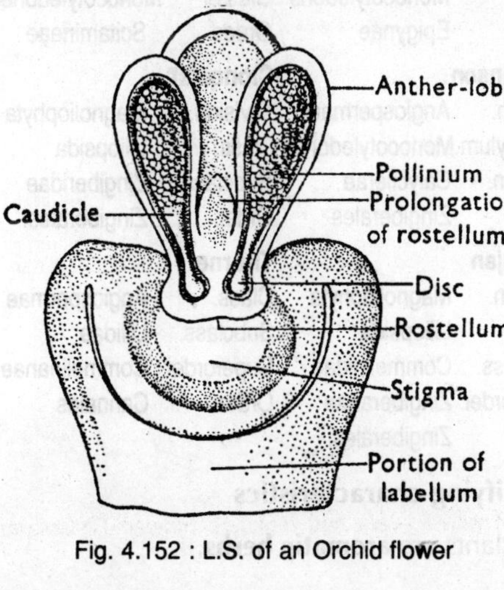

Fig. 4.152 : L.S. of an Orchid flower

Common plants

1. *Acampe papillosa* Lindl.;
2. *Cypripedilium elegans* Reichb;
3. *Dendrobium nobile* Lindl.;
4. *Didymoplexis pallens* Griff.;
5. *Geodorum densiflorum* (Lamk.) Schlecht.;
6. *Habenaria digitata* Lindl.;
7. *Orchis spathulata* Reichb. f ;
8. *Vanda tessellata* (Roxb.) ex G. Don (Syn. *V. roxburghii* R. Br.) etc.

Key to Genera

I. Terrestrial herbs with underground rhizomes; leaves elliptical, plicate, sheaths forming a pseudostem; lip white with pink lines or streaked with purple or red .... *Geodorum.*

II. Terrestrial herbs, leafless, saprophytic, with tuberous roots; inflorescence a scanty raceme of white flowers .... *Didymoplexis.*

III. Terrestrial 1-leaved tuberous herbs; leaf broadly cordate or orbicular, strongly plicate, lying on the surface of the ground; flowers on scapes arising from the tuber .... *Pogonia.*

IV. Terrestrial herbs with succulent glabrous stem; leaves membranous, often withering at flowering times; flowers small, in racemes or spikes .... *Zeuxine.*

V. Epiphytic herbs with leafy stems; leaves thickly coriaceous or fleshy, flat, keeled or terete; flowers large often showy in dense racemes or sometimes solitary; petals like sepals; lip large, base saccate or spurred; caudicle very broad ... *Vanda.*

VI. Epiphytic herbs; leaves coriaceous, flat, often keeled; flowers small in racemes, corymbs, umbels or panicles; petals like sepals; lip small, sessile, forms a cylindric or saccate spur; caudicle very narrow .... *Saccolabium.*

VII. Epiphytic herbs with pseudo-bulbs; leaves sessile, bases sheathing; flowers large and showy, solitary or in racemes; petals like sepals; lip sessile or clawed at the base; anther 2-celled; pollinia 4, usualy slightly coherent ... *Dendrobium.*

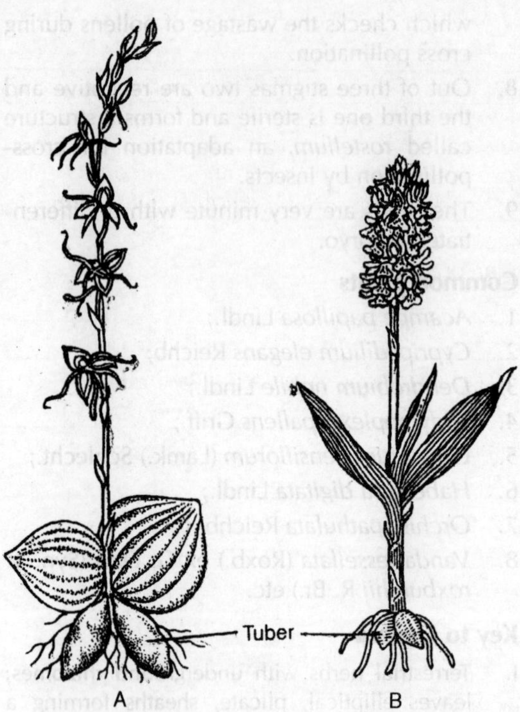

Fig. 4.144 : A. *Habenaria diphyla*, and B. *Orchis mascula*

VIII. Epiphytic shrub; stem terete or angled; leaves coriaceous or fleshy or 0; pollinia granular ... . *Vanilla.*

Plants of economic importance

Different plants of this family are used variously. These are :

A. **Food**

1. *Platenthera sussannae* (Syn. *Habenaria sussannae*). The tubers of this plant are edible.
2. *Orchis latifolia* and *Cymbidium aloifolium*. The 'salep' of commerce is extracted from the dry tuberous roots and is used as a farinaceous food. It is also used in silk industry.

B. **Flavouring material**

1. *Vanilla fragrans* (Syn. *V. planifolia*) (Vanilla). The dry fruits used to extract 'vanilla' of commerce. It is used for flavouring of various food products like chocolate and confectionery.

C. **Medicinal**

1. *Eulophia epidendrae*. The tubers of this plant are used as vermifuge.
2. *Geodrum densiflorum*. The roots are used as insecticide.
3. *Platenthera sussannae*. The tubers are used to cure the bullae on palms.
4. *Vanda tessellata (Syn. Vanda roxburghii).* The roots are used in the treatment of rheumatism and scorpion-bite.

D. **Dye**

1. *Calanthe veratrifolia*. The leaves contain a glycoside, which on hydrolysis yield a blue dye known as 'indigo blue'.

E. **Ornamental plants**

Different species of the following genera are cultivated for their beautiful and attractive colouration. These are :

1. *Aerides.*
2. *Bletilla.*
3. *Cattleya.*
4. *Coelogyne.*
5. *Cypripedium.*
6. *Cymbidium.*
7. *Dendrobium.*
8. *Epidendrum.*
9. *Epipactis.*
10. *Paphiopedium* and
11. *Vanda.*

2. ZINGIBERACEAE
(Zinger Family)

The family is represented by about 46 genera and 1275 species, distributed throughout the tropics and subtropics.

The type genus of this family is ***Zingiber***.

Systematic position

Bentham & Hooker		**Engler & Prantl**	
Division.	Angiospermia	Subdivision.	Angiospermae
Class.	Monocotyledons	Class.	Monocotyledoneae
Series.	Epigynae	Order.	Scitamineae
Hutchinson		**Cronquist**	
Phylum.	Angiospermae	Division.	Magnoliophyta
Subphylum.	Monocotyledones	Class.	Liliopsida
Division.	Calyciferae	Subclass.	Zingiberidae
Order.	Zingiberales	Order.	Zingiberales.
Takhtajan		**Thorne**	
Division.	Magnoliophyta	Class.	Angiospermae
Class.	Liliopsida	Subclass.	Liliidae
Subclass.	Commelinidae	Superorder.	Commelinanae
Superorder.	Zingiberanae	Order.	Cannales
Order.	Zingiberales.		

Identifying characteristics

1. Plants are **aromatic herbs**.

Zingiberaceae : Distinctive features

Common characteristics	Exceptions
Vegetative characteristics	
Plant. Tall and perennial herbs with rhizomatous stem.	Rhizome tuberous – *Curcuma*. Stem aerial – *Globba bulbifera*.
Leaves. Alternate or opposite, simple, radicle or cauline, with sheathing leaf bases. Upper part of sheath bears ligular outgrowth.	
Reproductive characteristics	
Inflorescence. Spike (*Curcuma*), raceme (*Globba*) or dense head (*Costus*).	Flowers solitary – *Gastrochilus*.
Flower. Bracteate, medianly zygomorphic, bisexual, trimerous and epigynous.	
Perianth. Tepals 6 (3+3), usually distinguished into outer calyx and inner corolla.	
Androecium. Stamens 6 in 2 trimerous whorls (in outer whorl-anterior member is suppressed and the other 2 form petaloid staminodes; in inner whorl–posterior median member is fertile and the other 2 form petaloid staminode, those united together and form labellum, the landing stage of the insect). The fertile stamen has a slender and deeply grooved filament and 2-celled anther dehiscing vertically.	
Gynoecium. Carpel 3, ovary inferior, 3-celled with many ovules in each cell on axile placentation.	Ovary 1 celled – *Globba*. Placentation parietal – *Globba*.
Fruits. Capsule or berry.	
Seeds. Generally arillate, round or angular with copious mealy or hard endosperm and straight embryo.	

2. Leaves **simple** and **ligulate**.
3. Inflorescence solitary, racemose or panicle.
4. Perianth 6 (3+3), differentiated into calyx and corolla.
5. **Stamens 6, of which 1 is fertile and the rest 5 form either petaloid staminodes** or **abortive**. Two petaloid staminodes of the inner whorl united together and form **labellum**.
6. Fruit **generally capsule**.

Floral formula

$\circ|\circ$ ⚥ $P_{(3+3)}\ A_6\ \bar{G}_{(3)}$

Common plants

1. *Costus speciosus* (Koenig ex Retz.) Smith;
2. *Curcuma aromatica* Salisb.;
3. *C. amada* Roxb.;
4. *Globba bulbifera* Roxb.;
5. *Hedychium coronarium Koen.*;
6. *Kaempferia rotunda* L.;
7. *Zingiber officinale* Rosc.

Key to Genera

I. Ovary 1-celled; stem leafy; flowers yellow; herbs with creeping rhizomes and erect stems; buds often replaced by bulbils .... *Globba*.

II. Ovary 3-celled with axile placentation :

1. Lateral staminodes broad
   - i. Connective not spurred at the base; corolla tube long and slender
     - (a) Filaments short; connectives broad crested; stem short with a few leaves, or no aerial stem ... .*Kaempferia*.

(b) Filaments long; connectives not crested; stem elongated, leafy ... .*Hedychium.*

ii. Connective spurred at the base; stemless herbs; leaves often very large; flowering spike surmounted by a coma of coloured bracts ... *Curcuma.*

2. Lateral staminodes small or 0 or adnate to the lip, Flowering spikes dense cone-like, but not surmounted by a coma of bracts –

i. Filaments short, not petaloid .... *Zingiber.*

ii. Filaments petaloid bearing the anther on its middle ....*Costus.*

iii. Filaments short, with or without petaloid crest .... *Amomum.*

3. Flowers in racemes or panicles at the top of tall leafy stems .... *Alpinia.*

Plants of economic importance

The members of this family are useful for various purposes such as spices, ornamentals and also have other uses :

A. **Spices**

Some members which yield spices are :

1. *Amomum aromaticum* (Bengal Cardamon, Elach, Elachi).
2. *A. subulatum* (Nepal Cardamon).
3. *Alpinia officinarum.*
4. *A. galangal* Sw.
5. *Curcuma amada* (Amada).
6. *C. anguistifolia* (East Indian arrowroot, Tikhur).
7. *C. longa* (Turmeric, Halud).
8. *Elettaria cardamomum* var. *minor* (Chota-elachi).
9. *Zingiber officinale* (Ginger, Ada).

B. **Ornamental plants**

Some members cultivated as ornamentals are :

1. *Alpinia speciosa* (Shell flower).
2. *Cautleya spicata.*
3. *Globba racemosa.*
4. *Hedychium coronarium* and *H. gardnerianum* are commonly called ginger lily.
5. *Kaempfera rotunda* (Bhui champa).
6. *Roscoea alpina* and *R. purpurea.*

C. **Plants of other uses**

1. *Curcuma longa* (Turmeric). The rhizome yields a yellow dye used in cooking (kitchen spice) and also in cosmetics.
2. *C. zedoaria* (Zedoary). The tubers yield zedoary used as a tonic or perfume.

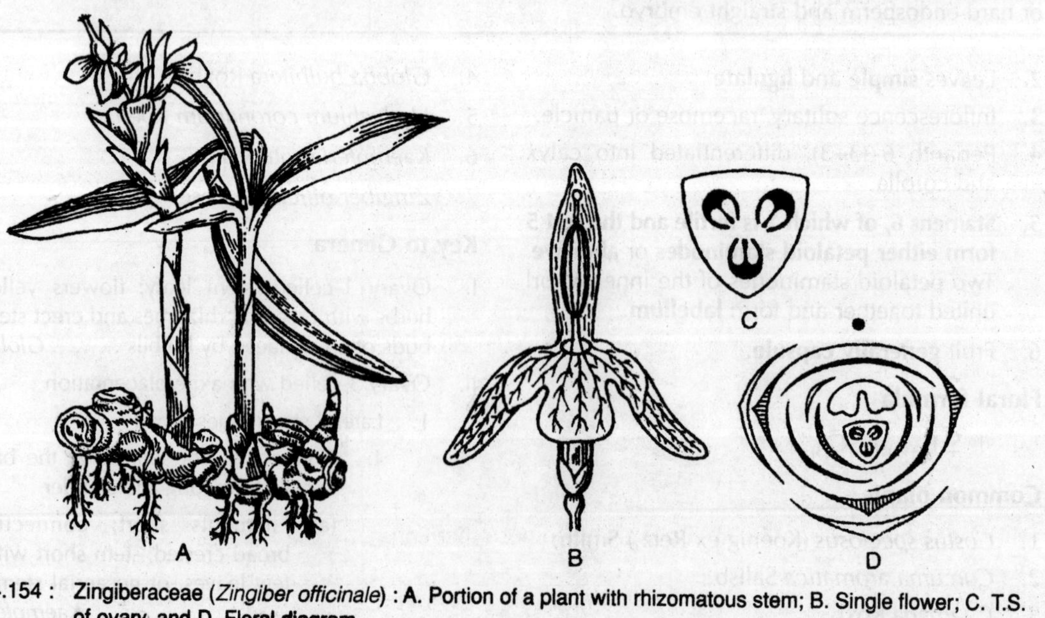

Fig. 4.154 : Zingiberaceae (*Zingiber officinale*) : A. Portion of a plant with rhizomatous stem; B. Single flower; C. T.S. of ovary, and D. Floral diagram

3. *Hedychium spicatum* (Kafur-kachri). The plant is the source of 'Kafur-kachri', the ingredient of scented powder.
4. *Zingiber officinale* (Ada). The ginger oil obtained from the rhizome is used in perfumery.

3. CANNACEAE

(Canna Family)

The family is represented by a single genus *Canna* (monotypic) with about 10 species, distributed in tropics and substropics.

The type genus of the family is ***Canna***.

Systematic position

Bentham & Hooker		**Engler & Prantl**	
Division.	Angiospermia	Subdivision.	Angiospermae
Class.	Monocotyledons	Class.	Monocotyledoneae
Series.	Epigynae	Order.	Scitamineae

Hutchinson		**Cronquist**	
Phylum.	Angiospermae	Division.	Magnoliophyta
Subphylum.	Monocotyledones	Class.	Liliopsida
Division.	Calyciferae	Subclass.	Zingiberidae
Order.	Zingiberales	Order.	Zingiberales.

Takhtajan		**Thorne**	
Division.	Magnoliophyta	Class.	Angiospermae
Class.	Liliopsida	Subclass.	Liliidae
Subclass.	Commelinidae	Superorder.	Commelinanaea
Superorder.	Zingiberanae	Order.	Cannales
Order.	Cannales.		

Identifying characteristics

1. Plants are **perennial herbs.**
2. Leaves **simple without ligule.**
3. Inflorescence **scorpioid cyme.**
4. Flowers **bisexual, zygomorphic, epigynous** and **asymmetrical.**
5. Perianth 6 (3+3), outer whorl sepaloid and

Cannaceae : Distinctive features

Common characteristics	Exceptions
Cannaceae : Vegetative characteristics	
Plant. Perennial aromatic herbs with creeping horizontal rhizomatous stem.	
Leaves. Alternate, simple, large, lanceolate with distinct midrib and sheathing leafbase without ligule.	
Reproductive characteristics	
Inflorescence. Spike or panicle.	
Flower. Zygomorphic, bisexual, bracteate, large and coloured (showy).	
Perianth. Six members are present in 2 whorls; outer whorl sepaloid, free, imbricate and persistent and inner whorl petaloid, united at the base, imbricate.	
Androecium. Stamens 6 in 2 trimerous whorls (in outer whorl– anterior member is suppressed and the other 2 form petaloid staminodes; in inner whorl– $1/2$ of the posterior median member is fertile and rest $1/2$ form the petaloid staminode and the other 2 form petaloid staminode, of which one is larger and form labellum, the landing stage of the insect), sometimes reduced to 4, all briefly basally connate.	
Gynoecium. Carpel 3, syncarpous, ovary inferior, warty, 3-chambered with many ovules in each chamber in 2 series on axile placentation; style petaloid, stigma oblique.	
Fruits. Capsule with a warty or weakly spinose pericarp, crowned by persistent calyx.	
Seeds. Many, rounded, with hard perisperm, scanty endosperm and a straight embryo; often arillate.	

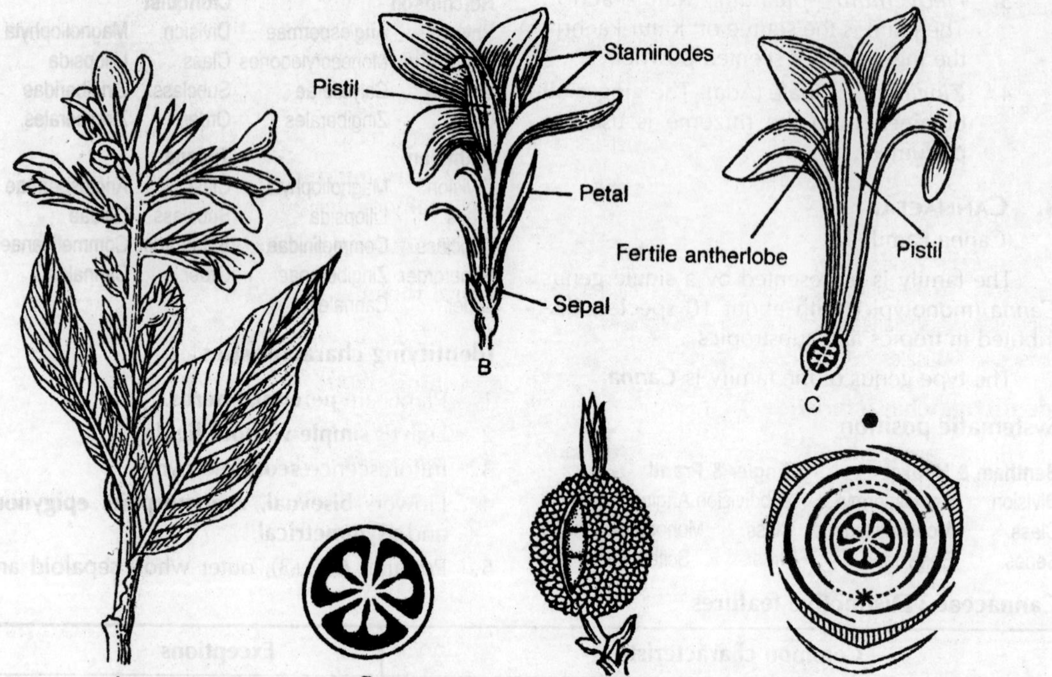

Fig. 4.155 : Cannaceae (*Canna indica*) : A. Portion of a flowering plant; B. Single flower; C. V.S. of a flower; D. T.S. of ovary; E. Single fruit with warty wall and persistant calyx, and F. Floral diagram

free but inner whorl petaloid and connate at the base.

6. **Stamens 6, of which posterior median member of the inner whorl forms petaloid staminode with ½ anther lobe fertile.** Out of the remaining 5, one (anterior member of the outer whorl) becomes abortive and the rest form petaloid staminode.
7. **Fruit a loculicidal capsule with warty wall crowned by persistent calyx.**

Floral formula

Asym. ⚥ $P_{(3+3)}$ A_6 $\overline{G}_{(3)}$

Common plants

1. *Canna indica* L.
2. *C. edulis* Ker-Gawl.
3. *C. orientalis* Rosc.

Plants of economic importance

This family is of limited economic importance. Some members are important for their ornamental value.

A. **Ornamental plants**

1. *Canna indica* and *C. orientalis.* Both the plants and their varieties are cultivated in the garden for beautification.

B. **Food**

1. *Canna edulis.* Rhizome is used in the preparation of starchy food.

C. **Vegetables**

1. *Canna orientalis.* The tuberous rhizome is used as vegetables.

4. MUSACEAE
(Banana Family)

The family is represented by about 5 genera and 150 species, widely distributed in tropics.

The type genus of this family is ***Musa.***

Systematic position

Bentham & Hooker		Engler & Prantl	
Division.	Angiospermia	Subdivision.	Angiospermae
Class.	Monocotyledons	Class.	Monocotyledoneae
Series.	Microspermae	Order.	Scitamineae

Hutchinson		**Cronquist**	
Phylum.	Angiospermae	Division.	Magnoliophyta
Subphylum.	Monocotyledones	Class.	Liliopsida
Division.	Calyciferae	Subclass.	Zingiberidae
Order.	Zingiberales	Order.	Zingiberales.
Takhtajan		**Thorne**	
Division.	Magnoliophyta	Class.	Angiospermae
Class.	Liliopsida	Subclass.	Liliidae
Subclass.	Commelinidae	Superorder.	Commelinanae
Superorder.	Zigiberanae	Order.	Cannales
Order.	Musales.		

Identifying characteristics

1. Plants are generally **perennial herbs** with **rhizomatous stem**.
2. Leaves simple with sheathing leaf base, those generally aggregated together and form a **false stem**.
3. **Flowers** are **epigynous, zygomorphic** and generally **bisexual**, develop on **mixed spadix inflorescence**.
4. Perianth 6 in two whorls, **petaloid**.
5. Stamens 6, 5 fertile and 1 sterile or wanting.
6. **Ovary inferior, carpel 3**, 3-chambered with many ovules on axile placentation.
7. **Fruit berry** or **capsule**.

Floral formula

$\%\ ⚥\ P_{(3+3)}\ A_{6 \text{ or } 5}\ \overline{G}_{(3)}$

Common plants

1. *Heliconia* sp.
2. *Musa balbisiana* L.;
3. *Ravenala madagascariensis* Gmel. – Panthopadap (Traveller's tree);
4. *Strelitzia reginae* Banks – Bird-of-paradise.

Musaceae : Distinctive features

Common characteristics	Exceptions
Vegetative characteristics	
Plant. Large perennial herbs with rhizomatous stems.	Tree – *Ravenala madagascariensis.*
Leaves. Simple, large, glabrous, lanceolate or oblong, entire with thick and prominent midrib, and stiff sheathing leaf bases. The sheathing leaf bases arranged spirally and form the pseudostem or false-stem.	
Reproductive characteristics	
Inflorescence. *Mixed spadix* (Individual flowers are arranged cymosely and is subtended by large spathe. The cymose inflorescences are arranged racemosely and finally they are covered by spathe), which develops singly and terminally from the rhizome and emerges at the top of the false stem.	
Flower. Zygomorphic, usually bisexual, epigynous, commonly large and showy.	
Perianth. Tepals 6 (3+3), petaloid or differentiated into calyx and corolla, united or free.	
Androecium. Stamens 6, of which the posterior 1 is rudimentary (staminode) or wanting, filaments free and filiform, anthers linear, 2-celled and introrse.	6 stamens are fertile – *Ravenala.* 6 fertile but posterior one is smaller than others – *Ensete.*
Gynoecium. Carpels 3, syncarpous; ovary inferior, 3-celled with numerous anatropous ovules in each cell on axile placentation; style filiform and stigma 3-lobed.	
Fruits. Berry.	Capsule – *Ravenala.*
Seeds. Numerous, with thick and hard testa containing mealy endosperm and straight embryo.	

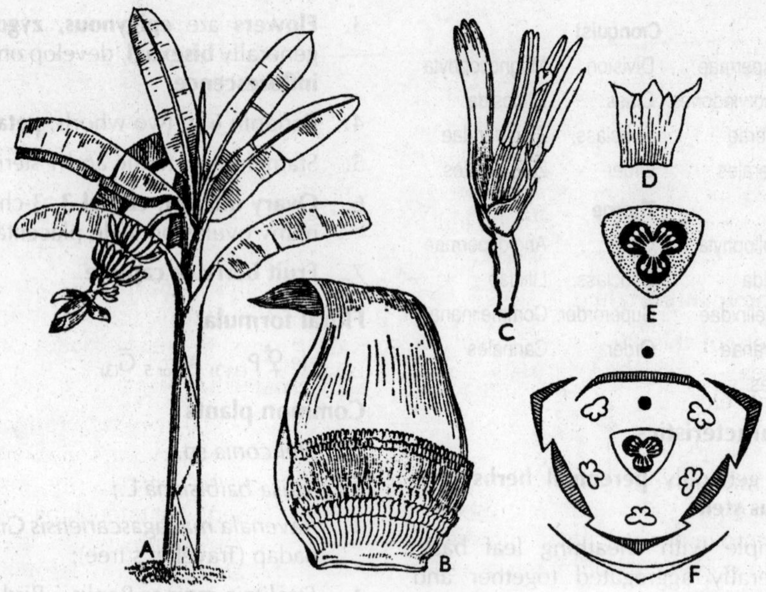

Fig. 4.156 : Musaceae (*Musa balbisiana*) : A. Single plant, B. A spathe enclosing flowers arranged in two rows; C. Single flower; D. Calyx; E. Ovary in T.S., and F. Floral diagram

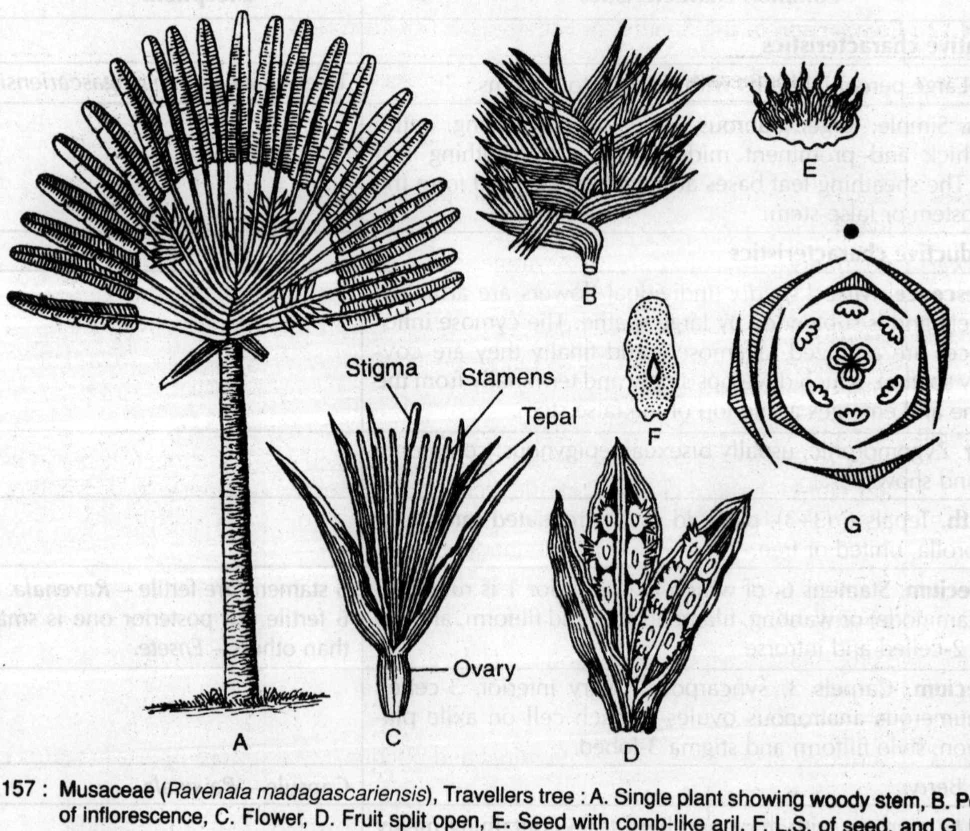

Fig 4.157 : Musaceae (*Ravenala madagascariensis*), Travellers tree : A. Single plant showing woody stem, B. Portion of inflorescence, C. Flower, D. Fruit split open, E. Seed with comb-like aril, F. L.S. of seed, and G. Floral diagram

Key to Genera

I. Sepals free; petals long and narrow; stamens 6 perfect; stem tree-like, marked with annular leaf-scars; leaves distichous, very large; fruit capsule .... *Ravenala.*

II. Sepals united into a split spathaceous tube; leaves spiral, oblong; stamens 5 perfect; fruit a berry .... *Musa.*

Plants of economic importance

The members of this family are useful for various purposes — vegetables, fruits, fibre, ornamental plants etc.

A. **Vegetables**

1. *Musa balbisiana* L. The unripe fruits, inflorescence stalk and inflorescence (after removing bracts, style with stigma and part of perianth) are taken as vegetables.

B. **Fruits**

1. *Musa paradisiaca* L. subsp. *sapientum* L. commonly called Martaman, Champa, Kanthali etc. The ripe berries are used as fruits.
2. *M. acuminata* (Singapuri, Basrai, Red banana etc.). The ripe berries are used as fruits.

C. **Fibre**

1. *Musa textilis* Nee. The sheathing leaf bases yield 'Manila hemp' or 'Abaca', the fibre is used for cordage and also used to prepare Abaca cloth.

D. **Ornamental plants**

1. *Ravenala madagascariensis* J. F. Gmel. The watery latex present in the aerial stem is used by the travellers in the deserts for quenching their thirst, collect by removing the leaves. The plant is also grown for beautification in the garden.
2. *Heliconia* sp. and *Strelitzia reginae* are also costly ornamental plants.

Table 4.12 : Comparison of the families of Scitamineae (Zingiberales)

Characters	Zingiberaceae	Cannaceae	Musaceae
1. Plant	Aromatic herbs with rhizome.	Herbs with rhizomes.	Large herbs with rhizome, rarely trees.
2. Leaves	Leaves simple and with ligule.	Leaves simple and without ligule.	Leaves simple, large and without ligule.
3. Flower	Medianly zygomorphic, bisexual.	Asymmetrical, bisexual.	Zygomorphic, bi- or unisexual.
4. Perianth	Six (3+3), generally outer sepaloides and inner petaloides.	Six (3+3), outer sepaloides and inner petaloides.	Six, petaloid, similar or differentiated into calyx and corolla.
5. Stamens	Stamens 6 of which 1 stamen is fertile and the rests are either abortive or form petaloid staminode.	Stamens 6, of which $1/2$ anther lobe of 1 stamen is fertile and rests are either abortive or form petaloid staminode.	Stamens 6, of which the posterior 1 is rudimentary (staminode) or wanting and rest 5 are fertile.
6. Ovary	3-celled with many ovules on axile placentation.	3-celled with 2 series of ovules on axile placentation.	Usually 3-celled with many ovules on axile placentation.
7. Style	Long, slender which passes through the furrow of the fertile stamen.	Style petaloid.	Simple, not petaloid.
8. Stigma	Variously developed.	Terminal and oblique.	More or less lobed or capitate.
9. Fruit	Capsule or berry.	A warty capsule crowned by persistent calyx.	Berry or capsule.

E. **Other uses**

1. *Musa acuminata* (Singapuri, Basrai, Red banana etc.) and other species. Half riped banana is used in the preparation of banana-chips in South India.
2. *Musa ensete*. The rhizome is rich in starch and used as food after cooking.
3. *Musa balbisiana* and other species. Leaves of banana plants are used as dish and both leaf and elongated sheathing leaf bases are also used in ritual purposes.

Phylogeny and systematics

Bentham and Hooker placed the natural order Scitaminae under the series Epigyne of the class Monocotyledons; which includes the tribes (family by many taxonomists) Musaceae, Zingiberaceae and Cannaceae. They have included the Marantaceae within the Cannaceae. Later on, most of the taxonomists divided the Bentham and Hooker's Scitamineae into 4 distinct families, Musaceae, Zingiberaceae, Cannaceae and Marantaceae. Engler and Prantl placed these families under the order Scitamineae just before the order Microspermae, due to the elaboration of flowers, zygomorphy and epigyny. However, Hutchinson splitted the family Musaceae into separate families like Musaceae, Strelitziaceae and Lowiaceae. He placed all the families under a new taxon Zingiberales in place of Scitamineae. So, the order Zingiberales consists of 6 families Musaceae, Strelitziaceae, Lowiaceae, Zingiberaceae, Cannaceae and Marantaceae was placed next to Bromeliaceae under the division Calyciferae of subphylum Monocotyledones.

Engler and Prantl included *Ravenala, Strelitzia* and *Heliconia* in the family Musaceae, but Hutchinson (1959, 69) placed them in separate family Strelitziaceae. Later, Takhtajan (1969, 80) and Cronquist (1968, 81) placed *Strelitzia* and *Heliconia* in two distinct families Strelitziaceae and Heliconiaceae.

5. Amaryllidaceae

(Amaryllis Family)

The family is represented by about 50 genera and 870 species, distributed mainly in the tropical and subtropical regions of the world. In India, the family consists of 9 genera and about 60 species — commonly found mainly in Himalayan regions.

The type genus of the family is ***Amaryllis***.

Systematic position

Bentham & Hooker		**Engler & Prantl**	
Division.	Angiospermia	Subdivision.	Angiospermae
Class.	Monocotyledons	Class.	Monocotyledoneae
Series.	Epigynae	Order.	Liliales.
Order.	Amaryllidales		
Hutchinson		**Cronquist**	
Phylum.	Angiospermae	Division.	Magnoliophyta
Subphylum.	Monocotyledones	Class.	Liliopsida
Division.	Corolliferae	Subclass.	Liliidae
		Order.	Liliales.
Takhtajan		**Thorne**	
Division.	Magnoliophyta	Class.	Angiospermae
Class.	Liliopsida	Subclass.	Liliales
Subclass.	Liliidae	Superorder.	Lilianae
Superorder.	Lilianae	Order.	Orchidales
Order.	Amaryllidales.		

Identifying characteristics

1. Commonly perennial **herbs with bulbs**.
2. Inflorescence **commonly umbellate cyme**.
3. Flowers **bisexual, epigynous, actinomorphic** or **medianly zygomorphic**.

Amaryllidaceae : Distinctive features

Common characteristics	Exceptions
Vegetative characteristics	
Plant. Perennial herbs with bulbous or sometimes with rhizomatous stems.	Tree – *Fourcroya*.
Leaves. Distichous, radical, simple.	
Reproductive characteristics	
Inflorescence. Umbellate cyme, subtended by spathe-like bracts and borne on a leafless scape.	Flowers solitary – *Zephyranthes* and many species of *Galanthus*.

Contd.

Contd.

Common characteristics	Exceptions
Flower. Actinomorphic or zygomorphic, bisexual, hypogynous.	Unisexual – *Smilax*. Zygomorphic – *Anigozanthon*.
Perianth. Tepal 6 (3+3), petaloid, united below into a short or long tube, rarely free; imbricate or valvate.	Tepals free – *Leucojum, Galanthus*. Gamophyllous with a corona – *Pancratium, Narcissus* and some species of *Crinum*.
Androecium. Stamens 6 in two whorls, epiphyllous, anther 2-celled, introrse and basifixed or versatile and dehiscing vertically.	Stamens 12 to 18 – *Gethyllis*. Filaments united at the base and form staminal cup – *Hymenocallis, Eucharis* and *Pancratium*.
Gynoecium. Carpel 3, syncarpous; ovary inferior, 3-celled with many anatropous ovules are arranged in 2 series on axile placentation. Style slender with 3-lobed or capitate stigma.	Ovary superior – *Tribonanthes*. Ovary 1-celled with 3 parietal placentae – *Alstroemeria, Leontochir*.
Fruits. Usually capsule or rarely berry.	
Seeds. Few to many; endosperm fleshy and embryo small.	

4. Perianth 6 (3+3), corona often present.
5. **Stamens 6, epipetalous**; anthers versatile and introrse.
6. Ovary **inferior**, carpel 3, **syncarpous**, 3-chambered.
7. Fruit **capsule** but rarely berry.

Floral formula

○|○ or $\oplus$ ⚥ $\widehat{P_{(3+3)}\ A_{6(3+3)}}\ \bar{G}_{(3)}$

Phylogeny and systematics

The family Amaryllidaceae is closely related to Liliaceae.

Bentham and Hooker placed the family Amaryllidaceae in the series Epigynae under the class Monocotyledons. Hutchinson placed Amaryllidaceae under a separate monotypic order Amaryllidales near to Liliales within the subphylum Monocotyledones. On the other hand, Engler

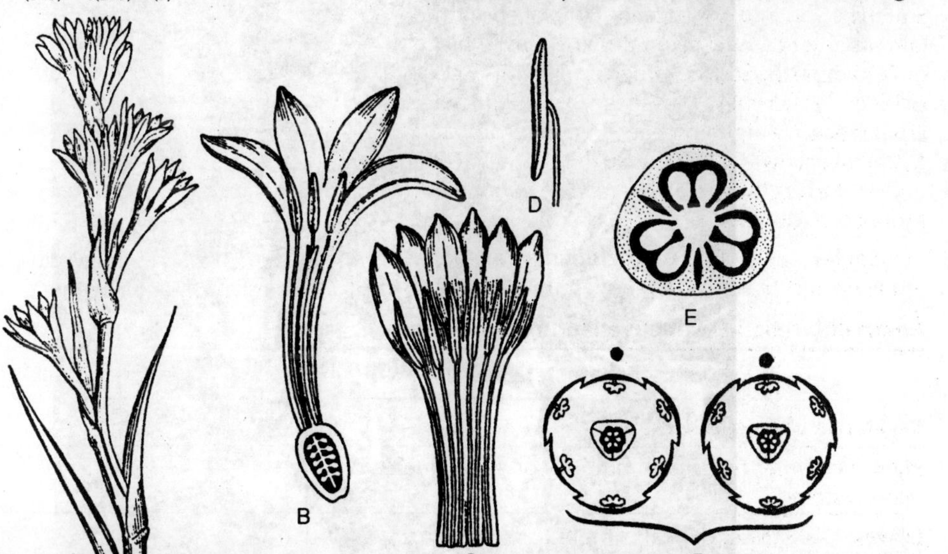

Fig. 4.158 : Amaryllidaceae (*Polianthes tuberosa*) : A. Apical region of a flowering shoot; B. V.S. of a flower; C. Epiphyllous stamens with perianth; D. Single stamen; E. T.S. of ovary, and F. Floral diagram of a pair of flowers at the axil of a bract

and Prantl placed both the families under the order Liliflorae of class Monocotyledoneae.

Common plants

1. *Crinum latifolium* Thumb.;
2. *Pancratium verecumdum* Ait,;
3. *P. triflorum* Roxb.;
4. *Polianthes tuberosa* L.;
5. *Zephyranthes grandiflora* Lindl.

Key to Genera

I. Leaves large, thick, fleshy, densely clustered at the top of a short stem, spiny at the tip and armed along the edge :
   Inflorescence laxly paniculate; stamens shorter than perianth ....*Furcraea.*

II. Leaves unarmed, thin, flat or somewhat fleshy :
   Rootstock a tunicated bulb; flowers white or pink, large, showy

   i. Filamets attached to perianth but not connate
      Scapes 1-flowered .... *Zephyranthes.*
      Scapes umbellate .... *Crinum.*
      Scapes a long spike, periath somewhat curved .... *Polianthes.*

   ii. Filaments united by an intervening petaloid membrane .... *Pancratium.*

Plants of economic importance

Many plants of this family are important mainly for their ornamental value although they have some other uses.

A. **Ornamental plants**

1 *Amaryllis belladonna* (Amaryllis). The plant is cultivated for its beautiful flower. It was introduced from South Africa.

2. *Crinum asiaticum, C. defixum* and *C.*

Fig. 4.159 : Amaryllidaceae (*Crinum asiaticum*)

latifolium. The plants are commonly known as Crinum lily (white flower with red strips or red tinged) and are cultivated in the gardens.

3. *Eucharis grandiflora* (Amazon lily). The plant having large showy, fragrant and white flowers, is very much attractive.
4. *Haemanthus coccineus* (Blood lily). The plants are grown in pots for their attractive red flowers in a dense head subtended by red bracts.
5. *Hippeastrum vittatum*. The plant is attractive for its white-striped red flowers. It is commonly grown in the garden.
6. *Polianthes tuberosa* (Tuberose, Rajanigandha). The plants are very popular for its attractive leathery, waxy-white flowers developed on elongated spike inflorescence. The flowers are commonly used in various occasions especially in marriage ceremony.
7. *Zephyranthes ajax, Z. atamasco, Z. candida* and other species. The plants develop single flower on each scape with attractive colours. The plants are commonly grown in the borders of a design in the garden.

B. **Other uses**

1. *Pancratium zeylanicum*. The plants contain **lycorine**, an alkaloid which is highly effective and even causes death by paralysis.

6. LILIACEAE
(Lily Family)

Liliaceae is a large family, consists of about 254 genera and 4,075 species commonly found to grow in most parts of the world. The family is represented in India by 35 genera and about 189 species, occurs mainly in Himalayas.

The type genus of the family is ***Lilium***.

Systematic position

Bentham & Hooker		**Engler & Prantl**	
Division.	Angiospermia	Subdivision.	Angiospermae
Class.	Monocotyledons	Class.	Monocotyledoneae
Series.	Coronarieae	Order.	Liliales
Hutchinson		**Cronquist**	
Phylum.	Angiospermae	Division.	Magnoliophyta
Subphylum.	Monocotyledones	Class.	Liliopsida
Division.	Calyciferae	Subclass.	Liliidae
Order.	Juncaginales	Order.	Liliales.
Takhtajan		**Thorne**	
Division.	Magnoliophyta	Class.	Angiospermae
Class.	Liliopsida	Subclass.	Liliidae
Subclass.	Liliidae	Superorder.	Lilianae
Superorder.	Lilianae	Order.	Liliales
Order.	Liliales.		

Liliaceae : Distinctive features

Common characteristics	Exceptions
Vegetative characteristics	
Plant. Perennial herbs with bulbous (*Lilium*) and rhizomatous (*Polygonatum*) stem.	Annual herbs – *Asparagus*. Climber – *Smilax, Gloriosa*. Tree – *Cordyline, Dracaena, Yucca*. Cladodes present – *Ruscus aculeatus* and *Asparagus*.
Leaves. Simple, radicle (*Allium* and *Lilium*) or cauline, alternate or whorled (*Paris* and *Trillium*), mostly lamellate, sometimes fleshy (*Aloe*), parallel veined.	Compound – *Paris polyphylla*. Opposite – *Scolyopus, Gloriosa*. Reticulate – *Smilax*. Leaves reduced to minute scales – *Asparagus* and *Ruscus*.
Reproductive characteristics	
Inflorescence. Generally racemose, sometimes cymose.	Huge terminal panicle – *Yucca*. Solitary axillary – *Gloriosa*. Solitary terminal – *Lilium* and *Tulipa*.

Contd.

Contd.

Common characteristics	Exceptions
Flower. Actinomorphic, bisexual, hypogynous.	Zygomorphic – *Hemerocallis.* Unisexual – *Smilax, Ruscus, Lomandra.*
Perianth. Tepals 6 (3+3), free or united, petaloid.	
Androecium. Stamens 6 (3+3), anthers extrorse or introrse.	Stamens 3 (inner whorl suppressed) – *Ruscus.*
Gynoecium. Carpel 3, syncarpous, ovary superior, 3-celled with few to many anatropous ovules in each cell on axile placentation; style united or separated; stigma 3 or 1 with 3 lobes.	Carpel 2 – *Maianthemum.* Ovary partially inferior – *Mondo.*
Fruits. Capsule.	Berry – *Smilax* and *Polygonatum.*
Seeds. Many, with fleshy or bony endosperm and small embryo.	

Identifying characteristics

1. Commonly **herbs** or **undershrubs with underground stem**.
2. Leaves are either **radicle** or **cauline**.
3. Flowers generally **bisexual, regular, hypogynous** and trimerous.
4. Perianths 6 (3+3), commonly petaloid.
5. Stamens 6, anther **basifixed, extrorse** or **introrse.**
6. Ovary **superior, syncarpous,** 3-chambered with many ovules on axile placentation.
7. Fruit **capsule** or **berry**.

Floral formula

$\oplus$ ⚥ $P_{6(3+3)}\ A_{6(3+3)}\ \underline{G}_{(3)}$

Phylogeny and systematics

The family Liliaceae is closely allied to Amaryllidaceae and has been separated by the

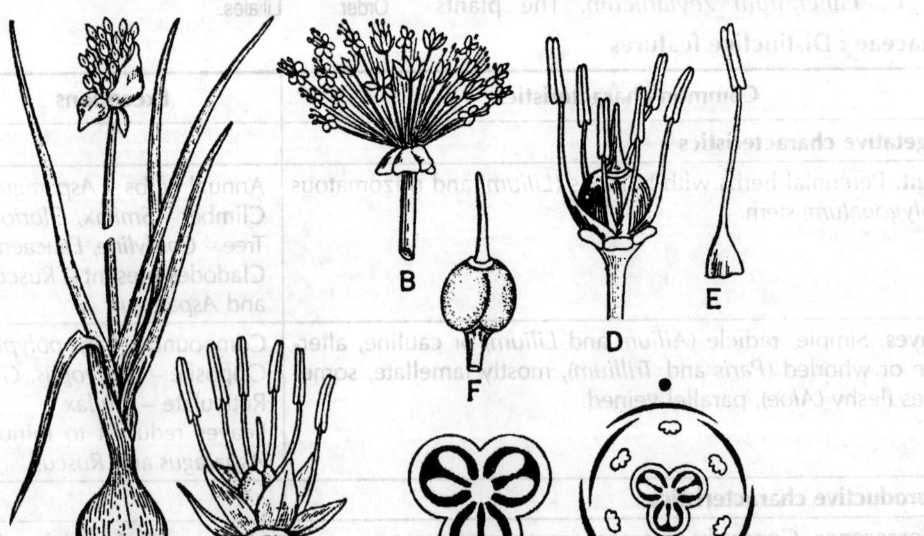

Fig. 4.160 : Liliaceae (*Allium cepa*) : A. Single plant with bulb and inflorescence; B. Inflorescence; C. Single flower; D. Flower, after removal of perianth lobes; E. Single stamen; F. Pistil; G. T.S. of ovary, and H. Floral diagram

position of ovary which is superior in the former and inferior in the latter. For the above Cronquist (1968, 81) merged Liliaceae with Amaryllidaceae. Based on the floral structure, the family Liliaceae is also related to some other families like Juncaceae, Pontederiaceae and Commelinaceae.

Bentham and Hooker placed the family in the series Coronarieae due to inner petaloid perianth, under the class Monocotyledons. Engler and Prantl placed the family under the order Liliflorae next to Farinosae of class Monocotyledoneae. Later, Hutchinson placed the family under the order Liliales, division Corolliferae of of subphylum Monocotyledones. Takhtajan and Cronquist also followed Hutchinson and placed it under Liliales.

Bentham and Hooker divided the family Liliaceae into 20 tribes. Later, Hutchinson (1934) removed the tribes Luzuriageae, Dracaeneae and Allieae from Liliaceae and ranked them in different families like Philesiaceae, Agavaceae and Amaryllidaceae.

Common plants

1. *Aloe vera* (L.) Burm f.;
2. *Allium cepa* L.;
3. *A. sativum* L.;
4. *Asparagus racemosus* Willd.;
5. *Asphodelus tenuifolius* Cav.;
6. *Gloriosa superba* L.;
7. *Iphigenia indica* Kunth;
8. *Dracaena terniflora* Jack. etc.
9. *Smilax zeylanica* L.

Key to Genera

I. Shrubs with perennial stems above ground; fruit berry-like :
   1. Stems without green leaves, usually straggling, leaves replaced by linear or acicular cladodes with minute scales ... *Asparagus.*
   2. Stems leafy –
      i. Leaves 3-5-nerved, reticulately veined; leaf-base with 2 tendrils ...... *Smilax.*
      ii. Leaves parallel veined, numerous, usually clustered at the apex of an erect caudex; each cell of ovary 1-ovuled .... *Dracaena.*
      iii. Leaves thick, succulent, spinose, fruit berry, globose, scarlet when ripe .... *Aloe.*

II. Climbing herb with leafy stems; leaves with a long, tendril-like apex; flowers large, showy .... *Gloriosa.*

III. Herbs with tunicated bulbs and characteristic odour, strong or pungent; leaves radical, fistular or flat enveloping the scape below; flowers umbellate or capitate at the top of the scape, at first enclosed in an involucre of bracts ...... *Allium.*

Plants of economic importance

The family is very important in respect of food, medicines, fibres, ornamentation and also other uses :

C. **Vegetable**

1. *Allium cepa* (Onion, Piyaz). Entire young plants are used as vegetable and also for flavouring other vegetables.
2. *Asparagus officinalis* (Garden asparagus). The fleshy shoots are used as vegetable.
3. *Asparagus racemosus* and *Chlorophytum arundinacea*. The roots are used as food and also in the preparation of jelly and jam.

D. **Spice**

1. *Allium sativum* (Garlic, Rasun). The mature bulbs are used as spice in different preparation during cooking.

E. **Medicines**

1. *Allium cepa* (Onion, Piyaz). The fresh juice of bulb has bacteriocidal properties. It also possesses diuretic, stimulant and expectorant properties.
2. *A. sativum* (Garlic, Rasun). The bulb yields an essential oil used as a tonic, vermifuge and as a stimulant to the stomach.
3. *Aloe vera* (Aloe, Ghritakumari). It is one of the constituents of many laxative preparations. It is also used in the treatment of fissures, piles and inflammations. The homeopathic medicine *'Aloe'* prepared from the plant is used during acute diarrhoea.
4. *Asparagus racemosus* (Satamuli). The oil extracted from the roots is used for rheumatic and nervous disorders.

5. *A. filicinus.* The extract of the root is an astringent (to constipate or compress) and also used in the tonic.
6. *A. gonocladus.* It is used in the treatment of skin diseases.
7. *Colchicum luteum.* Seeds and corms are used in liver diseases and also in rheumatism.
8. *Fritillaria cirrhosa.* The dry bulbs are used to control bronchitis, tuberculosis and asthma.
9. *F. roylei.* The bulb is used as an antipyretic and expectorant.
10. *Gloriosa superba* (Ulat-chandal). The leaf extract is used to control lice (sing. louse, a parasitic insect). The tubers are used to enhance labour pain.
11. *Hemerocallis fulva* (Orange day lily). The flowers are eaten to purify blood and for the treatment of all kinds of pains including labour pain of human being.
12. *Iphigenia indica.* The corm of this plant is used in the treatment of colic pain and headache.
13. *Paris polyphylla.* The rhizome of this plant has anthelmintic (any drug used to destroy parasites–especially intestinal worms like tapeworms, roundworms etc.) properties.
14. *Schoenocaulon officinale.* The plant has the insecticidal properties.
15. *Scilla hyacinthiana* (South Indian squill). The extract of the bulb is used as cardiac stimulant and also as expectorant.
16. *Smilax china, S. glabra* and *S. ovalifolia.* The root extract is used in the preparation of the drug 'Sarsaparilla' used for the remedy of skin diseases.
17. *Urginea indica* (Indian squill). The bulb is used to control rheumatism, skin diseases and also as cardiac stimulant.

F. **Fibres**

1. *Phormium tenax.* The fibres extracted from the leaves are used for twine, tow-line, etc.
2. *Yucca filamentosa.* The fibres extracted from the leaves are used for cordage.

G. **Ornamentals**

1. *Agapanthus africanus* (African lily).
2. *Asparagus plumosa* (Asparagus fern).
3. *Dracaena terniflora* (Dracena plant).
4. *Gloriosa superba* (Malabar glory lily).
5. *Hemerocallis fulva* (Orange day lily).
6. *Lilium candidum* and *L. gesneriana.*
7. *Rusca aculeatus* (Buchers broom).
8. *Tulipa gesneriana* and *T. suaveolens.*
9. *Yucca aloifolia* and *Y. gloriosa.*

H. **Other uses**

1. *Colchicum luteum* (meadow saffron). The important laboratory reagent 'Colchicine' is extracted from the large corm used to break the spindle fibre during cell division.
2. *Iphigenia indica.* The flowers yield a red dye.
3. *Phormium tenax.* The oil extracted from the seeds are used in paints and varnish.

Table 4.13 : Comparison of Amaryllidaceae and Liliaceae

Characteristics	Amaryllidaceae	Liliaceae
1. Plants	Usually bulbs are predominant; rhizomes are rarely present.	Underground stems are either bulbs, rhizomes or tubers.
2. Inflorescence	Usually umbellate cyme.	Racemose, rarely umbellate cyme.
3. Corona	Corona often present.	Corona absent.
4. Anther	Versatile, introrse.	Basifixed, both introrse and extrorse types.
5. Ovary	Inferior.	Superior or rarely semi-inferior.
6. Endosperm	Fleshy.	Both fleshy and cartilaginous.

7. Commelinaceae

(Spiderwort Family)

The family Commelinaceae is represented by 40 genera and 650 species found to grow mostly in the tropical and subtropical regions of the world. The family is represented in India by 10 genera and about 500 species available in tropical Himalayas and also in Southern and Western regions.

The type genus of the family is ***Commelina***.

Systematic position

Bentham & Hooker		**Engler & Prantl**	
Division.	Angiospermia	Subdivision.	Angiospermae
Class.	Monocotyledons	Class.	Monocotyledoneae
Series.	Coronarieae	Order.	Farinosae

Hutchinson		**Cronquist**	
Phylum.	Angiospermae	Division.	Magnoliophyta
Subphylum.	Monocotyledones	Class.	Liliopsida
Division.	Calyciferae	Subclass.	Commelinidae
Order.	Commelinales	Order.	Commelinales.

Takhtajan		**Thorne**	
Division.	Magnoliophyta	Class.	Angiospermae
Class.	Liliopsida	Subclass.	Liliidae
Subclass.	Commelinidae	Superorder.	Commelinanae
Superorder.	Commelinanae	Order.	Commelinales
Order.	Commelinales.		

Identifying characteristics

1. Plants **herb**, either erect or creeping, at marshy places.
2. **Leaves simple**, alternate **with sheathing leaf base**.

Commelinaceae : Distinctive features

Common characteristics	Exceptions
Vegetative characteristics	
Plant. Herbs with swollen nodes, either erect or creeping of marshy places.	Climber – *Streptolirion* and *Spatholirion* Epiphyte – *Cochliostema*.
Leaves. Alternate, simple, with prominent sheath; parallel veined.	
Reproductive characteristics	
Inflorescence. Monochasial cyme or solitary, enclosed by boat-shaped leaf sheath or leafy bract.	
Flower. Irregular, sometimes zygomorphic, bisexual or polygamous, hypogynous, often enclosed in spathes.	Small cleistogamous flowers are borne on rhizome branches under the soil – *Commelina*.
Perianth. Tepals 6 (3+3), outer sepaloid and inner petaloid, sepals generally free; petals generally free and commonly blue in colour.	
Androecium. Stamens 6 (3+3) in two whorls, often reduced to staminodes, anthers extrorse or introrse, basifixed, filaments hairy or smooth.	All 6 stamens are fertile – *Cyanotis, Pollia, Tradescantia*. 2 stamens fertile – *Cochliostema*. 3 stamens are fertile – *Pollia* and *Commelina*. 5 stamens are fertile – *Floscopa*. Staminodes absent – *Collisea*.
Gynoecium. Carpel 3, syncarpous, ovary superior, 3-celled with solitary or few orthotropous ovules in each cell on axile placentation; style terminal; stigma generally capitate or trifid.	Ovary 2-celled – *Floscopa*.
Fruits. Capsule.	
Seeds. Angled, albuminous with small embryo.	

3. **Inflorescence** cymose or solitary often **enclosed by leaf sheath or bract.**
4. **Perianth** 6 (3+3) **outer sepaloid** and **inner petaloid.**
5. **Stamens 6** (3+3).
6. Carpels 3, ovary superior.
7. **Fruit capsule.**

Floral formula

$\circ|\circ$ ⚥ $\overbrace{P_{(3+3)}\ A_{(3+3)}}\ \underline{G}_{(3)}$

Phylogeny and systematics

The family Commelinaceae is related to Pontederiaceae and Alismaceae.

Bentham and Hooker placed the family under the series Coronarieae of class Monocotyledons. Engler and Prantl placed the family under the 8th series Farinosae of Monocotyledoneae. Hutchinson placed the family Commelinaceae in a separate order Commelinales. Both Takhtajan and Cronquist followed Hutchinson.

Hutchinson interpreted that the order Commelinales has been derived from Alismatales and Butomales, but Takhtajan (1980) interpreted that the order Commelinales is, perhaps, derived from Liliales.

Common plants

1. *Amischophacelus axillaris* (L.) Rao & Kamm.;
2. *Commelina benghalensis* L.;
3. *C. paludosa* Blume;
4. *Cyanotis tuberosa* (Roxb.) Schultes f.;
5. *Murdannia nudiflora* (L.) Brenan;
6. *M. spirata* (L.) Brueckner.

Key to Genera

I. Stamens 3 perfect, and 1-3 staminodes :
   1. Cymes solitary, enclosed in spathaceous bract; one petal larger than others and often clawed .... *Commelina.*
   2. Cymes paniculate, not enclosed in spathes; petals all equal .... *Murdannia.*

II. Stamens 6 perfect :
   Petals united below in a tube .... *Cyanotis.*

Plants of economic importance

The members of this family have little economic importance. A few are grown as ornamental plants and some have different uses.

A. **Ornamental plants**
   1. *Tradescantia brateata* and *T. virginiana*, the spiderworts are grown in the garden for beautification.
   2. *Rhoeo discolor.*
   3. *Zebrina pendula* (Wandering jew).

B. **Other uses**
   1. *Commelina benghalensis.* The rhizome is used as vegetable in some places.
   2. *C. diffusa.* The leaves are used as vegetable and the entire plant as fodder.

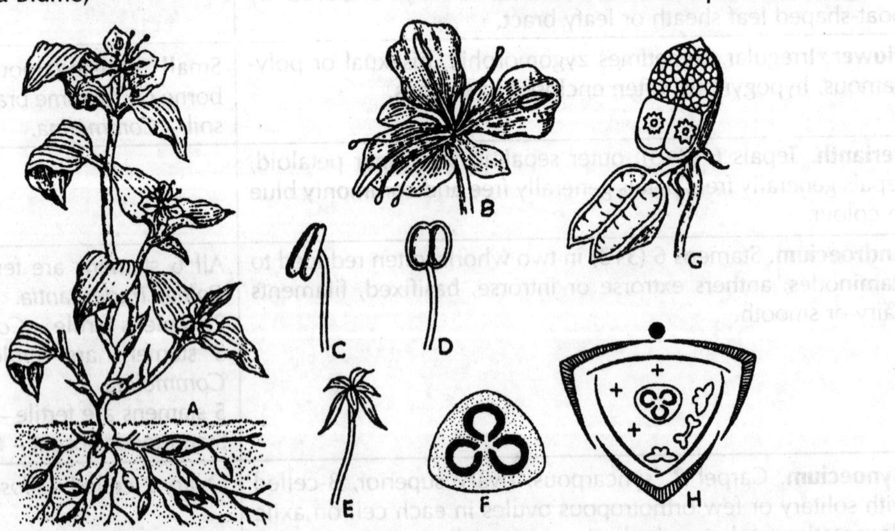

Fig. 4.161 : Commelinaceae (*Commelina benghalensis*) : A. Plant with underground cleistogamous and aerial flowers; B. Single flower; C and D. Fertile stamens; E. Sterile stamen; F. T.S. of ovary; G. Dehiscence of fruit, and H. Floral diagram

3. *C. longifolia*. The entire plant is used as fodder.

8. Palmae (Arecaceae)

(Palm Family)

The family Palmae consists of 217 genera and 2,500 species, distributed widely in tropical and subtropical regions of the world. About 28 genera and 93 species are found in India, mostly in Assam and Eastern Himalayas.

The type genus of the family is ***Areca***.

Systematic position

Bentham & Hooker		**Engler & Prantl**	
Division.	Angiospermia	Subdivision.	Angiospermae
Class.	Monocotyledons	Class.	Monocotyledoneae
Series.	Calycinae	Order.	Principes

Hutchinson		**Cronquist**	
Phylum.	Angiospermae	Division.	Magnoliophyta
Subphylum.	Monocotyledones	Class.	Liliopsida
Division.	Calyciferae	Subclass.	Arecidae
Order.	Palmales	Order.	Arecales.

Takhtajan		**Thorne**	
Division.	Magnoliophyta	Class.	Angiospermae
Class.	Liliopsida	Subclass.	Liliidae
Subclass.	Arecidae	Superorder.	Arecanae
Superorder.	Arecanae	Order.	Arecales
Order.	Arecales.		

Identifying characteristics

1. Plants are **usually tree, with unbranched (caudex) stem,**
2. Leaves are **pinnately** or **palmately-lobed,** present at the apex of the stem in the form of a crown,

Palmae : Distinctive features

Common characteristics	Exceptions
Vegetative characteristics	
Plant. Commonly trees with an erect, unbranched cylindrical stem, having short internodes covered with leaf-scars and have crown of leaves at the top.	Stem branched – *Hyphaene theboica* and *Calamus*. Stem with distinct internodes – *Calamus*. Climbers – *Calamus, Plectocomia, Korthalsia*.
Leaves. Simple, palmate (*Borassus*) or pinnate (*Phoenix*), often very large with well-developed petiole and a broad sheathing base, plicate in bud.	
Reproductive characteristics	
Inflorescence. Simple (*Borassus*) or compound spadix covered by 1 or more woody spathe called 'cymba'.	
Flower. Commonly unisexual, regular, hypogynous, arranged in close spiral on inflorescence axis, small actinomorphic sessile, monoecious (*Areca*) or dioecious (*Phoenix*).	Develops flower once in lifetime – *Corypha*. Bisexual – *Livistona* and *Licuala*.
Perianth. Tepals 6 (3+3), free, generally green, persistent, more or less fleshy or leathery; imbricate or valvate.	
Androecium. Stamens 6, in 2 whorls; filaments free or united; anther 2-celled.	Stamens 3 – *Nipa*. Stamens many – *Caryota, Pinanga*.
Gynoecium. Carpel 3, free or united, ovary superior, 3-celled with 1 ortho- or anatropous ovule in each cell on axile placentation; style short.	Carpel 1 – *Plectocomia*. 1-celled – *Cocos nucifera*. Carpels partly free – *Nypa*. Carpel completely free – *Phoenix*.
Fruits. Fibrous drupe or berry.	
Seeds. Albuminous with minute embryo.	

3. Inflorescence – **simple** or **compound spadix**,
4. Flower – small, actinomorphic, **generally unisexual**,
5. Perianths – **six (6), free, persistant**,
6. Ovary **superior, 3-chambered**,
7. Fruit berry or **drupe**.

Floral formula

Male flower : ⊕⚥ $P_{6(3+3)}$ $A_{6(3+3)}$

Female flower : ⊕♀ $P_{6(3+3)}$ $\underline{G}_{(3)}$

Phylogeny and systematics

The family is closely related with Araceae and difference lies on habit. It is also related with the families like Cyclanthaceae and Pandanaceae.

Bentham and Hooker placed this natural order under the series Calycine (because both perianth whorls are sepaloid and tough) of class Monocotyledons. Engler and Prantl put the family Palmae under a separate monotypic taxon Principes (far from Araceae and Lemnaceae) in the class Monocotyledoneae. Hutchinson placed the family under the monotypic order Palmales between Agavales and Pandanales in the subphylum Monocotyledones. Takhtajan and Cronquist placed the family under the order Arecales.

Common plants

1. *Areca catechu* L. – Betlenut palm (Supari),
2. *Borassus flabellifer* L. – Palmyra palm (Tal),
3. *Cocos nucifera* L. – Coconut palm (Narikel),
4. *Corypha elata* Roxb. – Tolipot palm,
5. *Calamus rotang* L. – Cane palm (Bet),
6. *Phoenix dactylifera* L. – Date palm (Khajur),
7. *P. paludosa* Roxb. – a mangrove plant,
8. *P. sylvestris* Roxb. – Wild date palm (Jangli khajur),
9. *Nypa fruticans* Wurmb. – a mangrove plant,
10. *Roystonea regia* (Kunth) Cook – Royal palm.

Key to Genera

I. Leaves flabelliform, fan-shaped or orbicular :
   1. Monocarpic palms, flowering once and then dying; flowers bisexual ...... *Corypha.*
   2. Polycarpic palms, flowering year after year –
      i. Flowers bisexual –
      Ovary of 3 triangular carpels .... *Licuala.*
      Ovary of 3 globose carpels .... *Livistona.*

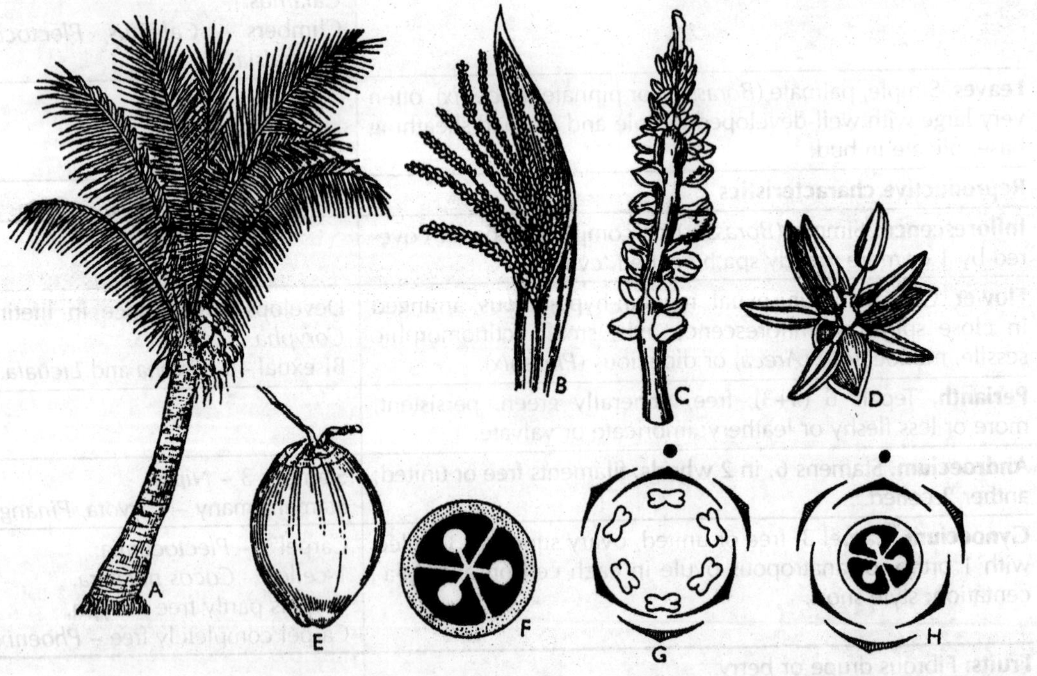

Fig. 4.162 : Palmae (*Cocos nucifera*) : A. Plant with hanging fruits; B. Inflorescence; C. Single rachilla; D. Single staminate flower; E. Fruit with persistent calyx; F. T.S. of ovary; G. Floral diagram of staminate flower, and H. Floral diagram of pistillate flower

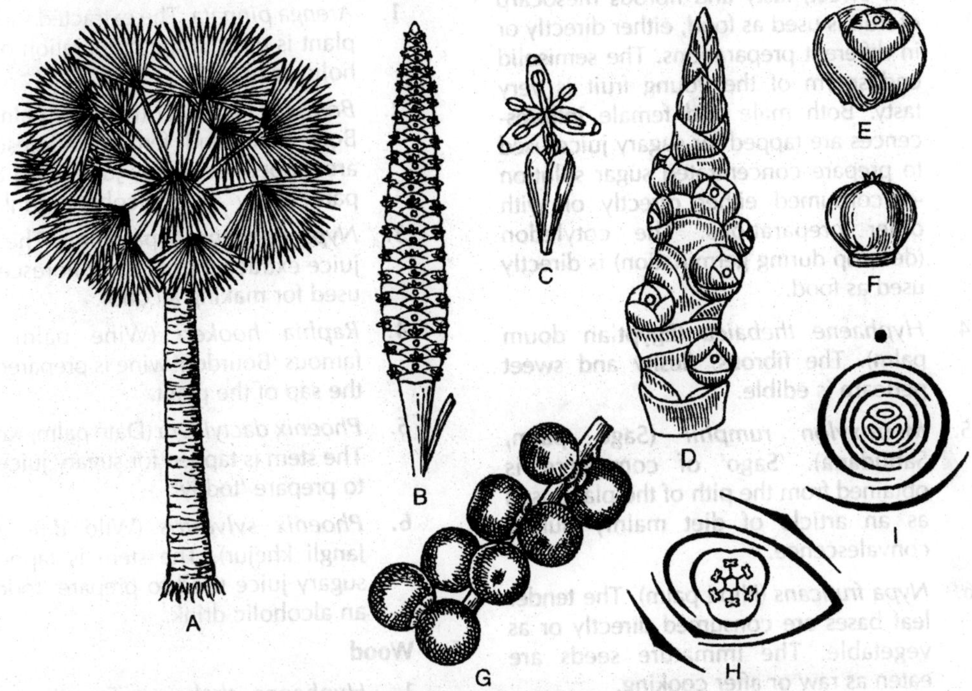

Fig. 4.163 : Palmae (*Borassus flabellifer*) : A. Single plant; B. Part of staminate spadix; C. Single staminate flower; D. Part of pistillate spadix; E. Single pistillate flower; F. Same after removal of perianth; G. Bunch of fruits; H and I. Floral diagram of staminate, and pistillate flower respectively

ii. Flowers unisexual; male flowers small, female flowers very large ... *Borassus*.

II. Leaves pinnate :

1. Fruit not covered with scales –
   i. Leaves twice pinnatisect ... *Caryota*.
   ii. Leaves completely or partially once pinnatisect –
      a. Dwarf palms of tidal swamps, branching (buried) .... *Nypa*.
      b. Tall palms with erect unbranched stems; fruit large ovoid nut with a fibrous husk .... *Cocos*.
      c. Tall or short palms with pinnatisect leaves, segments lanceolate or ensiform with margins induplicate; flowers dioecious, small, yellowish; fruit yellow or bright-red to blue-black .... *Phoenix*.
      d. Slender palms with erect annulate stems; leaves pinnatisect, distal segments confluent; flowers monoecious on branched intrafoliar spadices, male flowers minute, female much larger; perianth accrescent ... *Areca*.
2. Fruits covered with imbricating scales; stem climbing, rarely erect, armed ... *Calamus*.

Plants of economic importance

Economically, the family is very important and considered as second one among the monocotyledonous families. The plants are useful for food, medicine, oil, liquors, woods, ornamentation and also have other uses :

A. **Food**

1. *Arenga pinnata*. The extracted sap of the plant is used in the preparation of palm sugar. The tender leaf bases and young stem pith are edible and also used as vegetable.
2. *Cocos nucifera* (Coconut palm, Narikel). The young liquid endosperm is used as refreshing drink; the mature endosperm is used as food directly or added with different preparations and also used in the preparation of sweets. The ripe dried endosperm is also the source of vegetable oil used in South India and as hair oil throughout India.

3. *Borassus flabellifer* (Palmyra palm, Tal). The sweet, tasty and fibrous mesocarp of fruit is used as food, either directly or in different preparations. The semisolid endosperm of the young fruit is very tasty. Both male and female inflorescences are tapped for sugary juice used to prepare concentrated sugar solution — consumed either directly or with other preparations. The cotyledon (develop during germination) is directly used as food.
4. *Hyphaene thebaica* (Egyptian doum palm). The fibrous, fleshy and sweet pericarp is edible.
5. *Metroxylon rumphii* (Sago palm, Sabudana). 'Sago' of commerce is obtained from the pith of the plant used as an article of diet mainly during convalescence.
6. *Nypa fruticans* (Nipa palm). The tender leaf bases are consumed directly or as vegetable. The immature seeds are eaten as raw or after cooking.
7. *Phoenix dactylifera* (Date palm, Khejur). The fruits are edible, and have high food value. The fruits are used in confectionary and also in bakery. The stem is tapped for sugary juice used to prepare a tasty concentrated sugar solution after boiling for an appropriate time. It is also used in textile industry for sizing of cotton yarns.
8. *Phoenix sylvestris* (Wild date palm, Jangli khejur). The fruits are edible but commercially are not as important as *P. dactylifera*. The stem is tapped for sugary juice used to prepare a tasty concentrated sugar solution after boiling for appropriate time.

B. **Medicine**

1. *Nypa fruticans* (Nipa palm). The powder of burnt leaves and roots are used to get relief from toothache.

C. **Oil**

1. *Elaeis guineensis* (Oil palm). The mesocarp of the fruit is the source of palm oil (edible oil).

D. **Wines and other liquors**

1. *Arenga pinnata*. The extracted sap of the plant is used in the preparation of alcoholic drinks.
2. *Borassus flabellifer* (Palmyra palm, Tal). Both male and female inflorescences are tapped for sugary juice used to prepare 'toddy' — an alcoholic drink.
3. *Nypa fruticans* (Nipa palm). The sweet juice extracted from the inflorescence is used for making alcohol.
4. *Raphia hookeri* (Wine palm). The famous 'Bourdon' wine is prepared from the sap of the plant.
5. *Phoenix dactylifera* (Date palm, Khejur). The stem is tapped for sugary juice used to prepare 'toddy'.
6. *Phoenix sylvestris* (Wild date palm, Jangli khejur). The stem is tapped for sugary juice used to prepare 'toddy' — an alcoholic drink.

E. **Wood**

1. *Hyphaene thebaica* (Egyptian doum palm). The trunk is strong and heavy — used as wood.
2. *Borassus flabellifer* (Palmyra palm, Tal). The mature stems are divided into 4-6 longitudinal pieces and used in the construction of houses, bridges, etc.

F. **Ornamental plants**

Many plants are cultivated in the garden for their beautiful appearance. These are :

1. *Arenga saccharifolia* (Gomti palm).
2. *Bactris major.*
3. *Caryota urens* (Fish-tail palm).
4. *Chamaerops humilis* (European fan-palm).
5. *Heterospathe elata.*
6. *Livistona chinensis* (Chinese fan-palm).
7. *Roystonea regia* (Syn. *Oreodoxa regia*) (Royal palm), etc.

G. **Other uses**

1. *Areca catechu* (Betel nut palm, Supari). The seeds are used with betel leaves after making slice or small pieces.
2. *Arenga pinnata*. The fibres extracted from the petiole are used for making brushes.

3. *Borassus flabellifer* (Palmyra palm, Tal). The leaves are used for thatching and in the preparation of fan, hat, mat, basket etc. The mature plant, after removing the apical region and the inner soft content, is used as boat. The leaf base is the source of fibre used for making brushes etc.
4. *Calamus rotang* (Rattan palms or Cane) and other species (*C. latifolius, C. extensus* etc.). The plants are the source of common cane used in the preparation of chair, table, walking sticks, baskets etc.
5. *Ceroxylon andicola*. The wax produced from this plant is used in different industries like candle, boot-polish, etc.
6. *Cocos nucifera* (Coconut palm).
   i. The *oil cake* used as fodder, also eaten by people in some places.
   ii. The *fibrous mesocarp* is used in the preparation of ropes, door-mats, cushions, mattress etc.
   iii. The *woody endocarp* is used for decorating articles.
   iv. The *leaves* are used in making mat, basket and also for thatching.
   v. The *midrib of leaflets* is used to prepare broom.
   vi. The *stem* is used as pillar and also as fuel.

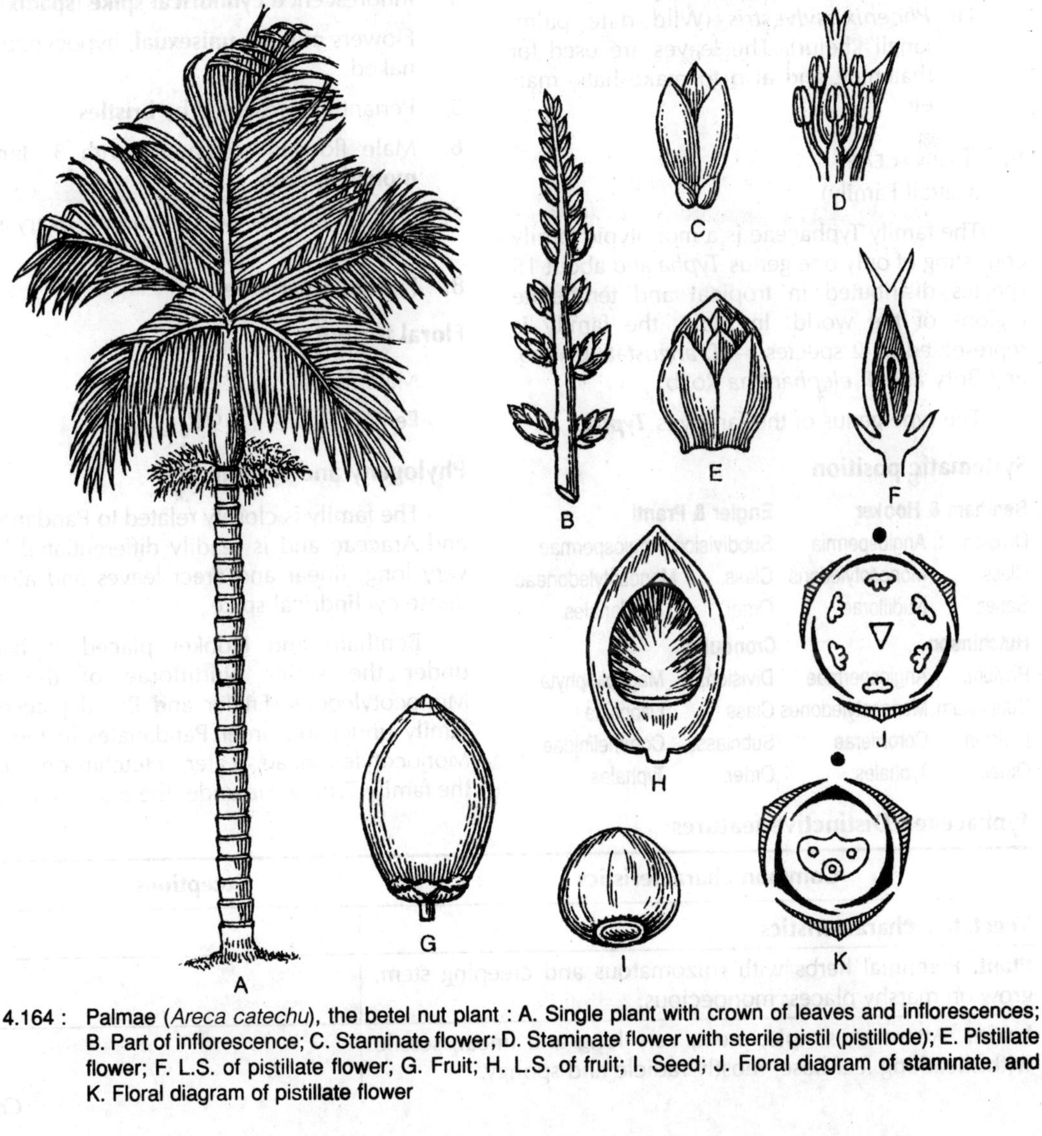

Fig. 4.164 : Palmae (*Areca catechu*), the betel nut plant : A. Single plant with crown of leaves and inflorescences; B. Part of inflorescence; C. Staminate flower; D. Staminate flower with sterile pistil (pistillode); E. Pistillate flower; F. L.S. of pistillate flower; G. Fruit; H. L.S. of fruit; I. Seed; J. Floral diagram of staminate, and K. Floral diagram of pistillate flower

7. *Copernicia ceribera*. The leaves exude 'canauba' wax which is used in candle and boot polish industry.
8. *Elaeis guineensis* (Oil palm). The palm oil extracted from the mesocarp is used in the manufacture of candles, soaps and lubricants. It is also used as fuel for internal combustion engines.
9. *Nypa fruticans* (Nipa palm). The sweet juice extracted from the inflorescence is used for making sugar, vinegar etc.
10. *Phoenix dactylifera* (Date palm, Khejur). The leaves are used to make hats, mats etc.
11. *Phoenix sylvestris* (Wild date palm, Jangli khejur). The leaves are used for thatching and also to make hats, mats etc.

9. TYPHACEAE
(Cattail Family)

The family Typhaceae is a monotypic family consisting of only one genus *Typha* and about 15 species distributed in tropical and temperate regions of the world. In India, the family is represented by 2 species — *T. angustata* Chaub. and Bory and *T. elephantina* Roxb.

The type genus of the family is ***Typha***.

Systematic position

Bentham & Hooker		**Engler & Prantl**	
Division.	Angiospermia	Subdivision.	Angiospermae
Class.	Monocotyledons	Class.	Monocotyledoneae
Series.	Nudiflorae	Order.	Pandanales

Hutchinson		**Cronquist**	
Phylum.	Angiospermae	Division.	Magnoliophyta
Subphylum.	Monocotyledones	Class.	Liliopsida
Division.	Corolliferae	Subclass.	Commelinidae
Order.	Typhales	Order.	Typhales.

Takhtajan		**Thorne**	
Division.	Magnoliophyta	Class.	Angiospermae
Class.	Liliopsida	Subclass.	Liliidae
Subclass.	Aridae	Superorder.	Commelinanae
Superorder.	Typhanae	Order.	Typhales
Order.	Typhales.		

Identifying characteristics

1. Perennial herbs with **creeping rhizomatous stem;**
2. Leaves **simple**, erect, **very long**, linear, **sessile with sheathing leaf-base**, mostly radicle and spongy.
3. Inflorescence **cylindrical spike** (**spadix**).
4. Flowers minute, **unisexual**, hypogynous and naked.
5. Perianth represented by **bristles**.
6. Male flowers commonly with 3 stamens, **monadelphous**.
7. Female flowers with 1 carpel, ovary **superior** situated on **gynophore**.
8. Fruit **minute nutlet**.

Floral formula

Male flower : $\oplus$ ♂ $P_0\,A_{(1-7)}$

Female flower : $\oplus$ ♀ $P_0\,\underline{G}_1$

Phylogeny and systematics

The family is closely related to Pandanaceae and Araceae and is readily differentiated by its very long, linear and erect leaves and also the dense cyclindrical spike.

Bentham and Hooker placed Typhaceae under the series Nudiflorae of the class Monocotyledons. Engler and Prantl placed the family under the order Pandanales in the class Monocotyledoneae. Later, Hutchinson placed the family Typhaceae under the order Typhales in

Typhaceae : Distinctive features

Common characteristics	Exceptions
Vegetative characteristics	
Plant. Perennial herbs with rhizomatous and creeping stem, grow on marshy places; monoecious.	
Leaves. Alternate, simple, erect, very long and linear, sessile with sheathing leaf-base, mostly radicle and spongy.	

Contd.

Contd.

Common characteristics	Exceptions
Reproductive characteristics	
Inflorescence. Cylindrical spike (spadix).	
Flower. Very small, unisexual, hypogynous and naked; male flowers are situated on the upper region and the female flowers on the lower region of the spadix. Both male and female flowers are subtended by one caduceus bract-like spathe.	
Male flower.	
Perianth. Represented by bristles.	
Androecium. Stamens usually 3, rarely 1-7, monadelphous.	
Female flower.	
Perianth. Represented by bristles.	
Gynoecium. Carpel 1, ovary superior–situated on gynophore, 1-celled with a single pendulous ovule; style 1, usually filiform; stigma 1; base of the gynophore bears many silky hairs.	
Fruits. A minute nutlet with persistent style.	
Seeds. Albuminous (endosperm mealy) with a straight embryo.	

the division Corolliferae of subphyla Monocotyledones. Recently Takhtajan (1980) placed the family under Typhales and placed it in subclass Arecidae under the class Liliopsida. Cronquist placed the family under the order Typhales like Takhtajan, but placed it in the subclass Commelinidae next to Hydatallales under the class Liliopsida. Recently, Taktajan (1997) placed the family under the order Typhales, superorder Typhanae and subclass Aridae under the class Liliopsida (Monocots).

Common plants

1. *Typha angustata* Chaub. & Bory;
2. *T. elephantina* Roxb. ;
3. *T. latifolia* L.

Plants of Economic importance

Economically, the family is not so important, but the long leaves of the 2 species — *T. angustata* and *T. elephantina,* are woven into flat sheet-like structures and used in thatching of huts, boats, temporary shades etc. The pollens of *T. angustata* are sometimes made into cakes and mostly used by the poor people.

10. ARACEAE

(Arum Family)

The family Araceae contains 104 genera and 3,040 species, grows mostly in the tropical regions of the world, out of which 24 genera and 136 species are found in India.

The type genus of this family is ***Arum***.

Systematic position

Bentham & Hooker

Division.	Angiospermia
Class.	Monocotyledons
Series.	Nudiflorae

Engler & Prantl

Subdivision.	Angiospermae
Class.	Monocotyledoneae
Order.	Spathiflorae

Hutchinson

Phylum.	Angiospermae
Subphylum.	Monocotyledones
Division.	Corolliferae
Order.	Arales

Cronquist

Division.	Magnoliophyta
Class.	Liliopsida
Subclass.	Arecidae
Order.	Arales.

Takhtajan

Division.	Magnoliophyta
Class.	Liliopsida
Subclass.	Arecidae
Superorder.	Aranae
Order.	Arales.

Thorne

Class.	Angiospermae
Subclass.	Liliidae
Superorder.	Aranae
Order.	Arales

Identifying characteristics

1. Mostly **perennial herbs with rhizome, corms** or **tubers**,
2. Leaves **radical** or **cauline**,
3. Flower bisexual or unisexual in **spadix inflorescence**,
4. Androecium – **stamens 2, 4** or **6**,
5. Gynoecium – **carpel 1** or **more, syncarpous**,
6. Fruit – **berry**.

Floral formula

Bisexual flower : ⊕ ⚥ $P_{3+3 \text{ or } 2+2}$ $A_{3+3 \text{ or } 2+2 \text{ or more or less}}$ $\underline{G}_{(3) \text{ or } 1}$

Male flower : ⊕ ♂ $P_{3+3 \text{ or } 2+2 \text{ or } 0}$ $A_{2\text{-}9}$

Female flower : ⊕ ♀ $P_{3+3 \text{ or } 2+2 \text{ or } 0}$ $\underline{G}_{1\text{-}(3)}$

Phylogeny and systematics

The family Araceae is closely related to Palmae (Arecaceae). It is also related with some members of the family Liliaceae.

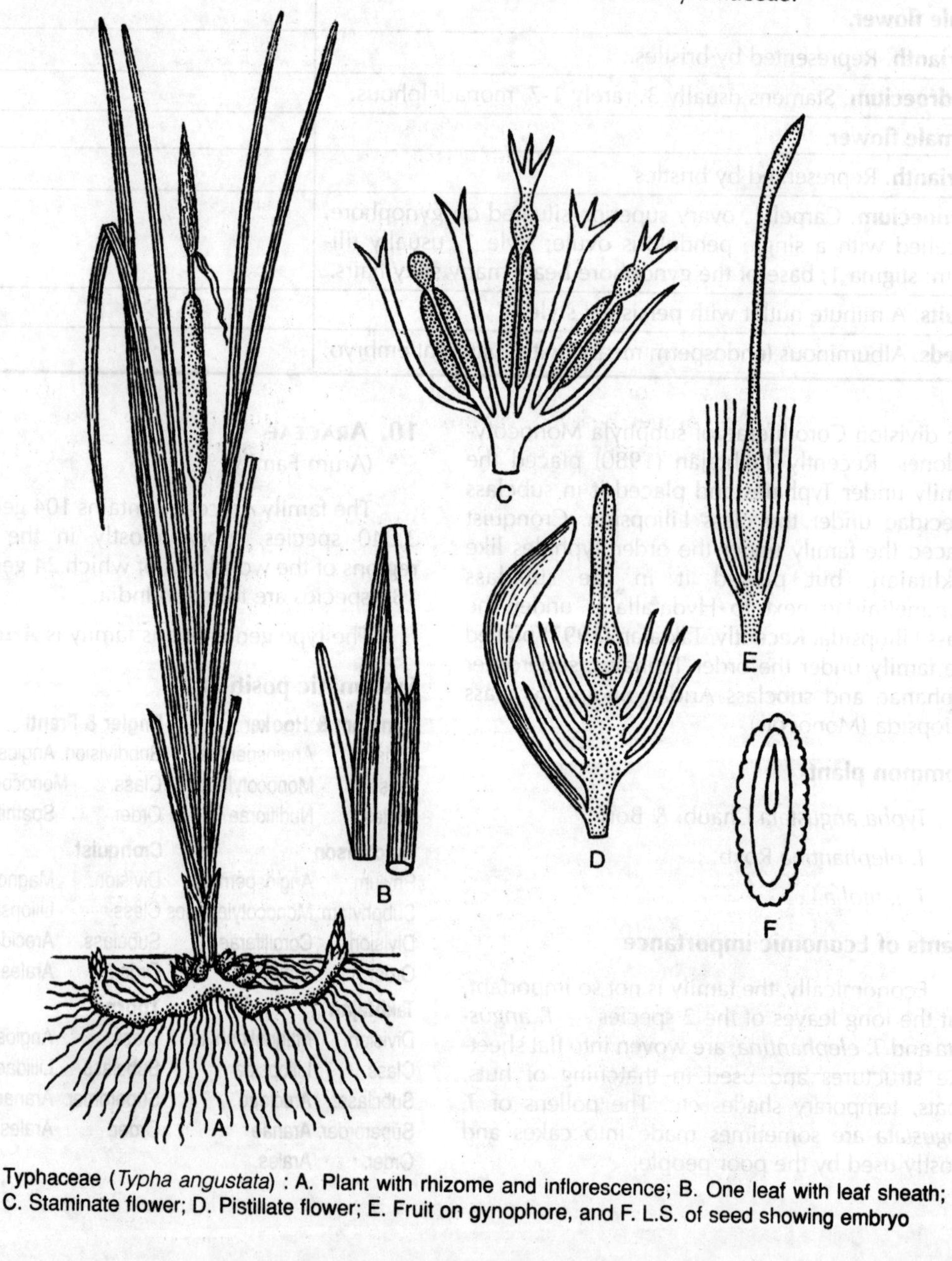

Fig. 4.165 : Typhaceae (*Typha angustata*) : A. Plant with rhizome and inflorescence; B. One leaf with leaf sheath; C. Staminate flower; D. Pistillate flower; E. Fruit on gynophore, and F. L.S. of seed showing embryo

Araceae : Distinctive features

Common characteristics	Exceptions
Vegetative characteristics	
Plant. Commonly terrestrial herbs, sometimes shrubby with corms, rhizomes and tubers; sometimes climbers; often with acrid watery latex.	Free floating aquatic herb – *Pistia*.
Leaves. Radicle, simple or compound (*Arisaema*), often broad, long-petioled and commonly net-veined.	Leaf single – *Arisaema*. Alternate and cauline – *Pothos*. Long and narrow with parallel veins – *Acorus calamus*.
Reproductive characteristics	
Inflorescence. Spadix subtended by a spathe. The spathe may be sepaloid i.e. green (*Colocasia*), petaloid (*Richardia*) or brilliant scarlet (*Anthurium*).	
Flower. Small, bisexual (*Pothos* and *Acorus*) or unisexual (*Arisaema*), closely crowded on the same or different (*Arisaema*) spadix. Perianth absent in unisexual flower.	Bisexual and naked – *Calla*.
Perianth. In bisexual flower tepals are represented by 4-6 scales those are either free (*Acorus*) or united (*Spathiphyllum*).	Tepals united into a cup – *Anadendrum*.
Androecium. Stamens 2, 4 or 8; opposite the perianth segments, filaments free, anther 2-celled, introrse, open by slits or pores. Sometimes staminodes are present in the female flowers.	Stamen 1 – *Biarum* and *Arisarum*. Filaments basally connate – *Arisaema*; united throughout their length forming *synandrium* – *Alocacia, Colocasia*.
Gynoecium. Carpels 3, syncarpous, ovary superior, 1-many locules with few to many ovules in each cell on parietal, axile, basal or apical placentation; stigma 1 or more.	Carpel l – *Pistia, Arum*. Placentation parietal (*Arum*), axile (*Pothos*); basal (*A isaema*); apical (*Lasia*).
Fruits. Commonly berry.	
Seeds. Albuminous, embedded in mucilaginous pulp or exalbuminous; embryo minute and straight.	

Engler considered the family to be derived from Palmae (Arecaceae). Later, Hutchinson considered that they have been derived from the stocks ancestral to the Liliaceae. Still later,

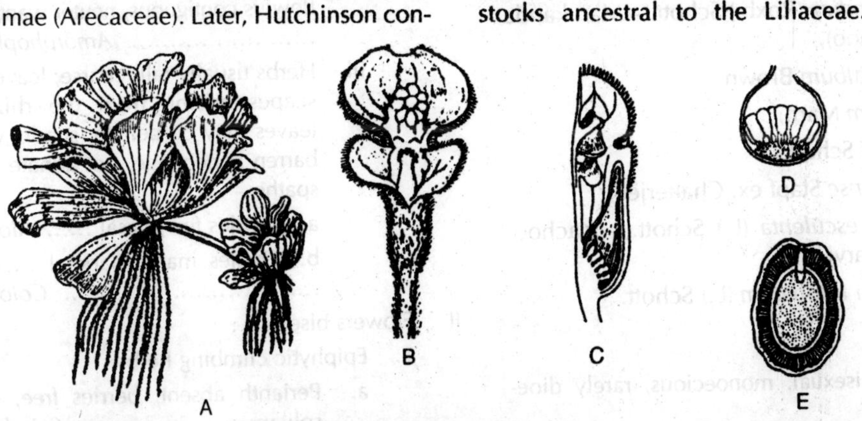

Fig. 4.166 : Araceae (*Pistia stratiotes*) : **A. Plant with stolon; B. Spadix; C. Spadix in L.S. showing staminate and pistillate flowers; D. Ovary in L.S., and E. Seed in L.S.**

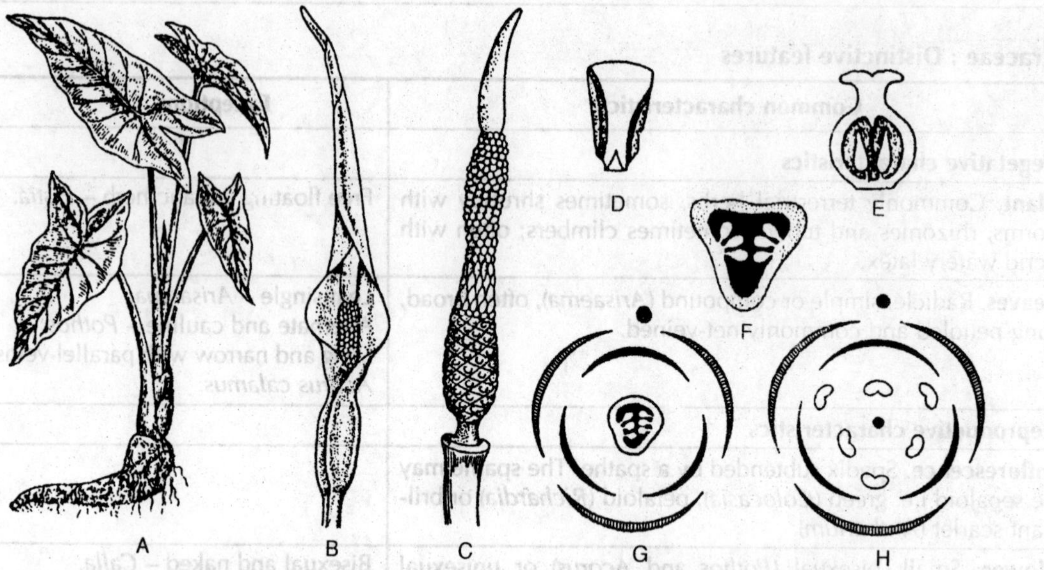

Fig. 4.167 : Araceae (*Alocasia indica*) : A. Single plant; B. Inflorescenece (Spadix); C. Spadix without spathe; D. Anther, E. L.S. of ovary, F. T.S. of ovary; G. Floral diagram of pistillate flower, and H. Floral diagram of staminate flower

Takhtajan stated that they have a common origin with Arecales and Cyclanthales from immediate ancestors of Liliaceae.

Bentham and Hooker placed the family Araceae as a Natural Order Aroideae under the 5th series Nudiflorae, together with Lemnaceae in the Class Monocotyledons. Later, Engler placed the family in the 7th order Spathiflorae under the class Monocotyledoneae. But, Hutchinson placed the family in a separate order Arales close to Typhales in the subphylum Monocotyledones. Takhtajan and also Cronquist placed the family under the order Arales associated with Lemnaceae.

Common plants

1. *Alocasia indica* (Roxb.) Schott. – Man kanda (Man kachoo),
2. *Arisaema album* Brown,
3. *A. costatum* Mart.,
4. *A. griffithii* Schott.,
5. *A. sikkimense* Stapf ex. Chatterjee,
6. *Colocasia esculenta* (L.) Schott. – Kachoo, ghuina or arvi,
7. *Typhonium trilobatum* (L.) Schott.

Key to genera

I. Flowers unisexual, monoecious, rarely dioecious :

1. Aquatic plants

i. Floating stemless herbs; leaves sessile in a rosette-like tuft ....*Pistia.*

ii. Submerged aquatic or marsh herbs; leaves tufted from a buried rootstock ....*Cryptocoryne.*

2. Terrestrial herbs

i. Tuberous; leaves entire, 3-5-lobed, cordate or hastate or pedate; flowers and leaves present together .... *Typhonium.*

ii. Tuberous; leaves solitary, 3-partite, segment pinnatisect; flowers appearing without leaves, male and female flowers contiguous, neuter wanting ... . *Amorphophallus.*

iii. Herbs usually tall, coarse; leaves and scapes arising from the rhizome; leaves may be peltate; spadix with a barren appendage, not adnate to the spathe –

a. Ovules few; basal ...... *Alocasia.*

b. Ovules many, parietal .... *Colocasia.*

II. Flowers bisexual :

1. Epiphytic climbing herbs

a. Perianth absent, berries free, ovule solitary .... *Scindapsus.*

b. Perianth of 4-6 segments, ovule solitary, plant unarmed .... *Pothos.*

2. Stout spinous aquatic or marsh herbs .... *Lasia.*

Plants of economic importance

The members of this family are important in respect of food, medicines and ornamentals :

A. **Food**

1. *Alocasia indica, A. cucullata.* The plants are cultivated for fleshy root stocks used as vegetables.
2. *Amorphophallus campanulatus* (Elephant foot). The large corm of this plant is used after boiling or used as vegetable.
3. *Colocasia esculenta* (Syn. *Colocasia antiquorum*). The plant is commercially cultivated for the starchy tuber used as food. The tubers are used in the production of industrial alcohol. The young leaves are used as vegetable.
4. *Monstera deliciosa* (Ceriman). The ripe fruits are edible.
5. *Remusatia vivipara* and *Xanthosoma niger.* Both corm and leaves of the plants are used as vegetable.

B. **Medicine**

1. *Acorus calamus.* The rhizome is useful in chronic stomach troubles and dyspepsia. The aromatic volatile oil is used as insecticide against bugs, lice, etc.
2. *Alocasia macrorrhiza.* The juicy extract of the stem is used to relieve pain from scorpion attack.
3. *Amorphophallus campanulatus* (Elephant foot). The large corm of this plant is recommended for dysentery and piles.

C. **Ornamental plants**

1. *Acorus calamus* var. *variegatus.* The plants are grown in the garden for their ornamented leaves.
2. *Alocasia indica* var. *metallica* and *A. lowii.* The plants are grown in the garden for beautification.
3. *Arisaema atrorubens.* The plants are cultivated in the garden for their beautiful and attractive spathe.
4. *Caladium bicolor* and *C. picturatum.* The plants are cultivated in the garden for their thin coloured leaves.
5. *Colocasia esculenta.* Some varieties (Elephant's ear) are grown in garden and also in tubs for their ornamental leaves.
6. *Dieffenbachia picta.* The plants are grown in tubs for their spotted and variegated leaves.
7. *Pistia stratiotes* (Water lettuce). These are floating herbs grown in aquaria, and also in garden water.
8. *Pothos aureus* (Money plant). Some varieties are climbers having variegated leaves, cultivated in the garden.

11. ALISMATACEAE

(Water plantain Family)

The family is represented by 16 genera and 100 species — cosmopolitan in distribution. In India, the family is represented by 7 genera and about 10 species found to grow throughout the plains.

The type genus of the family is ***Alisma.***

Systematic position

Bentham & Hooker		**Engler & Prantl**	
Division.	Angiospermia	Subdivision.	Angiospermae
Class.	Monocotyledons	Class.	Monocotyledoneae
Series.	Apocarpeae	Order.	Helobiae

Hutchinson		**Cronquist**	
Phylum.	Angiospermae	Division.	Magnoliophyta
Subphylum.	Monocotyledones	Class.	Liliopsida
Division.	Calyciferae	Subclass.	Alismatidae
Order.	Alismalales	Order.	Alismatales.

Takhtajan		**Thorne**	
Division.	Magnoliophyta	Class.	Angiospermae
Class.	Liliopsida	Subclass.	Liliidae
Subclass.	Alismatidae	Superorder.	Alismatanae
Superorder.	Alismatanae	Order.	Alismatales
Order.	Alismatales.		

Identifying characteristics

1. **Aquatic** or **marshy herbs.**
2. **Inflorescence on scape.**
3. Flower unisexual or bisexual.
4. Perianths 6 in 2 whorls, **outer sepaloid.**
5. Stamens 6 to many.
6. Carpels few to many, **apocarpous.**
7. Fruits are **etaerio of achene.**

Alismataceae : Distinctive features

Common characteristics	Exceptions
Vegetative characteristics	
Plant. Annual or perennial, aquatic or marshy, erect herbs, with usually a stought rhizome.	
Leaves. Generally radical or clustered at the nodes, simple with parallel venation, entire with long petioles, sagittate or hastate.	Venation reticulate – *Sagittaria guayanensis.*
Reproductive characteristics	
Inflorescence. Panicle or raceme or cymose umbel on scape.	
Flower. Actinomorphic, bisexual, bracteate, trimerous and hypogynous.	Unisexual – *Sagittaria*. Polygamous – *Limnophyton.*
Perianth. Tepals 6 (3+3), free, outer sepaloid and inner petaloid, imbricate.	
Androecium. Stamens 6-many, free.	Stamens numerous – *Sagittaria.*
Gynoecium. Carpel 3 to 6 or more, apocarpous, ovary superior, 1-celled with 1 to many basal anatropous ovules on marginal or superficial (*Butomus*) placentation.	Carpel 3 – *Wisneria;* Carpel about 18 in 1 whorl – *Sagittaria.* Carpel numerous – *Sagittaria, Ranalisma.*
Fruits. Etaerio of achene.	
Seeds. Minute, exalbuminous (non-endospermic) with straight, curved or horse-shoe shaped embryo.	

Floral formula

$\oplus ⚥ P_{3+3} A_{6-\alpha} \underline{G}_{3-\alpha}$

Phylogeny and systematics

The family Alismataceae is closely related to Ranunculaceae and also with Commelinaceae. Takhtajan (1959, 69) concluded that the order Alismales is linked with Nymphaeales. Further, he pointed out that there is no essential difference between the flowers of the Alismatales and Cabombaceae. Cronquist (1968) also had the same opinion.

Bentham and Hooker placed the family under the series Apocarpae. Later, Engler placed it under the order Helobiae. Still later, Hutchinson considered the family as a member under the order Alismatales. The placement of the family as made by Hutchinson was also accepted by Cronquist and Takhtajan. But Takhtajan named the order Alismales instead of Alismatales.

Common plants

1. *Alisma plantago* L. (Water plantain),
2. *Butomopsis lanceolata* Kunth.,
3. *Limnophyton obtusifolium* Miq.,
4. *Sagittaria sagittifolia* L. (Arrow-head),
5. *S. guayanensis* H.B.K.

Key to genera

I. Flowers hermaphrodite
   1. Fruit dehiscent of 6–7 follicles, stamens 8–12; ovules many .... *Butomopsis.*
   2. Fruit indehiscent of 3 or more achenes; stamens 6–9, ovule solitary ...... *Alisma.*

II. Flowers polygamous
   1. Unisexual flower, receptacle globose .... *Sagittaria.*
   2. Bisexual flower, receptacle flat .... *Limnophyton.*

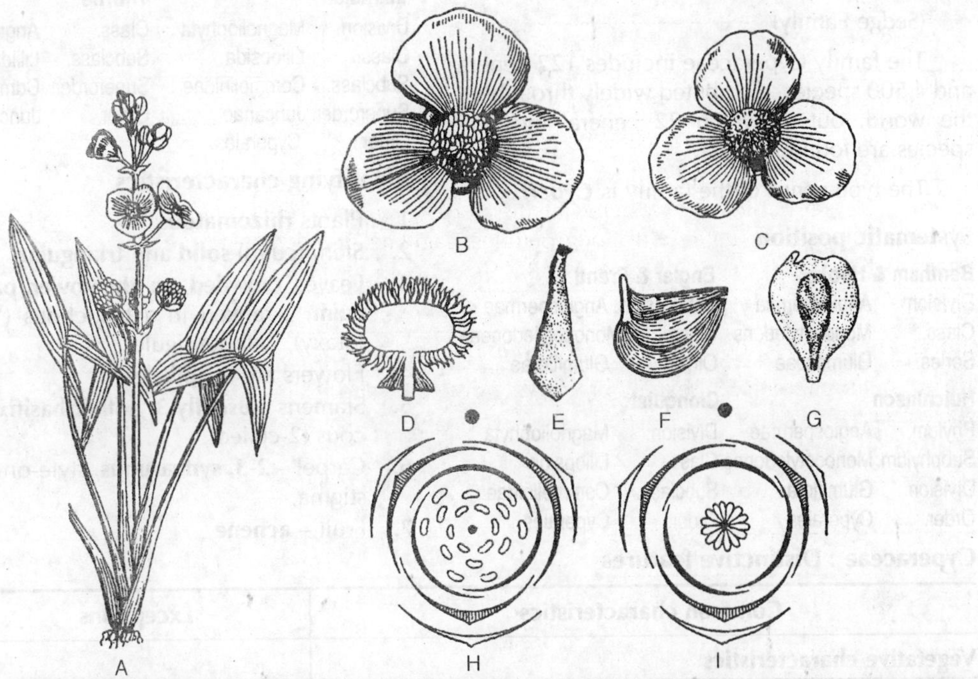

Fig. 4.168 : Alismataceae (*Sagittaria sagittifolia*) : A. Single plant; B. Staminate flower; C. Pistillate flower; D. V.S. of pistillate flower; E. Pistil; F. Fruit; G. V.S. of fruit; H. Floral diagram of staminate flower, and I. Floral diagram of pistillate flower

Plants of economic importance

Economically, the family is of little importance. Tubers of some species of *Sagittaria* are edible. A few species of *Sagittaria* and *Alisma* are grown in aquaria.

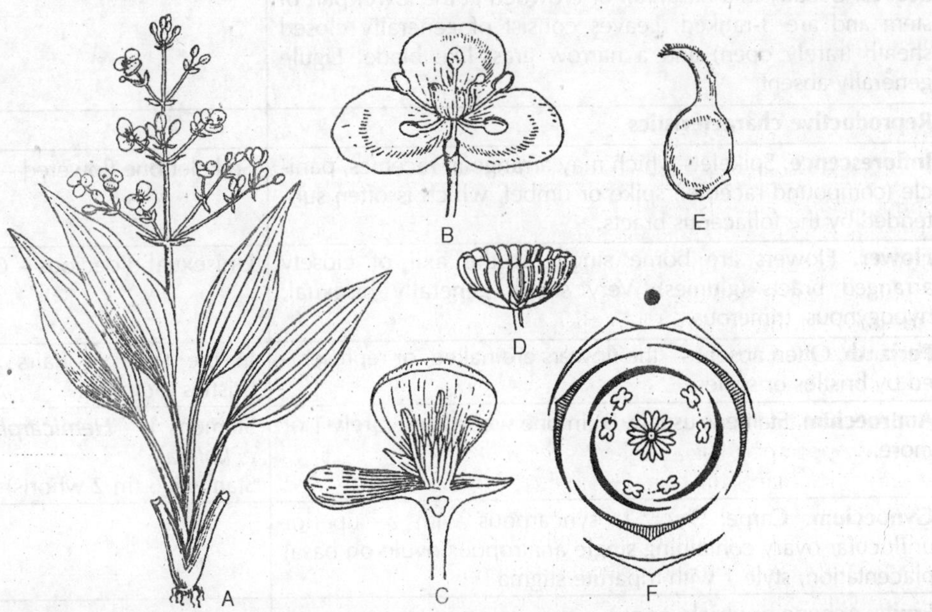

Fig. 4.169 : Alismataceae (*Alisma plantago*) : A. Single plant; B. Single flower; C. L.S. of a flower; D. Fruit; E. Pistil, and F. Floral diagram

12. Cyperaceae

(Sedge Family)

The family Cyperaceae includes 122 genera and 4,500 species, distributed widely throughout the world, out of which 22 genera and 405 species are found in India.

The type genus of the family is ***Cyperus***.

Systematic position

Bentham & Hooker

Division.	Angiospermia
Class.	Monocotyledons
Series.	Glumaceae

Engler & Prantl

Subdivision.	Angiospermae
Class.	Monocotyledoneae
Order.	Glumiflorae

Hutchinson

Phylum.	Angiospermae
Subphylum.	Monocotyledones
Division.	Glumiflorae
Order.	Cyperales

Cronquist

Division.	Magnoliophyta
Class.	Liliopsida
Subclass.	Commelinidae
Order.	Cyperales

Takhtajan

Division.	Magnoliophyta
Class.	Liliopsida
Subclass.	Commelinidae
Superorder.	Juncanae
Order.	Cyperales.

Thorne

Class.	Angiospermae
Subclass.	Liliidae
Superorder.	Commelinanae
Order.	Juncales

Identifying characteristics

1. Plants **rhizomatous**,
2. Stem (culm) **solid** and **triangular**,
3. Leaves **crowded on the lower part of the culm**; 3-ranked in a tristichous ($^1/_3$rd phyllotaxy), without ligule,
4. Flowers are in **spikelets**,
5. Stamens – **usually 3**, anther **basifixed**, dithecous (2-celled),
6. Carpel – **2-3, syncarpous**, style-one with 2-3 stigma,
7. Fruit – **achene**.

Cyperaceae : Distinctive features

Common characteristics	Exceptions
Vegetative characteristics	
Plants. They are grass-like, mostly perennial, sometimes annual herbs.	
Root. Fibrous in nature, those develop from sympodial rhizome.	
Stem (culm). Solid throughout, triangular, generally unbranched below the inflorescence.	Stem hollow – *Scirpus articulatus.*
Leaves. Usually in a basal tuft or crowded at the lower part of stem and are 3-ranked. Leaves consist of generally closed sheath (rarely open) and a narrow grass-like blade. Ligule generally absent.	
Reproductive characteristics	
Inflorescence. Spikelets which may arrange in racemes, panicle (compound raceme), spike or umbel, which is often subtended by the foliaceous bracts.	Spikelet one flowered – *Diplacrum.*
Flower. Flowers are borne singly in the axil of closely arranged bracts (glumes). Very small, generally bisexual, hypogynous, trimerous.	Unisexual dioecious – *Carex.*
Perianth. Often absent — the flowers are naked, or represented by bristles or scales.	In the form of hairs (*Eriophorum*), bristles (*Scirpus*).
Androecium. Stamens usually 3 (in one whorl) and rarely 1 or more.	Stamens 1 – *Hemicarpha* and *Bisboeckolera.* Stamens 6 (in 2 whorls) – *Gahnia.*
Gynoecium. Carpel 3 or 2, syncarpous with a superior unilocular ovary containing single anatropous ovule on basal placentation, style 1 with tripartite stigma.	
Fruits. Achene or utricle or nut.	
Seeds. Seeds with fleshy or mealy endosperm and minute embryo.	

Floral formula

$\oplus$ ⚥ $P_0\ A_3\ \underline{G}_{(2)\ or\ (3)}$

Phylogeny and systematics

The family is closely related with Gramineae (Poaceae). The chemical similarities between the two families (Cyperaceae and Gramineae) indicate their phyletic unity than convergence (Cronquist, 1981) and the Cyperaceae have been derived from Juncaceae (Takhtajan and Cronqiust).

Bentham and Hooker placed the family Cyperaceae in the last (7th) series Glumaceae under the class Monocotyledons. Engler includes Cyperaceae under the order Glumiflorae along with Gramineae. Hutchinson and later Takhtajan placed the families under separate orders Cyperales and Graminales or Poales. But Cronquist placed both the families under a single order Cyperales.

Common plants

1. *Carex plebeja* Clarke;
2. *C. speciosa* Kunth;
3. *Cyperus rotundus* L. (Mutha grass);
4. *C. iria* L.;
5. *C. kyllingia* Endl. (Syn. *Kyllinga monocephala* Rottb.
6. *C. tagetum* Roxb.;
7. *Juncellus inundatus* Clarke;
8. *Scirpus articulatus* L.
9. *Fimbristylis ferruginea* (L.) Vahl.

Key to genera

I. Ovary and fruit enclosed in a utricle ... *Carex.*

II. Ovary and fruit not enclosed in a utricle –

1. Flowering glumes distichous, scales or bristles 0
   - i. Style 2-fid; rachilla of spikelets deciduous; perennial herbs; stem trigonous; leaves narrow, radical ...... *Kyllinga.*
   - ii. Style 3-fid; rachilla of spikelets persistent; fruit trigonous; perennial or annual herbs; leaves all towards the

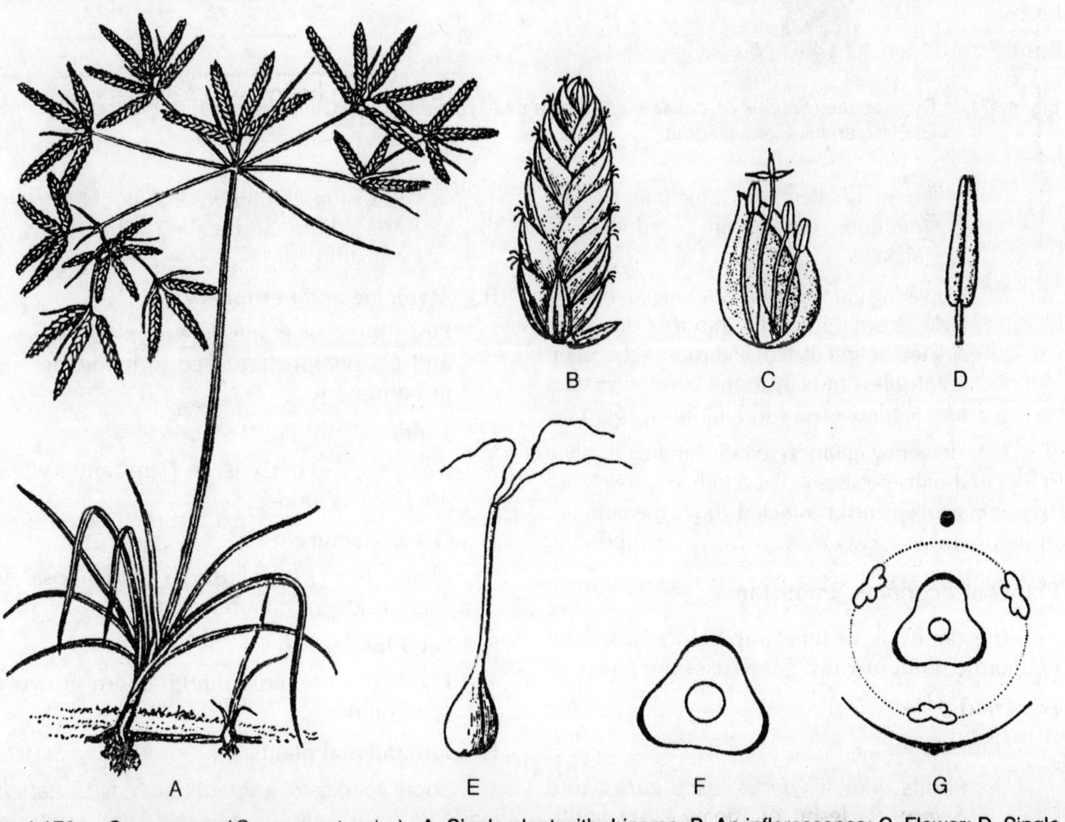

Fig. 4.170 : Cyperaceae (*Cyperus rotundus*) : A. Single plant with rhizome; B. An inflorescence; C. Flower; D. Single stamen; E. Pistil; F. T.S. of ovary, and G. Floral diagram

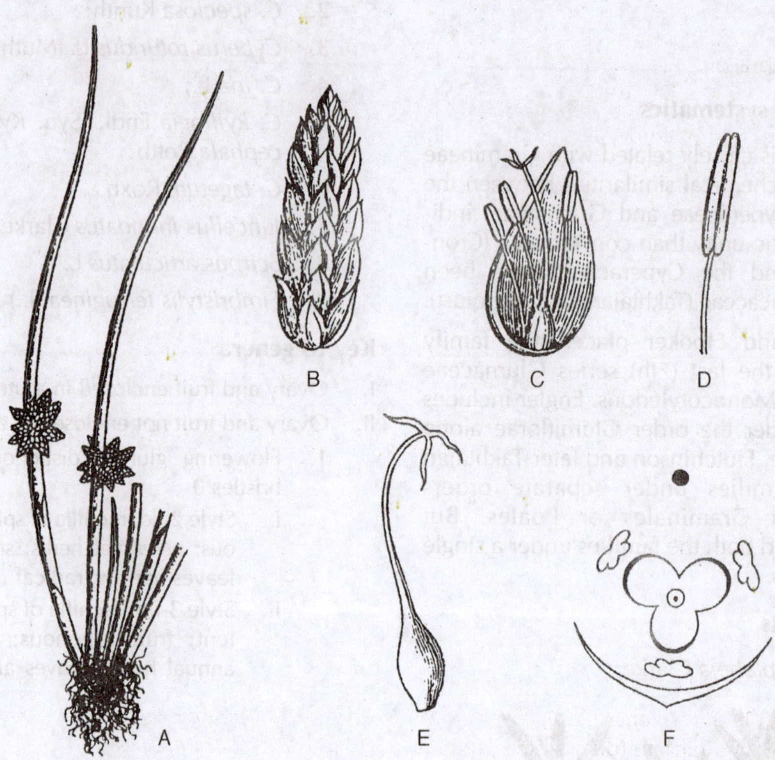

Fig. 4.171 : Cyperaceae (*Scripus articulatus*) : A. Single plant with rhizome; B. Inflorescence; C. Flower; D. Stamen; E. Pistil, and F. Floral diagram

base of stem or sometimes a few cauline, occasionally reduced to sheaths ....*Cyperus.*

2. Flowering glumes spirally arranged; base of style passing gradually into fruit and constricted or articulated; glabrous herbs, often tall, leafless or leafy at the base only; stem terete, trigonous or triquetrous ... *Scirpus.*

3. Flowering glumes spirally arranged, style usually persistent, flat and hairy; style-base swollen and constricted above the nut .... *Fimbristylis.*

Plants of economic importance

The family is of little importance from the economic point of view. Some uses are given :

A. **Food**

Edible tuber :

i. Plants like *Cyperus esculentus* and *Scirpus grossus* var. *kysoor* are cultivated for their tubers.

ii. In China, the tubers of *Eleocharis dulcis* are edible and called Chinese water chestnut.

B. **Medicine** and **perfumery**

Dry tubers of plants like *Cyperus scarious* and *C. rotundus* are used in medicine; also in perfumery.

C. **Fodder**

Many species of *Cyperus, Carex* and *Kyllinga* are used as fodder.

D. **Green manure**

Plants like *Rhynchospora corymbosa* are used as green manure.

E. **Sand binder**

Plants like *Remirea maritima* are grown as sand binder.

F. **Ornamental plants**

Some species of *Cyperus* such as *C. flabelliformis, C. alternifolius* etc. are grown as ornamental plants.

13. Gramineae (Poaceae)

(Grass Family)

The family Gramineae is one of the largest and economically most important family in Plant Kingdom. It includes about 656 genera and more than 9,975 species, distributed widely throughout the world. About 239 genera and 1,180 species are found in India.

The type genus of the family is ***Poa***.

Systematic position

Bentham & Hooker		**Engler & Prantl**	
Division.	Angiospermia	Subdivision.	Angiospermae
Class.	Monocotyledons	Class.	Monocotyledoneae
Series.	Glumaceae	Order.	Glumiflorae
Hutchinson		**Cronquist**	
Phylum.	Angiospermae	Division.	Magnoliophyta
Subphylum.	Monocotyledones	Class.	Liliopsida
Division.	Glumiflorae	Subclass.	Commelinidae
Order.	Graminales	Order.	Cyperales.

Gramineae : Distinctive features

Common characteristics	Exceptions
Vegetative characteristics	
Plants. Plants are annual, biennial or most frequently perennial herbs.	Tree – *Bambusa, Dendrocalamus.*
Root. Tuft of fibrous and adventitious.	
Rhizome. Present in perennial grasses formed by the lower internode of stem.	
Stem (culm). Erect, prostrate or creeping, usually cylindrical and usually with hollow internodes (fistular) and solid nodes.	Internodes solid – *Zea, Sorghum, Saccharum.*
Leaves. Simple, linear to lanceolate, rarely ovate, with sheathing leaf base, half alternate (distichous), ligulate.	Ligule absent in some species of *Echinochloa.*
Reproductive characteristics	
Inflorescence. The basic unit of inflorescence is spikelet. It may be stalked or sessile. Several spikelets arrange on branches as racemes (*Paspalum*), panicle (*Avena*) or spikes (*Triticum*).	
Spikelet. Each spikelet consists of a very short axis (rachilla) bearing a number of bracts, bracteoles and flower. The two lowermost bracts are sterile and termed **empty glumes**, the third one is flowering and called **lemma**. There is a bracteoles called **palea** placed opposite the flowering glume (lemma). Each spikelet may be one- or many-flowered. Lemma bears a flower in its axil.	
Flower. Small, bisexual, hypogynous and zygomorphic, chasmogamous.	Monoecious (*Zea, Coix*) or dioecious (*Spinifex littoreus*).
Perianth. Perianth is represented generally by two minute or fleshy and hyaline structure, the **lodicules.**	3 lodicules present in *Bambusa.*
Androecium. Stamens 3 or 6 (arranged in two whorls i.e., 3+3), free, anthers 2-celled and versatile.	Stamen 1 in *Uniola*, 2 in *Coleanthus*, numerous in *Pariana.* Androecium monadelphous in *Neohouzeaua, Ochlandra.*
Gynoecium. Carpel 1, ovary superior, unilocular, containing single anatropous ovule on basal placentation, style 2, stigma–feathery.	Style 3 in *Bambusa.*
Fruits. Mostly 1-seeded caryopsis.	Berry in some *Bambusa,* Utricle in *Eleusine.*

Takhtajan		**Thorne**	
Division.	Magnoliophyta	Class.	Angiospermae
Class.	Liliopsida	Subclass.	Liliidae
Subclass.	Commelinidae	Superorder.	Commelinanae
Superorder.	Poanae	Order.	Restionales
Order.	Poales.		

Identifying characteristics

1. Stem is cylindrical and **generally with hollow internodes** (**fistular**),
2. Leaves with distinct **sheathing leaf bases** those enclose the internodes, **ligulate**, phyllotaxy – alternate and **distichous**,
3. Inflorescence – **spikelet**,
4. Flowers small, **generally bisexual**, hypogynous and zygomorphic,
5. Stamens 6 (3+3) or 3, **anther – versatile**,
6. Carpel 1, style usually two, **stigma feathery**,
7. Fruit – **caryopsis**.

Floral formula

$\oplus ⚥ P_{0 \text{ or } 2\text{-}3} A_{3 \text{ or } 3+3} \underline{G}_{1 \text{ or } (3)}$

Phylogeny and systematics

The family is closely related with Cyperaceae. The chemical similarities between the two families (Cyperaceae and Gramineae) indicate their phyletic unity than convergence (Cronquist, 1981). The Gramineae is considered to be a very advanced group of angiosperm evolved from primitive liliaceous ancestor.

Bentham and Hooker placed the family Gramineae under the series Glumaceae along with the family Cyperaceae. Engler included Cyperaceae under the order Glumiflorae along with Gramineae. Hutchinson and Takhtajan placed the two families under separate orders Cyperales and Graminales or Poales respectively. But Cronquist placed both the families under a single order — Cyperales.

Common plants

1. *Bambusa arundinacea* (Retz.) Willd.;
2. *Brachiaria reptans* (L.) Gard.;
3. *Chloris barbata* SW.;
4. *Chrysopogon aciculatus* Trin;
5. *Cynodon dactylon* Pers. (Bermuda grass or Durba grass);
6. *Dactyloctenium aegyptium* (L.) Beauv.;
7. *Eragrostis tenella* (L.) Beauv.;
8. *Eleusine indica* (L.) Gaertn.;
9. *Imperata cylindrica* (L.) Beauv.;
10. *Oplismenus burmannii* (Retz.) Beauv.;
11. *Paspalidium flavidum* (Retz.) Camus;
12. *Saccharum spontaneum* L.

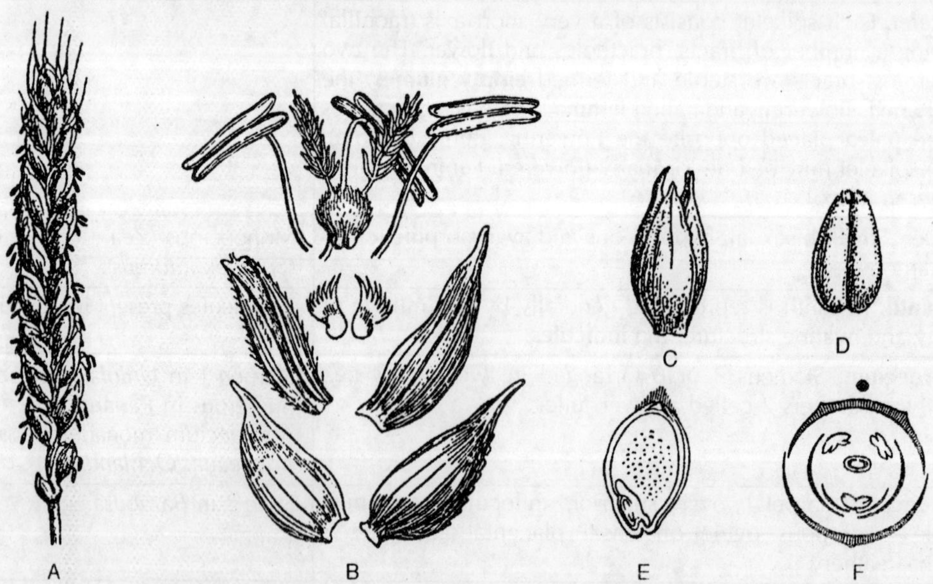

Fig. 4.172 : Gramineae (*Triticum aestivum*) : A. Single inflorescence; B. Flower dissected out; C. Fruit (caryopsis) enclosed in glumes; D. Single fruit; E. L.S. of grain, and F. Floral diagram

Key to genera

I. Herbs, flowers not dioecious :

1. Spikelets 2-flowered, upper bisexual, lower male or neuter –
   - i. Mature spikelets surrounded by an involucre of bristles which fall off along with the spikelets... *Pennisetum.*
   - ii. Spikelets subtended by an involucre of bristles .... *Setaria.*
   - iii. Spikelets not subtended by bristles; large grass; stem solid; leaves broad, flat .... *Thysanolaena.*
   - iv. Slender grasses, stems creeping below, branching; leaves ovate-lanceolate, thin, flat; ligules a ridge of long hairs .... *Oplismenus.*
   - v. Annual or perennial grasses; leaves broad or narrow, flat, or rarely plicate; ligules 0 or hairy; glume II not fimbriate, glume III with a palea or quite empty .... *Panicum.*
   - vi. Annual or perennial grasses; flowers (spikelets) in digitate spikes; lowest glume minute but usually present; glume III with a minute palea .... *Digitaria.*
2. Spikelets 1-flowered
   - i. Spikelets articulated on their pedicels and deciduous from them
     - a. Glumes I, II minute ...... *Oryza.*
     - b. Floating grass, diffusely branched, rooting in dense masses at the nodes; glumes I and II absent .... *Hygroryza.*
   - ii. Spikelets usually in pairs –
     - a. *Spikelets all similar –*

       Racemes of spikelets in compound panicles

       Panicle spiciform, silky, rachis tough; erect perennial grasses, stem leafy, internode solid, leaves narrow .... *Imperata.*

       Panicle much-branched, silky, rachis fragile, spikelets awnless; tall perennial grasses .... *Saccharum.*
     - b. *Spikelets dissimilar –*

       Spikelets in alternating pairs or the lower solitary –

       Sessile spikelet usually many, inflorescence long .... *Andropogon.*
   - iii. Male and female spikelets in different spikes –
     - a. Fruiting spikelets crowded on a spongy rachis, the grain exposed; a tall annual cultivated grass ... . *Zea.*
     - b. Fruiting spikelet enclosed in the stony polished nut-like bract ... .*Coix.*
3. Spikelets 2- or more-flowered; annual or perennial cultivated grasses; leaves flat; spikelet with twisted awns .... *Avena.*
4. Spikelets 3-7-flowered, in decompound panicles; tall perennial grasses with long leaves .... *Phragmites.*
5. Spikelets many-flowered, not pedicellate, with long silky hairs; outer glumes shorter than the lowest flowering glume; grain very minute; annual or perennial glabrous grasses, stems erect, leaves narrow .... *Eragrostis.*
6. Spikelets 2-seriate and secured –

   Spikelets not clustered, 1-flowered, awnless, spikes digitate; perennial creeping grasses, leaves narrow, flat ...... *Cynodon.*
7. Spikelets 3-12-flowered in numerous digitate or whorled spikes .... *Eleusine.*
8. Spikelets inserted in notches or pits of a simple rachis
   - i. Spikelets solitary at the nodes of the spikes; plant of spikelets tangential to the rachis; annual or biennial cultivated grass .... *Triticum.*
   - ii. Spikelets 2 or more, collected in fascicles at the nodes of the spikes; an annual cultivated grass ... *Hordeum.*

II. Shrubs or trees, rarely climbing; stem-sheaths broad, blade often triangular; leaves short-petioled

1. Pericarp crustaceous, not adnate to seed ....*Dendrocalamus.*
2. Pericarp thin, adnate to seed ... *Bambusa.*

Plants of economic importance

The Gramineae is the most valuable family of the entire plant kingdom. They yield cereal food useful to mankind and fodder for the animals — both domestic and wild. Besides cereals, they yield a number of products useful in our day-to-day life. These are sugar, paper, volatile and edible oils, fibres, alcohol, beverages, thatching and building materials, adhesives, plastics etc.

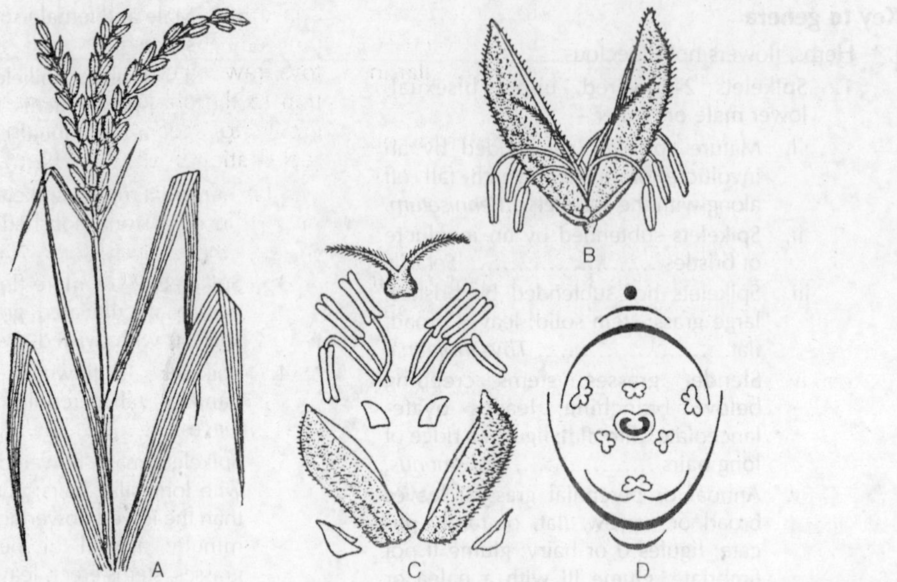

Fig. 4.173 : Gramineae (*Oryza sativa*) : A. Portion of a plant with inflorescence; B. Single flower; C. Flower dissected out, and D. Floral diagram

A. **Cereals**

The commonly used cereals are :

1. *Avena sativa* (Oats, Jai).
2. *Eleusine coracana* (Finger millet, Ragi).
3. *Hordeum vulgare* (Barley, Job, Jawa).
4. *Oryza sativa* (Paddy, Dhan, Chaul).
5. *Panicum miliaceum* (Common millet, Cheena, Sama).
6. *Pennisetum typhoides* (Pearl millet, Bajra).
7. *Secale cereale* (Rye).
8. *Sorgham glauca* (Italian millet, Kakni).
9. *S. vulgare* (Jowar).
10. *Triticum aestivum* (Wheat, Gom, Gehun).
11. *Zea mays* (Maize, Bhutta, Makka).

B. **Fodder**

Several grasses are used as fodder by domestic animals. Some — *Poa, Festuca, Agrostis* etc. are cultivated commercially as forage. Name of some grasses used as fodder are :

1. *Cenchrus ciliaris.*
2. *C. pennisetiformis.*
3. *Chrysopogon gryllus.*
4. *C. montanus.*
5. *Cynodon dactylon.*
6. *Dactyloctenium aegypticum.*
7. *Heteropogon contortus.*
8. *Lolium multiflorum.*
9. *L. perenne.*
10. *Phalaris aquatica.*
11. *P. canariensis.*
12. *Poa annua.*
13. *P. pratensis.*

C. **Sugar**

Saccharum officinarum (Sugarcane, Akh, Ganna). Sugar and 'molasses' are prepared from the juicy sap of the stem.

D. **Paper**

Saccharum officinarum (Sugarcane, Akh, Ganna). The baggase (parts left after extraction of juicy sap) is used in the preparation of straw-board.

E. **Volatile and edible oil**

Some grasses are important for the source of important volatile oil used as mosquito repellent, and also used in perfumery and in medicines. These are :

1. *Cymbopogon citratus* (Lemon grass oil from leaves).

2. *C. nardus* (Citronella oil). The leaves are used to extract Lemon grass oil used in phenyle, thereby it is very popular in the market (white phenyle).
3. *C. martinii* var. *motia* (Palmarosa oil from the leaves).
4. *C. martinii* var. *sofia* (Ginger grass oil).
5. *Vetiveria žizanoides* Nash. (Syn. *Andropogon muricatus* Retz.) Vetiver oil or Khus oil is obtained from the roots and rhizomes.
6. *Oryza sativa*. The edible rice bun oil is extracted from the husks.

F. **Bamboos**

These are tall arborescent grasses consisting of about 30 genera and 550 species; grow mostly in humid tropical regions. Some indigenous genera are *Bambusa, Arundinaria, Cephalostachyum, Melocanna, Gigantochloa* and *Dendrocalamus,* etc.

They are used in various purposes like the construction of building, ladder, walking sticks, umbrella handles, musical instruments, mats, toys, raw material in paper industry etc. In addition to the above importance the bamboos are grown to check soil erosion and also for soil reclamation in eroded regions :

1. *Arundinaria racemosa*. The plants are used as fodder for cattle. The leaves are the favourite food of elephants.
2. *Dendrocalamus arundinaceus*. The plant grows very fast and is used as a good source of raw material in paper industry.
3. *Phyllostachys* and *Arundinaria*. The plants are grown on pots and used for decoration of houses.
4. *Phragmites karka* (Common reed) and *Arundo donax* (Reed). Stem of both the plants are used for making musical pipes and many other things.
5. *Thysanolaena maxima* (Broom grass). The inflorescence (panicle) is used for making soft brooms.

Table 4.14 : Comparison of Gramineae and Cyperaceae

Characteristics	Gramineae	Cyperaceae
1. Plant	Annual or perennial herbs, rarely shrubs or tree.	Mostly perennial or sometimes annual herbs.
2. Stem	Stem cylindrical and generally with hollow internodes.	Stem triangular and solid.
3. Leaves	Leaves simple, ligulate, districhous, leaf-sheath open.	Leaf simple, without ligule, tristichous, leaf sheath closed.
4. Glumes and bracts	Empty glumes, flowering glumes and palea are present. Involucral bracts absent.	Glumes, palea etc. are absent. Involucral bract present.
5. Perianth	Perianth is represented by 2-3 lodicules.	Perianth is either absent or represented by bristles or scales.
6. Stamens	Stamens 3 or 6 (3+3); anther versatile.	Stamens usually 3, sometimes 1; anthers basifixed.
7. Gynoecium	Carpel 1 or 3 (of which two are abortive), Styles 2; stigma 2, hairy or feathery.	Carpels 2 or 3. Style 1; stigmas 2-3, plumose.
8. Fruit	Fruit caryopsis.	Fruit achene.
9. Embryo	Embryo small, straight and endosperm is starchy.	Embryo small, endosperm is mealy or fleshy.
10. Cotyledon	Cotyledon does not come out during germination.	Apical part of cotyledon comes out during germination.

G. **Fibres**

Saccharum spontaneum and *Erianthus munja* (Syn. *Saccharum munja*). The fibres extracted from the leaves are used for making ropes.

H. **Alcohol**

Saccharum officinarum (Sugarcane, Akh, Ganna). The byproduct of the sugar industry is the *molasses* used in the manufacture of industrial alcohol.

I. **Medicinal**

Secale cereale (Rye). The plant is cultivated to develop a disease, the Ergot disease of Rye (Pathogen : *Claviceps purpurea*) on the inflorescence. After infection the ovary becomes converted into a hard horney structure, the sclerotium. It contains many alkaloids of which Ergometrine is important, used to control the bleeding of mother during child birth (consult Chapter 4 (Vol. I), page 477, for more information).

J. **Thatching and building materials**

Culms (stems) of different species of *Bambusa* such as *B. tulda, B. vulgaris* are used either entirely or by splitting in the construction of houses.

Outline of George Bentham and Joseph Dalton Hooker's System (Genera Plantarum, 1862-1883)

Class : Dicotyledons

Subclass. 1. Polypetalae

Series 1. THALAMIFLORAE (Petals and stamens hypogynous)

Cohort 1. Ranales
Natural orders. Ranunculaceae, Dilleniaceae, Calycanthaceae, Magnoliaceae, Anonaceae, Menispermaceae, Berberideae, Nymphaeaceae.

Cohort 2. Parietales
Natural orders. Sarraceniaceae, Papaveraceae, Cruciferae, Cappari-deae, Resedaceae, Cistineae, Violarieae, Canellaceae, Bixineae.

Cohort 3. Polygalineae
Natural orders. Pittosporeae, Tremandreae, Polygaleae, Vochysiaceae.

Cohort 4. Caryophyllinae
Natural orders. Frankaniaceae, Caryophylleae, Portulacaceae, Tamariscineae.

Cohort 5. Guttiferales
Natural orders. Elatineae, Hypericineae, Guttiferae, Ternstroemiaceae, Dipterocarpeae, Chlaenaceae.

Cohort 6. Malvales
Natural orders. Malvaceae, Sterculiaceae, Tiliaceae.

Series 2. DISCIFLORAE, Stamens usually definite, inserted upon or inside or outside of a development of the floral axis which forms a ring or cushion at the base of the ovary, or is broken up into glands. Ovary superior.

Cohort 7. Geraniales
Natural orders. Lineae, Humariaceae, Malpighiaceae, Zygophyllaceae, Geraniaceae, Rutaceae, Simarubeae, Ochnaceae, Burseraceae, Meliaceae, Chailletiaceae.

Cohort 8. Olacales
Natural orders. Olacineae, Ilicineae, Cyrillaceae.

Cohort 9. Celastrales
Natural orders. Celastrineae, Stackhousieae, Rhamneae, Ampelideae.

Cohort 10. Sapindales
Natural orders. Sapindaceae, Sabiaceae, Anacardiaceae.

Ordines anomali. Coriarieae, Moringeae

Series 3. CALYCIFLORAE, Petals and stamens perigynous. Ovary often more or less enclosed by the development of the floral axis, sometimes inferior.

Cohort 11. Rosales
Natural orders. Connaraceae, Leguminosae, Rosaceae, Saxifrageae, Crassulaceae, Droseraceae, Hamamelideae, Bruniaceae, Halorageae.

Cohort 12. Myrtales
Natural orders. Rhizophoreae, Combretaceae, Myrtaceae, Melastomaceae, Lythraceae, Onagraceae.

Cohort 13. Passiflorales
Natural orders. Samydaceae, Loaseae, Turneraceae, Passifloreae, Cucurbitaceae, Begoniaceae, Datisceae.
Cohort 14. Ficoidales
Natural orders. Cacteae, Ficoideae.
Cohort 15. Umbellales
Natural orders. Umbelliferae, Araliaceae, Cornaceae.

Subclass. 2. Gamopetalae

Series 1. INFERAE. Ovary inferior, Stamens as many as the corolla-lobes, rarely fewer.
Cohort 1. Rubiales
Natural orders. Caprifoliaceae, Rubiaceae
Cohort 2. Asterales
Natural orders. Valerianeae, Dipsaceae, Calycereae, Compositae.
Cohort 3. Campanales
Natural orders. Stylideae, Goodenoveae, Campanulaceae.

Series 2. HETEROMERAE. Ovary generally superior. Stamens as many as the corolla-lobes, or more, epipetalous or free from the corolla. Carpels more than two.
Cohort 4. Ericales
Natural orders. Ericaceae, Vaccineae, Monotropeae, Epacrideae, Diapensiaceae, Lennoaceae.
Cohort 5. Primulales
Natural orders. Plumbagineae, Primulaceae, Myrsineae.
Cohort 6. Ebenales
Natural orders. Sapotaceae, Ebenaceae, Styraceae.

Series 3. BICARPELLATAE. Ovary generally superior, Stamens alternate with the corolla-lobes and equal in number or fewer. Carpels usually two.
Cohort 7. Gentianales
Natural orders. Oleaceae, Salvadoraceae, Apocynaceae, Asclepiadaceae, Loganiaceae, Gentianaceae.
Cohort 8. Polemoniales
Natural orders. Polemoniaceae, Hydrophyllaceae, Boraginaceae, Convolvulaceae, Solanaceae.
Cohort 9. Personales
Natural orders. Scrophulariaceae, Orobanchaceae, Lentibulariaceae, Columelliaceae, Gesneriaceae, Bignoniaceae, Pedaliaceae, Acanthaceae.
Cohort 10. Lamiales
Natural orders. Myoporineae, Selagineae, Verbenaceae, Labiatae.
Ordines anomali : Plantagineae.

Subclass. 3. Monochlamydeae

Series 1. CURVEMBRYEAE. Embryo curved round the generally mealy albumen. Ovules generally solitary, Flowers usually hermaphrodite. Stamens equal in number to or fewer than the perianth segments.
Natural orders. Nyctagineae, Illecebraceae, Amaranthaceae, Chenopodiaceae, Phytolaccaceae, Batideae, Polygonaceae.

Series 2. MULTIOVULATAE AQUATICAE. Submerged herbs. Ovary syncarpous, Ovules numerous.
Natural orders. Podostemaceae.

Series 3. MULTIOVULATAE TERRESTRIS. Terrestrial herbs or shrubs, Ovary syncarpous, ovules numerous.
Natural orders. Nepenthaceae, Cytinaceae, Aristolochiaceae.

Series 4. MICROEMBRYEAE. Embryo very small in a copious albumen, Ovary syncarpous or apocarpous. Ovules usually solitary.
Natural orders. Piperaceae, Chloranthaceae, Myristicaceae, Monimiaceae.

Series 5. DAPHNALES. Ovary usually monocarpellary, Ovules solitary or in pairs. Generally trees or shrubs with hermaphrodite flowers. Perianth sepaloid in one or two series.
Natural orders. Laurineae, Proteaceae, Thymelaeaceae, Penaeaceae, Elaeagnaceae.

Series 6. ACHLAMYDOSPOREAE. Ovary unilocular, 1-3-ovuled, Seeds devoid of testa. Albumen naked, Perianth sepaloid or petaloid.
Natural orders. Loranthaceae, Santalaceae, Balanophoreae.

Series 7. UNISEXUALES. Flowers unisexual, Ovary syncarpous or monocarpellary. Ovules solitary or in pairs. Seeds albuminous or exalbuminous, Perianth sometimes absent.
Natural orders. Euphorbiaceae, Balanopseae, Urticaceae, Platanaceae, Leitneriaceae, Juglandaceae,

Myricaceae, Casuarinaceae, Cupuliferae.

Series 8. ORDINES ANOMALI. Near the last series but not closely allied to any other Order. Flowers often unisexual.

Natural orders. Salicineae, Lacistemaceae, Empetraceae, Ceratophyllaceae.

Class : Gymnospermae

Natural orders. Gnetaceae, Coniferae, Cycadaceae.

Class : Monocotyledons

Series 1. MICROSPERMAE, At least the inner perianth petaloid, Ovary inferior, generally unilocular with three parietal placentae. Seeds very small, numerous and exalbuminous.

Natural orders. Hydrocharitaceae, Burmanniaceae, Orchidaceae.

Series 2. EPIGYNAE. At least the inner perianth petaloid. Ovary with a few exceptions inferior. Albumen copious.

Natural orders. Scitamineae, Bromeliaceae, Haemodoraceae, Irideae, Amaryllideae, Taccaceae, Dioscoreaceae.

Series 3. CORONARIEAE. At least the inner perianth petaloid, Ovary free, Albumen copious.

Natural orders. Roxburghiaceae, Liliaceae, Pontederiaceae, Philydraceae, Xyridaceae, Mayaceae, Commelinaceae, Rapataceae.

Series 4. CALYCINAE. Perianth small, sepaloid, somewhat stiff or herbaceous. Ovary free, Albumen copious.

Natural orders. Flagellariaceae, Juncaceae, Palmae.

Series 5. NUDIFLORAE. Perianth 0 or reduced to scales or setae. Ovary superior, carpels solitary or, if many, syncarpous, 1-many-ovuled. Seeds usually albuminous.

Natural orders. Pandaneae, Cyclanthaceae, Typhaceae, Araceae, Lemnaceae.

Series 6. APOCARPAE. Perianth in 1 or 2 series or 0, Ovary superior, Carpels solitary or free. Seeds exalbuminous.

Natural orders. Triuridaceae, Alismaceae, Naiadaceae.

Series 7. GLUMACEAE. Flowers in heads or spikelets, subtended by generally imbricated bracts. Perianth small, scale-like, or glumaceous, or 0. Ovary 1-ovuled or divided into 1-ovuled loculi, Seeds albuminous.

Natural orders. Eriocauleae, Centrolepideae, Restiaceae, Cyperaceae, Gramineae.

NOTES ON SOME PLANTS WITH THEIR ATTRACTIVE FEATURES

1. **Oldest living plant** : This was thought to be *bristlecone pines* (***Pinus longaeva/P. aristata***), living for nearly 5,600 years, high in the White Mountains of California and the Snake Range of eastern Nevada (Fig. 1.103).

2. **Oldest living fossil** : Oldest living fossil or the plant genera that lived during ancient times and still survives on earth today is *maiden hair tree* (***Ginkgo biloba***). Leaf imprints of an ancestral species of ***Ginkgo*** resembling the present-day ***Ginkgo biloba*** (Fig. 4.174) have been found abundantly in sedimentary rocks of the Jurassic and Triassic Periods (135-210 million years ago) when Dinosaurs roamed the earth.

3. **World's most massive living things** : Ancient bristlecone pines and creosote bush rings, the world's record for longevity went to the magnificent *giant sequoia* (***Sequoiadendron giganteum***) of the western slopes of the Sierra Nevada. The greatest authenticated age of a giant sequoia, derived from counting annual rings on a cut stump, was nearly 3,200 years. The largest tree, named General Sherman, is 272 feet (83 m) tall with a massive trunk 35 feet (11 m) in diameter and 109 feet (33 m) in circumference at the base. Even more remarkable is the fact that at a point 120 feet (36 m) in the air the trunk of General Sherman is still 17 feet (5 m) in diameter. It has been estimated to contain over 600,000 board feet of timber, enough to build 120 average-sized houses. In fact, a single giant sequoia may contain more wood than is found on several acres of some of the finest virgin timberland in the Pacific Northwest. The trunk of General Sherman alone weighs nearly 1,400 tons. By way of comparison, this is roughly equivalent to 15 adult blue whales, 10 diesel-electric train locomotives, or 25 average-sized military battle tanks.
   Another conifer species called the *Montezuma bald cypress* (***Taxodium mucronatum***) occasionally grows into a huge tree. One enormous specimen of this tree grows in the Churchyard of Santa Maria de Tule near Oaxaca, Mexico, called "El Gigante" by the locals. It is one of the most massive of all living things with a trunk circumference of 140 feet (43 m), larger than the General Sherman giant sequoia. With a diameter of 50 feet (15 m), the trunk of this spectacular tree is literally the size of a house. This gigantic tree was once thought to be

Fig. 4.174 : *Ginkgo biloba* : Plant bearing fruits

[Courtesy : Dr. M. R. Mazumdar]

10,000 years old, but botanists now consider it to be a youngster of only 1,500 to 2,000 years old.

4. **World's largest tree crown** : The Indian *banyan* (***Ficus benghalensis***) has enormous trunks up to 100 feet (30 m) or more in circumference, but do not grow as tall (Fig. 4.7, Page 301). It also has the record for the world's largest (spreading) tree crown, with 1,000 pillar-like prop roots supporting massive limbs that cover four acres of land.

5. **World's largest living organisms** : Many species of soil fungi have their hyphae intimately attached to the roots of forest trees in a symbiotic association called mycorrhizae ("fungus-roots"). Some mycorrhizal associations can be enormous. A single individual of ***Armillaria bulbosa*** has been discovered that spreads over more than 30 acres of forest soil in northern Michigan and may be one of the **world's largest living organisms**. Some scientists speculate that it was spawned by a single spore thousands of years ago. Another ***Armillaria*** in Washington was recently found to consist of a subterranean mycelial network with erect, above-ground mushrooms covering more than a thousand acres of forest soil.

6. **World's tallest tree** : The world's record for the tallest tree goes to another cone-bearing tree native to California, the coast *redwood* (***Sequoia sempervirens***). In fact, the tallest living redwood on record stands 367 feet (112 m), 62 feet (19 m) taller than the Statue of Liberty.

7. **World's heaviest and hardest wood** : Certainly one of the world's heaviest and hardest ironwoods is the Caribbean tree called *lignum vitae* (***Guaiacum officinale***), with a specific gravity of 1.37. The name *lignum vitae* means "wood of life," owing to the medicinal properties of the sweet-smelling resin. The density and high resin content of the wood make it extremely resistant to friction and abrasion and account for its remarkable self-lubrication properties. Under certain conditions it actually wears better than iron. In fact, the highly-prized wood was used for end grain thrust blocks which lined the propeller shafts of steamships. [Note: The **Guinness Book of World Records** lists the South African black ironwood (***Olea laurifolia***) as the heaviest wood with a specific gravity of 1.49. This is rather doubtful since the specific gravity of pure cell wall material is 1.5 (i.e., without any cellular structure), and samples of ***Olea laurifolia*** tested only weighed in at about 1.11.]

8. **World's Softest and lightest wood** : The tropical American *balsawood tree* (***Ochroma pyramidale***) is one of the world's softest and lightest woods with a specific gravity of only 0.19. The wood is used in making models of aircraft, dolls etc.

9. **World's smallest flowering plant** : Each plant is shaped like a microscopic **green football with a flat top**. An average individual plant of the Asian species ***Wolffia globosa***, or the equally minute Australian species ***W. angusta***, is small enough to pass through the eye of an ordinary sewing needle, and 5,000 plants could easily fit into a thimble (Fig. 4.175). It is difficult to say which is the smaller of the two, but perhaps ***W. globosa*** may be slightly smaller. An average individual plant is 0.6 mm long (1/42 of an inch) and 0.3 mm wide (1/85th of an inch). It weighs about 150 micrograms (1/190,000 of an ounce), or the approximate weight of 2-3 grains of table salt. One plant is 165,000 times shorter than the tallest

A

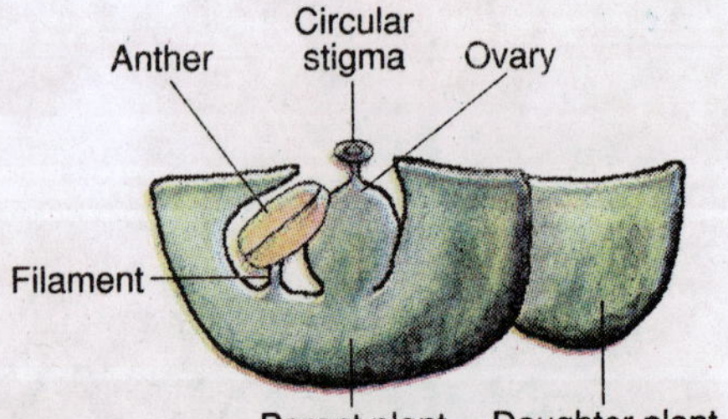

B

Fig. 4.175 : A. About 5,000 plants of *Wolffia* in a thimble, B. Lateral view of flowering *Wolffia globosa* showing the dorsal floral cavity

Australian eucalyptus (***Eucalyptus regnans***) and seven trillion times lighter than the most massive giant sequoia (***Sequoiadendron giganteum***). Another mind-boggling comparison is the size or volume of a single *Wolffia* plant: roughly intermediate between a water molecule and the planet earth. ***Arceuthobium minutissimum***, a stem parasite on *Pinus*, is the most minute dicot plant which is endemic in India.

10. **Smallest fruit** : *Wolffia* also produces the world's smallest fruit, although this record is not as yet recognised in the **Guinness Book**. Each one-seeded fruit is about the size of a cuboidal grain of ordinary table salt (0.3 mm long) and weighs about 70 micrograms (1/400,000 of an ounce). It is roughly five billion times lighter than a 900 pound squash.

11. **Largest fruit** : According to '**Cucurbits**', the official newsletter of the World Pumpkin Confederation, a 1993 record-breaking **pumpkin weighed in at 836 pounds** and a giant squash tipped the scales at just over 700 pounds. One year later at the "gourd lumose" in Port Elgin, Ontario, the reign of the pumpkin was broken by a **900 pound squash**. There have been other unofficial records for pumpkins exceeding 900 pounds. As per the official World Pumpkin Confederation, the pumpkin must be cream-yellow to orange; if it is green to grey or mottled in color it must be a squash. Many pumpkins are varieties of ***Cucurbita pepo***, although the largest pumpkins probably come from ***C. maxima***.

12. **Largest tree-bearing fruit** : The *jackfruit* (***Artocarpus heterophyllus***) produces massive seed-bearing fruits on its branches (Fig. 4.176). Native to the Indo-Malaysian region, this tree is grown throughout the tropics for its pulpy, edible fruit. It belongs to the same genus as the famous breadfruit (***A. altilis***). According to Charles Heiser (**Seed to Civilisation**, 1973), jackfruits may reach nearly three feet (0.9 m) and weigh up to 75 pounds (34 kg), thus making them the largest tree-bearing fruit on earth. Of course, the undisputed record for the world's largest fruit is a 1,337 pound pumpkin, a member of the gourd family (Cucurbitaceae).

13. **Largest sperms** : Cycads also have the record for the world's largest sperm. A single sperm cell from *Zamia roezlii*, an interesting cycad endemic to rain forests of Colombia, is about 0.4 mm in length and is visible to the unaided eye. In fact, it is almost as large as one entire *Wolffia* plant. It consists of several spiral bands of 20,000 to 40,000 cilia at one end

Fig. 4.176 : A jackfruit (*Artocarpus heterophyllus*)

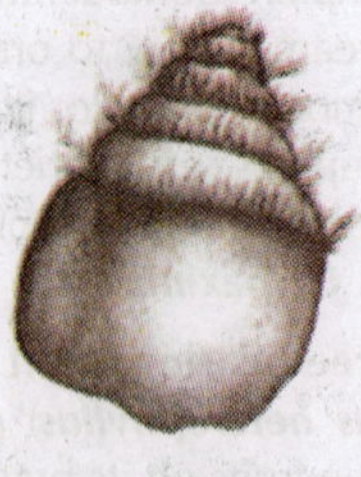

Fig. 4.177 : A sperm of *Zamia roezlii*

(Fig. 4.177). The pulsating beat of these cilia may help to propel the sperm through the pollen tube on its journey to fertilise the egg.

14. **Largest vegetable** : You define a botanical vegetable as an edible part of a plant that clearly excludes seed-bearing fruits, and then there are a number of possible contenders for this coveted record. Some of the top contenders for this record are the blades of large brown algae called kelp, and the tender leaves of the *horseradish tree* (***Moringa oleifera***), not to be confused with the true horseradish of the mustard family (***Armoracia lapathifolia***). Perhaps a more logical contender for this record is the massive subterranean yams of the genus ***Dioscorea***, some of which may **weigh over 120 pounds (54 kg)**.

15. **Smallest seed** : Certain epiphytic orchids of the tropical rain forest produce the world's *smallest seeds* weighing only 35 millionths of an ounce. They are dispersed into the air like minute dust particles or single-celled spores, eventually coming to rest in the upper canopy of rain forest trees.

16. **Largest seed** : The world's largest seed comes from the *coco-de-mer palm* (***Lodoicea maldivica***), native to the Seychelles Archipelago in the Indian Ocean. Although it belongs to a different genus from true coconut palms (***Cocos***), this enormous seed is often called the "**double**

Fig. 4.179 : Immature fruits of the Seychelles Island Palm (*Lodoicea maldivica*)

coconut." (Fig. 4.179 and 4.180) A single seed may be 12 inches (30 cm) long, nearly three feet (0.9 m) in circumference and weigh 40 pounds (18 kg). It should be noted here that the largest seed does not have the largest embryo. In fact, palm seeds are mostly composed of endosperm tissue and generally have relatively small embryos.

17. **Largest seed embryo** : According to C. R. Gunn (**World Guide To Tropical Drift Seeds and Fruits**, 1976), the record for the largest seed embryo goes to ***Mora oleifera*** (Fabaceae), a large tree that grows in tidal marshlands and estuaries along the Pacific coast of tropical America. In Costa Rica, this tree often forms nearly pure stands just behind the mangrove swamps. Seeds of ***M. oleifera*** may be up to 7 inches (18 cm) long and up to 5 inches (12.5 cm) wide (Fig. 4.178). Another species (***M. excelsa***) has slightly smaller seeds. Like other exalbuminous legume seeds, the two cotyledons com-

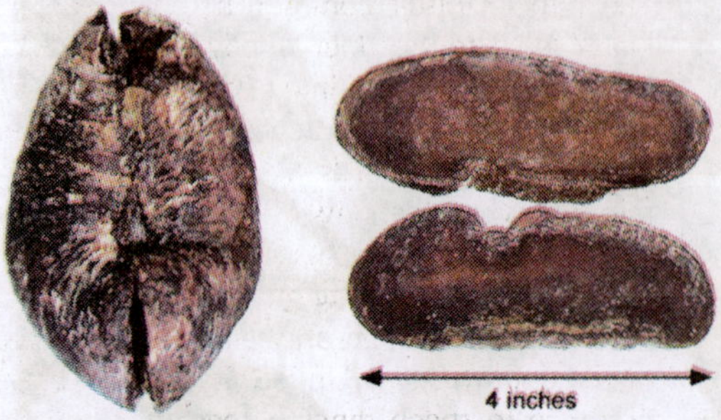

Fig. 4.178 : Seed and colyledons of *Mora oleifera*

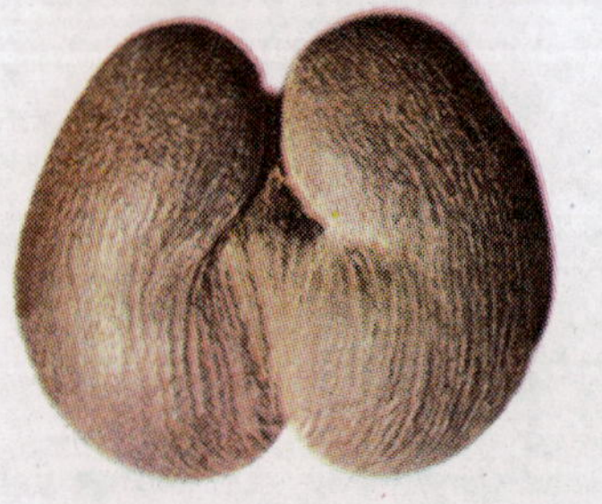

Fig. 4.180 : Seed of the Seychelles Island Palm (*Lodoicea maldivica*)

prise most of the seed. Since the cotyledons are part of the embryo, this species is certainly a strong contender for the record of world's largest seed. The seeds float in ocean current with their two large cotyledons connected or separated. Dried cotyledons washed up on beaches superficially resemble the shells of a bivalve mollusc.

18. **Largest Leaves** : In addition to the largest seeds, palms also have the record for the largest leaves. The *raffia palm* (*Raphia regalis*) of tropical Africa has huge pinnate leaves up to 80 feet (24 m) long. The leaves of the *Amazonian palm* (*Manicaria saccifera*) are nearly 30 feet (9 m) long, and have been listed by some authors as the longest undivided leaf of any plant. However, according to Chuck Hubbuch of Fairchild Tropical Garden, Coral Gables, Florida, the leaf is typically divided shallowly at the tip and is not truly entire.

    The golf ball-sized fruits of Manicaria palms, called "sea coconuts," commonly wash ashore on beaches throughout the Caribbean and southern Florida. There are two additional candidates for the record of longest undivided leaf: *Marojejya darianii*, a palm native to Madagascar with a leaf up to 5 meters (16.5 ft) in length that is divided only once at the tip; and *Johannesteijsmannia altifrons*, a palm native to Thailand with a leaf up to 4 meters (13 ft) long that is completely undivided.

19. **Largest flower** : The flower with the world's largest bloom is the ***Rafflesia arnoldii***. This rare flower is found in the rainforests of Indonesia (Fig. 4.181). It can grow to be 3 feet (0.9 m) across and weigh up to 15 pounds (7 kg). It is a parasitic plant, with no visible leaves, roots, or stem. It attaches itself to a host plant to obtain water and nutrients. When in bloom, the *Rafflesia* emits a repulsive odour, similar to that of rotting meat. This odour attracts insects that pollinate the plant.

Fig. 4.181 : Single flower of *Rafflesia arnoldii*

20. **Largest inflorescence** : Another enormous flower found in Indonesia is the *Amorphophallus titanum*, or **Titan arum**. It is also known as the "corpse flower" for its unpleasant odour. Like the *Rafflesia*, the Titan emits the smell of rotting flesh to attract pollinators. *Technically, the Titan arum is not a single flower.* It is a cluster of many tiny flowers, called an inflorescence. The Titan arum has the largest unbranched inflorescence of all flowering plants. The inflorescence can reach heights of 7-12 feet(2-4 m) and weigh as much as 170 pounds (77 kg).

    But, according to Charles E. Hubbuch, Director of Plant Collections at the Fairchild Tropical Garden in Coral Gables, Florida, the talipot palm (*Corypha umbraculifera*) of India has the largest inflorescence of any plant. The huge inflorescence may be 10 meters (33 feet) tall with millions of flowers.

21. **Largest Hitchhiking Fruit** : Seed-bearing fruits that cling to the bodies of animals are very effective methods of seed dispersal in the plant kingdom. There are literally hundreds of plant species with hitchhiking seed pods, but some of the largest are from the North American devil's claw (***Proboscidea***). The devil's claw fruit is technically a drupaceous capsule with a woody inner part surrounded by a fleshy layer. The rather sinister common name of "devil's claw" refers to the inner woody capsule which splits open at one end into two curved horns or claws (Fig. 4.182). Each capsule contains about 40 black seeds which are gradually released when the claws split apart. They are also called "elephant tusks" and readily cling to the hooves of grazing animals or your shoes if you happen to step on them. In some areas of the southwestern United States they are a nuisance to sheep ranchers because they get entangled in the fleece.

Fig. 4.182 : Seed of *Proboscidea parviflora* (a double-clawed variety).

In his fascinating book, **Plants and Planet** (1974), Anthony Huxley (son of Julian Huxley) eloquently describes the hitchhiking pods as "hookers." The fresh green pods (and dried black seed capsules) were important items in the cultures of many Red Indian tribes of the southwestern United States, and are still used for food and in basketry. The striking seed pods of the domesticated, white-seeded cultivar ***P. parviflora*** var. ***hohokamiana*** have claws up to 15 inches (38 cm) long.

Multiclawed forms have also been selected by native Americans because the horns split into 3-4 claws. The long claws are soaked in water and split into leathery strands. They provide durability and intricate black designs in tightly woven baskets made from sun-bleached yucca leaves (often ***Yucca elata***). The devil's claw is also known as "unicorn plant" — referring to the large, hornlike fruit before is has split open.

The seed capsules of devil's claws are clearly adapted for hitchhiking on the hooves of large grazing animals; however, with the exception of introduced livestock and people (and possibly desert bighorn sheep), there are few native North American animals living within the present range of devil's claws that are capable of dispersing these large hitchhikers. It is possible that the range of some large North American grazers, such as antelope, bison, deer and elk once overlapped the range of devil's claws thousands of years ago. It is also possible that devil's claw dispersal by grazing mammals may be a North and South American anachronism, or an occurrence that is out of its proper time in history. During the past one million years of the Pleistocene Epoch, the Americas were rich in large mammals (such as giant ground sloths) which are now extinct. Assuming devil's claw plants existed over 600,000 years ago, were some of these ancient mammals the true carriers of these hitchhiker pods ? A similar anachronism occurs in the New World tropics of Central and South America, where the natural dispersal agent for some large seed pods are unknown.

5 Plant Anatomy

INTRODUCTION

The word **anatomy** has derived its origin from *ana*, meaning asunder (into pieces), and *temnein*, meaning—to cut. The term is also known as **internal morphology**. It is the branch of botany which deals with the internal structures and organization of the plant body. The term **histology** (Greek *histos* meaning cobweb) deals with the minute internal structures like cellular arrangements, tissues and tissue systems. Specialised study of the cell belongs to **cytology** which is now treated as an independent branch of biology. Anatomy is really concerned with both minute and gross internal structures. Thus histology is covered by anatomy.

The modern studies on anatomy started with the invention of the microscope in the later part of the sixteenth century. Long before Christ and Greek Civilisation ancient Hindus of India focussed their attention to the subject. They defined five regions in of the plant body, viz. *tvac* (skin), *mansa* (soft tissues or bast), *asthi* (wood or bone), *majja* (pith), and *snayu* (fibres in the bast). *Mansa* was further divided into *valkala* (bark) and *sakara* (basts). The history of biological studies started from the time of Aristotle in the fourth century B.C. Theophrastus (372-286 B.C.), a pupil of Aristotole (384-322 B.C.), said — "Plants are made of *phloios* (bark), *zylon* (wood) and *metra* (pith), when pith is present". Thus the very commonly used terms xylem and phloem were coined by Theophrastus.

Actual interest started with the invention of the microscope. Robert Hooke (1635-1703) in England examined a piece of bottle cork under a microscope improvised by himself and noted the cells. His observations were published in *Micrographia* in 1665. N. Grew (1641-1712) published his work in 1682 under the title — *The Anatomy of Plants,* where he discussed the gross structures of plants, and compared the vascular skeleton of the plants with that of the animals. The commonly used terms of anatomy like parenchyma, vessels, cortex, etc. were introduced by him. M. Malphigi (1628-1694) discovered the spiral vessels and the stomata. His pioneering works on vascular plants came out in his '*Anatome Plantarum*' in 1675.

GENERAL STRUCTURE OF THE PLANT BODY

Growth of the plant body starts generally from the morphologically unicellular zygote, which gradually develops into the embryo and then to the whole plant. In spite of great diversities existing in the vascular plants as regards their size, structure and form, there is a fundamental uniformity. The higher plant body generally has an **axis**, the lower part of which is the underground **root** and the continuous upper aerial part is the **stem**. The root-stem axis bears appendages, which are of three types — the **leaves, emergences** and **hairs** or **trichomes**. The leaves are characteristic of the stems and absent in roots. They have the connections with the stem vasculature. Thus the root, stem and leaves are the three fundamental organs of the plant body.

THE AXIS

Starting from the periphery the plant axis has three zones or regions – the **epidermis, cortex,** and **stele**. The epidermis is the outermost layer forming the skin. It is usually single-layered due to anticlinal (at right angles to the axis) division of the mother cells and gives *protection* to the plant body. The next internal zone to epidermis is the cortex. It is multi-layered and is composed of different types of tissues performing functions like support, storage, etc. The innermost layer of the cortex is the **endodermis**. The central core zone is the stele or central cylinder, which remains encircled by the cortex.

Stele includes the **vascular bundles**, consisting of complex tissues like xylem and phloem, and other simple tissues. The vascular bundles perform the function of translocation of water and solutes as well as give mechanical support to the plant. Xylem translocates water along with dissolved mineral matters absorbed from the soil by roots and also forms the mechanical tissue system inside the plant. Phloem, on the other hand, is responsible for the translocation of prepared food matters from the leaves to the different parts of the body. A lateral meristem, called

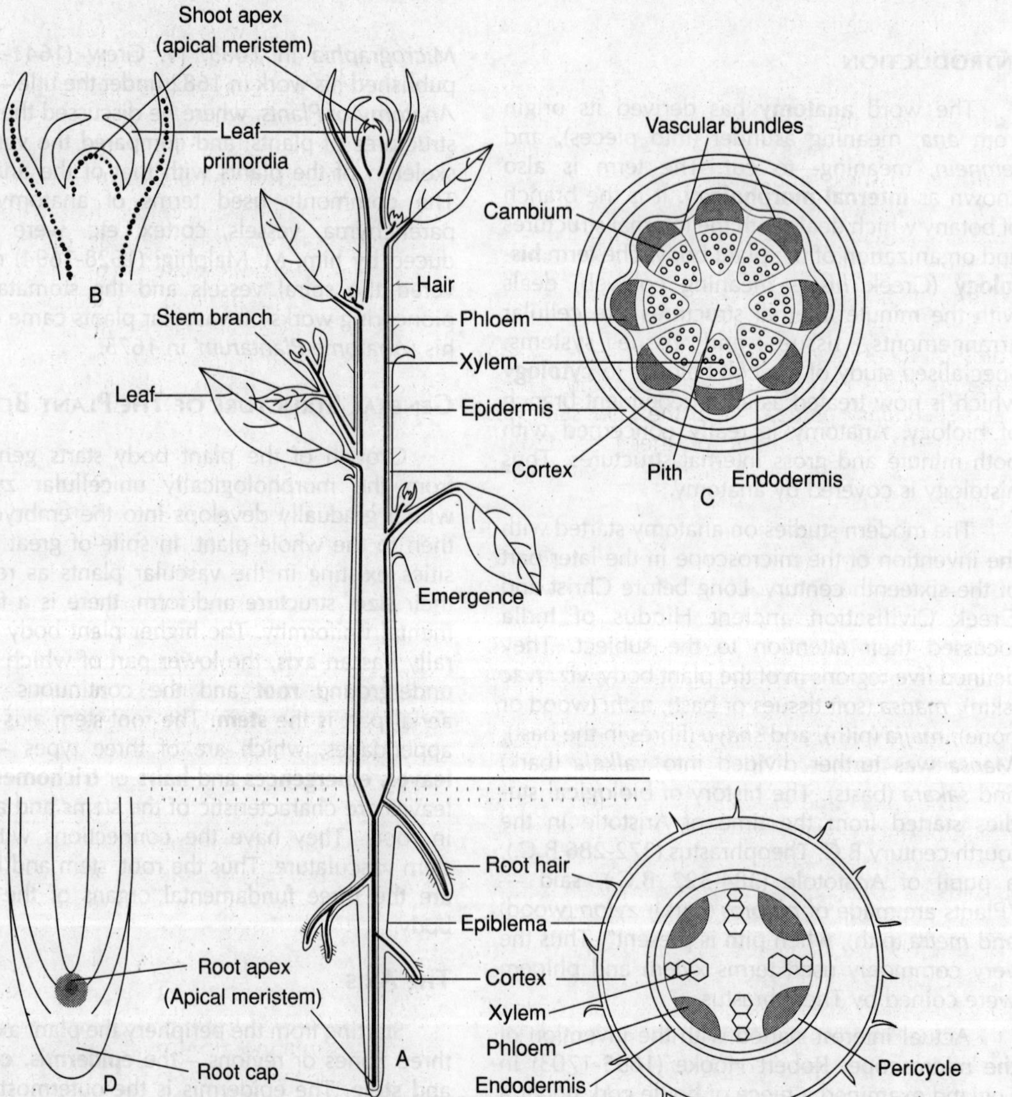

Fig. 5.1 : The structural plan of a vascular plant : A. The plant showing the parts and appendages, B. L.s. of the shoot apex, C. T.s. of stem, D. L.s. of the root apex, E. T.S. of root

cambium, may be present in between the xylem and phloem, as in the stems of dicotyledons and gymnosperms. Xylem and phloem remain arranged radially in the roots and occur on the same radius in other organs. The stele may form a solid rod, a hollow cylinder, etc. Very often they enclose a central, soft parenchymatous pith or medulla, which is continuous in-between the vascular bundles looking like rays radiating from the pith — called **pith rays** or **medullary rays.** In between the vascular bundles and endodermis there occurs one or more layers of non-conducting cells forming what is known as **pericycle.**

All the tissues forming the plant body are derived from the **apical meristem** present at the tips or apices of the axis. These tissues are called **primary tissues**, and the body composed of primary tissues is the **primary body**. During primary growth, the axis grows in length, leaves, axillary buds and other structures develop originating the branch systems. Cambium in-between primary xylem and phloem in the vascular bundles of dicotyledons and gymnosperms divide and produce new tissues — **secondary xylem** and **secondary phloem** — in the later stages of growth. After formation of these secondary tissues a new meristem, called **phellogen** or **cork cambium**, arises in the peripheral portion of the stem. This lateral meristem cuts **phellem** or **cork cells** with suberised wall on the outer side and parenchymatous **phelloderm** or **secondary cortex** on the inner side. As a result, the axis grows in girth or thickness. All these tissues, formed later by the lateral meristems, are known as **secondary tissues**; and the growth in thickness due to addition of secondary tissues is referred to as **secondary growth**.

Thus plant body is composed of morphologically distinct organs like stem, root and leaves, each performing specific functions. Every organ, again, is a collection of **tissue-systems** which carry on restricted functions. A tissue-system is made up of a number of contiguous **tissues**. A tissue is, in fact, a collection of **cells** formed in response to a basic division of labour. So the ultimate units of the plant body are the cells. Thus it is customary to begin studies on anatomy with the cell.

The Cell

The term *cellula* was first used by Robert Hooke in 1665 to the wall-surrounded small cavities in cork. Later he observed cells containing juice in other plant tissues. *The cell is defined as the structural and functional unit of the living body*. Subsequently, the substance within the cell termed **protoplasm,** was discovered. In 1880 Hanstein used the term **protoplast** to indicate the protoplasmic content in a single cell i.e., the unit of protoplasm. He favoured the use of the term protoplast instead of the term cell, but his suggestion did not receive universal acceptance. In plants, the term cell includes the protoplast together with the wall.

The *cell wall* was, for a long time, regarded as a non-living excretion of the living cell matter, but recently it is evident that organic unity exists between the protoplast and the cell wall, especially in young cells, and that they together form a single biological unit of structure and function.

Again, all the living cells in higher plants are interconnected by thin strands of cytoplasm known as **plasmodesmata**. Typically, they are cylindrical holes through the cell walls. Each hole is lined by a cylinder of cell membrane, the lumen of which is filled with cytoplasm and it often contains a minute tubule, known as the **desmotuble**.

In 1831, Robert Brown discovered the **nucleus** in an epidermal cell of an orchid. In 1846 Hugo von Mohl distinguished between the protoplasm and the cell sap, and Kolliker (1862) introduced the term **cytoplasm**.

So far as the structural organisation is concerned, all organisms belong to either of the two categories — **the prokaryotes**, including bacteria and blue-green algae, and the **eukaryotes**, comprising all other plants and animals. The French microbiologist Edward Chatton first used these terms in 1928.

Prokaryotes (Greek **pro**: before; **karyon**: nucleus) lack a membrane-bound definite nucleus. The cellular contents are differentiated into two distinct regions, the cytoplasm and the nucleoid, the latter consisting of the nuclear material containing fine DNA fibrils. The prokaryotic cell size ranges from 1-10 μm. The usual components occurring in prokaryotic cells are — the plasma membrane, ribosomes, nucleoid and the ground substance, the cytosol. Excepting the mycoplasma, a cell wall surrounding the cell membrane is usually present in all prokaryotic cells. Mitotic division is not found in prokaryotes and they divide by binary fission or other means.

In the eukaryotic (Greek **eu**: true; **karyon**: nucleus) cell, on the other hand, the nuclear matter is bound by a double membrane called the nuclear envelope. The cell size ranges from 10-100 μm. Eukaryotic cells divide mitotically — involving the microtubules. The cytoplasm is compartmentalised by membranous systems. These two fundamentally different kinds of cellular organisations have been recognised only from the early 1960s.

With the discovery of the electron microscope and the great progress in biochemical studies, a lot of knowledge has been accumulated on the structure and function of the cell.

The eukaryotic plant cells are of many types and can be divided into two broad groups – the **living cells** which are responsible for performing all the metabolic activities of the plant, and the **dead cells** which are metabolically inactive and function either as mechanical support or for conduction of fluids throughout the plant body. So, a living parenchymatous cell is the convenient choice for the discussion of a typical plant cell. A typical plant cell consists of two primary components, the **cell wall** and the **protoplast**. The cell wall completely surrounds the protoplast and provides mechanical support for it. It is composed of two layers: the outer layer is called the **primary wall**, whilst the inner layer, nearer to the protoplast and usually composed of several sub-layers, is known as the **secondary wall**. The cementing substance in-between the primary walls of adjacent cells is known as the **middle lamella**.

The starting point in the life of every individual is a single cell. If it continues as such the organism remains **unicellular**, as in many lower plants like yeasts, diatoms, desmids, etc.; but in large majority of plants the cell goes on dividing, so, ultimately, a distinct **multicellular** body is formed. In fact, the cellular complexity of an organism is based on the systematic position of the plant in the line of evolution.

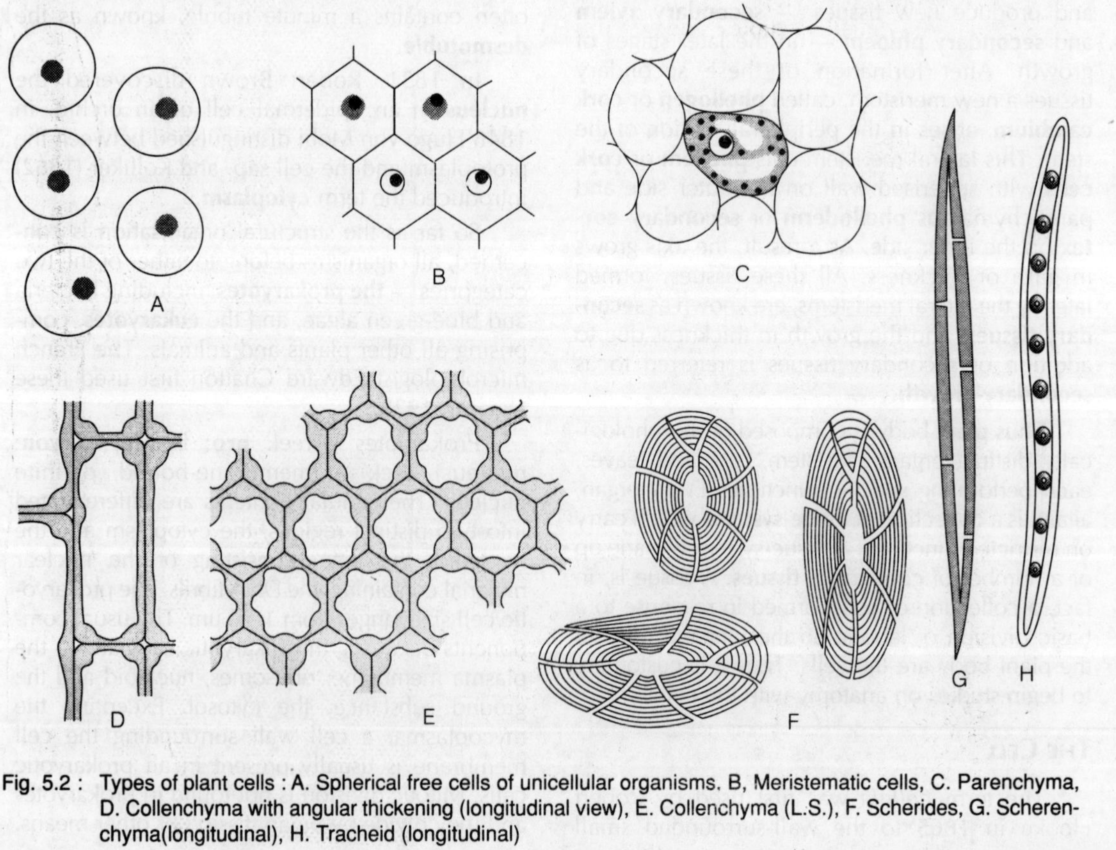

Fig. 5.2 : Types of plant cells : A. Spherical free cells of unicellular organisms. B, Meristematic cells, C. Parenchyma, D. Collenchyma with angular thickening (longitudinal view), E. Collenchyma (L.S.), F. Scherides, G. Scherenchyma(longitudinal), H. Tracheid (longitudinal)

The cells exhibit wide diversities as regards their shape, size and arrangement. With growth and differentiation, cells attain different shapes. Two chief types, however, are *polyhedral* with diameters more or less same and *elongate*. All transitional forms between the two extremes are possible. In case of young cells of same size and age the shape becomes polyhedral due to mutual contact and compression.

PROTOPLASM

Protoplasm is the essence of life. Huxley described it as 'the physical basis of life'. All the characteristics of living beings like metabolism, irritability, growth and reproduction are the external manifestations of the internal changes of the protoplasm. The protoplasmic mass inside a single cell is called the protoplast. So it is the unit of protoplasm and is the metabolically active component of the cell. It consists of three parts — the **cytoplasm**, the **nucleus** and the **vacuole**. The ground substance which resides immediately beneath the cell wall and surrounds the vacuole is the cytoplasm. The young cell contains very small vacuoles and, as the cell enlarges, the vacuoles coalesce to form a large one. With the increase in cell size the protoplasmic content does not increase in proportion to the vacuole size and, as a result, a thin cytoplasmic layer surrounds the vacuole. This thin layer is known as the **primordial utricle**. The vacuole is bound by a membrane known as the **tonoplast** separating the vacuolar sap or cell sap from the cytoplasm. The cytoplasm in its periphery is, again, bound by a membrane known as the **cell membrane** or **plasma membrane**.

The soluble portion of the cytoplasm in which the cell inclusions remain embedded is known as the **cytosol**. The extensive membranous system which traverses the entire cytoplasm is called the **endoplasmic reticulum** (ER). It consists of a ramified network of interconnected, membrane-bound tubules and vesicles.

Essential organelles occurring in all eukaryotic cells are the **mitochondria**, which are more numerous and somewhat smaller and serve as the centres of intracellular oxidation. They are generally spherical to ellipsoidal, and are commonly about 1 μm thick and 1-3 μm long. Like

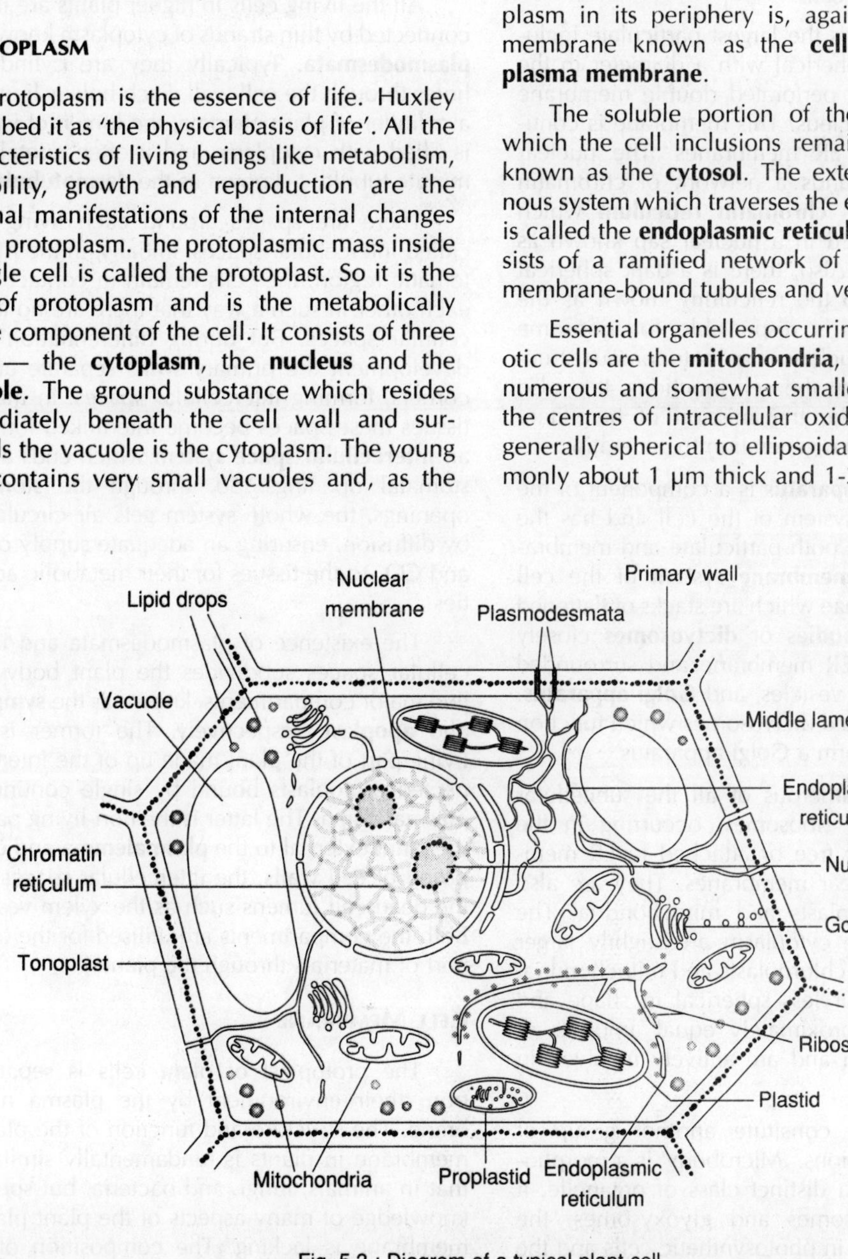

Fig. 5.3 : Diagram of a typical plant cell

plastids they are bound by a pair of membranes and are also semi-autonomous in function.

Plastids are the second largest semi-autonomous cell inclusions enclosed by a double membrane. They are confined to the plant kingdom. They may be coloured or colourless. They are typically lens-shaped with a diameter of up to 10 μm. There are about 200-400 such plastids per plant cell.

The **nucleus** is the largest particulate inclusion, which is spherical with a diameter in the range 6-8 μm. A perforated double membrane surrounds the nucleus. This membrane is continuous with the ER membranes. The nuclear membrane surrounds a network of chromatin matter known as **chromatin reticulum** which remains embedded in a nuclear sap known as the **karyolymph**. Also, there is a dark spherical body attached to the reticulum known as the **nucleolus** which is responsible for ribosome formation. The nucleus is an autonomous structure. It controls all the cell functions. It is also responsible for the transmission of hereditary characteristics from generation to generation.

The **Golgi apparatus** is a component of the endomembrane system of the cell and has the property of being both particulate and membranous. The **endomembrane system** of the cell consists of **cisternae** which are stacks of flattened vesicles, **Golgi bodies** or **dictyosomes** closely associated with ER membrane and surrounded by tiny spherical vesicles, and **Golgi apparatus**. The inter-associated dictyosomes which function synchronously, form a Golgi apparatus.

The most numerous of all the subcellular particles are the ribosomes, occurring in the cytoplasm, either free or attached to ER membranes and nuclear membranes. They are also found in chloroplasts and mitochondria. The ribosomes in the cytoplasm are slightly larger than those in the chloroplasts and mitochondria. They are more or less spherical in shape and made up of approximately equal amounts of RNA and protein and are actively involved in protein synthesis.

Microbodies consitute another group of particulate inclusions. 'Microbody' is a morphological term for a distinct class of organelle. It includes peroxisomes and glyoxysomes, the former occurring in photosynthetic cells and the latter restricted to cells of the endosperm or cotyledons of fat-storing seeds. The microbodies are roughly spherical and single-membrane structures with a diameter of 0.2 -1.5 μm.

Another debatable particulate inclusions are the **spherosomes** which are restricted to the cells of lipid-storing tissue. They are also single-membrane spherical structures with a diameter of 0.4-3 μm.

All the living cells in higher plants are interconnected by thin strands of cytoplasm known as **plasmodesmata**. Typically they are cylindrical holes through the cell wall. Each hole is lined by a cylinder of plasmalemma, the lumen of which is filled with cytoplasm and it often contains a minute tubule — known as the **desmotubule**.

There are spaces around each living cell called intercellular spaces. Initially, in the meristematic region, the cells remain in contact with each other in such a way that there are no intercellular spaces. But during differentiation and development the primary walls separate at the corners, forming intercellular spaces. In mature tissues these spaces become interlinked forming an **intercellular space system** which ends at the stomatal openings. So, through the stomatal openings, the whole system gets air circulation by diffusion, ensuring an adequate supply of O_2 and CO_2 to the tissues for their metabolic activities.

The existence of plasmodesmata and intercellular spaces subdivides the plant body into two major compartments, known as the **symplast** and **apoplast**, respectively. The former is the living part of the plant made up of the interconnected protoplasts bound by single continuous plasmalemma. The latter is the non-living part of the plant external to the **plasmalemma** and composed of cell walls, the intercellular spaces and the dead cell lumens such as the xylem vessels. Both the compartments are utilised for the transport of materials through the plant.

CELL MEMBRANE

The protoplast of plant cells is separated from their environment by the plasma membrane. The structure and function of the plasma membrane in plants is fundamentally similar to that in animals, fungi, and bacteria, but specific knowledge of many aspects of the plant plasma membrane is lacking. The composition of the lipid components and the properties of the pro-

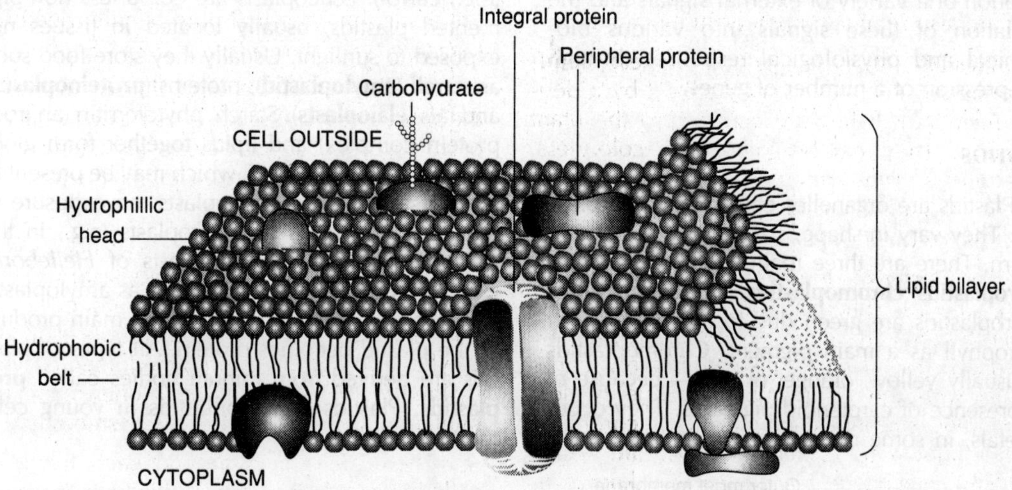

Fig. 5.4 : Three dimensional structure of the fluid-mosaic model of membrane

teins vary from membrane to membrane, on which membrane's unique functional characteristics depend. The plant cell membrane has a number of important functions, such as to mediate the transport of solutes into and out of the protoplast, to coordinate the synthesis and assembly of cell wall microfibrils, and to translate hormonal and environmental signals involved in the control of growth and differentiation.

The first important hypothesis of the structure of biological membranes was proposed by H. Davson and J. Danielli in 1935. This hypothesis says that membranes contain a continuous hydrocarbon phase contributed by the lipid components of the membrane. The hypothesis was later modified by J. D. Robertson into the **Unit-Membrane Hypothesis**. The unit membrane was proposed to consist of a bilayer of mixed polar lipids, with their hydrophobic tails oriented inward to form a continuous hydrocarbon phase and their hydrophilic heads oriented outward. Each surface was thought to be coated with a monomolecular layer of protein molecules with the polypeptide chains in extended form. It was suggested that the unit membrane is about 8.0 to 9.0 nm thick and that the lipid bilayer is about 6.0 to 7.0 nm thick.

The most satisfactory model of membrane structure at present is the dynamic **fluid-mosaic model** put forward by Singer and Nicholson in 1972. This model postulates that amphipathic lipids (e.g., phospholipids, glycolipids, sterols) are arranged in a bi-layer to form a fluid, liquid-crystalline matrix or core, in which individual molecules can move laterally. This movement gives the bi-layer fluidity, flexibility and characteristically high electrical resistance and relative impermeability to highly polar molecules. Globular proteins form a mosaic on the matrix. They may be partially embedded on either side or may completely penetrate the membrane. This mosaic is not fixed or static, because the proteins are free to diffuse laterally in two dimensions. The fluid-mosaic model accounts satisfactorily for many features and properties of biological membranes :

(i) it explains the widely different protein content per unit area,

(ii) it explains the variation in the thickness of different types of membranes,

(iii) it explains the electrical properties and permeability of membranes, and

(iv) it also explains the movement of some proteins in the cell membrane.

Functions

The plasma membrane has at least three major functions: (i) it is highly specialised for the transport of a variety of substance into and out of the cell, (ii) it has an important role in the synthesis and assembly of cell wall microfibrils, (iii) the plasma membrane is charged with the reception of a variety of external signals and the translation of these signals into various biochemical and physiological responses through de-repression of a number of genes.

PLASTIDS

Plastids are organelles characteristic of plant cells. They vary in shape, size, and pigmentation pattern. There are three main types of plastids – **chloroplastids**, **chromoplastids**, and **leucoplastids**. Chloroplastids are green due to the presence of chlorophyll as a main pigment. Chromoplastids are usually yellow, orange, or red — because of the presence of carotenoid pigments. They occur in petals, in some ripe fruits, and in some roots (e.g., carrot). Leucoplasts are colourless non-pigmented plastids, usually located in tissues not exposed to sunlight. Usually they store food such as starch (**amyloplastid**), proteins (**proteinoplasts**), and fats (**elaioplasts**). Starch, phytoferritin (an iron-protein complex) and lipids together form globules called **plastoglobuli**, which may be present in all types of plastids. Leucoplasts, on exposure to light, are converted to chloroplasts (e.g., in the potato tuber). The proteinoplasts of *Helleborus corsicus* were observed to initiate as amyloplasts; later the starch dis-appears and the main product of the plastid becomes protein. All types of plastids are derived from minute bodies called **proplastids**. Plastids and proplastids in young cells multiply by division.

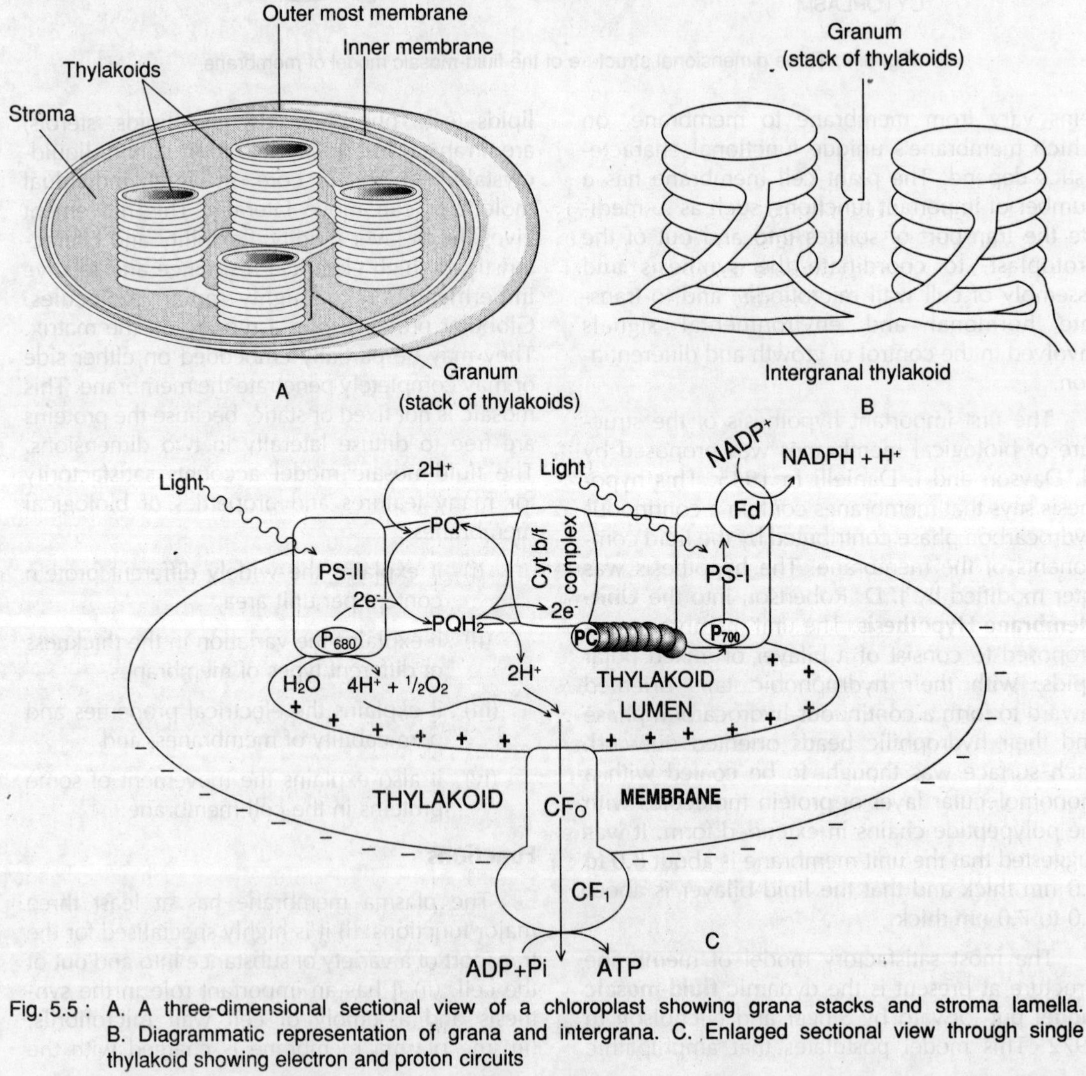

Fig. 5.5 : A. A three-dimensional sectional view of a chloroplast showing grana stacks and stroma lamella. B. Diagrammatic representation of grana and intergrana. C. Enlarged sectional view through single thylakoid showing electron and proton circuits

The chloroplast usually has a characteristic lens shape with a length of 4-10 μm. There are three major structural parts of the chloroplast : (i) an envelope consisting of two membranes with an enclosed space, (ii) the mobile stroma containing the soluble enzymes for CO_2 fixation, protein synthesis, and starch storage, and (iii) the highly organised internal lamellar membranes stacked together to form a **grana** which performs the function of light energy capture. Each stack is a disc or sac-like vesicle and is termed **thylakoid** (in Greek, sac-like) by Wilhelm Menke (1960) (Fig. 5.5).

The outer envelope is 50-80 Å thick. It is differentially permeable and regulates the passage of sugars and other carbon intermediates, NADPH/$NADP^+$, ATP and ADP in and out of the chloroplast. Synthesised starch is retained inside as storehouse of carbon skeleton for further use.

The stroma is a proteinaceous matrix containing about 50% ribulose 1, 5-bisphosphate carboxylase/oxygenase (Rubisco). There are small droplets rich in lipids, particularly plastoquinone close to the thylakoid membranes. These droplets are called **plastoglobuli**. The starch grains remain close to the thylakoids and are usually ovoid and up to 1.5 μm in length. The stroma also contains **ribosomes (70S)** attached to the outer surface of the thylakoid membranes. It also contains 20-50 identical histone-free **circular DNA** molecules with a circumference of 36-62 μm. A few chloroplast proteins are coded by chloroplast DNA. The large subunit of Rubisco is known to be coded by a chloroplast gene while the small subunit is coded by a nuclear gene.

The higher plant chloroplasts contain two types of thylakoid, the large or **stroma thylakoid** and the small or **grana thylakoid**. There may be 40-60 grana per chloroplast and usually 5-20 thylakoids per granum. Each thylakoid encloses an internal space called **thylakoid lumen** (Fig. 5.5). The membranes between the grana are perforated. The thylakoid system constitutes a single, complex cavity separated from stroma by the thylakoid membrane system. Built into the thylakoid membranes are pigment systems and electron carriers which carry out the light reaction of photosynthesis.

MITOCHONDRIA

Mitochondria are called the **power houses** of cell as they serve as sites of ATP synthesis. Generally, a plant cell contains several hundred simple mitochondria. They are spherical, ellipsoidal or rod-like structures. Mitochondria are specifically stained by the dye Janus green B. The mitochondria move freely in the streaming cytoplasm and appear to divide and coalesce; they undergo considerable changes in shape, being globular, cylindrical or branched.

Mitochondria have a double-membrane envelope. There is an aqueous matrix of solutes, soluble enzymes, and the mitochondrial glucose encircled by the inner membrane which is larger than the outer one and remains infolded to form sac-like **cristae** of variable shape and number, usually with a narrow neck (Fig. 5.6). There is an inter-membrane space in-between the two membranes. The mitochondrial matrix is finely granular and highly proteinaceous. The matrix contains a fibrillar region with DNA known as **nucleoid**. It also contains ribosomes and calcium containing granules. The ribosomes have a sedimentation coefficient of 70S. The ribosomes are composed of a large subunit (50S) and a small subunit (30S). Higher plant mitochondrial DNA is a circular molecule of 30μm length with a molecular weight of 70 x 10^6.

RIBOSOMES

The term ribosome was first introduced by Roberts in 1958. They are ribonucleoprotein particles and are essential machinery for protein biosynthesis. Green plant cells have 80S ribosome in the cytoplasm and 70S (sedimentation coefficient) ribosomes in the mitochondria and chloroplastids. The diameter of 80S ribosome is 20-22 nm, and the particle weight is about 4.0 megadaltons. The ribosomes are composed of two subunits, one of which is larger than the other (Fig. 5.7). The larger subunit is 60S and the smaller one is 40S. Plant ribosomes contain 25S and 16S rRNA.

The ribosomal RNAs are highly folded, and double helical stems alternate with single-stranded loops. The small subunit fits over the ends of the large subunit in such a way as to form a tunnel between the particles. This tunnel could accommodate the m-RNA, tRNA and protein synthesis factors. The shape of the ribosomes mainly depends on the concentration of Mg^{2+} ions in the medium. Removal of Mg^{2+} causes dissociation and unfolding of the subunits.

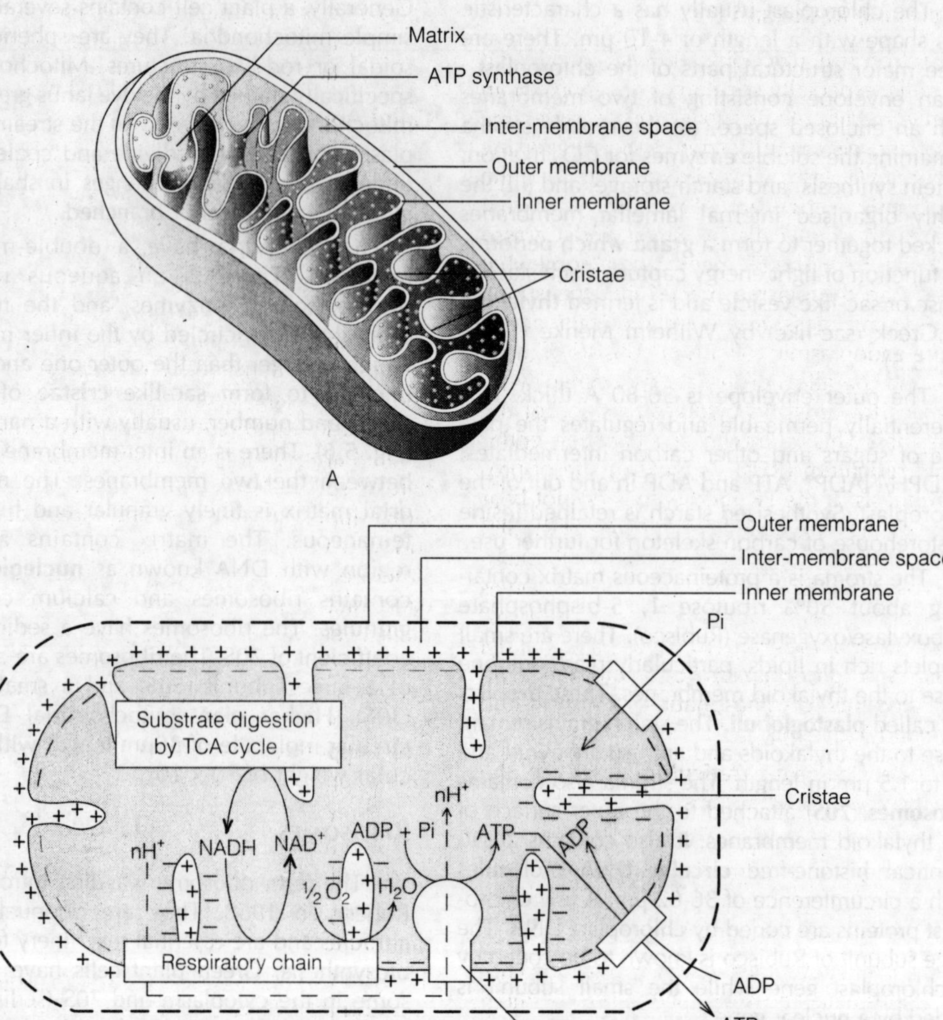

Fig. 5.6 : A. Three dimensional sectional view of a mitochondrion showing infoldings (cistae) of the inner membrane and the F_1-ATP synthase built in it. B. A scheme of mitochondrial electron transport mechanism

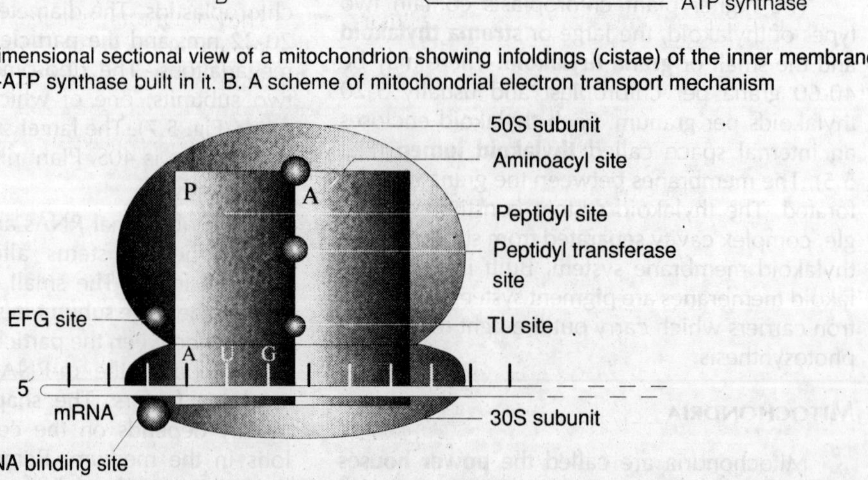

Fig. 5.7 : Three-dimensional diagrammatic view of a ribosome showing apparent locations of its different active sites

Endoplasmic Reticulum

Electron microscopic studies reveal that plant cells, especially secretory cells, contain an extensive membrane network of interconnecting **tubules** and **cisternae** (flattened sacs) traversing the entire cytoplasm. This network is termed the **endoplasmic reticulum** (ER) and was first described in plant cells by Buvat and Carasso (1957) (Fig. 5.8). The ER membrane is somewhat thinner than the plasma membrane, measuring 5-6 nm in thickness. In the metabolically active cell, the endoplasmic reticulum consists of interconnected parallel cisternae associated with ribosomes towards the cytoplasmic face of the membranes. This is known as the **rough endoplasmic reticulum (RER)** — a major site of protein synthesis. If the ER membranes do not bear ribosomes they are called **smooth-ER (SER)**. SER serves as a major site of lipid synthesis and membrane assembly. RER, on the other hand, serves as the working bench for the synthesis of proteins. The ER membranes are rich in phosphatidyl choline and phosphatidyl ethanolamine and lesser amounts of phosphatidyl inositol and phosphatidyl glycerol.

The endoplasmic reticulum has got the following important functions :

i. It functions as a communication or transport system throughout the cell, mediating the intracellular transport of small and large molecules.
ii. The ER tubules often end at plasmodesmata and thus play a role in intercellular communication.
iii. It plays an important role in cellular metabolism.
iv. It is the principal site of membrane synthesis and thus contributes to the formation of other organelles.

Golgi Apparatus

The Golgi apparatus was discovered in 1898 by the Italian scientist Camillo Golgi. It consists of a system of stacks of flat, smooth-walled, circular **cisternae** or **saccules** made up of unit membrane (Fig. 5.9). Each stack is termed **Golgi body** or **dictyosome**. Each dictyosome contains 5 – 8 cisternae. The margins of the cisternae may be vesicular, fenestrated or tubular in shape. The vesicles are budded-off from the cisternal margins. In some cases rod-shaped elements in parallel arrangement have been seen in the very narrow spaces between the individual cisternae. Golgi apparatus also contains coated or smooth surfaced vesicles of various types. It is a polar structure. The proximal pole (forming face) of each dictyosome remains associated with the nuclear envelope of ER in a characteristic manner. The opposite pole (maturing face), however, becomes more like the plasma membrane. The inter-associated dictyosomes together form the **Golgi apparatus**. The number of dictyosomes per Golgi apparatus ranges from 0–25,000.

The Golgi apparatus is a component of the endomembrane system of the cell and appears as intermediate between ER and cell membrane.

Golgi apparatus performs the following main functions :

1. The Golgi bodies are concerned with secretion processes. It functions in the packaging of materials for export to the cell's exterior. They are mainly involved in the secretion of sugars (in nectar secretion), polysaccharides

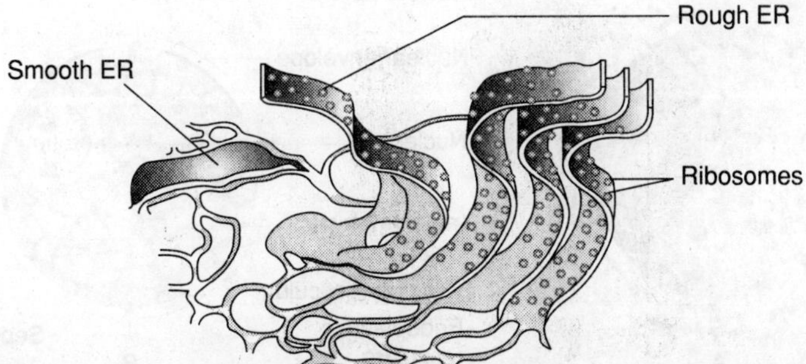

Fig. 5.8 : Structure of the endoplasmic reticulum

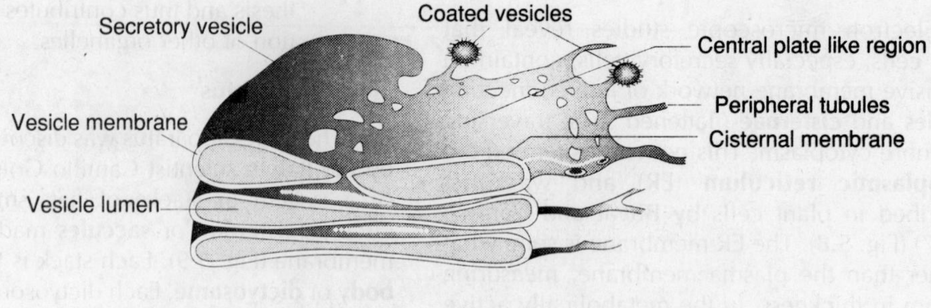

Fig. 5.9 : Generalised structure of Golgi complex (dictyosomes)

(cell-wall material, mucilage), and polysaccharide protein complexes (certain mucilages).

2. They are also responsible for the formation of a new plasma membrane to support growth or to replace the lost one.

Cytoskeleton

Throughout the cytosol there is a three-dimensional network of filamentous proteins called the **cytoskeleton**. It gives the scaffold structure for the positioning and movement of the organelles. In plant cells the cytoskeleton consists of **microtubules, microfilaments,** and **intermediate filaments** of fixed diameter and variable lengths. The microtubules are hollow filaments of 25 nm diameter, microfilaments are solid, having 7 nm diameter, and the intermediate filaments are helically coiled fibres of 1 nm diameter. Microtubules and microfilaments are made up of globular proteins. The protein polymers of tubulin are the components of microtubules.

Microfilaments are composed of globular actins or G-actins each of which is composed of a single polypeptide. Intermediate filaments are composed of linear polypeptide monomers of various types like **lamins, keratins** etc.

The cytoskeleton is involved in a variety of processes like intracellular movements such as cytoplasmic streaming, mitotic movements, secretory vesicle transport, cell plate formation, and cellulose micro-fibril deposition etc.

Nucleus

The nucleus is usually more or less spherical, though nuclei with other shapes have also been observed. The nucleus is surrounded by a nuclear envelope and contains the nuclear matrix called **nucleoplasm** and one or more **nucleoli** (Fig. 5.10).

The nucleoplasm contains chromatin reticulum consisting of deoxyribonucleic acid (DNA) and proteins. The DNA and protein complex in the chromosomes having affinity to basic dyes is called **chromatin**. In the interphase the chromosomes are uncoiled and can not be distinguished and form a network called

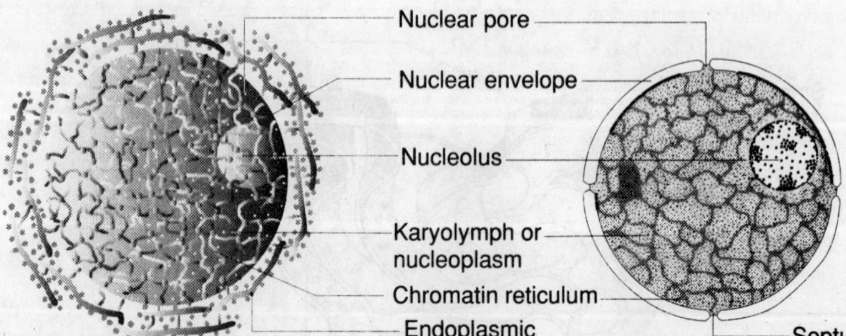

Fig. 5.10 : Structure of interphase nucleus : A. 3D structure, B. Line drawing

the **Chromatin reticulum**. At this phase some condensed and deeply stained chromatin masses are often seen in the reticulum. This chromatin is called **heterochromatin**. In contrast the less coiled and less stained masses of chromatin are called **euchromatin**. The nuclear membrane or **karyotheca** surrounds the nuclear content. Under electron microscope the nuclear membrane shows a special perinuclear cisternae of the cell endomembrane system with an inner and an outer membrane enclosing a lumen and traversed by **pores**.

In higher eukaryotes, the membrane disappears in late prophase and reappears around the daughter nuclei during telophase. The nuclear membrane separates the nucleus from the cytoplasm and regulates the nucleo-cytoplasmic interactions. Each nuclear membrane is about 75 to 90Å thick and lipoprotein in nature. A perinuclear space separates the outer and inner membranes of the nucleus. This intermembrane space is known as **perinuclear cisternae**. The inner membrane holds the nuclear content itself. The outer membrane is rough due to presence of ribosomes attached to it. Very often the outer nuclear membrane is continuous with the lumen or inner cavity of rough ER.

The nuclear membranes have a bi-layer structure similar to that of other biological membranes. The nuclear envelope is perforated by pores. The pores are enclosed by circular and cylindrical structures called annuli. The pores and annuli are collectively called the **nuclear pore complex**.

The nucleoli are very dense, granular, and fibrillar in structure, and are not bound by a membrane. They are often seen associated with some chromatin. They contain RNA, DNA, and proteins. The nucleus carries the information for the cellular proteins in its DNA molecule. The RNAs (mRNA, tRNA and rRNA) are transcribed on the DNA template. During protein synthesis mRNA is transcribed and transported to the cytoplasmic ribosomes where the synthesis of proteins takes place. Ribosomal RNAs take part in the ribosome formation whereas the tRNAs act as adaptor molecules in protein synthesis.

VACUOLES

The **vacuole** is a compartment within the plant cell containing an aqueous solution and separated from the cytoplasm by a membrane known as the **tonoplast** (Fig. 5.3). In the cells of multicellular plants, vacuoles are the most conspicuous compartments of the endomembrane system. They occupy more than 90% of the volume of most mature plant cells. Vacuole contains a variety of organic and inorganic substances, such as sugars, proteins, organic acids, phosphatides, tannins, flavonoid pigments, and calcium oxalate. Some substances may occur in solid or even in crystalline form.

Young meristematic cells possess many minute vacuoles. With growth and differentiation of a cell the vacuoles enlarge and fuse and ultimately form a large central one (Fig. 5.11). Various openions exist regarding the vacuole formation :

1. From the preexisting ones which multiply by fission, and, after cell division, each daughter cell receives a number of small vacuoles.
2. Vacuoles are formed *de novo* by attraction of water to a certain region of the cytoplasm followed by the formation of tonoplast around it.
3. From Golgi vescicles.
4. By the dilatation of ER derived cisternae or vescicles.

Tonoplast resembles the plasma membrane but reacts somewhat differently to the stains used in electron microscopy.

Vacuoles are involved in water metabolism, solute accumulation, turgor generation, and all

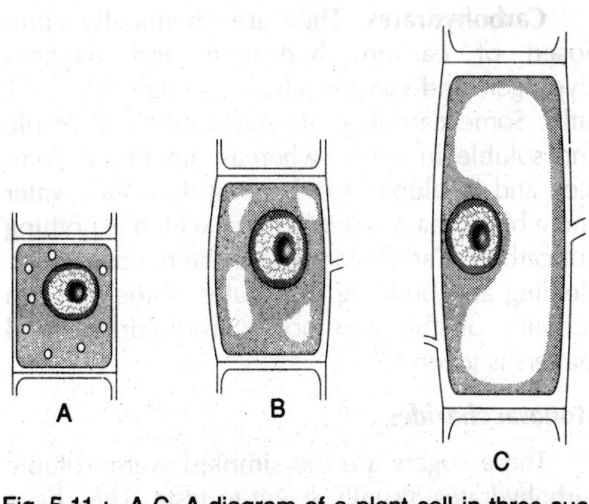

Fig. 5.11 : A-C. A diagram of a plant cell in three successive stages of development. The small vacuoles enlarge and coalscece

related functions. Vacuoles are responsible for different hydrolytic activities and solute accumulation.

Amorphous materials and crystals frequently occur as deposits in vacuoles. The cells of flower petals contain anthocyanin pigments in solution in the vacuolar sap.

Protoplasmic Inclusions or Ergastic Substances

Many types of organic and inorganic solid particles, as well as many substances like oils, gums, resins, etc., are frequently present within the protoplast, either in the cytoplasm or in the vacuole. These substances represent the (1) food products, such as starch and aleurone grains; (2) waste products, such as different types of crystals; and (3) other substances of unknown function, such as rubber, mucilages, tannins, latex, alkaloids etc. Some of these substances such as starch occur in most plants; others are characteristic of certain large or small groups of plants, and are lacking in others. Ergastic substances are grouped as (i) *Reserve* materials, (ii) *Secretory* materials and (iii) *Excretory* materials.

Reserve Materials

Reserve materials are the stored food matters. They usually remain stored in the underground stems and roots, buds, fruits and seeds. Reserve materials again are of the following types, according to their chemical nature — (a) *Carbohydrates,* (b) *Nitrogenous matters,* and (c) *Fats* and *oils.*

Carbohydrates. They are chemically composed of carbon, hydrogen and oxygen. Hydrogen and oxygen always remain in 2 : 1 ratio. Some carbohydrate molecules are simple and soluble in water, whereas others are complex and insoluble. On heating they lose water and a black mass is left behind, which is nothing but carbon. Carbohydrates are the major energy-yielding and building molecules of the cell. An account of the common carbohydrate food matters is given :

Monosaccharides

These sugars are the simplest water-soluble carbohydrates, usually sweet to taste. They have 3 to 8, usually 5 or 6, carbon atoms. The common formula of these sugars is $(CH_2O)n$, in which n is 3 or greater. They contain only one aldehyde or ketone as the functional group. They are reserve foods in many monocotyledons and also in some dicotyledons like sugar beet, carrot etc. Many of them are strong reducing agents. Six-carbon sugars are called hexoses. Common hexoses are glucose and fructose. Of them the former is abundant in all plants. The chemical formula of glucose is $C_6H_{12}O_6$. Fructose or fruit sugar with the same formula, $C_6H_{12}O_6$, occurs mainly in fruits. Two other common hexose sugars of plants are mannose and galactose.

Disaccharides

Disaccharides consist of two component monosaccharide sugars. The most abundant disaccharide in the plant kingdom is sucrose or cane sugar. Sucrose is quite valuable commercially and is extracted mainly from sugarcane stems and beet roots. On hydrolysis it yields one molecule of glucose and one molecule of fructose. Maltose is another disaccharide sugar commonly found in the germinating seeds during digestion of starch. It yields two molecules of glucose on hydrolysis.

Trisaccharides

Trisaccharide sugar molecules are composed of three component sugars. The most widespread trisaccharide in plants is raffinose, which occurs in very small quantities. On hydrolysis it yields three molecules of hexoses — glucose, fructose, and galactose.

Polysaccharides

Polysaccharides are complex molecules of higher molecular weight with multiple monosaccharide units held together by glycosidic linkages.

Inulin: It is a polysaccharide that forms a powder-like compound and occurs in colloidal condition in the cell-sap of the vacuoles of plants like *Dahlia,* Artichoke etc. of the family Compositae. It is precipitated by treatment with alcohol or glycerine. When blocks of root tubers of *Dahlia* are allowed to remain in alcohol or glycerine for some time and the sections are examined under the microscope, beautiful fan-shaped crystals are found inside the cells (Fig. 5.12). On hydrolysis, inulin yields fructose only. Inulin solution turns yellowish-brown when treated with acid phloroglucin. On precipitation, the crystals are easily recognised.

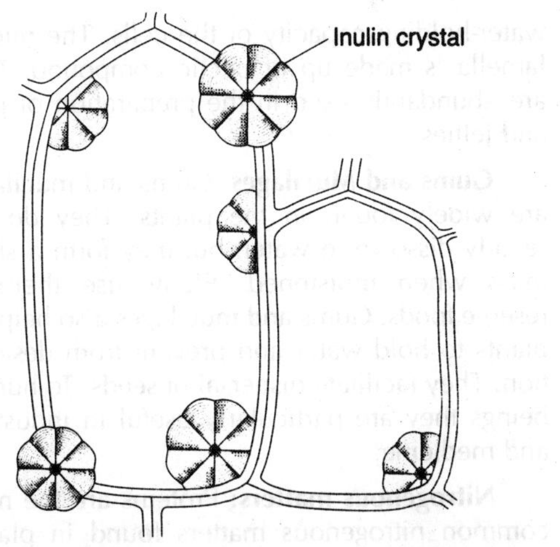

Fig. 5.12 : Root tuber cells of *Dahlia* showing inulin crystals

Starch grains: Starch grains are the common insoluble carbohydrates. They are the most important reserve foods of the plants and occur in almost all the green plants excepting some algae. Starch grains may occur in all parts of the plants, but they are more abundant in the storage organs, cereal endosperms, fruits and seeds. They are very complex chemically with the formula, $(C_6H_{10}O_5)n$ where the value of n is not clearly known. The building blocks of the molecule are unknown number of hexose sugars and is, thus, a polysaccharide. It is the stored food and is hydrolysed enzymatically into component hexoses when required.

There are two types of starch grains: **assimilatoty** starch and **reserve** starch. Assimilatory starch grains are temporary bodies formed by the chloroplasts in daytime during photosynthesis. They are granular in appearance and temporary in duration, as they are reconverted into sugar with nightfall.

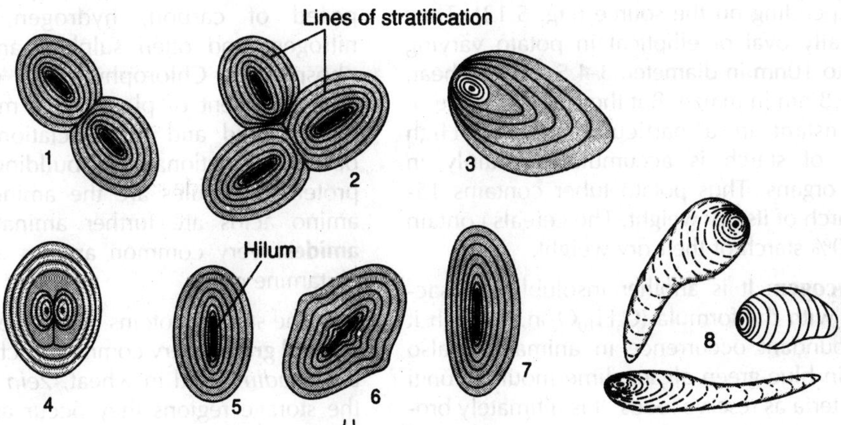

Fig. 5.13 : Starch grains : 1 and 2 compound starch grains in potato, 3. Simple starch grain, 4. Half compound starch grain, 5-7. Simple starch grains from *Pisum* cotyledon, 8. Banana starch grains, 9. Starch grains *in situ*

Surplus sugars are transported **to the sink** and there they are converted into **permanent** starch grains to be stored up for future use by the colourless leucoplasts—the **amyloplasts**. These are reserve starch. They occur profusely in storage organs like roots and underground stems and in seeds and fruits.

These starch grains are of diverse size and shape. Some of them are large enough to be seen even with the naked eye, whereas most of them are microscopic. Every starch grain has a shiny refractive point, known as **hylum**, which is the starting point. Starch molecules are deposited in layers around the hilum giving rise to a distinct stratified appearance. The layers are referred to as **lines of stratifications** (Fig. 5.13). The starch grains are regarded as sphero-crystals consisting of layers of sugar units arranged in a regular space-lattice.

Starch grains vary considerably in shape and size, depending on the source (Fig. 5.13). They are usually oval or elliptical in potato varying from 7 to 10nm in diameter, 3-4.5 nm in wheat, 1.2 to 1.8 nm in maize. But the shape is more or less constant in a particular species. High amount of starch is accumulated mainly in storage organs. Thus potato tuber contains 15-30% starch of its dry weight. The cereals contain 50 to 70% starch of their dry weight.

Glycogen: It is another insoluble polysaccharide with the formula $(C_6H_{10}O_5)n$. Though it is of abundant occurrence in animals, it also occurs in blue-green algae, slime moulds, fungi and bacteria as reserve foods. It is ultimately broken down to glucose by enzyme action and utilised as such. It gives reddish coloration with iodine solution.

Hemicellulose: The reserve cellulose is called *hemicellulose*. Cellulose $[(C_6H_{10}O_5)_n]$, is another polysaccharide which is much more complex than starch. The cell wall is primarily made up of cellulose. It is insoluble in water but readily soluble in alkalis, and is also indigestible. In the seeds of many plants like date palm, ivory nut palm etc., another insoluble carbohydrate resembling cellulose is deposited as extra layers making the wall very thick and hard. This is reserve cellulose or *hemicellulose.*

Pectins: Pectins are called fruit jellies and are present mainly in many fruits. They are readily soluble in water. Pectins can increase the water-holding capacity of the cells. The middle lamella is made up of pectic compound. They are abundantly used in the preparation of jams and jellies.

Gums and Mucilages: Gums and mucilages are widely found in the plants. They do not readily dissolve in water, but they form a slimy mass when moistened. Plants use them as reserve foods. Gums and mucilages also help the plants to hold water and prevent from desiccation. They facilitate dispersal of seeds. To human beings they are particularly useful in industries and medicine.

Nitrogenous matters: Proteins are the most common nitrogenous matters found in plants. They are the most complex organic macromolecules made up of amino acids. They are required in large amounts in the regions of active growth. They serve as an essential component of protoplasm. Chemically, the proteins are composed of carbon, hydrogen, oxygen and nitrogen, and often sulphur, and, sometimes, phosphorus. Chlorophyll, the most essential green pigment of plants, is a magnesium-containing lipid and in association of protein it becomes functional. The building blocks of the protein molecules are the amino acids. If the amino acids are further aminated they form **amides**. Very common amides are asparagine, glutamine etc.

The stored proteins inside the cell are called **proteid grains**. Very common such proteid grains are *gliadin* found in wheat, *zein* in corn etc. In the storage regions they occur as grains called **aleurone grains**. Many plant proteins have been found to occur in crystalline or amorphous forms.

In cereals, a layer of cells just beneath the seed coat, composed of pericarp and testa, remains filled with aleurone grains and called **aleurone layer**. In pulses they remain as small grains mixed with comparatively larger starch grains. More complex aleurone grains are found in the endosperm of castor seed in which the grain has a protein matrix. In that matrix there occurs a large crystalline body, called **crystalloid**, and a small round body, called **globoid**. The crystalloid is nitrogenous and the globoid is double phosphate of calcium and magnesium.

Fats and Oils (= **Triglycerides** or **triglycerols**): Fats and oils are also very important

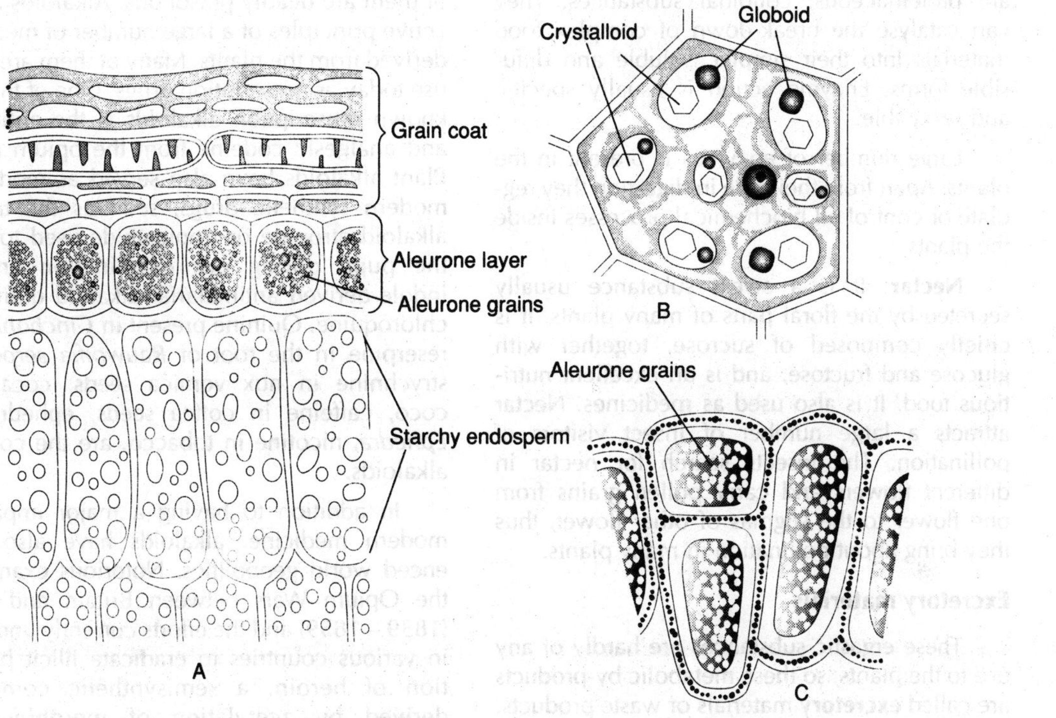

Fig. 5.14 : Aleurone grains in different plant materials : A. In *Triticum*, B. In *Ricinus*, and C. In *Zea*

energy-storage reserve materials of plants along with carbohydrates and proteins. They are composed of carbon, hydrogen and oxygen having much more of carbon and hydrogen and relatively very low oxygen. Chemically, triglycerides are esters of glycerol with three fatty acid molecules. Due to occurrence of relatively low oxygen and high carbon and hydrogen — fats and oils release much more energy than carbohydrates. Fats which are liquids at room temperature are called oils. They are lighter than water, their specific gravity ranges from 0.875 to 0.970, and are insoluble in water. The physical characteristics of a triglyceride are determined by the chain length of the constituent fatty acids — whether the fatty acids are saturated or unsaturated. The fats contain saturated fatty acids whereas the oils contain unsaturated fatty acids. Fats and oils dissolve in solvents like chloroform, ether, acetone, benzene etc. Oils are present in all living cells, often in form of droplets in the cytoplasm.

Triglycerides are esters of fatty acids and glycerol. Principal fatty acids are oleic acid, palmitic acid, capric acid, stearic acid, linoleic acid etc. They are abundantly present in storage organs, in seeds, embryos and meristematic tissues. They are formed in the cytosol and also by special fat-forming plastids called **elaioplasts**. Fats usually occur in the plants as oils. These oils are non-volatile and, hence, are called **fixed** oils. Some oils are also used as medicines, like chaulmoogra oil used in leprosy treatment and croton oil used as one of the most strong purgatives.

Besides their utility as source of food, fats are also used in many industries. If fatty acids are saponified when treated with caustic soda or caustic potash — soap is formed.

Secretory Materials

The secretory materials are not directly concerned with the nutrition and growth of the plants. They are often present in special sacs or glands. The secretory materials are of the following types :

Enzymes: Enzymes are the biocatalysts secreted by the protoplasm of many cells. These are proteinaceous, colloidal substances. They can catalyse the break-down of complex food materials into their simpler, soluble and diffusible forms. Enzyme action is usually specific and reversible.

Large number of enzymes is present in the plants. Apart from their role in digestion they regulate or control all biochemical processes inside the plants.

Nectar: It is a sweet substance usually secreted by the floral parts of many plants. It is chiefly composed of sucrose, together with glucose and fructose, and is an excellent nutritious food. It is also used as medicines. Nectar attracts a large number of insect visitors of pollination. The insects search for nectar in different flowers and carry pollen grains from one flower to the stigmas of other flower, thus they bring about pollination in many plants.

Excretory materials

These ergastic substances are hardly of any use to the plants, so these metabolic by-products are called **excretory** materials or waste products. Most of these substances are very useful to mankind. As the rate of metabolism in plants is less the waste products formation is also lesser in quantity. Moreover, the plants, unlike animals, have no definite excretory systems, excretory matters accumulate somewhere inside the plant-body. The plants, however, eliminate the waste products to some extent through leaf-falling and through fruits, feeds and barks of trees. Otherwise the accumulation of waste products would create toxicity and interfere with the activities of the living cells.

The common excretory materials are :

Alkaloid: The term alkaloid, coined in 1819 in Halle, Germany, by a pharmacist, Carl Meissner, finds its origin in the Arabic name ***al-qali***, the plant from which soda was first **isol**ated. Alkaloids were originally defined as pharmacologically active, nitrogen-containing **basic** compounds of plant origin. Thus alkaloids are nitrogenous waste products found in many plants. Chemically, they are composed of carbon, hydrogen, oxygen and nitrogen. They occur in storage organs like roots and seeds, in bark, leaves, wood and other parts. Alkaloids are usually insoluble in water, but readily soluble in alcohol. They have bitter taste, and many of them are deadly poisonous. Alkaloids are the active principles of a large number of medicines derived from the plants. Many of them are still in use today as prescription drugs. One of the best-known prescription alkaloids is the antitussive and analgesic codeine from the opium poppy. Plant alkaloids have also served as models for modern synthetic drugs, such as the tropane alkaloid atropine or tropicamide used to dilate the pupil during eye examination and the indole-derived antimalarial alkaloid quinine or chloroquine. Quinine present in *Cinchona* bark, reserpine in the root of *Rauwolfia serpentina*, strychnine in nux vomica seeds, cocaine in coco, caffeine in coffea seeds, ephedrine in *Ephedra*, nicotine in tobacco, are the common alkaloids.

In addition to having a major impact on modern medicine, alkaloids have also influenced world geopolitics. Notorious example is the Opium Wars between Britain and China (1839 – 1859) and the efforts currently underway in various countries to eradicate illicit production of heroin, a semisynthetic compound derived by acetylation of morphine, and cocaine, a natu-rally occurring alkaloid of coca plant.

The physiologically active alkaloids participate in plant chemical defenses. It is supported by their wide range of physiological effects on animals and by the antibiotic activities many alkaloids possess. Various alkaloids also are toxic to insects or function as feeding deterrents.

Tannins: 'Tannins' is a collective term for a variety of plant polyphenols used in the tanning of raw hides to produce leather. They are widely distributed in plants and occur in especially high amounts in the bark of certain trees (e.g., oak) and in galls. These are usually bitter and astringent substances. They occur in the cell-sap of the vacuoles, in the cell wall, in bark and woody parts of plants. Tannins are abundantly present in many fruits like myrobalans, in leaves, seeds, bark and also in pathological galls. The percentage of tannin is usually high in unripe fruits.

Tannins are classified as **condensed** and **hydrolysable** tannins. The condensed tannins are flavonoid polymers and are thus products of phenylpropanoid metabolism. The hydrolysable tannins consist of glycosylated **gallic acids**. Many of these gallic acids are linked to hexose molecules. Gallic acid in plants is formed mainly from shikimate.

The phenolic groups of the tannins bind so tightly to proteins — by forming hydrogen bonds with the –NH groups of peptides — that these bonds cannot be cleaved by digestive enzymes. In the tanning process, tannin binds to the collagen of animal hides and thus produces leather which is able to withstand the attack of degrading microorganisms. Tannins turn blue black in colour in contact with iron salts. Thus they are extensively used for manufacture of ink. Some of them have medicinal value as well.

Tannins have unpleasant taste. They bind with the enzymes of saliva and thus the animals are discouraged from eating plants containing tannin. Tannins also react with enzymes of the herbivore digestive tract. For these reasons tannins are very effective in protecting leaves from being eaten by animals. So it contributes to the defense mechanism of the plants.

Latex: Latex is usually a milky or often watery fluid present in some families of angiosperms. It is an emulsion of various substances like proteins, gums, resins, alkaloids, etc. remain suspended in an aqueous background. Dumb-bell-shaped starch grains are frequently noticed in latex. It is useful to the plants in healing of wounds. This milky fluid is very valuable as the chief commercial source of rubber. The majority of plants yielding rubber belong to the families Moraceae, Euphorbiaceae and Apocynaceae. *Hevea brazilensis* of family Euphorbiaceae needs special mention as a source of rubber. Latex of papaya contains papain, the enzyme helpful in digestion of protein food. The so-called 'Cow-tree' *Brosimum galactodendron* of family Moraceae of Venezuala yields the latex which is taken by the natives as a substitute of milk.

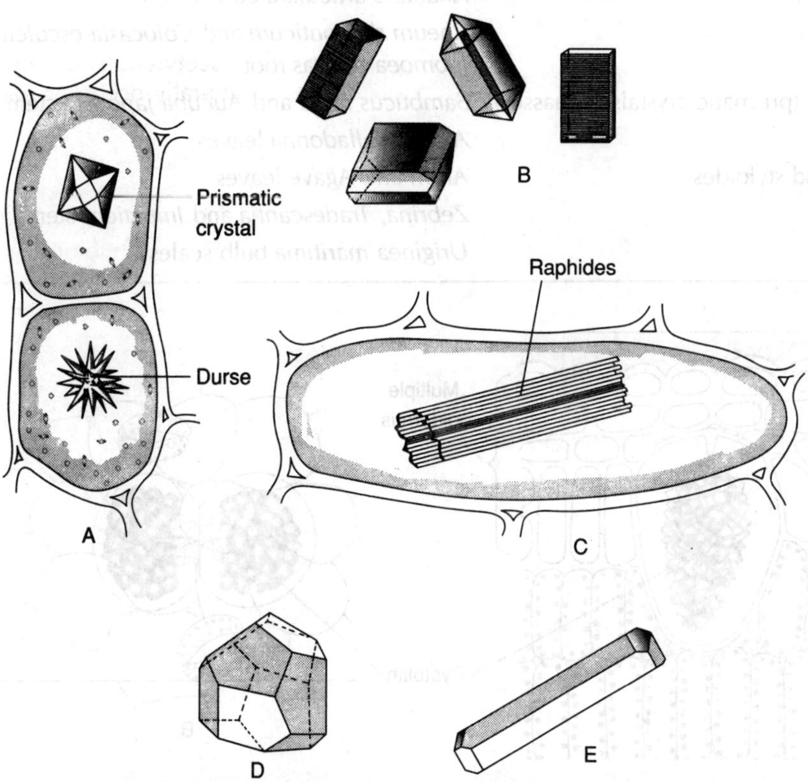

Fig. 5.15 : Crystals of various shapes : A. Two parenchyma cells containing solitary prismatic crystal and druse, B. Variously shaped prismatic crystals, C. A bundle of raphides in a parenchymatous cell, D. A crystal with pentagonal faces, E. Another elongated prismatic crystal

Mineral Crystals: Many plants deposit in their cells crystals of various chemical nature consisting mostly of salts of calcium, chiefly calcium oxalate, and silicon dioxide. Crystals of many organic substances, such as berberin, saponin etc., are frequent. Any part of the plant body may contain crystals, still they are more abundant in the pith, cortex, and phloem. There are many structural forms of crystals found in plant cells (Fig. 5.15). They may be solitary, rhombohedral, prismatic, pyramidal etc. Sheaf-like bundles of long acicular crystals known as **raphides**, and clustered crystals in spheroidal masses called **druses** are most common. The individual crystals of druses may be raphides. Solitary raphides, small prismatic crystals, and minute crystals called **crystal sand**. Usually one form of crystal may occur in a given cell. Two or more types may be found in the some cases. Many inorganic crystals appear to involve organic matter in the course of their formation. The larger crystals of xylem often show this condition, and druses frequently contain a prominent organic center.

Table 5.1 : Distribution of different types of crystals

Crystal Type	Plant Material
Rectangular or pyramidal prisms	*Citrus, Begonia, Hyoseyamus niger, Vicia sativa* leaves
	Pistacia palaestina bundle-sheath cells.
Druses (spheroidal aggregates of prismatic crystals)	*Datura stramonium* and *Ruta graveolens* leaves
	Opuntia ficus-indica stem
	Anabasis articulara cortex
	Rheum rhaponticum and *Colocasia esculenta* rhizome
	Ipomoea batatas root
Crystal sand (prismatic crystals in masses)	*Sambucus nigra* and *Aucuba japonica* stem
	Atropa belladonna leaves
Raphides and styloides	*Arum* and *Agave* leaves
	Zebrina, Tradescantia and *Impatiens* stems
	Uriginea maritima bulb scales

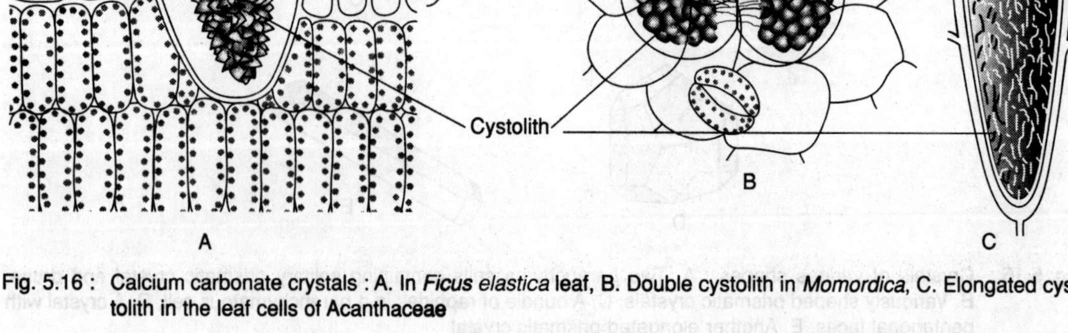

Fig. 5.16 : Calcium carbonate crystals : A. In *Ficus elastica* leaf, B. Double cystolith in *Momordica*, C. Elongated cystolith in the leaf cells of Acanthaceae

Calcium carbonate crystals usually occur in the epidermis impregnating a portion of cellulose cell wall in many plants. In rubber or banyan these Ca-carbonate crystals aggregate on a peg like projection of the cell wall looking like a bunch of grapes. The epidermis here is multiseriate and the crystal-aggregation takes place in some enlarged cells of the innermost layer. Such a crystal-aggregate is called **cystolith** (Fig. 5.16).

Essential oils: These are volatile oils found in many parts of the plant. They usually occur in special glands. The fragrant essence of many flowers and leaves, rinds of some fruits, bark and wood of some trees and even of some seeds are due to presence of essential oils. These oils are advantageous to the plants in bringing about pollination and dispersal of fruits and seeds by attracting insect visitors. They are used in manufacture of perfumes, soaps, etc, and also in medicine, e.g., eucalyptus oil from *Eucalyptus globulus* of family Myrtaceae and valerian from *Valeriana officinalis* of family Valeianaceae. Essential oils also have antiseptic value.

Resins: A resin is a polymer with a rigid three-dimensional network of phenol-formaldehyde repeating units. They occur in special glands or ducts usually in combination with gums and essential oils. They are water insoluble but soluble in many solvents like ether and alcohol. Resins are of three types. Pure resins which are secreted alone are usually solid and brittle, called **hard resins**. They readily dissolve in alcohol and are used in paint and varnish industry as waterproofing and glossy agent. Lacquer and shellac are the examples Resins mixed with essential oils are aromatic and more or less liquid in nature, known as **oleo-resins**, e.g., turpentines and Canadabalsam. The third type is **gum resins** which are mixtures of resins and gums. They are usually milky in appearance and collected by making an injury on the plant body. Many resins are used in medicine.

Gums. Gums are metamorphosed, unorganized, amorphous products of organized cell-wall materials. Gums may also be formed from starch. The process of gum formation is called **gummosis**. During this process disintegration of the walls starts in the primary wall and proceeds towards the innermost lamella of the secondary cell-wall. It also occurs in the bark. Gums are water soluble. They are exuded from the plants in natural course or through the injuries as mucilaginous masses. They are of great commercial value as adhesives, and are also used in other industries. Gumarabic of *Acacia senegal* and other *Acacia* species is formed in the bark. It is used in medicine as an emulsifier.

THE CELL WALL

Most of the plant cells have tough and semi-rigid cell wall which is non-living in nature. The presence of this wall in plant cells distinguishes them from animal cells. The cell wall was actually noticed by the early workers in the seventeenth century long before the protoplasm was recognized and since then it received considerable attention by many researchers. Various techniques of investigations such as chemical, physical, and morphological have been used. These investigations have been aided by organic chemistry, X-ray analysis, light, polarized and electron microscopy. The commercial importance of cellulose had given new impetus to the investigators to study the cell-wall.

The cell wall materials are synthesized and deposited by the protoplast. The cell walls are responsible for the shape of plant. They protect the protoplasts from adverse external influences and give mechanical stability to the cells. The walls of plant cells are analogous to the skeletons of animals. They also form a protective barrier against viral, bacterial and fungal attack. So, they function as the skeleton as well as the skin of plants. The cell walls control the growth rate of the plant cell and thus of the whole plant. So the cell walls are fundamentally involved in several aspects of plant biology, including the morphology, growth and development of plant cells. The cell wall of each cell merges with the walls of adjacent cells, giving the tissue physical coherence and strength.

The walls are not simply rigid and static structures. The cell walls expand as the cells

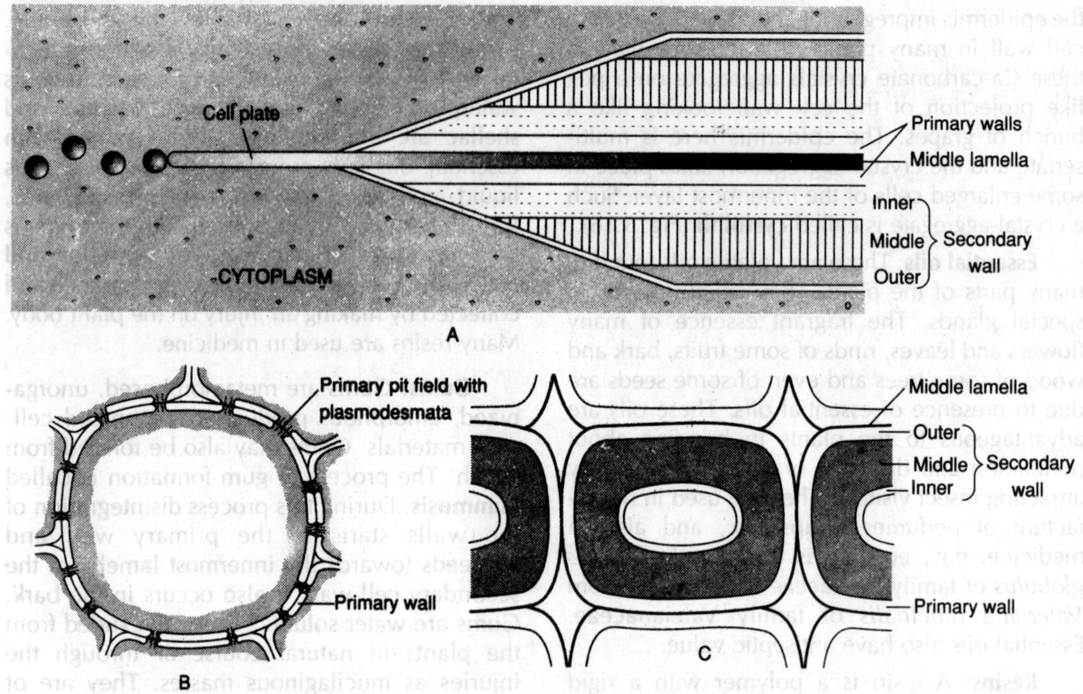

Fig. 5.17 : The cell wall and its different layers : A. Diagrammatic representation showing the formation of cell wall, origin of cell plate, its transformation into middle lamella and deposition of primary and secondary walls, B. Primary cell wall of a parenchymatous cell showing primary pit fields with plasmodesmata, C. Diagrammatic cross section of a cell showing primary and secondary wall layers

grow, and new components must be deposited onto the existing wall structure. With differentiation the walls change in size, shape and chemical composition.

The cell wall is formed during the cell division and it is comparatively thin in young cells than in fully mature cells. In mature cells the cell walls have the three general layers — the ***middle lamella***, a thin ***primary wall*** and a thicker ***secondary wall***. The primary walls are laid down by the still-growing undifferentiated cells. The primary wall is transformed into a secondary wall when the cells stop growing. The cell wall is composed of cellulose, hemicellulose, pectic polysaccharide, structural protein and lignin. Structurally, the primary cell wall is composed of cellulose fibers embedded in an amorphous mixture of polysaccharides and glycoproteins. Lignin is a characteristic component of secondary walls.

The middle lamella : The *middle lamella* is the cementing substance that holds the individual cells together to form the tissues and, accordingly, it is found between the primary cell walls of adjacent cells. It is an amorphous and colloidal substance and remains inactive in plane polarised light. In mechanical tissues it may also fill the intercellular spaces. In such tissues the middle lamella becomes quite obscure due to secondary depositions. At such times, it becomes difficult to separate the middle lamella from the primary walls on the either side. Esau (1953) proposed the term ***compound middle lamella*** for such three-ply structure. The middle lamella consists mainly of pectic substances such as calcium and magnesium pectate which can dissolve in strong acids. The enzyme pectinase and chemical reagents which digest pectins disintegrate tissues into individual loose cells — a procedure called ***maceration***.

The primary wall : The primary wall is the first product of cell wall synthesised by the protoplast. During cell enlargement the primary wall remains relatively thin and elastic. Thickening and rigidity comes only after the completion of cell enlargement. In many cells, primary wall is the only cell wall as the middle lamella is not the wall proper. This wall is optically active. It gets more and more stretched during the growth of the cell. Later on, secondary wall materials are deposited on the primary wall. The primary walls of two adjacent cells are present on either side of the middle lamella. It is composed of cellulose, hemicellulose, polysaccharides and many other pectic substances. Electron microscopic study reveals that during the development of middle lamella and primary wall, small openings are left at many places in walls between the adjacent cells. These openings are known as ***primary pit fields***. Through these pits cytoplasmic continuity between the neighbouring cells is maintained. These cytoplasmic connections are called ***plasmodesmata***.

Secondary cell wall (S-wall) : It develops on the inner surface of the primary wall, mainly when the cell growth has ceased. It is usually made up of a matrix of cellulose, hemicelluloses and polysaccharides. During maturation, lignin, suberin, waxes, tannins and calcium carbonate are also deposited on the secondary wall giving rise to different irreversible modifications. At the onset of secondary wall formation the cell wall becomes much less flexible and, finally, almost inelastic. That is why cell elongation ceases with the onset of secondary wall formation. However, the uniform layering of the secondary wall materials is discontinuous and may be localised to certain areas to form some special patterns. The secondary cell wall is optically active. It gives very high mechanical strength to the plants. In the majority of tracheids and fibres the secondary walls have three layers — the outer layer (S_1), the central layer (S_2), and the inner layer (S_3) (Fig. 5.18). The central layer usually is the thickest layer.

A non-cellulosic *tertiary wall* may also develop on the inner side of the secondary wall particularly in the tracheary elements. Frey-Wyssling (1976) suggests that in addition to the inner layer of the secondary wall an innermost layer may also be formed in some cases. According to him, this layer should be termed ***tertiary lamella*** consisting of a ***membranogenic stratum*** and a ***warty stratum***.

Secondary walls are usually present in the non-living cells like sclereids, fibres, tracheary elements etc. On the other hand, the xylem parenchyma and ray cells, though living, possess secondary walls.

Fine Structure of the Cell wall

Cellulose is the principal scaffolding component of all plant cell walls. It is the most abundant plant polysaccharide and exists in the form of ***microfibrils***, which are paracrystalline assemblies of several dozen (1→4) β-D-glucan chains hydrogen-bonded to one another along their length. In higher plants each microfibril contains 36 individual cellulose chains, but that of algae can form either large, round cables or flattened ribbons of several hundred chains. Angiospermic microfibrils are 5 to 12 nm wide in the electron microscope. Each (1→4) β-D-glucan chain may be about 2 to 3 μm long, but individual chains begin and end at different places within the microfibril which is analogous to a spool of thread that consists of thousands of individual cotton fibers, each about 2 to 3 cm long.

By electron diffraction, it has been observed the (l→4) β-D-glucan chains of cellulose are arranged parallel to one another with all of the reducing ends of the chains remain in the same direction. Recently, bacterial cellulose was shown to be synthesised by the adding of glucose units to the non-reducing ends of glucan chains.

Callose is another type of wall material. It is made by a few cell types at specific stages of wall development, such as in growing pollen tubes and in the cell plates of dividing cells. It differs from cellulose in consisting of (l→3) β-D-glucan chains, which can form helical duplexes and triplexes. It is also made in response to wounding or to attempted penetration by invading fungal hyphae.

Each cellulose molecule is 0.8 nm in diameter. The long chain molecules are again arranged in bundles to form ***elementary fibril***. Each elementary fibril has a diameter of about 3 nm and contains 40-100 cellulose molecules. Both the cellulose molecules and the fibrils are ribbon-like structures. Elementary fibrils contain some smaller units called ***micelles*** or ***crystallites*** which are small aggregations of cellulose molecules that lie parallel to one another. The elementary fibrils are again arranged in bundles called ***microfibrils*** (Fig. 5.18). Each microfibril is 25 nm in diameter and contains about 2,000 cellulose molecules. Microfibrils are aggregated into ***macrofibrils***. Each macrofibril has a diameter of 0.4μ and contains about 500,000 cellulose molecules.

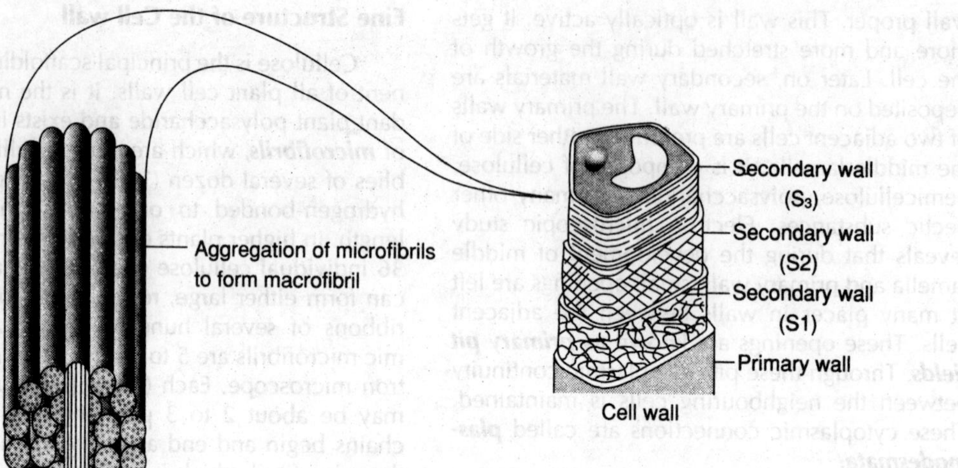

Fig. 5.18 : Paracrystalline arrays of several dozen. 1, 4 β-D-glucan chains tightly linked by numerous hydrogen bonds in cellose microfibrils

Cross-linking glycans, which are often called "hemicelluloses", interlock the cellulosic scaffold. The term hemicellulose is used for all materials extracted from the cell wall with molar solutions of alkali, regardless of the structure. They are a class of polysaccharide that links the microfibrills together to form a network.

There are two major cross-linking glycans found in primary cell walls of flowering plants *xyloglucans* (XyGs) and ***glucuronoarabinoxylans*** (GAXs).

The XyGs consist of linear chains of (1→4) β-D-glucan with numerous α-D-xylose (Xyl) units linked at regular sites to the 0-6 position of the glucose (Glc) units. Some of the xylosyl units are substituted further with α-L-Arabinose or β-D-Galactose, depending on the species, and sometimes the Galactose is substituted further with α-L-Fuctose.

The XyGs are constructed in block-like unit structures containing 6 to 11 sugars, the proportions of which vary among tissues and species. The XyGs can form three major structural variations. All of the non-commelinoid monocots and most of the dicots are fucogalacto-XyGs. The fundamental structure is composed of nearly equal amounts of XXXG and XXFG (Fig. 5.19).

But variations may occur in which α-L-Arabinose is added at some places along the glucan chain. In Solanaceae and peppermint

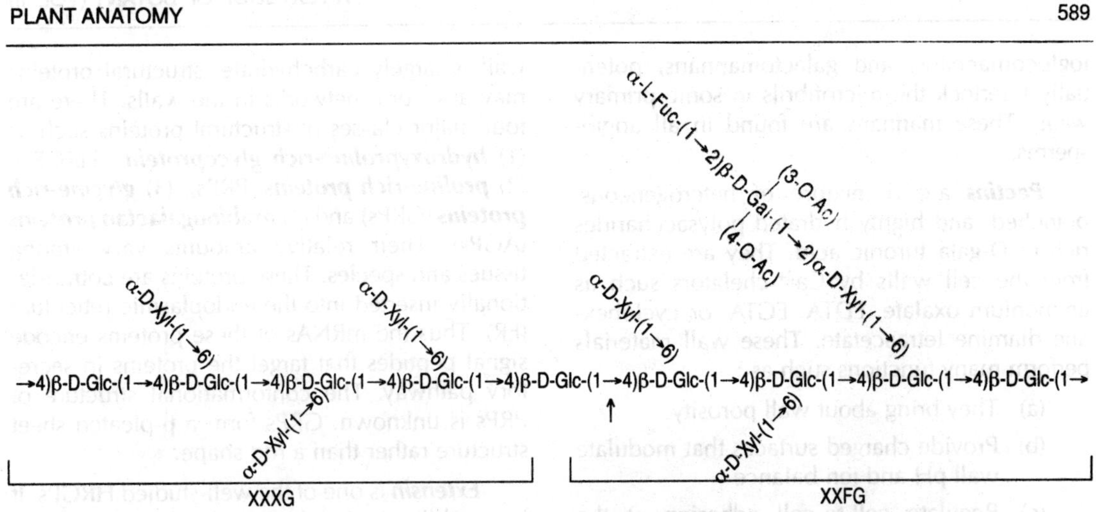

Fig. 5.19 : Fucogalactoxyloglucans in which α-D-xylosyl units are added to three contiguous glucosyl units of the backbone to produce a heptasaccharide unit (XXXG or XXFG) structure

arabino-XyGs are formed in which only two of every four glucosyl units contains a xylose unit, and the xylosyl units are substituted with either one or two α-L-Arabinose units to produce a mixture of AXGG, XAGG, and AAGG subunits. The commelinoid monocots also contain small amounts of XyG, but these include random additions of xylosyl units. All angiosperms also contain at least small amounts of GAXs, but their structure may vary considerably with respect to the degree of substitution and position of attachment of α-L-Ara residues.

In the commelinoid monocots, where GAXs are the major cross-linking polymers, the Arabinose units are invariably on the O-3 position. The α-L-Ara units are more commonly found at the O-2 position where XyG is the major cross-linking glycan.

A third major cross-linking glycan, called "mixed-linkage" (l→3), (l→4) β-D-glucans, is found in the cereals. It distinguishes the members of Poales from the commelinoid species.

These unbranched polymers consist of 90% cellotriose and cellotetraose units in a ratio of about 2 : 1 and connected by (l→3) β-D-linkages. These units form corkscrew-shaped about 50 residues long polymers which are intervened by oligomers of four or more contiguous (1→4) Glc units.

Other, much less abundant non-cellulosic polysaccharides, such as glucomannans, galac-

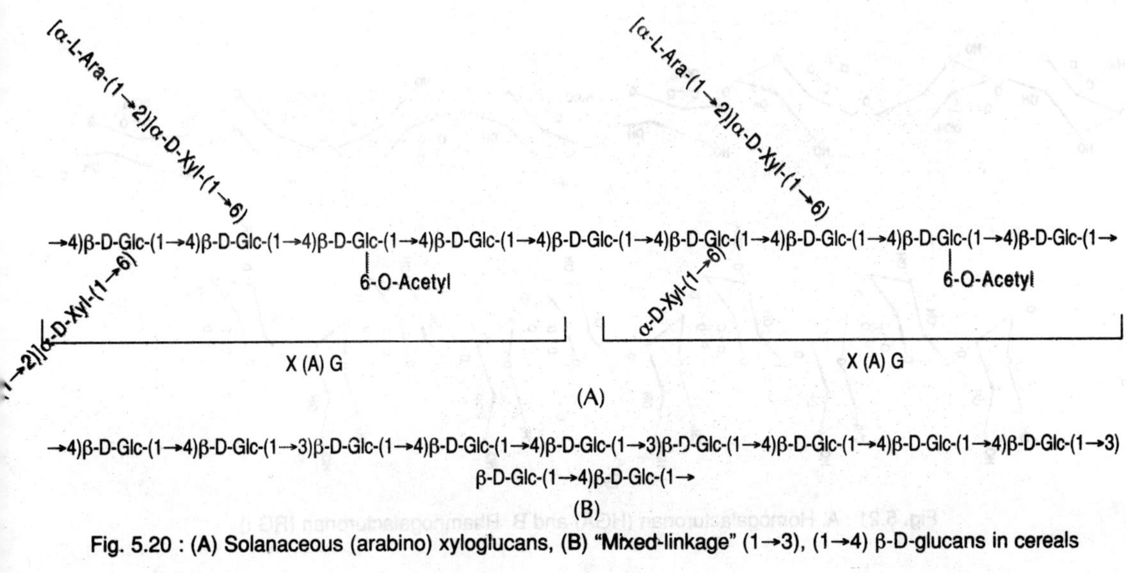

Fig. 5.20 : (A) Solanaceous (arabino) xyloglucans, (B) "Mixed-linkage" (1→3), (1→4) β-D-glucans in cereals

toglucomannans, and galactomannans, potentially interlock the microfibrils in some primary walls. These mannans are found in all angiosperms.

Pectins are a mixture of heterogeneous, branched, and highly hydrated polysaccharides rich in D-galacturonic acid. They are extracted from the cell walls by Ca^{2+}-chelators such as ammonium oxalate, EDTA, EGTA, or cyclohexane diamine tetraacetate. These wall materials perform many functions such as :

(a) They bring about wall porosity

(b) Provide charged surfaces that modulate wall pH and ion balance

(c) Regulate cell-to-cell adhesion at the middle lamella

(d) Serves as recognition molecules to alert plant cells from the presence of pathogens, insects and symbiotic organisms

(e) Cell wall enzymes bind to the charged pectin network to function at local regions

Two fundamental constituents of pectins are ***homogalacturonan*** (HGA) and ***rhamnogalacturonan I*** (RG I).

HGAs are homopolymers containing as many as 200 GalA (α-D-galacturonic acid) units and are about 100 nm long. RG I is a rod-like heteropolymer of repeating (1→2) α-L-Rha-(1→4) α-D-GalA disaccharide units.

Although the scaffold structure of the cell wall is largely carbohydrate, structural proteins may also form networks in the walls. There are four major classes of structural proteins such as (1) ***hydroxyproline-rich glycoproteins*** (HRGPs), (2) ***proline-rich proteins*** (PRPs), (3) ***glycine-rich proteins*** (GRPs) and (4) ***arabinogalactan proteins*** (AGPs). Their relative amounts vary among tissues and species. These proteins are cotranslationally inserted into the endoplasmic reticulum (ER). Thus the mRNAs of these proteins encode signal peptides that target the proteins in secretory pathway. The conformational structure of PRPs is unknown. GRPs form a β-pleated sheet structure rather than a rod shape.

Extensin is one of the well-studied HRGPs. It is a rod-like molecule encoded by a multigene family. It consists of repeating Ser-$(Hyp)_4$ and Try-Lys-Try sequences that are important for secondary and tertiary structure.

AGPs are named proteoglycans as they may contain more than 95% carbohydrate. These proteins form a broad class of molecules located in Golgi derived vesicles, the plasma membrane, and the cell wall. The exact function of AGPs is still unknown.

Aromatic substances are present in the non-lignified walls of commelinoid species. These moncots along with certain members of Chinopodiaceae (such as sugar-beet and spinach) contain significant amounts of aromatic substances in their non-lignified cell walls. The major components of these aromatics are ***hydroxycinnamic acids*** such as ferulic acids and *p*-coumaric acids.

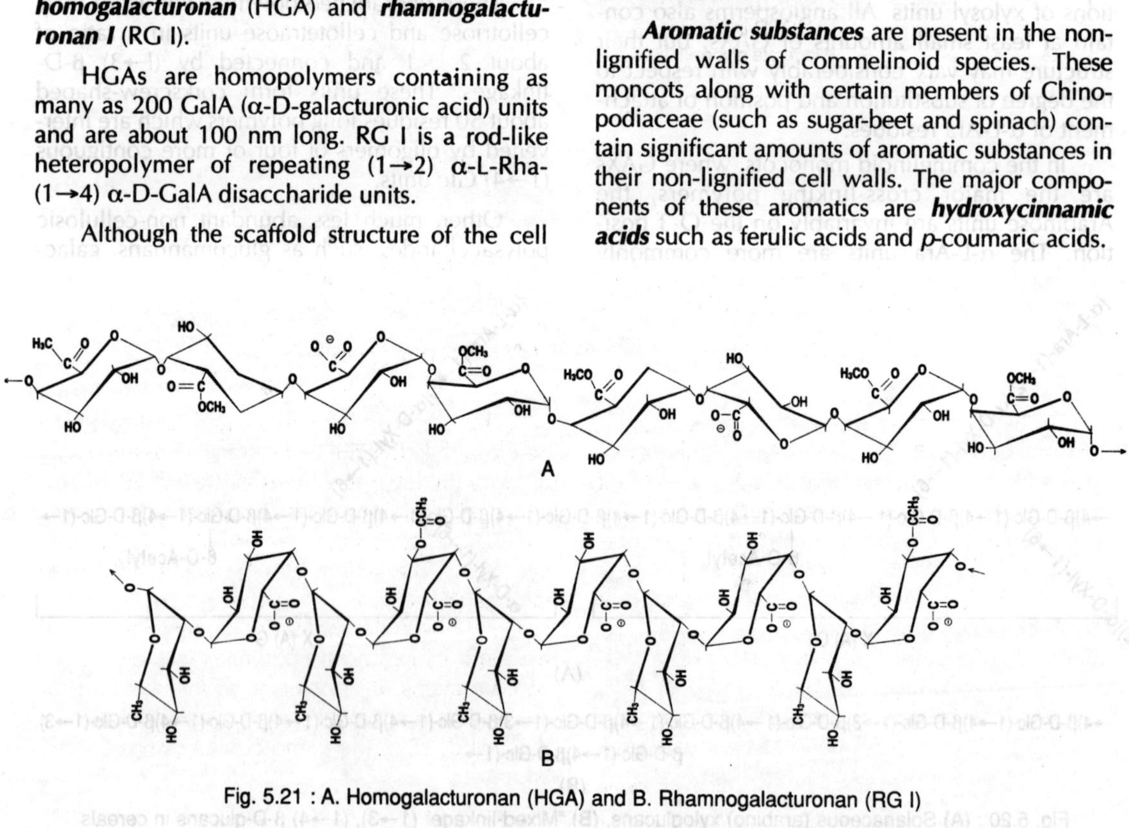

Fig. 5.21 : A. Homogalacturonan (HGA) and B. Rhamnogalacturonan (RG I)

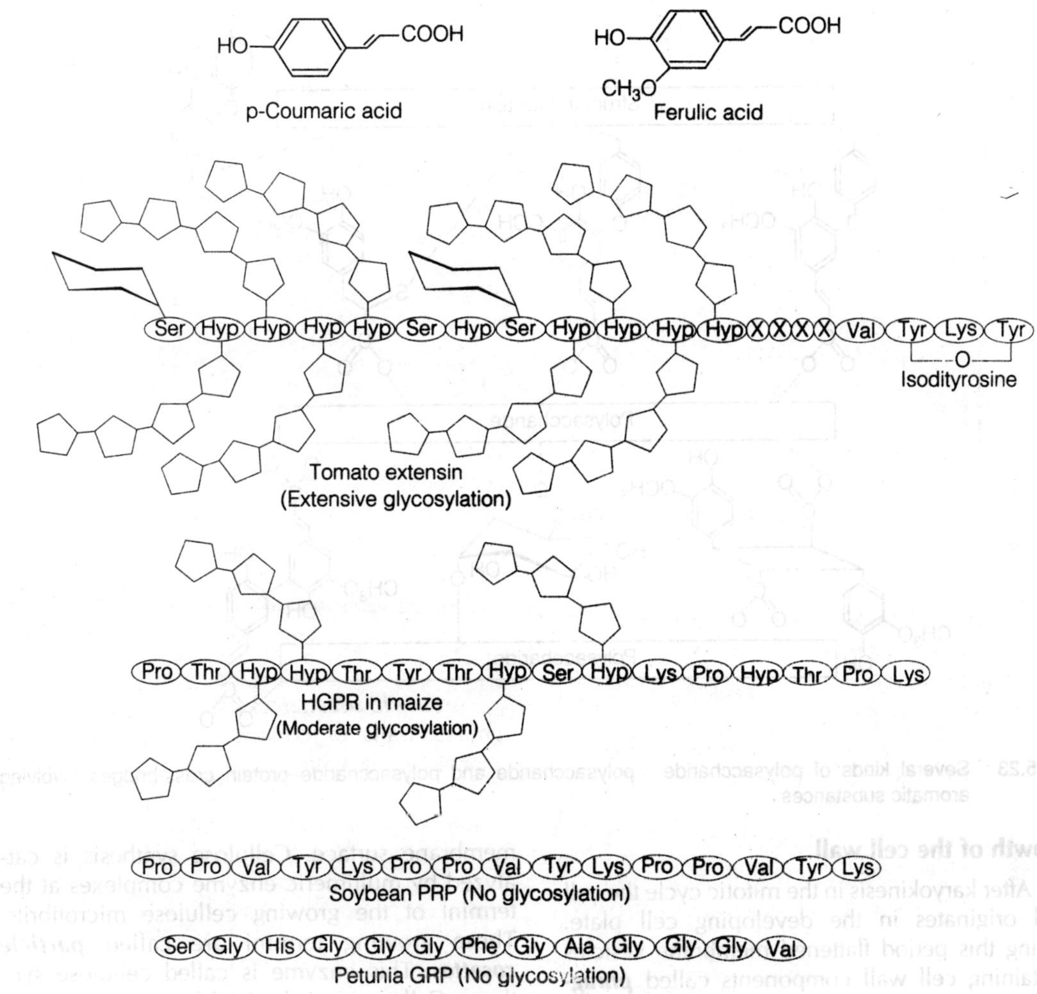

Fig. 5.22 : Extension molecules found in many flowering plants

The walls of most dicots and non-commelinoid monocots contain about equal amounts of XyGs and cellulose. These kinds of wall are called **Type I walls**. The commelinoid monocots have **Type II walls** which contain GAXs (glucuronoarabinoxylans) as principal interlocking polymers instead of XyGs. These walls are pectin poor, but contain α-L-GlcA units on GAX. In Type I walls, the cellulose-XyG framework is embedded in a protein matrix that controls wall porosity. Type II walls have very little structural protein compared with dicots and other monocots. There are extensive interconnecting networks of phenylpropanoids.

The polymers remain soluble until they can be cross-linked at the cell surface. Different types of cross-linking possibilities exist such as, hydrogen-bonding, ionic-bonding with Ca^{2+} ions, covalent ester linkages, ether linkages, and van der Waals' interactions etc. Assembly of the wall materials occur in an aqueous environment. Water is one of the major components of the cell wall and maintains the polymers in their proper conformations. It also serves as the medium for enzyme action as well as allows the passage of ions and signalling molecules through the apoplast.

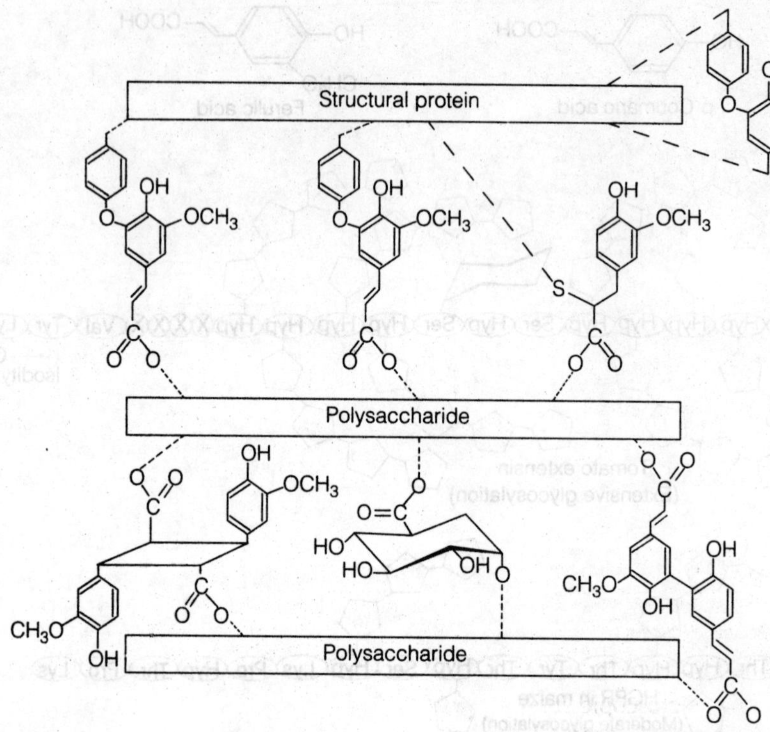

Fig. 5.23 : Several kinds of polysaccharide : polysaccharide and polysaccharide–protein cross-bridges involving aromatic substances

Growth of the cell wall

After karyokinesis in the mitotic cycle the cell wall originates in the developing cell plate. During this period flattened membrane vesicles containing cell wall components called ***phragmosomes*** arrange themselves across the cell. This structure of cytoskeletal array is called ***phragmoplast***. The plant Golgi apparatus is a factory for the synthesis, processing, and targeting of glycoproteins. The Golgi-derived vesicles fuse with the growing cell plate and ultimately the edges of the vesicle fuse with the plasma membrane dividing the cell into two. The non-cellulosic polysaccharides are also synthesised in the Golgi apparatus, packaged into vesicles and fuse with the growing cell plate. There is an enzyme system inside the Golgi apparatus that interconverts the nucleotide sugars *de novo*, which serve as substrates for polysaccharide synthesis.

Cellulose microfibrils are assembled at the plasma membrane surface. Cellulose and callose molecules are synthesized at the outer plasma membrane surface. Cellulose synthesis is catalyzed by multimeric enzyme complexes at the termini of the growing cellulose microfibrils. These growing termini are called ***particle rosettes***. This enzyme is called cellulose synthase. Callose is synthesised by an enzyme that does not require Ca^{2+} as cofactor. This enzyme is distinct from cellulose synthase. It is very difficult to purify the enzyme cellulose synthase in active form. Cellulose synthase genes have been cloned from plants.

The primary wall of many cells has a lamellate structure. The orientation of the microfibrils in the primary wall differs in different cell types as well as in different parts of the wall of a single cell. There are two classical concepts regarding cell wall extension growth — (1) growth in ***intussusception*** i.e. the new microfibrils are laid down between the existing microfibrils of the expanding cell wall, and (2) growth in ***apposition*** where new microfibrils are laid down upon the existing one forming a new layer.

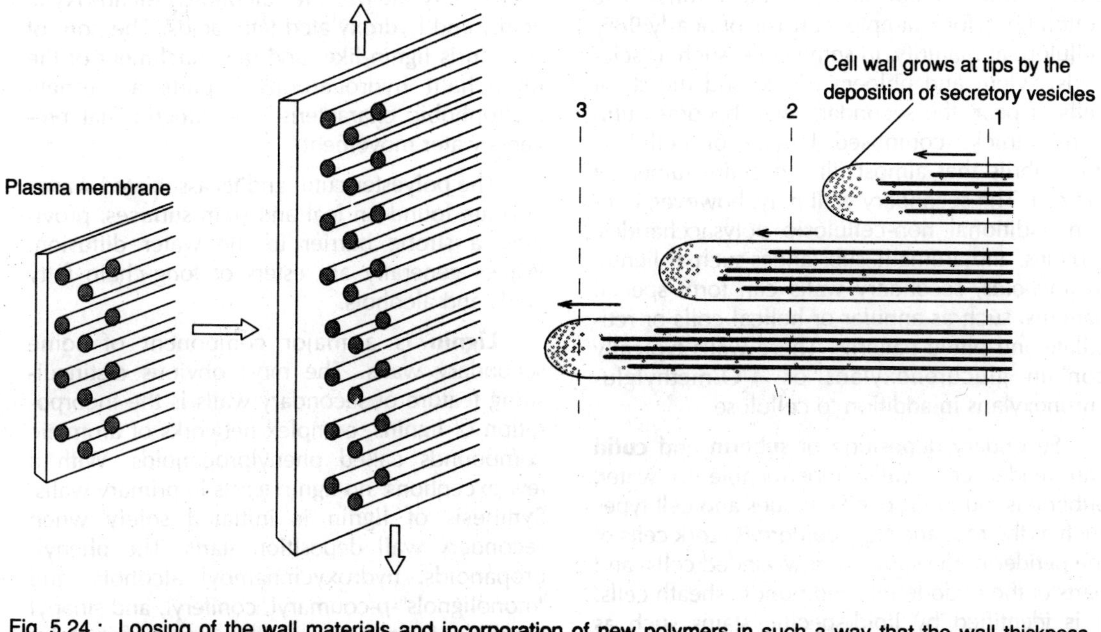

Fig. 5.24 : Loosing of the wall materials and incorporation of new polymers in such a way that the wall thickness remains constant during cell expansion

Based on electron microscopic studies, new theories have been developed regarding the primary wall growth. Frey-Wyssling and Stecher (1951) suggested the ***mosaic growth***. This explains that the microfibrils in certain wall areas become loose enzymatically and, as a result of turgor pressure, this region is stretched out and then newly synthesised microfibrils are deposited in the gaps caused by the strain. Growth hormones play a regulatory role in the process of extension growth. The acid-growth hypothesis postulates that auxin-dependent acidification of the cell wall promotes wall-extensibility and cell growth.

Another concept is the ***multinet growth hypothesis***. According to this theory the thickening and increase in surface area, of the primary wall is brought about by the axial displacement and reorientation of the crossed cellulose microfibrils as elongation proceeds.

According to the multinet growth hypothesis the fibrils are added to the growing cell wall by apposition, whereas in view of the mosaic growth concept it takes place also in intussusception. The wall matrix (hemicelluloses and pectins), however, is continuously secreted not only into the lamllae adjacent to the cytoplasm but also into the outer layers.

At present, two kinds of enzymes — ***xyloglucan endotransglycosylase*** (XET) and ***expansins*** have been evaluated as having wall-loosening activities. Xet carries out transglycosylation of XyG in which one chain of XyG is cleaved and reattached to the non-reducing terminus of another XyG chain. Expansins catalyze breakage of hydrogen bonds between cellulose and the load-bearing cross-linking glycans.

Secondary walls are laid down after the growth of the primary wall. Secondary walls

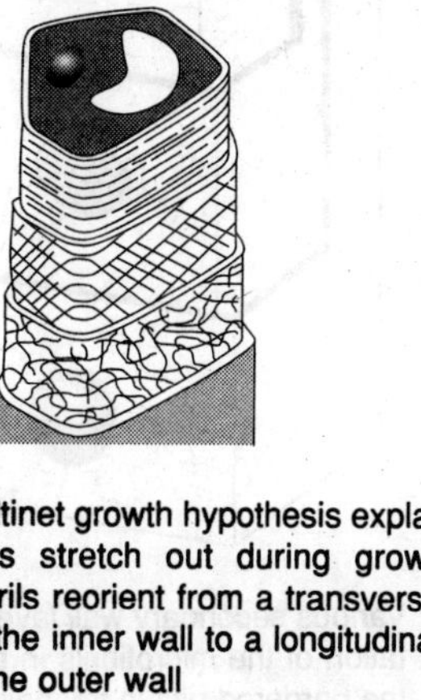

Fig. 5.25 : The multinet growth hypothesis explains that as walls stretch out during growth, the microfibrils reorient from a transverse direction on the inner wall to a longitudinal direction at the outer wall

often exhibit elaborate specialisations. The cotton fiber, for example, consists of nearly 98% cellulose at maturity. In some cells, such as sclereids, xylem and phloem fibers, and the stone cells of pear, the secondary wall becomes uniformly thick, composed largely of cellulose microfibrils that almost fill the entire lumen of the cell. The secondary wall may, however, contain additional non-cellulosic polysaccharides, proteins, and aromatic substances such as lignin. In tracheids, secondary walls can form special patterns, such as annular or helical coils or reticulate and pitted patterns. These walls typically contain glucuronoxylans or 4-O-methylglucuronoxylans in addition to cellulose.

Secondary deposition of **suberin** and **cutin** can render cell walls impermeable to water. Suberin is found in specific tissues and cell types such as the root and stem epidermis, cork cells of the periderm, the surfaces of wounded cells, and parts of the endodermis and bundle sheath cells. It is identified by lipid-specific stains such as Sudan IV, which detects the presence of long-chain fatty acids and alcohols, dicarboxylic acids, and hydroxylated fatty acids. The core of suberin is lignin-like, and the attachment of the long-chain hydrocarbons imparts a strongly hydrophobic characteristic to suberin that prevents water movement.

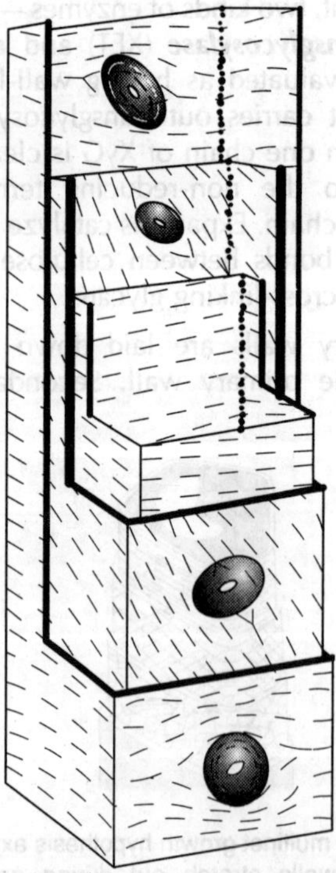

Fig. 5.26 : Various secondary wall layers and the orientation of the microfibrils in them and around the bordered pits in tracheid

The polyester cutin and its associated waxes also are found on leaf and stem surfaces, providing a strong barrier to the water diffusion. Waxes generally are esters of long-chain fatty acids and alcohols.

Lignin is a major component of some secondary walls. The most obvious distinguishing feature of secondary walls is the incorporation of lignins, complex networks of aromatic compounds called phenylpropanoids. With a few exceptions, no lignin exists in primary walls. Synthesis of lignin is initiated solely when secondary wall deposition starts. The phenylpropanoids, hydroxycinnamoyl alcohols, and "monolignols"-*p*-coumaryl, coniferyl, and sinapyl alcohols form most of the lignin networks. Lignins are detected in tissue sections by specific stains, such as the Mäule reagent (mixture of $KMnO_4$ and ammonia), acid fuchsin and the Wiesner reagent (phloroglucinol).

THICKENINGS OF THE CELL WALLS

The orientations of the microfibrils and the micelle in the secondary walls differs in different species, in the cells of different organs, in the different layers of the same wall, and sometimes even in the lamellae of the same layer. In the walls of vessels, tracheids and fibres there are three layers in the secondary wall. In the outer and inner layers the microfibrils are oriented almost horizontal and in the middle layer it is almost longitudinal to the long axis of the cell.

The deposition of the secondary wall materials is not uniform. It is localised in certain regions leading to the formation of different types of sculptures or designs. Thickening gives strength to the cell wall to resists different types of external and internal pressures.

Various sculptures or designs usually found on the inner surface of the cell wall are — (a) ***annular***, (b) ***spiral***, (c) ***scalariform***, (d) ***reticulate***, and (e) ***pitted***.

In *annular* or *ring-like* thickening, the secondary wall materials are disposed in the protoxylem elements in the form rings at regular intervals, e.g., annular vessels of xylem (Fig. 5.27A & B).

In ***spiral*** or ***helical thickening***, the deposition of secondary wall materials takes place in the form of spiral bands (Fig. 5.27C & D). The helical thickening may be single, or more than one helix and may be present in a single element. Again, the helices may be loose or dense. It is also found in protoxylem elements. Ontogeneti-cally the annular elements precede the helical elements.

In the ***scalariform type***, secondary thickening materials are disposed like the steps of a ladder — both in longitudinal as well as transverse directions (Fig. 5.27E & F). Actually, in the later-formed tracheary elements, the helical bands become joined in certain areas, giving rise to a ladder-like thickening.

Reticulate type of thickening results from irregular deposition of secondary wall materials in the form of a network (Fig. 5.27G). This type of tracheary elements are formed still later, in the ontogenetic stage. When the openings of the network are elongated in a direction perpendicular to the longitudinal axis of the element, it is termed ***scalariform-reticulate***.

In ***Pitted*** elements the secondary cell wall is interrupted only unthickened, small pits. Such elements are termed ***pitted elements*** (Fig. 5.27H & I). This type of thickening is found in the ontogenetically most advanced elements. It is the characteristic of the late primary and secondary xylem.

OTHER SCULPTURINGS ON THE WALL

(a) ***Trabeculae*** : These are rod-like or bar-like thickened projections of the wall which traverse the cell lumen extending across from one tangential wall to the other. These structures generally occur in radial rows in the wood elements. They are frequently present in the tracheids of the secondary wood of conifers.

(b) ***Crassulae*** : These are linear or crescent-shaped thickenings of intercellular materials and primary walls which occur along the outer and lower margins of pit pairs partially encircling them. These are visible through the secondary walls in stained preparations. They represent the borders of primary pit fields of young cells from which they have developed. Crassulae (Fig. 5.28) are also termed 'bars of Sanio' or 'rims of Sanio' after Sanio who extensively worked on wood of conifers in the nineteenth century.

(c) ***Wart structure*** : Certain wart-like structures are observed in the inner surface of the secondary wall of conifer tracheids and of fibers and vessels of certain dicots (Fig. 5.29). These structures consist of the remnants of the protoplast or lignin precursors. The diameter of these structures varies between 0.1μ and 0.5μ. They develop during the end of the differentiation.

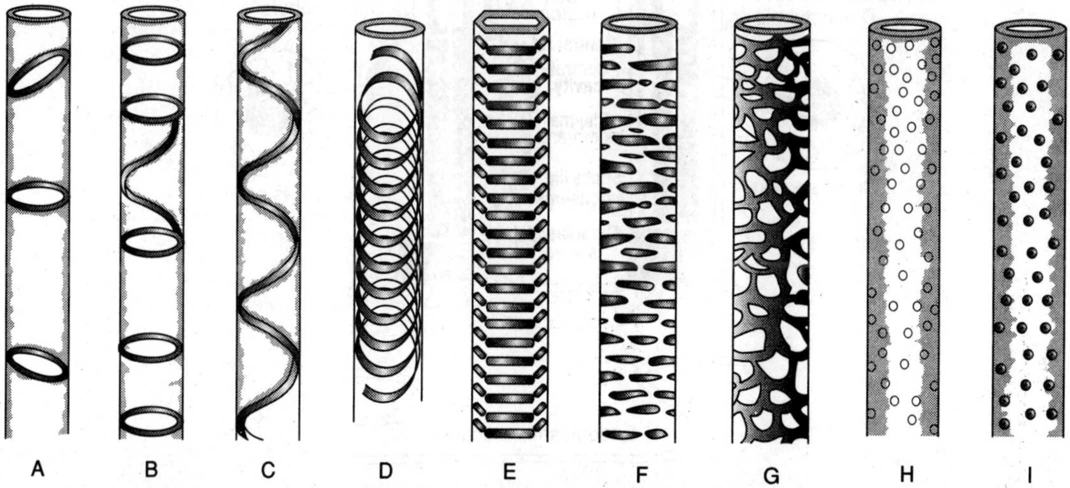

Fig. 5.27 : Different types of localized thickenings of the cell wall in longitudinal views : A & B. Annular, C and D. Spiral, E and F. Scalariform, G. Reticulate, H. Simple pitted, I. Bordered pitted

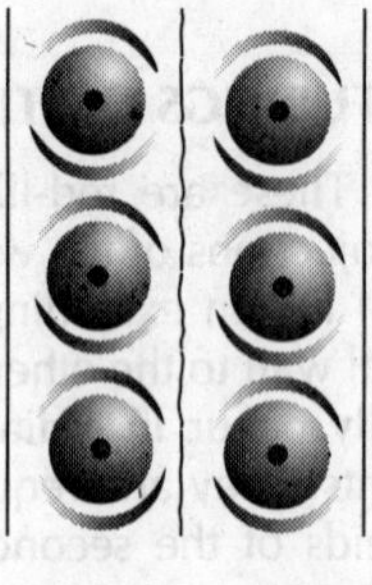

Fig. 5.28 : Crassulae in tracheids of *Pinus*

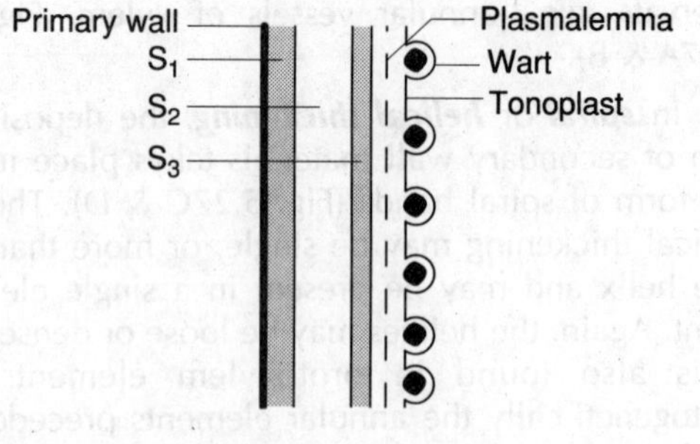

Fig. 5.29 : Diagram of a wall section showing warty structures

Primary Pit Fields

During cell growth the primary walls of young cells stretch and increase in surface area and thickness. But the growth of the wall is not uniform. Certain areas of the wall remain thin. These thin portions are termed ***primary pit fields*** or ***primary pits*** or ***primordial pits***. Presence of ***plasmodesmata*** is the characteristic feature of the primary pits. It makes protoplasmic continuity of neighbouring cells. They are ubiquitous in all living cells of higher plants. As a result, the protoplasts of the whole plant unite to form a single unit called ***symplast***. The cell wall continuum, consisting of the cell walls, intercellular spaces, and the lumen of non-living cells together form the ***apoplast***.

Plasmodesmata

Plasmodesmal structure is a tube of plasma membrane surrounding a strand of modified endoplasmic reticulum with particulate material between them. There is a cell-to-cell tubule of the plasma membrane surrounding a cylindrical strand of tightly furled endoplasmic reticulum (ER) called the ***desmotubule*** or ***appressed ER***. A thin central rod occupies the centre of the

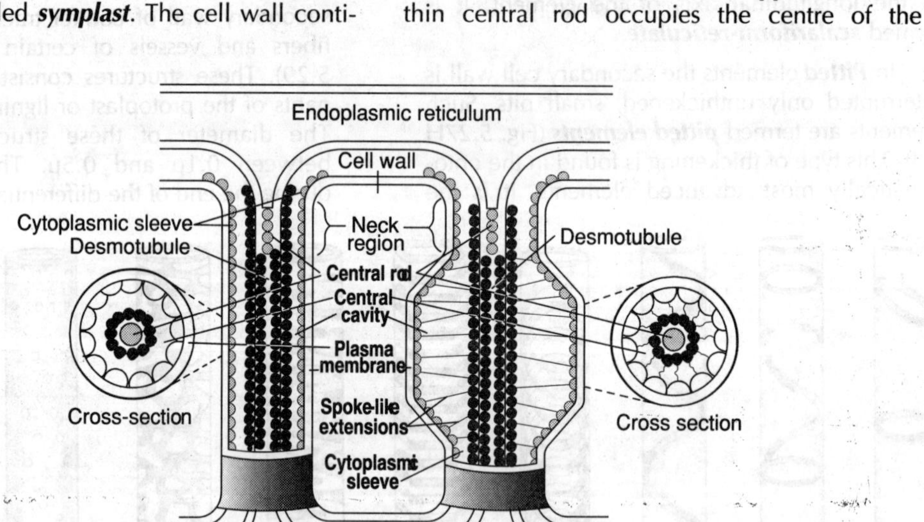

Fig. 5.30 : A model of plasmodesmal structure. The plasma membrane and endoplasmic reticulum are continuous cell to cell. The ER is tightly furled to form the desmotubule

desmotubule. The space between the desmotubule and the inner face of the plasma membrane is called ***cytoplasmic sleeve*** (or cytoplasmic annulus in cross-section). There remains a slightly raised annulus of cell wall material at either end of the plasmodesmata, called the ***collar***. There is a restricted neck region and a widened central cavity inside the wall. Plasmodesmata are often constricted at either or both ends (Fig. 5.30). During plasmolysis the protoplast withdraws itself from the wall in all places except where plasmodesmata are present. The diameter of the lumen of plasmodesmata is about 30-60 nm. The external diameter of the desmotubule is 16-20 nm.

Both the cytoplasmic sleeve and the desmotubule have been considered as possible pathways for cell-to-cell water and solute transport. Instead of being hollow, the desmotubule is essentially a solid strand of lipid. The central rod is composed of lipid polar groups and a few proteins that physically occupy the inner core of the tightly furled lipid bilayer. Much of the cytoplasmic annulus is occupied by particulate proteinaceous material. These particles are associated with both the outer surface of the desmotubule and the inner surface of the plasma membrane. In cross-section 7 to 9 gaps occur between the particles. The distance across the gaps is 2 to 3 nm. These gaps are the physical basis for the molecular sieving characteristic of cell-to-cell transport.

Plasmodesmata occur between most living cells of the plant. They are usually absent in interfaces between sporophyte and gametophyte, maternal tissues and embryos, and between guard cells and epidermal cells. Plasmodesmata usually occur in groups, but sometimes they may be more or less evenly distributed over the entire wall. Grouped plasmodesmata usually occur in the primary pit fields as has been demonstrated in cambial cells. Plasmodesmata are easily found in the endosperm of *Phoenix, Aesculus,* and *Diospyros* and in the cotyledons of some plants. The frequency of plasmodesmata varies substantially, ranging from about 0.1 to 10 μm^{-2} of cell wall area.

Function : Intercellular transport is the main function of plasmodesmata. This was immediately evident at the time of their discovery by Eduard Tangl in 1879. Plasmodesmatal function is usually evaluated by electrical coupling or by cell-to-cell movement of fluorescent tracers (dye coupling). They also play an important role in the relay of stimuli.

Origin : Primary plasmodesata form during cell division, whereas secondary plasmodesmata may be added across existing walls. However, secondary plasmodesmata can form through preexisting walls. Secondary plasmodesmata may also be formed in genetically distinct graft unions (Fig. 4.31), host-parasite connections, organs fused postgenitally (e.g., some carpels) etc. Plasmodesmata formed during cytokinesis are randomly distributed. They are typically grouped into pit fields in the fully elongated cell. Secondary plasmodesmata formation has been observed in the graft union between *Vicia* and *Helianthus*. The stages of formation is illustrated in the following figure.

Initially, the cells of the stock and scion remain separated from each other by debris from the cut surfaces of both of them. As the fusion process is advanced the ER gradually become adnated to the plasma membrane followed by localised thinning of the walls to the points

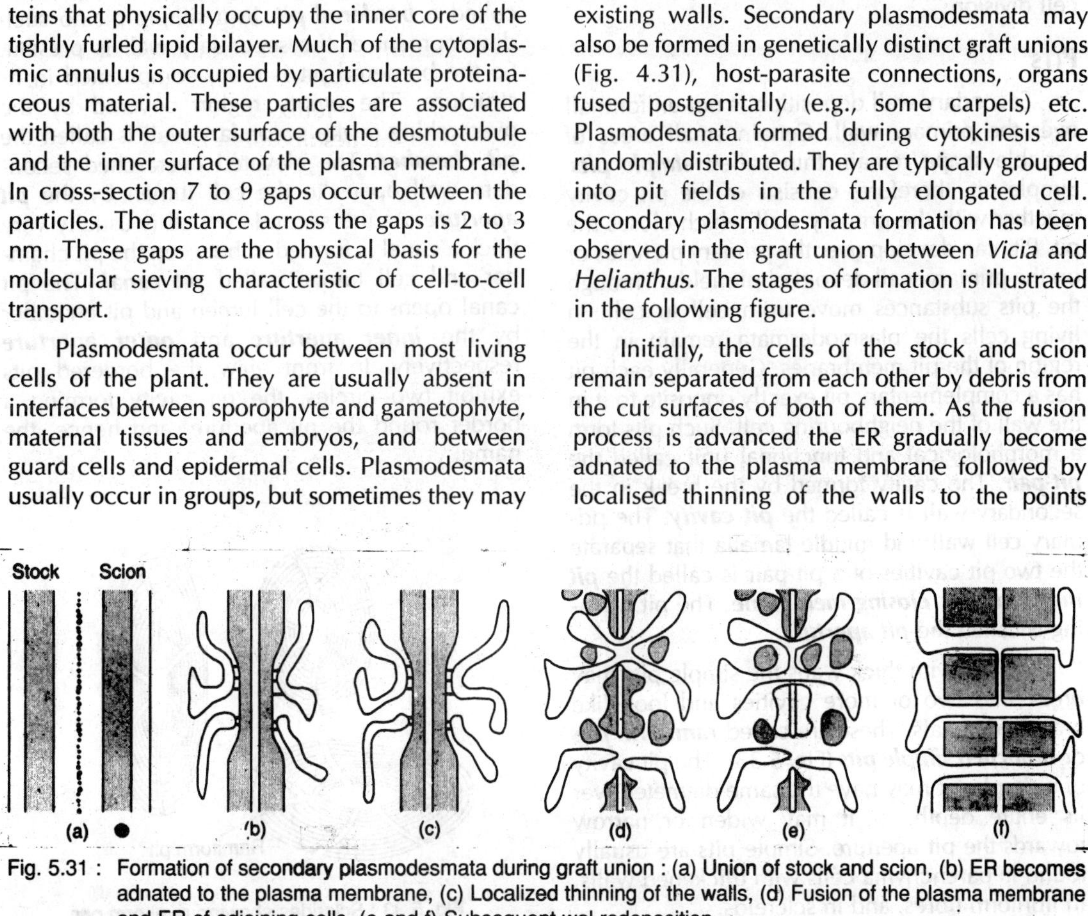

Fig. 5.31 : Formation of secondary plasmodesmata during graft union : (a) Union of stock and scion, (b) ER becomes adnated to the plasma membrane, (c) Localized thinning of the walls, (d) Fusion of the plasma membrane and ER of adjoining cells, (e and f) Subsequent wal redeposition

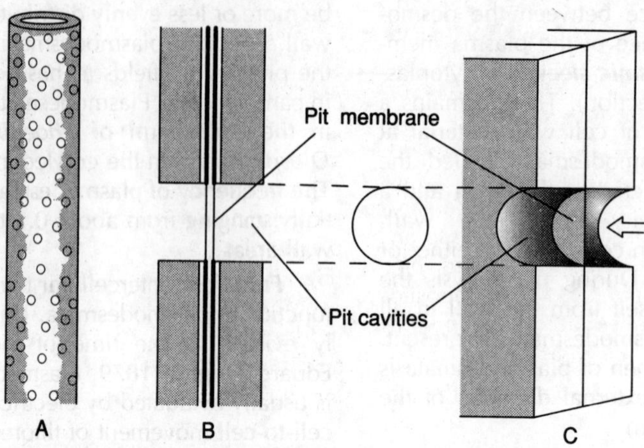

Fig. 5.32 : Simple pits on vessel wall : A. A simple-pitted vessel in front view, B. Sectional view of a simple pit, C. Perspective diagram of the same

where the plasma membrane and ER fuse. Subsequently, wall redeposition occurs much as it does during thickening of the cell plate after cell division.

PITS

Secondary wall deposition is not uniform all over the primary wall. Certain small areas of variable shape remain thin called ***simple pits***. Simple pit, therefore, consists of the pit cavity together with the primary wall which closes the pit. Pits can develop over the primary pit fields or on the primary wall devoid of pit fields. Through the pits substances move from cell to cell. In living cells the plasmodesmata remain in the region of the pit membranes. Generally each pit has a complementary pit exactly opposite to it in the wall of the neighbouring cell. Such pits form a morphological and functional unit called the ***pit-pair***. The cavity formed by the break in the secondary wall is called the ***pit cavity***. The primary cell wall and middle lamella that separate the two pit cavities of a pit-pair is called the ***pit membrane*** or ***closing membrane***. The pit opening is called the ***pit aperture***.

In cells with thick walls the simple pits may break into two or more cavities and look like branched canals. These are called ***ramiform pits*** or ***branched simple pits*** (Fig. 5.33). The pit cavity of a simple pit may have the same diameter over its entire depth, or it may widen or narrow towards the pit aperture. Simple pits are usually found in parenchyma cells with thickened walls, in libriform fibres, and in sclereids.

The second type, where the secondary wall arches or overhangs from both sides over the pit cavity leaving a narrow pore in its centre, is called a ***bordered pit***. In a simple pit no such development of the secondary wall is present. So the bordered pit is more complicated in its structure. The empty region covered by the overarching of the secondary wall is called the ***pit chamber***. The elevated overarched secondary wall opens to the cell lumen by the ***pit aperture***. As the secondary wall is usually very thick a canal is formed in between the pit chamber and cell lumen called ***pit canal***. The pit canal opens to the cell lumen and pit chamber by the ***inner aperture*** and ***outer aperture*** respectively. In front view the bordered pits exhibit two circles, the pit cavity forming a border round the pit aperture, and hence, the name.

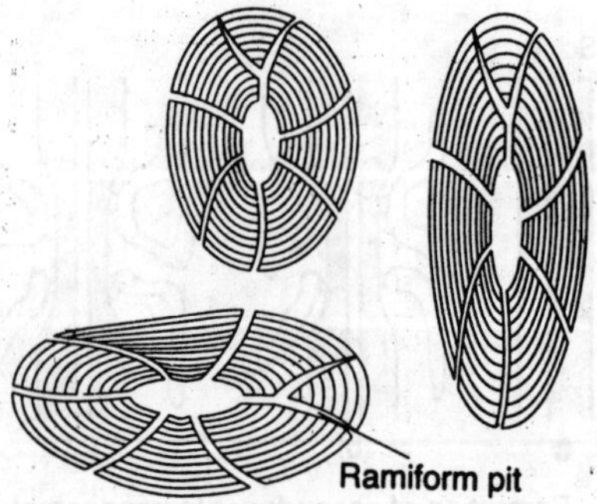

Fig. 5.33 : Sclerides showing ramiform pits

Generally, each pit remains in pair as complementary pits opposite to each other occur in the neighbouring walls. A pair of simple pits forms a ***simple pit-pair*** (Fig. 5.35B). A pair of bordered pits forms a ***bordered pit-pair*** (Fig. 5.35C) and a pair of a simple pit and a bordered pit forms a ***half-bordered pit-pair*** (Fig. 5.35D). If any pit is not accompanied by complementary pit in the adjacent wall or if it is opposite to an intercellular space it is termed a ***blind pit***. Sometimes two or more pits are found opposite to a large pit. It is called ***unilateral compound*** pit.

In some bordered pit-pairs (Fig. 5.34) the pit membrane becomes thickened disc-like in its central portion and is termed the ***torus***, the diameter of which is more than that of the aperture.

Other than torus, the pit membrane shows many types of thickening. The torus outline in *Cedrus* is fringed, which is a taxonomic characteristic. In the tracheids of many conifers the pit membrane around the torus, i.e. the ***margo*** (Fig. 5.35E) is porous. The pit membrane is usually flexible. When the torus is in the median position, water can move from one tracheid to another. Under certain conditions, the torus may be pushed against one of the

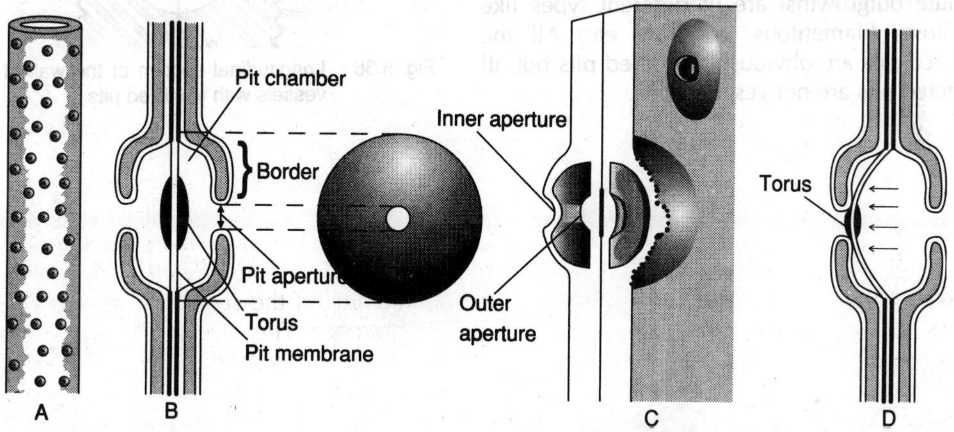

Fig. 5.34 : Diagrammatic view of bordered pits : A. A vessel with bordered pits in front view, B. Vertical sectional view of a bordered pit pair, C. Perspective diagram of the same, D. Torus closing one of the pit apertures to prevent water flow

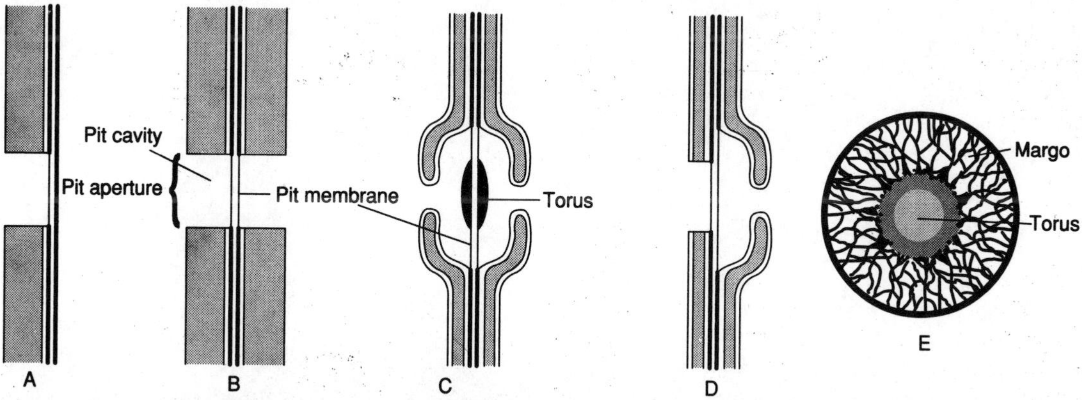

Fig. 5.35 : Structure of pits and pit pairs : A. Simple pit, B. Simple pit pair, C. Bordered pit pair, D. Half-bordered pit pair, E. Porous of membrane and torus of *Pinus*

pit apertures to occupy the lateral position and prevent the water movement. If the membrane flexibility is lost and the torus blocks the pit aperture permanently the pit is called ***aspirated pit***. This is found in most of the tracheids of the late wood and all of them in the heart-wood.

In some dicotyledons thin, simple, or branched sculpturings of various shape are formed on the secondary wall or around the pit aperture. These pits are called ***vestured pits*** or ***sieve-pits*** (Fig. 5.36). These pits appear as porous or net-like in top view. This type of pit is considered as advanced type as it is found in phylogenetically more advanced xylem. The vestures (minute outgrowths) are of different types like papillose, filamentous, coralloid etc. All the vestured pits are obviously bordered pits but all bordered pits are not vestured pits.

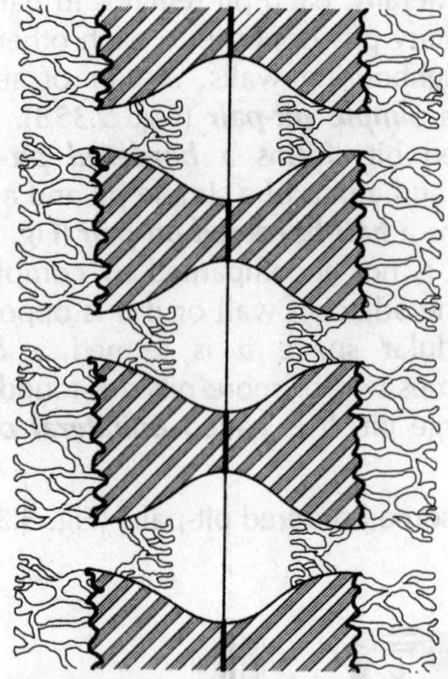

Fig. 5.36 : Longitudinal section of the wall of adjacent vessels with vestured pits

THE TISSUES

A tissue is defined as *a collection of cells of same origin* and *same methods of development performing a particular function in a harmonious way*. The cells in a tissue may be of same structural features (simple tissue) or of different structural complexities (complex tissue).

The plant tissues are classified on the basis of different characteristics viz., the position in plant body, types of constituent cells, functions, the methods of development and origin etc. However, on the basis of stages of development the tissues are grouped into — (i) *meristematic*, and (ii) *permanent*.

Meristematic tissue

Meristematic tissue or ***meristem*** is defined as the tissue in mature plant body, the cells of which continuously divide for indefinite period to add new cells continuously to the plant body. The meristem is obviously of embryonic origin and remains active throughout the grand period of growth.

Structure : Meristem cells are undifferentiated, isodiametric, and are usually small and compact. But intercellular spaces may appear in the meristem derived cells. The meristematic cells possess large prominent nuclei, dense cytoplasm with small scattered vacuoles throughout the cytoplasm or no vacuoles at all. Ergastic substances are absent. Plastids remain in proplastid stage; mitochondria are very feebly developed and the endoplasmic reticulum is rudimentary in comparison to mature cells. Cell wall is thin, homogeneous with cellulosic composition. Meristems with the above mentioned characteristics are termed as ***eumeristems***.

However, the meristematic cells of vascular cambium are fusiform in shape, often with conspicuous vacuoles. Thick walls with primary pit fields are found in some meristematic cells. Starch and tannins may also be present as ergastic matters.

Function : Meristems divide to increase the number of cells in the plant body to bring about growth and development of the plant body as a whole.

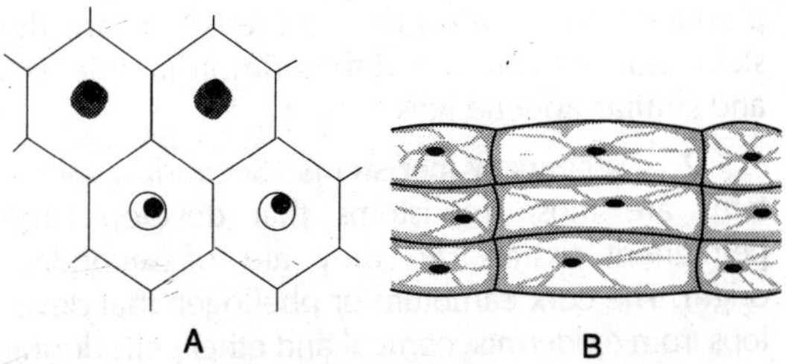

Fig. 5.37 : Cells of meristematic tissue : A. Typical cells of meristematic tissue. B. Cambium cells in transverse section

Distribution : Meristems occur at the growing tips of the main and lateral shoots and roots. They are also found at the bases of internodes intermingled with mature tissues. Again, secondary meristems are present parallel to the stem surface where secondary growth in thickness occurs e.g., the cambium of the vascular bundle and the phellogen (cork-cambiun) in the cortex.

The cells of meristems that remain active throughout are called ***initiating cells*** and the cells derived from them are called ***derivatives***. The derivatives gradually enlarge, change their shape, and ultimately become mature with definite shape and become specialised for a particular function. This process of maturation is referred to as ***differentiation*** through which permanent tissues originate. Sometimes, cells of the permanent tissue may regain the power of division (e.g., secondary meristem) the phenomenon is called ***dedifferentiation***.

Classification : Meristems are classified on the basis of development, origin, position in the plant body, function and plane of division.

CLASSIFICATION ON THE BASIS OF DEVELOPMENT

Promeristem or Primordial meristem – It is the very fundamental stage of meristem action. It initiates new organs and tissues. it is also known as **embryonic meristem**. It comprises the apical initials and their immediate derivatives. As soon as the cells of promeristem begin to change in shape, size, wall and cytoplasm characteristics, they do not remain a part of the promeristem.

CLASSIFICATION ON THE BASIS OF ORIGIN

1. **Primary Meristem** : The cells of which meristem originate directly from the embryonic cells and remain meristematic throughout the plant life is called primary meristem. From the primary meristems primary parts of the plant are produced. The main primary meristems are the stem and root apices and the primordia of leaves and similar appendages.

2. **Secondary meristems** : Secondary meristems are those meristems that develop from permanent tissues. So they are of secondary origin. The cork cambium or phellogen that develops from epidermis, cortical and other cells during secondary increase in thickness is an example of secondary meristem. The secondary meristems add new cells to the primary body forming supplementary tissues during secondary growth or serve in protection and repair of wounds.

CLASSIFICATION ON THE BASIS OF POSITION IN THE PLANT BODY

1. **Apical meristem** : The meristem which occurs at the shoot and root apices of the main and lateral branches is called apical meristem. Apical meristem includes the promeristem and the meristematic zone behind it i.e., primary meristem, in which three basic meristems — the protoderm, the procambium, and ground meristem — of the tissue system can be distinguished.

Increase in length of the plant axis is mainly achieved by the apical meristems. A single apical cell is found in the apical meristem of pteridophytes; but, in case of higher plants, a group of cells constitute the apical meristem, called *apical initials* or *apical cells.*

2. **Intercalary meristem** : These meristems remain in the intercalary position between the permanent tissues. During development the apical meristems go ahead and certain portions of it remain in between the permanent cells and finally become embedded between masses of permanent tissues. Intercalary meristems are found in different organs of plants like the leaf bases in *Pinus*, internode bases in the stem of many grasses and *Equisetum* or at the base of the node as in *Mentha* sp. etc.

The main axis and its branches increase in length by the activity of this type of meristem.

The lifespan of the intercalary meristem is short. They are converted into permanent tissues after a short period of time.

3. **Lateral Meristem** : These are laterally situated meristems i.e. parallel to the surface of the plant body and composed of a single layer

TISSUES

- I. Meristematic
  - Based on origin
    - 1. Primary
    - 2. Secondary
  - Based on function
    - 1. Protoderm
    - 2. Procambium
    - 3. Ground meristem
  - Based on position
    - 1. Apical
    - 2. Intercalary
    - 3. Lateral
  - Based on plane of division
    - 1. Mass meristem
    - 2. Plate meristem
    - 3. Rib meristem
- II. Permanent
  - 1. Simple
    - (a) Parenchyma
    - (b Collenchyma
    - (c) Sclerenchyma
      - Fibres
      - Sclereids
  - 2. Complex
    - Xylem
      - (a) Tracheary element
        - (i) Tracheids
        - (ii) Tracheae
      - (b) Xylem parenchyma
      - (c) Xylem fibre
      - (d) Phloem fibre
    - Phloem
      - (a) Sieve elements
        - (i) Sieve cells
        - (ii) Sieve tubes
      - (b) Companion cells
      - (c) Phoem parenchyma
  - 3. Special
    - Internal secretory structures
      - (a) Glands and ducts
      - (b) Laticifers
        - (i) Latex cells
        - (ii) Latex vessels
    - External secretory structures
      - (a) Glandular hairs
      - (b) Nectars
      - (c) Hydathodes

Fig. 5. 38 : Outline classification of tissues

of rectangular cells which divide mainly *periclinally* and produce secondary permanent tissues. The fascicular cambium and the phellogen or cork cambium are the examples of this type of meristem. By the activity of the lateral meristem the plant body increases in girth or diameter.

CLASSIFICATION ON BASIS OF FUNCTION

On functional basis the meristems are classified into :

1. **Protoderm** : Protoderm is the outermost cell-layer of the plant body the cells of which divide periclinally to give rise to epidermis. Multiple epidermis in *Ficus*, of course, is formed due to tangential divisions.

2. **Procambium** : From the cambium elongated tapering cells are derived in the growing region to constitute procambium. Those derived cell clusters are called procambial strands which occur in a ring in case of dicotyledonous stem. From the procambial strands the vascular bundles are formed, consisting of xylem, phloem and cambium. The strands remain scattered in monocotyledonous stems. In roots only one procambial strand is found in the centre. The procambium strands gradually increase giving rise to pericyle in some stems.

3. **Fundamental or ground meristem** : Excepting the procambium and protoderm the rest of the meristematic tissue forms the ground meristem. It is gradually differentiated into cortex, medullary rays and the pith.

CLASSIFICATION ON BASIS OF PLANE OF DIVISION

Based on the plane of cell division meristems are classified as mass meristem, plate meristem, and rib meristem.

1. **Mass meristem** : Cells of this type of meristem divide in three- or in all planes to produce a mass of cells. Early stages of many embryo development, the endosperm development in many plants, pith and cortex development in most plants and developing sporangia etc. are examples of mass meristems.

2. **Plate meristem** : In this type of meristem the cells mainly divide anticlinally in two planes. As a result there appears a plate-like increase in area. Single cell-layered plate meristem forms epidermis and 2 to several cell-layered plate meristem is responsible for the development of leaf blade.

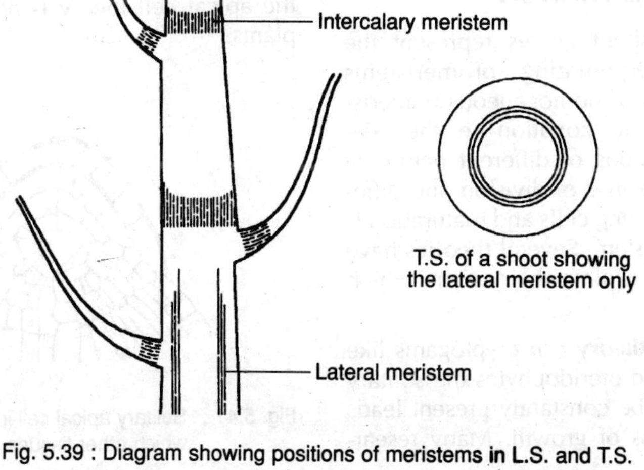

Fig. 5.39 : Diagram showing positions of meristems in L.S. and T.S.

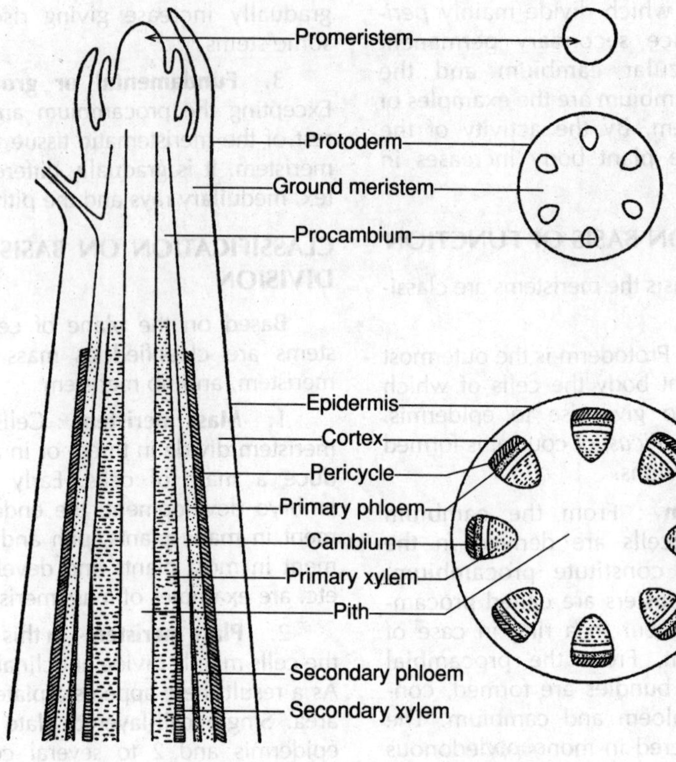

Fig. 5.40 : Diagrammatic representations of meristems in plant body and their gradual differentiation in L.S. with corresponding transverse views in T.S.

3. **Rib meristem** : In rib meristem continuous anticlinal division takes place to form columns or rows of cells. By the activity of this meristem young roots and pith and cortex of young stems are formed.

THEORIES OF STRUCTURAL DEVELOPMENT AND DIFFERENTIATION

The root and shoot apices represent the first-formed self-perpetuating promeristems consisting of a mass of homogeneous eumeristematic cells. A distinct zonation i.e. the existence of distinct regions of different nature of cells with different planes of division and different positions of initiating cells and maturation is noted in the promeristem. Several theories have been put forward to explain their origin, which are :

1. **Apical cell theory** : In cryptogams like algae, bryophytes and pteridophytes the solitary apical cell found to be constantly present leads to the whole process of growth. Many researchers believe that the same condition prevails in all higher plants. On that assumption apical cell theory put forward by Hofmeister and supported by Nageli in 1878. Subsequently it was found that the complex apices of gymnosperms and angiosperms could not be interpreted by this apical theory. So, soon it became consistent that the apical cell theory is not applicable to seed plants.

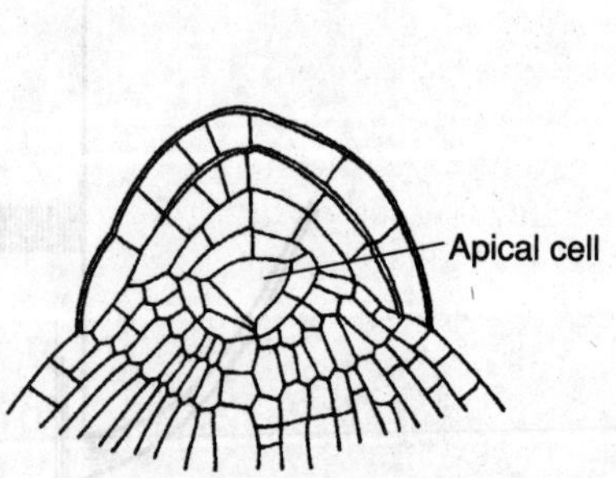

Fig. 5.41 : Solitary apical cell in the fern root apex from which other tissues originate

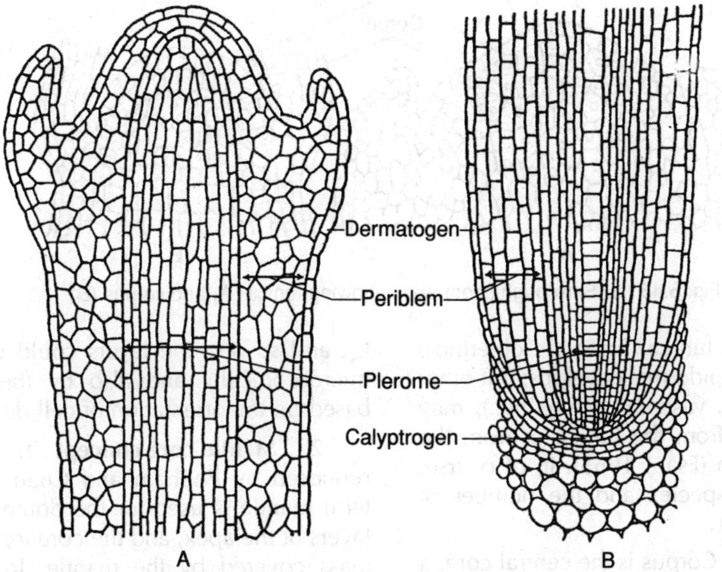

Fig.5.42 : L.S. through shoot apex (A) and root apex (B) showing histogens

2. **Histogen theory** : The older 'apical cell theory' was replaced by the *histogen theory* proposed by Hanstein (1870) and supported by Strasburger (1868). The theory states that the main plant body arises from a mass of meristem of considerable depth consisting of three zones or histogens – the *dermatogen,* the *plerome* and the *periblem.* This theory is concerned with the shoot and root apices.

(i) *The dermatogen* : Dermatogen is the outermost single cell-layer the cells of which usually divide by radial walls and give rise to epidermis. In some cases the cells of this layer divide by tangential as well as radial walls giving rise to *multiple* epidermis e.g. *Ficus* leaf.

(ii) *The plerome* : Plerome is the massive central core of the apices consisting of cells extended in longitudinal direction. The cells of this region divide in all planes giving rise to the central cylinder i.e. the stele consisting of pericycle, primary vascular tissues, medullary rays and pith. The plerome remains covered by the dermatogen and the periblem.

(iii) *The periblem* : Periblem is the intermediate zone lying in between the dermatogen and plerome. This zone is composed of isodiametric cells. From this region primary cortex is formed. The innermost layer of the cortex differentiates into the endodermis.

Hanstein's histogen theory, though it is applicable in root apex, is still criticised in case of shoot apex of angiosperm mainly for the following two reasons : (a) sharp distinction between periblem and plerome in the apical meristems is absent, (b) the origin of different regions of the plant body from the sharply-defined histogens cannot be demonstrated.

3. **Tunica-corpus theory** : Third theory of structural development and differentiation is the Tunica Corpus Theory propounded by Schmidt (1924) and supported by Foster (1949). The theory is based on the observations on angiosperm shoot apices. According to this theory, two tissue zones occur in the apical meristem, viz. (i) the *tunica,* and (ii) the *corpus.* In fact, in lower vascular plants like pteridophytes and even in some gymnosperms a sharp distinction between tunica and corpus is lacking. But in angiosperm, however, the demarcation between the two zones is prominent.

(i) *The tunica* : Tunica is the outer covering zone composed of one or more peripheral cell layers enclosing a central mass of tissue. The cells of tunica divide mainly anticlinally bringing about surface growth. The tunica cells are smaller than the corpus cells. Each tunica layer develops from a small group of separate initials. The number of tiers of initials is equal to the

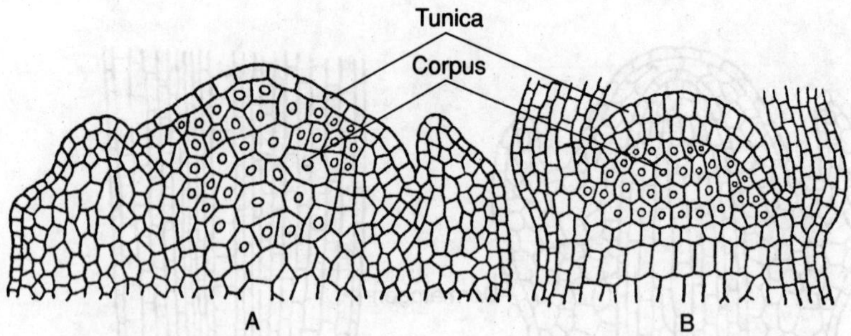

Fig. 5.43 : L. S. through shoot apex showing tunica (A) and corpus (B)

number of layers of tunica. From the outermost layer of the tunica epidermis develops; but other tissues (cortex, pith, vascular tissues etc.), may also be developed from the tunica or from the corpus or from both (Esau, 1965). It, of course, depends on plant species and the number of tunica layers present.

(ii) *The corpus* : Corpus is the central core, a mass of cells surrounded externally by tunica layers. The two regions differ in structure and appearance due to varying rates of cell division and methods of growth. The corpus cells are larger and divide in various planes, so that a mass of irregularly arranged cells is formed bringing about increase in volume. The corpus originates from its own initials located beneath those of tunica. The corpus zone gives rise to the pith, the vascular bundles and a part of cortex.

Other theories to explain the structural organisation of the shoot apex are discussed :

1. **Histogenic layer concept** : The concept was proposed by H. Dermen in 1947. Instead of naming the different layers of the apical meristerm he named the different layers as L_1, L_2, L_3, L_4, and so on. The layers could be recognised microscopically and also by the assumptions based on the orientation of cell divisions.

2. **Mantle core concept** : This concept was proposed by Popham and Chan in 1950. The term mantle is used for the dome-shaped outer layers of the apex, and the core for the inner cell mass covered by the mantle. In this concept mantle is used for tunica and the core for corpus.

Newman (1961) classified vegetative shoot apex into three types :

3. **Monoplex** : Where there is only one initial dividing periclinally. This type is found in many vascular cryptogams.

4. **Simplex** : In this type the apex is composed of several initials in a single cell layer. The initials divide periclinally as well as anticlinally. This type is found in most gymnosperms.

5. **Duplex** : In this type several initials in many layers are present at the apex. The initials of the surface layer divide anticlinally while cells of the inner layer divide in many planes. This type is found in most angiosperms.

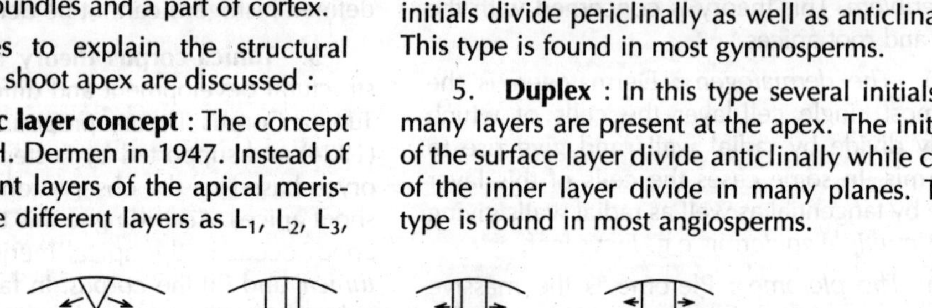

Fig. 5.44 : Diagram showing the different types of meristematic derivatives. The arrows indicate the displacement from the meristem derived cells

VEGETATIVE SHOOT APEX

The shoot apex proper is considered as the region of initiation of the primary organisation of the shoot in which the processes of unlimited growth take place. This region is situated immediately above the uppermost leaf primordium. In size and shape it varies enormously, but usually the shoot apex is more or less convex in longitudinal section. The apical meristem widens considerably before the initiation of the leaf and it again becomes narrow after the appearance of the leaf primordium, thus exhibiting a rhythmic change in shape. It is the region where initiation of the primary organisation of the shoot begins. The term ***plastochron*** has been used for the period between the initiation of two successive leaves. There are great variations in the shape and size of the shoot apices among spermatophytes. The apex usually appears more or less convex in a median longitudinal section.

The shoot apex size also varies among the seed plants. In some grasses the diameter is found to be 90 μ and in dicots it varies between 130-200 μ. In *Musa* it is 280 μ, but in *Nymphaea* and in some members of Palmae the width of the apex varies between 500-700 μ. In gymnosperms these variations are still greater.

In Pteridophyta a single or more apical initials are distinguishable from others and they give rise to all the cells of the apex. The shoot apices of conifers are relatively narrow and conical in *Ginkgo* and in the cycads they are little bit broader and flat. The shoot apices of some monocotyledons (e.g. *Elodea*) and dicotyledons (*Hippuris, Myriophyllum*) are narrow and elongated with round tips. Again, in many dicotyledons (*Drimys, Hibiscus syriacus* etc.) the shoot apex is concave. The apical cell is usually pyramidal in shape in Psilotales, *Equisetum* and leptosporangiate ferns. It divides to produce the cells on all sides. The eusporangiate ferns, on the other hand, possess two to four apical initials.

Angiosperm shoot apex : A number of theories have been enunciated regarding the growth

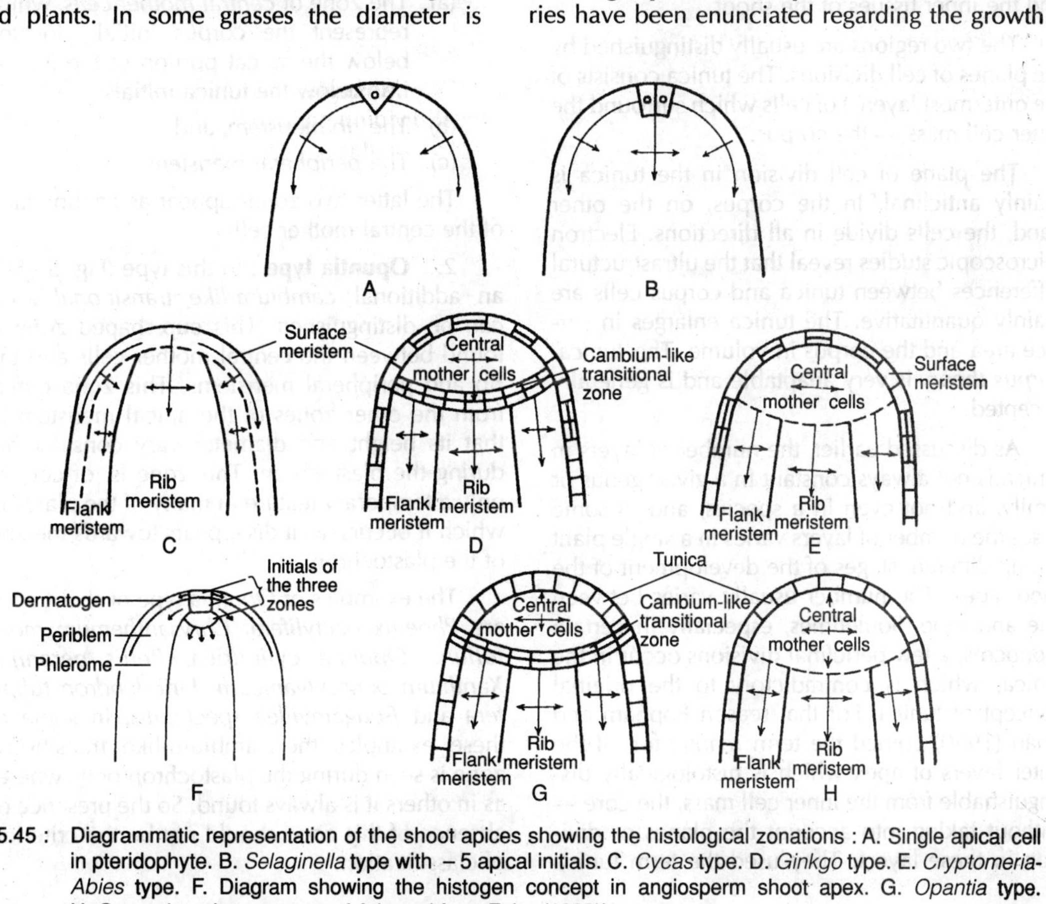

Fig.5.45 : Diagrammatic representation of the shoot apices showing the histological zonations : A. Single apical cell in pteridophyte. B. *Selaginella* type with 5 apical initials. C. *Cycas* type. D. *Ginkgo* type. E. *Cryptomeria-Abies* type. F. Diagram showing the histogen concept in angiosperm shoot apex. G. *Opantia* type. H. General angiosperm type. { Adapted from Fahn (1982)}

and differentiation of the angiosperm shoot apex. Initially the *histogen theory* of Hanstein was put forward, which distinguishes three zones in the shoot apex of angiosperms : the *dermatogen*, a *plerome*, and the *periblem*. He stated that the dermatogen, periblem and plerome develop from independent groups of initials, which act as direct histogens. The epidermis develops from the dermatogen, the cortex, and the leaf internal tissues from the periblem, and the central cylinder from the plerome. The theory was accepted for a long time, but later on it was disproved by the facts that (a) in most seed plants the periblem and plerome are not distinguishable, and (b) no predetermination of the mature tissues could be traced in the various initials.

In 1924, Schmidt postulated the tunica-corpus theory for the angiosperm shoot apex development, which divides the apex into two regions — the *tunica* and the *corpus*. The theory states that there is no constant relationship between the particular initials of the promeristem and the inner tissues of the shoot.

The two regions are usually distinguished by the planes of cell divisions. The tunica consists of the outermost layer(s) of cells which surround the inner cell mass — the corpus.

The plane of cell division in the tunica is mainly anticlinal. In the corpus, on the other hand, the cells divide in all directions. Electron microscopic studies reveal that the ultrastructural differences between tunica and corpus cells are mainly quantitative. The tunica enlarges in surface area and the corpus in volume. The *tunica-corpus theory* is very adaptable and is generally accepted.

As discussed earlier, the number of layers in tunica is not always constant in a given genus or family, and not even in a species, and in some cases the number of layers varies in a single plant during different stages of the development of the shoot apex. The number usually varies between one and nine. Sometimes, especially in certain monocots, a few periclinal divisions occur in the tunica, which is contradictory to the original concept of tunica. For that reason Popham and Chan (1950) coined the term *mantle* for all the outer layers of apex which is histologically distinguishable from the inner cell mass, the core — without taking into account the planes or division in these layers. They used the term tunica for those layers in which only anticlinal cell divisions take place.

In the entire tunica there are two cytologically recognisable zones. One of them is the central apical zone consisting of one or few *initials* which are usually larger and have larger nuclei and vacuoles than the other tunica cells and which, therefore, are also more lightly stai-ning. The second zone is the region on the sides of the apex between the initials and the leaf primordia. It consists of smaller, more darkly staining cells which divide more frequently and among which periclinal division may occur close to the primordia. The tunica is far more homogeneous than the corpus.

In angiosperms there are two principally recognisable types of corpus based on internal arrangement (Popham, 1952) :

1. **Usual angiosperm type** : In this type (Fig. 5.45H) 3 main zones can be distinguished in the corpus :

(a) The zone of *central mother cells*, which represent the corpus initials, located below the apical portion of the tunica, i.e., below the tunica initials;

(b) The *rib meristem*, and

(c) The *peripheral meristem*.

The latter two zones appear as continuation of the central mother cells.

2. **Opuntia type** : In this type (Fig. 5.45G) an additional *cambium-like transitional zone* can be distinguished. This cup-shaped zone is found between the central mother cells and the rib and peripheral meristems. This zone differs from the other zones of the apical meristem in that its height and diameter vary considerably during the plastochron. This zone is, of course, only a temporary feature in many of the plants in which it occurs, as it disappears towards the end of the plastochron.

The examples of *Opuntia* type of shoot apex are *Phoenix dactylifera, Chrysanthemum morifolium, Opuntia cylindrica, Bellis perennis, Xanthium pennsylvanicum, Liriodendron tulipifera* and *Bougainvillea spectabilis*. In some of these examples the cambium-like transitional zone is seen during the plastochron only, whereas in others it is always found. So the presence or absence of this zone should not be the criterion of classification.

Usually the rib meristem contains rows of cells narrowing towards the apex. The cells of this zone usually divide horizontally, but diagonal divisions may also occur. The series of cells can be distinguished. With the increased cell size the vacuoles also enlarge.

Johnson and Tolbert (1960) coined the term ***metrameristem*** to include the initials of the tunica and corpus. It can be concluded that the epidermis and the derived structures originate from the outermost layer of the tunica. The leaf primordia, the cortex, all or part of the procambium, and sometimes also the outer region of the pith are developed from the peripheral meristem.

Functional role of the apical meristematic zones

In the shoot apex two parallel cytohistological zones are generally distinguished :

(a) A central apical zone which includes the initials of the tunica and of the corpus and in which division is considered to occur rarely; and

(b) A peripheral zone to which much mitotic activity is observed.

The central apical zone is active only when the vegetative apex becomes reproductive (Buvat, 1952). This view is based on Plantefol's (1947) theory on phyllotaxy. The mitotically inactive central apical zone is termed *méristème d'attente* (waiting meristem).

The results of the study of natural and colchicine-induced polyploid chimeras favoured the concept that divisions do occur in this zone.

But the histological investigations indicate that cell divisions in the central zone of mother cells take place. Ball (1960) observed that the cells in the surface layer including those at the summit of the apex frequently divide. It was also supported by the DNA synthesis data with radiolabelled precursors. West and Gunckel (1968) with cytohistochemical studies found no major difference in RNA synthesis between the central apical cells and that of the flanking zones.

Lyndon (1976) concluded that in most of the plants there exists a gradient in the rate of cell division and growth from a minimum at the summit of the apex to a maximum in the region of leaf initiation.

Gymnosperm shoot apex : Recent cytohistological studies indicate that the gymnosperm shoot apex contains a complicated arrangement of groups of cells which are characterised by their size and nucleus, differential staining pattern, relative wall thickness, and the frequency and orientation of the plane of cell division. The cell division in the surface of the gymnosperm shoot apex is anticlinal and periclinal. This top layer represents the initiation zone and has been termed *surface meristem.* At the summit of the apex the cells divide periclinally and these cells are the *apical initials.* A distinct zone of *central mother cells* is the striking feature of the gymnosperm shoot apex. The cells of this zone are relatively larger, polyhedral, irregularly arranged and thick walled in the angles of the cells.

Electron microscopic studies have revealed that the apical initials and the central mother cells of *Pinus banksiana* contain numerous lipid bodies and their nuclei contain very little amount of heterochromatin. Both types of cells contain numerous vacuoles and large light-stained nuclei. These features of the apical mother cells and the central mother cells help to the appearance of the cytohistological zonation of the shoot apex.

Other apical regions also develop as a result of the diagonal and horizontal divisions of the central mother cells along the sides and the base of the central mother cell zone. Thus *peripheral* or *flank meristem* is formed laterally and the *rib meristem* zone at the base.

Based on the structure and development, Popham (1952) distinguished three main types of gymnosperm shoot apex :

1. ***Cycas* type** : In *Cycas* type (Fig. 5.45C) there are three meristematic zones :

(a) *The surface meristem* : The cells of this meristematic zone divide anticlinally, periclinally and diagonally. They are not uniform in size and the apical initials are distinguished in the centre of this zone in the seedling stage. Epidermis and other meristematic regions arise from this zone.

(b) *The rib meristem* : This meristem is situated centrally in the apex below the surface layer. Cells of this meristem are arranged in vertical rows. They divide periclinally, anticlinally and diagonally near the base. Pith develops from this zone.

(c) *The peripheral meristem* : The cells of this meristem are elongated. Cell division takes place within the zone itself. Further addition of cells to this zone takes place from the surface layer and from the periphery of the rib meristem. The derived cells of this zone differentiate into the cortex, the procambium and the leaf primordia.

2. **Ginkgo type** : In *Ginkgo* type (Fig. 5.45D) there are five meristematic zones as :

(a) *The surface meristem* : This zone remains at the summit of the apex. Cells of this zone are large, polyhedral and irregularly arranged; they divide in various planes.

(b) *The zone of central mother cells* : Beneath the surface meristem this zone is located centrally.

(c) *The rib meristem* : Just beneath the cambium like transitional zone the rib meristem is distinguished. The cells of these zones are usually arranged in rows. From this meristem pith of the stem develops.

(d) *The peripheral meristem* : The peripheral meristem forms a cylinder surrounding the rib meristem and is a continuation of the cambium-like transitional zone. In this zone, the number of cells increases by division in the meristem itself as well as by the addition of cells from the surface meristem and the cambium-like transitional zone.

(e) *The cambium-like transitional zone* : The cambium-like transitional zone is cup-shaped, relatively narrow and characterized by frequent cell divisions. This zone forms a transitional zone between the central mother cells and the rib and flank meristems. Most of the divisions are periclinal in relation to the central mother cells. So the cells are added to the zones below it. The cortex, leaf primordia, procambium and in some cases the outer region of the pith develop from this zone.

3. **The *Cryptomeria* – *Abies* type** : In this type (Fig. 5.45E) four meristematic zones are distinguished. Excepting the cambium-like transitional zone all other zones are remaning same like that of *Ginkgo type.* In *Pinus montana, Sequoia, Metasequoia glyptostroboides, Abies concolor, Taxus baccata, Ephedra altissima* and *Cryptomeria japonica.* This type shoot apex is found.

Shoot apex of pteridophytes : In pteridophyte shoot apex there are one or more readily distinguishable initials which — by division and redivision — give rise to all the remaining cells of the apex. Where there is a solitary initial at the apex it is called the **apical cell** and, if the number is more, they are called **apical initials.** In *Selaginella* (Fig. 5.45B) apical initials occupy the shoot apex, whereas solitary apical cell is found in *Equisetum,* in some ferns and in the members of Psilotales.

The single apical cell is generally tetrahedral in shape and its base is directed towards the apex surface. The apical cell divides in such a way that the new cells are formed on all sides except the surface.

The apical initials of pteridophytes are usually four-sided, but in *Salvinia* and *Azolla* and sometimes in *Selaginella,* they are only three-sided. In *Salvinia* and *Azolla* new cells are added on three sides while in *Selaginella* they are formed on two sides only. There is a general belief that in ferns the species with solitary apical cell are evolutionarily more advanced than those which possess apical initials in their apex.

Branching : Branches on the shoot originates at the apex. In many pteridophytes and in a few

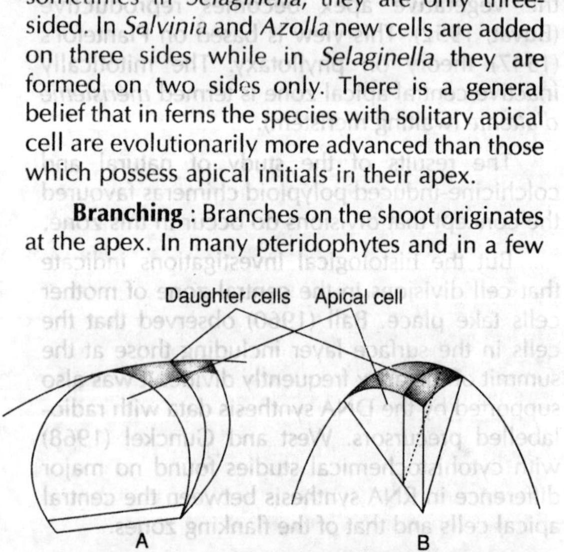

Fig. 5.46 : Apical cell and derived daughter cells and the pattern of cell division in pteridophytes : A. Apical cell showing the planes of division parallel to two faces only. B. Tetrahedral apical cell with base directed towards the surface of the apex and the divisional planes are parallel to other three faces

angiosperms branching is initiated by the equal division into two of the solitary apical cell or group of apical cells in such a way so that two apices are formed leading to a mode of ***dichotomous branching***. In remaining pteridophytes and in most seed plants lateral axillary buds are initiated in the axils of the leaf primordia. From these lateral axillary buds branches develop later. Accessory buds may also originate from part of the peripheral meristem of the axillary buds. Some branches, of course, may initiate extra-axillary buds from which tendrils may develop.

Reproductive apex : After a certain period of vegetative growth, under either the influence of the environmental factors or by the influence of internal factors or by both, the vegetative apex is replaced by the reproductive apex which produces a flower or an inflorescence. The classical concept is that the ***flower is a modified shoot***. It is generally thought that that at the time of flowering the vegetative shoot apex, terminal or lateral, undergoes various physiological and histological changes and becomes directly transformed into a reproductive apex. This may develop either a flower or an inflorescence. Like the vegetative shoots there are plastochronic size fluctuations in the floral apex also during successive stages of development of bracts, perianth, stamens, and carpels. Some floral apices may retain the cytohistological zonation of the vegetative apex. There is a progressive metamorphosis of the vegetative apex leading to the flower bud formation. According to Plantefol (1947) the *anneau initial* functions during vegetative growth in production of foliage leaves, the sepals and even the petals. Stamens and carpels derive from the *méristème d'attente*.

However, most authors do not support the opinion of the existence of a "waiting meristem" which gets active only when the vegetative apex is transformed into the reproductive apex. They consider the apex as a whole to enter a new phase of development.

The main function of the shoot apex is to promote longitudinal growth of the axis, while that of the floral apex is to produce a meristematic mantle with a large surface area from which the parts of a flower(s) develop. Inside this meristematic envelope there is a rib meristem consisting of relatively large, vacuolated cells. It has been shown that the transition from vegetative to reproductive apex is gradual.

The first observable change is the increase of mitotic index on the boundary between the central mother cell zone and the rib meristem zone. Gradually, this mitotic activity spreads into the central mother cell zone where the cells decrease in size and become rich in protoplasts. In this way all the cells above the rib meristem are added to the tunica, the cells of which are more or less isodiametric and are relatively small. Following these changes, mitotic activity and growth almost ceases in the cells of the rib meristem and the pith below it. Thus in the apex a parenchymatous pith surrounded by meristematic cells develops. Eventually only the flower parts or the bracts, the axillary branches of the inflorescence, and the flowers themselves develop from these meristematic cells. Many cytohistological studies on the transition of the vegetative apex to reproductive state have been made. Cytoplasmic basic histone proteins, RNA, and total protein level increases in all zones of the apex during vegetative to reproductive transitional stages.

During the transitional stages following overall events occur in the apical meristem of *Sinapis alba* when they are treated with proper photoperiods :

(a) Rise in the mitotic index and reaching a maximum in about 30h at the onset of the long day
(b) Initiation of DNA synthesis and its culmination at the 38th hour
(c) Increase in the size of nucleolus reaching to a maximum at about the 54th hour
(d) Cell volume starts to increase and culminates at the 62nd hour
(e) Further rise in the mitotic index and culminates at 62nd hour.

Floral induction is thought to occur during the first burst in mitotic activity in coordination with the second peak of the same. The cytohistological changes during floral induction are obviously preceded and accompanied by physiological and biochemical changes. The apical dominance of the main apex is lost with the formation of the inflorescence. The number of ribosomes and Golgi bodies is increased. The endomembrane system is also changed.

In this stage of development the apex ceases to elongate and is terminated by an inflorescence or by solitary flower. In banana, pineapple, etc., however, the elongation takes place by the activity of the rib meristem.

The root apex

From the root promeristem different tissues of the root are formed. It has been observed that in certain Pteridophytes, like the members of Polypodiaceae, Ophioglossaceae and *Equisetum*, the entire root develops from a single tetrahedral apical cell, while in others, e.g. the Marattiaceae, the root is developed from a group of initials. From its upper three sides the solitary tetrahedral cell cuts off cells to form the body of the root and from its base root cap is formed.

In *Azolla*, a division of the apical cell parallel to the distal surface produces the precursor cell of the root cap. All subsequent divisions of the apical cell occur at the three proximal faces. During the early stage of root development the apical cell actively divide and afterwards the apical cell becomes progressively more vacuolated and stops division.

In seed plants there is a larger group of initials in the median region of the root apex. In *Pseudotsuga* there is a central group of permanent initials with three groups of temporary initials (mother cells) on its periphery. The meristem of the vascular cylinder develops from the first group of temporary initials, that of the cortex from the second, and the ***columella*** from the third. The columella is a group of cells in longitudinal rows and forms the longitudinal axis of the root cap. By periclinal cell division on its periphery the columella adds cells to the root cap. The protoderm develops from the young cortex. In *Abies procera* there are two groups of temporary initials — one of which gives rise to the central cylinder and the other to the columella, from which the root cap and cortex develop.

In root apex, protoderm, meristem of the cortex and meristem of the vascular cylinder are derived from the promeristem. The meristems of the different tissue systems can be traced, in the root apex, at various distances from the central cells. In *Brassica* the initials of the various tissue systems are of **closed type** as they get separated immediately adjacent to the central cells. These initials represent the mother cells of the vascular cylinder, the cortex and the common initials of the protoderm and root cap. In *Zea* and *Triticum*, they represent common initials of the cortex and protoderm and separate initials of the vascular cylinder and of the root cap. The special initials of the root cap are termed ***calyptrogen***.

In other species, however, the meristems of the different tissue systems are of ***open type*** as they finally become distinct only some distance away from the central cells. This type is found in *Helianthus* where there are common initials for the cortex meristem, root cap, and protoderm. In *Allium* the meristems of all the tissue systems appear on the periphery of the central cells.

Recently it has been found that a group of cells at the pole of the stele and cortex with a very low mitotic activity appears in the form of a hemisphere or a disc, termed the ***quiescent centre***. The mitotically active cells lying over the surface of the quiescent centre are regarded as temporary initials. *In vitro* culture of isolated quiescent centre shows that it has got the capacity to regenerate a new root independently.

There are many views as to why a quiescent centre appears in the root :

(a) Hormones may be synthesised in the quiescent centre and may stimulate division at low concentration and inhibit at high concentration.

(b) There is a competition between cells for supplies of nutrients or hormones.

(c) The root cap controls quiescence in the particular portion of the meristem.

(d) The pressure exerted by the rapidly dividing neighbouring cells causes the inactivity in the quiescent centre.

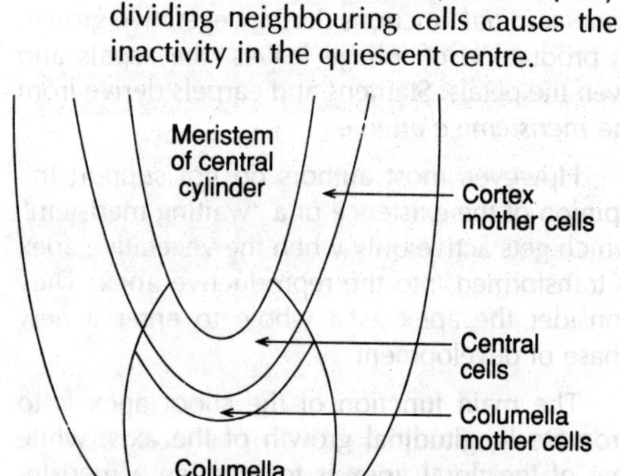

Fig. 5.47 : Diagrammatic representation of the root apical meristem showing different zones. (Redrawn after Mauseth, 1988)

The Körper-Kappe Theory

The theory was put forward by Schüepp (1917). It is similar to the tunica-corpus theory of the shoot apex. It is based on differences in the planes of cell division. The theory says that the cells in the root apex divide in a pattern called T-divisions. The outer region of the root apex is the Kappe. The cells of this region divide first horizontally. The lower daughter cell then divides longitudinally, i.e. at right angles to the plane of the first division. Thus the planes of the two divisions form a T in a median longitudinal section of the root. The inner region of the apex is the körper. In this region the T is inverted, i.e. the second division takes place in the upper daughter cell. This type of division has been found among the members of Gramineae and Fagaceae.

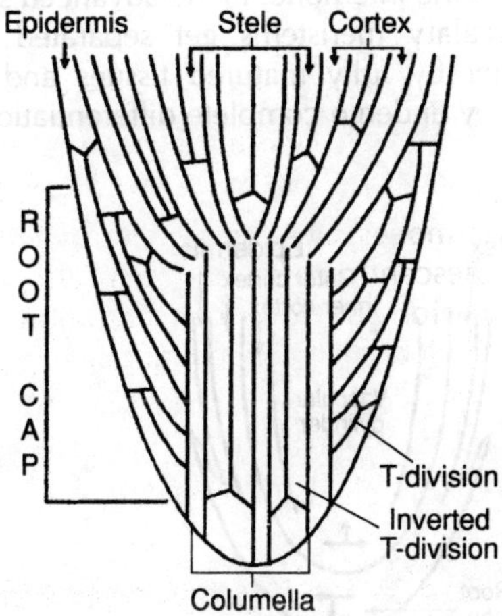

Fig. 5.48 : Diagrammatic representation showing the root apex meristem according to the Körper-Kappe theory

Root apices in angiosperms

Dicots : Dicot root apices are of the following three types according to the number of initials present :

(a) *Common type* : This type is found in majority of dicotyledonous plants. In this type, three groups of initials are found at the apex – outer (dermatogen), middle (periblem), and inner (plerome). The dermatogen forms the epidermis and the cap, the periblem forms the cortex, and the plerome gives rise to the central cylinder.

(b) *Ranunculus type* : The members of Ranunculaceae, Juglandaceae, Salicaceae, Casuarinaceae, Leguminosae etc. exhibit this type of root apex. In this type there is a single row of initials which, ultimately, gives rise to various zones of the root including the root cap. Some of the cells of the root cap differentiate into the epidermis.

(c) *Casuarina type* : This type of root apex development is evident in the members of the families Proteaceae, Casuarinaceae and in some members of Leguminosae. In this type, generally two rows of initials are observed at the apex. One of the layers gives rise to the stele and the other to the cortex and root cap. The epidermis develops from the outermost layer of the cortex. In certain members of the families Juglandaceae, Rosaceae, Leguminosae, Tiliaceae etc. one of these two rows of initials gives rise to the stele and the inner cortex, while the other gives rise to outer cortex and the root cap. The epidermis arises from the outermost layer of the cortex.

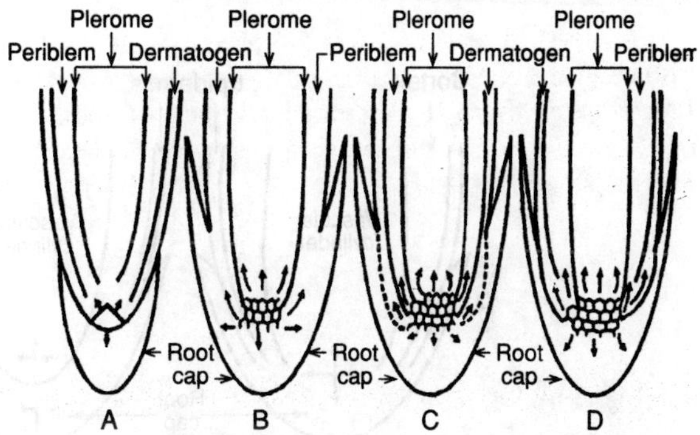

Fig. 5.49 : Diagrammatic representation of root apices : A. Pteridophytic type with solitary apical cell. B. As found in gymnosperms. C. Common dicotyledonous type. D. Monocot (maize) type

Monocots : In addition to the above discussed types in dicots there is a fourth type in monocotyledonous plants. The first type is found in *Zephyranthes* sp. The second type is very rare and found in *Allium sativum, Aloe vera, Amaryllis, Eucharis* etc. The third type is found in *Haemanthus coccineus*. In the fourth additional type, there are four rows of initials giving rise to the root cap, the epidermis, the cortex and the stele, independently. The root cap developing row of initials is called the ***calyptrogen*** which produces layers of cells outside to form a cap-like structure.

Differences between the shoot apex and the root apex

1. In contrast to that of the shoot apex, the apical meristem of the root apex generates cells not only toward the axis but also away from it to form the root cap.
2. The apical meristem of the root is sub-terminal in position as the root cap remains terminal.
3. The root apex forms no branches, and no lateral structures like the leaves. For that reason the root apex shows no change in morphology.
4. The root apex grows uniformly as it never produces nodes and internodes.
5. The root branches arise endogenously beyond the region of most active growth.

INTERCALARY MERISTEMS

Initial extension growth is established by the primary meristems. Further elongation of the shoot axis is caused by the uniform meristematic activity in the internodes. The mode of extension growth is distinguished between two types of meristems in developing internodes — ***uninterrupted meristems*** and ***intercalary meristems.*** A progressive confinement of the uninterrupted meristem to the upper region of the internode is observed and it remains continuous with the sub-apical meristematic region of the stem. Cell maturation within the internode is acropetal.

The nodes mature first. The internodes mature gradually in basipetal direction. The intercalary meristems are isolated meristematic zones which arise due to disjunction from the sub-apical meristematic regions. In most plants the least degree of differentiation is observed at the base of the internode. In the advanced stages the intercalary meristems get separated from each other by fully matured tissues and ultimately they undergo complete differentiation.

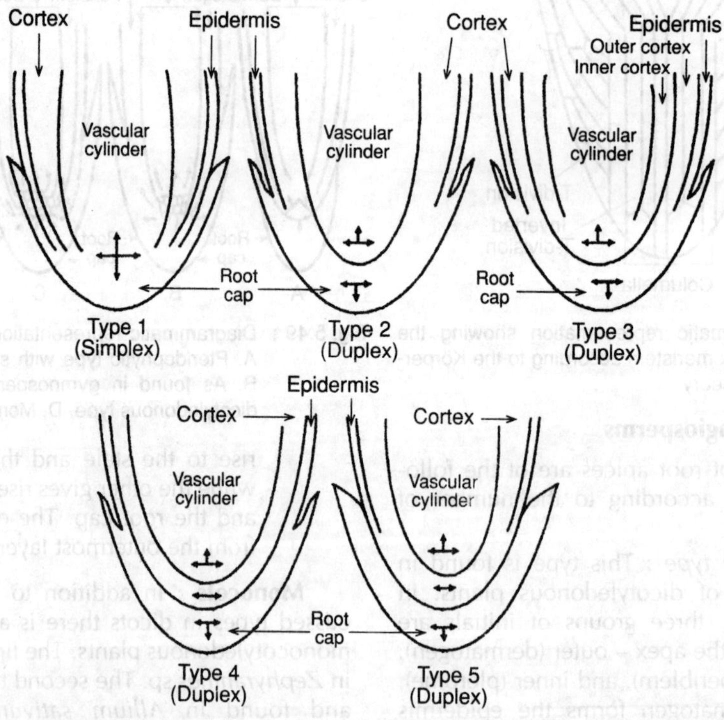

Fig. 5.50 : Diagrammatic types of root apical meristem in longitudinal section. Arrows indicate the direction of displacement of the meristem derived cells

The intercalary meristems are found in the stems of grasses and some other monocotyledons, some species of the Caryophyllaceae, Polygonaceae, Chenopodiaceae and *Equisetum*, in the inflorescence axis of certain plants, in the leaves of many monocotyledons and in the gynophore of *Arachis*.

The intercalary meristem of many grasses is termed rib meristem as the cells of the meristem divide to form parallel series of cells. The derived cells enlarge to contribute to the length of the internode. There is no fixed position of the intercalary meristems of grass internodes but their position changes with the elongation of the internode. During the early phases the intercalary meristem remains active throughout the internode but with the development of lacunae in most grasses, this activity becomes restricted to only the joint regions. The joints can regain the meristematic activity even in mature stems.

Cambium is a lateral meristem. In perennial plants the continuity of the vascular systems between the young parts of the root and shoot is brought about by secondary conducting tissues produced by this vascular cambium.

PERMANENT TISSUES

The meristematic cells gradually differentiate and their divisions get arrested at G_1 phase of the cell cycle (= G_0) and ultimately become mature or ***permanent***. Thus the permanent tissue is composed of cells in which division has been stopped. Of course, all living plant cells retain the potentialities of further division and growth. That is why they are called ***totipotent***.

Principally on the basis of constituent cells, permanent tissues are classified into two groups — (a) ***simple*** and (b) ***complex***. Simple tissues are composed of same type of cells, and thus are *homogeneous* in nature. The complex tissues, on the other hand, are heterogeneous, being composed of different types of cells. The third group of permanent tissue is the (c) ***secretory*** or ***special tissue***. In higher plants, tissues are very seldom homogeneous from morphological or physiological point of view.

SIMPLE TISSUES

There are three types of simple permanent tissues in plants. They are (i) ***parenchyma***, (ii) ***collenchyma*** and (iii) ***sclerenchyma***.

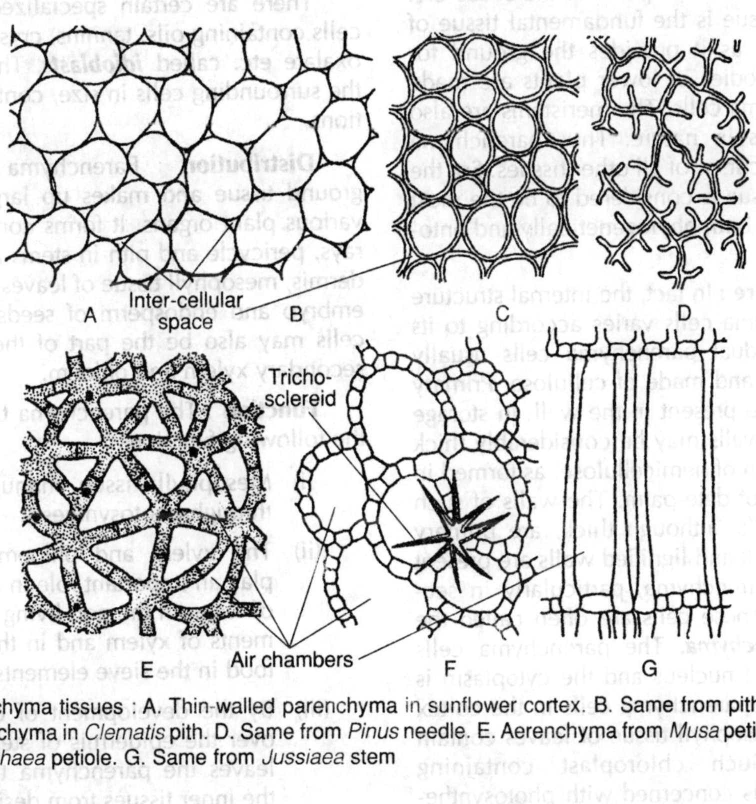

Fig. 5.51 : Parenchyma tissues : A. Thin-walled parenchyma in sunflower cortex. B. Same from pith. C. Thick-walled parenchyma in *Clematis* pith. D. Same from *Pinus* needle. E. Aerenchyma from *Musa* petiole. F. Same from *Nymphaea* petiole. G. Same from *Jussiaea* stem

PARENCHYMA

The term parenchyma refers to tissues which shows little specialisation and concerned with various physiological functions of the plant. The cells have the power of wound recovery and regeneration. A small group of parenchyma cells or even a single cell may be cultured and differentiated into a whole flowering plant in synthetic medium. Phylogenetically, the tissue is considered to be primitive as the multicellular plants of the lower groups consist only of parenchyma. Ontogenetically, parenchyma is also considered primitive as its cells are morphologically alike to those of meristems.

It is the most common simple tissue of the plants and occupies major portions of the plant body. Plant life begins with the formation of the parenchymatous tissue by the embryo. The tissue is composed usually of isodiametric cells with intercellular spaces (Fig. 5.51A). The cells have functional protoplasts. This tissue is ubiquitous in all plants and generally occupies the softer portions of the plant body like epidermis, cortex, pith, mesophyll tissue, pulp of the fleshy fruits, embryo and endosperm of the seeds etc. Parenchyma tissue is the fundamental tissue of the plant body as it provides the ground for other tissues. Bodies of lower plants are made up of parenchyma cells. The meristems are also parenchymatous in nature. Thus parenchyma tissue is the precursor of all other tissues. So, the parenchyma tissue is considered to be the most primitive tissue, both phylogenetically and ontogenetically.

Cell structure : In fact, the internal structure of the parenchyma cells varies according to its function. Individual parenchyma cells usually are thin walled and made of cellulose. Primary pit fields may be present in the wall. In storage region the cell walls may be considerably thick due to deposition of hemicellulose, as formed in the endosperm of date-palm. The walls of such parenchyma cells, although thick, are primary walls. Often thick and lignified walls are present in the xylem parenchyma, particularly in secondary xylem. Those cells are often called the ***sclerified parenchyma***. The parenchyma cells possess a distinct nucleus and the cytoplasm is vacuolated. The parenchyma cells in the cortex of stems and mesophyll tissue of leaves contain chloroplasts. Such chloroplast containing parenchyma cells concerned with photosynthesis are called ***chlorenchyma***. Usually the non-green parenchyma cells contain leucoplasts.

The shape of the parenchyma cells is variable, though they are generally described as isodiametric. On isolation they are more or less spherical but *in vivo* they take different shapes as various forces act upon them due to growth. In fact, they are usually polyhedral with 14 sides or facets. Parenchyma cells may be oblong and arranged in parallel as in the palisade cells of the mesophyll tissue, in the medullary rays etc. They may be variously lobed or folded as found in spongy cells of the mesophyll tissue. Star-like or stellate parenchyma cells are found in the stems with well developed air spaces of *Scirpus, Juncus* etc. and in the mesophyll tissue of *Canna*.

Intercellular spaces within the parenchyma cells arise either by the splitting apart of the cell walls or by breakdown of the cells. Intercellular spaces are larger in many aquatic plants and form a continuous system throughout the plant body to give buoyancy and to withstand the mechanical stress in the aquatic environment. Such a tissue is often called ***aerenchyma***.

There are certain specialized parenchyma cells containing oils, tannins, crystals of calcium oxalate etc. called ***idioblasts***. They differ from the surrounding cells in size, contents and functions.

Distribution : Parenchyma tissue is the ground tissue and makes up large portions of various plant organs. It forms cortex, medullary rays, pericycle and pith in stems and roots, epidermis, mesophyll tissue of leaves, pulp of fruits, embryo and endosperm of seeds. Parenchyma cells may also be the part of the primary and secondary xylem and phloem.

Function : The parenchyma tissue performs the following functions :

(i) Mesophyll tissue manufactures food through photosynthesis.

(ii) The xylem and phloem parenchyma play an important role in the movement of water in the non-living tracheary elements of xylem and in the transport of food in the sieve elements of phloem.

(iii) By the development of cutinised wall over the epidermis of stems, roots and leaves the parenchyma tissue protects the inner tissues from desiccation.

(iv) Aerenchyma in aquatic plants gives buoyancy, helps in gaseous exchange and withstands mechanical stress in aquatic environment.

(v) Parenchyma cells also help in secretion of various useful products like oils, nectar, resin etc.

(vi) Parenchyma cells store food and water.

(vii) Parenchyma tissue plays an important role in healing of wounds and regeneration.

Origin : The parenchyma of cortex and the pith, of the mesophyll of leaves, and of the floral parts originate from the ground meristem and protoderm. The primary and secondary xylem and phloem parenchyma are formed by the procambium and vascular cambium, respectively. The phelloderm originates from phellogen or cork cambium during secondary growth of the mature organs.

COLLENCHYMA

Collenchyma is another primary permanent simple tissue consisting of elongated living cells with uneven cellulosic walls. It is the supporting tissue of the plant. Ontogenetically, it develops from certain elongated cells resembling procambium which are formed in the very early stages of differentiation of the meristem. Younger cells of tissue show more extensibility and plasticity than the older ones which are less plastic, harder and more brittle. Collenchyma cells may also contain chloroplasts and carry out photosynthesis. In some cases collenchyma cells store tannins as secondary metabolites. The tissue is actually thick-walled parenchyma specialised to give mechanical support to the plants.

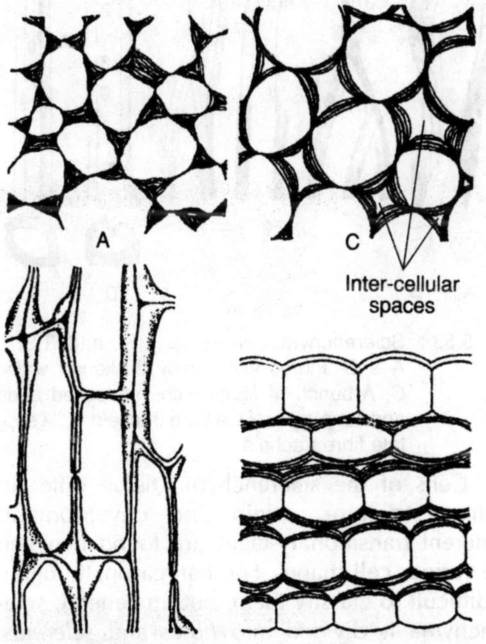

Fig. 5.52 : Different forms of collenchyma : A. Angular in T.S. B. Same in L.S. C. Lacunate in T.S. D. Lamellar in T.S.

Cell structure : Collenchyma cells are elongated in a plane parallel to the long axis of the organ in which they occur. The cells are living with vacuolated protoplast. Chloroplasts may also be present. The most distinctive characteristic of the collenchyma cells is the cell wall which becomes unevenly thickened. There are different types of deposition, but commonly, the thickenings are confined to the corners of the cells. Often the degrees of deposition may be so much pronounced that cells look circular in cross-section.

The size and shape of the collenchyma cells vary greatly. The shorter ones are more or less like parenchyma and the longer ones resembling the fibres with overlapping tapering ends. Individual collenchyma cell may often attain lengths of 2 mm. They are usually polygonal in cross-section. The cell wall is made up of cellulose, hemicellulose and pectic materials, and contains much water (about 67% of the fresh weight). The amount of pectin and hemicellulose in cell walls is high. Like parenchyma it can undergo reversible changes and retain the capacity of cell division. Primary pit fields can be distinguished in the walls of collenchyma cells. Collenchyma cells may or may not contain intercellular spaces.

On the basis of the thickening of the cell wall and arrangement of cells three main types of collenchyma are recognised :

(i) *Angular collenchyma* : It is the most common type where the deposition is localized at the corners or angles of the cells (Fig. 5.52A & B). The cells of this type of collenchyma are compact **and** irregularly arranged. This type of **collen**chyma is found in the stems of ***Datura, Dahlia, Cucurbita, Solanum tuberosum,*** ***Atropa belladona***, etc. and in the **peti-**

oles of the leaves of *Vitis, Begonia, Coleus, Cucur-bita, Beta, Morus*, etc.

(ii) *Lacunate collenchyma* : In this second type, the thickenings appear at the intercellular-space-facing parts of the cells (Fig. 5.52C). This type of collenchyma is also called **tubular collenchyma**. It is found in the petioles of *Salvia, Malva, Althaea, Asclepias* and in the members of Compositae.

(iii) *Plate or Lamellar collenchyma* : In this type of collenchyma, cells are compactly arranged without intercellular spaces (Fig. 5.52D). Thickenings occur in various patterns mainly on the tangential walls of the cells. This type of collenchyma tissue is found in stems of *Sambucus nigra* and *Rhamnus*, etc. and in the petiole of *Cochlearia armoracia*.

Distribution : Collenchyma occurs in the peripheral positions i.e. immediately beneath the epidermis in stems, leaves, floral parts, fruits and roots of dicotyledonous plants mainly. Collenchyma is not found in the stems of many monocots. It occurs as continuous band of solitary or many cell-layers thick hypodermis under the epidermis. It may also occur as isolated patches in the hypodermal region. Occasionally, the entire epidermis may be collenchymatous. In stems the collenchyma may also form a complete cylinder or remain in longitudinal strips. Collenchyma may occur in the cortex of light exposed roots. The collenchyma occurs on one or both sides of the vascular bundles and along the margins of the leaf blade in dicots.

Functions : The collenchyma tissue performs the following functions :

(i) Collenchyma cells give mechanical rigidity to the growing plant organs.

(ii) Due to the presence of collenchyma cells the plant parts or organs become flexible and shows plasticity.

(iii) Photosynthesis takes place in the chloroplast containing Collenchyma cells.

Origin : Ontogenetically, collenchyma cells develop from certain elongated cells resembling procambium which are formed in the very early stages of differentiation of the meristem. Sometimes less specialised collenchyma originates from the ground meristem. Some researchers believe that the collenchyma originates jointly with the vascular tissues from the procambium.

SCLERENCHYMA

Sclerenchyma is the third type of simple permanent tissue. Cells of this tissue are longer. Cell walls of this tissue are very thick, hard and lignified with very little water due to secondary deposition. The lumen is almost obliterated. At maturity, cells of the sclerenchyma tissue become dead. Simple pits or slightly bordered pits are found over the walls.

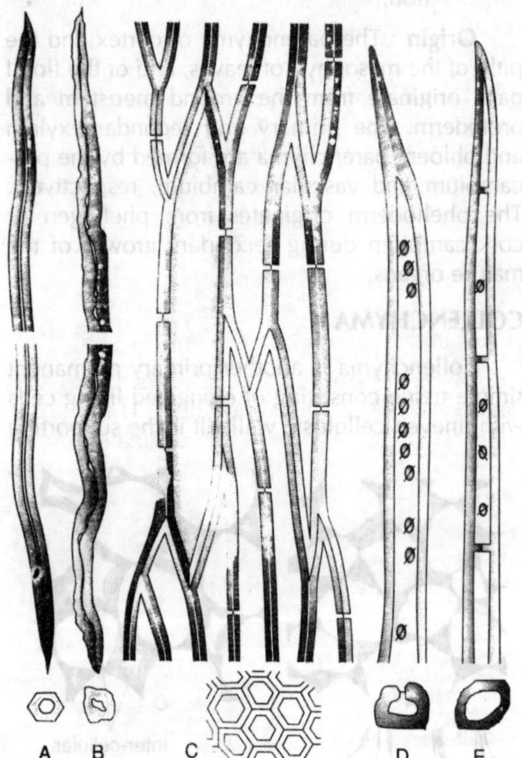

Fig. 5.53 : Sclerenchyma fibres in L.S. and T.S. : A & B. Fibres with highly thickened walls. C. A bunch of fibres with interlocked ends and simple pits. D. A fibre tracheid. E. A septate fibre tracheid

Cells of the sclerenchyma tissue differ in shape, structure, origin and development. Different transitional stages are found between the various cell shapes. For that reason, though it is difficult to classify them, but, in general, sclerenchyma is divided into *fibres* and *sclereids*. Fibres are generally elongated cells and sclereids as short cells. The former originate from meristematic cells whereas the latter develop from parenchyma cells due to secondary deposition of the wall materials.

Fibres : These are very long and narrow cells with tapered and sometimes branched ends. Walls are uniformly thickened and highly lignified. The pits are small, round or slit-like in outline. The cell lumen is small due to much thickening of the secondary wall and may be continuous or septate. The ends may be blunt or branched with interlocked arrangements. The fibres are always dead at maturity. They are angular in outline after cross section. In certain cases the fibre walls are cellulosic and non-lignified. Some fibres have mucilagenous walls.

Distribution : Fibres remain distributed in different organs of the plant body. In the leaflets of *Cycas* they occur singly as idioblasts. They may occur in separate strands in the cortex or as bundle caps associated with vascular bundles, or in the xylem and the phloem as wood and bast fibres.

Fibres are classified into two groups based on their positions in the plant body – (a) **xylary** fibres and (b) **extraxylary** fibres.

Xylary fibres : These are also known as ***intraxylary*** fibres or ***wood*** fibres. These are an integral part of the xylem and originate from the same meristematic tissues. Xylary fibres show variations in their shape, size, wall thickness and pitting pattern. The pits are simple or bordered in nature. These fibres possess lignified secondary walls. There are two main types of xylary fibres – *libriform fibres* and *fibre-tracheids*, based on the wall-thickness, type and amount of pits.

Libriform fibres : The term is derived from the word liber meaning inner bark as they resemble phloem fibres and are usually longer than the tracheids. Walls of these fibres are extremely thick with reduced simple pits. Sometimes in libriform fibres the pit canal becomes elongated and the inner pit aperture becomes slit-like. The inner pit apertures of a pit-pair are usually at right angles to each other.

Fibre-tracheids : These are intermediate between libriform fibres and tracheids. Cell walls of the fibre-tracheids are of medium thickness with bordered pits. The inner pit opening is slit-like. These structures are regarded as reduced tracheids. Like libriform fibres the fibre-tracheids may be septate (*septate fibres*). ***Gelatinous*** or ***mucilaginous fibre*** is observed in the secondary xylem of dicotyledonous plants. The innermost layer of the secondary wall of such fibres is made of β-cellulose called ***G-layer*** which after absorbing water almost covers the entire cell lumen. This less compact and porous G-layer irreversibly shrinks on drying.

Some elongated cells with thin secondary walls and living protoplasts occur in the secondary xylem, which may be confused with the living libriform fibres and fibre-tracheids. These elongated cells are called ***substitute fibres***.

Extraxylary fibres : The fibres present anywhere in the plant body other than xylem tissue are called ***extraxylary fibres*** or ***bast fibres***. They may remain distributed in the cortex, pericycle and phloem. These fibres are usually long with tapered, blunt or branched ends. The cell walls of these fibres are thick, lignified or non-lignified with simple or slightly bordered pits. Extraxylary fibres usually may form isolated strands or continuous bands in the cortex and pericycle. They may remain as caps above the vascular bundles. These fibres usually occur as patches in monocotyledonous leaves.

There are certain fibres termed ***septate fibres*** which are found in the xylem and the phloem even of the same species (e.g. *Vitis*). These fibres are characterized by the presence of partition walls (septa) and protoplasts with plasmodesmatal connections. The septum consists of a middle lamella and two primary wall-like layers and does not fuse with the fibre wall. In longitudinal section the fibre tips are viewed as pointed. These fibres may store starch, oil droplets, resins and some times calcium oxalate. Secondary xylem of many dicotyledons possesses septate fibres. Non-vascular septate fibres are found in some monocotyledons.

Functions : Presence of the fibres in the different parts of the plant body is mainly to give mechanical strength and rigidity to the plant body as well as to withstand strains and stresses.

Origin : Ontogenetically, fibres originate from different meristems such as procambium, cambium, ground meristem and protoderm. Cambium derived fibres develop from the elongated fusiform initials and do not elongate further during maturation. But those derived from the short initials elongate greatly at the time of maturation. Of course, the elongation is very gradual and may take a few months.

Evolution : Evolutionarily, fibres have developed from tracheids as many transitional forms between these two types of elements are found in some angiosperms. During the course of evolution the tracheid wall has become thickened, the number of pits and the size of the pit chamber have been reduced and ultimately the bordered pits disappeared.

Commercial Fibres : Commercially there are two types of fibres – **hard fibres** and **soft fibres**. The hard fibres are stiff and lignified as found in the leaves of *Agave, Yucca, Musa textilis,* etc. Soft fibres are extraxylary fibres. They are soft and flexible and may be lignified or non-lignified as found in jute, flax, hemp, ramie etc. The cotton fibres which are also known as **surface fibres** as they are produced from the testa of seeds represent the most important commercial fibres.

According to their use the fibres may be grouped as :

(a) *Textile fibres* (used in the manufacture of cloths; e.g., cotton, flax, ramie and hemp),

(b) *Cordage fibres* (used in making different types of cords or ropes; e.g., jute, cotton, hemp, flax, *Musa, Agave* etc.),

(c) *Brush fibres* (used in the manufacture of brushes and brooms; e.g., *Agave*, fibres from the Palmae and the inflorescences of *Sorghum vulgare* etc.), and

(d) *Filling fibres* (used in stuffing furnitures, mattresses, life-belts etc.; e.g., *Ceiba pentandra*, cotton, jute, *Tillandsia usneoides*).

Sclereids : Sclereids or sclerotic cells are non-prosenchymatous, isodiametric or irregular in shape. They normally become dead at maturity. They occur as hard masses of cells within soft parenchyma tissue in many different places in the plant body and are much shorter than true fibres in length. They are the major components in the shell of walnuts and seed coats of pea and many other plants. In many cases, sclereids may be readily distinguished from the surrounding cells by their shape, size and wall thickness, which are called ***idioblasts***. In some cases the sclereids are of very peculiar shapes. The sclereid walls generally possess reniform simple pits with branched canals. The hard and thick walls are lignified and also may be cutinised or suberised.

Distribution : Sclereids are abundantly present in cortex, phloem, pith, mesophyll tissue

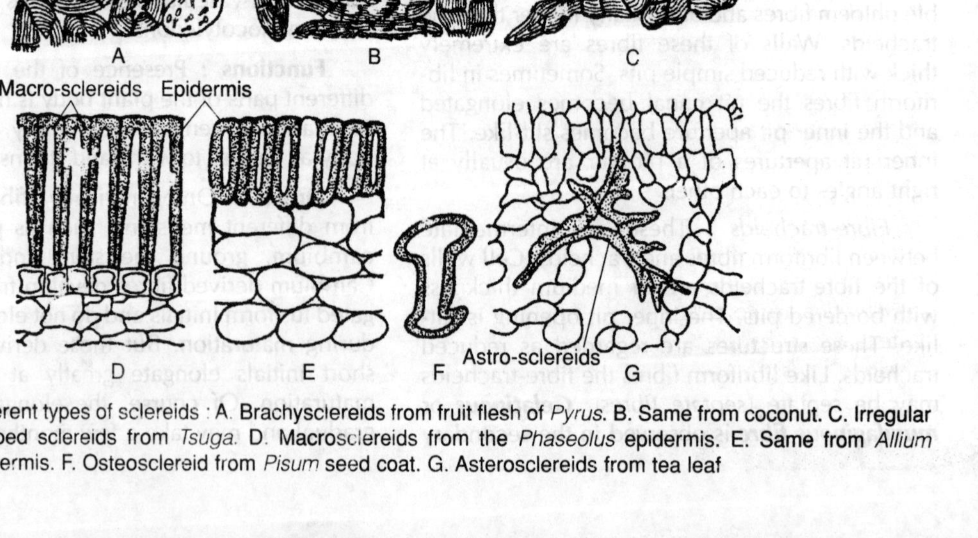

Fig. 5.54 : Different types of sclereids : A. Brachysclereids from fruit flesh of *Pyrus*. B. Same from coconut. C. Irregular shaped sclereids from *Tsuga*. D. Macrosclereids from the *Phaseolus* epidermis. E. Same from *Allium* epidermis. F. Osteosclereid from *Pisum* seed coat. G. Asterosclereids from tea leaf

etc. as idioblasts or a cluster of cells. They are found in fruit pericarp of *Pyrus, Psidium* etc. They also occur in the seed coats of many plants either singly or in groups.

Types : According to the shape, size and nature of wall thickening the sclereids may be of the following types :

(a) ***Brachysclereids*** or ***stone cell*** : These are more or less isodiametric in appearance. They are also called ***stone cells*** or ***grit cells*** as they give gritty texture of the fruit flesh of many plants (e.g., *Pyrus, Psidium* etc.). Brachysclereids are usually found in the phloem, the cortex and the bark of stems (Fig. 5.54A, B).

(b) ***Macrosclereids*** : These are rod-shaped columnar sclereids. They often form a continuous palisade like epidermal layer in the testa of seeds in Leguminosae (Fig. 5.54D, E). Macrosclereids also occur in the pulp of *Malus sylvestris*.

(c) ***Osteosclereids*** : These are bone- or spool-shaped sclereids. They are also columnar in arrangement. The ends of these sclereids are enlarged, lobed, or somewhat branched. Such sclereids are mainly found in seed coats (e.g., *Pisum*) and leaves of certain dicotyledons (e.g., *Pisum, Hakea* etc.) (Fig. 5.54F).

(d) ***Astrosclereids*** : These are branched and often star-shaped, mainly found in leaves and stems of many dicotyledonous plants like *Thea, Nymphaea, Trchodendron* etc. (Fig. 5.54G).

(e) ***Trichosclereids*** : This fifth type of sclereids are very elongated, hair-like, and always single branched sclereids.

Origin : Sclereids usually develop from parenchymatous small thin-walled initials by secondary thickening of the cell wall. During brachysclereid development the inner surface of the wall decreases and pits start to develop on the outside of the secondary wall. In the early stages of development of the branched sclereid, it begins to branch and ultimately acquires the form of the mature sclereid. The branches of the sclereid penetrate into the intercellular spaces. The secondary wall becomes thickly deposited in numerous concentric layers with the formation of branched pits. The mode of development of all types of sclereids is common excepting the degree of pitting. The probable cause of such sclerification may be physiological disturbances or ageing.

Table 5.2 : Differences between sclereids and fibres

Scleids	Fibres
1. Sclereids are developed from parenchyma cells due to secondary growth in thickness and lignin deposition on their walls.	These are developed from meristematic cells. The cell walls are lignified.
2. These are mostly isodiametric but may also show various shapes.	Fibres are elongated with tapering ends the usually interlock with each other.
3. Depending on shapes, they are termed brachy-, macro-, astro-, trichosclereids etc.	There is no major variation in shape which may be septate or asepted.
4. The sclereids have long pits with rounded pit aperture.	Fibres have pits with slit-like pit aperture.
5. Ramiform pits are present on brachysclereids.	Ramiform pits are absent on the fibre walls.
6. Sclereids are found in fruits, barks, pith, cortex, mesophyll tissue, seed coat etc. and rarely in association of xylem and phloem.	Fibres are present in leaves, stems, roots and fruits, in cortex, pith, xylem and phloem, bundle caps, pericycle etc.
7. Sclereids are responsible for mechanical rigidity and elasticity.	Fibres give more mechanical strength to the plant body.

Table 5.3 : Differences between parenchyma, collenchyma and sclerenchyma

Parenchyma	Collenchyma	Sclerenchyma
1. Parenchyma originate from protoderm and ground meristem.	Collenchyma originate from procambium like cells in the ground meristem.	Sclerenchyma originates from protoderm procambium and ground meristem.
2. Cells are living.	Cells are living.	Cells are dead.
3. Cells are usually isodiametric but various shapes are also found.	Cells are generally elongated.	Cells are isodiametric (sclereids) and elongated (fibres).
4. The cell wall is thin, uniformly thickened and cellulosic.	The cell walls are unevenly thick with more thickenings at the corners and composed of pectin and other substances.	The cell walls are uniformly thick and composed of lignin and other substances.
5. The cell walls are primary in nature without sculpturing.	The cell walls are primary in nature with no sculpturing.	The cell walls are formed secondarily with various sculpturing like annular, spiral, reticulate etc.
6. The walls show plasticity.	Cells walls are usually plastic.	Cell walls are elastic.
7. There are primary pit fields on the cell wall.	Pits are rare on cell wall.	Simple and bordered pits are present.
8. Parenchyma tissue have extensive intercellular spaces.	Very less intercellular spaces are present with occasional interlocking ends.	Intercellular spaces are absent and the ends of the fibres may be interlocked.
9. Parenchyma forms a ground to hold collenchyma and sclerenchyma.	It remains embedded in parenchyma.	It remains embedded in parenchyma.
10. The main function is storage.	Storage and mechanical rigidity are the main functions.	It is mechanical tissue.

COMPLEX TISSUES

Complex tissues are composed of two or more types of cells. Xylem and Phloem are complex tissues. They together comprise the vascular tissue system, the main function of which is conduction of water and minerals from the root xylem to the leaves and prepared food from the leaves to the different parts of the plant body.

Xylem : This complex tissue is composed of both living and non-living cells. The main components of the tissue are :

(a) ***Tracheary elements*** (vessels and tracheids),

(b) ***Xylem parenchyma***, and

(c) ***Xylary fibres*** or ***wood fibres***.

Xylem may be *primary* and *secondary* depending on the origin. The former one is derived from the procambium whereas the latter is derived from the vascular cambium during secondary growth. The vessel elements of the primary xylem, which differentiate and mature earlier during ontogeny, are known as the ***protoxylem*** and those which mature later are called ***metaxylem***.

(a) ***Tracheary elements*** : Tracheary elements are the main conducting elements. There are two types of tracheary elements : (i) ***tracheids*** and (ii) ***vessels*** or ***tracheae***. These elements are non-living. Instead of water conduction they give mechanical support to the plant.

(i) *Tracheids* : The tracheids are elongated thick-walled cells with tapering ends and remain parallel to the long axis of the organ where they occur. The ends may also be obtuse or rounded, and chisel-like or oblique. It is dead and devoid of protoplast at maturity. The cell lumen is comparatively larger. The hard and lignified cell wall contains bordered pits and annular, spiral, scalariform and reticulate thickenings. In cross section, the cell outline appears angular or polyhedral, or sometimes round. The tracheids remain in the tissue one above the other with overlapping ends. The transverse walls are not obliterated, rather persist with many perforations and the thickening is same as that in the lateral walls. They make communications with surrounding cells through pits. The pits remain on the lateral walls. Bordered pits are present between the adjacent tracheids. The pit pairs between tracheid and xylem parenchyma may commonl be simple, bordered or half-bordered (Fig. 5.55). In case of half bordered pit the border is formed on the tracheid cell wall and simple pitting on the parenchyma cell wall. Tracheids originate from procambium and vascular cambium.

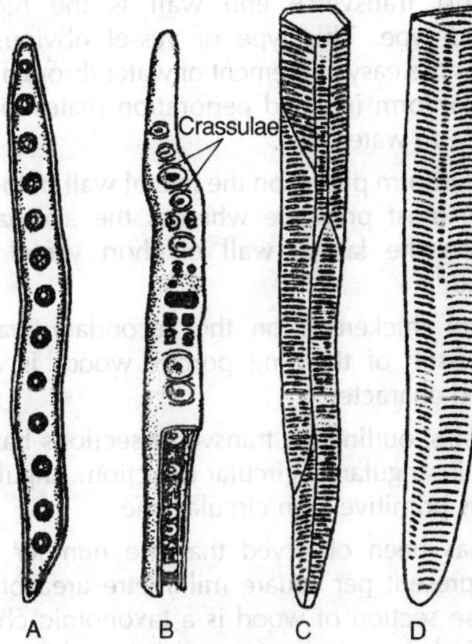

Fig. 5.55 : Tracheids : A. Bordered pitted tracheid. B. Bordered pits with crassulae on a portion of a tracheid. C & D. Parts of tracheids with scalariform thickening

Distribution : Tracheids are found in the primary and secondary xylems of the vascular plants. They predominantly occur in pteridophytes, gymnosperms and primitive angiosperms.

Functions : The functions of tracheid are conduction of water and mechanical support. Very rarely the tracheids store water.

Origin of tracheid : In primary xylem tracheids originate from procambium and in secondary xylem they originate from cambium ring. A single fusiform initial forms a tracheid.

Ontogeny of tracheid : Generally, vascular cryptogams have very long tracheids and gymnosperms have tracheids of intermediate length. Different types of secondary wall are found in tracheids the annular, helical, scalariform, reticulate and circular bordered pitted. Ontogenetically the annular elements precede the helical elements. The helical bands again became joined at certain areas forming scalariform elements. Phylogenetically pitted elements are most advanced. Tracheids with annular thickening have been reported from one of the oldest fossils *Cooksonia*, a leafless vascular plant from upper Silurian deposits (Stewart, 1983). *Baragawanathia* another very early leafy vascular plant — also contained the annularly thickened tracheids.

Phylogenetically, vessels have evolved from tracheids and are polyphyletic in origin. Vessels are found in pteridophyta (*Selaginella, Equisetum* and fern like *Actinopteris, Pteridium, Regnellidium* and *Marsilea*) and in gymnosperm (*Ephedra, Welwitschia* and *Gnetum*). Apart from angiosperms, the vessels found in the above mentioned genera of pteridophyta and gymnosperm are considered as anomalies.

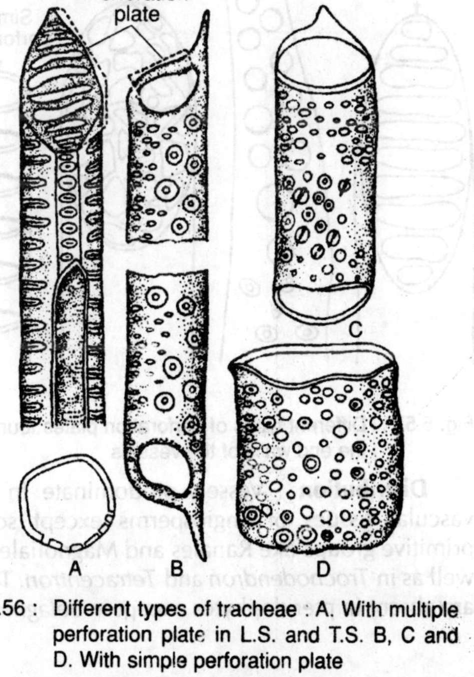

Fig. 5.56 : Different types of tracheae : A. With multiple perforation plate in L.S. and T.S. B, C and D. With simple perforation plate

Vessels : Vessels or tracheae are also thick-walled, lignified non-living members of the xylem tissue. They remain arranged in vertical rows and form a tube-like structure with their perforated end walls. They also have numerous pits on their lateral walls. Water conduction is their main function.

The vessels are dead at maturity. Each vessel is an elongated cylindrical structure. Sometimes their diameter is greater than their length. The perforations on their end walls are large. The vessel elements are parallel with the long axis of the organ in which they occur. The length of the vessel varies and it may be as long as three metres (e.g., *Fraxinus*).

The perforated end walls of vessel members are called ***perforation plates***, which may also remain on the lateral walls. So their positions are usually terminal, but sometimes sub-terminal or lateral. The pattern of perforations may be (a) ***simple*** with single large pore at the end (e.g., *Quercus*); (b) ***scalariform*** with several parallel elongated transverse pores (e.g., *Liriodendron*); (c) ***foraminate*** with several more or less circular pores in groups (e.g., *Ephedra*; and (d) ***reticulate*** with a network of small pores (e.g., Rhoeo).

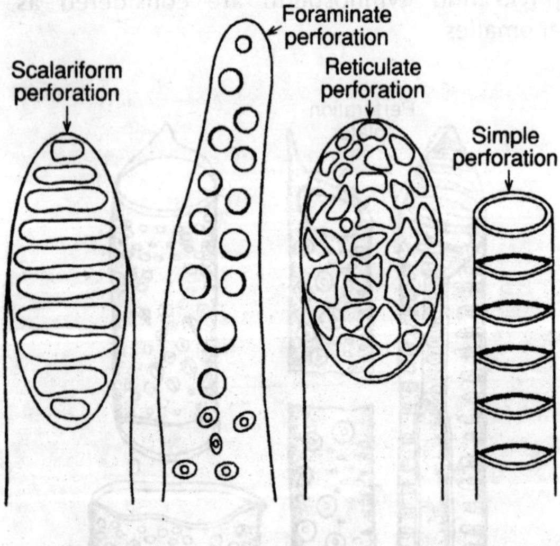

Fig. 5.57 : Different types of perforation plates found at the end walls of the vessels

Distribution : Vessels predominate in the vascular tissues of angiosperms except some primitive groups like Ranales and Magnoliales as well as in *Trochodendron* and *Tetracentron*. They are absent in pteridophytes except in *Selaginella, Equisetum, Pteridium* etc. They are also absent in gymnosperms except Gnetales. They are present in the xylem of both primary and secondary bodies of angiosperms.

Origin and phylogeny of vessel : Vessels of primary xylem originate from procambium and that of secondary xylem develop from cambium. The vessels have evolved from usually long and narrow tracheids. So, the long vessels are primitive, while the short and wide vessels are advanced.

Vessels with long inclined ends are considered as primitive as it is present in tracheid-like vessels and those with transverse end walls are advanced. Scalariform perforation plate in a very inclined end wall of long vessel is considered as primitive. Simple perforation with circular rim in an almost transverse end wall is the most advanced type. This type of vessel obviously facilitates the easy movement of water through it. The scalariform inclined perforation plates give resistance to water flow.

Scalariform pitting on the lateral wall of long vessel is most primitive whereas the alternate pitting on the lateral wall of short vessel is advanced.

Spiral thickening on the secondary wall, characteristic of the ring porous wood, is an advanced character.

Vessels' outlines in transverse sections have evolved in angular to circular direction. Angular outline is primitive than circular one.

It has been observed that the number of vessels present per square millimetre area of a transverse section of wood is a taxonomic character. High frequency of vessel is an advanced character.

Vessels may be present as single or in groups. These groupings and arrangements are taxonomic characters. The groups may be arranged in radial, oblique, or tangential lines. Solitary vessel represents the primitive condition.

(b) ***Xylem parenchyma*** : The parenchyma cells that occur as elements of the xylem tissue is termed as xylem parenchyma.

Distribution : Xylem parenchyma occurs in the primary and secondary xylem. In the latter they are present as axial and ray parenchyma. They are also known as ***wood parenchyma***.

Structure and arrangement : These cells may be oval, round, rectangular or square, elongated and sometimes irregular in shape. The cell wall usually consists of thin primary wall. Sometimes the wall becomes thick due to lignin deposition over the primary cell wall. Pits when occur in-between two parenchyma cells are simple. The pit pairs between the parenchyma and tracheary elements may be simple, half-bordered and bordered. Reserve foods in xylem parenchyma are starch and fat. Crystals and tannins are also found in these cells. The presence of chlorophyll is also reported in some herbs and deciduous trees.

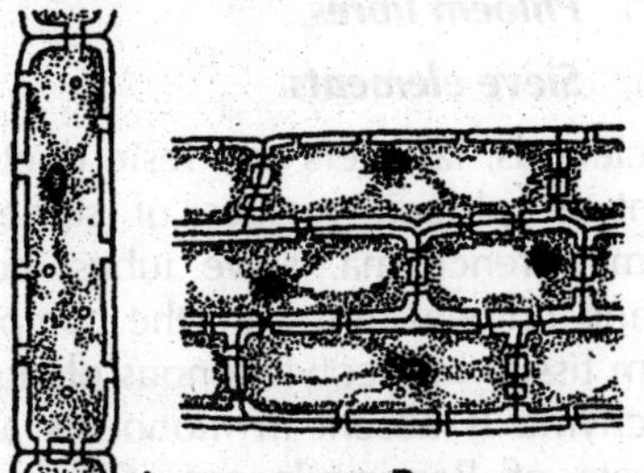

Fig. 5.58 : Xylem parenchyma : A. Parenchyma. B. Ray cells

The xylem parenchyma cells are oriented vertically or horizontally. The vertical orientation forms parenchyma strands that are more common in secondary xylem. Sometimes parenchyma protrudes into vessels to form ***tylosis***.

Function : Xylem parenchyma cell performs the following functions :

1. It helps in the transport of water and mineral salts.
2. It stores reserve food in the form of starch and fat, and ergastic substances e.g., oils, gums, resin, tannins, silica bodies, crystals etc.
3. The thick walled lignified parenchyma gives mechanical support to some extent.

Ontogeny and phylogeny : In primary xylem the xylem parenchyma originates from procambium. In secondary xylem the ray parenchyma cells originate from the ray initials of the cambium. The axial parenchyma along with tracheary elements and fibres originate from fusiform initials of cambium. The fusiform initials of cambium normally divide periclinally. In these cells anticlinal division also occur to keep pace with the growth of the stem in girth.

Phylogeny of ray parenchyma : Heterocellular multiseriate and uniseriate rays are the characteristics of the primitive wood. The multiseriate rays have uniseriate wings. In advanced type of wood, however, either multiseriate or uniseriate ray is present (homocellular), that is one has been lost during the course of evolution. The multiseriate rays are reduced in size and number. During evolution, the erect cells are lost and so there is a tendency for loss of heterogeneity. The advanced woods exhibit extremely short uniseriate wings on multiseriate rays.

In radial longitudinal section the ray parenchyma is observed to consist of square (isodiametric), erect (upright or vertically elongate) and procumbent (radially elongate) cells. The square cells are morphologically equivalent to erect cells. The ray parenchyma may be homocellular and heterocellular. In the former case rays are composed either of square cells or erect cells, or procumbent cells, or erect and square cells. In the latter case, however, the rays consist of both square and procumbent cells or of erect and procumbent cells. The rays may be uniseriate or biseriate or multiseriate with uniseriate wings (tapered ends) as observed in tangential longitudinal section (TLS). The outlines of rays appear to be fusiform in TLS.

Phylogeny of axial parenchyma : The fusiform initial of cambium gives rise to axial (vertical) parenchyma. Axial parenchyma may occur independently or remain associated with vessels. So, depending on the presence of this parenchyma, the timbers may be ***apotracheal*** (vessels without axial parenchyma) and ***paratracheal*** (vessels with axial parenchyma).

The common apotracheal timbers may be (i) ***diffuse*** (isolated strands of axial parenchyma), (ii) ***diffuse-in-aggregates*** (axial parenchyma occurs as aggregates), (iii) ***banded*** (axial parenchyma appears as narrow or wide bands) and (iv) ***marginal*** (axial parenchyma occurs either at the beginning of growth ring or at the end).

The common paratracheal forms are (i) ***scanty*** (discontinuous parenchymatous sheath surrounding a vessel), (ii) ***vasicentric*** (continuous parenchymatous sheath surrounding the vessels), (iii) ***abaxial*** (with more vasicentric parenchyma on abaxial side of vessel), (iv) ***adaxial*** (with more parenchymatous width on adaxial side of vessel), (v) ***aliform*** (with winged vasicentric parenchyma) and (vi) ***confluent*** (with continuous band of extended vasicentric parenchyma).

Phylogenetically, the primitive woods are usually devoid of parenchyma. If present, it is of diffuse type. The advanced wood exhibits a tendency towards grouping of axial parenchyma to form diffuse-in-aggregate type. The wide band of apotracheal parenchyma is more advanced than narrow band of parenchyma.

(c) **Xylem fibre** : The fibre components of xylem are known as xylary fibres or wood fibres. These fibres are of two types – libiriform fibre and fibre-tracheid (discussed under sclerenchyma). The libriform fibre is longer and thick-walled with simple pits. Xylary fibres may be septate. In the tension wood the libiriform fibre and fibre tracheid are of gelatinous type. The xylary fibres are responsible for mechanical support. They may remain living for a longer period and store reserve food.

Phylogeny of xylem fibre : Fibres originate from procambium in primary xylem whereas that of secondary xylem develops from fusiform initial of cambium. Phylogenetically, the fibres have evolved from tracheids. During evolution the wall thickness has increased, the length has decreased and the bordered pits became reduced in size. In dicotyledonous secondary xylem the evolution has occurred in the sequence of tracheid – fibre-tracheid – libiriform fibre. The tracheids have bordered pits; the pit border is diminished in fibre-tracheids and disappeared in libiriform fibres.

Phloem : Phloem is a complex vascular tissue through which photosynthates move from green tissues to the different parts of the plant. Sometimes it adds mechanical strength to the plant body. It is one of the component tissues of the vascular bundle. It remains alternately arranged with the xylem (*radial*) in the roots. In the stem, however, the phloem is usually external to the xylem. It may surround a central core of xylem (i.e., *haplostele*) or discrete strands of xylem remain surrounded by phloem (i.e., *mixed protostele*). Two cylinders of phloem may occur on both sides of xylem (i.e., *amphiphloic siphonostele*). In some cases the central phloem strand may be encircled by xylem (i.e., *amphivasal*, e.g., *Dracaena*). In most dicotyledonous stem phloem strand occurs external to xylem (e.g., *collateral vascular bundle*). In some species of Cucurbitaceae, Asclepiadaceae, Apocynaceae, Solanaceae etc. there are both external and internal phloem tissues. They may also be termed as *abaxial* and *adaxial* phloem respectively. The internal phloem is also termed as *intraxylary phloem*. Phloem strands embedded in the secondary xylem is termed as *included* or *interxylary* phloem.

Phloem is mainly composed of :

(a) Parenchyma termed ***phloem parenchyma***,

(b) Specialized parenchyma cells known as ***companion cells*** and ***albuminous cells***,

(c) ***Phloem fibres***,

(d) ***Sieve elements***.

Sclereids, laticifers and resin ducts are also present in phloem tissues of some species. Phloem parenchyma, sieve tubes, companion cells and phloem fibres are the composition of phloem tissue in dicotyledonous plants. Phloem parenchyma is absent in monocots and a few members of Ranunculaceae. Sieve cells and albuminous cells are present in gymnosperm and vascular cryptogams.

(a) **Phloem parenchyma** : The parenchyma other than albuminous and companion cells in phloem is called phloem parenchyma.

Distribution : Phloem parenchyma is the component of both primary and secondary phloem. In secondary phloem they remain as axial phloem parenchyma and phloem rays.

Structure and arrangement : Phloem parenchyma cells are rectangular or rounded in transverse section. It appears oblong with rounded or tapered ends in longitudinal section. The cell walls are thin and non-lignified with numerous pit fields through which plasmodesmatal connections are established with companion cells and sieve elements. Sometimes sclerified and thick-walled parenchyma cells occur which are inactive. Phloem parenchyma cells with wall ingrowths are known as ***transfer cells***. Phloem parenchyma may store starch, fats, resins, tannins etc.

In primary phloem the parenchyma cells remain parallel to the long axis of xylem, whereas, in secondary phloem, the axial parenchyma are parallel but the ray parenchyma are perpendicular to the long axis of the associated xylem.

Function : Phloem parenchyma cells perform the following functions :

1. Phloem parenchyma cells perform the function of storage of fat, starch, resin tannins etc.
2. Transfer cells are responsible for short distance transport of photosynthates
3. Phellogen derived from phloem parenchyma forms the periderm to protect the inner tissues
4. The inactive lignified cells add mechanical strength

Phylogeny : Procambium gives rise to phloem parenchyma of primary phloem. In secondary phloem the axial phloem parenchyma and phloem rays are developed from fusiform initial and ray initial of cambium, respectively. Some parenchymal cells originate from a common mother cell of sieve elements. In many dicotyledonous stems the parenchyma of protophloem often differentiates into fibre in later stages of development, i.e., when the protophloem elements become functionless.

The phloem parenchyma and fibre of the secondary phloem bear no phylogenetic trend in phloem evolution. The distribution and morphology of them may be of comparative value.

(b) **Companion cell** : These cells with dense cytoplasm remain associated with the sieve tubes. Both of them originate from the same mother cell. Companion cells load photosynthates into the sieve tubes.

Structure and arrangement : The cells are vertically elongated and somewhat angular in cross-section. They are usually shorter in length or may be as long as the associated sieve tubes. The uniformly thick cell wall possesses many sieve areas on the sieve tube side and primary pit fields on the opposite side. There are well-developed plasmodesmata in the sieve areas and primary pit fields. In some companion cells wall materials deposit on the inner side of the primary wall to form transfer cell.

The companion cells have prominent elongated or lobed nuclei. The cells abundantly contain Golgi apparatus, endoplasmic reticulum, mitochondria, ribosomes, plastids etc. In some companion cells P-proteins are found.

The sieve tube and companion cell **remain** strongly attached after their formation **from** the same mother cell to form a complex known as **SE/CC complex**. They cannot be separated by the usual maceration technique. Companion cells vary in number in relation to a single sieve tube. Usually, the number is one or two and occasionally up to five or several. Companion cells are absent in primary phloem. They are also absent in some primitive woody dicotyledons, gymnosperms and pteridophytes. In gymnosperm the associated parenchyma with sieve cells are termed as ***albuminous cells***.

Function : Following are the functions of the companion cell :

1. The companion cells are mainly related with the loading of the sieve tubes with sucrose. When the SE/CC complex is symplastically isolated the sucrose is absorbed actively by the companion cells from the apoplast of the minute veins. From the companion cells the sugars then pass through the plasmodesmata into the sieve elements. But when there is protoplasmic continuity with the mesophyll cells through plasmodesmata sucrose is loaded through polymer trapping mechanism. The sieve tube becomes non-functional when the associated companion cell dies.
2. They are the active sites of protein synthesis.

Phylogeny : The companion cell and sieve tube have got the common origin. The mother cell naturally remains in the primary and secondary phloem. The parent cell itself originates either from procambium or from cambium respectively. The mother cell divides by unequal longitudinal division and the smaller cell differentiates into the companion cell.

Companion cells remain associated with sieve tubes. Three categories of companion cells are observed :

1. Companion cell may be much shorter than the accompanying sieve tube;
2. It may be almost as long as the sieve tube, and
3. It may be as long as the sieve tube, but septate.

According to many workers the first type is the primitive type. The remaining two types are advanced. Reduced number of companion cells in the phloem indicates the advancement.

Albuminous cell : The parenchyma cells associated with the sieve cells are called albuminous cells. Each albuminous cell has a prominent nucleus and dense cytoplasm.

Structure and arrangement : Albuminous cells are vertically oblong and may be of the same length as that of sieve cells or shorter. There are symplastic connections between the two types of cells. The albuminous cells contain starch-free and protein-rich cytoplasm and stain deeply with cytoplasmic stains. These cells occur at the margins of rays.

Origin : Albuminous cells are present in both primary and secondary phloem. Therefore, their origin differs. In primary phloem they develop either from procambium derived phloem rays or from phloem parenchyma. In the secondary phloem of *Ephedra*, the albuminous cells originate from the fusiform initials of vascular cambium.

Function : There is a morphological as well as functional relationship between the albuminous cell and sieve cell even though they differ ontogenetically. When the sieve cell is non-functional the associated albuminous cell becomes dead. The possible function of the albuminous cell is to help in the conduction of proteins.

(c) **Phloem fibre** : The phloem fibres are the ***extraxylary fibres***. They are also called ***bast fibres*** or ***bass fibre*** or ***basswood*** or ***bast wood fibres***.

Structure and arrangement : The fibres are elongated cells and may be very long with tapering ends interlocked with other fibres. Its thick cell wall is usually lignified. But in *Linum* phloem fibre wall is made of cellulose. Simple pits with linear or round apertures are found in the fibre wall. Slightly bordered pits may rarely occur. The phloem fibre of *Vitis* is septate. The septate fibres store starch, oils, resins, calcium oxalate crystals etc. The fibres remain parallel to the long axis. In cross section they appear as isolated or scattered strands, as continuous or irregular bands, as clusters over the phloem strand and may form cylinders of tangential sheets encircling the inner tissues.

Function : The phloem fibres perform the following functions :

1. The phloem fibres with their interlocked ends give mechanical strength to the plant.
2. They protect the inner tissues.
3. Septate fibres may store starch, oils, resins etc.

Phylogeny : The phloem fibres occur in both primary and secondary phloem and, therefore, their origins differ. The primary phloem fibre originates from procambium whereas the secondary phloem fibre originates from cambium. The fusiform initial of cambium gives rise to fibre.

(d) **Sieve elements** : The conducting elements of the phloem are referred to as sieve elements that are characterized by the presence of sieve areas and absence of nuclei from mature protoplasts. These include ***sieve cell*** and ***sieve tube***. The sieve cells do not contain sieve plates. They are found in Pteridophytes and Gymnosperms.

(i) *Sieve tube* : It is the main solute conducting element in angiospermic phloem. It is living but enucleate and arranged longitudinally. Ribosomes and dictyosomes are also absent from mature protoplast. The sieve tubes have sieve plates and sieve areas on their transverse end walls. The sieve tube consists of longitudinally files of cells that are connected with each other through sieve areas on their transverse end walls.

Table 5.4 : Differences between albuminous cell and companion cell

Albuminous cell	Companion cell
1. It is present in the phloem of conifers and *Ginkgo*.	It is present in the phloem of angiosperm.
2. Ontogenetically, it is not related to the sieve cell.	It is ontogenetically related to sieve tube.
3. There is close morphological and physiological relationship with sieve cell.	Close morphological and physiological relationship exists with the sieve tubes.
4. It is not so strongly attached with sieve cell.	It attaches strongly with the sieve tube and cannot be separated after maceration.

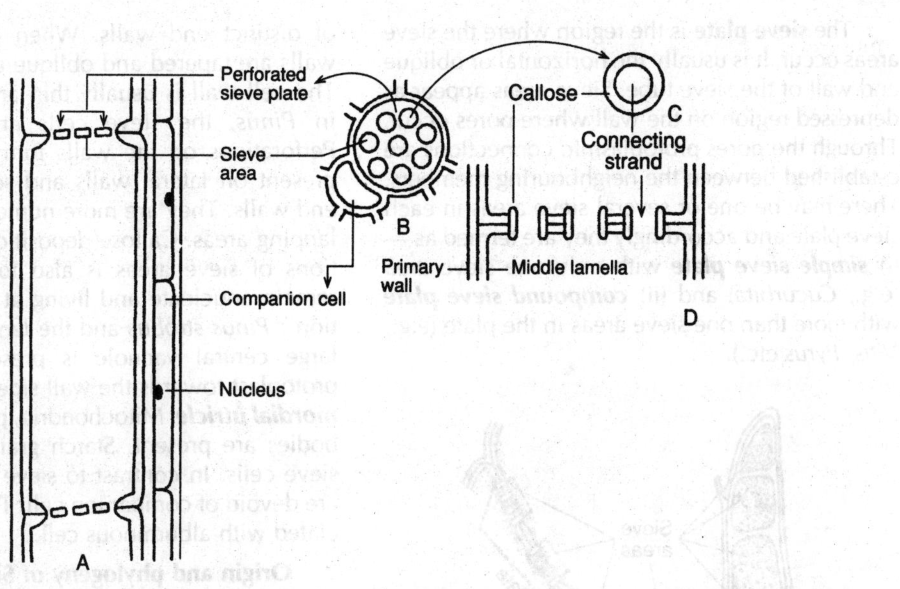

Fig. 5.59 : Diagrammatic representation of the sieve tube and companion cell in L.S. (A) and T.S. (B). C. A single sieve area. D. Vertical section of sieve plate

The cell wall of the sieve tube may be thin or thick and is usually primary. The wall is composed mainly of cellulose and pectin. The sieve tubes of *Smilax hispida*, and *Neptunia oleracea* contain nucleus.

The young sieve tube contains prominent nucleus, abundant dictyosomes, ribosomes, endoplasmic reticulum, plastids, mitochondria and other cell organelles. Sieve tubes accumulate starch which stains brownish red with iodine instead of blue. Discrete proteinaceous slime bodies are found in the young sieve tubes in the form of filament, tubule, granule or crystal, called ***P-protein***. It occurs in all dicotyledonous species without any exception and is rare in monocotyledons. They are absent in gymnosperms excepting *Ephedra* and pteridophytes. The P-proteins are synthesised in the cytoplasm and occupies the peripheral position. In the stained preparation of sieve tubes P-proteins accumulate at the transverse end walls of the tubes and plug the sieve plate pores. This plug is termed ***slime plug***.

Plastids occurring in the sieve tube protoplast may be either ***S-type*** or ***P-type***, depending on the nature of reserve food. Starch accumulating type is called S-type whereas protein accumulating type is called P-type plastid. S-type plastids are found in Bataceae, Polygonaceae, Plumbaginaceae etc. The ultrastructure of the plastids of the sieve tubes is a taxonomic characteristic. These ultrastructural details of sieve-element-plastids are, now-a-days, applied to characterise some higher taxa.

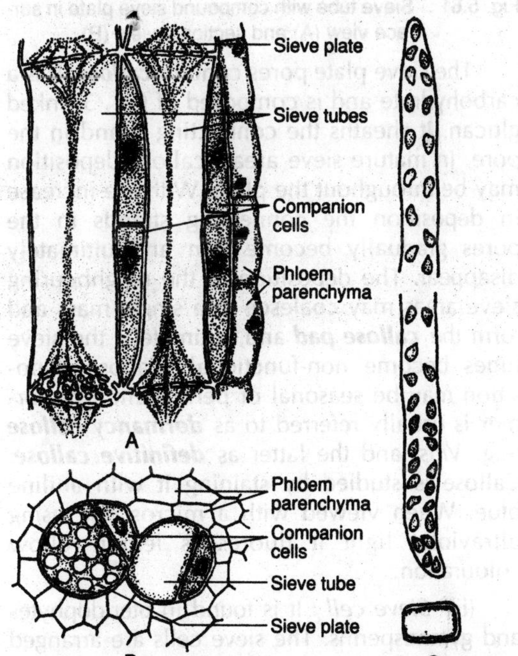

Fig. 5.60 : Sieve elements : A. Sieve tubes in L.S. B. Same in T.S. C. Sieve cell in L.S. and T.S.

The **sieve plate** is the region where the sieve areas occur. It is usually the horizontal or oblique end wall of the sieve tube. Sieve areas appear as depressed region on the wall where pores occur. Through the pores protoplasmic connections are established between the neighbouring members. There may be one or several sieve areas in each sieve plate and accordingly they are termed as — (i) ***simple sieve plate*** with only one sieve area (e.g., *Cucurbita*) and (ii) ***compound sieve plate*** with more than one sieve areas in the plate (e.g., *Vitis, Pyrus* etc.).

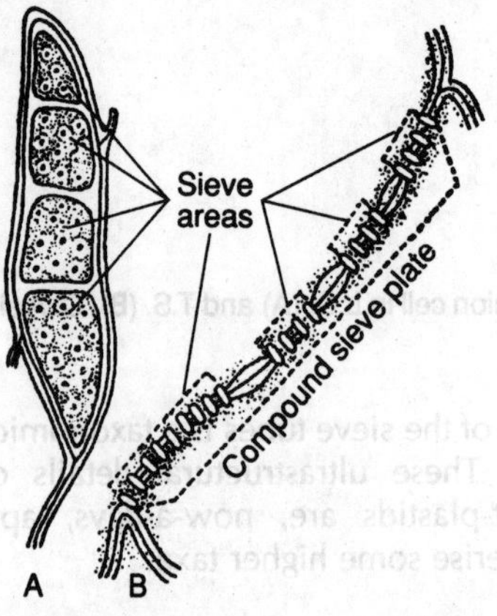

Fig. 5.61 : **Sieve tube with compound sieve plate in surface view (A) and sectional view (B)**

The sieve plate pores contain ***callose***. It is a carbohydrate and is composed of β-1, 3-linked glucan. It sheaths the connecting strand in the pore. In mature sieve areas, callose deposition may be throughout the plate. With the increase in deposition the connecting strands in the pores gradually become thin and ultimately disappear. The deposition of the neighbouring sieve areas may coalesce to a single mass and form the ***callose pad*** and, ultimately, the sieve tubes become non-functional. Callose deposition may be seasonal or permanent. The former is usually referred to as ***dormancy callose*** (e.g. *Vitis*) and the latter as ***definitive callose***. Callose is studied by staining it with aniline blue. When viewed with a microscope using ultraviolet light it fluoresces lemon-yellow colouration.

(ii) *Sieve cell* : It is found in pteridophytes and gymnosperms. The sieve cells are arranged longitudinally, but, unlike sieve tubes, not one above the other in a series. The cells are oblong and tapered at their ends. They are often devoid of distinct end walls. When present, the end walls are tapered and oblique and may overlap. The cell wall is usually thin and cellulosic but, in *Pinus,* the sieve cells are thick walled. Perforations on the walls form the sieve areas present on lateral walls and sometimes on the end walls. They are more numerous in the overlapping areas. Callose deposition in the perforations of sieve areas is also found. Sieve cells remain anucleate and living at maturity (exception : *Pinus strobus* and the family Taxaceae). A large central vacuole is present pushing the protoplast towards the wall side forming the ***primordial utricle***. Mitochondria, plastids and slime bodies are present. Starch grains are absent in sieve cells. In contrast to sieve tube, sieve cells are devoid of companion cell. They remain associated with albuminous cell.

Origin and phylogeny of Sieve tube : Sieve tubes in primary phloem originate from procambium and that of the secondary phloem originate from cambium. The mother cells divide longitudinally to form two daughter cells, one of which forms the companion cell and the other develops into sieve tube. The differentiating sieve tube increases in length and, at maturity, it contains mitochondria, plastids, P-protein and endoplasmic reticulum. The nucleus disappears except in *Taxus, Neptunia oleracea* etc. Sieve areas originate at the sieve plate, the common transverse wall of the sieve tubes.

Phylogenetically, long sieve tubes with numerous sieve areas are considered as primitive whereas the shorter ones are advanced. The cambial initials, during the course of evolution, tend to become shortened and this caused the formation of short sieve tubes. Usually, the cambial initial undergoes transverse septation before the differentiation of sieve tube leading to the formation of short sieve tubes. Small pores in the sieve plate are regarded as primitive. The simple horizontal sieve plate with single sieve area is an advanced feature.

Origin and phylogeny of sieve cell : The mother cells of sieve cells are slender, short cylindrical to oblong with tapering ends and numerous primary pit fields on their lateral walls. During differentiation, the mother cells elongate and the cell wall becomes thick. Sieve areas develop in the position of pit-fields. Plasmodesmata appear in the sieve areas and callose develops surrounding them.

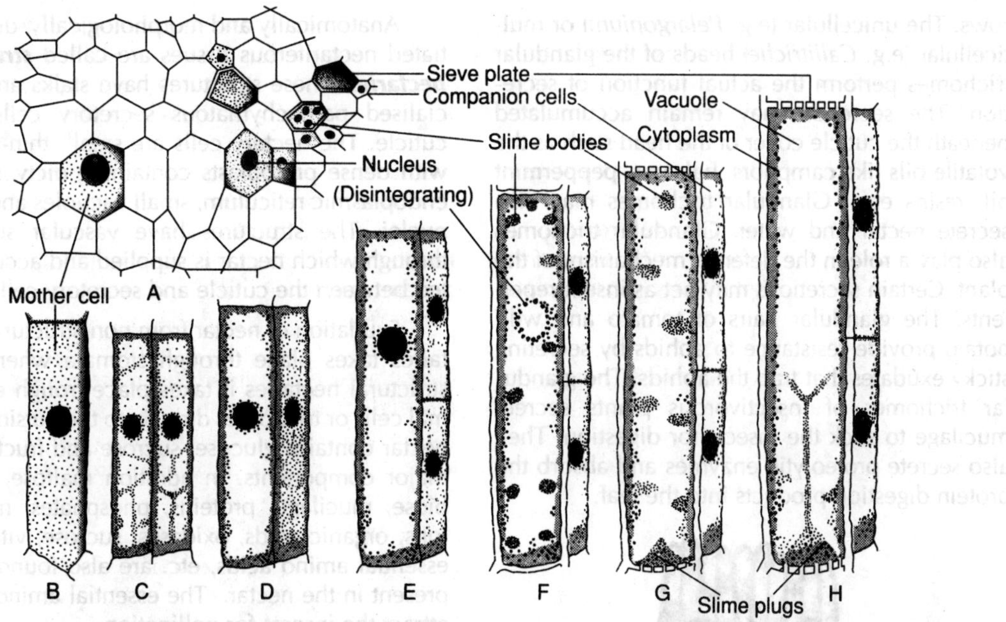

Fig. 5.62 : Differentiation of sieve tube : A. T.S. through phloem showing stages. B–H. L.S. through phloem showing different stages

Sieve cell predominates in pteridophytes and gymnosperms. It is reported among angiosperms in *Austrobaileya scandens* and *Sorbus aucuparia*. Sieve cells are considered as primitive. They are not arranged in axial files. Moreover, the structure of end walls is similar to the lateral walls.

SPECIAL TISSUES

The group of cells usually of different origin concerned in secretion of oils, nectar, resins, gums, mucilage, latex etc. is generally termed as ***secretory tissues***. This categorisation is based on physiological functions. The secretory cells may be any part of the plant body as a solitary cell, or group of cells, or a whole tissue like pith, cortex, xylem phloem etc.

Secretion means the process of release of by-products of metabolism from a cell. The secretory substances may also be stored in insoluble forms within the cell. These substances have either a special physiological function or may be secreted or stored as wastes. The removal of waste products of metabolism is defined as ***excretion***. Strictly, the secretory products take part in the metabolism like hormones and enzymes. At the same time the process of cell wall deposition, cutinisation, wax deposition, suberisation etc. are also the examples of secretion. On the other hand, examples of excretory products are terpenes, saponins, rubber, tannins, crystals etc. Of course, there is no sharp demarcation between the two terms secretion and excretion.

The secretory bodies vary greatly in structure and position. They are mainly of two types — (i) the ***secretory cells*** in which the secretory substances are formed and exuded cell outside e.g., simple glandular trichromes, nectaries, multicellular glands with vascular tissues etc., and (ii) the structures in which the secretory substances are stored and released after the breakdown of the cells. The secretory structures may be external and internal.

EXTERNAL SECRETORY STRUCTURES

External surfaces of the plant bear several secretory structures of epidermal origin or epidermal derivatives or emergences from deeper tissues. These include glands, glandular trichomes, nectaries, osmophores, hydathodes and salt glands.

Glandular trichomes : The epidermal surfaces of leaves and flowers of many plants contain glandular trichomes and glands. The glandular trichomes consist of a unicellular or multicellular stalk with a head above. The cells of multicellular stalk may be arranged in several rows. The unicellular (e.g. *Pelargonium*) or multicellular (e.g. *Callitriche*) heads of the glandular trichomes perform the actual function of secretion. The secretion may remain accumulated beneath the cuticle cover of the head such as the volatile oils like camphors, balsams, peppermint oil, resins etc. Glandular trichomes may also secrete nectar and water. Glandular trichomes also play a role in the defence mechanism of the plant. Certain secretions may act as insect repellents. The glandular hairs of tomato and wild potato provide resistance to aphids by secreting sticky exudates that trap the aphids. The glandular trichomes of insectivorous plants secrete mucilage to stick the insects for digestion. They also secrete proteolytic enzymes and absorb the protein digestion products into the leaf.

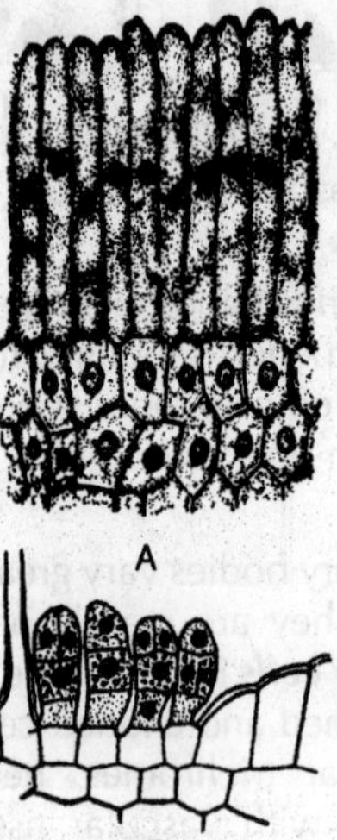

Fig. 5.63 : Nectary in sectional view. A. Nectary of *Euphorbia pulcherima*. B. Extrafloral nectary on the stipule of *Vicia* sp.

Nectary : Nectary is defined as a gland or any floral part that secretes nectar to attract the pollinators. Sometimes, it may occur on the vegetative organs and is not directly concerned with pollination. These structures are epidermal outgrowths or deeply sunken or at the level of the epidermis of the organ. The nectariferous tissues may not form any anatomically differentiated structure known as ***non-structural nectaries*** (e.g., *Dracaena reflexa* leaves, bracts of *Sansevieria zeylanica*, tepals of *Cattleya percivaliana* etc.). Histochemically, the cells of the non-structural nectaries can be distinguished. These cells show high acid phosphatase activity like the structural ones.

Anatomically and morphologically, differentiated nectariferous tissues are called ***structural nectaries***. These structures have stalks and specialised parenchymatous secretory cells with cuticle. The nectary cells are small, thin-walled with dense protoplasts containing dictyosomes, endoplasmic reticulum, small vacuoles and large nuclei. The structures have vascular supplies through which nectar is supplied and accumulated between the cuticle and secretory cells.

Exudation of nectar from non-structural nectaries takes place through stomata whereas in structural nectaries it takes place trough epidermal cells or trichomes directly to the outside. The nectar contains glucose, sucrose and fructose as major components. In addition maltose, melobiose, mucilage, proteins, phosphates, mineral ions, organic acids, oxidases, sucrose, vitamins, essential amino acids, etc. are also found to be present in the nectar. The essential amino acids attract the insects for pollination.

The extra floral nectaries are very common in dicots and much less in monocots. It rarely occurs in Poaceae (e.g., *Andropogon* and *Eragrostis*). In dicots it is present on all the organs.

Osmophores : Osmophores are some special areas on floral organs, which differ in structure from the neighbouring cells and produce fragrance to attract insects. The fragrance of a flower is due to volatile low terpenes which are not exuded by the glandular trichome. These substances at an appropriate temperature diffuse out of the cell in gaseous form through the cell

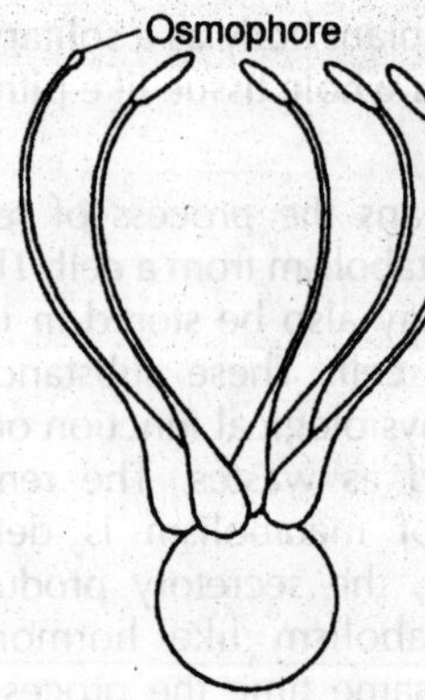

Fig. 5.64 : Osmophores in the flower of *Ceropegia loranthiflora*

wall and cuticles to produce fragrance. The fragrance producing cells differ markedly from the adjacent normal cells. The osmophores secrete terpenes as the main fragrant materials. In some species belonging to Araceae the fragrant substance may contain amines and ammonia in addition to terpenes.

The osmophores appear as flaps, cilia or brushes and can be stained with neutral red. Osmophores are present in Asclepiadaceae, Araceae, Orchidaceae, Aristolochiaceae and Burmaniaceae. The osmophores of *Ceropegia* consists of an epidermal layer and two rows of isodiametric cells situated below the epidermis. The epidermis contains dense cytoplasm and the isodiametric cells are filled with starch grains. The starch disappears after the emission of fragrant materials.

INTERNAL SECRETORY STRUCTURES

The internal secretory structures may be composed of a single cell or of groups of cells. They may occur throughout a tissue (e.g., oils or enzymes) or may be localised in distribution. Example : Castor oil obtained from the endosperm of Ricinus; groundnut oil extracted from the cotyledons of *Arachis*; the source of palm oil in the mesocarp of the fruit of *Elaeis guiensis*; the seed of *Carthemus tinctorius* yields safflower oil etc. Sometimes resin or oil-secreting idioblasts are formed, e.g., oleo-resin cells are present in the ground tissue of the rhizome of *Zingiber officinales*, the oil cells secreting the aromatic oil occur in the phloem of *Cinnamomum zeylanicum*.

Glands and ducts : Glands and ducts are formed by a group of cells or even by a solitary cell readily distinguishable from the neighbouring cells. These structures secrete or accumulate a specific substance. Cells of glands and ducts are thin-walled with dense protoplasm and sometimes occur as layer surrounding a cavity, known as *secretory cavity*. The cavity may be spherical or elongated to a tube-like structure and termed as *gland* or *duct*, respectively.The secretion product is discharged and accumulated within this cavity. The cavity may originate – (i) schizogenously, (ii) lysigenously and (iii) schizolysigenously. *Schizogenous glands* are formed by the dissolution of middle lamella separating apart the cells to form cavity (e.g., oil glands of *Eucalyptus*, the secretory ducts of *Rhus glabra*, resin duct of *Pinus* etc.) The schizogenous cavity is lined by a layer of intact parenchyma cells, termed *epithelium*. Lysigenous glands originate by lysis of a few cells forming the cavity (e.g., glands on leaves and fruits of *Citrus*). Schizolysigenous glands arise through schizogeny and lysis (e.g., *Eugenia caryophyllata*, the source of clove oil). These schizolysigenous glands are devoid of definite boundaries.

Laticifers : Laticifers are the specialised cell or row of cells that secrete the milky or watery fluid termed ***latex***. The term encompasses the various structures like ***latex cell, latex vessel, latex duct, latex tube*** and ***laticiferous duct***. The laticiferous duct is a tubular cavity into which latex is secreted and remains stored. A single vegetative cell may be converted into a simple or branched latex cell. The latex ducts are also modifications of the vegetative cells into aseptated, elongated and branched structures. The latex vessel is usually an anastomosing tubular

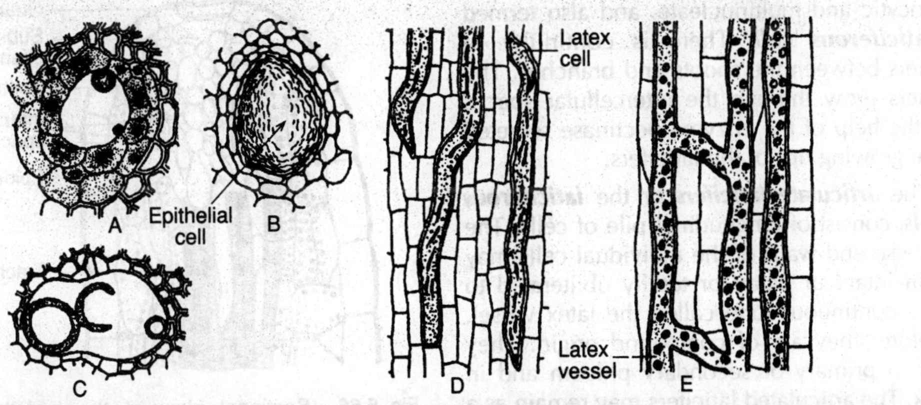

Fig. 5.65 : A. Resin duct of *Pinus* : B. Lysigenous oil cavity of *Citrus* fruit. C. Schizolysogenous protoxylem lacuna of maize. D. Non-articulate latex vessel of *Euphorbia pilulifera*. E. Latex vessel of *Carica papaya*

structure. It may also be simple or unbranched. The latex vessel is formed as a result of enlargement and fusion of a group of cells. Based on origin the laticifers may be ***simple*** or ***compound***. The simple laticifer is derived from a single cell whereas the compound laticifer originates from a longitudinal pile of cells.

The cell wall of the laticifer is non-lignified but thicker than the adjacent cells. Of course, the latex cell tip is thin-walled. The cell walls of laticifers grow in apposition and are composed of cellulose, hemicellulose and pectin.

Latex is produced within the latex vessels or cells. It is usually white and milky (e.g., *Euphorbia, Asclepias, Lactuca* etc.), yellow and brown (e.g., *Cannabis*), orange and sometimes colourless and clear (e.g., *Morus, Nerium* etc.). It contains many substances like sugars, proteins, alkaloids, oils, mineral salts, organic acids, terpenes, resins, rubber etc. The latex of *Euphorbia milii* contains dumb-bell shaped starch grains. The latex of *Carica papaya* contains the proteolytic enzyme papain. The latex of *Asclepias syriaca* contains the enzyme pectinase. The latex of some Euphorbia species is rich in vitamin B_1.

The laticifers, where present, may remain distributed throughout the plant body or may be confined to certain tissues. Laticifers may be ***non-articulated*** and ***articulated***. The former, which is derived from the enlargement of a single cell, has the potentiality of unlimited and rapid growth, and elongates to form a long unbranched latex tube (e.g., *Vinca, Cannabis, Urtica* etc.). In some plants (e.g., *Euphorbia, Nerium* etc.) the non-articulated latex tubes may be branched. The non-articulated laticifers are coenocytic and multinucleate, and also termed as ***laticiferous cell***. There is continuity of laticifers between the shoots and branches. The laticifers grow through the intercellular spaces with the help of the enzyme pectinase secreted by the growing tips of the laticifers.

The ***articulate laticifers*** or the ***laticiferous vessels***, consist of longitudinal pile of cells. The transverse end walls of the individual cells may remain intact or partly or totally obliterated to form a continuous tube called the latex vessel. Therefore, they are of compound origin. They occur in primary or secondary phloem and in cortex. The articulated laticifers may remain as a single chain of cells without anastomosis — ***articulated non-anastomosing laticifer*** (e.g., *Convolvulus, Allium, Musa* etc.). They may also form a complex anastomosing system called ***articulated anastomosing laticifers*** (e.g., *Lactuca, Papaver, Carica papaya* etc.). The enzyme cellulase is found in the latex of articulated laticifers suggesting that it may be involved in the lysis of common transverse walls during development.

Latex occurs in 900 genera distributed in 20 families, mostly in dicotyledons (e.g., Apocynaceae, Asclepiadaceae, Compositae, Euphorbiaceae, Papavaraceae etc.) and in a few families of monocotyledons (e.g., Araceae, Musaceae and Liliaceae). The different types of latex are of great economic value. The opium, a medicinally important alkaloid, is obtained from *Papaver somniferum*. The most important latex is rubber whose principal source is *Hevea brasiliensis*. The species of *Palaquium* yields gutta-percha. The latex of *Achras sapota* yields chicle, from which chewing gum is made.

Hydathodes : Hydathodes are ***water stomata, water pores*** or ***water glands*** on leaf tips and margins. Water exudation through hydathodes takes place in slowly transpiring well-watered plants. Cells of the hydathode are small, thin-walled with dense cytoplasm and are devoid of chloroplasts. The intercellular space system within the tissue is very extensive. Such tissue is called ***epithem***. There is xylem supply to the epithem and water moves through the intercellular spaces of the epithem to the outside through the pores. This process of water exudation is called ***guttation***. In some cases the cells of the hydathode secrete water and these hydathodes

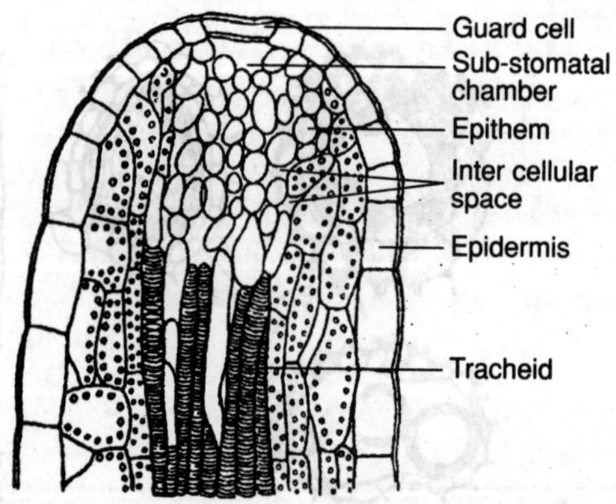

Fig. 5.66 : Sectional view of the hydathode from *Primula* sp.

are sometimes considered as ***glands***. Water stomata are modified stomata that are incapable of closing. Therefore, hydathodes are of two types (i) *epidermal hydathode* and (ii) *epithem hydathode*. The epidermal hydathodes secrete ions and minerals along with water. They are now referred to as salt glands. Epithem hydathodes, on the other hand, are generally referred to as hydathodes or water stomata.

Hydathode occurs in many families like Poaceae, Araceae, Ranunculaceae, Papavaraceae etc. including pteridophytes (*Equisetum*). They may occur over the entire surface of the leaf or may be restricted to margins or tips and even at the tips of tendrils (e.g., *Vitis vinifera*).

A typical hydathode consists of (a) water pore, (b) epithem, and (c) tracheids.

Water pore : Each hydathode opens to the exterior through one or more pores termed water pores which are generally regarded as modified stomata with guard cells. Each guard cell is chloroplast-free with a large nucleus, many mitochondria and a large vacuole. The guard cells are generally devoid of closing mechanics and, therefore, the pores always remain open with the exception in *Impatiens* and *Tropaeolum* where the water pores show diurnal periodicity of opening and closing.

The water pore internally opens to a chamber termed ***sub-stomatal chamber*** lined by one or two layers of large isodiametric cells easily distinguishable from neighbouring epithem cells (Fig. 5.66).

Epithem : Below the sub-stomatal chamber this parenchymatous tissue remains. Individual cells of this tissue are thin-walled, lack chloroplasts and some cells contain some small vacuoles. The epithem cells of *Taraxacum officinale, Papaver rhoeas* develop wall protuberances which are called **transfer cells**. Extensive intercellular spaces occur in the epithem tissue except *Crassula argentata*.

Tracheid : The tracheids of the xylem supply terminate at the intercellular spaces of the epithem and they even reach the sub-stomatal chamber (e.g., *Primula vulgaris*) or even they may extend up to the water pore (e.g., *Apomogeton distachyus*). Usually 1 to 3 veins fuse with each other before reaching the epithem. The terminal tracheids usually have spiral, scalariform or annular thickenings.

Function : The main function of hydathode is water exudation in the liquid form. A *Colocasia* leaf may exude 10 to 100 ml of water per night. Root pressure is the main cause of guttation. The exuded fluid may be pure water or it may contain ammonium phosphate, magnesium chloride, calcium nitrate etc. in low quantities. The occurrence of amino acids, sugars and water-soluble vitamins in the fluid are reported in the leaves of *Zea mays* and *Colocasia antiquarum.*

MECHANICAL TISSUE SYSTEM IN PLANTS

The tissue system that gives mechanical supports to the whole plant and their growing organs against different external and internal forces is called ***mechanical tissue system***. It is variously termed as *stereome* by Haberlandt (1918) and *stereids* by Schwendener (1874). Plant organs are subjected to various strains and stresses like bending and shearing stresses, stretching due to presence of large fruits, bending due to natural calamities like storm, heavy snow etc.

The stem has to withstand compression due to heavy weight of the large number of branches and leaves in the canopy. The branches, again, have to withstand bending as they are oriented obliquely or horizontally. The fruit stalks tend to be extended due to weight of fruits and the roots are also subjected to extension when the stem bends due to strong wind. Cell walls of all types of cells provide mechanical strength and rigidity to the plant. The woody plants achieve the structural stability and strength by the cell walls which contribute 95% of the dry weight of the wood.

The mechanical strength of non-lignified walls is due to the orientation of cellulose microfibrils. It is the major component of paper, cotton etc. Lignification in the walls of wood gives further strength. When cellulose is absent other wall polysaccharides may form microfibrils to give strength. The mechanical strength is more in the direction parallel to the microfibrils in both types of cell walls. Due to secondary growth several layers of different microfibrillar orientation are developed. Therefore, the cell wall is strong enough to resist forces from any direction. In stem the microfibrils of the thin-walled parenchymatous cells remain oriented transversely on the vertical walls so as to bend without break. In roots the microfibrils shows helical orientation to resist extensions.

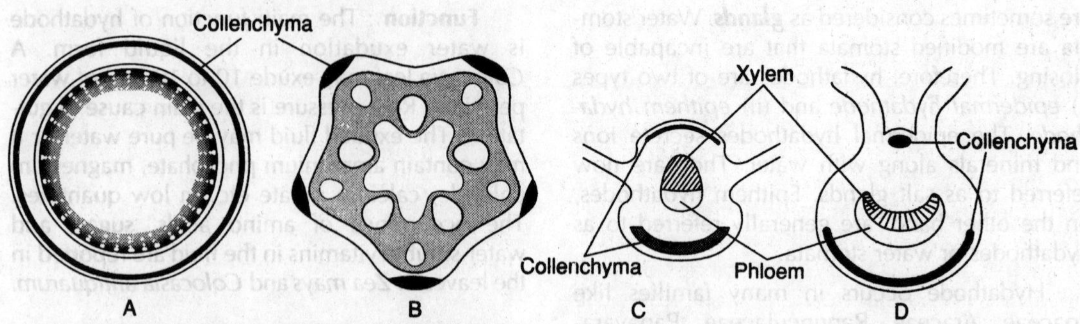

Fig. 5.67 : Diagrammatic representation of the distribution of collenchyma in t.s. of stems and leaves : A. *Sambuscus* stem. B. *Cucurbita* stem. C and D. Mid veins of leaves

Collenchyma and sclerenchyma cells, however, give maximum mechanical strength. Collenchyma walls get thickened at the corners or at the tangential walls due to the deposition of pectin, cellulose, hemicellulose, and protein. As the collenchyma cells are living they can regulate the deposition and orientation of wall materials according to the needs of the developing

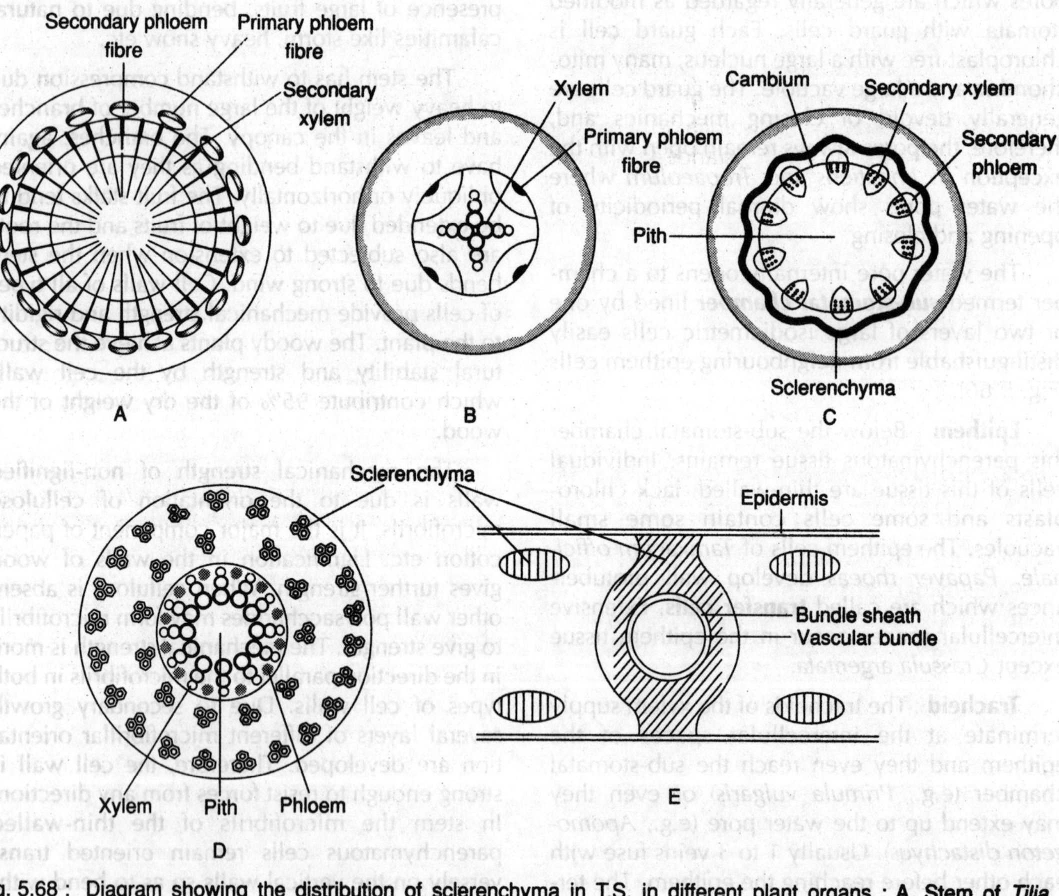

Fig. 5.68 : Diagram showing the distribution of sclerenchyma in T.S. in different plant organs : A. Stem of *Tilia*, B. Sclerenchyma in *Phaseolus* root. C. *Tinospora* stem. D. Scleren-chyma patches in stilt root of *Pandanus*. E. In monocotyledonous leaf

organs. The collenchyma cells are also elastic due to the presence of hydrated pectin on the walls.

In some plants the branches and stems show special adaptation to resist the gravitational force by the formation of ***reaction wood***. It differs in structure and location from ordinary wood. In conifers, this wood is located at the lower side of branches and stems to resist compression while in dicots it is present on the upper side of the branches to resist tension.

Principles governing the construction of mechanical tissue systems : The principle governing the construction of mechanical tissue system is the orientation of the mechanical tissues to obtain the maximum mechanical rigidity and elasticity with minimum expenditure of materials. This principle is observed in roots, stems, fruit stalks, branches etc.

Mechanical cells are developed to resist different strains and stresses. The resistances are designated as ***inflexibility*** (resistance to lateral bending), ***inextensibility*** (resistance to stretching), ***incompressibility*** (resistance to compression) and ***shearing stress*** (resistance to shearing action).

The principles of the construction and distribution of the mechanical tissues to form a system are :

Inflexibility : Both the terrestrial arboreal plants as well as the herbs are encountered by forces to bend laterally. The plants with arborial habit resist such forces by the formation of heartwood at the centre whereas the herbaceous plants resist through the formation of mechanical cells. All the aerial organs are the inflexible organs as they face lateral bending strain due to application of lateral forces like high wind. For inflexibility of plants the mechanical cells remain distributed in different organs like 'I'–girders as railway lines, and also that are used in the construction of bridges, buildings etc. The girder consists of two upper and lower horizontal plates again connected by a vertical plate (Fig. 5.69). In cross-sectional view it resembles the English capital letter 'I'. The horizontal plates are called ***flanges*** and the connecting vertical plate is called the ***web***. If a load is appplied at the middle of I-girder a bending strain is encountered. The upper flange tends to be compressed and the lower one tends to be extended. The flanges only suffer and resist the strains, the web on the other hand neither subjected to compression nor extension strains. This vertical line is termed as ***null-line*** as it is neutral regarding the strains.

A single I-girder can resist the lateral force in a single plane from both directions only. To resist such lateral forces from different directions several I-girders remain oriented in a circle perpendicular to the long axis. I-girders are arranged in such a way that their null-lines intercept at a common point forming a composite I-girder that can resist lateral forces from all directions. This simple I-girder formation is the characteristic of many plants and thus they resist bending forces either singly or in suitable combinations. This I-girder formation is found in fibrovascular bundles where bands of sclerenchyma are present above and below the actual vascular strand. These sclerenchymatous bands represent the upper and lower flanges and the vascular strand as the web. The fibrovascular bundles are present in the monocot leaves, e.g., *Cyperus*, where they are arranged in parallel. In the monocot stem, the vascular bundles are completely encircled by sclerenchyma termed bundle sheath.

In the cylindrical stems and other organs the mechanical cells remain at the periphery. In the hypodermis, collenchyma and sclerenchyma cells are found in dicot and monocot stems respectively. They are present either continuously or as isolated patches forming the flanges of composite I-girder. In some members of Compositae sclerenchymatous bundle caps are present. In case of rectangular or square stems (e.g., *Leonurus, Leucas* etc.) the mechanical collenchyma cells are present at the four corners to form the flanges and the pith acts as the web. Thus, diagonally opposite I-girders are formed. In *Nyctanthes* inverted vascular bundles are present at the four corners to act as flanges. In *Zea mays* stem sclerenchymatous bundle sheath is present and the stele is atactostele.

Inextensibility : The underground anchoring organs like rhizomes, roots, etc. have to face longitudinal tensions. The fruit stalks, the climbers and lianes also have to withstand the longitudinal tension, as they are to bear the weight of fruits and their own weight, respectively, when they hang over the supporting object. Many rooted hydrophytes have to be longitudinally stretched due to water currents. The mechanical tissues in the centre of the hydrophytic stems (e.g.,

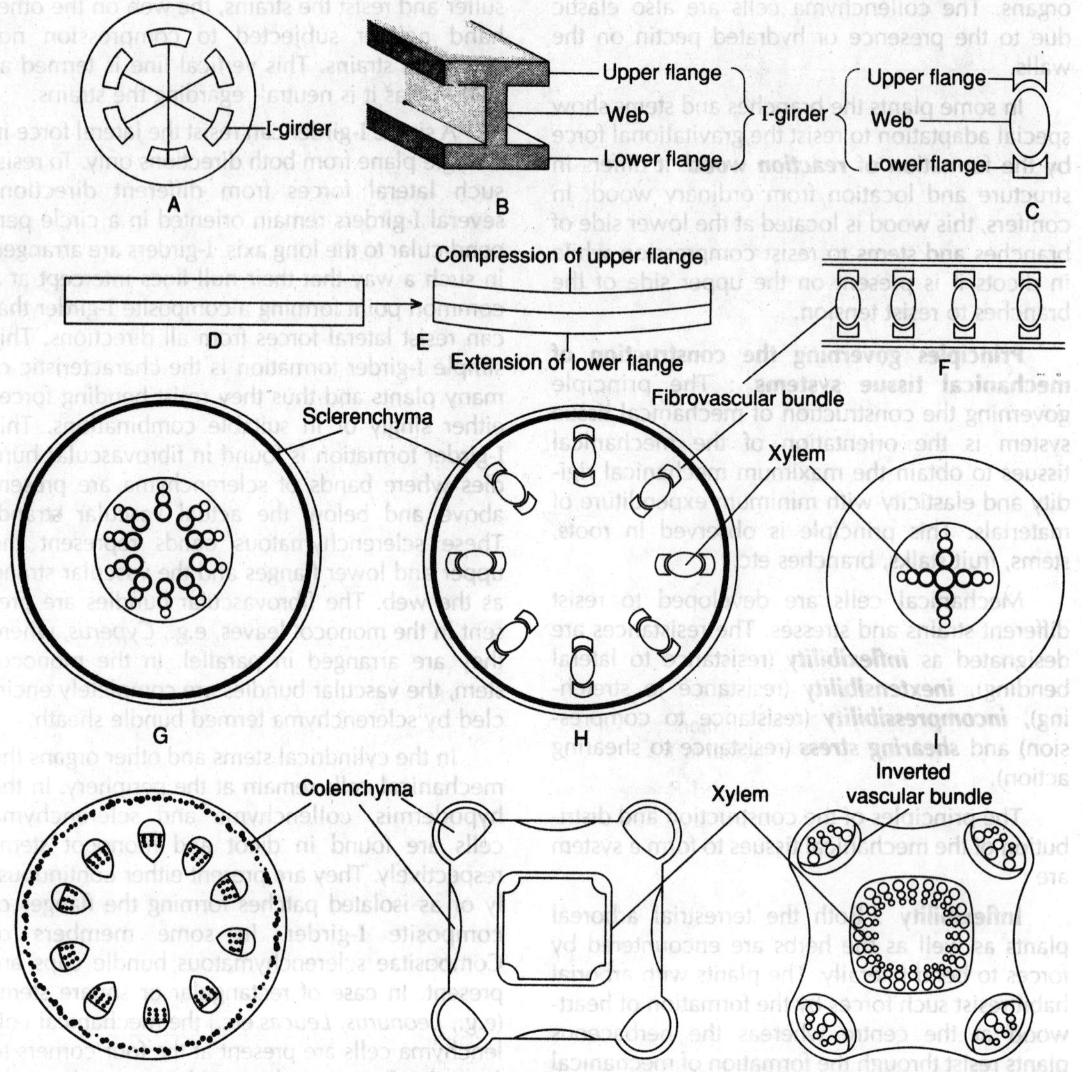

Fig. 5.69 : Principles of composite I-girder formation in different plants : A. A Composite I-girder, B. A portion of I-girder or railway line, C. A fibrovascular bundle compared with I-girder, D & E. I-girders before and after bending respectively, F. Parallel distribution of fibrovascular bundles as mechanical tissues, G. Diagrammatic representation of the mechanical tissues in stilt root of maize, H. Same in the stem of *Cyperus*, I. Mechanical tissue system in dicot root, J. Same in *Helianthus* stem, K. In *Leonurus* stem, L. In *Nyctanthes* stem

Potamogeton lanceolatus), in the fruit stalks (e.g., *Cucurbita*) and in the lianes (e.g., *Dioscorea*) help to resist longitudinal stretching. The centrally located mechanical tissues to resist longitudinal tension can be compared with electrical cable where rigidity is obtained from the central axile strands made up of metallic wires. This principle is observed in roots as well as in other inextensible organs. In roots the stele is small in comparison to the other external tissues and xylem and sclerenchyma are condensed within the stele.

Incompressibility : The stems are encountered by the longitudinal compression developed by the weight of the canopy. This situation can be compared with a heavy load on the top of a cylindrical axis where the axis is subjected to longitudinal compression. The metallic rods remain condensed at the centre of the axis like the effective aggregation of the mechanical tissues at the centre of the stems.

The subterranean and submerged organs of plants are encountered by radial compression or crushing pressure by the surrounding medium like the soil and water, respectively. Mechanical cells in these organs remain at the periphery to resist such radial compression.

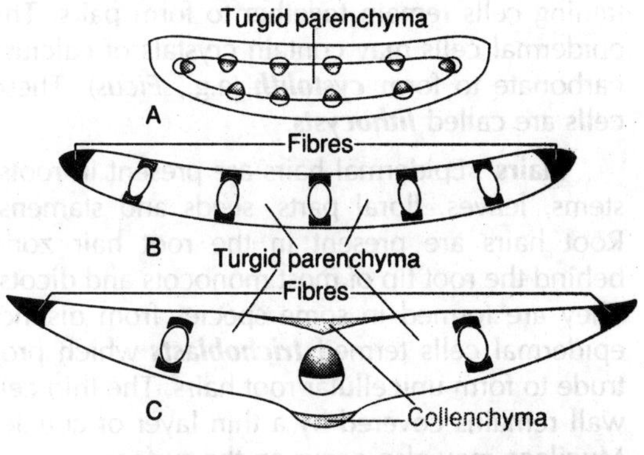

Fig. 5.70 : Diagrammatic representation of the distribution of fibres, collenchyma and turgid parenchyma : A. In succulent leaf of *Gasteria* (Liliaceae) in T.S.. B. In monocotyledonous leaf and C. In a dicot leaf in T.S.

The stilt roots of *Zea mays* and *Pandanus* the mechanical cells are distributed as such so that they can resist both compression and extension forces. The stilt roots on the wind blowing side have to withstand longitudinal extension and those present on the opposite side have to resist longitudinal compression. Thus the same root is periodically compressed and extended. In these roots in addition to centrally aggregated mechanical cells which resist longitudinal extension, sclerenchyma cells are present in the periphery to resist compression. *Zea mays* roots have peripheral sheet of sclerenchyma in the cortex (Fig. 5.69G) whereas *Pandanus* has developed isolated patches of sclerenchyma in the periphery of the cortex.

Shearing stress : The flat organs like leaves are often encountered by violent shearing stresses due to strong wind or water current. Such force acts at right angles to the surface of the leaves causing laceration. Dicot leaves are mechanically more resistant against such stress. The I-girders present in them are firmly held together by a large number of cross-ties in the form of vein network (Fig. 5.70C). In monocot leaves parallel I-girders formed by fibrovascular bundles are present (Fig. 5.70B). Sclerotic strands are also present at the hypodermal region and leaf margins and the leaf blades are usually cuticularised to withstand shearing stress.

TISSUE SYSTEMS

Plant tissues of varied nature and functions are aggregated to form ***tissue systems***. According to Sachs (1875) there are three tissue systems in higher plant body viz., ***epidermal tissue system, ground or fundamental tissue system*** and ***vascular tissue system***. The epidermal tissue system is the outer covering layer in a primary body thus protecting the inner tissues. The vascular tissue system consists of the vascular bundles made of the complex tissues like xylem and phloem the principal conducting tissues in plants. The ground or fundamental tissue system is the rest in which the vascular tissue system remains embedded. The meristematic tissues are the precursors of the three tissue systems—the epidermal tissue system developing from the protoderm, the vascular tissue system from the procambium and the ground tissue system from the ground meristem.

	Meristematic region	**Permanent region**	**Tissue system**
Promeristem	→ Protoderm →	Epidermis →	Epidermal tissue system
	→ Ground meristem →	Cortex, Pericycle, Medullary rays, Pith →	Ground meristem
	→ Procambium →	Vascular bundles →	Vascular tissue system

EPIDERMIS

The covering cell layer(s) in the whole plant forms the ***epidermis*** which remains in direct contact with the environment. Therefore, this particular tissue adjusts itself with its surroundings because it has to protect the inner tissues from any adverse external factors like high temperature, desiccation, mechanical injury, pathogenic infections etc. The epidermis may persist throughout the plant life or it may, in some cases, be replaced by periderm when the older epidermis becomes inactive.

Origin : The outermost layer of apical meristem gives rise to the epidermis, which is variously named as protoderm, dermatogen and tunica by different authors. In cryptogams, however, it originates from solitary initial cell along with cortex and stele. The periblem gives rise to epidermis along with root cap in gymnospermous root, whereas in dicots the epidermis develops from dermatogen. The periblem in monocot roots develops the epidermis and the cortex.

Structure : Generally the epidermis is single layer of cells. ***Multiple-epidermis*** is the feature of *Ficus* and *Nerium* leaves and orchid root. The term ***velamen*** is given to the multiple epidermis of the orchid root. The multiple-epidermis is derived by the periclinal division of the epidermal initials.

Epidermal cell walls are usually thin, but thick-walled lignified epidermal cells occur in some gymnosperms. The outer surface of the epidermal cell wall may usually be cuticularised or waxy, giving rise to the characteristic surface features of the epidermis, which very often becomes a taxonomic characteristic. The cuticle protects plants from dessication as it is impervious to water. It also prevents the plant from various pathogens. Sometimes wax may be deposited on the surface of the cuticle to form a white powdery lining on various fruits (e.g., the plum, grapes etc.) and on leaves. Wax is deposited on the inside surface of the pitcher of *Nepanthes* in the form of overlapping scales to stick to the feet of the insects.

Epidermal cells are horizontally flattened and remain parallel to surface of the organ. In surface view the epidermal cell walls show sinuous outline in dicot leaves and straight or sinuous outline in monocot leaves. Usually the intercellular space is absent among the epidermal cells.

Chloroplastids are absent in epidermal cells except the specialised guard cells of stomata. Sometimes silicon may be deposited in the epidermal cells walls or in their lumens. Trichome cell walls may be silicified. Silicon is deposited in the bracts of rice, in the marginal trichomes of oat, in the leaves of *Cyperus, Avena* etc. In *Avena sativa* the epidermal cells at the intercostals position of the internodes cork cells and silica containing cells remain together to form pairs. The epidermal cells may contain crystals of calcium carbonate to form ***cystolith*** (e.g., *Ficus*). These cells are called ***lithocysts***.

Hairs : Epidermal hairs are present in roots, stems, leaves, floral parts, seeds and stamens. Root hairs are present in the root hair zone behind the root tip of most monocots and dicots. They are formed in some species from distinct epidermal cells termed ***trichoblasts*** which protrude to form unicellular root hairs. The thin cell wall remains covered by a thin layer of cuticle. Mucilage may also occur on the surface.

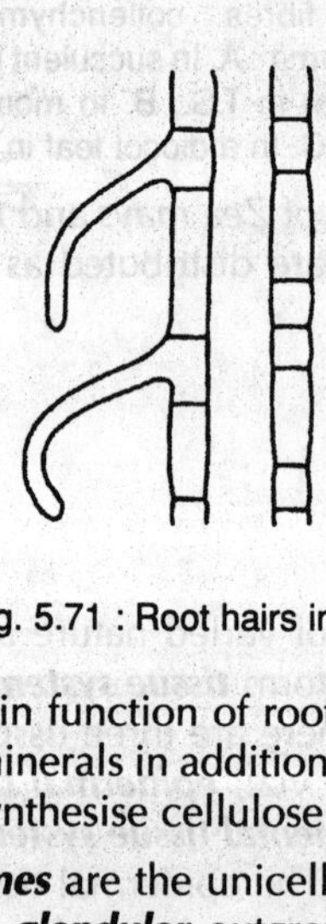

Fig. 5.71 : Root hairs in *Cyperus*

The main function of root hairs is to absorb water and minerals in addition to anchorage. The root hairs synthesise cellulose at their tips.

Trichomes are the unicellular or multicellular ***hairy*** or ***glandular*** outgrowths of epidermal cells. Single linear row or several rows of cells constitute the multicellular trichome. The glandular ones are secretory in function. The hairy trichomes give protection and also prevent water loss. The trichomes are also typified as ***stellate*** (star like) or ***dendroid*** (a miniature tree form, e.g., *Verbascum*). They may occur in tufts (e.g., *Hamamelis*). Hairy projections in flower petals are termed ***papillae***.

Most trichomes possess thin and cellulosic or lignified cell walls. The glandular trichomes

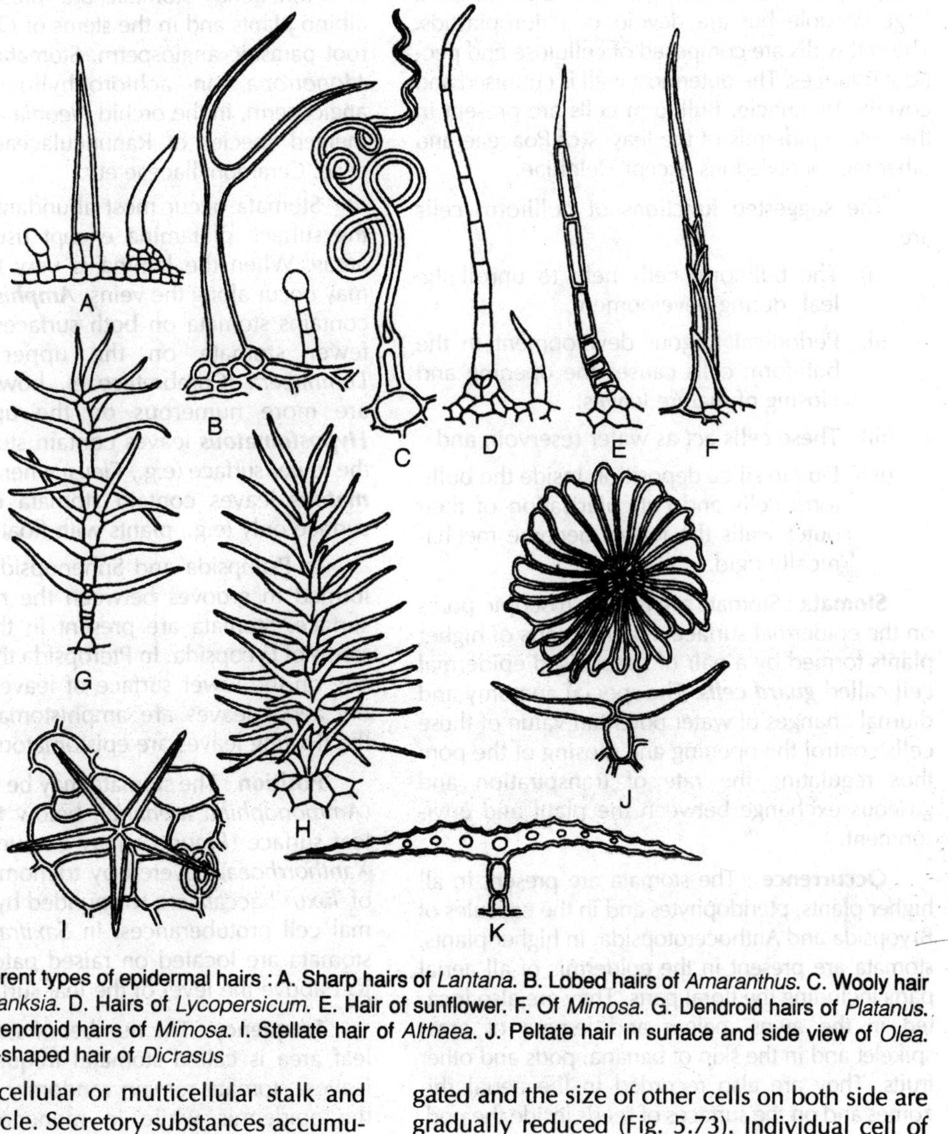

Fig. 5.72 : Different types of epidermal hairs : A. Sharp hairs of *Lantana*. B. Lobed hairs of *Amaranthus*. C. Wooly hair of *Banksia*. D. Hairs of *Lycopersicum*. E. Hair of sunflower. F. Of *Mimosa*. G. Dendroid hairs of *Platanus*. H. Dendroid hairs of *Mimosa*. I. Stellate hair of *Althaea*. J. Peltate hair in surface and side view of *Olea*. K. T-shaped hair of *Dicrasus*

consist of unicellular or multicellular stalk and head with cuticle. Secretory substances accumulate beneath the cuticle (e.g., oils, resins, camphor, peppermint etc.) Sometimes **scaly** outgrowth with a short stalk and head consisting of discoid plate of cells may be formed (e.g., leaves of *Olea*). Some of the epidermal cells of *Mesembryanthemum* may enlarge where water accumulates. These specialised epidermal cells are termed as ***vesicles*** or ***bladders***.

Bulliform cells : It is a group of outer larger epidermal cells found in monocot leaves easily distinguishable from other epidermal cells by their fan-shaped appearance in cross-section. The median cell of the group is the most elongated and the size of other cells on both side are gradually reduced (Fig. 5.73). Individual cell of

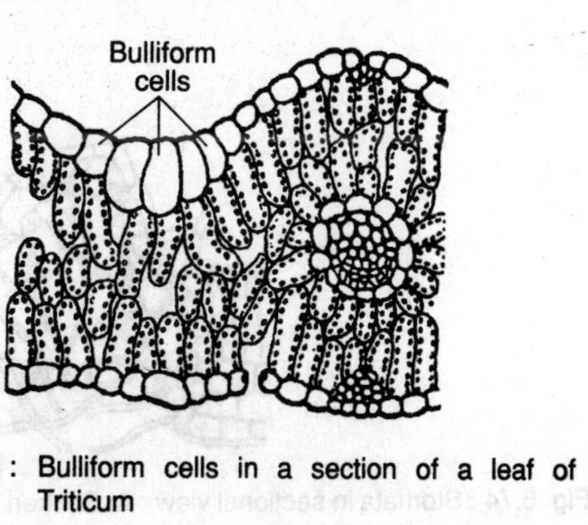

Fig. 5.73 : Bulliform cells in a section of a leaf of Triticum

the group is thin-walled, hyaline and possesses a large vacuole but are devoid of chloroplastids. The cell walls are composed of cellulose and pectic substances. The outermost wall is cutinised and covered by cuticle. Bulliform cells are present in the outer epidermis of the leaves of Poaceae and other monocotyledons except Helobiae.

The suggested functions of bulliform cells are :

(i) The bulliform cells help to unroll the leaf during development,

(ii) Periodical turgour development in the bulliform cells causes the opening and closing of mature leaves,

(iii) These cells act as water reservoir, and

(iv) Due to silica deposition inside the bulliform cells and cuticularisation of their outer walls the leaves become mechanically rigid.

Stomata : Stomata are the microscopic pores on the epidermal surface of aerial parts of higher plants formed by a pair of specialised epidermal cell called ***guard cells***. The special anatomy and diurnal changes of water potential value of these cells control the opening and closing of the pore thus regulating the rate of transpiration and gaseous exchange between the plant and environment.

Occurrence : The stomata are present in all higher plants, pteridophytes and in the capsules of Bryopsida and Anthocerotopsida. In higher plants, stomata are present in the epidermis of all aerial parts including the floral parts. They are also located in the awns, paleas and lemma of grass spikelet and in the skin of banana, pods and other fruits. They are also recorded in the aerial rhizomes and on the surfaces of seeds inside the pod. Non-functional stomata are present in petals, albino plants and in the stems of *Orobanche* — a root parasitic angiosperm. Stomata are absent in *Monotropa*, an achlorophyllous saprophytic angiosperm, in the orchid *Neottia* and in the submerged species of Ranunculaceae, Nymphaeaceae, Ceratophyllaceae etc.

Stomata occur most abundantly throughout the surface of lamina except usually the vein areas. When the lamina is very thick, stomata may occur along the veins. ***Amphistomatous*** leaf contains stomata on both surfaces with usually fewer stomata on the upper surface. In *Lumnitzera* (Combretaceae), however, stomata are more numerous on the upper surface. ***Hypostomatous*** leaves contain stomata only on the lower surface (e.g., *Ficus*) whereas the ***epistomatous*** leaves contain stomata on the upper surface only (e.g., plants with floating leaves).

In Psilopsida and Sphenopsida stomata are formed in grooves between the ridges of stem surface. Stomata are present in the leaves and stems of Lycopsida. In Pteropsida they occur usually on the lower surface of leaves. In *Marsilea* the aerial leaves are amphistomatous whereas the floating leaves are epistomatous.

Position : The stomata may be located in pits (*Ammonophilia arenaria*), below the epidermal leaf surface (*Pinus*) or may be sunken (*Nerium, Xanthorrhoea*) covered by trichomes. The pores of *Taxus baccata* are surrounded by large epidermal cell protuberances. In *Saxifraga stolonifera* stomata are located on raised patches and project above the level of the leaf surface.

Frequency : The number of stomata per unit leaf area is called stomatal frequency. In dicot leaves, stomata remain randomly distributed in the epidermis while in monocots and gym-

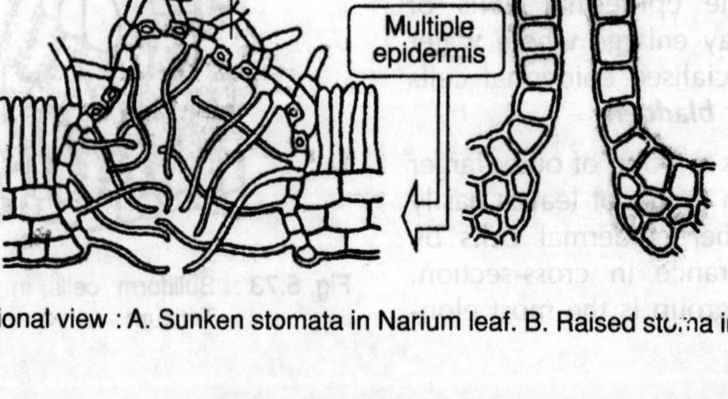

Fig. 5.74 : Stomata in sectional view : A. Sunken stomata in Narium leaf. B. Raised stoma in the peduncle of Cucurbita

nosperms stomata are usually arranged in rows. Stomatal frequency varies depending on the plant types, age and the portion of the leaf. Environmental factors like CO_2 concentration, temperature and light intensity also affect stomatal frequency. Generally, in higher stomatal frequency, the stomata are smaller in size. The area occupied by the total stomatal pores may be as much as five percent of the total leaf area, but normally the value remains below two percent.

As the stomatal frequency varies, the term stomatal index was introduced by Salisbury in 1928. The stomatal index is calculated as :

$$\text{Stomatal index} = \frac{\text{Stomatal frequency}}{\text{(Number of stomata + Number of epidermal cells per unit leaf area)}} \times 100$$

Stomatal index is found to be constant within the leaves of a plant.

Structure of stoma : The term ***stoma*** meaning mouth was first coined by deCandole in 1827. The stoma constitutes the pore and two guard cells. The surrounding epidermal cells in association with the guard cells are the subsidiary cells and they together constitute the ***stomatal complex*** or ***stomatal apparatus***. The stomatal pore opens below to the stomatal cavity or substomatal chamber.

Subsidiary cell : These are specialised epidermal cells which are distinguished by their shape and size from the guard cells and the rest of the epidermal cells. They are also called ***accessory cells***. Subsidiary cells are generally smaller in size than the normal epidermal cells. In *Equisetum* the subsidiary cells overarch the guard cells. In grasses no plasmodesmata or pit were observed between the subsidiary cells and guard cells, but they exist in the leaves of *Vicia faba* and *Nicotiana*.

Guard cell : Typical stomata of dicots consist of two kidney-shaped or semilunar guard cells. In grasses the guard cells tend to be more elongated or dumbbell shaped. Guard cells contain a few chloroplasts, whereas their neighbouring epidermal cells seldom do. There is report that the stoma of *Azolla pinnata* (Sen, 1983) consists of a unicelled and binucleate guard cell with a pore, and the stoma is associated with one or more subsidiary cell(s) on its proximal side.

Each kidney-shaped guard cell has four sides — thick ventral side facing the pore, thin dorsal side towards the subsidiary cell, upper lateral side facing the atmosphere and the lower lateral side facing the stomatal cavity. The ventral walls of the two guard cells are heavily thickened and are not fused in contrast to dorsal wall. A pair of heavily cuticularised lips protrudes from the upper edge of the ventral wall to guard the pore. The dumb-bell shaped guard cells of grasses are attached to each other by their thick-walled middle portion and connected by their bulbous thin-walled ends. The guard cells of *Zea mays* have thick upper and lower lateral walls while the dorsal and ventral walls are thin. In most cases the guard cells remain fused with the other epidermal cells by their walls only. But in *Pinus* and *Equisetum* the guard cells are enveloped by subsidiary cells.

Usually the guard cell wall is composed of cellulose microfibril and pectin excepting some ferns and gymnosperm where the guard cell walls contain lignin. The guard cell walls of *Ophioglossum* contain β-1, 3-glucan. The guard cell and subsidiary cell walls of grasses and *Equisetum* contain silicon. The cellulose microfibrils remain oriented radially in semilunar guard cell called ***radial micellation***. When a guard cell expands by taking up water, it cannot increase much in diameter as the microfibrils do not stretch much along their length. Therefore, because two guard cells are attached to each other at both ends, they bend outward when they swell, which opens the stomata. In the dumb-bell shaped guard cell walls the cellulose microfibrils have the axial arrangement in the middle of the guard cell while the bulbous ends have radial micellation.

The guard cells usually contain chloroplastids. The number of chloroplasts per guard cell ranges from two (e.g., *Anthoceros*) to as many as hundred or more (e.g., *Polypodium vulgare*). The guard cell chloroplasts are poorly developed with rudimentary grana compared to the mesophyll cell chloroplasts. Generally the plastids of guard cells are laden with starch grains but they lack in many species of Liliaceae, Amaryllidaceae etc.

The guard cells contain dense cytoplasm with single nucleus and all other cell organelles. The guard cell vacuoles are smaller in comparision to other cells. There are no plasmodesmatal connections between mature guard cell and the surrounding cells. The endomembrane system —

along with the subcellular particles involved in the synthesis of proteins and cell wall materials — are found to be present in the early stages of development of guard cells.

Origin : The stomata originate from the stomatal meristemoid cells of protoderm. These isodiametric cells possess a single conspicuous nucleus with dense cytoplasm. During development the stomatal meristemoids originate acropetally (e.g., stem of *Psilotum*), basipetally (e.g., Lycopodiales, Convolvulaceae, Rubiaceae etc.) or simultaneously (e.g., *Erythrina*). The differentiation of a stomatal meristemoids leading to the formation of stomata varies in different plant groups. In *Allium cepa* an elongated meristemoid cell of the protoderm divides unequally to produce a large and a small cell. The latter is the guard cell mother cell, which, by longitudinal division, forms two small sister guard cells. With maturity, the middle lamella between the two sister guard cells are lost enzymatically. As a result, the pore is formed. Gradually, the ventral surfaces of the guard cells become thick and non-elastic. Due to osmotic entrance of water the guard cells swell in length only caused by radial micellation. The elastic dorsal surface will stretch more than the thick non-elastic ventral side. As a result, the two guard cells gradually separate from each other at the pore region and become reniform.

The number of subsidiary cells in the stomatal complex varies in different plants. Subsidiary cells are absent in the stomata of *Allium*. A single meristemoid gives rise to the stoma.

Following are different ontogenetic types of stomata in plants (Pant, 1965) :

1. **Mesogenous type** : In this ontogenetic type of stomata the subsidiary cells and guard cells of a stomatal complex are derived from a single meristemoid cell. In *Lactuca viscosa* the stomatal meristemoid cell have four cutting faces. There are three anticlinal cutting faces in *Beta vulgaris, Datura stramonium, Nicotiana* etc. and two parallel cutting faces in Rubiaceae, *Casuarina, Gnetum gnemon, Ricinus communis* etc. In *Schizaea* a single annular subsidiary cell originates by a curved periclinal division of the stomatal meristemoid.

2. **Perigenous type** : In this type all the subsidiary cells of a stomatal complex are derived from protoderm cell. The subsidiary cell initials lie adjacent to guard cell mother cell. The guard cell mother cell divides once to produce two guard cells. *Lycopodium, Ephedra, Trapa, Cuscuta* etc. are the examples of this type.

3. **Mesoperigenous type** : In this developmental type the subsidiary cells are derived from different types of mother cells. The guard cell mother cell and one subsidiary cell are formed from a single stomatal meristemoid whereas the other subsidiary cells are derived from the neighbouring cells adjacent to guard cell mother cell. The first division of the meristemoid and the partition of the guard cell mother cell may be (i) parallel (e.g., *Tetracentron sinensis*), or (ii) at right angles (e.g, some members of Onagraceae and Caryophyllaceae), or (iii) at other angles to each other (e.g., *Ranunculus*).

There are two ontogenetic stomatal patterns in gymnosperm called ***haplocheilic*** and ***syndetocheilic*** (Florin, 1933).

Haplocheilic stomata : In this type the protoderm or stomatal meristemoid cell directly gives rise to guard cells and the neighbouring protoderm cells differentiate into subsidiary cells. Therefore, the subsidiary cells and the guard cells of the stomatal complex are not derived from the same stomatal meristemoid and

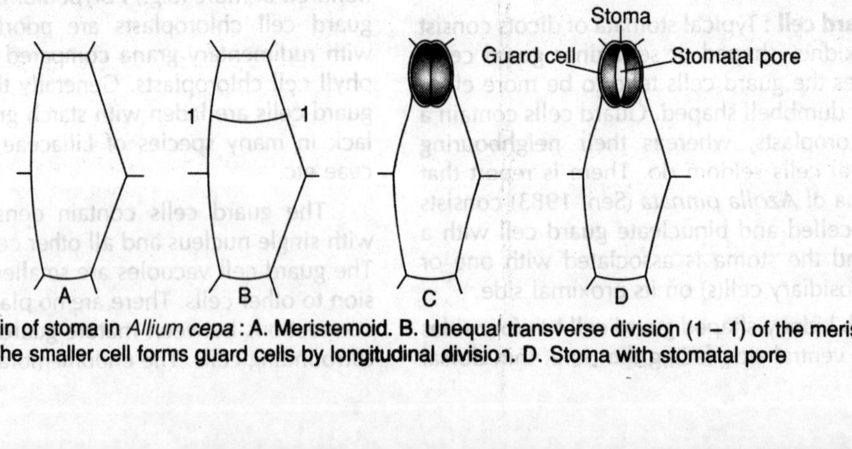

Fig. 5.75 : Origin of stoma in *Allium cepa* : A. Meristemoid. B. Unequal transverse division (1 – 1) of the meristemoid. C. The smaller cell forms guard cells by longitudinal division. D. Stoma with stomatal pore

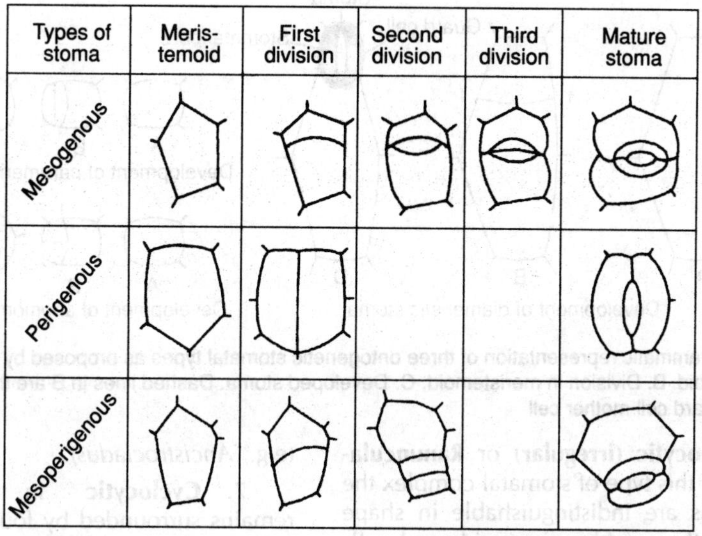

Fig. 5.76 : Diagrammatic representation of ontogenetic types of stomata

are termed as **perigene** (e.g., Cycadales, Pteridospermopsida, Coniferopsida, *Ephedra* etc.).

Syndetocheilic stomata : In syndetocheilic type the protoderm or stomatal meristemoid cells divide to form guard cell and subsidiary mother cells. Thus subsidiary cells are derived from the same stomatal meristemoid that also produces guard cell mother cell and are termed as ***mesogene*** (e.g., Cycadeoidales, *Gnetum, Welwitschia* etc.).

Later on, it has become evident that the protoderm cells cannot form trichomes or stomata directly. They are committed to form the meristemoid of either trichomes or stomata. So neither a trichome nor a stoma can be differentiated directly from protodermal cells by-passing the meristemoid stage. Thus there are no perigenous stomata instead they are either mesogenous or mesoperigenous.

Payne (1979) classified stomatal complex on the basis of mode of division of guard cell mother cell with respect to the orientation of the walls formed by it. The orientation of the wall that separates the two guard cells may be perpendicular, parallel or have no special orientation with respect to the wall that cuts the GMC from stomatal meristemoid.

Following three types are recognised :

1. **Diameristic** (*Dia* means across and *meros* means part) : In this particular type the stomatal meristemoid cuts off the GMC at the distal end, which again divides by perpendicular wall to the previous wall, i.e. at right angles to the wall that forms the GMC. The subsidiary cells may be mesogene (e.g., Lamiaceae) or mesoperegene (e.g., *Allium*, Caryophyllaceae).

2. **Parameristic** (*Para* means beside and *meros* means part) : In this type the stomatal meristemoid cuts off the GMC which then divides by a wall parallel to the preceding one. The subsidiary cells may be mesogene (e.g., Cruciferae, Rubiaceae) or mesoperegene (e.g., *Liriodendron*).

3. **Anomomeristic** (*Anomalos* means irregular and *meros* means part) : In this type the stomatal meristemoid cuts off the GMC that further divides by a wall without any special orientation i.e., at any angle with respect to the previous wall and, thus, the stoma lies at any angle to the wall that forms the GMC. The subsidiary cell is usually single and mesogene (e.g., Ranunculaceae).

The stomatal cavity or substomatal chamber underneath each stomatal complex is formed by the natural extension of existing air spaces and dissolution of some of the mesophyll cells.

Types of stomata : The following types of stomata are commonly found in different plant groups :

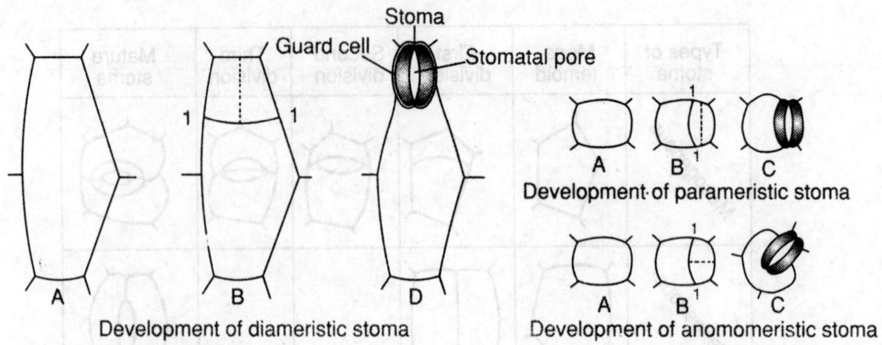

Fig. 5.77 : Diagrammatic representation of three ontogenetic stomatal types as proposed by Pyne (1979) : A. A meristemoid. B. Division in meristemoid. C. Developed stoma. Dashed lines in B are the sequences of division of guard cell mother cell

1. **Anomocytic (irregular)** or **Ranunculaceous type** : In this type of stomatal complex the subsidiary cells are indistinguishable in shape and size from the neighbouring epidermal cells and, at the same time, the number and arrangement of subsidiary cells may not be definite (e.g., *Cucurbita*, Capparidaceae, Malvaceae, Papaveraceae etc.).

2. **Anisocytic (unequal)** or **Cruciferous type** : These stomata remain encircled by three subsidiary cells of which one is distinctly smaller or larger in size than the other two. The subsidiary cells may not be very distinct from ordinary epidermal cells (e.g., *Petunia, Solanum, Sedum, Nicotiana* etc.).

3. **Paracytic (parallel)** or **Rubiaceous type** : This type of stoma usually has two subsidiary cells parallelly oriented to the long axis of guard cells and the stomatal pore (e.g., *Phaseolus, Arachis, Psoralea* etc.).

4. **Hemiparacytic** : There is a single subsidiary cell parallely placed to the long axis of the pore in the stomatal complex. This subsidiary cell may be long or short in length in contrast to the guard cells (e.g., *Tetracentron*).

5. **Diacytic (cross-walled)** or **Labiatous** or **Caryophyllaceous type** : In this type of stomatal complex there are two large subsidiary cells. The common walls of them remain perpendicular to the long axis of the guard cells as well as that of the stomatal aperture (e.g., *Hygrophila, Dianthus* etc.).

6. **Actinocytic** : In this stomatal complex the two guard cells remain surrounded by four or more radially arranged elongated subsidiary cells (e.g., *Ancistrocladus*).

7. **Cyclocytic** : In this type the stoma remains surrounded by four or more subsidiary cells arranged in a narrow ring around the guard cells (e.g., *Lumnitzera, Laguncularia* etc.).

8. **Tetracytic** : Tetracytic type shows four subsidiary cells surrounding the guard cells. Of them, two subsidiary cells remain on the polar sides and the other two on the lateral sides of the guard cells (e.g., *Rhoeo*).

9. **Hexacytic** : In hexacytic type six subsidiary cells surround the guard cells. Two of them are situated on the two polar sides and rest four subsidiary cells are found to occur on the two lateral sides in parallel to the long axis of the guard cells and the aperture.

10. **Graminaceous** : Metcalfe and Chalk (1950) described this type of stomata found in monocot leaves — it is the characteristic of the families Gramineae and Cyperaceae. In this type the guard cells are dumb-bell-shaped or osteate whose two attached ends are bulbous and the middle parts are much narrower and straight resembling dumb-bells. In the narrow portion the cell wall is highly thickened whereas the bulbous ends are relatively thin-walled. The subsidiary cells occur parallel to the long axis of the pore. The subsidiary cells of *Zea mays* are triangular-shaped.

Stebbins and Khush (1961), however, have reported four types of stomatal complex in monocotyledons :

1. **First type** : In this type the guard cells remain surrounded by 4 subsidiary cells — two at the polar sides and two at the lateral sides

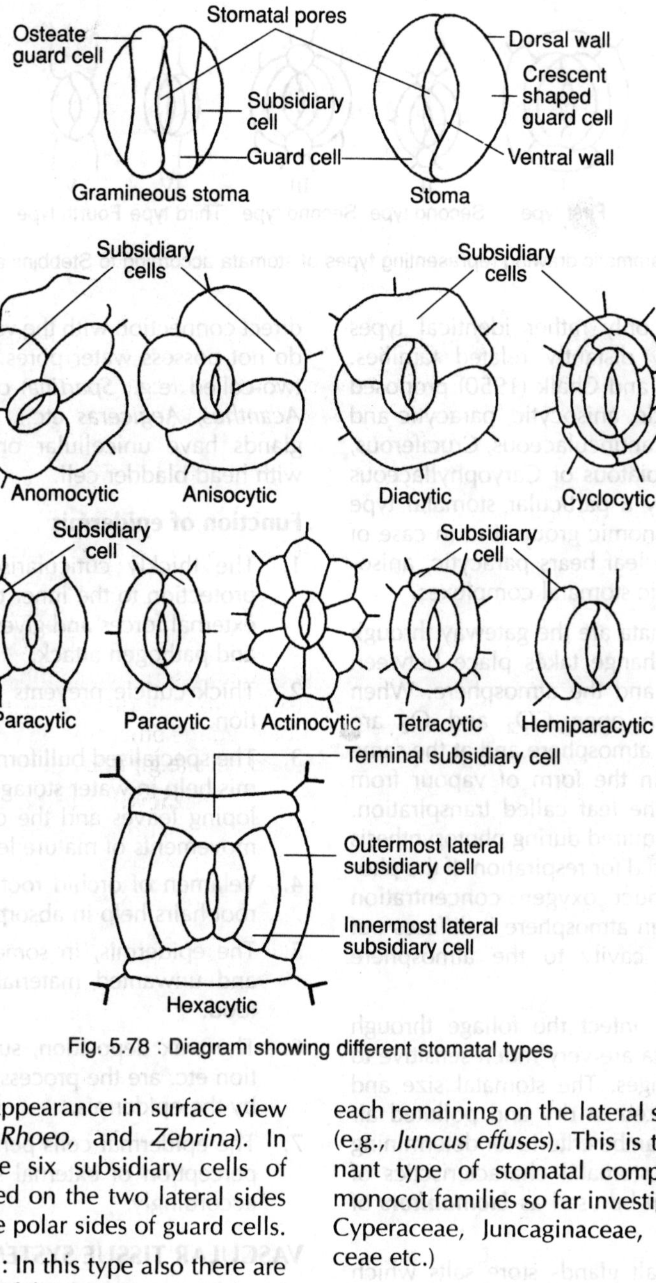

Fig. 5.78 : Diagram showing different stomatal types

giving a square like appearance in surface view (e.g., *Tradescantia, Rhoeo,* and *Zebrina*). In *Commelina* there are six subsidiary cells of which four are situated on the two lateral sides and the rest two at the polar sides of guard cells.

2. **Second type** : In this type also there are 4–6 subsidiary cells. Of them two are round and smaller in size and each lies at each pole. The remaining four are arranged on the lateral sides (e.g., *Pandanus haerbachii,* members of Palmae and Cyclanthaceae). *Caryota* and *Calamus* of Plamae, however, have two subsidiary cells only on the two lateral sides of the guard cells.

3. **Third type** : The two guard cells of a stoma are surrounded by two subsidiary cells each remaining on the lateral sides of guard cells (e.g., *Juncus effuses*). This is the most predominant type of stomatal complex found in 24 monocot families so far investigated (Gramineae, Cyperaceae, Juncaginaceae, Typhaceae, Liliaceae etc.)

4. **Fourth type** : In this type the stomatal complex is formed without any subsidiary cell. This is also very common type of stomatal complex found in the families Amaryllidaceae, Iridaceae, Orchidaceae, Agavaceae etc.

Stomata are usually described as Ranunculaceous, Cruciferous, Rubiaceous and Labiatous or Caryophyllaceous (Vesque, 1889) but it does not literally mean that these types are confined to

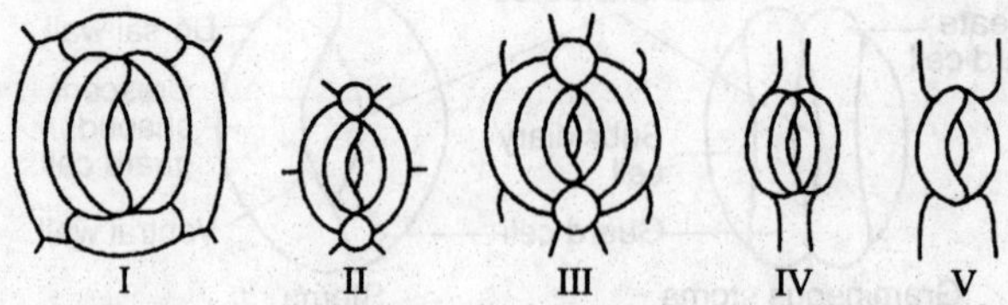

Fig. 5.79 : Diagrammatic drawing representing types of stomata according to Stebbins and Khush (1961)

respective families only rather identical types also occur in very distantly related families. Therefore, Metcalfe and Chalk (1950) proposed the terms anomocytic, anisocytic, paracytic and diacytic instead of Ranunculaceous, Cruciferous, Rubiaceous and Labiatous or Caryophyllaceous respectively. Usually a particular stomatal type characterises a taxonomic group. But in case of *Anopyxis*, the same leaf bears paracytic, anisocytic and anomocytic stomatal complexes.

Function : Stomata are the gateway through which gaseous exchange takes place between the plants' interior and the atmosphere. When the stomata remain open CO_2 and O_2 are exchanged with the atmosphere and at the same time water is lost in the form of vapour from internal tissues of the leaf called transpiration. Carbon dioxide is required during photosynthesis and oxygen is required for respiration. If the photosynthetic by-product oxygen concentration becomes greater than atmosphere it diffuses out from the stomatal cavity to the atmosphere through the stomata.

Pathogens may infect the foliage through stomata. The stomata are very much sensitive to environmental changes. The stomatal size and frequency are reduced in dry and polluted air and these changes can be criteria for determining air pollution. Thus stomatal characterisctics of certain fern taxa can be used as ***biomonitors*** of air-pollution.

Salt glands : Salt glands store salts which may be exuded by guttation flow. These glands play an important role in the regulation of mineral contents of plants. They occur in the epidermis of several angiospermic families like Plumbaginaceae, Tamariaceae, Acanthaceae, Chenopodiaceae, Primulaceae, and Gramineae etc. They predominantly occur in mangrove plants (e.g., *Avicennia, Aegiceras* etc.). In contrast to epithem hydathodes these glands have no direct connection with the vascular elements and do not possess water pores. Salt glands may be two-celled (e.g., *Spartina*) or many-celled (e.g., *Acanthus, Aegiceras* etc.). In *Artiplex* the salt glands have unicellular or multicellular hairs with head bladder cell.

Function of epidermis

1. The thickly cuticularised epidermis gives protection to the inner tissues from different external forces and gives resistance to insect and pathogen attack.
2. Thick cuticle prevents plants from dessication.
3. The specialised bulliform cells of the epidermis help in water storage, unrolling of developing leaves and the opening and closing movements of mature leaves.
4. Velamen of orchid root and epiblema with root hairs help in absorption.
5. The epidermis, in some cases, store water and unwanted materials (salt glands) and food.
6. The wax deposition, suberisation, cutinisation etc. are the processes of secretion done by the epidermis.
7. The epidermal cells perform the function of perception of external stimuli and respond accordingly.

VASCULAR TISSUE SYSTEM

The vascular tissue system consists of the complex tissues, ***xylem*** and ***phloem***, which constitute discrete conducting strands called ***vascular bundles***. Each bundle is the isolated unit of the longitudinal strands of conducting tissues consisting essentially of xylem and phloem, frequently with a sheath of thick walled cells or other interspersed cells. These are usually primary in nature. In spite of the mechanical support

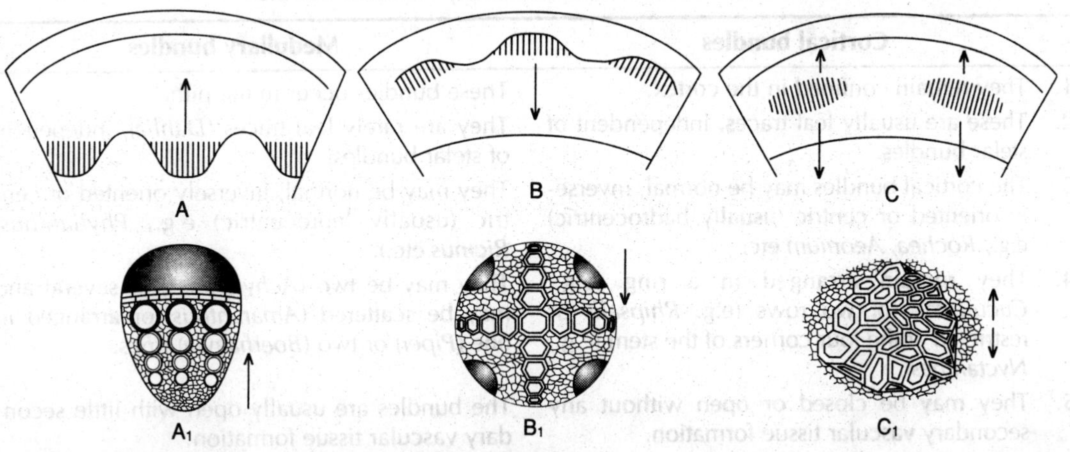

Fig. 5.80 : Diagrams showing order of development of primary vascular tissues in cross-section : A & A_1. Centrifugal development with endarch xylem. B & B_1. Centripetal development with exarch xylem. C & C_1. Both centripetal and centrifugal development with mesarch xylem

the vascular bundles primarily function in ***conduction*** — xylem for conduction of water with dissolved mineral matters, and phloem for conduction of prepared food matters in solution.

Occurrence : Every organ of the vascular plants possesses vascular bundles. In stems either they remain scattered throughout the ground tissue or may remain arranged in a compact ring in the stem axis to form the stele. In some abnormal cases bundles may also be formed in the cortex or pith to form ***cortical*** (e.g., members of Cactaceae, Oleaceae) and ***medullary bundles*** (e.g., *Dahlia, Piper* etc.), respectively. In roots the xylems and phloems remain radially arranged at the centre.

Origin : Vascular bundles originate from primary meristem. According to Hanstein and Schmidt ***plerome*** gives rise to the vascular bundles whereas the ***corpus*** gives rise to the non-vascular tissues. The procambial cells divide longitudinally and the peripheral derivatives form the phloem while the inner ones form the xylem (Haberlandt, 1914). The first cells of the xylem initials mature into ***protoxylem*** whose development may be centripetal or centrifugal, or both, in relation to the centre of the axis of the organ and are termed as ***exarch, endarch*** and ***mesarch,*** respectively. Exarch bundles are found in roots, endarch bundles are found in stems, while the mesarch bundles are found in stems of *Tmesipteris* and ferns. The xylem tissue formed later is termed as ***metaxylem***. The first matured phloem is termed ***protophloem*** and it develops acropetally in relation to the centre of the axis. The lately formed phloem is termed ***metaphloem.***

Types of vascular bundle : According to the different types of arrangements of the xylem and phloem tissues, vascular bundles are of different types. Following types of vascular bundes are found :

1. **Collateral bundle** : When xylem and phloem tissues remain on the same radius with reference to the centre point of the organ the phloem being external to the xylem — this type of vascular bundles are called collateral vascular bundle. Again, depending on the presence or absence of the cambium, the vascular bundles are of the following types :

(a) *Closed collateral bundle* : In this type of vascular bundles cambium is absent in-between xylem and phloem (Fig. 5.81) and, therefore, there is no further growth of the xylem and phloem tissues. Stems having this type of bundle do not have normal secondary growth as in monocotyledonous stems. Such closed vascular bundles usually remain enclosed within sclerenchymatous bundle sheath.

(b) *Open collateral bundle* : When cambium is present in between the xylem and phloem tissues it is called ***open collateral vascular bundle*** (Fig. 5.81). This cambium is called ***fascicular cambium***. The bundles can increase in diameter by normal secondary growth by the activity of the fascicular cambium. In dicotyledonous stem this type of vascular bundle is found.

Table 5.5 : Differences between cortical and medullary vascular bundles

	Cortical bundles	Medullary bundles
1.	They remain confined in the cortex.	These bundles occur in the pith.
2.	These are usually leaf traces, independent of stelar bundles.	They are rarely leaf traces (*Dahlia*), indepenent of stelar bundles.
3.	The cortical bundles may be normal, inversely oriented or centric (usually hadrocentric) e.g., *Rochea, Aeonium* etc.	They may be normal, inversely oriented or centric (usually leptocentric) e.g., *Phyllanthus, Ricinus* etc.).
4.	They remain arranged in a ring (e.g., *Cucurbita*), in two rows (e.g. *Rhipsalis* or restricted at the four corners of the stem (e.g., *Nyctanthes*).	They may be two (*Achyranthes*) or several and may be scattered (*Amaranthus*) or arranged in one (*Piper*) or two (*Boerhaavia*) rings.
5.	They may be closed or open without any secondary vascular tissue formation.	The bundles are usually open with little secondary vascular tissue formation.
6.	Translocation is their main function.	They give mechanical support to the stem.

Table 5.6 : Differences between protoxylem and metaxylem

	Protoxylem	Metaxylem
1.	The first formed xylem is the protoxylem.	The lately developed xylem is the metaxylem.
2.	In root it is exarch and in stem it is endarch.	It is just opposite of the protoxylem.
3.	The protoxylem vessels usually has annular, spiral and rarely reticulate wall thickenings.	It has spiral, reticulate and pitted secondary wall.
4.	The protoxylem vessels are of narrower diameter.	The metaxylem vessels have wider diameter.
5.	The elements of protoxylem usually get lost during the development of the organ.	The metaxylem elements do not get lost even after the development of the organ.
6.	No tylosis formation is observed.	Sometimes tylosis is formed.
7.	It is less compact as it contains few tracheids and vessels.	It appears more compact due to the presence of high proportion of lignified tracheids, vessels and fibres.

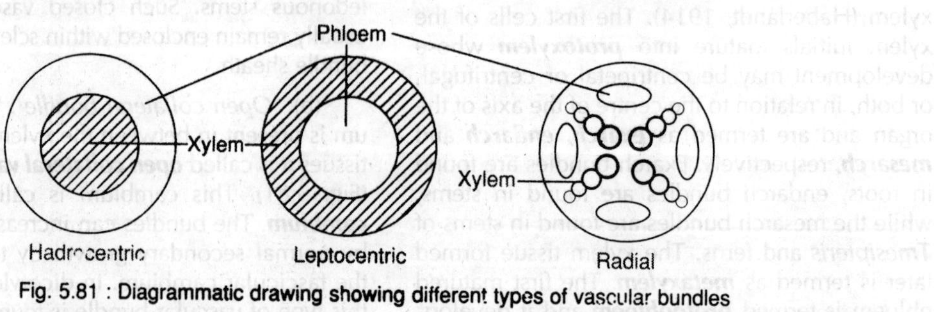

Fig. 5.81 : Diagrammatic drawing showing different types of vascular bundles

Table 5.7 : Differences between vascular bundles of monocot and dicot stem

	Vascular bundle of monocot stem	Vascular bundle of dicot stem
1.	The vascular bundles remain scattered throughout the ground tissue, the smaller ones being restricted towards the periphery.	The vascular bundles of more or less uniform size are arranged in a ring surrounding a central pith.
2.	Each vascular bundle is encircled by sclerenchymatous bundle sheath.	Bundle seath is lacking. Medullary parenchyma occurs on the side of the vascular bundle.
3.	There is no bundle cap.	In some cases sclerenchymatous bundle cap is present.
4.	Vascular bundles are conjoint, collateral or leptocentric and closed.	Each vascular bundle is conjoint, collateral or bicollateral and open.
5.	The protoxylem and metaxylem are arranged in the form of capital 'Y'.	The protoxylem and metaxylem remain arranged in the same radius.
6.	Phloem is situated in between and above the metaxylem.	Phloem is situated on the outer side of xylem.
7.	Usually a protoxylem cavity is formed.	There is no protoxylem cavity.
8.	There is no secondary growth as a cambium is absent.	Secondary growth is occurred by fascicular cambium.

2. **Bicollateral bundle** : If phloem tissue is situated on both the peripheral and central side of the xylem separated by strips of cambia in between is called **bicollateral vascular bundle** (Fig. 5.81). So the sequence of vascular tissues in bicollateral bundles from the periphery is outer phloem, outer cambium, xylem, inner cambium and inner phloem. These bundles are obviously open type as two strips of cambia are present but the secondary growth is achieved only by the outer cambium as in *Cucurbita* stem.

3. **Concentric bundle** : When either of the vascular tissues surrounds the other or, *vice versa*, it is known as ***concentric bundle***. This type of vascular bundles are of the following types :

(a) *Amphivasal bundle* : In this type of bundle the xylem encircles the central strand of phloem and is also called ***leptocentric vascular bundle*** (Fig. 5.81). This type of bundle is found in *Dracaena, Yucca* etc.

(b) *Amphicribral bundle* : In this type the phloem encircles the central strand of xylem. It is also known as ***hadrocentric bundle*** (Fig. 5.81) as found in Selaginella. The concentric bundles either amphivasal or amphicribal are closed, as there is no cambium in between the two tissues.

4. **Radial vascular bundle** : In this type of vascular bundle the primary xylem and phloem strands remain separated from each other by non-vascular tissues and are situated on alternate radii from the central point of the organ (Fig. 5.81). These bundles are the characteristic of roots. There is no primary cambium in this type of bundle and the secondary thickening may occur by the secondary cambium originating during secondary growth in dicotyledonous root only. In dicot roots there are usually 4–6 arcs of xylem in contrast to monocot roots where there are more than six xylem arcs.

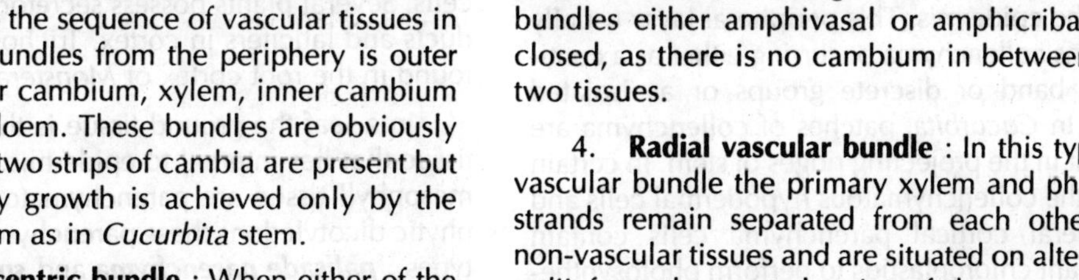

Fig. 5.82 : Diagrammatic drawing showing the different arrangements protoxylem in roots in cross section

GROUND OR FUNDAMENTAL TISSUE SYSTEM

This tissue system includes all the tissues excepting the epidermis and vascular bundles. Therefore, it is the largest tissue system beginning from the layer next to epidermis and continuing up to the centre point of the organ. Ground tissue system is heterogeneous in nature including diverse type of cells specialised for different types of function. The ground tissue outside the stele is called is the ***cortex*** or ***extrastellar ground tissue*** and that inside the stele is called ***intrastellar ground tissue*** or ***pith***. Both the ground tissues are further differentiated into different specialized zones to perform certain particular functions. Ground tissues are derived from the ground meristem of the embryo.

Cortex : The tissue internal to the epidermis and external to the vascular cylinder is the ***cortex*** consisting of various types of cells. The cortex is usually composed of thin walled parenchyma cells only (e.g., *Salicornia, Pelargonium* etc.). Sometimes the peripheral layers of it, immediately beneath the epidermis, form ***hypodermis*** with other types of cells of different origin from the epidermis. This peripheral layer usually contains collenchyma occurring either as a continuous band or discrete groups or as isolated strips. In *Cucurbita*, patches of collenchyma are present in the projecting ridges of stem. In certain cases the collenchymatous hypodermal cells and peripheral cortical parenchyma cells contain abundant chloroplastids to perform photosynthesis and are known as ***chlorenchyma***. In the cortex other types of cells like sclereids (e.g., *Trochodendron*), laticifers (e.g., *Calotropis*) etc. may also occur. The cortical parenchyma may contain starch, crystals etc.

In monocot stems all the tissues but vascular bundles and epidermis constitute the ground tissue without distinction between cortex and pith. In many genera of Liliaceae (e.g., *Asparagus*) and Amaryllidaceae, however, there is differentiation of cortex and stele with distinct endodermis and pericycle. In certain cases the outer layers of the cortex may be sclerified parenchyma or sclerenchyma. In *Asparagus* sclerenchymatous pericycle is present. The remaining portion of ground tissue is parenchymatous in which the vascular bundles are embedded. The vascular bundles in monocot stem usually remain scattered throughout the ground tissue and this type of distribution is termed as **atactostele**. But in many genera like *Coix, Triticum, Oryza* etc. they are arranged more or less in a ring.

The cortex of the root is mainly composed of parenchyma cells. In many monocots sclerenchyma develops in the root cortex in addition to parenchyma. Schizogenous and lysigenous intercellular spaces are frequently formed in the root cortex of some members of Gramineae and Cyperaceae. In many hydrophytes the air containing root cortical cells with regular arrangement form the ***aerenchyma***.

In *Iris, Smilax, Phoenix,* and *Citrus*, epiphytic orchids etc. uniseriate or multiseriate exodermis is present forming a kind of hypodermis. The exodermal cell walls are thickened with suberin and lignin to form casparian strips to protect the inner delicate cortical cells. The exodermal cells remain living even after maturity.

Collenchyma cells rarely found in root cortex of monocots e.g., *Monstera*. The root cortex is usually very wide and the cells often contains starch, crystals etc. The aerial roots of many epiphytes contain chloroplastids in the cortical cells. Several plants possess secretory cells, resin ducts and laticifers in cortex. Trichosclereids are found in the root cortex of *Monstera*.

In a leaf the ground tissue is the mesophyll tissue that lies internal to epidermis. The cells of mesophyll tissue are parenchymatous. In mesophytic dicotyledons these parenchyma are of two types : ***palisade parenchyma*** and ***spongy parenchyma***.

The palisade parenchyma cells are cylindrical, elongated or rod shaped packed densely in one or more rows. Each palisade cell contains numerous chloroplasts. The palisade cells usually remain on the adaxial side of the leaf beneath the epidermis. Leaf of *Thymelaea* is an exception where there is abaxial palisade only. Again in *Dianthus, Atriplex,* and *Artemisia* etc. palisade parenchyma is present on both adaxial and abaxial sides whereas the strips of spongy parenchyma occur in the middle. A leaf is said to be ***isobilateral*** when the palisade parenchyma occurs on both adaxial and abaxial surface. When the palisade parenchyma occurs only on the adaxial surface of the leaf, it is said to be ***dorsiventral***. Monocotyledonous leaves belong to this category. The cylindrical leaves of *Hakea* contain palisade tissue along the periphery.

The spongy mesophyll occurs in the abaxial surface of a dorsiventral leaf and may be of various shapes. They may be elongated resembling the palisade or spherical or irregular. The cells contain abundant chloroplastids and are arranged in such a manner that there occur large intercellular spaces.

Endodermis : The innermost layer of the cortex or the outermost layer of stele is the ***endodermis***. It is the delimiting layer of cortex from the stele. This layer is inconspicuous in aerial stems except in *Piper* where casparian strips are also observed. Endodermis is prominent in underground stems. It usually surrounds the entire stele and in certain exceptional polystelic cases it encircles individual vascular strand (e.g., *Nymphaea*). The endodermal cells of many dicotyledonous stems contain abundant starch and such endodermis is called **starch sheath** which in some stems may be transformed into endodermis having casparian strips.

An endodermis consists of a layer of compactly set living cells. Lignin and suberin, in addition to cellulose, are deposited on the transverse and radial walls of endodermal cells as strips or bands called ***Casparian strip*** (Fig. 5.83A).

In the roots also the endodermis is uniseriate with compactly arranged cells with casparian strips in young cells. The casparian strip is a narrow or wide band of suberin running around the radial and cross walls of the endodermal cells. In roots (e.g., *Iris*) without secondary growth suberin deposition takes place as lamellae over the whole inner primary wall of the endodermal cells including the casparian strips. Cellulose is then deposited centripetally on the inner side of suberin lamella to make the radial walls of the endodermal cells considerably thick, which may be further thickened by lignification. Suberin deposition starts initially in the endodermal cells adjacent to the phloem strands. It then spreads radially to the other adjacent endodermal cells except those opposite to the protoxylem. These thin-walled cells are known as ***passage cells*** (Fig. 5.83E) through which water and dissolved substances move from cortex to xylem vessels. The casparian strips prevent such passage of materials. Plasmodesmatal connections exist between cortical and endodermal cells but it is absent in the region of casparian strips.

Function : The endodermis performs the following functions :

1. It accumulates starch, protein granules, fats, and tannins in some cases.
2. The suberised cell wall of endodermis acts as a barrier for the movement of heavy metals into or out of vascular tissues.

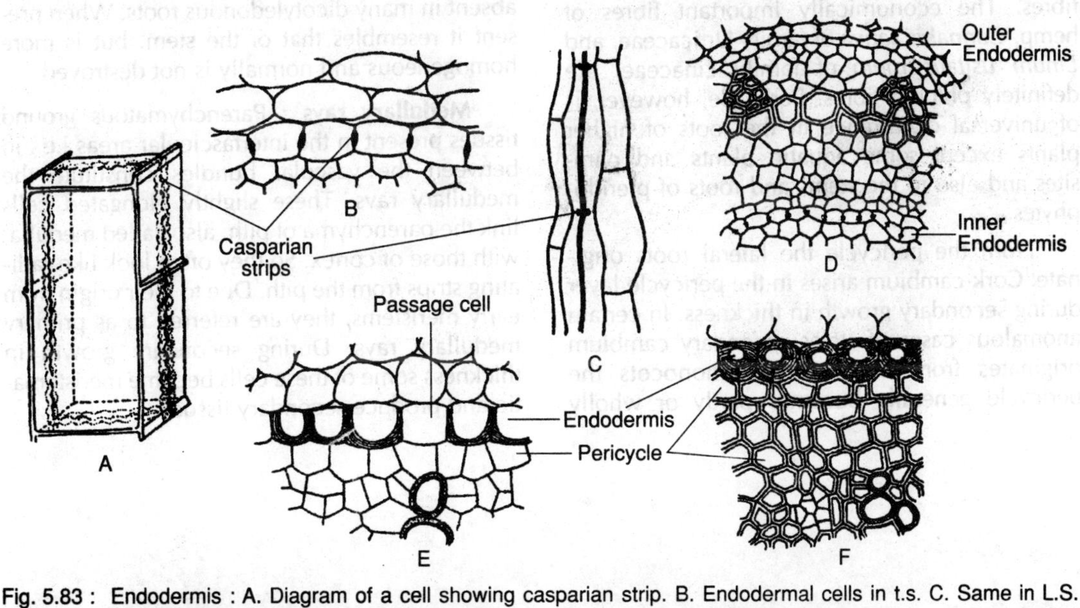

Fig. 5.83 : Endodermis : A. Diagram of a cell showing casparian strip. B. Endodermal cells in t.s. C. Same in L.S. D. Transverse section of the rhizome of *Marsilea* showing the outer and inner endodermis. E. Thick-walled endodermis in root of maize with passage cell. F. Same in the root of *Smilax*

3. It secretes water into the adjacent xylem vessels as it possesses more positive water potential value than the neighbouring cortical cells.
4. It controls the conduction of water between the stele and cortex in rhizomes.
5. It forms a barrier for the pathogens as the cells of the endodermis contain large amounts of benzoquinones, napthaquinones and anthroquinones that inhibit the growth of fungi and bacteria.
6. In *Cocculus, Paederia foetida* stem the cork cambium originates from endodermis.
7. Endodermis may behave as a meristem termed ***proendodermis*** that divides to form a part of the cortex.

Pericycle : The pericycle is the region consisting of one or more layers of cells immediately inner to the endodermis and surrounding the vascular tissues. It is regarded as the limiting layers of stele forming the outermost part of intrastellar ground tissues. The other portions of the stellar ground tissue are the pith and medullary rays. The pericycle typically consists of parenchymatous cells as found in all roots and pteridophytes. Heterogeneous pericycle with parenchyma and discrete bands of sclerenchyma fibres may also be formed. But modern anatomists have shown that most of the so-called pericyclic fibres of stems are really phloem fibres. The economically important fibres of hemp, *Cannabis sativa* of family Urticaceae, and *Linum usitatissimum* of family Linaceae, are definitely phloem fibres. Pericycle, however, is of universal occurrence in the roots of higher plants except some aquatic plants and parasites and also in the stems and roots of pteridophytes.

From the pericycle the lateral roots originate. Cork-cambium arises in the pericycle layer during secondary growth in thickness. In certain anomalous cases of steles secondary cambium originates from this layer. In monocots the pericycle generally becomes partly or wholly sclerified. The pericycle may also contain laticiferous tissues and secretory ducts.

Pericycle is usually single-layered in the roots of angiosperms but in some cases like monocotyledons (e.g., *Smilax, Agave* of family Liliaceae, palms and some grasses) and dicotyledons like *Morns* of family Moraceae, it may be multilayered. The pericycle of gymnosperms is typically multiseriate.

Pith : Pith is the main internal ground tissue located at the centre of the organs. Very often it is quite large parenchymatous tissue with profuse intercellular spaces. The pith cells are usually isodiametric and sometimes remain arranged in longitudinal series due to their development from rib meristems. The cellulosic thin-walled cells usually contain colourless leucoplasts. The pith cells may store starch, fatty substances and often crystals and tannins. Secretory tissues are frequently present. Fibres and sclereids may also rarely occur. Sometimes the outer pith cells are smaller with thicker walls and dense protoplasm and thus become morphologically distinct from the rest. This distinct outer region has been called ***perimedullary zone*** or ***medullary sheath***. In many herbaceous plants the pith is destroyed during rapid elongation accompanied by radial expansion of the stem. In the family Cucurbitaceae and many grasses ***hollow pith*** with broken walls lining the cavity may be formed. Pith is absent in many dicotyledonous roots. When present it resembles that of the stem, but is more homogeneous and normally is not destroyed.

Medullary rays : Parenchymatous ground tissues present in the interfascicular areas i.e., in between the vascular bundles constitute the medullary rays. These slightly elongated cells link the parenchyma of pith, also called medulla, with those of cortex. So they often look like radiating strips from the pith. Due to their origin from early meristems, they are referred to as primary medullary rays. During secondary growth in thickness some of these cells become meristematic and produce secondary tissues.

ROOT-STEM TRANSITION

The change in orientation of vascular tissues from exarch radial type in root to endarch and conjoint collateral type in stem is designated as root-stem transition.

The continuity between the vascular strands of roots and stems is maintained but the orientational change takes place through splitting, twisting and inversion of xylem strands at the very short transition zone which is neither a root nor a stem. Transition occurs at the hypocotyl region either gradually or abruptly. There exist two main concepts on the nature of root-stem transition.

According to Eames and MacDaniels (1947) there are three types of root-stem transitions in dicot and single type in monocot. In the diagram (Fig. 5.84) the row of circles at the base represents the transverse sections of roots while the uppermost row is the transverse sections of stems. The three intermediate rows are the successive transverse sections through transition zone.

Type I (Fig. 5.84A) : In this type there are four strands of xylem and phloem in the root and stem. The phloem remains in same position in both root and stem. Each xylem strand of root forks by radial division into two branches and gradually at the upper region they swing laterally to opposite directions by 180° and join the

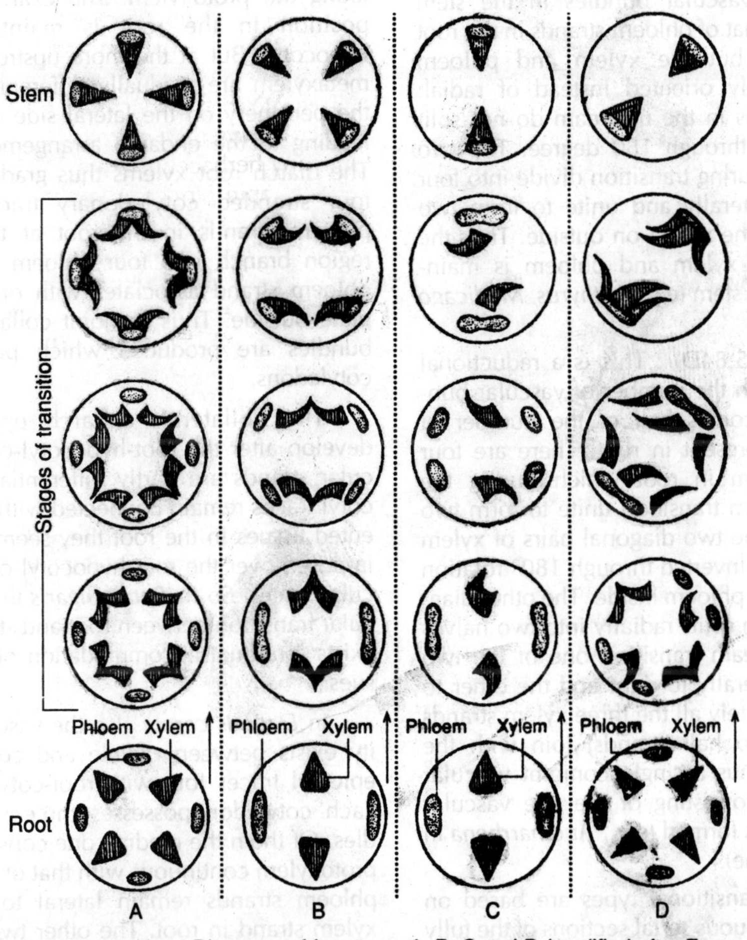

Fig. 5.84 : Root-stem transition : Diagrams of four types A, B, C and D (modified after Eames and MacDaniels)

phloem strand in between. Thus the two xylem halves are inverted and become conjoint and endarch in stem. In this way the radial bundles and exarch xylem become conjoint collateral with endarch (e.g., *Mirabilis, Fumaria* and *Dipsacus*) xylem orientation.

Type II (Fig.5.84B) : In this second type the number of bundles in stem is twice than in root. In root there are two alternate xylem and two phloem strands. Phloem in its upstream becomes elongated and dumb-bell shaped due to constriction at the middle. In the still upper region the constriction becomes deep and ultimately the phloem strand is divided into two. Thus the number is doubled to four in the stem. The two xylem strands also split and become inverted like type I. The four xylem halves thus produced join with the phloem inside to become conjoint (e.g., *Cucurbita, Tropaeolum, Acer* and *Phaseolus*).

Type III (Fig.5.84C) : In this type, like type I, the number of vascular bundles in the stem remain same as that of phloem strands in the root after transition, but the xylem and phloem become conjointly oriented instead of radial. The xylem strands in the upstream do not split but get inverted through 180 degree. The two phloem strands during transition divide into four strands, swing laterally and unite to form two strands that join the xylem on outside. Thus the same number of xylem and phloem is maintained in root and stem (e.g., *Lathyrus, Medicago* and *Phoenix*).

Type IV (Fig.5.84D) : This is a reductional transition in which the number of vascular bundles in stem becomes half of the number of phloem strands present in root. There are four strands of phloem in root which during the course of root-stem transition unite to form two strands. One of the two diagonal pairs of xylem does not split but inverted through 180° rotation and joins with the phloem inside. The other diagonal pair of xylem splits radially into two halves and during upstream transition one of the two halves swings laterally to right and the other to the left and ultimately all the three xylem strands (one full and two half strands) join with the phloem inside. Thus a single conjoint vascular bundle of stem consisting of the five vascular elements of root is formed (e.g., *Anemarrhena* in the family Liliaceae).

The above transitional types are based on the study of continuous serial sections of the fully differentiated transitional zone. A completely different picture is offered by the developing transitional zone. According to Esau (1965) during ontogeny, vascular continuity is established between root and cotyledon through hypocotyls for the first time. After differentiation the epicotyl traces are superimposed over the root-hypocotyl-cotyledon traces and the tissues present between them are mutually accommodated. This view is now considered as recent concept of root-stem transition. The vascular transition of *Beta vulgaris* and *Daucus carota* has been cited based on ontogeny in favour of this recent concept.

There is an epicotyl shoot in between the two cotyledons of the *Beta vulgaris* seedling and the hypocotyl and root on the lower side. Anatomically, the root shows two exarch xylems and two phloem strands arranged alternately in radial fashion. The cotyledon traces consist of two vascular bundles that are partially fused along the protoxylem. The exarch protoxylem position in the root is maintained up to hypocotyl. But at the more upstream level two metaxylem are gradually differentiated towards the periphery on the lateral side of protoxylem leading to the endarch arrangement of xylem. The diarch root xylems thus gradually become four stranded cotyledonary traces. The two phloem strands in the root at the hypocotyl region branch into four phloem strands. Each phloem strand associates with one metaxylem plate outside. Thus conjoint collateral vascular bundles are produced which persists in the cotyledons.

The collateral, endarch epicotyl traces develop after the root-hypocotyl-cotyledon vascular strands are partly differentiated. The epicotyl traces remain connected with similarly oriented tissues in the root they seem to be superimposed over the root-hypocotyl-cotyledon vascular traces. So in *Beta vulgaris* there is no vascular transition between root and stem. But there exists a mutual accommodation of vascular tissues.

In *Daucus carota* also the vascular continuity exists between radicle and cotyledon. The epicotyl traces join with root-cotyledon traces. Each cotyledon possesses three vascular bundles. Of them the median one consists of exarch protoxylem continuous with that of the root. Two phloem strands remain lateral to the median xylem strand in root. The other two lateral vascular bundles are collateral with endarch xylem. These two bundles originate from the diarch xylem of root. So the primary vascular tissues of cotyledon and root remain unaltered. The epicotyl traces develop later that are conjoint and collateral and they ultimately join with radicle-hypocotyl-cotyledon traces.

In conclusion, it may be concluded that the ontogenetic studies of vascular transition of *Beta* and *Daucus* reveals no root-stem transition as illustrated by Eames and MacDaniels (1947) in the earlier common concept. The modern concept, which is based on developmental studies, reveals that during ontogeny vascular connections are established between radicle and cotyledon via hypocotyl. The epicotyl traces are developed later from procambial strand, which ultimately join with the fully differentiated radicle-hypocotyl-cotyledon unit and the tissues between the traces are accommodated mutually.

Significance

In the transition zone the internal tissue arrangement is completely different from that of root and stem. Thus, this region does not belong to any of those categories. Several interpretations are available to describe the structure and evolutionary significance of the transition zone.

The old and most common concept is that the seedling has a unit vascular system which is morphologically equivalent to root and stem. The anatomical differences in this zone are due to splitting, twisting, rotation, inversion etc. According to this concept the protoxylem, metaxylem and phloem differentiate in the same positions where they occur in the mature state. At the transition zone, these vascular elements are differentiated in such a way that they can maintain the continuity of unit vascular system.

According to the other concept there is double origin of the vascular system. There are two discontinuous parts of the vascular system in the seedling. One is the radicle–hypocotyl part and the other is the cotyledon part. These two parts of the vascular system are joined at the upper region of hypocotyl. The root and shoot tips of the seedling have their own meristem which forms the root and stem, respectively. During development, vascular connections are established between radicle and cotyledon via hyopocotyl. The epicotyl traces are joined with the radicle-hypocotyl-cotyledon unit and tissues between the traces are mutually accommodated (Esau, 1965).

The radial arrangement of xylem and phloem is usually considered as primitive whereas the collateral arrangement of them is regarded as advanced. Therefore, the xylem and phloem arrangements at the upstream transition zone represent the different evolutionary stages which culminate into collateral arrangements in the stem.

THE PRIMARY TISSUES

The higher vascular plants possess a large variation in the arrangement and structure of primary tissues in their roots, stems and leaves.

Primary tissues in dicotyledonous stem

Anatomically, the stem possesses the dermal tissue system which forms the epidermis, vascular tissue system which forms the vascular bundles and the ground tissue system which includes rest of the tissues found in the stem. All these primary tissues show a great diversity in their structure and arrangement. A brief account of the tissues found in mature dicot stems are discussed below in basipetal order.

Epidermis : Epidermis is the peripherial protective cell layer in the stem the details of which is discussed in the chapter 'Tissue System'. The compactly arranged cells of this layer are more or less tubular or rectangular with thin primary walls. The outer cell surface of this layer is often covered by a layer of cuticle. In some cases the stem contains chlorenchyma cells to perform the function of photosynthesis and, in that case, the epidermis is interrupted by the presence of stomata. Glandular and non-glandular hairy outgrowths are formed on the epidermis.

Cortex : The tissue in-between the epidermis and the stele is the cortex which may contain various types of cells. It may contain thin-walled parenchyma cells only (e.g., *Salicornia, Pelargonium* etc.) or other types of cells immediately below the epidermis to form the hypodermis. The cells of this layer differ in origin from the epidermis and structure from the cortical cells below. The hypodermis usually contains peripheral collenchyma cells either as a continuous band or in groups. Groups of collenchyma occur at the ridges of *Cucurbita* stem. The photosynthesizing stems contain abundant chloroplastids in the peripheral collenchyma and cortical parenchyma. The hypodermis contains sclereids in *Trochodendron*, oleoresins in *Zingiber*, laticifers in *Calotropis*. The parenchymatous cells of the cortex may contain starch, crystals etc.

Endodermis : It is the inner delimiting layer of cortex. This layer is inconspicuously demarketed in aerial stems with exception in *Piper* where casparian strips are also observed. But it is usually prominent in underground stems. The entire vascular cylinder usually remains surrounded by the endodermis but in polystele it encircles the individual vascular bundles as in *Nymphaea*. Endodermis in some cases develops casparian strips and *Phaseolus* stem contains abundant starch to form the starch sheath. Endodermal cells deposit lignin and suberin in addition to cellulose on their transverse and radial walls as strips or bands to form ***casparian strip*** or ***band***.

Function of endodermis : Endodermis performs the following functions :

1. Fat, starch, protein granules and tannins are accumulated in endodermal cells.
2. The casparian strips of the endodermal cells restrict the movement of water, iron, lead and copper in either direction.
3. The endodermal cells form a barrier for pathogens as it contains large amount of quinones that inhibit the growth of fungi and bacteria.
4. In *Cocculus, Paederia foetida* stem the cork cambium or phellogen originates from endodermis.
5. Endodermis may behave as meristem termed ***proendodermis*** the cells of which divide to form a part of the cortex.

Stele : The central core of the vascular and ground tissue systems encircled by the endodermis is the stele. The parenchymatous pith or medulla forms the ground tissue. The pith cells sometimes may be lignified and pitted. The vascular tissues lie in between cortex and pith in dicots as a continuous or split cylinder. The vascular bundles are conjoint and collateral in which the xylem and phloem lie on the same radii and the phloem remains at the external side. The vascular bundles remain separated by interfascicular parenchyma. In some members of the families Solanaceae, Cucurbitaceae, Apocynaceae etc. phloem is present on both the inner and ouer side of xylem. It is called ***bicollateral vascular bundle***. The internal phloem is also termed ***intraxylary phloem***. The phloem strand may also occur as isolated strands in close contact on the inner side of xylem of a vascular bundle (e.g., *Calotropis, Strychnos*). In dicot stem the collateral and bicollateral bundles are always open with cambium in-between xylem and phloem and all the bundles remain arranged more or less in a ring and lie within a common endodermis to form the ***monostele*** structure. But in *Nymphaea* ***polystele*** is formed due to enclosure of each vascular strand by an endodermis. In some genera, medullary and cortical bundles may also occur.

Primary tissues in monocotyledonous stem

Monocotyledonous stems also contain the dermal, vascular and ground tissues systems. The dermal tissue system forms the epidermis, vascular tissue system consists of vascular bundles and the ground tissue system consists mainly of parenchyma. A brief account of the tissues found in mature monocot stems are discussed in basipetal order :

Epidermis : The epidermis of monocot stem is more or less like that of dicots. In contrast to dicot stem, the monocot stem epidermis is usually devoid of glandular and non-glandular trichomes.

Ground tissue : In monocot stems all the tissues except the epidermis and vascular bundles are described as ground tissue as there is no distinction between cortex and pith. In certain genera of Liliaceae (e.g., *Asparagus*) and

Amaryllidaceae, however, there is distinction between cortex and stele with distinct endodermis and sclerenchymatous pericycle. The external layers of the ground tissues adjacent to epidermis may be composed of sclerified parenchyma or sclerenchyma.

The vascular bundles are usually scattered throughout the ground tissue to form the atactostele. In *Coix, Triticum, Oryza* etc. they are, however, arranged more or less in a ring. Individual vascular bundle may be leptocentric or amphivasal (e.g., *Dracaena, Aloe* etc.) or collateral and closed (e.g., *Zea, Asparagus* etc.). In the collateral bundles the metaxylem and protoxylem vessels are arranged in such a way that they appear in the form of the capital 'Y' or 'V' or 'U' in cross section. The large metaxylem vessel elements form the arms of the alphabets and the protoxylem compose the rest. In some genera, protoxylem cavity or lacuna or canal is formed due to disintegration of the protoxylem. The phloem remains in between the two arms of Y- or V- or U-shaped xylem. A sclerenchymatous bundle sheath usually encircles each vascular bundle.

Arrangements of primary tissues in roots

A cross-section of root reveals the following tissues :

Epiblema : The dermal tissue system is called epiblema or rhizodermis as it differs from that of the shoot in origin, structure and function. The cells of this layer are elongated, thin-walled with a thin cuticle on the outer cell wall and devoid of intercellular spaces. Sometimes mucilage may occur superficially. In the long lasting parts of the root system the cells of the epiblema may become thick-walled due to the deposition of lignin, cutin or suberin.

The most notable feature of the epiblema is the presence of unicellular and unbranched root hairs that are projections of certain cells of this layer. They perform the function of anchorage as well as absorption of water and mineral salts from the soil solution. The root hairs are formed from some special epidermal cells that can be distinguished from the rest by their smaller size and dense cytoplasm. These specialised cells are known as ***trichoblasts*** or ***piliferous cells***. For that reason the epiblema is also termed as ***piliferous layer***. Root hairs are usually short-lived except in some species of Asteraceae. Multicellular root hairs are observed in the aerial adventitious root of *Kalancho fedtschenkoi*.

The epiblema is typically single-layered with exception in the aerial roots of certain epiphytes where it is multi-layered. This multiple epidermis is specially termed as ***velamen***, the cells of which are derived from the dermatogen. This layer contains several layers of dead cells externally forming a sheath around the root. The cell walls of the compactly arranged cells are strengthened by spiral or reticulate bands of lignin. The walls also contain many primary pit fields.

The innermost layer of the velamen is known as ***exodermis***. This layer demarcates the thick-walled cells of velamen from thin-walled cells of cortex. The exodermis is considered as the outermost layer of the cortex as it is derived from the periblem and not from the dermatogen that gives rise to velamen. In exodermis long and short cells remain arranged alternately of them the smaller ones are thin-walled and are known as **passage cells**. The longer cells are thickened on their tangential and radial walls only. Thus exodermis may be comparable with endodermis of root. Uniseriate or multiseriate exodermis is also observed in *Iris, Smilax, Phoenix,* and *Citrus* etc. just beneath the epidermis.

Velamen protects the delicate thin-walled cortical cells from desiccation. The velamen and exodermis are almost impermeable to water and solutes.

Cortex : The root cortex is mainly parenchymatous with schizogenous and lysigenous intercellular spaces and in many monocotyledons sclerenchyma may develop in addition to parenchyma. In hydrophytes the root cortical cells are very regularly arranged to form aerenchyma. The root cortex is usually very wide and the cells often contains starch, crystals etc. Collenchyma rarely occurs in roots cortex (e.g., *Monstera*) and the cortical cells of many epiphytes contain chloroplastids. Several plants possess secretory cells, resin ducts and laticifers in cortex. Trichosclereids are found in the root cortex of *Monstera*.

Endodermis : The innermost layer of the cortex is the endodermis that encircles the stele. The cells of this uniseriate layer are compact and contain casparian strips in their radial and cross walls when young. The strip may be narrow, thread-like or may be as wide as the wall itself. In

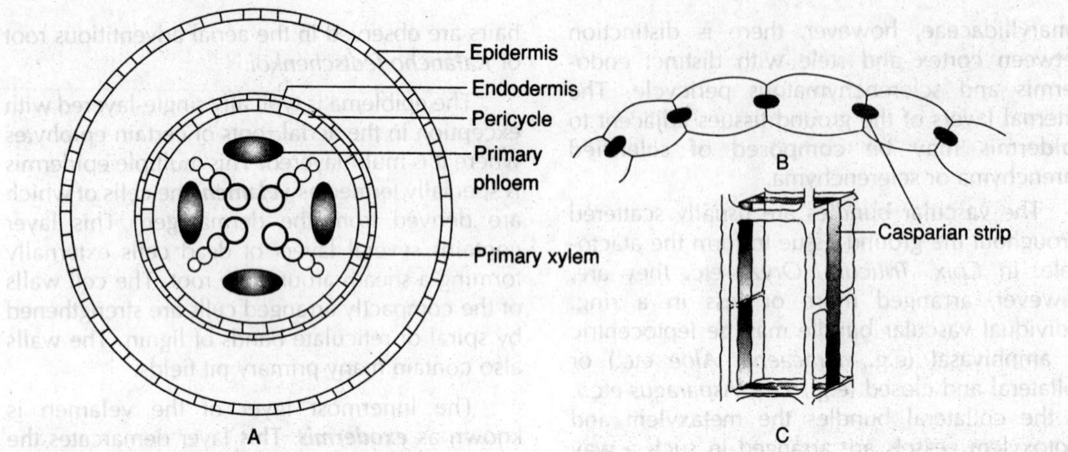

Fig. 5.85 : A. Shematic diagram of root in T.S. B. Endodermal cells with Casparian strips, C. Anenlarged single endodermal cell shwing Casparian strip

roots devoid of secondary growth suberin deposits as lamellae over the whole inner primary wall of the endodermal cells including the casparian strips and afterwards cellulose is deposited centripetally on the inside of suberin lamella (e.g., *Iris*). Thus the endodermal cells attain considerable thickness on their radial walls. The inner radial and tangential walls become further thickened due to lignification. Suberisation starts in the endodermal cells adjacent to the phloem followed by the deposition on adjacent endodermal cells. In the protoxylem adjacent endodermal cells thickening is either delayed or there is no thickening except the casparian strip. These cells are known as ***passage cells*** which may remain thin-walled throughout or may become thick-walled later. Through the passage cells water and mineral matters get entry into the xylem from the cortical cells. The main functions of the root endodermis are more or less same as that of stem.

Stele : The stele consists of radial vascular bundles and pith. Just beneath the endodermis and external to the vascular tssues there is a cell layer called ***pericycle***. This layer is usually uniseriate but sometimes may be multiseriate (e.g., *Agave, Smilax, Salix* etc.) containing thin-walled parenchyma and occasionally fibres. The pericycle is sometimes termed as pericambium as it retains its meristematic activity. It can give rise to the lateral root primordia, vascular cambium that giveş rise to secondary vascular tissues in dicots and the phellogen – the cork meristem. Laticifers and secretory ducts may sometimes be present in pericycle. Sclerified pericycle cells may be found in monocots.

The primary xylem and phloem remain alternately and radially arranged beneath the pericycle. The xylem contains thick-walled, lignified tracheary elements that mature centripetally. Protoxylem lies on the periphery and metaxylem towards the center and, therefore, the xylem is exarch. Centripetal differentiation is also observed in phloem. In many dicots the xylem strands extend towards the centre to form a star-shaped solid central core. In most monocotyledons, however, there is large parenchymatous or sclerenchymatous pith in the center. The number of xylem strands with protoxylem groups varies from 1 to many (Fig. 5.82) and termed as ***monarch*** (e.g., *Trapa natans*), ***diarch*** (e.g., *Beta, Daucus, Raphanus* etc.), ***triarch*** (e.g., *Pisum*), ***tetrach*** (e.g., *Vicia, Ranunculus* etc.), ***pentarch*** and ***polyarch*** (e.g., monocotyledonous root) respectively. Usually the number of xylem arcs is constant for a plant but exceptions are noted in *Libocedrus decurrens* where the root may be diarch, triarch, pentarch and hexarch. The aerial root of *Tinospora cordifolia* also has tri-, tetra-, or pentarch xylem. In polyarch roots the number of xylem strands may be as high as 100 or more (e.g., Palmae). In *Triticum* a single large metaxylem vessel occupies the centre and remains separated from the peripheral protoxylem by parenchyma. A number of meta-xylem vessels remain arranged in a circle around the pith as in *Iris, Zea* etc.

Arrangements of tissues in leaves

Leaves also consist of three main tissue systems — the dermal tissue system, the ground tissue system, and the vascular tissue system. The dermal tissue system consists of the upper and lower epidermis, the ground tissue system comprises the mesophyll and the vascular tissue system is represented by the vascular bundles. In cross-section a leaf shows — (i) the upper epidermis; (ii) lower epidermis, and (iii) mesophyll tissue in-between with (iii) embedded vascular bundles.

Epidermis : It is the uniseriate outermost layer. In *Ficus, Piper*, and *Nerium* the epidermis is multiseriate consisting of 2 – 16 cell layers which is a xerophytic characteristic and protects the inner tissues from desiccation. In cases where both the epidermises are multiseriate the upper one consists of more layers than the lower one. In multiple lower epidermis a sub-stomatal cavity is present (e.g., *Piper*) in the mesophyll tissue beneath stoma. In *Nerium* the lower epidermis is multiseriate. The stomata are present in stomatal chambers only where the epidermis is uniseriate. These stomata are called ***sunken stomata***. Stomata may occur either on the upper epidermis (e.g., *Nymphaea*) or on the lower epidermis (e.g., *Ficus*) or on both surfaces. They may remain at the same level of the other epidermal cells or may be raised above (e.g., *Hakea*) or sunken (e.g., *Nerium*).

Usually the epidermal cells are achlorophyllous except the guard cells. In the aquatic plants like *Ranunculus, Phyllospadix* etc. chloroplastids occur in the epidermal cells. Glandular or non-glandular hairs, crystals, cystoliths etc. may occur in the epidermis. The epidermal cell walls usually contain cutin, occasionally lignin and wax.

Mesophyll : It is the parenchymatous ground tissue of leaf lying internal to the epidermis. In dorsiventral leaves these parenchyma are of two types : **palisade parenchyma** and **spongy parenchyma**.

In leaf t.s. the palisade cells appear as elongated cylindrical or rod structures densely packed in one or more rows. In l. s. they appear as separated or slightly attached spherical structures. These cells are laden with chloroplasts. Sometimes a hypodermis remains in between the epidermis and palisade cells. Though palisade cells occur on the dorsal side only, the leaves of *Thymelaea* possess palisade cells on ventral side only. In the leaves of *Dianthus, Atriplex*, and *Artemisia* etc. palisade parenchyma is present on both sides and the spongy parenchyma remains in between in the form of strips. If the palisade parenchyma occurs on the dorsal surface only

Table 5.8 : Differences between dicotyledonous stem and root

Dicotyledonous Stem	Dicotyledonous Root
1. Epidermis has cuticular covering, stomata, hairs, scales etc.	Epiblema is devoid of cuticular layer and stomata but contains unicellular root hairs.
2. It contains comparatively narrower parenchymatous cortex with collenchymatous hypodermis.	Cortex is usually parenchymatous but wider than stem.
3. The stele is siphonostele, dictyostele.	The stele is radial.
4. Pericycle is multiseriate consisting of sclerenchyma and parenchyma.	Pericycle is uniseriate consisting of parenchyma only.
5. The vascular bundles are conjoint collateral or bicollateral, open and arranged more or less in a ring.	The vascular bundle is radial.
6. Protoxylem is endarch and phloem occurs on the periphery or on both inner and outer side of xylem.	Protoxylem is exarch and phloem alternates with xylem which may be monarch to hexarch.
7. Pith is well developed.	Pith is either ill-developed or absent.
8. Intrastelar secondary growth occurs with the help of cambium ring that is partly primary and partly secondary.	Intrastelar secondary growth occurs with the help of wavy cambium ring that is secondary in origin.

the leaf is termed as ***dorsiventral*** whereas a leaf is said to be ***isobilateral*** when the spongy parenchyma occurs on both surfaces. Usually dicot leaves are dorsiventral and monocot leaves are isobilateral. The cylindrical leaves of *Hakea* have palisade tissue situated all round the periphery.

The spongy parenchyma occurs in the ventral surface of a dorsiventral leaf and may be variously shaped — elongated, spherical or irregular. The chloroplastid-rich spongy cells are arranged in such a manner that there occur large intercellular spaces among the cells to give a spongy appearance.

Vascular Bundles : Vascular bundles are situated at the center (mid rib) with inverse orientation of vascular tissues in contrast to stem. Xylem occurs on adaxial side and phloem on the abaxial side. Vascular bundles are sheathed by parenchyma cells morphologically distinguishable from the adjacent mesophyll cells in some monocotyledons like *Cyperus, Saccharum, Zea* etc. These parenchyma cells are usually larger, thick-walled and may contain fewer chloroplasts. This parenchyma sheath is termed as ***bundle sheath***. These cells are concerned in conduction and storing of food materials. The bundle sheath is regarded either as endodermis due to occasional occurrence of casparian strips or as starch sheath as they sometimes contain starch.

Table 5.9 : Differences between dicotyledonous stem and monocotyledonous stem

Dicotyledonous Stem	Monocotyledonous Stem
1. Anatomically, dicot stem is differentiated into epidermis, cortex and stele.	It is differentiated into epidermis and cortex in which vascular bundles remain scattered.
2. The epidermis bears multicellular hairs.	The epidermis is devoid of hairs but may contain stomata.
3. The cortex is parenchymatous with peripheral collenchymatous hypodermis.	The cortical parenchymatous ground tissue has peripheral hypodermis containing sclerenchyma or sclerified parenchyma.
4. The innermost layer of cortex forms the endodermis.	No endodermis is formed as the stele is atactostele.
5. Next to endodermis there occurs pericycle inside.	There is no pericycle.
6. The stele is siphonostelic dictyostele in which the vascular bundles are arranged in a ring around the central pith.	The stele is atactostele in which the vascular bundles are scattered throughout the ground tissue and thus devoid of pith.
7. The vascular bundles are more or less uniform in size.	The vascular bundles are of different sizes, the smaller ones being restricted to the periphery.
8. Each vascular bundle is conjoint, collateral and open.	Each vascular bundle is conjoint, collateral and closed.
9. Sclerenchymatous bundle cap occurs over each vascular bundle.	Each vascular bundle is encircled by a sclerenchymatous bundle sheath.

Table 5.10 : Differences between monocotyledonous stem and root

Monocotyledonous Stem	Monocotyledonous Root
1. The stem is differentiated into epidermis and ground tissue with scattered vascular bundles in it.	It is differentiated into epiblema, cortex and stele.
2. Epidermis is uniseriate with cuticle and stomata.	Epiblema is uniseriate or multiseriate with unicellular root hairs but lacks cuticle and stomata.

Contd.

Contd. 5.10

Monocotyledonous Stem	Monocotyledonous Root
3. Ground tissue is parenchymatous with hypodermis containing sclerenchyma or sclerified parenchyma.	Cortex is parenchymatous with sclerenchymatous (stilt root of maize) or collenchymatous (*Monstera*).
4. The stele is atactostele without endodermis.	The stele is radial encircled by endodermis.
5. There is no pericycle.	Pericycle is uniseriate.
6. The vascular bundles are leptocentric, conjoint, collateral, closed and numerous.	The vascular bundles are radial and xylem is polyarch.
7. Protoxylem is endarch.	Protoxylem is exarch.
8. Protoxylem and metaxylem form the shape of the capital 'Y'.	Protoxylem and metaxylem remain in the same line.
9. A protoxylem cavity is present beneath the protoxylem.	No such cavity is formed.
10. Pith is absent.	Large pith is present in the centre.

ANATOMY OF SOME DICOTYLEDONOUS STEMS

Sunflower stem

Transverse sections through the internode of a young sunflower stem (*Helianthus annuus*) exhibits the following tissue arrangements from the periphery to the centre (Fig. 5.86 & 87).

Epidermis : It is the outermost uniseriate zone, consisting of tubular cells joined end to end without intercellular spaces but interrupted by stomata. The cells are living, achlorophylous and vacuolate. The outer walls of the epidermal cells have cuticle to check water loss. Numerous multicellular hairs develop on the epidermis.

Cortex : It forms the extrastelar ground tissues beneath the epidermis. Cortex is differentiated into three zones :

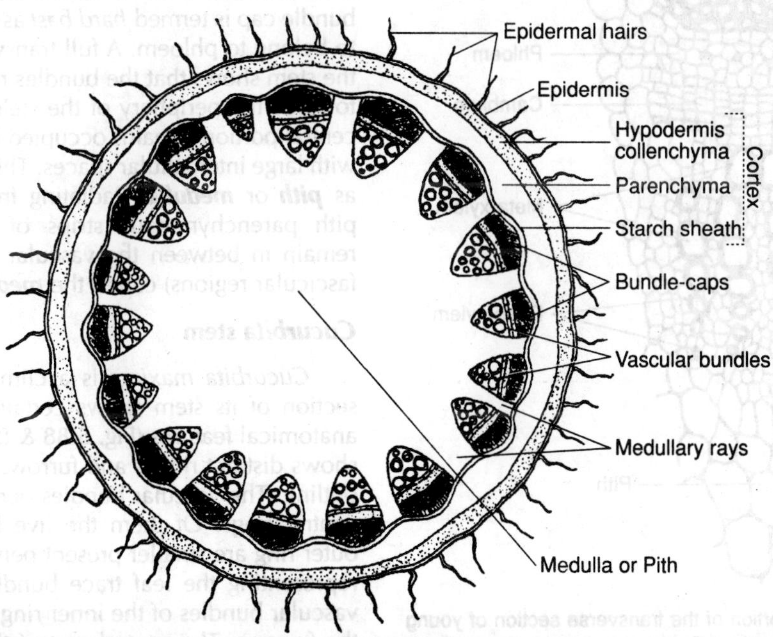

Fig. 5.86 : Diagrammatic anatomical view of young stem of sunflower in transverse section

1. *Peripheral collenchymatous hypodermis* : Cells of this zone are usually angular collenchyma and form a continuous band. It gives mechanical support to the growing stem.

2. *Parenchymatous zone* : Next to hypodermis a few layers of thin-walled parenchyma cells with conspicuous intercellular spaces occur. A few glands with hollow cavity surrounded by small epithelial cells are present here and there.

3. *Starch sheath* : This is the innermost wavy layer of cortex made of compactly set barrel-shaped parenchyma cells containing abundant starch grains. This layer is homologous to the endodermis in root.

Stele : The central core of tissues encircled by the starch sheath constitutes the stele. It is a dissected siphonostele consisting of vascular bundles arranged in a ring and the intrastelar ground tissue.

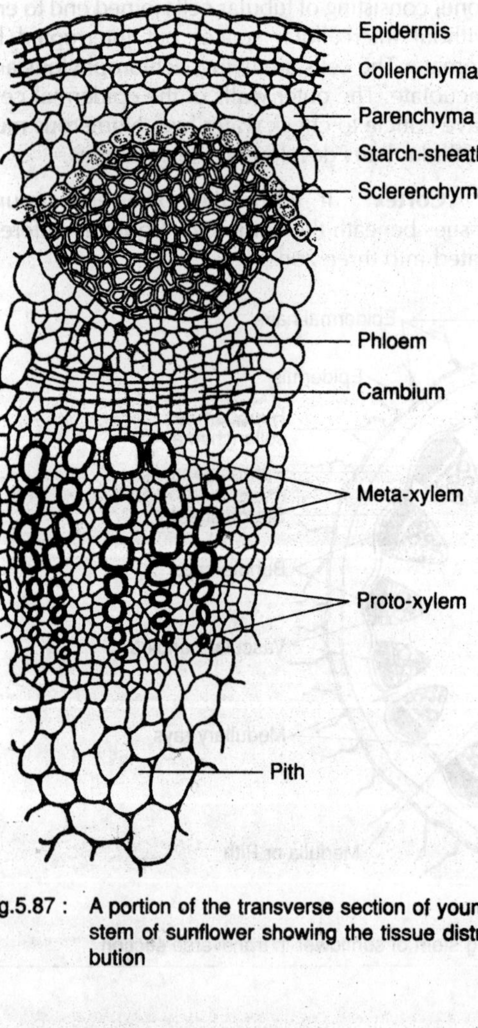

Fig.5.87 : A portion of the transverse section of young stem of sunflower showing the tissue distribution

Vascular bundles : The vascular bundles are conjoint, collateral and open with internal xylem and external phloem on the same radius. A strip of lateral cambium is present in-between xylem and phloem.

The xylem is *endarch* showing centrifugal mode of differentiation from the procambium. The protoxylem elements like narrower tracheids and tracheae with annular and spiral thickenings occur towards the centre. Metaxylem with pitted and other types of thickenings remains on the periphery. Xylem parenchyma cells are smaller than other parenchyma cells of the stele.

The phloem is composed of sieve elements (SE), companion cells (CC) and phloem parenchyma. The sieve tubes and companion cells collectively from the SE-CC complex.

The cambium of the vascular bundle known as *fascicular cambium* appears to consist of two or three layers of fusiform cells which appear rectangular in cross-section. This tissue is responsible for growth in thickness.

Intrastelar ground tissues : There exists a patch of sclerenchyma on the top of every vascular bundle which is called ***bundle cap***. This bundle cap is termed *hard bast* as it is considered to belong to phloem. A full transverse section of the stem shows that the bundles remain in a ring towards the periphery of the stele and the large central portion remains occupied by parenchyma with large intercellular spaces. This part is known as ***pith*** or ***medulla***. Radiating from the central pith parenchymatous strips of ground tissue remain in between the vascular bundles (interfascicular regions) called the ***medullary rays***.

Cucurbita stem

Cucurbita maxima is a climber. Transverse section of its stem shows certain characteristic anatomical features (Fig. 5.88 & 89). The section shows distinct ridges and furrows giving a wavy outline. The vascular bundles occur in two concentric rings. Of them the five bundles of the outer ring are smaller present beneath the ridges representing the leaf trace bundles. Five larger vascular bundles of the inner ring occur beneath the furrows. The central part of the stem is hollow due to disintegration of pith. The t.s. of

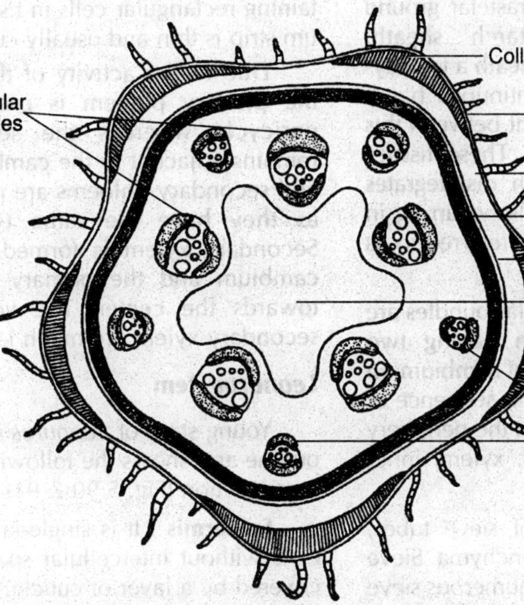

Fig. 5.88 : Diagrammatic view of the transverse section of a young *Cucurbita* stem

Cucurbita stem shows the following tissue distribution from periphery to the center :

Epidermis : It is the uniseriate outer layer made of compactly-set tubular cells with cuticular outer covering. The cells of this layer are living with vacuolate protoplast. Numerous multicellular hairs are present on the epidermis. Stomata may be present in the young stem.

Cortex : It is differentiated into three zones :

1. *Peripheral collenchymatous hypodermis* : The thickness of the hypodermis is not uniform. It is most reduced beneath the furrows and even interrupted by some parenchyma cells containing abundant chloroplasts just beneath the epidermis where stomata are present. Conspicuous collenchymatous patches are formed at the ridges. Chloroplasts may also be present in the collenchyma cells.

2. Internal to the hypodermis there are a few layers of parenchyma cells which are larger in size having intercellular spaces among them. These cells contain lesser number of chloroplasts.

3. The innermost layer of cortex is the *starch sheath* the cells of which are barrel-shaped and larger than other cells of cortex. These cells contain abundant starch grains.

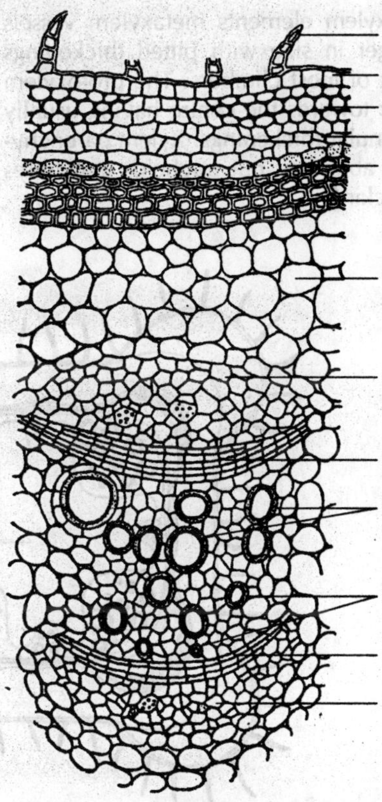

Fig. 5.89 : The cellular diagram of a portion of young *Cucurbita* stem in T.S. to show the tissue distribution

Stele : It is the central cylinder core containing the vascular bundles and intrastelar ground tissues encircled by the starch sheath. Immediately beneath the starch sheath a few layers of sclerenchyma form a continuous band. Little parenchyma cells are present between this band and the vascular bundles. These tissues together form the pericycle. Pith disintegrates early in *Cucurbita* stem. The cells occurring in between the vascular bundles are referred to as internal parenchyma.

Vascular bundles : The vascular bundles are conjoint, bicollateral and open having two patches of phloem and two strips of cambium on either side of xylem. Therefore, the sequence of tissues in the vascular bundle from the periphery is outer phloem, outer cambium, xylem, inner cambium and inner phloem.

The phloem is composed of sieve tubes, companion cells and phloem parenchyma. Sieve tubes are quite conspicuous with numerous sieve plates. The outer phloem is more massive than the inner one.

Of the xylem elements metaxylem vessels are very larger in size with pitted thickenings characteristic of most climbers. The protoxylem vessels occur towards the centre having usually spiral and annular thickenings. Xylem parenchyma cells are abundant but tracheids and fibres are few or lacking.

The outer cambium is many-layered containing rectangular cells in t.s. The inner cambium strip is thin and usually curved.

Due to the activity of the outer cambium the primary phloem is pressed towards the pericycle whereas the secondary phloem remains adjacent to the cambium. The primary and secondary phloems are not distinguishable as they have the same type of elements. Secondary xylem is formed internal to outer cambium and the primary xylem is pressed towards the center. The vessel elements in secondary xylem are much larger.

Leonurus stem

Young stem of *Leonurus* is almost square in outline and shows the following tissues in transverse section (Fig. 5.90 & 91) :

Epidermis : It is single-layered with tubular cells without intercellular spaces but externally covered by a layer of cuticle. Multicellular hairs and stomata are present in the epidermis.

Cortex : The cortex in this stem is comparatively shallow but differentiated into three zones – the hypodermis, parenchymatous middle zone and the starch sheath. The collenchymatous hypodermis is densely aggregated at the corners of the stem. These diagonally placed collenchymatous patches form the I-girders to give mechanical support to the stem. These patches extend from the

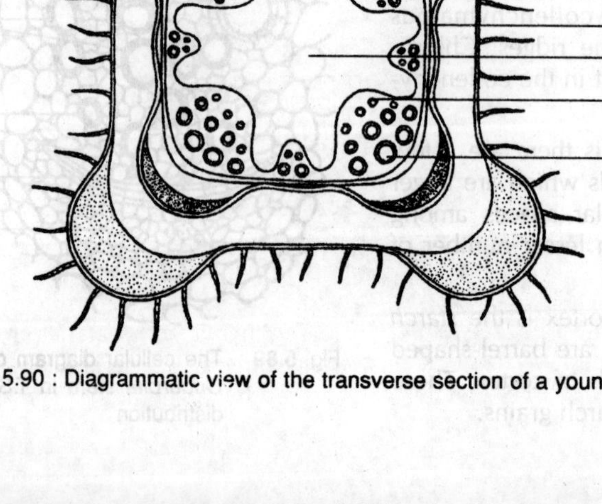

Fig. 5.90 : Diagrammatic view of the transverse section of a young Leonurus stem

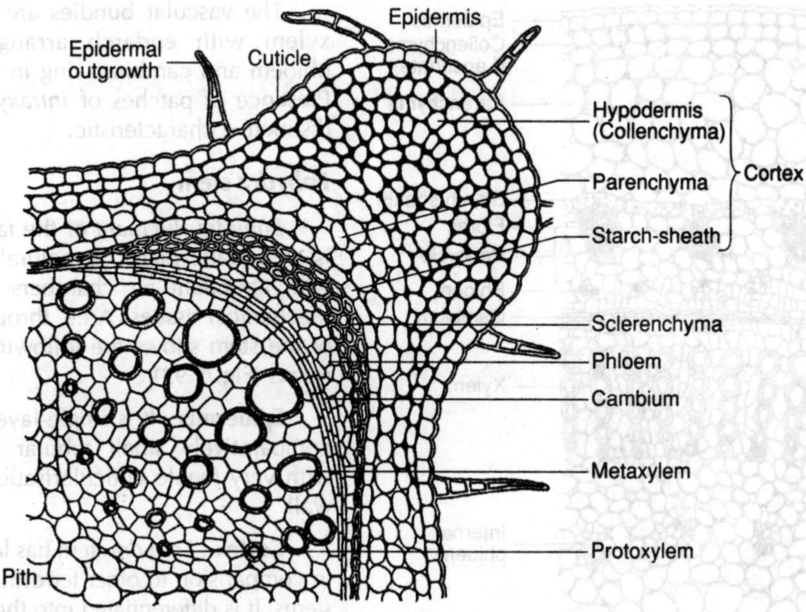

Fig. 5.91 : The cellular diagram of a portion of young *Leonurus* stem in t.s. to show the tissue distribution

corners to some extent but do not form the continuous band of hypodermis. The innermost layer of the cortex is the starch sheath consisting of barrel-shaped compactly arranged cells with abundant starch grains. In between the hypodermis and starch sheath a few layers of thin-walled parenchyma cells remain. The cells of this layer contain abundant chloroplasts and appear as a green belt under the microscope.

Stele : The central core of tissues made of vascular strands and intrastelar ground tissues encircled by the starch sheath is the stele. Internal to the starch sheath a few layers of sclerenchyma form a continuous band. It may be called pericycle or perivascular tissue. Large parenchymatous pith present at the center presses the vascular bundles towards the periphery. The primary medullary rays become indistinct due to secondary growth in thickness.

The vascular bundles are ***conjoint, collateral*** and ***open*** and remain arranged more or less in a ring. The phloem remains towards the periphery having sieve tubes, companion cells, and phloem parenchyma. Internal to the phloem there is a strip of cambium, made of a few layers of fusiform cells appearing more or less rectangular in outline. Internal to the layer of cambium xylem occurs with usual tracheary elements—tracheids and tracheae, parenchyma and fibres. Xylem is endarc, i.e., protoxylem remains towards centre and metaxylem towards circumference.

Calotropis stem

A transverse section through the internode a young stem of *Calotropis procera* of family Asclepiadaceae shows the following arrangement of tissues (Fig. 5.92) :

Epidermis : Epidermis is uniseriate composed of tubular cells with cuticularized outer walls. There may be further deposition of wax external to the outer wall of the epidermis. Stomata are present here and there.

Cortex : It is differentiated into **hypodermis, parenchymatous zone** and the **starch sheath**. The hypodermis is composed of two or three layers of lacunate collenchyma. The intermediate zone consists of large parenchyma cells with intercellular spaces and the starch sheath is made of barrel-shaped compactly arranged cells with starch grains with starch grain accumulation inside.

Stele : The central core of tissues consists of compactly arranged vascular bundles with intervening ray cells and intrastelar ground tissues. The pericycle is made of a few cell layers of parenchyma with patches of cellulosic fibres here and there. The central portion of the stele is occupied by large parenchymatous pith.

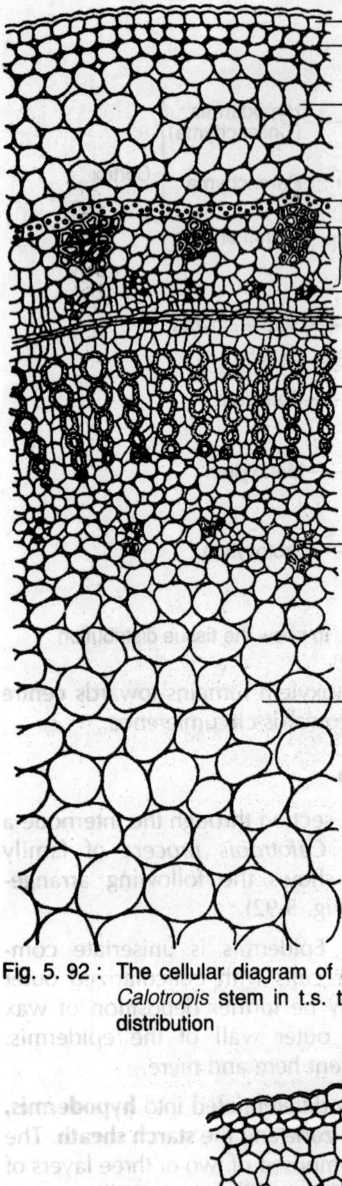

Fig. 5. 92 : The cellular diagram of a portion of young *Calotropis* stem in t.s. to show the tissue distribution

The vascular bundles are made of internal xylem with endarch arrangement, external phloem and cambium ring in-between the two. Presence of patches of *intraxylary* phloem is a distinctive characteristic.

Enhydra stem

Enhydra fluctuant of the family Compositae is an aquatic plant. So naturally its stem is soft with abundant air chambers and rudimentary mechanical tissues. A t.s. through the internode of the stem shows the following tissue arrangements (Fig. 5.93) :

Epidermis : It is single-layered composed of comparatively small tubular thin-walled cells with very feeble cuticularisation on the outside wall.

Cortex : *Enhydra* stem has less massive cortex in comparision to other terrestrial dicotyledonous stems. It is differentiated into three zones :

1. A few layers of parenchymatous ***hypodermis*** with comparatively thicker walls forming a continuous band. The absence of peripheral collenchyma is thus a notable feature.
2. Internal to the hypodermis there are a few layers of large parenchymatous cells having abundant spaces and conspicuous ***air chambers.*** These air spaces and air chambers give buoyancy to the whole plant apart from normal aeration. The parenchyma cells contain chloroplastids in them.
3. The innermost layer of cortex is as usual the ***starch sheath,*** the barrel shaped compactly

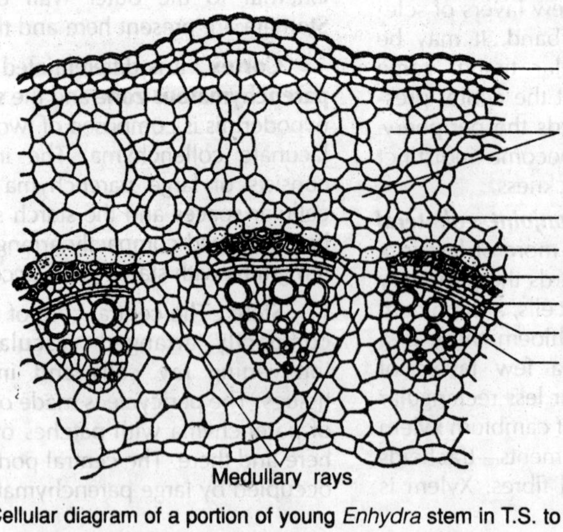

Fig. 5.93 : Cellular diagram of a portion of young *Enhydra* stem in T.S. to show the tissue distribution

arranged cells of which store starch grains inside.

Stele : The central cylindrical core of tissues constitutes the ***stele*** consisting of the vascular bundles arranged in a ring and intrastelar ground tissues. There is a ***bundle cap*** or hard bast against every vascular bundle as a patch of sclerenchyma similar to sunflower stem but much smaller in size. These sclerenchyma patches together with intervening parenchyma form the pericycle, though doubt remains regarding the origin of the fibres. The pith is hollow as it disintegrates leaving a large cavity at the central region. Some parenchyma cells occur in the interfascicular regions.

The vascular bundles remain arranged in a ring. They are conjoint, collateral and open. The amount of phloem tissue is rather small in comparision to xylem with the usual components like sieve tubes, companion cells and phloem parenchyma. A strip of cambium occurs in between the xylem and phloem. Xylem is composed of a few tracheary elements including protoxylem and metaxylem with endarch orientation and a few xylem parenchyma cells.

ANATOMY OF SOME MONOCOTYLEDONOUS STEMS

Anatomical features of stem of some monocotyledonous plants are discussed in the next page :

Maize

Zea mays belongs to the family Poaceae and its stem in t.s. shows three tissue systems — epidermis, ground tissue and the vascular bundles. No stele is formed (Fig. 5.94 & 95).

Epidermis : It is uniseriate layer of cells consisting of small compactly arranged cells. The outer walls are cuticularised and stem hairs are absent.

Ground tissue : It is not well-differentiated. Next to epidermis a few layers of sclerenchyma cells form the ***hypodermis***. It is usually continuous but parenchyma may intervene here and there. Rest of the ground tissue internal to the hypodermis is composed of thin-walled parenchyma cells with profuse intercellular spaces. The vascular bundles remain scattered throughout the ground tissue without forming any stele.

Stele : The vascular bundles are not surrounded by any endodermis or starch sheath to form the central cylindrical core of tissues or stele. Still this type of arrangement of vascular bundles throughout the ground tissue is called ***atactostele***. The peripheral vascular bundles are comparatively smaller in size.

The vascular bundles are conjoint, collateral and closed (Fig. 5.96). Xylem components are aligned in the form of the letter Y. The two wider metaxylem vessels with pitted thickening remain at the two arms of the Y and one or two narrower protoxylem vessels with spiral or annular

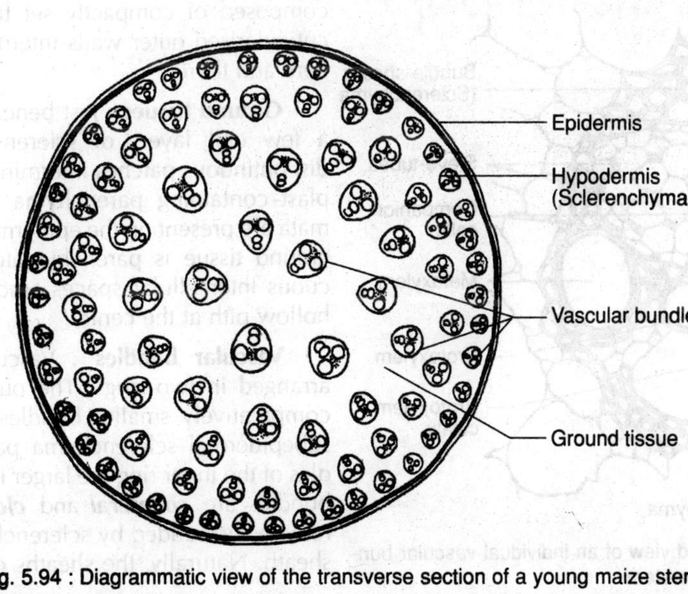

Fig. 5.94 : Diagrammatic view of the transverse section of a young maize stem

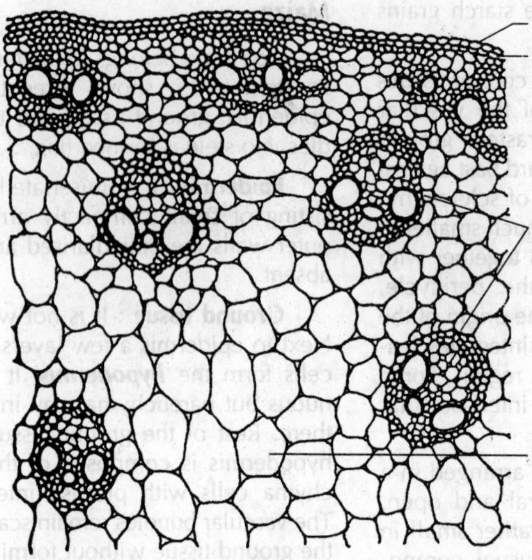

Fig. 5.95 : The cellular diagram of a portion of young maize stem in T.S. to show the tissue distribution

thickening remain at the base of the Y. At the vicinity of the protoxylem a few tracheids remain. With the advancement of the development the xylem elements become more and more lignified and the lowermost protoxylem disintegrates to form a lacuna or cavity known as ***protoxylem cavity***.

The small phloem tissue consisting of only sieve tubes and companion cells remains in-between the two arms of the Y. Phloem parenchyma is absent. The individual vascular

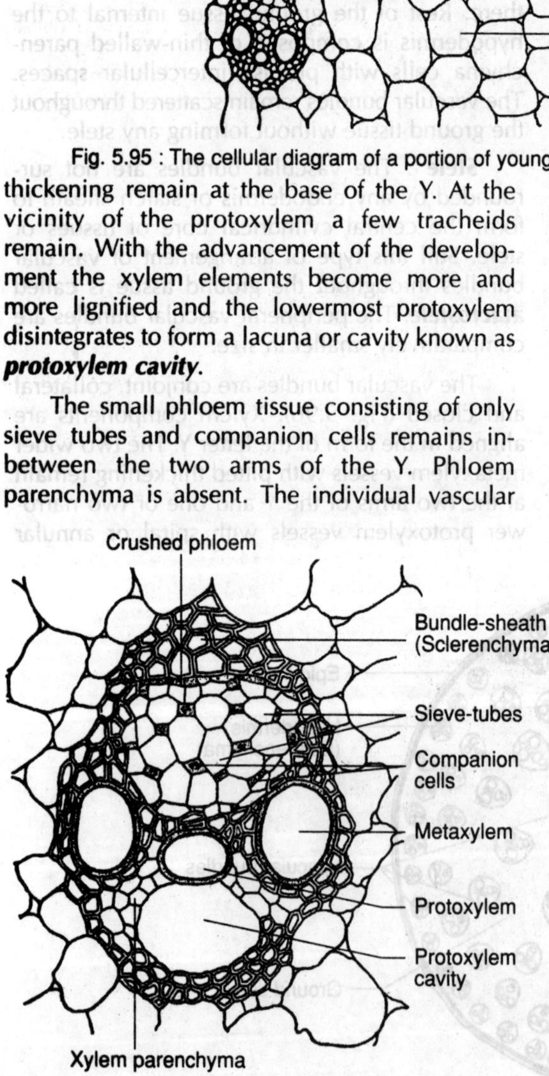

Fig. 5.96 : An enlarged view of an individual vascular bundle of maize stem

bundles remain surrounded by sclerenchymatous bundle sheath. The ***bundle sheaths*** of small peripheral bundles may coalesce with the sclerenchymatous hypodermis. As a result, those peripheral bundles appear to be embedded in the sclerenchymatous sheath.

Wheat

In wheat (*Triticum vulgare*) the internode is hollow and a T.S. through this portion shows following anatomical features (Fig. 5.97 & 98) :

Epidermis : The single-layered epidermis is composed of compactly set tubular cells with cuticularised outer walls interrupted by stomata here and there.

Ground tissues : Just beneath the epidermis a few cell layers of sclerenchyma occur as discontinuous patches intermingled with chloroplast–containing parenchyma cells where stomata are present on the epidermis. The rest of the ground tissue is parenchymatous with conspicuous intercellular spaces among the cells and hollow pith at the centre.

Vascular bundles : Vascular bundles are arranged in two rings. The outer ring contains comparatively smaller bundles associated with subepidermal sclerenchyma patches. The bundles of the inner ring are larger in size. Individual bundles are ***collateral*** and ***closed*** and usually remain surrounded by sclerenchymatous bundle sheath. Naturally, the sheaths of the bundles of

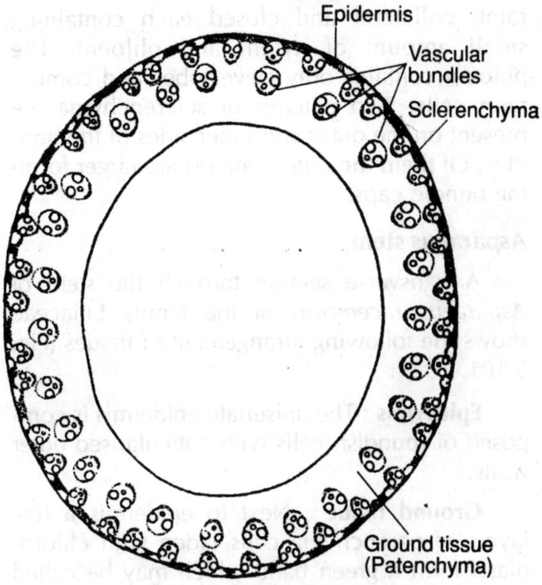

Fig. 5.97 : Diagrammatic view of the transverse section of a young *Triticum* stem

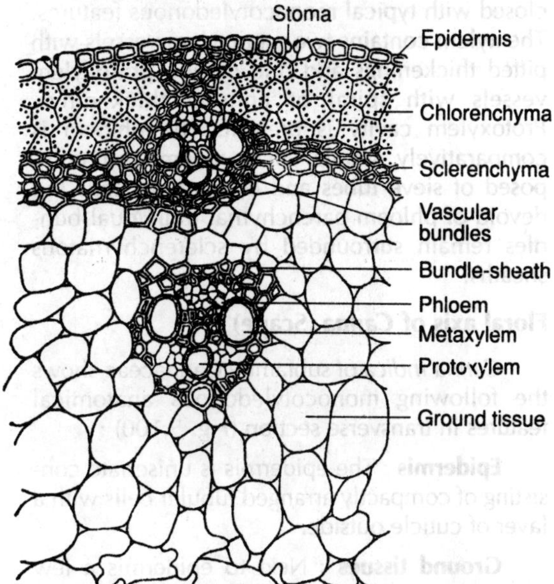

Fig. 5.98 : Cellular diagram of a portion of young wheat stem in T.S. to show the tissue distribution

the outer ring fuse with the subepidermal sclerenchyma patches.

Stem of an Aquatic monocotyledon

Leptochloa chinensis of family Poaceae is an aquatic grass. Transverse section of it shows the following arrangement of the tissues (Fig. 5.99).

Epidermis : The uniseriate epidermis is composed of a row of compactly arranged cells with a thin layer of cuticle outside.

Ground tissues : Just beneath the epidermis two or three layers of sclerenchyma cells form the hypodermis next to which parenchymatous ground tissue with lots of intercellular spaces occur. Another band of 1–2 cell layer of sclerenchyma is present again in the ground tissue. The central portion of the stem is hollow.

Vascular bundles : The vascular bundles are scattered throughout the ground tissue. The peripheral bundles are smaller in size.

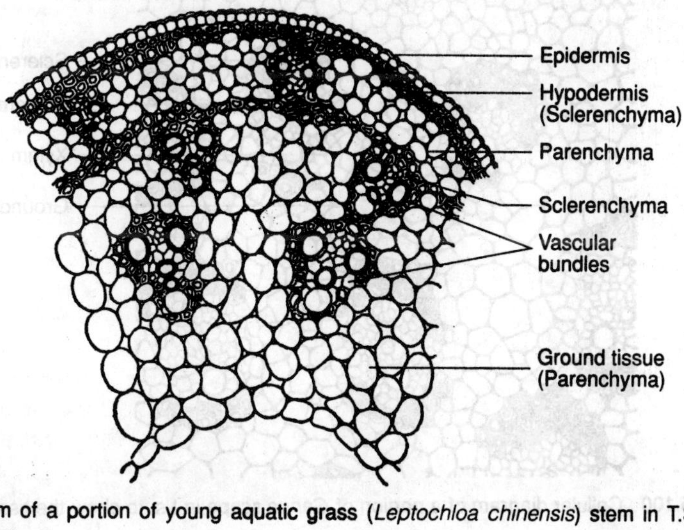

Fig. 5.99 : Cellular diagram of a portion of young aquatic grass (*Leptochloa chinensis*) stem in T.S. to show the tissue distribution

Individual bundles are conjoint, collateral and closed with typical monocotyledonous features. The xylem contains two metaxylem vessels with pitted thickenings and one or two protoxylem vessels with spiral or annular thickenings. Protoxylem cavity is present. The phloem is comparatively smaller than xylem and composed of sieve tubes and companion cells but devoid of phloem parenchyma. Individual bundles remain surrounded by sclerenchyrnatous sheaths.

Floral axis of *Canna* (Scape)

Canna indica of subfamily Cannaceae shows the following monocotyledonous anatomical features in transverse section (Fig. 5.100) :

Epidermis : The epidermis is uniseriate consisting of compactly arranged tubular cells with a layer of cuticle outside.

Ground tissues : Next to epidermis a few parenchymatous cell layers form a small cortex, internal to which is a band of chloroplast-containing cells referred to as ***chlorophyllous tissue***. Sclerenchyma patches remain here and there attached to the chlorophyllous tissue. The remaining portion is the parenchymatous ground tissue bearing the scattered vascular bundles.

Vascular bundles : The bundles are conjoint, collateral and closed each containing small amount of xylem and phloem. The phloem contains only sieve tubes and companion cells. Two patches of sclerenchyma are present on the outer and inner sides of the bundles. Of them the outer patches are larger forming bundle caps.

Asparagus stem

A transverse section through the stem of *Asparagus racemosus* of the family Liliaceae shows the following arrangement of tissues (Fig. 5.101, 102) :

Epidermis : The uniseriate epidermis is composed of roundish cells with cuticularised outer walls.

Ground tissue : Next to epidermis a few layers of parenchyma cells laden with chloroplasts form a green band which may be called the cortex. Internal to this green band a starch sheath is present. Beneath the starch sheath a band of sclerenchyma of few cells thick remains encircling the ground tissue with scattered vascular bundles. The center is occupied by thin-walled comparatively larger parenchyma cells with distinct intercellular spaces.

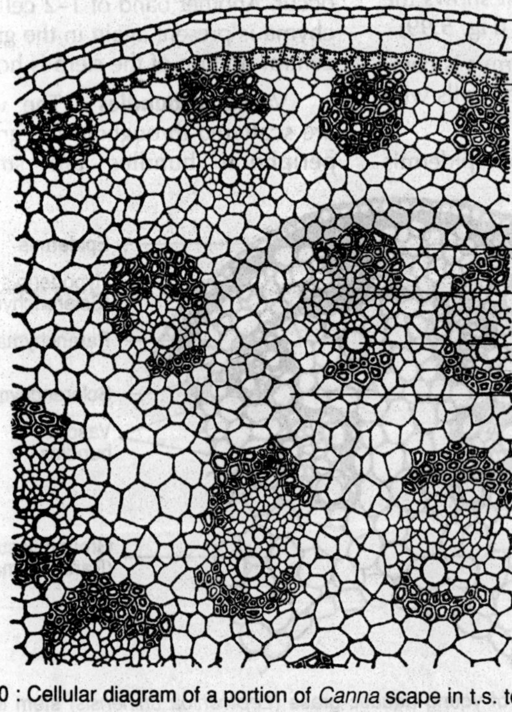

Fig. 5.100 : Cellular diagram of a portion of *Canna* scape in t.s. to show the tissue distribution

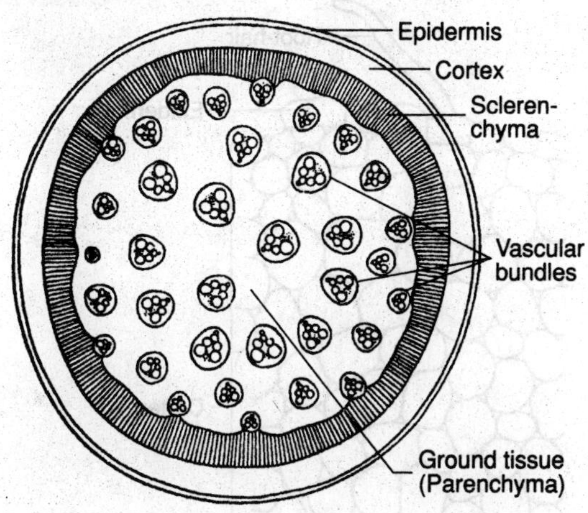

Fig. 5.101 : Diagrammatic view of the transverse section of a young *Asparagus* stem

Vascular bundles : Of the scattered vascular bundles the central ones are larger. The peripheral bundles may touch the sclerenchymatous band. The vascular bundles are conjoint, collateral and closed. The xylem occurs in the form of the letter V, the metaxylem vessels form the arms and protoxylem forms the base of the letter V. Phloem remains within the arms of V. It is composed of sieve tubes and companion cells and phloem parenchyma is absent.

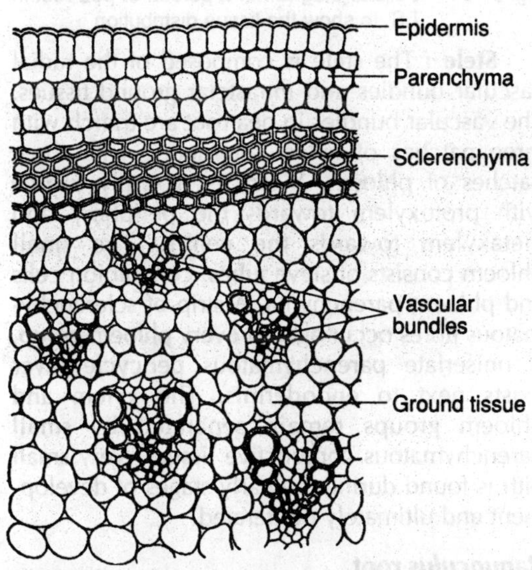

Fig. 5.102 : Cellular diagram of a portion of *Asparagus* stem in T.S. to show the tissue distribution

ANATOMICAL STUDY OF SOME SELECTED ROOTS

DICOTYLEDONOUS ROOTS

Gram root

A transverse section of *Cicer arietinum* of subfamily Papilionaceae shows the following arrangement of tissues (Fig. 5.103) :

Epidermis : Epidermis or *epiblema* or *piliferous layer* is uniseriate outer boundary layer consisting of tubular cells with cuticle on the outer walls. Some epiblema cells at a particular zone of the root protrude to form long unicellular root hairs. This zone is called root hair zone and located next to the root tip.

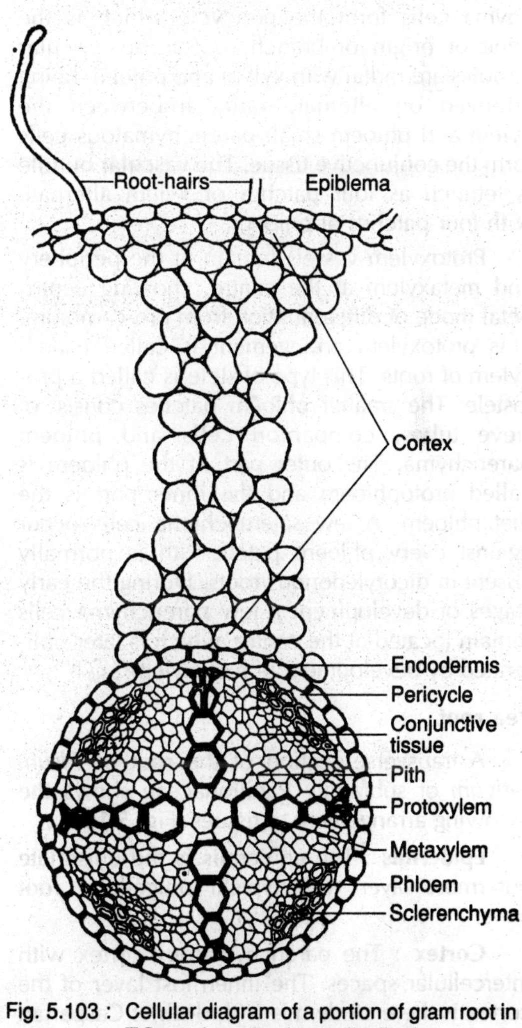

Fig. 5.103 : Cellular diagram of a portion of gram root in T.S. to show the tissue distribution

Cortex : Next to epidermis there is a homogeneous massive parenchymatous zone called the cortex consisting of parenchyma cells with conspicuous intercellular spaces. The cells contain large amount of leucoplasts. Their main function is to translocate water and minerals absorbed by the root hairs to the conducting elements. The innermost layer of cortex is the endodermis consisting of compactly arranged barrel-shaped cells forming a distinct cell layer surrounding the stele. The cells of this layer possess Casparian strips on their radial walls.

Stele : The stele or the central core of tissue is precisely demarcated from the cortex by the endodermis. The stele is composed of vascular tissues and intrastelar ground tissues. Internal to the endodermis a layer of thin-walled parenchyma cells form the pericycle which is the zone of origin of branch roots. The vascular bundles are radial with xylem and phloem being arranged on alternate radii. In-between the xylem and phloem small parenchymatous cells form the conjunctive tissue. The vascular bundle is tetrarch as four patches of xylem alternate with four patches of phloem.

Protoxylem vessels remain at the periphery and metaxylem at the centre, showing centripetal mode of differentiation from pro-cambium. This protoxylem arrangement is called exarch xylem of roots. This type of stele is called a protostele. The smaller phloem patches consist of sieve tubes, companion cells and phloem parenchyma. The outer part of the phloem is called protophloem and the inner part is the metaphloem. A few sclerenchyma cells occur against every phloem patch. Pith is normally absent in dicotyledonous roots. During the early stages of development a few parenchyma cells remain located at the centre, which is later obliterated by development of metaxylem.

Pea root

A transverse section of the root of *Pisum sativum* of subfamily Papilionaceae shows the following arrangement of tissues (Fig. 5.104) :

Epidermis : The epidermis is the uniseriate outermost layer with typical unicellular root hairs.

Cortex : The parenchymatous cortex with intercellular spaces. The innermost layer of the cortex is the endodermis with distinct Casparian strips.

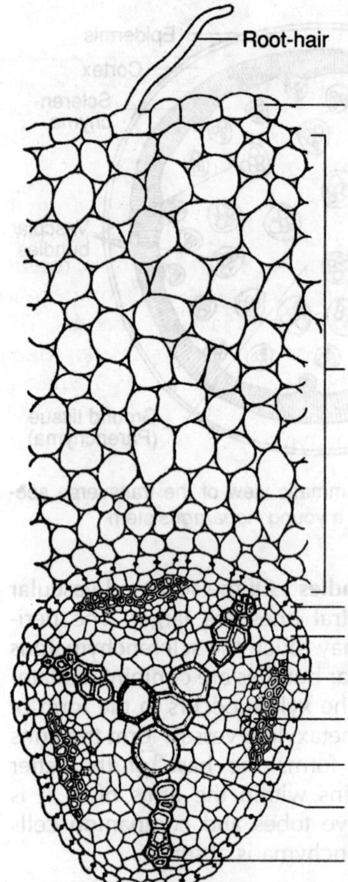

Fig. 5.104 : Cellular diagram of a portion of pea root in T.S. to show the tissue distribution

Stele : The stele is composed of the radial vascular bundles and intrastelar ground tissues. The vascular bundles in pea root are triarch with three patches of xylem alternating with three patches of phloem. Xylem is typically *exarch* with protoxylem towards the periphery and metaxylem towards the centre. The small phloem consists of sieve tubes, companion cells and phloem parenchyma. A strip of sclerenchymatous fibres occur against every phloem group. A uniseriate parenchymatous pericycle layer exists next to endodermis. The xylem and phloem groups remain separated by small parenchymatous conjunctive tissues. Very small pith is found during the early stages of development and ultimately obliterated.

***Ranunculus* root**

A transverse section through the root of *Ranunculus* sp. of family Ranunculaceae exhibits the following arrangement of tissues (Fig. 5.105) :

Epidermis : The single layered epidermis is composed of a row of living comparatively small tubular cells, but disorganised protoplast may be found in some cells of the layer.

Cortex : The cortex is very massive encircling the stele. It consists of (i) a narrow outer zone next to epidermis, called exodermis, resembling the hypodermis of stems, and (ii) a broad inner zone occurring internal to exodermis. The exodermis is composed of compactly arranged smaller cells with scanty intercellular spaces. The broad inner zone consists of larger starch containing parenchyma cells with conspicuous intercellular spaces. The innermost layer of cortex forms the endodermis. It is made of a row

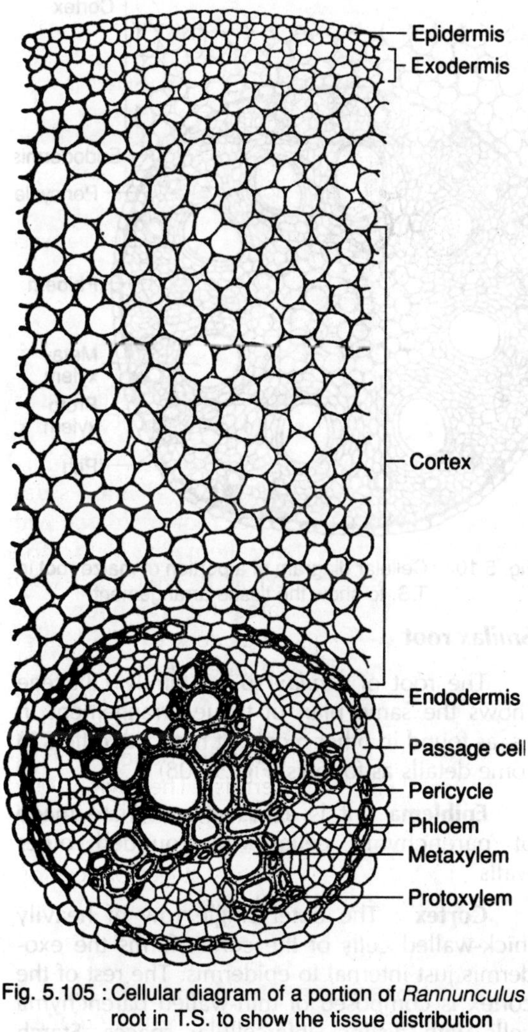

Fig. 5.105 : Cellular diagram of a portion of *Rannunculus* root in T.S. to show the tissue distribution

of barrel-shaped cells with Casparian strips and further secondary depositions. Only in certain cells of the layer facing against the protoxylem secondary deposition does not take place and their walls remain thin. These cells are called the ***passage cells***.

Stele : The central core of tissues forms the stele. The outermost boundary of the stele is the uniseriate parenchymatous *pericycle*. Vascular bundles are radially arranged. Four to five strands exarch xylem alternate with the same number of small patches of phloem strands. Parenchymatous conjunctive tissue separates the xylem and phloem.

MONOCOTYLEDONOUS ROOTS

Arum root

A transverse section through the root of arum (*Colocasia* sp. of family Araceae) shows the following anatomical features (Fig. 5.106) :

Epidermis : The uniseriate epiblema is composed of a row of compactly arranged tubular cells without intercellular spaces. Some cells of the layer protrude to form the root hairs.

Cortex : The massive cortex mainly consists of unspecialised parenchyma with profuse schizogenously formed spaces. In older roots the outer zone of the cortex next to epiblema forms the exodermis the cells of which become suberized when the epiblema is decayed and thus protects the internal tissues in absence of epiblema.The innermost layer of cortex forms the endodermis consisting of barrel-shaped closely set cells with conspicuous Casparian strips. The radial walls of the endodermal cells become secondarily thickened except some cells against the protoxylem groups to form passage cells or transfusion cells for ready diffusion of absorbed sap into the xylem stream.

Stele : The central cylindrical core of tissues encircled by the endodermis forms the stele. It consists of radially, arranged vascular strands and intrastelar ground tissues. Internal to the endodermis there is a uniseriate parenchymatous cell-layer called the pericycle. There is alternate orientation of a good number of exarch xylem and phloem strands. This type of stele is referred to as polyarch. Phloem is composed of sieve tubes, companion cells and phloem parenchyma. The outer protophloem and inner metaphloem can hardly be distinguished. Small amount of parenchymatous conjunctive tissue

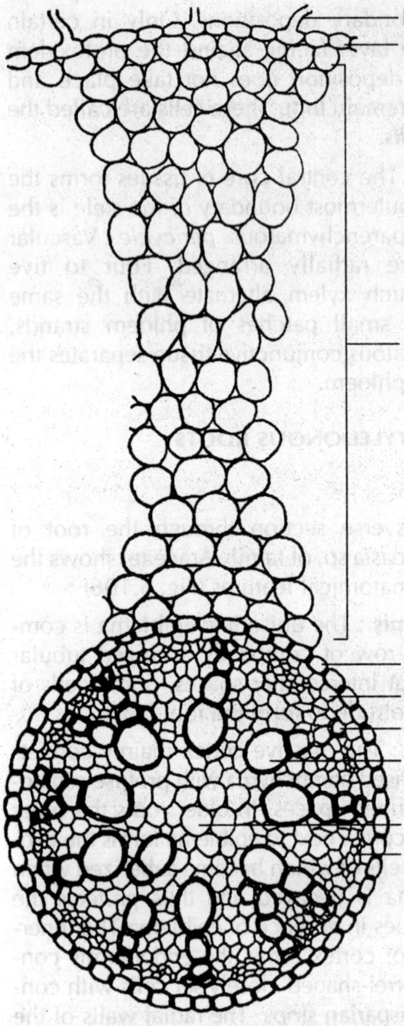

Fig. 5.106 : Cellular diagram of a portion of *Colocasia* (arum) root in T.S. to show the tissue distribution

separates the xylem and phloem patches. The central portion of the stele is occupied by large pith.

Maize root

The anatomical features of the maize root (*Zea mays* of family Poaceae) are almost similar to that of arum, so far as the epiblema and cortex are concerned. Formation of exodermis in the outer zone of cortex is common in almost all monocotyledonous roots. The innermost layer of the cortex forms the endodermis composed of thick-walled cells. As the secondary and tertiary layers are deposited on the walls of the endodermal cells the Casparian strips are not recognised.

Internal to the endodermis uniseriate pericycle is present but it is partly sclerenchymatous. The stele is polyarch with large number of exarch xylem and phloem strands. Xylem parenchyma cells undergo sclerosis and thus become thick-walled. A large pith made of loosely-arranged parenchyma cells containing abundant starch grains occupies the center (Fig. 5.107).

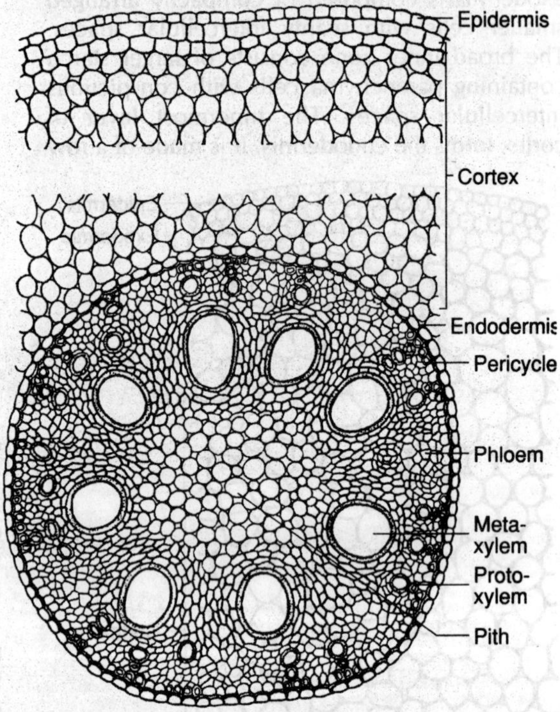

Fig. 5.107 : Cellular diagram of a portion of maize root in T.S. to show the tissue arrangement

Smilax root

The root of *Smilax* sp. of family Liliaceae shows the same plan of tissue arrangement in t.s. as found in other monocot roots but differs in some details as follows (Fig. 5.108) :

Epiblema : It is as usual uniseriate made of parenchyma cells with rounded outer walls.

Cortex : The outer single row of heavily thick-walled cells of the cortex forms the exodermis just internal to epidermis. The rest of the cortex is composed of thin-walled parenchyma cells with distinct intercellular spaces. Starch

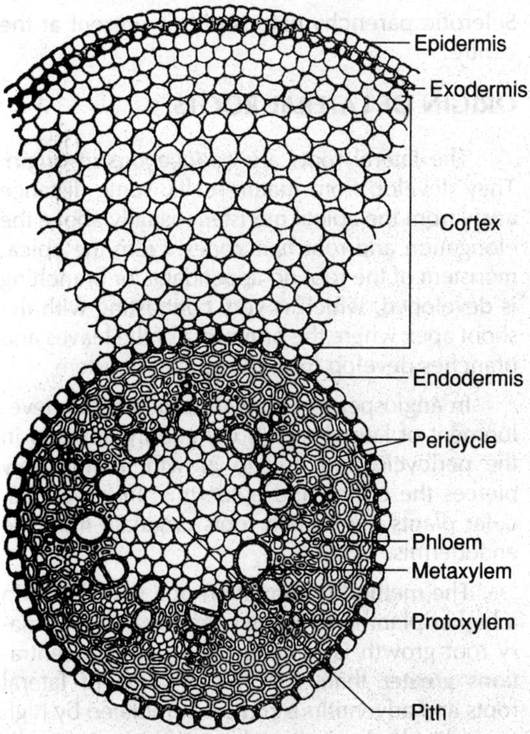

Fig. 5.108 : Cellular diagram of a portion of *Smilax* root in T.S. to show the tissue arrangement

grains are abundantly present in the cortical cells. The innermost layer of the cortex forms the endodermis the cells of which are thick-walled due to secondary deposition on their radial and inner walls.

Stele : The stele is polyarch consisting of a large number of xylem and phloem strands with alternate arrangement. The pericycle, unlike that of other roots, is multiseriate and sclerenchymatous. The outer protophloem elements are smaller than the inner metaphloem elements. Pith at the centre is fairly large and is made of thick-walled parenchyma.

Orchid root

The epiphytic orchid (*Vanda* spp. of family Orchidaceae) roots have a spongy outer tissue to absorb moisture from the atmosphere, known as velamen, instead of uniseriate epidermis (Fig. 5.109). The tissue consists of a few layers of compactly arranged dead and empty cells which

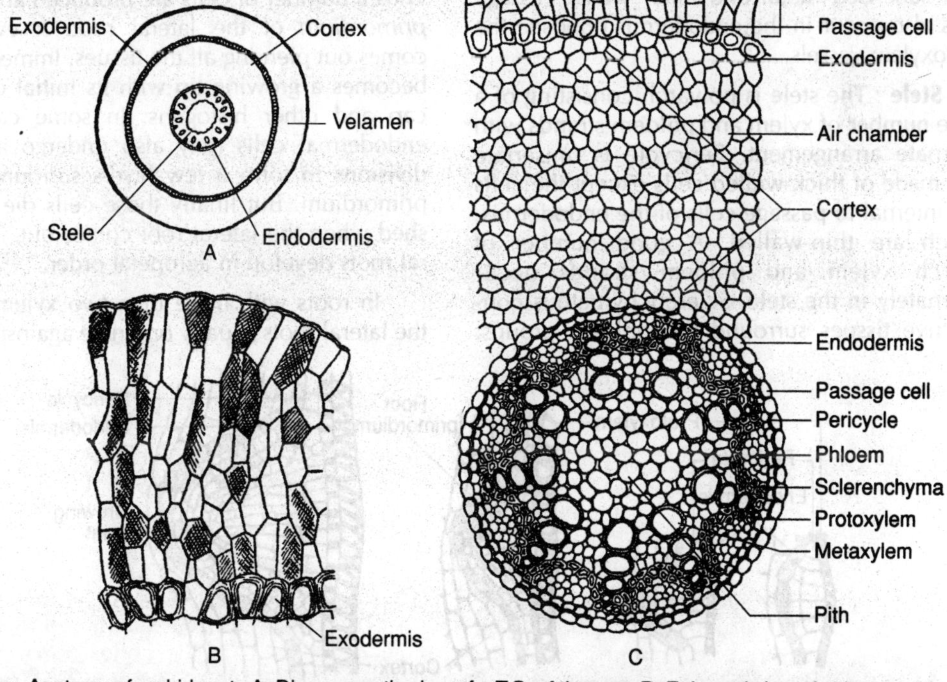

Fig. 5.109 : Anatomy of orchid root : A. Diagrammatic view of a T.S. of the root. B. Enlarged view of velamen with exodermis. C. Cellular diagram of a portion of the root in T.S. to show the tissue arrangement

often form a silvery outer coat. The variously thickened porous cell walls of the tissue work like a sponge. The spirally or reticulately arranged fibres form the supporting ribs. In dry weather the cells remain filled with air, and in moist weather they can absorb moisture. Under water-saturated condition gaseous exchange takes place through some special structures, called *pneumatodes,* consisting of groups of cells with dense spiral wall thickenings. Velamen originates from protoderm, and hence may be interpreted as a typical *multiseriate epidermis,* specially adapted to epiphytic condition. The outermost layer of velamen is known as *limiting layer.*

Cortex : The outermost layer of cortex consists of a row of thick-walled cells forms the exodermis. The cells of this layer deposit suberin on their outer and lateral walls except certain of them remain unthickened to form *passage* cells. Through these cells water, absorbed by the velamen, moves inside. The main bulk of the cortex occurs in-between exodermis and endodermis and composed of a few layers of chloroplast containing parenchyma cells with intercellular spaces. The root looks green due to these cells. The innermost layer of cortex is the *endodermis* with suberised radial and inner walls. Passage cells also occur in the endodermis opposite the protoxylem vessels.

Stele : The stele is polyarch consisting of a large number of xylem and phloem strands with alternate arrangement. Pericycle is uniseriate and made of thick-walled cells except the cells just internal to passage cells of the endodermis, which are thin-walled. A good number of exarch xylem and phloem strands occur alternately in the stele. Sclerenchymatous conjunctive tissues surround the phloem groups. Sclerotic parenchymatous pith is present at the centre.

ORIGIN OF LATERAL ROOTS

The lateral roots are *endogenous* in origin. They develop from mature cells, some distance away from the apical meristem usually above the elongation and root hair zones. From the apical meristem of the root no appendage for branching is developed, which is very contrasting with the shoot apex where the primordia of the leaves and branches develop from the apical meristem.

In angiosperms and gymnosperms the development of lateral root is commonly initiated in the pericycle and due to growth it ultimately pierces the cortex and epidermis. In lower vascular plants the branch roots originate from the endodermis.

The method of formation of a lateral root in a higher plant is as follows. Although the primary root growth is inhibited by auxin concentrations greater than 10^{-8} M, initiation of lateral roots and adventitious roots is stimulated by high auxin levels. Auxin stimulates a few mature cells of pericycle, usually opposite a protoxylem group, to divide periclinally and anticlinally. Thus a number of cells are produced and form a *primordium* of the lateral root. It eventually comes out piercing all the tissues. Immediately it becomes a growing tip with its initial cells, the cap and other histogens. In some cases, the endodermal cells may also undergo anticlinal divisions to form a few layers surrounding the primordium. But finally those cells die and are shed when the lateral root comes out. The lateral roots develop in acropetal order.

In roots with more than two xylem strands the lateral roots usually originate against the pro-

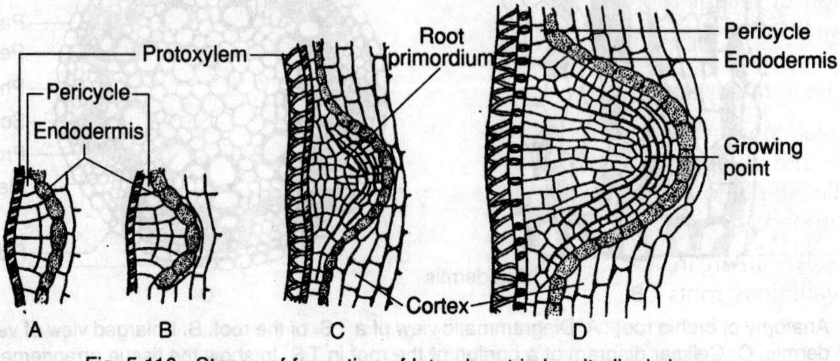

Fig. 5.110 : Stages of formation of lateral root in longitudinal views

toxylem groups to form vertical rows. Therefore, the number of lateral roots becomes equal to that of the xylem strands. But in diarch roots, the primordia develop at each side of the phloem group also, so that the number of lateral roots formed is double that of xylem strands.

The lateral roots show all the characteristics of the primary root, having four distinct anatomical regions. The lateral branches form an extensive root system ramifying through the soil particles. Many of the lateral roots continue normal growth, whereas some remain undeveloped, or may even be lost. In some fleshy roots like carrot additional lateral roots may arise at the bases of original ones when the latter perish.

ORIGIN OF ADVENTITIOUS ROOTS

The roots which originate from non-root tissues are called adventitious roots. The mode of adventitious root development is diverse. They may develop independently or in association with buds either in the young or mature roots. They usually originate *endogenously* from the dormant primordia or form the newly formed primordia. Exogenous development of adventitious roots cannot be ruled out. Adventitious root primordia may develop from epidermis, different internal regions, even from the tissues of leaf margins and petioles, e.g., *Begonia, Kalanchoe* etc.

Adventitious root formation takes place by the activity of a group of cells forming the primordium near the periphery of vascular system in case of young stem, and near the vascular cambium in case of mature stem. The cells forming the primordium are derived from interfascicular parenchyma, medullary rays, or from vascular rays. These dividing cells develop into root apical in a manner somewhat analogous to the formation of lateral roots. It is often stated that the pericycle is the seat of origin of adventitious roots, but the very existence of pericycle in a stem has been rather doubtful.

The developing adventitious root lies close to the xylem and phloem strands of the stem and thus vascular supply to the newly formed root is easily established.

The role of auxin in the initiation of lateral and adventitious roots can be realised from different *Arabidopsis* mutants, named *alf* (aberrant lateral root formation) series. In *alf*1 mutant endogenous auxin level increases more or less 17-fold causing extreme proliferation of lateral and adventitious roots.

In *alf*4 mutant no lateral root is formed and no lateral-root primordia were observed microscopically. This mutant is insensitive to exogenous application of IAA.

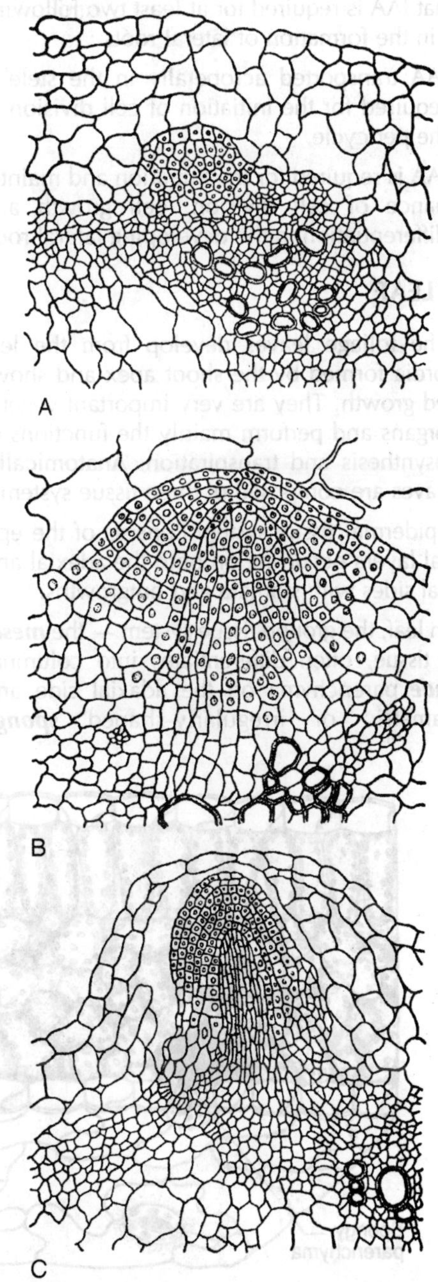

Fig.5.111 : Stages of formation of adventitious root from the tomato stem

Another mutant *alf*3 is defective in the development of lateral-root primordia into mature lateral roots. The primary root remains covered with arrested lateral-root primordia. The arrested growth can be revived by application of exogenous IAA.

In these mutants of *Arabidopsis* it is suggested that IAA is required for at least two following steps in the formation of lateral roots :

1. IAA transported acropetally in the stele is required for the initiation of cell division in the pericycle.
2. IAA is required for the initiation and maintenance of cell division, cell growth and differentiation in the developing lateral root.

THE LEAF

The foliage leaves develop from the leaf primordia formed by the shoot apex and shows limited growth. They are very important vegetative organs and perform mainly the functions of photosynthesis and transpiration. Anatomically, the leaves are composed of three tissue systems.

Epidermal tissue system consists of the epidermal layers occurring on both the adaxial and abaxial sides with stomata and outgrowths.

In leaf, the ground tissue system — the ***mesophyll*** tissue, often differentiates into columnar ***palisade*** parenchyma on the adaxial side and isodiametric or irregularly-shaped ***spongy*** parenchyma on the abaxial side. A leaf having such differentiation in mesophyll tissue is referred to as ***dorsiventral***, the characteristic of dicotyledons. Whereas a leaf with undifferentiated mesophyll tissue, commonly found in monocotyledons, is known as an ***isobilateral leaf***. Mesophyll tissue is provided with conspicuous air spaces for the gaseous exchange with the atmosphere.

The vascular tissue system is composed of ***collateral*** and ***closed*** vascular bundles with the xylem always on the upper side and phloem on the lower. In leaves the vascular tissue forms a network throughout the lamina intermingled with the ground tissue. The main vascular supply through the petiole into the lamina usually forms the midrib bearing branches and branchlets resulting in either reticulate or parallel venation. The ultimate branches are very small and their ends bend into minute specialised photosynthetic areas known as ***vein islets*** or they may just extend into the mesophyll. Bundle-ends vary considerably in the leaves, but usually it consists of a single tracheid with a single sieve element representing xylem and phloem, respectively, surrounded by a parenchymatous bundle sheath. In extreme cases the veinlet may be made of a single spiral tracheid without any phloem cell.

Each vascular bundle usually remains surrounded by a circle of parenchymatous cells called ***bundle sheath*** or ***border parenchyma***, which may or may not contain chloroplasts. The

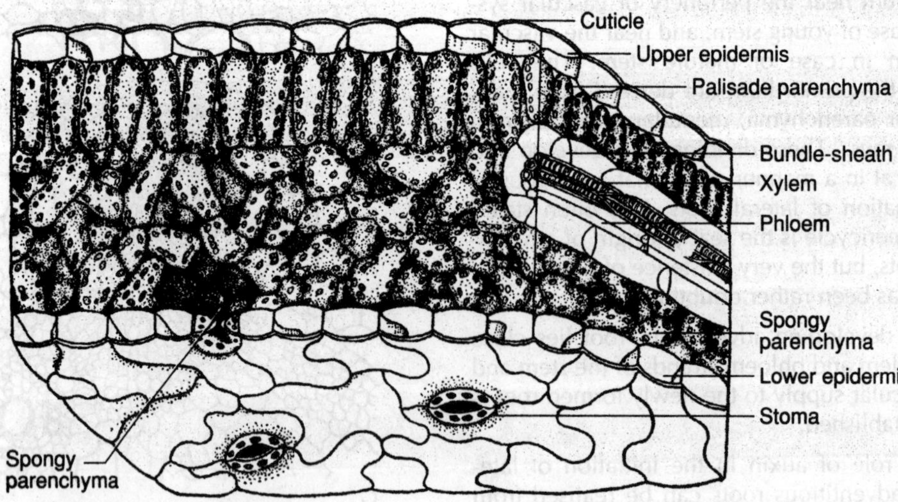

Fig. 5.112 : Diagrammatic view in three dimensions of the vertical section of a dorsiventral leaf showing the arrangement of tissues

monocotyledonous leaves often have two bundle sheaths, an outer chloroplast containing parenchymatous sheath and an inner achlorophyllous thick-walled sheath. In some dicotyledonous leaves the bundle sheath may extend up to the epidermis, either on one or on both sides of the leaf, and are termed ***bundle sheath extensions.*** According to Fann the bundle sheath is regarded as an endodermis as Casparian strips in some cases have been observed. It may also be considered as ***starch sheath*** as often parenchyma cells of the sheath contain starch.

Ontogeny

From the shoot apex promeristem ***leaf primordia*** appear as small rounded protrusions just below the apex. The meristematic cells at the tip of the primordium constitute the apical meristem of the leaf from which a leaf is developed. Initially, apical growth takes place and soon by intercalary growth further expansion takes place to give the proper shape of the leaf. During this development the basal part of the primordium remains unexpanded which finally gives rise to the petiole. At the early stages of the lamina development two strips of ***marginal meristems*** formed along the two surfaces of the leaf axis composing of files of ***marginal initials.*** They form the protoderm by anticlinal divisions. The cells beneath the marginal initials are ***submarginal initials*** which divide in all planes and produce the internal tissues of the leaf often differentiating into three layers—adaxial, middle and abaxial layers. From the procambial strands of the shoot apex a median procambium develops. Whenever the marginal growth stops, further growth of the lamina takes place by anticlinal divisions to increase the surface area of the leaf without changing the thichness of the cell layers.

The two protoderm layers produce upper and lower epidermis. Adaxial layer of submarginal initials produce the palisade parenchyma whereas the abaxial layer produce the lower spongy parenchyma. The middle layer forms the middle spongy parenchyma. The vascular bundles are formed from the procambium.

A developmental scheme of the leaf is shown :

Marginal initials → Upper protoderm → Upper epidermis

Marginal initials → Submarginal initials → Adaxial layer → Palisade prenchyma

Marginal initials → Submarginal initials → Procambium → Vascular bundles

Marginal initials → Submarginal initials → Middle layer → Middle spongy parenchyma

Marginal initials → Submarginal initials → Abaxial layer → Spongy parenchyma

Marginal initials → Lower protoderm → Lower epidermis

DORSIVENTRAL LEAVES

Anatomical features of some common dorsiventral leaves are discussed :

Mango leaf

A vertical section through the leaf lamina of *Mangifera indica* of the family Anacardiaceae shows this arrangement of tissues (Fig. 5.113) :

Epidermis : There are uniseriate epidermises on both upper and lower surfaces composed of a row of compactly arranged tubular cells. The outer walls of epidermal cells possess thin cuticle. The upper epidermis possesses more thickned cuticle. Stomata occur on the lower epidermis.

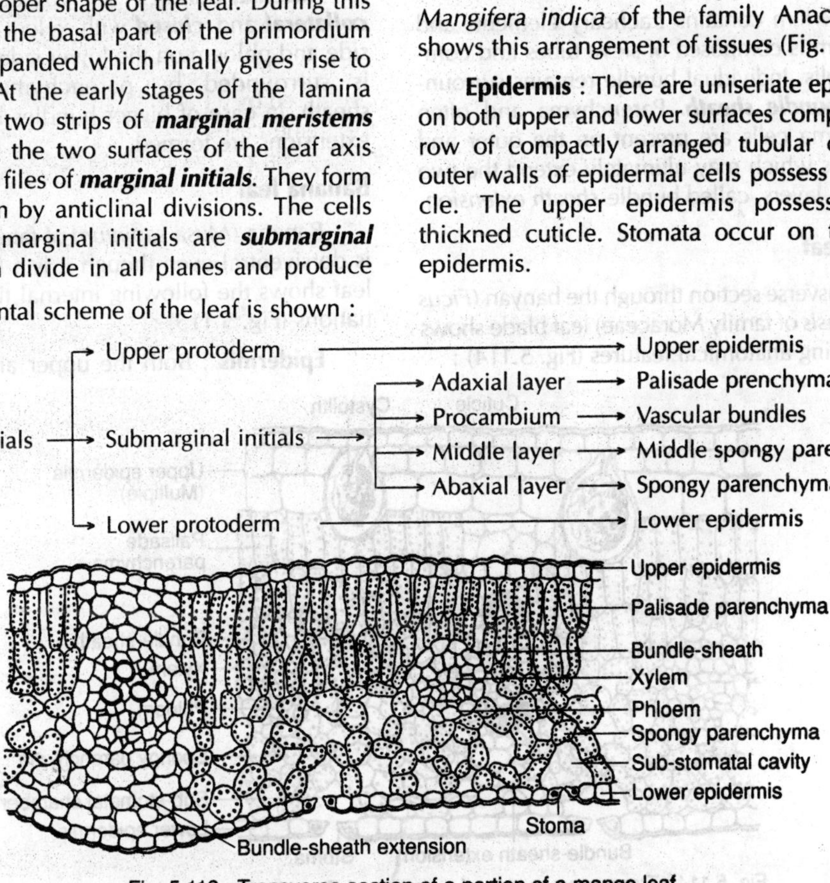

Fig. 5.113 : Transverse section of a portion of a mango leaf

Mesophyll : The ground tissue is the ***mesophyll tissue*** which is differentiated into ***palisade*** and ***spongy*** parenchyma. The columnar palisade cells with scanty intercellular spaces remain beneath the upper epidermis. They are paralelly arranged and remain at right angles to the leaf surface. Large number of chloroplasts is present in peripheral orientation inside the palisade cells. There are two layers of palisade cells. The ***spongy*** cells occur adjacent to the lower epidermis. They are very loosely arranged with conspicuous intercellular spaces. These cells contain comparatively lesser number of chloroplasts — so the lower surface of the leaf is pale green in colour.

Vascular bundles : The vascular bundles are ***collateral*** and ***closed*** and remain distributed in the mesophyll tissue. The size of the bundle varies depending on its position in the leaf. Actually the veins and veinlets represent the vascular bundles. A bigger bundle is composed of xylem towards the upper epidermis and phloem towards the lower side.

The xylem contains tracheary elements and the phloem is composed of sieve tubes and companion cells. Individual bundle remains surrounded by ***bundle sheath***. Parenchyma and often collenchyma cells are present on the outer and inner sides which may ultimately extend the two epidermal layers, called bundle *sheath extension.*

Banyan leaf

A transverse section through the banyan (*Ficus benghalensis* of family Moraceae) leaf blade shows the following anatomical features (Fig. 5.114) :

Epidermis : The upper epidermis is typically multiseriate lithocysts here and there containing well-developed calcium carbonate crystals, called ***cystoliths***. The lower epidermis is uniseriate. The outer layer of upper multiple epidermis as well as the lower epidermis is made of compactly arranged tubular cells with cutinised outer walls with a layer of cuticle which is more thick on the upper side. Stomata occur on the lower epidermis.

Mesophyll : The mesophyll tissue is differentiated into ***palisade*** and ***spongy*** parenchyma. The palisade zone consists of two or three layers of chloroplast-rich columnar cells oriented at right angles to the leaf surface. These are the principal photosynthetic cells. The spongy cells occurring towards lower side of the leaf are isodiametric, and often irregular in shape with profuse intercellular spaces. The number of chloroplasts is naturally much smaller here in comparison to palisade cells.

Vascular bundles : The vascular bundles are ***collateral*** and ***closed*** with xylem on the upper side and phloem on the lower side. Each bundle is surrounded by parenchymatous bundle sheath. In case of bigger bundles bundle sheath extensions are formed.

Banana leaf

Banana (*Musa sapintum* of family Musaceae) is dorsiventral even though it is a monocot. The leaf shows the following internal tissue differentiations (Fig. 5.115) :

Epidermis : Both the upper and lower epi-

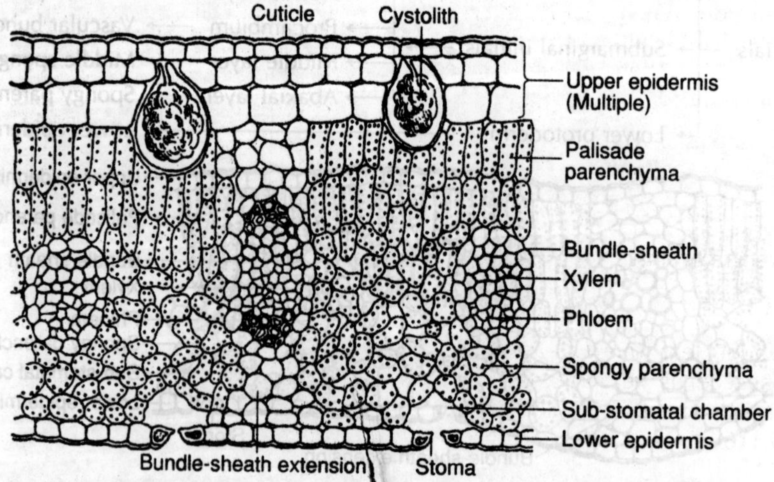

Fig. 5.114 : Transverse section of a portion of *Ficus benghalensis* (banyan) leaf

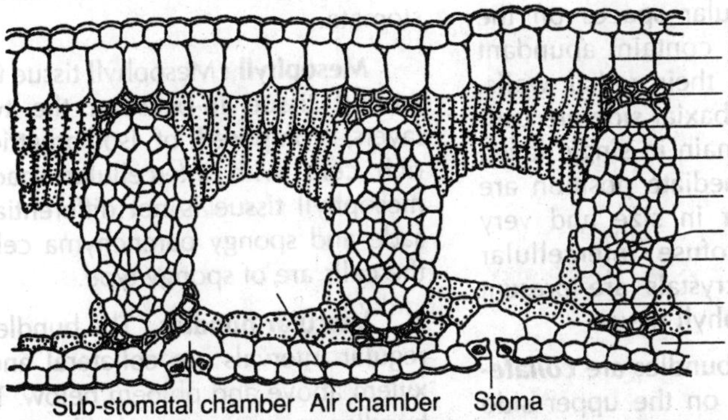

Fig. 5.115 : Transverse section of a portion of a *Musa* leaf

dermal layers are composed of compactly-arranged more or less rounded cells with cuticularized outer walls. The lower epidermis possesses slightly sunken stomata.

Internal to both the epidermal layers, one or two layers of larger thin-walled parenchyma cells occur. Those on the upper side are comparatively larger in size and are markedly different from the mesophyll cells. These subepidermal layers often designated as hypodermis. These cells are devoid of chlorophyll and perform the function of water-storage.

Mesophyll : It is differentiated into ***palisade*** and ***spongy*** parenchyma cells. The former occur towards upper side in two or three layers. These cells are oblong and columnar in shape and very tightly arranged with scanty intercellular spaces. The spongy cells are comparatively much smaller and more or less isodiametric in shape. Series of large air chambers occur at regular intervals towards abaxial side. When young stellate cells are differentiated, they ultimately disintegrate and form the cavities. Therefore, these cavities originate lysigenously.

Vascular bundles : The poorly-developed vascular bundles are ***collateral*** and ***closed*** with scanty xylem. Sclerenchyma patches occur on the upper and lower sides of the bundles to provide mechanical strength against shearing stresses.

Oleander leaf

The oleander leaf (*Nerium oleander* of family Apocynaceae) is xerophytic and shows the following arrangement of tissues in transverse section (Fig. 5.116) :

Epidermis : Both the upper and lower epidermal layers are multiseriate. They are composed of a few layers of isodiametric cells without intercellular spaces. There is a thick layer of cuticle on the epidermal surface. The sunken stomata on the lower epidermis are located in a depression known as stomatal pit. Large number of hairs is present in the bordering layer of the depression. This whole stomatal apparatus reduces the rate of transpiration.

Mesophyll : The mesophyll tissue consists of ***palisade*** and ***spongy*** parenchyma cells. But in this leaf palisade, parenchyma cells are found internal to both upper and lower epidermis and spongy parenchyma cells remain in-between. Three or more layers of palisade

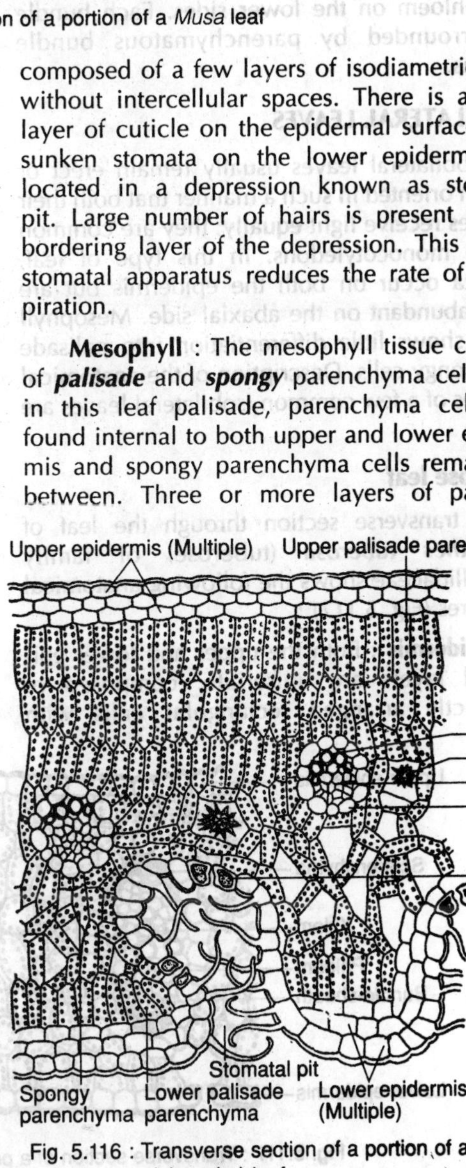

Fig. 5.116 : Transverse section of a portion of a *Nerium* (oleander) leaf

parenchyma form a compact zone of columnar cells with scanty intercellular spaces on the adaxial side. These cells contain abundant chloroplasts usually along their radial walls. The palisade cells on the abaxial side are usually smaller in size and remain in single layer. The spongy cells in intermediate position are isodiametric, much smaller in size and very loosely arranged with profuse intercellular spaces. Calcium oxalate crystals are present here and there in the mesophyll tissue.

Vascular bandles : The bundles are ***collateral*** and ***closed*** with xylem on the upper side and phloem on the lower sides. Each bundle is surrounded by parenchymatous bundle sheaths.

ISOBILATERAL LEAVES

Isobilateral leaves usually remain erect or remain oriented in such a manner that both their surfaces receive light equally; they are common in the monocotyledons. In this type of leaf, stomata occur on both the epidermis but are more abundant on the abaxial side. Mesophyll tissue shows little differentiation into palisade and spongy cells. Description of the anatomical features of a few common isobilateral leaves are given :

Tuberose leaf

A transverse section through the leaf of *Polianthes tuberosa* (tuberose) of family Amaryllidaceae shows the following anatomical structures (Fig. 5.117) :

Epidermis : Both the upper and lower epidermal layers are uniseriate, composed of compactly arranged rectangular cells with rounded cuticularised outer walls and contain stomata.

Mesophyll : Mesophyll tissue forms the main ground tissue in-between the two epidermal layers, composed of isodiametric chloroplast-rich cells with intercellular spaces. Here the mesophyll tissue is not differentiated into palisade and spongy parenchyma cells; instead all the cells are of spongy type.

Vascular bundles : The bundles occurring at regular intervals are collateral and closed with xylem above and phloem below. The individual bundles remain surrounded by a row of achlorophyllous parenchymatous cells forming the ***bundle sheath***. A few sclerenchyma cells are present on the upper and lower ends of the bundles inside the bundle sheath.

Maize leaf

A transverse section through maize (*Zea mays* of family Poaceae) leaf blade shows the following anatomical features (Fig. 5.118) :

Epidermis : Both upper and lower epidermal layers are uniseriate and composed of more or less oval cells with a cuticle layer on the leaf surface. The upper epidermis contains large and empty bulliform cells. Stomata are present on both the epidermal layers.

Mesophyll : The mesophyll tissue is not differentiated into palisade and spongy cells. It is composed of rather compactly-arranged isodiametric cells.

Vascular bundles : The vascular bundles are collateral and closed, arranged in parallel lines. Most of the bundles in series are small. Larger bundles occur at regular intervals. Small bundles

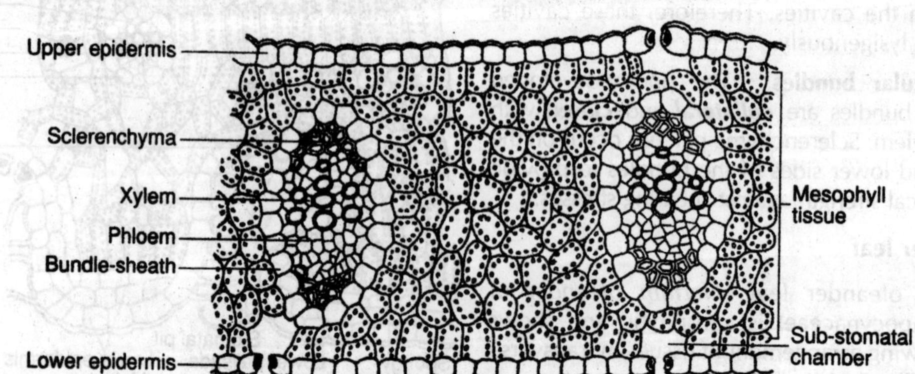

Fig. 5.117 : Transverse section of a portion of a *Polianthes* (tuberose) leaf

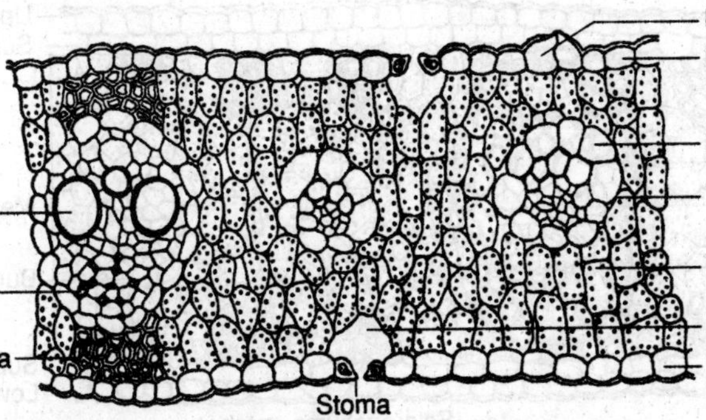

Fig. 5.118 : Transverse section of a portion of a *Zea* (maize) leaf

have xylem on the upper side and phloem on lower side surrounded by bundle sheaths. The bundle sheath cells contain plastids which are agranal or with very feebly developed grana with starch grains. Maize is a C-4 plant where the Hatch and Slack pathway operates to concentrate CO_2 in the bundle sheath cells to minimise photo-respiration. Inside the bundle sheath cells Calvin cycle is operated. It is assumed that this layer serves as a temporary storage tissue, apart from conducting the products of photosynthesis to the phloem. Xylem consists of tracheary elements, and phloem tissue contains sieve tubes and companion cells. Sclerenchyma cells occur in patches on both edges of the bundles external to the bundle sheath to give mechanical strength to the leaf.

Leaf of Bamboo

A transverse section through the leaf blade of bamboo reveals the following anatomical features (Fig. 5.119) :

Epidermis : There are upper and lower epidermal layers. The upper one contains a numerous conspicuous bulliform cells, whereas the lower one bears stomata and a thick cuticle layer on its outer surface. Silicon is deposited on the walls of the epidermal layers. Stiff sharply pointed hairs are also present.

Mesophyll : The mesophyll tissue composed of compactly-arranged cells and forms the ground tissue, which shows no differentiation into palisade and spongy cells. The cells are less elongated and palisade type. Some distinct cavities are present here and there.

Vascular bundles : The vascular bundles are collateral and closed arranged in lines. Xylem remains on the upper side and phloem on the lower side. The larger bundles contain more prominent xylem and phloem surrounded by a *bundle sheath*. Patches of sclerenchyma cells on the two sides of the larger bundles external to the bundle sheath occur to give mechanical strength to the leaf.

Leaf of Datepalm

A transverse section through the leaf blade of date-palm (*Phoenix sylvestris* of family Palmae) shows the following structure (Fig. 5.120) :

Epidermis : Epidermal layers on both upper and lower surfaces are composed of compactly-arranged tubular cells with a deposition of

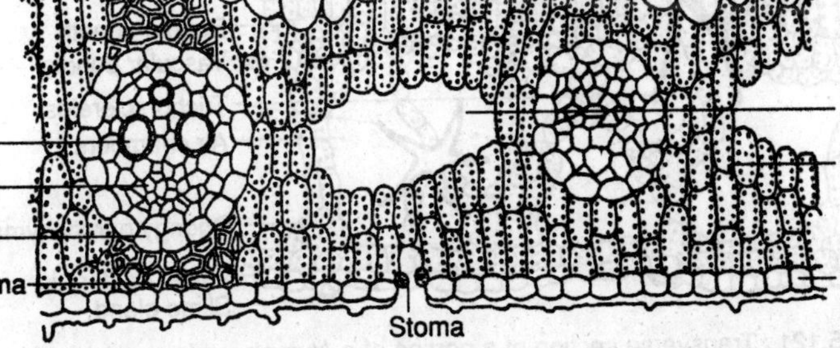

Fig. 5.119 : Transverse section of a portion of a *Bambusa* (bamboo) leaf

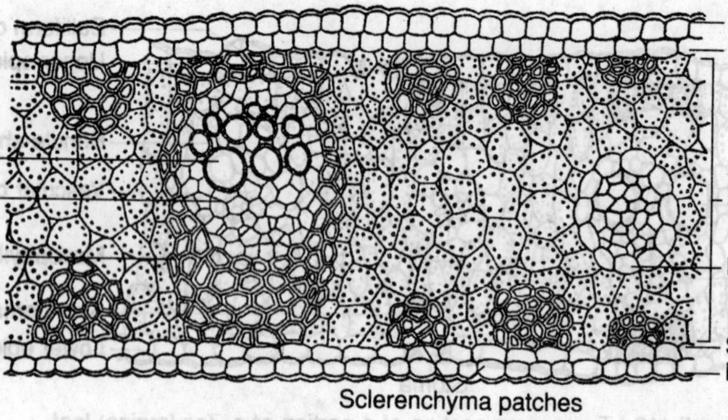

Fig. 5.120 : Transverse section of a portion of a *Phoenix sylvestris* (datepalm) leaf

cuticle layer on the outer surface. Stomata are present on both the surfaces. A subepidermal hypodermis consisting of parenchyma cells with scanty chloroplastids occurs just internal to both the epidermal layers.

Mesophyll : The mesophyll tissue is not differentiated into spongy and palisade cells. The cells of this zone are more or less isodiametric with small intercellular spaces. Sclerenchyma patches occur more or less in parallel series just internal to both the upper and lower hypodermis in such a way that the opposite patches may form I-girders for withstanding shearing stress.

Vascular bundles : The vascular bundles are collateral and close, arranged in parallel series. Each bundle has xylem on the upper and phloem on the lower side. The larger bundles have patches of heavily thick-walled sclerenchyma on their two edges, whereas the smaller ones remain surrounded by achlorophyllous parenchymatous *bundle sheath*.

Floating leaf of Waterlily

A transverse section through the leaf lamina of waterlily (*Nymphaea stellata* of family Nyphaeaceae) reveals the following anatomical characteristics (Fig. 5.121). The main feature of this type of leaf is that the vascular and mechanical tissues in it are very much reduced with well-developed air-chambers due to aquatic adaptations.

Epidermis : The internal tissues of the leaf blade are bounded by the upper and lower uniseriate epidermal layers. They are composed of compactly arranged cells. Stomata are restricted on the upper side only.

Mesophyll : The mesophyll tissue of the leaf is differentiated into ***palisade*** and ***spongy*** parenchyma cells. The palisade zone is composed of a few layers of chloroplastid containing columnar cells occur towards the adaxial side forming the palisade. The spongy cells with irregular outline are present towards lower epider-

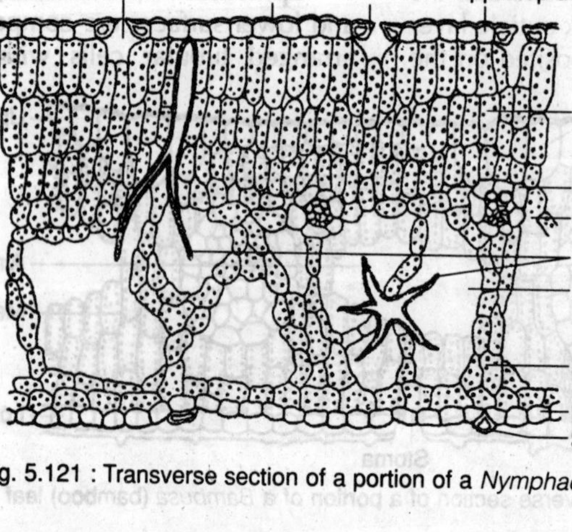

Fig. 5.121 : Transverse section of a portion of a *Nymphaea* (waterlily) leaf

mis. Large air chambers are present in this spongy zone of the mesophyll. Elongated sclerotic trichosclereid cells commonly called 'internal hairs' are frequently present. Very often their ends are branched

Vascular bundles : The vascular bundles are very much reduced composed of xylem and phloem. Each bundle is surrounded by parenchymatous *bundle sheath.*

Anatomy of a Succulent leaf

A transverse section through the leaf blade of a succulent leaf with halophytic adaptations reveals the following anatomical features (Fig. 5.122) :

Epidermis : Both the epidermal layers are uniseriate composed of compactly-arranged tubular cells with a cuticle layer on the outer surface.

Mesophyll : A row of columnar palisade parenchyma cells occurs beneath the epidermis forming more or less a compact band throughout.

Central zone : The central part of the leaf is made of well-developed water-storage tissue, consisting of large parenchymatous cells with conspicuous vacuoles and intercellular spaces.

Vascular bundles : Vascular bundles are very much reduced with very scanty mechanical elements located just beneath the palisade layer.

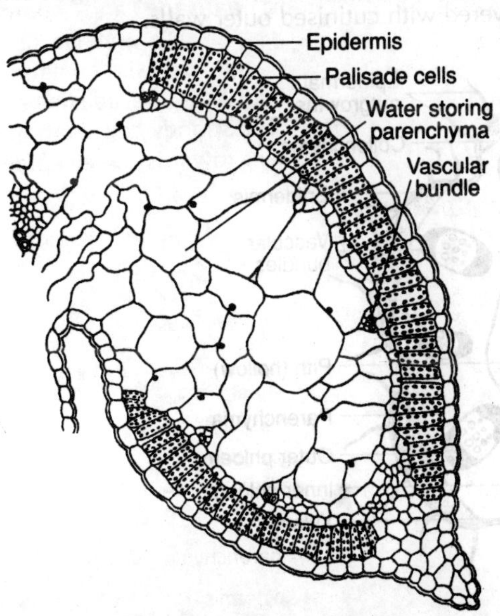

Fig. 5.122 : Transverse section of half lamina of a *Solsola* (succulent) leaf

Cylindrical leaf of Onion

Onion leaves are cylindrical in shape and are more or less circular in cross-sectional outline. A transverse section through the cylindrical leaf of onion (*Allium* cepa of family Liliaceae) shows the following anatomical features (Fig. 5.123) :

Epidermis : The single uniseriate circular epidermis is composed of small more or less rounded cells with strongly cuticularised outer walls. It is interrupted here and there by slightly-depressed stomata.

Mesophyll : The mesophyll tissue consists of palisade and spongy parenchyma cells. Just internal to epidermis there are usually two layers of columnar palisade cells with abundant chloroplasts forming something like a green belt. Internal to this band there occur a few layers of isodiametric spongy cells, with scanty chloroplastids. The next zone towards the centre is occupied by larger thin walled achlorophyllous parenchymatous cells with conspicuous intercellular spaces. The central part of the cylinder is hollow.

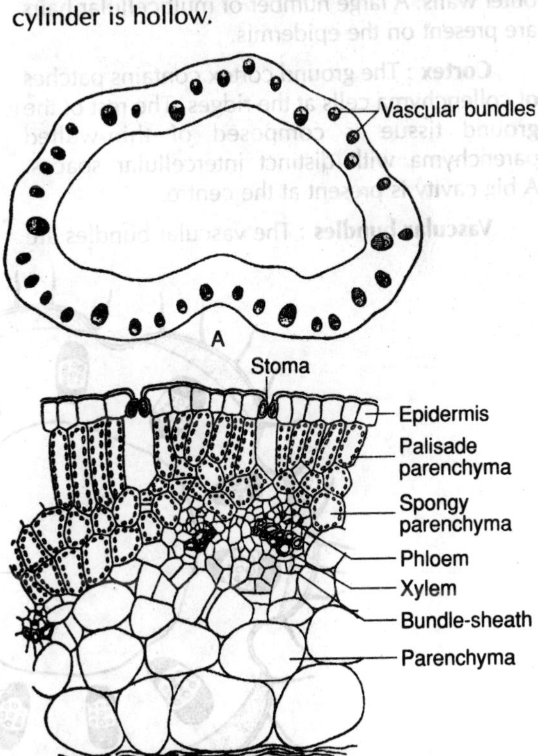

Fig. 5.123 : Transverse section of leaf of *Allium* (onion) : A. Diagrammatic view in T.S., B. Cellular drawing of a portion of the leaf in T.S.

Vascular bundles : The bundles remain scattered in the lower part of the mesophyll. They are ***collateral*** and ***closed*** with xylem towards the centre and phloem towards the periphery. Each bundle remains surrounded by parenchymatous *bundle sheath.*

THE PETIOLE

The petiole connects the lamina with the base and thus holds the lamina in proper orientation on the axis so that the lamina may receive maximum amount of light and air. Anatomically the petiole resembles the stem.

Cucurbita

A transverse section through the petiole of *Cucurbita* shows the following anatomical features (Fig. 5.124) :

It is wavy in outline having distinct ridges and furrows and a large cavity at the centre.

Epidermis : The single layered epidermis is made of a row of tubular cells with cuticularised outer walls. A large number of multicellular hairs are present on the epidermis.

Cortex : The ground cortex contains patches of collenchyma cells at the ridges. The rest of the ground tissue is composed of thin-walled parenchyma with distinct intercellular spaces. A big cavity is present at the centre.

Vascular bundles : The vascular bundles are arranged in a ring. Individual bundles are conjoint, bicollateral and open.

Sunflower

The cross-sectional outline of the petiole of sunflower (*Helianthus annus* of family Compositae) is concave at one side and convex at the other. The tissue arrangement is as given (Fig. 5.125) :

Epidermis : The single layered epidermis consists of tubular cells and a large number of epidermal hairs.

Cortex : In the ground cortex there occurs a collenchymatous hypodermis, broken by chlorenchyma here and there, internal to epidermis. The rest of ground tissue is made of isodiametric parenchyma cells with distinct intercellular spaces.

Vascular bundles : A number of vascular bundles of different sizes occur in the form of a band. Some bundles are small and some are quite large. Every bundle is conjoint and collateral.

Piper betle

The cross sectional outline of the petiole of *Piper betle* of family Piperaceae is more or less heart-shaped with a concave groove at one side and convex surface at the other (Fig. 5.126).

Epidermis : The epidermis is single cell-layered with cutinised outer walls.

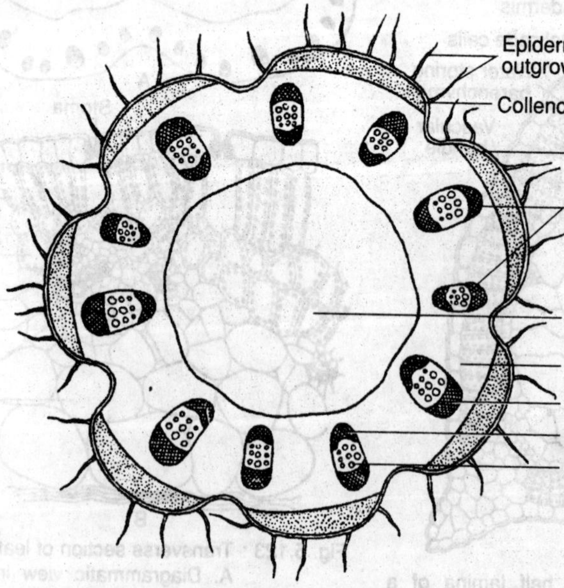

Fig. 5.124 : Transverse section of petiole of *Cucurbita* (diagrammatic)

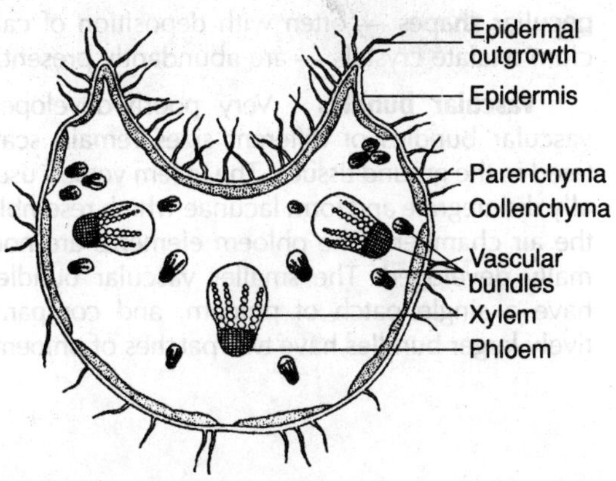

Fig. 5.125 : Transverse section of petiole of *Helianthus* (sunflower)

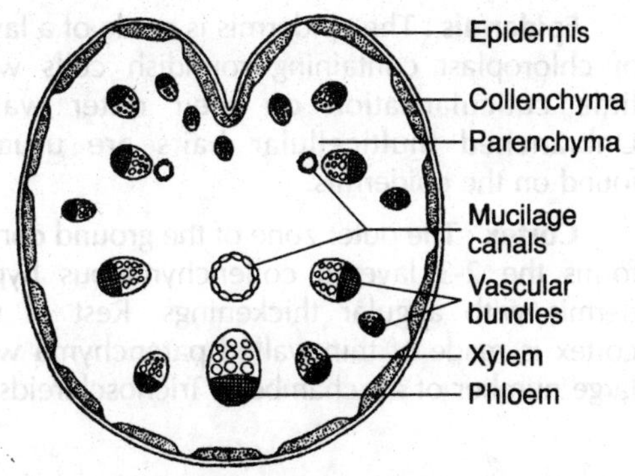

Fig. 5.126 : Transverse section of petiole of *Piper betle* (diagrammatic)

Cortex : The outer zone of the ground cortex forms the collenchymatous hypodermis interrupted here and there by small patches of chlorenchyma. The rest zone is distinctly parenchymatous. Many large and small mucilage canals remain scattered throughout the ground tissue. The larger canals remain at the center.

Vascular bundles : Vascular bundles of different sizes occur in the form of an arc. The vascular bundles are conjoint and collateral. The xylem remains towards the centre and the phloem remains towards the periphery.

Waterlily

The petiole of waterlily (*Nymphaea stellata* of the family Nymphaeaceae) are characterised by large air chambers, branched trichosclerides or internal hairs with calcium oxalate deposition, and scattered vascular bundles with very poorly developed xylem elements. Transverse section of the petiole of waterlily shows the following anatomical features (Fig. 5.127) :

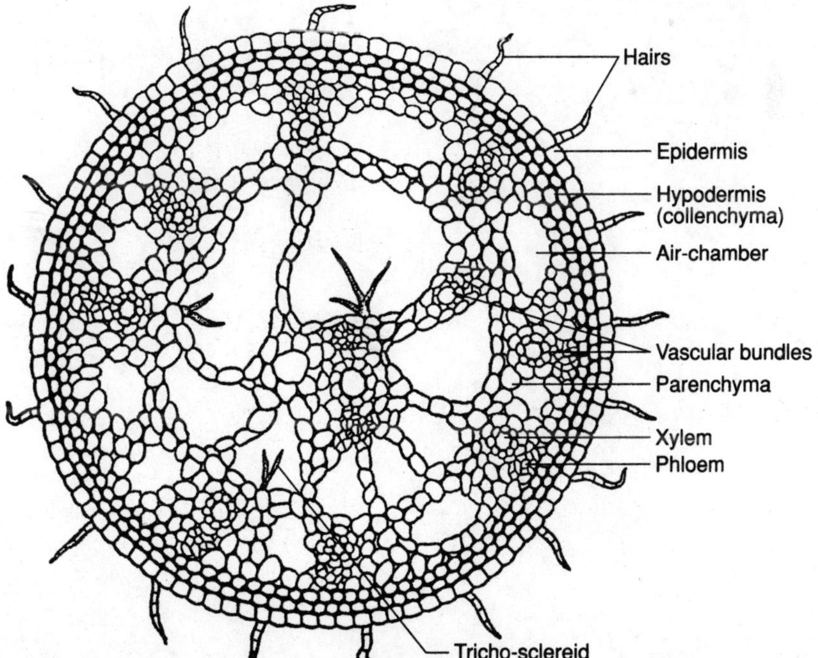

Fig. 5.127 : Transverse section of petiole of *Nymphaea* (waterlily)

Epidermis : The epidermis is made of a layer of chloroplast containing roundish cells with little cuticularisation on their outer walls. Unbranched multicellular hairs are usually found on the epidermis.

Cortex : The outer zone of the ground cortex forms the 2-3 layered collenchymatous hypodermis with angular thickenings. Rest of the cortex is made of thin-walled parenchyma with large number of air chambers. Trichosclereids of peculiar shapes — often with deposition of calcium oxalate crystals — are abundantly present.

Vascular bundles : Very poorly-developed vascular bundles of different sizes remain scattered in the ground tissues. The xylem vessels usually disintegrate and form lacunae which resemble the air chambers. The phloem elements are normally developed. The smaller vascular bundles have a single patch of phloem, and comparatively larger bundles have two patches of phloem.

SECONDARY GROWTH IN THICKNESS

The primary plant body, originates from the apical meristem i.e., the fundamental parts of the plant body are produced there and grow in length and thickness. Further growth in thickness is achieved in the regions which become quite distant from the apices in the dicotyledons and gymnosperms. This further growth is called ***secondary growth in thickness***. This growth in thickness starts by the activities of the lateral meristems like ***cambium*** and ***phellogen***, also called ***cork cambium***. The derived tissues are known as ***secondary tissues***. The cambium gives rise to the secondary vascular tissues and the phellogen or cork cambium gives rise to the periderm.

SECONDARY GROWTH IN DICOTYLEDONOUS STEMS

Growth in Intrastelar region

In dicotyledonous stems, distinct secondary growth in thickness is the usual occurrence as their vascular bundles are conjoint, collateral and open. This vascular cambium is also known as a ***fascicular cambium***. There are certain types of secondary growth in thickness in dicotyledonous stems.

In one type, where vascular bundles remain arranged in a ring with narrow medullary rays, secondary growth occurs in the fascicular as well as interfascicular regions. Due to the differentiation of interfascicular parenchymatous cells of the medullary rays into interfascicular cambium, a complete ***cambium ring*** is produced. The secondary vascular tissues are added on either side of the cambium ring. Thus a continuous cylinder of secondary vascular tissues is formed. This method of secondary growth is commonly found in erect stems like the members of compositae and also in many dicotyledonous stems.

In another type, where the primary vascular bundles remain arranged in a continuous cylinder, the secondary vascular tissues are also added in the same form. This type is found in *Tilia, Nicotiana*, etc.

In the third type — that prevails in climbers like *Aristolochia* — broad medullary rays present in-between the vascular bundles are converted into interfascicular cambium and goes on producing only parenchyma cells, and, as a result, a conti-nuous ring of secondary vascular tissues is not formed. Thus the secondary vascular tissues occur in discontinuous strands. In climbers like *Cucurbita*, secondary growth is confined to the individual bundles and does not extend to the interfascicular regions. Besides these, anomalous secondary growth occurs in a good number of dicotyledonous plants.

The process of secondary growth in thickness in a typical dicotyledonous stem is discussed below. It starts in the intrastelar region and then extends to the extrastelar region.

Cambium

Cambium is a lateral meristem the cells of which are in a layer or strips or cylinder at stelar or extrastelar region with chiefly periclinal mode of cell division.

Cambium is formed from the procambium strand which remains undifferentiated during the progressive process of maturation leading to the formation of the primary body and ultimately occurs between xylem and phloem. It is called ***fascicular cambium*** (fascicle means bundle). During the course of secondary growth a few living parenchymatous cells of the medullary rays, which remain in the G_0 stage of the cell cycle, resume the meristematic activity and, thus, new strips of meristems are formed in the same line of circumference with the strips of fascicular cambia. Thus a continuous ***cambium ring*** or ***cylinder*** is formed. The newly formed cambium strips in-between the vascular bundles are known as ***interfascicular*** cambia.

The fascicular cambium consists of spindle-shaped elongated ***fusiform initials***, and comparatively smaller and isodiametric ***ray initials***. The former ones produce the secondary tissues forming the ***vertical*** or ***longitudinal system*** parallel to the long axis such as all the elements of xylem and phloem except the horizontally arranged ray cells. The xylem and phloem rays forming the ***transverse*** or ***horizontal system*** are produced by the ray initials.

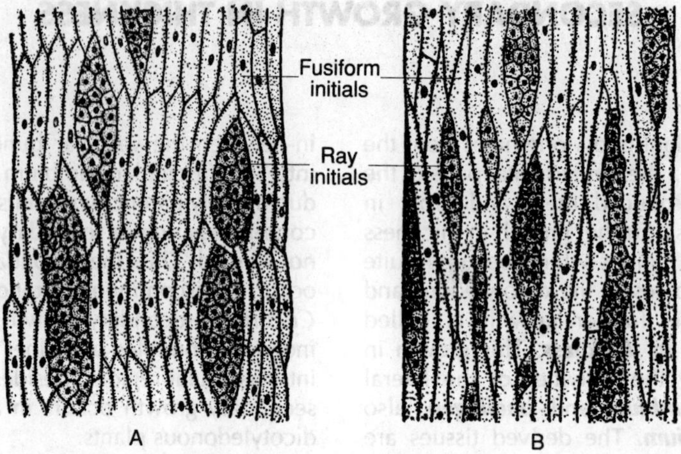

Fig. 5.128 : Fusiform initials and ray initials of cambium in longitudinal views : A. Storied cambium. B. Non-storied cambium

Initially, the cambium remains single-layered in thickness. With the onset of the active secondary growth the cambium cells indistinguishable with their nascent derivatives together form a zone, referred to as ***cambium zone***. The cells of this zone possess conspicuous vacuoles and primary pit-fields with plasmodesmatal connections.

In tangential views the cambium cells appear in two ways. In one way, the fusiform initials appear in horizontal tiers in which the lengths of the fusiform cells are equal and their ends remain more or less at the same level without overlapping. This type of cambium cells is known as ***storied*** or ***stratified cambium***. In other way, the comparatively longer fusiform initials do not appear in horizontal tiers but overlap at their ends. This is known as ***non-storied*** or ***non-stratified*** cambium. In all highly developed dicotyledons the cambium is of storied type.

The plane of division of the cambium cells is tangential adding ***secondary xylem*** and ***secondary phloem*** to the internal and external sides, respectively. Each cambium initial divides into two daughter derivatives, one of which becomes either a xylem mother cell or a phloem mother cell depending on its position, while the other one remains meristematic. The xylem or phloem mother cell is finally differentiated through further division into secondary xylem or secondary phloem elements. Thus the cambium cells continue secondary growth and perpetuate themselves as active for a longer period. Usually, the cambium produces more xylem than phloem thus pushing all the tissues external to the cambium ring more and more towards the periphery. Due to this mode of growth of secondary tissues

Table 5.12 : Difference between procambium and cambium

	Procambium	Cambium
1.	It produces the primary vascular tissues i.e., primary xylem and phloem during differentiation.	It produces secondary vascular tissues and non-vascular tissues in some cases.
2.	It originates from promeristem.	It originates from procambium.
3.	The procambial cells take deep stain.	The cambial cells do not stain deeply.
4.	It consists of single type of cell.	It consists of fusiform initials and ray initials.
5.	In radial longitudinal section (RLS) the procambial cells appear as elongated with tapered ends.	In RLS the cambial cells are elongated and the two flat ends.

the conjoint primary xylem and primary phloem are gradually pushed apart from each other as well as from the cambium. Usually, in the plants growing in tropical regions cambium is active throughout the year. In temperate regions cambial activity may cease with the onset of unfavourable winter and enter into a dormant state till favourable summer conditions are available.

Secondary Xylem

Secondary xylem constitutes the major portion of the secondary vascular tissue in woody plants performing a number of important functions, viz., conduction of water and dissolved mineral matters, mechanical support, storage of food etc. Secondary xylem is mostly made of thick-walled cells consisting of both vertical and horizontal systems as derived from the fusiform and ray initials of the cambium. The vertical system consists of the tracheary elements, tracheae and tracheids, xylem parenchyma and fibres. The horizontal system consists only of xylem rays arranged at right angles to the long axis. The living ray cells, xylem parenchyma, pith cells, living cells of phloem and cortex together form a ***symplastic system*** through plasmodesmatal connections.

The secondary xylem elements are more or less similar to those occurring in primary xylem with abundant larger pitted vessels and shorter fusiform xylem parenchyma cells meant for storage of starch or fat. Tannins, mineral crystals etc., are also frequently present in these cells. Secondary wall may be absent in xylem parenchyma, if present, the pits between them and tracheary elements are bordered, half-bordered or even simple. Secondary xylem parenchyma may occur abundantly in some species or may be reduced in number or entirely absent. The characteristics of the secondary xylem parenchyma have been considered as a

Table 5.13 : Difference between Apotracheal parenchyma and Paratracheal parenchyma

Apotracheal parenchyma	Paratracheal parenchyma
1. The axial parenchyma of secondary xylem is situated independent of vessel.	The axial parenchyma of secondary xylem is distinctly associated with the vessel.
2. It may be : (a) Diffuse: parenchyma cells appear as single cell or as small uniseriate band throughout growth ring (*Quercus*). (b) Banded: parenchyma cells appear as concentric rings (*Hicoria*). (c) Boundary : parenchyma appears at the beginning or end of the growth ring (*Magnolia, Ceratonia* etc.).	It may be : (a) Scanty : the parenchyma cells do not form a continuous sheath surrounding the vessel (*Acer*). (b) Vasicentric : the parenchyma cells form a continuous sheath around the vessel of different width (*Tamarix*). (c) Aliform : vasicentric parenchyma extends laterally as wings (*Acacia*).

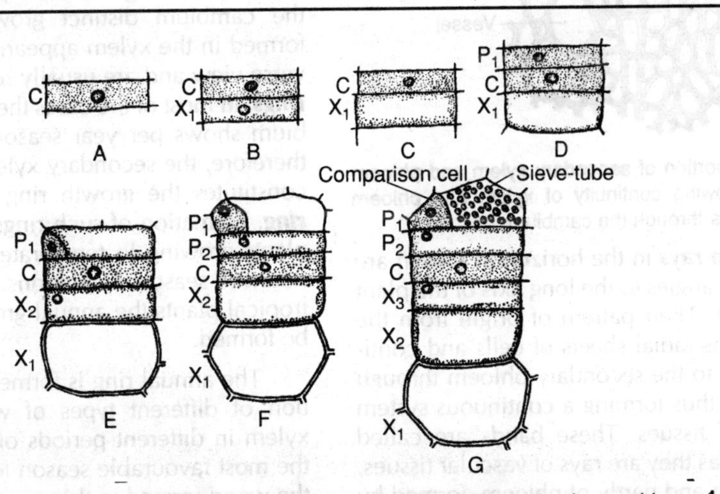

Fig. 5.129 : Stages of formation of secondary xylem and phloem by the cambium

taxonomic characteristic in recent years. If they occur in association with the vessels, they are called ***paratracheal*** and, if they occur independently, they are called ***apotracheal***. The former are of different types based on their distribution. The most common type is the ***vesicentric paratracheal parenchyma*** which forms entire sheaths, appearing circular or elliptical in cross-section, around the vessels. Apotracheal parenchyma is considered as primitive and, through its different evolutionary types, advanced to the paratracheal forms culminating in highest developed vasicentric type.

The secondary xylem fibres have thick walls with reduced bordered pits. The two main types of fibres are the ***fibre tracheids***, and the ***libriform fibres***. The former have pits with poorly developed borders, whereas the latter possess simple pits resembling the phloem fibres. Both these fibres are the important mechanically supporting elements of xylem.

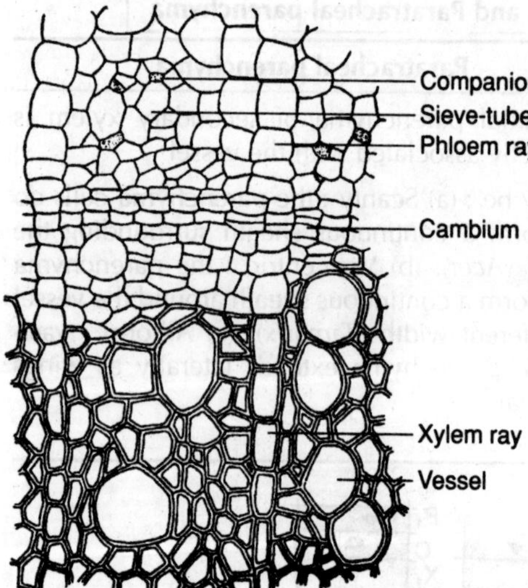

Fig. 5.130 : A portion of secondary xylem and phloem showing continuity of xylem and phloem rays through the cambium

The ***xylem rays*** in the horizontal system are placed at right angles to the long axis of the plant body or organ. Their pattern of origin from the ray initials is as radial sheets of cells and continues as bands to the secondary phloem through the cambium, thus forming a continuous system of conducting tissues. These bands are called ***vascular rays***, as they are rays of vascular tissues, partly of xylem and partly of phloem, formed by the cambium. They establish transverse communications in the living cells of the vascular tissues. They also make connections with ***intercellular space system*** of the whole plant body for gaseous exchange with intercellular spaces of their own. Apart from storage function the ray cells also perform the function of conduction of water from the xylem to the cambium and phloem, and that of food from the phloem to the cambium and living cells of xylem.

The parenchymatous ray cells in dicotyledons may be uniseriate or multiseriate. The ray cells in angiosperms are always living and shorter than other ray cells. Ray parenchyma cells may be oriented ***upright*** or ***procumbent***. In the former type the ray cells remain oriented vertically with their long axes, and, in the latter type, the ray cells orient their long axes horizontally. If the secondary xylem contains both the types it is said to be ***heterogeneous***; whereas if it consists of only single type, it is called ***homogeneous***. Both uniseriate and multiseriate ray cells may be homogeneous or heterogeneous. Uniseriate vertically elongated ray parenchyma and multiseriate heterogeneous ray parenchyma are considered to be primitive and homogeneous ray parenchyma to be advanced. In almost all gymnosperms the rays, are uniseriate. The ray cells may be of single type, usually living with comparatively thicker walls called ***homocellular*** rays. In gymnosperms, tracheids may be present in addition to the living parenchyma in xylem rays called ***heterocellular rays***. The dead ***ray tracheids*** have lignified walls with bordered pits.

Annual Rings : Due to periodical activity of the cambium distinct growth layers may be formed in the xylem appearing as rings in transverse view and are usually referred to as ***growth rings***. In most of the cases the activity of the cambium shows per year seasonal periodicity and, therefore, the secondary xylem formed in a year constitutes the growth ring and called ***annual ring***. Formation of such rings is characteristic in plants growing in temperate climates with pronounced seasonal variations. In tropical and subtropical plants the annual growth rings may also be formed.

The annual ring is formed due to the formation of different types of wood or secondary xylem in different periods of the year. Spring is the most favourable season for plant growth and the wood formed in this season is less dense and

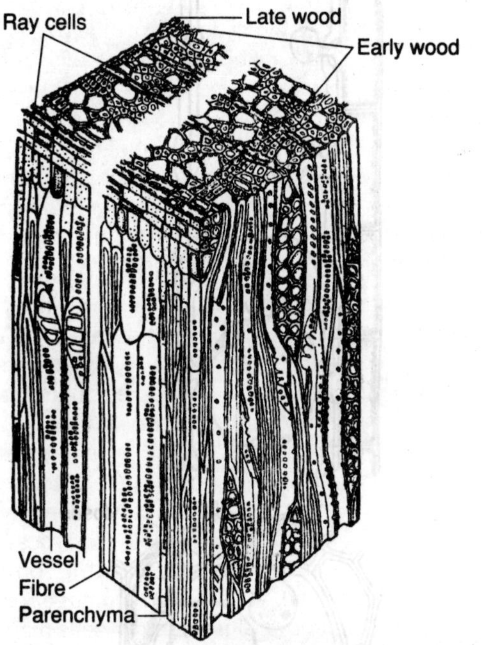

Fig. 5.131 : Block diagram of secondary xylem showing early and late woods

is made of vessels with wider diameters. Due to the rapid rate of photosynthesis in this favorable season there appears a high demand of water for photosynthate transport and, therefore, larger and wider vessels in secondary xylem are formed in that season. This is called the ***early wood*** or ***spring wood***.

In summer, the rate of photosynthesis gradually decreases mainly due to high temperature and water scarcity, and thus the secondary xylem formed in this season is dense and made of narrower, compact and lignified vessel elements. This part is called ***late wood*** or ***summer wood***.

The terms spring wood and summer wood are well suited in case of temperate plant where this type of growth pattern coincides with the seasons. But in some plants, however, the formation of late wood and summer wood is not strictly seasonal even in temperate climates. They are designated simply as early wood and late wood. Winter is the resting season when cambium shows minimum activity. In-between the early wood and late wood of one season a sharp line of demarcation exists. Thus concentric annual rings each consisting of inner early wood and outer late wood are observed in the transverse view of the trunks of temperate deciduous and evergreen plants. In some tropical trees distinct annual rings are formed due to periodic cambial activity in alternate wet and dry seasons. Well-marked rings are found in the important timber yielding plants like *Tectona grandis* (teak), *Dalbergia sissoo* (sissoo), *Albizzia lebbeck* (sirish), *Salmalia malabarica* (silk cotton), etc.

The seasonal activity of the cambium is influenced by adverse climatic conditions like drought, defoliation and diseases. In such situation, wood development is temporarily stopped and growth is resumed later in the same season. As a result, additional layers, called ***false annual rings***, are formed. Thus more than two false

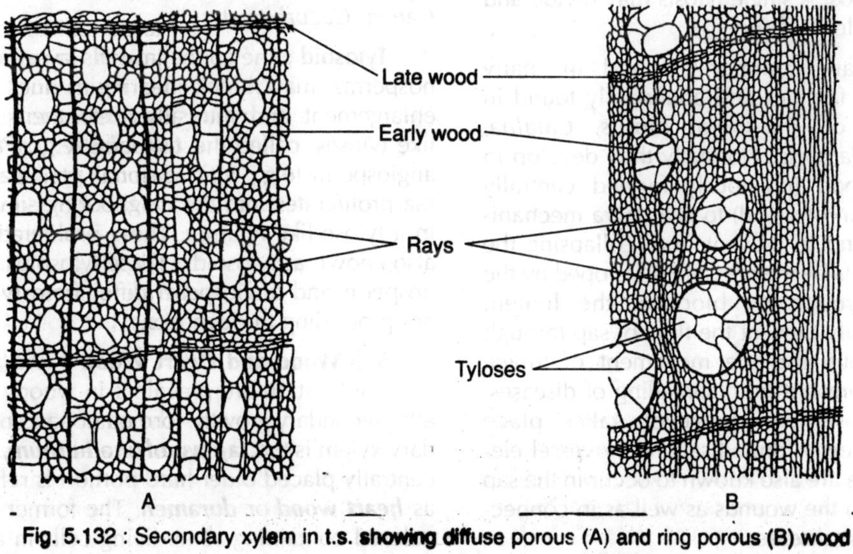

Fig. 5.132 : Secondary xylem in t.s. showing diffuse porous (A) and ring porous (B) wood

rings may be formed. This is known as ***multiple annual rings***.

The early wood is more conspicuous than the late wood due to the formation of wider and larger vessels. The vessels in the early wood are of unequal diameter and the largest ones show a ring-like arrangement in transverse section. This type of wood is called ***ring porous wood***. Sometimes the vessels of uniform diameter are found to be distributed uniformly throughout the wood. But when there is a gradient of distribution in size from smaller to larger ones throughout the growth ring it is called ***diffuse porous wood***. Phylogenetically ring-porous wood is considered more advanced than the diffuse porous wood as intermediate forms are observed between them.

Tyloses : Tyloses (singular : tylosis) are the balloon-like protrusions into the tracheary elements of the secondary xylem in certain plants. These are developed by the adjacent xylem parenchyma and ray parenchyma cells. Such ingrowths may be developed in the primary xylem also. Tyloses are the enlarged protrusions of the pit membranes of the half-bordered pits from a living parenchyma cell into a vessel or a tracheid lumen. Often they are so large that the lumen of the vessel or tracheid is almost blocked. A tylosis contains a nucleus with a certain amount of cytoplasm coming from the living cell. Starch, resin and other matters may also be present in it. The initial thin walls of tyloses may undergo lignification later leading to the formation of sclereids. A single tylosis may divide and form a multiple structure.

Tyloses are commonly found in many angiospermic families and abundantly found in the species of *Tinospora, Morus, Catalpa, Quercus, Juglans* etc. Normally they develop in the heart wood (previously formed centrally located portion of wood) to give extra mechanical support and to prevent from collapsing the vessel elements by the pressure developed by the secondary growth by blocking the lumen. Tyloses not only prevent the flow of sap through the vessels but also the movement of fungal hyphae and prevent from spreading of diseases. The sap movement, however, takes place through the newly formed secondary vessel elements. Tyloses are also known to occur in the sap wood beneath the wounds as well as in connection with some diseases.

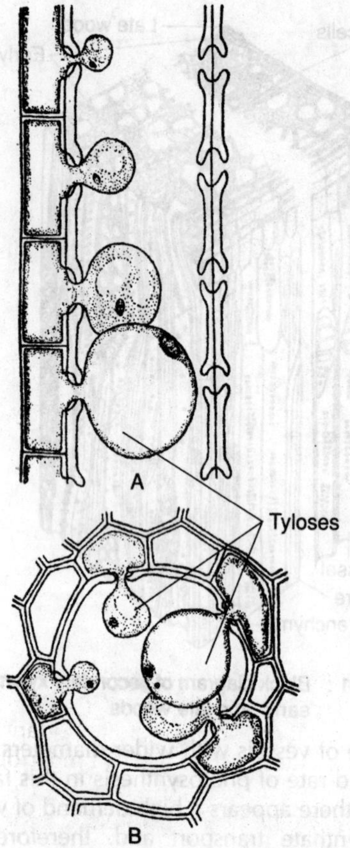

Fig. 5.133 : Secondary xylem in T.S. showing diffuse porous (A) and ring porous (B) wood

Occurrence of tyloses has also been reported in some herbaceous plants like *Coleus, Canna, Cucurbita* etc.

Tylosoid : The epithelial cells in certain gymnosperms may block the resin duct by the enlargement and intrusions into them looking like tyloses, called the ***tylosoids*** (e.g., *Pinus*). In angiosperm (e.g., *Vitis, Bombax* etc.) parenchyma proliferates into the neighboring sieve tubes in a tylose-like manner. These proliferations are also known as tylosoids. These tylosoids of gymnosperm and angiosperm differ from tyloses in not protruding through pits.

Sap Wood and Heart Wood : Two types of wood formation are prevalent in woody species after secondary growth – the outer nascent secondary xylem is the ***sap wood*** or ***alburnum***, and the centrally placed older hard portion is referred to as ***heart wood*** or ***duramen***. The former is light-colored containing some living cells in addition to the tracheary elements and fibres. This part actually carries on the water and solute conduction in addition to mechanical support and storage of food. The hard central dark colored heart wood mainly consists of dead elements representing merely a solid mechanical column. The sap wood is gradually transformed into heart wood through certain changes like :

1. The loss of protoplasts in the living cells.
2. Reduced water contents.
3. Withdrawal of the food matters from the cells.
4. Formation of tyloses to block the lumens of the tracheary elements.
5. Considerable lignification of the walls of the parenchyma cells.
6. Infiltration of substances like gums, resins, oils and particularly tannins and colouring matters on both the walls and lumen of the elements.

Finally, the heart wood becomes more compact, strong and dark-colored. The heart wood is commercially more valuable than the sap wood but the latter is functionally more important. The heart wood yields good quality of timber due to its durability and resistance to decay. The dye haematoxylin is obtained from the heart wood of *Haematoxylum campechianum*.

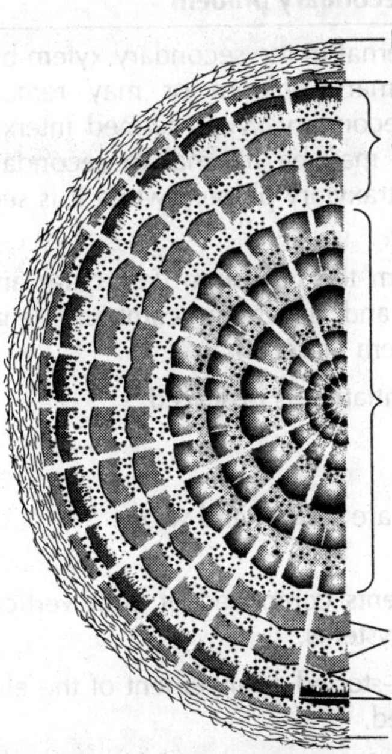

Fig. 5.134 : Secondary growth in stem in T.S. showing sap wood and heart wood

The living xylem elements of sap wood in course of time become dead and finally converted to heartwood. However, in *Sequoia sempervirens* the ray parenchyma of heart wood may remain alive for hundred years. The ratio between the quantity and the degree of difference of heart wood and sap wood varies with different conditions of growth. In *Abies, Picea* etc. the heart woods are not well differentiated. *Taxus* and *Morus* possess thin sap wood while *Fagus* and *Acer* contain thick sap wood.

Secondary phloem

Due to the tangential divisions of the cambium cells secondary phloem elements are produced on the outer side. Usually the amount of secondary phloem tissue is comparatively less than that of the secondary xylem tissue. In majority of the dicotyledons the primary phloem is crushed due to pressure originated by the secondary growth in thickness and subsequently the secondary phloem takes over the physiological functions for a considerable period.

The secondary phloem elements again are oriented in vertical and horizontal systems. The sieve tubes, companion cells, phloem parenchyma and fibers together form the vertical system. They are the derivatives of the fusiform initials of the cambium, chiefly responsible for vertical conduction of food matters in different directions. The ray parenchyma cells formed by the ray initials are the only components of the horizontal system. These ray parenchyma cells may also remain arranged in storied and non-storied form like the secondary xylem.

The secondary phloem is more complex than the primary phloem though both are fundamentally similar on the basis of elemental analysis. The secondary phloem elements remain arranged in radial rows. The sieve tubes and phloem fibers are usually more numerous. The sieve plates are usually simple, but compound plates may also be present in some plants. The companion cells and the sieve tubes together form the sieve elements as they are derived from the same mother cell. Single companion cell may accompany the sieve tube along its entire length or a few companion cells may be associated with it.

In gymnosperms the vertical system consists of sieve cells, phloem parenchyma and albuminous cells. In dicotyledons, phloem parenchyma are very numerous in the secondary phloem, but with maturity their number decreases. These parenchyma cells may be elongated with pointed ends or may be short rectangular or oblong. They play a role in vertical conduction of solute as well as storage of starch, crystals and other ergastic substances. Abundant presence of sclerenchyma as discrete tangential bands, or as isolated patches, or even singly is the characteristic of secondary phloem. Patches of fibers may also be present in regular alteration with the bands of sieve tubes, companion cells and parenchyma. Sclereids are one of the components of the secondary phloem in association with the fibres, the number of which increases in the non-functional phloem through sclerosis of phloem parenchyma. Elongated fibre-like elements formed by sclerosis of parenchyma cells have been called ***fibre-sclereids***. The fibres and the sclereids give the mechanical strength. Many fibres of commercial importance belong to this secondary phloem.

Phloem Rays : From the ray initials of the cambium phloem rays originate and constitute the ***horizontal system*** of the phloem. The phloem ray cells are the components of this system. They establish connections with the xylem rays and other living cells of the vascular tissues through the cambium. They are responsible for horizontal transport of photosynthates to cambium and other living cells and water from xylem into them. The phloem rays and xylem rays adja-

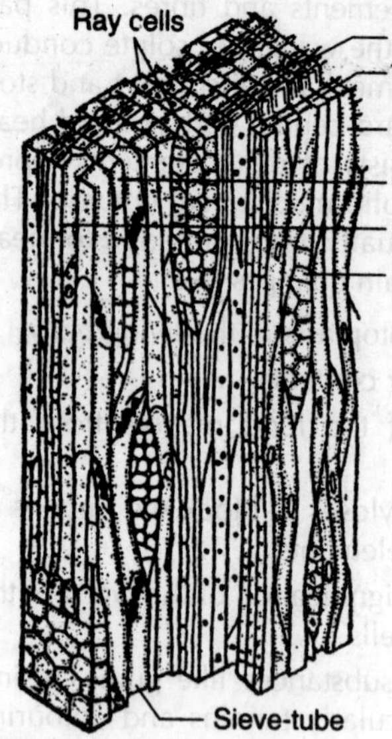

Fig. 5.135 : Block diagram of secondary phloem elements with the cambium

Table 5.14 : Difference between primary and secondary phloem

Primary phloem	Secondary phloem
1. It is usually present external to the primary xylem. It may also be present internal to the primary xylem called intraxylary phloem.	It is formed external to the secondary xylem but internal to primary phloem or may remain embedded in secondary xylem termed interxylary phloem. It may be internal to secondary xylem called intraxylary phloem when it is secondary in origin.
2. It originates from procambium of the apical meristem.	It originates from the cambium (fascicular and interfascicular) and accessory cambium (e.g., intraxylary phloem of *Tecoma*).
3. Protophloem and metaphloem are distinguished in it, based on mode of differentiation.	No such differentiation is observed.
4. The sieve tubes are larger with narrow lumen.	The sieve tubes are shorter and wider.
5. The sieve elements remain parallel to the long axis of the organ.	The sieve elements comprised of both vertical and horizontal systems.
6. Storied and non-storied arrangement of the elements cannot be distinguished.	Storied and non-storied arrangement of the elements is observed.
7. No growth rings are formed.	In some dicotyledonous trees, faint growth rings are observed.

cent to the cambium are of equal size. With maturity of the plant phloem ray cells gradually increase in width. They may be (a) ***uniseriate*** and parenchymatous, as in conifers, or (b) ***multiseriate*** and homogeneous, composed of either upright or procumbent cells, or (c) ***heterogeneous***, made of both upright and procumbent cells.

No growth ring is formed in secondary phloem. The elements of secondary phloem remain arranged in regular tangential series and thus exhibit very faint concentric rings. But unlike the secondary xylem there is no seasonal variation in the growth pattern and, therefore, no line of demarcation is observed between the elements formed in consecutive seasons. In some species of dicotyledons, however, very faint rings are observed due to differences in the cells produced at the beginning and end of the season. Tangential bands of fibres have also been observed in some plants. But, in any way, they do not indicate the age of secondary phloem.

GROWTH IN EXTRASTELAR REGION

Periderm

Continued secondary growth in the intrastelar region by the activity of the cambium cylinder large amount of pressure is exerted on the extrastelar tissues, which migrate ultimately to the epidermis. As a result, the epidermis gets more stretched and ultimately tends to rupture exposing the internal tissues to the outside. In this situation, to protect the inner tissues, a new dermal tissue is formed secondarily, called ***periderm***, in the extrastelar region. It consists of three tissues :

1. A meristem known as ***phellogen*** or ***cork cambium***
2. The phellogen derived cells on the outer side, called ***phellem*** or ***cork cells***
3. The phellogen derived cells on the inner side, called ***phelloderm***.

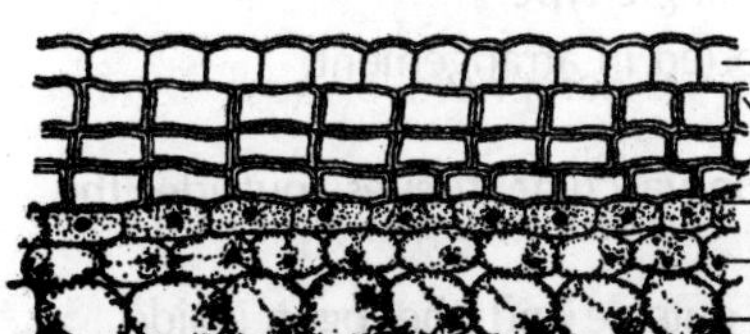

Fig. 5.136 : Diagram showing periderm formation

The phellogen or cork cambium is a secondary meristem arising from the living cells staying in the G_0 stage of the cell cycle. It originates as a single layer of initiating cells either (a) in the subepidermal portion, or (b) in the epidermis itself, (c) in the deep-seated layers of cortex or (d) may even extend up to the phloem.

Phellogen consists of only one type of compactly arranged initials which appear rectangular in shape and flattened radially in transverse section. The cells divide tangentially producing new tissues in both centrifugal and centripetal directions. The derivatives usually remain arranged in radial rows.

The amount of phellem produced by the phellogen on the outer side is much more than that of phelloderm on the inner side. The cork cells or phellem are uniformly produced being arranged in distinct radial rows as they originate from a tangentially dividing meristem. The cork cells are compactly arranged and have no intercellular spaces. After differentiation they become dead and become air-filled and accumulate coloring matters. The primary walls of

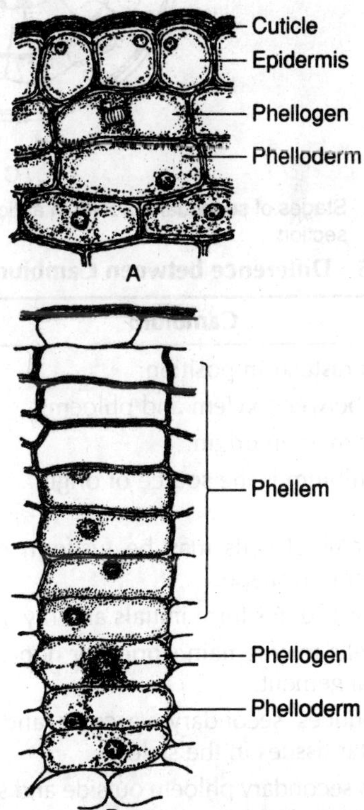

Fig. 5.137 : Stages of formation of periderm

these cells are mostly made of cellulose, partly suberin and lignin. A thick layer of suberin is deposited next to the primary walls. So the cork cells become impervious to water and gases. In *Eucalyptus* the cork cells are extremely thick-walled and the cell lumens are completely filled up with resinous and other materials. But usually they remain radially elongated, thin-walled empty cells. In *Betula* cork cell layer is peeled off like sheets of paper and the periderm serves as an effective secondary protective tissue against desiccation and mechanical injuries due to suberisation of the compactly arranged cork cells.

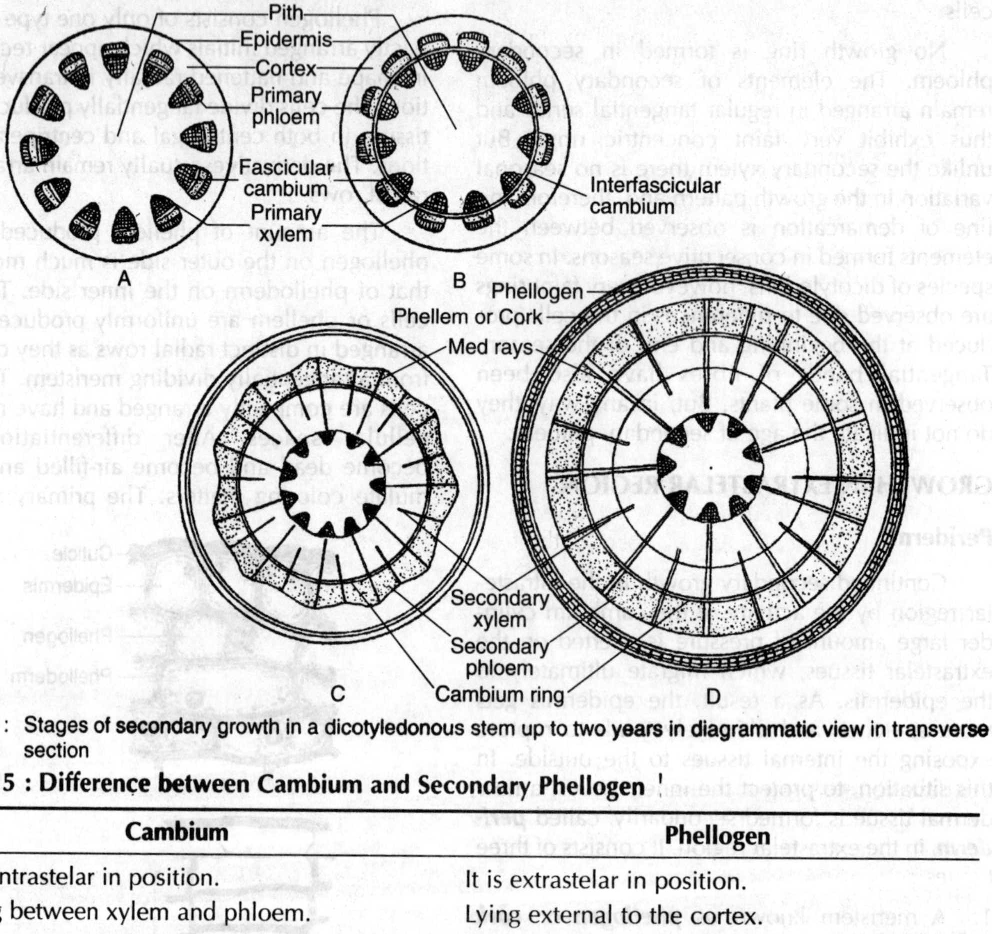

Fig. 5.138 : Stages of secondary growth in a dicotyledonous stem up to two years in diagrammatic view in transverse section

Table 5.15 : Difference between Cambium and Secondary Phellogen

	Cambium	Phellogen
1.	It is intrastelar in position.	It is extrastelar in position.
2.	Lying between xylem and phloem.	Lying external to the cortex.
3.	It is primary in origin.	It is secondary in origin.
4.	Procambium is the source of origin.	Epidermis, cortex or even phloem are the sources of origin.
5.	The cambial cells may be fusiform or isodiametric in shape.	The cork cambial cells may be polygonal or rectangular.
6.	It consists of fusiform initials and ray initials.	The cells are of single type.
7.	The cells may remain storied or non-storied in arrangement.	The cells are storied in arrangement.
8.	It produces secondary vascular and non-vascular tissues in the stele.	It produces non-vascular tissues outside the stele.
9.	It adds secondary phloem outside and secondary xylem inside.	It adds phellem outside and phellogen inside.
10.	It is also known as vascular cambium.	It is also known as cork cambium.

The phelloderm cells on the inner side resemble those of cortex. These cells are living, more or less isodiametric with intercellular spaces arranged in definite radial rows. They may photosynthesise and store food.

In conclusion, it can be said that secondary growth to increase in girth in a dicotyledonous stem is initiated in the intrastelar region by the activity of the cambium ring formed by the joining of the fascicular and interfascicular cambia and to cope up with pressure generated in the intrastelar region, periderm is formed in the extrastelar periphery.

Bark

In large trees, due to continued formation of secondary tissues in the intrastelar region the periderm may not be adequate to withstand the inwardly generated pressure. In that case additional layers of periderm are formed successively in the deeper regions of the cortex, pericycle and even phloem. The outwardly cut-off phellem or cork cells by the phellogen ultimately get devoid of water and food supply and eventually become dead. All these dead tissues lying outside the active phellogen constitute the ***bark*** of the trees. The term bark is loosely used and the term ***rhytidome*** is used for the outer bark covering all the tissues external to the innermost phellogen. Actually, the term bark is given to all tissues external to the vascular cambium.

Due to the formation of successive layers of periderm in the deeper regions of the stem the bark is formed in concentric rings surrounding the entire stem, which is known as ***ring bark***. In some plants the periderm is formed as overlapping scale-like layers, known as ***scale bark***. In *Eucalyptus, Platanus*, etc. the bark is intermediate between these two types, where the outer layers of the bark peel off in the form of sheets.

Lenticels

The secondary growth in thickness makes the epidermis to be replaced by the periderm in performing the protective function. The suberised walls of the dead cork cells are partly impervious to gases and thus for gaseous exchange between the internal living cells and the atmosphere some lens-shaped pores develop on the surface of the stem. These pores are known as ***lenticels***. Only a few plants, mostly climbers, do not form lenticels though periderm is formed.

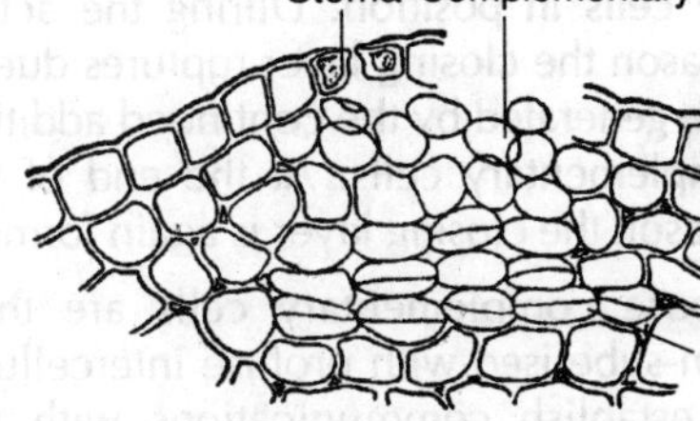

Fig. 5.139 : Early stages of formation of lenticel

Lenticel formation takes place for the first time just beneath the stomata present in the epidermis. They usually start to form either (a) just before the commencement of periderm formation, or (b) simultaneously with the periderm initiation, or (c) after the initiation of the periderm. The parenchyma cells adjacent to the substomatal chamber lose chlorophyll and divide in various planes to form a mass of colourless loose

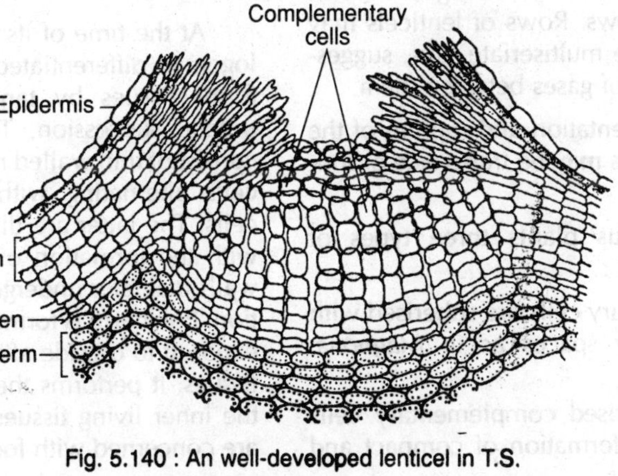

Fig. 5.140 : An well-developed lenticel in T.S.

cells. Sooner or later normal phellogen is differentiated in the adjoining deeper region, which further produces loose cells on the outer side. The whole loose cell mass formed by the divisions in the sub-stomatal parenchyma as well as in phellogen on the outside is known as ***complementary cells***.

With the continued addition of the complementary cell mass the epidermis is protruded above and finally ruptured, thus appearing like so many raised spots. The outermost cells at the atmospheric interface often die and are replaced by the cells derived from the phellogen. Often mass of compact cells, known as closing cells, alternate with loose complementary cells. The ***closing cells*** together form a layer referred to as ***closing layer*** which helps to keep the loose complementary cells in position. During the active growing season the closing layer ruptures due to the pressure generated by the continued addition of the complementary cells. At the end of the growing season the closing layer is again formed.

The loose complementary cells are thin-walled, non-suberised with profuse intercellular spaces to establish communications with the internal tissues. Like the stomata the lenticels are thus mainly concerned with gaseous exchange between the atmosphere and the internal tissues.

In some cases, lenticels are formed in the stomata free regions. After the formation of the cork cells by the phellogen for a while loose complementary cells are produced in localised areas which ultimately protrude above and rupture through the cork giving rise to lenticels. The number of lenticels formed in stem is variable. They may remain scattered or arranged in vertical or longitudinal rows. Rows of lenticels may occur opposite to the multiseriate rays, suggesting free interchange of gases between them.

Based on the orientation and rupture of the epidermis the lenticels may be transverse or longitudinal.

In dicotyledonous plants three types of lenticels are observed :

1. The complementary cells are suberised with little intercellular spaces as in *Magnolia, Pyrus, Salix*, etc.
2. Loose, non-suberised complementary cells followed by the formation of compact and suberized complementary cells as in *Tilia, Quercus*, etc.
3. Alternately arranged loose non-suberised complementary cells and compact suberised cells to form a multilayered complementary tissue as in *Betula, Fagus*, etc.

Rhytidome

Rhytidome is a special type of bark composed of successive layers of periderm as well as either the cortical parenchyma or secondary phloem. In *Robinia* periderm formation continues successively by the development of phellogen in the deeper regions of stem giving rise to periderm bands due to the death of the peripheral ones. The cork cells get suberised and the successive periderm bands enclose either cortical tissues or secondary phloem. All the periderm bands together with the enclosed cortex or secondary phloem and all the tissues present external to the innermost phellogen are collectively referred to as ***rhytidome***. The term bark is applied to all the tissues present external to the vascular cambium. Sometimes the term ***outer bark*** is applied to rhytidome and the living part of the bark inside the rhytidome is referred to as ***inner bark***.

Polyderm

Polyderm is a special protective tissue consisting of twenty or more alternating layers of uniseriate suberised cells and multiseriate non-suberised cells. The peripheral cells are dead and the inner cells including the suberised cells contain living protoplasts. It is found in underground stems and roots of Rosaceae, Onagraceae, Myrtaceae and Hypericaceae.

At the time of its formation, a special phellogen is differentiated at the pericycle, which forms tissues by tangential divisions in centripetal succession. The newly formed tissues consist of thin walled non-suberised parenchyma cells alternating with uniseriate endodermoid cells. The latter are differentiated into cork cells with the formation of casparian strips on their walls that later undergo more extensive suberisation. Polyderm is formed at the pericycle and it is exposed to outside after the death of the cortical tissues. It performs the function of protection of the inner living tissues. The non-suberised cells are concerned with food storage.

Table 5.16 : Difference between Periderm and Polyderm

	Periderm	Polyderm
1.	It is formed on the aerial stems and roots.	It is formed on the underground stems and roots.
2.	It is composed of phellem, phellogen and phelloderm.	It is composed of alternating layers of uniseriate endodermoid cells and multiseriate parenchyma.
3.	Suberin and sometimes lignin and wax are deposited.	Suberin is deposited on endodermoid cells.
4.	Casparian strips are not formed.	Casparian strips are formed before suberisation in endodermoid cells.
5.	The site of origin may be epidermis, hypodermis, cortex or phloem.	The site of origin is pericycle.
6.	The phellogen forms phellem outside and phelloderm inside.	The phellogen forms polyderm outside.
7.	Phellem cells differentiate into cork.	Endodermoid cells become cork due to suberization.
8.	The phellem and phelloderm are responsible for protection and storage of food respectively.	The outermost dead cells and non-suberized cells are responsible for protection and storage of food, respectively.
	Example : *Alistonia, Vitis, Clematis* etc.	Example : Rosaceae, Onagraceae etc.

SECONDARY GROWTH IN CERTAIN DICOTYLEDONOUS STEMS

Aristolochia

Aristolochia is a liane (woody climber) and its secondary growth in stem differs from the normal woody plants. The most striking differences are the formation of only parenchymatous medullary rays by the interfascicular cambium, and vascular tissues in the fascicular regions.

The primary structure of the *Aristolochia* stem shows following characteristics (Fig. 5.141) :

Epidermis : It is uniseriate with cuticularised outer walls.

Cortex : It is differentiated into outer collenchymatous hypodermis, middle parenchymatous portion and the innermost starch sheath. Both the collenchyma and parenchyma cells contain chloroplasts. Internal to the starch

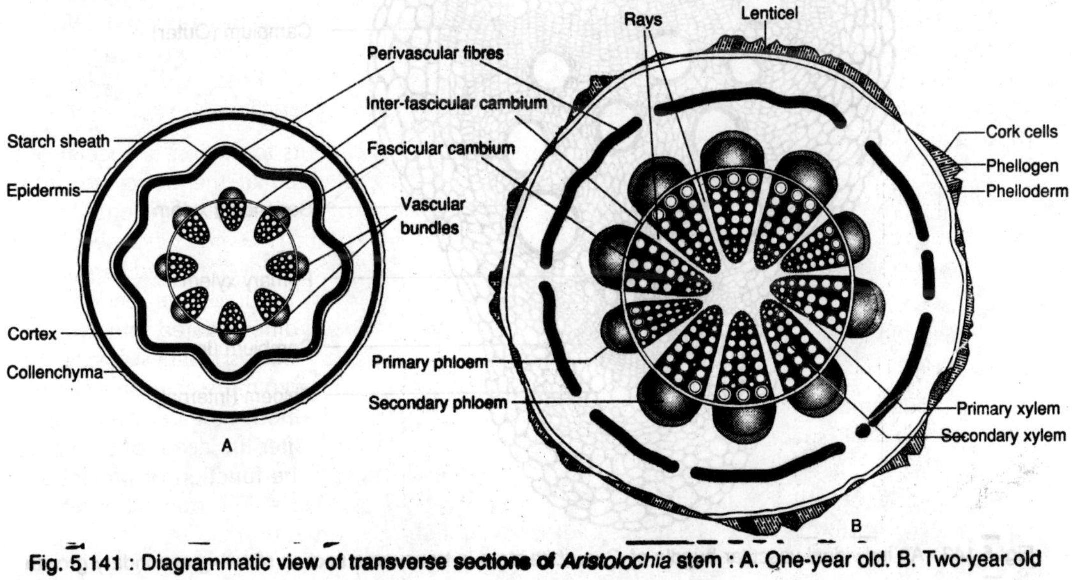

Fig. 5.141 : Diagrammatic view of transverse sections of *Aristolochia* stem : A. One-year old. B. Two-year old

sheath a continuous band of sclerenchyma occurs, which are ***perivascular fibres***, or may be said to form pericycle together with the adjoining parenchyma cells.

Stele : Broad medullary rays composed of parenchyma cells occur between the vascular bundles. The central portion of the stem is occupied by a large parenchymatous pith. The vascular bundles are arranged in a ring. The bundles are conjoint, collateral and open.

With onset of secondary growth in thickness a strip of parenchymatous cells of the broad medullary rays in the line of *fascicular* cambium become meristematic to form the *interfascicular* cambium. Both of them join up to form the cambium ring. Only the fascicular cambium produces secondary xylem and secondary phloem on the inner and outer sides, respectively, in individual vascular bundles and thus the primary xylem and primary phloem are pushed apart from each other. The secondary xylem has the usual elements arranged in vertical and horizontal systems. Annual rings are formed as a result of seasonal activities of the cambium. The primary phloem on the outer side gets crushed due to the pressure. Bands of sieve elements alternate with the bands of parenchyma in the secondary phloem. On the whole, the vascular bundles increase enormously in size due to continued activity of the fascicular cambium.

The interfascicular cambium simply produces parenchyma cells on the outer and inner sides. Thus the medullary rays become increasingly more broad and long.

The central pith gets more and more reduced in size. The continuous cylinder of sclerenchyma is disrupted and ruptured here and there, usually in the region of medullary rays. The adjoining parenchyma cells fill up the gaps thus formed. The ruptured bands of sclerenchyma gradually undergo sclerosis and are ultimately transformed into sclereids. The hypodermal collenchymatous band also breaks due to pressure. The parenchyma cells of the cortex fill up the gaps thus produced.

The epidermis also ruptures and periderm develops in the subepidermal layers. Phellogen is formed in patches to produce a thick layer of cork cells on the outer side and considerable phelloderm on the inner side. Lenticels are formed here and there on the stem surface (Fig. 5.142).

Cucurbita

The process of secondary growth in thickness is more or less similar to that of *Aristolochia*.

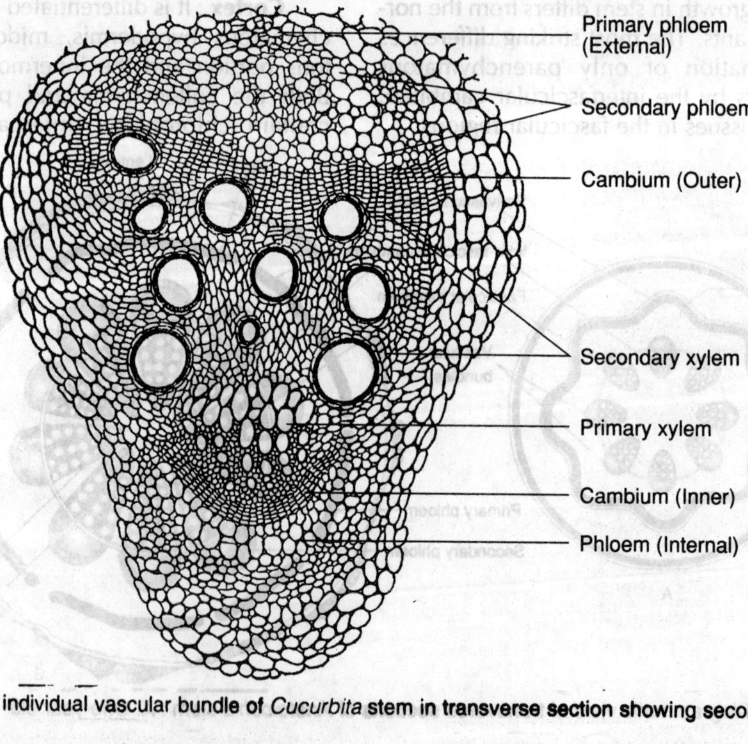

Fig. 5.142 : An individual vascular bundle of *Cucurbita* stem in transverse section showing secondary growth

But as this stem is not so woody, the cambial activities remain confined to the individual vascular bundles. Moreover, the perivascular fibrous band is not ruptured here.

Due to the tangential divisions of the cells of the outer cambium secondary xylem elements are produced inside and secondary phloem elements on the outside. The vessels are usually pitted and enormously large in size. So the characteristic radial arrangement and positions of other tissues are considerably disturbed. The secondary phloem elements are hardly distinguishable from their primary counterparts. The phloem consists of broad sieve tubes, narrow companion cells and parenchyma cells of intermediate size. Fibres and ray cells are absent. The sieve plates are simple and have large pores and connecting strands. The inner cambium strip remains inactive.

SECONDARY GROWTH IN MONOCOTYLEDONOUS STEM

Normally, the vascular bundles of the monocotyledonous stems are closed ones. Thus due to absence of the cambium they lack secondary growth in thickness and the vascular system is wholly composed of primary tissues. The bundles remain irregularly scattered in the ground tissues, forming an atactostele. Normal secondary growth is absent in monocotyledonous stems. However, some monocotyledons belonging to the family Liliaceae, mainly the arborescent, woody ones like *Dracaena, Yucca, Cordyline, Agave, Aloe* etc. exhibit anomalous secondary growth in thickness, which is discussed in the anomalous growth section.

SECONDARY GROWTH IN DICOTYLEDONOUS ROOT

The roots of some herbaceous dicotyledons and of all gymnosperms and woody dicotyledons show secondary growth in thickness. The tissues of secondary origin in the dicotyledonous roots are basically similar to those of the stem, but the way of formation is different. In dicot roots there are limited number of radially arranged vascular bundles with exarch xylem. Pith is very little or absent. A group of parenchyma cells beneath each phloem patch become meristematic and thus form strips of cambium which divide and produce secondary xylem inside and secondary phloem outside. The cells of the uniseriate pericycle external to each protoxylem group start dividing to form a few layers. The first-formed cambium internal to each phloem patch now extends both ways and joins with the innermost derivatives of the pericycle and finally a continuous wavy ring of cambium is formed. The cambial cells produce much more xylem at a higher rate than phloem and thus the wavy cambium cylinder ultimately becomes circular.

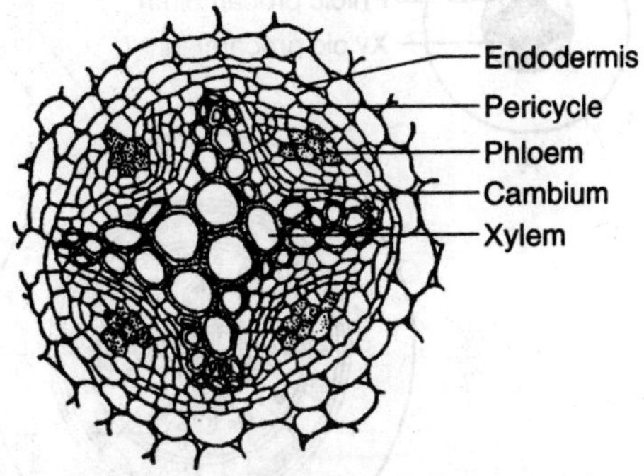

Fig. 5.143 : Early stages of secondary growth in a dicotyledonous root showing formation of the wavy cambium cylinder.

Due to continued secondary growth the primary xylem is pushed towards the centre and the primary phloem is pushed towards the periphery. Finally, the centrally located exarch primary xylems are fused to give a star-like appearance in transverse section. The strands of secondary vascular tissues remain collaterally arranged like those of the stem. The sieve elements of the primary phloem often get crushed. The cambial cells originating from the pericycle against protoxylem groups function as ray initials and produce broad bands of vascular rays which are also called medullary rays. These rays — running between xylem and phloem through the cambium are characteristic of the roots (Fig. 5.144).

To cope up with the pressure generated by the secondary growth, the periderm is formed in the outer region. Phellogen arises outside the pericycle. It produces cork cells on the outer side and some phelloderm inside. The pressure due to secondary growth ruptures the endodermis and cortex, which are ultimately sloughed-off. Lenticels may occur as transversely elongated rough areas.

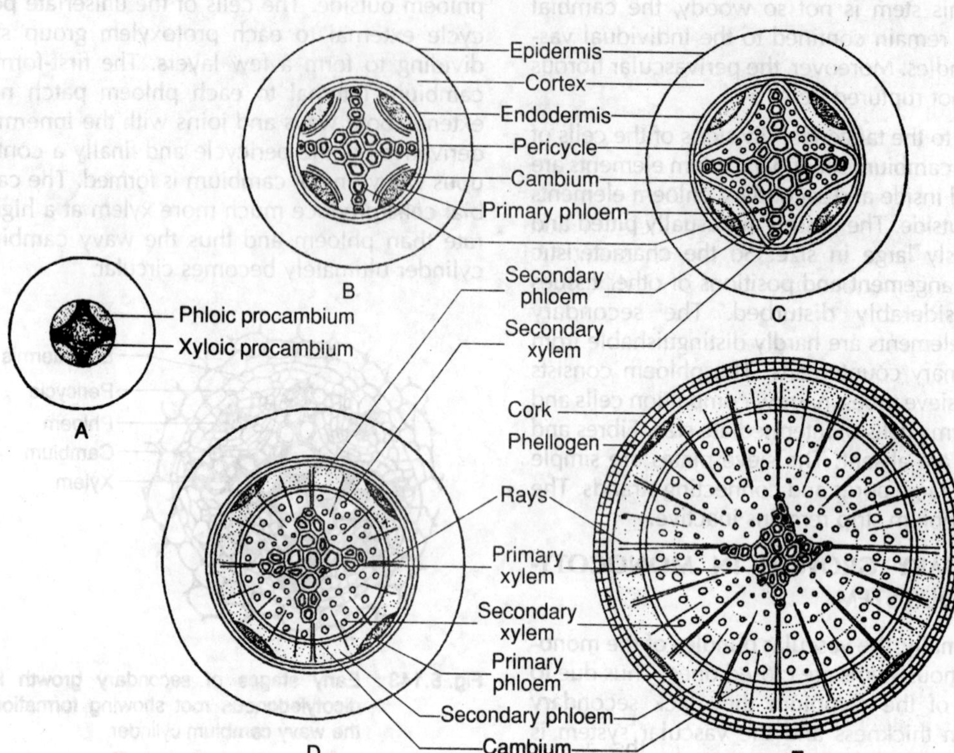

Fig. 5.144 : Diagrammatic view of the stages of secondary growth in thickness in dicotyledonous root

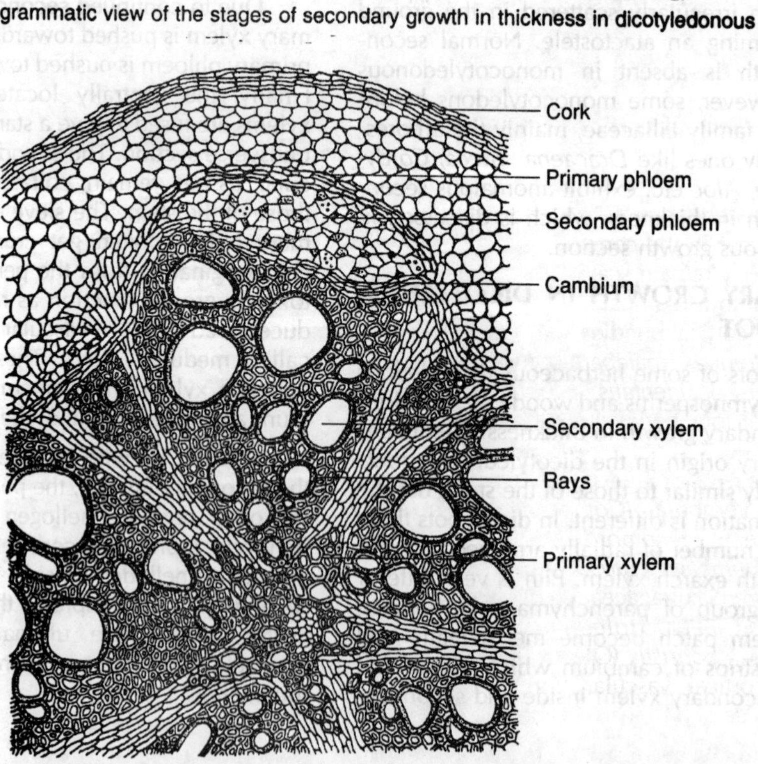

Fig. 5.145 : Cellular view of a portion of a *Tinospora* root showing secondary growth in thickness

Anomalous structures

The arrangements of tissues exhibited by the majority of the plants are considered as ***normal***, variations from which found in minority are considered as ***anomalous***. The anomalous structures found in the primary body are defined as primary anomalous structures, which occur in both monocotyledons and dicotyledonous stem, and may persist throughout the life of the plant. Following are the common anomalous structures found in stems :

Occurrence of atactostele in dicotyledonous stem : Normal arrangement of the vascular bundles in the dicotyledonous stem is more or less in a ring. In *Thalictrum* (Ranunculaceae), *Podophyllum peltatum* (Podophyllaceae), *Papaver orientale* and *P. somniferum* (Papaveraceae), *Bougainvillea* (Nyctaginaceae), *Piper betle* and *Peperomia langsdorfii* (Piperaceae), *Nymphaea* (Nymphaeaceae) etc. the vascular bundles remain irregularly scattered throughout the ground tissue.

Occurrence of arranged vascular bundles in monocotyledonous stem : In monocot stems the vascular bundles are scattered in the ground tissue. But in *Coix, Triticum, Oryza* (Poaceae), *Tamus communis* (Dioscoriaceae) etc. the vascular bundles are more or less arranged in one or two rings.

Occurrence of cortical bundles : In dicotyledonous stem the vascular bundles remain encircled by the endodermis to form the stele. But in some dicotyledonous stems accessory vascular bundles occur in the cortex in addition to stellar bundles. They remain scattered in the cortex and are referred to as cortical bundles. In most of the species the cortical bundles are the leaf trace bundles. In *Bombax malabaricum* (Bombacaceae), *Calycanthus occidentalis, Chimonanthus* (Calycanthaceae) four pericyclic bundles (leaf traces) occur with internal phloem and external xylem. In *Campanula pyramidalis* (Campanulaceae), *Crepis, Senecio, Vernonia (Compositae), Cotyledon* (Crassulaceae), *Momordica* (Cucurbitaceae), *Armeria, Plumbago europaea* (Plumbaginaceae), *Peltiphyllum* (Saxifragaceae), petiole of *Eryngium campestre, Siler trilobum* (Umbelliferae), *Nyctanthes arbortristis* (Oleaceae) four inversely oriented cortical vascular bundles occur with xylem outside and phloem inside.

Occurrence of medullary bundle : The extra vascular bundles may be found in the pith of some dicotyledonous stem in addition to the normal ring of bundles, which are referred to as medullary bundles. In *Acanthus* (Acanthaceae) collateral and inversely oriented medullary bundles occur. In *Achyranthes aspara* (Amaranthaceae) two medullary bundles at internode and in *Cyathula prostrata* four medullary bundles are observed.

Occurrence of internal bundles : This type of bundles is found below the normal ring of bundles but above the medullary bundles. These bundles are referred to as internal bundles. In *Rumex crispus, R. orientalis* etc. this type of bundles are observed.

Occurrence of internal phloem or intraxylary (perimedullary, medullary) phloem : The phloem that occurs at the periphery of the pith in the form of a continuous cylinder or strands is referred to as intraxylary phloem or internal phloem or medullary phloem. In *Wrightia, Vinca, Thevetia, Landolphia* (Apocynaceae); *Calotropis* (Asclepiadaceae); *Convolvulus, Ipomaea, Evolvulus* (Convolvulaceae); *Acanthus, Barleria* (Acanthaceae); *Solanum, Nicotiana* (Solanaceae) etc. such type of bundles are found.

Absence of vessels in angiosperm : Normally the xylem of angiosperms contains vessels but in certain members of the families Tetracentraceae (*Tetracentron*), Trochodendraceae (*Trochodendron*) and Winteraceae (*Drymis, Pseudowintera, Bubbia, Zygogynum*) vessels are absent.

Occurrence of polystele in dicotyledons : Usually the vascular bundles of dicotyledonous stem remain encircled by a common endodermis to form monostele, but variations occur when each vascular bundle becomes encircled by an endodermis and thus forms a polystelic structure. The nodes of *Nymphaea* (Nymphaeaceae) and *Parnassia palustris* (Parnassiaceae) form polystele. *Dianthera americana* forms seven steles and each stele is represented by a vascular bundle completely enclosed by an endodermis. One stele is located at the centre and the rest towards the periphery.

Occurrence of separate xylem and phloem bundles : Normally, xylem and phloem together form the conjoint bundles. In some abnormal cases, vascular bundles may consist of either only xylem or phloem in addition to the normal bundles. In *Cuscuta* fully developed phloem strands occurs interspersed with other normal collateral bundles. Incomplete phloem bundles are found in *Rumex crispus, Ricinus communis, Xanthium strumarium, Antigonon leptopus, Achyranthes aspera, Mirabilis jalapa, Boerhaavia diffusa* etc.

REVIEW OF CERTAIN ANOMALOUS TYPES

Monocot stem

Some monocotyledons belonging to the family Liliaceae have arborescent habit with woody stems like *Dracaena, Yucca, Cordyline, Agave, Aloe* etc. They exhibit an anomalous secondary growth in thickness producing a cambium ***which is abnormal both in position and function***.

After the cease of primary growth a ***cambium ring*** of a few cells thick outside the vascular bundles in the cortex is formed. The cambial cells are fusiform or rectangular in shape. Instead of forming phloem and xylem on the outer and inner sides, as in normal condition, the cambial cells go on adding secondary tissues on the inner side first, and, later, little amount of secondary tissues are added on the outer side as well.

From the secondary parenchymatous tissues formed on the inner side differentiate into oval-shaped ***leptocentric*** vascular bundles. These parenchyma cells, in which the secondary vascular bundles remain embedded, are called the ***conjunctive tissue***. The cells of this tissue are radially arranged due to their origin by tangential divisions of the cambial cells. So they may be easily distinguished from the irregularly arranged parenchyma of the primary ground tissue. They may be thin-walled or thick-walled.

A few cambium derived cells divide longitudinally (anticlinal divisions followed by periclinal divisions) and even haphazardly to form xylem and phloem of the secondary bundles. These bundles differ from the primary bundles in presence of small amount of phloem and in absence of annular and spiral protoxylem elements. The secondary bundles are mostly ***amphivasal***. Some of them may be collateral as well.

The small amount of phloem consists of short sieve tubes, companion cells and phloem parenchyma. The xylem is made of only tracheids, usually with scalariform thickening and small amount of xylem parenchyma with lignified walls. The cambium adds small amount of parenchymatous cells on the outer side. The primary bundles are comparatively larger. They are also mostly amphivasal or rarely collateral ones.

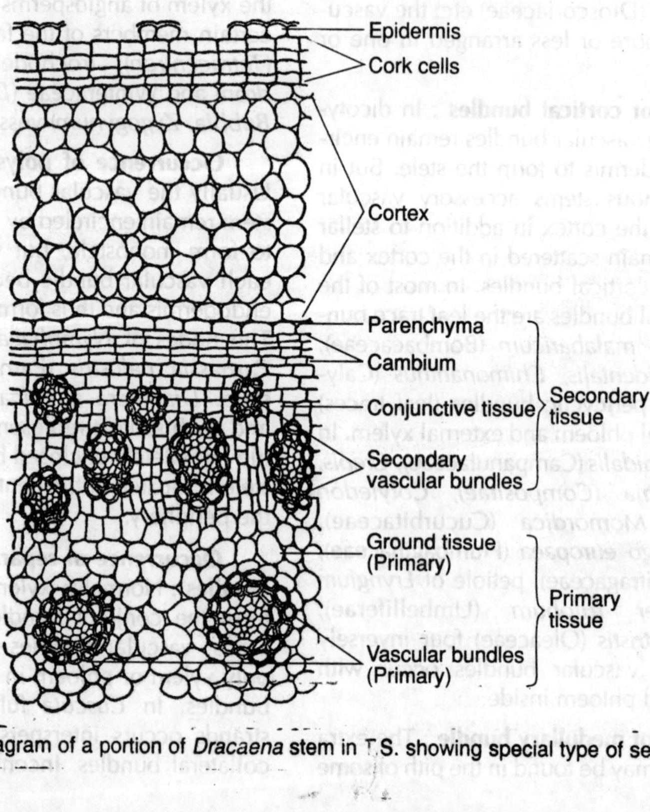

Fig. 5.146 : Cellular diagram of a portion of *Dracaena* stem in T.S. showing special type of secondary growth

The secondary bundles are more numerous and the peripheral ones are smaller in size.

In *Dracaena* typical periderm formation by the activity of phellogen is not found but ***storied cork cells*** are formed by the continuous periclinal division of the outermost cortical cells followed by suberisation of their walls. This storied cork cells form a special type of protective layer outside the stem. In *Aloe*, coconut and royal palm periderm is formed in the same manner as in the dicotyledons.

Considerable thickening is often noticed at the bases of many palms. This is not due to the activity of a cambium cylinder, but it is due long-term primary growth. The apical meristem produces only a central column of parenchyma and vascular strands. Another primary meristem occurs just beneath the leaf primordia forming a mantle-like tissue zone, known as ***primary thickening meristem.*** The cells of this meristem divide periclinally to form rows of cells which differentiate into ***ground parenchyma*** traversed by procambial strands. From the procambial strands vascular bundles develop. By the continued divisions of the ground parenchyma cells further increase in thickness takes place. Thus the primary thickening meristem produces the main bulk of the stem tissues, viz., ground parenchyma and most of the vascular bundles.

The structural anomaly may be directly influenced \by external environmental conditions. The woody climbers or lianes and the storage organs show structural anomaly as well as abnormal secondary growth in thickness as they are different from the normal ones — from morphological and physiological points of view. At the same time, there are many cases of anomalous condition which are not influenced by the environment and appear to be merely cases of variations of the design. Haberlandt has called these two cases as ***adaptive*** and ***non-adaptive*** anomalies of secondary growth in thickness, respectively.

Dicot stem

Cambium — Normal in Position but Abnormal in Activity

(a) In some lianes like *Bignonia* and other members of family Bignoniaceae, the cambium

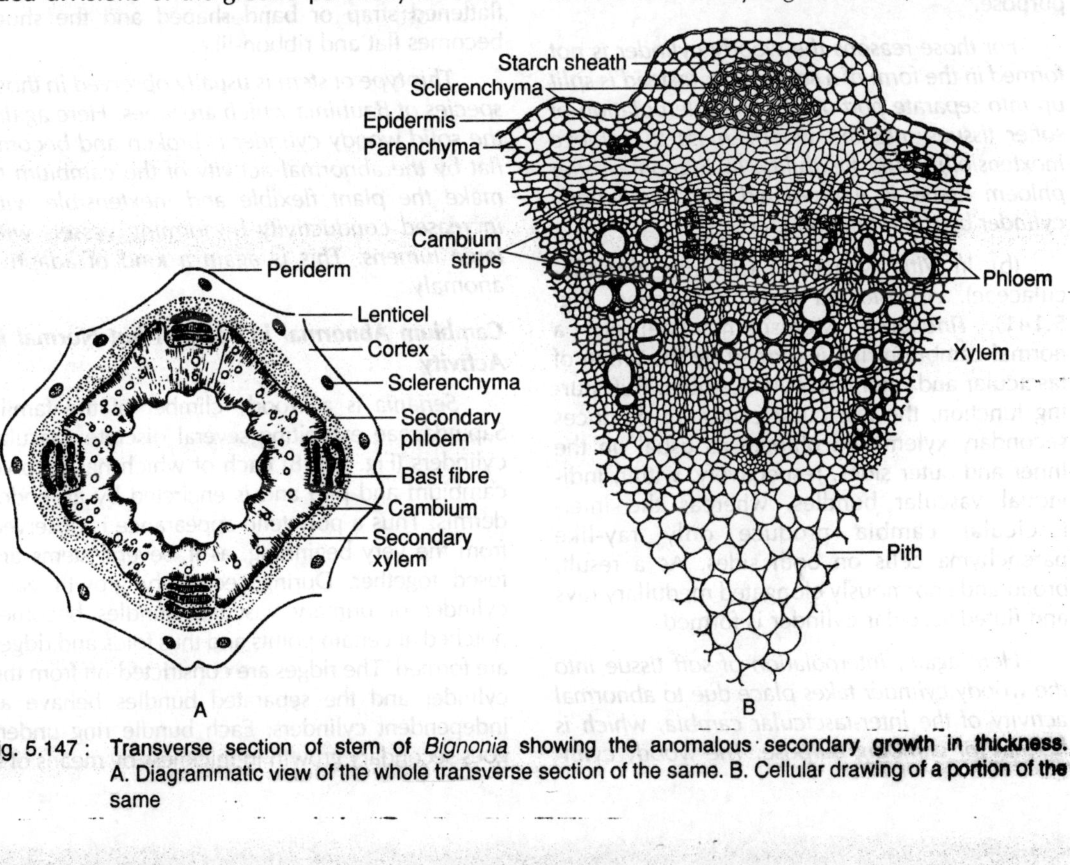

Fig. 5.147 : Transverse section of stem of *Bignonia* showing the anomalous secondary growth in thickness. A. Diagrammatic view of the whole transverse section of the same. B. Cellular drawing of a portion of the same

is normal in position and activity for the first time, but immediately it cuts off different proportions of xylem and phloem at four alternating points arranged in form of a cross. In one set of alternating points the cambium produces less amount of secondary xylem and much amount of secondary phloem, and vice versa in other set of alternating points. As a result the woody cylinder appears to have four longitudinal grooves alternating with four ridges. The ridges are wider than the grooves. The cambium breaks up into a number of strips. The wider ones remain on the top of the ridges while the narrower ones remain at the bases of the grooves. As a result a peculiar structure with ridged and furrowed xylem cylinder is formed. In the secondary phloem formed in the groove parallel strips of **bast fibres** are formed.

This is a kind of adaptive anomaly. *Bignonia* is a woody climber (liane) and so its stem is twining and thus is inextensible as it hangs freely and has to bear its own weight. The stem is also flexible as it has to withstand high velocity of wind and again subject to radial compression due to unequal growth in thickness for twining purpose.

For those reasons the woody cylinder is not formed in the form of a solid mass, instead is split up into separate portions by the interpolation of softer tissues into the woody mass to achieve inextensibility and flexibility. In Bignonia softer phloem tissue is interpolated into the woody cylinder by the formation grooves.

(b) In *Vitis* (Vitaceae), *Clematis* (Rannunculaceae), *Aristolochia* (Aristolochiaceae, Fig. 5.141), *Tinospora* (Menispermaceae) etc. a normal cambium ring is formed by the union of fascicular and inter-fascicular cambia. But, during function, the fascicular cambium produces secondary xylem and phloem as usual on the inner and outer sides, respectively, within individual vascular bundles, whereas the interfascicular cambia produce only ray-like parenchyma cells on both sides. As a result, broad and enormously elongated medullary rays and fluted vascular cylinder is formed.

Here again, interpolation of soft tissue into the woody cylinder takes place due to abnormal activity of the inter-fascicular cambia, which is helpful for climbing purpose. The woody cylinder is broken by wide rays and the sclerenchymatous cylinder encircling the vascular bundles is ruptured and the resulting gaps are filled by parenchyma cells. So, this is again an example of adaptive type of anomaly.

(c) In *Bauhinia divaricata, B. sericella* and *Prestonia macrocarpa* the stem is flat, ribbon-like at maturity having the flattened strap shaped stele. Secondary growth starts by the activity of a normal cambium ring producing secondary xylem and phloem towards inner and outer sides, respectively, as usual. At that early stage the vascular cylinder as well as the stem is more or less round with uniform amount of secondary xylem consisting of small tracheary elements surrounding the pith.

After such a period of secondary growth the cambium becomes more active at two opposite sides, and at alternate opposite sides it becomes less active. The nature of the secondary xylem produced by the alternating segments of cambium is different. In the less active region the secondary xylem shows small tracheary element, whereas that of adjacent more active region exhibits vessels with larger diameter. With such continued growth the stele gradually becomes flattened strap or band shaped and the shoot becomes flat and ribbon-like.

This type of stem is usually observed in those species of Bauhinia which are vines. Here again, the solid woody cylinder is broken and become flat by the abnormal activity of the cambium to make the plant flexible and inextensible with increased conductivity by forming vessels with large lumens. This is again a kind of adaptive anomaly.

Cambium Abnormal in Position but Normal in Activity

Serjania is a woody climber of the family Sapindaceae exhibiting several discrete vascular cylinders (Fig. 5.148) each of which has its own cambium and pith and is encircled by an endodermis. Thus a polystelic appearance is observed from the very beginning, as if, several stems are fused together. During secondary growth each cylinder of primary vascular bundles becomes notched at certain points and thus folds and ridges are formed. The ridges are constricted-off from the cylinder and the separated bundles behave as independent cylinders. Each bundle ring undergoes secondary growth in thickness by means of a ring of cambium present in them. These cambia, though unusual in position, normally give rise to internal xylem and external phloem. At maturity each vascular cylinder exhibits bast fibers and wood enclosing parenchymatous pith at the centre. Certain amount of parenchyma also occurs between the vascular cylinders.

With age and consequent development of periderm the compound structure becomes more marked. In fact, there is a large central stele, surrounded by a few peripheral ones. The secondary xylem is composed of vessels and tracheids, on the inner side, and a few layers of phloem on the outer. A sclerenchymatous cylinder occurs encircling the vascular cylinders. It is again notched and wavy in outline with less amount of tissue at the notches.

In Serjania and other members of the family again, instead of forming single solid woody cylinder the stele is broken into many — even in the procambial stage to make the stem compatible with twining habit. The vascular cylinders remain encircled by soft parenchymatous tissue, which helps to make the stem tough, inextensible but flexible. So, this peculiarity is obviously adaptive in nature.

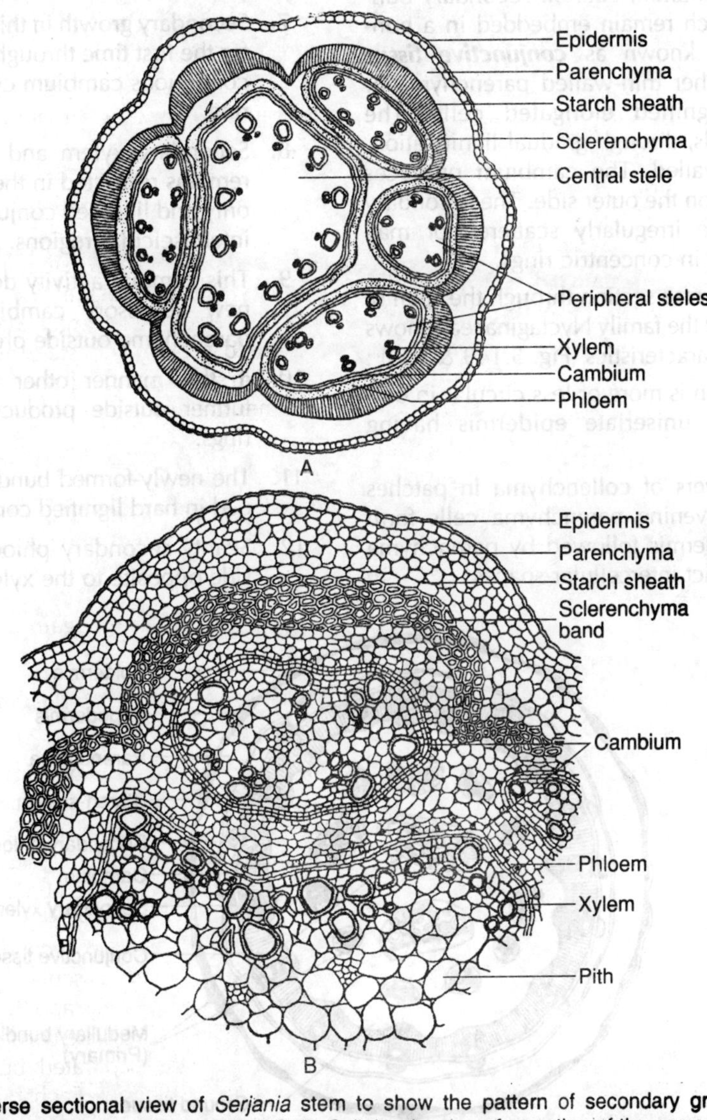

Fig. 5.148 : Transverse sectional view of *Serjania* stem to show the pattern of secondary growth in thickness : A. Diagrammatic view of the whole T.S. B. Cellular drawing of a portion of the same in T.S.

Anomaly due to the formation of Accessory Cambium

In the families Nyctaginaceae, Amaranthaceae, Chenopodiaccae, etc., different types of anomaly is observed. The collateral and open vascular bundles normally appear either in a ring or they remain irregularly scattered in the ground tissues called ***medullary bundles.*** Cambial activity in the primary bundles commences but soon ceases.

This is followed by the formation a secondary accessory cambium encircling the primary bundles. This cambium cuts off secondary bundles inside, which remain embedded in a non-vascular tissue, known as ***conjunctive tissue*** consisting of either thin-walled parenchyma or thick-walled lignified elongated cells. The parenchyma cells, through gradual lignification, become thick-walled. The cambium produces very little tissue on the outer side. The secondary bundles may be irregularly scattered or may remain arranged in concentric rings.

(a) A transverse section through the stem of *Boerhaavia* sp. of the family Nyctaginaceae shows the following characteristics (Fig. 5.149 & 150) :

1. The section is more or less circular in outline with uniseriate epidermis having stomata.
2. A few layers of collenchyma in patches with intervening parenchyma cells form the hypodermis followed by parenchyma with distinct intercellular spaces.
3. The stele includes primary vascular bundles in three rings.
4. Two fairly large bundles occur at the central region surrounded by quite a few comparatively much smaller ones forming a loose ring.
5. The outermost ring consists of pretty large number of much smaller bundles occurring just beneath the pericycle.
6. The central bundles are collateral and open but cambial activity remains confined to individual bundles.
7. Secondary growth in thickness commences for the first time through the formation of a continuous cambium cylinder in the outer ring.
8. Secondary xylem and phloem formation remains restricted in the fascicular regions only and lignified conjunctive tissue in the interfascicular regions.
9. This cambial activity declines soon and a new accessory cambium arises in the parenchyma outside phloem.
10. In this manner other cambia may arise further outside producing a few growth rings.
11. The newly-formed bundles remain embedded in hard lignified conjunctive tissue.
12. Scanty secondary phloem is formed outside opposite to the xylem vessels.

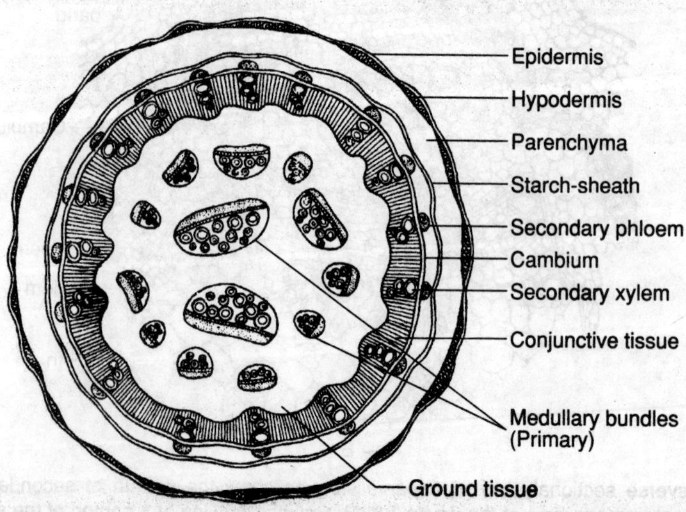

Fig. 5.149 : Diagrammatic view of the transverse section of *Boerhaavia* stem

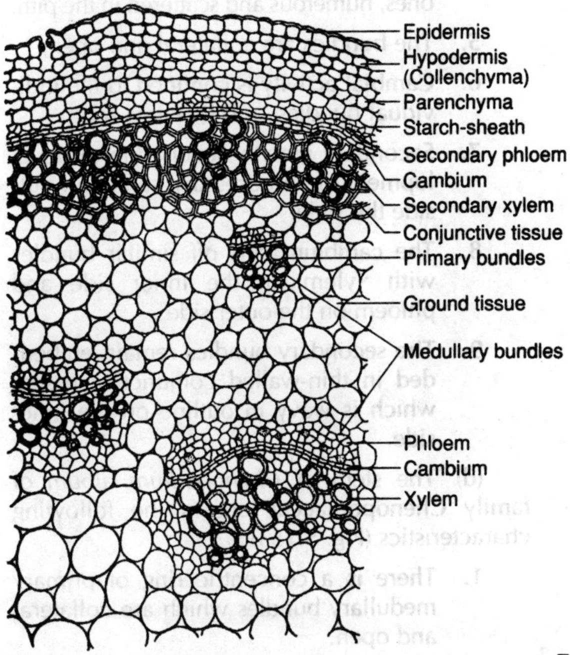

Fig. 5.150 : **Cellular drawing of a portion of *Boerhaavia* stem in T.S.**

(b) Transverse section of stem of *Mirabilis jalapa* of the same family shows the following characteristics (Fig. 5.151) :

1. The section is rather quadrangular in outline.
2. It more or less resembles that of *Boerhaavia*, but the medullary bundles are more numerous.
3. Those occurring towards periphery are smaller in size and more numerous, whereas those at the centre are larger and more scattered.
4. Secondary growth is initiated by the formation of secondary accessory cambium originating in the same manner as in *Boerhaavia*.
5. This cambium cuts off secondary tissues, usually secondary xylem elements on the inner side which remain embedded in the conjunctive tissue.
6. Secondary phloem elements are occasionally formed.
7. The cambium produces very little secondary tissues on the outer side.

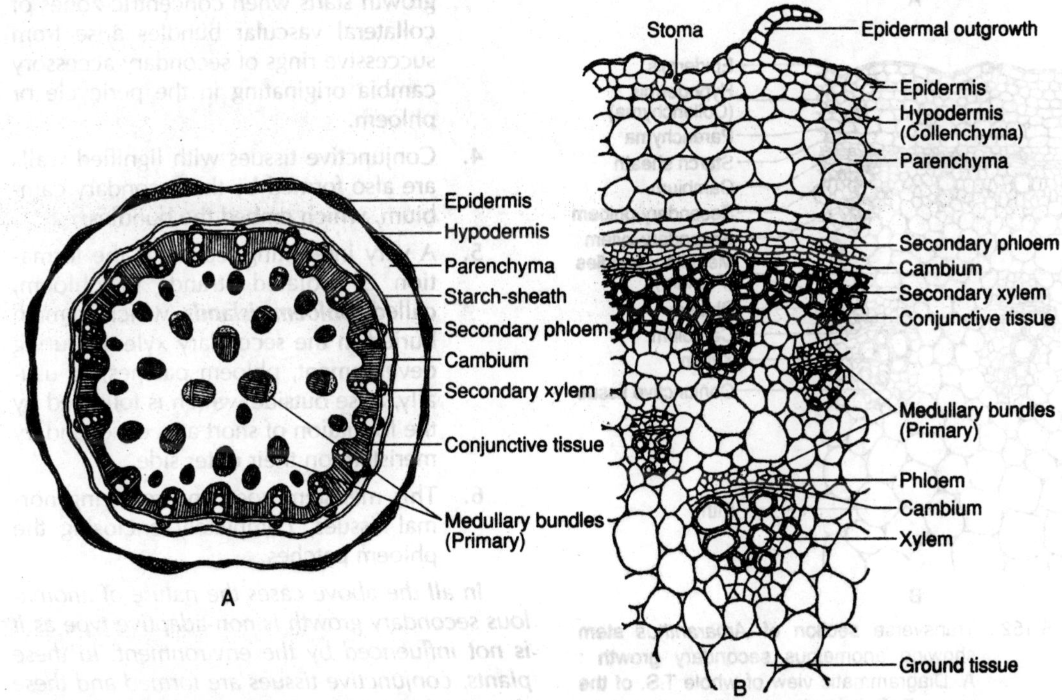

Fig. 5.151 : **A. Diagrammatic view of the whole transverse section of *Mirabilis jalapa* stem in T.S. B. Cellular drawing of a portion of the same in T.S.**

(c) The transverse section of the stem of *Amaranthus* of the family Amaranthaceae shows the following characteristics (Fig. 5.152) :

1. The transverse section is more or less circular in outline with uniseriate epidermis.
2. The ground tissue is well-differentiated.
3. Collenchyma cells occur in the hypodermal region interrupted by chlorophyll-containing parenchyma cells here and there.
4. The vascular bundles are medullary ones, numerous and scattered in the pith.
5. The bundles are collateral and open.
6. Cambial activity is confined to the individual bundles, and it ceases soon.
7. Secondary growth occurs due to development of an accessory cambium outside the stele.
8. The cambium cuts off similar bundles with xylem on the inner side and phloem on the outer side.
9. The secondary bundles remain embedded in thin-walled conjunctive tissue, which is wavy in outline on the inner side.

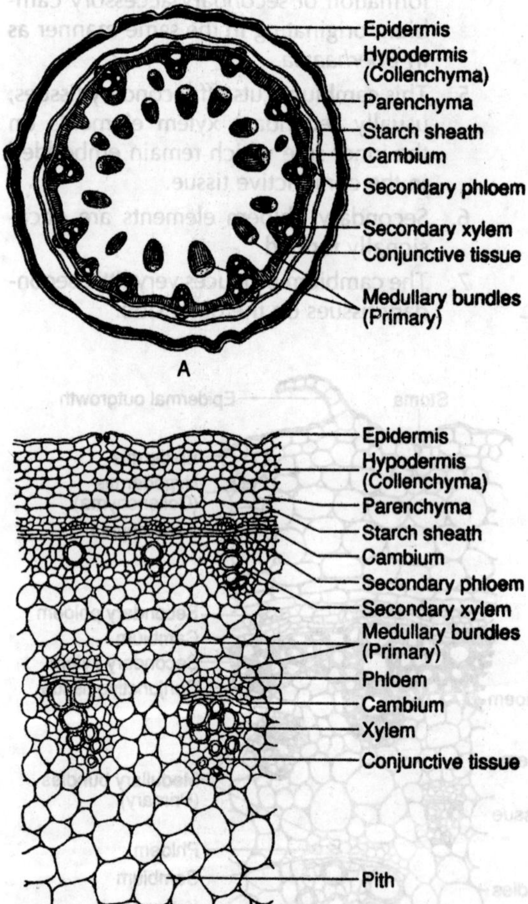

Fig. 5.152 : Transverse section of *Amaranthus* stem showing anomalous secondary growth : A. Diagrammatic view of whole T.S. of the same. B. Cellular drawing of a portion of the same in T.S.

(d) The stem of *Chenopodium album* of family Chenopodiaceae shows the following characteristics (Fig. 5.153) :

1. There is a concentric ring of primary medullary bundles which are collateral and open.
2. They secondarily grow little in thickness.
3. Then a peculiar anomalous secondary growth starts when concentric zones of collateral vascular bundles arise from successive rings of secondary accessory cambia originating in the pericycle or phloem.
4. Conjunctive tissues with lignified walls are also formed by the secondary cambium, which embed the bundles.
5. A very interesting feature is the formation of isolated strands of phloem, called ***phloem islands*** which remain buried in the secondary xylem. During development, phloem patches, as usually, arise outside, which is followed by the formation of short arcs of secondary meristem on their outer side.
6. This meristem goes on producing normal tissues, completely enclosing the phloem patches.

***In all the above cases the nature of anomalous secondary growth is non-adaptive type as it is not influenced by the environment.** In these plants, conjunctive tissues are formed and these tissues give mechanical strength in the usual manner.*

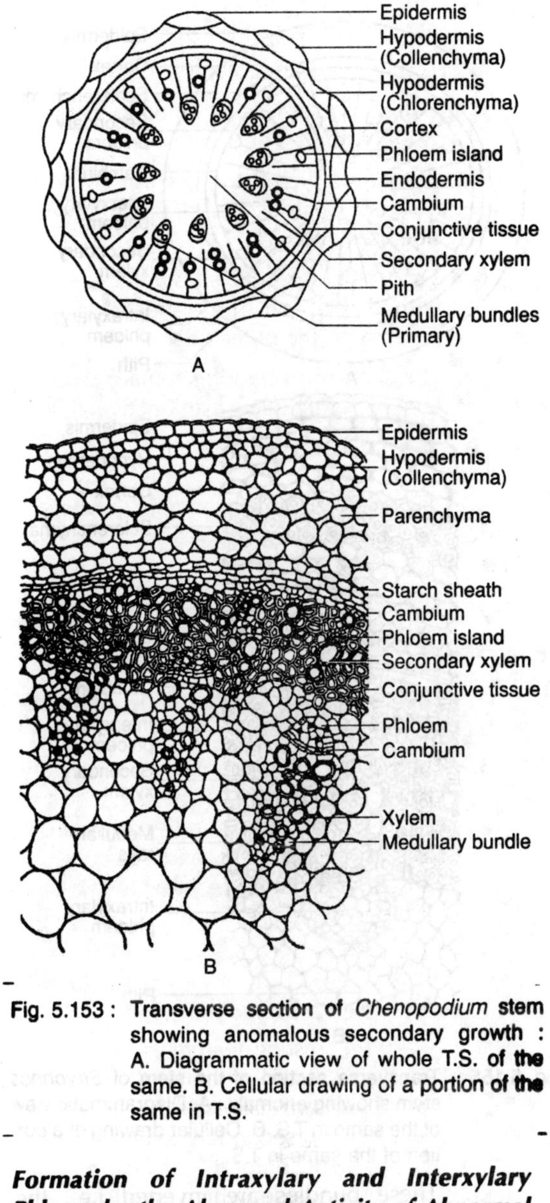

Fig. 5.153 : Transverse section of *Chenopodium* stem showing anomalous secondary growth : A. Diagrammatic view of whole T.S. of the same. B. Cellular drawing of a portion of the same in T.S.

Formation of Intraxylary and Interxylary Phloem by the Abnormal Activity of Abnormal Accessory Cambium

(a) In *Campsis radicans* (=*Tecoma radicans*) and *C. grandiflora* (=*Tecoma andrepens*) anomaly is produced during secondary growth (Fig. 5.154) :

1. In this genus, the normal cambium ring forms the typical xylem and phloem inside and outside, respectively.
2. Later accessory cambia are formed in two arcs at the outer margin of pith and on the inner side of the normal vascular cylinder.
3. These cambia produces secondary xylem and phloem in inverse order i.e., xylem outside and phloem inside.
4. As a result, two arcs of vascular bundles are formed at the margin of pith, showing inverse orientation of xylem and phloem in contrast to normal bundles.
5. The two patches of internal phloem, thus formed, gradually crush the pith.
6. So, these phloem patches here are ***intraxylary*** and secondary in origin.

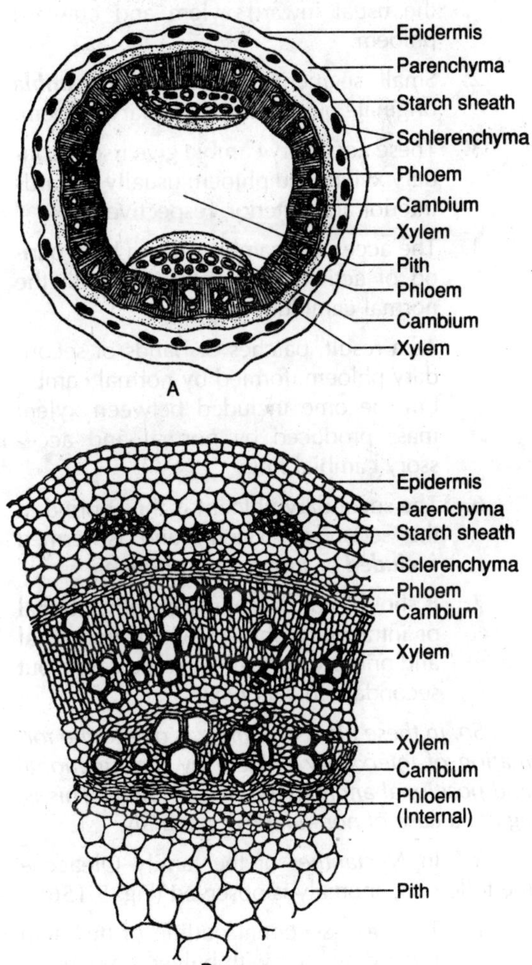

Fig. 5.154 : Transverse section of the stem of *Tecoma* stem showing anomaly : A. Diagrammatic view of the same in T.S. B. Cellular drawing of a portion of the same in T.S.

7. In the developing stem, some parenchyma cells are found between the wood formed by normal and accessory cambia.

(b) In some members of family Combretaceae — *Combretum, Entada,* and *Strychnos* of family Loganiaceae anomalous structures result from occurrence of unusual type of phloem which remains embedded in secondary xylem. This type is known as ***interxylary*** or ***included phloem.***

In *Strychnos, Thunbergia* etc. anomaly is produced (Fig. 5.155) :

1. The normal cambium initially gives rise the usual inward xylem and outward phloem.
2. Small segments of ***accessory cambia*** originate outside the secondary phloem.
3. These accessory cambia give rise secondary xylem and phloem usually towards interior and exterior, respectively.
4. The accessory cambia, after a brief period of activity, gradually join with the normal cambium ring.
5. As a result, patches or bands of secondary phloem, formed by normal cambium, become included between xylem mass produced by normal and accessory cambial ring.
6. The secondary phloem is thus buried in the secondary xylem, and is termed ***included*** or ***interxylary phloem.***
7. It should not be confused with internal or intraxylary phloem which is normal and primary in origin in some cases, but secondary in *Tecoma.*

So, in these plants anomaly is due to the formation of interxylary phloem by the functional and positional anomaly of the cambium. This is, again, a case of non-adaptive anomaly.

(c) In *Nyctanthes* of the family Oleaceae the following anomaly is observed (Fig. 5.156) :

1. The cross-sectional outline of the stem is quadrangular with bulged corners.
2. Here, apart from normal vascular bundles occurring more or less in a ring, there are four cortical bundles at the bulged corners of the stem.

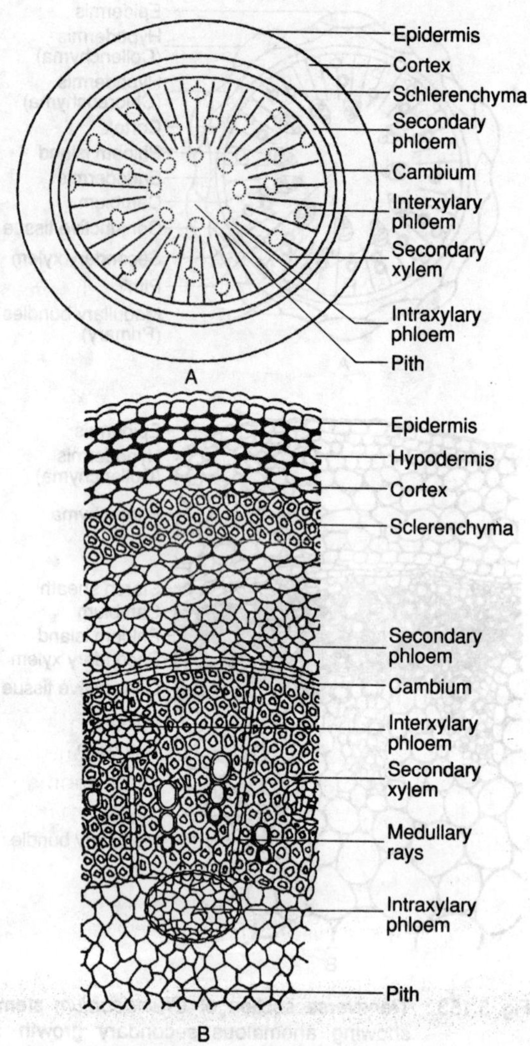

Fig. 5.155 : Transverse section of the stem of *Strychnos* stem showing anomaly : A. Diagrammatic view of the same in T.S. B. Cellular drawing of a portion of the same in T.S.

3. These bundles are inverted i.e., the phloem remains towards the centre and the xylem towards the periphery.
4. The cambium in these leaf trace bundles shows anomalous function.
5. It adds xylem outside and phloem inside. They are obviously leaf trace bundles.
6. The central vascular bundles are rather compact collateral and open ones with intervening patches of parenchyma in form of rays.

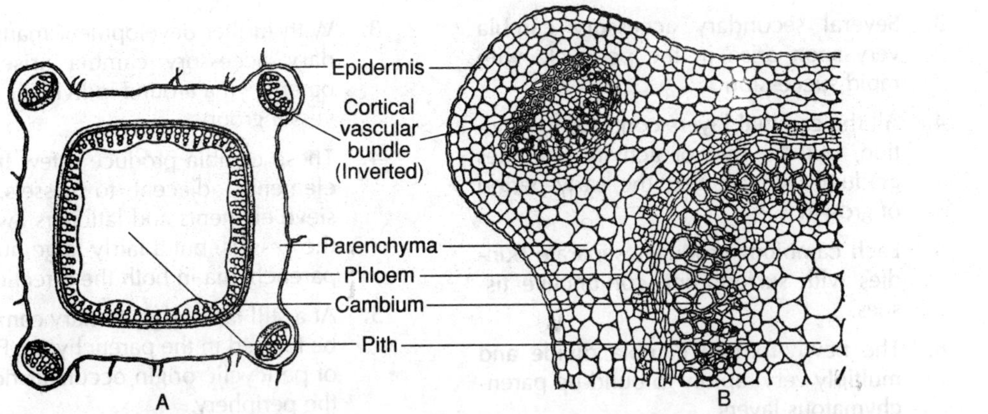

Fig. 5.156 : Transverse section of the stem of *Nyctanthes* stem showing leaf trace bundles : A. Diagrammatic view of the same in T.S. B. Cellular drawing of a portion of the same in T.S.

7. The secondary growth in this central vascular cylinder is normal.

Dicot root

In a number of dicotyledons, fleshy roots are developed to serve as organs of storage. Anatomical anomalies are often noticed in these organs which develop considerable amount of storage parenchyma. Again, due to these anomalous cambial activities, the conducting elements are formed with suitable arrangements for translocation. Massive development of storage parenchyma is noticed in many cases in the region of cortex or secondary phloem. Therefore, many fleshy roots exhibit a notable deviation from the normal structure in the comparatively poor development of woody tissues.

(a) In the fusiform underground tap roots of radish (*Raphanus sativus* of family Cruciferae) the storage tissue mainly originates by proliferation of parenchyma of pith region. Massive storage parenchyma may also be produced from the secondary xylem parenchyma and xylem rays. The vascular elements — with some poorly developed fibres — remain arranged in concentric rings here.

(b) The fleshy nature of the conical tap root of carrot (*Daucus carota* of family Umbelliferae) is also due to massive development of storage parenchyma in the phloem and xylem.

(c) In napiform root of sugar beet (*Beta vulgaris*) of family Chenopodiaceae a striking anomalous form of secondary growth is observed (Fig. 5.157) :

1. In this root several successive accessory cambial layers are formed.
2. The first cambium layer forms a ring of vascular bundles near the primary xylem.

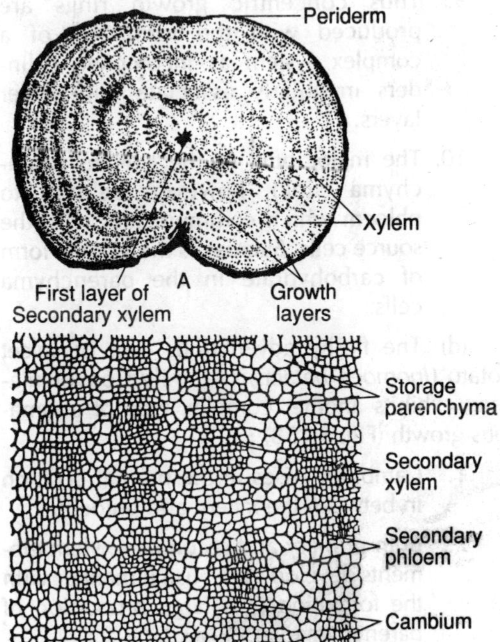

Fig. 5.157 : Transverse section of the root of sugar beet showing anomalous secondary growth : A. Diagrammatic view of the same in T.S. B. Cellular drawing of a portion of the same in T.S.

3. Several secondary accessory cambia very soon arise in the pericycle with rapid succession.
4. All the cambial layers continue to function, but the rate of activity becomes gradually slower after the early period of growth.
5. Each cambium produces separate bundles with surrounding conjunctive tissues.
6. The newly-formed cambia divide and multiply very rapidly to build-up parenchymatous layers.
7. As a result, alternate layers of proliferated pericycle and vascular elements are formed. The bundles are largely parenchymatous with very scanty lignified elements.
8. Growth continues in the bundles by cambial activity and by proliferation of xylem and phloem parenchyma.
9. Thus concentric growth rings are produced with the formation of a complex system as incomplete cylinders irregularly associate with other layers.
10. The major bulk of the storage parenchyma (sink), however, belongs to phloem which communicates with the source cells. Sucrose is the storage form of carbohydrate in the parenchyma cells.

(d) The fleshy adventitious root of sweet potato (*Ipomoea batatas*) of family Convolvulaceae exhibits another complex type of anomalous growth (Fig. 5.158) :

1. Cambium arises in the normal position in between the xylem and phloem.
2. Both primary and secondary xylem elements develop in the usual manner with the formation of a large proportion of parenchyma.
3. With further development many secondary accessory cambia arise in the parenchyma around individual vessel or vessel groups.
4. These cambia produce a few tracheary elements adjacent to vessels, a few sieve elements and laticifers away from the vessels, but a fairly large number of parenchyma in both the directions.
5. At a still later stage, tertiary cambia may be formed in the parenchyma. Periderm of pericyclic origin occurs as normal in the periphery.

In all these roots secondary growth occurs mainly for the storage purpose with formation of scanty conducting tissues but large amounts of storage parenchyma in concentric rings. So, the anomaly in these cases is adaptive in nature.

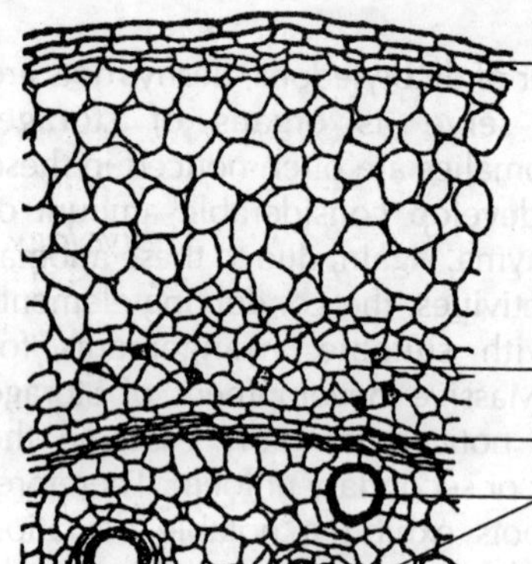

Fig. 5.158 : Cellular drawing of a portion of *Ipomoea batatas* root

Selected Readings

Andrews HN 1961. *Studies in Paleobotany.* John Wiley and Sons, New York.

Arnold CA 1947. *An Introduction to Palebotany.* McGraw-Hill. New York.

Bhattacharya K, Majumder MR and Gupta Bhattachariya, S. 2006. *A Textbook of Palynology.* New Central Book Agency (P) Ltd., Kolkata.

Bhatnagar SP and Moitra A. 1996. *Gymnosperms.* New Age India (P) Ltd., Pub. New Delhi.

Bhojwani SS, Bhatnagar SP 1992. *Embryology of Angiosperms.* Vikas Publ. House Pvt. Ltd., New Delhi.

Bierhorst DW 1971. *Morphology of Vascular Plants.* Macmillan, New York.

Bold HC, Alexopoulos CJ and Delevoryas T. 1980. *Morphology of Plants and Fungi,* 4th ed. Harper and Row, New York.

Carlquist S. 1961. *Comparative Plant Anatomy.* Holt, Rinehart and Winston, New York.

Coulter JM and Chamberlain CJ. 1917. *Morphology of Gymnosperms.* University Chicago Press, Chicago.

Cronquist A 1981. *A Integrated System of Classification of Flowering Plants,* Columbia University Press, New York.

Cronquist A 1988. *The Evolution and Classification of Flowering Plants,* 2nd ed., New York Botanical Garden, New York.

Datta SC. 1985. *Systematic Botany,* Wiley Eastern, New Delhi.

Cutter EG 1971. *Plant Anatomy.* Part I : Cells and Tissues. Edward Arnold, London.

Delevoryas T 1962. *Morphology and Evolution of Fossil Plants.* Holt, Rinehart and Winston, New York.

Eames AJ and MacDaniels LH. 1947. *An Introduction to Plant Anatomy.* McGraw Hill, New York and London.

Easu K. 1965. *Plant Anatomy.* 2nd ed. John Wiley & Sons, New York.

Engler A. (H. Melchior, ed.) 1964. *Syllabus der pflanzenfamilien,* 12th ed. Vol. 2. Gebruder Borntozaeger, Berlin.

Fahn A. 1982 *Plant Anatomy,* Aditya Books (P) Ltd., New Delhi.

Gifford EM and Foster AS. 1989. *Morphology and Evolution of Vascular Plants.* 3rd ed. W. H Freeman and Company, New York.

Hutchinson J. 1973. *The Families of Flowering Plants,* 3rd ed. Oxford University Press, Oxford.

Judd WS, Cambell CS, Kellogg EA, Stevens PF and Donoghue MJ. 2002. *Plant Systematics : A Phylogenetic Approach,* 2nd ed., Sinauer Associates Inc, USA.

Lawrance GHM. 1964 *Taxonomy of Vascular Plants.* The Macmillan Company, N.Y.

Maheswari P. 1950. *An Introduction to the Embryology of Angiosperms.* McGraw Hill Book Co. India, New Delhi.

Mauseth JD. 1988. *Plant Anatomy.* The Benjamin/ Cummings Pub. Com. Inc.

Metcalfe CR. and Chalk L. 1950. *Anatomy of the Dicotyledons,* Vol. I and Vol. II, Clarendon Press, Oxford.

Meyen SV. 1987. *Fundamentals of Palaeobotany.* Chapman and Hall, London and New York.

Naik VN. 1984. *Taxonomy of Angiosperms,* Tata McGraw Hill, New Delhi.

Radford AE. 1986. *Fundamentals of Plant Systematics,* Harper and Row, Pubs. Inc. New York.

Roy P. 2006. *Plant Anatomy.* New Central Book Agency (P) Ltd., Kolkata.

Shiranna KR and Johri BM. 1985. *The Angiosperm Pollen : Structure and Function.* Wiley Eastern, New Delhi.

Simpson MG. 2006. *Plant Systematics.* Elsevier Inc.

Singh G. 2004. *Plant Systematics. Theory and Practice,* 2nd ed., Oxford and IBH, New Delhi.

Sivarajan VV. 1984. *Introduction to Principles of Plant Taxanomy.* Oxford and IBH, New Delhi.

Sneath PHA and Sokal RR. 1973. *Numerical Taxonomy. The Principles and Practice of Numerical Classification.* Freeman, San Francisco.

Sporne KR. 1974. *The Morphology of Gymnosperms. The Structure and Evolution of Primitive Seed Plants.* 2nd ed. Hutchinson University Library, London.

Stewart WN and Rottwell GW. 1993. Paleobotany and the Evolution of Plants. Cambridge University Press, London.

Stuessy TF. 1990. Plant Taxonomy : *The Systematic Evoluation of Comparative Data,* Columbia University Press, New York.

Takhtagarn AL. 1997. *Diversity and Classification of Flowering Plants.* Columbia University Press, New York.

Taylor TN. 1981. *Paleobotay : An Introduction to Fossil Plant Biology.* McGraw Hill, New York.

Thomas BA and Spicer RA. 1986. *The Evolution and Palaeobiology of Land Plants,* Croom Helm, London and Sydney.

Thorne RF. 1976. A Phylogenetic Classification of the Angiospermae. *Evol. Biol.* 9 : 35-106.

Thorne RF. 1992. Classification and Geography of the Flowering Plants. *Bot. Review.* 58 : 225-345.

Index

Some exciting gifts are up for grabs with this book!!

A TEXTBOOK OF BOTANY

[Vol. II]

Bhattacharya, Hait and Ghosh

FREE Gifts Rs. 1000/-

Dear Readers

By now you must have gone through this book in the entirety. So we invite your assessment on this title with absolute impartiality and candid observations. After all, we publishers blindly depend on your benchmarking. Friends, please fill up the open spaces below with CAPITAL LETTERS. You may add some extra sheets of paper if the space provided is not sufficient for you.

i. What are the important topics of your syllabus which are not adequately covered in this book?

ii. In which chapters and on what topics the treatment of the subject-matter by the author is not satisfactory and why?

iii. Have you come across the typos, factual inaccuracies, inconsistencies, and other forms of mistakes in the book? If yes, then please specify the mistakes and page numbers.

iv. Name top three books on the same subject, which, in your consideration, are better than this book. Could you provide some reasons?

Your 1st Preference: ....

Your 2nd Preference: ....

Your 3rd Preference: ....

v. Any suggestion to improve the quality of illustrations, diagrams etc.?

vi. How did you like the cover design? Any suggestions on colours?

vii. Suggestions for further improvement of production standard:

viii. Who recommended you this book:

☐ Teacher ☐ Friend ☐ Parents ☐ Bookseller ☐ Librarian

ix. Name of the person who has recommended:

Designation: ....

Address: ....

Phone No.: ....

e-mail: ....

x. Name and address of the bookseller from whom you have purchased this book:

....

....

....

xi. Name and address of the library where you would like to find the book:

....

....

xii. Name and address of your institution (Please mention the University or Board):

....

....

xiii. Your name, e-mail, and complete postal address:

....

....

....

xiv. A few other titles published by us that you would like to read:

....

....

The best assessment will be rewarded half-yearly. The award will be in the form of our publications, as decided by the Editorial and Marketing Operations Division, amounting to Rs. 1000/- (Rupees one thousand) only.

....

Please mail the filled up coupon at your earliest to :

Editorial & Marketing Operations Division

NEW CENTRAL BOOK AGENCY (P) Ltd

8/1 Chintamoni Das Lane, Kolkata 700 009

West Bengal, INDIA

Ph. No.: 033 + 22418411, 22418573, 22419094

Fax: 033 + 22419573, 22413852

e-mail: ncbapvtltd@eth.net, ncba@cal2.vsnl.net.in

Website: www.centralgroup.com